T0215755

Lecture Notes in Computer Science   9216

Commenced Publication in 1973
Founding and Former Series Editors:
Gerhard Goos, Juris Hartmanis, and Jan van Leeuwen

Rosario Gennaro · Matthew Robshaw (Eds.)

Advances in Cryptology – CRYPTO 2015

35th Annual Cryptology Conference
Santa Barbara, CA, USA, August 16–20, 2015
Proceedings, Part II

 Springer

Editors
Rosario Gennaro
City College of New York
New York, NY
USA

Matthew Robshaw
Impinj, Inc.
Seattle, WA
USA

ISSN 0302-9743   ISSN 1611-3349 (electronic)
Lecture Notes in Computer Science
ISBN 978-3-662-47999-5   ISBN 978-3-662-48000-7 (eBook)
DOI 10.1007/978-3-662-48000-7

Library of Congress Control Number: 2015944435

LNCS Sublibrary: SL4 – Security and Cryptology

Printed on acid-free paper

Springer-Verlag GmbH Berlin Heidelberg is part of Springer Science+Business Media
(www.springer.com)

Preface

CRYPTO 2015, the 35th Annual International Cryptology Conference, was held August 16–20, 2015, on the campus of the University of California, Santa Barbara. The event was sponsored by the International Association for Cryptologic Research (IACR) in cooperation with the UCSB Computer Science Department.

The program of CRYPTO 2015 reflects significant advances and trends in all areas of cryptology. Seventy-four papers were included in the program; this two-volume proceedings contains the revised versions of these papers. The program also included two invited talks: Shai Halevi on 'The state of cryptographic multilinear maps' and Ed Felten on 'Cryptography, Security, and Public Safety: A Policy Perspective'. The paper "Integral Cryptanalysis on Full MISTY1" by Yosuke Todo was selected for both the best paper award and the award for the best paper authored by a young researcher.

This year we received a record number of submissions (266), and in an effort to accommodate as many high-quality submissions as possible, the conference ran in two parallel sessions.

The papers were reviewed by a Program Committee (PC) consisting of 40 leading researchers in the field, in addition to the two co-chairs. Each PC member was allowed to submit two papers. Papers were reviewed in a double-blind fashion, with each paper assigned to three reviewers (four for PC-authored papers). During the discussion phase, when necessary, extra reviews were solicited.

We would like to sincerely thank the authors of all submissions—those whose papers made it into the program and those whose papers did not. Our deep appreciation also goes out to the PC members, who invested an extraordinary amount of time in reviewing papers, and to the many external reviewers who significantly contributed to the comprehensive evaluation of the submissions. A list of PC members and external reviewers follows. Despite all our efforts, the list of external reviewers may contain errors or omissions; we apologize for that in advance.

We would like to thank Tom Ristenpart, the general chair, for working closely with us throughout the whole process and providing the much-needed support at every step, including artfully creating and maintaining the website and taking care of all aspects of the conference's logistics—particularly the novel double-track arrangements.

As always, special thanks are due to Shai Halevi for providing his tireless support of the *websubrev* software, which we used for the whole conference planning and operation, including paper submission and evaluation, interaction among PC members, and communication with the authors. Alfred Hofmann and his colleagues at Springer provided a meticulous service for the timely production of this volume.

Finally, we would like to thank Qualcomm, NSF, and Microsoft for sponsoring the conference, and Cryptography Research for their continuous support.

August 2015

Rosario Gennaro
Matthew Robshaw

CRYPTO 2015

The 35th IACR International Cryptology Conference

University of California, Santa Barbara, CA, USA
August 16–20, 2015

Sponsored by the *International Association for Cryptologic Research*

General Chair

Thomas Ristenpart   Cornell Tech, New York, USA

Program Chairs

Rosario Gennaro   The City College of New York, USA
Matthew Robshaw   Impinj, USA

Program Committee

Michel Abdalla	École Normale Supérieure and CNRS, France
Masayuki Abe	NTT Labs, Japan
Paulo Barreto	University of Sao Paulo, Brazil
Colin Boyd	University of Science and Technology, Norway
Zvika Brakerski	Weizmann Institute of Science, Israel
Emmanuel Bresson	Airbus Cybersecurity, France
Anne Canteaut	Inria, France
Dario Catalano	Università di Catania, Italy
Nishanth Chandran	Microsoft Research, India
Melissa Chase	Microsoft Research, USA
Joan Daemen	ST Microelectronics, Belgium
Orr Dunkelman	University of Haifa, Israel
Karim ElDefrawy	HRL Laboratories, USA
Dario Fiore	IMDEA Software Institute, Spain
Steven Galbraith	Auckland University, New Zealand
Sanjam Garg	University of California, Berkeley, USA
Carmit Hazay	Bar-Ilan University, Israel
Tetsu Iwata	Nagoya University, Japan
Stas Jarecki	University of California, Irvine, USA
Thomas Johansson	Lund University, Sweden
Lars R. Knudsen	Technical University of Denmark

Gregor Leander	Ruhr-Universität Bochum, Germany
Allison B. Lewko	Columbia University, USA
Huijia (Rachel) Lin	University of California, Santa Barbara, USA
Mitsuru Matsui	Mitsubishi Electric, Japan
Sarah Meiklejohn	University College London, UK
Daniele Micciancio	University of California, San Diego, USA
Steve Myers	Indiana University, USA
Bryan Parno	Microsoft Research, USA
Giuseppe Persiano	Università di Salerno, Italy
Thomas Peyrin	Nanyang Technological University, Singapore
Josef Pieprzyk	Queensland University of Technology, Australia
Axel Poschmann	NXP Semiconductors, Germany
Bart Preneel	KU Leuven, Belgium
Mariana Raykova	SRI International, USA
Carla Ràfols	Ruhr-Universität Bochum, Germany
Palash Sarkar	Indian Statistical Institute, India
Nigel Smart	University of Bristol, UK
Franois-Xavier Standaert	Université catholique de Louvain, Belgium
John Steinberger	Tsinghua University, China

Additional Reviewers

Divesh Aggarwal	Harry Bartlett	Ran Canetti
Shashank Agrawal	Georg Becker	Angelo De Caro
Shweta Agrawal	Christof Beierle	David Cash
Martin Albrecht	Sonia Belaid	Debrup Chakraborty
Mehrdad Aliasgari	Mihir Bellare	Eshan Chattopadhyay
Prabhanjan Ananth	Fabrice Benhamouda	Binyi Chen
Elena Andreeva	Guido Bertoni	Jie Chen
Kazumaro Aoki	Nir Bitansky	Mahdi Cheraghchi
Daniel Apon	Olivier Blazy	Céline Chevalier
Benny Applebaum	Celine Blondeau	Chongwon Cho
Frederik Armknecht	Florian Boehl	Joo Yeon Cho
Hassan Asghar	Sonia Bogos	Ashish Choudhury
Gilad Asharov	Jonathan Bootle	Michele Ciampi
Gilles Van Assche	Joppe Bos	Ran Cohen
Nuttapong Attrapadung	Christina Boura	Dana Dachman-Soled
Jean-Philippe Aumasson	Elette Boyle	Hani T. Dawoud
Shi Bai	Cerys Bradley	Ed Dawson
Josep Balasch	Anne Broadbent	Yi Deng
Foteini Baldimtsi	Andre Chailloux	Claus Diem
Achiya Bar-On	Christian Cachin	Itai Dinur
Joshua Baron	Seyit Camtepe	Yevgeniy Dodis

Alexandre Duc
Leo Ducas
Stefan Dziembowski
Oriol Farràs
Sebastian Faust
Serge Fehr
Joan Feigenbaum
Ben Fisch
Marc Fischlin
Christopher Fletcher
Georg Fuchsbauer
Thomas Fuhr
Eiichiro Fujisaki
Marc Fyrbiak
Romain Gay
Ran Gelles
Craig Gentry
Hossein Ghodosi
Kristian Gjøsteen
Florian Gopfert
Vincent Grosso
Jian Guo
Divya Gupta
Shai Halevi
Brett Hemenway
Nadia Heninger
Javier Herranz
Ryo Hiromasa
Shoichi Hirose
Viet Tung Hoang
Justin Holmgren
Naofumi Homma
Yan Huang
Vincenzo Iovino
Yuval Ishai
Zahra Jafargholi
Tibor Jager
Abhishek Jain
Jrmy Jean
Anthony Journault
Saqib A. Kakvi
Pierre Karpman
Elham Kashefi
Aniket Kate
Jonathan Katz
Stefan Katzenbeisser

Marcel Keller
Nathan Keller
Carmen Kempka
Sotirios Kentros
Dmitry Khovratovich
Dakshita Khurana
Aggelos Kiayias
Hyun-Jin (Tiffany) Kim
Susumu Kiyoshima
Miroslav Knezevic
Markulf Kohlweiss
Ilan Komargodski
Venkata Koppula
Luke Kowalczyk
Thorsten Kranz
Ranjit Kumaresan
Junichiro Kume
Eyal Kushilevitz
Tatsuya Kyogoku
Thijs Laarhoven
Mario Lamberger
Joshua Lampkins
Martin Mehl Lauridsen
Tancrède Lepoint
Gaëtan Leurent
Anthony Leverrier
Benoit Libert
Fuchun Lin
Zhen Liu
Steve Lu
Atul Luykx
Anna Lysyanskaya
Vadim Lyubashevsky
Mohammad Mahmoody
Antonio Marcedone
Daniel Masny
Alexander May
Willi Meier
Carlos Aguilar Melchor
Florian Mendel
Bart Mennink
Peihan Miao
Eric Miles
Brice Minaud
Kazuhiko Minematsu
Ilya Mironov

Rafael Misoczki
Payman Mohassel
Amir Moradi
Pawel Morawiecki
Paz Morillo
Nicky Mouha
Pratyay Mukherjee
Sean Murphy
Michael Naehrig
Preetum Nakkiran
Chanathip Namprempre
Mara Naya-Plasencia
Phong Nguyen
Jesper Buus Nielsen
Ivica Nikolic
Ventzi Nikov
Svetla Nikova
Ryo Nishimaki
Luca Nizzardo
Adam O'Neill
Miyako Ohkubo
Olya Ohrimenko
Tatsuaki Okamoto
Claudio Orlandi
Rafail Ostrovsky
Carles Padro
Jiaxin Pan
Omer Paneth
Saurabh Panjwani
Alain Passelègue
Valerio Pastro
Arpita Patra
Michaël Peeters
Roel Peeters
Chris Peikert
Christopher Peikert
Olivier Pereira
Thomas Peters
Duong Hieu Phan
Krzysztof Pietrzak
Benny Pinkas
Oxana Poburinnaya
David Pointcheval
Joop van de Pol
Antigoni Polychroniadou
Christopher Portmann

Contents – Part II

Contents – Part I

Integrity

Assumptions

Hash Functions and Stream Cipher Cryptanalysis

Implementations

Multiparty Computation I

A Simpler Variant of Universally Composable Security for Standard Multiparty Computation

Ran Canetti[1,2], Asaf Cohen[3], and Yehuda Lindell[3](✉)

[1] Boston University, Boston, USA
canetti@tau.ac.il
[2] Tel-Aviv University, Tel Aviv, Israel
[3] Bar-Ilan University, Ramat Gan, Israel
asafc@me.com, lindell@biu.ac.il

Abstract. In this paper, we present a simpler and more restricted variant of the universally composable security (UC) framework that is suitable for "standard" two-party and multiparty computation tasks. Many of the complications of the UC framework exist in order to enable more general tasks than classic secure computation. This generality may be a barrier to entry for those who are used to the stand-alone model of secure computation and wish to work with universally composable security but are overwhelmed by the differences. The variant presented here (called simplified universally composable security, or just SUC) is closer to the definition of security for multiparty computation in the stand-alone setting. The main difference is that a protocol in the SUC framework runs with a fixed set of parties, and machines cannot be added dynamically to the execution. As a result, the definitions of polynomial time and protocol composition are much simpler. In addition, the SUC framework has authenticated channels built in, as is standard in previous definitions of security, and all communication is done via the adversary in order to enable arbitrary scheduling of messages. Due to these differences, not all cryptographic tasks can be expressed in the SUC framework. Nevertheless, standard secure computation tasks (like secure function evaluation) can be expressed. Importantly, we show that for every protocol that can be represented in the SUC framework, the protocol is secure in SUC if and only if it is secure in UC. Therefore, the UC composition theorem holds and any protocol that is proven secure under SUC is secure under the general framework (with some technical changes to the functionality definition). As a result, protocols that are secure in the SUC framework are secure when an a priori unbounded number of concurrent executions of the protocols take place (relative to the same fixed set of parties).

1 Introduction

1.1 Background

The framework of universally composable security (UC) provides very strong security guarantees. In particular, a protocol that is UC secure maintains its

---

The full version of this work can be found on the *IACR Cryptology ePrint Archive* [9]. This work was supported by THE ISRAEL SCIENCE FOUNDATION (grant No. 189/11).

R. Gennaro and M. Robshaw (Eds.): CRYPTO 2015, Part II, LNCS 9216, pp. 3–22, 2015.
DOI: 10.1007/978-3-662-48000-7_1

security properties when run together with many other arbitrary secure and insecure protocols. To be a little more exact, if a protocol π UC securely realizes some ideal functionality $\mathcal{F}$, then π will "behave just like $\mathcal{F}$" in whatever arbitrary computational environment it is run. This security notion matches today's computational and network settings and thus has become the security definition of choice in many cases.

One of the strengths of the UC framework is that it is possible to express almost any cryptographic task as a UC ideal functionality, and it is possible to express almost any network environment within the UC framework (e.g., authenticated and unauthenticated channels, synchronous and asynchronous message delivery, fair and unfair protocol termination, and so on). Unfortunately, this generality and power of expression comes at the price of the UC formalization being very complicated. It is important to note that many of these complications exist in order to enable general cryptographic tasks to be expressible within the framework. For example digital signatures involve local computation alone, and also have no a priori polynomial bound on how many signatures will be generated (by an honest party) since the adversary can determine this. This is very different from standard "secure computation tasks" that involve an a priori known number of interactions between the honest parties.

In this paper, we present a simpler and more restricted variant of the universally composable security (UC) framework; we call this framework simple UC, or SUC for short. Our simplified framework suffices for capturing classic secure computation tasks like secure function evaluation, mental poker, and the like. However, it does not capture more general tasks like digital signatures, and has a more rigid network model (e.g., the set of parties is a priori fixed and authenticated channels are built into the framework). These restrictions make the formalization much simpler, and far closer to the classic stand-alone definition of security which many are more familiar with. Importantly, our simplifications are with respect to the expressibility of the framework and not the security guarantees obtained. Thus, we can prove that any protocol that is expressed and proven secure in the SUC framework is automatically secure also in the full UC framework (relative to an appropriately modified ideal functionality). This means that it is possible to work in the simpler SUC framework, and automatically obtain security in the full UC framework. In Sect. 3, we provide an illustrative example demonstrating that it is significantly more simple to work in the SUC model than in the full UC model.

Remark: We assume familiarity with the ideal/real model paradigm and the standard definitions of security for multiparty computation; see [3] and [15, Chapter 7] for a detailed treatment on these definitions. In addition, we assume that the reader has basic familiarity and understanding of the notion of UC security. This paper is not intended as a tutorial of the UC framework.

1.2   An Informal Introduction to Universally Composable Security

We begin by informally outlining the framework for universally composable security [4,7]. The framework provides a rigorous method for defining the security of cryptographic tasks, while ensuring that security is maintained under concurrent general composition. This means that the protocol remains secure when run concurrently with arbitrary other secure and insecure protocols. Protocols that fulfill this definition of security are called universally composable.

As in other general definitions (e.g., [1,3,15,16,25,27]), the security requirements of a given task (i.e., the functionality expected from a protocol that carries out the task) are captured via a set of instructions for a "trusted party" that obtains the inputs of the participants and provides them with the desired outputs (in one or more iterations). We call the algorithm run by the trusted party an ideal functionality. Since the trusted party just runs the ideal functionality, we do not distinguish between them. Rather, we refer to *interaction between the parties and the functionality*. Informally, a protocol securely carries out a given task if no adversary can gain more from an attack on a real execution of the protocol, than from an attack on an ideal process where the parties merely hand their inputs to a trusted party with the appropriate functionality and obtain their outputs from it, without any other interaction. In other words, it is required that a real execution can be *emulated* in the above ideal process (where the meaning of *emulation* is described below). We stress that in a real execution of the protocol, no trusted party exists and the parties interact amongst themselves.

In order to prove the universal composition theorem, the notion of emulation in the UC framework is considerably stronger than in previous ones. Traditionally, the model of computation includes the parties running the protocol, plus an adversary $\mathcal{A}$ that potentially corrupts some of the parties. In the setting of concurrency, the adversary also has full control over the scheduling of messages (i.e., it fully determines the order that messages sent between honest parties are received); thus, the model is inherently asynchronous. Emulation means that for any adversary $\mathcal{A}$ attacking a real protocol execution, there should exist an "ideal process adversary" or simulator $\mathcal{S}$, that causes the outputs of the parties in the ideal process to be essentially the same as the outputs of the parties in a real execution. In the universally composable framework, an additional adversarial entity called the environment $\mathcal{Z}$ is introduced. This environment generates the inputs to all parties, reads all outputs, and in addition interacts with the adversary in an arbitrary way throughout the computation. (As is hinted by its name, $\mathcal{Z}$ represents the external environment that consists of arbitrary protocol executions that may be running concurrently with the given protocol.) A protocol is said to UC-securely compute a given ideal functionality $\mathcal{F}$ if for any "real-life" adversary $\mathcal{A}$ that interacts with the protocol there exists an "ideal-process adversary" $\mathcal{S}$, such that *no environment $\mathcal{Z}$ can tell whether it is interacting with $\mathcal{A}$ and parties running the protocol, or with $\mathcal{S}$ and parties that interact with $\mathcal{F}$ in the ideal process. (In a sense, here $\mathcal{Z}$ serves as an "interactive distinguisher" between a run of the protocol and the ideal process with access to $\mathcal{F}$.) Note that the definition requires the "ideal-process adversary" (or simulator) $\mathcal{S}$ to interact with $\mathcal{Z}$ throughout the computation. Furthermore, $\mathcal{Z}$ cannot be "rewound".

The following *universal composition theorem* is proven in [4,7]: Consider a protocol π that operates in a *hybrid* model of computation where parties can communicate as usual, and in addition have ideal access to an unbounded number of copies of some ideal functionality $\mathcal{F}$. (This model is called the $\mathcal{F}$-hybrid model.) Furthermore, let ρ be a protocol that UC-securely computes $\mathcal{F}$ as sketched above, and let π^ρ be the "composed protocol". That is, π^ρ is identical to ρ with the exception that each interaction with the ideal functionality $\mathcal{F}$ is replaced with a call to (or an activation of) an appropriate instance of the protocol ρ. Similarly, ρ-outputs are treated as values provided by the functionality $\mathcal{F}$. The theorem states that in such a case, π and π^ρ have essentially the same input/output behaviour. Thus, ρ behaves just like the ideal functionality $\mathcal{F}$, even when composed concurrently with an arbitrary protocol π. This implies the notion of concurrent general composition. A special case of the composition theorem states that if π UC-securely computes some ideal functionality $\mathcal{G}$ in the $\mathcal{F}$-hybrid model, then π^ρ UC-securely computes $\mathcal{G}$ from scratch.

In order to model dynamic settings, the UC formulation enables programs to dynamically generate other programs and dynamically determine their code, and a control function must be defined to determine what operations are allowed and not allowed. This model provides great flexibility, and enables one to model almost any conceivable setting. However, this also adds considerable complexity to the definition, in part due to subtleties that arise with respect to polynomial time, and with respect to the communication rules [18,19,22].

1.3 The SUC Framework

The SUC framework is designed to be as similar as possible to the stand-alone definitions of secure multiparty computation (cf. [3,15]), with the addition of an interactive environment as is required for proving concurrent general composition [23]. In this section we outline the SUC definition, and discuss the main differences between it and the full UC framework.

An Outline of the SUC Framework. The SUC framework was designed by starting with the stand-alone model of secure computation, and adding the seemingly minimal changes required to obtain security under concurrent general composition for standard secure computation tasks, without many of the complications of the UC framework. Thus, in the SUC framework a *fixed* set of parties interact with each other and/or with an ideal functionality (depending on whether an execution is real, ideal or hybrid). An adversary may corrupt some subset of the parties, in which case it sees their state and controls them in the standard way depending on whether it is semi-honest or malicious. As in the UC framework, an environment machine $\mathcal{Z}$ interacts with the adversary throughout the computation and serves as an "interactive distinguisher" between a real execution of the protocol and an ideal execution.

In order to model the fact that the adversary controls all message scheduling, the parties (and any ideal functionality) are connected in a *star configuration* via a router machine. The router queues all communication, and forwards messages only when instructed by the adversary. The adversary sees all the messages sent, and delivers or blocks these messages at will. We note that although the adversary may block messages, it cannot modify messages sent by honest parties (i.e., the communication lines are ideally authenticated). Thus messages sent by a party can arrive in a different order or not arrive at all, but cannot be forged unless the adversary has corrupted the sending party. In order to model the fact that inputs sent to ideal functionalities are private, the SUC framework defines that any message between the parties and the ideal functionality is comprised of a public header and private content. The public header contains any information that is public and thus revealed to the adversary (e.g., the type of message is being sent or what its length is), whereas the private content contains information that the adversary is not supposed to learn.

Composition is defined by replacing the Turing machine code for sending a message to an ideal functionality by the Turing machine code of the protocol that realizes the functionality. Thus, subroutines are executed internally as in the sequential modular composition modeling in [3], unlike the modeling in the full UC framework where subprotocols are invoked as separate ITMs.

The Main Differences Between UC and SUC

Defining Polynomial Time. In the UC framework, machines can be dynamically added to the computation through the mechanism of an external write instruction. Thus, bounding the running time of a single machine by a polynomial does not guarantee that the overall computation is bounded by polynomial time. For example, consider an execution with a single machine that generates a copy of itself and halts. Clearly, each machine is polynomial time. However, an execution of this machine will generate an infinite series of machines and will thus never halt. This makes defining polynomial time in this setting difficult. The definition in the UC framework states that a machine M runs in polynomial time if it runs at most $p(\tilde{n})$ steps where p is a polynomial, and $\tilde{n}$ is the length of the input tape of M plus the security parameter, *minus* the length of all the inputs M provides to other machines. It can be shown that under this definition, the overall execution is bounded by a polynomial, and pathological examples like the one provided above are ruled out.

In the SUC framework, machines cannot generate other machines, and the set of all machines running is fixed ahead of time. Thus, the aforementioned challenges do not arise. We can therefore define polynomial time in the more standard way by simply requiring that each machine run in $p(|x| + n)$ steps, where $|x|$ is the length of its input and n is the security parameter.

Authentication versus Unauthenticated Channels. The basic UC framework has plain, unauthenticated channels; authenticated channels are obtained via an ideal functionality $\mathcal{F}_{\text{AUTH}}$ that provides message authentication. However, almost all secure computation protocols rely on authenticated channels and this is the modeling used in [3,15]. We therefore adopt authenticated channels as the default in SUC, thus simplifying the description of protocols (formally, the real model of computation in the SUC framework corresponds to the $\mathcal{F}_{\text{AUTH}}$-hybrid model of computation in the UC framework). Although this is mainly an **aesthetic** difference, it makes protocol descriptions much more simple.

Defining Composition. The dynamic generation of machines in the UC framework also adds complications regarding defining composition. For example, security under composition is only guaranteed to hold for *subroutine respecting protocols*, which places limitations on the input/output interface of machines with other machines; see [4, Sect. 5.1]. These difficulties arise since when a party calls a subroutine in the UC framework, the subroutine machine is a distinct machine. In order to simplify this issue, in the SUC framework a subroutine call is simply a call to a local routine on the same machine, *exactly* as in the formulation of sequential modular composition in [3].

We stress that although the number of parties in an SUC protocol is a priori fixed, security is guaranteed under composition even when an *unbounded* number of instances of the protocol are run concurrently. This is obtained via the SUC/UC composition theorem.

Expressibility. As we have mentioned, there are cryptographic tasks that can be modeled in the UC framework, but not in the SUC framework. One class of examples is non-interactive cryptographic primitives like digital signatures, encryption, pseudorandom functions and so on. These cannot be modeled in the SUC framework since any interaction with an ideal functionality requires communication that goes via the router and thus its scheduling is controlled by the adversary. This does not model the real-world behavior of local computation for these primitives. Another example is that of protocols in synchronous networks that guarantee output to all parties. This is not possible since the adversary controls the scheduling and thus it is inherently asynchronous. In addition, the adversary can always block messages. Despite this, the SUC framework suffices for modeling any interactive protocol between parties in the most common model of communication for the concurrent setting where the adversary has full control over all message scheduling.

The UC Security of SUC Protocols. We define a transformation $T_P : SUC \rightarrow UC$ that translates SUC-protocols to UC-protocols, and a transformation ϕ that translates ideal functionalities from the SUC framework to the UC framework. We prove that a protocol π SUC-securely computes some ideal functionality $\mathcal{F}$ if and only if $T_P(\pi)$ UC-securely computes $\phi(\mathcal{F})$. SUC composition is derived as a result. The implication is that one may build secure computation protocols in SUC and automatically derive UC security without working with the complex structures of the UC framework. Composition of SUC and UC protocols can also be done freely. Since SUC is less expressive than UC, it is not possible to express every functionality in SUC. SUC cannot replace UC, but is intended as a convenient *interface* to the UC framework that offers the same security standard, and can simplify the process of proving UC security of protocols.

Organization. Due to lack of space in this extended abstract, the proof of equivalence between UC and SUC security is not included, and can be found in the full version [9]. Although this proof is crucial to this work, our main contribution is a simple model that can be used. As such, a presentation of the framework, and a demonstration of why it is easier to use – as can be found in Sect. 3 – covers the main goals.

1.4 Related Work

There has been considerable work in refining the UC framework and solving all the subtleties that arise in the fully dynamic and concurrent setting [17,18,20]. In addition, there have been other frameworks developed to capture the same setting of dynamic concurrency as that of the UC framework [19,21,22,24,27]. However, all of these attempt to capture the same generality of the UC framework in alternative ways. In this work, we make no such attempt and our aim is to capture concurrency for more restricted tasks and obtain a simpler definition. Due to its simplicity, our work can also act as a bridge for connecting the full UC framework with alternative formalisms like [26]. A similar attempt at providing a simplified framework, but without a proof of equivalence, also appeared in [28, Ch. 4].

2 The Simpler UC Model and Definition

In this section, we present a simpler variant of universally composable security that is suitable for standard multiparty computation tasks. It does not have the generality and expressibility of the full-fledged UC framework, but suffices for classic secure computation tasks where a set of parties compute some function of their inputs (a.k.a. secure function evaluation). It also suffices for reactive computations where parties give inputs and get outputs in stages.

2.1 Preliminaries

We denote the security parameter by n. A function $\mu : \mathbb{N} \to [0, 1]$ is negligible if for every polynomial $p(\cdot)$ there exists a value $n_0 \in \mathbb{N}$ such that for every $n > n_0$ it holds that $\mu(n) < 1/p(n)$. All entities (parties, adversary, etc.) are interactive Turing machines (ITM); each such machine has an input tape, an output tape, an incoming communication tape, an outgoing communication tape, and a security parameter tape. If the machine is probabilistic then it also has a random tape. The value written on the security parameter tape is in unary.

We say that a machine is polynomial time if it runs in time that is polynomial in the sum of the lengths of the values that are written on its input tape during its execution plus the security parameter (note that in reactive computations there may be many inputs). Thus, we require that there exists a polynomial $q(\cdot)$ so that for any series of inputs $x_1, x_2, ..., x_\ell$ written on the machine's input tape throughout its lifetime, it always halts after at most $q(n+|x_1|+|x_2|+\cdots+|x_\ell|) = q\left(n + \sum_{j=1}^{\ell} |x_j|\right)$ steps. This is equivalent to saying that each machine receives 1^n as its first input, and n is polynomial in the sum of the lengths of all its inputs.

It is important to note that even if the inputs are short (e.g., constant length), a polynomial-time party can still run in time that is polynomial in the security parameter in every invocation. In order to see this, observe that $\left(n + \sum_{j=1}^{\ell} |x_j|\right)^2 > \sum_{j=1}^{\ell} n \cdot |x_j|$ and thus a machine that runs in time n^c in every invocation is polynomial-time by taking $q(n + \sum_{j=1}^{\ell} |x_j|) = (n + \sum_{j=1}^{\ell} |x_j|)^{2c}$.

Interactive Turing Machines. The formal definition of interactive Turing machines (ITMs) can be found in the full version.

2.2   The Communication and Execution Models

We consider a network where the adversary sees all the messages sent, and delivers or blocks these messages at will. We note that although the adversary may block messages, it cannot modify messages sent by honest parties (i.e., the communication lines are ideally authenticated). We consider a completely asynchronous point-to-point network, and thus the adversary has full control over when messages are delivered, if at all. We now formally specify the communication and execution model. This general model is the same for the real, ideal and hybrid models; we will describe below how each of the specific models are derived from the general communication and execution model.

Communication. In each execution there is an environment $\mathcal{Z}$, an adversary $\mathcal{A}$, participating parties $P_1, \ldots, P_m$, and possibly an ideal functionality $\mathcal{F}$. The parties, adversary and functionality are "connected" in a star configuration, where all communication is via an additional *router machine* that takes instructions from the adversary (see Fig. 1). Formally, this means that the outgoing communication tape of each machine is connected to the incoming communication tape of the router, and the incoming communication tape of each machine is connected to the outgoing communication tape of the router. (For this to work, we define the router so that it has one incoming and one outgoing tape for every other entity in the network except the environment). As we have mentioned, the adversary has full control over the scheduling of all message delivery. Thus, whenever the router receives a message from a party it stores the message and forwards it to the adversary $\mathcal{A}$. Then, whenever the adversary wishes to deliver a message, it sends it to the router who then checks that this message has been stored. If yes, it delivers the message to the designated recipient and erases it, thereby ensuring that every message is delivered only once. If no, the router just ignores the message. If the same message is sent more than once, then the router will store multiple copies and will erase one every time it is delivered.

Observe that $\mathcal{A}$ can only influence when a message is delivered but cannot modify its content. This therefore models authenticated channels, which is standard for secure computation. By convention, a message x from a party P_i to P_j will be of the form (P_i, P_j, x); after P_i writes this message to its outgoing communication tape, the router receives it and checks that the correct sending party identifier P_i is written in the message; if yes, it stores it and works as

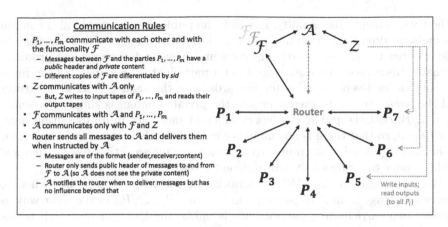

Fig. 1. The communication model and rules

above (sending only to the P_j designated in the message); if no, it ignores the message. Observe that this means that P_j also knows who sent the message to it. In addition, we assume that the set of parties is fixed and known to all.[1]

The above communication model is the same regarding the communication between the functionality $\mathcal{F}$ and the parties and adversary, with two differences. First, the different copies of $\mathcal{F}$ are differentiated by a unique session identifier sid for each copy. Specifically, each message sent to the ideal functionality has a session identifier sid. When, the "main ideal functionality" receives a message, it first checks if there exists a copy of the ideal functionality with that sid. If not, then it begins a new execution of the actual ideal functionality code with that sid, and executes the functionality on the given message. If a copy with that sid does already exist, then that copy is invoked with the message. Likewise, any message sent from a copy of the ideal functionality to a party is sent together with the sid identifying that copy.

The second difference is that any message between the parties and the ideal functionality is comprised of a **public header** and **private content**. The public header contains any information that is public and thus revealed to the adversary, whereas the private content contains information that the adversary is not supposed to learn. For example, in a standard two-party computation functionality where $\mathcal{F}$ computes $f(x, y)$ for some function f (where x is P_1's input and y is P_2's input), the inputs x and y sent by the parties to $\mathcal{F}$ are private. The output from $\mathcal{F}$ to the parties may be public or private, depending on whether this output is supposed to remain secret (say from an eavesdropping adversary between two honest parties) even after the computation.[2] A more interesting example

---

[1] Observe that in contrast to the full UC model, a protocol party here cannot write to the input tapes of other parties. All communication between protocol parties is via the router.

[2] If one of the parties is corrupted then $f(x, y)$ is always learned by the adversary. However, if both are honest, then it may or may not be learned depending on how one defines it.

is the commitment functionality, in which the public header would also contain the message type (i.e., "commit" or "reveal"), since we typically do not try to hide whether the parties are running a commitment or decommitment protocol. Formally, upon receiving a message from a participating party P_i for the functionality or vice versa, the router forwards only the sender/receiver identities and the *public header* to the adversary; the private content is simply not sent.[3] We remark that the public headers of different messages in an execution must be *different*, so that there is no ambiguity regarding the adversary's instructions to the router (formally, the router ignores any new message that has an identical public header to a previously sent different message).

We stress that in the SUC framework, the adversary determines when to deliver a message from $\mathcal{F}$ to participating parties $P_1, \ldots, P_m$ in the same way as between two participating parties. This is unlike the UC framework where the adversary has no such power. In the UC model ideal functionalities are invoked as subroutine machines, and the protocol parties of the *main instance* communicate with the invoked *sub-protocol machine* directly via the input and output tapes, without passing through the adversary. Thus, the class of functionalities that can be expressed in SUC is more restricted. Specifically, we cannot guarantee fairness in the SUC framework, nor model local computation via an ideal functionality (e.g., as is used to model digital signatures in the UC framework).

Finally, the environment $\mathcal{Z}$ communicates with the adversary directly and not via the router. This is due to the fact that it cannot send messages to anyone apart from the adversary; this includes the ideal functionality $\mathcal{F}$. However, differently to all other interaction between parties, the environment $\mathcal{Z}$ can write inputs to the honest parties' input tapes and can read their output tapes (we do not call this "communication" in the same sense since it is not via the communication tapes). The adversary $\mathcal{A}$ itself can send messages in the name of any corrupted party (see Sect. 2.3 below), and can send messages to $\mathcal{Z}$ and $\mathcal{F}$ (the fact that it *can* communicate with $\mathcal{F}$ is useful for relaxing functionalities to allow some adversarial influence; see [4,7]). The adversary $\mathcal{A}$ cannot "directly" communicate with the participating parties.

Execution. An execution of a set of machines connected as above and communicating according to the above rules proceeds as follows. All machines are initialized to have the same value 1^n on their security parameter tapes. Then, the environment is given an initial input $z \in \{0, 1\}^*$ and is the first to be "activated".

In the concurrent setting, and unlike the classic stand-alone setting for secure computation, there are no synchronous rounds in which all parties send messages, compute their next message, and then send it. Rather, the adversary is given full control over the scheduling of messages sent. In order to model this but still to have a well-defined execution model, an execution is modeled by a series of activations of machines one after another, where the order of activations is determined by the adversary. As we have stated, the environment $\mathcal{Z}$ is activated first. In any activation of the environment, it may write to the input tapes of any

---

[3] In order to formalize this, every ideal functionality $\mathcal{F}$ has an associated public-header function $H_{\mathcal{F}}(x)$ that defines the public-header portion of the input x.

of the participating parties $P_1, \ldots, P_m$ that it wishes to, and read their output tapes. In addition, it can send a message to the adversary by writing on its outgoing communication tape. When it halts, the adversary is activated next. In any activation of the adversary, it may read all messages written to entities' outgoing communication tapes (apart from the private content sent between a party and $\mathcal{F}$), carry out any local computation, and write a message on its outgoing communication tape to $\mathcal{Z}$. It then completes its activation by doing one of the following:

1. Instructing the router to deliver a message to any single party that it wishes (including messages between the parties and $\mathcal{F}$). In this case the router is activated next to deliver the message. After the router has delivered the message the recipient party (or $\mathcal{F}$) is activated.
2. Sending a direct message to $\mathcal{F}$ (this type of communication is not via the router). In this case $\mathcal{F}$ is activated next.
3. Sending a direct message to $\mathcal{Z}$. In this case $\mathcal{Z}$ is activated next.

If the activated machine is $\mathcal{F}$ or $\mathcal{Z}$, it reads the message from $\mathcal{A}$, runs a local computation and then sends a response to $\mathcal{A}$, in which case $\mathcal{A}$ is activated next. Otherwise, the activated party ($P_1, \ldots, P_m$ or $\mathcal{F}$) can read the message on its incoming communication tape, carry out any local computation it wishes, and write any number of messages to its outgoing communication tape to the router; its activation ends when it halts. The router is activated next and sends all of the messages that it received to $\mathcal{A}$. The adversary is then once again activated, and so on. One technicality is that the adversary may wish to activate a party to whom no message has previously been sent. This makes most sense at the beginning of a protocol execution where a party already has input but has not yet been sent any messages. Since the adversary is not generally allowed to communicate to parties, it cannot activate such a party since there are no messages to deliver. We therefore allow the adversary to deliver an "empty message" to a party to activate it whenever it wishes. The execution ends when the environment writes a bit to its output tape (the fact that the environment's output is just a single bit is without loss of generality, as shown in [4,7]).

We stress that the ideal functionality has no input on its input tape and never writes to its output tape; it only communicates with the participating parties and the adversary (Fig. 2).

2.3 Corruptions and Adversarial Power

As in the standard model of secure computation, the adversary is allowed to corrupt parties. In the case of **static** adversaries the set of corrupted parties is fixed at the onset of the computation. In the **adaptive** case the adversary corrupts parties at will throughout the computation. In the static corruption case, the environment $\mathcal{Z}$ is given the set of corrupted parties at the onset of the computation. In the active corruption case, whenever the adversary corrupts a party, $\mathcal{Z}$ is notified of the corruption immediately. The adversary is allowed to corrupt parties whenever it is activated. (Formally, the adversary sends a

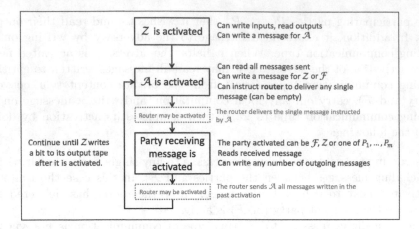

Fig. 2. The execution flow and order of activations

(corrupt, P_i) message first to P_i via the router, and P_i returns its full internal state to the adversary. Then, by convention, the adversary is required to send the corrupt message to $\mathcal{Z}$ who is activated at the end of the corruption sequence.)

We also distinguish between malicious and semi-honest adversaries: If the adversary is malicious then corrupted parties follow the arbitrary instructions of the adversary. In the semi-honest case, even corrupted parties follow the prescribed protocol and the adversary only gets read access to the internal state of the corrupted parties. In the case of a *malicious* adversary, we stress that the adversary can send any message that it wishes in the name of a corrupted party. Formally, this means that the router delivers any message in the name of a corrupted party at the request of the adversary. Observe that in the case of *adaptive malicious corruptions*, any messages that were sent by a party (to another party or to the ideal functionality) before it was corrupted but were not yet delivered may be modified arbitrarily by the adversary. This follows from the fact that from the point of corruption the router delivers any message requested by the adversary. This mechanism assumes that the router is notified whenever a party is corrupted.

We stress that unlike in the full UC model, here it is not possible to "partially corrupt" a party. Rather, if a party is corrupted, then the adversary learns everything. This means that we cannot model, for example, the *forward security* property of key exchange that states that if a party's session key is stolen in one session, then this leaks nothing about its session key in a different session (since modeling this requires corrupting one session of the key exchange and not another). For the same reason, it is not possible to model *proactive security* in the SUC framework [11].

2.4   The Real, Ideal and Hybrid Models

We are now ready to define the real, ideal and hybrid models. These are all just special cases of the above communication and execution models:

- The real model with protocol π: In the real model, there is no ideal functionality and the (honest) parties send messages to each other according to the specified protocol π. We denote the output bit of the environment $\mathcal{Z}$ after a real execution of a protocol π with environment $\mathcal{Z}$ and adversary $\mathcal{A}$ by SUC-REAL$_{\pi,\mathcal{A},\mathcal{Z}}(n,z)$, where z is the input to $\mathcal{Z}$.
- The ideal model with $\mathcal{F}$: In the ideal model with $\mathcal{F}$ the parties follow a *fixed ideal-model protocol*. According to this protocol, the parties send messages only to the ideal functionality but never to each other. Furthermore, these messages are the inputs that they read from their input tapes, and nothing else (unless they are corrupted and the adversary is malicious, in which case they can send anything to $\mathcal{F}$). In addition, they write any message received back from the ideal functionality to their output tapes. That is, the ideal-model protocol instructs a party upon activation to read any new input on its input tape and send it unmodified to $\mathcal{F}$ as an outgoing message, and to read all incoming messages (from $\mathcal{F}$) on its incoming message tape and write them unmodified to its output tape. This then ends the party's activation. We denote the output of $\mathcal{Z}$ after an ideal execution with ideal functionality $\mathcal{F}$ and adversary $\mathcal{S}$ (denoted by $\mathcal{S}$ since it is actually a "simulator") by SUC-IDEAL$_{\mathcal{F},\mathcal{S},\mathcal{Z}}(n,z)$, where n and z are as above. We stress that in the ideal model, the adversary/simulator $\mathcal{S}$ interacts with $\mathcal{Z}$ in an online way; in particular, it cannot rewind $\mathcal{Z}$ or look at its internal state. In addition, in keeping with the general communication model all messages between the parties and $\mathcal{F}$ are delivered by the adversary.[4]
- The hybrid model with π and $\mathcal{F}$: In the hybrid model, the parties follow the protocol π as in the real model. However, in addition to regular messages sent to other parties, π can instruct the parties to send messages to the ideal functionality $\mathcal{F}$ and also instructs them how to process messages received from $\mathcal{F}$. We stress that the messages sent to $\mathcal{F}$ may be any values specified by π and are not limited to inputs like in the ideal model. We denote the output of $\mathcal{Z}$ from a hybrid execution of π with ideal calls to $\mathcal{F}$ by SUC-HYBRID$_{\pi,\mathcal{A},\mathcal{Z}}^{\mathcal{F}}(n,z)$, where $\mathcal{A}, \mathcal{Z}, n, z$ are as above. When $\mathcal{F}$ is the ideal functionality we call this the $\mathcal{F}$-hybrid model.

In all models, there is a fixed set of participating parties $P_1, \ldots, P_m$, where each party has a unique party identifier. Observe that we formally consider a *single* ideal-functionality type $\mathcal{F}$, and not multiple different ones.[5] This is not a limitation even though protocols often use multiple different subprotocols (e.g., commitment, zero knowledge, and oblivious transfer). This is because one can define a single functionality computing multiple subfunctionalities. Thus, formally we consider one. When defining protocols and proving security, it is

---

[4] The fact that the adversary delivers these messages and thus message delivery is not guaranteed frees us from the need to explicitly deal with the "early stopping" problem of protocols run between two parties or amongst many parties where only a minority may be honest. This is because the adversary can choose which parties receive output and which do not, even in the ideal model.

[5] This is not to be confused with multiple copies of the same functionality $\mathcal{F}$ which is included in the model.

customary to refer to multiple functionalities with the understanding that this is formally taken care of as described.

2.5   The Definition and Composition Theorem

We are now ready to define SUC security, and to state the composition theorem. Informally, security is defined as in the classic stand-alone definition of security by requiring the existence of an ideal-model simulator for every real-model adversary. However, in addition, the simulator must work for every environment, as in the aforementioned communication and execution models. The environment behaves as the interactive distinguisher, and therefore we say that a protocol π SUC-securely computes a functionality if the environment outputs 1 with almost the same probability in a real execution of π with $\mathcal{A}$ as in an ideal execution with $\mathcal{F}$ and $\mathcal{S}$. Recall that the SUC-IDEAL and SUC-REAL notation denotes the output of $\mathcal{Z}$ after the respective executions.

Balanced Environments. A balanced environment is an environment for which at any point in time during the execution, the overall length of the inputs given to the parties of the main instance of the protocol is at most n times the length of the input to the adversary [7]. As in the full UC framework, we require balanced environments in order to prevent unnatural situations where the input length and communication complexity of the protocol is arbitrarily large relative to the input length and complexity of the adversary. In such case no PPT adversary can deliver even a fraction of the protocol communication. The definition of UC security considers only balanced environments, and we adopt this same convention.

Definition 1. *Let π be a protocol for up to m parties and let $\mathcal{F}$ be an ideal functionality. We say that π SUC-securely computes $\mathcal{F}$ if for every probabilistic polynomial-time real-model adversary $\mathcal{A}$ there exists a probabilistic polynomial-time ideal-model adversary $\mathcal{S}$ such that for every probabilistic polynomial-time balanced environment $\mathcal{Z}$ and every constant $d \in \mathbb{N}$, there exists a negligible function $\mu(\cdot)$ such that for every $n \in \mathbb{N}$ and every $z \in \{0,1\}^*$ of length at most n^d,*

$$\left| \Pr\left[\text{SUC-IDEAL}_{\mathcal{F},\mathcal{S},\mathcal{Z}}(n,z) = 1\right] - \Pr\left[\text{SUC-REAL}_{\pi,\mathcal{A},\mathcal{Z}}(n,z) = 1\right] \right| \le \mu(n).$$

The SUC composition theorem is essentially the same as the UC composition theorem: secure protocols "behave like" ideal functionalities when run in arbitrary environments. See the full version for a formal statement of the theorem.

3   An Example – Proving in the UC vs SUC Models

In this section, we demonstrate the difference between proving security in the full UC framework and in the SUC framework. We consider the classic commitment functionality $\mathcal{F}_{\text{COM}}$, due to its relative simplicity. We also consider realizing the $\mathcal{F}_{\text{ZK}}$ functionality in the $\mathcal{F}_{\text{COM}}$-hybrid model, since existing protocols "gloss over" the details of using the composition theorem correctly.

3.1   Differences in Defining the Ideal Functionality for Commitments

Before describing the functionality, we need to introduce the *delayed output* terminology, which is a convention that appears in the full UC framework. Quoting from [6, Sect. 6.2]: *"we say that an ideal functionality $\mathcal{F}$ sends a delayed output v to party P if it engages in the following interaction: Instead of simply outputting v to P, $\mathcal{F}$ first sends to the adversary a message that it is ready to generate an output to P. If the output is public, then the value v is included in the message to the adversary. If the output is private then v is not mentioned in this message. Furthermore, the message contains a unique identifier that distinguishes it from all other messages sent by $\mathcal{F}$ to the adversary in this execution. When the adversary replies to the message (say, by echoing the unique id), $\mathcal{F}$ outputs the value v to P."*

We now consider the definition of secure commitments. For simplicity, we consider the *single* commitment functionality (typically, the multiple commitment functionality is used, but this even further complicates the definition). This is the definition that appears in [5, Sect. 7.3.1]:

FIGURE 1 (Functionality $\mathcal{F}_{COM}$ for the Full UC Framework)

1. Upon receiving an input (Commit, sid, x) from C, verify that $sid = (C, R, sid')$ for some R, else ignore the input. Next, record x and generate a public delayed output (Receipt, sid) to R. Once x is recorded, ignore any subsequent Commit inputs.
2. Upon receiving an input (Open, sid) from C, proceed as follows: If there is a recorded value x then generate a public delayed output (Open, sid, x) to R. Otherwise, do nothing.
3. Upon receiving a message (Corrupt-committer, sid) from the adversary, output a Corrupted value to C, and send x to the adversary. Furthermore, if the adversary now provides a value x', and the Receipt output was not yet written on R's tape, then change the recorded value to x'.

The Ideal Commitment Functionality $\mathcal{F}_{COM}$

In contrast, in the SUC framework the functionality description is far simpler. Before writing the functionality, we introduce a convention that was used in [12] for the public headers and private contents in functionalities. The "operation labels" (e.g., ,, Receipt, etc.) and the session identifiers are by convention (and unless explicitly stated otherwise) part of the public header, and the rest of the message constitutes the private contents. In addition, we parameterize the functionality by some $m = \text{poly}(n)$, which means that all commitment values are of length m. This is needed since SUC parties have a fixed polynomial running time, and so a receiver who does not receive input to the commitment functionality cannot process arbitrarily long strings. Note that all known UC commitment schemes work in this way (i.e., they are either commitments to bits, fixed-length strings, or group elements, etc.). Thus, this definition matches existing constructions.[6] We have:

FIGURE 2 (Functionality $\mathcal{F}_{\text{COM}}$ for the SUC Framework)

$\mathcal{F}_{\text{COM}}$ runs with length parameter m, as follows:

1. Upon receiving an input (Commit, sid, x) from C, verify that $x \in \{0,1\}^m$ and that $sid = (C, R, sid')$ for some R, else ignore the input. Next, record x and send (Receipt, sid) to R. Once x is recorded, ignore any subsequent Commit inputs.
2. Upon receiving an input (Open, sid) from C, proceed as follows: If there is a recorded value x then send (Open, sid, x) to R. Otherwise, do nothing.

The Ideal Commitment Functionality $\mathcal{F}_{\text{COM}}$

Explaining the Differences Between the Functionalities. In the full UC framework, it is necessary to refer to public delayed outputs, since honest parties write their inputs locally to ideal functionalities; to be more exact, an ideal call is a subroutine invocation. Thus, in interactive scenarios, it is necessary for the ideal functionality to explicitly communicate with the adversary to ask permission to send the receipt, and so on. Due to the fact that this is tiresome to describe each time, the convention of a "delayed output" was introduced. In contrast, in the SUC framework, since the adversary automatically controls all delivery, it suffices to naturally send messages. However, this does come at the price of explicitly stating which parts of the messages are public (and seen by the adversary when it delivers) and which parts are private. Nevertheless, by our convention, this is typically simple.

A more significant difference arises in the context of corruption. In the full UC model, an ideal functionality is modeled as a *subroutine* of the main protocol instance. Therefore, parties "send" messages/inputs to an ideal functionality $\mathcal{F}$ by writing them directly on the input tape of $\mathcal{F}$. This means that the adversary cannot change the contents of such a written message, even in the case that the party is corrupted before the input was effectively used. In real protocols, it is often possible for the adversary to make such a change. (For example, consider the case that the honest party sends its first message and is corrupted before it is delivered. In this case, the adversary can choose not to deliver that message and instead send a new message in its place for the corrupted party, possibly using a different input. Thus, this has to also be possible in the ideal model.) This forces such treatment to be explicitly defined in the ideal functionality. In contrast, in the SUC framework, this issue does not arise at all. This is because all messages, *including inputs to an ideal functionality and messages in a real protocol*, are treated in the same way and sent via the router. By the way the

---

[6] We remark that it is also possible to define $\mathcal{F}_{\text{COM}}$ so that S inputs x and R inputs $1^{|x|}$. This ensures that R can run in time that is polynomial in the length of the committed value. We chose the formulation of a fixed m since it more closely models how UC commitments are typically constructed.

router is defined, an adversary can choose not to deliver messages to an ideal functionality in the same way that it can choose not to deliver messages in a real protocol.

3.2 Proving Security of Commitment Protocols and Zero Knowledge Protocols

In this section, we consider the problem of constructing UC commitments in the CRS model, and then zero knowledge protocols using UC commitments. This is the standard way of working; see [10,12], and see [14] for a more recent work following the same paradigm. The authors of [14] claim security of their zero knowledge protocol by referring to the proof of security of zero knowledge from commitments that appears in [10]. However, this proof is much closer to the SUC framework and does not take into account a number of issues that must be considered in the (current version of the) full UC model. We describe *some* of the additional issues that need to be taken into account in order to prove the full UC security of the zero knowledge protocol from full UC commitments. For the sake of concreteness, when considering polynomial time, we refer specifically to the constructions in [14].

Before proceeding, denote the commitment protocol of [14] by Π_{COM}, the CRS functionality by $\mathcal{F}_{\mathrm{CRS}}$, and the zero knowledge protocol of [10,14] by Π_{ZK}. Protocol Π_{ZK} works by running the classic zero knowledge Hamiltonicity protocol of Blum [2], while using UC commitments. Actually, since many commitments are needed with respect to the same CRS, the multiple commitment functionality $\mathcal{F}_{\mathrm{MCOM}}$ is used but for simplicity we will ignore this here. Note that the commitment protocol Π_{COM} in [14] uses a fully-homomorphic encryption scheme denoted Q_{ENC} and a CCA-secure encryption scheme ENC_{CCA}.

Proof of Polynomial-Time. One of the requirements of the UC composition theorem is that all the protocols involved are polynomial time. The mentioned proofs do not formally prove that the protocols are polynomial time. In the SUC model, the fact that Π_{COM} in [14] is polynomial time is immediate, and simply follows from the fact that the Q_{ENC} and ENC_{CCA} encryption schemes run in polynomial time (since in each invocation each party trivially runs in time that is polynomial in the security parameter and input; see Sect. 2.1 for why this suffices in the SUC framework). However, in order to prove that Π_{COM} in [14] is polynomial time in the full UC framework, one needs to first pad the input of each party in Π_{COM} with sufficient tokens, so that it runs in time that is polynomial in the length of its (padded) input minus the length of the inputs/messages that it sends to $\mathcal{F}_{\mathrm{CRS}}$. If $\mathcal{F}_{\mathrm{CRS}}$ is assumed to be a local functionality (e.g., secure setup), then this is not difficult since the only input to $\mathcal{F}_{\mathrm{CRS}}$ is the pair (CRS, sid). However, if $\mathcal{F}_{\mathrm{CRS}}$ is implemented via coin-tossing using a local $\mathcal{F}_{\mathrm{CRS}}$ functionality (as suggested in the JUC [13] solution to achieving independent CRS invocations per protocol), then the number of tokens needed to be provided is different. Essentially, a different $\mathcal{F}_{\mathrm{CRS}}$ ideal functionality has to be defined for each of these cases. (The reason that a different ideal functionality is needed is that the functionality defines the length of the input, which depends on the number of tokens needed.)

Consider next the case of constructing Π_{ZK} using $\mathcal{F}_{\text{COM}}$. These zero-knowledge protocols make multiple calls to the commitment functionality. The number of calls to $\mathcal{F}_{\text{COM}}$, and thus the length of the input written by the parties in Π_{ZK} to $\mathcal{F}_{\text{COM}}$, differs *significantly* when the zero-knowledge is based on Hamiltonicity versus when it is based on 3 coloring. The proof of polynomial-time complexity must take into account that for Hamiltonicity, for a graph with n nodes, $O(n^3)$ calls to $\mathcal{F}_{\text{COM}}$ are made (repeating n times where in each time a matrix of size $O(n^2)$ is committed to). However, the size of the graph depends on the Karp reduction of the statement being proven to Hamiltonicity, and this must also be counted. This bound must then be included in the ideal functionality for $\mathcal{F}_{\text{COM}}$, since the actual length of the input includes these tokens. Notice, however, that the number of tokens needed in 3 coloring will be different, and so the definition of $\mathcal{F}_{\text{COM}}$ can actually depend on the implementation of $\mathcal{F}_{\text{ZK}}$ as used by Π_{COM}. To make this even more complex, if $\mathcal{F}_{\text{COM}}$ uses $\mathcal{F}_{\text{CRS}}$ as described above, then the number of token further depends on whether $\mathcal{F}_{\text{CRS}}$ is a local functionality or derived by some type of coin-tossing protocol.

We are not aware of any research paper whose focus is protocol construction that relates to the issue of defining the number of tokens–equivalently how much to pad the input–when defining the functionality, and proving that the protocol is polynomial time as defined in the full UC framework.

Subroutine Respecting Protocols. The UC composition theorem demands that protocols are subroutine respecting; see [6]. Informally speaking, this means that subroutines only accept messages from other parties or subsidiaries of the subroutine instance. In addition, upon the first activation, the adversary receives notification of the code and SID of the instance. Since these are messages sent to the adversary, they need to be dealt with by the adversary in the proof of security. To the best of our knowledge, the adversary's treatment of these notifications are typically not described.

Corruptions. In the full UC framework, the protocol specification has to include what the parties should do upon receiving a Corrupt message. This is due to the fact that the UC framework enables great flexibility in dealing with corruptions (and thus can model partial corruptions, proactive corruptions, and so on). In contrast, in the SUC model, a party is either honest or fully corrupted, and in the latter case the adversary obtains full control of the party. Although describing what a party should do upon corruption is not complicated, it is once again an example of a detail that needs to be addressed, but is to the best of our knowledge omitted in current protocol specifications.

Order of Activations. In the full UC framework, the order of activations depends on the adversary and on the protocol, and is derived from the order of external write calls made by the machines in the system. Each machine can only write one external message (be it input to a subroutine, output, or a regular message) per activation, and by writing the message it passes the execution to the receiving machine. This means that multiple invocation patterns are possible, yielding multiple case analyses in the proof. In addition, when writing the proof, one must distinguish between the different types of messages (writing to an ideal functionality is fundamentally different to sending a message to another party). Both of these complicate the presentation and make it harder for one writing the proof to be exact. In contrast, in the SUC model, one of our aims was to make the order of activations the same in all models (real, ideal and hybrid) and to use the same method for all types of messages. (The only exception is the parties' inputs written by the environment and their outputs read by the environment.) Thus, the scheduling of activations and the terminology with respect to messages is always the same (under full control of the adversary), simplifying the presentation.

Conclusions – Current UC Research and UC/SUC Proofs. We are not aware of *any written proof* in the UC framework that actually takes these details into account. Rather, researchers writing protocols in the UC framework do not specify the number of tokens needed in order to be polynomial time (which is the most serious issue), do not describe what the adversary should do with invocation messages, do not consider the varying order of activations, and so on. Essentially, researchers today write their proofs as if they are working in something similar to the SUC framework. The main contribution of this paper can therefore be viewed as a *justification of the soundness* of working in this way. In addition, we provide an *exact* model that can be used, instead of handwaving away the full UC details. Finally, our proof that SUC protocols are actually UC secure (with the appropriate adjustments) means that for the standard interactive secure computation tasks, nothing is lost by working with our simpler model.

References

1. Beaver, D.: Foundations of secure interactive computing. In: Feigenbaum, J. (ed.) CRYPTO 1991. LNCS, vol. 576, pp. 377–391. Springer, Heidelberg (1992)
2. Blum, M.: How to prove a theorem so no one else can claim it. In: Proceedings of the International Congress of Mathematicians, pp. 1444–1451, USA
3. Canetti, R.: Security and composition of multiparty cryptographic protocols. J. cryptology **13**(1), 143–202 (2000)
4. Canetti, R.: Universally composable security: a new paradigm for cryptographic protocols. In: In the 42nd FOCS, pp. 136–145 (2001)
5. Canetti, R.: Universally composable security: a new paradigm for cryptographic protocols. Cryptology ePrint Archive, Report 2000/067 (revision of 13 December 2005)
6. Canetti, R.: Universally composable security: a new paradigm for cryptographic protocols. Cryptology ePrint Archive, Report 2000/067 (revision of 16 July 2013)
7. Canetti, R.: Universally composable security: a new paradigm for cryptographic protocols. Cryptology ePrint Archive, Report 2000/067 (revised 13 December 2005 and re-revised April 2013)
8. Canetti, R.: Obtaining universally compoable security: towards the bare bones of trust. In: Kurosawa, K. (ed.) ASIACRYPT 2007. LNCS, vol. 4833, pp. 88–112. Springer, Heidelberg (2007)

9. Canetti, R., Cohen, A., Lindell, Y.: A simpler variant of universally composable security for standard multiparty computation (full version). Cryptology ePrint Archive, Report 2014/553 (2014)
10. Canetti, R., Fischlin, M.: Universally composable commitments. In: Kilian, J. (ed.) CRYPTO 2001. LNCS, vol. 2139, pp. 19–40. Springer, Heidelberg (2001)
11. Canetti, R., Herzberg, A.: Maintaining security in the presence of transient faults. In: Desmedt, Y.G. (ed.) CRYPTO 1994. LNCS, vol. 839, pp. 425–438. Springer, Heidelberg (1994)
12. Canetti, R., Lindell, Y., Ostrovsky, R.,Sahai, A.: Universally composable two-party and multi-party secure computation. In: In the 34th STOC, pp. 494–503 (2002). Reference is to page 13 of Cryptology ePrint Archive Report 2002/140 (version of 14 July 2003)
13. Canetti, R., Rabin, T.: Universal Composition with Joint State. In: Boneh, D. (ed.) CRYPTO 2003. LNCS, vol. 2729, pp. 265–281. Springer, Heidelberg (2003)
14. Damgård, I., Polychroniadou, A., Rao, V.: Adaptively Secure UC constant round multi-party computation. Cryptology ePrint Archive, Report 2014/830 (2014)
15. Goldreich, O.: Foundations of Cryptography: Volume 2 - Basic Applications. Cambridge University Press, Cambridge (2004)
16. Goldwasser, S., Levin, L.A.: Fair computation of general functions in presence of immoral majority. In: Menezes, A., Vanstone, S.A. (eds.) CRYPTO 1990. LNCS, vol. 537, pp. 77–93. Springer, Heidelberg (1991)
17. Hofheinz, D., Müller-Quade, J., Steinwandt, R.: Initiator-resilient universally composable key exchange. In: Snekkenes, E., Gollmann, D. (eds.) ESORICS 2003. LNCS, vol. 2808, pp. 61–84. Springer, Heidelberg (2003)
18. Hofheinz, D., Müller-Quade, J., Unruh, D.: Polynomial runtime and composability. IACR Cryptology ePrint Archive, report 2009/23 (2009)
19. Hofheinz, D., Shoup, V.: GNUC: a new universal composability framework. IACR Cryptology ePrint Archive, report 2011/303 (2011)
20. Katz, J., Maurer, U., Tackmann, B., Zikas, V.: Universally composable synchronous computation. In: Sahai, A. (ed.) TCC 2013. LNCS, vol. 7785, pp. 477–498. Springer, Heidelberg (2013)
21. Küsters, R.: Simulation-based security with inexhaustible interactive turing machines. In: CSFW, pp. 309–320 (2006)
22. Küsters, R., Tuengerthal, M.: The IITM model: a simple and expressive model for universal composability. IACR Cryptology ePrint Archive, report 2013/25 (2013)
23. Lindell, Y.: General composition and universal composability in secure multi-party computation. J. Cryptology **22**(3), 395–428 (2009). An extended abstract appeared in the 44th FOCS, pp. 394–403 (2003)
24. Lynch, N.A., Segala, R., Vaandrager, F.W.: Compositionality for probabilistic automata. In: Amadio, R.M., Lugiez, D. (eds.) CONCUR 2003. LNCS, vol. 2761, pp. 208–221. Springer, Heidelberg (2003)
25. Micali, S., Rogaway, P.: Secure computation. In: Feigenbaum, J. (ed.) CRYPTO 1991. LNCS, vol. 576, pp. 392–404. Springer, Heidelberg (1992)
26. Micciancio, D., Tessaro, S.: An Equational Approach to Secure Multi-Party Computation. In: ITCS 2013, pp. 355–372 (2013)
27. Pfitzmann, B., Waidner, M.: Composition and integrity preservation of secure reactive systems. In: In 7th ACM Conference on Computer and Communication Security, pp. 245–254 (2000)
28. Wikström, D.: On the Security of Mix-Nets and Hierarchical Group Signatures. Ph.D. thesis (2005)

Concurrent Secure Computation via Non-Black Box Simulation

Vipul Goyal[1]([✉]), Divya Gupta[2], and Amit Sahai[2]

[1] Microsoft Research India, Bengaluru, India
vipul@microft.com
[2] University of California, Los Angeles and Center for Encrypted Functionalities,
Los Angeles, USA
{divyag,sahai}@cs.ucla.edu

Abstract. Recently, Goyal (STOC'13) proposed a new non-black box simulation techniques for fully concurrent zero knowledge with straight-line simulation. Unfortunately, so far this technique is limited to the setting of concurrent zero knowledge. The goal of this paper is to study what can be achieved in the setting of concurrent secure computation using non-black box simulation techniques, building upon the work of Goyal. The main contribution of our work is a secure computation protocol in the fully concurrent setting with a straight-line simulator, that allows us to achieve several new results:

- We give first positive results for concurrent blind signatures and verifiable random functions in the plain model *as per the ideal/real world security definition*. Our positive result is somewhat surprising in light of the impossibility result of Lindell (STOC'03) for black-box simulation. We circumvent this impossibility using non-black box simulation. This gives us a quite natural example of a functionality in concurrent setting which is impossible to realize using black-box simulation but can be securely realized using non-black box simulation.
- Moreover, we expand the class of realizable functionalities in the concurrent setting. Our main theorem is a positive result for concurrent secure computation as long as the ideal world satisfies the *bounded pseudo-entropy condition* (BPC) of Goyal (FOCS'12). The BPC requires that in the ideal world experiment, the total amount of information learnt by the adversary (via calls to the ideal functionality) should have "bounded pseudoentropy".

D. Gupta, A. Sahai—Research supported in part from a DARPA/ONR PROCEED award, a DARPA/ARL SAFEWARE award, NSF Frontier Award 1413955, NSF grants 1228984, 1136174, 1118096, and 1065276, a Xerox Faculty Research Award, a Google Faculty Research Award, an equipment grant from Intel, and an Okawa Foundation Research Grant. This material is based upon work supported by the Defense Advanced Research Projects Agency through the U.S. Office of Naval Research under Contract N00014-11-1-0389. The views expressed are those of the author and do not reflect the official policy or position of the Department of Defense, the National Science Foundation, or the U.S. Government.

D. Gupta—Work done in part while interning at Microsoft Research India.

R. Gennaro and M. Robshaw (Eds.): CRYPTO 2015, Part II, LNCS 9216, pp. 23–42, 2015.
DOI: 10.1007/978-3-662-48000-7_2

- We also improve the round complexity of protocols in the single-input setting of Goyal (FOCS'12) both qualitatively and quantitatively. In Goyal's work, the number of rounds depended on the length of honest party inputs. In our protocol, the round complexity depends only on the security parameter, and is completely independent of the length of the honest party inputs.

Our results are based on a non-black box simulation technique using a new language (which allows the simulator to commit to an Oracle program that can access information with bounded pseudoentropy), and a simulation-sound version of the concurrent zero-knowledge protocol of Goyal (STOC'13). We assume the existence of collision resistant hash functions and constant round semi-honest oblivious transfer.

1   Introduction

Secure computation protocols enable a set of mutually distrustful parties to securely perform a task by interacting with each other. Traditional security notions for secure computation [21,49] were defined for the *stand-alone setting* where security holds only if a single protocol session is executed in isolation. In today's connected world (and especially over internet), many instances of these protocols may be executing concurrently. In such a scenario, a protocol that is secure in the classical stand-alone setting may become completely insecure [5,37]. Ambitious efforts have been made to generalize the results for the stand-alone setting, starting with concurrently-secure zero-knowledge protocols [7,14,34,45,47].

However, in the plain model, the effort to go beyond the zero-knowledge functionality were, unfortunately, less than fully satisfactory. In fact, for the plain model far reaching unconditional impossibility results were shown in a series of works [1,5,8,19,24,37,38]. Two notable exceptions giving positive results in the plain model are the works on *bounded* concurrency [36,43,44] (where there is an a-priori fixed bound on the total number of concurrent sessions in the system and the protocol in turn can depend on this bound), and, the positive results for a large class of functionalities in the so called "single input" setting [24]. In this setting, there is a server interacting with multiple clients concurrently with the restriction that the server (if honest) is required to use the *same* input in all sessions. There is a large body of literature on getting concurrently secure computation in weaker models such as using a super-polynomial time simulator, or a trusted setup. A short survey of these works is given later in this section. We emphasize that in this work, we are interested in concurrently secure computation protocols with no trusted set up assumptions where the security holds according to standard ideal/real paradigm.

An intriguing functionality that cannot be realized in the fully concurrent setting by these results is blind signatures in the plain model. The blind signature functionality, introduced by [11], allows users to obtain unforgeable signatures on messages of their choice without revealing the message being signed to the signer (blindness property). The question of whether a concurrently-secure protocol for this functionality can be constructed as per the ideal/real model simulation paradigm has been open so far. Moreover, given the impossibility result for concurrent blind signatures for black box simulation by Lindell [37], it is clear that we need to use non-black box techniques. Until recently, no non-black box technique was known which applies to full concurrency with polynomial time simulation. However, Goyal [25] recently proposed new non-black box simulation techniques for (fully) concurrent zero-knowledge with straight line simulation. Unfortunately, the result of Goyal is limited to the setting of concurrent zero-knowledge. We ask the question: *Can we construct non-box black techniques for (fully) concurrent secure computation, building upon the work of Goyal [25]?*

Our Contributions. The main contribution of our work is a secure computation protocol in the fully concurrent setting with a straight-line simulator, that allows us to achieve several new results. In short, we expand the class of realizable functionalities in the concurrent setting and give the first positive results for concurrent blind signatures and verifiable random functions in the plain model *as per the ideal/real world security definition.* Moreover, the round complexity of our protocol depends only on the security parameter and hence, improves the round complexity of [24] both qualitatively and quantitatively. Finally, our work can be seen as a *unifying framework*, which essentially subsumes all the previous work on positive results for concurrent secure computation achieving polynomial time simulation based security in the plain model. For detailed description of our results, see Sect. 1.1.

Other Models. In order to circumvent the above mentioned impossibility results in the plain model, there has been quite some work studying various trust assumptions such as common reference string (CRS) model and tamper proof hardware tokens [3,10,32]. Another interesting line of work has studied weaker security definitions [16,39,42,46] while still remaining in the plain model, and most notably obtains positive results in models like super polynomial time simulation [6,9,17,46] and input indistinguishable security [17,39].

Note that these trust assumptions and these relaxed notions of security are sometimes restrictive and are not applicable to many situations. We again emphasize that the focus of this work is concurrent secure computation in the plain model achieving *polynomial* time simulation. In the plain model, there are point to point authenticated channels between the parties, but there is no global trusted third party.

What Goes Wrong in Concurrent Setting in Plain Model? A well established approach to constructing secure computation protocols is to use the GMW compiler: take a semi-honest secure computation protocol and "compile" it with zero-knowledge arguments. The natural starting point in the concurrent setting is to follow the same principles: somehow compile a semi-honest secure computation protocol with a *concurrent* zero-knowledge protocol (actually compile with concurrent *non-malleable* zero-knowledge [5]). Does such an approach (or minor variants) already give us protocols secure according to the standard ideal/real world definition in the plain model?

There is a fundamental problem with this approach which poses a key *bottleneck* in a number of previous works (see [17, 24, 26, 28–30]). All known concurrent zero-knowledge simulators in the fully concurrent setting work by rewinding the adversarial parties. Such an approach is highly problematic for secure computation in the concurrent setting, where the adversary controls the scheduling of the messages of different sessions. For instance, consider the following scenario: Due to nesting of sessions by the adversary, a rewinding based simulator may need to execute some sessions more than once. Since the adversary can choose a different input in each execution (e.g. based on transcript so far), the simulator would have to query the ideal functionality for than once. However, for any session, the simulator is allowed at most one query! Indeed, such problems are rather inherent as indicated by various impossibility results [5, 38].

Trying to solve this bottleneck of "handling extra queries" in various ways has inspired a number of different works which revolve around a unified theme: first construct a protocol where the simulator requires multiple queries per session in the ideal world, and then, somehow manage to either eliminate or answer these extra queries by exploiting some property of the specific setting in question. Examples of these include Resettable and Stateless computation [29, 30], Multiple Ideal Query model [26–28], Single-Input setting [24], Leaky Ideal Query model [26], etc[1].

Indeed, as is natural to expect, there are limitations on how much one can achieve using the above paradigm of constructing protocols. A very natural question that arises is *whether there exists a different approach which allows us to construct concurrent secure computation protocols in the plain model without the need of additional output queries?* Moreover, if such a different approach does exist, we know that due to impossibility results [1, 5, 8, 19, 24, 37, 38], there will be some limitations on the scope of its applicability. This leads to some more natural questions. *What all can we achieve using this approach? In particular, can we expand the class of realizable functionalities in the concurrent setting? Can we improve the parameters (e.g. round complexity) of the protocols which exist in the plain model?*

1.1   Our Results

The key contribution of this work is a new way of approaching the problem of concurrent secure computation in the plain model facilitated by recent advances in concurrent non-black box simulation [25]. We give a protocol with non-black box and straightline simulator. Since, very informally, our simulator does not rely on rewinding at all, we are able to avoid the key bottleneck of additional output queries to the ideal functionality during the rewinds.

However, our simulator has to overcome a number of additional obstacles not present in [25]. Note that unlike secure computation, an adversary in concurrent zero-knowledge does not receive any outputs. Dealing with the outputs given to the adversary in each session is a key difficulty we have to overcome.

---

[1] For a detailed survey of these works, see our full version.

In particular, one might think that a straightline simulator for concurrent zero-knowledge should give a concurrently secure computation protocol trivially for all functionalities and in particular for concurrently secure oblivious transfer. Note that this *cannot* be true given unconditional impossibility results for oblivious transfer. For more on such technical hurdles, please refer to the technical overview (Sect. 1.2).

Informally stated, our main theorem is a general positive result for concurrent secure computation as long as the ideal world satisfies our so called *bounded pseudo-entropy condition* (BPC). Very informally, the bounded pseudoentropy condition requires that in the ideal world experiment, the total amount of information learnt by the adversary (via calls to the trusted party) should have "bounded pseudoentropy". The origin of the bounded pseudoentropy condition comes from a conjecture of Goyal [24]. More precisely, the bounded pseudoentropy condition says the following:

Definition 1 (Bounded Pseudoentropy Condition (BPC)). *An ideal world experiment satisfies bounded pseudoentropy condition if there exists $B \in \mathbb{N}$ and a PPT algorithm T such that for all $m = m(n)$ concurrent sessions, for all adversarial input vectors I (where an element of the vector represents the input of the adversary in that session), there exists a set S of possible output vectors such that the following conditions are satisfied*

- *All valid output vectors corresponding to the input vector I of the adversary are contained in S. Observe that for a given I, for different honest party input vectors, the output vectors may be different. We require that any such output vector be contained in S. Furthermore, $|S| \le 2^B$.*
- *For every $O \in S$, $T(I, O) = 1$, and for every $O \notin S$, $T(I, O) = 0$. That is, the set S is efficiently recognizable.*

Intuitively, this condition says the following: The adversary might be scheduling an unbounded polynomial number of sessions and gaining information from each of the outputs obtained. However for any vector of adversarial inputs, the number of possible output vectors is bounded (and hence so is the information that adversary learns). Further note that this condition places a restriction only on the ideal world experiment, which consists of the functionality being computed and the honest party inputs. There is no restriction on the ideal world adversary, which may follow any (possibly unbounded state) polynomial time strategy.

It can be seen that in concurrent zero-knowledge, as well as, in the bounded concurrency setting, the BPC is satisfied. Also note that the class of ideal worlds which satisfy BPC is significantly more general compared to the single input setting of [24]. For a formal proof of this claim, refer to Sect. 2. In our work, we prove the following main theorem.

Theorem 1. *Assume the existence of collision resistant hash functions and constant-round semi-honest oblivious transfer. If the ideal world for the functionality $\mathcal{F}$ satisfies the bounded pseudoentropy condition in Definition 1, then for any constant ϵ, there exists a $O(n^\epsilon)$ round real world protocol Π which securely realizes the ideal world for functionality $\mathcal{F}$.*

To understand the power of our result, a positive result for all ideal worlds satisfying BPC allows us to get the following "concrete" results:

- **Resolving the Bounded Pseudoentropy Conjecture.** Goyal [24] considered the so called "single input setting" and obtained a positive result for many functionalities in the plain model. Goyal further left open the so called bounded pseudoentropy conjecture which if resolved would give a more general and cleaner result (see [24] for the exact statement).

  Our BPC is inspired from this conjecture (and can be seen as one way of formalizing it). Thus, Theorem 1 allows us to resolve the bounded pseudoentropy conjecture in the positive. Our positive result for the BPC subsumes most known positive results for concurrent secure computation in the plain model such as for zero-knowledge [34,45,47], bounded concurrent computation [36,43,44], and the positive results in the single input setting [24].

- **Improving the Round Complexity of Protocols in the Single Input Setting.** The round complexity of the construction of Goyal [24] in the single input setting was a large polynomial depending not only upon the security parameter but also on the length of the input and the nature of the functionality. For example, for concurrent private information retrieval, the round complexity would depend multiplicatively of the number of bits in the database and the security parameter. Our construction only has n^ϵ rounds, where n is the security parameter. Therefore, we obtain a significant qualitative improvement in the round complexity for protocols in the single input setting.

- **Expanding the Class of Realizable Functionalities, and, Getting Blind Signatures.** The blind signature functionality is an interesting case in the paradigm of secure computation both from theoretical as well as practical standpoints. The question of whether concurrent blind signatures (secure as per the ideal/real model simulation paradigm) exist is currently unresolved. Lindell [36,38] showed an impossibility result for concurrent blind signature based on *black-box simulation*. This result has also been used as a motivation to resort to weaker security notions (such as game based security) or setup assumptions in various subsequent works (see e.g., [15,18,20,31,33,41]). We show that a positive result for BPC directly implies a construction of concurrent blind signatures *secure in the plain model as per the standard ideal/real world security notion*. Prior to our work, the only known construction of concurrently secure blind signatures was according to the weaker game based security notion due to Hazay et al. [31].

  This implies that concurrent blind signatures is a "natural" example of a functionality which is impossible to realize using black-box simulation but can be securely realized using non-black box simulation in the concurrent setting.[2] The only previous such example known [29] was for a reactive (and arguably rather contrived) functionality. Another concrete (and related) example of a new functionality that can be directly realized using our techniques is that of a secure verifiable random function.

---

[2] Previous separations between the power of black-box and non-black box simulation are known only if we place additional constraints on the design of the real world protocol (e.g., it should be public coin, or constant rounds, etc.).

It would also be interesting to see what our approach yields in the plain model for different settings and security notions where the previous rewinding based approach has been useful (such as resettable computation, super-polynomial simulation, etc.). We leave that as future work.

1.2 Our Techniques

Our protocol and analysis for the concurrent secure computation is admittedly quite complex and we face a number of hurdles on the way. Below, we try to sketch the main difficulties and our ideas to circumvent them at a high level.

To construct concurrent secure computation, we roughly follow the [21] strategy of first constructing an appropriate zero-knowledge protocol, and then "somehow compiling" a semi honest secure computation protocol using that. In our concurrent setting, in order to avoid the multiple output queries per session, we need a concurrently secure protocol for zero-knowledge with a *straightline* simulator. Recently, the first such protocol was given by Goyal [25] based on non-black box techniques[3].

Another property of the zero-knowledge protocol which is crucial for compilation is *simulation-soundness*. Our first (and arguably smaller) technical hurdle is to construct a simulation-sound version of Goyal's protocol. This is necessary because the simulator would rely on the soundness of the proofs given by the adversary while simulating the proofs where it is acting as the prover. Another issue is that in our protocol for concurrent secure computation, the adversary is allowed to choose the statement proved till a very late stage in the protocol. Hence, we need simulation-soundness to hold even when the statements to prove are being chosen adaptively by the adversary. We note that this issue is somewhat subtle to deal with. Our construction of simulation-sound concurrent zero-knowledge relies on the following ingredients: Goyal's concurrent simulation strategy, a robust non-malleable commitment scheme [35], and a special language to be used in the universal arguments. The final construction along with a description of the main ideas is given in Sect. 3.

The next (and arguably bigger) difficulty is the following. In secure computation, the adversary receives an output in each session (this is unlike the case of zero-knowledge). It turns that it is not clear how to handle these outputs while performing non-black box simulation. Note that some such challenge is inherent in the light of the long list of general impossibility results known [5,38]. Before we describe the challenge faced in detail, it would be helpful to recall how the non-black box techniques based on [2] work at a high level.

---

[3] Before this, all the (fully) concurrent zero-knowledge protocols were based on rewinding techniques, while, the construction of [2] (which had a non-rewinding simulator) worked only in the bounded concurrent setting. The main result in [25] was the first public-coin concurrent zero-knowledge protocol where the non-rewinding nature of the simulation technique was not crucial. However in the current work, we would crucially exploit the fact that the simulation strategy was straightline.

- **Non-black Box Technique.** In each session, the simulator has to commit to a program Π, which has to generate the adversary's random string r in that session. In the transcript between the commitment to Π and r, there may be messages of other sessions, which Π has to regenerate. Even if the program Π consists of the entire state of the simulator and the adversary at the point of the commitment, it runs into a problem in the case of secure computation (where the adversary is getting non-trivial output in each session).
- **Key Challenge.** Note that to reach from the commitment of Π to the message r, the simulator makes use of some external information: namely the outputs it learns by querying the ideal functionality as it proceeds in the simulation. This information, however, is not available with the program Π (since the simulator may query the ideal functionality *after* the program Π was committed to). Also, note that the number of outputs learnt could be any unbounded polynomial. Hence, it is not clear how to regenerate the transcript.

The first obvious solution, which does not work, is to allow the program Π to take inputs of unbounded length. This would allow the simulator to pass all the outputs obtained to the program Π. But now the soundness of the protocol seems to be completely compromised. On the other hand, if Π does not receive all the outputs, it cannot regenerate the transcript!

To resolve this issue, we use the idea of "Oracle programs" due to Deng, Goyal, and Sahai [13]. The program Π, while running, is allowed to make any (polynomially unbounded) number of queries (to be answered by the simulator) as long as the response to each query is information theoretically fixed by the query. The soundness is still preserved: an adversarial prover still cannot communicate any information about the verifier's random string r to Π. However, the program Π can still access a potentially unbounded length string using such an "Oracle interface".

Unfortunately, *the above idea is still not sufficient for our purpose*: the outputs given by the ideal functionality are not fixed given the adversary's input in the session. Here we rely on the fact that we are only considering the ideal worlds which satisfy the bounded pseudoentropy condition. Very roughly, it is guaranteed that the entire output vector has only bounded pseudoentropy (B), given the input of the adversary. Moreover, given the adversary's input vector, all possible output vectors are efficiently testable by the PPT algorithm $\mathcal{T}$. In other words, for every vector of queries, there is only a bounded (although potentially *exponential*) number of response vectors accepted by $\mathcal{T}$. We allow the program Π to make any number of queries such that the response vector is accepted by $\mathcal{T}$. More details regarding our precise language for non-black box simulation may be found in Fig. 1. This idea allows the simulator to supply the entire output vector (learnt from the ideal functionality) to Π while still preserving soundness. The soundness proof relies on the fact that the queries only allow for communication of up to B-bit string to Π, which is still not sufficient for communicating the string r.

Finally, there are additional challenges due to the requirement of straightline extraction. Towards that end, we rely on input indistinguishable computation introduced by Micali, Pass, and Rosen [39]. Challenges also arise with performing hybrid arguments in the setting where the code of the simulator itself is committed (because of non-black box simulation). The full construction along with the main ideas is given in Sect. 4.

Other Related Work: Though Goyal et al. [25] gave the first protocol for concurrent zero-knowledge with a straightline simulator, recently, Chung et al. [12] gave a *constant round* concurrent zero-knowledge protocol for uniform adversaries based on a new assumption of P-certificates, which is also straightline simulatable. Their protocol represents an exciting idea which opens an avenue for getting *constant round* concurrently secure computation protocols (albeit for uniform adversaries only, and, based on a new assumption). We believe that our techniques could also be applicable in constructing concurrent secure computation protocols using the protocol of [12].

2 Concurrently Secure Computation: Our Model

In this section, we begin by giving a brief sketch of our model. For formal description (building upon the model of [38]) of our model, see full version. In this work, we consider a malicious, static and probabilistic polynomial time adversary that chooses whom to corrupt before the execution of the protocol and controls the scheduling of the concurrent executions. Additionally, the adversary can choose the inputs of different sessions adaptively. We denote the security parameter by n. We give a real world/ideal world based security definition. There are k parties $Q_1, Q_2, \ldots, Q_k$, where each party may be involved in multiple sessions with possibly interchangeable roles. In the ideal world, there is a trusted party for computing the desired two-party functionality $\mathcal{F} : \{0,1\}^{r_1} \times \{0,1\}^{r_2} \to \{0,1\}^{s_1} \times \{0,1\}^{s_2}$. Let the total number of executions be $m = m(n)$. Note that there is no a-priori bound on the number of sessions m and the adversary can start any (possibly unbounded) polynomial number of sessions. On the other hand, in the real world there is no trusted party and the two parties involved in a session, say P_1 and P_2, execute a two party protocol Π for computing $\mathcal{F}$. Our security definition requires that any adversary in the real model can be *emulated* by an adversary in the ideal model.

2.1 Our Result and Its Applications

As mentioned in the introduction, our main result (see Theorem 1, Sect. 1.1) is a general positive result for concurrent secure computation as long as the ideal world satisfies the bounded pseudo-entropy condition (Definition 1, Sect. 1.1).

Next, we show that our theorem not only subsumes the positive results of [24] in the single input setting but also improves the round complexity.

Comparing Our Results with [24]. In [24], Goyal showed that if the ideal world satisfies the "key technical property" (KTP), then there exists a real world protocol which securely realizes this ideal world. The key technical property, taken verbatim from [24], is as follows:

Definition 2 (Key Technical Property (Definition 3, [24])). *The key technical property (KTP) of an ideal world experiment requires the existence of a PPT predictor $\mathcal{P}$ satisfying the following conditions. For all sufficiently large n, there exists a bound D such that for all adversaries and honest party inputs,*

$$\left|\left\{j\ \mathcal{P}(\{I[\ell]\}_{\ell\leq j}, \{O[\ell]\}_{\ell<j}) \neq O[j]\right\}\right| < D$$

For the ideal worlds which satisfy KTP, [24] gave a $O(n^3 D^2)$ round secure protocol which realizes the functionality, where D is the parameter in Definition 2.

In our full version, we prove the following lemma:

Lemma 1. *If an ideal world experiment satisfies the key technical property (Definition 2), then it also satisfies the bounded pseudoentropy condition (Definition 1).*

As mentioned before, the round complexity of Goyal [24] is $O(n^3 D^2)$ which is a polynomial in security parameter n as well as D (which depends upon length of single input as well as nature of functionality). Our Theorem 1 and Lemma 1 imply a quantitative and qualitative improvement in round complexity. This leads to lower round protocols for applications like private database search, secure set intersection, computing k^{th} ranked element etc. For details see the full version.

Moreover, [24] only gave a positive result for functionalities with hardness free ideal world, i.e. in the ideal world the trusted party is not required to perform any cryptographic operations. There is no such restriction in our setting. In fact, we show that blind signatures and verifiable random functions satisfy the bounded pseudoentropy condition. More interestingly, they do not satisfy the key technical property. We next describe our results for these functionalities.

Blind Signatures. Blind signatures, introduced by [11], allow users to obtain signatures on messages of their choice without revealing the message being signed to the signer (blindness property). In addition, they also need to satisfy the unforgeability property of the digital signature schemes. In this work, we give the following positive result for concurrent blind signatures.

Theorem 2. *Assume the existence of collision resistant hash functions and constant-round semi-honest oblivious transfer. Then for any constant ϵ, there exists a $O(n^\epsilon)$ round secure protocol which realizes the ideal world for concurrent blind signature functionality.*

We prove this theorem by using unique signatures [22] as the underlying signature scheme and showing that blind signatures satisfy the bounded pseudoentropy condition when the underlying signature scheme is unique. (Note that Lindell's black box impossibility result also holds in this setting.) A signature scheme is said to be *unique* if for each public key and each message, there exists at most one valid signature which verifies.

We can model blind signature as a two party computation between the signer and the user for the circuit for generating signatures. Note that the circuit will have the verification key vk hardcoded. At the end of the protocol, the user outputs a valid signature σ if obtained, and signer always outputs $\perp$. Now we show that this functionality satisfies BPC for $B = 0$ and T algorithm which is same as the signature verification algorithm. Note that if the adversary is playing the role of the user, its output is unique and is completely determined by its input message since vk is fixed by the function being computed. If the adversary is playing the role of the signer, its output is always $\perp$. Hence, set S will contain only one output vector, which is information theoretically fixed by the adversary inputs and the ideal world experiment (which fixes the verification keys for all the sessions). The algorithm T simply verifies the user's signatures w.r.t. corresponding vk and ensures that signer's outputs are $\perp$.

Finally note that blind signatures will not satisfy the key technical property. Consider the case when the adversary is acting as the user in all the sessions. By the unforgeability property of the scheme, any PPT predictor which receives k valid input/output (message/signature) pairs *cannot* predict the signature on the next message with non-negligible probability. Also, note that blind signatures will not satisfy the generalized key technical property discussed in the full version [23] for the same reason.

Verifiable Random Functions. Verifiable random functions (VRFs) were introduced by Micali, Rabin, and Vadhan [40]. They combine the properties of pseudo-random functions with the verifiability property. Intuitively, they are pseudo-random functions with a public key and proofs for verification. Along with pseudo-randomness, they are required to satisfy *uniqueness*, i.e., given the public key, for any input x, there is a unique y which can verify. In this work, we show the following:

Theorem 3. *Assume the existence of collision resistant hash functions and constant-round semi-honest oblivious transfer. Then for any constant ϵ, there exists a $O(n^\epsilon)$ round concurrent real world protocol which realizes the ideal world experiment for verifiable random functions.*

We again prove this theorem by showing that VRFs satisfy BPC for $B = 0$ and T algorithm which is same as verification algorithm. Here, we again rely on the uniqueness property. Finally, note that VRFs too will not satisfy the key technical property due to pseudo-randomness guarantee. For details see the full version.

3 Our Simulation-Sound Non-Black Box Zero-Knowledge Protocol

Constructing a family of polynomially many zero-knowledge protocols which are *simulation-sound* with respect to each other under (unbounded polynomially many) concurrent executions is one of the difficulties in constructing protocols for fully concurrent multi-party computation (MPC). Simulation-soundness, introduced by Sahai [48], means that the soundness of each of the proofs given by the adversary should hold even when the adversary is getting unbounded polynomial number of simulated proofs. To avoid the problem of providing multiple outputs due to a rewinding based simulator for concurrent MPC, we need to construct simulation-sound zero-knowledge protocols which are straight-line simulatable. Note that Pass [43] also gave a construction of polynomially many protocols which are concurrent zero-knowledge and simulation-sound w.r.t. each other in the restricted setting of bounded concurrency. In this work, we construct such simulation-sound zero-knowledge protocols building upon the non-black box public coin concurrent zero-knowledge protocol of Goyal [25].

First, we give a brief overview of [25]. Some of the text has been taken verbatim from [25]. One of the main technical ideas in [25] is to have $N = n^\epsilon$ non-black box slots, for any constant ϵ (each consisting of a commitment to a machine and a verifier challenge string). Each slot is followed by a universal argument (UA) execution. Any of the UA's in a session may be picked for simulation. If a UA is picked for simulation, to make the analysis go through, the simulator could choose of any of the previously completed slots and prefer the slots which are computationally lighter. In a UA execution, the prover proves that in one of the completed slots, the machine committed successfully outputs the verifier challenge string. Other main idea was to have encrypted executions of the UAs (using its public coin property) to hide the location of the convincing UA executions in the transcript. Finally there is an execution of a witness-indistinguishable argument of knowledge (WIAOK), where the prover proves that either the statement $x \in L$ or there exists a decryption of one of the UAs which is accepting. In the subsequent discussion, we will refer to the part of the protocol with non black box slots and encrypted UAs as the *preamble phase* and last phase as the WIAOK phase.

Two main ideas are required to transform the above described protocol into simulation-sound zero-knowledge protocols, which can then be used to construct protocols for concurrent MPC. Firstly, observe that unless the parties have identities it is impossible to construct a simulation-sound protocol because a man-in-the-middle attack cannot be prevented. Hence, we focus on a setting where each party has a unique identity of n bits. Let NMCom be a k-robust identity-based non-malleable commitment scheme. Now, after the preamble phase of the protocol, the prover with identity id gives a non-malleable commitment to the witness under its identity id. More precisely, the prover, having witness w to $x \in L$, gives a commitment $c = \mathsf{NMCom}(w)$ under his identity id. In the final WIAOK phase, the prover proves that either there exists a w such that $c = \mathsf{NMCom}(w)$ and $w \in R_L(x)$ or one of the UA executions was convincing. We will be able to prove the simulation-soundness of our protocol using the non-malleability and k-robustness of NMCom. Note that (as described later) our protocol will be simulation-sound even when the adversary is allowed to choose the statements to be proven adaptively till the point when he gives this non-malleable commitment.

Secondly, in our UA executions we will use a special generalized language Λ (see Fig. 1) for the UA executions. Here, along with [13] kind of queries decommit($\cdot$) whose response is information theoretically fixed given the query itself, we will also have a second kind of queries, which we will denote by output($\cdot$). Note that though the responses of these queries is not information theoretically fixed, they have a bounded pseudoentropy. Next, we give some intuition about the use of these oracle queries.

The language Λ is defined w.r.t. an algorithm T and bound B with the following property: For any vector $\boldsymbol{x}$ (of possibly unbounded polynomial length) there exists a set S containing vectors $\boldsymbol{y}$ such that $|S| \leq 2^B$ and for all $\boldsymbol{y}' \notin S, T(\boldsymbol{x}, \boldsymbol{y}') = 0$. Now the language Λ is defined as follows:

We say that $(h, z, r) \in \Lambda$ if there exists an oracle program Π s.t. $z = \text{COM}(h(\Pi))$ and there exist strings $y_1 \in \{0, 1\}^{\leq |r| - B - n}$, $y_2 \in \{0, 1\}^{\leq n^{\log \log n}}$ and $y_3 \in \{0, 1\}^{\leq n^{\log \log n}}$ with the following properties. The oracle program Π takes y_1 as input and outputs r within $n^{\log \log n}$ steps. Program Π can make two kinds of calls to the oracle

1. Produce a query of the form decommit(str) and expecting (r) with str $= \text{COM}(r)$ in return such that the tuple (str, r) is guaranteed to be found in the string y_2 (as per a suitable encoding of y_2). Thus, such oracle calls by Π can be answered using y_2.
2. Produce a query of the form output(x) and expecting y in return, such that the tuple (x, y) is guaranteed to be found in the string y_3 (as per a suitable encoding of y_3). Thus, such oracle calls by Π can be answered using y_3.

If the program Π makes a query that cannot be answered by strings y_2 or y_3, Π aborts and we have that $(h, z, r) \notin \Lambda$. Also, let $\boldsymbol{x}$ denote the vector containing all the output($\cdot$) queries made by Π (throughout its execution) and $\boldsymbol{y}$ be the corresponding responses, then Π aborts if $T(\boldsymbol{x}, \boldsymbol{y}) = 0$ and we have that $(h, z, r) \notin \Lambda$.

Fig. 1. Our language for zero-knowledge with non-black-box simulation

Intuition Behind the Oracle Queries output($\cdot$) in Language Λ. The algorithm T and the bound B are introduced to capture the information learnt by the adversary. When only concurrent sessions of zero-knowledge are running, there is no information passed to the adversary, hence we can have T to reject all outputs and still be able to simulate the view of the adversary. This notion will be important for the concurrent executions of multiparty computation because the adversary learns non-trivial information from calls to the trusted party. In particular, it learns the output of the function in each session. We will use the oracle queries output($\cdot$) to communicate the information learnt from the trusted party to the adversary in the ideal world. But still to get our positive result, we will need to bound the amount of information learnt by the adversary. The bound B will be the number of bits of information passed on to the adversary. This is intuitively captured by the condition that there are only 2^B vectors of oracle responses which might be accepted by T. Looking ahead, the description of T will depend on the functionality being computed.

Formal Protocol Description. Let $\mathrm{COM}(\cdot)$ denote a non-interactive perfectly binding commitment scheme. Whenever we need to be explicit about the randomness, we denote by $\mathrm{COM}(s; r)$ a commitment to a string s computed with randomness r. Unless stated otherwise, all commitments in the protocol are executed using this commitment scheme. Let NMCom be the k-robust non-malleable commitment scheme, where k is a parameter computed later. Let $\mathsf{len} = n^2 + B + \eta$, where B and η are parameters computed later.

The common input to P and V is the security parameter n. The input to P is x in the language $L \in NP$, and a witness w to $x \in L$. Let id be the n bit identity of the prover. Our protocol $\langle P, V \rangle$ or $c\mathcal{ZK}_{\mathsf{id}}$, where id is the identity of the prover, proceeds as follows: Parts of the protocol have been taken verbatim from [25].

1. The verifier V chooses a random collision resistant hash function h from a function family $\mathcal{H}$ and sends it to P.
2. For $i \in [n^6]$, the protocol proceeds as follows:[4]
   - The prover P computes $z_i = \mathrm{COM}(h(0))$ and sends it to V.
   - The verifier V selects a challenge string $r_i \xleftarrow{\$} \{0,1\}^{\mathsf{len}}$ and sends it to the prover P. The above two messages (consisting of the prover commitment and the verifier challenge) are referred to as a "slot".
   - The prover P and the verifier V will now start a three-round public coin universal argument (of knowledge) [4] where P proves to V that *there exists $j \leq i$, s.t., $\tau_j (= (h, z_j, r_j))$ is in the language Λ* (see Fig. 1).
     The three messages of this UA protocol are called as the *first UA message*, *verifier UA challenge*, and, the *last UA message*.
     Observe that the UA does not just refer to the slot immediately preceding it but rather has a choice of using *any of the slots that have completed* in the protocol so far.
   - The prover computes the first UA message and sends a *commitment* to this message to the verifier. The honest prover will simply commit to a random string of appropriate size.
   - The verifier now sends the UA challenge message.
   - The prover computes the last UA message and again sends only a *commitment* to this message to the verifier. The honest prover will simply commit to a random string of appropriate size.
3. The prover declares the statement $x \in L$ and commits to the witness w using the non-malleable commitment scheme NMCom under prover's identity id. *Note that a cheating prover can adaptively choose the statement x here.*
4. Finally, the prover proves the following statement to V using WIAOK
   1. The value committed to in Step 3 is a value w such that it is a valid witness to $x \in L$, (i.e. $w \in R_L(x)$), *or*
   2. There exists i such that the i-th UA execution was "convincing". That is, there exists an $i \in [n^6]$ such that there exists an opening to the prover first and last UA messages such that an honest verifier would have accepted the transcript of the UA execution.

---

[4] Note that the round complexity of our protocol can be made n^ϵ using standard techniques involving "scaling down" the security parameter.

An honest prover simply commits to the witness for $x \in L$ in Step 3 and uses the first part of the statement to complete the witness-indistinguishable argument of knowledge protocol.

Observe that a witness to the second part of the above statement would be the opening of the commitments to the UA first and last messages. Hence, the size of the witness is fixed and depends only upon the communication complexity of the 3-round UA system being used.

Remark 1. We call the Steps 1 and 2 of the protocol as non-black box *preamble*, step 3 as the NMCOM phase and step 4 as the WIAOK phase.

Parameter k. We set k to be the round complexity of WIAOK. Hence, we set $k = 3$.

Parameter B. Note that the parameter B in len is same as the one in Fig. 1, i.e. the parameter specified for algorithm T in the description of language Λ.

Setting the Parameter η. Let η be the sum of the following: prover's maximum communication complexity in different primitives used in the protocol described above, and communication complexity of NMCom. More precisely, we set

$$\eta = \max(c_z, c_{UA1}, c_{UA2}, c_{WIAOK}, c_{NMCom,S}) + c_{NMCom,R},$$

where c_z is the length of the slot begin message z, c_{UA1} is the length of the UA first message, c_{UA2} is the length of the UA last message, c_{WIAOK} is the prover's communication complexity in the final WIAOK execution, $c_{NMCom,S}$ is the sender's communication complexity in NMCom and $c_{NMCom,R}$ is the receiver's communication complexity in NMCom.

Looking ahead, (very informally) while proving the simulation-soundness of the above protocol, different parts of the protocol will be taken externally and NMCom given by the adversary will be exposed to an external receiver, etc. Hence, different parts of the protocol will be given externally to the machine committed by the simulator as part of the string y_1 in Λ.

Note that the entire $\langle P, V \rangle$ protocol is run w.r.t. to language Λ having a specific algorithm T and bound B. We will prove that the security properties hold for any such T and bound B when η is chosen as above. Next, we prove the soundness of the protocol for any fixed value of B. Then we will prove the simulation-soundness of the protocol. Our ZK simulator will not use the oracle queries of the type output($\cdot$). Later on our MPC simulator will make a non-trivial use of these oracle queries.

The proof of security of simulation-sound non-black box zero-knowledge protocol proceeds along the lines discussed in the introduction (see Sect. 1.2). We give a detailed formal proof of security in the full version.

4   Concurrently Secure Computation: Our Protocol

In this section, we will describe our protocol Σ for concurrently secure computation for ideal world experiments which satisfy the bounded pseudoentropy condition (Definition 1) for some parameter $B \in \mathbb{N}$ and algorithm T.

Common input: Let $\text{COM}(\cdot)$ be a non-interactive perfectly binding commitment scheme. The functionality $f_{\text{com}_1,\text{com}_2}$ is parameterized by two commitments com_1 and com_2 under $\text{COM}(\cdot)$, which are the common inputs to the functionality and the parties P_1^{iic} and P_2^{iic}.
Inputs: Let (z_1, td_1) and (z_2, td_2) be the inputs of P_1^{iic} and P_2^{iic} respectively.

Computation: Party P_1^{iic} sends its input (z_1, td_1) and party P_2^{iic} sends its input (z_2, td_2) to the trusted functionality $f_{\text{com}_1,\text{com}_2}$.
If td_1 is a *valid* opening of com_1 to bit 1, $f_{\text{com}_1,\text{com}_2}$ sends z_2 to P_1^{iic}, otherwise it sends $\perp$.
Similarly, if td_2 is a *valid* opening of com_2 to bit 1, $f_{\text{com}_1,\text{com}_2}$ sends z_1 to P_2^{iic}, otherwise it sends $\perp$.

Fig. 2. The functionality $f_{\text{com}_1,\text{com}_2}$

Our Construction. In order to describe our construction, we first recall the notation associated with the primitives that we use in our protocol. Let $\text{COM}(\cdot)$ denote the commitment function of a non-interactive perfectly binding commitment scheme. Let $\langle P, V \rangle$ denote the simulation-sound non-black box concurrent zero-knowledge protocol as described in Sect. 3 with length of challenge strings modified to be $\text{len} = n^2 + B + \theta$, where θ is a parameter computed later. Let $\langle P_1^{\text{iic}}, P_2^{\text{iic}} \rangle$ be the constant round protocol for input indistinguishable computation [17,39]. Let NMCom be the k-robust non-malleable commitment scheme, where k is a parameter computed later. Further, let $\langle P_{\text{wi}}, V_{\text{wi}} \rangle$ denote a witness indistinguishable argument and let $\langle P_1^{\text{sh}}, P_2^{\text{sh}} \rangle$ denote a constant round *semi-honest* two party computation protocol $\langle P_1^{\text{sh}}, P_2^{\text{sh}} \rangle$ that securely computes $\mathcal{F}$ in the stand-alone setting as per the standard definition of secure computation.

Let P_1 and P_2 be two parties with inputs x_1 and x_2. Let n be the security parameter. Protocol $\Sigma = \langle P_1, P_2 \rangle$ proceeds as follows:

I. Non-Black Box Simulation Phase

1. $P_1 \Rightarrow P_2$: P_1 and P_2 engage in the *preamble* phase of $\langle P, V \rangle$ where P_1 is the prover. Next, in the NMCOM phase, P_1 creates a non-malleable commitment com_1 to bit 0, i.e. $\text{com}_1 = \text{NMCom}(0)$ and sends com_1 to P_2. P_1 and P_2 now engage in the WIAOK phase where P_1 proves that either (1) com_1 is a commitment to 0 , or (2) there exists i such that the i-th UA execution in the preamble phase was "convincing".

2. $P_2 \Rightarrow P_1$: P_2 now acts symmetrically. P_1 and P_2 engage in the *preamble* phase of $\langle P, V \rangle$ where P_2 is the prover. Next, P_2 creates a non-malleable commitment com_2 to bit 0, i.e. $\text{com}_2 = \text{NMCom}(0)$ to bit 0 and sends com_2 to P_1. P_1 and P_2 now engage in the WIAOK phase where P_2 proves that either (1) com_2 is a commitment to 0 , or (2) there exists i such that the i-th UA execution in the preamble phase was "convincing".

Informally speaking, the purpose of this phase is to aid the simulator in obtaining a "trapdoor" to be used during the simulation of the other two phases of the protocol.

II. Input Indistinguishable Computation Phase. Intuitively speaking, in this phase, the parties "commit" to their inputs and random coins (to be used in the final secure computation phase) by engaging in a execution of $\langle P_1^{\text{iic}}, P_2^{\text{iic}} \rangle$ for the functionality $f_{\text{com}_1, \text{com}_2}$ described in Fig. 2. More precisely, P_1 and P_2 engage in an execution of $\langle P_1^{\text{iic}}, P_2^{\text{iic}} \rangle$ for the functionality $f_{\text{com}_1, \text{com}_2}$ where P_1 plays the role of P_1^{iic}, while P_2 plays the role of P_2^{iic} as follows:

1. P_1 first samples a random string r_1 (of appropriate length, to be used as P_1's randomness in the execution of $\langle P_1^{\text{sh}}, P_2^{\text{sh}} \rangle$ in Phase III) and uses input $z_1 = x_1 \| r_1$ and $\text{td}_1 = \bot$ in execution of $\langle P_1^{\text{iic}}, P_2^{\text{iic}} \rangle$ for $f_{\text{com}_1, \text{com}_2}$.
2. $P_2 \Rightarrow P_1 : P_2$ now acts symmetrically. P_2 first samples a random string r_2 (of appropriate length, to be used as P_2's randomness in the execution of $\langle P_1^{\text{sh}}, P_2^{\text{sh}} \rangle$ in Phase III) and uses input $z_2 = x_2 \| r_2$ and $\text{td}_2 = \bot$ in execution of $\langle P_1^{\text{iic}}, P_2^{\text{iic}} \rangle$ for $f_{\text{com}_1, \text{com}_2}$.

Informally speaking, the purpose of this phase is to aid the simulator in extracting the adversary's input and randomness with the help of the trapdoor obtained in the previous phase. As we will show later, an adversary will never be able to input a valid trapdoor.

III. Final Secure Computation Phase.[5] In this phase, P_1 and P_2 engage in an execution of $\langle P_1^{\text{sh}}, P_2^{\text{sh}} \rangle$ where P_1 plays the role of P_1^{sh}, while P_2 plays the role of P_2^{sh}. Since $\langle P_1^{\text{sh}}, P_2^{\text{sh}} \rangle$ is secure only against semi-honest adversaries, parties first run a coin-flipping protocol to enforce that the coins of each party are truly random. We then compile the semi-honest $\langle P_1^{\text{sh}}, P_2^{\text{sh}} \rangle$ with $\langle P_{\text{wi}}, V_{\text{wi}} \rangle$ to ensure correct behavior on part of each party. More precisely, after sending each protocol message, a party also gives a proof using $\langle P_{\text{wi}}, V_{\text{wi}} \rangle$ that the message generated is consistent with the transcript so far and the input used in the previous phase. More precisely, this phase proceeds as follows:

1. $P_1 \leftrightarrow P_2 : P_1$ samples a random string r_2' (of same length as r_2) and sends it to P_2. Similarly, P_2 samples a random string r_1' (of same length as r_1) and sends it to P_1. Let $r_1'' = r_1 \oplus r_1'$ and $r_2'' = r_2 \oplus r_2'$. Now, r_1'' and r_2'' are the random coins that P_1 and P_2 will use during the execution of $\langle P_1^{\text{sh}}, P_2^{\text{sh}} \rangle$.
2. Let q be the number of rounds in $\langle P_1^{\text{sh}}, P_2^{\text{sh}} \rangle$, where one round consists of a message from P_1^{sh} followed by a reply from P_2^{sh}. Let transcript $T_{1,j}$ (resp., $T_{2,j}$) be defined to contain all the messages exchanged between P_1^{sh} and P_2^{sh} before the point P_1^{sh} (resp., P_2^{sh}) is supposed to send a message in round j. For $j = 1, \ldots, q$:
   (a) $P_1 \Rightarrow P_2 :$ Compute $\beta_{1,j} = P_1^{\text{sh}}(T_{1,j}, x_1, r_1'')$ and send it to P_2. P_1 and P_2 now engage in an execution of $\langle P_{\text{wi}}, V_{\text{wi}} \rangle$, where P_1 proves the following statement:
      i. *either* there exist values $\hat{x}_1$, $\hat{r}_1$ and $\hat{\text{td}}_1$ such that (a) the $f_{\text{com}_1, \text{com}_2}$ is *valid* with respect to the value $\hat{z}_1 = \hat{x}_1 \| \hat{r}_1$ and $\hat{\text{td}}_1$ and (b) $\beta_{1,j} = P_1^{\text{sh}}(T_{1,j}, \hat{x}_1, \hat{r}_1 \oplus r_1')$
      ii. *or*, the non-malleable commitment com_1 is a commitment to bit 1.
   (b) $P_2 \Rightarrow P_1 : P_2$ now acts symmetrically.

---

[5] Part of the text in this phase has been taken verbatim from [17].

This completes the description of the protocol $\Sigma = \langle P_1, P_2 \rangle$. Note that Π consists of several instances of WI, such that the proof statement for each WI instance consists of two parts. Specifically, the second part of the statement states that prover committed to bit 1 in the non-black box simulation phase. In the sequel, we will refer to the second part of the proof statement as the *trapdoor* condition. Further, we will call the witness corresponding to the first part of the statement as *real* witness and that corresponding to the second part of the statement as the *trapdoor* witness.

Setting the Parameters k and θ. We will set k to be the maximum round complexity among UA, WIAOK, $\langle P_1^{\text{iic}}, P_2^{\text{iic}} \rangle$ and $\langle P_1^{\text{sh}}, P_2^{\text{sh}} \rangle$. We will set θ to be the sum of the following: a party's maximum communication complexity in different primitives used in the protocol described above (excluding when it acts as a verifier in $\langle P, V \rangle$), and communication complexity of NMCom. More precisely,

$$\theta = \max(c_z, c_{\text{UA1}}, c_{\text{UA2}}, c_{\text{WIAOK}}, c_{\text{WI}}, c_{\text{IIC}}, c_{\text{TPC}}, c_{\text{NMCom},S}) + c_{\text{NMCom},R},$$

where c_z is the length of the message z (the slot begin message), c_{UA1} is the length of the UA first message, c_{UA2} is the length of the UA last message, c_{WIAOK} is the prover's communication complexity in the final WIAOK execution, c_{WI} is the prover's communication complexity in WI, c_{IIC} is the communication complexity of any party in $\langle P_1^{\text{iic}}, P_2^{\text{iic}} \rangle$, c_{TPC} is the total communication complexity of the semi-honest two party computation $\langle P_1^{\text{sh}}, P_2^{\text{sh}} \rangle$ for the functionality $\mathcal{F}$, $c_{\text{NMCom},S}$ is the sender's communication complexity in NMCom and $c_{\text{NMCom},R}$ is the receiver's communication complexity in NMCom. Looking ahead, while proving the security of the above protocol, different parts of the protocol will be taken externally and NMCom given by the adversary will be exposed to external receiver, etc. Hence, all of these will be given externally to the machine committed by the simulator as part of the string y_1 in Λ.

The proof of Theorem 1 proceeds along the lines discussed in the introduction (see Sect. 1.2). For a complete proof refer to the full version of the paper.

References

1. Agrawal, S., Goyal, V., Jain, A., Prabhakaran, M., Sahai, A.: New impossibility results for concurrent composition and a non-interactive completeness theorem for secure computation. In: Safavi-Naini, R., Canetti, R. (eds.) CRYPTO 2012. LNCS, vol. 7417, pp. 443–460. Springer, Heidelberg (2012)
2. Barak, B.: How to go beyond the black-box simulation barrier. In: FOCS (2001)
3. Barak, B., Canetti, R., Nielsen, J.B., Pass, R.: Universally composable protocols with relaxed set-up assumptions. In: FOCS (2004)
4. Barak, B., Goldreich, O.: Universal arguments and their applications. In: IEEE Conference on Computational Complexity (2002)
5. Barak, B., Prabhakaran, M., Sahai, A.: Concurrent non-malleable zero knowledge. In: FOCS (2006)
6. Barak, B., Sahai, A.: How to play almost any mental game over the net - concurrent composition via super-polynomial simulation. In: FOCS (2005)

7. Canetti, R., Kilian, J., Petrank, E., Rosen, A.: Black-box concurrent zero-knowledge requires $\tilde{\Omega}$ (log n) rounds. In: STOC (2001)
8. Canetti, R., Kushilevitz, E., Lindell, Y.: On the limitations of universally composable two-party computation without set-up assumptions. In: Biham, E. (ed.) EUROCRYPT 2003, vol. 2656, pp. 68–86. Springer, Heidelberg (2003)
9. Canetti, R., Lin, H., Pass, R.: Adaptive hardness and composable security from standard assumptions. In: FOCS (2010)
10. Canetti, R., Lindell, Y., Ostrovsky, R., Sahai, A.: Universally composable two-party and multi-party secure computation. In: STOC (2002)
11. Chaum, D.: Blind signatures for untraceable payments. In: Chaum, D., Rivest, R.L., Sherman, A.T. (eds.) Blind Signatures for Untraceable Payments, pp. 199–203. Springer, New York (1982)
12. Chung, K.M., Lin, H., Pass, R.: Constant-round concurrent zero knowledge from p-certificates. In: FOCS (2013)
13. Deng, Y., Goyal, V., Sahai, A.: Resolving the simultaneous resettability conjecture and a new non-black-box simulation strategy. In: FOCS (2009)
14. Dwork, C., Naor, M., Sahai, A.: Concurrent zero-knowledge. In: STOC (1998)
15. Fischlin, M.: Round-optimal composable blind signatures in the common reference string model. In: Dwork, C. (ed.) CRYPTO 2006. LNCS, vol. 4117, pp. 60–77. Springer, Heidelberg (2006)
16. Garay, J.A., MacKenzie, P.D.: Concurrent oblivious transfer. In: FOCS (2000)
17. Garg, S., Goyal, V., Jain, A., Sahai, A.: Concurrently secure computation in constant rounds. In: Pointcheval, D., Johansson, T. (eds.) EUROCRYPT 2012. LNCS, vol. 7237, pp. 99–116. Springer, Heidelberg (2012)
18. Garg, S., Gupta, D.: Efficient round optimal blind signatures. In: Nguyen, P.Q., Oswald, E. (eds.) EUROCRYPT 2014. LNCS, vol. 8441, pp. 477–495. Springer, Heidelberg (2014)
19. Garg, S., Kumarasubramanian, A., Ostrovsky, R., Visconti, I.: Impossibility results for static input secure computation. In: Safavi-Naini, R., Canetti, R. (eds.) CRYPTO 2012. LNCS, vol. 7417, pp. 424–442. Springer, Heidelberg (2012)
20. Garg, S., Rao, V., Sahai, A., Schröder, D., Unruh, D.: Round optimal blind signatures. In: Rogaway, P. (ed.) CRYPTO 2011. LNCS, vol. 6841, pp. 630–648. Springer, Heidelberg (2011)
21. Goldreich, O., Micali, S., Wigderson, A.: How to play any mental game or a completeness theorem for protocols with honest majority. In: STOC (1987)
22. Goldwasser, S., Ostrovsky, R.: Invariant signatures and non-interactive zero-knowledge proofs are equivalent. In: Brickell, E.F. (ed.) CRYPTO 1992. LNCS, vol. 740, pp. 228–245. Springer, Heidelberg (1993)
23. Goyal, V.: Positive results for concurrently secure computation in the plain model. IACR Cryptology ePrint Archive 2011 (2011)
24. Goyal, V.: Positive results for concurrently secure computation in the plain model. In: FOCS (2012)
25. Goyal, V.: Non-black-box simulation in the fully concurrent setting. In: STOC (2013)
26. Goyal, V., Gupta, D., Jain, A.: What information is leaked under concurrent composition? In: Canetti, R., Garay, J.A. (eds.) CRYPTO 2013, Part II. LNCS, vol. 8043, pp. 220–238. Springer, Heidelberg (2013)
27. Goyal, V., Jain, A.: On concurrently secure computation in the multiple ideal query model. In: Johansson, T., Nguyen, P.Q. (eds.) EUROCRYPT 2013. LNCS, vol. 7881, pp. 684–701. Springer, Heidelberg (2013)

28. Goyal, V., Jain, A., Ostrovsky, R.: Password-authenticated session-key generation on the internet in the plain model. In: Rabin, T. (ed.) CRYPTO 2010. LNCS, vol. 6223, pp. 277–294. Springer, Heidelberg (2010)
29. Goyal, V., Maji, H.K.: Stateless cryptographic protocols. In: FOCS (2011)
30. Goyal, V., Sahai, A.: Resettably secure computation. In: Joux, A. (ed.) EUROCRYPT 2009. LNCS, vol. 5479, pp. 54–71. Springer, Heidelberg (2009)
31. Hazay, C., Katz, J., Koo, C.-Y., Lindell, Y.: Concurrently-secure blind signatures without random oracles or setup assumptions. In: Vadhan, S.P. (ed.) TCC 2007. LNCS, vol. 4392, pp. 323–341. Springer, Heidelberg (2007)
32. Katz, J.: Universally composable multi-party computation using tamper-proof hardware. In: Naor, M. (ed.) EUROCRYPT 2007. LNCS, vol. 4515, pp. 115–128. Springer, Heidelberg (2007)
33. Kiayias, A., Zhou, H.-S.: Concurrent blind signatures without random oracles. In: De Prisco, R., Yung, M. (eds.) SCN 2006. LNCS, vol. 4116, pp. 49–62. Springer, Heidelberg (2006)
34. Kilian, J., Petrank, E.: Concurrent and resettable zero-knowledge in poly-loalgorithm rounds. In: STOC (2001)
35. Lin, H., Pass, R.: Non-malleability amplification. In: STOC (2009)
36. Lindell, Y.: Bounded-concurrent secure two-party computation without setup assumptions. In: STOC (2003)
37. Lindell, Y.: General composition and universal composability in secure multi-party computation. In: FOCS (2003)
38. Lindell, Y.: Lower bounds and impossibility results for concurrent self composition. J. Cryptology 21(2), 200–249 (2008)
39. Micali, S., Pass, R., Rosen, A.: Input-indistinguishable computation. In: FOCS (2006)
40. Micali, S., Rabin, M.O., Vadhan, S.P.: Verifiable random functions. In: FOCS (1999)
41. Okamoto, T.: Efficient blind and partially blind signatures without random oracles. In: Halevi, S., Rabin, T. (eds.) TCC 2006. LNCS, vol. 3876, pp. 80–99. Springer, Heidelberg (2006)
42. Pass, R.: Simulation in quasi-polynomial time, and its application to protocol composition. In: Biham, E. (ed.) EUROCRYPT 2003, vol. 2656, pp. 160–176. Springer, Heidelberg (2003)
43. Pass, R.: Bounded-concurrent secure multi-party computation with a dishonest majority. In: STOC (2004)
44. Pass, R., Rosen, A.: Bounded-concurrent secure two-party computation in a constant number of rounds. In: FOCS (2003)
45. Prabhakaran, M., Rosen, A., Sahai, A.: Concurrent zero knowledge with logarithmic round-complexity. In: FOCS (2002)
46. Prabhakaran, M., Sahai, A.: New notions of security: achieving universal composability without trusted setup. In: STOC (2004)
47. Richardson, R., Kilian, J.: On the concurrent composition of zero-knowledge proofs. In: Stern, J. (ed.) EUROCRYPT 1999. LNCS, vol. 1592, pp. 415–431. Springer, Heidelberg (1999)
48. Sahai, A.: Non-malleable non-interactive zero knowledge and adaptive chosen-ciphertext security. In: FOCS (1999)
49. Yao, A.C.C.: How to generate and exchange secrets (extended abstract). In: FOCS (1986)

Concurrent Secure Computation with Optimal Query Complexity

Ran Canetti[1,2], Vipul Goyal[3]([✉]), and Abhishek Jain[4]

[1] Boston University, Boston, USA
canetti@bu.edu
[2] Tel Aviv University, Tel Aviv, Israel
[3] Microsoft Research, Bangalore, India
vipul@microsoft.com
[4] Johns Hopkins University, Baltimore, USA
abhishek@cs.jhu.edu

Abstract. The multiple ideal query (MIQ) model [Goyal, Jain, and Ostrovsky, Crypto'10] offers a relaxed notion of security for concurrent secure computation, where the simulator is allowed to query the ideal functionality *multiple times per session* (as opposed to just once in the standard definition). The model provides a quantitative measure for the degradation in security under concurrent self-composition, where the degradation is measured by the number of ideal queries. However, to date, all known MIQ-secure protocols guarantee only an overall *average* bound on the number of queries per session throughout the execution, thus allowing the adversary to potentially fully compromise some sessions of its choice. Furthermore, [Goyal and Jain, Eurocrypt'13] rule out protocols where the simulator makes only an adversary-independent constant number of ideal queries per session.

We show the first MIQ-secure protocol with worst-case per-session guarantee. Specifically, we show a protocol for any functionality that matches the [GJ13] bound: The simulator makes only a *constant* number of ideal queries in *every* session. The constant depends on the adversary but is independent of the security parameter.

As an immediate corollary of our main result, we obtain the first password authenticated key exchange (PAKE) protocol for the fully concurrent, multiple password setting in the standard model with no set-up assumptions.

1 Introduction

General feasibility results for secure computation were established nearly three decades ago in the seminal works of [14,33]. However, these results only promise

R. Canetti — Supported by the Check Point Institute for Information Security, ISF grant 1523/14, and NSF Frontier CNS 1413920 and 1218461 grants.

A. Jain — Work done in part while visiting Microsoft Research, India, and at Boston University and MIT, where the author was supported in part by NSF 1218461 and DARPA FA8750-11-2-0225. Presently supported in part by a DARPA Safeware grant.

© International Association for Cryptologic Research 2015
R. Gennaro and M. Robshaw (Eds.): CRYPTO 2015, Part II, LNCS 9216, pp. 43–62, 2015.
DOI: 10.1007/978-3-662-48000-7_3

security for a protocol if it is executed in isolation, "unplugged" from any network activity. In particular, these results are not suitable for the Internet setting where multiple protocol executions may occur *concurrently* under the control of a common adversary.

A Brief History of Concurrent Security. Towards that end, an ambitious effort to understand and design concurrently secure protocols kicked into gear with early works such as [10,15], and later the study of the *concurrent zero knowledge* setting [7,11,23,30,32]. For other functionalities and in more general settings, however, far-reaching impossibility results were established [1,3,6,8,13, 18,24]. These results refer to the "plain model" where the participating parties have no trusted set-up, and hold even if the parties have access to pairwise authenticated communication and a broadcast channel.

Two main lines of research have emerged in order to circumvent these impossibility results. The first concerns with the use of *trusted setup assumptions* such as a common random string, strong public key infrastructure or tamper-proof hardware tokens (see, e.g. [2,5,22]).

The second line of research is dedicated to the study of weaker security definitions that allow for positive results in the plain model, without additional trust assumptions. The most notable examples of this include security w.r.t. super-polynomial time simulation [4,9,12,28,31] and input-indistinguishable computation [12,26]. One main drawback in this line of research is that it is not always clear by "how much" is the definition of security relaxed, or in other words "how much security" is being lost due to concurrent attacks.

The Multiple Ideal Query Model and Its Applications. The multiple ideal query model (or, the MIQ model in short) of Goyal, Jain and Ostrovsky [21] takes a different approach to the problem of quantifying the security loss. In this model, the simulator is allowed to query the ideal functionality *multiple times per session* (as opposed to just once in the standard definition). On the technical side, allowing the simulator multiple queries indeed facilitates proofs of security in a concurrent setting. On the conceptual side, this model allows for a natural quantification of the "security loss" incurred by concurrent attack: the more ideal queries, the weaker the security guarantee. Furthermore, the effect of multiple ideal queries strongly depends on the task at hand, thus allowing for more fine-tuned notions of security for a given problem or setting.

One functionality where this approach proved very effective is that of password-based key exchange (namely the two-party function that outputs a secret random value to both parties if the inputs provided by the two parties are equal). When the number of queries made by the simulator per session is a constant, the security guarantees of the MIQ model actually imply fully concurrent password-based authenticated key exchange (see [16,17,21]). This fact was exploited by Goyal et al. [21] to get the first concurrent PAKE in the plain model — albeit with the significant restriction that the *same* password is to be used as input in every session. This restriction results from a weakness in their modeling and analysis - a weakness that we overcome in this work.

The Central Question: How Many Queries? So, how to best bound the number of ideal queries made by the simulator? Intuitively, if we allow a large number of queries, then the security guarantee may quickly degrade and become meaningless; in particular, if enough queries are allowed, then the adversary may be able to completely learn the inputs of the honest parties. On the other hand, if the number of allowed queries is very small (say only $1 + \epsilon$ per session) then the security guarantee is very close to that of the standard definition.

To exemplify this further, consider 1-out-of-m OT. Here, as long as λ, the simulator's query complexity, is smaller than m, MIQ provides meaningful security which degrades gracefully with λ. More generally, the remaining security for any session i in concurrently secure computation of function f is proportional to the "level of unlearnability" of $f(\cdot, x_i)$ after q queries, where x_i is the secret input of the honest party in session i. Password-based key exchange is an extreme case of an unlearnable function. Ideally, we would like to bring λ as close as possible to 1.

Prior Work: Average Case vs. Worst Case Guarantees. The best positive result in the MIQ model is due to Goyal, Gupta, and Jain [19] (improving upon [21]). They provide a construction where the number of ideal queries in a session are $(1 + \frac{\log^6 n}{n})$, where n is the security parameter. However, this is only an *average-case* guarantee over the sessions that provides very weak security. In particular, it does not preclude the ideal adversary from making an arbitrarily large number of queries in some chosen sessions (while keeping the number of queries low in the other sessions). In cases of interest, such as the PAKE functionality or the above oblivious polynomial evaluation functionality, this means that the security in some sessions may be *completely compromised*!

Furthermore, Goyal and Jain [20] recently proved an unconditional lower bound on the number of ideal queries per session. Specifically, they show that there exists a two-party functionality that cannot be securely realized in the MIQ model with any (adversary independent) constant number of ideal queries per session. A natural and important question is thus what is the best worst-case bound we can give on the number of ideal queries asked per session?

1.1   Our Results

In this work, we fully settle the question of worst-case number of per session ideal queries in the context of general function evaluation. Our main result is stated below.

Theorem 1.1 (Main Result (Informally Stated)). *Under standard cryptographic assumptions, for every PPT functionality f, there exists a protocol in the MIQ model where the simulator makes only a constant number of ideal queries in every session. The aforementioned constant is dependent upon the adversary, and, in particular on the number of sessions (rather than being universal).*

If the number of concurrent sessions being executed by the adversary is n^c, then the constant in the above theorem will be derived from c.

We stress that due to the worst-case guarantee of our result, we are able to achieve, for the *first* time in the study of the MIQ model, meaningful security for *all sessions*, which is much closer to standard security for secure computation. Interestingly, our protocol is the same as the [19] protocol. Still, we provide a significantly better analysis of its security. We stress that prior to this work, no approach for obtaining a worst-case bound on the ideal query complexity was known.

Our upper bound tightly matches the lower bound of Goyal and Jain [20] which rule out protocols where the simulator makes a constant number of ideal queries per session for any universal constant. Taken together, this fully resolves the central problem in the study of the MIQ problem: a (adversary dependent) constant number of ideal queries per session is both necessary and sufficient for simulation. Thus, our work can be viewed as the *final step* in understanding the simulator query complexity of the MIQ model.

Fully Concurrent PAKE Without Setup. Say that a password-based key exchange protocol is *fully concurrent* if it remains secure in a setting where unboundedly many executions of the protocol run concurrently, on potentially different passwords. An immediately corollary of our main result is the resolution of the long standing open problem of designing a fully concurrent PAKE protocol in the standard model and with no setup assumptions.

1.2   Technical Overview

Simulator Query Complexity and Precise Simulation. The question of simulator query complexity in the MIQ model is intimately connected to the notion of precise simulation introduced by Micali and Pass [25]. Recall that traditional simulator strategies allow for the simulator's running time to be an arbitrary polynomial factor of the (worst-case) running time of the real adversary. The notion of precise simulation concerns with the study of how low this polynomial can be. This idea is, in fact, much more general and can also be used in the context of resources other than running time, such as memory, etc. Thus, in the most general sense, the goal of precise simulation is to develop simulation strategies whose resource utilization is "close" to the resource utilization of the real adversary.

As observed in [21], the study of simulator query complexity in the MIQ model can also be cast as a precise simulation problem by viewing the trusted party queries as the resource of the simulator. Therefore, advances in precise simulation strategies go hand in hand with improvements in the simulator query complexity in the MIQ model. Indeed, prior works in the MIQ model [19,21] have relied upon sophisticated precise simulation strategies in order to obtain their positive results. We note, however, that till date, all precise simulation strategies only focus on minimizing the *total cost* of the simulator across all the sessions. Indeed, this is why these works only yield an *average-case* bound on the simulator query complexity.

In this work, we are interested in minimizing the *worst-case* simulator query complexity per session. In other words, we are interested in simulation strategies that guarantee **local precision for every session**.

Our Approach in a Nutshell. Towards that end, our starting observation is that the problem of bounding the simulator query complexity per session can be reduced to bounding the number of times the output message of a session appears in the entire simulation transcript.[1] In other words, we need a precise (concurrent) simulation strategy where the output message of every session appears only a constant number of times across the *entire* simulation transcript.[2] For this purpose, we revisit existing precise simulation strategies. Concretely, we show that a slight variant of the "sparse" rewinding strategy of Goyal, Gupta and Jain [19] (that we henceforth refer to as the GGJ simulation strategy) satisfies our desired property. We prove this by a novel, purely combinatorial analysis. Our final secure computation protocol remains essentially identical to those in the prior works in the MIQ model.

We now give an overview of the steps involved in our proof. Say that we wish to analyze the number of queries in session i. Consider the specific point in the protocol execution of session i where, the simulator actually makes a query to the ideal functionality: call this point p_i (for example, this may be the 5th message of the protocol execution in session i). This means that whenever the simulator reaches the point p_i (in the overall concurrent execution), it will have to call the trusted functionality for session i to compute the next outgoing message. Thus, now the problem reduces to *simply counting* how many times the point p_i occurs in the entire rewinding schedule. Observe that in each thread of execution, point p_i only occurs once. However, there could be multiple threads of execution resulting because of rewinding. Therefore, p_i may also occur multiple times in the rewinding schedule.

While a direct (full) analysis of the GGJ rewinding strategy [19] turns out to be complex, we are able to break it down into three different steps. Each step builds upon the previous one, with the final step yielding us the desired bound on the simulator query complexity. Below, we provide an informal overview of each of the three steps and refer the reader to the later sections for details.

Step 1. Lazy-KP with *Static* Scheduling: We first consider the warm-up case when scheduling of messages by the adversary is static. This means that the ordering of the messages of different sessions is decided by the adversary ahead of time and is fixed (and does not change upon rewinding by the simulator). Further, instead of directly analyzing the GGJ simulator [19], here we will analyze the query complexity of the (simpler) "lazy-KP" simulator [23,29,30] for the case where the simulator uses a splitting factor of n for rewinding. That is, during simulation, each thread is divided into n equal parts, and, each resulting part is rewound individually (resulting in different threads of execution).

---

[1] More concretely, we wish to bound the first message in the protocol where the simulator is forced to query the trusted party in order to obtain the function output.

[2] Note that the output message of a session may appear more than once in the simulation transcript if the simulator employs rewinding.

In this case, we are able to prove that the simulator makes at most $O(1)$ queries to the ideal functionality in any given session. This is done by relying on the following fact. Say that the point p_i does *not* occur in a given thread. Then, since the adversary only employs static scheduling, this would mean that the point p_i also cannot occur in any threads resulting from rewinding this thread. Thus, the proof reduces to a counting argument on the number of threads resulting from rewinding the part of the main thread containing p_i. If d is the depth of recursion for our recursive rewinding schedule, then we are able to show that there are at most $O(2^d)$ threads containing point p_i. However, the depth d will be a constant for lazy-KP simulation with splitting factor n.

Step 2. Lazy-KP with *Dynamic* Scheduling: Now we analyze a general adversary that may dynamically change the ordering of the messages across different sessions upon being rewound. Hence, different threads of execution may have different ordering of the messages. We shall continue to analyze the lazy-KP simulation strategy with splitting factor n.

In this case, we prove that the simulator makes at most $O(\log(n))$ queries to the ideal functionality in any given session. The key difficulty in this case is that even if a given thread does *not* contain the point p_i, the threads resulting from its rewinding may still have p_i. Hence, it seems hard to rule out the possibility that p_i may show up in a large number of threads throughout the simulation.

To overcome this problem, we rely on the following fact: once the point p_i is seen in the main thread of execution, it cannot occur in any thread arising out of the main thread *after* that point. We also observe that *before* this point is seen in the main thread, there seems hope to rule out its occurrence in a "large" number of look ahead threads. This relies on the symmetry of the main and the look-ahead threads, and, on the fact that this point has roughly equal probability of occurring first in the main thread vs occurring first in any given look ahead thread. This step of the proof is more involved than the first step and we refer the reader to Sect. 4 for details.

Step 3. *Sparsifying* the Lazy-KP Simulation: In the final step, we analyze the *sparse* rewinding strategy of [19]. Very roughly speaking, the sparse rewinding strategy of [19] aims to rewind the adversary in "as few places as possible" while still solving all the sessions. More specifically, there is a cost associated with creating each look ahead, and, the goal of the rewinding strategy is to solve all sessions while minimizing the cost.

The sparse rewinding strategy of [19] builds upon the lazy-KP simulator with splitting factor n. Very roughly, [19] pick a subset of the total threads resulting out of the lazy-KP simulation, and choose to execute only the threads in the subset (while ignoring the remaining threads by aborting them at their start). In more detail, at each level of recursion, [19] randomly chooses $\frac{\text{polylog}(n)}{n}$ fraction of the total threads and execute them while ignoring the rest. Interestingly, Goyal et al. [19] show that, if one uses protocols with somewhat higher round complexity, all the session will still be solved even though most of the look-ahead threads are never executed.

The key idea of our final step is to leverage this sparsification in order the reduce the number of queries from $O(\log(n))$ from the previous step to $O(1)$. Recall from above that if we were to use the full lazy-KP simulation, the point p_i would have occurred at $O(\log(n))$ places in the entire simulation. However, now, in the GGJ rewinding strategy, it will occur only $O(1)$ times because most of the threads will never be executed. More details are given in Sect. 5.

2   Our Model

Let n denote the security parameter. We consider malicious, static adversaries that choose whom to corrupt before the start of any protocol. We work in the static input setting, i.e., we assume that the inputs of the honest parties in all sessions are fixed at the beginning. We do not require fairness.

Ideal Model. In the ideal world experiment, there is a trusted party for computing the desired two-party functionality f. Let there be two parties P_1 and P_2 that are involved in multiple, say $m = m(n)$, evaluations of f. Let $\mathcal{S}$ denote the adversary. The ideal world execution (parametrized by λ) proceeds as follows.

I. Inputs: P_1 and P_2 obtain a vector of m inputs, denoted $\vec{x}$ and $\vec{y}$ respectively. The adversary is given auxiliary input z, and chooses a party to corrupt. Without loss of generality, we assume that the adversary corrupts P_2. The adversary receives the input vector $\vec{y}$ of the corrupted party.

II. Session Initiation: The adversary initiates a new session by sending a start-session message to the trusted party. The trusted party then sends (start-session, i) to P_1, where i is the index of the session.

III. Honest Parties Send Inputs to Trusted Party: Upon receiving the message (start-session, i) from the trusted party, P_1 sends (i, x_i) to the trusted party, where x_i denotes its input for session i.

IV. Adversary Sends Input to Trusted Party and Receives Output: At any point, the adversary may send a message $(i, \ell, y'_{i,\ell})$ to the trusted party for any $y'_{i,\ell}$ of its choice. It receives back $(i, \ell, f(x_i, y'_{i,\ell}))$ where x_i is the input value that P_1 previously sent to the trusted party for session i. For any i, the trusted party accepts at most λ tuples indexed by i from the adversary.

V. Adversary Instructs Trusted Party to Answer Honest Party: When the adversary sends a message of the type (output, i, ℓ) to the trusted party, the trusted party sends $(i, f(x_i, y'_{i,\ell}))$ to P_1, where x_i and $y'_{i,\ell}$ denote the respective inputs sent by P_1 and adversary for session i.

VI. Outputs: The honest party P_1 always outputs the values $f(x_i, y'_{i,\ell})$ that it obtained from the trusted party. The adversary may output an arbitrary efficient function of its auxiliary input z, input vector $\vec{y}$ and the outputs obtained from the trusted party.

The ideal execution of a function $\mathcal{F}$ with security parameter n, input vectors $\vec{x}, \vec{y}$ and auxiliary input z to $\mathcal{S}$, denoted $\mathsf{Ideal}_{\mathcal{F},\mathcal{S}}(n, \vec{x}, \vec{y}, z)$, is defined as the output pair of the honest party and $\mathcal{S}$ from the above ideal execution.

Definition 2.1 (λ-Ideal Query Simulator). *Let $\mathcal{S}$ be a non-uniform proba-bilistic (expected) PPT machine representing the ideal-model adversary. We say that $\mathcal{S}$ is a λ-ideal query simulator if it makes at most λ output queries per session in the above ideal experiment.*

Real Model. Let Π be a two-party protocol for computing $\mathcal{F}$. Let $\mathcal{A}$ denote a non-uniform probabilistic polynomial-time adversary that controls either P_1 or P_2. The parties run concurrent executions of the protocol Π, where the honest party follows the instructions of Π in each execution i using input x_i. The scheduling of all messages is controlled by the adversary. At the conclusion of the protocol, an honest party computes its output as prescribed by the protocol. Without loss of generality, we assume the adversary outputs exactly its entire view of the execution of the protocol.

The real concurrent execution of Π with security parameter n, input vectors $\vec{x}$, $\vec{y}$ and auxiliary input z to $\mathcal{A}$, denoted $\mathsf{Real}_{\Pi,\mathcal{A}}(n,\vec{x},\vec{y},z)$, is defined as the output pair of the honest party and $\mathcal{A}$, resulting from the above real-world process.

Definition 2.2 (λ-Secure Concurrent Computation in the MIQ Model). *A protocol Π is said to λ-securely realize a functionality $\mathcal{F}$ under concurrent self composition in the MIQ model if for every real model non-uniform PPT adversary $\mathcal{A}$, there exists a non-uniform (expected) PPT λ-ideal query simulator $\mathcal{S}$ such that for all polynomials $m = m(n)$, every pair of input vectors $\vec{x} \in X^m$, $\vec{y} \in Y^m$, every $z \in \{0,1\}^*$,*

$$\{\mathsf{Ideal}_{\mathcal{F},\mathcal{S}}(n,\vec{x},\vec{y},z)\}_{n\in\mathbb{N}} \stackrel{c}{\equiv} \{\mathsf{Real}_{\Pi,\mathcal{A}}(n,\vec{x},\vec{y},z)\}_{n\in\mathbb{N}}$$

3 Framework for Concurrent Extraction

The Setting. Consider the following two-party computation protocol $\Pi = (P_1, P_2)$:

- **Stage 1:** First, P_1 and P_2 interact in the commit phase of an execution of an extractable commitment scheme $\langle C, R \rangle$ (described below) where P_2 acts as the committer, committing to a random string, and, P_1 acts as the receiver.
- **Stage 2:** At the end of the commitment protocol, P_1 sends a special message msg to P_2.

Now, consider the scenario where P_1 and P_2 are interacting in multiple con-current executions of Π. Suppose that P_2 is corrupted. Our goal is to design a simulator algorithm $\mathcal{S}$ that satisfies the following two properties:

- **Extraction in all Sessions:** $\mathcal{S}$ must successfully extract the value committed by adversarial P_2^* in each execution of Π.
- **Minimize the Query Parameter:** Let λ denote the upper bound on the number of times the special message msg_s of any session s appears in the entire simulation transcript. We refer to λ as the *query parameter*. Then, the goal of $\mathcal{S}$ is to minimize the query parameter.

In the next subsection, we describe the extractable commitment scheme $\langle C, R \rangle$ from [30]. Later, in Sects. 4 and 5, we analyze the "lazy-KP" rewinding strategy [23,29,30] and the "sparse" rewinding strategy of Goyal, Gupta and Jain (GGJ) [19].

3.1 Extractable Commitment Protocol $\langle C, R \rangle$

Let $\text{COM}(\cdot)$ denote the commitment function of a non-interactive perfectly binding string commitment scheme. Let $\ell = \omega(\log n)$. Let $N = N(n)$ which will be determined later depending on the extraction strategy. The commitment scheme $\langle C, R \rangle$ between the committer C and the receiver R is described as follows.

Commit Phase: This consists of two stages, namely, the Init stage and the Challenge-Response stage, described below:

INIT: To commit to a n-bit string σ, C chooses $(\ell \cdot N)$ independent random pairs of n-bit strings $\{\alpha_{i,j}^0, \alpha_{i,j}^1\}_{i,j=1}^{\ell, N}$ such that $\alpha_{i,j}^0 \oplus \alpha_{i,j}^1 = \sigma$ for all $i \in [\ell], j \in [N]$. C commits to all these strings using COM, with fresh randomness each time. Let $B \leftarrow \text{COM}(\sigma)$, and $A_{i,j}^0 \leftarrow \text{COM}(\alpha_{i,j}^0)$, $A_{i,j}^1 \leftarrow \text{COM}(\alpha_{i,j}^1)$ for every $i \in [\ell], j \in [N]$.

CHALLENGE-RESPONSE: For every $j \in [N]$, do the following:

- Challenge: R sends a random ℓ-bit challenge string $v_j = v_{1,j}, \ldots, v_{\ell,j}$.
- Response: $\forall i \in [\ell]$, if $v_{i,j} = 0$, C opens $A_{i,j}^0$, else it opens $A_{i,j}^1$ by sending the decommitment information.

Open Phase: C opens all the commitments by sending the decommitment information for each one of them. R verifies the consistency of the revealed values. This completes the description of $\langle C, R \rangle$.

Notation. We introduce some terminology that will be used in the remainder of this paper. We refer to the committed value σ as the *preamble secret*. A slot$_i$ of the commitment scheme consists of the i'th Challenge message from R and the corresponding Response message from C. Thus, in the above protocol, there are N slots.

4 Lazy-KP Extraction Strategy

In this section, we discuss the "lazy-KP" rewinding strategy [23,29,30] with a "splitting factor" of n. We note that the idea of using a large splitting factor was first used in [27].

For this strategy, we will first prove that $\lambda = \mathcal{O}(1)$ for *static* adversarial schedules. Next, we will prove that for *dynamic* schedules, $\lambda = \mathcal{O}(\log n)$. In both of these results, the constants in $\mathcal{O}$ depend on number of sessions started by the concurrent adversary.

Lazy-KP Simulator. The rewinding strategy of the lazy-KP simulator is specified by the Lazy-KP-SIMULATE procedure. Very roughly, the simulator divides the current thread (given as input) into n equal parts and then rewinds each part individually and recursively. The input to the Lazy-KP-SIMULATE procedure consists of a triplet $(\ell, \mathsf{hist}, \mathcal{T})$. The parameter ℓ denotes the adversary's messages to be explored, the string hist is a transcript of the *current* thread of execution, and $\mathcal{T}$ is a table containing the contents of all the adversary's messages explored so far (to extract the preamble secrets and for sending the Stage 2 special message in protocol Π in any session).

The simulation is performed by invoking the procedure Lazy-KP-SIMULATE with appropriate parameters. Let $m = \mathsf{poly}(n)$ denote the number of concurrent sessions in the adversarial schedule. Then, the Lazy-KP-SIMULATE procedure is invoked with input $(m(N+1), \emptyset, \emptyset)$, where $m(N+1)$ is the total number of adversary's messages in a schedule of m sessions. The Lazy-KP-SIMULATE procedure is described in Fig. 1. Note that here (similar to [27]) we divide each thread into n parts. In other words, we consider a splitting factor of n. For every

---

Lazy-KP-SIMULATE($\ell, \mathsf{hist}, \mathcal{T}$):

Bottom level ($\ell = 1$):

- Run P_1's algorithm to choose the next message α_1 and feed P_2^* with $(\mathsf{hist}, \alpha_1)$. Let α_2 be the answer of P_2^*.
- Output $((\alpha_1, \alpha_2), \alpha_2)$.

Recursive step ($\ell > 1$):

1. Initialize $\widetilde{\mathsf{hist}} = \emptyset$, $\widetilde{\mathcal{T}} = \emptyset$.
2. For every $i \in [n]$:
    (a) Compute $(\widetilde{\mathsf{hist}}_{i,1}, \widetilde{\mathcal{T}}_{i,1}) \leftarrow$ Lazy-KP-SIMULATE $\left(\ell/n, \left(\mathsf{hist}, \widetilde{\mathsf{hist}}\right), \left(\mathcal{T}, \widetilde{\mathcal{T}}\right)\right)$.
    (b) Compute $(\widetilde{\mathsf{hist}}_{i,2}, \widetilde{\mathcal{T}}_{i,2}) \leftarrow$ Lazy-KP-SIMULATE $\left(\ell/n, \left(\mathsf{hist}, \widetilde{\mathsf{hist}}\right), \left(\mathcal{T}, \widetilde{\mathcal{T}}\right)\right)$.
    (c) Update $\widetilde{\mathsf{hist}} = (\widetilde{\mathsf{hist}}, \widetilde{\mathsf{hist}}_{i,1})$ and $\widetilde{\mathcal{T}} = (\widetilde{\mathcal{T}}, \widetilde{\mathcal{T}}_{i,1}, \widetilde{\mathcal{T}}_{i,2})$.
3. Output $(\widetilde{\mathsf{hist}}, \widetilde{\mathcal{T}})$.

Fig. 1. Lazy-KP Simulator with splitting factor n. Even though the messages in $\{\widetilde{\mathsf{hist}}_{i,2}\}$ do not appear in the output, some of them do appear in $\widetilde{\mathcal{T}}$.

session s consisting of an execution of Π, the goal of the simulator is to find two instances of any slot $i \in [N]$ of the commitment protocol $\langle C, R \rangle$ where the simulator's challenges are different and adversary responds with a valid response to each challenge. Note that in this case, the simulator can extract the preamble secret of $\langle C, R \rangle$ from the two responses of the adversary. On the other hand, if the simulation reaches Stage 2 in Π at any time, without having extracted the preamble secret from the adversary, then it gives up the simulation and outputs $\perp$. In this case, we say the simulator *gets stuck*.

It follows from [29] that the lazy-KP simulator (as described above) gets stuck with only negligible probability.

4.1 Terminology for Concurrent Simulation

We introduce some terminology and definitions regarding concurrent simulation that will be used in the rest of the paper.

Execution Thread. Consider any adversary that starts $m = \text{poly}(n)$ number of concurrent sessions of Π. In order to extract the preamble secret in every session, the simulator creates multiple execution threads, where a thread of execution is a simulation of (part of) the protocol messages in the m sessions. We differentiate between the following:

Main Thread vs Look-ahead Thread: The *main thread* is a simulation of a complete execution of the m sessions, and this is the execution thread that is output by the simulator. In addition, from any execution thread, the simulator may create other threads by rewinding the adversary to a previous state and continuing the execution from that state. Such a thread is called a *look-ahead thread*. Note that a look-ahead thread can be created from another look-ahead thread.

Complete vs Partial Thread: We say that an execution thread T is a *complete* thread if it shares a prefix with the main thread: it starts where the main thread starts, and, continues until it is terminated by the simulator. Other threads that start from intermediary points of the simulation are called *partial* threads. Note that by definition, the main thread is a complete thread. In general, a complete thread may consist of various partial threads. Various complete threads may overlap with each other. For simplicity of exposition, unless necessary, we will not distinguish between complete and partial threads in the sequel.

Simulation Transcript. The simulation transcript is the set of all the messages between the simulator and the adversary during the simulation of all the concurrent sessions. In particular, this includes the messages that appear on the main thread as well as all the look-ahead threads.

Simulation Index. Consider $m = \text{poly}(n)$ concurrent executions of Π. Let $M = m(2N+2)$, where $2N+2$ is the round complexity of Π. Then, a simulation index i denotes the point where the i'th message (out of a maximum of M messages) is sent on any complete execution thread in the simulation transcript.

Note that a simulation index i may appear *multiple* times over various threads in the simulation transcript. However, a simulation index i can appear at most once on any given thread (complete or partial). In particular, every simulation index $i \in [M]$ appears on the main thread (unless the main thread is aborted prematurely). Further, if a look-ahead thread T was created from a thread at simulation index i, then only simulation indices $j > i$ can appear on T.

Static vs Dynamic Scheduling. Consider the concurrent execution of $m = \text{poly}(n)$ instances of Π. Recall that the adversary controls the scheduling of the protocol messages across the m sessions. We say that a concurrent schedule is *static* if the scheduling of the protocol messages is decided by the adversary ahead of time and does not change upon rewindings. Thus, in a static schedule, protocol messages appear in the *same* order on every complete thread. In particular, for every $i \in [M]$, every instance of a simulation index i in the simulation transcript corresponds to the *same* message index $j \in [2N+2]$ of the *same* session s (out of the m sessions). However note that the actual content of the j'th message may differ on every execution thread.

We say that a concurrent schedule is *dynamic* if at any point during the execution, the adversary may decide which message to schedule next based on the protocol messages received so far. Therefore, in a dynamic schedule, the ordering of messages may be *different* on different execution threads in the simulation. In particular, each instance of a simulation index i may correspond to a *different* message j_i of a *different* session s_i.

Recursion Levels. We define recursion levels of simulation and count the number of threads at each recursion level for the lazy-KP simulator. We say that the main thread is at recursion level 0. Note that the Lazy-KP-SIMULATE divides the main thread of execution into n parts and executes each part twice. This results in $2n$ execution threads, n of which are part of the main thread, while the remaining n are look-ahead threads. All of these $2n$ threads are said to be at recursion level 1. Now, each of these threads at recursion level 1 is divided into n parts and each part is executed twice. This creates $2n$ threads at recursion level 2. Since there are $2n$ threads at recursion level 1, in total, we have $(2n)^2$ threads at recursion level 2. (Again, out of these $(2n)^2$ threads, $2n^2$ threads actually lie on the $2n$ threads at level 1.) This process is continued recursively. At recursion level ℓ, there are $(2n)^\ell$ threads. Since there are $m(2N+2)$ messages across the m sessions, the depth of recursion is a constant c', where $c' = c + \log(2N+2)$ when $m = n^c$. Then, at recursion level c', there are $(2n)^{c'}$ threads.

Sibling Threads. Consider Fig. 2 where a thread T at some recursion level ℓ is divided into $n = 4$ parts, which leads to the creation of 8 threads at recursion level $\ell + 1$. Each pair of threads (T_i, T'_i) that are started from the same point are referred to as *sibling* threads.

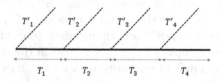

Fig. 2. One recursion step for splitting factor 4. Every T_i and T'_i are sibling threads.

4.2   Analysis of λ for Static Schedules

We start by analyzing the lazy-KP extraction strategy for static schedules. Let $\lambda_{\mathsf{lazy\text{-}KP}}$ denote the query parameter for the lazy-KP simulator.

Theorem 4.1. *For any constant c and any concurrent execution of $m = n^c$ instances of Π where the scheduling of messages is static, $\lambda_{\mathsf{lazy\text{-}KP}} = 2^{c'}$, where $c' = c + \log(2N + 2)$.*

In order to prove Theorem 4.1, we use the following lemma that follows by a simple counting argument (the proof is deferred to the full version).

Lemma 4.2. *For any constant c and any concurrent execution of $m = n^c$ instances of Π, the simulation transcript generated by the lazy-KP simulator is such that every simulation index $i \in [M]$ appears $2^{c'}$ times, where $c' = c + \log(2N + 2)$.*

Consider any session s. From the definition of static scheduling, we have that for every $j \in [2N+2]$, if the j'th message of session s appears at simulation index i on any thread, then *every* instance of simulation index i in the simulation transcript corresponds to the j'th message of session s. Now, from Lemma 4.2, since each simulation index appears $2^{c'}$ times in the simulation transcript, we have that the special message of every session s appears $2^{c'}$ times in the simulation. Thus, we have that $\lambda_{\mathsf{lazy\text{-}KP}} = 2^{c'}$ for static schedules.

4.3   Analysis of λ for Dynamic Schedules

Theorem 4.3. *For any polynomial $m = \mathsf{poly}(n)$, for any concurrent execution of m instances of Π (with possibly dynamic scheduling of messages), $\lambda_{\mathsf{lazy\text{-}KP}} = \mathcal{O}(\log n)$ except with negligible probability.*

Proof of Theorem 4.3. Fix any session s out of the $m = n^c$ sessions. Note that the special message msg_s of session s appears exactly once on the main thread. Let i_{main} denote the simulation index where msg_s appears on the main thread. Now, we will count:

1. The number of times msg_s appears in the simulation transcript *before* i_{main}. Let δ_1 denote this number.
2. The number of times msg_s appears in the simulation transcript at i_{main} or *after* i_{main}. Let δ_2 denote this number.

Thus, the total number of times msg_s appears in the simulation transcript is $\delta_1 + \delta_2$. It suffices to prove that $\delta_1 + \delta_2 = \mathcal{O}(\log n)$.

Let $i_1, \ldots, i_k$ be the *distinct* simulation indices where msg_s appears in the simulation transcript. Let $i_1, \ldots, i_k$ be ordered, i.e., for every $\ell \in [k-1]$, $i_\ell < i_{\ell+1}$. Let $k_1 \leq k$ be such that $i_{k_1} < i_{\mathsf{main}}$ and $i_{k_1+1} \geq i_{\mathsf{main}}$.

Lemma 4.4. *For any $\ell \in [k]$, the probability that msg_s does not appear on the main thread at simulation index i_ℓ is at most $(1 - \frac{1}{c'})$.*

Proof. Consider the simulation index i_1. From Lemma 4.2, we have that i_1 appears on $2^{c'}$ threads in the simulation transcript. Let $T[i_1] = T_1, \ldots, T_{2^{c'}}$ denote these threads. Now, let q be such that the special message msg_s appears at simulation index i_1 on q of these $2^{c'}$ threads. Let $T^*[i_1] = T_1^*, \ldots, T_q^*$ denote these q threads. Let T_{main} denote the main thread. Then, we have that:

$$\Pr\left[T_{\text{main}} \in T^*[i_1]\right] = \frac{q}{2^{c'}} \tag{1}$$

To see this, recall that the Lazy-KP-SIMULATE procedure uses uniformly random coins on each execution thread, and follows the same strategy. Thus, the view of the adversary is indistinguishable on each thread. In particular, if p is the probability that a message α appears on a thread T and m' appears on its sibling thread T' with, then with probability $p - \text{negl}(n)$, m' appears on T and m appears on T'. (This is the "symmetry" property for threads in the lazy-KP simulation.) Therefore, Eq. 1 follows.

From Eq. 1, we have that:

$$\Pr\left[T_{\text{main}} \notin T^*[i_1]\right] = 1 - \frac{q}{2^{c'}}$$

Note that the above probability is maximum when $q = 1$. Hence, we have that:

$$\Pr[\text{msg}_s \text{ does not occur on main thread at } i_1] \leq 1 - \frac{1}{2^{c'}}. \tag{2}$$

Now, consider simulation index i_2. Again, from Lemma 4.2, we have that i_2 appears on $2^{c'}$ threads. Let $T[i_2]$ denote the set of these threads. Now, note that msg_s cannot appear on the look-ahead threads $T \in T^*[i_1] \cap T[i_2]$. Thus, following Eq. 2, we have that:

$$\Pr[\text{msg}_s \text{ does not occur on main thread at } i_2] \leq 1 - \frac{1}{2^{c''}}.$$

where $c'' \leq c'$. Continuing the same argument, we have that for every $\ell \in [k-1]$,

$$\Pr[\text{msg}_s \text{ is not on main thread at } i_{\ell+1}] \leq \Pr[\text{msg}_s \text{ is not on main thread at } i_\ell]$$

Thus, for every i_ℓ, we have that the probability that msg_s does not occur on main thread at i_ℓ is at most $1 - \frac{1}{c'}$.

Computing δ_1. Now, note that $(1 - \frac{1}{c'})^t = \text{negl}(n)$ for $t = \omega(\log n)$. Therefore, we have that $k_1 = \mathcal{O}(\log n)$. Now, since each of the simulation indices $i_1, \ldots, i_{k_1}$ appears $2^{c'}$ times in the simulation transcript, we have that:

$$\delta_1 \leq 2^{c'} \mathcal{O}(\log n) \tag{3}$$

Computing δ_2. We now compute the value of γ_2. Towards this, let us suppose that for every simulation index $i \in [\ell]$, the Lazy-KP-SIMULATE procedure runs all threads starting from simulation index i in *parallel*. That is, Lazy-KP-SIMULATE performs one step of execution on each of these threads. It then performs the next execution step on each of these threads, and so on. Note that this is without loss of generality since the Lazy-KP-SIMULATE procedure runs all such threads *independently*.

Now, we first observe that msg_s cannot appear on a look-ahead thread that starts at a simulation index $i > i_{\mathsf{main}}$. Thus, to compute δ_2, we only need to consider the look-ahead threads that started at simulation indices $i < i_{\mathsf{main}}$ and did not finish before reaching i_{main}. Let T_{good} denote the set of such threads.

By using Lemma 4.2, we can claim that $|T_{\mathsf{good}}| \leq 2^{c'}$. Then, assuming the worst case where msg_s appears on each thread $T \in T_{\mathsf{good}}$, we have that $\delta_2 \leq 2^{c'}$

5 GGJ Extraction Strategy

In this section, we discuss the GGJ extraction strategy [19] and analyze the query complexity parameter for the same. Unlike [19] that used a splitting factor of 2, we will work with n as the splitting factor. For this strategy, we will prove that for every concurrent schedule of polynomial number of sessions, the query parameter $\lambda = \mathcal{O}(1)$. Here, the constant in $\mathcal{O}$ depends on the number of concurrent sessions.

Overview. Roughly speaking, the GGJ rewinding strategy can be viewed as a "stripped down" version of the lazy-KP simulation strategy. In particular, unlike lazy-KP that executes *every* thread at every recursion level, here we only execute a small fraction of them. The actual threads that are to be executed are chosen uniformly at random, at every level. It is shown in GGJ that by slightly increasing the round complexity – (roughly) $N = n^2$ from $N = n$, executing a $\frac{\mathsf{polylog}n}{N}$ fraction of threads at every level is sufficient to extract the preamble secret in every session.

We describe the GGJ rewinding strategy in two main steps:

1. We first describe an algorithm Sparsify that essentially selects which threads to execute in the lazy-KP recursion tree (Sect. 5.1).
2. Next, we describe the actual GGJ simulation procedure GGJ-SIMULATE that is essentially the same as the Lazy-KP-SIMULATE strategy, except that it only executes the threads selected by Sparsify (Sect. 5.2).

5.1 The Sparsification Procedure

We first describe the lazy-KP simulation tree and give a coloring scheme for the same. Next, we describe the Sparsify algorithm that takes the lazy-KP simulation tree as input and outputs a "trimmed" version of it that will correspond to the GGJ simulation tree.

Lazy-KP Simulation Tree. Let $m = n^c$ be the total number of concurrent sessions of Π started by an adversary $\mathcal{A}$. Then, the Lazy-KP-SIMULATE strategy for $\mathcal{A}$ can be described by a $2n$-ary tree $\mathsf{Tree}_{\mathsf{lazy\text{-}KP}}$ of constant depth c' where $c' = c + \log(2N + 2)$. The nodes in $\mathsf{Tree}_{\mathsf{lazy\text{-}KP}}$ are colored *white* or *black* as per the following strategy:

– The root node is colored white.
– Consider the $2n$ child nodes of any parent node. The odd numbered nodes are colored white and the even numbered nodes are colored black.

Let us explain our coloring strategy. The root node (which is colored white) corresponds to the main thread of execution. Each black colored node Node corresponds to a look-ahead thread that was forked from the thread corresponding to node Parent(Node). A white colored node Node (except the root node) corresponds to a thread T' that is a part of the thread T corresponding to node Parent(Node).

Figure 3 denotes the lazy-KP simulation tree for splitting factor $n = 2$ with white boxes representing white nodes and grey boxes representing black nodes.

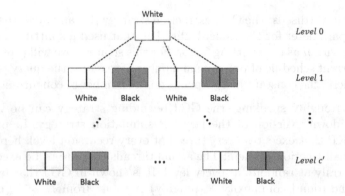

Fig. 3. The lazy-KP simulation tree for splitting factor 2.

Node Labeling. To facilitate the description of the GGJ simulation strategy, we first describe a simple tree node labeling strategy for $\mathsf{Tree}_{\text{lazy-KP}}$. The root node is labeled 1. The i'th child (out of $2n$ children) of the root node is labeled $(1, i)$. More generally, consider a node Node at level $\ell \in [c']$. Let path be its label. Then the i'th child of Node is labeled (path, i).

Below, whenever necessary, we shall refer to the nodes by their associated labels.

The Sparsify Procedure. Let p be such that $\frac{1}{p} = \frac{\text{polylog}(n)}{N}$. The Sparsify function transforms the lazy-KP simulation tree $\mathsf{Tree}_{\text{lazy-KP}}$ into a "sparse" tree $\mathsf{Tree}_{\text{sp}}$ in the following manner.

Let the root node correspond to level 0 and the leaf nodes correspond to level c'. The Sparsify procedure starts at level 0 and traverses down $\mathsf{Tree}_{\text{lazy-KP}}$, stopping at level c'. It performs the following steps at every level $\ell \in [c']$:

1. Choose $\frac{1}{p}$ fraction of the total black nodes at level ℓ, uniformly at random. Let B_ℓ denote the set of these nodes.
2. Delete from $\mathsf{Tree}_{\text{lazy-KP}}$, every black node Node at level ℓ that is not present in set B_ℓ. Further, delete the entire subtree of Node from $\mathsf{Tree}_{\text{lazy-KP}}$.

The resultant tree is denoted as $\mathsf{Tree}_{\text{sp}}$. Looking ahead, we will describe the GGJ rewinding strategy as essentially a modification of Lazy-KP-SIMULATE in that it only executes the threads corresponding to the nodes in $\mathsf{Tree}_{\text{sp}}$.

5.2   The GGJ-Simulate Procedure

The rewinding strategy of the GGJ simulator is specified by the GGJ-SIMULATE procedure. The input to the GGJ-SIMULATE procedure is a tuple (path, ℓ, hist, $\mathcal{T}$). The parameter path denotes the label of the node in $\mathsf{Tree_{sp}}$ that is to be explored, ℓ denotes the number of adversary's messages to be explored (on the thread corresponding to the node labeled with path), the string hist is a transcript of the *current* thread of execution, $\mathcal{T}$ is a table containing the contents of all the adversary's messages explored so far (to extract the preamble secrets and for sending the Stage 2 special message in Π in any session).

The simulation is performed by invoking the procedure GGJ-SIMULATE with appropriate parameters. Let $m = \mathsf{poly}(n)$ denote the number of concurrent sessions in the adversarial schedule. Then, the GGJ-SIMULATE procedure is invoked with input $(1, m(N+1), \emptyset, \emptyset)$, where $m(N+1)$ is the total number of adversary's messages in a schedule of m sessions. The GGJ-SIMULATE procedure is described in Fig. 4. Note that unlike [19], where each thread is recursively divided into two parts, here we divide each thread into n parts. In other words, we consider a splitting factor of n. For every session s consisting of an execution of Π, the goal of the simulator is to find two instances of any slot $i \in [N]$ of the commitment protocol $\langle C, R \rangle$ where the simulator's challenges are different and adversary responds with a valid response to each challenge. Note that in this case, the simulator can extract the preamble secret of $\langle C, R \rangle$ from the two responses of the adversary. On the other hand, if the simulation reaches Stage 2 in Π at any time, without having extracted the preamble secret from the adversary, then it gives up the simulation and outputs $\perp$. In this case, we say the simulator *gets stuck*.

It is implicit in [19] that the GGJ simulator (as described above) gets stuck with only negligible probability when $N - \mathcal{O}(n^2)$. We now analyze the query parameter λ_{GGJ} for the GGJ simulation strategy. A formal proof is deferred to the full version.

Theorem 5.1. *For every constant c, every $m = n^c$ number of concurrent executions of Π, the query parameter $\lambda_{\mathsf{GGJ}} = \mathcal{O}(1)$, where the constant depends on c.*

Proof (Sketch). Fix any session s. We will show that the special message msg_s can appear at most $\mathcal{O}(1)$ times at each recursion level RL_ℓ. Then, since there are only a constant number of recursion levels, it will follow that $\lambda_{\mathsf{GGJ}} = \mathcal{O}(1)$.

Towards that end, lets fix a recursion level ℓ. First recall from Theorem 4.3 that for the lazy-KP simulation strategy, $\lambda_{\mathsf{lazy\text{-}KP}} = \mathcal{O}(\log n)$. In particular, this implies that at every recursion level ℓ in the lazy-KP simulation, msg_s for a session s appears on at most $\mathcal{O}(\log n)$ threads. Using the tree terminology as introduced earlier, we have that msg_s appears on (the threads corresponding to) at most $\mathcal{O}(\log n)$ black nodes at level ℓ in $\mathsf{Tree_{lazy\text{-}KP}}$. Now, recall that at every level ℓ, the Sparsify procedure selects only $\frac{1}{p} = \frac{\mathsf{polylog} n}{N}$ fraction of black nodes, uniformly at random, and deletes the rest of the black nodes. Using Chernoff bound, we can then show that the probability that Sparsify selects $\omega(1)$ black nodes containing msg_s is negligible.

GGJ-SIMULATE(path, ℓ, hist, $\mathcal{T}$):

Bottom level ($\ell = 1$):

- Run P_1's algorithm to choose the next message α_1 and feed P_2^* with (hist, α_1).
  Let α_2 be the answer of P_2^*.
- Output $((\alpha_1, \alpha_2), \alpha_2)$.

Recursive step ($\ell > 1$):

1. Initialize $\widetilde{\text{hist}} = \emptyset$, $\widetilde{\mathcal{T}} = \emptyset$.
2. For every $i \in [n]$:
   - If node (path, $2i - 1$) $\notin$ Tree$_{\text{sp}}$, set $\widetilde{\text{hist}}_{i,1} = \emptyset$, $\widetilde{\mathcal{T}}_{i,1} = \emptyset$.
     Else, compute:
     $$\left(\widetilde{\text{hist}}_{i,1}, \widetilde{\mathcal{T}}_{i,1}\right) \leftarrow \text{GGJ-SIMULATE}\left((\text{path}, 2i - 1), \ell/n, \left(\text{hist}, \widetilde{\text{hist}}\right), \left(\mathcal{T}, \widetilde{\mathcal{T}}\right)\right).$$
   - If node (path, $2i$) $\notin$ Tree$_{\text{sp}}$, set $\widetilde{\text{hist}}_{i,2} = \emptyset$, $\widetilde{\mathcal{T}}_{i,2} = \emptyset$.
     Else, compute:
     $$\left(\widetilde{\text{hist}}_{i,2}, \widetilde{\mathcal{T}}_{i,2}\right) \leftarrow \text{GGJ-SIMULATE}\left((\text{path}, 2i), \ell/n, \left(\text{hist}, \widetilde{\text{hist}}\right), \left(\mathcal{T}, \widetilde{\mathcal{T}}\right)\right).$$
   - Update $\widetilde{\text{hist}} = (\widetilde{\text{hist}}, \widetilde{\text{hist}}_{i,1})$ and $\widetilde{\mathcal{T}} = (\widetilde{\mathcal{T}}, \widetilde{\mathcal{T}}_{i,1}, \widetilde{\mathcal{T}}_{i,2})$.
3. Output $(\widetilde{\text{hist}}, \widetilde{\mathcal{T}})$.

Fig. 4. GGJ Simulator with splitting factor n. Even though the messages in $\{\widetilde{\text{hist}}_{i,2}\}$ do not appear in the output, some of them do appear in $\widetilde{\mathcal{T}}$.

6 From Concurrent Extraction to Concurrent Secure Computation

Theorem 6.1. *Assuming 1-out-of-2 oblivious transfer, for any efficiently computable functionality f there exists a protocol Π that $\mathcal{O}(1)$-securely realizes f in the MIQ model.*

We construct such a protocol by following the exact recipe of [19,21]. We note that the works of [19,21] show how to compile a semi-honest secure computation protocol Π_{sh} for any functionality f into a new protocol Π that securely realizes f in the MIQ model. The core ingredient of their compiler is a concurrently extractable commitment $\langle C, R \rangle$: if there exists a concurrent simulator for $\langle C, R \rangle$ with query parameter λ, then the resultant (compiled) protocol Π λ-securely realizes f.

In order to prove Theorem 6.1, we construct such a protocol Π by simply plugging in our $O(n^2)$-round extractable commitment scheme in the construction of [19,21]. Then, it follows from Theorem 5.1 that protocol Π $\mathcal{O}(1)$-securely realizes f in the MIQ model, where the constant in $\mathcal{O}$ depends on c, where n^c is the number of sessions opened by the concurrent adversary.

Fully Concurrent PAKE in the Plain Model. Consider the PAKE functionality: it takes a password as input from each party, and, if they match, outputs a randomly generated key to both of them. The above protocol, when executed for the PAKE functionality gives a PAKE construction in the MIQ model where the simulator makes a constant number of queries per session in the ideal world. We then plug in Lemma 7 in [21] which shows that a PAKE construction in the MIQ model for a constant number of queries implies a concurrent PAKE as per the definition of Goldreich and Lindell [16] (with the modification that the constant in big O is adversary dependent). Put together, this gives us a construction of concurrent password-authenticated key exchange in the plain model.

References

1. Agrawal, S., Goyal, V., Jain, A., Prabhakaran, M., Sahai, A.: New impossibility results on concurrently secure computation and a non-interactive completeness theorem for secure computation. In: CRYPTO (2012)
2. Barak, B., Canetti, R., Nielsen, J., Pass, R.: Universally composable protocols with relaxed set-up assumptions. In: FOCS (2004)
3. Barak, B., Prabhakaran, M., Sahai, A.: Concurrent non-malleable zero knowledge. In: FOCS (2006)
4. Barak, B., Sahai, A.: How to play almost any mental game over the net - concurrent composition using super-polynomial simulation. In: Proc. 46th FOCS (2005)
5. Canetti, R., Lindell, Y., Ostrovsky, R., Sahai, A.: Universally composable two-party and multi-party secure computation. In: STOC (2002)
6. Canetti, R., Fischlin, M.: Universally composable commitments. In: Kilian, J. (ed.) CRYPTO 2001. LNCS, vol. 2139, p. 19. Springer, Heidelberg (2001)
7. Canetti, R., Kilian, J., Petrank, E., Rosen, A.: Black-box concurrent zero-knowledge requires $\tilde{\Omega}$ ($\log n$) rounds. In: STOC, pp. 570 579 (2001)
8. Canetti, R., Kushilevitz, E., Lindell, Y.: On the limitations of universally composable two-party computation without set-up assumptions. In: Eurocrypt (2003)
9. Canetti, R., Lin, H., Pass, R.: Adaptive hardness and composable security in the plain model from standard assumptions. In: FOCS (2010)
10. Dolev, D., Dwork, C., Naor, M.: Nonmalleable cryptography. SIAM J. Comput. 30(2), 391–437 (electronic) (2000), preliminary version in STOC 1991
11. Dwork, C., Naor, M., Sahai, A.: Concurrent zero-knowledge. In: STOC, pp. 409–418 (1998)
12. Garg, S., Goyal, V., Jain, A., Sahai, A.: Concurrently secure computation in constant rounds. In: Pointcheval, D., Johansson, T. (eds.) EUROCRYPT 2012. LNCS, vol. 7237, pp. 99–116. Springer, Heidelberg (2012)
13. Garg, S., Kumarasubramanian, A., Ostrovsky, R., Visconti, I.: Impossibility results for static input secure computation. In: Safavi-Naini, R., Canetti, R. (eds.) CRYPTO 2012. LNCS, vol. 7417, pp. 424–442. Springer, Heidelberg (2012)
14. Goldreich, O., Micali, S., Wigderson, A.: How to play any mental game. In: STOC (1987)
15. Goldreich, O., Krawczyk, H.: On the composition of zero-knowledge proof systems. SIAM J. Comput. 25(1), 169–192 Feburary 1996. http://epubs.siam.org/sam-bin/dbq/article/22068, preliminary version appeared in ICALP1990

16. Goldreich, O., Lindell, Y.: Session-key generation using human passwords only. In: Kilian, J. (ed.) CRYPTO 2001. LNCS, vol. 2139, p. 408. Springer, Heidelberg (2001)
17. Goldreich, O., Lindell, Y.: Session-key generation using human passwords only. J. Cryptology 19(3), 241–340 (2006)
18. Goyal, V.: Positive results for concurrently secure computation in the plain model. In: FOCS (2012)
19. Goyal, V., Gupta, D., Jain, A.: What information is leaked under concurrent composition? In: Canetti, R., Garay, J.A. (eds.) CRYPTO 2013, Part II. LNCS, vol. 8043, pp. 220–238. Springer, Heidelberg (2013)
20. Goyal, V., Jain, A.: On concurrently secure computation in the multiple ideal query model. In: Johansson, T., Nguyen, P.Q. (eds.) EUROCRYPT 2013. LNCS, vol. 7881, pp. 684–701. Springer, Heidelberg (2013)
21. Goyal, V., Jain, A., Ostrovsky, R.: Password-authenticated session-key generation on the internet in the plain model. In: Rabin, T. (ed.) CRYPTO 2010. LNCS, vol. 6223, pp. 277–294. Springer, Heidelberg (2010)
22. Katz, J.: Universally composable multi-party computation using tamper-proof hardware. In: Naor, M. (ed.) EUROCRYPT 2007. LNCS, vol. 4515, pp. 115–128. Springer, Heidelberg (2007)
23. Kilian, J., Petrank, E.: Concurrent and resettable zero-knowledge in polyloalgorithm rounds. In: STOC (2001)
24. Lindell, Y.: Bounded-concurrent secure two-party computation without setup assumptions. In: STOC, pp. 683–692. ACM (2003)
25. Micali, S., Pass, R.: Local zero knowledge. In: STOC (2006)
26. Micali, S., Pass, R., Rosen, A.: Input-indistinguishable computation. In: FOCS (2006)
27. Pandey, O., Pass, R., Sahai, A., Tseng, W.-L.D., Venkitasubramaniam, M.: Precise concurrent zero knowledge. In: Smart, N.P. (ed.) EUROCRYPT 2008. LNCS, vol. 4965, pp. 397–414. Springer, Heidelberg (2008)
28. Pass, R.: Simulation in quasi-polynomial time, and its application to protocol composition. In: Eurocrypt (2003)
29. Pass, R., Tseng, W.L.D., Venkitasubramaniam, M.: Concurrent zero knowledge, revisited. J. Cryptology 27(1), 45–66 (2014)
30. Prabhakaran, M., Rosen, A., Sahai, A.: Concurrent zero knowledge with logarithmic round-complexity. In: FOCS (2002)
31. Prabhakaran, M., Sahai, A.: New notions of security: achieving universal composability without trusted setup. In: STOC (2004)
32. Richardson, R., Kilian, J.: On the concurrent composition of zero-knowledge proofs. In: Stern, J. (ed.) EUROCRYPT 1999. LNCS, vol. 1592, p. 415. Springer, Heidelberg (1999)
33. Yao, A.C.C.: How to generate and exchange secrets. In: FOCS (1986)

Constant-Round MPC with Fairness and Guarantee of Output Delivery

S. Dov Gordon[1(✉)], Feng-Hao Liu[2], and Elaine Shi[3]

[1] Department of Computer Science, George Mason University, Fairfax, USA
crypto@dovgordon.com
[2] Department of Computer and Electrical Engineering and Computer Science,
Florida Atlantic University, Boca Raton, USA
fenghao.liu@fau.edu
[3] Department of Computer Science, Cornell University, Ithaca, USA
runting@gmail.com

Abstract. We study the round complexity of multiparty computation with fairness and guaranteed output delivery, assuming existence of an honest majority. We demonstrate a new lower bound and a matching upper bound. Our lower bound rules out any two-round fair protocols in the standalone model, even when the parties are given access to a common reference string (CRS). The lower bound follows by a reduction to the impossibility result of virtual black box obfuscation of arbitrary circuits.

Then we demonstrate a three-round protocol with guarantee of output delivery, which in general is harder than achieving fairness (since the latter allows the adversary to force a fair abort). We develop a new construction of a threshold fully homomorphic encryption scheme, with a new property that we call "flexible" ciphertexts. Roughly, our threshold encryption scheme allows parties to adapt flexible ciphertexts to the public keys of the non-aborting parties, which provides a way of handling aborts without adding any communication.

1 Introduction

Secure multi-party computation (MPC) allows mutually distrusting parties to securely compute a function on their inputs with several desired properties, including: **correctness** (honest parties should not receive a wrong output), and **privacy** (corrupted parties cannot learn anything beyond the prescribed output). In addition to these two basic properties, one might further require **fairness** (corrupted parties receive their output only if all honest parties receive output), or the stronger **guarantee of output delivery** (corrupted parties cannot prevent honest parties from receiving their output). Alternatively, a relaxed

---

S. Dov Gordon — This work was done when the author was a research scientist at Applied Communication Sciences.

F.-H. Liu — The work was done when the author was a postdoc at the University of Maryland.

R. Gennaro and M. Robshaw (Eds.): CRYPTO 2015, Part II, LNCS 9216, pp. 63–82, 2015.
DOI: 10.1007/978-3-662-48000-7_4

security notion is often used, called **security with abort** – it is possible that the attacker can prevent the honest parties from receiving output. All of these requirements can be formalized in an Ideal/Real paradigm [5,13], which provides a nice way to analyze security.

In this work, we explore the *round complexity* required for achieving these various properties. For the setting of *security with abort*, we already understand the round complexity fairly well – Asharov et al. [1] constructed a 3-round protocol (in the common reference string model) under the learning with error (LWE) assumption; Garg et al. [10] constructed a 2-round protocol for general computation (in the CRS model) using indistinguishable obfuscation; it is well-known that one-round protocols are in general not possible.

However, for protocols with *fairness* and *guarantee of output delivery*, our understanding of round complexity is still incomplete. Regarding *feasibility*, everything is well understood: if there is no honest majority, Cleve proved [6] that fair MPC for general computation is not possible. In the setting of an honest majority, we know that we can always achieve fairness [4], and, assuming a broadcast channel, we can always guarantee output delivery [7]. However, the optimal round complexity for this setting (of an honest majority) is still unknown[1]. Asharov et al. [1] show that their basic protocol can be extended to achieve security with guarantee of output delivery (and thus fairness) in 5 rounds, assuming there exists an honest majority. By slightly modifying the multi-key FHE protocol of Lopez-Alt, Tromer, and Vaikuntanathan [18], we can obtain a 5-round protocol with guarantee of output delivery, assuming there exists an honest majority. This is the best known round complexity for achieving fairness for any $t < N/2$. For the lower-bounds, Gennaro et al. [11] showed that there are functionalities that cannot be computed (fairly) by 2-round protocols, even in the CRS model. Recently, Garg et al. [10] claimed that their 2-round protocol (in the CRS model) also achieves fairness, but the claim contradicts the lower-bound of [11], as well as the (stronger) lower bound we present here.

1.1 Our Results

Our main two results are matching upper and lower bounds for three-round multiparty computation with guaranteed output delivery with security against a malicious minority of parties. More specifically:

- We show that 2-round, fair MPC for general functions is impossible, even if there is an honest majority. We strengthen the impossibility result of Gennaro et al. [11], demonstrating impossibility even when a fail-stop adversary corrupts only a single party. Both our result and the result of Gennaro et al. extend to the CRS model. (Sect. 3.)
- There exists a 3-round MPC with guaranteed output delivery for general functions in the CRS model, secure against a minority of semi-honest fail-stop adversaries. The security relies on the learning with errors (LWE) assumption. (Sect. 4.)

---

[1] For more restricted corruption settings, we do know how to construct 2-round protocols. See related work for more discussions.

- If parties have access to an authenticated broadcast channel[2], then the above 3-round protocol can be upgraded to one that is secure against malicious adversaries, without any additional rounds. (Sect. 5.)
- Additionally, we show that security of the two-round protocol by Garg et al. [10] can be based on witness encryptions for general NP statements, which is weaker than indistinguishable obfuscation for general circuits[3] as presented in their work. Together with an idea in Sect. 1.2, we can construct a three-round *fair* protocol (but not guarantee of output delivery) based on witness encryptions for general NP statements. Due to space limit, we present the results in the full version of this paper [14].

In summary, 2-round general fair MPC is not possible, and 3-round general MPC with guarantee of output delivery can be constructed under a falsifiable assumption. Guarantee of output delivery implies fairness (by definition), and thus 3-round fair MPC can also be constructed under the same falsifiable assumption.

All of our positive results are UC-secure [5]. Our protocols, along with those appearing in the prior work of Garg et al. [10] and Asharov et al. [1], require a CRS.

1.2 Overview of Our Techniques

Impossibility of Fairness in Two Rounds. We show that a two round, fair, polynomial-time protocol for general functions yields a construction of virtual black box (VBB) secure program obfuscation for P/Poly, in contradiction of the well-known impossibility result of Barak et al. [2].

Consider a symmetric 3-ary functionality $f(x_1, x_2, x_3)$ that interpret x_1 as a circuit C, ignores x_3 and outputs $C(x_2)$. Suppose there exists a two-round fair protocol π that computes f with fairness, then we make the following observations. We assume the three parties are Alice, Bob, and Charlie.

- If the adversary (only) corrupts Alice and instructs her to abort in the second round, then after Bob and Charlie send their messages in the second round, the adversary can learn the output $C(x_2)$.
- By the property of fairness, Bob and Charlie must be able to learn the output $C(x_2)$, since the adversary in the above case has learned the output.
- It follows that Alice's second message is redundant. Whether she sends her second message or not, the other parties can compute the outcome.

Using the above observations, we can construct a program obfuscator for general circuits: we view Alice's first message as the obfuscation of C. To evaluate $C(x)$, we just simulate Bob and Charlie with Bob's input x. Since Alice's first

---

[2] An authenticated broadcast channel enables a party to send a message to all other parties, ensuring that each party knows both the identity of the sender and that all other parties have received the same message.

[3] Indistinguishable obfuscation for general circuits is a stronger assumption. We know that indistinguishable obfuscation for general circuits implies witness encryption for general NP statements, but the other way is unclear.

message is independent of the other parties' inputs, we can rewind Bob and Charlie and compute $C(x)$ repeatedly on arbitrary values of x.

We note that Garg et al. state (without proof) that their two-round protocol achieves fairness [10], and one can see why this mistake might have been made. Their protocol works by collapsing some protocol with greater round complexity into a two-round protocol through the use of obfuscation, and they state that if the underlying protocol is fair, then the resulting two round protocol will also be fair. Speaking very roughly, in their construction, each party sends a commitment to their input and their randomness in round one, and in round two, they each send obfuscations of the next message functions from the underlying fair protocol. They then each finish the protocol locally, using the obfuscated programs to generate the correct protocol messages. At first glance, it would seem that this preserves fairness, because if a party aborts in round two, the other parties can simply generate next-messages as though he aborted, and, by the fairness of the underlying protocol, fairness should be preserved. In fact, this misses the following subtlety. If a party aborts in round two, he still receives all of the obfuscated programs, and can still compute the output of the function: this is equivalent to aborting in the very last round of the underlying fair protocol. On the other hand, because he never sent his obfuscated next-message programs, the other parties will be forced to treat him as though he aborted in round one of the underlying protocol, perhaps replacing his input with some default value. In particular, then, it could be that the malicious party learns $f(x_1, \ldots, x_N)$ while the other parties learn $f(\bot, x_2, \ldots, x_N)$.

Fairness in Three Rounds. The construction of Garg et al. can be modified slightly to get a three-round fair protocol, as we now outline. However, we note that there is no clear way to guarantee output delivery without increasing the round complexity; our main technical result is a new protocol for achieving guaranteed output delivery in three rounds.

To achieve fairness in three rounds, we can start with the protocol of Garg et al., but instead of sending obfuscations of the next message functions that compute the underlying secure computation, the parties will send obfuscations that compute an $N/2$-out-of-N secret sharing of the output. They then add one additional round to reconstruct the output. Now, if the adversary aborts in round two, even though he learns all of the next message functions, he still cannot recover the output (since there is an honest majority). If he aborts in round three, the honest parties already have enough shares to reconstruct the output on their own.

In general, we can compile any fair protocol into one that guarantees output delivery, assuming a broadcast channel [7], but we cannot necessarily preserve the round complexity. In the particular protocol just described, note that the obfuscated programs sent in round two have commitments to the parties' inputs embedded inside of them. If a party aborts in round two, the other parties would need to replace their obfuscations, embedding a commitment to some default input value in place of the aborting party's true input. But this will incur additional communication rounds.

Guarantee of Output Delivery. Before we describe our protocol, we first give an overview the approach by Asharov et al. [1]. Asharov et al. proposed a new primitive called *Threshold Fully Homomorphic Encryption* (TFHE), which is essentially a distributed version of fully homomorphic encryption (FHE). For their TFHE, there is a joint public key pk^* whose secret key is shared among all parties, i.e. $\mathsf{sk}^* = \mathsf{sk}_1 + \mathsf{sk}_2 + \cdots + \mathsf{sk}_N$. (There is also an evaluation key, but we omit it for simplicity of exposition). The keys $(\mathsf{pk}^*, \mathsf{sk}^*)$ constitute an FHE key pair, so the encryption and evaluation algorithms can remain the same as those used in the original FHE scheme. To decrypt, parties need to run a *threshold decryption protocol*, since the secret key is shared among all parties.

Using the TFHE scheme, their basic three round protocol has the following structure: (1) in the first round parties establish a joint public key pk^*; (2) in the second round parties output an encryption of their inputs, i.e. $\mathsf{Enc}_{\mathsf{pk}^*}(x_i)$; (3) in the third round, parties perform the homomorphic operations (for computing f) to obtain an evaluated ciphertext C^*, and then run the threshold decryption protocol to decrypt C^*. Asharov et al. [1] presented a simple idea to make the basic protocol fair (and to guarantee output delivery) in the first round, the parties also secret share their inputs and all random coins. If any party aborts in the second or third round, the honest majority would reconstruct his states and resume the protocol. (Note that if a party aborts in the first round, he is simply ignored). This approach will add two additional rounds for the worst case.

We note that their construction uses an N-out-of-N sharing of the secret key sk^*, so they require all parties in order to decrypt. This means, if any party aborts, the other parties need to reconstruct his view to resume. Thus, these two additional rounds seem inherent if we follow this approach. To get a 3-round protocol, we need a new approach. In particular we propose and construct a new variant of TFHE with more fine-grained features. Using it as a building block, we are able to get around the barriers mentioned above. We highlight our new ideas below.

Instead of establishing a "fixed" joint public key, our new TFHE uses $\mathsf{pk}_{[N]} = \{\mathsf{pk}_i\}_{i \in [N]}$ as the public keys, where pk_i is contributed by party P_i. Then with $\{\mathsf{pk}_1, \ldots, \mathsf{pk}_N\}$, P_i can encrypt the input x_i and produce a **flexible** ciphertext C_i. We introduce a new algorithm $\mathsf{TransCT}(C; S)$ that transforms a flexible ciphertext C into C', where C' is with respect to the public keys $\mathsf{pk}_S = \{\mathsf{pk}_j : j \in S\}$. Intuitively, a *flexible* ciphertext is one that is not yet committed to a set of public keys, and a *transformed* one commits to some pk_S and can be homomorphically evaluated. Finally, our threshold decryption protocol works when there is an honest majority of parties (as opposed to the previous one which requires all parties).

Using the new TFHE, our protocol has the following structure: (1) in the first round, parties generate $\{\mathsf{pk}_1, \ldots, \mathsf{pk}_N\}$; (2) in the second round, each party output a flexible ciphertext $C_i = \mathsf{Enc}(x_i)$; (3) let S be the parties that did not abort in the second round. Now each party transforms the ciphertexts to C'_i with respect to pk_S and performs the homomorphic evaluation for computing f. Then they perform the threshold decryption to obtain the output.

Intuitively, if a party aborts in the first round, then he is simply ignored. If he aborts in the second round, he is also ignored: since the other parties output flexible ciphertexts, these can be transformed to a public key representing the set of non-aborting parties. Those remaining parties can then proceed to perform the homomorphic computation. Finally, if a party aborts at the end, then it is too late – our threshold decryption algorithm only requires an honest majority of parties. We describe our three round protocol in Sect. 4.3.

Constructing TFHE. Our construction of TFHE is a distributed variant of the FHE scheme by Gentry, Sahai, and Waters [12]. We inherit from their scheme that our TFHE does not need the evaluation keys that Asharov et al. required, which allows for a cleaner presentation. We outline some of the technical aspects of our construction here, after we recall the GSW construction. The public key in their construction is a matrix $\mathbf{B}$ and a vector $b = \mathbf{B}s + e$ of the LWE form; the secret key is the LWE secret s. To encrypt a bit m, the algorithm generates a random 0-1 matrix $\mathbf{R}$, and outputs $C = \mathsf{Flatten}\left(m \cdot I_D + \mathsf{BitDecomp}(\mathbf{R} \cdot b \parallel \mathbf{R} \cdot \mathbf{B})\right)$, where I_D is the identity matrix. (We will define $\mathsf{BitDecomp}$ and $\mathsf{BitDecomp}^{-1}$ in Sect. 2, but, essentially, these functions act as their names suggest, decomposing a field element into a binary representation, and building a field element from a binary string.) To decrypt, the algorithm takes row β (where roughly $2^\beta >$ some noise bound) and parses the row into $(C_{\beta,1}, C_{\beta,2}) \in \mathbb{Z}_q^\ell \times \mathbb{Z}_q^{n\cdot\ell}$. (The parameters ℓ, n, q will be set in the scheme. Here for exposition, we can omit them.) Then it outputs

$$\left\lfloor \frac{\mathsf{BitDecomp}^{-1}(C_{\beta,1}) - \langle \mathsf{BitDecomp}^{-1}(C_{\beta,2}), s \rangle}{2^\beta} \right\rceil.$$

The homomorphic evaluation of the GSW scheme is surprisingly simple and beautiful! For addition it is $C + C'$ and for multiplication $C \cdot C'$.

As we discussed, our TFHE does not immediately determine a public key with respect to all parties, as done by Asharov et al. [1]. Instead, we set the public parameter (CRS) to be $\mathbf{B}$, and let each party P_i output $\mathsf{pk}_i = b_i = \mathbf{B}s_i + e_i$. Note that each $(\mathbf{B}, b_i)$ is a GSW public key. The next challenge is how to generate flexible ciphertexts. A first natural idea would be: for P_i to generate a flexible ciphertext on some message m, P_i encrypts m under all GSW-type public keys $\{(\mathbf{B}, b_i)\}_{[N]}$ to get $C = (C_1, \ldots, C_N)$. To transform C with respect to a set S, we simply output $\{C_i\}_{i \in S}$. However, this is not secure since it allows every party, independently, to decrypt P_i's ciphertext. A next idea would be P_i encrypts m under the key $(\mathbf{B}, b_i)$ corresponding to his public key, and encrypts 0 for other keys $\{(\mathbf{B}, b_j)\}_{j\neq i}$. The transform algorithm works the same. Intuitively, semantic security holds since P_i does not encrypt m under other people's keys. However, it is not clear how to jointly evaluate two ciphertexts from two parties, since the essential messages are encrypted under two different GSW public keys.

Our new idea to solve such challenge modifies C_j's for $j \neq i$: instead of generating $\mathsf{Enc}(0)$'s under $\{(\mathbf{B}, b_i)\}_{[N]\setminus\{i\}}$, P_i outputs some *hints* for the transformation algorithm, but such hints will not hurt security. More specifically, we have the following design: P_i generates $C_i = \mathsf{Flatten}\left(m \cdot I_D + \mathsf{BitDecomp}(\mathbf{R} \cdot b_i \parallel \mathbf{R} \cdot \mathbf{B})\right)$

and $C_j = \mathsf{BitDecomp}(\mathbf{R} \cdot \boldsymbol{b}_j \parallel \mathbf{0})$, for $j \neq i$, where the same $\mathbf{R}$ is used for all $\{C_j\}_{j \in [N]}$. Since each C_j only decreases the entropy of $\mathbf{R}$ by $|\mathbf{R} \cdot \boldsymbol{b}_j|$, we can still use a leftover-hash-lemma style approach to argue that m is hidden.

Then given a set S (including i), we can compute $C_S = \sum_{j \in S} C_j$. By unfolding the equation, we can see:

$$
C_S = \left(m \cdot I_D + \mathsf{BitDecomp}\left(\mathbf{R} \cdot \left(\sum_{j \in S} \boldsymbol{b}_j\right) \parallel \mathbf{R} \cdot \mathbf{B}\right) \right),
$$

which is of the form $\mathsf{Enc}(m)$ under the GSW public key $(\mathbf{B}, \sum_{j \in S} \boldsymbol{b}_j)$! This means any flexible ciphertext, after being transformed, results in an encryption under the GSW public key. Therefore, ciphertexts from different parties can be jointly computed after transformed to ones with respect to the same set S.

Our threshold decryption protocol needs to work for any set S of participants such that $|S| > [N/2]$. So the parties should distribute the secret $\boldsymbol{s}_i$'s to all the other parties using a threshold secret sharing scheme. The challenging part is to design a *one-round* protocol. We use the fact that the decryption algorithm of the GSW scheme is essentially computing inner product (of a publicly known vector and the secret key), and Shamir's secret sharing scheme is highly compatible with inner product computation. In particular, each party P_i shares $\boldsymbol{s}_i$ into $(\boldsymbol{p}_i(1), \boldsymbol{p}_i(2), \ldots, \boldsymbol{p}_i(N))$ and sends $\boldsymbol{p}_i(j)$ to P_j, where $\boldsymbol{p}$ is a vector of polynomials for Shamir's shares. To compute $w = \langle \boldsymbol{u}, \sum_{j \in S} \boldsymbol{s}_j \rangle$ for some publicly known vector $\boldsymbol{u}$ (think of it as part of a ciphertext), each party can output $w_i = \langle \boldsymbol{u}, \sum_{j \in S} \boldsymbol{p}_j(i) \rangle$. Then it is not hard to see that these w_i's form shares of w, so after receiving a majority of shares each party can run the reconstruction without interaction!

Finally, we need to handle an additional technicality to deal with noise of evaluated ciphertexts, as pointed out by Asharov et al. [1]. Intuitively, an evaluated ciphertext $\mathsf{Enc}(f(x))$ might contain noise that is related to the original input x, so we need to add additional smudging noise to eliminate any such link. In the decryption protocol of the work [1], each party adds independent small noise to the output. However, this method will not work for our case because in our reconstruction procedure, these noise values are multiplied by the Lagrange coefficient, which can be too large. To solve this issue, we let each party P_i secret share some small noise η_i into $(r_i(1), \ldots, r_i(N))$ and send the shares to the other parties (where r_i is a random polynomial for the shares). Then each party P_i adds $\sum_{j \in S} r_j(i)$ to their output. By the linearity of the Shamir's sharing scheme, this is equivalent to adding $\sum_{j \in S} \eta_j$ to the original reconstructed output value. In Sect. 4.2 we go through this construction in detail. The new TFHE may be of independent interests.

1.3 Related Work

There is a long line of work studying the round complexity of secure computation, both in the semi-honest and malicious models, the two-party and multi-party settings, the honest majority and honest minority settings, and even in a variety of other models. We will not aim to survey all of this work, but mention what we know to be the best round complexity in the most relevant settings.

Constant round protocols have been known since Yao's original two-round construction for the two-party, semi-honest setting [21], and Beaver et al.'s constant round protocol for the setting of a malicious minority [3]. In the two-party, malicious setting, Katz and Ostrovsky give a five-round protocol and demonstrate that this is tight [16]. There are several works demonstrating constant round protocols in the multiparty, malicious majority setting ([17,19] Of course, with a malicious majority (including the two-party case), fairness is unachievable, so these results are in the security-with-abort model, and are not directly relevant to our own work.

In the multiparty setting with a malicious minority, the best known round complexity is achieved by the two-round protocol of Garg et al. [10], but, as we outlined above, their result does not ensure fairness. For $t < N/5$ corruptions, Damgård and Ishai give a three-round protocol with a guarantee of output delivery [8], though they require private point-to-point channels, and establishing these would add at least one additional round. For $t < N/2$, the exact round complexity of their protocol is a bit hard to discern, but it is greater than four (and we believe more); in this domain, the five-round protocol of Asharov et al. [1], which also guarantees output delivery, is the best known. For $t = 1$ corruption, Ishai et al. [15] showed that $N \geq 5$ parties are sufficient to securely compute general functionalities with guarantee of output delivery. The work [15] also showed 2-round protocols (guarantee of output delivery) for general functionalities in the server-client model, with a more restricted corruption pattern (e.g. one corrupted client and coalitions of $t < N/3$ servers). In the semi-honest, two-party setting, Yao's original construction already achieves two-rounds.

Very recently and independent of this paper, Mukherjee and Wichs [] constructed 2-round protocols (in the CRS model) that achieve security with abort against any number of corruptions. In the setting of an honest majority, their protocol can be easily modified to achieve guarantee of output delivery in 3 rounds, assuming private communication channels, and in 4 rounds without private communication channels.

Gennaro et al. [11] provide a lower bound on the round complexity of fair protocols whenever $1 < t < N/2$. Our lower-bound strengthens theirs, ruling out even a fail-stop adversary that corrupts a single party.

2   Preliminaries

In this section, we present basic vector operations. Due to space limit, we describe the security definitions for MPC and the LWE assumptions in the full version of this paper [14].

2.1 Elementary Vector Operations

We define a number of vector/matrix operations that we describe below. Let a, b be vectors of dimension k. Let $\ell = \lfloor \log q \rfloor + 1$ for some modulus q. Note that the operations we describe are also defined over matrices, operating *row by row* on the matrix, and that all arithmetic is over $\mathbb{Z}_q$.

$\mathsf{BitDecomp}(a) = $ the $k \cdot \ell$ dimensional vector $(a_{1,0}, \ldots, a_{1,\ell-1}, \ldots, a_{k,0}, \ldots a_{k,\ell-1})$
  where $a_{i,j}$ is the jth bit in the binary representation of a_i, with bits ordered
  from least significant to most significant.
$\mathsf{BitDecomp}^{-1}(a')$ For $a' = (a_{1,0}, \ldots, a_{1,\ell-1}, \ldots, a_{k,0}, \ldots a_{k,\ell-1})$, let
$\mathsf{BitDecomp}^{-1}(a') = \left(\sum_{j=0}^{\ell-1} 2^j a_{1,j}, \ldots, \sum_{j=0}^{\ell-1} 2^j a_{k,j} \right)$, but defined even when a'
  isn't binary.
$\mathsf{Flatten}(a') = \mathsf{BitDecomp}\left(\mathsf{BitDecomp}^{-1}(a')\right)$
$\mathsf{Powersof2}(b) = (b_1, 2b_1, 4b_1, \ldots, 2^{\ell-1}b_1, \ldots, b_k, \ldots 2^{\ell-1}b_k)$.

3 Impossibility Result

In this section, we are going to show that it is impossible to construct a two-round secure protocol for general multi-party computation with fairness, even with an honest majority of players. Our impossibility results holds in the *standalone* model, even with non-rushing fail-stop adversaries with access to a CRS. Our result strengthens that of Gennaro, as it holds even for adversaries corrupting only a single party, while their result cannot rule out the case where $t = 1$.

  We assume that the players have both point to point channels and a public broadcast channel, but they do not have private point-to-point channels – an eavesdropper can listen to all channels.[4] We note that our three round protocol from Sect. 4 can be collapsed into a three round protocol if we give the users access to a PKI of the appropriate form, so the assumption of non-private channels in our lower-bound is natural.[5] A more formal proof follows.

Theorem 1. *Let $\mathcal{C}$ be a family of circuits, and let Π be a polynomial-time, 2-round, 3-party secure protocol for computing $U(C, x, 0) = C(x)$ for any $C \in \mathcal{C}$, with fairness in the standalone model. Then there exists a virtual black-box obfusctor for general circuits[6].*

Proof. We describe a VBB obfuscator $\mathcal{O}$ for all circuits in $\mathcal{C}$. Before doing that, we define some notation and our next message functions. We let $\mathcal{M}$ denote the set of valid messages in the secure computation. We let $\perp \in \mathcal{M}$ denote a special abort symbol, and we let $\emptyset$ denote the empty transcript (before any messages

---

[4] Our lower bound holds even when the eavesdropper only listens to some channels.

[5] Although we allow the eavesdropping adversary to corrupt two private channels at once, we do not allow it to corrupt the parties themselves, so we do still maintain an honest majority. However, there is still room to consider a weaker model where the eavesdropper can only listen to a single channel.

[6] We describe the definition of VBB obfuscation in the full version of this paper [14].

have been sent). A *partial incoming transcript*, is either $\emptyset$ (if no messages have been sent yet), or of the form $(\mathcal{M}, \mathcal{M})$, where each message is received from one of the two other parties in the first round of the protocol. A *partial outgoing transcript* is of the same form, but represents the two messages sent by a single party in the first round, each going to one of the other parties. A *full incoming transcript* is of the form $((\mathcal{M}, \mathcal{M}), (\mathcal{M}, \mathcal{M}))$, where the first pair of messages are those received in the first round, and the second pair are those received in the second round. We define the following set of circuits.

$\pi_{i,j}(x, \tau, r)$: for parties $i, j \in \{1, 2, 3\}$, on input value x, partial incoming transcript τ and randomness r, the circuit outputs i's next message to j.

$\pi_{\mathsf{out}}(x, \tau, r_2)$: the circuit computes P_2's output in the secure computation, given input x, full incoming transcript τ and randomness r_2.

The VBB obfuscation of circuit C is as follows. $\mathcal{O}(C)$ chooses randomness r_1 and computes $\alpha_2 = \pi_{12}(C, \emptyset, r_1)$, $\alpha_3 = \pi_{13}(C, \emptyset, r_1)$. Note that they are the first-round messages from P_1 to P_2 and P_3. Then the obfuscator outputs the following circuit $\Gamma_{\alpha_2,\alpha_3}(x; r_2, r_3)$, which, on input x and randomness r_2, r_3, performs the following computations:

- $\gamma^{(1)} = \pi_{32}(0, \emptyset, r_3)$; $\beta = \pi_{23}(x, \emptyset, r_2)$. (The relevant first round messages.)
- $\gamma^{(2)} = \pi_{32}(0, (\alpha_3, \beta), r_3)$. (The relevant second round message.)
- Output $\pi_{\mathsf{out}}(x, ((\alpha_2, \gamma^{(1)}), (\bot, \gamma^{(2)})), r_2)$. ($\mathsf{P}_2$'s output, given his full incoming transcript.)

Basically, the circuit simulates $\mathsf{P}_2, \mathsf{P}_3$'s messages when P_1 sends out α_2, α_3 and then aborts in the second round Fig. 1

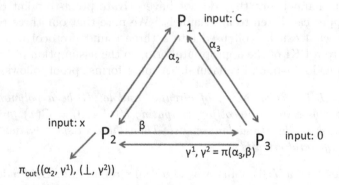

Fig. 1. A depiction of the messages used in the circuit $\Gamma_{\alpha_2,\alpha_3}$. α messages are sent by party P_1, β messages by P_2, and γ messages by P_3. A subscript i indicates that the recipient is P_i, and a superscript indicates a round number. Since we do not need all protocol messages, we drop subscripts and superscripts where we can.

We claim that for any $C \in \mathcal{C}$, $\Gamma_{\alpha_2,\alpha_3}$ is a secure VBB obfuscation of C. Efficiency of $\Gamma_{\alpha_2,\alpha_3}$ follows from the fact that the secure computation is polynomial-time. Correctness follows from the fairness of the underlying secure computation protocol – we note that by the correctness of the protocol, P_1 can learn the output after he sees all the incoming messages, regardless of whether or not he aborts in the second round. Thus by fairness of the protocol, P_2 should also receives the output, regardless of whether or not P_1 aborts in the second round. Thus, given the transcript $(x, ((\alpha_2, \gamma^{(1)}), (\bot, \gamma^{(2)})))$, P_2 can compute the output.

To prove the VBB property, recall that we need to prove that for any adversary $\mathcal{A}_\mathcal{O}$, for any circuit C, there exists a simulator $\mathcal{S}_\mathcal{O}$ and a negligible function ϵ such that

$$|\Pr[\mathcal{A}_\mathcal{O}(\Gamma_{\alpha_2,\alpha_3}) = 1] - \Pr[\mathcal{S}_\mathcal{O}^C(1^{|C|}) = 1]| < \epsilon(|C|)$$

By the security of the underlying secure computation, we know there exists an ideal-world simulator, which we will denote by $\mathcal{S}_E$, that simulates the view of an eavesdropper who listens to the channels between P_1 and P_2, and between P_1 and P_3. We will denote by $\mathcal{S}_E^{(1)}$ the result of running $\mathcal{S}_E$ and restricting the output to the partial outgoing transcript, i.e. the first round messages sent from P_1. Then, $\mathcal{S}_\mathcal{O}$ gets $(\tilde{\alpha}_2, \tilde{\alpha}_3) \leftarrow \mathcal{S}_E^{(1)}$, and constructs the circuit $\Gamma_{\tilde{\alpha}_2,\tilde{\alpha}_3}$, as described above; we note that neither C nor r_1 are needed, once $\tilde{\alpha}_2$ and $\tilde{\alpha}_3$ are computed. Finally, $\mathcal{S}_\mathcal{O}$ outputs $\mathcal{A}_\mathcal{O}(\Gamma_{\tilde{\alpha}_2,\tilde{\alpha}_3})$.

Suppose that this does not meet the above security requirement. It follows that there exists a distinguisher $\mathcal{D}$ that distinguishes between a real world execution of the protocol and the ideal simulation of $\mathcal{S}_E$. This follows immediately, because the only difference between the true obfuscation and the simulated obfuscation is the way in which α_2 and α_3 are generated. Therefore, on input transcript τ, $\mathcal{D}$ simply takes the messages α_2, α_3 that constitute the first round messages sent from P_1 to P_2 and P_3 respectively, and he completes the construction of $\Gamma_{\alpha_2,\alpha_3}$ himself. $\mathcal{D}$ then runs $\mathcal{A}_\mathcal{O}$ on the resulting circuit and determines from the output whether τ was simulated.

Remark. The lower bound proof can be extended to rule out protocols in the CRS model, using the same idea. The obfuscated circuit will now embed crs as a common reference string, and in the security proof, the simulator will simulate the string, i.e. $(\tilde{crs}, \tilde{\alpha}_2, \tilde{\alpha}_3) \leftarrow \mathcal{S}_E^{(1)}$.

Two Round Feasibility with a PKI. Note that the proof breaks down if the parties have access to private channels, including in the scenario where they have access to a PKI. This is because we need *both* first-round messages sent from P_1 in order to simulate round two of the protocol. In particular, without access to α_3, we could not correctly simulate $\gamma^{(2)}$ (as sent from P_3 to P_2 in round two), and therefore we could guarantee the correct output of the obfuscation. The only way to gain access to both α_2 and α_3 is either by eavesdropping on multiple channels, or by corrupting two parties, but this latter approach would violate our assumption of an honest majority. Indeed, as we mentioned previously, and as we will see later, our construction in Sect. 4 can be collapsed to two rounds if we have access to a PKI, with public keys of a particular form. It is still an open question whether a two-round protocol with guaranteed output delivery is possible, given access only to private channels.

4   Towards Fairness and Guarantee of Output Delivery

The previous section shows that two-round fair protocols are in general impossible. As discussed in the introduction, we can construct a three-round fair protocol by adding one more round to the protocol by Garg et al. [10]; yet it is unclear how to construct three-round protocols with guarantee of output delivery. In this section, we present our main contribution – we construct a new threshold FHE scheme, which extends the notion of threshold FHE by Asharov et al. [1] with enriched features. We elaborate on these below.

4.1   New Threshold Fully Homomorphic Encryption Scheme

As discussed in the introduction, our TFHE introduces a new idea of *flexible* and *transformed* ciphertexts that play an important role in our 3-round MPC construction. Here we first present the syntax: a threshold fully homomorphic encryption scheme (TFHE) is basically a homomorphic encryption scheme, with the difference that the key generation and decryption are N-party protocols instead of algorithms. We will consider protocols defined in terms of some common parameter pp.

- TFHE.Gen(pp) **(Key Generation Protocol).** Initially each party holds some parameter pp. At the conclusion of the protocol, each party P_i for $i \in [N]$ publishes a public key pk_i, and keeps a private key sk_i.
- TFHE.Dec$_S(C; v)$ **(Threshold Decryption Protocol).** Let S be a set in $[N]$, and $v = \{v_i : i \in S\}$ be some secret values, each held by one party. The protocol is run among parties $\{P_i : i \in S\}$. Initially each party holds a secret input $v[i]$, a secret key sk_i, and receives a ciphertext C as the public input. At the end, the parties in the set can compute the decrypted message m. Intuitively, the secret input v is used for smudging the noise.

    Note: in the setting with honest majority, we assume that $|S| \geq [N/2] + 1$. For simplicity, we assume the input ciphertext C has already been transformed to one that corresponds to the set of public keys $pk_S = \{pk_i : i \in S\}$. See the syntax below for further exposition.
- TFHE.Enc$_i$(pp, $pk_1, \ldots, pk_N; m$) **(Encryption Algorithm).** Let parties $\{P_i\}_{i \in [N]}$ participate in the protocol, and $\{pk_i\}_{i \in [N]}$ be the set of their public keys. The encryption algorithm is non-interactive and run by party P_i. The algorithm takes inputs the public parameter, the public keys $\{pk_i\}_{i \in [N]}$, a message m, and computes a ciphertext C.

    We implicitly require that the ciphertexts here are **flexible** in the sense that they do not *commit* to a particular public key/secret key yet; in particular, we can use the algorithm below to transform a flexible ciphertext into one that corresponds to a set of public keys.
- TFHE.TransCT$(C; S)$ **(Ciphertext Transform Algorithm).** The algorithm takes inputs a *flexible* ciphertext C (from the above encryption algorithm), and a set $S \subseteq [N]$ and outputs a *transformed* ciphertext C_S. The ciphertext can be thought as one under the set of joint keys: $pk_S = \{pk_i : i \in S\}$.

– TFHE.Eval$(f, C_1, \ldots, C_t; S)$ **(Evaluation Algorithm).** The evaluation algorithm is non-interactive. A party P_i (can be any party) receives inputs a function $f : \{0,1\}^t \to \{0,1\}$, flexible ciphertexts $C_1, \ldots, C_t$, and a set $S \subseteq [N]$. He computes an evaluated ciphertext C'_S with respect to the set S, which can be thought as an evaluated ciphertext under the joint public key pk_S defined as above.

We summarize the main differences between our TFHE and that of the prior work [1].

1. Our key generation does not output a *joint* public key. Instead each party will only output their own public key pk_i. Then parties can run the Eval algorithm to homomorphically compute on the ciphertexts under pk_S for some set S decided later. As pointed out in the introduction, this is an important feature.
2. The construction of the prior work requires all parties to participate in the decryption protocol (in the non-interactive case). Here we allow a subset of parties to run the protocol; moreover, we allow a "threshold" type of decryption where a majority of parties can decrypt the ciphertext.

These new features play an important role: intuitively, when a party generates a ciphertext, he does not know who else might abort. The flexibility of ciphertexts handles this problem – the parties can generate ciphertexts first, and later on decide a set of public key (namely $\mathsf{pk}_S = \{\mathsf{pk}_i : i \in S\}$), so that the flexible ciphertexts can be transformed with respect to pk_S. Then the parties can perform homomorphic computation with respect to pk_S and run the threshold decryption algorithm.

Similar to the work [1], we do not define the security of TFHE on its own. The reason is similar: requiring that the above protocols securely realize some ideal key-generation and decryption functionalities is unnecessarily restrictive. Instead, we will show that our TFHE scheme is secure directly in the context of our implementation of general MPC in Sect. 4.3.

4.2   Construction of Our New TFHE

Following the intuition in the introduction, we describe our construction.

Common Parameter. All parties receive the common parameter pp of the form: let N be the number of parties, $L = \mathrm{poly}(\kappa)$ be the maximum depth of the circuits supported by the TFHE evaluation algorithm. Then we choose a modulus q of $\mathrm{poly}(L, N)$ bits, lattice dimension parameter $n = n(L, N)$, and error distribution $\chi = \chi(\kappa, L, N)$ appropriately for LWE security against 2^κ known attacks. Also, choose parameter $m = m(\kappa, L) = O((n + N)\log q)$. Let the distribution χ be B_χ-bounded (i.e. with overwhelming probability, a sample from χ has the absolute value less than B_χ). Let $\ell = \lceil \log q \rceil + 1$, $D = (n+1) \cdot \ell$, and $B_{\mathsf{smug}} \in \mathbb{Z}$ be an integer bound, satisfying the following relations:

$$\frac{(D+1)^L \cdot N \cdot B_\chi}{B_{\mathsf{smug}}} = \mathsf{negl}(\kappa), \quad B_{\mathsf{smug}} < q/8.$$

Then $\mathsf{pp} = (n, q, \chi, m, B_\chi, B_{\mathsf{smug}}, \mathbf{B})$ where $\mathbf{B}$ is sampled uniformly from $\mathbb{Z}_q^{m \times n}$.

TFHE.Gen(pp): This is a two-round protocol among N parties.

- **(Round 1):** Each party P_i samples a random vector $\boldsymbol{s}_i \in \mathbb{Z}_q^n$, and computes $\boldsymbol{b}_i = \mathbf{B} \cdot \boldsymbol{s}_i + \boldsymbol{e}_i$ where $\boldsymbol{e}_i \leftarrow \chi^m$. Then P_i broadcasts $\mathsf{pk}_i = \boldsymbol{b}_i$, and keep $\boldsymbol{s}_i$ secretly.
- **(Round 2):** Each party P_i secret shares $\boldsymbol{s}_i$ using the Shamir Secret Sharing Scheme with threshold $[N/2] + 1$. Let $\boldsymbol{p}_i$ denote the random polynomial vector (of degree $[N/2]+1$) generated by P_i where $\boldsymbol{p}_i(0) = \boldsymbol{s}_i$ (This is how the Shamir Secret Sharing works). P_i sends $\boldsymbol{p}_i(j)$ to P_j for $j \in [N]$. At the end, P_i sets $\mathsf{sk}_i = (\boldsymbol{p}_1(i), \boldsymbol{p}_2(i), \ldots, \boldsymbol{p}_N(i))$.

  Note that although we do not assume secure point-to-point channels, sending private message in the second round is achievable – everyone can send a public key in the first round, and later on every party encrypts the outgoing messages. For simplicity, we just assume there are secure point-to-point channels available in the second round.

TFHE.Dec$_S(C; \boldsymbol{v})$: Let $\boldsymbol{v}$ be a vector of $|S|$ numbers (error terms), where $\boldsymbol{v}[i]$ (the element indexed by i) is held by party P_i for $i \in S$; let C be a ciphertext. For $S \subseteq [N]$ such that $|S| \geq N/2$, this is a one-round protocol among parties $\{P_i : i \in S\}$. For simplicity, we assume that C is a transformed ciphertext that corresponds to pk_S.

- Each party P_i parses C as a matrix in $\mathbb{Z}_q^{D \times D}$. Then he picks the β-th row, C_β, where $\beta = \lfloor \log_2(q/2) \rfloor$. Note that $2^\beta \in (q/4, q/2]$. Then parse $C_\beta = (C_{\beta,1}, C_{\beta,2})$ where $C_{\beta,1} \in \mathbb{Z}_q^\ell$, $C_{\beta,2} \in \mathbb{Z}_q^{n \cdot \ell}$. Then he computes $\boldsymbol{z}_i = \sum_{j \in S} \boldsymbol{p}_j(i)$ and broadcasts $w_i = \langle \mathsf{BitDecomp}^{-1}(C_{\beta,2}), \boldsymbol{z}_i \rangle + \boldsymbol{v}[i]$.
- At the end, each party picks an arbitrary subset $T \subseteq S$ such that $|T| = [N/2] + 1$. Then they compute $w = \sum_{k \in T} \mu_k(0) w_k$, where μ_k is the Lagrange polynomial. Finally they output $\left\lfloor \frac{\mathsf{BitDecomp}^{-1}(C_{\beta,1}) - w}{2^\beta} \right\rceil$.

TFHE.Enc$_i(\mathsf{pp}, \mathsf{pk}_1, \ldots, \mathsf{pk}_N; m)$: This is the i-th party's encryption algorithm. Let $m \in \{0, 1\}$ be the input message.

- The algorithm parses pp as a matrix $\mathbf{B} \in \mathbb{Z}_q^{m \times n}$, $\mathsf{pk}_j = \boldsymbol{b}_j \in \mathbb{Z}_q^m$ for $j \in [N]$. Then it samples a random matrix $\mathbf{R} \in \{0, 1\}^{D \times m}$, and computes $\mathbf{W}_j = \mathsf{BitDecomp}(\mathbf{R} \cdot \boldsymbol{b}_j \parallel 0^{D \times n})$ for $j \neq i$. It computes $\mathbf{W}_i = \mathsf{Flatten}\left(m \cdot I_D + \mathsf{BitDecomp}(\mathbf{R} \cdot \boldsymbol{b}_i \parallel \mathbf{R} \cdot \mathbf{B})\right)$, where I_D is the identity matrix of dimension $D \times D$. It outputs $C = (\mathbf{W}_1, \ldots, \mathbf{W}_N)$.

TFHE.TransCT$(C; S)$:

- The algorithm parses C as a N matrices $(\mathbf{W}_1, \ldots, \mathbf{W}_N)$. It outputs $C_S = \sum_{j \in S} \mathbf{W}_j$.

TFHE.Eval$(f, C_1, \ldots, C_t; S)$:

- For simplicity, we assume that all the ciphertexts $C_1, \ldots, C_t$'s are transformed to ones that correspond to pk_S (otherwise we can apply the above TFHE.TransCT first). We then observe that actually a transformed ciphertext is of the same form of the GSW scheme [12] where the public key is $(\sum_{k \in S} \boldsymbol{b}_k \parallel \mathbf{B})$. Thus, we can run exactly the same evaluation as the GSW scheme! More specifically, we represent f as a circuit (with all NAND gates). Then we can homomorphically compute $NAND(C, C')$ by outputting $\mathsf{Flatten}(I_D - C \cdot C')$. See the work [12] for detailed explanation.

With out setting of parameters, we can argue that flexible ciphertexts do not leak the underlying messages to the other parties (and so do the transformed ciphertexts, since they can be obtained deterministically from flexible ciphertexts). This can be shown formally using the lemma below in a strait-forward way as done by the work [12]. See their work [12] for further exposition[7].

Lemma 1 (Implicit in [20]). *Let n, m, χ, q be parameters such that the $LWE_{n,q,\chi}$ holds, and N be some polynomial. Then for $m = O((n + N) \log q)$, for any vectors $\boldsymbol{b}_1, \boldsymbol{b}_2, \ldots, \boldsymbol{b}_{N-1} \in \mathbb{Z}_q^m$, then the distribution described as above $(\mathbf{B}, \boldsymbol{b}, \mathbf{R} \cdot (\mathbf{B} \| \boldsymbol{b}), \mathbf{R} \cdot (\boldsymbol{b}_1 \| \ldots \| \boldsymbol{b}_{N-1}))$ is computationally indistinguishable from $(\mathbf{B}, \boldsymbol{u}, \mathbf{U}, \mathbf{R} \cdot (\boldsymbol{b}_1 \| \ldots \| \boldsymbol{b}_{N-1}))$, where $\mathbf{B}$ is uniform over $\mathbb{Z}_q^{m \times n}$, $\boldsymbol{u}$ is uniform over $\mathbb{Z}_q^m$, $\mathbf{U}$ is uniform over $\mathbb{Z}_q^{D \times (n+1)}$, and $\mathbf{R}$ is uniform over $\{0,1\}^{D \times m}$, $D = (n+1) \cdot \ell$, $\ell = \lceil \log q \rceil + 1$.*

4.3 Three-Round MPC with Guarantee of Output Delivery

Now we are ready to present our new three-round MPC for general functions using the new TFHE we have developed in the previous section. We first present a simpler case that considers MPC for polynomial-time deterministic boolean function f (where all parties receive the same bit). Moreover, the security holds against static *semi-malicious* fail-stop attackers[8] corrupting less than half of the parties. In Sect. 5, we discuss how to handle general cases using standard techniques.

Remark 1. Our protocol only needs a public broadcast channel[9]. For simplicity of presentation, we make the following two assumptions. First, there are secure point-to-point channels available. Second, when a party distributes shares to the

---

[7] Our setting of parameters is slightly different from that of the work [12], so our parameters in the lemma are slightly different. The analysis is essentially identical.

[8] Basically, a semi-malicious attacker is one whose behavior follows the protocol with *some* input and randomness he must know. Protocols that achieve such security can be upgraded to malicious security without adding a coin flipping round (c.f. Sect. 5). See the full version of this paper [14] for further details about the notion and its advantage.

[9] In the semi-malicious setting, this can be easily implemented by reliable public point-to-point channels, where an eavesdropper can listen to the all channels but cannot modify the messages.

other parties, he must either send messages to all parties or send messages to no one. These assumptions are not necessary, and we sketch how to achieve them in our protocol using the broadcast channel. We observe that our protocol will only use the secure channels to distribute shares in the second round. So in the first round everyone can publish a public key, and then in the second round, everyone broadcasts encryptions of the shares (under different parties' public keys). This can implement the secure channels, and ensure that parties will either abort (not broadcast at all) or distribute messages to all the other parties.

Our Construction. Let $f : \{0,1\}^{(\ell_{in})^N} \to \{0,1\}$ be a function computed by a depth L circuit, where ℓ_{in} is the input length of each party.

Input: Each party P_i holds some input $x_i \in \{0,1\}^{\ell_{in}}$. The parties share the public parameter pp as described in the TFHE scheme. (pp can be viewed as the common reference string. The generation of pp depends on L, since we need the TFHE to support circuits up to depth L).

The Protocol:

- **Round 1:** The parties execute the first round of the TFHE.Gen(pp). If anyone aborts in this round, then he is simply ignored. Let $S_1 \subseteq [N]$ be the set of non-aborting parties at this round. At the end of this round, each party holds all $\{pk_i\}_{i \in S_1}$.
- **Round 2:** The parties execute the following procedures at the same time:
  - The (currently non-aborting) parties execute the second round of the TFHE.Gen(pp).
  - For $i \in S_1$, P_i broadcasts an encryption of his input using the algorithm TFHE.Enc$_i(x_i)$ (encrypt it bit-by-bit). Note that these are a flexible ciphertexts.
  - Each P_i samples a uniformly random error term from $\eta_i \leftarrow [-B_{smug}, B_{smug}]$, and compute random Shamir secret shares (with the same threshold $[T/2] + 1$). Denote the polynomial as r_i (note that $r_i(0) = \eta_i$). Then each P_i sends $r_i(j)$ to party P_j for $j \neq i$.

  Let $C_i = (C_{i,1}, C_{i,2}, \ldots, C_{i,\ell_{in}})$ be the broadcasted ciphertexts from P_i, and $(r_i(1), \ldots, r_i(N))$ be the shares from P_i to the other parties.

  If anyone aborts at this round, either not sending the second round of TFHE.Gen(pp), the ciphertexts, or the shares of error terms, then he (and his input) are again ignored. Let $S_2 \subseteq S_1$ be the set of non-aborting parties.
- **Round 3:** Now each non-aborting party in S_2 first transforms the ciphertexts he received to ones that correspond to pk_{S_2}. Let $\{C_{j,k}\}_{j \in S_2, k \in [\ell_{in}]}$ be the broadcasted ciphertexts. For $i \in S_2$, P_i first computes $C_{j,k}^{S_2} = $ TFHE.TransCT$(C_{j,k}; S_2)$ for $j \in S_2$, $k \in [\ell_{in}]$.

  Let f^{S_2} be the residual function where the inputs of $[N] \setminus S_2$ are replaced with the default values. P_i homomorphically computes the residual function, i.e. $C^* = $ TFHE.Eval$(f^{S_2}, \{C_{j,k}^{S_2}\}_{j \in S_2, k \in [\ell_{in}]})$.

  Then each P_i computes $v_i = \sum_{k \in S_2} r_k(i)$. Finally, they run the threshold decryption TFHE.Dec$_{S_2}(\{C^*; v_{S_2}\})$, where v_{S_2} denotes the vector of the following set $\{v_j : j \in S_2\}$.

Recall that the protocol TFHE.Dec handles situations when parties abort. In this round, parties broadcast some messages, and a majority of them is sufficient to recover the output.

Theorem 2. *Let f be any deterministic functionality with N inputs and one output. Let pp be parameters sampled according to the choice as the TFHE above, and the corresponding LWE assumption holds. Then the above protocol π UC-realizes the ideal functionality $\mathcal{F}_f$ with guarantee of output delivery, in the presence of any static (semi-malicious) fail-stop adversary who corrupts less than $\lceil N/2 \rceil$ parties.*

As explained in the introduction, the transformed ciphertexts $\{C_{j,k}^{S_2}\}_{j \in S_2, k \in [\ell_{in}]}$ are GSW ciphertexts under the public key $(\mathbf{B}, \sum_{i \in S_2} \boldsymbol{b}_i)$. Therefore, by applying the evaluation algorithm, C^* is a ciphertext of the output y. Each party in our threshold decryption protocol, as explained, outputs a share of y by computing some inner product with the shares (and substraction). Thus, the correctness holds.

To prove security, we need to construct a simulator $\mathcal{S}$ that generates the views of the honest parties. We sketch the construction: the simulator simulates the public parameter faithfully, and generates the messages in each round as follows. Let $\mathcal{I}$ be the set of corrupted set.

- **(First round).** $\mathcal{S}$ simulates the public keys of honest parties' by random vectors $\boldsymbol{u}_i$ for $i \notin \mathcal{I}$.
- **(Second round).** $\mathcal{S}$ simulates the encrypted ciphertexts by TFHE.Enc(0), and simulates the error terms and shares of secret keys by sending random values (or vectors).
- **(Third round).** $\mathcal{S}$ then reads the witness tapes of the adversary to get secret keys and inputs from the corrupted parties. He sets the aborting parties' inputs to be the default value, and then queries the ideal functionality to receive the output y. From the output y and the secret keys of the corrupted parties, $\mathcal{S}$ then figures out consistent outputs of the honest parties.

Intuitively, the LWE assumption guarantees that the simulation in round 1 is indistinguishable, and Lemma 1 guarantees that TFHE.Enc(0) is indistinguishable form the encryptions in the real world. The last step is the most challenging, and we will further explain the ideas in the appendix.

In the full version of this paper [14], we present the detailed analyses of correctness and security with further exposition.

5 Variants and Generalizations

In this section, we discuss variants of our basic protocol in the following aspects: (1) how to handle functionalities with longer inputs, (2) how to handle randomized functionalities, (3) how to compile a protocol that is secure against semi-malicious adversaries into one that is secure against malicious adversaries, and (4) how to reduce one round by using a PKI setup. These issues can be handled using standard techniques as presented in the work of Asharov et al. [1]. We highlight the ideas and refer curious readers to their work for further details.

Functions with Longer Outputs. Let $f : \{0,1\}^{(\ell_{in})^N} \to \{0,1\}^{\ell_{out}}$ be an N-ary functionality. We consider ℓ_{out} boolean functionalities $\left\{ f_i : \{0,1\}^{(\ell_{in})^N} \to \{0,1\} \right\}_{i \in [\ell_{out}]}$ where each f_i outputs the i-th bit of f. Let π_i be the protocol computing f_i as we described in Sect. 4. To compute f, we simply run $\pi_1, \ldots, \pi_{\ell_{out}}$ in parallel, and we treat an abort in any one of the execution as an abort in all executions. To argue that the resulting protocol is secure against an arbitrary semi-malicious adversary, we also require the adversary to include proofs, in the form of witnesses written to their witness tape, of input-consistency across the parallel executions. This is to enforce that the adversary is using the same inputs for all the subprotocols. Below we will describe a compiler that upgrades the protocol to one against malicious adversaries.

Randomized Functionalities. Our basic MPC protocol only considers deterministic functionalities where all the parties receive the same output. It can be generalized to handle with randomized functionalities and individual outputs via a standard transformation. Basically in this transformation, instead of computing some randomized function $f(x_1, \ldots, x_N; r)$, the parties compute the deterministic function $f'\left((x_1, r_1), \ldots, (x_N, r_N)\right) = f\left(x_1, \ldots, x_N; \oplus_{i \in [N]} r_i\right)$. This transformation does not add additional rounds.

Semi-malicious Security to Malicious Security. Our basic MPC protocol is only secure in the semi-malicious setting. Asharov et al. [1] presents a simple and general round-preserving compiler from semi-malicious to fully malicious security using UC NIZKs [9] in the CRS model. In particular, in each round, the attacker must prove (in zero-knowledge) that it is following the protocol consistently with *some* setting of the random coins. In particular, we present the theorem of Asharov et al. [1]:

Theorem 3 ([1]). *There is a generic round-preserving compiler such the following holds. Let $\mathcal{F}$ be an N-ary functionality and π be an N-party protocol. Suppose π t-securely computes F against semi-malicious fail-stop adversaries with guarantee of output delivery (or fairness), then the compiled protocol π' t-securely computes $\mathcal{F}$ against malicious adversaries with guarantee of output delivery (or fairness, respectively) in the CRS, $\mathsf{F_{ZK}}$, and authenticated broadcast-hybrid model. Moreover, π' has the same round complexity as π.*

Together with Theorem 2, we are able to achieve the following corollary:

Corollary 1. *Assume that the LWE assumption holds and UC-NIZK exists. Then there exists a three-round MPC in the CRS and authenticated broadcast hybrid model, with a guarantee of output delivery, and providing security against a malicious adversary that corrupts less than half of the parties.*

Two Rounds with PKI. We recall that in the first round of our protocol, each party just publishes some public key $b_i = \mathbf{B} \cdot s_i + e_i$, which is independent of the input. If there is an additional setup *public-key infrastructure* (PKI), then we can move the first round to the PKI. Thus the entire MPC execution would consist only of the remaining two rounds. The resulting PKI is very simple and does not require a trusted party for setup; we just need a trusted party to choose a CRS, and then each party can choose its own public key individually (possibly maliciously). Moreover, the PKI can be reused for many MPC executions of arbitrary functions f with arbitrary inputs.

The security analysis is exactly the same as that of our original three-round protocol in the CRS model, just by noting that the first round there consists of broadcast message, which does not depend on the inputs of the parties (and hence we can think of it as a public key). In the malicious case, the parties need to provide a zero-knowledge proof of knowing some randomness of their public keys registered in the PKI. This is similar to our original protocol (without PKI) where the parties need to provide a zero-knowledge proof of knowing some randomness of their first round messages.

Acknowledgments. This research is partially supported by an NSF grant CNS-1314857, a Sloan Fellowship, and Google Research Awards.

References

1. Asharov, G., Jain, A., López-Alt, A., Tromer, E., Vaikuntanathan, V., Wichs, D.: Multiparty computation with low communication, computation and interaction via threshold FHE. In: Pointcheval, D., Johansson, T. (eds.) EUROCRYPT 2012. LNCS, vol. 7237, pp. 483–501. Springer, Heidelberg (2012)
2. Barak, B., Goldreich, O., Impagliazzo, R., Rudich, S., Sahai, A., Vadhan, S.P., Yang, K.: On the (im)possibility of obfuscating programs. In: Kilian, J. (ed.) CRYPTO 2001. LNCS, vol. 2139, p. 1. Springer, Heidelberg (2001)
3. Beaver, D., Micali, S., Rogaway, P.: The round complexity of secure protocols (extended abstract). In: 22nd ACM STOC, pp. 503–513. ACM Press, May 1990
4. Ben-Or, M., Goldwasser, S., Wigderson, A.: Completeness theorems for non-cryptographic fault-tolerant distributed computation (extended abstract). In: 20th ACM STOC, pp. 1–10. ACM Press, May 1988
5. Canetti, R.: Universally composable security: a new paradigm for cryptographic protocols. In: 42nd FOCS, pp. 136–145. IEEE Computer Society Press, October 2001
6. Cleve, R.: Limits on the security of coin flips when half the processors are faulty (extended abstract). In: Proceedings of the 18th Annual ACM Symposium on Theory of Computing, May 28–30, 1986, Berkeley, California, USA, pp. 364–369 (1986)
7. Cohen, R., Lindell, Y.: Fairness versus guaranteed output delivery in secure multiparty computation. In: Sarkar, P., Iwata, T. (eds.) ASIACRYPT 2014, Part II. LNCS, vol. 8874, pp. 466–485. Springer, Heidelberg (2014)
8. Damgård, I.B., Ishai, Y.: Constant-round multiparty computation using a black-box pseudorandom generator. In: Shoup, V. (ed.) CRYPTO 2005. LNCS, vol. 3621, pp. 378–394. Springer, Heidelberg (2005)

9. De Santis, A., Di Crescenzo, G., Ostrovsky, R., Persiano, G., Sahai, A.: Robust non-interactive zero knowledge. In: Kilian, J. (ed.) CRYPTO 2001. LNCS, vol. 2139, p. 566. Springer, Heidelberg (2001)
10. Garg, S., Gentry, C., Halevi, S., Raykova, M.: Two-round secure MPC from indistinguishability obfuscation. In: Lindell, Y. (ed.) TCC 2014. LNCS, vol. 8349, pp. 74–94. Springer, Heidelberg (2014)
11. Gennaro, R., Ishai, Y., Kushilevitz, E., Rabin, T.: On 2-round secure multiparty computation. In: Yung, M. (ed.) CRYPTO 2002. LNCS, vol. 2442, p. 178. Springer, Heidelberg (2002)
12. Gentry, C., Sahai, A., Waters, B.: Homomorphic encryption from learning with errors: conceptually-simpler, asymptotically-faster, attribute-based. In: Canetti, R., Garay, J.A. (eds.) CRYPTO 2013, Part I. LNCS, vol. 8042, pp. 75–92. Springer, Heidelberg (2013)
13. Goldreich, O.: Foundations of Cryptography: Basic Applications, vol. 2. Cambridge University Press, Cambridge (2004)
14. Gordon, S.D., Liu, F.-H., Shi, E.: Constant-round mpc with fairness and guarantee of output delivery. Cryptology ePrint Archive, report 2015/371 (2015)
15. Ishai, Y., Kushilevitz, E., Paskin, A.: Secure multiparty computation with minimal interaction. In: Rabin, T. (ed.) CRYPTO 2010. LNCS, vol. 6223, pp. 577–594. Springer, Heidelberg (2010)
16. Katz, J., Ostrovsky, R.: Round-optimal secure two-party computation. In: Franklin, M. (ed.) CRYPTO 2004. LNCS, vol. 3152, pp. 335–354. Springer, Heidelberg (2004)
17. Katz, J., Ostrovsky, R., Smith, A.: Round efficiency of multi-party computation with a dishonest majority. In: Biham, E. (ed.) EUROCRYPT 2003. LNCS, vol. 2656, pp. 578–595. Springer, Heidelberg (2003)
18. López-Alt, A., Tromer, E., Vaikuntanathan, V.: On-the-fly multiparty computation on the cloud via multikey fully homomorphic encryption. In: Karloff, H.J., Pitassi, T. (eds.) 44th ACM STOC, pp. 1219–1234. ACM Press, New York (2012)
19. Pass, R.: Bounded-concurrent secure multi-party computation with a dishonest majority. In: Babai, L. (ed.) 36th ACM STOC, pp. 232–241. ACM Press, New York (2004)
20. Regev, O.: On lattices, learning with errors, random linear codes, and cryptography. In: Gabow, H.N., Fagin, R. (eds.) 37th ACM STOC, pp. 84–93. ACM Press, New York (2005)
21. Yao, A.C.-C.: How to generate and exchange secrets (extended abstract). In: 27th FOCS, pp. 162–167. IEEE Computer Society Press, October 1986

Zero-Knowledge

Statistical Concurrent Non-malleable
Zero-Knowledge from One-Way Functions

Susumu Kiyoshima$^{(\boxtimes)}$

NTT Secure Platform Laboratories, Tokyo, Japan
kiyoshima.susumu@lab.ntt.co.jp

Abstract. *Concurrent non-malleable zero-knowledge* (CNMZK) protocols are zero-knowledge protocols that are secure even when the adversary interacts with multiple provers and verifiers simultaneously. Recently,
the first *statistical* CNMZK argument for $\mathcal{NP}$ was constructed by Orlandi
et al. (TCC'14) under the DDH assumption.

In this paper, we construct a statistical CNMZK argument for $\mathcal{NP}$
assuming only the existence of one-way functions. The security is proven
via black-box simulation, and the round complexity is $\mathsf{poly}(n)$. Under
the existence of collision-resistant hash functions, the round complexity
can be reduced to $\omega(\log n)$, which is essentially optimal for black-box
concurrent zero-knowledge.

1 Introduction

Zero-knowledge (ZK) *proofs* and *arguments* are protocols that enable the prover
to convince the verifier of the correctness of a mathematical statement while
providing *zero additional knowledge*. This "zero additional knowledge" property
is formalized by using the *simulation paradigm*: An interactive proof or argument
is said to be zero-knowledge if for any adversarial verifier there exists a *simulator*
that can output a simulated view of the adversary. In the original definition of
the ZK property, the adversary interacts with a single prover at a time. Thus,
the original definition guarantees the ZK property in the stand-alone setting.

Non-malleable zero-knowledge (NMZK) [6] and *concurrent zero-knowledge*
(CZK) [7] are security notions that guarantee the ZK property in the concurrent
setting. Specifically, NMZK guarantees the ZK property in the setting where the
adversary concurrently interacts with a honest prover in the *left session* and a
honest verifier in the *right session*, and CZK guarantees the ZK property in the
setting where the adversary concurrently interacts with unbounded number of
honest provers.

As a security notion that implies both NMZK and CZK, Barak et al. [1] proposed *concurrent non-malleable zero-knowledge* (CNMZK). CNMZK guarantees
the ZK property in the setting where the adversary concurrently interacts with
many provers in the left sessions and many verifiers in the right sessions. In particular, it guarantees that receiving proofs in the left session does not help the
adversary to give proofs in the right sessions—that is, it guarantees that if the

© International Association for Cryptologic Research 2015
R. Gennaro and M. Robshaw (Eds.): CRYPTO 2015, Part II, LNCS 9216, pp. 85–106, 2015.
DOI: 10.1007/978-3-662-48000-7_5

adversary can prove some statements in the right sessions while receiving proofs in the left sessions, the adversary could prove the same statements *even without receiving proofs in the left sessions*. In the definition of CNMZK, this guarantee is formalized as the existence of a *simulator-extractor* that can simulate the adversary's view in the left and right sessions while extracting witnesses from the adversary in the simulated right sessions.

The first CNMZK argument was constructed by Barak et al. [1]. Subsequently, a computationally efficient construction was shown by Ostrovsky et al. [21]. The first CNMZK *proof* was constructed by Lin et al. [16], and a variant of their protocol was shown to be secure with adaptively chosen inputs by Lin and Pass [14]. Additionally, a CNMZK argument that is secure with "fully" adaptively chosen inputs was recently constructed by Venkitasubramaniam [26].

Very recently, Orlandi et al. [20] constructed the first *statistical* CNMZK argument—that is, a CNMZK argument such that the view simulated by the simulator-extractor is statistically indistinguishable from the adversary's view. Statistical CNMZK is clearly of great interest since it guarantees quite strong security in the concurrent setting. However, statistical CNMZK is hard to achieve, and the existing techniques of computational CNMZK protocols seem to be insufficient for constructing statistical CNMZK protocols (see Sect. 2.1).

On statistical CNMZK protocols, an important open question is what hardness assumption is needed for constructing them. The statistical CNMZK argument of Orlandi et al. [20] was constructed under the DDH assumption (or the existence of dense cryptosystems). Thus, it is already known that statistical CNMZK protocols can be constructed under standard assumptions. However, since it is known that the existence of one-way functions is sufficient for constructing both statistical ZK protocols and computational CNMZK protocols [1,10], it is important to study the following question.

Can we construct statistical concurrent non-malleable zero-knowledge protocols by assuming only the existence of one-way functions?

1.1   Our Result

In this paper, we answer the above question affirmatively.

Theorem 1. *Assume the existence of one-way functions. Then, there exists a statistical concurrent non-malleable zero-knowledge argument for $\mathcal{NP}$ with round complexity $\mathsf{poly}(n)$. Furthermore, if there exists a family of collision-resistant hash functions, the round complexity can be reduced to $\omega(\log n)$.*

The round complexity of our statistical CNMZK argument—$\mathsf{poly}(n)$ rounds when only the existence of one-way functions is assumed and $\omega(\log n)$ rounds when the existence of a family of collision-resistant hash functions is assumed—is the same as the round complexity of the known statistical CZK arguments [9]. Thus, our result closes the gap between statistical CNMZK arguments and statistical CZK arguments. Furthermore, since the security of our statistical CNMZK protocol is proven via black-box simulation, the logarithmic round complexity of our hash-function-based protocol is essentially tight due to the lower bound on black-box CZK protocols [3].

2 Techniques

2.1 Previous Techniques

Before explaining our technique, we explain the difficulty of constructing statistical CNMZK protocols by using the techniques of existing computational CNMZK protocols [1,16].

We first recall the protocols of [1,16]. The definition of CNMZK requires the existence of a simulator-extractor that simulates the adversary's view while extracting the witnesses for the statements proven by the adversary in the simulated view. To satisfy this definition, protocols need to satisfy the following properties: (i) the proofs in the left sessions can be simulated for the adversary; (ii) even when the adversary receives simulated proofs in the left sessions, the witnesses can be extracted from the adversary in the right sessions. In the protocol of [1,16], the simulatability of the left sessions is guaranteed by requiring the verifier to commit to a random trapdoor by using a *concurrently extractable commitment scheme* CECom [17]. Since the committed values of CECom can be extracted by a rewinding extractor even in the concurrent setting, the proofs in the left sessions can be simulated by extracting the trapdoors from CECom. On the other hand, the witness-extractability of the right sessions is guaranteed by requiring the prover to commit to the witness with a non-malleable commitment scheme NMCom [6] and additionally designing the protocols so that the following hold.

1. When the adversary receives honest proofs in the left sessions, the committed value of the NMCom commitment is indeed a valid witness in every accepted right session.
2. When the proofs in the left sessions are switched to the simulated ones, the committed values of the NMCom commitments do not change in the right sessions due to the non-malleability of NMCom.

It follows from these that even when the adversary receives simulated proofs in the left sessions, the committed value of the NMCom commitment is a witness for the statement in every accepted right session. Therefore, the witnesses can be extracted in the right sessions by extracting the committed values of the NMCom commitments.

As mentioned above, the techniques of [1,16] alone seem to be insufficient for constructing statistical CNMZK protocols. This is because the techniques of [1,16] requires the prover to commit to the witness by using NMCom, which is only computationally hiding.[1] Since in the simulation the committed values of NMCom need to be switched to another values (e.g., 0^n) in the left sessions, the simulated view can be only computational indistinguishable from the real view.

---

[1] NMCom need to be *non-malleable w.r.t. commitment* [6], which roughly says that the committed value of the commitment that the man-in-the-middle adversary gives is independent of the committed value of the commitment that adversary receives. Since the definition of non-malleability w.r.t. commitment is meaningless when the committed values cannot be uniquely determined, NMCom cannot be statistically hiding.

Recently, Orlandi et al. [20] constructed a statistical CNMZK protocol by modifying the CNMZK protocol of [1] with *mixed non-malleable commitment scheme* MXNMCom. MXNMCom is parametrized by a string and is either statistically hiding or non-malleable depending on the string.[2] Very roughly speaking, Orlandi et al. circumvent the above problem by switching the parameter string of MXNMCom in the security proof—when proving the statistical indistinguishability of the simulation, the string is set so that MXNMCom is statistically hiding, and when proving the non-malleability, the string is set so that MXNMCom is non-malleable. The use of MXNMCom, however, requires assumptions that are stronger than the existence of one-way functions (such as the DDH assumption or the existence of dense cryptosytems). Thus, the technique of Orlandi et al. cannot be used to construct statistical CNMZK protocols from one-way functions.

2.2 Our Technique

Since the techniques of [1, 16] cannot be used for statistical CNMZK protocols because the committed values of NMCom need to be switched during the simulation, one potential strategy for statistical CNMZK is to construct a protocol such that the adversary's view can be simulated *without switching the committed value of* NMCom *(and of any other computationally hiding commitment)*. However, when the simulator commits to the same value in NMCom as a honest prover, it is not clear how non-malleability of NMCom can be used in the security proof. Below, we show that the CNMZK property can be shown even in this case *if we use a stronger variant of* NMCom.

A key technical tool in our technique is *CCA-secure commitment schemes* [4], which is a stronger variant of (concurrent) non-malleable commitment schemes. Roughly speaking, CCA security guarantees that the scheme is hiding even against adversaries that have access to the *committed-value oracle*, which receives concurrent commitments from the adversary and returns their committed values to the adversary. (In non-malleability, the oracle receives only parallel commitments from the adversary and returns the committed values only after the adversary finishes the interaction with the committer.) Several CCA-secure commitment schemes were constructed from one-way functions [4, 8, 12, 15]; furthermore, although CCA security itself does not provide any extractability, all of these schemes satisfy concurrent extractability as well.

Using CCA-secure commitment schemes, we construct the following protocol as a starting point.

Stage 1. (V commits to trapdoor)
1. The verifier V chooses random $r_V \in \{0,1\}^n$ and commits to r_V by using a statistically binding commitment scheme Com, which can be constructed from one-way functions [11, 18]. Let (r_V, d) be the decommitment.

---

[2] Specifically, Orlandi et al. [20] used the scheme such that (i) when the string is sampled from a uniform distribution, the scheme is statistically hiding and (ii) when the string is taken from another (computationally indistinguishable) distribution, the scheme is non-malleable.

2. V commits to (r_V, d) by using CCA-CECom, where CCA-CECom is a CCA-secure commitment scheme that is also concurrent extractable [4,8,12,15].

Stage 2. (P proves $x \in L$ or knowledge of trapdoor) The prover P proves that it knows a witness for $x \in L$ or a valid decommitment (r_V, d) of the Com commitment that V gives in Stage 1. P proves this statement by using a statistical witness-indistinguishable argument of knowledge sWIAOK, which can be constructed from one-way functions by instantiating Blum's Hamiltonian-cycle protocol with the statistically hiding commitment scheme of [10].

In this protocol, the verifier's view can be statistically simulated by a simulator that extracts (r_V, d) from CCA-CECom and uses it as a witness in sWIAOK. (Note that this simulator executes Stage 1 honestly; thus, even if computationally hiding commitment schemes are used as building blocks in CCA-CECom, the simulator commits to the same values by using them as a honest prover.) Also, intuitively this protocol seems to be CNMZK from the following reason.

- The CCA security of CCA-CECom guarantees that the trapdoors of the right sessions are hidden from the adversary even when the trapdoors of the left sessions are extracted and returned to the adversary.
- Then, since the simulated proofs are generated in the left sessions by extracting the trapdoors, the trapdoors in the right sessions are hidden from the adversary even when the adversary receives simulated proofs in the left sessions.
  Thus, even when the adversary receives the simulated proofs in the left sessions, the adversary cannot "cheat" in the right sessions, and therefore witnesses for the statements must be extractable from sWIAOK in the right sessions.

Of course, to formally show the statistical CNMZK property, we need to show a simulator-extractor that statistically simulates the adversary's view and also extracts witnesses for the statements in the right sessions.

As the simulator-extractor, we consider the following $\mathcal{SE}$.

1. First, $\mathcal{SE}$ simulates the view of the adversary $\mathcal{A}$ by executing the following simulator $\mathcal{S}$: Simulator $\mathcal{S}$ internally invokes $\mathcal{A}$ and interacts with it in the left and right sessions honestly except that in each left session, $\mathcal{S}$ extracts (r_V, d) by using the concurrent extractor of CCA-CECom and uses it as a witness in sWIAOK.

2. After simulating the view of $\mathcal{A}$ as above, $\mathcal{SE}$ extracts witnesses from the right sessions by doing the following for each right session. First, $\mathcal{SE}$ rewinds $\mathcal{S}$ until the point just before $\mathcal{S}$ sends the challenge message of sWIAOK to $\mathcal{A}$.[3] Then, $\mathcal{SE}$ repeatedly executes $\mathcal{S}$ from this point with flesh randomness until it obtains another accepted transcript of sWIAOK. After obtaining another accepted transcript, $\mathcal{SE}$ extracts a witness by using the argument-of-knowledge property of sWIAOK.

---

[3] Since $\mathcal{S}$ rewinds $\mathcal{A}$ during the concurrent extraction of CCA-CECom, $\mathcal{S}$ may send the challenge message of sWIAOK of a right session to $\mathcal{A}$ multiple times. Here, $\mathcal{SE}$ rewinds $\mathcal{S}$ until the point just before $\mathcal{S}$ sends it to $\mathcal{A}$ on the "main thread."

It is easy to see that $\mathcal{SE}$ statistically simulates the real view of $\mathcal{A}$. Thus, it remains to show that $\mathcal{SE}$ extracts witnesses for the statements in the right sessions.

To show the witness extractability of $\mathcal{SE}$, a natural approach is to follow the above-mentioned approach of [1,16] and show the following.

1. When $\mathcal{A}$ receives honest proofs in the left sessions, a witness for the statement is extracted from the sWIAOK proof in every accepted right session.
2. When the honest proofs in the left sessions are switched to the simulated ones, the value extracted from sWIAOK does not change in every accepted right session.

Note that here we argue about the extracted values instead of the committed values. At first sight, it seems that this is not a big difference and it seems that the above can be shown by using an argument similar to the one used in [1,16].

However, this approach does not work. In particular, we cannot show the second part—that is, we cannot show that the extracted values remain to be the same when the honest proofs in the left sessions are switched to the simulated ones. To see this, observe the following. Since the witnesses used in sWIAOK are switched in the simulated proofs, we need to use the witness indistinguishability of sWIAOK of the left sessions. However, since $\mathcal{A}$ is rewound during the witness extraction of the sWIAOK proofs of the right sessions, if the left and the right sessions are scheduled so that the sWIAOK proofs of the left sessions are executed in parallel with the sWIAOK proofs of the right sessions, the sWIAOK proofs of the left sessions are also rewound, and thus we cannot use their witness indistinguishability.[4]

Thus, we instead use the following approach. Informally, the above approach does not work because the honest proofs and the simulated proofs are "too different." We thus introduce a hybrid experiment in which $\mathcal{A}$ receives *hybrid proofs* in the left sessions, where a hybrid proof is generated by extracting (r_V, d) by brute force and using it as a witness in sWIAOK. (Notice that the only difference between the hybrid proofs and the simulated proofs is how the trapdoors are extracted.) We then show that (i) witnesses for the statements are extracted in the right sessions when $\mathcal{A}$ receives hybrid proofs in the left sessions, and (ii) when hybrid proofs are switched to the simulated ones, the extracted values do not change. In particular, our analysis proceeds as follows.

- First, we show the second part, i.e., we show that the values extracted in the right sessions do not change when the proofs in the left sessions are switched from the hybrid proofs to the simulated ones. Since the only difference between the hybrid proofs and the simulated ones is how the committed values of the

---

[4] If we use the robust extraction technique [8], for each left session there exists a rewinding strategy that allows us to extract witnesses from the right sessions without rewinding sWIAOK of this left session. However, since what we want to show is that the values extracted in the right sessions *by the rewinding strategy that $\mathcal{SE}$ uses* are unchanged, the robust extraction technique cannot be used here (unless there exists a rewinding strategy that allows us to extract witnesses from the right sessions without rewinding the sWIAOK proof of *every* left session).

CCA-CECom commitments are extracted (by brute-force or by the concurrent extractability), we can show this by using the concurrent extractability of CCA-CECom. We note however that there is a subtlety since CCA-CECom in the left sessions can be rewound not only by the concurrent extractor of CCA-CECom but also by the extractor of sWIAOK. Nonetheless, by carefully using a standard technique (the "good prefix" argument), we can show that the concurrent extractor of CCA-CECom works even in this case.

- Next, we show that in the hybrid experiment, witnesses for the statements are extracted from the right sessions. Since the simulated proofs can be efficiently generated given access to the committed-value oracle of CCA-CECom, at first sight it seems that this follows directly from the CCA security of CCA-CECom and argument-of-knowledge property of sWIAOK—if a witness for the statement is not extracted, (r_V, d) must be extracted, and thus we can break the CCA security of CCA-CECom. However, there are two problems.

  1. Since CCA-CECom in the left sessions can be rewound during the witness extraction of sWIAOK of the right sessions, the hybrid experiment cannot be emulated even given access to the committed-value oracle of CCA-CECom. Hence, the CCA-secure commitments in the right sessions may not be hiding in the hybrid experiment.

  2. Since the adversary obtains hybrid proofs, which are generated in super-polynomial time, the argument-of-knowledge property of sWIAOK may not hold in the hybrid experiment. We note that although existing CCA-secure commitment schemes provides *robustness*, which guarantees that arbitrary "small"-round protocol remains secure even when adversaries have access to the committed-value oracle, we cannot use robustness here since CCA-CECom in the left sessions can be rewound during the witness extraction of sWIAOK of the right sessions and therefore the hybrid experiment cannot be emulated even given access to the committed-value oracle.

  Because of these problems, we cannot use the security of CCA-CECom directly in the analysis. Thus, instead of using existing CCA-secure commitment schemes in a modular way, we directly use their building blocks in the protocol and directly use their proof technique in the analysis. (In particular, we use the robust concurrent extraction technique of [8] and a one-one CCA-secure commitment scheme of [13].) The proof techniques of existing CCA-secure commitment schemes are strong enough to solve the above problems, and thus we can show that witnesses for the statements are extracted in the hybrid experiment.

From the above two, it follows that even when $\mathcal{A}$ receives simulated proofs in the left session, valid witnesses are extracted in right sessions. This completes the overview of our technique.

3   Definitions

In this section, we sketch the definitions used in this paper. The formal definitions are given in the full version.

3.1   Statistical Concurrent Non-malleable Zero-Knowledge Arguments

The definition of (statistical) concurrent non-malleable zero-knowledge [1,20] is closely related to the definition of simulation extractability of [22]. Let $\langle P, V \rangle$ be an interactive argument for a language $L \in \mathcal{NP}$. For any man-in-the-middle adversary $\mathcal{A}$, let us consider a probabilistic experiment in which $\mathcal{A}$ participates in the following left and right interactions. In the left interaction, $\mathcal{A}$ interacts with a honest prover P of $\langle P, V \rangle$ and verifies the validity of statements $x_1, \ldots, x_m$ using identities $\mathsf{id}_1, \ldots, \mathsf{id}_m$. In the right interaction, $\mathcal{A}$ interacts with a honest verifier V of $\langle P, V \rangle$ and proves the validity of statements $\widetilde{x}_1, \ldots, \widetilde{x}_m$ using identities $\widetilde{\mathsf{id}}_1, \ldots, \widetilde{\mathsf{id}}_m$. The statements proven in the left interaction, $x_1, \ldots, x_m$, are given to P and $\mathcal{A}$ prior to the experiment. In contrast, the statements proven in the right interaction, $\widetilde{x}_1, \ldots, \widetilde{x}_m$, and the identities used in the left and the right interactions, $\mathsf{id}_1, \ldots, \mathsf{id}_m$ and $\widetilde{\mathsf{id}}_1, \ldots, \widetilde{\mathsf{id}}_m$, are chosen by $\mathcal{A}$ during the experiment. Then, roughly speaking, $\langle P, V \rangle$ is *statistical concurrent non-malleable zero-knowledge* (statistical CNMZK) if for any adversary $\mathcal{A}$, there exists a PPT machine called the *simulator-extractor* that can statistically simulate the view of $\mathcal{A}$ in the above experiment while extracting witnesses for the statements proven by $\mathcal{A}$ in the accepted right interactions that use different identities from the left interactions.

3.2   Concurrently Extractable Commitment Schemes

Roughly speaking, a commitment scheme is *concurrently extractable* if there exists a PPT extractor such that for any adversarial committer that concurrently commits to many values by using the scheme, the extractor can extract the committed value from the adversarial committer in every valid commitment.[5]

Micciancio et al. [17] showed a $\omega(\log n)$-round concurrently extractable commitment CECom (Fig. 1), which is an abstraction of the preamble stage of the concurrent zero-knowledge protocol of [25] and can be constructed from one-way functions. The extractor of CECom performs the extraction by rewinding the adversarial committer according to the rewinding strategy of [23,25]—the extractor internally invokes the adversarial committer C^* and interacts with C^* as honest receivers on the "main thread"; at the same time, the extractor rewinds the main thread and generates "look-ahead threads" on which the extractor interacts with C^* again as honest receivers with flesh randomness; then, at the end of each commitment on each thread, the extractor extracts the committed values by using the information collected on the other threads.

Robust Concurrent Extraction.   On the concurrently extractable commitment scheme CECom of [17], Goyal et al. [8] showed a very useful lemma called the *robust concurrent extraction lemma*. Roughly speaking, this lemma states that even when the adversarial committer additionally participates in an external

---

[5]   A commitment is *valid* if there exists a value to which it can be decommitted.

CECom can be seen as concurrent executions of the extractable commitment scheme ExtCom of [24], which consists of three stages—commit, challenge, and reply—and can be constructed from one-way functions.

Commit phase. The committer C and the receiver R receive common input 1^n and parameter ℓ. (In [17], $\ell = \omega(\log n)$.) To commit to $v \in \{0, 1\}^n$, the committer C commits to v concurrently ℓ times by using ExtCom as follows.
  1. C and R execute commit stage of ExtCom ℓ times in parallel.
  2. For each $j \in [\ell]$ in sequence, C and R do the following.
     (a) R sends the challenge message of ExtCom for the j-th session.
     (b) C sends the reply message of ExtCom for the j-th session.
Decommit phase. C sends v to R and decommits all the ExtCom commitments.

Fig. 1. Concurrently extractable commitment CECom [17].

protocol, the committed values can be extracted from the adversarial committer *without rewinding the external protocol* as long as the round complexity of the external protocol is "small." In particular, the lemma guarantees that the robust concurrent extraction is possible as long as $\ell - O(k \cdot \log n) = \omega(\log n)$, where ℓ is the parameter of CECom and k is the round complexity of the external protocol. (Thus, we need to set $\ell := \omega(\log n)$ when $k = O(1)$ and set $\ell := \mathsf{poly}(n)$ when $k = \mathsf{poly}(n)$.)

In this work, we cannot use the lemma in a black-box way since in the security analysis we use a specific property of the extractor shown in [8]. In particular, in our security analysis, it is important that the extractor of [8] performs the extraction by generating the main thread and the look-ahead threads as in the rewinding strategies of [23, 25].

3.3 (One-one) CCA-secure Commitment Schemes

We recall the definition of (one-one) CCA security and κ-robustness of commitment schemes [4, 13, 15].

(One-one) CCA Security. Roughly speaking, a tag-based commitment scheme $\langle C, R \rangle$ (i.e., a commitment scheme that takes an n-bit string—a *tag*—as an additional input) is *CCA-secure* if it is hiding even against adversary $\mathcal{A}$ that interacts with the following *committed-value oracle*: The committed-value oracle $\mathcal{O}$ interacts with $\mathcal{A}$ as an honest receiver in many concurrent sessions of the commit phase of $\langle C, R \rangle$ using tags chosen adaptively by $\mathcal{A}$; at the end of each session, if the commitment of this session is invalid or has multiple committed values, $\mathcal{O}$ returns $\bot$ to $\mathcal{A}$; otherwise, $\mathcal{O}$ returns the unique committed value to $\mathcal{A}$.

If $\langle C, R \rangle$ is CCA secure only against adversaries that interact with the *one-session committed-value oracle*, which is the same as the committed-value oracle except that it interacts with the adversary only in a single session, $\langle C, R \rangle$ is *one-one CCA secure.*

κ-*Robustness.* Roughly speaking, a tag-based commitment scheme is κ-*robust* if for any adversary $\mathcal{A}$ and any ITM B, the joint output of a κ-round interaction between $\mathcal{A}^{\mathcal{O}}$ and B can be simulated without $\mathcal{O}$ by a PPT simulator. Intuitively, κ-robustness guarantees that the security of any κ-round protocol (say, the hiding property of a κ-round commitment scheme) holds even against the adversary that interacts with $\mathcal{O}$.

The Scheme We Use. From a result shown in [8], we can obtain a constant-round κ-robust one-one CCA-secure commitment scheme for every constant $\kappa \in \mathbb{N}$ from one-way functions. In [8], Goyal et al. constructed a $\omega(\log n)$-round CCA-secure commitment scheme from one-way functions. This scheme has $\omega(\log n)$ rounds because CECom with parameter $\ell = \omega(\log n)$ is used as a building block. The reason why ℓ is set to be $\omega(\log n)$ is that in the security analysis, the committed values of CECom need to be extracted when polynomially many CECom commitments are concurrently executed. In the setting of *one-one* CCA security, however, the security analysis works even if the committed values of CECom are extractable only when a single CECom commitment is executed; hence, we can set $\ell := O(1)$. For completeness, we give the protocol and the proof of one-one CCA security in the full version.

4   Our Statistical Concurrent Non-malleable ZK Argument

We show that a statistical concurrent non-malleable zero-knowledge argument can be constructed from any statistically hiding commitment scheme.

Theorem 2. *Assume the existence of statistically hiding commitment schemes with round complexity $R_{\mathsf{SH}}(n)$. Then, there exists an $\omega(R_{\mathsf{SH}}(n)\log n)$-round statistical concurrent non-malleable zero-knowledge argument* sCNMZK.

Since poly(n)-round statistically hiding commitment schemes can be constructed from one-way functions [10] and constant-round ones can be constructed from a family of collision-resistant hash functions [5,19], our main theorem (Theorem 1) follows from Theorem 2.

Proof (of Theorem 2). In sCNMZK, we use the following building blocks, all of which can be constructed from $R_{\mathsf{SH}}(n)$-round statistically hiding commitment schemes (or one-way functions, which can be obtained from statistically hiding commitment schemes).

- Two-round statistically binding commitment scheme $\mathsf{Com}_{\mathsf{SB}}$ [11,18].
- Constant-round 4-robust one-one CCA-secure commitment scheme $\mathsf{CCACom}^{1:1}$ (see Sect. 3.3).
- Four-round witness-indistinguishable proof of knowledge WIPOK, which is a parallel version of Blum's Hamiltonian-cycle protocol [2].
- $(R_{\mathsf{SH}}(n)+2)$-round statistical witness-indistinguishable argument of knowledge sWIAOK, which is a parallel version of Blum's Hamiltonian-cycle protocol that is instantiated with a $R_{\mathsf{SH}}(n)$-round statistically hiding commitment scheme $\mathsf{Com}_{\mathsf{SH}}$.

Input. The common input is statement $x \in L$ and identity $\mathsf{id} \in \{0,1\}^n$. The prover's private input is witness $w \in \mathbf{R}_L(x)$.

Stage I. (V commits to trapdoor)

1. V chooses random $r_V \in \{0,1\}^n$ and commits to r_V by using $\mathsf{Com_{SB}}$. Let (r_V, d) be the decommitment of this commitment.
2. V commits to (r_V, d) by using CECom.

Stage II. (V proves knowledge of trapdoor)

1. P chooses random $r_P \in \{0,1\}^n$ and commits to r_P by using $\mathsf{CCACom}^{1:1}$ with tag id.
2. V commits to 0^n by using CECom.
3. P decommits the $\mathsf{CCACom}^{1:1}$ commitment of Stage II-1 to r_P.
4. V proves the following by using WIPOK:
   - the committed value of the CECom commitment of Stage I-2 is a valid decommitment of the $\mathsf{Com_{SB}}$ commitment of Stage I-1, or
   - the committed value of the CECom commitment of Stage II-2 is r_P.

Stage III. (P proves $x \in L$ or knowledge of trapdoor)

1. P proves the following by using sWIAOK.
   - $x \in L$, or
   - There exists (r'_V, d') such that (r'_V, d') is a valid decommitment of the $\mathsf{Com_{SB}}$ commitment of Stage I-1.

Fig. 2. Statistical concurrent non-malleable zero-knowledge argument sCNMZK.

$\omega(R_{\mathsf{SH}}(n) \log n)$-round concurrently extractable commitment scheme CECom, which is the scheme of [17] with parameter $\ell = \omega(R_{\mathsf{SH}}(n) \log n)$. From the robust concurrent extraction lemma [8], we can extract the committed values from any adversarial committer even when it additionally participates in any $O(R_{\mathsf{SH}}(n))$-round external protocol.

Protocol sCNMZK is shown in Fig. 2. Roughly speaking, soundness can be proven as follows. Assume that an adversary breaks the soundness. From the witness extractability of sWIAOK, a valid decommitment (r'_V, d') of the $\mathsf{Com_{SB}}$ commitment of Stage I can be extracted from this adversary in Stage III. Furthermore, from the hiding property of CECom and the witness indistinguishability of WIPOK, it can be shown that (r'_V, d') can be extracted even when Stage I is simulated by extracting r_P in Stage II-1 and using it in Stage II-2 and II-4. Then, since Stage 2 is now simulated without using the decommitment of the $\mathsf{Com_{SB}}$ commitment of Stage 1, we can derive a contradiction by breaking the hiding property of $\mathsf{Com_{SB}}$ or CECom by using (r'_V, d'). The formal proof is given in the full version.

In the following, we prove the statistical CNMZK property.

Simulator-Extractor $\mathcal{SE}$. Recall that to prove the statistical CNMZK property, we need to show a simulator-extractor that simulates the view of the adversary $\mathcal{A}$ and also extracts a witness in every accepted right session. We construct our simulator-extractor step by step. First, we construct a super-polynomial-time simulator $\hat{S}$ that simulates the view of $\mathcal{A}$ but does not extract witnesses in the right seasons. Next, we construct a super-polynomial-time simulator-extractor $\hat{\mathcal{SE}}$ that simulates the view of $\mathcal{A}$ by executing $\hat{S}$ and then extracts the witnesses by rewinding $\hat{S}$. Finally, we construct a polynomial-time simulator-extractor $\mathcal{SE}$ that emulates the execution of $\hat{\mathcal{SE}}$ in polynomial time.

Remark 1. In the following, we use the hat symbol in the names of simulators and simulator-extractors if they run in super-polynomial time (e.g., $\hat{S}$ and $\hat{\mathcal{SE}}$). Also, we use the tilde symbol in the names of the messages of sCNMZK if they are the messages of the right sessions (e.g., $\tilde{r}_V$ and $\tilde{r}_P$); if necessary, we use subscript to denote the index of the session.

Super-Polynomial-Time Simulator $\hat{S}$. First, we show the simulator $\hat{S}$, which simulates the view of $\mathcal{A}$ in super-polynomial time as follows. $\hat{S}$ internally invokes $\mathcal{A}$ and interacts with $\mathcal{A}$ as provers and verifiers in the following way.

- In each left session, $\hat{S}$ interacts with $\mathcal{A}$ in the same way as a honest prover except for the following. In Stage I-2, $\hat{S}$ extracts the committed value (r_V, d) of the CECom commitment by brute force. (If the committed value is not uniquely determined, (r_V, d) is defined to be $(\bot, \bot)$.) In Stage III, $\hat{S}$ checks whether (r_V, d) is a valid decommitment of the $\mathsf{Com_{SB}}$ commitment of Stage I-1; if so, $\hat{S}$ gives a sWIAOK proof by using (r_V, d) as a witness; otherwise, $\hat{S}$ terminates with output fail.
- In each right session, $\hat{S}$ interacts with $\mathcal{A}$ in the same way as a honest verifier.

Finally, $\hat{S}$ outputs the view of internal $\mathcal{A}$. Notice that $\hat{S}$ does not rewind $\mathcal{A}$.

Super-Polynomial-Time Simulator-Extractor $\hat{\mathcal{SE}}$. Next, we show the simulator-extractor $\hat{\mathcal{SE}}$, which simulates the view of $\mathcal{A}$ in super-polynomial time and also extracts witnesses in every accepted right session as follows. First, $\hat{\mathcal{SE}}$ simulates the view of $\mathcal{A}$ by executing $\hat{S}$. We call this execution of $\hat{S}$ the WI-*main thread*. Next, for each $i \in [m]$, if the i-th right session is accepted on the WI-main thread and uses a different identity from every left session, $\hat{\mathcal{SE}}$ extracts a witness from this session as follows.

- $\hat{\mathcal{SE}}$ rewinds the WI-main thread until the point just before the challenge message of sWIAOK of the i-th right session is sent. Then, from this point, $\hat{\mathcal{SE}}$ executes $\hat{S}$ again with flesh randomness (i.e., interacts with $\mathcal{A}$ as $\hat{S}$ does with flesh randomness). $\hat{\mathcal{SE}}$ repeats this rewinding until it obtains another accepting transcript of the i-th right session. We call each execution of $\hat{S}$ in this step a WI-*auxiliary thread* .
- After obtaining two accepting transcripts of the i-th right session (one is on the WI-main thread and the other is on an WI-auxiliary thread), $\hat{\mathcal{SE}}$ extracts a witness from sWIAOK by using the witness extractability of sWIAOK. If $\hat{\mathcal{SE}}$ fails to extract a witness for $\tilde{x}_i \in L$ (the statement proven in the i-th right session), $\hat{\mathcal{SE}}$ terminates with output fail$_{\mathsf{WI}}$. Otherwise, let $\tilde{w}_i$ be the extracted witness.

If the i-th right session is not accepted or uses the same identity as a left session, define $\widetilde{w}_i \stackrel{\text{def}}{=} \perp$. The output of $\hat{\mathcal{SE}}$ is $(\mathsf{view}, \{\widetilde{w}_i\}_{i \in [m]})$, where view is the view of $\mathcal{A}$ on the WI-main thread.

Polynomial-Time Simulator-Extractor $\mathcal{SE}$. Finally, we show the simulator-extractor $\mathcal{SE}$, which emulates the execution of $\hat{\mathcal{SE}}$ in polynomial time as follows. First, $\mathcal{SE}$ emulates the WI-main thread in polynomial time as follows.

- $\mathcal{SE}$ internally invokes $\mathcal{A}$ and interacts with $\mathcal{A}$ as $\hat{\mathcal{S}}$ does except that in each left session, $\mathcal{SE}$ extracts (r_V, d) by using the concurrent extractability of CECom. Recall that a concurrent extraction of CECom involves the generation of a main thread and many look-ahead threads. We call the main thread generated during the concurrent extraction of CECom the CEC-*main thread*, and call the look-ahead threads generated during the concurrent extraction of CECom the CEC-*auxiliary threads.*[6]

Next, for each $i \in [m]$, if the i-th right session is accepted on the emulated WI-main thread and uses a different identity from every left session, $\mathcal{SE}$ emulates WI-auxiliary threads as follows.

- $\mathcal{SE}$ rewinds the emulation of the WI-main thread until the point just before the challenge message of sWIAOK of the i-th right session is sent on the CEC-main thread. Then, from this point, $\hat{\mathcal{SE}}$ emulates the WI-main thread again with flesh randomness (i.e., generates the rest of CEC-main thread and CEC-auxiliary threads with flesh randomness). $\mathcal{SE}$ repeats this rewinding until it obtains another accepted transcript of the i-th right session on an emulated WI-auxiliary thread.

Let $(\mathsf{view}, \{\widetilde{w}_i\}_{i \in [m]})$ be the output of the emulated $\hat{\mathcal{SE}}$. Then, $\mathcal{SE}$ outputs $(\mathsf{view}, \{\widetilde{w}_i\}_{i \in [m]})$.

Analysis of Poly-Time Simulator-Extractor $\mathcal{SE}$.

To prove the statistical CNMZK property, we show that $\mathcal{SE}$ statistically simulates the view of $\mathcal{A}$ and also extracts witnesses for the statements in the right sessions.

Lemma 1. *The view of $\mathcal{A}$ simulated by $\mathcal{SE}$ is statistically indistinguishable from the view of $\mathcal{A}$ in the real experiment. Furthermore, except with negligible probability, $\mathcal{SE}$ outputs witnesses for the statements proven by $\mathcal{A}$ in the accepted right sessions that use different identities from the left sessions.*

Proof (sketch). In this proof, we use the following claim, which states that the super-polynomial-time simulator-extractor $\hat{\mathcal{SE}}$ statistically simulates the view of $\mathcal{A}$ and also extracts the witnesses from the right sessions.

---

[6] Note that the WI-main thread is also a CEC-main thread.

Claim 1. *The view of A simulated by $\hat{S\mathcal{E}}$ is statistically indistinguishable from the view of A in the real experiment. Furthermore, except with negligible probability, $\hat{S\mathcal{E}}$ outputs witnesses for the statements proven by A in the accepted right sessions that use different identities from the left sessions.*

Before proving this claim, we finish the proof of Lemma 1. Given Claim 1, we can prove Lemma 1 by showing that the output of $S\mathcal{E}$ is statistically indistinguishable from that of $\hat{S\mathcal{E}}$. This indistinguishability can be shown by observing the following.

- In $S\mathcal{E}$, the emulation of $\hat{S\mathcal{E}}$ is perfect if in every left session that reaches Stage III, the value extracted by the concurrent extractability of CECom is equal to the value that would be extracted by brute force.
- In every such left session, the value extracted by the concurrent extractability of CECom is indeed equal to the value that would be extracted by brute force. This is because the CECom commitment in Stage I-2 is valid in every such left session except with negligible probability, which in turn is because of the soundness of WIPOK and the hiding property of $\mathsf{CCACom}^{1:1}$.

We note that there is a subtlety since the concurrent extraction of CECom itself is rewound in $S\mathcal{E}$ when the witnesses are extracted from the right sessions. The formal proof is given in the full version. □

Analysis of Super-Poly-Time Simulator-Extractor $\hat{S\mathcal{E}}$.

It remains to prove Claim 1, which states that (i) super-polynomial-time simulator-extractor $\hat{S\mathcal{E}}$ statistically simulates the real view of A and (ii) $\hat{S\mathcal{E}}$ also extracts a valid witness from every accepted right session in the simulated view.

Proof (of Claim 1). First, we show that $\hat{S\mathcal{E}}$ statistically simulates the real view of A. Since $\hat{S\mathcal{E}}$ simulates the view of A by executing $\hat{S}$, it suffices to show that the output of $\hat{S}$ is statistically indistinguishable from the real view of A. In $\hat{S}$, each left session is simulated by extracting (r_V, d) from the CECom commitment in Stage I-2 and giving a sWIAOK proof in Stage III with witness (r_V, d). Hence, the indistinguishability follows from the statistical witness indistinguishability of sWIAOK and the following claim.

Claim 2. *In $\hat{S}$, the following holds except with negligible probability: In every left session that reaches Stage III, the CECom commitment in Stage I-2 of this session is valid and its committed value is a valid decommitment of the $\mathsf{Com_{SB}}$ commitment of Stage I-1.*

We do not prove Claim 2, since it is implied by the claim that we prove later (Claim 5).

Next, we show that $\hat{S\mathcal{E}}$ extracts a valid witness from every accepted right session except with negligible probability. Since $\hat{S\mathcal{E}}$ outputs $\mathsf{fail_{WI}}$ when it fails to extract a witness in an accepted right session, it suffices to show that $\hat{S\mathcal{E}}$ outputs $\mathsf{fail_{WI}}$ only with negligible probability. Assume for contradiction that there exists

$\tilde{i}^* \in [m]$ such that $\hat{\mathcal{SE}}$ outputs $\mathsf{fail_{WI}}$ during the witness extraction of the $\tilde{i}^*$-th right session with non-negligible probability. Then, let us consider the following hybrid simulator-extractor $\hat{\mathcal{SE}}_{\tilde{i}^*}$.

– $\hat{\mathcal{SE}}_{\tilde{i}^*}$ is the same as $\hat{\mathcal{SE}}$ except that $\hat{\mathcal{SE}}_{\tilde{i}^*}$ tries to extract a witness only from the $\tilde{i}^*$-th right session (and therefore rewinds the WI-main thread only from the challenge message of sWIAOK of the $\tilde{i}^*$-th right session).

Clearly, $\hat{\mathcal{SE}}_{\tilde{i}^*}$ outputs $\mathsf{fail_{WI}}$ with non-negligible probability. Then, we reach a contradiction roughly as follows.

Step 1. First, we show that in $\hat{\mathcal{SE}}_{\tilde{i}^*}$, the probability that $\tilde{r}_V$ is extracted as a witness during the witness extraction of the $\tilde{i}^*$-th right session is non-negligible, where $\tilde{r}_V$ is the value chosen by the verifier in Stage I-1 of the $\tilde{i}^*$-th right session.

Step 2. Next, we define a sequence of hybrid simulator-extractors. The first hybrid is the same as $\hat{\mathcal{SE}}_{\tilde{i}^*}$, and we gradually modify the $\tilde{i}^*$-th right session so that it is independent of $\tilde{r}_V$ in the last hybrid.

Step 3. Finally, we show that even in the last hybrid, the probability that $\tilde{r}_V$ is extracted during the witness extraction of the $\tilde{i}^*$-th right session is non-negligible. Since the $\tilde{i}^*$-th right session is independent of $\tilde{r}_V$ in the last hybrid, we reach a contradiction.

Details are given below.

Step 1. Prove that $\hat{\mathcal{SE}}_{\tilde{i}^*}$ extracts $\tilde{r}_V$. We first prove the following claim.

Claim 3. *Let $\tilde{r}_V$ be the value chosen by the verifier in Stage I-1 of the $\tilde{i}^*$-th right session. If $\hat{\mathcal{SE}}_{\tilde{i}^*}$ outputs $\mathsf{fail_{WI}}$ with non-negligible probability, then in $\hat{\mathcal{SE}}_{\tilde{i}^*}$ the probability that $\tilde{r}_V$ is extracted during the witness extraction of the $\tilde{i}^*$-th right session is non-negligible.*

Proof. Assume for contradiction that $\tilde{r}_V$ is extracted during the witness extraction of the $\tilde{i}^*$-th right session with at most negligible probability. Then, since we assume that $\hat{\mathcal{SE}}_{\tilde{i}^*}$ outputs $\mathsf{fail_{WI}}$ with non-negligible probability, the following occurs in $\hat{\mathcal{SE}}_{\tilde{i}^*}$ with non-negligible probability:

– $\hat{\mathcal{SE}}_{\tilde{i}^*}$ obtains two accepting transcript of the $\tilde{i}^*$-th right session (and therefore that of sWIAOK) such that the commit-messages of sWIAOK are the same,[7] but
– from these two transcript, $\hat{\mathcal{SE}}_{\tilde{i}^*}$ fails to extract any witness from sWIAOK (either a witness for $\tilde{x}_{\tilde{i}^*} \in L$ or a valid decommitment of the Stage I-1 commitment).

We first show that when the above occurs, the two accepting sWIAOK transcripts are *admissible* except with negligible probability, where a pair of accepted transcripts of sWIAOK are admissible if their commit-messages are the same but their challenge-messages are different. Toward this end, it suffices to show that $\hat{\mathcal{SE}}_{\tilde{i}^*}$ chooses the same challenge-message of sWIAOK on two WI-auxiliary threads with at most negligible probability. This can be shown as follows.

---

[7] Recall that WIPOK consists of three stages: commit, challenge, and response.

– From a standard argument, we can show that the expected number of rewinding of the WI-main thread is 1 in $\hat{\mathcal{SE}}_{\tilde{i}^*}$.[8] Thus, the probability that $\hat{\mathcal{SE}}_{\tilde{i}^*}$ rewinds the WI-main thread more than $2^{n/2}$ times is at most $2^{-n/2}$. Furthermore, under the condition that $\hat{\mathcal{SE}}_{\tilde{i}^*}$ rewinds the WI-main thread at most $2^{n/2}$ times, the probability that $\hat{\mathcal{SE}}_{\tilde{i}^*}$ chooses the same challenge-message on two WI-auxiliary threads is at most $2^{n/2} \cdot 2^{-n} = 2^{-n/2}$. Thus, the probability that $\hat{\mathcal{SE}}_{\tilde{i}^*}$ chooses the same challenge-message in two WI-auxiliary thread is at most $2^{-n/2} + 2^{-n/2} = \mathsf{negl}(n)$.

Thus, with non-negligible probability $\hat{\mathcal{SE}}_{\tilde{i}^*}$ obtains two admissible transcripts of sWIAOK from which no witness can be computed.

We then reach a contradiction as follows. Since sWIAOK is a parallel version of Blum's Hamiltonian-cycle protocol, if no witness is extracted from two admissible transcripts of sWIAOK, a $\mathsf{Com_{SH}}$ commitment in the commit-messages is decommitted to two different values in the transcripts. Thus, we derive a contradiction by breaking the binding property of $\mathsf{Com_{SH}}$ using $\hat{\mathcal{SE}}_{\tilde{i}^*}$. A problem is that since $\hat{\mathcal{SE}}_{\tilde{i}^*}$ runs in super-polynomial time, the *computational* hiding property of $\mathsf{Com_{SH}}$ may not hold in $\hat{\mathcal{SE}}_{\tilde{i}^*}$. To overcome this problem, we consider hybrid simulator-extractor $\mathcal{SE}_{\tilde{i}^*}$ that emulates the execution of $\hat{\mathcal{SE}}_{\tilde{i}^*}$ in polynomial time. Specifically, $\mathcal{SE}_{\tilde{i}^*}$ emulates $\hat{\mathcal{SE}}_{\tilde{i}^*}$ in the same way as $\mathcal{SE}$ emulates $\hat{\mathcal{SE}}$ (i.e., by using the concurrent extractability of CECom instead of the brute-force extraction) except for the following.

– During the emulation of the WI-main thread, the value (r_V, d) is extracted in Stage I-2 of each left session by using the *robust* concurrent extractability of CECom so that the commit-message of sWIAOK of the $\tilde{i}^*$-th right session is not rewound.

As in the proof of Lemma 1, we can show that $\mathcal{SE}_{\tilde{i}^*}$ statistically emulates the execution of $\hat{\mathcal{SE}}_{\tilde{i}^*}$. Thus, with non-negligible probability, $\mathcal{SE}_{\tilde{i}^*}$ obtains two valid decommitments of a $\mathsf{Com_{SH}}$ commitment (in the commit-messages of sWIAOK of the $\tilde{i}^*$-th right session) such that decommitted values are different. Then, since $\mathcal{SE}_{\tilde{i}^*}$ runs in polynomial time and since the commit-messages of sWIAOK (and therefore the $\mathsf{Com_{SH}}$ commitment) of the $\tilde{i}^*$-th right session is not rewound in $\mathcal{SE}_{\tilde{i}^*}$,[9] we can break the binding property of $\mathsf{Com_{SH}}$. Thus, we reach a contradiction. □

---

[8] For any prefix ρ of the transcript up until the challenge message of sWIAOK of the i-th right session, let p_ρ be the probability that the i-th right session is accepted when the prefix of the transcript is ρ. Then, we have $\mathrm{E}\left[T_i \mid \mathsf{prefix}_\rho\right] = p_\rho \cdot 1/p_\rho = 1$, where T_i is the random variable representing the number of rewinding of the WI-main thread and prefix_ρ is the event that the prefix of the transcript is ρ. Thus, we have $\mathrm{E}\left[T_i\right] = \sum_\rho \mathrm{E}\left[T_i \mid \mathsf{prefix}_\rho\right] \Pr\left[\mathsf{prefix}_\rho\right] = 1$.

[9] Note that the commit-messages of sWIAOK of the $\tilde{i}^*$-th right session appear only on the WI-main thread.

Step 2. Introduce hybrid simulator-extractor. Next, we introduce hybrid simulator-extractors. To clarify the exposition, we first define a sequence of hybrid simulators by gradually modifying $\hat{S}$ and then define the hybrid simulator-extractors by using them. Below, when we refer to a particular stage of sCNMZK, we always means the corresponding stage of sCNMZK in the $\tilde{i}^*$-th right session.

Hybrid simulator h-$\hat{S}_0$ is identical with $\hat{S}$.

Hybrid simulator h-$\hat{S}_1$ is the same as h-$\hat{S}_0$ except that $\tilde{r}_P$ is extracted by brute force in Stage II-1 and the committed value of the CECom commitment in Stage II-2 is switched from 0^n to $\tilde{r}_P$.

Hybrid simulator h-$\hat{S}_2$ is the same as h-$\hat{S}_1$ except that in Stage II-4, the WIPOK proof is computed by using a witness for the fact that the committed value of the CECom commitment of Stage II-2 is $\tilde{r}_P$.

Hybrid simulator h-$\hat{S}_3$ is the same as h-$\hat{S}_2$ except that in Stage I-2, the committed value of the CECom commitment is switched from $(\tilde{r}_V, \tilde{d})$ to $(0^{|\tilde{r}_V|}, 0^{|\tilde{d}|})$.

Hybrid simulator h-$\hat{S}_4$ is the same as h-$\hat{S}_3$ except that in Stage I-1, the committed value of the $\mathsf{Com}_{\mathsf{SB}}$ commitment is switched from $\tilde{r}_V$ to 0^n.

Then, for each $k \in \{0, \ldots, 4\}$, hybrid simulator-extractor h-$\hat{\mathcal{SE}}_k$ is defined as follows.

Hybrid simulator-extractor h-$\hat{\mathcal{SE}}_k$ is the same as $\hat{\mathcal{SE}}_{\tilde{i}^*}$ except that the execution of $\hat{S}$ is replaced with that of h-$\hat{S}_k$. The output of h-$\hat{\mathcal{SE}}_k$ is the value extracted during the witness extraction of the $\tilde{i}^*$-th right session.

Note that the value $\tilde{r}_V$ is not used anywhere in h-$\hat{\mathcal{SE}}_4$.

Step 3. Prove that $\tilde{r}_V$ is extracted in every hybrid. Finally, we show that $\tilde{r}_V$ is extracted with non-negligible probability in each hybrid. First, we consider h-$\hat{\mathcal{SE}}_1$.

Claim 4. *Let $\tilde{r}_V$ be the value chosen by the verifier in Stage I-1 of the $\tilde{i}^*$-th right session. If $\hat{\mathcal{SE}}_{\tilde{i}^*}$ outputs $\mathsf{fail}_{\mathsf{WI}}$ with non-negligible probability, then in h-$\hat{\mathcal{SE}}_1$ the probability that $\tilde{r}_V$ is extracted during the witness extraction of the $\tilde{i}^*$-th right session is non-negligible.*

Proof. In this proof, we use intermediate hybrid simulator-extractors in which the CECom commitment in Stage II-2 of the $\tilde{i}^*$-th right session is gradually modified. Again, we first introduce hybrid simulators. Recall that a CECom commitment consists of $\ell = \omega(R_{\mathsf{SH}}(n) \log n)$ ExtCom commitments. Then, the intermediate hybrid simulators h-$\hat{S}_{0:0}, \ldots, h$-$\hat{S}_{0:\ell}$ are defined as follows.

Hybrid simulator h-$\hat{S}_{0:0}$ is the same as h-$\hat{S}_0$ except that $\tilde{r}_P$ is extracted by brute force in Stage II-1 of the $\tilde{i}^*$-th right session.

Hybrid simulator h-$\hat{S}_{0:k}$ $(k \in [\ell])$ is the same as h-$\hat{S}_{0:k-1}$ except that the committed value of the k-th ExtCom commitment in the CECom commitment of Stage II-2 is switched from 0^n to $\tilde{r}_P$ in the $\tilde{i}^*$-th right session.

Then, for each $k \in \{0, \ldots, \ell\}$, hybrid simulator-extractor $h\text{-}\hat{\mathcal{SE}}_{0:k}$ is defined as follows.

Hybrid simulator-extractor $h\text{-}\hat{\mathcal{SE}}_{0:k}$ is the same as $h\text{-}\hat{\mathcal{SE}}_0$ except that the execution of $h\text{-}\hat{S}_0$ is replaced with that of $h\text{-}\hat{S}_{0:k}$.

Note that $h\text{-}\hat{\mathcal{SE}}_{0:\ell}$ is identical with $h\text{-}\hat{\mathcal{SE}}_1$.

Below, we show that for every $k \in [\ell]$, the output of $h\text{-}\hat{\mathcal{SE}}_{0:k-1}$ and that of $h\text{-}\hat{\mathcal{SE}}_{0:k}$ are indistinguishable. (Recall that the outputs of $h\text{-}\hat{\mathcal{SE}}_{0:k-1}$ and $h\text{-}\hat{\mathcal{SE}}_{0:k}$ are the value extracted in the i^*-th right session.) Since the probability that $\tilde{r}_V$ is extracted in $h\text{-}\hat{\mathcal{SE}}_{0:0}$ is non-negligible from Claim 3, this suffices to prove Claim 4.

Roughly speaking, we show this indistinguishability as follows. Since $h\text{-}\hat{\mathcal{SE}}_{0:k-1}$ and $h\text{-}\hat{\mathcal{SE}}_{0:k}$ differ only in the committed values of a ExtCom commitment, we use the hiding property of the ExtCom commitment to show the indistinguishability. A problem is that we cannot use it directly since $h\text{-}\hat{\mathcal{SE}}_{0:k-1}$ and $h\text{-}\hat{\mathcal{SE}}_{0:k}$ run in super-polynomial time. To overcome this problem, we observe that the only super-polynomial computations in $h\text{-}\hat{\mathcal{SE}}_{0:k-1}$ and $h\text{-}\hat{\mathcal{SE}}_{0:k}$ are the brute-force extraction of $\mathsf{CCACom}^{1:1}$ in the i^*-th right session and those of CECom in the left sessions. Based on this observation, we first show that the execution of $h\text{-}\hat{\mathcal{SE}}_{0:k-1}$ and $h\text{-}\hat{\mathcal{SE}}_{0:k}$ can be emulated in polynomial-time by using the one-session committed-value oracle $\mathcal{O}$ of $\mathsf{CCACom}^{1:1}$ and the concurrent extractability of CECom. We then combine the 4-robustness of $\mathsf{CCACom}^{1:1}$ with the hiding property of ExtCom (which has only four rounds) to argue that the output of $h\text{-}\hat{\mathcal{SE}}_{0:k-1}$ and that of $h\text{-}\hat{\mathcal{SE}}_{0:k}$ are indistinguishable. To formally implement this idea, we need to make sure that the ExtCom commitment and the $\mathsf{CCACom}^{1:1}$ commitment are not rewound during the concurrent extraction of CECom. Details are given below.

First, we introduce hybrid simulator-extractors $h\text{-}\mathcal{SE}^{\mathcal{O}}_{0:k-1}$ and $h\text{-}\mathcal{SE}^{\mathcal{O}}_{0:k}$, where $\mathcal{O}$ is the one-session committed-value oracle of $\mathsf{CCACom}^{1:1}$. Hybrid $h\text{-}\mathcal{SE}^{\mathcal{O}}_{0:k}$ (resp., $h\text{-}\mathcal{SE}^{\mathcal{O}}_{0:k-1}$) emulates $h\text{-}\hat{\mathcal{SE}}_{0:k}$ (resp., $h\text{-}\hat{\mathcal{SE}}_{0:k-1}$) in the same way as $\mathcal{SE}$ emulates $\hat{\mathcal{SE}}$ except for the following.

– During the emulation of the WI-main thread, the value (r_V, d) is extracted in Stage I-2 of each left session by using the robust concurrent extractability so that the $\mathsf{CCACom}^{1:1}$ commitment of Stage II-1 and the k-th ExtCom commitment of the CECom commitment of Stage II-2 are not rewound in the i^*-th right session. In addition, in the i^*-th right session, the committed value of $\mathsf{CCACom}^{1:1}$ is extracted by forwarding the commitment to $\mathcal{O}$. Note that the $\mathsf{CCACom}^{1:1}$ commitment in the i^*-th right session is not rewound and therefore it can be forwarded to $\mathcal{O}$.

Next, we show that for each $h \in \{k-1, k\}$, the output of $h\text{-}\hat{\mathcal{SE}}_{0:h}$ and that of $h\text{-}\mathcal{SE}^{\mathcal{O}}_{0:h}$ are indistinguishable. This can be proven in a similar way to Lemma 1. In particular, we can use the same argument if we use the following claim instead of Claim 2.

Claim 5. *In $h\text{-}\hat{S}_{0:h}$ for each $h \in \{k-1, k\}$, the following holds except with negligible probability: In every left session that reaches Stage III, the* CECom *commitment in Stage I-2 of this session is valid and its committed value is a valid decommitment of the* $\mathsf{Com_{SB}}$ *commitment of Stage I-1.*

Note that since h-$\hat{\mathcal{S}}_{0:0}$ is identical to $\hat{\mathcal{S}}$, Claim 5 implies Claim 2.

Proof (of Claim 5). Let us say that a left session is *bad* if it reaches Stage III and either the CECom commitment in Stage I-2 is invalid or its committed value is not a valid decommitment of the $\mathsf{Com_{SB}}$ commitment in Stage I-1; a left session is *good* if it is not bad. What we want to prove is that every left session is good except with negligible probability.

Roughly speaking, the proof proceeds as follows. From the soundness of WIPOK, if a left session is bad, then in Stage II-2 of this left session, the committed value of the CECom commitment is r_P, which is the committed value of the CCACom$^{1:1}$ commitment of Stage II-1; thus, before r_P is decommitted to in Stage II-3, we can obtain r_P by extracting the committed value from CECom in Stage II-2. This itself does not contradict to the hiding property of CCACom$^{1:1}$ since h-$\hat{\mathcal{S}}_{0:h}$ runs in super-polynomial time in the brute-force extraction of CECom and CCACom$^{1:1}$. Thus, we again replace the brute-force extraction with the concurrent extraction of CECom and an oracle access to the one-session committed-value oracle $\mathcal{O}$ of CCACom$^{1:1}$, and use the one-one CCA-security of CCACom$^{1:1}$ instead of its hiding property. Here, since we want to use the one-one CCA-security of CCACom$^{1:1}$, we perform the concurrent extraction of CECom so that the CCACom$^{1:1}$ commitment in a left session and the CCACom$^{1:1}$ in the $\tilde{i}^*$-th right session are not rewound. Details are given below.

Assume for contradiction that there exists $h \in \{k-1, k\}$ such that in h-$\hat{\mathcal{S}}_{0:h}$, a left session is bad with non-negligible probability. (Here, the indices of the left sessions are determined by the order in which Stage III begins; the reason why we define the indices in this way will become clear later.) Then, there exists $i^* \in [m]$ such that in h-$\hat{\mathcal{S}}_{0:h}$, the first $(i^* - 1)$ left sessions are good except with negligible probability but the i^*-th left session is bad with non-negligible probability. Note that from the soundness of WIPOK, when the i^*-th left session is bad, the committed value of the CECom commitment in Stage II-2 is r_P in the i^*-th left session except with negligible probability, where r_P is the value committed to in Stage II-1 of the i^*-th left session. In the following, we use BAD to denote the event that the i^*-th left session is bad, and use CHEAT to denote the event that the committed value of the CECom commitment in Stage II-2 is r_P in the i^*-th left session. Then, let us consider the following hybrids.

Hybrid simulator h-$\hat{\mathcal{S}}_{0:h:0}$ is the same as h-$\hat{\mathcal{S}}_{0:h}$. From our assumption, BAD occurs in h-$\hat{\mathcal{S}}_{0:h:0}$ with non-negligible probability. Thus, from the above argument, CHEAT occurs in h-$\hat{\mathcal{S}}_{0:h:0}$ with non-negligible probability.

Hybrid simulator h-$\hat{\mathcal{S}}_{0:h:1}$ is the same as h-$\hat{\mathcal{S}}_{0:h:0}$ except that h-$\hat{\mathcal{S}}_{0:h:1}$ terminates just before Stage III of the i^*-th left session begins. Clearly, BAD and CHEAT also occur in h-$\hat{\mathcal{S}}_{0:h:1}$ with non-negligible probability.

Hybrid simulator h-$\mathcal{S}_{0:h:1}^{\mathcal{O}}$ emulates h-$\hat{\mathcal{S}}_{0:h:1}$ in polynomial time as follows.
  – At the beginning, a random left session s is chosen. (Here, we guess that session s is the i^*-th left session.)

- In every left session, in Stage I-2, the committed value (r_V, d) is extracted by the robust concurrent extractor of CECom in such a way that the $\mathsf{CCACom}^{1:1}$ commitment of left session s and the $\mathsf{CCACom}^{1:1}$ commitment of the $\tilde{i}^*$-th right session are not rewound. In addition, in the $\tilde{i}^*$-th right session, the committed value of $\mathsf{CCACom}^{1:1}$ is extracted by forwarding the commitment to $\mathcal{O}$.
- In left session s, the committed value is also extracted in Stage II-2 by the robust concurrent extractor of CECom without rewinding the $\mathsf{CCACom}^{1:1}$ commitment of the $\tilde{i}^*$-th right session.

Note that when Stage III of a left session is executed, the CECom commitment in Stage I-2 of that session is valid except with negligible probability (since that session is one of the first $(i^* - 1)$ left sessions and therefore it is good except with negligible probability). Thus, the values extracted from the concurrent extractor are equal to the values that would be extracted by brute force except with negligible probability; therefore, $h\text{-}\mathcal{S}^{\mathcal{O}}_{0:h:1}$ statistically emulates $h\text{-}\hat{\mathcal{S}}_{0:h:1}$, and BAD and CHEAT occur in $h\text{-}\mathcal{S}^{\mathcal{O}}_{0:h:1}$ with non-negligible probability.

Note that session s is the i^*-th left session with non-negligible probability. Then, since CHEAT occurs in $h\text{-}\mathcal{S}^{\mathcal{O}}_{0:h:1}$ with non-negligible probability, r_P is extracted from the CECom commitment in Stage II-2 of session s with non-negligible probability, where r_P is the value committed to in Stage II-1 of session s. Then, since the $\mathsf{CCACom}^{1:1}$ commitment of session s is not rewound in $h\text{-}\mathcal{S}^{\mathcal{O}}_{0:h:1}$, we can break the one-one CCA security of $\mathsf{CCACom}^{1:1}$. Thus, we reach a contradiction. □

Thus, for each $h \in \{k-1, k\}$, the outputs of $h\text{-}\hat{\mathcal{SE}}_{0:h}$ and $h\text{-}\mathcal{SE}^{\mathcal{O}}_{0:h}$ are indistinguishable.

To show that the outputs of $h\text{-}\hat{\mathcal{SE}}_{0:k-1}$ and $h\text{-}\hat{\mathcal{SE}}_{0:k}$ are indistinguishable, it remains to prove that the outputs of $h\text{-}\mathcal{SE}^{\mathcal{O}}_{0:k-1}$ and $h\text{-}\mathcal{SE}^{\mathcal{O}}_{0:k}$ are indistinguishable. This can be shown as follows. Observe that $h\text{-}\mathcal{SE}^{\mathcal{O}}_{0:k-1}$ and $h\text{-}\mathcal{SE}^{\mathcal{O}}_{0:k}$ differ only in the k-th ExtCom commitment of the CECom commitment of the $\tilde{i}^*$-th right session, and this ExtCom commitment is not rewound in $h\text{-}\mathcal{SE}^{\mathcal{O}}_{0:k-1}$ and $h\text{-}\mathcal{SE}^{\mathcal{O}}_{0:k}$. In addition, $h\text{-}\mathcal{SE}^{\mathcal{O}}_{0:k-1}$ and $h\text{-}\mathcal{SE}^{\mathcal{O}}_{0:k}$ run in polynomial time given oracle access to the one-session committed-value oracle $\mathcal{O}$ of $\mathsf{CCACom}^{1:1}$. Thus, from the hiding property of ExtCom and the 4-robustness of $\mathsf{CCACom}^{1:1}$, the output of $\mathcal{SE}^{\mathcal{O}}_{0:k-1}$ and that of $h\text{-}\mathcal{SE}^{\mathcal{O}}_{0:k}$ are indistinguishable.

Thus, we conclude that the probability that $\tilde{r}_V$ is extracted in $h\text{-}\hat{\mathcal{SE}}_1$ is non-negligible. This concludes the proof of Claim 4. □

By using essentially the same argument as in the proof of Claim 4, we can show that $\tilde{r}_V$ is extracted with non-negligible probability also in $h\text{-}\hat{\mathcal{SE}}_2$, $h\text{-}\hat{\mathcal{SE}}_3$, and $h\text{-}\hat{\mathcal{SE}}_4$.

Concluding the Proof of Claim 1. In $h\text{-}\hat{\mathcal{SE}}_4$, the $\tilde{i}^*$-th right session is independent of $\tilde{r}_V$, and therefore the probability that $\tilde{r}_V$ is extracted is negligible. However, we show above that this probability is non-negligible. Thus, we reach a contradiction.

This concludes the proof of Theorem 2. □

References

1. Barak, B., Prabhakaran, M., Sahai, A.: Concurrent non-malleable zero knowledge. In: FOCS, pp. 345–354 (2006)
2. Blum, M.: How to prove a theorem so no one else can claim it. In: International Congress of Mathematicians, pp. 1444–1451 (1987)
3. Canetti, R., Kilian, J., Petrank, E., Rosen, A.: Black-box concurrent zero-knowledge requires (almost) logarithmically many rounds. SIAM J. Comput. 32(1), 1–47 (2002)
4. Canetti, R., Lin, H., Pass, R.: Adaptive hardness and composable security in the plain model from standard assumptions. In: FOCS, pp. 541–550 (2010)
5. Damgård, I., Pedersen, T.P., Pfitzmann, B.: Statistical secrecy and multibit commitments. IEEE Trans. Inf. Theory 44(3), 1143–1151 (1998)
6. Dolev, D., Dwork, C., Naor, M.: Nonmalleable cryptography. SIAM J. Comput. 30(2), 391–437 (2000)
7. Dwork, C., Naor, M., Sahai, A.: Concurrent zero-knowledge. J. ACM 51(6), 851–898 (2004)
8. Goyal, V., Lin, H., Pandey, O., Pass, R., Sahai, A.: Round-efficient concurrently composable secure computation via a robust extraction lemma. In: Dodis, Y., Nielsen, J.B. (eds.) TCC 2015, Part I. LNCS, vol. 9014, pp. 260–289. Springer, Heidelberg (2015)
9. Goyal, V., Moriarty, R., Ostrovsky, R., Sahai, A.: Concurrent statistical zero-knowledge arguments for NP from one way functions. In: Kurosawa, K. (ed.) ASIACRYPT 2007. LNCS, vol. 4833, pp. 444–459. Springer, Heidelberg (2007)
10. Haitner, I., Nguyen, M.-H., Ong, S.J., Reingold, O., Vadhan, S.P.: Statistically hiding commitments and statistical zero-knowledge arguments from any one-way function. SIAM J. Comput. 39(3), 1153–1218 (2009)
11. Håstad, J., Impagliazzo, R., Levin, L.A., Luby, M.: A pseudorandom generator from any one-way function. SIAM J. Comput. 28(4), 1364–1396 (1999)
12. Kiyoshima, S.: Round-efficient black-box construction of composable multi-party computation. In: Garay, J.A., Gennaro, R. (eds.) CRYPTO 2014, Part II. LNCS, vol. 8617, pp. 351–368. Springer, Heidelberg (2014)
13. Kiyoshima, S., Manabe, Y., Okamoto, T.: Constant-round black-box construction of composable multi-party computation protocol. In: Lindell, Y. (ed.) TCC 2014. LNCS, vol. 8349, pp. 343–367. Springer, Heidelberg (2014)
14. Lin, H., Pass, R.: Concurrent non-malleable zero knowledge with adaptive inputs. In: Ishai, Y. (ed.) TCC 2011. LNCS, vol. 6597, pp. 274–292. Springer, Heidelberg (2011)
15. Lin, H., Pass, R.: Black-box constructions of composable protocols without set-up. In: Safavi-Naini, R., Canetti, R. (eds.) CRYPTO 2012. LNCS, vol. 7417, pp. 461–478. Springer, Heidelberg (2012)
16. Lin, H., Pass, R., Tseng, W.-L.D., Venkitasubramaniam, M.: Concurrent non-malleable zero knowledge proofs. In: Rabin, T. (ed.) CRYPTO 2010. LNCS, vol. 6223, pp. 429–446. Springer, Heidelberg (2010)
17. Micciancio, D., Ong, S.J., Sahai, A., Vadhan, S.P.: Concurrent zero knowledge without complexity assumptions. In: Halevi, S., Rabin, T. (eds.) TCC 2006. LNCS, vol. 3876, pp. 1–20. Springer, Heidelberg (2006)
18. Naor, M.: Bit commitment using pseudorandomness. J. Cryptol. 4(2), 151–158 (1991)

19. Naor, M., Yung, M.: Universal one-way hash functions and their cryptographic applications. In: STOC, pp. 33–43 (1989)
20. Orlandi, C., Ostrovsky, R., Rao, V., Sahai, A., Visconti, I.: Statistical concurrent non-malleable zero knowledge. In: Lindell, Y. (ed.) TCC 2014. LNCS, vol. 8349, pp. 167–191. Springer, Heidelberg (2014)
21. Ostrovsky, R., Pandey, O., Visconti, I.: Efficiency preserving transformations for concurrent non-malleable zero knowledge. In: Micciancio, D. (ed.) TCC 2010. LNCS, vol. 5978, pp. 535–552. Springer, Heidelberg (2010)
22. Pass, R., Rosen, A.: New and improved constructions of non-malleable cryptographic protocols. In: STOC, pp. 533–542 (2005)
23. Pass, R., Tseng, W.-L.D., Venkitasubramaniam, M.: Concurrent zero knowledge, revisited. J. Cryptol. 27(1), 45–46 (2012)
24. Pass, R., Wee, H.: Black-box constructions of two-party protocols from one-way functions. In: Reingold, O. (ed.) TCC 2009. LNCS, vol. 5444, pp. 403–418. Springer, Heidelberg (2009)
25. Prabhakaran, M., Rosen, A., Sahai, A.: Concurrent zero knowledge with logarithmic round-complexity. In: FOCS, pp. 366–375 (2002)
26. Venkitasubramaniam, M.: On adaptively secure protocols. In: Abdalla, M., De Prisco, R. (eds.) SCN 2014. LNCS, vol. 8642, pp. 455–475. Springer, Heidelberg (2014)

Implicit Zero-Knowledge Arguments and Applications to the Malicious Setting

Fabrice Benhamouda, Geoffroy Couteau, David Pointcheval$^{(\boxtimes)}$, and Hoeteck Wee

ENS, CNRS, INRIA, and PSL, Paris, France
{fabrice.benhamouda,geoffroy.couteau,david.pointcheval,
hoeteck.wee}@ens.fr

Abstract. We introduce *implicit zero-knowledge* arguments (iZK) and simulation-sound variants thereof (SSiZK); these are lightweight alternatives to zero-knowledge arguments for enforcing semi-honest behavior. Our main technical contribution is a construction of efficient two-flow iZK and SSiZK protocols for a large class of languages under the (plain) DDH assumption in cyclic groups in the common reference string model. As an application of iZK, we improve upon the round-efficiency of existing protocols for securely computing inner product under the DDH assumption. This new protocol in turn provides privacy-preserving biometric authentication with lower latency.

Keywords: Hash proof systems · Zero-knowledge · Malicious adversaries · Two-party computation · Inner product

1 Introduction

Zero-Knowledge Arguments (ZK) enable a prover to prove the validity of a statement to a verifier without revealing anything else [13,30]. In addition to being interesting in its own right, zero knowledge has found numerous applications in cryptography, most notably to simplify protocol design as in the setting of secure two-party computation [28,29,46], and as a tool for building cryptographic primitives with strong security guarantees such as encryption secure against chosen-ciphertext attacks [19,41].

In this work, we focus on the use of zero-knowledge arguments as used in efficient two-party protocols for enforcing semi-honest behavior. We are particularly interested in round-efficient two-party protocols, as network latency and round-trip times can be a major efficiency bottleneck, for instance, when a user wants to securely compute on data that is outsourced to the cloud. In addition, we want to rely on standard and widely-deployed cryptographic assumptions. Here, a standard interactive zero-knowledge argument based on the DDH assumption would require at least three flows; moreover, this overhead in round complexity is incurred each time we want to enforce semi-honest behavior via zero knowledge. To avoid this overhead, we could turn to non-interactive zero-knowledge proofs

R. Gennaro and M. Robshaw (Eds.): CRYPTO 2015, Part II, LNCS 9216, pp. 107–129, 2015.
DOI: 10.1007/978-3-662-48000-7_6

(NIZK). However, efficient NIZK would require either the use of pairings [32] and thus stronger assumptions and additional efficiency overhead, or the use of random oracles [6, 23].

We would like to point out that, contrary to some common belief, there is no straightforward way to reduce the number of rounds of zero-knowledge proofs "à la Schnorr" [42] by performing the first steps (commitment and challenges) in a preprocessing phase, so that each proof only takes one flow subsequently. Indeed, as noticed by Bernhard-Pereira-Warinsky in [9], the statement of the proof has to be chosen before seeing the challenges, unless the proof becomes unsound.

On the Importance of Round-Efficiency. In addition to being an interesting theoretical problem, improving the round efficiency is also very important in practice. If we consider a protocol between a client in Europe, and a cloud provider in the US, for example, we expect a latency of at least 100ms (and even worse if the client is connected with 3 g or via satellite, which may induce a latency of up to 1s [14]). Concretely, using Curve25519 elliptic curve of Bernstein [10] (for 128 bits of security, and 256-bit group elements) with a 10 Mbps Internet link and 100 ms latency, 100 ms corresponds to sending 1 flow, or 40,000 group elements, or computing 1,000 exponentiations at 2 GHz on one core of current AMD64 microprocessor[1], hence 4,000 exponentiations on a 4-core microprocessor[2]. As a final remark on latency, while speed of networks keeps increasing as technology improves, latency between two (far away) places on earth is strongly limited by the speed of light: there is no hope to get a latency less than 28 ms between London and San Francisco, for example.

Our Contributions. In this work, we introduce *implicit Zero-Knowledge Arguments* or iZK and simulation-sound variants thereof or SSiZK, lightweight alternatives to (simulation-sound) zero-knowledge arguments for enforcing semi-honest behavior in two-party protocols. Then, we construct efficient two-flow iZK and SSiZK protocols for a large class of languages under the (plain) DDH assumption in cyclic groups without random oracles; this is the main technical contribution of our work. Our SSiZK construction from iZK is very efficient and incurs only a small additive overhead. Finally, we present several applications of iZK to the design of efficient secure two-party computation, where iZK can be used in place of interactive zero-knowledge arguments to obtain more round-efficient protocols.

While our iZK protocols require an additional flow compared to NIZK, we note that eliminating the use of pairings and random oracles offers both theoretical and practical benefits. From a theoretical stand-point, the DDH assumption in cyclic groups is a weaker assumption than the DDH-like assumptions used in Groth-Sahai pairing-based NIZK [32], and we also avoid the theoretical pitfalls associated with instantiating the random oracle methodology [5, 16]. From a practical stand-point, we can instantiate our DDH-based protocols over a larger

---

[1] According to [20], an exponentiation takes about 200,000 cycles.

[2] Assuming exponentiations can be made in parallel, which is the case for our iZKs.

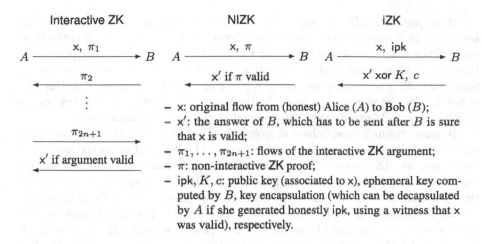

- x: original flow from (honest) Alice (A) to Bob (B);
- x': the answer of B, which has to be sent after B is sure that x is valid;
- $\pi_1, \ldots, \pi_{2n+1}$: flows of the interactive ZK argument;
- π: non-interactive ZK proof;
- ipk, K, c: public key (associated to x), ephemeral key computed by B, key encapsulation (which can be decapsulated by A if she generated honestly ipk, using a witness that x was valid), respectively.

Fig. 1. Enforcing semi-honest behavior of Alice (A)

class of groups. Concrete examples include Bernstein's Curve25519 [10] which admit very efficient group exponentiations, but do not support an efficient pairing and are less likely to be susceptible to recent breakthroughs in discrete log attacks [4,31]. By using more efficient groups and avoiding the use of pairing operations, we also gain notable improvements in computational efficiency over Groth-Sahai proofs. Moreover, additional efficiency improvements come from the structure of iZK which makes them *efficiently batchable*. Conversely, Groth-Sahai NIZK cannot be efficiently batched and do not admit efficient SS-NIZK (for non-linear equations).

New Notion: Implicit Zero-Knowledge Arguments. iZK is a two-party protocol executed between a prover and a verifier, at the end of which both parties should output an ephemeral key. The idea is that the key will be used to encrypt subsequent messages and to protect the privacy of a verifier against a cheating prover. Completeness states that if both parties start with a statement in the language, then both parties output the same key K. Soundness states that if the statement is outside the language, then the verifier's ephemeral output key is hidden from the cheating prover. Note that the verifier may not learn whether his key is the same as the prover's and would not be able to detect whether the prover is cheating, hence the soundness guarantee is *implicit*. This is in contrast to a standard ZK argument, where the verifier would "explicitly" abort when interacting with a cheating prover. Finally, zero-knowledge stipulates that for statements in the language, we can efficiently simulate (without the witness) the joint distribution of the transcript between an honest prover and a malicious verifier, together with the honest prover's ephemeral output key K. Including K in the output of the simulator ensures that the malicious verifier does not gain additional knowledge about the witness when honest prover uses K in subsequent interaction, as will be the case when iZK is used as part of a bigger protocol.

More precisely, iZK are key encapsulation mechanisms in which the public key ipk is associated with a word x and a language $i\mathscr{L}$. In our case, x is the flow[3] and $i\mathscr{L}$ the language of valid flows. If x is in $i\mathscr{L}$, knowing a witness proving so (namely, random coins used to generate the flow) enables anyone to generate ipk together with a secret key isk, using a key generation algorithm iKG. But, if x is not in $i\mathscr{L}$, there is no polynomial-time way to generate a public key ipk for which it is possible to decrypt the associated ciphertexts (*soundness*).

To ensure semi-honest behavior, as depicted in Fig. 1, each time a player sends a flow x, he also sends a public key ipk generated by iKG and keeps the associated secret key isk. To answer back, the other user generates a key encapsulation c for ipk and x, of a random ephemeral key K. He can then use K to encrypt (using symmetric encryption or pseudo-random generators and one-time pad) all the subsequent flows he sends to the first player. For this transformation to be secure, we also need to be sure that c (and the ability to decapsulate K for any ipk) leaks no information about random coins used to generate the flow (or, more generally, the witness of x). This is ensured by the *zero-knowledge* property, which states there must exist a trapdoor (for some common reference string) enabling to generate a public key ipk and a trapdoor key itk (using a trapdoor key algorithm iTKG), so that ipk looks like a classical public key and itk allows to decapsulate any ciphertext for ipk.

Overview of Our iZK and SSiZK Constructions. We proceed to provide an overview of our two-flow iZK protocols; this is the main technical contribution of our work. Our main tool is Hash Proof Systems or Smooth Projective Hash Functions (SPHFs) [18]. We observe that SPHFs are essentially "honest-verifier" iZK; our main technical challenge is to boost this weak honest-verifier into full-fledged zero knowledge, without using pairings or random oracles.

Informally speaking, a smooth projective hash function on a language $\mathscr{L}$ is a sort of hash function whose evaluation on a word $C \in \mathscr{L}$ can be computed in two ways, either by using a *hashing key* hk (which can be seen as a private key) or by using the associated *projection key* hp (which can be seen as a public key). On the other hand, when $C \notin \mathscr{L}$, the hash of C cannot be computed from hp; actually, when $C \notin \mathscr{L}$, the hash of C computed with hk is statistically indistinguishable from a random value from the point of view of any individual knowing the projection key hp only. Hence, an SPHF on $\mathscr{L}$ is given by a pair (Hash, ProjHash) with the requirements that, when there is a witness w ensuring that $C \in \mathscr{L}$, Hash(hk, $\mathscr{L}, C$) = ProjHash(hp, $\mathscr{L}, C, w$), while when there is no such witness (i.e. $C \notin \mathscr{L}$), the smoothness property states that $H = $ Hash(hk, $\mathscr{L}, C$) is random and independent of hp. In this paper, as in [26], we consider a weak form of SPHFs, where the projection key hp can depend on C.

Concretely, if we have an SPHF for some language $\mathscr{L}$, we can set the public key ipk to be empty ($\bot$), the secret key isk to be the witness w, the ciphertext c to be the projection key hp, and the encapsulated ephemeral key K would be the hash value. (Similar connections between SPHF and zero knowledge were made

---

[3] In our formalization, actually, it is the flow together all the previous flows. But we just say it is the flow to simplify explanations.

in [1,12,25,26]). The resulting iZK would be correct and sound, the soundness coming from the smoothness of the SPHF: if the word C is not in $\mathscr{L}$, even given the ciphertext $c = \mathsf{hp}$, the hash value K looks random. However, it would not necessarily be zero-knowledge for two reasons: not only, a malicious verifier could generate a malformed projection key, for which the projected hash value of a word depends on the witness, but also there seems to be no trapdoor enabling to compute the hash value K from only $c = \mathsf{hp}$.

These two issues could be solved using either Trapdoor SPHF [7] or NIZK of knowledge of hk. But both methods require pairings or random oracle, if instantiated on cyclic or bilinear groups. Instead we construct it as follows:

First, suppose that a projection key is well-formed (i.e., there exists a corresponding hashing key). Then, there exists an *unbounded* zero-knowledge simulator that "extracts" a corresponding hashing key and computes the hash value. To boost this into full-fledged zero knowledge with an efficient simulator, we rely on the "OR trick" from [22]. We add a random 4-tuple (g', h', u', e') to the CRS, and build an SPHF for the augmented language $C \in \mathscr{L}$ or (g', h', u', e') is a DDH tuple. In the normal setup, (g', h', u', e') is not a DDH tuple with overwhelming probability, so the soundness property is preserved. In the trapdoor setup, $(g', h', u', e') := (g', h', g'^r, h'^r)$ is a random DDH tuple, and the zero-knowledge simulator uses the witness r to compute the hash value.

Second, to ensure that the projection key is well-formed, we use a second SPHF. The idea for building the second SPHF is as follows: in most SPHF schemes, proving that a projected key hp is valid corresponds to proving that it lies in the column span of some matrix Γ (where all of the linear algebra is carried out in the exponent). Now pick a random vector tk: if hp lies in the span of Γ, then $\mathsf{hp}^\mathsf{T}\mathsf{tk}$ is completely determined given $\Gamma^\mathsf{T}\mathsf{tk}$; otherwise, it is completely random. The former yields the projective property and the latter yields smoothness, for the SPHF with hashing key hk and projection key $\mathsf{tp} = \Gamma^\mathsf{T}\mathsf{tk}$. Since the second SPHF is built using the transpose Γ^T of the original matrix Γ (defining the language $\mathscr{L}$), we refer to it as a "transpose SPHF". As it turns out, the second fix could ruin soundness of the ensuing iZK protocol: a cheating prover could pick a malformed $\Gamma^\mathsf{T}\mathsf{tk}$, and then the hash value $\mathsf{hp}^\mathsf{T}\mathsf{tk}$ computed by the verifier could leak additional information about his witness hk for hp, thereby ruining smoothness. To protect against the leakage, we would inject additional randomness into hk so that smoothness holds even in the presence of leakage from the hash value $\mathsf{hp}^\mathsf{T}\mathsf{tk}$. This idea is inspired by the 2-universality technique introduced in a very different context of chosen-ciphertext security [18].

Finally, to get simulation-soundness (i.e., soundness even if the adversary can see fake or simulated proofs), we rely on an additional "OR trick" (mixed up with an idea of Malkin et al. [40]): we build an SPHF for the augmented language $C \in \mathscr{L}$, or (g', h', u', e') is a DDH tuple (as before), or $(g', h', \mathcal{W}_1(C), \mathcal{W}_2(C))$ is not a DDH tuple (with $\mathcal{W}_k$ a Waters function [45], $\mathcal{W}_k(m) = v_{k,0}\prod_{i=1}^{|m|} v_{k,i}^{m_i}$, when $m = m_1\|\ldots\|m_{|m|}$ is a bitstring, the $v_{k,0}, \ldots, v_{k,|m|}$ are random group elements, and C is seen as a bitstring, for $k = 1,2$). In the security proof, with non-negligible probability, $(g'', h'', \mathcal{W}_1(C), \mathcal{W}_2(C))$ is a non-DDH tuple for simulated proofs, and a DDH tuple for the soundness challenge, which proves simulation-soundness.

Organization. First, we formally introduce the notion of *implicit zero-knowledge proofs* (iZK) in Sect. 2. Second, in Sect. 3, we discuss some difficulties related to the construction of iZK from SPHF and provide an intuition of our method to overcome these difficulties. Next, we show how to construct iZK and SSiZK from SPHF over cyclic groups for any language handled by the generic framework [7], which encompasses most, if not all, known SPHFs over cyclic groups. This is the main technical part of the paper. Third, in Sect. 4, we indeed show a concrete application of our iZK constructions: the most efficient 3-round two-party protocol computing inner product in the UC framework with static corruption so far. We analyze our construction and provide a detailed comparison with the Groth-Sahai methodology [32] and the approach based on zero-knowledge proofs "à la Schnorr" [42]. In addition, as proof of concept, we show in the full version [8] that iZK can be used instead of ZK arguments to generically convert any protocol secure in the semi-honest model into a protocol secure in the malicious model. This conversion follows the generic transformation of Goldreich, Micali and Wigderson (GMW) in their seminal papers [28,29]. While applying directly the original transformation with Schnorr-like ZK protocols blows up the number of rounds by a multiplicative factor of at least three (even in the common reference string model), our conversion only adds a small constant number of rounds. Eventually, in the full version [8], we extend our construction of iZK from SPHF to handle larger classes of languages described by computational structures such as circuits or branching programs.

Additional Related Work. Using the "OR trick" with SPHF is reminiscent of [2]. However, the methods used in our paper are very different from the one in [2], as we do not use pairings, but consider weaker form of SPHF on the other hand.

A recent line of work has focused on the cut-and-choose approach for transforming security from semi-honest to malicious models [34,35,37–39,43,44] as an alternative to the use of zero-knowledge arguments. Indeed, substantial progress has been made towards practical protocols via this approach, as applied to Yao's garbled circuits. However, the state-of-the-art still incurs a large computation and communication multiplicative overhead that is equal to the security parameter. We note that Yao's garbled circuits do not efficiently generalize to arithmetic computations, and that our approach would yield better concrete efficiency for natural functions F that admit compact representations by arithmetic branching programs. In particular, Yao's garbled circuits cannot take advantage of the structure in languages handled by the Groth-Sahai methodology [32], and namely the ones defined by multi-exponentiations: even in the latter case, Groth-Sahai technique requires pairings, while we will be able to avoid them.

The idea of using implicit proofs (without the zero-knowledge requirement) as a lightweight alternative to zero-knowledge proofs also appeared in an earlier work of Aiello, Ishai and Reingold [3]. They realize implicit proofs using conditional disclosure of secrets [27]. The latter, together with witness encryption [24] and SPHFs, only provide a weak "honest-verifier zero-knowledge" guarantee.

Recently, Jarecki introduced the concept of conditional key encapsulation mechanism [36], which is related to iZK as it adds a "zero-knowledge flavor" to SPHFs by allowing witness extraction. The construction is a combination of SPHF and zero-knowledge proofs "à la Schnorr". Contrary to iZK, it does not aim at reducing the interactivity of the resulting protocol, but ensures its covertness.

Witness encryption was introduced by Garg et al. in [24]. It enables to encrypt a message M for a word C and a language $\mathscr{L}$ into a ciphertext c, so that any user knowing a witness w that $C \in \mathscr{L}$ can decrypt c. Similarly to SPHFs, witness encryption also only has this "honest-verifier zero-knowledge" flavor: it does not enable to decrypt ciphertext for words $C \notin \mathscr{L}$, with a trapdoor. That is why, as SPHF, witness encryption cannot be used to construct directly iZK.

2   Definition of Implicit Zero-Knowledge Arguments

2.1   Notations

Since we will now be more formal, let us present the notations that we will use. Let $\{0,1\}^*$ be the set of bitstrings. We denote by PPT a probabilistic polynomial time algorithm. We write $y \leftarrow A(x)$ for 'y is the output of the algorithm A on the input x', while $y \xleftarrow{\$} A(x)$ means that A will additionally use random coins. Similarly, $X \xleftarrow{\$} \mathcal{X}$ indicates that X has been chosen uniformly at random in the (finite) set $\mathcal{X}$. We sometimes write st the state of the adversary. We define, for a distinguisher A and two distributions $\mathcal{D}_0, \mathcal{D}_1$, the advantage of A (i.e., its ability to distinguish those distributions) by $\mathsf{Adv}^{\mathcal{D}_0,\mathcal{D}_1}(A) = \mathrm{Pr}_{x \in \mathcal{D}_0}[A(x) = 1] - \mathrm{Pr}_{x \in \mathcal{D}_1}[A(x) = 1]$. The qualities of adversaries will be measured by their successes and advantages in certain experiments $\mathsf{Exp}_{\mathcal{A}}^{\mathsf{sec}}$ or $\mathsf{Exp}_{\mathcal{A}}^{\mathsf{sec}-b}$: $\mathsf{Succ}^{\mathsf{sec}}(\mathcal{A}, \mathfrak{K}) = \mathrm{Pr}[\mathsf{Exp}_{\mathcal{A}}^{\mathsf{sec}}(1^{\mathfrak{K}}) = 1]$ and $\mathsf{Adv}^{\mathsf{sec}}(\mathcal{A}, \mathfrak{K}) = \mathrm{Pr}[\mathsf{Exp}_{\mathcal{A}}^{\mathsf{sec}-1}(1^{\mathfrak{K}}) = 1] - \mathrm{Pr}[\mathsf{Exp}_{\mathcal{A}}^{\mathsf{sec}-0}(1^{\mathfrak{K}}) = 1]$ respectively, where $\mathfrak{K}$ is the security parameter, and probabilities are over the random coins of the challenger and of the adversary.

2.2   Definition

Let $(i\mathscr{L}_{\mathsf{crs}})_{\mathsf{crs}}$ be a family of NP languages, indexed by a common reference string crs, and defined by a witness relation $i\mathcal{R}_{\mathsf{crs}}$, namely $i\mathscr{L} = \{x \in i\mathcal{X}_{\mathsf{crs}} \mid \exists iw, i\mathcal{R}_{\mathsf{crs}}(x, iw) = 1\}$, where $(i\mathcal{X}_{\mathsf{crs}})_{\mathsf{crs}}$ is a family of sets. crs is generated by some polynomial-time algorithm $\mathsf{Setup}_{\mathsf{crs}}$ taking as input the unary representation of the security parameter $\mathfrak{K}$. We suppose that membership to $i\mathcal{X}_{\mathsf{crs}}$ and $i\mathcal{R}_{\mathsf{crs}}$ can be evaluated in polynomial time (in $\mathfrak{K}$). For the sake of simplicity, crs is often implicit.

To achieve stronger properties (namely simulation-soundness in Sect. 3.4), we sometimes also assume that $\mathsf{Setup}_{\mathsf{crs}}$ can also output some additional information or trapdoor $\mathcal{T}_{\mathsf{crs}}$. This trapdoor should enable to check, in polynomial time, whether a given word x is in $i\mathscr{L}$ or not. It is only used in security proofs, and is never used by the iZK algorithms.

An iZK is defined by the following polynomial-time algorithms:

- icrs $\overset{\$}{\leftarrow}$ iSetup(crs) generates the (normal) common reference string (CRS) icrs (which implicitly contains crs). The resulting CRS provides statistical soundness;
- (icrs, $\mathcal{T}$) $\overset{\$}{\leftarrow}$ iTSetup(crs)[4] generates the (trapdoor) common reference string icrs together with a trapdoor $\mathcal{T}$. The resulting CRS provides statistical zero-knowledge;
- (ipk, isk) $\overset{\$}{\leftarrow}$ iKG$^\ell$(icrs, x, iw) generates a public/secret key pair, associated to a word x $\in i\mathscr{L}$ and a label $\ell \in \{0,1\}^*$, with witness iw;
- (ipk, itk) $\overset{\$}{\leftarrow}$ iTKG$^\ell$(icrs, $\mathcal{T}$, x) generates a public/trapdoor key pair, associated to a word x $\in \mathcal{X}$ and a label $\ell \in \{0,1\}^*$;
- (c, K) $\overset{\$}{\leftarrow}$ iEnc$^\ell$(icrs, ipk, x) outputs a ciphertext c of a value K (an ephemeral key), for the public key ipk, the word x, and the label $\ell \in \{0,1\}^*$;
- $K \leftarrow$ iDec$^\ell$(icrs, isk, c) decrypts the ciphertext c for the label $\ell \in \{0,1\}^*$, and outputs the ephemeral key K;
- $K \leftarrow$ iTDec$^\ell$(icrs, itk, c) decrypts the ciphertext c for the label $\ell \in \{0,1\}^*$, and outputs the ephemeral key K.

The three last algorithms can be seen as key encapsulation and decapsulation algorithms. Labels ℓ are only used for SSiZK and are often omitted. The CRS icrs is often omitted, for the sake of simplicity.

Normally, the algorithms iKG and iDec are used by the user who wants to (implicitly) prove that some word x is in $i\mathscr{L}$ (and we often call this user the prover), while the algorithm iEnc is used by the user who wants to (implicitly) verify this (and we often call this user the verifier), as shown in Figs. 1 and 3. The algorithms iTKG and iTDec are usually only used in proofs, to generate simulated or fake implicit proofs (for the zero-knowledge property).

2.3   Security Requirements

An iZK satisfies the four following properties (for any (crs, $\mathcal{T}_{crs}$) $\overset{\$}{\leftarrow}$ Setup$_{crs}(1^\Re)$):

- **Correctness.** The encryption is the reverse operation of the decryption, with or without a trapdoor: for any icrs $\overset{\$}{\leftarrow}$ iSetup(crs) or with a trapdoor, for any (icrs, $\mathcal{T}$) $\overset{\$}{\leftarrow}$ iTSetup(crs), and for any x $\in \mathcal{X}$ and any $\ell \in \{0,1\}^*$,
  - if x $\in i\mathscr{L}$ with witness iw, (ipk, isk) $\overset{\$}{\leftarrow}$ iKG$^\ell$(icrs, x, iw), and (c, K) $\overset{\$}{\leftarrow}$ iEnc$^\ell$(ipk, x), then we have $K =$ iDec$^\ell$(isk, c);
  - if (ipk, itk) $\overset{\$}{\leftarrow}$ iTKG$^\ell$($\mathcal{T}$, x) and (c, K) $\overset{\$}{\leftarrow}$ iEnc$^\ell$(ipk, x), then we have $K =$ iTDec$^\ell$(itk, c).

---

[4] When the CRS is word-dependent, i.e., when the trapdoor $\mathcal{T}$ does only work for one word x* previously chosen, there is a second argument: (icrs, $\mathcal{T}$) $\overset{\$}{\leftarrow}$ iTSetup(crs, x*). Security notions are then slightly different. See details in the full version [8].

- **Setup Indistinguishability.** A polynomial-time adversary cannot distinguish a normal CRS generated by iSetup from a trapdoor CRS generated by iTSetup. More formally, no PPT can distinguish, with non-negligible advantage, the two distributions:

$$\{\text{icrs} \mid \text{icrs} \xleftarrow{\$} \text{iSetup(crs)}\} \qquad \{\text{icrs} \mid (\text{icrs}, \mathit{IT}) \xleftarrow{\$} \text{iTSetup(crs)}\}.$$

- **Soundness.** When the CRS is generated as icrs $\xleftarrow{\$}$ iSetup(crs), and when x $\notin \mathscr{L}$, the distribution of K is statistically indistinguishable from the uniform distribution, even given c. More formally, if Π is the set of all the possible values of K, for any bitstring ipk, for any word x $\notin i\mathscr{L}$, for any label $\ell \in \{0,1\}^*$, the two distributions:

$$\{(c, K) \mid (c, K) \xleftarrow{\$} \text{iEnc}^\ell(\text{ipk}, \text{x})\} \quad \{(c, K') \mid (c, K) \xleftarrow{\$} \text{iEnc}^\ell(\text{ipk}, \text{x}); K' \xleftarrow{\$} \Pi\}$$

are statistically indistinguishable (iEnc may output $(\perp, K)$ when the public key ipk is not well formed).

- **Zero-Knowledge.** For any label $\ell \in \{0,1\}^*$, when the CRS is generated using $(\text{icrs}, \mathit{IT}) \xleftarrow{\$} \text{iTSetup}^\ell(\text{crs})$, for any message x$^* \in i\mathscr{L}$ with the witness iw*, the public key ipk and the decapsulated key K corresponding to a ciphertext c chosen by the adversary, either using isk or the trapdoor itk, should be indistinguishable, even given the trapdoor IT. More formally, we consider the experiment $\text{Exp}^{\text{iZK-zk-}b}$ in Fig. 2. The iZK is (statistically) zero-knowledge if the advantage of any adversary $\mathcal{A}$ (not necessarily polynomial-time) for this experiment is negligible.

We defined our security notion with a "composable" security flavor, as Groth and Sahai in [32]: soundness and zero-knowledge are statistical properties, the only computational property is the setup indistinguishability property. This is slightly stronger than what is needed, but is satisfied by our constructions and often easier to use.

$\text{Exp}^{\text{iZK-zk-}b}(\mathcal{A}, \text{crs}, \mathfrak{K})$	$\text{Exp}^{\text{iZK-ss-}b}(\mathcal{A}, \text{crs}, \mathfrak{K})$
$(\text{icrs}, \mathit{IT}) \xleftarrow{\$} \text{iTSetup(crs)}$	$(\text{icrs}, \mathit{IT}) \xleftarrow{\$} \text{iTSetup(crs)}$
$(\ell, \text{x}^*, \text{iw}, \text{st}) \xleftarrow{\$} \mathcal{A}(\text{icrs}, \mathit{IT})$	$(\ell^*, \text{x}^*, \text{ipk}, \text{st}) \xleftarrow{\$} \mathcal{A}^{\mathcal{O}}(\text{icrs})$
if $i\mathcal{R}(\text{x}^*, \text{iw}) = 0$ **then return** random bit	$(c, K) \xleftarrow{\$} \text{iEnc}^\ell(\text{ipk}, \text{x}^*)$
if $b = 0$ **then** $(\text{ipk}, \text{isk}) \xleftarrow{\$} \text{iKG}^\ell(\text{icrs}, \text{x}^*, \text{iw}^*)$	**if** $b = 0$ **then** $K' \leftarrow K$
else $(\text{ipk}, \text{itk}) \xleftarrow{\$} \text{iTKG}^\ell(\mathit{IT}, \text{x}^*)$	**else** $K' \xleftarrow{\$} \Pi$
$(c, \text{st}) \xleftarrow{\$} \mathcal{A}(\text{st}, \text{icrs}, \mathit{IT}, \text{ipk})$	$b' \xleftarrow{\$} \mathcal{A}^{\mathcal{O}}(\text{st}, c, K')$
if $b = 0$ **then** $K \leftarrow \text{iDec}^\ell(\text{isk}, c)$	**if** $\exists$itk, $(\ell^*, \text{x}^*, \text{ipk}, \text{itk}) \in L \cup L'$ **then**
else $K \leftarrow \text{iTDec}^\ell(\text{itk}, c)$	**return** random bit
return $\mathcal{A}(\text{st}, K)$	**if** $\text{x}^* \in i\mathscr{L}$ **then return** random bit
	return b'

Fig. 2. Experiments $\text{Exp}^{\text{iZK-zk-}b}$ for zero-knowledge of iZK, and $\text{Exp}^{\text{iZK-ss-}b}$ for simulation-soundness of SSiZK

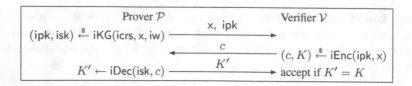

Fig. 3. Three-round zero-knowledge from iZK for a word $x \in i\mathscr{L}$ and a witness iw

We also consider stronger iZK, called simulation-sound iZK or SSiZK, which satisfies the following additional property:

- **Simulation Soundness.** The soundness holds (computationally) even when the adversary can see simulated public keys and decryption with these keys. More formally, we consider the experiment $\mathsf{Exp}^{\mathtt{iZK\text{-}ss\text{-}}b}$ in Fig. 2, where the oracle $\mathcal{O}$, and the lists L and L' are defined as follows:
  - on input (ℓ, x), $\mathcal{O}$ generates $(\mathsf{ipk}, \mathsf{itk}) \overset{\$}{\leftarrow} \mathsf{iTKG}(\mathsf{icrs}, \mathbb{T}, x)$, stores $(\ell, x, \mathsf{ipk}, \mathsf{itk})$ in a list L, and outputs ipk;
  - on input (ipk, c), $\mathcal{O}$ retrieves the record $(\ell, x, \mathsf{ipk}, \mathsf{itk})$ from L (and aborts if no such record exists), removes it from L, and adds it to L', computes $K \leftarrow \mathsf{iTDec}^{\ell}(\mathsf{icrs}, \mathsf{itk}, c)$, and outputs K.

The iZK is (statistically) simulation-sound if the advantage of any adversary $\mathcal{A}$ (not necessarily polynomial-time) for this experiment is negligible.

Remark 1. An iZK for some language $i\mathscr{L}$ directly leads to a 3-round zero-knowledge arguments for $i\mathscr{L}$. The construction is depicted in Fig. 3 and the proof is provided in the full version [8]. If the iZK is additionally simulation-sound, the resulting zero-knowledge argument is also simulation-sound.

Remark 2. For the sake of completeness, in the full version [8], we show how to construct iZK from either NIZK or Trapdoor SPHFs. In the latter case, the resulting iZK is not statistically sound and zero-knowledge but only computationally sound and zero-knowledge. In both cases, using currently known constructions over cyclic groups, strong assumptions such as the random oracle model or pairings are needed.

3   Construction of Implicit Zero-Knowledge Arguments

Let us first recall the generic framework of SPHFs [7] for the particular case of cyclic groups, and when the projection key hp can depend on the word C, as it is at the core of our construction of iZK. Second, we explain in more details the limitations of SPHFs and the fact they cannot directly be used to construct iZK (even with a concrete attack). Third, we show how to overcome these limitations to build iZK and SSiZK.

3.1 Review of the Generic Framework of SPHFs over Cyclic Groups

Languages. Let $\mathbb{G}$ be a cyclic group of prime order p and $\mathbb{Z}_p$ the field of integers modulo p. If we look at $\mathbb{G}$ and $\mathbb{Z}_p$ as the same ring $(\mathbb{G}, +, \bullet)$, where internal operations are on the scalars, many interesting languages can be represented as subspaces of the vector space $\mathbb{G}^n$, for some n. Here are some examples.

Example 3 (DDH or ElGamal Ciphertexts of 0). Let g and h be two generators of $\mathbb{G}$. The language of DDH tuples in basis (g, h) is

$$\mathscr{L} = \{(u, e) \in \mathbb{G}^2 \mid \exists r \in \mathbb{Z}_p, \, u = g^r \text{ and } e = h^r\} \subseteq \mathbb{G}^2,$$

where r is the witness. It can be seen as the subspace of $\mathbb{G}^2$ generated by (g, h). We remark that this language can also be seen as the language of (additive) ElGamal ciphertexts of 0 for the public key $\mathsf{pk} = (g, h)$. $\quad\square$

Example 4 (ElGamal Ciphertexts of a Bit). Let us consider the language of ElGamal ciphertexts of 0 or 1, under the public key $\mathsf{pk} = (g, h)$:

$$\mathscr{L} := \{(u, e) \in \mathbb{G}^2 \mid \exists r \in \mathbb{Z}_p, \exists b \in \{0, 1\}, \, u = g^r \text{ and } e = h^r g^b\}.$$

Here $C = (u, e)$ cannot directly be seen as an element of some vector space. However, a word $C = (u, e) \in \mathbb{G}^2$ is in $\mathscr{L}$ if and only there exists $\boldsymbol{\lambda} = (\lambda_1, \lambda_2, \lambda_3) \in \mathbb{Z}_p^3$ such that:

$$u = g^{\lambda_1} \, (= \lambda_1 \bullet g) \qquad\qquad e = h^{\lambda_1} g^{\lambda_2} \, (= \lambda_1 \bullet h + \lambda_2 \bullet g)$$
$$1 = u^{\lambda_2} g^{\lambda_3} \, (= \lambda_2 \bullet u + \lambda_3 \bullet g) \quad 1 = (e/g)^{\lambda_2} h^{\lambda_3} \, (= \lambda_2 \bullet (e - g) + \lambda_3 \bullet h),$$

because, if we write $C = (u, e) = (g^r, h^r g^b)$ (with $r, b \in \mathbb{Z}_p$, which is always possible), then the first three equations ensure that $\lambda_1 = r$, $\lambda_2 = b$ and $\lambda_3 = -rb$, while the last equation (right bottom) ensures that $b(b - 1) = 0$, i.e., $b \subset \{0, 1\}$, as it holds that $(h^r g^b/g)^b h^{-rb} = g^{b(b-1)} = 1$.

Therefore, if we introduce the notation $\hat{C} = \theta(C) := (u \, e \, 1 \, 1) \in \mathbb{G}^4$, then the language $\mathscr{L}$ can be defined as the set of $C = (u, e)$ such that $\hat{C}$ is in the subspace of $\mathbb{G}^4$ generated by the rows of the following matrix

$$\Gamma := \begin{pmatrix} g & h & 1 & 1 \\ 1 & g & u & e/g \\ 1 & 1 & g & h \end{pmatrix}. \qquad\qquad\square$$

Example 5 (Conjunction of Languages). Let g_i and h_i (for $i = 1, 2$) be four generators of $\mathbb{G}$, and $\mathscr{L}_i$ be (as in Example 3) the languages of DDH tuples in bases (g_i, h_i) respectively. We are now interested in the language $\mathscr{L} = \mathscr{L}_1 \times \mathscr{L}_2 \subseteq \mathbb{G}^4$, which is thus the conjunction of $\mathscr{L}_1 \times \mathbb{G}^2$ and $\mathbb{G}^2 \times \mathscr{L}_2$: it can be seen as the subspace of $\mathbb{G}^4$ generated by the rows of the following matrix

$$\Gamma := \begin{pmatrix} g_1 & h_1 & 1 & 1 \\ 1 & 1 & g_2 & h_2 \end{pmatrix}. \qquad\qquad\square$$

This can also be seen as the matrix, diagonal by blocks, with Γ_1 and Γ_2 the matrices for $\mathscr{L}_1$ and $\mathscr{L}_2$ respectively.

More formally, the generic framework for SPHFs in [7] considers the languages $\mathscr{L} \subseteq \mathcal{X}$ defined as follows: There exist two functions θ and Γ from the set of words $\mathcal{X}$ to the vector space $\mathbb{G}^n$ of dimension n, and to set $\mathbb{G}^{k \times n}$ of $k \times n$ matrices over $\mathbb{G}$, such that $C \in \mathscr{L}$ if and only if $\hat{C} := \theta(C)$ is a linear combination of the rows of $\Gamma(C)$. From a witness w for a word C, it should be possible to compute such a linear combination as a row vector $\boldsymbol{\lambda} = (\lambda_i)_{i=1,\dots,k} \in \mathbb{Z}_p^{1 \times k}$:

$$\hat{C} = \theta(C) = \boldsymbol{\lambda} \bullet \Gamma(C). \tag{1}$$

For the sake of simplicity, because of the equivalence between w and $\boldsymbol{\lambda}$, we will use them indifferently for the witness.

SPHFs. Let us now build an SPHF on such a language. A hashing key hk is just a random column vector $\mathsf{hk} \in \mathbb{Z}_p^n$, and the associated projection key is $\mathsf{hp} := \Gamma(C) \bullet \mathsf{hk}$. The hash value of a word C is then $H := \hat{C} \bullet \mathsf{hk}$, and if $\boldsymbol{\lambda}$ is a witness for $C \in \mathscr{L}$, this hash value can also be computed as:

$$H = \hat{C} \bullet \mathsf{hk} = \boldsymbol{\lambda} \bullet \Gamma(C) \bullet \mathsf{hk} = \boldsymbol{\lambda} \bullet \mathsf{hp} = \mathsf{proj}H,$$

which only depends on the witness $\boldsymbol{\lambda}$ and the projection key hp. On the other hand, if $C \notin \mathscr{L}$, then $\hat{C}$ is linearly independent from the rows of $\Gamma(C)$. Hence, $H := \hat{C} \bullet \mathsf{hk}$ looks random even given $\mathsf{hp} := \Gamma(C) \bullet \mathsf{hk}$, which is exactly the *smoothness* property.

Example 6. The SPHF corresponding to the language in Example 4, is then defined by:

$$\mathsf{hk} = (\mathsf{hk}_1, \mathsf{hk}_2, \mathsf{hk}_3, \mathsf{hk}_4)^\mathsf{T} \xleftarrow{\$} \mathbb{Z}_p^4$$

$$\mathsf{hp} = \Gamma(C) \bullet \mathsf{hk} = (g^{\mathsf{hk}_1} h^{\mathsf{hk}_2}, g^{\mathsf{hk}_2} u^{\mathsf{hk}_3} (e/g)^{\mathsf{hk}_4}, g^{\mathsf{hk}_3} h^{\mathsf{hk}_4})$$

$$H = \hat{C} \bullet \mathsf{hk} = u^{\mathsf{hk}_1} e^{\mathsf{hk}_2} \qquad \mathsf{proj}H = \boldsymbol{\lambda} \bullet \mathsf{hp} = \mathsf{hp}_1^r \cdot \mathsf{hp}_2^b \cdot \mathsf{hp}_3^{-rb}.$$

For the sake of clarity, we will omit the C argument, and write Γ, instead of $\Gamma(C)$.

3.2 Limitations of Smooth Projective Hash Functions

At a first glance, as explained in the introduction, it may look possible to construct an iZK from an SPHF for the same language $\mathscr{L} = \mathit{i}\mathscr{L}$ as follows:

- iSetup(crs) and iTSetup(crs) outputs the empty CRS icrs $:=\perp$;
- iKG(icrs, x, iw) outputs an empty public key ipk $:=\perp$ together with the secret key isk $:= (x, iw)$;
- iEnc(ipk, x) generates a random hashing key $\mathsf{hk} \xleftarrow{\$} \mathsf{HashKG}(\mathsf{crs}, x)$ and outputs the ciphertext $c := \mathsf{hp} \leftarrow \mathsf{ProjKG}(\mathsf{hk}, \mathsf{crs}, x)$ together with the ephemeral key $K := H \leftarrow \mathsf{Hash}(\mathsf{hk}, \mathsf{crs}, x)$;
- iDec(isk, c) outputs the ephemeral key $K := \mathsf{proj}H \leftarrow \mathsf{ProjHash}(\mathsf{hp}, \mathsf{crs}, x, \mathsf{iw})$.

This construction is sound: if $x \notin \mathscr{L}$, given only $c = \mathsf{hp}$, the smoothness ensures that $K = H$ looks random. Unfortunately, there seems to be no way to compute K from only c, or in other words, there does not seem to exist algorithms iTKG and iTDec.

Example 6 is not Zero-Knowledge. Actually, with the SPHF from Example 6, no such algorithm iTKG or iTDec (verifying the zero-knowledge property) exists. It is even worse than that: a malicious verifier may get information about the witness, even if he just has a feedback whether the prover could use the correct hash value or not (and get the masked value or not), in a protocol such as the one in Fig. 1. A malicious verifier can indeed generate a ciphertext $c = \mathsf{hp}$, by generating hp_1 honestly but by picking hp_2 and hp_3 uniformly at random. Now, a honest prover will compute $\mathsf{proj}H = \mathsf{hp}_1^r \mathsf{hp}_2^b \mathsf{hp}_3^{-rb}$, to get back the ephemeral key (using iDec). When C is an encryption of $b = 1$, this value is random and independent of H, as hp_2 and hp_3 have been chosen at random, while when $b = 0$, this value is the correct $\mathsf{proj}H$ and is equal to H. Thus the projected hash value $\mathsf{proj}H$, which is the ephemeral output key by the honest prover, reveals some information about b, part of the witness.

If we want to avoid such an attack, the prover has to make sure that the hp he received was built correctly. Intuitively, this sounds exactly like the kind of verifications we could make with an SPHF: we could simply build an SPHF on the language of the "correctly built" hp. Then the prover could send a projection key for this new SPHF and ask the verifier to XOR the original hash value H with the hash value of this new SPHF. However, things are not that easy: first this does not solve the limitation due to the security proof (the impossibility of computing H for $x \notin i\mathscr{L}$) and second, in the SPHF in Example 6, all projection keys are valid (since Γ is full-rank, for any hp, there exists necessarily a hk such that $\mathsf{hp} = \Gamma \bullet \mathsf{hk}$).

3.3 iZK Construction

Let us consider an SPHF defined as in Sect. 3.1 for a language $i\mathscr{L} = \mathscr{L}$. In this section, we show how to design, step by step, an iZK for $i\mathscr{L}$ from this SPHF, following the overview in Sect. 1. At the end, we provide a summary of the construction and a complete proof. We illustrate our construction on the language of ElGamal ciphertexts of bits (Examples 4 and 6), and refer to this language as "our example". We suppose a cyclic group $\mathbb{G}$ of prime order p is fixed, and that DDH is hard in $\mathbb{G}^5$.

We have seen the limitations of directly using the original SPHF are actually twofold. First, SPHFs do not provide a way to compute the hash value of a word outside the language, with just a projection key for which the hashing key is not known. Second, nothing ensures that a projection key has really been derived from an actually known hashing key, and in such a bad case, the projected hash value may leak some information about the word C (and the witness).

---

[5] The construction can be trivially extended to DLin, or any MDDH assumption [21] though.

To better explain our construction, we first show how to overcome the first limitation. Thereafter, we will show how our approach additionally allows to check the validity of the projection keys (with a non-trivial validity meaning). It will indeed be quite important to notice that the projection keys coming from our construction (according to one of the setups) will not necessarily be valid (with a corresponding hashing key), as the corresponding matrix Γ will not always be full rank, contrary to the projection keys of the SPHF in Example 6. Hence, the language of the valid projection keys will make sense in this setting.

Adding the Trapdoor. The CRS of our construction is a tuple $\mathsf{icrs} = (g', h', u' = g'^{r'}, e' = h'^{s'}) \in \mathbb{G}^4$, with g', h' two random generators of $\mathbb{G}$, and

- r', s' two random distinct scalars in $\mathbb{Z}_p$, for the normal CRS generated by iSetup, so that (g', h', u', e') is not a DDH tuple;
- $r' = s'$ a random scalar in $\mathbb{Z}_p$, for the trapdoor CRS generated by $\mathsf{iTSetup}$, with $\mathcal{T} = r'$ the trapdoor, so that (g', h', u', e') is a DDH tuple.

Then, we build an SPHF for the augmented language $\mathscr{L}_t$ defined as follows: a word $C_t = (C, u', e')$ is in $\mathscr{L}_t$ if and only if either C is in the original language $\mathscr{L}$ or (u', e') is a DDH tuple. This new language $\mathscr{L}_t$ can be seen as the disjunction of the original language $\mathscr{L}$ and of the DDH language in basis (g', h'). Construction of disjunctions of SPHFs were proposed in [2] but require pairings. In this article, we use an alternative more efficient construction without pairing[6]. Let us show it on our example, with $C_t = (C, u', e')$. We set $\hat{C}_t := (g'^{-1}, 1, 1, 1, 1, 1, 1)$ and $\Gamma_t(C_t) \in \mathbb{G}^{(k+3) \times (n+3)}$ as

$$
\Gamma_t(C_t) := \left(
\begin{array}{c|c}
1 & \Gamma(C) \\
\hline
\begin{array}{c|cc} g' & 1 & 1 \end{array} & \hat{C} = \theta(C) \\
\hline
\begin{array}{c|cc} 1 & g' & h' \end{array} & 1 \ldots \quad 1 \\
\hline
\begin{array}{c|cc} g' & u' & e' \end{array} & 1 \ldots \quad 1
\end{array}
\right)
= \left(
\begin{array}{ccc|cccc}
1 & 1 & 1 & g & h & 1 & 1 \\
1 & 1 & 1 & 1 & g & u & e/g \\
1 & 1 & 1 & 1 & 1 & g & h \\
\hline
g' & 1 & 1 & u & e & 1 & 1 \\
1 & g' & h' & 1 & 1 & 1 & 1 \\
g' & u' & e' & 1 & 1 & 1 & 1
\end{array}
\right). \tag{2}
$$

Let us show the language corresponding to Γ_t and $\hat{C}_t$ is indeed $\mathscr{L}_t$: Due to the first column of Γ_t and the first element of $\hat{C}_t$, if $\hat{C}_t$ is a linear combination of rows of Γ_t with coefficients $\boldsymbol{\lambda}_t$ (i.e., $\hat{C}_t = \boldsymbol{\lambda}_t \bullet \Gamma_t$), one has $\lambda_{t,4} + \lambda_{t,6} = -1$, and thus at least $\lambda_{t,4}$ or $\lambda_{t,6}$ is not equal to zero.

- If $\lambda_{t,6} \neq 0$, looking at the second and the third columns of Γ_t gives that:

$$
\lambda_{t,5} \bullet (g', h') + \lambda_{t,6} \bullet (u', e') = (1, 1), \text{ i.e., } (u', e') = (g'^{\lambda_{t,5}/\lambda_{t,6}}, h'^{\lambda_{t,5}/\lambda_{t,6}}),
$$

  or in other words (u', e') is a DDH tuple in basis (g', h');
- if $\lambda_{t,4} \neq 0$, looking at the last four columns of Γ_t gives that: $\lambda_{t,4} \bullet \hat{C} = \lambda_{t,4} \bullet (u, e, 1, 1)$ is a linear combination of rows of Γ, hence $\hat{C}$ too. As a consequence, by definition of $\mathscr{L}$, $C \in \mathscr{L}$.

---

[6] Contrary to [2] however, our matrix Γ_t depends on the words C_t, which is why we get this more efficient construction.

Now, whatever the way the CRS is generated (whether (u', e') is a DDH tuple or not), it is always possible to compute $\mathsf{proj}H$ as follows, for a word $C \in \mathscr{L}$ with witnesses r and b:

$$\mathsf{proj}H = \boldsymbol{\lambda}_t \bullet \mathsf{hp} \qquad \boldsymbol{\lambda}_t = (\boldsymbol{\lambda}, -1, 0, 0) = (r, b, -rb, -1, 0, 0)$$

When the CRS is generated with the normal setup, as shown above, this is actually the only way to compute $\mathsf{proj}H$, since (u', e') is not a DDH tuple and so $\hat{C}_t$ is linearly dependent of the rows of Γ_t if and only if $C \in \mathscr{L}$. On the opposite, when the CRS is generated by the trapdoor setup with trapdoor r', we can also compute $\mathsf{proj}H$ using the witness r': $\mathsf{proj}H = \boldsymbol{\lambda}_t' \bullet \mathsf{hp}$ with $\boldsymbol{\lambda}_t' = (0, 0, 0, 0, r', -1)$.

However, the latter way to compute $\mathsf{proj}H$ gives the same result as the former way, only if $\mathsf{hp}_{t,5}$ and $\mathsf{hp}_{t,6}$ involve the correct value for hk_1. A malicious verifier could decide to choose random $\mathsf{hp}_{t,5}$ and $\mathsf{hp}_{t,6}$, which would make $\boldsymbol{\lambda}_t' \bullet \mathsf{hp}$ look random and independent of the real hash value!

Ensuring the Validity of Projection Keys. The above construction and trapdoor would provide zero-knowledge if we could ensure that the projection keys hp (generated by a potentially malicious verifier) is valid, so that, intuitively, $\mathsf{hp}_{t,5}$ and $\mathsf{hp}_{t,6}$ involve the correct value of hk_1. Using a zero-knowledge proof (that hp derives from some hashing key hk) for that purpose would annihilate all our efforts to avoid adding rounds and to work under plain DDH (interactive ZK proofs introduce more rounds, and Groth-Sahai [32] NIZK would require assumptions on bilinear groups). So we are left with doing the validity check again with SPHFs.

Fortunately, the language of valid projection keys hp can be handled by the generic framework, since a valid projection key hp is such that: $\mathsf{hp} = \Gamma_t \bullet \mathsf{hk}$, or in other words, if we transpose everything $\mathsf{hp}^\mathsf{T} - \mathsf{hk}^\mathsf{T} \bullet \Gamma_t^\mathsf{T}$. This is exactly the same as in Eq. (1), with $\hat{C} \leftrightarrow \mathsf{hp}^\mathsf{T}$, $\Gamma \leftrightarrow \Gamma_t^\mathsf{T}$ and witness $\boldsymbol{\lambda} \leftrightarrow \mathsf{hk}^\mathsf{T}$. So we can now define a smooth projective hash function on that language, where the projection key is called transposed projection key tp, the hashing key is called transposed hashing key tk, the hash value is called transposed hash value $\mathsf{t}H$ and the projected hash value is called transposed projected hash value $\mathsf{tproj}H$.

Finally, we could define an iZK, similarly to the one in Sect. 3.2, except, ipk contains a transposed projection key tp (generated by the prover from a random transposed hashing key tk), and c contains the associated transposed projected hash value $\mathsf{tproj}H$ in addition to hp, so that the prover can check using tk that hp is valid by verifying whether $\mathsf{tproj}H = \mathsf{t}H$ or not.

An Additional Step. Unfortunately, we are not done yet, as the above modification breaks the soundness property! Indeed, in this last construction, the prover now learns an additional information about the hash value H: $\mathsf{tproj}H = \mathsf{hk}^\mathsf{T}\mathsf{tp}$, which does depend on the secret key hk. He could therefore choose $\mathsf{tp} = \hat{C}_t^\mathsf{T}$, so that $\mathsf{tproj}H = \mathsf{hk}^\mathsf{T}\hat{C}_t^\mathsf{T} = \hat{C}_t\mathsf{hk}$ is the hash value $H = K$ of C under hk.

We can fix this by ensuring that the prover will not know the extended word $\hat{C}_t$ on which the SPHF will be based when he sends tp, using an idea similar to the 2-universality property of SPHF introduced by Cramer and Shoup in [18]. For that purpose, we extend Γ_t and make $\hat{C}_t$ depends on a random scalar $\zeta \in \mathbb{Z}_p$ chosen by the verifier (and included in c).

Detailed Construction. Let us now formally show how to build an iZK from any SPHF built from the generic framework of [7], following the previous ideas. We recall that we consider a language $\mathscr{L} = i\mathscr{L}$, such that a word $\mathsf{x} = C$ is in $i\mathscr{L}$, if and only if $\hat{C} = \theta(C)$ is a linear combination of the rows of some matrix $\Gamma \in \mathbb{G}^{k \times n}$ (which may depend on C). The coefficients of this linear combination are entries of a row vector $\boldsymbol{\lambda} \in \mathbb{Z}_p^{1 \times k}$: $\hat{C} = \boldsymbol{\lambda} \bullet \Gamma$, where $\boldsymbol{\lambda} = \boldsymbol{\lambda}(\mathsf{iw})$ can be computed from the witness iw for x.

The setup algorithms iSetup(crs) and iTSetup(crs) are defined as above (page 13). We define an extended language using the generic framework:

$$\theta_t(\mathsf{x}, \zeta) = \hat{C}_t = (g'^{-1}, 1, \ldots, 1, g'^{-\zeta}, 1, \ldots, 1) \qquad \in \mathbb{G}^{1 \times (2n+6)}$$

$$\Gamma_t(\mathsf{x}) = \left(\begin{array}{c|c} \Gamma_t'(\mathsf{x}) & 1 \\ \hline 1 & \Gamma_t'(\mathsf{x}) \end{array} \right) \qquad \in \mathbb{G}^{(2k+6) \times (2n+6)},$$

where $\Gamma_t'(\mathsf{x})$ is the matrix (initially called $\Gamma_t(\mathsf{x})$ in Eq. (2), 1 is the matrix of $\mathbb{G}^{(2k+3) \times (2n+3)}$ with all entries equal to 1, and ζ is a scalar used to ensure the prover cannot guess the word $\hat{C}_t$ which will be used, and so cannot choose $\mathsf{tp} = \hat{C}_t$. As explained above, this language corresponds to a 2-universal SPHF for the disjunction of the language of DDH tuples (g', h', u', e') and the original language $\mathscr{L}$. We write:

$$\boldsymbol{\lambda}_t(\zeta, \mathsf{iw}) = (\boldsymbol{\lambda}(\mathsf{iw}), -1, 0, 0, \zeta\boldsymbol{\lambda}(\mathsf{iw}), -\zeta, 0, 0)$$
$$\boldsymbol{\lambda}_t(\zeta, \mathit{\Pi}) = (0, \ldots, 0, r', -1, 0, \ldots, 0, \zeta r', -\zeta) \qquad \text{with } \mathit{\Pi} = r',$$

so that:

$$\hat{C}_t = \begin{cases} \boldsymbol{\lambda}_t(\zeta, \mathsf{iw}) \bullet \Gamma_t(\mathsf{x}) & \text{if } (g', h', u', e') \text{ is a DDH tuple, with witness } \mathit{\Pi} \\ \boldsymbol{\lambda}_t(\zeta, \mathit{\Pi}) \bullet \Gamma_t(\mathsf{x}) & \text{if } \mathsf{x} \in i\mathscr{L} \text{ with witness iw.} \end{cases}$$

The resulting iZK construction is depicted in Fig. 4. This is a slightly more efficient construction that the one we sketched previously, where the prover does not test anymore explicitly tprojH, but tprojH (or tH) is used to mask K. Thus, tprojH no more needs to be included in c.

Variants. In numerous cases, it is possible to add the trapdoor in a slightly more efficient way, if we accept to use word-dependent CRS. While the previous construction would be useful for security in the UC framework [15], the more efficient construction with a word-dependent CRS is enough in the stand-alone setting. Independently of that improvement, it is also possible to slightly reduce the size of hp, by computing ζ with an entropy extractor, and so dropping it from hp. Details for both variants are given in the full version [8].

iSetup(crs)	iTSetup(crs)
$(g', h') \xleftarrow{\$} \mathbb{G}^{*2}$	$(g', h') \xleftarrow{\$} \mathbb{G}^{*2}$
$(r', s') \xleftarrow{\$} \mathbb{Z}_p^2 \setminus \{(a, a) \mid a \in \mathbb{Z}_p\}$	$r' \xleftarrow{\$} \mathbb{Z}_p$
$(u', e') \leftarrow (g'^{r'}, h'^{s'}) \in \mathbb{G}^2$	$(u', e') \leftarrow (g'^{r'}, h'^{r'}) \in \mathbb{G}^2$
$\text{icrs} \leftarrow (g', h', u', e')$	$\text{icrs} \leftarrow (g', h', u', e'); \mathit{IT} \leftarrow r'$
return icrs	**return** (icrs, IT)
iKG(icrs, x, iw)	iTKG(icrs, x, IT)
$\text{tk} \xleftarrow{\$} \mathbb{Z}_p^{2k+6}$	$\text{tk} \xleftarrow{\$} \mathbb{Z}_p^{2k+6}$
$\text{ipk} := \text{tp} \leftarrow \Gamma_t(\text{x})^{\mathsf{T}} \bullet \text{tk} \in \mathbb{G}^{2n+6}$	$\text{ipk} := \text{tp} \leftarrow \Gamma_t(\text{x})^{\mathsf{T}} \bullet \text{tk} \in \mathbb{G}^{2n+6}$
$\text{isk} := (\text{x}, \text{tk}, \text{iw})$	$\text{itk} := (\text{x}, \text{tk}, \mathit{IT})$
return (ipk, isk)	**return** (ipk, itk)
iEnc(icrs, ipk, x)	$H \leftarrow \theta_t(\text{x}, \zeta) \bullet \text{hk} \in \mathbb{Z}_p$
$\text{tp} \leftarrow \text{ipk}; \text{hk} \xleftarrow{\$} \mathbb{Z}_p^{2n+6}; \zeta \xleftarrow{\$} \mathbb{Z}_p$	$K \leftarrow H \cdot \text{tproj}H \in \mathbb{G}$
$\text{hp} \leftarrow \Gamma_t(\text{x}) \bullet \text{hk} \in \mathbb{Z}_p^{2k+6}$	$c := (\zeta, \text{hp})$
$\text{tproj}H \leftarrow \text{hk}^{\mathsf{T}} \bullet \text{tp} \in \mathbb{G}$	**return** (K, c)
iDec(icrs, isk, c)	iTDec(icrs, itk, c)
$(\text{x}, \text{tk}, \text{iw}) \leftarrow \text{isk}$	$(\text{x}, \text{tk}, \mathit{IT}) \leftarrow \text{itk}$
$(\zeta, \text{hp}) \leftarrow c$	$(\zeta, \text{hp}) \leftarrow c$
$tH \leftarrow \text{hp}^{\mathsf{T}} \bullet \text{tk} \in \mathbb{Z}_p$	$tH \leftarrow \text{hp}^{\mathsf{T}} \bullet \text{tk} \in \mathbb{Z}_p$
$\text{proj}H \leftarrow \lambda_t(\zeta, \text{iw}) \bullet \text{hp} \in \mathbb{G}$	$\text{trap}H := \lambda_t(\zeta, \mathit{IT}) \bullet \text{hp} \in \mathbb{G}$
return $K := \text{proj}H \cdot tH \in \mathbb{G}$	**return** $K := \text{trap}H \cdot tH \in \mathbb{G}$

Fig. 4. Construction of iZK

3.4 SSiZK Construction

Our SSiZK construction is similar to our iZK construction, except that, in addition both iSetup and iTSetup add the CRS icrs, a tuple $(v_{k,i})_{i=0,\ldots,2\mathfrak{K}}^{k=1,2}$ of group elements constructed as follows: for $i = 0$ to $2\mathfrak{K}$ (with $\mathfrak{K}$ the security parameter): $r_i' \xleftarrow{\$} \mathbb{Z}_p, v_{1,i} \leftarrow g'^{r_i'}, v_{2,i} \leftarrow h'^{r_i'}$. We also define the two Waters functions [45] $\mathcal{W}_k : \{0, 1\}^{2\mathfrak{K}} \rightarrow \mathbb{G}$, as $\mathcal{W}_k(m) = v_{k,0} \prod_{i=1}^{2\mathfrak{K}} v_{k,i}^{m_i}$, for any bitstring $m = m_1 \| \ldots \| m_{2\mathfrak{K}} \in \{0, 1\}^{2\mathfrak{K}}$. Finally, the CRS is also supposed to contain a hash function $\mathcal{H} : \{0, 1\}^* \rightarrow \{0, 1\}^{2\mathfrak{K}}$ drawn from a collision-resistant hash function family $\mathcal{HF}$.

Next, the language $\mathscr{L}_t$ is further extended by adding 3 rows and 2 columns (all equal to 1 except on the 3 new rows) to both the sub-matrices $\Gamma_t'(\text{x})$ of $\Gamma_t(\text{x})$, where the 3 new rows are:

$$\begin{pmatrix} 1 & 1 & 1 & 1 \ldots 1 & g' & h' \\ 1 & 1 & 1 & 1 \ldots 1 & u'' & e'' \\ g' & 1 & 1 & 1 \ldots 1 & g' & 1 \end{pmatrix} \in \mathbb{G}^{3 \times (n+5)},$$

with $u'' = \mathcal{W}_1(\mathcal{H}(\ell, \text{x}))$ and $e'' = \mathcal{W}_2(\mathcal{H}(\ell, \text{x}))$. The vector $\hat{C}_t$ becomes $\hat{C}_t = (g^{-1}, 1, \ldots, 1, g^{-\zeta}, 1, \ldots, 1)$ (it is the same except for the number of 1's).

Due to lack of space, the full matrix is depicted in the full version [8], where the security proof can also be found. The security proof requires that $\mathsf{Setup_{crs}}$ also outputs some additional information or trapdoor $\mathcal{T}_{\mathsf{crs}}$, which enables to check, in polynomial time, whether a given word x is in $i\mathscr{L}$ or not.

Here is an overview of the security proof. Correctness, setup indistinguishability, and zero-knowledge are straightforward. Soundness follows from the fact that (g', h', u'', e'') is a DDH-tuple, when parameters are generated by iSetup (and also iTSetup actually), and so $(g', 1)$ is never in the subspace generated by (g', h') and (u'', e'') (as $h' \neq 1$), hence the corresponding language $\mathscr{L}_t$ is the same as for our iZK construction. Finally, to prove simulation-soundness, we use the programmability of the Waters function [33] and change the generation of the group elements $(v_{k,i})$ so that for the challenge proof (generated by the adversary) (g', h', u'', e'') is not a DDH-tuple, while for the simulated proofs it is a DDH-tuple. Then, we can change the setup to $\hat{\mathsf{iSetup}}$, while still being able to simulate proofs. But in this setting, the word $\hat{C}_t$ for the challenge proof is no more in $\mathscr{L}_t$, and smoothness implies simulation-soundness.

4   Application to the Inner Product

In case of biometric authentication, a server $\mathcal{S}$ wants to compute the Hamming distance between a fresh user's feature and the stored template, but without asking the two players to reveal their own input: the template y from the server side and the fresh feature x from the client side. One can see that the Hamming distance between the ℓ-bit vectors x and y is the sum of the Hamming weights of x and y, minus twice the inner product of x and y. Let us thus focus on this private evaluation of the inner product: a client $\mathcal{C}$ has an input $x = (x_i)_{i=1}^{\ell} \in \{0,1\}^{\ell}$ and a server $\mathcal{S}$ has an input $y = (y_i)_{i=1}^{\ell} \in \{0,1\}^{\ell}$. The server $\mathcal{S}$ wants to learn the inner product $\mathsf{IP} = \sum_{i=1}^{\ell} x_i y_i \in \{0, \ldots, \ell\}$, but nothing else, while the client $\mathcal{C}$ just learns whether the protocol succeeded or was aborted.

Semi-Honest Protocol. $\mathcal{C}$ can send an ElGamal encryption of each bit under a public key of her choice and then $\mathcal{S}$ can compute an encryption of $\mathsf{IP} + R$, with $R \in \mathbb{Z}_p$ a random mask, using the homomorphic properties of ElGamal, and sends this ciphertext. $\mathcal{C}$ finally decrypts and sends back $g^{\mathsf{IP}+R}$ to $\mathcal{S}$ who divides it by g^R to get g^{IP}. Since IP is small, an easy discrete logarithm computation leads to IP.

Malicious Setting. To transform this semi-honest protocol into one secure against malicious adversaries, we could apply our generic conversion presented in the full version [8]. Here, we propose an optimized version of this transformation for this protocol. We use the ElGamal scheme for the encryption $\mathcal{E}_{\mathsf{pk}}$, where pk is a public key chosen by $\mathcal{C}$ and the secret key is $\mathsf{sk} = (\mathsf{sk}_j)_{j=1}^{\log p}$, and the Cramer-Shoup scheme [17] for commitments Com, of group elements or multiple group elements with randomness reuse, where the public key is in the CRS. The CRS additionally contains the description of a cyclic group and a generator g of this group. The construction is presented on Fig. 5. First, the client commits to her

$$\mathcal{C} \xrightarrow{\quad \mathsf{pk}, (c_i = \mathcal{E}_{\mathsf{pk}}(g^{x_i}))_{i=1}^{\ell} \quad} \mathcal{S}$$
$$\xleftarrow{\quad \prod_{i=1}^{\ell} c_i^{y_i} \cdot \mathcal{E}_{\mathsf{pk}}(g^R) \equiv \mathcal{E}_{\mathsf{pk}}(g^{\mathsf{IP}+R}) \quad}$$
$$\xrightarrow{\quad g^{\mathsf{IP}+R} \quad}$$

$$\mathcal{C} \xrightarrow{\quad \mathsf{pk}, \mathsf{Com}\,((g^{\mathsf{sk}_j})_{j=1}^{\log p}), (c_i = (u_i, e_i) = \mathcal{E}_{\mathsf{pk}}(g^{x_i}))_{i=1}^{\ell}, \mathsf{ipk}_{\mathcal{C}} \quad} \mathcal{S}$$
$$\xleftarrow{\quad \mathsf{Com}\,((g^{y_i})_{i=1}^{\ell}, g^R, g^{R'}, \prod u_i^{y_i}, \prod e_i^{y_i}), (\hat{u}, \hat{e}), \mathsf{ipk}_{\mathcal{S}}, c_{\mathcal{C}} \quad}$$
$$\xrightarrow{\quad g^{R\cdot \mathsf{IP}+R'} \cdot K_{\mathcal{S}}, c_{\mathcal{S}} \quad}$$

Fig. 5. Semi-honest and malicious protocols for secure inner product computation

secret key (this is the most efficient alternative as soon as $n \gg \ell$) and sends encryptions $(c_i)_{i \leq n}$ of her bits. Then, the server commits to his inputs $(y_i)_i$ and to two random integers (R, R'), computes the encryption $(\hat{u}, \hat{e})$ of $g^{R\cdot \mathsf{IP}+R'}$), re-randomized with a randomness ρ, masked by an iZK to ensure that the c_i's encrypt bits under the key pk whose corresponding secret key sk is committed (masking one of the two components of an ElGamal ciphertext suffices). The client replies with $g^{R\cdot \mathsf{IP}+R'}$, masked by a SSiZK (this is required for UC security) to ensure that the $\mathsf{Com}(g^{y_i})$ contains bits, and that the masked ciphertext has been properly built. The server then recovers $g^{R\cdot \mathsf{IP}+R'}$, removes R and R', and tries to extract the discrete logarithm IP. If no solution exists in $\{0, \ldots, \ell\}$, the server aborts. This last verification avoids the 2-round verification phase from our generic compiler: if the client tries to cheat on $R \cdot \mathsf{IP} + R'$, after removing R and R', the result would be random, and thus in the appropriate range with negligible probability ℓ/p, since ℓ is polynomial and p is exponential. We prove in the full version [8] that *the above protocol is secure against malicious adversaries in the UC framework with static corruptions, under the plain DDH assumption, and in the common reference string setting.*

Efficiency and Comparison with Other Methodologies. In the full version [8], we provide a detailed analysis of our inner product protocol in terms of complexity. Then, we estimate the complexity of this protocol when, instead of using iZK, the security against malicious adversaries in the UC model is ensured by using the Groth-Sahai methodology [32] or Σ-protocols. In this section, we sum up our comparisons in a table. The notation $>$ indicates that the given complexity is a lower bound on the real complexity of the protocol (we have not taken into account the linear blow-up incurred by the conversion of NIZK into SS-NIZK), and $\gg$ indicates a very loose lower bound. We stress that with usual parameter, an element of $\mathbb{G}_2$ is twice as big as an element of $\mathbb{G}_1$ (or $\mathbb{G}$) and the number of rounds in the major efficiency drawback (see Sect. 1). The efficiency improvement of iZK compared to NIZK essentially comes from their "batch-friendly" nature.

Moreover, our iZKs do not require pairings, which allows us to use more efficient elliptic curves than the best existing curves for the Groth-Sahai

Proofs	Pairings	Exponentiations	Communication	Rounds
Σ-proofs	0	38ℓ	20ℓ	5
GS proofs	$>14\ell$	$\gg 28\ell(\mathbb{G}_1) + 6\ell(\mathbb{G}_2)$	$>11\ell(\mathbb{G}_1) + 10\ell(\mathbb{G}_2)$	3
iZK (this paper)	0	67ℓ	21ℓ	3

methodology. With a reasonable choice of two curves, one without pairing and one with pairing, for 128 bits of security, we get the following results: (counting efficiency as a multiple of the running time of an exponentiation in $\mathbb{G}_1$).

Curve\Efficiency	Pairings	Exponentiations in $\mathbb{G}_1$	Exponentiations in $\mathbb{G}_2$
Curve25519 [10]	no pairings	1	✗
[11]	≈ 8	≈ 3	≈ 6

Acknowledgments. This work was supported in part by the CFM Foundation, ANR-14-CE28-0003 (Project EnBid), and the European Research Council under the European Community's Seventh Framework Programme (FP7/2007-2013 Grant Agreement no. 339563 – CryptoCloud).

References

1. Abdalla, M., Benhamouda, F., Blazy, O., Chevalier, C., Pointcheval, D.: SPHF-friendly non-interactive commitments. In: Sako, K., Sarkar, P. (eds.) ASIACRYPT 2013, Part I. LNCS, vol. 8269, pp. 214–234. Springer, Heidelberg (2013)
2. Abdalla, M., Benhamouda, F., Pointcheval, D.: Disjunctions for hash proof systems: New constructions and applications. Cryptology ePrint Archive, Report 2014/483 (2014). http://eprint.iacr.org/2014/483
3. Aiello, W., Ishai, Y., Reingold, O.: Priced oblivious transfer: how to sell digital goods. In: Pfitzmann, B. (ed.) EUROCRYPT 2001. LNCS, vol. 2045, p. 119. Springer, Heidelberg (2001)
4. Barbulescu, R., Gaudry, P., Joux, A., Thomé, E.: A heuristic quasi-polynomial algorithm for discrete logarithm in finite fields of small characteristic. In: Nguyen, P.Q., Oswald, E. (eds.) EUROCRYPT 2014. LNCS, vol. 8441, pp. 1–16. Springer, Heidelberg (2014)
5. Bellare, M., Boldyreva, A., Palacio, A.: An uninstantiable random-oracle-model scheme for a hybrid-encryption problem. In: Cachin, C., Camenisch, J.L. (eds.) EUROCRYPT 2004. LNCS, vol. 3027, pp. 171–188. Springer, Heidelberg (2004)
6. Bellare, M., Rogaway, P.: Random oracles are practical: a paradigm for designing efficient protocols. In: Ashby, V. (ed.) ACM CCS 1993, pp. 62–73. ACM Press, November 1993
7. Benhamouda, F., Blazy, O., Chevalier, C., Pointcheval, D., Vergnaud, D.: New techniques for SPHFs and efficient one-round PAKE protocols. In: Canetti, R., Garay, J.A. (eds.) CRYPTO 2013, Part I. LNCS, vol. 8042, pp. 449–475. Springer, Heidelberg (2013)

8. Benhamouda, F., Couteau, G., Pointcheval, D., Wee, H.: Implicit zero-knowledge arguments and applications to the malicious setting. Cryptology ePrint Archive, Report 2015/246 (2015). http://eprint.iacr.org/2015/246

9. Bernhard, D., Pereira, O., Warinschi, B.: How not to prove yourself: pitfalls of the fiat-shamir heuristic and applications to helios. In: Wang, X., Sako, K. (eds.) ASIACRYPT 2012. LNCS, vol. 7658, pp. 626–643. Springer, Heidelberg (2012)

10. Bernstein, D.J.: Curve25519: new diffie-hellman speed records. In: Yung, M., Dodis, Y., Kiayias, A., Malkin, T. (eds.) PKC 2006. LNCS, vol. 3958, pp. 207–228. Springer, Heidelberg (2006)

11. Beuchat, J.-L., González-Díaz, J.E., Mitsunari, S., Okamoto, E., Rodríguez-Henríquez, F., Teruya, T.: High-speed software implementation of the optimal ate pairing over barreto–naehrig curves. In: Joye, M., Miyaji, A., Otsuka, A. (eds.) Pairing 2010. LNCS, vol. 6487, pp. 21–39. Springer, Heidelberg (2010)

12. Blazy, O., Pointcheval, D., Vergnaud, D.: Round-optimal privacy-preserving protocols with smooth projective hash functions. In: Cramer, R. (ed.) TCC 2012. LNCS, vol. 7194, pp. 94–111. Springer, Heidelberg (2012)

13. Brassard, G., Chaum, D., Crépeau, C.: Minimum disclosure proofs of knowledge. J. Comput. Syst. Sci. **37**(2), 156–189 (1988). http://dx.doi.org/10.1016/0022-0000(88)90005-0

14. Brodkin, J.: Satellite internet faster than advertised, but latency still awful, Feb 2013. http://arstechnica.com/information-technology/2013/02/satellite-internet-faster-than-advertised-but-latency

15. Canetti, R.: Universally composable security: a new paradigm for cryptographic protocols. In: 42nd FOCS, pp. 136–145. IEEE Computer Society Press, October 2001

16. Canetti, R., Goldreich, O., Halevi, S.: The random oracle methodology. J. ACM **51**(4), 557–594 (2004). revisited. http://doi.acm.org/10.1145/1008731.1008734

17. Cramer, R., Shoup, V.: A practical public key cryptosystem provably secure against adaptive chosen ciphertext attack. In: Krawczyk, H. (ed.) CRYPTO 1998. LNCS, vol. 1462, p. 13. Springer, Heidelberg (1998)

18. Cramer, R., Shoup, V.: Universal hash proofs and a paradigm for adaptive chosen ciphertext secure public-key encryption. In: Knudsen, L.R. (ed.) EUROCRYPT 2002. LNCS, vol. 2332, p. 45. Springer, Heidelberg (2002)

19. Dolev, D., Dwork, C., Naor, M.: Non-malleable cryptography (extended abstract). In: 23rd ACM STOC, pp. 542–552. ACM Press, May 1991

20. ECRYPT II: eBATS. http://bench.cr.yp.to/results-dh.html

21. Escala, A., Herold, G., Kiltz, E., Ràfols, C., Villar, J.: An algebraic framework for diffie-hellman assumptions. In: Canetti, R., Garay, J.A. (eds.) CRYPTO 2013, Part II. LNCS, vol. 8043, pp. 129–147. Springer, Heidelberg (2013)

22. Feige, U., Lapidot, D., Shamir, A.: Multiple non-interactive zero knowledge proofs based on a single random string (extended abstract). In: 31st FOCS, pp. 308–317. IEEE Computer Society Press, October 1990

23. Fiat, A., Shamir, A.: How to prove yourself: practical solutions to identification and signature problems. In: Odlyzko, A.M. (ed.) CRYPTO 1986. LNCS, vol. 263, pp. 186–194. Springer, Heidelberg (1987)

24. Garg, S., Gentry, C., Sahai, A., Waters, B.: Witness encryption and its applications. In: Boneh, D., Roughgarden, T., Feigenbaum, J. (eds.) 45th ACM STOC, pp. 467–476. ACM Press, June 2013

25. Gennaro, R., Lindell, Y.: A framework for password-based authenticated key exchange. In: Biham, E. (ed.) EUROCRYPT 2003. LNCS, vol. 2656, pp. 524–543. Springer, Heidelberg (2003). http://eprint.iacr.org/2003/032.ps.gz

26. Gennaro, R., Lindell, Y.: A framework for password-based authenticated key exchange. ACM Trans. Inf. Syst. Secur. **9**(2), 181–234 (2006)
27. Gertner, Y., Ishai, Y., Kushilevitz, E., Malkin, T.: Protecting data privacy in private information retrieval schemes. In: 30th ACM STOC, pp. 151–160. ACM Press, May 1998
28. Goldreich, O., Micali, S., Wigderson, A.: How to play any mental game or A completeness theorem for protocols with honest majority. In: Aho, A. (ed.) 19th ACM STOC, pp. 218–229. ACM Press, May 1987
29. Goldreich, O., Micali, S., Wigderson, A.: How to prove all NP-statements in zero-knowledge and a methodology of cryptographic protocol design. In: Odlyzko, A.M. (ed.) CRYPTO 1986. LNCS, vol. 263, pp. 171–185. Springer, Heidelberg (1987)
30. Goldwasser, S., Micali, S., Rackoff, C.: The knowledge complexity of interactive proof systems. SIAM J. Comput. **18**(1), 186–208 (1989)
31. Granger, R., Kleinjung, T., Zumbrägel, J.: Breaking '128-bit Secure' supersingular binary curves. In: Garay, J.A., Gennaro, R. (eds.) CRYPTO 2014, Part II. LNCS, vol. 8617, pp. 126–145. Springer, Heidelberg (2014)
32. Groth, J., Sahai, A.: Efficient non-interactive proof systems for bilinear groups. In: Smart, N.P. (ed.) EUROCRYPT 2008. LNCS, vol. 4965, pp. 415–432. Springer, Heidelberg (2008)
33. Hofheinz, D., Kiltz, E.: Programmable hash functions and their applications. J. Cryptol. **25**(3), 484–527 (2012)
34. Huang, Y., Katz, J., Evans, D.: Efficient secure two-party computation using symmetric cut-and-choose. In: Canetti, R., Garay, J.A. (eds.) CRYPTO 2013, Part II. LNCS, vol. 8043, pp. 18–35. Springer, Heidelberg (2013)
35. Ishai, Y., Kushilevitz, E., Lindell, Y., Petrank, E.: Black-box constructions for secure computation. In: Kleinberg, J.M. (ed.) 38th ACM STOC, pp. 99–108. ACM Press, May 2006
36. Jarecki, S.: Practical covert authentication. In: Krawczyk, H. (ed.) PKC 2014. LNCS, vol. 8383, pp. 611–629. Springer, Heidelberg (2014)
37. Lindell, Y.: Fast cut-and-choose based protocols for malicious and covert adversaries. In: Canetti, R., Garay, J.A. (eds.) CRYPTO 2013, Part II. LNCS, vol. 8043, pp. 1–17. Springer, Heidelberg (2013)
38. Lindell, Y., Pinkas, B.: An efficient protocol for secure two-party computation in the presence of malicious adversaries. In: Naor, M. (ed.) EUROCRYPT 2007. LNCS, vol. 4515, pp. 52–78. Springer, Heidelberg (2007)
39. Lindell, Y., Pinkas, B.: Secure two-party computation via cut-and-choose oblivious transfer. In: Ishai, Y. (ed.) TCC 2011. LNCS, vol. 6597, pp. 329–346. Springer, Heidelberg (2011)
40. Malkin, T., Teranishi, I., Vahlis, Y., Yung, M.: Signatures resilient to continual leakage on memory and computation. In: Ishai, Y. (ed.) TCC 2011. LNCS, vol. 6597, pp. 89–106. Springer, Heidelberg (2011)
41. Naor, M., Yung, M.: Public-key cryptosystems provably secure against chosen ciphertext attacks. In: 22nd ACM STOC, pp. 427–437. ACM Press, May 1990
42. Schnorr, C.-P.: Efficient identification and signatures for smart cards. In: Brassard, G. (ed.) CRYPTO 1989. LNCS, vol. 435, pp. 239–252. Springer, Heidelberg (1990)
43. Shelat, A., Shen, C.: Two-output secure computation with malicious adversaries. In: Paterson, K.G. (ed.) EUROCRYPT 2011. LNCS, vol. 6632, pp. 386–405. Springer, Heidelberg (2011)
44. Shelat, A., Shen, C.H.: Fast two-party secure computation with minimal assumptions. In: Sadeghi, A.R., Gligor, V.D., Yung, M. (eds.) ACM CCS 13, pp. 523–534. ACM Press, November 2013

45. Waters, B.: Efficient identity-based encryption without random oracles. In: Cramer, R. (ed.) EUROCRYPT 2005. LNCS, vol. 3494, pp. 114–127. Springer, Heidelberg (2005)
46. Yao, A.C.C.: How to generate and exchange secrets (extended abstract). In: 27th FOCS, pp. 162–167. IEEE Computer Society Press, October 1986

Impossibility of Black-Box Simulation Against Leakage Attacks

Rafail Ostrovsky[1], Giuseppe Persiano[2], and Ivan Visconti[2]([✉])

[1] University of California, Los Angeles, USA
rafail@cs.ucla.edu
[2] Università di Salerno, Fisciano, Italy
{pino.persiano,visconti}@unisa.it

Abstract. In this work, we show how to use the positive results on succinct argument systems to prove impossibility results on leakage-resilient black-box zero knowledge. This recently proposed notion of zero knowledge deals with an adversary that can make leakage queries on the state of the prover. Our result holds for black-box simulation only and we also give some insights on the non-black-box case. Additionally, we show that, for several functionalities, leakage-resilient multi-party computation is impossible (regardless of the number of players and even if just one player is corrupted).

More in details, we achieve the above results by extending a technique of [Nielsen, Venturi, Zottarel – PKC13] to prove lower bounds for leakage-resilient security. Indeed, we use leakage queries to run an execution of a communication-efficient protocol in the head of the adversary. Moreover, to defeat the black-box simulator we connect the above technique for leakage resilience to security against reset attacks.

Our results show that the open problem of [Ananth, Goyal, Pandey – Crypto 14] (i.e., continual leakage-resilient proofs without a common reference string) has a negative answer when security through black-box simulation is desired. Moreover our results close the open problem of [Boyle et al. – STOC 12] for the case of black-box simulation (i.e., the possibility of continual leakage-resilient secure computation without a leak-free interactive preprocessing).

Keywords: Zero knowledge · MPC · Resettability · Succinct arguments · Impossibility results · Black-box vs non-black-box simulation

1 Introduction

The intriguing notion of a zero-knowledge proof introduced by Goldwasser, Micali and Rackoff [31] has been for almost three decades a source of fascinating open questions in Cryptography and Complexity Theory. Indeed, motivated by new real-world attacks, the notion has been studied in different flavors (e.g., non-interactive zero knowledge [8], non-malleable zero knowledge [21], concurrent zero knowledge [23], resettable zero knowledge [16]) and each of them required

© International Association for Cryptologic Research 2015
R. Gennaro and M. Robshaw (Eds.): CRYPTO 2015, Part II, LNCS 9216, pp. 130–149, 2015.
DOI: 10.1007/978-3-662-48000-7_7

extensive research to figure out the proper definition and its (in)feasibility. Moreover all such real-world attacks have been considered also for the natural generalization of the concept of zero knowledge: secure computation [30].

Leakage Attacks. Leakage resilience deals with modeling real-word attacks where the adversary manages through some physical observations to obtain side-channel information on the state (e.g., private input, memory content, randomness) of the honest player (see, for example, [42]). Starting with the works of [25,34,35,41] leakage resilience has been a main-stream research topic in Cryptography, and recently the gap between theory and practice has been significantly reduced [22,40,43].

The notions of leakage-resilient zero knowledge [28] (LRZK) and secure multi-party computation [10] (LRMPC) have been also considered. Despite the above intensive research on leakage resilience, LRZK and LRMPC are still rich of interesting open problems.

1.1 Previous Work and Open Problems

Leakage Resilience vs. Tolerance. The first definition for leakage-resilient zero knowledge (LRZK, in short) was given by Garg et al. in [28]. In their definition, the simulator is allowed to make leakage queries in the ideal world. This was justified by the observation that an adversary can, through leakage queries, easily obtain some of the bits of the witness used by the prover in the real world. Clearly, these bits of information can not be simulated, unless the simulator is allowed to make queries in the ideal model. Therefore the best one can hope for is that a malicious verifier does not learn anything from the protocol beyond the validity of the statement being proved and the leakage obtained from the prover. This formalization of security has been extensively studied by Bitansky et al. in [6] for the case of universally composable secure computation [15]. Similar definitions have been used in [9,11,12,36].

In [28], constructions for LRZK in the standard model and for non-interactive LRZK in the common reference string (CRS) model were given. The simulator of [28] for LRZK asks for a total of $(1 + \epsilon) \cdot l$ bits in the ideal world, where l is the number of bits obtained by the adversarial verifier. Thus the simulator is allowed to obtain more bits than the verifier and this seems to be necessary as Garg et al. show that it is impossible to obtain a simulator that ask for less than l bits in the ideal world. Very recently, Pandey [39] gave a constant-round construction for LRZK under the definition of [28].

Nowadays, *leakage tolerance* is the commonly accepted term for the security notion used in [6,28,39] as it does not prevent a leakage attack but only guarantees that a protocol does not leak more than what can be obtained through leakage queries. Bitansky et al. [7] obtained UC-secure continual leakage tolerance using an input-independent leak-free preprocessing phase.

Open Problems: Leakage Resilience with Leak-Free Encoding. The motivation to study leakage-tolerant Cryptography is based on the observation that a private input can not be protected in full from a leakage query. However this notion is quite extreme and does not necessarily fit all real-world scenarios. Indeed, it is commonly expected that an adversary attacks the honest player during the execution of the protocol, while they are connected through some communication channel. It is thus reasonable to assume that a honest player receives his input in a preliminary phase, before having ever had any interaction with the adversary. Once this input is received, the honest player can encode it in order to make it somewhat intelligible from leakage queries but still valid for the execution of a protocol. This encoding phase can be considered leak-free since, as stressed before, the honest player has never been in touch with the adversary[1]. Later on, when the interaction with the adversary starts, leakage queries will be possible but they will affect the current state of the honest player that contains an encoding of the input. The need of a leak-free phase to protect a secret from leakage queries was considered also in [26,32,33].

The above realistic scenario circumvents the argument that leakage tolerance is the best one can hope for, and opens the following challenging open questions:

Open Question 1: *"Assuming players can encode their inputs during a leak-free phase, is it possible to construct LRZK argument/proof systems?"*

Open Question 2: *"Assuming players can encode their inputs during a leak-free phase, is it possible to construct protocols for leakage-resilient Multi-Party Computation (LRMPC)?"*

Leakage Resilience Assuming the Existence of a CRS. Very recently, Ananth et al. [1], showed that in the CRS (common reference string) model it is possible to have an interactive argument system that remains *non-transferable* even in presence of continual leakage attacks. More precisely, in their model a prover encodes the witness in a leak-free environment and, later on, the prover runs the protocol with a verifier using the encoded witness. During the execution of the protocol, the adversarial verifier is allowed to launch leakage queries. Once the protocol has been completed, the prover can refresh (again, in a leak-free environment) its encoded witness and then it can play again with the verifier (under leakage attacks). Non-transferability means that an adversarial verifier that mounts the above attack against a honest prover does not get enough information to later prove the same statement to a honest verifier. The main contribution of [1] is the construction of an encoding/refreshing mechanism and a protocol for non-transferable arguments against such continual leakage attacks. They left explicitly open the following open problem (see page 167 of [1]): is it possible to obtain non-transferable arguments/proofs that remain secure against continual leakage attacks without relying on a CRS? This problem has similarities with Open Problem 1. Indeed, zero knowledge (without a CRS) implies non-transferability and therefore solving Open Problem 1 in the positive and with continual leakage would solve the problem opened by [1] in

---

[1] Moreover such a phase can be run on a different device disconnected from the network, running an operating system installed on some read-only disk.

a strong sense since non-transferability would be achieved through zero knowledge, and this goes even beyond the security definition of [1][2]. However, as we will show later we will give a negative answer to Open Problem 1 for the case of black-box simulation. Even in light of our negative results, the open problem of [1] remains open as one might be able to construct leakage resilient non-black-box zero knowledge (which is clearly non-transferable) or leakage resilient witness hiding/indistinguishable proofs (that can still be non-transferable since non-malleable proofs can be achieved with non-malleable forms of WI as shown in [37]).

Leakage Resilience Assuming Leak-Free Preprocessing. In [10], Boyle et al. proposed a model for leakage-resilient secure computation based on the following three phases:

1. a leak-free interactive preprocessing to be run only once, obliviously w.r.t. inputs and functions;
2. a leak-free stand-alone input-encoding phase to be run when a new input arrives (and of course after the interactive preprocessing), obliviously w.r.t. functions to be computed later;
3. an on-line phase where parties, on input the states generated during the last executions of the input-encoding phases, and on input a function f, run a protocol that aims at securely computing the output of f.

In the model of [10] leakage attacks are not possible during the first two phases but are possible in any other moment, including the 3rd phase and in between phases.

Reference [10] showed (a) the impossibility of leakage-resilient 2-party computation and, more in general, of n-party LRMPC when $n-1$ players are corrupted; (b) the feasibility of leakage-resilient MPC when the number of players is polynomial and a constant fraction of them is honest.

The positive result works for an even stronger notion of leakage resilience referred to as "continual leakage" that has been recently investigated in several papers [13, 14, 19, 20, 24]). Continual leakage means that the same input can be re-used through unbounded multiple executions of the protocol each allowing for a bounded leakage, as long as the state can be refreshed after each execution. Leakage queries are allowed also during the refreshing.

Boyle et al. explicitly leave open (see paragraph "LR-MPC with Non-Interactive Preprocessing" on page 1240 of [10]) the problem of achieving their results without the preprocessing (i.e., Open Question 2) and implicitly left open the case of zero-knowledge arguments/proofs. (i.e., Open Question 1) since when restricting to the ZK functionality only, the function is known in advance and therefore their impossibility for the two-party case does not directly hold.

We notice that the result of [1] does not yield a continual leakage-resilient non-transferable proof system for the model of [10]. Indeed, while the preprocessing of [10] can be used to establish the CRS needed by [1], the refresh of the

---

[2] Their definition does not require zero knowledge.

state of [1] requires a leak-free phase that is not available in the model of [10]. We finally stress that the construction of [1] is not proved to be LRZK.

However the interesting open question in the model of [10] consists in achieving continual LRZK *without* an interactive preprocessing. Indeed, if an interactive preprocessing is allowed, continual LRZK can be trivially achieved as follows. The preprocessing can be used to run a secure 2-party computation for generating a shared random string. The input-encoding phase can replace the witness with a non-interactive zero-knowledge proof of knowledge (NIZKPK). The on-line phase can be implemented by simply sending the previously computed NIZKPK. This trivial solution would allow the leakage of the entire state, therefore guaranteeing continual leakage (i.e., no refresh is needed).

Impossibility Through Obfuscation. In the model studied by Garg et al. [28], the simulator is allowed to see the leakage queries issued by the adversarial verifier (and not the replies) and, based on these, it decides his own leakage queries in the ideal model. Nonetheless, the actual simulator constructed by [28] does not use this possibility; such a simulator is called *leakage-oblivious*. In our setting (in which the simulator is not allowed to ask queries) leakage-oblivious simulators are very weak: an adversarial verifier that asks the query for function $R(x, \cdot)$ applied to the witness w (here R is the relation associated to $\mathbb{NP}$ language L and x is the common input) cannot be simulated. Notice though that in the model we are interested in, the leak-free encoding phase might invalidate this approach since the encoded witness could have a completely different structure and therefore could make R evaluate to 0. Despite this issue (that is potentially fixable), the main problem is that in our setting the simulator can read the query of the adversarial verifier and could easily answer 1 (the honest prover always has a valid witness). Given the recent construction of circuit obfuscators [27], one could then think of forcing simulators to be leakage-oblivious by considering an adversary that obfuscates its leakage queries. While this approach has a potential, we point out that our goal is to show the impossibility under standard assumptions (e.g., the existence of a family of CRHFs).

The Technique of Nielsen et al. [36]. We finally discuss the very relevant work of Nielsen et al. [36] that showed a lower bound on the size of a secret key for leakage-tolerant adaptively secure message transmission. Nielsen et al. introduced in their work a very interesting attack consisting in asking a collision-resistant hash of the state of a honest player through a leakage query. Then a succinct argument of knowledge is run through leakage queries in order to ask the honest player to prove knowledge of a state that is consistent with the previously sent hash value. As we will discuss later, we will extend this technique to achieve our main result. The use of CRHFs and succinct arguments of knowledge for impossibility of leakage-resilience was also used in [18] but in a very different context. Indeed in [18] the above tools are used to check consistency with the transcript of played messages with the goal of proving that full adaptive security is needed in multi-party protocols as soon as some small amount of leakage must be tolerated.

1.2   Our Results

In this paper we study the above open questions and show the following results.

Black-Box LRZK Without CRS/Preprocessing. As a main result, we show that, if a family of collision-resistant hash functions exist, then black-box LRZK is impossible for non-trivial languages if we only rely on a leak-free input-encoding phase (i.e., without CRS/preprocessing). More in details, with respect to the works of [1,10], our results shows that, by removing the CRS/preprocessing, not only non-transferable continual black-box LRZK is impossible, but even ignoring non-transferability and continual leakage, the simple notion of 1-time black-box LRZK is impossible. Extending the techniques of [36], we design an adversarial verifier V* that uses leakage queries to obtain a very small amount of data compared to the state of the prover and whose view cannot be simulated in a black-box manner. The impossibility holds even knowing already at the input-encoding phase which protocol will be played later.

Overview of Our Techniques. We prove the above impossibility result by extending the previously discussed technique of [36]: the adversary will attack the honest player without running the actual protocol at all! Indeed, the adversary will only run an execution of another (insecure) protocol in its head, using leakage queries to get messages from the other player for the "virtual" execution of the (insecure) protocol.

More in details, assuming by contradiction the existence of a protocol (P, V) for a language $L \not\subseteq$ BPP, we show an adversary V* that first runs a leakage query to obtain a collision-resistant (CR) hash $\tilde{w}$ of the state $\hat{w}$ of the prover. Then it takes a communication-efficient (insecure) protocol $\Pi = (\Pi.P, \Pi.V)$ and, through leakage queries, V* runs in its head an execution of Π playing as a honest verifier $\Pi.V$, while the prover P will have to play as $\Pi.P$ proving that the hash is a good one: namely, it corresponds to a state that would convince a honest verifier V on the membership of the instance in L. We stress that this technique was introduced in [36].

Notice that in the real-world execution P would convince V* during the "virtual" execution of Π since P runs as input an encoded witness that by the completeness of (P, V) convinces V.

Therefore a black-box simulator will have to do the same without having the encoding of a witness but just relying on rewinding capabilities. To show our impossibility we extend the technique of [36] by making useless the capabilities of the simulator. This is done by connecting leakage resilience with resettability. Indeed we choose Π not only to be communication efficient on $\Pi.P$'s side (this helps so that the sizes of the outputs of leakage queries will correspond to a small portion of the state of P), but also to be a resettable argument of knowledge (and therefore resettably sound). Such arguments of knowledge admit an extractor Π.Ext that works even against a resetting prover $\Pi.P^*$ (i.e., such an adversary in our impossibility will be the simulator Sim of (P, V)).

The existence of a family of CR hash functions gives not only the CR hash function required by the first leakage query but also the communication-efficient resettable argument of knowledge for NP. Indeed we can use Barak's public-coin universal argument [3] that enjoys a *weak* argument of knowledge property when used for languages in NEXP. Instead when used for NP languages, Barak's construction is a *regular* argument of knowledge with a black-box extractor. We can finally make it extractable also in presence of a resetting prover by using the transformation of Barak et al. [4] that only requires the existence of one-way functions.

Summing up, we will show that the existence of a black-box simulator for (P, V) implies either that the language is in BPP, or that (P, V) is not sound or that the family of hash functions is not collision resistant.

The Non-Black-Box Case. Lower bounds in the case of non-black-box simulation are rare in Cryptography and indeed we can not rule out the existence of LRZK argument whose security is based on the existence of a non-black-box simulator. We will however discuss some evidence that achieving a positive result under standard assumptions requires a breakthrough on non-black-box simulation that goes beyond Barak's non-black-box techniques.

Impossibility of Leakage-Resilient MPC for Several Functionalities.
Additionally, we address Open Question 2 by showing that for many function-alities LRMPC with a leak-free input-encoding phase (and without an interactive preprocessing phase) is impossible. This impossibility holds regardless of the number of players involved in the computation and only assumes that one player is corrupted. It applies to functionalities that when executed multiple times keeping unchanged the input x_i of a honest player P_i, produce outputs delivered to the dishonest players that reveal more information on x_i than what a single output would reveal. Similar functionalities were studied in [17]. We also require outputs to be short.

Our impossibility is actually even stronger since it holds also in case the functionality and the corresponding protocol to be run later are already known during the input-encoding phase.

For simplicity, we will discuss a direct example of such a class of functionalities: a variation of Yao's Millionaires' Problem, where n players send their inputs to the functionality that will then send as output a bit b specifying whether player P_1 is the richest one.

High-Level Overview. The adversary will focus on attacking player P_1 that has an input to protect. The adversary can play in its head by means of a single leakage query the entire protocol selecting inputs and randomnesses for all other players, and obtaining as output of the leakage query the output of the function (i.e., the bit b). This "virtual" execution can be repeated multiple times, therefore extracting more information on the input of the player. Indeed playing multiple times and changing the inputs of the other players while the input of P_1 remains the same, it is possible to restrict the possible input of P_1 to a much smaller range of values than what can be inferred by a single execution.

The above attack will be clearly impossible to simulate since it would require the execution of multiple queries in the ideal world, but the simulator by definition can make only one query.

When running the protocol through leakage queries, we are of course assuming that authenticated channels do not need to be simulated by the adversary[3] since their management is transparent to the state of the players running the leakage-resilient protocol. This is already assumed in previous work like [10] since otherwise leakage-resilient authenticated channels would have been required, while instead [10] only requires an authenticated broadcast channel (see Sect. 3 of [10]).

We will give only a sketch of this additional simpler result.

2   Definitions

We will denote by "$\alpha \circ \beta$" the string resulting from appending β to α, and by $[k]$ the set $\{1, \ldots, k\}$. A polynomial-time relation R is a relation for which it is possible to verify in time polynomial in $|x|$ whether $R(x, w) = 1$. We will consider NP-languages L and denote by R_L the corresponding polynomial-time relation such that $x \in L$ if and only if there exists w such that $R_L(x, w) = 1$. We will call such a w a *valid witness for $x \in L$* and denote by $W_L(x)$ the set of valid witnesses for $x \in L$. We will slightly abuse notation and, whenever L is clear from the context, we will simply write $W(x)$ instead of $W_L(x)$. A *negligible* function $\nu(k)$ is a function such that for any constant $c < 0$ and for all sufficiently large k, $\nu(k) < k^c$.

We will now give all definitions required for the main result of our work, the impossibility of black-box LRZK. Since we will only sketch the additional result on LRMPC, we defer the reader to [10] for the additional definitions.

2.1   Interactive Proof Systems

An *interactive proof system* [31] for a language L is a pair of interactive Turing machines (P, V), satisfying the requirements of *completeness* and *soundness*. Informally, completeness requires that for any $x \in L$, at the end of the interaction between P and V, where P has on input a valid witness for $x \in L$, V rejects with negligible probability. Soundness requires that for any $x \notin L$, for any computationally unbounded P*, at the end of the interaction between P* and V, V accepts with negligible probability. When P* is only probabilistic polynomial-time, then we have an argument system. We denote by $\langle P, V \rangle(x)$ the output of the verifier V when interacting on common input x with prover P. Also, sometimes we will use the notation $\langle P(w), V \rangle(x)$ to stress that prover P receives as

---

[3] More in details, we are assuming that the encoded state of the player does not include any information useful to check if a message supposed to be from a player P_j is genuine.

additional input witness w for $x \in L$. We will write $\langle P(w; r_P), V(r_V) \rangle(x)$ to make explicit the randomness used by P and V. We will also write $V^*(z)$ to denote an adversarial verifier V^* that runs on input an auxiliary string z.

Definition 1 *[31]. A pair of interactive Turing machines* (P, V) *is an interactive proof system for the language L, if V is probabilistic polynomial-time and*

1. *Completeness: There exists a negligible function $\nu(\cdot)$ such that for every $x \in L$ and for every $w \in W(x)$* $\mathrm{Prob}[\langle P(w), V \rangle(x) = 1] \geq 1 - \nu(|x|)$.
2. *Soundness: For every $x \notin L$ and for every interactive Turing machines P^* there exists a negligible function $\nu(\cdot)$ such that* $\mathrm{Prob}[\langle P^*, V \rangle(x) = 1] \leq \nu(|x|)$.

If the soundness condition holds only with respect to probabilistic polynomial-time interactive Turing machines P^* then (P, V) is called an argument.

We now define the notions of reset attack and of resetting prover.

Definition 2 *[4]. A reset attack of a prover P^* on V is defined by the following two-step random process, indexed by a security parameter k.*

1. *Uniformly select and fix $t = \mathsf{poly}(k)$ random tapes, denoted by $r_1, \ldots, r_t$, for V, resulting in deterministic strategies $V^{(i)}(x) = V_{x,r_i}$ defined by $V_{x,r_i}(\alpha) = V(x, r_i, \alpha)$, where $x \in \{0,1\}^k$ and $i \in 1, \ldots, t$. Each $V^{(i)}(x)$ is called an* incarnation *of V.*
2. *On input 1^k, machine P^* is allowed to initiate $\mathsf{poly}(k)$-many interactions with V. The activity of P^* proceeds in rounds. In each round P chooses $x \in \{0,1\}^k$ and $i \in 1, \ldots, t$, thus defining $V^{(i)}(x)$, and conducts a complete session (a session is complete if is either terminated or aborted) with it.*

We call resetting prover *a prover that launches a reset attack.*

We now define proofs/arguments of knowledge, in particular considering the case of a prover launching a reset attack.

Definition 3 *[5]. Let R be a binary relation and $\epsilon : \{0,1\}^* \to [0,1]$. We say that a probabilistic polynomial-time interactive machine V is a knowledge verifier for the relation R with knowledge error ϵ if the following two conditions hold:*

Non-triviality: *There exists a probabilistic polynomial-time interactive machine P such that for every $(x, w) \in R$, with overwhelming probability an interaction of V with P on common input x, where P has auxiliary input w, is accepting.*

Validity (or Knowledge Soundness) with Negligible Error ϵ: *for every probabilistic polynomial-time machine P^*, there exists an expected polynomial-time machine Ext, such that and for every $x, aux, r \in \{0,1\}^*$, Ext satisfies the following condition: Denote by $p(x, aux, r)$ the probability (over the random tape of V) that V accepts upon input x, when interacting with the prover P^* who has input x, auxiliary-input aux and random-tape r. Then, machine Ext, upon input (x, aux, r), outputs a solution $w \in W(x)$ with probability at least $p(x, aux, r) - \epsilon(|x|)$.*

A pair (P, V) *such that* V *is a knowledge verifier with* negligible *knowledge error for a relation R and* P *is a machine satisfying the non-triviality condition (with respect to V and R) is called an argument of knowledge for the relation R. If the validity condition holds with respect to any (not necessarily polynomial- time) machine* P*, *then* (P, V) *is called a proof of knowledge for R. If the validity condition holds with respect to a polynomial-time machine* P* *launching a reset attack, then* (P, V) *is called a resettable argument of knowledge for R.*

In the above definition the extractor does not depends on the code of the prover (i.e., the same extractor works with all possible provers) Ext then the interactive argument/proof system is a *black-box* (resettable) argument/proof of knowledge.

The Input-Encoding Phase. Following previous work we will assume that the prover receives the input and encodes it running in a leak-free environment. This is unavoidable since otherwise a leakage query can cask for some bits of the witness and therefore zero knowledge would be trivially impossible to achieve, unless the simulator is allowed to ask leakage query in the ideal world (i.e., leakage tolerance). After this leak-free phase that we call input-encoding phase, the prover has a state consisting only of the encoded witness and is ready to start the actual leakage-resilient protocol.

Leakage-Resilient Protocol [39]. As in previous work, we assume that random coins are available only in the specific step in which they are needed. More in details, the prover P at each round of the protocol obtains fresh randomness r for the computations related to that round. However, unlike in previous work, we do not require the prover to update its state by appending r to it. We allow the prover to erase randomness and change its state during the protocol execution. This makes our impossibility results even stronger.

The adversarial verifier performs a leakage query by specifying a polynomial-sized circuit C that takes as input the current state of the prover. The verifier gets immediately the output of C and can adaptively decide how to continue. An attack of the verifier that includes leakage queries is called a leakage attack.

Definition 4. *Given a polynomial p, we say that an interactive argument/proof system* (P, V) *for a language* $L \in \mathbb{NP}$ *with a witness relation R, is* $p(|x|)$-*leakage-resilient zero knowledge if for every probabilistic polynomial-time machine* V* *launching a leakage attack on* P *after the input-encoding phase, obtaining at most* $p(|x|)$ *bits, there exists a probabilistic polynomial-time machine* Sim *such that for every* $x \in L$, *every* w *such that* $R(x, w) = 1$, *and every* $z \in \{0, 1\}^*$ *distributions* $\langle P(w), V^*(z) \rangle(x)$ *and* $Sim(x, z)$ *are computationally indistinguishable.*

The definition of standard zero-knowledge is obtained by enforcing that no leakage query is allowed to any machine and removing the input-encoding phase.

In the above definition the simulator does not depends on the code of the verifier (i.e., the same simulator works with all possible verifiers) Sim then the interactive argument/proof system is leakage-resilient *black-box* zero knowledge. We will denote by Sim^{V^*} an execution of Sim having oracle access to V*.

3   Impossibility of Leakage-Resilient Zero Knowledge

Here we prove that LRZK argument systems exist only for BPP languages.

Tools. In our proof we assume the existence of a communication-efficient argument system $\Pi = (\Pi.\mathsf{P}, \Pi.\mathsf{V})$ for a specific auxiliary $\mathbb{NP}$ language (to be defined later). Moreover we require such an argument system to be a resettable argument of knowledge. Specifically, we require that on common input x, $\Pi.\mathsf{P}$ sends $O(|x|^\epsilon)$ bits to $\Pi.\mathsf{V}$ for an arbitrarily chosen constant $\epsilon > 0$. We denote, with a slight abuse of notation, by $\Pi.\mathsf{P}$ the prover's next message function; that is, $\Pi.\mathsf{P}$ on input x, randomness $r_1, \ldots, r_{i-1}$ used in the previous $i-1$ rounds, fresh randomness r_i and verifier messages $v_1, \ldots, v_i$ received so far, outputs msg_i, the prover's i-th message. Similarly, we denote the verifier's next message function by $\Pi.\mathsf{V}$. Finally, we denote by $\Pi.\mathsf{Ext}$ the extractor that in expected polynomial time outputs a witness for $x \in L$ whenever a polynomial-time prover can make $\Pi.\mathsf{V}$ accept $x \in L$ with non-negligible probability.

Such a resettable argument of knowledge Π exists based on the existence of a family of collision-resistant hash functions. It can be obtained by starting with the public-coin universal argument of [3] that for $\mathbb{NP}$ languages is also an argument of knowledge. Then by applying the transformation of [4] that requires one-way functions, we have that the resulting protocol is still communication efficient, and moreover is a resettable argument of knowledge.

Theorem 1. *Assume the existence of a family of collision-resistant hash functions. If an $\mathbb{NP}$-language L admits an $(|x|^\epsilon)$-leakage-resilient black-box zero-knowledge argument system $\Pi_{LRZK} = (\mathsf{P}, \mathsf{V})$ for some constant $\epsilon > 0$ then $L \in \mathsf{BPP}$.*

Proof. For sake of contradiction, we assume that language $L \notin \mathsf{BPP}$ admits a $(|x|^\epsilon)$-leakage-resilient zero-knowledge argument system (P, V) with black-box simulator Sim for some constant $\epsilon > 0$. We now describe an adversarial verifier $\mathsf{V}^\star = \mathsf{V}^\star_{x,s,h,t}$, parameterized by input x, strings s and t, and function h from a family of collision-resistant hash functions. In the description of $\mathsf{V}^\star$, we let $\{F_s\}$ be a pseudorandom family of functions.

Our proof makes use of the auxiliary language Λ consisting of the tuples $\tau = (h, \tilde{w}, \mathbf{rand}^\mathsf{P}, \mathbf{rand}^\mathsf{V})$ for which there exists $\hat{w}$ such that $h(\hat{w}) = \tilde{w}$ and $\langle \mathsf{P}(\hat{w}; \mathbf{rand}^\mathsf{P}), \mathsf{V}(\mathbf{rand}^\mathsf{V})\rangle(x) = 1$. Clearly, $\Lambda \in \mathbb{NP}$. Let $\Pi = (\Pi.\mathsf{P}, \Pi.\mathsf{V})$ be a communication-efficient argument system for Λ. We assume wlog that the number of rounds of Π is 2ℓ (i.e., ℓ messages played by the verifier and ℓ messages played by the prover) where $\ell > 1$ and that the verifier speaks first.

1. At the start of the interaction between P and $\mathsf{V}^\star$ on an n-bit input x with $n = \mathsf{poly}(k)$, the state of P consists solely of the encoding $\hat{w}$ of the witness w for $x \in L$, where $|\hat{w}| = \mathsf{poly}(n)$.
2. $\mathsf{V}^\star$ issues leakage query Q_0 by specifying function h; as a reply, $\mathsf{V}^\star$ receives $\tilde{w} = h(\hat{w})$, a hash of the encoding of the witness used by P.

3. V* then selects randomness

$$\mathbf{rand} = (\mathbf{rand}^P, \mathbf{rand}^V, \mathbf{rand}_1^{\Pi.P}, \ldots, \mathbf{rand}_\ell^{\Pi.P}, \mathbf{rand}_1^{\Pi.V}, \ldots, \mathbf{rand}_\ell^{\Pi.V}, \mathbf{rand}_{\ell+1}^{\Pi.V})$$

by setting $\mathbf{rand} = F_s(\tilde{w} \circ x)$.

4. V* performs, by means of leakage queries, an execution of the protocol Π on common input $(h, \tilde{w}, \mathbf{rand}^P, \mathbf{rand}^V)$.

Specifically, for round $i = 1, \ldots, \ell$, V* computes

$$v_i = \Pi.V\left((h, \tilde{w}, \mathbf{rand}^P, \mathbf{rand}^V), \{\mathbf{msg}_j\}_{0 < j < i}, \{\mathbf{rand}_j^{\Pi.V}\}_{0 < j \le i}\right)$$

and issues leakage query Q_i for the prover's next message function

$$\Pi.P\left((h, \tilde{w}, \mathbf{rand}^P, \mathbf{rand}^V), \cdot, \{v_j\}_{0 < j \le i}, \{\mathbf{rand}_j^{\Pi.P}\}_{0 < j \le i}\right)$$

that is to be applied to the state $\hat{w}$ of prover P. In other words, the query computes the prover's i-th message $\mathbf{msg}_i$ of an interaction of protocol Π in which prover $\Pi.P$ (running on randomness $\mathbf{rand}_1^{\Pi.P}, \ldots, \mathbf{rand}_\ell^{\Pi.P}$) tries to convince verifier $\Pi.V$ (running on randomness $\mathbf{rand}_1^{\Pi.V}, \ldots, \mathbf{rand}_\ell^{\Pi.V}, \mathbf{rand}_{\ell+1}^{\Pi.V}$) that $(h, \tilde{w}, \mathbf{rand}^P, \mathbf{rand}^V) \in \Lambda$.

After receiving prover $\Pi.P$'s last message, V* computes $\Pi.V$'s output in this interaction:

$$b = \Pi.V((h, \tilde{w}, \mathbf{rand}^P, \mathbf{rand}^V), \mathbf{msg}_1, \ldots, \mathbf{msg}_\ell, \mathbf{rand}_1^{\Pi.V}, \ldots, \mathbf{rand}_{\ell+1}^{\Pi.V}).$$

5. If $b = 1$ then V* outputs t; otherwise, V* outputs $\perp$.

This concludes the description of V*.

Counting the Number of Bits Leaked. The total number of bits leaked is equal to the output of the first leakage query (i.e., the length in bits of a range element of the collision-resistant hash function) $|\tilde{w}| = k$ and the number of bits sent by the prover in Π which, for inputs of length n, is $O(n^{\epsilon'})$ for an arbitrarily constant $\epsilon' > 0$. Being $n = \mathsf{poly}(k)$, we have that the amount of leakage can be made smaller than n^ϵ for any $\epsilon > 0$.

Sim *Can Get* t *only by Succeeding in* Π, *Therefore Properly Answering to Leakage Queries.* We continue by observing that the output of the real game (i.e., when P and $V^*_{x,s,h,t}$ interact) is t. Therefore, Sim must output t when interacting with $V^*_{x,s,h,t}$ with overwhelming probability. Since Sim is a black-box simulator, and since all messages of $V^*_{x,s,h,t}$ except for the last one, are independent of t, the only way Sim can obtain t from $V^*_{x,s,h,t}$ is by replying with a value $\tilde{w}$ to the first leakage query and by replying to queries $Q_1, \ldots, Q_\ell$ so to define a transcript $\mathsf{Conv} = (v_1, \mathbf{msg}_1, \ldots, v_\ell, \mathbf{msg}_\ell)$ that for common input $(h, \tilde{w}, \mathbf{rand}^P, \mathbf{rand}^V)$ produces $1 = \Pi.V((h, \tilde{w}, \mathbf{rand}^P, \mathbf{rand}^V), \mathbf{msg}_1, \ldots, \mathbf{msg}_\ell, \mathbf{rand}_1^{\Pi.V}, \ldots, \mathbf{rand}_{\ell+1}^{\Pi.V})$.

By the security of the pseudorandom function, we can consider the same experiment except having that $\mathbf{rand} = \mathcal{R}(\tilde{w} \circ x)$ (computed by V* in step 3 of its description) where $\mathcal{R}$ is a truly random function (i.e., each time $\tilde{w} \circ x$ is new, $\mathbf{rand}$ is computed by sampling fresh randomness).

We denote by $\mathsf{Sim}_R^{V^*}$ the simulation in such a modified game. We can show (the proofs of the following lemmas are omitted for lack of space) that when considering $\mathsf{Sim}_R^{V^*}$, still t is given in output with overwhelming probability.

Lemma 1. *The output of* $\mathsf{Sim}_R^{V^*}$ *is computationally indistinguishable from the output of* Sim^{V^*}.

We can then show that $\mathsf{Sim}_R^{V^*}(x, z)$ outputs t also for some $x \notin L$.

Lemma 2. *If* $L \notin \mathsf{BPP}$ *then there exists some* $x \notin L$ *such that* $\mathsf{Sim}_R^{V^*}(x, z)$ *outputs* t *with probability greater than* $2/3$.

Let $x \notin L$ be a special statement such that $\mathsf{Sim}_R^{V^*}(x, z)$ outputs t with probability at least $2/3$ (such an x exists since we are assuming that $L \notin \mathsf{BPP}$). This means that Sim_R feeds V^* with a transcript of messages that with non-negligible probability produces t as output.

Let $\mathsf{time}_{\mathsf{Sim}_R}$ be the expected running time of Sim_R. Consider the strict polynomial-time machine Sim_{pR} that consists of running the first $3\mathsf{time}_{\mathsf{Sim}_R}$ steps of Sim_R.

We can prove the following lemma.

Lemma 3. *If* $L \notin \mathsf{BPP}$ *then there exists some* $x \notin L$ *such that* $\mathsf{Sim}_{pR}^{V^*}(x, z)$ *outputs* t *with probability greater than* $1/3$.

For notation purposes, we say that a query of Sim_{pR} to V^* belongs to the i-th session if it is a tuple $(h, \tilde{w}, \ldots)$ where $\tilde{w}$ is the i-th different value played by Sim_{pR} as first message of $\Pi.\mathsf{P}$ answering a leakage query of V^*. Let $\mathsf{time}_{\mathsf{Sim}_{pR}}$ be the strict polynomial corresponding to the running time of Sim_{pR}.

We can then prove the existence of a critical session i.

Lemma 4. *There exist* $x \notin L$ *and* $i \in [\mathsf{time}_{\mathsf{Sim}_{pR}}]$ *such that* $\mathsf{Sim}_{pR}^{V^*}$ *obtains* t *after answering to a query of the* i-th *session with non-negligible probability.*

Consider now the augmented simulator $\mathsf{Sim}_{pR}^{i^{V^*}}$ that works as $\mathsf{Sim}_{pR}^{V^*}$ except that V^* in the i-th session will only send h, while all other messages of V^* will be asked to an external oracle that plays as honest verifier of Π. Let $\mathsf{time}_{\Pi.\mathsf{Ext}}$ be the expected running time of $\Pi.\mathsf{Ext}$.

We can prove the following lemma.

Lemma 5. *There exist* $x \notin L$ *and* $i \in [\mathsf{time}_{\mathsf{Sim}_{pR}}]$ *such that the extractor* $\Pi.\mathsf{Ext}$ *of* Π *outputs a witness* $\hat{w}$ *for* $\tau = (h, \tilde{w}, \mathsf{rand}^\mathsf{P}, \mathsf{rand}^\mathsf{V}) \in \Lambda$ *with non-negligible probability and running in expected polynomial time. Moreover* $\mathrm{Prob}\left[\langle \mathsf{P}(\hat{w}), \mathsf{V}\rangle(x) = 1\right]$ *is non-negligible.*

We now show an adversarial prover P^* that violates the soundness of Π_{LRZK}. Let $\Pi.\mathsf{Ext}_p$ be the strict polynomial-time extractor that behaves precisely as $\Pi.\mathsf{Ext}$ (up to a given polynomial number of steps) as specified in the last part of the proof of Lemma 5.

P^* works as follows:

1. P^* picks at random $i \in [\text{time}_{\text{Sim}_{pR}}]$ and then runs $\Pi.\text{Ext}_p$ with respect to $\text{Sim}_{pR}^{i^{V^*}}$. If $\Pi.\text{Ext}_p$ does not give in output a state $\hat{w}$ as part of a witness proving that $\tau \in \Lambda$, then P^* aborts.
2. P^* then runs the honest prover P of Π_{LRZK} on input $\hat{w}$ for proving to a honest verifier V that $x \in L$ where x is the above special statement (i.e., $x \notin L$).

First of all, the running time of P^* is clearly polynomial since both the above steps take polynomial time. Then, we notice that by Lemma 5, both Step 1 and 2 correspond to runs without aborting with non-negligible probability. This is due to the fact that the extractor $\Pi.\text{Ext}_p$ fails only with negligible probability and that the extracted state $\hat{w}$ gives to a honest prover of (P, V) non-negligible probability to convince the verifier. Therefore P^* succeeds in proving a false statement to honest V with non-negligible probability.

We have proved that if $L \notin BPP$ then Π_{LRZK} can not be both LRZK and sound.

3.1 Discussion on Non-Black-Box LRZK

Since we have shown that LRZK is impossible when security is proved through black-box simulation, a natural question is whether non-black-box simulation can be useful to overcome this impossibility result.

The technique that we have shown for the black-box case is based on an adversarial verifier V^* that uses leakage queries to perform an execution of a resettably sound communication-efficient argument of knowledge Π against a honest prover. This makes the rewinding capabilities of the simulator ineffective therefore showing the impossibility of a black-box simulation.

However, the technique proposed by Barak in [2] allows for non-black-box straight-line simulation thus bypassing the difficulties to simulate a protocol where rewinds are useless. The construction and simulator proposed by Barak in [2] allows to get public-coin constant-round zero knowledge with a straight-line simulator, going therefore beyond the limits of black-box simulation [29]. It is also known that non-black-box simulation allows for resettably sound zero knowledge [4] where a prover can reset a verifier while the protocol still remains sound and zero knowledge. This is similar to the setting in which our black-box impossibility result holds. Indeed our adversarial verifier V^* is resilient to rewinds of the black-box simulator.

Having in mind the goal of overcoming the above impossibility result through non-black-box simulation, remember that in order to answer properly to the leakage queries of our adversarial verifier, a simulator either must simulate the execution of the universal argument[4] or must use a special trapdoor. Such a trapdoor must allow a honest prover of Π_{LRZK} to succeed in convincing a honest verifier that runs on input a randomness r. Such randomness is later revealed by

---

[4] Proving that Kilian's construction, analyzed in [2,3] as a 4-round public-coin universal argument, is zero-knowledge would be a major breakthrough.

V* only after seeing the short representation of the state $\tilde{w}$. Barak's construction does not allow to run the prover with an input different from a witness for $x \in L$, however, we next present a simple variant of it that does.

A Variation of Barak's Construction. Consider the following variant of Barak's protocol: (1) the verifier sends the description of a CRHF h; (2) the prover sends $h_w = h(\mathsf{Com}(w, u))$ to the verifier[5] where w is its private input, Com is the commitment function of a non-interactive commitment scheme and u is a random string; (3) the verifier sends a random string z; (4) the prover runs a witness indistinguishable universal argument proving that either $x \in L \vee h_w$ corresponds to the hash of a commitment of a machine M that in at most $n^{\log \log n}$ steps outputs z; the prover uses its private input w and u as witness in the universal argument.

Notice that the variation is really minimal: it just consists in asking the prover to use its private input when computing h_w. The impact of this variation is that the prover now can run successfully the protocol both when receiving as input a witness for $x \in L$ and also when receiving as input the code of the verifier.

The above small variation does not affect the zero-knowledge property (the proof is the same as Barak's), but allows the simulator to answer leakage queries of V* since the description of V* can be used as a legitimate encoded state that a prover can use in order to convince a verifier using a randomness r (again, such r is revealed by V* upon receiving through a leakage query the short representation of the state of the prover).

We stress that the discussion so far does not propose a LRZK protocol, rather it shows that the impossibility result given for the black-box case fails spectacularly when Barak's non-black-box techniques are considered.

Defeating Barak's Non-Black-Box Simulation Technique. While the above discussion seems to say that Barak's techniques could be used to design a LRZK protocol[6], we argue here that a breakthrough on non-black-box simulation is required in order to obtain a LRZK protocol. Notice that the above variation of Barak's construction allowed the prover to use a special trapdoor (the code of the verifier) instead of a witness to successfully run the protocol. Moreover, notice that the size of such a trapdoor is not bound by a fixed polynomial in the length of the common input since it depends on the size of the adversarial verifier. Instead there exists a constant $c > 0$ such that the length of a legitimate encoded witness of a LRZK protocol for a common input of length n is at most n^c. Therefore, let us consider an adversarial verifier that, just as in the impossibility proof for black-box LRZK, uses the leakage queries to execute a special protocol with a prover. In such a protocol, in addition to proving that the encoded state (that is consistent with the commitment already sent) makes the verifier accept, the prover also proves that the committed value is the hash of an

---

[5] Note that in Barak's protocol the prover uses 0^n instead of w.

[6] We stress that our work sticks with the use of standard/falsifiable assumptions.

encoded state of length at most n^c. Then the code of the adversarial verifier can not be used anymore as the simulation fails for adversarial verifiers whose code is longer than n^c. In other words, Barak's technique turns out to be insufficient. Additionally, the adversarial verifier might send a long vector of random strings $r_1, \ldots, r_\ell$ therefore asking the prover to prove in the universal argument that the verifier would have accepted the proof running with any of those ℓ randomnesses. Since ℓ can be greater than the upperbound on the encoded witness, there is no way to commit to a small machine that can predict all such strings.

In other words, we would need a non-black-box simulation technique that relies on standard assumptions and allows to construct a protocol where the trapdoor used by the simulator is of an a-priori fixed bounded size and can thus be given as input to the prover. Notice that it is exactly because of this limitation (or, rather, because of the lack of it) on the size of the trapdoor that the construction from [2] requires the use of a witness indistinguishable universal arguments instead of a witness-indistinguishable arguments of knowledge. In turn, this implies that the straight-line simulation of [2] can only be extended to bounded concurrency, leaving still unsolved the question of achieving constant-round concurrent zero knowledge under standard assumptions.

As a conclusion, as for many other lower bounds in zero knowledge, when taking into account non-black-box simulation, we can not rule out the existence of a non-black-box LRZK argument system, but at the same time we gave evidence that, to obtain such a result, new breakthroughs on non-black-box simulation are required.

4   Impossibility of LRMPC

We now use again the technique of running a protocol in the head of the adversary through leakage queries to show that LRMPC is impossible, therefore solving a problem opened in [10]. For this simpler result we give only a sketch of the proof and we defer for the additional definitions to [10]. We stress that the only variation here is that the interactive preprocessing does not take place (as required in the formulation of the open problem in [10]).

We can show that for many functionalities LRMPC with a leak-free input-encoding phase is impossible. The involved functionalities are the ones such that when they are run multiple times keeping unchanged the input x_i of a honest player P_i, the (short) outputs delivered to the dishonest players reveal more information on x_i than what a single output would reveal. Our impossibility requires just one dishonest player.

For simplicity we will now consider one such functionality: a variation of Yao's Millionaires' Problem, where n players $P_1, \ldots, P_n$ send their inputs to the functionality $\mathcal{F}$ and then $\mathcal{F}$ outputs to all players a bit b specifying whether P_1 is the richest one.

Theorem 2. *Consider the n-party functionality $\mathcal{F}$ that on input n k-bit strings $x_1, \ldots, x_n$ outputs to all players the bit $b = 1$ when $x_1 \geq x_j$ for $1 < j \in [n]$ and*

0 *otherwise. If at least one among* $P_2, \ldots, P_n$ *is corrupted and can get two bits as total output of leakage queries then there exists no LRMPC for* $\mathcal{F}$.

Proof. We will sketch the proof since the main ideas were already used in the proof of the impossibility of LRZK.

Assume by contradiction that there exists a secure multi-party protocol Π. Assume wlog that all players are honest except P_n. The adversary Adv controls P_n and works as follows.

1. It sends a leakage query that includes different encodings of the same value $x_2 = \cdots = x_n = 2^{k-1}$ for players $P_2, \ldots, P_n$; the leakage query asks for a "virtual" execution of the protocol where P_1 uses its state $\hat{x}_1$, and requires to give in output the output of P_n.
2. It repeats Step 1 changing the value to be used for the $n-1$ encodings of $P_2, \ldots, P_n$ (still a unique value for all of them) according to binary search (i.e., $2^{k-1} + 2^{k-2}$ if the previous output was 1 or 2^{k-2} otherwise).
3. Adv ends the protocol by giving in output the first two bits of the original (i.e., pre-encoding) input of P_1.

The communication complexity (from honest player to adversary) of this execution through leakage queries is the constant 2. Notice that the above leakage attack can be mounted with two queries each obtaining one bit as output, or with one single query obtaining two bits as output. As a result of the above leakage attack, Adv in the real world obtains the first two bits of x_1, the original input of P_1. Sim in the ideal world does not have such an information since it can perform only one query to $\mathcal{F}$, therefore getting at most one bit.

Acknowledgments. We thank the anonymous reviewers for their useful comments. The full version of this work appears in [38].

Part of this work was done while the second and third authors were visiting the Computer Science Department of UCLA.

This work has been supported by NSF grants 09165174, 1065276, 1118126 and 1136174, US-Israel BSF grant 2008411, OKAWA Foundation Research Award, IBM Faculty Research Award, Xerox Faculty Research Award, B. John Garrick Foundation Award, Teradata Research Award, and Lockheed-Martin Corporation Research Award. This material is based upon work supported by the Defense Advanced Research Projects Agency through the U.S. Office of Naval Research under Contract N00014 -11 -1-0392. The views expressed are those of the author and do not reflect the official policy or position of the Department of Defense or the U.S. Government.

References

1. Ananth, P., Goyal, V., Pandey, O.: Interactive proofs under continual memory leakage. In: Garay, J.A., Gennaro, R. (eds.) CRYPTO 2014, Part II. LNCS, vol. 8617, pp. 164–182. Springer, Heidelberg (2014)
2. Barak, B.: How to go beyond the black-box simulation barrier. In: 42nd Annual Symposium on Foundations of Computer Science, FOCS 2001, pp. 106–115. IEEE Computer Society (2001)

3. Barak, B.: Non-black-box techniques in cryptography. Ph.D. Thesis (2004). http://www.boazbarak.org/Papers/thesis.pdf
4. Barak, B., Goldreich, O., Goldwasser, S., Lindell, Y.: Resettably-sound zero-knowledge and its applications. In: 42nd Annual Symposium on Foundations of Computer Science, FOCS 2001, pp. 116–125. IEEE Computer Society (2001)
5. Bellare, M., Goldreich, O.: On defining proofs of knowledge. In: Brickell, E.F. (ed.) CRYPTO 1992. LNCS, vol. 740, pp. 390–420. Springer, Heidelberg (1993)
6. Bitansky, N., Canetti, R., Halevi, S.: Leakage-tolerant interactive protocols. In: Cramer, R. (ed.) TCC 2012. LNCS, vol. 7194, pp. 266–284. Springer, Heidelberg (2012)
7. Bitansky, N., Dachman-Soled, D., Lin, H.: Leakage-tolerant computation with input-independent preprocessing. In: Garay, J.A., Gennaro, R. (eds.) CRYPTO 2014, Part II. LNCS, vol. 8617, pp. 146–163. Springer, Heidelberg (2014)
8. Blum, M., De Santis, A., Micali, S., Persiano, G.: Non-interactive zero knowledge. SIAM J. Comput. **20**(6), 1084–1118 (1991)
9. Boyle, E., Garg, S., Jain, A., Kalai, Y.T., Sahai, A.: Secure computation against adaptive auxiliary information. In: Canetti, R., Garay, J.A. (eds.) CRYPTO 2013, Part I. LNCS, vol. 8042, pp. 316–334. Springer, Heidelberg (2013)
10. Boyle, E., Goldwasser, S., Jain, A., Kalai, Y.T.: Multiparty computation secure against continual memory leakage. In: Proceedings of the 44th Symposium on Theory of Computing Conference, STOC 2012, pp. 1235–1254. ACM (2012)
11. Boyle, E., Goldwasser, S., Kalai, Y.T.: Leakage-resilient coin tossing. In: Peleg, D. (ed.) Distributed Computing. LNCS, vol. 6950, pp. 181–196. Springer, Heidelberg (2011)
12. Boyle, E., Goldwasser, S., Kalai, Y.T.: Leakage-resilient coin tossing. Distrib. Comput. **27**(3), 147–164 (2014)
13. Boyle, E., Segev, G., Wichs, D.: Fully leakage-resilient signatures. J. Cryptol. **26**(3), 513–558 (2013)
14. Brakerski, Z., Kalai, Y.T., Katz, J., Vaikuntanathan, V.: Overcoming the hole in the bucket: Public-key cryptography resilient to continual memory leakage. In: 51th Annual IEEE Symposium on Foundations of Computer Science, FOCS 2010, pp. 501–510. IEEE Computer Society (2010)
15. Canetti, R.: Universally composable security: a new paradigm for cryptographic protocols. In: 42nd Annual Symposium on Foundations of Computer Science, FOCS 2001, pp. 136–145. IEEE Computer Society (2001)
16. Canetti, R., Goldreich, O., Goldwasser, S., Micali, S.: Resettable zero-knowledge (extended abstract). In: Proceedings of the Thirty-Second Annual ACM Symposium on Theory of Computing, STOC 2000, pp. 235–244. ACM (2000)
17. Dagdelen, Ö., Mohassel, P., Venturi, D.: Rate-limited secure function evaluation: definitions and constructions. In: Kurosawa, K., Hanaoka, G. (eds.) PKC 2013. LNCS, vol. 7778, pp. 461–478. Springer, Heidelberg (2013)
18. Damgård, I., Dupuis, F., Nielsen, J.B.: On the orthogonal vector problem and the feasibility of unconditionally secure leakage resilient computation. IACR Cryptology ePrint Archive 2014 (2014). http://eprint.iacr.org/2014/282
19. Dodis, Y., Haralambiev, K., López-Alt, A., Wichs, D.: Cryptography against continuous memory attacks. In: 51th Annual IEEE Symposium on Foundations of Computer Science, FOCS 2010, pp. 511–520. IEEE Computer Society (2010)
20. Dodis, Y., Lewko, A.B., Waters, B., Wichs, D.: Storing secrets on continually leaky devices. In: IEEE 52nd Annual Symposium on Foundations of Computer Science, FOCS 2011, pp. 688–697. IEEE (2011)

21. Dolev, D., Dwork, C., Naor, M.: Non-malleable cryptography (extended abstract). In: Proceedings of the 23rd Annual ACM Symposium on Theory of Computing, STOC 1991, pp. 542–552. ACM (1991)

22. Duc, A., Dziembowski, S., Faust, S.: Unifying leakage models: from probing attacks to noisy leakage. In: Nguyen, P.Q., Oswald, E. (eds.) EUROCRYPT 2014. LNCS, vol. 8441, pp. 423–440. Springer, Heidelberg (2014)

23. Dwork, C., Naor, M., Sahai, A.: Concurrent zero-knowledge. In: Proceedings of the Thirtieth Annual ACM Symposium on the Theory of Computing, STOC 1998, pp. 409–418. ACM (1998)

24. Dziembowski, S., Faust, S.: Leakage-resilient circuits without computational assumptions. In: Cramer, R. (ed.) TCC 2012. LNCS, vol. 7194, pp. 230–247. Springer, Heidelberg (2012)

25. Dziembowski, S., Pietrzak, K.: Leakage-resilient cryptography. In: 49th Annual IEEE Symposium on Foundations of Computer Science, FOCS 2008, pp. 293–302. IEEE Computer Society (2008)

26. Faust, S., Rabin, T., Reyzin, L., Tromer, E., Vaikuntanathan, V.: Protecting circuits from leakage: the computationally-bounded and noisy cases. In: Gilbert, H. (ed.) EUROCRYPT 2010. LNCS, vol. 6110, pp. 135–156. Springer, Heidelberg (2010)

27. Garg, S., Gentry, C., Halevi, S., Raykova, M., Sahai, A., Waters, B.: Candidate indistinguishability obfuscation and functional encryption for all circuits. In: 54th Annual IEEE Symposium on Foundations of Computer Science, FOCS 2013, pp. 40–49. IEEE Computer Society (2013)

28. Garg, S., Jain, A., Sahai, A.: Leakage-resilient zero knowledge. In: Rogaway, P. (ed.) CRYPTO 2011. LNCS, vol. 6841, pp. 297–315. Springer, Heidelberg (2011)

29. Goldreich, O., Krawczyk, H.: On the composition of zero-knowledge proof systems. SIAM J. Comput. 25(1), 169–192 (1996)

30. Goldreich, O., Micali, S., Wigderson, A.: How to play any mental game or A completeness theorem for protocols with honest majority. In: Proceedings of the 19th Annual ACM Symposium on Theory of Computing, STOC 1987, pp. 218–229. ACM (1987)

31. Goldwasser, S., Micali, S., Rackoff, C.: The knowledge complexity of interactive proof-systems (extended abstract). In: Proceedings of the 17th Annual ACM Symposium on Theory of Computing, STOC 1985, pp. 291–304. ACM (1985)

32. Goldwasser, S., Rothblum, G.N.: Securing computation against continuous leakage. In: Rabin, T. (ed.) CRYPTO 2010. LNCS, vol. 6223, pp. 59–79. Springer, Heidelberg (2010)

33. Goldwasser, S., Rothblum, G.N.: How to compute in the presence of leakage. In: 53rd Annual IEEE Symposium on Foundations of Computer Science, FOCS 2012, pp. 31–40. IEEE Computer Society (2012)

34. Ishai, Y., Sahai, A., Wagner, D.: Private circuits: securing hardware against probing attacks. In: Boneh, D. (ed.) CRYPTO 2003. LNCS, vol. 2729, pp. 463–481. Springer, Heidelberg (2003)

35. Micali, S., Reyzin, L.: Physically observable cryptography. In: Naor, M. (ed.) TCC 2004. LNCS, vol. 2951, pp. 278–296. Springer, Heidelberg (2004)

36. Nielsen, J.B., Venturi, D., Zottarel, A.: On the connection between leakage tolerance and adaptive security. In: Kurosawa, K., Hanaoka, G. (eds.) PKC 2013. LNCS, vol. 7778, pp. 497–515. Springer, Heidelberg (2013)

37. Ostrovsky, R., Persiano, G., Visconti, I.: Constant-round concurrent non-malleable zero knowledge in the bare public-key model. In: Aceto, L., Damgård, I., Goldberg, L.A., Halldórsson, M.M., Ingólfsdóttir, A., Walukiewicz, I. (eds.) ICALP 2008, Part II. LNCS, vol. 5126, pp. 548–559. Springer, Heidelberg (2008)
38. Ostrovsky, R., Persiano, G., Visconti, I.: Impossibility of black-box simulation against leakage attacks. IACR Cryptology ePrint Archive 2014 (2014). http://eprint.iacr.org/2014/865
39. Pandey, O.: Achieving constant round leakage-resilient zero-knowledge. In: Lindell, Y. (ed.) TCC 2014. LNCS, vol. 8349, pp. 146–166. Springer, Heidelberg (2014)
40. Standaert, F.-X., Malkin, T., Yung, M.: Does physical security of cryptographic devices need a formal study? (Invited Talk). In: Safavi-Naini, R. (ed.) ICITS 2008. LNCS, vol. 5155, pp. 70–70. Springer, Heidelberg (2008)
41. Standaert, F.-X., Malkin, T.G., Yung, M.: A unified framework for the analysis of side-channel key recovery attacks. In: Joux, A. (ed.) EUROCRYPT 2009. LNCS, vol. 5479, pp. 443–461. Springer, Heidelberg (2009)
42. Standaert, F., Pereira, O., Yu, Y., Quisquater, J., Yung, M., Oswald, E.: Leakage resilient cryptography in practice. In: Sadeghi, A., Naccache, D. (eds.) Towards Hardware-Intrinsic Security - Foundations and Practice. Information Security and Cryptography, pp. 99–134. Springer, Heidelberg (2010)
43. Yu, Y., Standaert, F., Pereira, O., Yung, M.: Practical leakage-resilient pseudo-random generators. In: Al-Shaer, E., Keromytis, A.D., Shmatikov, V. (eds.) Proceedings of the 17th ACM Conference on Computer and Communications Security, CCS 2010, pp. 141–151. ACM (2010)

Efficient Zero-Knowledge Proofs
of Non-algebraic Statements with Sublinear
Amortized Cost

Zhangxiang Hu[1], Payman Mohassel[2]([✉]), and Mike Rosulek[1]

[1] Oregon State University, Corvallis, USA
{huz,rosulekm}@eecs.oregonstate.edu
[2] Yahoo Labs, Sunnyvale, CA, USA
pmohassel@yahoo-inc.com

Abstract. We describe a zero-knowledge proof system in which a prover holds a large dataset M and can repeatedly prove NP relations about that dataset. That is, for any (public) relation R and x, the prover can prove that $\exists w : R(M, x, w) = 1$. After an initial setup phase (which depends only on M), each proof requires only a constant number of rounds and has communication/computation cost proportional to that of a *random-access machine (RAM)* implementation of R, up to polylogarithmic factors. In particular, the cost per proof in many applications is sublinear in $|M|$. Additionally, the storage requirement between proofs for the verifier is constant.

1 Introduction

Zero-knowledge (ZK) proofs are a fundamental concept in cryptography and are used as a building block in numerous applications. ZK proofs allow a prover with the knowledge of a witness w to prove statements of the form $\exists w : R(x, w) = 1$ to a verifier V, for a public NP statement R and a public input x. The *soundness* of such a proof guarantees that a malicious prover cannot prove a false statement to a verifier, and the *zero-knowledge* property guarantees that a malicious verifier cannot learn any information about the witness except for validity of the proved statement.

Since the conception of zero-knowledge proofs [GMR89], a large body of work has focused on design of efficient constructions that are practical enough for use in practice. But until recently, all such constructions were practical only for proving statements about certain algebraic structures such as proving knowledge of and relations for discrete logarithms, RSA public keys, and bilinear equations [Sch90, CDS94, CM99, GS08].

The recent work of [JKO13] proposes a new approach based on garbled circuits (GC) that is suitable for general-purpose statements represented as boolean circuits. This is particularly powerful for proving non-algebraic statements, e.g., proving knowledge of x such that $y = \mathsf{Sha256}(x)$ for a public value y. The construction is very efficient, only requiring a constant number of rounds and communication/computation cost that is similar to that of *semi-honest* 2PC based

© International Association for Cryptologic Research 2015
R. Gennaro and M. Robshaw (Eds.): CRYPTO 2015, Part II, LNCS 9216, pp. 150–169, 2015.
DOI: 10.1007/978-3-662-48000-7_8

on garbled circuits (i.e., Yao's protocol). Given the recent advances in design & implementation of circuit garbling techniques, these ZK proofs are scalable to statements with billions of gates.

Need for ZK Proof of RAM Programs. But the GC-based approach falls short when the statement being proven involves access to a large dataset committed by the prover. For instance, recall the problem solved by *zero-knowledge sets* [MRK03]: a prover commits to a set S in an initial phase and is later able to prove membership and non-membership statements $(x \in S, x \notin S)$ for any input x without revealing additional information.

A natural extension is to prove membership for a (possibly private) value x that satisfies a predicate p without leaking any additional information about x or the set S. For instance, the prover may need to prove knowledge of an $x \in S$ where $\mathsf{Sha256}(x) = y$ for a public y in order to prove inclusion of a password in a password-file. Furthermore, to improve on storage cost, the prover may want to store his set S in a *Bloom filter* [Blo70]. This would lead to major storage improvement, especially when considering the inevitable overhead caused by crypto for every bit of memory stored. Now, the prover needs to prove knowledge of an x where $\mathsf{Sha256}(x) = y$ and where the Bloom filter stores a bit 1 at each of the locations $H_1(x), \ldots, H_k(x)$ (the H_i's are the hash functions associated with the Bloom filter and can be public). Such a statement involves several hash evaluations and memory lookups. More generally, the prover may want to store its data in a data-structure of its own choice and still have efficient tools for proving statements about it.

In all of these scenarios, the statements being proven are naturally expressed as RAM programs whose running time is sublinear in the size of the large dataset. By comparison, directly applying a circuit-based approach (i.e., [JKO13]) would involve garbled circuits that are at least linear in the size of the large dataset.

Existing Solutions for RAM-ZK. One can combine the GC-based proof system of [JKO13] with the recent garbled RAM constructions [LO13, GHL+14] that directly garble RAM programs as opposed to circuits. But the existing constructions for garbled RAM are not efficient enough for practical use. In particular, one needs to perform cryptographic operations inside the garbled circuits for every step of RAM computation, which is a major bottleneck.

Finally, given that ZK proofs are a special case of secure two-party computation against malicious adversaries (*i.e.*, a malicious 2PC where one party, the verifier, has no input), we can obtain a solution by employing an efficient malicious 2PC for RAM programs [AHMR15] and not assigning one party any input. But for statistical security 2^{-s}, such a proof would be a factor of s more expensive than the semi-honest 2PC for the same RAM program, and the number of rounds would also be proportional to the running time of the RAM program.

1.1 Our Contribution

We propose a new solution for zero-knowledge proof of statements of the form $\exists w : R(M, x, w) = 1$ where R is a RAM program and M is its (large) memory.

Here, M is committed upfront by the prover and can in general remain private from the verifier. Our construction is constant-round, and incurs online computation and communication cost that is linear in the running time of the RAM program (upto a polylogarithmic factor), competitive with the best *semi-honest* 2PC for RAM programs ([GKK+12]), and hence sublinear in $|M|$ for many applications of interest. Sublinear-time 2PC is not possible in general when expressing the NP relation as a boolean circuit. Furthermore, in our protocol the verifier maintains only constant storage space between multiple proofs.

Our construction combines an Oblivious RAM [GO96] and garbled circuits, but it avoids the use of cryptographic operations inside the garbled circuits as in current garbled-RAM constructions. Unlike previous 2PC constructions based on RAM computation [GKK+12, AHMR15], our construction requires only a constant number of rounds of interaction. We discuss the construction in more detail next.

2   Overview of the Protocol

The JKO Protocol. Our starting point is the garbled-circuit-based ZK protocol of [JKO13], which we summarize here. To prove a statement $\exists w : R(x, w) = 1$ (for public R and x), the protocol proceeds as follows:

1. The verifier generates a garbled circuit computing $R(x, \cdot)$. Using a committing oblivious transfer, the prover obtains the wire labels corresponding to his private input w. Then the verifier sends the garbled circuit to the prover.
2. The prover evaluates the garbled circuit, obtaining a single garbled output (wire label). He commits to this garbled output.
3. The verifier opens his inputs to the committing oblivious transfer, giving the prover all garbled inputs. From this, the prover can check whether the garbled circuit was generated correctly. If so, the prover opens his commitment to the garbled output; if not, the prover aborts.
4. The verifier accepts the proof if the prover's commitment holds the output wire label corresponding to TRUE.

Security against a cheating prover follows from the properties of the circuit garbling scheme. Namely, the prover commits to the output wire label before the circuit is opened, so the *authenticity* property of the garbling scheme ensures that he cannot predict the TRUE output wire label unless he knows a w with $R(x, w) = \text{TRUE}$. Security against a cheating verifier follows from correctness of the garbling scheme. The garbled output of a *correctly generated* garbled circuit reveals only the output of the (plain) circuit, and this garbled output is not revealed until the garbled circuit was shown to be correctly generated.

Note that in this protocol, the prover evaluates the garbled circuit on an input which is completely known to him. This is the main reason that the garbled circuit used for evaluation can also be later opened and checked for correctness, unlike in the setting of cut-and-choose for general 2PC. Along the same lines, it was further pointed out in [FNO15] that the circuit garbling scheme need not satisfy the *privacy* requirement of [BHR12], only the *authenticity* requirement. Removing the privacy requirement from the garbling scheme leads to a non-trivial reduction in garbled circuit size.

Adapting to the ORAM Setting, Using Constant Rounds. We follow roughly the RAM-2PC paradigm of [GKK+12, AHMR15], with some important differences. Let Π be an Oblivious RAM program with memory $\widehat{M}$, that implements $R(M, x, \cdot)$.[1] We assume a trusted setup phase in which Π's memory $\widehat{M}$ and state st are initialized from M. The prover learns $\widehat{M}$, st, as well as a garbled encoding of these values (i.e., one wire label for each bit of memory & state); the verifier specifies the garbled encoding to be used (i.e., both wire labels for each bit). If we follow [GKK+12, AHMR15] strictly, we would have both parties repeatedly evaluate the next-memory-access circuit of Π, updating memory $\widehat{M}$, until it halts. However, this would result in a protocol with one round of interaction for each memory access of Π.

To see how to achieve the same effect in a constant number of rounds, imagine that when executing an ORAM program, the memory access pattern $\mathcal{I}$ is known in advance. Then it is possible to express the entire computation in a single circuit. The circuit includes many copies of the RAM program's next-memory-access circuit, but is wired together under the assumption that the memory accesses will be $\mathcal{I}$. For example, if $\mathcal{I}$ says that Π writes to some memory block at time 2, and later reads from the same memory block at time 10, then the memory-output wires of subcircuit copy #2 will be connected to the memory-input wires of subcircuit copy #10, and so on.

We can leverage this optimization in our setting because the prover knows all (plaintext) inputs to Π, including the contents of memory and the ORAM state. Hence, the prover can execute Π locally to determine the complete memory access pattern $\mathcal{I}$. Since Π is an oblivious RAM, its access pattern $\mathcal{I}$ leaks no information about the inputs/memory/state, so the prover can safely send $\mathcal{I}$ to the verifier. Using $\mathcal{I}$, the verifier constructs a *single* garbled circuit $C_{x,\mathcal{I}}$ as described above. To prevent the prover from lying about the access pattern $\mathcal{I}$, the circuit recomputes the memory access pattern of Π and compares it to (hard-coded) $\mathcal{I}$.

Hence, this setting admits a constant-round solution based on ORAM, but avoiding tools like garbled RAM [LO13, GHL+14] which incorporate expensive additional crypto circuitry into the garbled circuits.

Reusing M to Perform Many Proofs. We follow the approach of [AHMR15], where the prover stores the ORAM memory and ORAM state encoded as wire labels from the various garbled circuits. The idea is that these wire labels can be reused directly as inputs to subsequent circuits, avoiding oblivious transfers for garbled circuit input. However, some modifications are required to adapt this idea to our setting.

---

[1] We use M to refer to the logical RAM memory, and $\widehat{M}$ to refer to the physical ORAM memory.

After evaluating a garbled circuit, the prover holds a garbled output encoding of ORAM state & memory. The *authenticity* property of the garbling scheme guarantees that the prover knows at most one valid label per wire. As soon as the garbled circuit is opened, however, the prover learns both labels for each wire and authenticity is lost. The output wire labels are no longer useful for input to subsequent circuits, as the prover can now feed arbitrary garbled state/memory into subsequent garbled circuits. We need a mechanism to restore authenticity on all wire labels that may be later used (this includes the ORAM internal state as well as all memory locations that are read or written by the garbled circuit).

Say the two wire labels on some output wire are y_0 and y_1, and that the prover knows only y_b. Let us call y_0 and y_1 the *temporary* wire labels, since they will soon be discarded. The verifier chooses a random function h from a strongly universal hash family. Just before the garbled circuit is opened (clobbering wire-label authenticity), the parties perform a *private function evaluation (PFE)*, where the prover gives y_b, the verifier gives h, and the prover learns $h(y_b)$. After the PFE, the garbled circuit can be opened, revealing y_0 and y_1.

Define $y'_0 = h(y_0)$ and $y'_1 = h(y_1)$ to be the *permanent* wire labels for this wire. At the time of the PFE, the prover could not have guessed y_{1-b}, and so learned the output of h on some point that was not y_{1-b}. From strong universality of h, even if y_{1-b} is later revealed, $y'_{1-b} = h(y_{1-b})$ is still random from the prover's point of view. Hence the PFE "transfers" the authenticity guarantee from the temporary wire labels y_0, y_1 to the permanent ones y'_0, y'_1, preserving authenticity even after both of y_0, y_1 are revealed. Hence, y'_0, y'_1 are safe to use as input wire labels to a subsequent garbled circuit. We emphasize that all wire labels are used only in a single garbled circuit — we use the term "permanent" since these wire labels will be the long-term representation of the RAM program's memory between proof instances. (It may be many proof instances before a particular block of memory is next accessed.)

For technical reasons, the PFE needs to be committing with respect to the input h (so that the verifier can later "open" the h that was used). We suggest two efficient instantiations of committing-PFE for strongly universal families: one based on oblivious linear function evaluation (OLFE) [WW06] and one based on the string-select variant of OT presented in [KK12].

Note that all the PFE instances can be run in parallel hence, maintaining the constant round complexity of the overall protocol.

Eliminating the Verifier's Storage Requirement. As described so far, the verifier is required to keep track of two wire labels for each bit of $\widehat{M}$, at all times. We can decrease this burden somewhat by letting the verifier derive these wire labels from a PRF. Let s be a seed to a PRF. For simplicity, suppose a wire label encoding truth value b on the jth bit of the ith memory block, last accessed at time t, is chosen as $\mathsf{PRF}(s, i\|j\|t\|b)$. In the actual protocol, the choice of wire labels is slightly more complicated.

Using this choice of wire labels, the verifier need only remember the last-access time of each block of $\widehat{M}$. However, this is still storage proportional to $|\widehat{M}|$. To reduce the storage even further, we "outsource" the maintenance of these last-access times to the prover. Let $T[i]$ denote the last-access time of block i. We let the prover store the array T authenticated by a Merkle tree for which the verifier remembers only the root node.[2]

Whenever the verifier is about to garble a circuit, he must be reminded of $T[i]$ for each memory block i to be read by the RAM in its computation. We make the prover report each such $T[i]$ to the verifier, authenticating each value via the Merkle tree. The ORAM circuit performs some reads & writes in $\widehat{M}$, so T and the Merkle tree are updated accordingly, for each memory block that was accessed. Note that all accesses to the Merkle tree are done at the same time (in parallel), and similarly for the updates at the end of the execution.

Overall, accessing/updating the authenticated array T adds polylogarithmic (in $|\widehat{M}|$) communication/computation overhead and only a small constant number of rounds to the protocol. Instead of remembering two wire labels for each bit of $\widehat{M}$, the verifier need now remember only a PRF seed and the root of a Merkle tree.

3   Preliminaries

Throughout the paper, we let $k \in \mathbb{N}$ be the security parameter. We say a function $\epsilon : \mathbb{N} \to [0, 1]$ is negligible if for any polynomial p, there exists a large enough k' such that for all $k > k'$, $\epsilon(k) < 1/p(k)$. Also, for a integer n, we define $[n] = \{1, 2, \ldots, n\}$.

3.1   ZK Proofs and Other Standard Functionalities

Here we define the variant of ZK proofs that we achieve, as well as other standard ideal functionalities used in our protocol.

Zero-Knowledge Proofs: Roughly speaking, a zero-knowledge proof is an interactive protocol in which a party P (the prover) can prove to another party V (the verifier) that some NP statement x is true by using a valid witness w, leaking no information about w (except that the statement x is true).

More precisely, for any language $\mathcal{L} \in$ NP with some binary relation $\mathcal{R}_\mathcal{L}$, for all valid instances $x \in \mathcal{L}$, there exists a string w such that $\mathcal{R}_\mathcal{L}(x, w) = 1$. Otherwise, if $x \notin \mathcal{L}$, then for all string w we have $\mathcal{R}_\mathcal{L}(x, w) = 0$.

The ideal functionality $\mathcal{F}_{\mathsf{ZK}}^\mathcal{R}$ is defined in Fig. 1, which allows for many proofs to reference a common (secret) value M.

Commitment: The commitment functionality $\mathcal{F}_{\mathsf{com}}$ is described in Fig. 2. It allows a party to commit to a secret value at one time and reveal that value at a later time.

---

[2] More generally, T can be stored in any authenticated data structure that provides small storage for the verifier.

$\mathcal{F}_{\mathsf{ZK}}^{\mathcal{R}}$ is parametrized by a relation $\mathcal{R}$. It involves two parties: a prover P and a verifier V.

- Setup: On input (INIT, M) from P, if no previous INIT command has been given, then $\mathcal{F}_{\mathsf{ZK}}^{\mathcal{R}}$ stores M internally.
- Proof: On input $(\mathrm{PROVE}, sid, x, w)$ from P, if $\mathcal{R}(M, x, w) = 1$, output $(\mathrm{ACCEPT}, sid, x)$ to V.

Fig. 1. Ideal functionality $\mathcal{F}_{\mathsf{ZK}}^{\mathcal{R}}$ for zero-knowledge proofs of NP-relation $\mathcal{R}$

Let $\mathcal{M}$ denote the space of valid messages. $\mathcal{F}_{\mathsf{com}}$ receives input from party P and sends output message to party V. It consists of two phases: Commit and Open.

- Commit: On input (COMMIT, m) from P with $m \in \mathcal{M}$, if there is no value m already stored in memory, then $\mathcal{F}_{\mathsf{com}}$ stores m internally and outputs $\mathrm{COMMITTED}$ to party V.
- Open: On input OPEN from P, if value m exists in memory, then $\mathcal{F}_{\mathsf{com}}$ outputs (OPENED, m) to party V.

Fig. 2. Ideal functionality $\mathcal{F}_{\mathsf{com}}$ for commitment

- Initialization: $\mathcal{F}_{\mathsf{otc}}$ takes private input E (an $m \times 2$ array) from party V and the private input $\sigma \in \{0, 1\}^m$ from party P, then stores (E, σ) internally and output $\mathrm{COMMITTED}$.
- Transfer: On command $\mathrm{TRANSFER}$ from V, $\mathcal{F}_{\mathsf{otc}}$ sends $(\mathrm{TRANSFERRED}, E|_\sigma)$ to P.
- Open: On command OPEN from V, $\mathcal{F}_{\mathsf{otc}}$ sends (OPENED, E) to P.

Fig. 3. Ideal functionality $\mathcal{F}_{\mathsf{otc}}$ for committing oblivious transfer. Notation $E|_\sigma$ is defined in Sect. 3.4.

- Initialization: On input (INIT, N) from party V, $\mathcal{F}_{\mathsf{Aut}}$ initialize an array T of size N. For each $T[i]$, $i \in \{1, \ldots, N\}$, set $T[i] = 0$.
- Update: On input $(\mathrm{UPDATE}, id, data)$ from party V, set $T[id] = data$ and output $(\mathrm{UPDATED}, id, data)$ to both parties.
- Open: On input (ACCESS, id) from party V, where $id \in \{1, \ldots, N\}$, send $(\mathrm{ACCESSED}, id, T[id])$ to V.

Fig. 4. Ideal functionality $\mathcal{F}_{\mathsf{Aut}}$ for authenticated array access.

Committing Oblivious Transfer: The definition of committing oblivious transfer was first given by Kiraz and Schoenmakers [KS06]. In the general OT protocol, party V inputs a description of wire labels E and party P has input σ. After running oblivious transfer, P receives a garbled encoding of σ under the encoding E. See Sect. 3.4 for more details about the wire-label syntax used in the figure.

$\mathcal{F}_{\mathsf{cpfe}}$ is parametrized by a class of functions $\mathcal{H}$, with each $h \in \mathcal{H}$ having a common domain A.

- Evaluation: On input $h \in \mathcal{H}$ from party V and input $x \in A$ from party P, give output $h(x)$ to party P. Remember h internally.
- Open: On input OPEN from party V, give output h to party P.

Fig. 5. Ideal functionality $\mathcal{F}_{\mathsf{cpfe}}$ for committing private function evaluation.

The "committing" aspect of committing OT allows party V to reveal E at a later time. The ideal functionality $\mathcal{F}_{\mathsf{otc}}$ is defined in Fig. 3.

Authenticated Array: The functionality $\mathcal{F}_{\mathsf{Aut}}$ in Fig. 4 simply provides storage of an array, in which the party V has control over modifications. Such a functionality becomes interesting in our setting when it is realized by a protocol with minimal (constant) storage for party V. A simple approach is to use an authenticated Merkle-tree, with V storing only the root of the tree.

3.2 Committing Private Function Evaluation (of a Strongly Universal Family)

Private function evaluation (PFE) takes input h (a function) from a sender, input x from a receiver, and gives output $h(x)$ to the receiver. We define and use a committing variant of PFE in which the sender can later reveal the h that was used. The formal description is given in Fig. 5.

In our final protocol, we require committing PFE supporting a **strongly universal** class $\mathcal{H}$ of functions. Suppose each function h in $\mathcal{H}$ is of the form $h : A \rightarrow B$. Then $\mathcal{H}$ is strongly universal if for all distinct $a, a' \in A$ and all (possibly equal) $b, b' \in B$,

$$\Pr_{h \leftarrow \mathcal{H}}[h(a) = b \mid h(a') = b'] = 1/|B|$$

Below we suggest several efficient choices for PFE of strongly universal families:

Using 1-out-of-2 OT: Let X be an $n \times 2$ matrix of length-m strings. For such an X, define the function $h_X : \{0,1\}^n \rightarrow \{0,1\}^m$ via:

$$h_X(z) = \bigoplus_{i=1}^{n} X_{i,z_i}$$

Then the class $\mathcal{H} = \{h_X \mid X \in (\{0,1\}^m)^{n \times 2}\}$ is strongly universal.

A simple protocol for private function evaluation of $\mathcal{H}$ uses standard 1-out-of-2 oblivious transfer (of strings) in the following way: For $i = 1$ to n, the sender gives input $X_{i,0}$ and $X_{i,1}$ as input to an instance of OT. The receiver gives input z_i and obtains $r_i = X_{i,z_i}$. Finally the receiver outputs $r_1 \oplus \cdots \oplus r_n$.

Technically, this protocol is *not* a secure PFE for the family $\mathcal{H}$, because the receiver learns more than $h_X(z)$. In particular, the receiver learns various X_{i,z_i} values. However, the protocol suffices for our needs, by considering slightly relaxed definitions. Let $\mathcal{H}$ be a family of *pairs* of functions. We write $(h, \widehat{h}) \in \mathcal{H}$, where $h : A \to B$ and $\widehat{h} : A \to \widehat{B}$. Then we say that $\mathcal{H}$ is **modified strongly universal** if for all distinct $a, a' \in A$ and all (possibly equal) $b \in B$, $\widehat{b'} \in \widehat{B}$:

$$\Pr_{(h,\widehat{h}) \leftarrow \mathcal{H}} [h(a) = b \mid \widehat{h}(a') = \widehat{b'}] = 1/|B|$$

The family $\mathcal{H} = \{h_X\}$ we described above satisfies this definition, taking $\widehat{h}_X(z) = (X_{1,z_1}, \ldots, X_{n,z_n})$. That is, the value of $h_X(z')$ is distributed uniformly even after fixing the output of $\widehat{h}_X(z)$ for $z \neq z'$.

Then the protocol just described is secure for a variant of Fig. 5 in which an adversarial receiver obtains not $h(x)$ but $\widehat{h}(x)$. It should be clear that such a modified functionality suffices for our eventual usage of $\mathcal{F}_{\mathsf{cpfe}}$ when the family $\mathcal{H}$ is *modified* strongly universal. For simplicity we write our eventual ZK protocol in terms of the simpler $\mathcal{F}_{\mathsf{cpfe}}$ defined in Fig. 5.

Furthermore, when the underlying OT protocol is a committing OT, then the PFE protocol is also committing in a natural way (with the sender revealing all committed-OT inputs). We note that this protocol is essentially the "string-select oblivious transfer" protocol of [KK12] but without the final verification step which is not needed here.

Using OLFE: In a finite field $\mathbb{F}$, the class of functions of the form $x \mapsto ax + b$ is strongly universal (with $a, b \in \mathbb{F}$). A private function evaluation for this class therefore accepts $a, b \in \mathbb{F}$ from the sender, $x \in \mathbb{F}$ from the receiver, and gives output $ax + b$ to the receiver. Such a functionality is already known by the name of *oblivious linear function evaluation* (OLFE or OLE) [WW06].

The state of the art for malicious-secure OLFE is due to the general protocol of Ishai, Prabhakaran, and Sahai [IPS09] for evaluating arithmetic circuits in the OT-hybrid model. Since OLFE can be represented by an arithmetic circuit with just 2 gates, their construction yields an OLFE protocol with (amortized) constant number of field elements communicated per OLFE and computation roughly $O(\log k)$ field operations per OLFE.

The general construction of [IPS09] combines an outer MPC protocol among imaginary parties and an inner 2PC protocol between the real parties. It is easy to see that if the inner protocol is committing, so is the overall protocol.

3.3 Oblivious RAM Program

Oblivious RAM (ORAM) programs were first introduced by Goldreich & Ostrosvsky [GO96]. ORAM allows a client to hide its access pattern and data to the server. In this work we freely identify a RAM program Π with its deterministic *next-instruction* circuit. We use M to represent the logical memory of a RAM program and $\widehat{M}$ to indicate the physical memory array in Oblivious RAM program. We consider all memory to be split into **blocks**, where $M[i]$ denotes the ith block of M.

Without loss of generality, we assume that the RAM program is deterministic. Although constructions of oblivious RAM require randomness, we can allow the prover to provide that randomness as part of the witness w. Thus, we think of w as $w = w_{\mathsf{real}} \| r$, where w_{real} is the actual witness to the statement and r is randomness used by the ORAM. An honest prover will choose r uniformly so that the ORAM memory access sequence hides private information. Allowing a corrupt prover to choose r does not compromise soundness in practical ORAM constructions (e.g., [SvDS+13]) — it only affects the probability of an overflow error event (in which case we can have the ORAM circuit output FALSE).

Let the next-instruction circuit Π have syntax:

$$(inst, st, block) \leftarrow \Pi(st, \Sigma, block)$$

where Σ is external input, st is the ORAM state, $block$ is the memory blocks and $inst$ represents a RAM memory access instruction, which must have one of the following forms: $(read, i)$, $(write, i)$, or (HALT, z), where i is the index of a memory block.

The execution of an ORAM program Π on input (x, w) using memory $\widehat{M}$ is as follows:

$\underline{\mathrm{RAMEval}(\Pi, \widehat{M}, (x, w), st)}$
    $\mathcal{I} := \emptyset$
    $(inst, st, block) := \Pi(st, (x, w), \perp)$
    do until inst has the form $(\mathrm{HALT}, \mathrm{TRUE})$:
        $block := [\text{if } inst = (read, id) \text{ then } \widehat{M}[id] \text{ else } \perp]$
        $(inst, st, block) := \Pi(st, \perp, block)$
        if $inst = (write, id)$ then $\widehat{M}[id] := block$
        $\mathcal{I} := \mathcal{I} \| inst$
    output $\mathcal{I}$

Note that we have RAMEval output the access sequence $\mathcal{I}$. We say $\mathcal{I}$ is an **accepting access sequence** if the last instruction in $\mathcal{I}$ is $(\mathrm{HALT}, \mathrm{TRUE})$.

We assume a function Initialize with syntax:

$$(\widehat{M}, st) \leftarrow \mathsf{Initialize}(1^k, M)$$

This function returns the initial value of st and also the initialized physical memory array $\widehat{M}$ encoding the logical memory M.

The security definition of an oblivious RAM program Π requires that the memory access sequence $\mathcal{I}$ does not leak information about the data set M or witness w_{real}. More formally:

Definition 1. *We say that Π is a* **secure ORAM** *if there exists an efficent $\mathcal{S}$ such that, for all M, all $(\widehat{M}, st) \leftarrow \mathsf{Initialize}(1^k, M)$, all (x, w_{real}) such that $\mathcal{R}(M, x, w_{\mathsf{real}}) = 1$ and for all PPT $\mathcal{A}$, the following difference:*

$$\left| \Pr[\mathcal{A}(\mathcal{S}(1^k, |\widehat{M}|, x) = 1] - \Pr_r[\mathcal{A}(\mathit{RAMEval}(\Pi, \widehat{M}, (x, w_{\mathsf{real}} \| r), st)) = 1] \right|$$

is negligible in k.

Any RAM program can be converted into an oblivious one satisfying our definitions, using standard constructions [SvDS+13, CP13]. Note that $\mathcal{I}$ (the output of
RAMEval) contains only the memory *locations* and not the *contents* of memory.
Hence, we do not require the ORAM construction to encrypt/decrypt memory
contents.

3.4   Garbling Scheme

We assume some familiarity with standard constructions of garbled circuits. We
employ the abstraction of garbling scheme [BHR12] introduced by Bellare *et al.*,
but we use a slightly different syntax for our needs.

We represent a set of wire labels on m wires via a $m \times 2$ array W. For each
wire i, $W[i, 0] \in \{0, 1\}^k$ and $W[i, 1] \in \{0, 1\}^k$ are two wire labels that encode
FALSE and TRUE, respectively. For a truth value x, the corresponding wire labels
are defined as $W|_x = (W[1, x_1], \ldots, W[m, x_m])$.

Our protocol adopts the idea of [MGFB14, AHMR15] of re-using wire labels
between different garbled circuits. We require somewhat different syntax for the
garbling scheme in order to facilitate this reuse.

For our purposes, a garbling scheme consists of the following algorithms:

- $\mathsf{Gb}(1^k, f, E, D) \to F$. Takes as input a boolean circuit f, descriptions of input
  wire labels E and output wire labels D, and outputs a garbled circuit F.
- $\mathsf{En}(E, x) \to X = E|_x$. Takes as input description of input wire labels E, a
  plaintext input x and outputs a garbled input X. In our schemes, encoding is
  always done via $E|_x$.
- $\mathsf{Ev}(F, X) \to Y$. Takes as input a garbled circuit F and a garbled input X and
  returns a garbled output Y.
- $\mathsf{Chk}(f, F, E) \to D$ or $\perp$. Takes as input a boolean circuit, a (purported) garbled circuit F and input wire label desription E and outputs either D or an
  error indicator $\perp$.

The correctness and security condition of garbling scheme we require here
is slightly different from those given in [BHR12], but any garbling scheme that
meet the requirements in [BHR12] also works well for our definitions.

Definition 2. *A garbling scheme satisfies* **correctness** *if:*

1. For all circuits f, circuit-inputs x, and valid wire label descriptions E, D,

$$\mathsf{Chk}(f, F, E) = D \text{ whenever } F \leftarrow \mathsf{Gb}(1^k, f, E, D)$$

*2. For all circuits f, (possibly malicious) garbled circuits F and wire-label
descriptions E,*

$$\mathsf{Ev}(F, E|_x) = D|_{f(x)} \text{ whenever } \mathsf{Chk}(f, F, E) = D \neq \perp$$

Definition 3. *Let* $\mathcal{W}$ *denote the uniform distribution of* $m \times 2$ *matrices as described above. A garbling scheme has* **authenticity** *if for every circuit* f, *circuit-input* x, *and PPT algorithm* $\mathcal{A}$, *the following probability:*

$$\Pr[\exists y \neq f(x), \tilde{D} = D|_y : E \leftarrow \mathcal{W}, F \leftarrow \mathsf{Gb}(1^k, f, E, D), \tilde{D} = \mathcal{A}(F, E|_x)]$$

is negligible in k.

The above definition says that when given F and $E|_x$, there is no efficient adversary that can forge valid output wire labels $\tilde{D}$ such that $\tilde{D} \neq D|_{f(x)}$.

We emphasize that the garbling scheme we use here only requires only the authenticity property and not any privacy property. Hence, the protocol may use a more efficient and simpler garbling scheme (*e.g.*, the "privacy-free" constructions of [FNO15, ZRE15]).

4   Zero-Knowledge by Oblivious RAM

4.1   Notation and Helper Routines

ORAM components: Let $\mathcal{I}$ be an ORAM memory access sequence. We define $\mathsf{read}(\mathcal{I}) = \{i \mid (\mathrm{READ}, i) \in \mathcal{I}\}$, $\mathsf{write}(\mathcal{I}) = \{i \mid (\mathrm{WRITE}, i) \in \mathcal{I}\}$, and $\mathsf{access}(\mathcal{I}) = \mathsf{read}(\mathcal{I}) \cup \mathsf{write}(\mathcal{I})$; *i.e.*, the indices of blocks that are read/write/accessed in $\mathcal{I}$. If $S = \{s_1, \ldots, s_n\}$ is a set of memory-block indices, then we define $M[S] = (M[s_1], \ldots, M[s_n])$.

Let Π denote the next-instruction circuit of an ORAM. Given a zero-knowledge statement x and ORAM access sequence $\mathcal{I}$, we let circuit $C_{x,\mathcal{I}}$ denote the following circuit:

$$
\begin{aligned}
&C_{x,\mathcal{I}}(st, w, \widehat{M}[\mathsf{read}(\mathcal{I})]): \\
&\quad (inst, st, block) := \Pi(st, (x, w), \bot) \\
&\quad \text{for } i = 1 \text{ to } |\mathcal{I}| - 1: \\
&\quad\quad \text{if } \mathcal{I}[i] = (read, id) \text{ then:} \\
&\quad\quad\quad (st, inst, \bot) \leftarrow \Pi(st, \bot, \widehat{M}[id]) \\
&\quad\quad \text{if } \mathcal{I}[i] = (write, id) \text{ then:} \\
&\quad\quad\quad (st, inst, block) \leftarrow \Pi(\mathsf{st}, \bot, \bot) \\
&\quad\quad\quad \widehat{M}[id] = block \\
&\quad\quad \mathcal{I}' := \mathcal{I}' \| inst \\
&\quad z := [\mathcal{I} \stackrel{?}{=} \mathcal{I}'] \\
&\quad \text{return } (st, z, \widehat{M}[\mathsf{access}(\mathcal{I})])
\end{aligned}
$$

As described in Sect. 2, $C_{x,\mathcal{I}}$ is the circuit that will be garbled in the protocol. Note that both x and $\mathcal{I}$ are hard-coded into $C_{x,\mathcal{I}}$. Also, the circuit verifies that $\mathcal{I} = \mathcal{I}'$, and this entails checking the correctness of the witness since the final element of $\mathcal{I}$ is (HALT, TRUE).

Garbling Notation: The circuit $C_{x,\mathcal{I}}$ has 3 logical inputs and 3 logical outputs, and we must distinguish among them. When garbling the circuit via $F \leftarrow \mathsf{Gb}(C_{x,\mathcal{I}}, E, D, 1^k)$, we denote by E a description of input wire labels (*i.e.*, two labels per wire) and D a description of output wire labels. We write $E = E_{\mathsf{st}} \| E_{\mathsf{wit}} \| E_{\mathsf{mem}}$, denoting the corresponding input wire labels for state, witness, and memory blocks, respectively. We define $D = D_{\mathsf{st}} \| D_{\mathsf{z}} \| D_{\mathsf{mem}}$ similarly. When referring to a specific memory block i, we use notation $E_{\mathsf{mem},i}$ and $D_{\mathsf{mem},i}$.

We use X to denote the prover's garbled input, and Y to denote the prover's garbled output (*i.e.*, one label per wire). As above, we define X_{st}, X_{wit}, X_{mem}, Y_{st}, Y_{z}, Y_{mem}. Finally, we have the prover maintain an array R_{mem} at all times, containing the current wire labels for all of the ORAM memory $\widehat{M}$.

For an overview of the notation used in the protocol, see Fig. 6.

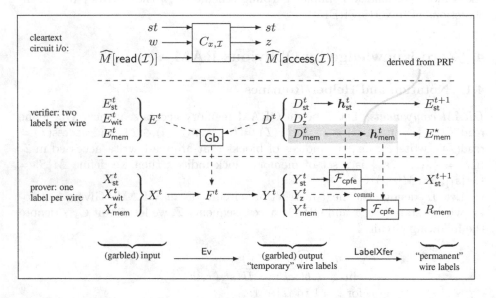

Fig. 6. Summary of variables and notation used in the protocol.

Temporary and Permanent Wire Labels. Recall from Sect. 2 that the output wire labels of a circuit are "temporary" in the sense that their authenticity is lost when the garbled circuit is opened. We use PFE to transfer the authenticity property of these temporary wire labels to a different set of "permanent" wire labels.

We transfer authenticity with the LabelXfer subprotocol, where Y is a list of "temporary" wire labels (*i.e.*, one label per wire), and $\boldsymbol{h}$ is a list of elements from a strongly universal hash family $\mathcal{H}$.

prot LabelXfer$(Y, \boldsymbol{h})$:
 for $i = 1$ to $|Y|$ (in parallel):
  V sends $Y[i]$ and P sends $\boldsymbol{h}[i]$ to an instance of $\mathcal{F}_{\mathsf{cpfe}}$
  P receives output $Z[i] := \boldsymbol{h}[i](Y[i])$
 P outputs Z

Note that all instances of $\mathcal{F}_{\mathsf{cpfe}}$ are run in parallel and hence the protocol remains constant-round given that $\mathcal{F}_{\mathsf{cpfe}}$ is itself constant-round.

Selecting Wire Labels. Now let's consider how the verifier generates wire labels for the circuit. Recall from Sect. 2 that the verifier uses a PRF to generate wire labels corresponding to the ORAM memory, in order to reduce storage.

Since permanent wire labels are derived by applying strongly universal functions to temporary wire labels, the verifier must also select strongly universal functions using the PRF to be able to reconstruct the choice of functions later.

Let s be the seed to a PRF. The verifier derives the *temporary* wire labels for a set S of memory block indices, last updated at time t, via the subroutine TempMemLabels. The verifier derives the choice of strongly universal functions via the subroutine GenH.

Finally, the verifier derives the *current, permanent* wire labels for a set S of memory block indices via the subprotocol PermMemLabels. Since each block may have been last accessed a different time, the authenticated array $\mathcal{F}_{\mathsf{Aut}}$ is referenced. For each block, the most recent temporary wire labels and strongly universal functions are reconstructed to derive the permanent wire labels.

func TempMemLabels(S, t):
$\quad D := \emptyset$
$\quad$ for $i \in S$:
$\quad\quad$ for $j \in \{1, \ldots, l\}, b \in \{0, 1\}$:
$\quad\quad\quad D_i[j, b] = \mathsf{PRF}(s, 0\|i\|j\|t\|b)$
$\quad\quad D := D\|D_i$
$\quad$ return D

func GenH(S, t):
$\quad \boldsymbol{h} = \emptyset$
$\quad$ for $i \in S$:
$\quad\quad$ for $j \in \{1, \ldots, l\}$:
$\quad\quad\quad \boldsymbol{h}_i[j] = \mathsf{PRF}(s, 1\|i\|j\|t)$
$\quad\quad \boldsymbol{h} := \boldsymbol{h}\|\boldsymbol{h}_i$
$\quad$ return $\boldsymbol{h}$

prot PermMemLabels(S):
$\quad E := \emptyset$
$\quad$ for all i in S (in parallel):
$\quad\quad$ send ($access, i$) to $\mathcal{F}_{\mathsf{Aut}}$
$\quad\quad$ receive $t_i := T[i]$
$\quad\quad D_i := \mathsf{TempMemLabels}(\{i\}, t_i)$
$\quad\quad \boldsymbol{h}_i := \mathsf{GenH}(\{i\}, t_i)$
$\quad\quad E_i := \boldsymbol{h}_i(D_i)$
$\quad\quad E := E\|E_i$
$\quad$ return E

When $\boldsymbol{h}$ is an array of functions and D is a matrix of wire labels, the notation $\boldsymbol{h}(D)$ refers to the matrix E whose entries are $E[j, b] = \boldsymbol{h}[j](D[j, b])$.

4.2   Detailed Protocol

Now we present the full protocol π. We refer to the prover as P and the verifier as V. The setup phase uses the initialization functionality $\mathcal{F}_{\mathsf{init}}$ defined in Fig. 7.

Setup: On input M for prover P, let N denote the number of blocks in the ORAM encoding of M. Then both parties do the following:

- Initialize: On command (INIT, M) from P and (INIT, D_{st}, D_{mem}), where M is logical ORAM memory, and D_{st} & D_{mem} are wire label descriptions, run $(st, \widehat{M}) \leftarrow \mathsf{Initialize}(1^k, M)$. Give output $(st, \widehat{M}, D_{st}|_{st}, D_{mem}|_{\widehat{M}})$ to P.
- Open: On command OPEN from V, give output (D_{st}, D_{mem}) to P.

Fig. 7. Ideal functionality $\mathcal{F}_{init}$ for initializing an ORAM program along with wire labels.

1. V picks random wire label descriptions D_{st}^0 and computes $D_{mem}^0 = \mathsf{TempMemLabels}([N], 0)$. V also chooses a random PRF seed $s \leftarrow \{0,1\}^k$.
2. P sends $(init, M)$ to $\mathcal{F}_{init}$; V sends $(init, D_{st}^0, D_{mem}^0)$ to $\mathcal{F}_{init}$. P receives output $(st, \widehat{M}, Y_{st}^0 = D_{st}^0|_{st}, Y_{mem}^0 = D_{mem}^0|_{\widehat{M}})$.
3. **[Transfer wire-label authenticity]:**[3]
   (a) V picks random vector h_{st}^0 of strongly universal functions and sets $E_{st}^1 = h_{st}^0(D_{st}^0)$. The parties perform subprotocol $\mathsf{LabelXfer}(Y_{st}^0, h_{st}^0)$, with P obtaining output $h_{st}^0(Y_{st}^0)$ which he stores as X_{st}^1.
   (b) V picks vector $h_{mem}^0 = \mathsf{GenH}([N], 0)$ and the parties perform subprotocol $\mathsf{LabelXfer}(Y_{mem}^0 h_{mem}^0)$. P receives output $h_{mem}^0(Y_{mem}^0)$ which he stores as R_{mem}.
   (c) V sends $open$ to $\mathcal{F}_{init}$, and P receives output (D_{st}^0, D_{mem}^0).
4. P sends $(init, N)$ to $\mathcal{F}_{Aut}$ to initialize authenticated array T (with $T[i] = 0$ for all i).

Proofs: On input (x, w) for the prover, let this be the tth such proof. The parties do the following:

5. **[ORAM Evaluation]:** P runs $\mathcal{I} \leftarrow \mathsf{RAMEval}(\Pi, \widehat{M}, x, w, st)$, then sends $(x, \mathcal{I})$ to V. V aborts if $\mathcal{I}$ is not an accepting access sequence. Note that $\mathsf{RAMEval}$ modifies $\widehat{M}$ for the prover.
6. **[Garbling the circuit]:** V generates a garbled circuit as follows:
   (a) V chooses input wire labels to the circuit as follows: E_{wit}^t are chosen randomly. E_{mem}^t are chosen as $E_{mem}^t \leftarrow \mathsf{PermMemLabels}(\mathsf{read}(\mathcal{I}))$. Recall that E_{st}^t has been set previously.
   (b) V chooses output wire labels D_z^t and D_{st}^t randomly, and chooses $D_{mem}^t = \mathsf{TempMemLabels}(\mathsf{access}(\mathcal{I}), t)$.
   (c) V sets $E^t = E_{st}^t \| E_{wit}^t \| E_{mem}^t$, sets $D^t = D_{st}^t \| D_z^t \| D_{mem}^t$, then invokes garbling algorithm $F^t \leftarrow \mathsf{Gb}(1^k, C_{x,\mathcal{I}}, E^t, D^t)$.
7. **[Evaluating garbled circuit]:**
   (a) The parties invoke $\mathcal{F}_{otc}$ with P giving input w and V giving input E_{wit}^t. P receives $X_{wit}^t = E_{wit}^t|_w$. Additionally, P finds X_{st}^t in its memory and sets $X_{mem}^t = R_{mem}[\mathsf{read}(\mathcal{I})]$.
   (b) V sends F^t to P, and P evaluates the garbled circuit $Y^t \leftarrow \mathsf{Ev}(F^t, X^t)$.

---

[3] This step could be easily incorporated into $\mathcal{F}_{init}$, but is written separately so that the remainder of the protocol has no edge-cases involving $t = 0$.

    (c) P commits to Y_z^t (a single wire label) under $\mathcal{F}_{\text{com}}$.

8. **[Transfer wire-label authenticity]:**

    (a) V picks random vector h_{st}^t of strongly universal functions and sets $E_{\text{st}}^{t+1} = h_{\text{st}}^t(D_{\text{st}}^t)$. The parties perform subprotocol $\textsf{LabelXfer}(Y_{\text{st}}^t, h_{\text{st}}^t)$, with P obtaining output $h_{\text{st}}^t(Y_{\text{st}}^t)$ which he stores as X_{st}^{t+1}.

    (b) V picks vector $h_{\text{mem}}^t = \textsf{GenH}(\text{access}(\mathcal{I}), t)$ and the parties perform subprotocol $\textsf{LabelXfer}(Y_{\text{mem}}^t, h_{\text{mem}}^t)$. P receives output $h_{\text{mem}}^t(Y_{\text{mem}}^t)$ which he stores as $R_{\text{mem}}[\text{access}(\mathcal{I})]$.

9. **[Check garbled circuit]:**

    (a) V sends *open* to the $\mathcal{F}_{\text{otc}}$-instance from time t, and P receives output E_{wit}^t.

    (b) V sends *open* to the PFE-instances used for the state wire labels in time $t - 1$. The prover thus obtains h_{st}^{t-1} and sets $E_{\text{st}}^t = h_{\text{st}}^{t-1}(D_{\text{st}}^{t-1})$.

    (c) For each $i \in \text{read}(\mathcal{I})$, verifier sends *open* to the PFE-instances used for memory block i in time $T[i]$. The prover thus obtains $h_{\text{mem},i}^{T[i]}$ and sets $E_{\text{mem},i}^t = h_{\text{mem},i}^{T[i]}(D_{\text{mem},i}^{T[i]})$.

    (d) The verifier sets $E^t = E_{\text{st}}^t \| E_{\text{wit}}^t \| E_{\text{mem}}^t$ and runs $D^t = \textsf{Chk}(C_{x,\mathcal{I}}, F^t, E^t)$. If the result is $\bot$, then V aborts. Otherwise, V opens his commitment to Y_z^t.

10. **[Check prover's output]:** V checks whether $Y_z^t = D_z^t|_{\text{TRUE}}$. If not, then V aborts the protocol. Otherwise, V outputs $(accept, t, x)$.

11. **[Update T]:** For all $i \in \text{access}(\mathcal{I})$ (in parallel), V sends $(update, i, t)$ to $\mathcal{F}_{\text{Aut}}$.

Other Discussion. Our protocol is written in a hybrid model with access to various setup functionalities. In particular, $\mathcal{F}_{\text{cpfe}}$ is a *reactive* functionality, and our protocol involves many $(O(|\widehat{M}|))$ instances of $\mathcal{F}_{\text{cpfe}}$ that remain "active" between ZK proofs. We have shown how the verifier's *inputs* to the $\mathcal{F}_{\text{cpfe}}$ instances can be derived from a PRF, eliminating the need to explicitly store them. However, when these $\mathcal{F}_{\text{cpfe}}$ instances are realized by concrete protocols, both parties are required to keep internal state between the PFE phase and opening phase. Hence, the verifier's *random coins* for the $\mathcal{F}_{\text{cpfe}}$-protocols should also be derived from a PRF. In that way, the verifier's entire view can be reconstructed as needed when it is time to *open* each $\mathcal{F}_{\text{cpfe}}$ instance.

4.3 Security Proof

Theorem 1. *The protocol π presented in Sect. 4.2 is a secure realization of the $\mathcal{F}_{\text{ZK}}^R$ functionality.*

Proof. We describe two simulators, depending on which party is corrupted.

Prover is corrupt: The primary role of the simulator in this case is to extract the witness from P. We construct the simulator in a sequence of hybrid interactions:

$\mathcal{H}_0$: Simulator plays the role of an honest verifier V (who has no input) and all ideal functionalities. In particular, the simulator obtains all of P's inputs to the ideal functionalities. This interaction is identical to the real interaction with π.

$\mathcal{H}_1$: Same as $\mathcal{H}_0$ except that instead of using a PRF, the simulated veri-fier chooses output wire labels D^t_{mem} and h^t_{mem} functions uniformly (in TempMemLabels and GenH). We have $\mathcal{H}_1 \approx \mathcal{H}_0$ by the security of the PRF.

$\mathcal{H}_2$: Same as $\mathcal{H}_1$ except that the simulator aborts in certain cases as follows. The simulator has initially generated $\widehat{M}$ and st (while simulating $\mathcal{F}_{\text{init}}$) and obtains w as P's input to $\mathcal{F}_{\text{otc}}$ in each step (7a). Hence, each time in step 7, the simulator executes $C_{x,\mathcal{I}}(st, w, \widehat{M}[\text{read}(\mathcal{I})]) \rightarrow (st, z, \widehat{M}[\text{access}(\mathcal{I})])$, updating its internal st and $\widehat{M}$.

In the LabelXfer subprotocols in steps (3) and (8), P is meant to provide his garbled output Y^t_{mem} and Y^t_{st} to the $\mathcal{F}_{\text{cpfe}}$ functionalities. Similarly, in step (7c), the prover is expected to commit to $Y^t_z|_{\text{TRUE}}$. In $\mathcal{H}_2$, the simulator artificially aborts if P provides a *valid* encoding $D^t|_y$ for y not equal to the simulated output of $C_{x,\mathcal{I}}$ at time t.

Now we claim that the simulator artificially aborts with only negligible probability (so $\mathcal{H}_1 \approx \mathcal{H}_2$) and that the prover's view of E^t during step (8) in time t can be simulated given only $E^t_{\text{mem}}|_{\widehat{M}[\text{read}(\mathcal{I})]}$ and $E^t_{\text{st}}|_{st}$. This follows essentially from the authenticity property of the garbling scheme and the strong-universal hashing property of $\mathcal{H}$.

Consider the LabelXfer subprotocol in step (3) (i.e., time $t = 0$). At this time, all wire labels in D^0 besides $D^0_{\text{mem}}|_{\widehat{M}}$ and $D^0_{\text{mem}}|_{st}$ are independent of the adversary's view by definition of the $\mathcal{F}_{\text{init}}$ functionality. Hence, the simulator artificially aborts with negligible probability during these steps. Conditioned on not aborting, the action of the strongly universal hash func-tions on the "wrong" wire labels of D^0 — and hence the value of the "wrong" input wire labels in E^1 — is distributed independently of P's view. Thus P's view in step (7) can be simulated given only the claimed subset of E^1. Induc-tively, the prover's view of E^t at the time of the LabelXfer steps depends only on the "expected" input wire labels. Hence, the simulator artificially aborts with negligible probability, due to the authenticity property of the garbling scheme. As above, conditioned on not aborting, the strong universal hashing property ensures that the prover's view of E^{t+1} depends only on the claimed subset of E^{t+1}.

$\mathcal{H}_3$: Same as $\mathcal{H}_2$ except that in step (2) the simulator sends P's input M to $\mathcal{F}^{\mathcal{R}}_{\text{ZK}}$. In step (10), if the simulated verifier does not abort, then the simulator sends (x, w_{real}) to $\mathcal{F}^{\mathcal{R}}_{\text{ZK}}$ (where w was extracted from the prover in step (7a)). We claim that the output of the ideal verifier always matches that of the simulated verifier. The simulated verifier accepts the proof if P has commit-ted to $D^t_z|_{\text{TRUE}}$. Provided that the simulator has not artificially aborted, then it must be that the simulated $C_{x,\mathcal{I}}$ has output $z = \text{TRUE}$. By the correctness of the RAM program, it must be that w_{real} is a valid witness for x.

Hence, the simulator implicit in $\mathcal{H}_3$ is our final simulator.

Verifier is Corrupt: In this case, the primary role of the simulator is to simulate its view without knowledge of the witness w. We note that the only information that needs to be simulated in each proof is the memory access sequence $\mathcal{I}$ and the opened commitment to output wire label Y_z^t. Again we proceed in a sequence of hybrid interactions.

$\mathcal{H}_0$: Simulator plays the role of an honest prover P (including M and witnesses w as input) and all ideal functionalities. Hence, the simulator obtains all of V's inputs to the ideal functionalities. This interaction is identical to the real interaction with π.

$\mathcal{H}_1$: Same as $\mathcal{H}_0$ except for the following changes. An honest prover computes D^t in step (9d) when the verifier decommits to certain inputs to ideal functionalities. Here we have the simulator perform the same computations, but as soon as possible given the ability to see the verifier's inputs to the functionalities. Hence, in step (7c), the simulator will know the entire contents of D^t. Instead of evaluating the garbled circuit to obtain garbled output Y_z^t, we have the simulator simply commit to $D_z^t|_{\text{TRUE}}$.

This commitment is only opened when the garbled circuit F^t is shown to be correct. Hence, $\mathcal{H}_0 \equiv \mathcal{H}_1$.

$\mathcal{H}_2$: Same as $\mathcal{H}_1$ except for the following changes. Note that in $\mathcal{H}_1$ the simulator uses secret values M and w only to generate the memory access sequence $\mathcal{I}$. All of the simulated prover's other inputs to ideal functionalities can be set to dummy values, as V gets no outputs. So in $\mathcal{H}_2$ we have the simulated prover generate $\mathcal{I}$ in step (5) using the ORAM simulator instead of actually executing the RAM program itself. We have $\mathcal{H}_1 \approx \mathcal{H}_2$ by the security of the ORAM.

The simulator implicit in $\mathcal{H}_2$ defines our final simulator, since it no longer requires the secret values M and w to operate.

This completes the security proof of our protocol.

Acknowledgements. We thank anonymous CRYPTO reviewers for helpful feedback. Zhangxiang Hu and Mike Rosulek were supported by NSF award CCF-1149647.

References

[AHMR15] Afshar, A., Hu, Z., Mohassel, P., Rosulek, M.: How to efficiently evaluate RAM programs with malicious security. In: Oswald, E., Fischlin, M. (eds.) EUROCRYPT 2015. LNCS, vol. 9056, pp. 702–729. Springer, Heidelberg (2015)

[BHR12] Bellare, M., Hoang, V.T., Rogaway, P.: Foundations of garbled circuits. In: Yu, T., Danezis, G., Gligor, V.D. (eds.) ACM CCS 12: 19th Conference on Computer and Communications Security, pp. 784–796. ACM Press, October 2012

[Blo70] Bloom, B.H.: Space/time trade-offs in hash coding with allowable errors. Commun. ACM **13**(7), 422–426 (1970)

[CDS94] Cramer, R., Damgård, I.B., Schoenmakers, B.: Proof of partial knowledge and simplified design of witness hiding protocols. In: Desmedt, Y.G. (ed.) CRYPTO 1994. LNCS, vol. 839, pp. 174–187. Springer, Heidelberg (1994)

[CM99]   Camenisch, J.L., Michels, M.: Proving in zero-knowledge that a number is the product of two safe primes. In: Stern, J. (ed.) EUROCRYPT 1999. LNCS, vol. 1592, p. 107. Springer, Heidelberg (1999)

[CP13]   Chung, K.-M., Pass, R.: A simple ORAM. Cryptology ePrint Archive, Report 2013/243 (2013). http://eprint.iacr.org/2013/243

[FNO15] Frederiksen, T.K., Nielsen, J.B., Orlandi, C.: Privacy-free garbled circuits with applications to efficient zero-knowledge. In: Oswald, E., Fischlin, M. (eds.) EUROCRYPT 2015. LNCS, vol. 9057, pp. 191–219. Springer, Heidelberg (2015)

[GHL+14] Gentry, C., Halevi, S., Lu, S., Ostrovsky, R., Raykova, M., Wichs, D.: Garbled RAM revisited. In: Nguyen, P.Q., Oswald, E. (eds.) EUROCRYPT 2014. LNCS, vol. 8441, pp. 405–422. Springer, Heidelberg (2014)

[GKK+12] Gordon, S.D., Katz, J., Kolesnikov, V., Krell, F., Malkin, T., Raykova, M., Vahlis, Y.: Secure two-party computation in sublinear (amortized) time. In: Yu, T., Danezis, G., Gligor, V.D. (eds.) ACM CCS 12: 19th Conference on Computer and Communications Security, pp. 513–524. ACM Press, October 2012

[GMR89] Goldwasser, S., Micali, S., Rackoff, C.: The knowledge complexity of interactive proof systems. SIAM J. Comput. 18(1), 186–208 (1989)

[GO96]   Goldreich, O., Ostrovsky, R.: Software protection and simulation on oblivious RAMs. J. ACM 43(3), 431–473 (1996)

[GS08]   Groth, J., Sahai, A.: Efficient non-interactive proof systems for bilinear groups. In: Smart, N.P. (ed.) EUROCRYPT 2008. LNCS, vol. 4965, pp. 415–432. Springer, Heidelberg (2008)

[IPS09] Ishai, Y., Prabhakaran, M., Sahai, A.: Secure arithmetic computation with no honest majority. In: Reingold, O. (ed.) TCC 2009. LNCS, vol. 5444, pp. 294–314. Springer, Heidelberg (2009)

[JKO13] Jawurek, M., Kerschbaum, F., Orlandi, C.: Zero-knowledge using garbled circuits: how to prove non-algebraic statements efficiently. In: Sadeghi, A.-R., Gligor, V.D., Yung, M. (eds.) ACM CCS 13: 20th Conference on Computer and Communications Security, pp. 955–966. ACM Press, November 2013

[KK12]   Kolesnikov, V., Kumaresan, R.: Improved secure two-party computation via information-theoretic garbled circuits. In: Visconti, I., De Prisco, R. (eds.) SCN 2012. LNCS, vol. 7485, pp. 205–221. Springer, Heidelberg (2012)

[KS06]   Kiraz, M., Schoenmakers, B.: A protocol issue for the malicious case of Yao's garbled circuit construction. In: 27th Symposium on Information Theory in the Benelux, pp. 283–290 (2006)

[LO13]   Lu, S., Ostrovsky, R.: How to garble RAM programs? In: Johansson, T., Nguyen, P.Q. (eds.) EUROCRYPT 2013. LNCS, vol. 7881, pp. 719–734. Springer, Heidelberg (2013)

[MGFB14] Mood, B., Gupta, D., Feigenbaum, J., Butler, K.: Reuse it or lose it: more efficient secure computation through reuse of encrypted values. In: ACM CCS (2014)

[MRK03] Micali, S., Rabin, M.O., Kilian, J.: Zero-knowledge sets. In: 44th Annual Symposium on Foundations of Computer Science, pp. 80–91. IEEE Computer Society Press, October 2003

[Sch90] Schnorr, C.-P.: Efficient identification and signatures for smart cards. In: Brassard, G. (ed.) CRYPTO 1989. LNCS, vol. 435, pp. 239–252. Springer, Heidelberg (1990)

[SvDS+13] Stefanov, E., van Dijk, M., Shi, E., Fletcher, C.W., Ren, L., Yu, X., Devadas, S.: Path ORAM: an extremely simple oblivious RAM protocol. In: Sadeghi, A.-R., Gligor, V.D., Yung, M. (eds.) ACM CCS 13: 20th Conference on Computer and Communications Security, pp. 299–310. ACM Press, November 2013

[WW06] Wolf, S., Wullschleger, J.: Oblivious transfer is symmetric. In: Vaudenay, S. (ed.) EUROCRYPT 2006. LNCS, vol. 4004, pp. 222–232. Springer, Heidelberg (2006)

[ZRE15] Zahur, S., Rosulek, M., Evans, D.: Two halves make a whole. In: Oswald, E., Fischlin, M. (eds.) EUROCRYPT 2015. LNCS, vol. 9057, pp. 220–250. Springer, Heidelberg (2015)

Theory

Parallel Hashing via List Recoverability

Iftach Haitner[1]([⊠]), Yuval Ishai[2], Eran Omri[3],
and Ronen Shaltiel[4]

[1] Tel Aviv University, Tel Aviv, Israel
iftachh@cs.tau.ac.il
[2] Technion, Haifa, Israel
yuvali@cs.technion.ac.il
[3] Ariel University, Ariel, Israel
omrier@ariel.ac.il
[4] Haifa University, Haifa, Israel
ronen@cs.haifa.ac.il

Abstract. Motivated by the goal of constructing efficient hash functions, we investigate the possibility of hashing a long message by only making parallel, non-adaptive calls to a hash function on short messages. Our main result is a simple construction of a collision-resistant hash function $h : \{0,1\}^n \to \{0,1\}^k$ that makes a polynomial number of parallel calls to a *random* function $f : \{0,1\}^k \to \{0,1\}^k$, for any polynomial $n = n(k)$. This should be compared with the traditional use of a Merkle hash tree, that requires at least $\log(n/k)$ rounds of calls to f, and with a more complex construction of Maurer and Tessaro [26] (Crypto 2007) that requires two rounds of calls to f. We also show that our hash function h satisfies a relaxed form of the notion of indifferentiability of Maurer et al. [27] (TCC 2004) that suffices for implementing the Fiat-Shamir paradigm. As a corollary, we get sublinear-communication non-interactive arguments for NP that only make two rounds of calls to a small random oracle.

An attractive feature of our construction is that h can be implemented by Boolean circuits that only contain parity gates in addition to the parallel calls to f. Thus, we get the first domain-extension scheme which is *degree-preserving* in the sense that the algebraic degree of h over the binary field is equal to that of f.

Our construction makes use of *list-recoverable codes*, a generalization of list-decodable codes that is closely related to the notion of randomness condensers. We show that list-recoverable codes are necessary for any construction of this type.

The first author was supported by ISF grant 1076/11, I-CORE grant 4/11, BSF grant 2010196, and Check Point Institute for Information Security. The second author was supported by ERC starting grant 259426, ISF grant 1709/14, and BSF grant 2012378. The third author was supported by ERC starting grants 259426 and 279559, and by ISF grant 544/13. The fourth author was supported by ERC starting grant 279559, BSF grant 2010120, and ISF grant 864/11.

R. Gennaro and M. Robshaw (Eds.): CRYPTO 2015, Part II, LNCS 9216, pp. 173–190, 2015.
DOI: 10.1007/978-3-662-48000-7_9

1   Introduction

In this work we consider the problem of extending the domain of cryptographic hash functions. We start by discussing the case of collision-resistant hash functions, and later address extensions to other types of hash functions.

A family $\left\{ g : \{0,1\}^v \to \{0,1\}^k \right\}$ of efficiently computable, length-decreasing functions is called *collision resistant* if given the description of a random g from the family, it is computationally infeasible to find a pair of distinct inputs s, s' such that $g(s) = g(s')$.

Collision-resistant hashing is a fundamental primitive in cryptography that has been the subject of a large body of work. Its applications span many areas, ranging from the commonly used "hash and sign" paradigm for practical digital signatures [10, 29] to cryptographic protocols such as sublinear-communication commitments [9, 21], succinct and efficiently verifiable arguments for NP [23, 30], and protocols that bypass black-box simulation barriers [1].

The existence of collision-resistant hash functions can be based on a variety of standard number theoretic or algebraic cryptographic assumptions, including the conjectured intractability of factoring, discrete logarithms, and lattice problems [8, 13, 25, 33]. Yet, the task of heuristically constructing highly efficient hash functions that can also be conjectured to have near-optimal security is quite challenging. In particular, this task is arguably more challenging than a similar task for other "symmetric" cryptographic primitives such as one-way functions [41], pseudorandom generators [5, 41], and universal one-way hash functions [31]. This intuition is supported by theoretical results that rule out the possibility of obtaining collision-resistant hash functions from any of these other symmetric primitives via a black-box construction [19, 36]. Practical collision attacks on commonly used hash functions such as MD5 [40] may also be viewed as an indication for the subtle nature of hash function design. Despite the above, there are many practical constructions of cryptographic hash functions that are conjectured to satisfy collision resistance as well as other useful properties. See [4] for a description of SHA-3, the winner of the recent NIST hash function competition, as well as an overview of other work on practical hash function design.

A common technique for building a hash function $g : \{0,1\}^v \to \{0,1\}^k$ that compresses a long input into a short output is by combining multiple invocations of a smaller hash function $f : \{0,1\}^{k_{in}} \to \{0,1\}^{k_{out}}$ in a way that supports a black-box reduction of the collision-resistance of g to that of f. This technique, known as *domain-extension*, is motivated by the possibility of carefully designing and analyzing an optimized implementation of f on some fixed input length, and then scaling up its efficiency and security advantages to apply to arbitrarily long inputs. It is sometimes the case that the collision-resistance of g relies on a stronger assumption on f than just collision-resistance. In fact, several domain-extension schemes assume f to be a completely random function (see e.g., [26, 34, 37, 38] and references therein, as well as [20] for discussion of the meaningfulness of such results). In the following we will use the term "domain-extension" in this broader sense. A simple domain-extension technique due to

Merkle [28] extends the domain of a hash function $f\colon \{0,1\}^{2k} \to \{0,1\}^k$ for short inputs into a hash function $g\colon \{0,1\}^{nk} \to \{0,1\}^k$ for long inputs by applying a tree of invocations of f whose leaves are k-bit input blocks and whose root is the output.

In this work we consider the question of minimizing the *parallel complexity* of domain-extension schemes. A natural measure of this complexity is the number of rounds of parallel calls to f. Ideally, one could hope to compute g by only making a single round of calls to f, where the input for each call is computed directly from the input for g, and the outputs of the calls are used to compute the output of g. The hash tree construction falls short of this goal, requiring at least $\lceil \log_2 n \rceil$ rounds. A more complex construction of [26] comes close to this goal, requiring only two rounds of calls to f.[1]

Our main result is a simple construction of a fully parallel (single-round) domain-extension scheme that realizes a collision-resistant $g : \{0,1\}^v \to \{0,1\}^k$ by making a polynomial number of parallel calls to a *random* function $f :$ $\{0,1\}^k \to \{0,1\}^k$, for any polynomial $v = v(k)$. The construction achieves a near-optimal level of security, requiring an attacker to make roughly $2^{k/2}$ calls to f in order to find a collision in g with high probability. However, this may come at the cost of a higher number of calls to f compared to traditional domain-extension schemes. See Sect. 7 for a more detailed discussion of the achievable parameters.

Our domain-extension scheme has the attractive feature that g can be implemented by Boolean circuits consisting only of parity gates in addition to the parallel calls to f. Thus, we get the first *degree-preserving* domain-extension scheme, in the sense that the algebraic degree of g over the binary field is equal to that of f. In contrast, in constructions that make two rounds of calls to f, the degree of g is at least quadratic in that of f. Low-degree hash functions are motivated by applications in the domain of secure computation, in which the cost of evaluating a function may depend on its algebraic degree. See [20] for further discussion.

Our construction makes use of *list-recoverable codes*, a generalization of list-decodable codes that is closely related to the notion of randomness condensers. We show that list-recoverable codes are necessary for any construction of this type. In the following we give a more detailed account of our results and the underlying techniques.

2   Parallel Domain-Extension

Recall that a domain-extension scheme for hash functions takes as input a fixed length hash function $f\colon \{0,1\}^{k_{in}} \to \{0,1\}^{k_{out}}$ and outputs a new hash function for much larger inputs, namely a function $g\colon \{0,1\}^v \to \{0,1\}^{k_{out}}$ for a given $v > k_{in}$. (For the sake of simplicity, we assume that the output length of g

---

[1] Their construction actually realizes the stronger goal of constructing a function g that is indistinguishable from a random function. See Sect. 8 for further discussion.

is k_{out}, rather than an additional parameter.) We consider the standard model in which the function f is provided to the construction after being chosen by some randomized process, and the function g uses f as a black-box (i.e., it is oblivious to the concrete implementation of f). The focus of this work is on *parallel* domain-extension: upon receiving an input $s \in \{0,1\}^v$, the function g first prepares n queries to the hash function f (which we will denote by $C(s) = (x_1, \ldots, x_n)$), and then the final output of g is obtained by computing some function h on the input s and the answers $f(x_1), \ldots, f(x_n)$.

Definition 1 (Parallel Domain-Extension Scheme). *Let k_{in}, k_{out}, v, n be integers, let $C \colon \{0,1\}^v \rightarrow \left(\{0,1\}^{k_{in}}\right)^n$ and let h be defined over $\{0,1\}^v \times \left(\{0,1\}^{k_{out}}\right)^n$. The* parallel domain-extension scheme *(C, h) is the oracle-aided function $g_{(C,h)}$ defined as follows. For $f \colon \{0,1\}^{k_{in}} \rightarrow \{0,1\}^{k_{out}}$, let $g^f_{(C,h)} \colon \{0,1\}^v \rightarrow \{0,1\}^{k_{out}}$ be defined by*

$$g^f_{(C,h)}(s) = h(s, f(C(s)_1), \ldots, f(C(s)_n)).$$

If the value of (C, h) is clear from the context, we refer to $g^f_{(C,h)}$ as g^f or g.

Such a construction should maintain the security of the underlying hash function (i.e., f). In particular, whenever f is chosen from a collision-resistant hash function family, the resulting function g should be collision-resistant as well. As a step towards this goal, it is common to consider the following intermediate goal: assume that f is a random function (which in particular is collision-resistant), and prove that the resulting function g is collision-resistant. In the following let $\mathcal{F}_{k_{in}, k_{out}}$ be the family of all functions mapping k_{in}-bit strings to k_{out}-bit strings.

Definition 2 (Collision-Resistance in the Random Oracle Model). *Let g be an oracle-aided function (i.e., deterministic algorithm), with an oracle mapping k_{in}-bit strings to k_{out}-bit strings. The function g is (ℓ, ε)-collision-resistant in the random oracle model, if for any ℓ-query adversary $\mathcal{A}$ (i.e., $\mathcal{A}$ makes ℓ oracle calls), it holds that $\Pr_{f \leftarrow \mathcal{F}_{k_{in}, k_{out}}} \left[(s_1, s_2) \leftarrow \mathcal{A}^f \colon s_1 \neq s_2 \wedge g^f(s_1) = g^f(s_2)\right] \leq \varepsilon$.*

An important goal is to come up with an *efficient* collision-resistant parallel domain-extension scheme, according to Definition 2, where efficiency can be measured in terms of circuit size, depth, or algebraic degree. This motivates schemes in which n (the number of queries) is as small as possible, and the functions C and h are efficiently computable.

Our main result is that if we take C to be a *list-recoverable code* (defined below), then for making the resulting scheme collision-resistant in the random-oracle model, it suffices to take h to be simply the XOR function applied to the n outputs of f.

It turns out that when used with (short) random oracle, our parallel domain-extension scheme also maintains other useful properties of the random oracle. While we cannot show our parallel domain-extension scheme to be indifferentiable from a random function in the sense of Maurer et al. [27], we show that it enjoys a weaker form of indifferentiability, which neither implies nor is implied by collision-resistance. This property will turn out to be sufficient for converting interactive proof system into non-interactive ones using the Fiat-Shamir paradigm [12].

Definition 3 (Weak Indifferentiability). *Let* $g \colon \{0,1\}^v \mapsto \{0,1\}^t$ *be an oracle-aided function, taking an oracle mapping* k_{in}-*bit strings to* k_{out}-*bit strings. The function* g *is* (ℓ, R, r)-*weak-indifferentiable from a random function, if for any two-oracle algorithm* D *making* ℓ *queries to the left-hand side oracle, and a single query to the right-hand side oracle, there exists a single-query algorithm* Sim *such that* $\Pr_{f_{Ideal} \leftarrow \mathcal{F}_{k_{in}, k_{out}}} \left[\mathsf{D}^{f_{Ideal}, g^{f_{Ideal}}} \in E \right] \leq R \cdot \Pr_{g_{Ideal} \leftarrow \mathcal{F}_{v,t}} \left[\mathsf{D}^{\mathrm{Sim}^{g_{Ideal}}, g_{Ideal}} \in E \right]$ *for any event* E. *The simulator* Sim *is of size* r, *i.e., it is implemented by a next message circuit of size* r *(i.e., a circuit that gets as input the past queries and the current one, and returns the answer to the current query).*

The main difference between Definition 3 and the standard notion of indifferentiability from [27], is that the above definition only requires domination between the real and emulated pair of systems, whereas [27] require statistical closeness. We also provide Sim the query parameter ℓ as a parameter, which makes our relaxed definition easier to realize. For simplicity, we have fixed the number of queries to the right-hand side oracle to one (both for the distinguisher and the simulator). It turns out that this type of security is achieved by our parallel domain-extension scheme and is sufficient for applying the Fiat-Shamir paradigm as described below. Concretely, we show that by taking C and h to be as above, the resulting function is weakly indifferentiable from a random function, with small (i.e., polynomial) parameters.

We now give some brief intuition for why the above weak indifferentiability property suffices for applying the Fiat-Shamir paradigm to simulate the verifier's challenge in 3-message public-coin interactive proofs. Let P be a malicious prover that convinces the verifier V with noticeable success probability. P may make ℓ queries to the real short-input oracle before sending his message to V. We can consider a distinguisher D that simulates P and then uses one query to the long-input oracle to see whether V accepts. D accepts iff V accepts. Now consider the behavior of D in the two experiments that appear in Definition 3. In the real experiment, the probability that D accepts is the success probability of P. In the ideal experiment, V uses an ideal (full length) oracle and so the success probability of D is bounded by the success probability when applying the Fiat-Shamir paradigm with an ideal hash function.

3   List-Recoverable Codes

Definition 4 (List-Recoverable Code). *Let* $\alpha \in [0,1]$. *A tuple* $x \in \left(\{0,1\}^k\right)^n$ *is*

- α-consistent *with a set* $T \subseteq \{0,1\}^k$, *if* $|\{i \colon x_i \in T\}| \geq \alpha \cdot n$.
- α-consistent *with sets* $T_1, \ldots, T_n \subseteq \{0,1\}^k$, *if* $|\{i \colon x_i \in T_i\}| \geq \alpha \cdot n$.

A function $C\colon \{0,1\}^v \to \left(\{0,1\}^k\right)^n$ *is* (α, ℓ, L)-list recoverable, *if for every set* $T \subseteq \{0,1\}^k$ *of size at most* ℓ, *there are at most* L *strings* $s \in \{0,1\}^v$ *such that* $C(s)$ *is* α-consistent *with* T. *It is* strongly (α, ℓ, L)-list recoverable, *if for every* $T_1, \ldots, T_n \subseteq \{0,1\}^k$ *each of size at most* ℓ, *there are at most* L *strings* $s \in \{0,1\}^v$ *such that* $C(s)$ *is* α-consistent *with* $T_1, \ldots, T_n$.

For $\alpha = 1$, *we omit* α *in the above notation. The strings in the image of* C *are referred to as* codewords, *and* C *has* distance β, *if every two codewords differ on at least* $\beta \cdot n$ *of the indices.*

The function C *has a size* r list-recovering algorithm, *if there exists a circuit of size* r *that given a set* $T \subseteq \{0,1\}^k$ *of size at most* ℓ *returns the full list of (at most* L) *strings that are* α-consistent *with* T.

The notion of strongly list-recoverable codes (explicitly defined in [15]) is a natural extension of the more standard *uniquely decodable codes* (captured by $\ell = L = 1$) and *list-decodable codes* (captured by $\ell = 1$ and $L > 1$). The reader is referred to [14] for a comprehensive treatment of list-decodable codes. In this paper we use the weaker notion of list-recoverable codes (with a single set T instead of a collection $T_1, \ldots, T_n$), as it turns out to be more natural for the applications we consider.[2] List-recoverable codes show up naturally in coding theory when one considers list-decoding of concatenated codes.[3] Conveniently, many list-decoding algorithms (e.g., [16,17,32,39]) solve the more general list-recovering problem, and list-decoding is achieved as a special case. The parameter regime that we consider is less standard in coding theory and is strongly related to *unbalanced expanders* and *randomness condensers*. We elaborate on this connection in [20].

In our construction we require codes that, in addition to having large distance and being list-recoverable, are also *well ordered*.

Definition 5 (Well-Ordered Codes). *A function* $C\colon \{0,1\}^v \to \left(\{0,1\}^k\right)^n$ *is* well ordered, *if for every* $s_1, s_2 \in \{0,1\}^v$ *(not necessarily distinct) and for every* $i \neq j$, $C(s_1)_i \neq C(s_2)_j$.

Constructions of list-recoverable codes in the literature typically have this property. Furthermore, a given function $C\colon \{0,1\}^v \to \left(\{0,1\}^k\right)^n$ can be converted into a function $\bar{C}\colon \{0,1\}^v \to \left(\{0,1\}^{k+\log n}\right)^n$ that is well ordered by defining $\bar{C}(s)_i = (C(s)_i, i)$. This transformation increases the alphabet of the code, but does not compromise the distance or list-recoverability. In our setting $\log n$ is

---

[2] Note that it is immediate that a strongly list recoverable code is also (weakly) list-recoverable, and that a weakly list-recoverable code with $L' = n \cdot L$ is strongly list recoverable. In our setting n is negligible compared to L and so the distinction between the two notions of list-recoverable code makes little difference.

[3] More precisely, if the inner code is list-decodable (rather than uniquely decodable) then to obtain a list-decodable code, the outer code needs to be list-recoverable (and not only list-decodable).

typically negligible compared to k and so the increase in alphabet size is immaterial. Hence, one can assume without loss of generality that a list-recoverable code is well ordered.

4 Parallel Domain-Extension via List-Recoverable Codes

We show that well-ordered, list-recoverable codes with large distance yield parallel domain-extension schemes that are collision-resistant in the random-oracle model, and furthermore are weak-indifferentiable from a random function. Specifically, this holds for any domain-extension scheme of the form $g^f(s) = \bigoplus_{i=1}^n f(C(s)_i)$, where C is such a list-recoverable code. Hereafter, we refer to this scheme as the *XOR parallel domain extension scheme*.

Theorem 1. *Let k_{in}, k_{out}, v be integers, $\alpha > 0$, and let $C\colon \{0,1\}^v \to \left(\{0,1\}^{k_{in}}\right)^n$ be a well-ordered, (α, ℓ, L)-list recoverable code of distance α. Define $h\colon \{0,1\}^v \times \left(\{0,1\}^{k_{out}}\right)^n \to \{0,1\}^{k_{out}}$ by $h(s, a_1, \ldots, a_n) = \bigoplus_{i=1}^n a_i$. Then $g_{(C,h)}$ is $(\ell, L^2/2^{k_{out}})$-collision-resistant in the random-oracle model.*

We remark that the collision-resistance of $g_{(C,h)}$ holds even if we only require that the function f it gets as oracle be L^2-wise independent. Thus, using codes with small L allows us to require less of the oracle.

Theorem 2. *Let k_{in}, k_{out}, v be integers, and let $C\colon \{0,1\}^v \to \left(\{0,1\}^{k_{in}}\right)^n$ be a well-ordered, (ℓ, L)-list recoverable code, with size r list-recovering algorithm, and let h be as in Theorem 1. Then $g_{(C,h)}$ is $(\ell, L, \hat{r})$-weak-indifferentiable from random function (from v bits to k_{out} bits), with $\hat{r} = O(r + \ell \cdot (k_{out} + k_{in}))$.*

Note that the weak-indifferentiablity of the scheme requires much less from the underlying code. In particular, it is not sensitive to the consistency parameter (allowing it to be 1) nor to the distance of the code, and hence does not imply collision-resistance. On the other hand, our application of this notion in the context of computationally sound arguments will require the list-recovering algorithm to be computationally efficient, a feature that is not needed for collision-resistance.

We prove Theorem 1 below. For the proof of Theorem 2, and proofs of the other theorems in this paper, see full version [20].

4.1 Proving Theorem 1

We show that an ℓ-query adversary is unlikely to find a collision in the above construction (i.e., find two elements $s_1 \neq s_2 \in \{0,1\}^v$, with $g^f(s_1) = g^f(s_2)$), when f is chosen at random from $\mathcal{F}$ — the set all functions mapping k_{in}-bit strings to k_{out}-bit strings.

Fix a code C of the type considered in Theorem 1 and an ℓ-query (without loss of generality, deterministic) adversary $\mathcal{A}$, and let $g = g_{(C,\text{xor})}$. The core of the argument is using the list-recoverability of C, and its distance, to bound the number of input pairs that $\mathcal{A}$ is able to try out. We use the following definition.

Definition 6 (Dangerous Pairs). *A pair* $(s_1, s_2) \in (\{0,1\}^v)^2$ *of distinct elements is* dangerous *w.r.t. a (query) set* Q *of elements in* $\{0,1\}^{k_{in}}$, *if* $C(s_1)_i, C(s_2)_i \in Q$ *for all* $1 \leq i \leq n$ *with* $C(s_1)_i \neq C(s_2)_i$.

We bound the number dangerous pairs w.r.t. an ℓ-size query set Q using the bound on the number of codewords that are α-consistent with Q.

Claim 3. *Let* (s_1, s_2) *be a dangerous pair w.r.t. a query set* Q, *then both* $C(s_1)$ *and* $C(s_2)$ *are* α-*consistent with* Q.

Proof. Assume that (s_1, s_2) is a dangerous pair w.r.t. a query set Q. Let $D = \{i: C(s_1)_i \neq C(s_2)_i\}$. Since the distance of C is α, it holds that $|D| \geq \alpha \cdot n$. Since (s_1, s_2) is a dangerous pair, $C(s_1)_i, C(s_2)_i \in Q$ for all $i \in D$, and hence, both $C(s_1)$ and $C(s_2)$ are α-consistent with Q.

Corollary 1. *There are at most* $\binom{L}{2}$ *dangerous pairs w.r.t. an* ℓ-*size query set.*

Proof. Since C is (α, ℓ, L)-list recoverable, there are at most L strings $s \in \{0,1\}^v$ such that $C(s)$ is α-consistent with an ℓ-size query set Q. Hence, by Claim 3, there are at most $\binom{L}{2}$ dangerous pairs w.r.t. Q.

For $f \in \mathcal{F}$, let $Q_{\mathcal{A},f}$ be the ℓ-size query set asked by $\mathcal{A}^f$. Corollary 1 yields that there are at most $\binom{L}{2}$ dangerous pairs w.r.t. $Q_{\mathcal{A},f}$. A straightforward union bound yields that a *non*-adaptive $\mathcal{A}$ (i.e., one that "writes" all its queries in advance) is unlikely to find a collision within the dangerous pairs w.r.t. $Q_{\mathcal{A},f}$. A slightly more involved argument yields the same bound also for adaptive adversaries. Specifically, we give the following bound (proof given below).

Claim 4. $\Pr_{f \leftarrow \mathcal{F}}[(s_1, s_2) \leftarrow \mathcal{A}^f : (s_1, s_2) \text{ is dangerous w.r.t. } Q_{\mathcal{A},f} \wedge g^f(s_1) = g^f(s_2)] \leq \binom{L}{2} \cdot 2^{-k_{out}}$.

On the other hand, it is immediate that $\mathcal{A}$ is unlikely to find a collision of a *non*-dangerous pair.

Claim 5. $\Pr_{f \leftarrow \mathcal{F}}[(s_1, s_2) \leftarrow \mathcal{A}^f : s_1 \neq s_2 \wedge (s_1, s_2) \text{ is non-}dangerous \text{ w.r.t. } Q_{\mathcal{A},f} \wedge g^f(s_1) = g^f(s_2)] = 2^{-k_{out}}$.

Proof. Since (s_1, s_2) is non-dangerous, it follows that $C(s_1)_i \neq C(s_2)_i$ for some $i \in [v]$, and without loss of generality $C(s_1)_i \notin Q_{\mathcal{A},f}$. Consider any fixing of all f queries but $C(s_1)_i$ that is consistent with the actual answers of f on the queries in $Q_{\mathcal{A},f}$. Since C is well ordered, this fixes $C(s_t)_j$ for all $t \in \{1, 2\}$ and $j \notin [n]$. The claim follows, since for each such fixing, it holds that

$$\Pr\left[g^f(s_1) = g^f(s_2)\right] = \Pr\left[f(C(s_1)_i) = \bigoplus_{j \in [n] \setminus \{i\}} f(C(s_1)_j) \oplus \bigoplus_{j \in [n]} f(C(s_2)_j)\right]$$

$$= 2^{-k_{out}}.$$

It follows that $\mathcal{A}^f$ finds a collision with probability at most $\left(\binom{L}{2}+1\right)\cdot 2^{-k_{out}} \leq L^2/2^{k_{out}}$, proving the first part Theorem 1.

Proving Claim 4. Recall that the ℓ-query adversary $\mathcal{A}$ in consideration may be adaptive, which means that it possibly selects its oracle queries based on the answers it received for previous queries. Our goal is to bound the probability that $\mathcal{A}$ finds a pair of codewords that is both dangerous (with respect to $Q_{\mathcal{A},f}$) and forms a collision.

To this end, we first introduce the following notations. Let $Q_{\mathcal{A},f}^{(j)} = \left\{q_{\mathcal{A},f}^{(1)}, \ldots, q_{\mathcal{A},f}^{(j)}\right\}$ be the set of first j queries made by $\mathcal{A}^f$. Let $E_{\mathcal{A},f}^{(j)}$ be the event that there exists a pair $(s_1, s_2) \in \left(\{0,1\}^v\right)^2$ of distinct elements that is dangerous w.r.t. $Q_{\mathcal{A},f}^{(j)}$ and $g^f(s_1) = g^f(s_2)$. Finally, denote by $d_{\mathcal{A},f}^{(j+1)}$ the number of pairs $(\hat{s}_1, \hat{s}_2)$ that are dangerous w.r.t. $Q_{\mathcal{A},f}^{(j+1)}$ and there exists $1 \leq i \leq n$ such that $C(\hat{s}_1)_i = q_{\mathcal{A},f}^{(j+1)} \neq C(\hat{s}_2)_i$.

We next bound the probability that after making the $j+1$ query, the adversary finds – for the first time – a pair that is both dangerous and colliding.

Claim 6. *For any* $1 \leq j < \ell$ *and* $d \in \mathbb{N}$, *it holds that*
$$\Pr_{f \leftarrow \mathcal{F}}\left[E_{\mathcal{A},f}^{(j+1)} \wedge \neg E_{\mathcal{A},f}^{(j)} \mid d_{\mathcal{A},f}^{(j+1)} = d\right] \leq \frac{d}{2^{k_{out}}}.$$

Proof. By simple rules of conditional probability, it suffices to prove $\Pr_{f \leftarrow \mathcal{F}}\left[E_{\mathcal{A},f}^{(j+1)} \mid \neg E_{\mathcal{A},f}^{(j)} \wedge d_{\mathcal{A},f}^{(j+1)} = d\right] \leq \frac{d}{2^{k_{out}}}$. For $E_{\mathcal{A},f}^{(j+1)}$ to occur, there needs to be a pair $(\hat{s}_1, \hat{s}_2)$ that is dangerous w.r.t. $Q_{\mathcal{A},f}^{(j+1)}$ and $g^f(\hat{s}_1) = g^f(\hat{s}_2)$. The condition that $E_{\mathcal{A},f}^{(j)}$ does not occur yields that if $(\hat{s}_1, \hat{s}_2)$ is dangerous w.r.t. $Q_{\mathcal{A},f}^{(j)}$, then $g^f(\hat{s}_1) \neq g^f(\hat{s}_2)$. Hence, for computing the probability that such a pair exists, one should only consider pairs that are dangerous w.r.t. $Q_{\mathcal{A},f}^{(j+1)}$ and are *not* dangerous w.r.t. $Q_{\mathcal{A},f}^{(j)}$.

Let $(\hat{s}_1, \hat{s}_2)$ be a pair that is dangerous w.r.t. $Q_{\mathcal{A},f}^{(j+1)}$ and not dangerous w.r.t. $Q_{\mathcal{A},f}^{(j)}$. Note that there exists a (single) $1 \leq i \leq n$ with $C(\hat{s}_1)_i = q_{\mathcal{A},f}^{(j+1)} \neq C(\hat{s}_2)_i$; the existences holds since otherwise, this pair is already a dangerous pair w.r.t. $Q_{\mathcal{A},f}^{(j)}$, and the uniqueness follows since C is well-ordered. We next compute the probability that $g^f(\hat{s}_1) = g^f(\hat{s}_2)$. Consider any fixing of all f queries but $C(s_1)_i$ that is consistent with the actual answers of f on the queries in $Q_{\mathcal{A},f}^{(j)}$ (specifically, $E_{\mathcal{A},f}^{(j)}$ does not occur and $d_{\mathcal{A},f}^{(j+1)} = d$ for such fixings). Since C is well ordered, this fixes $C(s_t)_j$ for all $t \in \{1,2\}$ and $j \notin [n]$. For each such fixing, it holds that

$$\Pr\left[g^f(\hat{s}_1) = g^f(\hat{s}_2)\right] = \Pr\left[f(C(\hat{s}_1)_i) = \bigoplus_{j \in [n]\setminus\{i\}} f(C(\hat{s}_1)_j) \oplus \bigoplus_{j \in [n]} f(C(\hat{s}_2)_j)\right]$$
$$= 2^{-k_{out}}.$$

By assumption, there are d such dangerous pairs. Hence, by a union bound, the claim follows.

Proof (Proof of Claim 4). Since $Q_{\mathcal{A},f} = Q_{\mathcal{A},f}^{(\ell)}$, it holds that $E_{\mathcal{A},f} := E_{\mathcal{A},f}^{(\ell)}$ is the event that there exists a pair $(\hat{s}_1, \hat{s}_2) \in (\{0,1\}^v)^2$ of distinct elements that is dangerous w.r.t. $Q_{\mathcal{A},f}$ and $g^f(\hat{s}_1) = g^f(\hat{s}_2)$. Clearly, the probability of $E_{\mathcal{A},f}$ upperbounds the probability that $\mathcal{A}$ outputs such a pair.

Evidently, $E_{\mathcal{A},f}^{(1)}$ can never occur, since no pair is dangerous w.r.t. a single query. Furthermore, $E_{\mathcal{A},f}^{(j')}$ for any $j' \leq j$ implies $E_{\mathcal{A},f}^{(j)}$. Hence, we have that

$$\Pr_{f \leftarrow \mathcal{F}}[E_{\mathcal{A},f}] = \sum_{j=1}^{\ell-1} \Pr_{f \leftarrow \mathcal{F}}[E_{\mathcal{A},f}^{(j+1)} \wedge \neg E_{\mathcal{A},f}^{(j)}] \leq \sum_{j=1}^{\ell-1} \mathop{\mathrm{E}}_{f \leftarrow \mathcal{F}} \frac{d_{\mathcal{A},f}^{(j+1)}}{2^{k_{out}}}, \tag{1}$$

where the inequality follows from Claim 6. By linearity of expectation, it holds that

$$\Pr_{f \leftarrow \mathcal{F}}[E_{\mathcal{A},f}] \leq 2^{-k_{out}} \cdot \mathop{\mathrm{E}}_{f \leftarrow \mathcal{F}} \sum_{j=1}^{\ell-1} d_{\mathcal{A},f}^{(j+1)} \leq 2^{-k_{out}} \cdot \binom{L}{2}. \tag{2}$$

The last inequality follows since

$$\sum_{j=1}^{\ell-1} d_{\mathcal{A},f}^{(j+1)} \leq \binom{L}{2} \tag{3}$$

for every $f \in \mathcal{F}$. To see that Eq. (3) holds, note that each pair that is dangerous w.r.t. $Q_{\mathcal{A},f}^{(j)}$ is also dangerous w.r.t. $Q_{\mathcal{A},f}^{(\ell)}$ (i.e., the set of all queries made by $\mathcal{A}$). Furthermore, each such dangerous pair (s, s') is only counted by a single $d_{\mathcal{A},f}^{(j)}$, i.e., for the first j in which Q_j contains all the queries $C(s)_i \neq C(s')_i$. Hence, Eq. (3) follows from Corollary 1.

5   Beyond Collision-Resistance

We suggest some applications of the XOR parallel domain-extension scheme described in [20] to parallel constructions of other cryptographic primitives in the random oracle model. These applications exploit both the collision-resistance and weak-indifferentiability properties of our construction. In this section we give a high level description of these applications and refer the reader to [20] for formal statements.

Fiat-Shamir Paradigm. We show that the XOR parallel domain-extension scheme can be used to implement the Fiat-Shamir paradigm for converting any three-message public-coin argument, which may possibly employ a random oracle, into a non-interactive (i.e., single-message) argument in the random oracle model.

We start by describing the Fiat-Shamir transformation when applied to three-message protocols. Let $\langle P, V \rangle$ be a public-coin three-message argument system for an NP language. Such a protocol has the following high level structure: (1) P send a v-bit message to V; (2) V sends a *random* k-bit challenge to P; (3) P responds to this challenge; (4) V decides whether to accept by applying an efficient predicate to the input and the protocol's transcript.

The Fiat-Shamir transformation makes P generate all three messages by applying a hash function $h \colon \{0,1\}^v \to \{0,1\}^k$ to the first message of P to simulate the random challenge. This paradigm is provably secure in the random oracle model, but requires the random oracle input length to be as long as P's first message. We then use the weak-indifferentiability property, as discussed in Sect. 2, to show that the resulting scheme is also secure when h is the hash function obtained by applying the XOR parallel domain-extension scheme to a random oracle $f \colon \{0,1\}^k \to \{0,1\}^k$. Namely, we create a Fiat-Shamir like transformation that uses parallel calls to a small random oracle.

Parallel Commitment with Local Decommitment. Next, we consider commitment schemes for strings $s \in \{0,1\}^v$ that support a sublinear-communication local decommitment of any bit from s. Intuitively, in such schemes we require that the sender be bound to the string it committed to, but we do not explicitly require that it hide s. Instead, we require that the communication of both the commitment to s and the decommitment of each bit s_i be sublinear in v. We observe that such a commitment scheme can be obtained by dividing s into $\sqrt{v}$ blocks of length $\sqrt{v}$ each and applying the XOR parallel domain-extension separately to each block. To decommit s_i, the sender reveals the entire block containing s_i, and the receiver applies the hash function to ensure consistency. In this scheme, both the sender and the receiver only make parallel calls to a small random oracle.

Two-Adaptive Sublinear Non-interactive Arguments. Finally, we combine the above two applications to obtain sublinear-communication non-interactive arguments for NP in the random oracle model, which does not require the oracle input length to be large. To this end, we first apply the three-message protocol of [22], which combines a probabilistically checkable proof (PCP) with a commitment scheme as above. Then, following [30], we apply the Fiat-Shamir transformation to make this argument non-interactive.

By using efficient PCP constructions (e.g., those from [2]) and applying the XOR parallel domain-extension in both steps of the process, we get the following corollary: every NP language that can be recognized by a non-deterministic Turing machine of running time $T(n)$ has a non-interactive argument of length $\tilde{O}(T^{1/2}(n))$ in the random oracle model, in which the prover and the verifier make only *two* rounds of calls to the oracle.

6   Necessity of List-Recoverability for Parallel Domain-Extension

It turns out that some form of list-recoverability is necessary for the collision-resistance of a parallel domain-extension. Let (C, h) be a domain-extension scheme, and assume that C is *not* $(1, \ell, L)$-list recoverable. Namely, there exists a set $T \subseteq \{0,1\}^{k_{in}}$ of size at most ℓ, for which there are (at least) $L + 1$ distinct elements $s_1, \ldots, s_{L+1} \in \{0,1\}^v$ such that for every $1 \leq i \leq L + 1$: $g^f(s_i) = h(s_i, f(x_1), \ldots, f(x_n))$ for $x_1, \ldots, x_n \in T$. Hence, by querying f only on the ℓ elements in T, an adversary obtains the required information for computing $g^f(s_1), \ldots, g^f(s_{L+1})$. This attack finds a collision in g^f if $L \geq 2^{k_{out}}$. Thus, $(1, \ell, L)$-list-recoverability with $L \leq 2^{k_{out}}$, is *necessary* for the collision-resistance of g in the random-oracle model. This is formally stated below.

Theorem 7. *Let k_{in}, k_{out}, v, n be integers. For every $C \colon \{0,1\}^v \to \left(\{0,1\}^{k_{in}}\right)^n$ and $h \colon \{0,1\}^v \times \left(\{0,1\}^{k_{out}}\right)^n \to \{0,1\}^{k_{out}}$ if C is not $(1, \ell, 2^{k_{out}})$-list-recoverable then $g_{(C,h)}$ is* not *$(\ell, 0.99)$-collision-resistant in the random oracle model.*

We note that there are codes of large minimal distance (such as the repetition code) for which the attack in the proof of Theorem 7 can be implemented in polynomial time, by using linear algebra.

Necessity of List-Recoverability with $L \approx 2^{\frac{k_{out}}{2}}$. Note that Theorem 7 discusses $L = 2^{k_{out}}$ while in Theroem 1 we require $L \leq 2^{\frac{k_{out}}{2}}$ to get a meaningful result. Is it possible to show that (ℓ, L) list-recoverability with $L \approx 2^{\frac{k_{out}}{2}}$ is also necessary for security? We give a partial answer to this question below.

Observe that the above attack allows the adversary to use ℓ queries into f and come up with $\binom{L+1}{2} \approx 2^{k_{out}}$ pairs $s \neq s'$, such that he can compute $g(s)$ and $g(s')$. In some natural settings, computing g on this number of pairs suffices to find a collision. For instance, this is the case if the function g is 4-wise independent.[4] There are codes C satisfying the properties requested in Theorem 1, with which the construction of Theorem 1 is 4-wise independent. This implies that Theorem 1 cannot be improved to imply security with $L > 2^{\frac{k}{2}}$.

Necessity of List-Recoverability with $\alpha < 1$. Theorem 7 shows that it is necessary that C is list-recoverable with $\alpha = 1$ in any parallel domain-extension scheme. In our construction, however, we use stronger codes with $\alpha < 1$, and we also require that the codes have large distance. The next theorem shows that this assumption is necessary in case h is the XOR function (as we chose in Theorem 1).

Theorem 8. *There exists $c > 0$ such that the following holds for every $\alpha < 1$, integers $k_{in} \geq c \cdot \log(\frac{v}{1-\alpha})$, k_{out}, $v \geq c \cdot \max\{k_{in}, k_{out}\}$, $c \cdot (\frac{v}{1-\alpha}) \leq n \leq 2^{k_{in}/2}$, $c \cdot$*

---

[4] g is 4-wise independent, if for every four distinct $s_1, s_2, s_3, s_4 \in \{0,1\}^v$ the random variables $g(s_1), g(s_2), g(s_3), g(s_4)$ are uniformly distributed and independent (over the random choice of the oracle f). For such g, the expectation of the random variable counting the number of pairs $s \neq s'$ such that $g(s) = g(s')$ is at least $\binom{L+1}{2}/2^{k_{out}}$ (which is large if $L \geq 2^{k_{out}/2}$). Moreover, 4-wise indpendence implies that the variance of the random variable above is small, and therefore, the number of collisions is with high probability, close to the expectation. This implies that the adversary obtains a collision with high probability.

$(\frac{n}{1-\alpha}) \leq \ell \leq 2^{k_{in}/4}$ and $L \geq \ell$. There exists a function $C \colon \{0,1\}^v \to \left(\{0,1\}^{k_{in}}\right)^n$ that is $(1, \ell, L)$-list recoverable, well ordered, and has distance α, and (yet) for $h(s, a_1, \ldots, a_n) = \bigoplus_{i=1}^n a_i$, the parallel domain-extension scheme $g_{(C,h)}$ is not $(O(\frac{n}{1-\alpha}), 0.99)$-collision-resistant in the random-oracle model.

Theorem 8 shows that there exist codes C which satisfy all the requirements of Theorem 1 with the single exception being that the list-recoverability parameter is taken to be one (rather than the distance α of the code). Yet, the resulting construction is insecure. In fact, there is a lot of slack in the counterexample, one can choose the parameters ℓ, L to be much more favorable than in Theorem 1, and still an adversary with only $O(\frac{n}{1-\alpha})$ queries can break the scheme with probability arbitrarily close to one.

It should be noted that the previous construction of [26] extends the domain of a random function by relying on a notion of "input-restricting families", which is equivalent to strongly list-recoverable codes with $\alpha = 1$.[5] Such input-restricting families were subsequently used in [11] for the purpose of extending the domain of MACs. The construction from [26] is not fully parallel, requiring two rounds of calls to the random oracle f. The example provided in Theorem 8 gives a formal explanation why the use of input-restricting families does not suffice for using a single round of calls, even if one is only interested in collision-resistance as in this work.

Intuitively, the issue is as follows. In order to break collision-resistance the adversary is only required to produce a distinct pair (s, s') of inputs such that $g(s) = g(s')$, and the adversary is not required to be able to compute $g(s)$. Loosely speaking, parallel domain-extension schemes in which C is $(\alpha = 1, \ell, L)$-list recoverable, have the property that after asking ℓ queries the adversary cannot come up with $t > L$ inputs $s_1, \ldots, s_t$ such that he can compute $g(s_1), \ldots, g(s_t)$. The example in Theorem 8 shows that there are $(1, \ell, L)$-list-recoverable codes, in which the adversary can a produce a collision (s, s') even though he did not query f on all the inputs required to compute $g(s), g(s')$ (and therefore is not controlled by list-recoverability with $\alpha = 1$). In Theorem 1 we show how to bypass this limitation by using list-recoverable codes with $\alpha < 1$. We hope that the introduction of this stronger combinatorial object to the area of domain-extensions may help to improve and simplify other tasks in this area.

7   Using Known Explicit List-Recoverable Codes

In this section we plug in list-recoverable codes with specific parameters to obtain concrete results. We use the Parvaresh-Vardy code [32] in the range of parameters analyzed by Guruswami, Umans and Vadhan [18].

Theorem 9 ([18]). For every $\alpha \geq 1/2$, $0 < \beta < 1$, and $k < v \in \mathbb{N}$, there exists a $\mathrm{poly}(v)$-time computable function $C \colon \{0,1\}^v \to \left(\{0,1\}^k\right)^n$ for $n =$

---

[5] This notion is also equivalent to certain unbalanced expander graphs, see discussion in [20].

$O(v \cdot k)^{\frac{1}{1-\beta}}$, that is well ordered, has distance α, and for every $L \leq 2^{\beta \cdot (k - 2 \log n)}$, it is (α, ℓ, L)-list recoverable with $\ell = \Omega(n \cdot L)$ and has a $\mathrm{poly}(v, \ell)$-size list-recovering algorithm. Furthermore, when viewed as a function $C \colon \mathbb{F}_2^v \to \mathbb{F}_2^{k \cdot n}$, every output bit can be expressed as a degree one polynomial in the input bits.

We remark that [18] give a more general trade-off of parameters as well as a tighter connection between the parameters. More specifically, the theorem of [18] is stated as a condenser, and the statement given here is using the interpretation of condensers as list-recoverable codes (see [20] for more details). The facts that the construction of [18] is well-ordered and has large distance are not explicitly stated in [18], but are easily verified from the actual construction. The list-recovering algorithm is also not explicitly stated but follows directly from the proof of [18]. Finally, the fact that the mapping can be seen as a collection of degree one polynomials over $\mathbb{F}_2$ also follows from the specific structure of the construction of [18], or more generally from the structure of the Parvaresh-Vardy code.[6]

We now plug this code into Theorem 1 and obtain concrete results. For simplicity, we assume here that the input and output length of the oracle (i.e., f) are the same, and denote both lengths by k. We consider powerful adversaries with $\ell = 2^{(\frac{1}{2} - \gamma) \cdot k}$ for a small constant $\gamma > 0$ and shoot for ε that is exponentially small in k. Plugging the code of [18] into the construction of Theorem 1, yields that for desired security ε, it suffices to take $L = c \cdot \varepsilon^{\frac{1}{2}} \cdot 2^{\frac{k}{2}}$ for some constant c. By the construction of [18] we can achieve this with $\ell = \Omega(L \cdot n) = \Omega(\varepsilon^{\frac{1}{2}} \cdot 2^{\frac{k}{2}} \cdot n)$. Namely, we can achieve $\varepsilon = 2^{-2\gamma k}$ for $\ell = 2^{(\frac{1}{2} - \gamma) \cdot k}$-query adversaries. Furthermore, ℓ can be taken to be $\Omega(2^{k/2} \cdot n)$ (that is larger than $2^{k/2}$) for any small constant $\varepsilon > 0$. This is best possible in the sense that with $2^{\frac{k}{2}} \cdot n$ queries to f, one can simulate a birthday attack against g, and find a collision.

Comparing to the standard Merkle-tree based domain-extension, the resulting construction does make significantly more oracle calls to the underlying small domain function. Specifically, our construction makes $n = O(v^2 \cdot k^2)$ calls, whereas the Merkle-tree construction makes $O(v/k)$ calls. We remark that if we were to use a random code (rather than an explicit one), then the number of calls decreases to $O(v/k)$ as is the case for Merkle trees. Furthermore, even when using explicit codes, if we settle for security against $2^{\beta k}$-query adversaries, the

---

[6] More precisely, the function C has the following form. It sets $v = v_1 \cdot v_2$ for some integers v_1, v_2. Given an input $x \in \{0,1\}^v$ it is interpreted as a vector in $\mathbb{F}_{2^{v_1}}^{v_2}$ which is in turn interpreted as the coefficients of a degree v_2 univariate polynomial $f(X)$ over $\mathbb{F}_{2^{v_1}}$. For every $i \in [n]$, $C(x)_i = (i, f_0(\alpha_i), \ldots, f_{m-1}(\alpha_i))$ where $\alpha_i \in \mathbb{F}_{2^{v_1}}$ is a constant that depends only on i, and for every $j \in [m]$, $f_j(X)$ is a univariate polynomial defined by $f_j = f^{h^j} \mod E$, where h is a parameter and E is some degree $v_2 + 1$ irreducible polynomial. Thus, the code is immediately seen to be well-ordered and to inherit distance from the Reed-Solomon code (that corresponds to $j = 1$). The analysis of [18] allows choosing h that is even. Note that for an even h, the identity $(x + y)^h = x^h + y^h$ holds in $\mathbb{F}_{2^{v_1}}$. It is standard that this implies that for every fixed $\alpha \in \mathbb{F}_{2^{v_1}}$, the map $f \to (f^h \mod E)(\alpha)$ is $\mathbb{F}_2$-linear. This indeed implies that viewing the function C as a map from $\mathbb{F}_2^v$ to $\mathbb{F}_2^{k \cdot n}$, it is a degree one mapping.

query complexity of our construction can be reduced to roughly $(v \cdot k)^{\frac{1}{1-\beta}}$, which roughly matches the Merkle-tree construction for v that is significantly larger than k (which is the interesting range of parameters). Parvaresh-Vardy codes allow for some other trade-offs between security and number of queries that we do not examine here.

The code C that we use can be evaluated by degree one polynomials over $\mathbb{F}_2$. This immediately gives a very efficient parallel implementation in the standard model of Boolean circuits with parity gates of fan-in 2. Such circuits can compute the code C with depth $\log_2 v$. Moreover, in this model, computing the final xor in our construction, can be done by circuits of depth $\log_2 n$. Thus, overall our final hash function g can be implemented by circuits whose depth is bigger than the depth of f by $\log_2 v + \log_2 n$. By our bounds on n, this quantity is roughly $3 \log_2 v$ for $2^{(\frac{1}{2}-\gamma) \cdot k}$-query adversaries with small $\gamma > 0$, and roughly $(2 + \beta) \cdot \log_2 v$ for small $\beta > 0$ and $2^{\beta k}$-query adversaries. We remark that future developments in the area of list-recoverable codes or randomness condensers may reduce n to $O(v/k)$. It is also natural to expect that random $\mathbb{F}_2$-linear codes (or even families of efficiently encodable LDPC codes that are used in practice) achieve this bound. However, this is not known at this point.

8 Additional Related Work

Extending the domain of collision-resistant hash functions is of great importance for many cryptographic applications that depend on collision-resistance. Classical construction paradigms for domain-extension are the Merkle hash tree [28] and the Merkle-Damgård paradigm [10,29]. Both paradigms are iterative, namely, use sequential calls to the underlying hash-function. More specifically, in both paradigms $n = O(v/k)$ calls are made to the primitive, where the former paradigm requires $\log(n)$ rounds of calls, and the latter paradigm requires n rounds. Indeed, the Merkle-Damgård paradigm realizes the much stronger task of extending a fixed domain hash function to a full-fledged hash function, i.e., one that can deal with input of any length. The Merkle-Damgård paradigm is extensively used in practice and was the subject of much theoretical research and extensions (see, e.g., [3,7,24]). Lower bounds on the security of these domain-extension techniques were obtained, e.g., in [37,38]. The construction of Shrimpton and Stam [35] was the first construction achieving optimal collision-resistance security in an inherently non-trivial way. Their construction only doubles the domain, and requires two rounds of calls.

Most relevant to our work is the work of Maurer and Tessaro [26], already discussed above. This work considers the more challenging problem of extending the domain of a random function. Specifically, given a random function f from k bits to k bits, they construct a function g from $m(k)$ bits to $\ell(k)$ bits, for arbitrary polynomials m, ℓ, such that g is indistinguishable from a random function. The latter is formalized by using the indifferentiability framework from [27], which implies collision-resistance as a special case. The main goal of [27] is to obtain near-optimal security, namely to guarantee security against attackers that make $2^{(1-\varepsilon)k}$ oracle queries to f, improving over previous works.[7] However, their construction also achieves a high level of parallelism, requiring only two rounds of calls to f. Compared to the construction from [27], our construction is considerably simpler, it is fully parallel (i.e., requires only one round of calls to f), and it preserves the algebraic degree of f (whereas the construction from [27] more than squares the degree). As discussed in Sect. 6 (below Theorem 8), these disadvantages of [26] seem inherent given the type of combinatorial object on which they rely.

Building on and extending the techniques of [26], Dodies and Steinberger [11] construct a domain-extension scheme for MACs that has security beyond the "birthday barrier". Finally, Canetti et al. [6] considered the related, but somewhat orthogonal, goal of amplifying the security of a collision-resistant hash function.

Acknowledgments. We thank Yevgeniy Dodis, Swastik Kopparty, Phil Rogaway, Atri Rudra and Stefano Tessaro for helpful discussions and pointers.

References

1. Barak, B.: How to go beyond the black-box simulation barrier. In: Proceedings of the 42nd Annual Symposium on Foundations of Computer Science (FOCS), pp. 106–115 (2001)
2. Ben-Sasson, E., Sudan, M.: Short pcps with polylog query complexity. SIAM J. Comput. **38**(2), 551–607 (2008)
3. Bertoni, G., Daemen, J., Peeters, M., Assche, G.V.: Sufficient conditions for sound tree and sequential hashing modes. Int. J. Inf. Sec. **13**(4), 335–353 (2014a)
4. Bertoni, G., Daemen, J., Peeters, M., Assche, G.V.: The making of KEC-CAK. Cryptologia **38**(1), 26–60 (2014b). doi:10.1080/01611194.2013.856818. http://dx.doi.org/10.1080/01611194.2013.856818
5. Blum, M., Micali, S.: How to generate cryptographically strong sequences of pseudo random bits. In: Proceedings of the 23th Annual Symposium on Foundations of Computer Science (FOCS), pp. 112–117 (1982)
6. Canetti, R., Rivest, R., Sudan, M., Trevisan, L., Vadhan, S.P., Wee, H.M.: Amplifying collision resistance: a complexity-theoretic treatment. In: Menezes, A. (ed.) CRYPTO 2007. LNCS, vol. 4622, pp. 264–283. Springer, Heidelberg (2007)
7. Coron, J.-S., Dodis, Y., Malinaud, C., Puniya, P.: Merkle-Damgård revisited: how to construct a hash function. In: Shoup, V. (ed.) CRYPTO 2005. LNCS, vol. 3621, pp. 430–448. Springer, Heidelberg (2005)
8. Damgård, I.B.: Collision free hash functions and public key signature schemes. In: Price, W.L., Chaum, D. (eds.) EUROCRYPT 1987. LNCS, vol. 304, pp. 203–216. Springer, Heidelberg (1988)
9. Damgård, I., Pedersen, T.P., Pfitzmann, B.: On the existence of statistically hiding bit commitment schemes and fail-stop signatures. J. Cryptol. **10**(3), 163–194 (1997)
10. Damgård, I.B.: A design principle for hash functions. In: Brassard, G. (ed.) CRYPTO 1989. LNCS, vol. 435, pp. 416–427. Springer, Heidelberg (1990)

---

[7] Note that in the context of collision-resistance, the birthday paradox implies that collisions can be found with high probability using $2^{k_{out}/2}$ oracle calls.

11. Dodis, Y., Steinberger, J.: Domain extension for MACs beyond the birthday barrier. In: Paterson, K.G. (ed.) EUROCRYPT 2011. LNCS, vol. 6632, pp. 323–342. Springer, Heidelberg (2011)
12. Fiat, A., Shamir, A.: How to prove yourself: practical solutions to identification and signature problems. In: Odlyzko, A.M. (ed.) CRYPTO 1986. LNCS, vol. 263, pp. 186–194. Springer, Heidelberg (1987)
13. Goldreich, O., Goldwasser, S., Halevi, S.: Public-key cryptosystems from lattice reduction problems. In: Kaliski Jr., B.S. (ed.) CRYPTO 1997. LNCS, vol. 1294, pp. 112–131. Springer, Heidelberg (1997)
14. Guruswami, V.: List Decoding of Error-Correcting Codes. Ph.D. thesis, Massachusetts Institute of Technology (2005)
15. Guruswami, V., Indyk, P.: Expander-based constructions of efficiently decodable codes. In: 42nd Annual Symposium on Foundations of Computer Science, pp. 658–667 (2001)
16. Guruswami, V., Rudra, A.: Explicit codes achieving list decoding capacity: error-correction with optimal redundancy. IEEE Trans. Inf. Theory 54(1), 135–150 (2008)
17. Guruswami, V., Sudan, M.: Improved decoding of reed-solomon and algebraic-geometry codes. IEEE Trans. Inf. Theory 45(6), 1757–1767 (1999)
18. Guruswami, V., Umans, C., Vadhan, S.P.: Unbalanced expanders and randomness extractors from Parvaresh-Vardy codes. J. ACM 56(4), 20:1–20:34 (2009)
19. Haitner, I., Hoch, J.J., Reingold, O., Segev, G.: Finding collisions in interactive protocols - tight lower bounds on the round and communication complexities of statistically hiding commitments. SIAM J. Comput. 44(1), 193–242 (2015a). Preliminary version in STOC'07
20. Haitner, I., Ishai, Y., Omri, E., Shaltiel, R.: Parallel hashing via list recoverability (2015b). www.cs.tau.ac.il/~iftachh/papers/CRHDomainExtension/CRH.pdf. Full version of this paper
21. Halevi, S., Micali, S.: Practical and provably-secure commitment schemes from collision-free hashing. In: Koblitz, N. (ed.) CRYPTO 1996. LNCS, vol. 1109, pp. 201–215. Springer, Heidelberg (1996)
22. Kilian, J.: A note on efficient zero-knowledge proofs and arguments (extended abstract).In: Proceedings of the 24th Annual ACM Symposium on Theory of Computing (STOC), pp. 723–732 (1992)
23. Kilian, J.: On the complexity of bounded-interaction and noninteractive zero-knowledge proofs. In: Proceedings of the 35th Annual Symposium on Foundations of Computer Science (FOCS), pp. 466–477 (1994)
24. Lucks, S.: Design principles for iterated hash functions. Technical report, Cryptology ePrint Archive (2004)
25. Lyubashevsky, V., Micciancio, D.: Generalized compact knapsacks are collision resistant. In: Bugliesi, M., Preneel, B., Sassone, V., Wegener, I. (eds.) ICALP 2006. LNCS, vol. 4052, pp. 144–155. Springer, Heidelberg (2006)
26. Maurer, U.M., Tessaro, S.: Domain extension of public random functions: beyond the birthday barrier. In: Menezes, A. (ed.) CRYPTO 2007. LNCS, vol. 4622, pp. 187–204. Springer, Heidelberg (2007)
27. Maurer, U.M., Renner, R.S., Holenstein, C.: Indifferentiability, impossibility results on reductions, and applications to the random oracle methodology. In: Naor, M. (ed.) TCC 2004. LNCS, vol. 2951, pp. 21–39. Springer, Heidelberg (2004)
28. Merkle, R.C.: A digital signature based on a conventional encryption function. In: Pomerance, C. (ed.) CRYPTO 1987. LNCS, vol. 293, pp. 369–378. Springer, Heidelberg (1988)

29. Merkle, R.C.: A certified digital signature. In: Brassard, G. (ed.) CRYPTO 1989. LNCS, vol. 435, pp. 218–238. Springer, Heidelberg (1990)
30. Micali, S.: Computationally sound proofs. SIAM J. Comput. **30**(4), 1253–1298 (2000). Preliminary version in FOCS 1994
31. Naor, M., Yung, M.: Universal one-way hash functions and their cryptographic applications. In: Proceedings of the 21st Annual ACM Symposium on Theory of Computing (STOC), pp. 33–43. ACM Press (1989)
32. Parvaresh, F., Vardy, A.: Correcting errors beyond the Guruswami-Sudan radius in polynomial time. In: 46th Annual IEEE Symposium on Foundations of Computer Science, FOCS 2005, pp. 285–294 (2005)
33. Peikert, C., Rosen, A.: Efficient collision-resistant hashing from worst-case assumptions on cyclic lattices. In: Halevi, S., Rabin, T. (eds.) TCC 2006. LNCS, vol. 3876, pp. 145–166. Springer, Heidelberg (2006)
34. Rogaway, P., Steinberger, J.P.: Constructing cryptographic hash functions from fixed-key blockciphers. In: Wagner, D. (ed.) CRYPTO 2008. LNCS, vol. 5157, pp. 433–450. Springer, Heidelberg (2008)
35. Shrimpton, T., Stam, M.: Building a collision-resistant compression function from non-compressing primitives. In: Aceto, L., Damgård, I., Goldberg, L.A., Halldórsson, M.M., Ingólfsdóttir, A., Walukiewicz, I. (eds.) ICALP 2008, Part II. LNCS, vol. 5126, pp. 643–654. Springer, Heidelberg (2008)
36. Simon, D.R.: Findings collisions on a one-way street: can secure hash functions be based on general assumptions? In: Nyberg, K. (ed.) EUROCRYPT 1998. LNCS, vol. 1403, pp. 334–345. Springer, Heidelberg (1998)
37. Stam, M.: Beyond uniformity: better security/efficiency tradeoffs for compression functions. In: Wagner, D. (ed.) CRYPTO 2008. LNCS, vol. 5157, pp. 397–412. Springer, Heidelberg (2008)
38. Steinberger, J., Sun, X., Yang, Z.: Stam's conjecture and threshold phenomena in collision resistance. In: Safavi-Naini, R., Canetti, R. (eds.) CRYPTO 2012. LNCS, vol. 7417, pp. 384–405. Springer, Heidelberg (2012)
39. Sudan, M.: Decoding of reed solomon codes beyond the error-correction bound. J. Complex. **13**(1), 180–193 (1997)
40. Wang, X., Yu, H.: How to break MD5 and other hash functions. In: Cramer, R. (ed.) EUROCRYPT 2005. LNCS, vol. 3494, pp. 19–35. Springer, Heidelberg (2005)
41. Yao, A.C.: Theory and applications of trapdoor functions. In: Proceedings of the 23th Annual Symposium on Foundations of Computer Science (FOCS), pp. 80–91 (1982)

Cryptography with One-Way Communication

Sanjam Garg[1]([✉]), Yuval Ishai[2], Eyal Kushilevitz[2], Rafail Ostrovsky[3],
and Amit Sahai[3]

[1] UC Berkeley, Berkeley, USA
sanjamg@berkeley.edu
[2] Technion, Haifa, Israel
[3] UCLA, Los Angeles, USA

Abstract. There is a large body of work on using noisy communication
channels for realizing different cryptographic tasks. In particular, it is
known that secure message transmission can be achieved unconditionally using only *one-way* communication from the sender to the receiver.
In contrast, known solutions for more general secure computation tasks
inherently require interaction, even when the entire input originates from
the sender.

We initiate a general study of cryptographic protocols over noisy channels in a setting where only one party speaks. In this setting, we show
that the landscape of what a channel is useful for is much richer. Concretely, we obtain the following results.

- **Relationships Between Channels.** The binary erasure channel
  (BEC) and the binary symmetric channel (BSC), which are known

S. Garg—Supported by NSF CRII Award 1464397.

Y. Ishai—Supported by BSF grant 2012378 and by the European Union's Tenth
Framework Programme (FP10/2010-2016) under grant agreement no. 259426 ERC-
CaC.

E. Kushilevitz—Supported by ISF grant 1709/14 and BSF grant 2012378.

R. Ostrovsky—Work supported in part by NSF grants 09165174, 1065276, 1118126
and 1136174, US-Israel BSF grant 2008411, OKAWA Foundation Research Award,
IBM Faculty Research Award, Xerox Faculty Research Award, B. John Garrick
Foundation Award, Teradata Research Award, and Lockheed-Martin Corporation
Research Award. This material is based upon work supported by the Defense
Advanced Research Projects Agency through the U.S. Office of Naval Research under
Contract N00014 -11 -1-0392. The views expressed are those of the author and do
not reflect the official policy or position of the Department of Defense or the U.S.
Government.

A. Sahai—Research supported in part from a DARPA/ONR PROCEED award,
a DARPA/ARL SAFEWARE award, NSF Frontier Award 1413955, NSF grants
1228984, 1136174, 1118096, and 1065276, a Xerox Faculty Research Award, a Google
Faculty Research Award, an equipment grant from Intel, and an Okawa Foundation Research Grant. This material is based upon work supported by the Defense
Advanced Research Projects Agency through the U.S. Office of Naval Research under
Contract N00014-11-1-0389. The views expressed are those of the author and do not
reflect the official policy or position of the Department of Defense, the National
Science Foundation, or the U.S. Government.

R. Gennaro and M. Robshaw (Eds.): CRYPTO 2015, Part II, LNCS 9216, pp. 191–208, 2015.
DOI: 10.1007/978-3-662-48000-7_10

to be securely reducible to each other in the interactive setting, turn
out to be qualitatively different in the setting of one-way communication. In particular, a BEC cannot be implemented from a BSC, and
while the erasure probability of a BEC can be manipulated in both
directions, the crossover probability of a BSC can only be manipulated
in one direction.

- **Zero-knowledge Proofs and Secure Computation of Deterministic Functions.** One-way communication over BEC or BSC is
sufficient for securely realizing any deterministic (possibly reactive)
functionality which takes its inputs from a sender and delivers its
outputs to a receiver. This provides the first truly non-interactive
solutions to the problem of zero-knowledge proofs.
- **Secure Computation of Randomized Functions.** One-way communication over BEC or BSC *cannot* be used for realizing general
randomized functionalities which take input from a sender and deliver
output to a receiver. On the other hand, one-way communication over
other natural channels, such as bursty erasure channels, can be used
to realize such functionalities. This type of protocols can be used for
distributing certified cryptographic keys without revealing the keys to
the certification authority.

1   Introduction

The seminal work of Wyner [Wyn75] demonstrated the usefulness of noise for
secure communication. Since then, there has been a large body of work on
basing various cryptographic primitives, such as key agreement and commitment [BBCM95,BBR88,Mau91,DKS99,WNI03,Wul09,RTWW11], on different
types of noisy communication channels.

In 1988, Crépeau and Kilian [CK88] showed that noise in a communication channel can be used to realize essentially everything a cryptographer could
wish for. In particular, they showed that any non-trivial *binary-symmetric channel* (BSC) can be used to realize *oblivious transfer* (OT) which is sufficient
for realizing two-party secure computation. (More efficient construction were
later considered in [KM01,SW02,IKO+11b].) Finally, Crépeau, Morozov and
Wolf [CMW04] generalized these results to arbitrary *discrete memory-less* channels. Other results towards characterizing the types of channels on which OT
can be based appeared in [Kil88,DKS99,DFMS04,Wul07,Wul09].

Following the work of Crépeau and Kilian [CK88], the entire body of research
on secure two-party computation over noisy channels requires parties to interact. In contrast, the present paper considers cryptographic protocols which only
use *one-way communication*, namely ones in which only one party speaks. There
has been a considerable amount of work on realizing information-theoretic secure
message transmission in this setting. These works are motivated not only by the
goal of achieving information-theoretic security, but also by the goal of efficiency;
see [BTV12] for discussion. Our goal is to extend this study to more general cryptographic tasks, including useful special cases of secure two-party computation
in which the input originates from only one party.

1.1 Our Model

We model a channel as an ideal functionality $\mathcal{C}$. This is done in order to capture the security properties of the channel in a clean way and in order to facilitate the use of composition theorems. A channel provides a communication medium between a *sender* and a *receiver*. The sender can invoke the channel $\mathcal{C}$ on an input of its choice. The channel "based on its nature" processes the input and outputs the processed value to the receiver. The correctness and secrecy requirements of a channel and the protocols we build on top of it can be specified in terms of UC security. For example, consider a binary erasure channel (BEC) parameterized by a probability $p \in (0, 1)$. For this channel, the sender inputs a bit $x \in \{0, 1\}$ and the channel outputs (for the receiver) x with a probability p and $\perp$ with a probability $1 - p$.[1] Even for this basic channel, stating the correctness and security properties is non-trivial. Correctness requires that if the sender sends x then the receiver outputs either x or $\perp$ with the right probability distribution. Security is a bit more involved; it requires that no malicious sender can figure out whether the receiver actually received the sent bit or not, and that a malicious receiver does not learn any partial information about the sent bit in the case of an erasure.

In this work, we consider various such channels. Two other channels that would be of great interest to us are the *binary symmetric channel* (BSC) and the *random oblivious transfer* (ROT) channel. A BSC is parameterized by a probability $p \in (\frac{1}{2}, 1)$. For this channel, the sent bit is transmitted correctly with probability p and is flipped with probability $1 - p$. An ROT channel takes as input two strings m_0 and m_1 from the sender and outputs either $(m_0, \perp)$ or $(\perp, m_1)$ to the receiver, with equal probability.

When considering protocols built on top of such channels, we distinguish between the weaker *semi-honest* model, where the sender follows the protocol but tries to learn information about the receiver's output from its random coins, and the *malicious model*, where the sender may send arbitrary information over the channel. When the sender follows the protocol, the receiver's output should be as specified by the functionality. When the sender deviates from the protocol, the security requirement uses the standard real-ideal paradigm, asserting that the sender's strategy can be simulated by a distribution over honest strategies. It is important to note, however, that in this case the standard definition of "security with abort" also allows the sender to make the protocol fail, as long as the receiver can detect this failure. By default, the term "secure" refers to the malicious model, though most of our negative results apply also to the semi-honest model.

1.2 Our Results

We initiate a general study of one-way secure computation (OWSC) protocols over noisy channels in a setting where only one party speaks. Surprisingly, the

---

[1] In the literature, p sometimes stands for the error probability, while in our paper it is the probability of the "no noise" event.

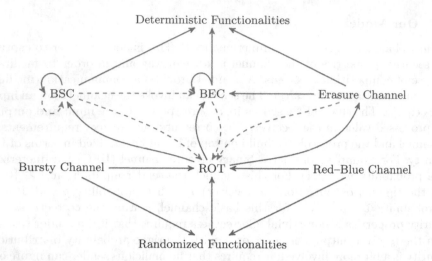

Fig. 1. Relationships among different kinds of channels and their applications. Solid arrows are used to denote a positive reduction, i.e. $A \to B$ implies that B can be constructed given A. On the other hand, dashed arrows indicate negative results, i.e. $A \dashrightarrow B$ implies that B cannot be constructed given A. Solid self-edge of BEC indicates that the transmission probability of a BEC can be manipulated in both directions. On the other hand, the solid and dashed self-edges of BSC respectively indicate that the probability of correct transmission of a BSC can be diminished (and brought closer to $\frac{1}{2}$) but cannot be amplified (Color figure online).

one-way setting is strikingly different from the interactive setting. In the interactive setting, all finite channels are either trivial, equivalent to secure message transmission, or equivalent to oblivious transfer. On the other hand, in the setting of OWSC, the landscape of what a channel is useful for is much richer. Specifically, we obtain the following results. All the implications have been summarized in Fig. 1.

- **Relationships Between Channels.** Binary erasure channel (BEC) and binary symmetric channel (BSC), which are known to be securely reducible to each other in the interactive setting, turn out to be qualitatively very different in the setting of one-way communication. In particular, we show that a BEC cannot be implemented given a BSC. Also, somewhat surprisingly, we show that while the erasure probability of a BEC can be manipulated in both directions the probability of correct transmission of a BSC can only be manipulated in one direction.
- **Deterministic Functions.** We show that both BEC and and BSC are sufficient for securely realizing any deterministic (possibly reactive) functionality that takes input from a sender and delivers its output to a receiver with only one-way communication. This provides the *first* truly non-interactive solution to the problem of zero-knowledge. We extend our results to the Generalized Erasure Channel (GEC) which is a generalization of BEC (see Sect. 3 for formal definition).

- **Randomized Functions.** We show that neither BEC nor BSC can be used (even assuming computational assumptions) for the task of realizing randomized functionalities which take input from a sender and deliver output to a receiver, in the setting of one-way communication. Nonetheless, one-way communications over natural channels, such as bursty erasure channels, can be used to realize such functionalities. This result is obtained by first constructing a random oblivious-transfer channel (ROT) and building on the techniques from [IPS08,IKO+11a]. This provides the first non-trivial feasibility result for secure-computation in a setting where only one party speaks.

1.3   Applications

One-way secure computation (OWSC) both for deterministic and randomized functionalities enable a number of applications for which there are no known solutions.

Truly Non-interactive Zero-Knowledge. Non-interactive zero-knowledge proof systems (NIZKs) [BFM90,FLS99] are a fundamental tool in cryptography with widespread applications. However, all known constructions rely on a common random string (or a random oracle)[2] and inherently fail to achieve useful features such as non-transferability or deniability [Pas03]. OWSC for deterministic functions provides the *first* truly non-interactive solution to the problem of zero-knowledge. This solution does not rely on a shared string between parties or a random oracle and achieves non-transferability and deniability properties. Furthermore, this solution achieves information theoretic and composable security.

Oblivious Certification of Cryptographic Keys. Public-key cryptography relies on the existence of certification authorities (like Verisign) who sign the public keys of different parties. All known implementations of this certification procedure rely on interaction. Our OWSC for randomized functionalities provides for the *first* candidate to realize this procedure with just one-way communication. More specifically, our protocol allows the certification authority to send a public-key secret-key pair along with a certificate on the public key with just one-way communication. We stress that in this setting the certification authority itself does not learn the secret key of the recipient party, as the randomness used in its generation is derived from the channel. However, if the certificate authority deviates from the protocol, the recipient may detect failure rather than output a pair of keys.

Fair Puzzle Distribution. Consider a Sudoku Puzzle competition where the organizer of the competition would like to generate signed puzzles for all the participants. However the participants do not trust the organizer and would like

---

[2] The result of Barak and Pass [BP04] is an exception to this. However they only achieve a weaker notion where security is only guaranteed against uniform provers. We, on the other hand, are interested in the standard notion of zero-knowledge.

their challenge Sudoku puzzles to be of the same difficulty. More specifically, we would like to have a mechanism that allows the competition organizer to provide independent puzzles of a pre-specified difficulty level (along with a signature on this puzzle) to each of the participants. The participants should be assured not only that the puzzles were generated independently from the correct distribution, but also that the organizers do not have an edge in solving the puzzles they generated (e.g., by generating random solved puzzles). There are no known solutions for this problem in a setting with just one-way communication. Our OWSC protocol for randomized functions gives the first such solution.

2   Preliminaries

Let λ denote a security parameter. We say that a function is *negligible* in λ if it is asymptotically smaller than the inverse of any fixed polynomial in λ. Otherwise, the function is said to be *non-negligible* in λ. We say that an event happens with *overwhelming* probability if it happens with probability $p(\lambda) = 1 - \nu(\lambda)$, where $\nu(\lambda)$ is a negligible function in λ. We use $[n]$ to denote the set $\{1, \ldots, n\}$.

Monotone Sets. Let $X_1, X_2 \ldots X_n$ be independent Bernoulli variables with $\Pr[X_i = 1] = p_i$. We define $Q_n = \{0, 1\}^n$ (the *n-cube*) and identify each element $a \in Q_n$ with the corresponding subset of $[n]$; i.e., $\{i \mid a_i = 1\}$. We define a probability measure $\Pr$ on Q_n by:

$$\Pr(a) = \prod_{i \in a} p_i \prod_{i \notin a} (1 - p_i) .$$

A set $A \subseteq Q_n$ is said to be a *monotone* if $a \in A$ and $a \subseteq b$ implies that $b \in A$.

Lemma 1 (Harris [Har60], Kleitman [Kle66]). If A and B are two monotone subsets of Q_n then A and B are *positively correlated*; namely,

$$\Pr[A \cap B] \geq \Pr[A] \Pr[B].$$

Chernoff Bounds. Let $X_1, X_2 \ldots X_n$ be independent Bernoulli variables with $\Pr[X_i = 1] = p_i$. Let $X = \sum_{i=1}^n X_i$ and μ be the expectation of X. Then,

$$\Pr(X \geq (1 + \delta)\mu) \leq e^{-\frac{\delta^2 \mu}{3}}, \text{ for } 0 < \delta < 1.$$

$$\Pr(X \leq (1 - \delta)\mu) \leq e^{-\frac{\delta^2 \mu}{2}}, \text{ for } 0 < \delta < 1.$$

3   Different Kinds of Channels

In this work, we model a channel as an ideal functionality $\mathcal{C}$. This is done in order to capture the security properties of a channel in a clean way. A channel provides a (one-way) communication medium between a *sender* and a *receiver*. The sender can invoke the channel $\mathcal{C}$ on an input of its choice. The channel "based on its nature", processes the input and outputs the processed value to the receiver. The correctness and secrecy requirements of a channel can be specified by a two-party functionality, which takes an input from the sender, generates some internal randomness, and delivers an output to the receiver. Our formulation of channel functionalities, as well as the security definition of protocols that build on top of them, follow the standard UC framework [Can05]. All of our positive results hold with statistical security, and some of our negative results apply also to the case of computational security. We will consider the following types of channels.

Binary Erasure Channel. The binary erasure channel (BEC) is perhaps the simplest non-trivial channel model considered in the literature. We denote this channel by $\mathcal{C}_{BEC}^p$. For this channel, the sender inputs a bit $x \in \{0,1\}$ and the channel outputs (to the receiver) x with a probability p and $\perp$ with a probability $1 - p$.

Binary Symmetric Channel. The binary symmetric channel (BSC) denoted by $\mathcal{C}_{BSC}^p$ (for $p > \frac{1}{2}$) is a channel in which the sender inputs a bit $x \in \{0,1\}$ and the channel outputs (for the receiver) x with a probability p and $1 - x$ with a probability $1 - p$.

Generalized Erasure Channel. The generalized erasure channel (GEC) is a generalization of the BEC, where k strings are sent by the sender and some subset of them, determined by a probability distribution $\mathcal{D}$, is erased. We denote this channel by $\mathcal{C}_{GEC}^{k,\ell,\mathcal{D}}$. Formally, the functionality takes as input k strings $x_1, \ldots, x_k \in \{0,1\}^\ell$ from the sender. It samples a string $s \subset \{0,1\}^k$ (which we call the *randomness of the channel*) according to the distribution $\mathcal{D}$. If $s_i = 1$ then set $y_i = x_i$ and, otherwise, $y_i = \perp$. The functionality outputs $y_1, \ldots, y_k$ to the receiver. We will consider the following special cases of the generalized erasure channel.

- *ℓ-Bit Random Oblivious Transfer.* The ℓ-bit random oblivious transfer channel (ℓ-ROT) denoted by $\mathcal{C}_{ROT}^\ell$ corresponds to the channel $\mathcal{C}_{GEC}^{2,\ell,\mathcal{D}_{2,OT}}$, where $\mathcal{D}_{2,OT}$ is the distribution that outputs a uniformly random value in $\{01, 10\}$. We also consider a p-biased ℓ-bit ROT channel denoted by $\mathcal{C}_{ROT}^{\ell,p}$ corresponds to the channel $\mathcal{C}_{GEC}^{2,\ell,\mathcal{D}_{2,p,OT}}$, where $\mathcal{D}_{2,p,OT}$ is the distribution that outputs 10 with probability p and 01 with a probability $1 - p$.
- *(k, ℓ, p)-Erasure Channel.* The (k, ℓ, p)-erasure channel corresponds to the channel $\mathcal{C}_{GEC}^{k,\ell,\mathcal{D}_{k,p}}$, where $\mathcal{D}_{k,p}$ is the distribution that outputs a k bit string s such that, for every $i \in [k]$, we have $s_i = 1$ with probability p and $s_i = 0$ with probability $1 - p$.
- *(k, ℓ)-Perfect Red-Blue Channel.* The (k, ℓ)-Perfect Red-Blue channel corresponds to the channel $\mathcal{C}_{GEC}^{k,\ell,\mathcal{D}_{k,RB}}$, where $\mathcal{D}_{k,RB}$ is any distribution such that each string in its output space (namely $\{0,1\}^k$) may be labeled either Red or Blue (or none) in a way that $\Pr[\text{Red} \cup \text{Blue}] = 1$, $\Pr[\text{Red}] = \Pr[\text{Blue}]$ and

$\forall r \in$ Red and $\forall s \subseteq r$ we have that $s \notin$ Blue and, similarly, $\forall b \in$ Blue and $\forall c \subseteq b$ we have that $c \notin$ Red.[3]

– $(k, \ell, \mu, \nu, \eta)$-*Statistical Red-Blue Channel.* The $(k, \ell, \mu, \nu, \eta)$-Statistical Red-Blue channel is a relaxed version of the Perfect Red-Blue Channel, that corresponds to the channel $C_{GEC}^{k,\ell,\mathcal{D}_{k,\mu,\nu,\eta}}$, where $\mathcal{D}_{k,\mu,\nu,\eta}$ is any distribution whose output space can be labelled Red and Blue such that (i) $\Pr[\text{Red}\cup\text{Blue}] \geq 1-\mu$, (ii) $|\Pr[\text{Red}] - \Pr[\text{Blue}]| \leq \nu$, (iii) $\Pr_{r\in\text{Red}}[\exists s \subseteq r$ such that $s \in$ Blue$] \leq \eta$, and (iv) $\Pr_{b\in\text{Blue}}[\exists c \subseteq b$ such that $c \in$ Red$] \leq \eta$.

– (k, ℓ, b)-*Perfect Bursty Channel.* This is an erasure channel where all b erasures appear in a "burst". Formally, the (k, ℓ, b)-Perfect bursty channel corresponds to the channel $C_{GEC}^{k,\ell,\mathcal{D}_{k,b}}$, where $\mathcal{D}_{k,b}$ is the distribution that outputs a k bit string such that all the bits are set to 1 besides the bits in locations $x + 1, x + 2, \ldots, x + b$ where x is chosen uniformly from $\{0, \ldots, k - b\}$.

– (k, ℓ, b, σ)-*Noisy Bursty Channel.* This is an erasure channel where erasures still appear in a "burst" but their number b' is normally distributed around b. Formally, the (k, ℓ, b, σ)-noisy bursty channel corresponds to the channel $C_{GEC}^{k,\ell,\mathcal{D}_{k,b,\sigma}}$ for typical $k \gg b$, where $\mathcal{D}_{k,b,\sigma}$ is the distribution that outputs a k bit string such that all the bits are set to 1 besides the bits in locations $x + 1, x + 2, \ldots, x + b'$ where b' is sampled from a gaussian and rounded to the closest non-negative integer $\leq k$ with mean b and standard deviation σ and then x is chosen uniformly from $\{0, \ldots, k - b'\}$.

4   Classification of Functionalities

Below we define the notion of one-way secure computation (OWSC) over a channel $\mathcal{C}$ (thought of as a non-reactive ideal functionality). We shall refer to such a OWSC scheme as $OWSC/\mathcal{C}$.

An $\mathsf{OWSC}^f/\mathcal{C}$ scheme for a function $f : X \to Y$ is a two-party protocol between Sender and Receiver and it follows the following format:

– Sender gets an input $x \in X$.
– Sender invokes the channel $\mathcal{C}$ (possibly multiple instances of the channel) with inputs of its choice. The channel, based on its nature, processes the input value and outputs it to the Receiver.
– Receiver carries out a local computation and outputs $f(x)$ or an error message.

Similarly, we can consider reactive functionality specified by a *stateful* function $f : \Sigma \times X \to \Sigma \times Y$. The Sender of a $\mathsf{OWSC}^f/\mathcal{C}$ scheme for a stateful function f obtains multiple inputs on the fly. On obtaining an input $x \in X$, Sender can invoke the channel $\mathcal{C}$ multiple times and in each execution the Receiver should either output y where $(\sigma', y) \leftarrow f(\sigma, x)$ (where $\sigma \in \Sigma$ is the current state and σ' is the state for the next execution) or an error message. The first execution of the protocol sets the state to ϵ.

---

[3] Here, again, we identify each $a \in \{0, 1\}^k$ with a subset of $[k]$ in the natural way.

The correctness and secrecy requirements of an OWSC scheme can be specified in terms of an ideal functionality. An $\mathsf{OWSC}^f/\mathcal{C}$ scheme for f is required to be a secure realization of the following function $\mathcal{F}_f$ in the $\mathcal{C}$-hybrid model.

- $\mathcal{F}_f$ accepts $x \in X$ from the Sender and outputs $f(x)$ to the receiver. If x is a special input **error**, then it outputs **error** to the Receiver.

We shall denote the security parameter by λ and require that the sender and the receiver in any scheme run in time polynomial in λ and the size of the circuit computing the function f. Further, for a scheme to be considered secure, we require that the simulation error be at most $2^{-\Omega(\lambda)}$.

Definition 1 (Completeness for Deterministic Functionalities). A channel $\mathcal{C}$ is said to be OWSC *complete for deterministic functionalities*, if for every deterministic function $f : X \rightarrow Y$ there exists a $\mathsf{OWSC}^f/\mathcal{C}$ scheme that is a UC-secure realization of the functionality $\mathcal{F}_f$ in the $\mathcal{C}$-hybrid model.

Definition 2 (Completeness for Randomized Functionalities). A channel $\mathcal{C}$ is said to be OWSC *complete for randomized functionalities*, if for every randomized function $f : X \rightarrow Y$ there exists a $\mathsf{OWSC}^f/\mathcal{C}$ scheme that is a UC-secure realization of the functionality $\mathcal{F}_f$ in the $\mathcal{C}$-hybrid model.

5   Reductions Among Channels

In this section, we study the relationships between different kinds of channels. Specifically:

- **Impossibility Results for** $\mathcal{C}_{ROT}$. One of the key channels of interest to us is the random oblivious transfer channel. We start by establishing (in Sect. 5.1) that this channel cannot be securely realized out of the most basic channels such as $\mathcal{C}_{BEC}$ (in fact, from any $\mathcal{C}_{GEC}^{k,\ell,\mathcal{D}_{k,p}}$, where $\mathcal{D}_{k,p}$ is the distribution that outputs a k bit string s such that, for every $i \in [k]$, we have $s_i = 1$ with probability p and $s_i = 0$ with probability $1 - p$) and $\mathcal{C}_{BSC}$. In full-version, we provide extensions of these results to the computational setting (but ruling out only protocols with negligible error rather than small noticeable error).
- **Positive Results for** $\mathcal{C}_{ROT}$. We consider a variety of more structured channels, such as the Red-Blue channel and the bursty channel, and give constructions of random oblivious transfer channel from such channels (Sect. 5.2).
- **Self-transformations for** $\mathcal{C}_{BEC}$ **and** $\mathcal{C}_{BSC}$. We move back to the basic channels ($\mathcal{C}_{BEC}$ and $\mathcal{C}_{BSC}$) and study additional properties of them. Although both these channels do not imply $\mathcal{C}_{ROT}^1$, they are of a very different nature. We show (in Sect. 5.3) that erasure probabilities of the $\mathcal{C}_{BEC}$ can be easily manipulated but the flipping probability of $\mathcal{C}_{BSC}$ is harder to manipulate. In particular, we show that, given a $\mathcal{C}_{BEC}$, we can construct another $\mathcal{C}_{BEC}$ with amplified or diminished erasure probabilities. On the other hand, given a $\mathcal{C}_{BSC}$, we can only construct another $\mathcal{C}_{BSC}$ with amplified flipping probability. In fact, diminishing the flipping probability turns out to be is impossible.

We remark that all the impossibility results (in this section) are stated in terms of the simulation based notion but hold even for a weaker game-based security notion. These stronger impossibility results are implied by the proofs and are not spelled out explicitly.

5.1 Impossibility Results for $\mathcal{C}_{ROT}$

In this subsection, we rule out the construction of $\mathcal{C}^1_{ROT}$ (random oblivious transfer) from the most basic channels such as $\mathcal{C}_{BEC}$ and $\mathcal{C}_{BSC}$. In particular, we show:

- $\mathcal{C}^{\ell'}_{ROT}$ (and, in fact, even biased-ROT) cannot be non-interactively securely realized from $\mathcal{C}^{k,\ell,\mathcal{D}_{k,p}}_{GEC}$.
- $\mathcal{C}^{p'}_{BEC}$ cannot be non-interactively securely realized from $\mathcal{C}^p_{BSC}$. It is easy to realize $\mathcal{C}^{\frac{1}{2}}_{BEC}$ from $\mathcal{C}^{\ell'}_{ROT}$. Hence, combining with the above result, we also conclude that $\mathcal{C}^{\ell'}_{ROT}$ cannot be non-interactively securely realized from $\mathcal{C}^p_{BSC}$.

The following theorem and its proof can be adapted to rule out even $\mathcal{C}^{\ell',q}_{ROT}$ for any constant q. We state the result and the proof in the simpler setting where $q = \frac{1}{2}$.

Theorem 1. $\exists\ \varepsilon \in (0,1)$ and $\ell' \in \mathbb{Z}^+$ such that $\forall k, \ell, p$, the channel $\mathcal{C}^{\ell'}_{ROT}$ cannot be ε-securely realized in the $\mathcal{C}^{k,\ell,\mathcal{D}_{k,p}}_{GEC}$ hybrid model even against semi-honest adversaries.

We start by giving some intuition for the case of binary erasure channel. The intuition extends to (k, ℓ, p)-erasure channels in a natural way. In any protocol for non-interactively realizing $\mathcal{C}^1_{ROT}$ the sender will need to encode both its inputs m_0, m_1 into its first message. Whether the receiver obtains m_0 or m_1 should depend solely on the random coins of the channel. In other words, erasure of certain bits (or more generally one combination from a list of possible choices) allows the receiver to obtain m_0 while erasure of another combination allows the receiver to learn m_1. The key issue is that a binary erasure channel erases each bit sent by the sender independently with a probability $1 - p$. Consider the scenario in which a receiver can obtain m_0 from the received bits. In this scenario, since each bit sent by the sender is treated independently we have that the receiver also obtains m_1 with a large enough probability, contradicting the security of the protocol. Arguing the last step formally is tricky and we rely on the Harris-Kleitman inequality for our argument. The full proof appears in the full-version.

Theorem 2. $\forall p \in (\frac{1}{2}, 1)$, $p' \in (0, 1)$ and protocol π, $\exists \varepsilon$ such that π does not ε-securely realize $\mathcal{C}^{p'}_{BEC}$ in the $\mathcal{C}^p_{BSC}$-hybrid model even against semi-honest adversaries.

We start by giving some intuition. Any protocol for non-interactively securely realizing $\mathcal{C}_{BEC}$ will need the sender to encode its input m into its first message. Whether the receiver obtains m or not should depend solely on the random coins of the channel. In other words when certain bits (or, more generally, one combination from a list of possible choices) is flipped then the receiver loses all information about m while flipping another combination allows the receiver to learn m completely. Consider a sequence of hybrid strings between a pair of strings on which the receiver outputs m and $\perp$ respectively. Among the hybrid strings there must exist two strings that differ in exactly one bit but are such that the receiver's output on the two differs completely. At this point, we argue that a change of just one bit cannot affect the receiver's best guess about the sent bit very dramatically, contradicting the security of the protocol. The key technical challenge of the proof lies in proving that this happens with a noticeable probability. The full proof appears in the full-version.

5.2   Positive Constructions for $\mathcal{C}_{ROT}$

We start by presenting a construction of a random oblivious transfer channel in Red-Blue channel hybrid model. Our construction provides a solution for any arbitrary Red-Blue channel and is inefficient. Furthermore, such a channel in its generality is not very natural. Therefore, we study natural examples of Red-Blue channels (and their approximate variants) and attempt at more efficient solutions.

We start by considering the basic setting of an arbitrary Red-Blue Channel and prove that it is sufficient to realize a random oblivious transfer channel.

Theorem 3. $\mathcal{C}_{ROT}^{\ell}$ *can be* $\max\{\mu, \nu, \eta\}$*-UC-securely realized (even against malicious adversaries) in the* $(k, \ell', \mu, \nu, \eta)$*-Red-Blue Channel hybrid model where* $\ell' = \ell \cdot 2^k$.

The proof appears in the full-version. Note that for the case of perfect Red-Blue Channel, we have that $\mu = \nu = \eta = 0$, and hence $\mathcal{C}_{ROT}^{\ell}$ can be perfectly-UC-securely realized in the (k, ℓ')-Perfect Red-Blue Channel hybrid model where $\ell' = \ell \cdot 2^k$.

Efficient Construction for ROT. We will start by considering the case of perfect bursty channel and show that it can be used to realize ROT. Recall that a (k, ℓ, b)-perfect bursty channel corresponds to the channel $\mathcal{C}_{GEC}^{k,\ell,\mathcal{D}_{k,b}}$, where $\mathcal{D}_{k,b}$ is the distribution that outputs a k bit string such that all the bits are set to 1 besides the "burst" of bits in locations $x+1, x+2, \ldots, x+b$ which are set to 0, where x is chosen uniformly from $\{0, \ldots, k-b\}$. In this setting we claim that:

Theorem 4. $\mathcal{C}_{ROT}^{\ell}$ *can be UC-securely realized (even against malicious adversaries) in the* (k, ℓ, b)*-perfect bursty channel hybrid model when* $b > \frac{k}{2}$ *or when* b *is odd.*

Proof. We start by giving the intuition. The key idea is to use Shamir's secret sharing (with shares of length ℓ) and secret share the first string in the first

$\Pi = \langle S, R \rangle$ **protocol with sender input** m_0, m_1

1. Let $\theta = t - \lfloor b/2 \rfloor$. Let $\{\alpha_1, \dots, \alpha_t\}$ be a θ-out-of-t Shamir's secret sharing of m_0. Similarly, let $\{\alpha_{t+1}, \dots, \alpha_k\}$ be a θ-out-of-t Shamir's secret sharing of m_1.
2. Send $(\alpha_1, \dots, \alpha_k)$ to the receiver.
3. Let the starting point of the burst in the symbols received by the receiver be i^*. If $i^* > \theta$ compute m_0 using the shares $\alpha_1, \dots, \alpha_\theta$ and output $(m_0, \perp)$; otherwise, output $(\perp, m_1)$ where m_1 is computed using the shares $\alpha_{k-\theta+1}, \dots, \alpha_k$.

Fig. 2. $\mathcal{C}_{ROT}^\ell$ in the (k, ℓ, b)-perfect bursty channel hybrid model, for odd b

half and the second string in the second half (with some appropriate threshold). Both when $b > \frac{k}{2}$ or when b is odd we will have an asymmetry in terms of the deletion pattern. If more terms from the first half are erased then the first string is deleted and, on the other hand, if more terms from the second half get erased then the second string is deleted. If k is odd then our construction will only give a biased-ROT but this bias can be corrected using the transformation from Sect. 7. Similarly, we note that in our construction we do not need the distribution over where the burst happens to be uniform. Our protocol can be very easily modified so that this restriction is not crucial. This would however only give biased ROT protocols and this bias will need to be corrected using the transformation from Sect. 7.

Next we give the construction for the case when b is odd. We assume, for simplicity, that k is even and $t = \frac{k}{2}$. The construction for the setting when k is odd or when b is not necessarily odd but $k > b/2$ are identical except that the parameters should be adjusted appropriately.

The construction appears in Fig. 2. Since b is odd, either in the first half or in the second half at least $\lceil b/2 \rceil$ of the strings are erased and hence that value remains hidden. On the other hand, in the other half the value can always be computed since at most $\lfloor b/2 \rfloor$ strings are deleted. The proof is identical to the case of Red-Blue Channel (proved in the full-version and is therefore omitted.

Channel with Imprecise Burst. Finally, we consider a bursty erasure channel where the size of burst is not precisely known but comes from roughly a discrete gaussian distribution. Recall that (k, ℓ, b, σ)-noisy bursty channel corresponds to the channel $\mathcal{C}_{GEC}^{k, \ell, \mathcal{D}_{k,b,\sigma}}$, where $\mathcal{D}_{k,b,\sigma}$ is the distribution that outputs a k bit string such that all the bits are set to 1 besides the bits in locations $x+1, x+2, \dots, x+b'$ where b' is sampled from a gaussian and rounded to the closest non-negative integer $\leq k$ with mean b and standard deviation σ and then x is chosen uniformly from $\{0, \dots, k - b'\}$.

Theorem 5. $\mathcal{C}_{ROT}^\ell$ *can be* $\frac{(1-\alpha)b}{k-(1+\alpha)b} + \frac{\sigma^2}{\alpha^2 b^2}$-*UC-securely realized in the* (k, ℓ, b, σ)-*noisy bursty channel hybrid model for any constant* $\alpha \in (0, 1)$.

Proof. We use the same construction as in Fig. 2 except the threshold parameter θ of the Shamir secret sharing. We set it up in a way so that it is possible to obtain m_0 if less than $(1-\alpha)b/2$ symbols are erased from the first half. Similarly secret sharing is done for the second half. By Chebyshev's inequality, the probability that the size of the burst, b', lies outside the range $\{(1-\alpha)b, \ldots, (1+\alpha)b\}$ is at most $\frac{\sigma^2}{\alpha^2 b^2}$ (if b' is too big the receiver may not learn any value, while if b' is too small it may learn both values). Assuming this does not happen, then the receiver gets only one of the sent values as long as the burst does not happen "in the middle" (i.e., $(1-\alpha)b/2$ symbols are erased from each half). The probability that the burst happens in the middle is at most $\frac{(1-\alpha)b}{k-(1+\alpha)b}$.

5.3 Self-transformations for $\mathcal{C}_{BEC}$ and $\mathcal{C}_{BSC}$

In this subsection, we show that any erasure channel can be used to construct a binary erasure channel with any desired erasure probability. On the other hand, the case of BSC is very different. The probability of correct transmission in a BSC channel can be reduced but cannot be increased. Formally,

Theorem 6. $\forall\ \mathcal{C}_{GEC}^{k,\ell,\mathcal{D}}$ *such that* $\mathcal{D}$ *is not a constant distribution,* $\exists\ p$ *such that* $\mathcal{C}_{BEC}^p$ *can be (perfectly) UC-securely realized (even against malicious adversaries) in the* $\mathcal{C}_{GEC}^{k,\ell,\mathcal{D}}$*-hybrid model.*

Theorem 7. $\forall p, p' \in (0,1)$ *and* $\epsilon > 1$, $\exists p'' \in [p', \epsilon p']$, *such that* $\mathcal{C}_{BEC}^{p''}$ *can be (perfectly) UC-securely realized (even against malicious adversaries) in the* $\mathcal{C}_{BEC}^p$ *hybrid model.*

Theorem 8. $\forall p \in (\frac{1}{2},1)$ *and* $t \in \mathbb{Z}^+$, *the channel* $\mathcal{C}_{BSC}^{p'}$ *can be (perfectly) UC-securely realized (even against malicious adversaries) in the* $\mathcal{C}_{BSC}^p$*-hybrid model where* $p' = \frac{1}{2} + 2^{t-1}\left(p - \frac{1}{2}\right)^t$.

Theorem 9. $\forall\ p, p' \in (\frac{1}{2},1), p' > p$ *and protocol* π, $\exists \varepsilon$ *such that* π *does not* ε*-securely realize* $\mathcal{C}_{BSC}^{p'}$ *in the* $\mathcal{C}_{BSC}^p$*-hybrid model even against semi-honest adversaries.*

Proofs of the above theorems appear in the full-version.

6 OWSC Scheme for Deterministic Functionalities

$\text{OWSC}^f/\mathcal{C}$ is a meaningful notion only for those deterministic functions f such that given a value y identifying if there exists an input x such that $y = f(x)$ is non-trivial (cannot be done in efficiently). This, in particular, rules out all functions with polynomial sized input domains. Furthermore, this notion is useful only in the setting of malicious adversaries because it is trivial to realize this notion in the setting of semi-honest adversaries.

We start by noting that a $\mathsf{OWSC}^f/\mathcal{C}$ scheme, for any deterministic function f, can be realized by using a $\mathsf{OWSC}^{\mathsf{zk}}/\mathcal{C}$ scheme for the zero-knowledge functionality. This can be achieved simply by having the sender send the output to the receiver and along with it prove in zero-knowledge, knowledge of an input x for which $f(x)$ yields the provided output. Here we implicitly assume that besides the channel $\mathcal{C}$ the sender also has access to an error free channel which can be implemented using $\mathcal{C}$ itself (with a negligible error). Formally,

Theorem 10. *For every deterministic function f, there exists a $\mathsf{OWSC}^f/\mathcal{C}$ scheme that is a UC-secure realization (even against malicious adversaries) of the functionality $\mathcal{F}_f$ in the $\mathcal{C}$-hybrid model where $\mathcal{C} \in \{\mathcal{C}_{GEC}^{k,\ell,\mathcal{D}}, \mathcal{C}_{BSC}^{p}\}$.*

As already mentioned, proving the above theorem reduces to the task of realizing a $\mathsf{OWSC}^{\mathsf{zk}}/\mathcal{C}$ scheme. In our construction, we will make use of oblivious ZK-PCPs (see definitions in full-version).

Lemma 2. *There exists a $\mathsf{OWSC}^{\mathsf{zk}}/\mathcal{C}$ scheme that is a UC-secure realization (even against malicious adversaries) of the zero-knowledge functionality in the $\mathcal{C}$-hybrid model where $\mathcal{C} \in \{\mathcal{C}_{GEC}^{k,\ell,\mathcal{D}}, \mathcal{C}_{BSC}^{p}\}$.*

We start by giving some intuition. The key idea is to use an erasure channel or a binary symmetric channel to send over multiple instances of independently chosen ZK-PCPs and observe the statistical gap that can be created only if valid proofs were sent. However, a number of difficulties arise in realizing this intuition, particularly in our construction from BSC. Below, we provide our construction from erasure channels. The more involved construction from binary symmetric channel is deferred to full-version.

Erasure Channels. We start by considering the case of binary erasure channels with error probability $\frac{1}{2}$; i.e., when $\mathcal{C} = \mathcal{C}_{BEC}^{\frac{1}{2}}$. It follows from Theorems 6 and 7 that any $\mathcal{C}_{GEC}^{k,\ell,\mathcal{D}}$ can be used to realize $\mathcal{C}_{BEC}^{\frac{1}{2}}$.[4] We give the protocol in Fig. 3.

Completeness. For every $i \in [k]$, using Chernoff bound, we have that:

$$\Pr\left[\Upsilon(\pi_i') \le \frac{n}{4}\right] \le e^{-\frac{n}{16}},$$

where $\Upsilon(\pi_i')$ denotes the number of occurrences of $\perp$ in π_i'.

Hence, except with negligible probability for each $i \in [k]$, R receives at least c. Given this the completeness of the protocol follows from the completeness of the oblivious ZK-PCP.

Soundness. We will construct an extractor E', that extracts valid witnesses from any cheating prover P^* that makes the honest verifier accept with non-negligible probability. We will first describe our extractor E' and then argue that it indeed works (with overwhelming probability).

---

[4] Theorem 7 only guarantees a channel $\mathcal{C}_{BEC}^{p'}$ with p' close enough to p. We will use the value $\frac{1}{2}$ for concreteness but any value close enough to $\frac{1}{2}$, say in the range $\frac{1}{2}$ to $\frac{51}{100}$, will suffice as well.

OWSC^{zk}/$\mathcal{C}^p_{BEC}$ protocol for language L

Common Input: $x \in \{0,1\}^\lambda$.
Auxiliary Input for prover P: w such that $(x,w) \in R_L$.
Parameters: Let (P_{oZK}, V_{oZK}) be any (c, ν)-oblivious ZK-PCP system (see full-version)(with $c \le \frac{n}{4}$ and $\nu \ge \frac{3}{4}$) with knowledge soundness κ. Let $\ell = \frac{\lambda}{\kappa}$.

- P samples proofs $\pi_1, \ldots, \pi_\ell$ from $P_{oZK}(\lambda, x, w)$ and sends $(\pi_1, \ldots, \pi_\ell)$ to V via the erasure channel $\mathcal{C}^p_{BEC}$.
- V receives $\pi'_1, \ldots, \pi'_\ell$ and for all $i \in [\ell]$ checks if $V_{oZK}(\pi'_i)$. It outputs accept if all the checks pass and reject otherwise.

Fig. 3. Realizing zero-knowledge from binary erasure channel

Our extractor E' proceeds as follows. Let $(\pi_1, \pi_2, \ldots, \pi_\ell)$ be the proofs generated by the cheating prover P^*. For every $i \in [\ell]$, E' obtains $y_i = E(x, \pi_i)$. If $\exists i^* \in [\ell]$ such that $y_{i^*} \in R(x)$ then output y_{i^*} (breaking ties arbitrarily). If no such i^* exists then output $\perp$.

Note that since our extractor E' failed to extract witness out of π_i for any $i \in [\ell]$ we have (by soundness of the ZK-PCP) that $\Pr[V_{oZK}(x, \pi'_i) = 0] \ge \kappa$, for every $i \in [\ell]$, where the probability is taken over the random choices of obtaining π'_i from π_i. Hence, if E' outputs $\perp$ then the verifier must also always reject, except with probability at most $\le (1 - \kappa)^\ell$, which is negligible for $\ell = \frac{\lambda}{\kappa}$.

Zero-Knowledge. We need to construct a simulator S' for our protocol. This construction follows immediately from the ν-zero-knowledge property of the oblivious ZK-PCP.

The full proof for the case of BSC appears in full-version.

7 $\mathcal{C}^\ell_{ROT}$ is OWSC Complete for Randomized Functionalities

In this section, we describe an OWSC scheme for any randomized function in the $\mathcal{C}_{ROT}$-hybrid model that uses only a *single* round of random OTs and no additional interaction. The functionalities considered here provide output to only one party. This result follows directly from [IPS08, Appendix B] and we include the construction and proof in the full-version for completeness (much of the text have been taken verbatim from [IPS08, Appendix B]). More efficient alternatives have been considered by [IKO+11a] however we consider the simplest feasibility result for our setting.

One technical difference in our setting compared to [IPS08] is in the underlying primitive from which the protocols are constructed. While the protocol in [IPS08] uses a regular 1-out-of-N OT protocol, in our case we only have access to a 1-out-of-2 ROT protocol and need to convert it to a 1-out-of-N ROT protocol. (Recall that the choice about which 1-out-of-N strings the receiver obtains is made by the channel in the ROT protocol.) This however can be done easily using standard techniques and a sketch of the construction has been provided in full-version.

Theorem 11. *For every randomized function* f, $\exists \ell$ *and a* $OWSC^f / C^\ell_{ROT}$ *scheme that is a UC-secure realization (even against malicious adversaries) of the functionality* $\mathcal{F}_f$ *in the* C^ℓ_{ROT}*-hybrid model.*

ϵ*-secure Variant.* We can also use the ϵ-UC realization of ROT (based on noisy bursty channel as in Theorem 5) in order to obtain a $\epsilon \cdot r$-UC realization of $OWSC^f$ where r is the number of ROT calls made inside our construction. r for our construction is a fixed polynomial in the security parameter λ, independent of the size of the function being computed.

Construction Using Biased-ROT. The above theorem is stated just for the case of C^ℓ_{ROT}-hybrid model. However we note that the same construction continues to work in the $C^{\ell,p}_{ROT}$-hybrid model, for any constant $p \in (0,1)$, with one small change. When using the $C^{\ell,p}_{ROT}$ channel, the input provided by the channel for the function evaluation will be biased. This issue can be resolved by using security parameter λ number of independent bits from the channel to obtain each bit for the functionality being evaluated. More specifically, each input bit for the functionality is obtained by taking the exclusive or of λ independent input bits. By the XOR Lemma, we claim that the obtained bits will be close to uniform.

Furthermore, when using the $C^{\ell,p}_{ROT}$-hybrid model, the construction itself does not depend on the precise value of the constant p. Hence, our construction is robust in the sense that it remains secure even if the adversary gets to specify the value of p (within some bounded range).

References

[Ajt10] Ajtai, M.: Oblivious RAMs without cryptogrpahic assumptions. In: Schulman, L.J. (ed.) 42nd Annual ACM Symposium on Theory of Computing, pp. 181–190. ACM Press, Cambridge (2010)

[BBCM95] Bennett, C.H., Brassard, G., Crepeau, C., Maurer, U.M.: Generalized privacy amplification. IEEE Trans. Inf. Theory **41**(6), 1915–1923 (1995)

[BBR88] Bennett, C.H., Brassard, G., Robert, J.-M.: Privacy amplification by public discussion. SIAM J. Comput. **17**(2), 210–229 (1988)

[BCR86] Brassard, G., Crépeau, C., Robert, J.-M.: Information theoretic reductions among disclosure problems. In: FOCS, pp. 168–173 (1986)

[BFM90] Blum, M., Feldman, P., Micali, S.: Proving security against chosen cyphertext attacks. In: Goldwasser, S. (ed.) CRYPTO 1988. LNCS, vol. 403, pp. 256–268. Springer, Heidelberg (1990)

[BGW88] Ben-Or, M., Goldwasser, S., Wigderson, A.: Completeness theorems for non-cryptographic fault-tolerant distributed computation. In: Proceedings of the 20th STOC, pp. 1–10. ACM (1988)

[BP04] Barak, B., Pass, R.: On the possibility of one-message weak zero-knowledge. In: Naor, M. (ed.) TCC 2004. LNCS, vol. 2951, pp. 121–132. Springer, Heidelberg (2004)

[BTV12] Bellare, M., Tessaro, S., Vardy, A.: Semantic security for the wiretap channel. In: Safavi-Naini, R., Canetti, R. (eds.) CRYPTO 2012. LNCS, vol. 7417, pp. 294–311. Springer, Heidelberg (2012)

[Can01] Canetti, R.: Universally composable security: a new paradigm for cryptographic protocols. Electronic Colloquium on Computational Complexity (ECCC) TR01-016 (2001). (Previous version "A unified framework for analyzing security of protocols" availabe at the ECCC archive TR01-016. Extended abstract in FOCS 2001)

[Can05] Canetti, R.: Universally composable security: a new paradigm for cryptographic protocols. Cryptology ePrint Archive, Report 2000/067 (2005). Revised version of [Can01]

[CK88] Crépeau, C., Kilian, J.: Achieving oblivious transfer using weakened security assumptions (extended abstract). In: FOCS, pp. 42–52 (1988)

[CMW04] Crépeau, C., Morozov, K., Wolf, S.: Efficient unconditional oblivious transfer from almost any noisy channel. In: Blundo, C., Cimato, S. (eds.) SCN 2004. LNCS, vol. 3352, pp. 47–59. Springer, Heidelberg (2005)

[DFMS04] Damgård, I.B., Fehr, S., Morozov, K., Salvail, L.: Unfair noisy channels and oblivious transfer. In: Naor, M. (ed.) TCC 2004. LNCS, vol. 2951, pp. 355–373. Springer, Heidelberg (2004)

[DKS99] Damgård, I.B., Kilian, J., Salvail, L.: On the (Im)possibility of basing oblivious transfer and bit commitment on weakened security assumptions. In: Stern, J. (ed.) EUROCRYPT 1999. LNCS, vol. 1592, p. 56. Springer, Heidelberg (1999)

[FLS99] Feige, U., Lapidot, D., Shamir, A.: Multiple noninteractive zero knowledge proofs under general assumptions. SIAM J. Comput. 29(1), 1–28 (1999)

[GMW87] Goldreich, O., Micali, S., Wigderson, A.: How to play ANY mental game. In: ACM (ed.) Proceedings of the 19th STOC, pp. 218–229. ACM (1987). (See [Gol04 Chap. 7] for more details)

[Gol04] Goldreich, O.: Foundations of Cryptography: Basic Applications. Cambridge University Press, Cambridge (2004)

[Har60] Harris, T.E.: A lower bound for the critical probability in a certain percolation process. Proc. Cambridge Phil. Soc. 56, 13–20 (1960)

[IKO+11a] Ishai, Y., Kushilevitz, E., Ostrovsky, R., Prabhakaran, M., Sahai, A.: Efficient non-interactive secure computation. In: Paterson, K.G. (ed.) EUROCRYPT 2011. LNCS, vol. 6632, pp. 406–425. Springer, Heidelberg (2011)

[IKO+11b] Ishai, Y., Kushilevitz, E., Ostrovsky, R., Prabhakaran, M., Sahai, A., Wullschleger, J.: Constant-rate oblivious transfer from noisy channels. In: Rogaway, P. (ed.) CRYPTO 2011. LNCS, vol. 6841, pp. 667–684. Springer, Heidelberg (2011)

[IPS08] Ishai, Y., Prabhakaran, M., Sahai, A.: Founding cryptography on oblivious transfer – efficiently. In: Wagner, D. (ed.) CRYPTO 2008. LNCS, vol. 5157, pp. 572–591. Springer, Heidelberg (2008)

[ISW03] Ishai, Y., Sahai, A., Wagner, D.: Private circuits: securing hardware against probing attacks. In: Boneh, D. (ed.) CRYPTO 2003. LNCS, vol. 2729, pp. 463–481. Springer, Heidelberg (2003)

[Kil88] Kilian, J.: Founding cryptography on oblivious transfer. In: STOC, pp. 20–31 (1988)

[Kle66] Kleitman, D.J.: Families of non-disjoint subsets. J. Combin. Theory **1**, 153–155 (1966)

[KM01] Korjik, V., Morozov, K.: Generalized oblivious transfer protocols based on noisy channels. In: Gorodetski, V.I., Skormin, V.A., Popyack, L.J. (eds.) MMM-ACNS 2001. LNCS, vol. 2052, p. 219. Springer, Heidelberg (2001)

[Liu] Liu, H.: M400 msci project - discrete isoperimetric inequalities

[Mau91] Maurer, U.M.: Perfect cryptographic security from partially independent channels. In: STOC, pp. 561–571 (1991)

[Mau02] Maurer, U.M.: Secure multi-party computation made simple. In: Cimato, S., Galdi, C., Persiano, G. (eds.) SCN 2002. LNCS, vol. 2576, pp. 14–28. Springer, Heidelberg (2003)

[Pas03] Pass, R.: On deniability in the common reference string and random oracle model. In: Boneh, D. (ed.) CRYPTO 2003. LNCS, vol. 2729, pp. 316–337. Springer, Heidelberg (2003)

[RTWW11] Wullschleger, J., Ranellucci, S., Tapp, A., Winkler, S.: On the efficiency of bit commitment reductions. In: Lee, D.H., Wang, X. (eds.) ASIACRYPT 2011. LNCS, vol. 7073, pp. 520–537. Springer, Heidelberg (2011)

[SW02] Stebila, D., Wolf, S.: Efficient oblivious transfer from any non-trivial binary-symmetric channel. In: 2002 IEEE International Symposium on Information Theory, Proceedings, p. 293 (2002)

[Wik13] Wikipedia. Binomial distribution (2013). Accessed 17 October 2013

[WNI03] Winter, A.J., Nascimento, A.C.A., Imai, H.: Commitment capacity of discrete memoryless channels. In: Paterson, K.G. (ed.) Cryptography and Coding 2003. LNCS, vol. 2898, pp. 35–51. Springer, Heidelberg (2003)

[Wul07] Wullschleger, J.: Oblivious-transfer amplification. In: Naor, M. (ed.) EUROCRYPT 2007. LNCS, vol. 4515, pp. 555–572. Springer, Heidelberg (2007)

[Wul09] Wullschleger, J.: Oblivious transfer from weak noisy channels. In: Reingold, O. (ed.) TCC 2009. LNCS, vol. 5444, pp. 332–349. Springer, Heidelberg (2009)

[Wyn75] Wyner, A.D.: The wire-tap channel. Bell Syst. Tech. J. **54**(8), 1334–1387 (1975)

(Almost) Optimal Constructions of UOWHFs from 1-to-1, Regular One-Way Functions and Beyond

Yu Yu[1,2,3]($\boxtimes$), Dawu Gu[1], Xiangxue Li[4], and Jian Weng[5]

[1] Department of Computer Science and Engineering,
Shanghai Jiao Tong University, Shanghai, China
{yyuu,dwgu}@sjtu.edu.cn

[2] State Key Laboratory of Cryptology, P.O. Box 5159, Beijing 100878, China

[3] State Key Laboratory of Information Security, Institute of Information
Engineering, Chinese Academy of Sciences, Beijing 100093, China

[4] Department of Computer Science and Technology,
East China Normal University, Shanghai, China
xxli@cs.ecnu.edu.cn

[5] College of Information Science and Technology,
Jinan University, Guangzhou, China
cryptjweng@gmail.com

Abstract. We revisit the problem of black-box constructions of universal one-way hash functions (UOWHFs) from several typical classes of one-way functions (OWFs), and give respective constructions that either improve or generalize the best previously known.

- For any 1-to-1 one-way function, we give an optimal construction of UOWHFs with key and output length $\Theta(n)$ by making a single call to the underlying OWF. This improves the constructions of Naor and Yung (STOC 1989) and De Santis and Yung (Eurocrypt 1990) that need key length $O(n \cdot \omega(\log n))$.
- For any known-(almost-)regular one-way function with known hardness, we give an optimal construction of UOWHFs with key and output length $\Theta(n)$ and a single call to the one-way function.
- For any known-(almost-)regular one-way function, we give a construction of UOWHFs with key and output length $O(n \cdot \omega(1))$ and by making $\omega(1)$ non-adaptive calls to the one-way function. This improves the construction of Barhum and Maurer (Latincrypt 2012) that requires key and output length $O(n \cdot \omega(\log n))$ and $\omega(\log n)$ calls.
- For any weakly-regular one-way function introduced by Yu et al. at TCC 2015 (i.e., the set of inputs with maximal number of siblings is of an n^{-c}-fraction for some constant c), we give a construction of UOWHFs with key length $O(n \cdot \log n)$ and output length $\Theta(n)$. This generalizes the construction of Ames et al. (Asiacrypt 2012) which requires an unknown-regular one-way function (i.e., $c = 0$).

Along the way, we use several techniques that might be of independent interest. We show that almost 1-to-1 (except for a negligible fraction) one-way functions and known (almost-)regular one-way functions are

© International Association for Cryptologic Research 2015
R. Gennaro and M. Robshaw (Eds.): CRYPTO 2015, Part II, LNCS 9216, pp. 209–229, 2015.
DOI: 10.1007/978-3-662-48000-7_11

equivalent in the known-hardness (or non-uniform) setting, by giving an optimal construction of the former from the latter. In addition, we show how to transform any one-way function that is far from regular (but only weakly regular on a noticeable fraction of domain) into an almost-regular one-way function.

1   Introduction

Informally, a family of compressing hash functions, denoted by $\mathcal{G}$, is called *universal one-way*, if given a random function $g \in \mathcal{G}$ and a random (or equivalently, any pre-fixed) input x, it is infeasible for any efficient algorithm to find any $x' \neq x$ satisfying $g(x) = g(x')$. The seminal result that one-way functions (OWFs) imply universal one-way hash functions (UOWHFs) [17] constitutes one of the central pieces of modern cryptography. Applications of UOWHFs include basing digital signatures [9] on minimal assumptions (one-way functions), Cramer-Shoup encryption scheme [4], statistically hiding commitment scheme [12,13], etc.

UOWHFs FROM ANY OWFS. The principle possibility result that UOWHFs can be based on any OWF was established by Rompel [17] (with some corrections given in [15,18]). However, Rompel's construction was quite complicated and extremely unpractical. In particular, for any one-way function on n-bit inputs it requires key length $\tilde{O}(n^{12})$ and output length $\tilde{O}(n^8)$. Haitner et al. [11] improved the construction via the notion of inaccessible entropy [13], and reduced key and output length to $\tilde{O}(n^7)$. Therefore, even the best known generic UOWHF constructions (based on arbitrary OWFs) are mainly of theoretical interest and are too inefficient to be of any practical use.

UOWHFs FROM SPECIAL OWFS. Another line of research focuses on more efficient (and nearly practical) constructions of UOWHFs from special structured OWFs. Naor and Yung gave an elegant "hash-then-truncate" construction of UOWHFs with key and output length $\Theta(n)$ which does a single call to any one-way permutation. However, for a slightly weaker primitive, namely, 1-to-1 one-way functions, the authors of [16] only gave a rather complicated construction. De Santis and Yung [19] gave an improved construction from any 1-to-1 OWF $f : \{0,1\}^n \to \{0,1\}^l$ as below:

$$\mathcal{G}_{1-1} \stackrel{def}{=} \{(h_{n-1}^n \circ \ldots \circ h_{l-2}^{l-1} \circ h_{l-1}^l \circ f) : \{0,1\}^n \to \{0,1\}^{n-1}, h_{i-1}^i \in \mathcal{H}_{i-1}^i, n \leq i \leq l \},$$

where "$\circ$" denotes function composition, each $\mathcal{H}_{i-1}^i$ denotes a family of pairwise-independent hash functions that compress i-bit strings into $(i-1)$ bits. Although $\mathcal{G}_{1-1}$ enjoys linear output length and a single function call, it requires[1] key length $O(\omega(\log n) \cdot n)$. In addition, the work of [19] also introduced a construction from

---

[1] A straightforward calculation suggests that $\mathcal{G}_{1-1}$ needs key length $O(l \cdot (l - n))$, and we know (see Fact 1) that every 1-to-1 one-way function implies another one-way function $f' : \{0,1\}^{n' \in \Theta(n)} \to \{0,1\}^{n' + \omega(\log n)}$ that is 1-to-1 except on a negligible fraction of inputs, which implies that the key length of [16,19] can be pushed to $O(\omega(\log n) \cdot n)$.

any known-regular[2] one-way function with key and output length $O(\omega(\log^2 n) \cdot n)$ and $O(\omega(1) \cdot \log n)$ adaptive calls, which was recently improved by Barhum and Maurer [3] to key and output length $O(\omega(\log n) \cdot n)$ and $O(\omega(1) \cdot \log n)$ non-adaptive calls. Based on unknown-regular one-way functions, Ames et al. [1] presented a more general construction with output length $\Theta(n)$, key length $O(\log n \cdot n)$ and $\tilde{O}(n)$ adaptive calls. We refer to Table 1 for a summary of previous constructions and a comparison to our work.

Table 1. A summary of existing constructions [1,3,16,19] and our work, where KR-OWF and UR-OWF are the shorthands for known-regular and unknown-regular one-way functions respectively, ε-hard KR-OWF additionally assumes that the hardness parameter ε of KR-OWF is known, and n^{-c}-WUR-OWF is the shorthand for weakly unknown-regular one-way functions (see Footnote 7 and formally Definition 9).

	Assumption	Output Length	Key Length	# of Calls	Type of Call
[16]	OWP	$\Theta(n)$	$\Theta(n)$	1	non-adaptive
[16,19]	1-to-1 OWF	$\Theta(n)$	$O(\omega(\log n) \cdot n)$	1	non-adaptive
[19]	KR-OWF	$O(\omega(\log^2 n) \cdot n)$	$O(\omega(\log^2 n) \cdot n)$	$O(\omega(\log n))$	adaptive
[3]	KR-OWF	$O(\omega(\log n) \cdot n)$	$O(\omega(\log n) \cdot n)$	$O(\omega(\log n))$	non-adaptive
[1]	UR-OWF	$\Theta(n)$	$O(\log n \cdot n)$	$\tilde{O}(n)$	adaptive
ours	1-to-1 OWF	$\Theta(n)$	$\Theta(n)$	1	non-adaptive
ours	ε-hard KR-OWF	$\Theta(n)$	$\Theta(n)$	1	non-adaptive
ours	KR-OWF	$O(\omega(1) \cdot n)$	$O(\omega(1) \cdot n)$	$O(\omega(1))$	non-adaptive
ours	n^{-c}-WUR-OWF	$\Theta(n)$	$O(\log n \cdot n)$	$\tilde{O}(n^{2c+1})$	adaptive

SUMMARY OF OUR CONSTRUCTIONS. In this paper, we give the following constructions from the respective aforementioned one-way functions. The first two constructions enjoy optimal parameters simultaneously and they are (almost) security-preserving[3], the third achieves parameters that are almost optimal up to an arbitrarily small super-constant factor $\omega(1)$ (e.g., $\log \log \log n$ or even less), and thus they all improve upon the respective known constructions. The fourth construction generalizes to beyond regular one-way functions (as introduced in [21]) with optimal output length $\Theta(n)$ and key length $O(n \cdot \log n)$.

1. For any 1-to-1 one-way function, we construct an optimal family of UOWHFs with key and output length $\Theta(n)$ and a single OWF call.

---

[2] A function f is regular if every image has the same number (say α) of preimages, and it is known- (resp., unknown-) regular if α is efficiently computable (resp., inefficient to approximate). More generally (as introduced in [21]), f is weakly unknown-regular if the fraction of x's with maximal $|f^{-1}(f(x))|$ (which is not necessarily efficiently computable) is noticeable. We stress that here "weakly" is used to describe "regularity" (rather than "one-way-ness" as in "weakly one-way functions").

[3] The security of the first UOWHF is essentially the same as the respective OWF, and the security of the second one is roughly a square root of its underlying OWF.

2. For any known-regular one-way function with known hardness, we give another optimal construction of UOWHFs with key and output length $\Theta(n)$ and a single call.
3. For any known-regular one-way function, we give a construction of UOWHFs with key and output length $O(\omega(1)\cdot n)$ and $\omega(1)$ non-adaptive calls.
4. For any one-way function f that is weakly unknown-regular on a noticeable fraction (i.e., n^{-c} for constant c) of domain [21], we give a construction of UOWHFs with key length $O(n\cdot\log n)$ and output length $\Theta(n)$.

ON THE (A)SYMMETRY TO PRGs. Our results further exhibit the inherent "black-box duality" [5,11,13] between UOWHFs and PRGs. Firstly, we abstract out a lemma about universal hashing (see Lemma 1) that is implicit in previous works [13,15,17] and plays a dual role in UOWHF constructions to the leftover hash lemma in PRG constructions. Secondly, constructions #2 and #3 above match the best known results about constructions of PRGs from known-regular OWFs (see [22]), namely, seed length $O(\omega(1)\cdot n)$ or even $\Theta(n)$ if the hardness of the underlying OWF is known. Thirdly, construction #4 is symmetric to the recent PRG construction [21] based on the same class of one-way functions with succinct key/seed length $O(n \cdot \log n)$. Finally (and perhaps more interestingly), construction #1 is asymmetric to the case of PRGs, where we do not know how to construct a linear seed length PRG from an arbitrary 1-to-1 one-way function[4].

ON THE EFFICIENCY, FEASIBILITY AND LIMITS. Constructions #1, #2 and #3 are practically relevant as most one-way function candidates turn out to be known-almost-regular or even 1-to-1. Goldreich, Levin and Nisan [8] showed how to base almost 1-to-1 (except for a negligible fraction) one-way functions on intractable problems such as RSA and DLP, and thus construction #1 enables to build optimal UOWHFs from those problems. A byproduct of construction #2 is the equivalence of almost 1-to-1 one-way functions and known-(almost-)regular one-way functions in certain (known-hardness or non-uniform) settings, where we give an optimal construction of the former from the latter. Moreover, unknown regular one-way functions further reduce the knowledge required about the underlying one-way functions, and the problem of basing cryptographic primitives (PRGs, UOWHFs, etc.) on weaker assumptions is of theoretic interests. It improves our understanding about the feasibility and limits of black-box reductions. In particular, Holenstein and Sinha [14], Barhum and Holenstein [2] showed that $\Omega(n/\log n)$ black-box calls to an arbitrary (including unknown-regular) one-way function is necessary to construct PRGs and UOWHFs, and the lower bound is matched by explicit constructions of PRGs [10] and UOWHFs [1] respectively. The recent work of [21] carried on this line of research even further by

---

[4] Given a 1-to-1 one-way function f, one might think of getting a PRG by hashing $f(U_n)$ into $n - s$ bits concatenated with $s+1$ hard-core bits of f, where $s \in \omega(\log n)$ is the necessary entropy loss due to the leftover hash lemma. This is in general not possible without knowing the exact hardness of the underlying f. See more discussions and the relaxed solutions to this problem by Goldreich [6, Sect. 3.5.1.3].

considering a more general class of one-way functions (which they call weakly unknown-regular one-way functions), namely, the underlying one-way function can have an arbitrary structure as long as the set of x with maximal number of siblings (i.e., x and x' are siblings of each other if $f(x) = f(x')$) is of noticeable fraction. The authors of [21] gave a construction of PRG with seed length $O(n \cdot \log n)$ from weakly unknown-regular OWFs. However, their analysis is quite ad-hoc (see Remark 2), and doesn't seem to generalize to UOWHFs. As an intermediate step of construction #4, we prove that "iterating such a one-way function (weakly regular on only a noticeable fraction) polynomially many times yields a one-way function that is almost-regular on an overwhelming fraction" and thus unify the approach to the two dual objects (i.e., PRGs and UOWHFs).

THE ROADMAP. We outline below the steps to build UOWHFs from the respective one-way function $f : \{0,1\}^n \rightarrow \{0,1\}^l$ introduced above. We note that the following assumptions (about output length) can be made without loss of generality: $l \in O(n)$ for 1-to-1 one-way functions and length-preserving-ness (i.e., $l = n$) for arbitrary one-way functions. More specifically, any 1-to-1 one-way function $f : \{0,1\}^n \rightarrow \{0,1\}^l$ implies a one-way function $f' : \{0,1\}^{n' \in \Theta(n)} \rightarrow \{0,1\}^{l' \in \Theta(n)}$ that is 1-to-1 except for a negligible fraction. Any one-way function f with $\alpha \leq |f^{-1}(y)| \leq \alpha \cdot \beta$ implies another length-preserving one-way function $f' : \{0,1\}^{n' \in \Theta(n)} \rightarrow \{0,1\}^{n'}$ with $\alpha' \leq |f'^{-1}(y)| \leq \alpha' \cdot \beta$ except for a negligible fraction, where the size of range β is preserved, and α' is efficiently computable if α is. We refer to [20] for a full proof.

BASED ON 1-TO-1 OWFS. We adapt Naor-Yung's elegant "hash-then-truncate" approach (for one-way permutation) to any 1-to-1 one-way function:

$$\mathcal{G}_1 \stackrel{def}{=} \{ (\text{trunc} \circ h \circ f) : \{0,1\}^n \rightarrow \{0,1\}^{n-s} , h \in \mathcal{H} \},$$

where $\mathcal{H}$ is a family of universal hash permutations on l bits, and $\text{trunc} : \{0,1\}^l \rightarrow \{0,1\}^{n-s}$ is a truncating function that outputs the first $n - s$ bits of input. We show that if f is a (t,ε)- 1-to-1 OWF then the resulting $\mathcal{G}_1$ is a $(t - n^{O(1)}, 2^{s+1} \cdot \varepsilon)$-UOWHF family with key and output length $\Theta(n)$ and shrinkage s (see Definitions 3 and 7 for formal definitions). The construction enjoys optimal parameters and somewhat counter-intuitively the security bound drops only by factor 2^s (which is optimal by [5]) rather than by 2^{l-n+s} (i.e., exponential in the number of bits truncated which would render the construction useless). We refer to the proof of Theorem 1 and Remark 1 for more technical details and further discussions.

BASED ON KNOWN-(ALMOST-)REGULAR ε-HARD OWFS. Given an almost-regular f (see Definition 6) which is known to be (t,ε)-one-way for some efficiently computable ε, we define the following function family

$$\mathcal{G}_2 \stackrel{def}{=} \{ g : \{0,1\}^n \rightarrow \{0,1\}^{n-s}, g(x) = (\text{trunc}(h(f(x))), h_1(x)), h \in \mathcal{H}, h_1 \in \mathcal{H}_1 \}$$

where $\mathcal{H}$ is a family of universal hash permutations, and let $\mathcal{H}_1$ and trunc be a family of universal hash functions and the truncating function (both with appropriate output sizes) respectively. We show that $\mathcal{G}_2$ is a UOWHF family with key and output length $\Theta(n)$ and shrinkage s. The rationale is that for any[5] $x \neq x'$ colliding on $g \in \mathcal{G}_2$ it either satisfies "$f(x) = f(x') \wedge h_1(x) = h_1(x')$" or "$f(x) \neq f(x') \wedge \mathsf{trunc}(h(f(x))) = \mathsf{trunc}(h(f(x')))$". The former is unconditionally bounded by universal hashing, and the latter is computationally bounded (and reducible to the one-way-ness of f). Interestingly, by abstracting out function $f'(x, h_1) \overset{def}{=} (f(x), h_1(x), h_1)$ from the above construction, we further show that f' is a one-way function that is 1-to-1 except for a negligible fraction. We refer to Theorem 2, Lemma 2 and Theorem 3 for the details.

BASED ON KNOWN-(ALMOST-)REGULAR OWFs. Next, we consider any known-(almost-)regular OWF f whose hardness parameter is ε unknown (i.e., ε is negligible but may not be efficiently computable). In this case, we run q independent copies of f, and we get a construction by making q non-adaptive calls with shrinkage $q \log n$, key and output length $O(q \cdot n)$, where $q \in \omega(1)$ can be any efficiently computable super-constant. The parallel repetition technique was also used in similar contexts (e.g., the construction of PRG from any known regular OWF [22]). We refer to Theorem 4 for the detailed construction and proof.

BASED ON A MORE GENERAL CLASS OF OWFs. We show iterating the class of one-way functions introduced in [21] sufficiently many times yields a one-way function f' that is almost-regular, and thus plugging this f' into the construction of Ames et al. [1] yields a construction of UOWHFs with output length $\Theta(n)$ and key length $O(n \cdot \log n)$.

2   Preliminaries

NOTATIONS AND DEFINITIONS. We use $[n]$ to denote set $\{1, \ldots, n\}$. We use capital letters (e.g., X, Y) for random variables, standard letters (e.g., x, y) for values, and calligraphic letters (e.g. $\mathcal{X}$, $\mathcal{Y}$) for sets. The support of a random variable X, denoted by $\mathsf{Supp}(X)$, refers to the set of values on which X takes with non-zero probability, i.e., $\{x : \Pr[X = x] > 0\}$. For a binary string $x = x_1 \ldots x_n$, denote by $x_{[t]}$ the first t bits of x, i.e., $x_1 \ldots x_t$. $x \| y$ refers the concatenation of x and y. We denote by $\mathsf{trunc} : \{0,1\}^n \rightarrow \{0,1\}^t$ a truncating function that outputs the first t bits of input, i.e., $\mathsf{trunc}(x) = x_{[t]}$. $|\mathcal{S}|$ denotes the cardinality of set $\mathcal{S}$. For function $f : \{0,1\}^n \rightarrow \{0,1\}^{l(n)}$, we use shorthand $f(\{0,1\}^n) \overset{def}{=} \{f(x) : x \in \{0,1\}^n\}$, and denote by $f^{-1}(y)$ the set of y's preimages under f, i.e., $f^{-1}(y) \overset{def}{=} \{x : f(x) = y\}$. We say f is length-preserving if $l(n) = n$. We use $s \leftarrow S$ to denote sampling an element s according to distribution S, and let $s \overset{\$}{\leftarrow} \mathcal{S}$ denote sampling s uniformly from set $\mathcal{S}$, and $y := f(x)$ denote value assignment. We use U_n and $U_{\mathcal{X}}$ to denote uniform distributions over $\{0,1\}^n$ and $\mathcal{X}$ respectively, and let $f(U_n)$ be the distribution induced by applying function

---

[5] More precisely, x is sampled at random and x' can be any distinct value (i.e., $x' \neq x$) efficiently computable from x and g.

f to U_n. For probabilistic algorithm A, we use $\mathsf{A}(x; r)$ to denote the output of A on input x and internal coin r. The min-entropy and max-entropy (see, e.g., [13]) of a random variable X, denoted by $\mathbf{H}_\infty(X)$ and $\mathbf{H}_0(X)$ respectively, are defined as:

$$\mathbf{H}_\infty(X) \overset{def}{=} \log \min_{x \in \mathsf{Supp}(X)} \frac{1}{\Pr[X = x]} \; ; \quad \mathbf{H}_0(X) \overset{def}{=} \log |\mathsf{Supp}(X)|.$$

We use '$+/-$' and '$\cdot$' for addition/subtraction and multiplication between field elements respectively. The zero element of any finite field is denoted by $\mathbf{0}$.

COLLISION PROBABILITY. We use $\mathsf{CP}(X)$ to denote the collision probability of X, i.e., $\mathsf{CP}(X) \overset{def}{=} \sum_x \Pr[X = x]^2$, and denote by $\mathsf{CP}(X|Z)$ the average collision probability of X conditioned on another (possibly correlated) random variable Z by

$$\mathsf{CP}(X|Z) \overset{def}{=} \mathbb{E}_{z \leftarrow Z} \left[\; \sum_x \Pr[X = x| \; Z = z]^2 \; \right].$$

SIMPLIFYING NOTATIONS. Parameters (e.g., ε, r) are said to be known if they are polynomial-time computable from the security parameter n. By notation $f : \{0,1\}^n \to \{0,1\}^l$ we refer to the ensemble of functions $\{f : \{0,1\}^n \to \{0,1\}^{l(n)}\}_{n \in \mathbb{N}}$. As slight abuse of notion, poly might be referring to the set of all polynomials or a certain polynomial, and h might be either a function or its description which will be clear from context. For example, in $h(y) \overset{def}{=} h \cdot y$ the first h denotes a function, the second h refers to a string (a finite field element) that describes the function (i.e., multiplying y by h).

Definition 1 (ρ-almost universal hashing). *A family of functions $\mathcal{H} = \{h : \{0,1\}^l \to \{0,1\}^t\}$ is ρ-almost universal if for any distinct $x_1, x_2 \in \{0,1\}^l$, it holds that*

$$\Pr_{h \overset{\$}{\leftarrow} \mathcal{H}} [h(x_1) = h(x_2)] \le \rho.$$

In the special case $\rho = 2^{-t}$, we say that $\mathcal{H}$ is universal.

Definition 2 (pairwise independent hashing). *A family of functions $\mathcal{H} = \{h : \{0,1\}^l \to \{0,1\}^t\}$ is pairwise independent if any distinct $x_1, x_2 \in \{0,1\}^l$ and any $v_1, v_2 \in \{0,1\}^t$ it holds that $\Pr_{h \overset{\$}{\leftarrow} \mathcal{H}} [h(x_1) = v_1 \wedge h(x_2) = v_2] = 2^{-2t}$.*

Definition 3 (one-way functions). *A sequence of functions $\{f : \{0,1\}^n \to \{0,1\}^{l(n)}\}_{n \in \mathbb{N}}$ is $(t(n), \varepsilon(n))$-one-way if f is polynomial-time computable and for any probabilistic algorithm A of running time $t(n)$*

$$\Pr_{x \overset{\$}{\leftarrow} \{0,1\}^n} [\mathsf{A}(1^n, f(x)) \in f^{-1}(f(x))] \le \varepsilon(n).$$

Hereafter we use simplified notation $f : \{0,1\}^n \to \{0,1\}^{l(n)}$ for the above one-way function, where $t(\cdot)$ and $1/\varepsilon(\cdot)$ are super-polynomial.

Definition 4 (a family of one-way functions). *A sequence of function family* $\mathcal{F} = \{\mathcal{F}_n\}_{n\in\mathbb{N}}$, *where* $\mathcal{F}_n = \{f_u : \{0,1\}^n \to \{0,1\}^{l(n)}, u \in \{0,1\}^{q(n)}\}$, *is* $(t(n),\varepsilon(n))$-*one-way if for any* $n \in \mathbb{N}$, $u \in \{0,1\}^{q(n)}$ *and* $x \in \{0,1\}^n$, *the value* $f_u(x)$ *can be computed in polynomial time, and for any probabilistic algorithm* A *of running time* $t(n)$, *we have that*

$$\Pr_{x\xleftarrow{\$}\{0,1\}^n;\ u\xleftarrow{\$}\{0,1\}^{q(n)}}[\ \mathsf{A}(1^n, u, f_u(x))\in f_u^{-1}(f_u(x))\ ] \ \leq\ \varepsilon(n).$$

We use shorthands $\mathcal{F} = \{f_u : \{0,1\}^n \to \{0,1\}^{l(n)}, u \in \{0,1\}^{q(n)}\}$ *for* $\{\mathcal{F}_n\}_{n\in\mathbb{N}}$.

Definition 5 (almost 1-to-1 functions). *A function* $f : \{0,1\}^n \to \{0,1\}^{l(n)}$ *is* $\varepsilon(n)$-*almost 1-to-1 if there exists a negligible function* $\varepsilon(n)$, *such that for every* $n \in \mathbb{N}$ *we have*

$$\Pr_{x\xleftarrow{\$}\{0,1\}^n}\ [\ \exists x' : x' \neq x \wedge f(x) = f(x')\ ] \leq \varepsilon(n).$$

In particular, f *is 1-to-1 if* $\varepsilon(n) \equiv 0$.

Definition 6 (almost regular functions). *For integer functions* $\alpha = \alpha(n)$ *and* $\beta = \beta(n)$, *a function* $f : \{0,1\}^n \to \{0,1\}^{l(n)}$ *is* α-*regular if for every* $n \in \mathbb{N}$ *and* $x \in \{0,1\}^n$ *we have*

$$|f^{-1}(f(x))| = \alpha.$$

f *is* $(\alpha,\ \alpha\cdot\beta)$-*almost regular if for every* $n \in \mathbb{N}$ *and* $x \in \{0,1\}^n$ *we have*

$$\alpha \ \leq \ |f^{-1}(f(x))| \ \leq \ \alpha \cdot \beta.$$

In particular, f *is known-(almost)-regular if* α *is polynomial-time computable, or otherwise it is called unknown-(almost)-regular. Standard "almost-regularity" for a* (t, ε)-*one-way function* f *refers to that* f *is* $(\alpha,\ \alpha\cdot\beta)$-*almost-regular for* $\beta = \mathsf{poly}(n)$ *or at most* $\beta = (1/\varepsilon)^{\Theta(1)}$ *for certain small constant* $0 < \Theta(1) < 1$.

Definition 7 (UOWHFs [16]). *A sequence of function family* $\mathcal{G} = \{\mathcal{G}_n\}_{n\in\mathbb{N}}$, *where* $\mathcal{G}_n = \{g_u : \{0,1\}^{\ell(n)} \to \{0,1\}^{\ell(n)-s(n)}, u \in \{0,1\}^{q(n)}, \ell \in \mathsf{poly}\}$, *is a family of* $(t(n),\varepsilon(n))$-*universal one-way hash functions if for every* $n \in \mathbb{N}$, $u \in \{0,1\}^{q(n)}$ *and* $x \in \{0,1\}^{\ell(n)}$, *the value* $g_u(x)$ *can be computed in polynomial time, and for every probabilistic algorithm* A *of running time* $t(n)$, *it holds that*

$$\Pr_{x\xleftarrow{\$}\{0,1\}^{\ell(n)};\ u\xleftarrow{\$}\{0,1\}^{q(n)};\ x'\leftarrow\mathsf{A}(1^n,x,u)}[\ x \neq x' \wedge g_u(x) = g_u(x')\ ] \ \leq\ \varepsilon(n).$$

The difference between input and output lengths (i.e., $s(n)$*) is called* **shrinkage**. *For succinctness, hereafter we will use shorthand* $\mathcal{G} = \{g_u : \{0,1\}^{\ell} \to \{0,1\}^{\ell-s}, u \in \{0,1\}^q\}$ *for* $\{\mathcal{G}_n\}_{n\in\mathbb{N}}$ *defined above.*

3 UOWHFs from 1-to-1 One-Way Functions

3.1 A Technical Lemma and Its Applications

We state below a folklore lemma about universal hashing which is symmetric to the leftover hash lemma.

Lemma 1 (The injective hash lemma [20]). *For any integers a, d, k and l satisfying $a \leq l$, let Y be any random variable over $\{0,1\}^l$ with $\mathbf{H}_0(Y) \leq a$, and let $\mathcal{H} \stackrel{def}{=} \{h : \{0,1\}^l \to \{0,1\}^{a+d}\}$ be a family of $(k \cdot 2^{-(a+d)})$-almost universal hash functions. Then, we have that*

$$\Pr_{y \leftarrow Y,\ h \stackrel{\$}{\leftarrow} \mathcal{H}} [\ \exists \tilde{y} \in \mathsf{Supp}(Y) : \ \tilde{y} \neq y \wedge h(\tilde{y}) = h(y)\ ] \leq k \cdot 2^{-d}.$$

Recall that $k = 1$ corresponds to the special case that $\mathcal{H}$ is universal.

We also mention the fact that the input and output lengths of a 1-to-1 one-way function $f : \{0,1\}^n \to \{0,1\}^{l(n)}$ can be assumed to be linearly related (i.e., $l(n) = O(n)$). For almost regular one-way functions, we can even assume that they are length-preserving (i.e., $l(n) = n$). We refer to [20] for the proof of Fact 1.

Fact 1. *For any $r_1 = r_1(n) \leq r_2 = r_2(n)$ and any efficiently computable $\kappa = \kappa(n) \in O(n)$, we have*

1. *Any 1-to-1 (t, ε)-one-way function $f : \{0,1\}^n \to \{0,1\}^l$ implies a $(t - n^{O(1)}, \varepsilon + \mathsf{poly}(n) \cdot 2^{-\kappa})$-one-way function $f' : \{0,1\}^{n' \in \Theta(n)} \to \{0,1\}^{(n'+\kappa) \in \Theta(n)}$ which is 1-to-1 except on a $(\mathsf{poly}(n) \cdot 2^{-\kappa})$-fraction of inputs, i.e.,*

$$\Pr_{x \stackrel{\$}{\leftarrow} \{0,1\}^{n'}} [\ \exists x' \in \{0,1\}^{n'} : x' \neq x \wedge f'(x) = f'(x')\ ] \leq \mathsf{poly}(n) \cdot 2^{-\kappa}$$

2. *Any $(2^{r_1}, 2^{r_2})$-almost regular (t, ε)-one-way function $f : \{0,1\}^n \to \{0,1\}^l$ implies a length-preserving $(t - n^{O(1)}, \varepsilon + \mathsf{poly}(n) \cdot 2^{-(r_1+\kappa)})$-one-way function $\bar{f} : \{0,1\}^{n' \in \Theta(n)} \to \{0,1\}^{n'}$ which is $(2^{\kappa+r_1}, 2^{\kappa+r_2})$-almost regular except on a $(\mathsf{poly}(n) \cdot 2^{-(r_1+\kappa)})$-fraction of inputs, i.e.,*

$$\Pr_{x \stackrel{\$}{\leftarrow} \{0,1\}^{n'}} [\ 2^{\kappa+r_1} \leq |\bar{f}^{-1}(\bar{f}(x))| \leq 2^{\kappa+r_2}\ ] \geq 1 - \mathsf{poly}(n) \cdot 2^{-(r_1+\kappa)}.$$

Therefore, we will assume in the remainder of the paper that the underlying 1-to-1 one-way function has linear output length (i.e., $l(n) = O(n)$) and that the almost-regular and weakly unknown-regular one-way functions are length-preserving (i.e., $l(n) = n$).

3.2   UOWHFs from 1-to-1 OWFs

For a 1-to-1 OWF $f : \{0,1\}^n \to \{0,1\}^l$, we define a cryptographic game between a challenger C and an inverter Inv. That is, C samples a random $y^* \xleftarrow{\$} \{0,1\}^l$ and sends it to Inv, and Inv wins the game iff he comes up with any x' satisfying $f(x') = y^*$. Note that even unbounded Inv wins this game with advantage no more than $2^{-(l-n)}$ (which is probability that $y^* \in f(\{0,1\}^n)$), and Fact 2 states that the chance to win is even smaller for computationally bounded Inv.

Fact 2. *For any 1-to-1 (t,ε)-one-way function $f : \{0,1\}^n \to \{0,1\}^l$ and any probabilistic algorithm Inv of running time t, it holds that*

$$\Pr_{y^* \xleftarrow{\$} \{0,1\}^l} [\, f(\mathsf{Inv}(y^*)) = y^* \,] \; \leq \; 2^{-(l-n)} \cdot \varepsilon.$$

Proof.

$$\Pr_{y^* \xleftarrow{\$} \{0,1\}^l} [f(\mathsf{Inv}(y^*)) = y^*] \leq \Pr_{y^* \xleftarrow{\$} \{0,1\}^l}[y^* \in f(\{0,1\}^n)] \cdot \Pr_{y^* \xleftarrow{\$} f(\{0,1\}^n)}[\, f(\mathsf{Inv}(y^*)) = y^* \,] \leq 2^{-(l-n)} \cdot \varepsilon.$$

Remark 1 (on the proof sketch of Theorem 1). We use a trick to prove Theorem 1. We show that any A that ε'-breaks the TCR of the constructed UOWHF implies an $\mathsf{Inv}^{\mathsf{A}}$ (of almost the same efficiency as A) that wins the above game (i.e., inverting f on a random $y^* \in \{0,1\}^l$) with advantage roughly $2^{n-l-s} \cdot \varepsilon'$. This may seem useless since $l-n$ can be $\Omega(n)$ or even $\mathsf{poly}(n)$. However, by Fact 2 this term (i.e., $2^{n-l-s} \cdot \varepsilon'$) is actually upper bounded by $2^{-(l-n)} \cdot \varepsilon$. The conclusion $\varepsilon' \leq 2^s \varepsilon$ immediately follows by cancelling the factor $(l-n)$. In other words, the security bound does not depend on the number of bits truncated (i.e., $l-n+s$), but only on shrinkage s, and it is tight due to [5].

Theorem 1 (UOWHFs from 1-to-1 OWFs). *Let $f : \{0,1\}^n \to \{0,1\}^{l \in O(n)}$ be any 1-to-1 (t, ε)-one-way function, let $\mathcal{H}$ be a family of permutations[6] over $\{0,1\}^l$ as follows:*

$$\mathcal{H} = \{h : \{0,1\}^l \to \{0,1\}^l \; , \; h(y) \overset{def}{=} h \cdot y, \quad where \;\; y \in GF(2^l), \;\; 0 \neq h \in GF(2^l) \; \},$$

let $\mathsf{trunc} : \{0,1\}^l \to \{0,1\}^{n-s}$ be a truncating function, where $s = s(n)$ is efficiently computable. Then, we have that

$$\mathcal{G}_1 \overset{def}{=} \{ \, (\mathsf{trunc} \circ h \circ f \,) : \{0,1\}^n \to \{0,1\}^{n-s} \, , \; h \in \mathcal{H} \, \}$$

is a family of $(t - n^{O(1)}, 2^{s+1} \cdot \varepsilon)$-UOWHFs with key and output length $\Theta(n)$, and shrinkage s.

---

[6] In fact, $\mathcal{H}$ constitutes a family of universal hash permutations. However, our proofs only use the concrete construction of $\mathcal{H}$ and benefit from its algebraic property over finite fields, rather than assuming a universal $\mathcal{H}$ plus a constructible property [13] (given any x and y there exists a PPT sampler to output $h \xleftarrow{\$} \{h \in \mathcal{H} : h(x) = y\}$).

Algorithm 1. $\mathsf{Inv}^{\mathsf{A}}$ that inverts f on input y^* using random coins (x, v).

Input: $y^* \xleftarrow{\$} \{0,1\}^l$

   Sample $x \xleftarrow{\$} \{0,1\}^n$
   if $f(x) = y^*$ then
      Output x and **terminate.**
   end if

   sample $h := (f(x) - y^*)^{-1} \cdot v$, where $v \xleftarrow{\$} \mathcal{V} = \{v \in \{0,1\}^l \setminus \{\mathbf{0}\} : v_{[n-s]} = \overbrace{0 \ldots 0}^{n-s}\}$
   {The above implies $h \xleftarrow{\$} \{h \in \mathcal{H} : h(f(x))_{[n-s]} = h(y^*)_{[n-s]}\}$ by the $GF(2^l)$ arithmetics. }
   $x' \leftarrow \mathsf{A}(x, h)$
   if $f(x') = y^*$ then
      Output x'
   else
      Output $\perp$
   end if
   Terminate

Proof. Suppose for contradiction that there exists a $\mathcal{G}_1$-collision finder A of running time t' that on input (x, h), breaks the target collision resistance with some non-negligible probability ε', i.e.,

$$\Pr_{x \xleftarrow{\$} \{0,1\}^n, h \xleftarrow{\$} \mathcal{H}}[x' \leftarrow \mathsf{A}(x,h) : x \neq x' \wedge h(f(x))_{[n-s]} = h(f(x'))_{[n-s]}] = \varepsilon' > 2^{s+1} \cdot \varepsilon$$

We define algorithm $\mathsf{Inv}^{\mathsf{A}}$ (that inverts f on input $y^* \xleftarrow{\$} \{0,1\}^l$ by invoking A) as in Algorithm 1. Define event $\mathcal{E}_{\mathsf{neq}} \overset{def}{=} (f(x) \neq y^*)$. We argue that $\mathsf{Inv}^{\mathsf{A}}$ inverts f with the following probability (see the rationale below)

$$\Pr_{y^* \xleftarrow{\$} \{0,1\}^l, x \xleftarrow{\$} \{0,1\}^n, v \xleftarrow{\$} \mathcal{V}}[f(\mathsf{Inv}^{\mathsf{A}}(y^*)) = y^*]$$

$$\geq \Pr_{x \xleftarrow{\$} \{0,1\}^n, y^* \xleftarrow{\$} \{0,1\}^l}[\mathcal{E}_{\mathsf{neq}}] \cdot \Pr_{x \xleftarrow{\$} \{0,1\}^n, y^* \xleftarrow{\$} \{0,1\}^l \setminus \{f(x)\}, v \xleftarrow{\$} \mathcal{V}}[f(\mathsf{Inv}^{\mathsf{A}}(y^*)) = y^* \mid \mathcal{E}_{\mathsf{neq}}]$$

$$\geq (1 - 2^{-l}) \cdot \Pr_{x \xleftarrow{\$} \{0,1\}^n, h \xleftarrow{\$} \mathcal{H}, x' \leftarrow \mathsf{A}(x,h), v \xleftarrow{\$} \mathcal{V}}[x \neq x' \wedge h(f(x))_{[n-s]} = h(f(x'))_{[n-s]} \wedge y^* = f(x')]$$

$$\geq (1 - 2^{-l}) \cdot \varepsilon' \cdot \Pr_{v \xleftarrow{\$} \mathcal{V}}[y^* = f(x') \mid \mathcal{E}_{\mathsf{neq}} \wedge x \neq x' \wedge h(f(x))_{[n-s]} = h(f(x'))_{[n-s]}]$$

$$= \frac{(1 - 2^{-l}) \cdot \varepsilon'}{|\mathcal{V}|} = \frac{(1 - 2^{-l}) \cdot \varepsilon'}{2^{l-n+s} - 1} > \frac{\varepsilon'/2}{2^{l-n+s}} > \varepsilon \cdot 2^{-(l-n)},$$

where the first inequality is straightforward (note that conditioned on $\mathcal{E}_{\mathsf{neq}}$ the sampling of x and y^* are uniform over $\{0,1\}^n$ and $\{0,1\}^l \setminus \{f(x)\}$ respectively), the second inequality follows from Claim 1, namely, conditioned on $\mathcal{E}_{\mathsf{neq}}$ it is equivalent to consider $(x, h, v) \xleftarrow{\$} \{0,1\}^n \times \mathcal{H} \times \mathcal{V}$ and then $y^* := f(x) - v \cdot h^{-1}$, and the third inequality is due to that A takes only x and h as input (i.e., independent of v). That is, conditioned on that A produces a valid $x' \neq x$ satisfying $h(f(x'))_{[n-s]} = h(f(x))_{[n-s]}$, we have by Claim 1 that string y^* is uniformly distributed over set $\mathcal{Y}^* \stackrel{def}{=} \{f(x) - v \cdot h^{-1}, v \in \mathcal{V}\}$. Note that the already fixed $f(x')$ is also an element of $\mathcal{Y}^*$ and thus y^* hits $f(x')$ with probability $1/|\mathcal{Y}^*| = 1/|\mathcal{V}|$. We complete the proof by reaching a contradiction to Fact 2.

Claim 1 (equivalent sampling). *Let the values h, v, x, y^* be sampled as in Algorithm 1, and conditioned on event $\mathcal{E}_{\mathsf{neq}} \stackrel{def}{=} (f(x) \neq y^*)$, it is equivalent to sample $(x, h, v) \stackrel{\$}{\leftarrow} \{0,1\}^n \times \mathcal{H} \times \mathcal{V}$ uniformly and independently and then determine $y^* := f(x) - v \cdot h^{-1}$.*

Proof of Claim 1. We know that (x, v) is uniformly sampled from $\{0,1\}^n \times \mathcal{V}$ by definition, and thus it suffices to show that "fix any (x, v), and conditioned on $y^* \neq f(x)$ (i.e., Y^* is uniform distributed over $\{0,1\}^l \setminus \{f(x)\}$), it holds that h is uniform over $\mathcal{H}$". This follows from that $v \neq \mathbf{0}$ ($\mathcal{V}$ excludes $\mathbf{0}$ by definition) and hence $h = (f(x) - Y^*)^{-1} \cdot v$ is uniform over $\{0,1\}^l \setminus \{\mathbf{0}\}$, namely, $h \stackrel{\$}{\leftarrow} \mathcal{H}$. Finally, for any given (x, h, v), one efficiently determines the value $y^* = f(x) - v \cdot h^{-1}$ due to the arithmetics over the finite field. □

4   UOWHFs from Known Regular OWFs

We proceed to the more general case that f is a known almost-regular function. Recall that by Fact 1 we can assume WLOG that the underlying almost regular one-way function is length-preserving. We first show a construction where the hardness parameter ε is known, and then remove the dependency on ε.

4.1   Compressing the Output Is Necessary but not Sufficient

We attempt to generalize the Naor-Yung approach for one-way permutations (and 1-to-1 one-way functions) to almost regular one-way functions by compressing (using $\mathsf{trunc} \circ h$) the output $Y = f(X)$ into $\mathbf{H}_\infty(Y) - s'$ bits for $s' \in O(\log(1/\varepsilon))$. However, this only gives a weak form of guarantee, as stated in Lemma 2 below, that given a random x it is infeasible for efficient algorithms to find any $f(x') \neq f(x)$ such that $\mathsf{trunc}(h(f(x'))) = \mathsf{trunc}(h(f(x)))$. Otherwise said, it does not rule out the possibility that one may easily find $x' \neq x$ satisfying $f(x') = f(x)$. Hence, compressing the output is only a useful intermediate step to obtain UOWHFs. Lemma 2 below further generalizes Theorem 1 to known-(almost-)regular functions, whose proof is similar to that of Theorem ref1-to-1-OWF (see [20]).

Lemma 2. *For any constant c, any efficiently computable $r = r(n)$ and $s' = s'(n)$, let $f : \{0,1\}^n \to \{0,1\}^n$ be any $(2^r, 2^r n^c)$-almost regular (length-preserving) (t, ε)-one-way function, let $\mathcal{H}$ be a family of permutations over $\{0,1\}^n$ as below*

$$\mathcal{H} = \{h : \{0,1\}^n \to \{0,1\}^n , h(y) \stackrel{def}{=} h \cdot y, \text{ where } y \in GF(2^n), \mathbf{0} \neq h \in GF(2^n) \},$$

let trunc $: \{0,1\}^n \rightarrow \{0,1\}^{n-r-c \cdot \log n - s'}$ *be a truncating function. Then, for any* $\tilde{\mathsf{A}}$ *of running time* $t - n^{O(1)}$ *(for some universal constant* $O(1)$*) we have that*

$$\Pr_{x \xleftarrow{\$} \{0,1\}^n,\ h \xleftarrow{\$} \mathcal{H},\ x' \leftarrow \tilde{\mathsf{A}}(x,h)} [\ f(x) \neq f(x') \ \wedge\ \mathsf{trunc}(h(f(x))) = \mathsf{trunc}(h(f(x')))\ ] \leq n^c \cdot 2^{s'+1} \cdot \varepsilon.$$

4.2   Known (Almost-)Regular OWFs with Known Hardness

We first give an optimal construction assuming that the inversion probability upper bound ε is known. Note that in addition to hashing the output $f(x)$ (as we did in Lemma 2), we also hash the input x to ensure that no distinct x' collides with x with respect to the resulting function.

Theorem 2 (UOWHFs from known-almost-regular ε-hard OWFs). *Let* $f : \{0,1\}^n \rightarrow \{0,1\}^n$ *be any* $(2^r, 2^r n^c)$*-almost regular (length-preserving)* (t, ε)*-one-way function as assumed in Lemma 2. Let shrinkage* $s = s(n)$ *be any efficiently computable function, and let* $\mathcal{H}$ *and* trunc *be as defined in Lemma 2 with* $s' = (s + \log(1/\varepsilon) - c \log n)/2$, *and let* $\mathcal{H}_1 = \{h_1 : \{0,1\}^n \rightarrow \{0,1\}^{r+c \log n + s' - s}\}$ *be a family of universal hash functions. Then, we have that*

$$\mathcal{G}_2 \overset{def}{=} \{\ g : \{0,1\}^n \rightarrow \{0,1\}^{n-s}\ ,\ g(x) \overset{def}{=} (g_1(x), h_1(x)),\ g_1 \in \mathcal{H}\ , h_1 \in \mathcal{H}_1\ \}$$

where $g_1 \overset{def}{=} (\mathsf{trunc} \circ h \circ f)$, *is a* $(t - n^{O(1)}, O(\sqrt{2^s \cdot n^c \cdot \varepsilon}))$*-universal one-way hash function family with key and output length* $\Theta(n)$.

Proof. Define shorthands $\mathcal{E}_1 \overset{def}{=} (x \neq x' \wedge f(x) = f(x') \wedge h_1(x) = h_1(x'))$ and $\mathcal{E}_2 \overset{def}{=} (f(x) \neq f(x') \wedge g_1(x) = g_1(x'))$. For any $\mathcal{G}_2$-collision finder A, we have

$$\Pr_{x \xleftarrow{\$} \{0,1\}^n,\ (h,h_1) \xleftarrow{\$} (\mathcal{H},\mathcal{H}_1),\ x' \leftarrow A(x,h,h_1)} [\ x \neq x' \ \wedge\ g(x) = g(x')\ ]$$

$$\leq \Pr_{x \xleftarrow{\$} \{0,1\}^n,\ (h,h_1) \xleftarrow{\$} (\mathcal{H},\mathcal{H}_1),\ x' \leftarrow A(x,h,h_1)} [\ \mathcal{E}_1 \vee \mathcal{E}_2\ ]$$

$$\leq \Pr_{x \xleftarrow{\$} \{0,1\}^n,\ h_1 \xleftarrow{\$} \mathcal{H}_1} [\ \exists\, x' \neq x \ \wedge\ f(x) = f(x') \ \wedge\ h_1(x) = h_1(x')\ ]$$

$$+ \Pr_{x \xleftarrow{\$} \{0,1\}^n,\ (h,h_1) \xleftarrow{\$} (\mathcal{H},\mathcal{H}_1),\ x' \leftarrow A(x,h,h_1)} [\ f(x) \neq f(x') \ \wedge\ g_1(x) = g_1(x')\ ]$$

$$\leq 2^{-(s'-s)} + n^c \cdot 2^{s'+1} \cdot \varepsilon = \sqrt{2^s \cdot n^c \cdot \varepsilon} + 2\sqrt{2^s \cdot n^c \cdot \varepsilon} = 3\sqrt{2^s \cdot n^c \cdot \varepsilon},$$

where the first inequality refers to that any collision on $g \in \mathcal{G}_2$ (for $x' \neq x$) must satisfy either $\mathcal{E}_1$ or $\mathcal{E}_2$ and the second inequality follows by a union bound. We already know by Lemma 2 that the second term is bounded by $n^c \cdot 2^{s'+1}\varepsilon$, and it thus remains to show that the first term is bounded by $2^{-(s'-s)}$. Conditioned on any $y = f(X)$ random variable X is a flat distribution on a set of size at most $2^r \cdot n^c$, so we apply Lemma 1 (setting $a = r + c \cdot \log n$, $d \geq s' - s$ and $k = 1$) to get

$$\Pr_{x \xleftarrow{\$} \{0,1\}^n,\ h_1 \xleftarrow{\$} \mathcal{H}_1} [\ \exists\ x' \neq x\ \wedge\ f(x) = f(x')\ \wedge\ h_1(x) = h_1(x')\ ]$$

$$= \mathbb{E}_{y \leftarrow f(U_n)} \left[\Pr_{x \xleftarrow{\$} f^{-1}(y),\ h_1 \xleftarrow{\$} \mathcal{H}_1} [\ \exists\ x' \neq x\ \wedge\ f(x) = f(x')\ \wedge\ h_1(x) = h_1(x')\ ] \right]$$

$$\leq \mathbb{E}_{y \leftarrow f(U_n)} [\ 2^{-(s'-s)}\ ] = 2^{-(s'-s)},$$

which completes the proof.

4.3   An Alternative Approach to Sect. 4.2

A neater (and perhaps more intuitive) approach is to construct an almost 1-to-1 one-way function f' (with input and output lengths $\Theta(n)$) based on f (stated as Theorem 3) and then plug f' into Theorem 1 (using f' in place of f)[7]. This statement is interesting in its own right as it implies that almost 1-to-1 one-way functions and known-(almost-)regular one-way functions (with known hardness) are equivalent. Taking a closer look at Theorem 3 we find that this almost 1-to-1 f' is also present (as an intermediate function) in construction $\mathcal{G}_2$ of Theorem 2 (except with slightly different length parameters). Lemmas 3 and 4 state the almost injectiveness and one-way-ness of f' respectively, for which we determine a judicious value for d (assuming knowledge about ε) in Theorem 3 to achieve injectiveness and one-way-ness simultaneously.

Theorem 3 (almost 1-to-1 OWF from almost-regular ε -hard OWF). *Let* $f : \{0,1\}^n \to \{0,1\}^n$ *be any* $(2^r, 2^r n^c)$*-almost regular (length-preserving)* (t,ε)*-one-way function as assumed in Lemma 2. For efficiently computable* $d = d(n) \in \mathbb{N}$, *define*

$$f' : \{0,1\}^n \times \mathcal{H}_1 \to \{0,1\}^n \times \{0,1\}^{r+c \cdot \log n + d} \times \mathcal{H}_1$$

$$'(x, h_1) \overset{def}{=} (f(x), h_1(x), h_1)$$

where $\mathcal{H}_1$ *is a family of universal hash functions from* n *bits to* $r + c \cdot \log n + d$ *bits. Then, for* $d = \frac{\log(1/\varepsilon) - c \cdot \log n - 3}{3}$ *we have that* f' *is* $2\sqrt[3]{\varepsilon \cdot n^c}$*-almost 1-to-1 and* $(t - O(n),\ 2\sqrt[3]{\varepsilon \cdot n^c})$*-one-way with input and output lengths* $\Theta(n)$.

Proof. The almost 1-to-1-ness and one-way-ness of f' follow from Lemmas 3 and 4 respectively by setting parameter $d = \frac{\log(1/\varepsilon) - c \cdot \log n - 3}{3}$.

Lemma 3 (f' is almost 1-to-1 [20]). *f' defined in Therorem 3 is* 2^{-d}*-almost 1-to-1.*

Lemma 4 (f' is one-way [20]). *f' defined in Therorem 3 is a* $(t - O(n),$ $\sqrt{2^{d+3} \cdot n^c \cdot \varepsilon})$*-one-way function.*

---

[7] Strictly speaking, we need to show that the construction works even if the underlying OWF is only 1-to-1 on an overwhelming fraction of inputs. The proof is given in [20].

4.4 UOWHFs from any Known (Almost-)Regular OWFs

REMOVING THE DEPENDENCY ON ε. Unfortunately, Theorem 2 doesn't immediately apply to an arbitrary regular function as in general we assume no knowledge about ε (other than that ε is negligible). To see the difficulty, check the proof of Theorem 2 where the security of the resulting UOWHF is bounded by the sum of two terms, i.e., $2^{-(s'-s)} + n^c \cdot 2^{s'+1} \cdot \varepsilon$. Without knowing ε, one may end up setting some super-polynomial $2^{s'}$ (to make the first term negligible) which kills the second term $n^c \cdot 2^{s'+1} \cdot \varepsilon$. Same problems arise in similar situations (e.g., construction of PRGs from regular OWFs [22]). A remedy for this is parallel repetition: run $q \in \omega(1)$ copies of f on $\boldsymbol{x} = (x_1, \ldots, x_q)$, apply hash-then-truncate (setting $s' = 2 \log n$) to every copy $f(x_i)$, which shrinks the entropies by $2q \log n$ bits and yields a bound $O(\varepsilon \cdot n^{c+2})$. Next, apply a single hashing to $\boldsymbol{x}$ that expands $q \cdot \log n$ bits (to yield another negligible term n^{-q}). This gives a family of UOWHFs with shrinkage $2q \log n - q \log n = q \log n$, and key and output length $O(q \cdot n)$ for any (efficiently computable) $q \in \omega(1)$. The proof is similar in spirit to that of Theorem 2 (see [20]).

Definition 8 (parallel repetition). *For any function $g : \mathcal{X} \to \mathcal{Y}$, we define its q-fold parallel repetition $g^q : \mathcal{X}^q \to \mathcal{Y}^q$ as*

$$g^q(x_1, \ldots, x_q) = (\, g(x_1)\, , \ldots, \, g(x_q)\,).$$

For simplicity, we use shorthand $\boldsymbol{x} \stackrel{def}{=} (x_1, \ldots, x_q)$ and thus $g^q(\boldsymbol{x}) = g^q(x_1, \ldots, x_q)$.

Theorem 4 (UOWHFs from any known almost-regular OWFs). *Let $f : \{0,1\}^n \to \{0,1\}^n$ be any $(2^r, 2^r n^c)$-almost regular (length-preserving) (t, ε)-one-way function as assumed in Lemma 2. Then, for any efficiently computable $q = q(n) = \omega(1)$, let $\mathcal{H}$ and trunc be as defined in Lemma 2 with $s' = 2 \log n$, and let $\mathcal{H}_1 = \{h_1 : \{0,1\}^{q \cdot n} \to \{0,1\}^{q(r+(c+1)\log n)}\}$ be a family of universal hash functions, we have that*

$$\mathcal{G}_3 \stackrel{def}{=} \{\, g : \{0,1\}^{qn} \to \{0,1\}^{qn - q\log n}\, , \, g(\boldsymbol{x}) \stackrel{def}{=} (g_1(\boldsymbol{x}), h_1(\boldsymbol{x})), \, h \in \mathcal{H}\, , h_1 \in \mathcal{H}_1\, \}$$

where $g_1 \stackrel{def}{=} (\text{trunc} \circ h \circ f)^q$, is a $(t - n^{O(1)}, n^{-q} + 2q \cdot n^{c+2} \cdot \varepsilon)$-universal one-way hash function family with key and output length $O(q \cdot n)$, and shrinkage $q \cdot \log n$.

5 Going Beyond Almost-Regular OWFs

Although (almost) optimal, our foregoing constructions need at least almost-regularity, i.e., the one-way function f satisfies $\alpha \le |f^{-1}(f(x))| \le \alpha \cdot \beta$ for all (or at least an overwhelming portion of) x, where α is efficiently computable and $\beta = \text{poly}(n)$ (or at most $\beta = O(\log(1/\varepsilon))$ for an $(\varepsilon^{-1}, \varepsilon)$-hard f). Complementary to our work, Ames et al. [1] gave an elegant construction from unknown-(almost-)regular one-way functions, namely, without knowledge about α, for which they pay a cost of much increased number of one-way function calls (i.e., $O(n/\log n)$) and key length $O(n \log n)$. In this section, we further weaken the assumption so that f can have an arbitrary structure (i.e., β is not bounded) as long as the fraction of x's with (nearly) maximal number of siblings is noticeable.

5.1   A More General Class of OWFs

The following class of one-way functions was introduced in [21] as a relaxation to unknown-(almost-)regular one-way functions.

Definition 9 (weakly unknown-regular OWFs [21]). *Let $f : \{0,1\}^n \to \{0,1\}^{l(n)}$ be a one-way function, and for every $n \in \mathbb{N}$, divide domain $\{0,1\}^n$ into sets $\mathcal{X}_1, \ldots, \mathcal{X}_n$ (i.e., $\mathcal{X}_1 \cup \ldots \cup \mathcal{X}_n = \{0,1\}^n$) such that $\mathcal{X}_j \stackrel{def}{=} \{x : 2^{j-1} \leq |f^{-1}(f(x))| < 2^j\}$, and define $\max = \max(n)$ to be the maximal subscript of the non-empty sets, i.e., $|\mathcal{X}_{\max}| > 0$ and $|\mathcal{X}_{\max+1} \cup \ldots \cup \mathcal{X}_n| = 0$. We say that f is **weakly unknown-regular** if there exists a constant c such that for all sufficiently large n:*

$$\Pr[U_n \in \mathcal{X}_{\max}] \geq n^{-c}. \tag{1}$$

Note that $\max(\cdot)$ can be arbitrary (not necessarily efficient) functions and thus unknown-regular one-way functions fall into a special case[8] for $c = 0$.

5.2   UOWHFs from Beyond Almost-Regular OWFs

We state below the main results of this section, namely, the fourth construction which is based on weakly unknown-regular one-way functions (see Definition 9).

Theorem 5. *Assume that f is a weakly unknown-regular one-way function on an n^{-c}-fraction of domain for constant c. Then, there exists an explicit construction of UOWHF family with output length $\Theta(n)$, key length $O(n \cdot \log n)$ by making $n^{2c+1} \cdot \omega(1)$ black-box calls to f.*

The main idea is to transform any weakly unknown-regular one-way function f into a family of functions $\mathcal{F} = \{f_u : u \in \{0,1\}^{O(n \log n)}\}$ such that $\mathcal{F}$ is almost regular and that it preserves the one-way-ness of f. $\mathcal{F}$ is constructed based on (the derandomized version of) the randomized iterate with a succinct description u. Finally, we sample a random $f_u \stackrel{\$}{\leftarrow} \mathcal{F}$ and plug it into the construction by Ames et al. to get the UOWHFs as desired. We refer to [20] for more details about the explicit construction.

Definition 10 (the randomized iterate [7,10]). *Let $n \in \mathbb{N}$, function $f : \{0,1\}^n \to \{0,1\}^n$, and let $\mathcal{H}$ be a family of pairwise-independent length-preserving hash functions over $\{0,1\}^n$. For $k \in \mathbb{N}$, $x_1 \in \{0,1\}^n$ and vector $\boldsymbol{h}^k = (h_1, \ldots, h_k) \in \mathcal{H}^k$, recursively define the ith randomized iterate by:*

$$x_1 \stackrel{f}{\longrightarrow} y_1 \stackrel{h_1}{\longrightarrow} x_2 \stackrel{f}{\longrightarrow} y_2 \stackrel{h_2}{\longrightarrow} \cdots \quad x_k \stackrel{f}{\longrightarrow} y_k \stackrel{h_k}{\longrightarrow}$$

$$y_i = f(x_i), \ x_{i+1} = h_i(y_i).$$

---

[8] In fact, our construction #4 only assumes a relaxed condition than (1), i.e., $\Pr[U_n \in \mathcal{X}_{\max-O(\log n)} \cup \ldots \cup \mathcal{X}_{\max}] \geq n^{-c}$, so that unknown-almost-regular one-way functions become a special case for $c = 0$.

We denote the ith iterate by function f^i, i.e., $y_i = f^i(x_1, \mathbf{h}^k)$, where $\mathbf{h}^k$ is possibly redundant as for $i \leq k + 1$ y_i only depends on $\mathbf{h}^{i-1}$.

*The **randomized version** refers to the case where $x_1 \overset{\$}{\leftarrow} \{0,1\}^n$ and $\mathbf{h}^k \overset{\$}{\leftarrow} \mathcal{H}^k$.*

*The **derandomized version** refers to that $x_1 \overset{\$}{\leftarrow} \{0,1\}^n$, $u \overset{\$}{\leftarrow} \{0,1\}^{q \in O(n \cdot \log n)}$, $\mathbf{h}^k := BSG(u)$, where $BSG : \{0,1\}^q \to \{0,1\}^{k \cdot \log|\mathcal{H}|}$ is a bounded-space generator that 2^{-2n}-fools every $(2n+1, k, \log|\mathcal{H}|)$-LBP (layered branching program), and $\log|\mathcal{H}|$ is the description length of $\mathcal{H}$ (e.g., $2n$ bits for concreteness).*

Remark 2 (on what is proven in [21]). The authors of [21] introduced weakly unknown-regular one-way functions from which they constructed a pseudorandom generator with seed length $O(n \cdot \log n)$ based on the randomized iterate. They showed that "every $k = n^{2c} \cdot \log n \cdot \omega(1)$ iterations are hard-to-invert", i.e., for any j it is hard to predict x_j given $y_{j+k} = f^{j+k}(x_1, BSG(u))$ and u. A PRG thus follows by outputting $\log n$ hardcore bits for every k iterations. In this paper, we first adapt their findings to show that $f_u(\cdot) = f^k(\cdot, BSG(u))$ constitutes a family of one-way functions, i.e., given $y_k = f_u(x_1)$ and u it is infeasible to find any x_1' such that $y_k = f^k(x_1', BSG(u))$. This is stated as Lemma 6. However, it is still insufficient to construct UOWHFs with the one-way-ness of f_u. We further show in Lemma 7 that a random $f_u \overset{\$}{\leftarrow} \mathcal{F}$ is almost regular (in a slightly weaker sense than Definition 6 but already suffices for our needs).

Following [21], we define the following event and recall some inequalities.

Definition 11. *For any $n, j \leq k \in \mathbb{N}$, define events*

$$\mathcal{E}_j' \overset{def}{=} \left((X_1, U_q) \in \{ (x_1, u) : y_j = f^j(x_1, BSG(u)) \in \mathcal{Y}_{\max} \} \right)$$

where $\mathcal{Y}_{\max} \overset{def}{=} \{y : 2^{\max-1} \leq |f^{-1}(y)| < 2^{\max}\}$, and (X_1, U_q) are uniform over $\{0,1\}^n \times \{0,1\}^q$. Note that by definition $\mathcal{Y}_{\max} = f(\mathcal{X}_{\max})$ (see Definition 9) and thus $\Pr[f(U_n) \in \mathcal{Y}_{\max}] \geq n^{-c}$.

Lemma 5 (Some inequalities from [20])

$$\mathsf{CP}(Y_k' \mid U_q) \leq k \cdot 2^{\max - n + 1} + 2^{-2n}, \tag{2}$$

$$\Pr[\mathcal{E}_1' \lor \mathcal{E}_2' \lor \ldots \lor \mathcal{E}_k'] \geq 1 - 2^{-k/n^{2c}} - 2^{-2n}, \tag{3}$$

where $Y_k' \overset{def}{=} f^k(X_1, BSG(U_q))$.

Lemma 6 ($\mathcal{F}$ is one-way [20]). *Assume that f is a (t, ε)-OWF that is weakly unknown-regular on an n^{-c} fraction of domain, define a family of functions*

$$\mathcal{F} \overset{def}{=} \{ f_u : \{0,1\}^n \to \{0,1\}^n, f_u(x) = f^k(x, BSG(u)), u \in \{0,1\}^{O(n \cdot \log n)} \} \tag{4}$$

where $\mathcal{H}, f^k$ and $BSG : \{0,1\}^{q \in O(n \cdot \log n)} \to \{0,1\}^{k \cdot \log|\mathcal{H}|}$ are as defined in Definition 10. Then, for any A of running time $t - n^{O(1)}$ it holds that

$$\Pr_{u \overset{\$}{\leftarrow} \{0,1\}^q, \ x \overset{\$}{\leftarrow} \{0,1\}^n} [A(u, f_u(x)) \in f_u^{-1}(f_u(x))] \leq \sqrt{2^8 \cdot k^4 \cdot n^{3c} \cdot \varepsilon} + 2^{-k/n^{2c}} + 2^{-2n}. \tag{5}$$

Lemma 7 ($\mathcal{F}$ is almost-regular). *Let $\mathcal{F} = \{f_u\}$ be as defined in Lemma 6. Then, for any $a \geq 0$ it holds that*

$$
\Pr_{u \xleftarrow{\$} \{0,1\}^q,\ x \xleftarrow{\$} \{0,1\}^n} [\ 2^{\max -a-1} \leq |f_u^{-1}(\ f_u(x)\ )| \leq 2^{\max +a+1}\ ] \geq 1 - \frac{k}{2^{a-2}} - \frac{1}{2^{k/n^{2c}}},
$$

(6)

where $u \in \{0,1\}^{q \in O(n \cdot \log n)}$ and $f_u(x) = f^k(x, BSG(u))$.

Proof. We define $\mathcal{S}_{low} \overset{def}{=} \left((X_1, U_q) \in \{(x,u) : 0 < |f_u^{-1}(f_u(x))| < 2^{\max -a-1}\} \right)$ and $\mathcal{S}_{up} \overset{def}{=} \left((X_1, U_q) \in \{(x,u) : |f_u^{-1}(f_u(x))| > 2^{\max +a+1}\} \right)$, where X_1 is uniform over $\{0,1\}^n$. The left-hand of (6) is lower bounded by $1 - \Pr[\mathcal{S}_{low}] - \Pr[\mathcal{S}_{up}]$ and thus it suffices to upper bound both $\Pr[\mathcal{S}_{low}]$ and $\Pr[\mathcal{S}_{up}]$. We have

$$
\Pr[\mathcal{S}_{low}] = \Pr[\mathcal{S}_{low} \wedge (\mathcal{E}_1' \vee \mathcal{E}_2' \vee \ldots \vee \mathcal{E}_k')] + \Pr[\mathcal{S}_{low} \wedge \neg(\mathcal{E}_1' \vee \mathcal{E}_2' \vee \ldots \vee \mathcal{E}_k')]
$$

$$
\leq \Pr[\bigvee_{j=1}^{k} (\mathcal{S}_{low} \wedge \mathcal{E}_j')] + \Pr[\neg(\mathcal{E}_1' \vee \mathcal{E}_2' \vee \ldots \vee \mathcal{E}_k')]
$$

$$
\leq \sum_{j=1}^{k} \Pr[\mathcal{S}_{low} \wedge \mathcal{E}_j'] + (2^{-k/n^{2c}} + 2^{-2n})
$$

$$
\leq k \cdot 2^{-a} + 2^{-k/n^{2c}} + 2^{-2n}
$$

where the first inequality is trivial, the second is by the union bound and (3), and the third is due to that for every $j \in [k]$ with shorthand $f_{u,j}(x) \overset{def}{=} f^j(x, BSG(u))$ it holds that

$$
\Pr[\mathcal{S}_{low} \wedge \mathcal{E}_j'] = \sum_{u} \Pr[U_q = u] \cdot \sum_{x:\ f_{u,j}(x) \in \mathcal{Y}_{\max} \wedge 0 < |f_u^{-1}(f_u(x))| < 2^{\max -a-1}} \Pr[X_1 = x | U_q = u]
$$

$$
\leq \sum_{u} \Pr[U_q = u] \cdot \sum_{x:\ f_{u,j}(x) \in \mathcal{Y}_{\max} \wedge 0 < |f_{u,j}^{-1}(f_{u,j}(x))| < 2^{\max -a-1}} \Pr[X_1 = x | U_q = u]
$$

$$
\leq \sum_{u} \Pr[U_q = u] \cdot |\mathcal{Y}_{\max}| \cdot 2^{\max -a-1} \cdot 2^{-n}
$$

$$
\leq 2^{n+1-\max} \cdot 2^{-n+\max -a-1} = 2^{-a}
$$

where the first inequality is due to Fact 3 (setting $f_1 = f_{u,j}$, $f_2 = f \circ h_{k-1} \circ \ldots \circ f \circ h_j$ and thus $\bar{f} = f_u$), the second follows from the fact that there are $|\mathcal{Y}_{\max}|$ possible values for $f_{u,j}(x) \in \mathcal{Y}_{\max}$ and every $f_{u,j}(x)$ has less than $2^{\max -a-1}$ preimages (by definition of $\mathcal{S}_{low}$), and the third is due to $|\mathcal{Y}_{\max}| \leq 2^{n+1-\max}$. Next we proceed to bounding the second term, i.e., $\Pr[\mathcal{S}_{up}] \leq k \cdot 2^{-a+1}$.

$$k \cdot 2^{\max - n + 1} + 2^{-2n} \geq \mathsf{CP}(\, Y_k' \mid U_q) = \mathbb{E}_{u \leftarrow U_q}\left[\sum_y \Pr[\, f_u(X_1) = y \mid U_q = u]^2\right]$$

$$> 2^{\max + a - n + 1} \cdot \mathbb{E}_{u \leftarrow U_q}\left[\sum_{y:\, |f_u^{-1}(y)| > 2^{\max + a + 1}} \Pr[\, f_u(X_1) = y \mid U_q = u]\right]$$

$$= 2^{\max + a - n + 1} \cdot \Pr[\mathcal{S}_{up}],$$

where the first inequality is by (2), and the second is due to that for any (y, u) satisfying $|f_u^{-1}(y)| > 2^{\max + a + 1}$ and it holds that

$$\Pr[\, f_u(X_1) = y \mid U_q = u] = \Pr[\, X_1 \in f_u^{-1}(y)\,] > 2^{-n} \cdot 2^{\max + a + 1} = 2^{\max + a - n + 1}.$$

It follows that $\Pr[\mathcal{S}_{up}] \leq (k \cdot 2^{\max - n + 1} + 2^{-2n})/2^{\max + a - n + 1} \leq k \cdot 2^{-a + 1}$ and hence completes the proof.

Fact 3. *Let $f_1 : \mathcal{X} \to \mathcal{Y}$ and $f_2 : \mathcal{Y} \to \mathcal{Z}$ be any functions, and let $\bar{f} \stackrel{def}{=} f_2 \circ f_1$. Then for any $t \in \mathbb{N}^+$ it holds that*

$$\{x : 0 < |\bar{f}^{-1}(\bar{f}(x))| < t\} \subseteq \{x : 0 < |f_1^{-1}(f_1(x))| < t\}.$$

Proof. Any x satisfying $0 < |\bar{f}^{-1}(\bar{f}(x))| < t$ implies $0 < |f_1^{-1}(f_1(x))| < t$.

Given that $\mathcal{F}$ is a family of unknown-(almost-)regular one-way functions with description length $O(n \cdot \log n)$, we just plug a random $f_u \in \mathcal{F}$ into the Ames et al.'s construction [1] to yield a family of UOWHFs with output length $\Theta(n)$ and key length $O(n \cdot \log n)$. We refer to a more complete version of this work [20], where we put together all the necessary technical details.

Acknowledgement. This research work was supported by the National Basic Research Program of China (Grant 2013CB338004). Yu Yu was supported by the National Natural Science Foundation of China Grant (Nos. 61472249, 61103221). Dawu Gu was supported by the National Natural Science Foundation of China Grant (Nos. 61472250, 61402286), the Doctoral Fund of Ministry of Education of China (No. 20120073110094) and the Innovation Program by Shanghai Municipal Science and Technology Commission (No. 14511100300). Xiangxue Li was supported by the National Natural Science Foundation of China (Nos. 61472472, 61272536) and Science and Technology Commission of Shanghai Municipality (Grant 13JC1403500). Jian Weng was supported by NSFC under Grant Nos. 61133014, 61472165 and 61272413, the Program for New Century Excellent Talents in University under Grant No. NCET-12-0680, and the Research Fund for the Doctoral Program of Higher Education of China under Grant No. 20100073110060.

References

1. Ames, S., Gennaro, R., Venkitasubramaniam, M.: The generalized randomized iterate and its application to new efficient constructions of UOWHFs from regular one-way functions. In: Wang, X., Sako, K. (eds.) ASIACRYPT 2012. LNCS, vol. 7658, pp. 154–171. Springer, Heidelberg (2012)

2. Barhum, K., Holenstein, T.: A cookbook for black-box separations and a recipe for UOWHFs. In: Sahai, A. (ed.) TCC 2013. LNCS, vol. 7785, pp. 662–679. Springer, Heidelberg (2013)
3. Barhum, K., Maurer, U.: UOWHFs from OWFs: trading regularity for efficiency. In: Hevia, A., Neven, G. (eds.) LatinCrypt 2012. LNCS, vol. 7533, pp. 234–253. Springer, Heidelberg (2012)
4. Cramer, R., Shoup, V.: Design and analysis of practical public-key encryption schemes secure against adaptive chosen ciphertext attack. SIAM J. Comput. **33**(1), 167–226 (2003)
5. Gennaro, R., Gertner, Y., Katz, J., Trevisan, L.: Bounds on the efficiency of generic cryptographic constructions. SIAM J. Comput. **35**(1), 217–246 (2005)
6. Goldreich, O.: Foundations of Cryptography: Basic Tools. Cambridge University Press, New York (2001)
7. Goldreich, O., Krawczyk, H., Luby, M.: On the existence of pseudorandom generators. SIAM J. Comput. **22**(6), 1163–1175 (1993)
8. Goldreich, O., Levin, L.A., Nisan, N.: On constructing 1-1 one-way functions. In: Goldreich, O. (ed.) Studies in Complexity and Cryptography. LNCS, vol. 6650, pp. 13–25. Springer, Heidelberg (2011)
9. Goldwasser, S., Micali, S., Rivest, R.L.: A digital signature scheme secure against adaptive chosen-message attacks. SIAM J. Comput. **17**(2), 281–308 (1988)
10. Haitner, I., Harnik, D., Reingold, O.: On the power of the randomized iterate. In: Dwork, C. (ed.) CRYPTO 2006. LNCS, vol. 4117, pp. 22–40. Springer, Heidelberg (2006)
11. Haitner, I., Holenstein, T., Reingold, O., Vadhan, S., Wee, H.: Universal one-way hash functions via inaccessible entropy. In: Gilbert, H. (ed.) EUROCRYPT 2010. LNCS, vol. 6110, pp. 616–637. Springer, Heidelberg (2010)
12. Haitner, I., Nguyen, M.H., Ong, S.J., Reingold, O., Vadhan, S.P.: Statistically hiding commitments and statistical zero-knowledge arguments from any one-way function. SIAM J. Comput. **39**(3), 1153–1218 (2009)
13. Haitner, I., Reingold, O., Vadhan, S.P., Wee, H.: Inaccessible entropy. In: Proceedings of the 41st ACM Symposium on the Theory of Computing. pp. 611–620 (2009)
14. Holenstein, T., Sinha, M.: Constructing a pseudorandom generator requires an almost linear Number of calls. In: Proceedings of the 53rd IEEE Symposium on Foundation of Computer Science. pp. 698–707 (2012)
15. Katz, J., Koo, C.Y.: On constructing universal one-way hash functions from arbitrary one-way functions. IACR Cryptology ePrint Archive (2005). http://eprint.iacr.org/2005/328
16. Naor, M., Yung, M.: Universal one-way hash functions and their cryptographic applications. In: Johnson, D.S. (ed.) Proceedings of the Twenty First Annual ACM Symposium on Theory of Computing, Seattle, Washington, pp. 33–43, 15–17 May 1989
17. Rompel, J.: One-way functions are necessary and sufficient for secure signatures. In: Proceedings of the Twenty Second Annual ACM Symposium on Theory of Computing, Baltimore, Maryland, pp. 387–394, 14–16 May 1990
18. Rompel, J.: Techniques for computing with low-independence randomness. Ph.D. thesis, Massachusetts Institute of Technology (1990). http://dspace.mit.edu/handle/1721.1/7582
19. De Santis, A., Yung, M.: On the design of provably-secure cryptographic hash functions. In: Damgård, I.B. (ed.) EUROCRYPT 1990. LNCS, vol. 473, pp. 412–431. Springer, Heidelberg (1991)

20. Yu, Y., Gu, D., Li, X., Weng, J.: (Almost) Optimal Constructions of UOWHFs from 1-to-1, Regular One-way Functions and Beyond. Cryptology ePrint Archive, Report 2014/393 (2014). http://eprint.iacr.org/2014/393/
21. Yu, Y., Gu, D., Li, X., Weng, J.: The randomized iterate, revisited - almost linear seed length PRGs from a broader class of one-way functions. In: Dodis, Y., Nielsen, J.B. (eds.) TCC 2015, Part I. LNCS, vol. 9014, pp. 7–35. Springer, Heidelberg (2015)
22. Yu, Y., Li, X., Weng, J.: Pseudorandom generators from regular one-way functions: new constructions with improved parameters. In: Sako, K., Sarkar, P. (eds.) ASIACRYPT 2013, Part II. LNCS, vol. 8270, pp. 261–279. Springer, Heidelberg (2013)

Signatures

Practical Round-Optimal Blind Signatures in the Standard Model

Georg Fuchsbauer[1]([✉]), Christian Hanser[2], and Daniel Slamanig[2]

[1] Institute of Science and Technology Austria, Klosterneuburg, Austria
georg.fuchsbauer@ist.ac.at
[2] IAIK, Graz University of Technology, Graz, Austria
{christian.hanser,daniel.slamanig}@iaik.tugraz.at

Abstract. Round-optimal blind signatures are notoriously hard to construct in the standard model, especially in the malicious-signer model, where blindness must hold under adversarially chosen keys. This is substantiated by several impossibility results. The only construction that can be termed theoretically efficient, by Garg and Gupta (EUROCRYPT'14), requires complexity leveraging, inducing an exponential security loss.

We present a construction of practically efficient round-optimal blind signatures in the standard model. It is conceptually simple and builds on the recent structure-preserving signatures on equivalence classes (SPS-EQ) from ASIACRYPT'14. While the traditional notion of blindness follows from standard assumptions, we prove blindness under adversarially chosen keys under an interactive variant of DDH. However, we neither require non-uniform assumptions nor complexity leveraging.

We then show how to extend our construction to partially blind signatures and to blind signatures on message vectors, which yield a construction of one-show anonymous credentials à la "anonymous credentials light" (CCS'13) in the standard model.

Furthermore, we give the first SPS-EQ construction under non-interactive assumptions and show how SPS-EQ schemes imply conventional structure-preserving signatures, which allows us to apply optimality results for the latter to SPS-EQ.

Keywords: (Partially) Blind signatures · Standard model · SPS-EQ · One-show anonymous credentials

1 Introduction

The concept of blind signatures [22] dates back to the beginning of the 1980s. A blind signature scheme is an interactive protocol where a user (or obtainer) requests a signature on a message which the signer (or issuer) must not learn. In

G. Fuchsbauer—Supported by the European Research Council, ERC Starting Grant (259668-PSPC).
C. Hanser—Supported by EU FP7 through project MATTHEW (GA No. 610436).
C. Hanser, D. Slamanig—Supported by EU FP7 through project FutureID (GA No. 318424).

© International Association for Cryptologic Research 2015
R. Gennaro and M. Robshaw (Eds.): CRYPTO 2015, Part II, LNCS 9216, pp. 233–253, 2015.
DOI: 10.1007/978-3-662-48000-7_12

particular, the signer must not be able to link a signature to the execution of the issuing protocol in which it was produced (*blindness*). Furthermore, it should even for adaptive adversaries be infeasible to produce a valid blind signature without the signing key (*unforgeability*). Blind signatures have proven to be an important building block for cryptographic protocols, most prominently for e-cash, e-voting and one-show anonymous credentials. In more than 30 years of research, many different (> 50) blind signature schemes have been proposed. The spectrum ranges from RSA-based (e.g., [19,22]) over DL-based (e.g., [2,41]) and pairing-based (e.g., [12,14]) to lattice-based (e.g., [44]) constructions, as well as constructions from general assumptions (e.g., [25,35,36]).

Blind Signatures and Their Round Complexity. Two distinguishing features of blind signatures are whether they assume a common reference string (CRS) set up by a trusted party to which everyone has access; and the number of rounds in the signing protocol. Schemes which require only one round of interaction (two moves) are called *round-optimal* [25]. Besides improving efficiency, round optimality also directly yields concurrent security (which otherwise has to be dealt with explicitly; e.g., [35,37]). There are very efficient round-optimal schemes [11,14,23] under interactive assumptions (chosen target one more RSA inversion and chosen target CDH, respectively) in the random oracle model (ROM), as well as under the interactive LRSW [39] assumption in the CRS model [32]. All these schemes are in the honest-key model, where blindness only holds against signers whose keys are generated by the experiment.

Fischlin [25] proposed a generic framework for constructing round-optimal blind signatures in the CRS model with blindness under malicious keys: the signer signs a commitment to the message and the blind signature is a non-interactive zero-knowledge (NIZK) proof of a signed commitment which opens to the message. Using structure-preserving signatures (SPS) [3] and the Groth-Sahai (GS) proof system [33] instead of general NIZKs, this framework was efficiently instantiated in [3]. In [12,13], Blazy et al. gave alternative approaches to compact round-optimal blind signatures in the CRS model which avoid including a GS proof in the final blind signature. Another round-optimal solution with comparable computational costs was proposed by Seo and Cheon [46] building on work by Meiklejohn et al. [40].

Removing the CRS. Known impossibility results indicate that the design of round-optimal blind signatures in the standard model has some limitations. Lindell [38] showed that concurrently secure (and consequently also round-optimal) blind signatures are impossible in the standard model when using simulation-based security notions. This can however be bypassed via game-based security notions, as shown by Hazay et al. [35] for non-round-optimal constructions.

Fischlin and Schröder [27] showed that black-box reductions of blind-signature unforgeability to non-interactive assumptions in the standard model are impossible if the scheme has three moves or less, blindness holds statistically (or computationally if unforgeability and blindness are unrelated) and protocol transcripts allow to verify whether the user is able to derive a signature. Existing constructions [30,31] bypass these results by making non-black-box use of the underlying primitives (and preventing signature-derivation checks in [31]).

Garg et al. [31] proposed the first round-optimal generic construction in the standard model, which can only be considered as a theoretical feasibility result. Using fully homomorphic encryption, the user encrypts the message sent to the signer, who evaluates the signing circuit on the ciphertext. To remove the CRS, they use two-round witness-indistinguishable proofs (ZAPs) to let the parties prove honest behavior; to preserve round-optimality, they include the first fixed round of the ZAP in the signer's public key.

Garg and Gupta [30] proposed the first efficient round-optimal blind signature constructions in the standard model. They build on Fischlin's framework using SPS. To remove a trusted setup, they use a two-CRS NIZK proof system based on GS proofs, include the CRSs in the public key while forcing the signer to honestly generate the CRS. Their construction, however, requires complexity leveraging (the reduction for unforgeability needs to solve a subexponential DL instance for every signing query) and is proven secure with respect to non-uniform adversaries. Consequently, communication complexity is in the order of hundreds of KB (even at a 80-bit security level) and the computational costs (not considered by the authors) seem to limit their practical application even more significantly.

Partially Blind Signatures. Partially blind signatures are an extension of blind signatures, which additionally allow to include common information in a signature. Many non-round-optimal partially blind signature schemes in the ROM are based on a technique by Abe and Okamoto [7]. The latter [42] proposed an efficient construction for non-round-optimal blind as well as partially blind signatures in the standard model. Round-optimal partially blind signatures in the CRS model can again be obtained from Fischlin's framework [25]. Round-optimal partially blind signatures in the CRS model are constructed in [13,40,46]. To date, there is—to the best of our knowledge—no round-optimal partially blind signature scheme that is secure in the standard model.

One-Show Anonymous Credentials Systems. Such systems allow a user to obtain a credential on several attributes from an issuer. The user can later selectively show attributes (or prove relations about attributes) to a verifier without revealing any information about undisclosed attributes. No party (including the issuer) can link the issuing of a credential to any of its showings, yet different showings of the same credential are linkable. An efficient implementation of one-show anonymous credentials is Microsoft's U-Prove [16].

Baldimtsi and Lysyanskaya [9] showed that the underlying signature scheme [15] cannot be proven secure using known techniques. To mitigate this problem, in [8] they presented a generic construction of one-show anonymous credentials in the vein of Brands' [15] approach from so-called blind signatures with attributes. They also present a scheme based on a non-round-optimal blind signature scheme by Abe [2] and prove their construction secure in the ROM.

Our Contribution

Blind Signatures and Anonymous Credentials. Besides Fischlin's generic *commit-prove* paradigm [25], there are other classes of schemes. For instance,

RSA and BLS blind signatures [11,14,23] follow a *randomize-derandomize* approach, which exploits the homomorphic property of the respective signature scheme. Other approaches follow the *commit-rerandomize-transform* paradigm, where a signature on a commitment to a message can be transformed into a rerandomized (unlinkable) signature on the original message [12,32]. Our construction is based on a new concept, which one may call *commit-randomize-derandomize-open* approach. It does not use non-interactive proofs at all and is solely based on the recent concept of structure-preserving signature schemes on equivalence classes (SPS-EQ) [34] and commitments. As we also avoid a trusted setup of the commitment parameters, we do not require a CRS. We do however prove our scheme secure under interactive hardness assumptions.

In SPS-EQ the message space is partitioned into equivalence classes and given a signature on a message anyone can *adapt* the signature to a different representative of the same class. SPS-EQ requires that after signing a representative a signer cannot distinguish between an adapted signature for a new representative of the same class and a fresh signature on a completely random message.

In our blind-signature scheme the obtainer combines a commitment to the message with a normalization element yielding a representative of an equivalence class (*commit*). She chooses a random representative of the same class (*randomize*), on which the signer produces a signature. She then adapts the signature to the original representative containing the commitment (*derandomize*), which can be done without requiring the signing key. The blind signature is the rerandomized (unlinkable) signature for the original representative plus an opening for the commitment (*open*). Our contributions to blind signatures are the following:

- We propose a new approach to constructing blind signatures in the standard model based on SPS-EQ. It yields conceptually simple and compact constructions and does not rely on techniques such as complexity leveraging. Our blind signatures are practical in terms of key size, signature size, communication and computational effort (when implemented with known instantiations of SPS-EQ [29], a blind signature consists of 5 bilinear-group elements).
- We provide the first construction of round-optimal partially blind signatures in the standard model, which follow straightforwardly from our blind signatures and are almost as efficient.
- We generalize our blind signature scheme to message vectors, which yields one-show anonymous credentials à la "anonymous credentials light" [8]. We thus obtain one-show anonymous credentials secure in the standard model (whereas all previous ones have either no security proof or ones in the ROM).

SPS-EQ. We give the first structure-preserving signatures on equivalence classes satisfying all security notions from [34] under non-interactive assumptions. (Unfortunately, the scheme does not have all the properties required for building blind signatures from it, for which we strengthen the notions from [34].)

Moreover, we show how any SPS-EQ scheme can be turned into a standard structure-preserving signature scheme. This transformation allows us to apply the optimality criteria by Abe et al. [4,5] to SPS-EQ. We conclude that the scheme from [29] is optimal in terms of signature size and verification complexity and that it cannot be proven unforgeable under non-interactive assumptions.

2 Preliminaries

A function $\epsilon \colon \mathbb{N} \to \mathbb{R}^+$ is *negligible* if $\forall c > 0 \; \exists k_0 \; \forall k > k_0 : \epsilon(k) < 1/k^c$. By $a \xleftarrow{R} S$ we denote that a is chosen uniformly at random from a set S. We write $A(a_1, \ldots, a_n; r)$ to make the randomness r used by a probabilistic algorithm $A(a_1, \ldots, a_n)$ explicit. If $\mathbb{G}$ is an (additive) group then $\mathbb{G}^*$ denotes $\mathbb{G} \setminus \{0_{\mathbb{G}}\}$.

Definition 1 (Bilinear Map). Let $(\mathbb{G}_1, +)$, $(\mathbb{G}_2, +)$, generated by P and $\hat{P}$, resp., and $(\mathbb{G}_T, \cdot)$ be cyclic groups of prime order p. We call $e \colon \mathbb{G}_1 \times \mathbb{G}_2 \to \mathbb{G}_T$ a *bilinear map (pairing)* if it is efficiently computable and the following holds:

Bilinearity: $e(aP, b\hat{P}) = e(P, \hat{P})^{ab} = e(bP, a\hat{P}) \quad \forall a, b \in \mathbb{Z}_p$.
Non-degeneracy: $e(P, \hat{P}) \neq 1_{\mathbb{G}_T}$, i.e., $e(P, \hat{P})$ generates $\mathbb{G}_T$.

If $\mathbb{G}_1 = \mathbb{G}_2$, then e is *symmetric* (Type-1) and *asymmetric* (Type-2 or 3) otherwise. For Type-2 pairings there is an efficiently computable isomorphism $\Psi \colon \mathbb{G}_2 \to \mathbb{G}_1$; for Type-3 pairings no such isomorphism is known. Type-3 pairings are currently the optimal choice in terms of efficiency and security trade-off [21].

Definition 2 (Bilinear-Group Generator). A bilinear-group generator is a polynomial-time algorithm BGGen that takes a security parameter 1^κ and outputs a bilinear group $\mathsf{BG} = (p, \mathbb{G}_1, \mathbb{G}_2, \mathbb{G}_T, e, P, \hat{P})$ consisting of groups $\mathbb{G}_1 = \langle P \rangle$, $\mathbb{G}_2 = \langle \hat{P} \rangle$ and $\mathbb{G}_T$ of prime order p with $\log_2 p = \kappa$ and a pairing $e \colon \mathbb{G}_1 \times \mathbb{G}_2 \to \mathbb{G}_T$. In this work we assume that BGGen is a *deterministic* algorithm.[1]

Definition 3 (Decisional Diffie-Hellman Assumption). Let BGGen be a bilinear-group generator that outputs $\mathsf{BG} = (p, \mathbb{G}_1, \mathbb{G}_2, \mathbb{G}_T, e, P_1 = P, P_2 = \hat{P})$. The DDH assumption holds in $\mathbb{G}_i$ for BGGen if for all probabilistic polynomial-time (PPT) adversaries $\mathcal{A}$ there is a negligible function $\epsilon(\cdot)$ such that

$$\Pr\left[\begin{matrix} b \xleftarrow{R} \{0,1\}, \; \mathsf{BG} = \mathsf{BGGen}(1^\kappa), \; r, s, t \xleftarrow{R} \mathbb{Z}_p \\ b^* \leftarrow \mathcal{A}(\mathsf{BG}, rP_i, sP_i, ((1 - b) \cdot t + b \cdot rs)P_i) \end{matrix} : b^* = b \right] - \frac{1}{2} \leq \epsilon(\kappa).$$

Definition 4 ((Symmetric) External Diffie-Hellman Assumption). The XDH and SXDH assumptions hold for BGGen if the DDH assumption holds in $\mathbb{G}_1$ and holds in both $\mathbb{G}_1$ and $\mathbb{G}_2$, respectively.

The next assumption is a static computational assumption derived from the SXDH version of the q-Diffie-Hellman inversion assumption [21].

Definition 5 (Co-Diffie-Hellman Inversion Assumption). Let BGGen be a bilinear-group generator that outputs $\mathsf{BG} = (p, \mathbb{G}_1, \mathbb{G}_2, \mathbb{G}_T, e, P_1 = P, P_2 = \hat{P})$. The co-DHI$^*_i$ assumption holds for BGGen if for every PPT adversary $\mathcal{A}$ there is a negligible function $\epsilon(\cdot)$ such that

$$\Pr\left[\mathsf{BG} = \mathsf{BGGen}(1^\kappa), \; a \xleftarrow{R} \mathbb{Z}_p^* : \tfrac{1}{a} P_i \leftarrow \mathcal{A}(\mathsf{BG}, aP_1, aP_2) \right] \leq \epsilon(\kappa).$$

---

[1] This is e.g. the case for BN-curves [10]; the most common choice for Type-3 pairings.

co-DHI$_1^*$ is implied by a variant of the decision linear assumption in asymmetric groups stating that given $(BG, (aP_j, bP_j)_{j \in [2]}, raP_2, sbP_2)$ for $a, b, r, s \xleftarrow{R} \mathbb{Z}_p^*$ it is hard to distinguish $T = (r + s)P_2$ from a random $\mathbb{G}_2$ element. (A co-DHI$_i^*$ solver could be used to compute $\frac{1}{a}P_1$ and $\frac{1}{b}P_1$, which enables to check whether $e(\frac{1}{a}P_1, raP_2) \, e(\frac{1}{b}P_1, sbP_2) = e(P_1, T)$.) This holds analogously for co-DHI$_2^*$.

Generalized Pedersen Commitments. These are commitments to a vector of messages $\boldsymbol{m} = (m_i)_{i \in [n]} \in \mathbb{Z}_p^n$ that consist of one group element. They are perfectly hiding and computationally binding under the discrete-log assumption.

Setup$_P(1^\kappa, n)$: Choose a group $\mathbb{G}$ of prime order p with $\log_2 p = \kappa$ and $n + 1$ distinct generators $(P_i)_{i \in [n]}, Q$ and output parameters $\mathsf{cpp} \leftarrow (\mathbb{G}, p, (P_i)_{i \in [n]}, Q)$ (which is an implicit input to the following algorithms).

Commit$_P(\boldsymbol{m}; r)$: On input a vector $\boldsymbol{m} \in \mathbb{Z}_p^n$ and randomness $r \in \mathbb{Z}_p$, output a commitment $C \leftarrow \sum_{i \in [n]} m_i P_i + rQ$ and an opening $O \leftarrow (\boldsymbol{m}, r)$.

Open$_P(C, O)$: On input $C \in \mathbb{G}$ and $O = (\boldsymbol{m}, r)$, if $C = \sum_{i \in [n]} m_i P_i + rQ$ then output $\boldsymbol{m} = (m_i)_{i \in [n]}$; else output $\perp$.

Remark 1. Setup$_P$ is typically run by a trusted party; it can however also be run by the receiver since commitments are perfectly hiding.

2.1   Structure-Preserving Signatures on Equivalence Classes

Structure-preserving signatures (SPS) [3,4,6,18] can sign elements of a bilinear group without requiring any prior encoding. In such a scheme public keys, messages and signatures consist of group elements only and the verification algorithm evaluates a signature by deciding group membership and evaluating pairing-product equations (PPEs).

The notion of SPS on equivalence classes (SPS-EQ) was introduced by Hanser and Slamanig [34]. Their initial instantiation turned out to only be secure against random-message attacks (cf. [28] and the updated full version of [34]), but together with Fuchsbauer [29] they subsequently presented a scheme that is unforgeable under chosen-message attack (EUF-CMA) in the generic group model.

The concept of SPS-EQ is as follows. Let p be a prime and $\ell > 1$; then $\mathbb{Z}_p^\ell$ is a vector space and we can define a projective equivalence relation on it, which propagates to $\mathbb{G}_i^\ell$ and partitions $\mathbb{G}_i^\ell$ into equivalence classes. Let $\sim_\mathcal{R}$ be this relation, i.e., for $M, N \in \mathbb{G}_i^\ell : M \sim_\mathcal{R} N \Leftrightarrow \exists s \in \mathbb{Z}_p^* : M = sN$. An SPS-EQ scheme signs an equivalence class $[M]_\mathcal{R}$ for $M \in (\mathbb{G}_i^*)^\ell$ by signing a representative M of $[M]_\mathcal{R}$. It then allows for switching to other representatives of $[M]_\mathcal{R}$ and updating the signature without access to the secret key. An important property of SPS-EQ is *class-hiding*, which roughly means that two message-signature pairs corresponding to the same class should be unlinkable.

Here, we discuss the abstract model and the security model of such a signature scheme, as introduced in [34].

Definition 6 (Structure-Preserving Signatures on Equivalence Classes). An SPS-EQ scheme SPS-EQ on $(\mathbb{G}_i^*)^\ell$ (for $i \in \{1, 2\}$) consists of the following PPT algorithms:

$\mathsf{BGGen}_\mathcal{R}(1^\kappa)$, a bilinear-group generation algorithm, which on input a security parameter κ outputs an asymmetric bilinear group BG.

$\mathsf{KeyGen}_\mathcal{R}(\mathsf{BG}, \ell)$, on input BG and vector length $\ell > 1$, outputs a key pair $(\mathsf{sk}, \mathsf{pk})$.

$\mathsf{Sign}_\mathcal{R}(M, \mathsf{sk})$, given a representative $M \in (\mathbb{G}_i^*)^\ell$ and a secret key sk, outputs a signature σ for the equivalence class $[M]_\mathcal{R}$.

$\mathsf{ChgRep}_\mathcal{R}(M, \sigma, \mu, \mathsf{pk})$, on input a representative $M \in (\mathbb{G}_i^*)^\ell$ of class $[M]_\mathcal{R}$, a signature σ on M, a scalar μ and a public key pk, returns an updated message-signature pair (M', σ'), where $M' = \mu \cdot M$ is the new representative and σ' its updated signature.

$\mathsf{Verify}_\mathcal{R}(M, \sigma, \mathsf{pk})$ is deterministic and, on input a representative $M \in (\mathbb{G}_i^*)^\ell$, a signature σ and a public key pk, outputs 1 if σ is valid for M under pk and 0 otherwise.

$\mathsf{VKey}_\mathcal{R}(\mathsf{sk}, \mathsf{pk})$ is a deterministic algorithm, which given a secret key sk and a public key pk outputs 1 if the keys are consistent and 0 otherwise.

An SPS-EQ scheme must satisfy *correctness*, *EUF-CMA security* and *class-hiding*.

Definition 7 (Correctness). An SPS-EQ scheme SPS-EQ on $(\mathbb{G}_i^*)^\ell$ is *correct* if for all $\kappa \in \mathbb{N}$, all $\ell > 1$, all key pairs $(\mathsf{sk}, \mathsf{pk}) \leftarrow \mathsf{KeyGen}_\mathcal{R}(\mathsf{BGGen}_\mathcal{R}(1^\kappa), \ell)$, all messages $M \in (\mathbb{G}_i^*)^\ell$ and all $\mu \in \mathbb{Z}_p^*$: $\mathsf{VKey}_\mathcal{R}(\mathsf{sk}, \mathsf{pk}) = 1$,

$$\Pr\left[\mathsf{Verify}_\mathcal{R}(M, \mathsf{Sign}_\mathcal{R}(M, \mathsf{sk}), \mathsf{pk}) = 1\right] = 1 \quad \text{and}$$
$$\Pr\left[\mathsf{Verify}_\mathcal{R}(\mathsf{ChgRep}_\mathcal{R}(M, \mathsf{Sign}_\mathcal{R}(M, \mathsf{sk}), \mu, \mathsf{pk}), \mathsf{pk}) = 1\right] = 1.$$

In contrast to standard signatures, EUF-CMA security is defined with respect to equivalence classes, i.e., a forgery is a signature on a message from an equivalence class from which no message has been signed.

Definition 8 (EUF-CMA). An SPS-EQ scheme SPS-EQ is *existentially unforgeable under adaptively chosen-message attacks*, if for all PPT algorithms $\mathcal{A}$ with access to a signing oracle $\mathcal{O}$, there is a negligible function $\epsilon(\cdot)$ such that:

$$\Pr\left[\begin{array}{l} \mathsf{BG} \leftarrow \mathsf{BGGen}_\mathcal{R}(1^\kappa), \\ (\mathsf{sk}, \mathsf{pk}) \leftarrow \mathsf{KeyGen}_\mathcal{R}(\mathsf{BG}, \ell), \quad : \quad \begin{array}{l} [M^*]_\mathcal{R} \neq [M]_\mathcal{R} \ \forall M \in \mathcal{Q} \ \wedge \\ \mathsf{Verify}_\mathcal{R}(M^*, \sigma^*, \mathsf{pk}) = 1 \end{array} \\ (M^*, \sigma^*) \leftarrow \mathcal{A}^{\mathcal{O}(\cdot, \mathsf{sk})}(\mathsf{pk}) \end{array}\right] \leq \epsilon(\kappa),$$

where $\mathcal{Q}$ is the set of queries that $\mathcal{A}$ has issued to the signing oracle $\mathcal{O}$.

Class-hiding is defined in [34] and uses the following oracles and a list $\mathcal{Q}$ to keep track of queried messages M.

$\mathcal{O}^{RM}$: Pick a message $M \xleftarrow{R} (\mathbb{G}_i^*)^\ell$, append it to $\mathcal{Q}$ and return M.

$\mathsf{BGGen}_{\mathcal{R}}(1^{\kappa})$: Generate a Type-3 bilinear group BG with order p of bitlength κ.

$\mathsf{KeyGen}_{\mathcal{R}}(\mathsf{BG}, \ell)$: On input BG and vector length $\ell > 1$, choose $(x_i)_{i \in [\ell]} \xleftarrow{R} (\mathbb{Z}_p^*)^{\ell}$, set $\mathsf{sk} \leftarrow (x_i)_{i \in [\ell]}$, $\mathsf{pk} \leftarrow (\hat{X}_i)_{i \in [\ell]} = (x_i \hat{P})_{i \in [\ell]}$ and output $(\mathsf{sk}, \mathsf{pk})$.

$\mathsf{Sign}_{\mathcal{R}}(M, \mathsf{sk})$: Given a representative $M = (M_i)_{i \in [\ell]} \in (\mathbb{G}_1^*)^{\ell}$ of class $[M]_{\mathcal{R}}$ and secret key $\mathsf{sk} = (x_i)_{i \in [\ell]}$, choose $y \xleftarrow{R} \mathbb{Z}_p^*$ and output $\sigma = (Z, Y, \hat{Y})$ with

$$Z \leftarrow y \sum_{i \in [\ell]} x_i M_i \qquad Y \leftarrow \tfrac{1}{y} P \qquad \hat{Y} \leftarrow \tfrac{1}{y} \hat{P}$$

$\mathsf{Verify}_{\mathcal{R}}(M, \sigma, \mathsf{pk})$: Given $M = (M_i)_{i \in [\ell]} \in (\mathbb{G}_1^*)^{\ell}$, $\sigma = (Z, Y, \hat{Y}) \in \mathbb{G}_1 \times \mathbb{G}_1^* \times \mathbb{G}_2^*$ and public key $\mathsf{pk} = (\hat{X}_i)_{i \in [\ell]}$, output 1 if the following hold and 0 otherwise:

$$\prod_{i \in [\ell]} e(M_i, \hat{X}_i) = e(Z, \hat{Y}) \qquad e(Y, \hat{P}) = e(P, \hat{Y})$$

$\mathsf{ChgRep}_{\mathcal{R}}(M, \sigma, \mu, \mathsf{pk})$: Given representative $M = (M_i)_{i \in [\ell]} \in (\mathbb{G}_1^*)^{\ell}$, $\sigma = (Z, Y, \hat{Y})$, scalar $\mu \in \mathbb{Z}_p^*$ and pk, return $\bot$ if $\mathsf{Verify}_{\mathcal{R}}(M, \sigma, \mathsf{pk}) = 0$. Otherwise pick $\psi \xleftarrow{R} \mathbb{Z}_p^*$ and return $(\mu M, \sigma')$ with $\sigma' \leftarrow (\psi \mu Z, \tfrac{1}{\psi} Y, \tfrac{1}{\psi} \hat{Y})$.

$\mathsf{VKey}_{\mathcal{R}}(\mathsf{sk}, \mathsf{pk})$: Given $\mathsf{sk} = (x_i)_{i \in [\ell]} \in (\mathbb{Z}_p^*)^{\ell}$ and $\mathsf{pk} = (\hat{X}_i)_{i \in [\ell]} \in (\mathbb{G}_2^*)^{\ell}$, output 1 if $x_i \hat{P} = \hat{X}_i \ \forall i \in [\ell]$ and 0 otherwise.

Scheme 1. EUF-CMA-secure construction of an SPS-EQ scheme

$\mathcal{O}^{RoR}(M, \mathsf{sk}, \mathsf{pk}, b)$: Given message M, key pair $(\mathsf{sk}, \mathsf{pk})$ and bit b, return $\bot$ if $M \notin \mathcal{Q}$. On the first valid call, record M and $\sigma \leftarrow \mathsf{Sign}_{\mathcal{R}}(M, \mathsf{sk})$; if later called on $M' \neq M$, return $\bot$. Pick $R \xleftarrow{R} (\mathbb{G}_i^*)^{\ell}$ and $\mu \xleftarrow{R} \mathbb{Z}_p^*$, set $(M_0, \sigma_0) \leftarrow \mathsf{ChgRep}_{\mathcal{R}}(M, \sigma, \mu, \mathsf{pk})$ and $(M_1, \sigma_1) \leftarrow (R, \mathsf{Sign}_{\mathcal{R}}(R, \mathsf{sk}))$ and return (M_b, σ_b).

Definition 9 (Class-Hiding). An SPS-EQ scheme SPS-EQ on $(\mathbb{G}_i^*)^{\ell}$ is called *class-hiding* if for all $\ell > 1$ and PPT adversaries $\mathcal{A}$ with oracle access to $\mathcal{O} \leftarrow \{\mathcal{O}^{RM}, \mathcal{O}^{RoR}(\cdot, \mathsf{sk}, \mathsf{pk}, b)\}$ there is a negligible function $\epsilon(\cdot)$ such that

$$\Pr \left[\begin{matrix} \mathsf{BG} \leftarrow \mathsf{BGGen}_{\mathcal{R}}(1^{\kappa}), \ b \xleftarrow{R} \{0,1\}, & b^* = b \ \wedge \\ (\mathsf{st}, \mathsf{sk}, \mathsf{pk}) \leftarrow \mathcal{A}(\mathsf{BG}, \ell), \ b^* \leftarrow \mathcal{A}^{\mathcal{O}}(\mathsf{st}, \mathsf{sk}, \mathsf{pk}) \end{matrix} : \mathsf{VKey}_{\mathcal{R}}(\mathsf{sk}, \mathsf{pk}) = 1 \right] - \frac{1}{2} \leq \epsilon(\kappa).$$

Fuchsbauer, Hanser and Slamanig [29] present an EUF-CMA-secure scheme, which we give as Scheme 1, and prove the following.

Theorem 1. *Scheme 1 is EUF-CMA secure against generic forgers and class-hiding under the DDH assumption.*

3 New Results on SPS-EQ

In the following, we present the first standard-model construction of SPS-EQ as modeled in [34]. We then introduce new properties to characterize SPS-EQ constructions, strengthening the notion of class-hiding. Finally, we show

---

BGGen$'_\mathcal{R}(1^\kappa)$: Output BG $\leftarrow$ BGGen$_\mathcal{R}(1^\kappa)$.

KeyGen$'_\mathcal{R}($BG$, \ell)$: Given BG and $\ell > 1$, output (sk, pk) $\leftarrow$ KeyGen$_\mathcal{R}($BG$, \ell + 2)$.

Sign$'_\mathcal{R}(M, sk)$: Given $M = (M_i)_{i \in [\ell]} \in (\mathbb{G}_1^*)^\ell$ and sk, choose $(R_1, R_2) \xleftarrow{R} (\mathbb{G}_1^*)^2$, compute $\tau \leftarrow$ Sign$_\mathcal{R}((M, R_1, R_2), sk)$ and output $\sigma \leftarrow (\tau, R_1, R_2)$.

Verify$'_\mathcal{R}(M, \sigma, pk)$: Given $M = (M_i)_{i \in [\ell]} \in (\mathbb{G}_1^*)^\ell$, signature $\sigma \leftarrow (\tau, R_1, R_2)$ and pk, return Verify$_\mathcal{R}((M, R_1, R_2), \tau, pk)$.

ChgRep$'_\mathcal{R}(M, \sigma, \mu, pk)$: Given $M = (M_i)_{i \in [\ell]} \in (\mathbb{G}_1^*)^\ell$, $\sigma \leftarrow (\tau, R_1, R_2)$, $\mu \in \mathbb{Z}_p^*$ and pk, run $((\tilde{M}, \tilde{R}_1, \tilde{R}_2), \tilde{\tau}) \leftarrow$ ChgRep$_\mathcal{R}((M, R_1, R_2), \tau, \mu, pk)$ and output $(\tilde{M}, \tilde{\sigma})$ with $\tilde{\sigma} \leftarrow (\tilde{\tau}, \tilde{R}_1, \tilde{R}_2)$ (or $\perp$ if ChgRep$_\mathcal{R}$ output $\perp$).

VKey$'_\mathcal{R}($sk, pk$)$: Return VKey$_\mathcal{R}($sk, pk$)$.

Scheme 2. Standard-model SPS-EQ construction from Scheme 1

how to turn any SPS-EQ construction into an SPS construction. This does not only provide a new, efficient standard-model SPS scheme derived from our SPS-EQ scheme; it also allows us to infer optimality of the SPS-EQ scheme from [29], (Scheme 1) and the impossibility of basing its EUF-CMA security on non-interactive assumptions.

3.1 A Standard-Model SPS-EQ Construction

Following the approach by Abe et al. [4], we construct from scheme SPS-EQ, given as Scheme 1, an SPS-EQ scheme SPS-EQ', given as Scheme 2, and prove that it satisfies EUF-CMA and class-hiding, both under non-interactive assumptions.

The scheme for ℓ-length messages is simply Scheme 1 with message space $(\mathbb{G}_1^*)^{\ell+2}$, where before each signing two random group elements are appended to the message. Scheme 2 features constant-size signatures ($4\,\mathbb{G}_1 + 1\,\mathbb{G}_2$ elements), has public keys of size $\ell + 2$ and still uses 2 PPEs for verification.

Unforgeability follows from a q-type assumption that states that Scheme 1 for $\ell = 2$ is secure against *random-message attacks*. (That is, no PPT adversary, given the public key and signatures on q random messages, can, with non-negligible probability, output a message-signature pair for an equivalence class that was not signed.) Class-hiding follows from class-hiding of Scheme 1. Both proofs can be found in the full version.

3.2 Perfect Adaption of Signatures

We now introduce new definitions characterizing the output distribution of ChgRep$_\mathcal{R}$, which lead to stronger notions than class-hiding. The latter only guarantees that given an *honestly* generated signature σ on M, the output $(\mu M, \sigma')$ of ChgRep$_\mathcal{R}$ for a random μ looks like a random message-signature pair.

This however does not protect a user against a signer when the user randomizes a pair obtained from the signer. We thus explicitly require that an adaption of any valid (not necessarily honestly generated) signature is distributed like a fresh signature.

Definition 10 (Perfect Adaption of Signatures). SPS-EQ on $(\mathbb{G}_i^*)^\ell$ *perfectly adapts signatures* if for all tuples $(\mathsf{sk}, \mathsf{pk}, M, \sigma, \mu)$ with

$$\mathsf{VKey}_\mathcal{R}(\mathsf{sk}, \mathsf{pk}) = 1 \qquad \mathsf{Verify}_\mathcal{R}(M, \sigma, \mathsf{pk}) = 1 \qquad M \in (\mathbb{G}_i^*)^\ell \qquad \mu \in \mathbb{Z}_p^*$$

$\mathsf{ChgRep}_\mathcal{R}(M, \sigma, \mu, \mathsf{pk})$ and $(\mu M, \mathsf{Sign}_\mathcal{R}(\mu M, \mathsf{sk}))$ are identically distributed.

We now show the relation between Definitions 9 and 10. The following is proven analogously to the proof of class-hiding of Scheme 1 in [29].

Proposition 1. *Let* SPS-EQ *be an SPS-EQ scheme on* $(\mathbb{G}_i^*)^\ell$, $\ell > 1$, *with perfect adaption of signatures. If* $M \xleftarrow{R} [M]_\mathcal{R}$ *is computationally indistinguishable from* $M \xleftarrow{R} (\mathbb{G}_i^*)^\ell$ *then* SPS-EQ *is class-hiding.*

Corollary 1. *If the DDH assumption holds in* $\mathbb{G}_i$ *then any SPS-EQ scheme on* $(\mathbb{G}_i^*)^\ell$ *satisfying Definition 10 is class-hiding (Definition 9).*

We note that the converse is not true, as witnessed by Scheme 2: it satisfies class-hiding, but the discrete logs of (R_1, R_2) contained in a signature σ have the same ratio as those of $(\tilde{R}_1, \tilde{R}_2)$ from the output of $\mathsf{ChgRep}_\mathcal{R}$.

Maliciously Chosen Keys. Whereas Definition 10 strengthens Definition 9 in that it considers maliciously generated signatures, the next definition strengthens this further by considering maliciously generated public keys. As there might not even be a corresponding signing key, we cannot compare the outputs of $\mathsf{ChgRep}_\mathcal{R}$ to those of $\mathsf{Sign}_\mathcal{R}$. We therefore require that $\mathsf{ChgRep}_\mathcal{R}$ outputs a random element that satisfies verification.

Definition 11 (Perfect Adaption Under Malicious Keys). SPS-EQ on $(\mathbb{G}_i^*)^\ell$ *perfectly adapts signatures under malicious keys* if for all tuples $(\mathsf{pk}, M, \sigma, \mu)$ with

$$\mathsf{Verify}_\mathcal{R}(M, \sigma, \mathsf{pk}) = 1 \qquad M \in (\mathbb{G}_i^*)^\ell \qquad \mu \in \mathbb{Z}_p^*$$

we have that $\mathsf{ChgRep}_\mathcal{R}(M, \sigma, \mu, \mathsf{pk})$ outputs $(\mu M, \sigma')$ such that σ' is a random element in the space of signatures, conditioned on $\mathsf{Verify}_\mathcal{R}(\mu M, \sigma', \mathsf{pk}) = 1$.

Proposition 2. *Scheme 1, from [29], satisfies both Definitions 10 and 11.*

Proof (sketch). For any $M \in (\mathbb{G}_1^*)^\ell$ and $\mathsf{pk} \in (\mathbb{G}_2^*)^\ell$, let $(x_i)_{i \in [\ell]}$ be s.t. $\mathsf{pk} = (x_i \hat{P})_{i \in [\ell]}$. A signature $(Z, Y, \hat{Y}) \in \mathbb{G}_1 \times \mathbb{G}_1^* \times \mathbb{G}_2^*$ satisfying $\mathsf{Verify}_\mathcal{R}(M, (Z, Y, \hat{Y}), \mathsf{pk}) = 1$ must be of the form $(Z = y \sum x_i M_i, Y = \frac{1}{y} P, \hat{Y} = \frac{1}{y} \hat{P})$ for some $y \in \mathbb{Z}_p^*$. $\mathsf{ChgRep}_\mathcal{R}$ outputs $\sigma' = (y\psi \sum x_i \mu M_i, \frac{1}{y\psi} P, \frac{1}{y\psi} \hat{P})$, which is a random element in $\mathbb{G}_1 \times \mathbb{G}_1^* \times \mathbb{G}_2^*$ satisfying $\mathsf{Verify}_\mathcal{R}(M, \sigma', \mathsf{pk}) = 1$. $\qquad \square$

3.3   From SPS-EQ to (Rerandomizable) SPS Schemes

We now show how *any* EUF-CMA-secure SPS-EQ scheme that signs equivalence classes of $(\mathbb{G}_i^*)^{\ell+1}$ with $\ell > 0$ can be turned into an EUF-CMA-secure SPS scheme signing vectors of $(\mathbb{G}_i^*)^{\ell}$. (We note that SPS schemes typically allow messages from $\mathbb{G}_1$ and/or $\mathbb{G}_2$, which is preferable when used in combination with Groth-Sahai proofs.) The transformation works by embedding messages $(M_i)_{i\in[\ell]} \in (\mathbb{G}_i^*)^{\ell}$ into $(\mathbb{G}_i^*)^{\ell+1}$ as $M' = ((M_i)_{i\in[\ell]}, P)$ and signing M'. To verify a signature σ on a message $(M_i)_{i\in[\ell]} \in (\mathbb{G}_i^*)^{\ell}$ under key pk, one checks whether $\mathsf{Verify}_{\mathcal{R}}(((M_i)_{i\in[\ell]}, P), \sigma, \mathsf{pk}) = 1$.

What we have done is to allow only one single representative of each class, namely the one with P as its last element, a procedure we call *normalization*. EUF-CMA of the SPS-EQ states that no adversary can produce a signature on a message from an unqueried class, which therefore implies EUF-CMA of the resulting SPS scheme.

Moreover, from any SPS-EQ with perfect adaption of signatures the above transformation yields a rerandomizable SPS scheme, since signatures can be rerandomized by running $\mathsf{ChgRep}_{\mathcal{R}}$ for $\mu = 1$ (Definition 10 guarantees that this outputs a random signature). This also means that the lower bounds for SPS over Type-3 groups given by Abe et al. in [4,5] carry over to SPS-EQ: any SPS must use at least 2 PPEs for verification and must have at least 3 signature elements, which cannot be from the same group. Moreover, EUF-CMA security of optimal (that is, 3-element-signature) SPS-EQ schemes cannot be reduced to non-interactive assumptions.

Finally, let us investigate the possibility of SPS-EQ in the Type-1 and Type-2 pairing setting and implied lower bounds. Class-hiding requires the DDH assumption to hold on the message space. This excludes the Type-1 setting, while in Type-2 settings the message space must be $(\mathbb{G}_1^*)^{\ell}$. In [6] Abe et al. identified the following lower bounds for Type-2 SPS schemes with messages in $\mathbb{G}_1$: 2 PPEs for verification and 3 group elements for signatures. The above transformation converts a Type-2 SPS-EQ into a Type-2 SPS, hence these optimality criteria apply to Type-2 SPS-EQ schemes as well.

Implications. Applying the above transformation to the SPS-EQ scheme from [29] (Scheme 1) yields a perfectly rerandomizable SPS scheme in Type-3 groups with constant-size signatures of unilateral length-ℓ message vectors and public keys of size $\ell + 1$. Scheme 1 is optimal as it only uses 2 PPEs and its signatures consist of 3 bilateral group elements. Hence, by [5] there is no reduction of its EUF-CMA security to a non-interactive assumption and the generic group model proof in [29] is the best one can achieve.

Applying our transformation to Scheme 2 yields a new standard-model SPS construction for unilateral length-ℓ message vectors in Type-3 groups. It has constant-size signatures (4 $\mathbb{G}_1 + 1$ $\mathbb{G}_2$ elements), a public key of size $\ell + 3$ and uses 2 PPEs for verification; it is therefore almost as efficient as the best known direct SPS construction from non-interactive assumptions in [4], whose signatures consist of 3 $\mathbb{G}_1 + 1$ $\mathbb{G}_2$ elements. Scheme 2 is partially rerandomizable [3], whereas the scheme in [4] is not.

244 G. Fuchsbauer et al.

4 Blind Signatures from SPS-EQ

We first present the abstract model for blind signature schemes. Security is defined by unforgeability and blindness and was initially studied in [36,43] and then strengthened in [26,45].

Definition 12 (Blind Signature Scheme). A blind signature scheme BS consists of the following PPT algorithms:

KeyGen$_{\mathsf{BS}}(1^\kappa)$, on input κ, returns a key pair $(\mathsf{sk}, \mathsf{pk})$. The security parameter κ is also an (implicit) input to the following algorithms.

$(\mathcal{U}_{\mathsf{BS}}(m, \mathsf{pk}), \mathcal{S}_{\mathsf{BS}}(\mathsf{sk}))$ are run by a user and a signer, who interact during execution. $\mathcal{U}_{\mathsf{BS}}$ gets input a message m and a public key pk and $\mathcal{S}_{\mathsf{BS}}$ has input a secret key sk. At the end $\mathcal{U}_{\mathsf{BS}}$ outputs σ, a signature on m, or $\perp$ if the interaction was not successful.

Verify$_{\mathsf{BS}}(m, \sigma, \mathsf{pk})$ is deterministic and given a message-signature pair (m, σ) and a public key pk outputs 1 if σ is valid on m under pk and 0 otherwise.

A blind signature scheme BS must satisfy *correctness*, *unforgeability* and *blindness*.

Definition 13 (Correctness). A blind signature scheme BS is *correct* if for all $\kappa \in \mathbb{N}$, all $(\mathsf{sk}, \mathsf{pk}) \leftarrow \mathsf{KeyGen}_{\mathsf{BS}}(1^\kappa)$, all messages m and $\sigma \leftarrow (\mathcal{U}_{\mathsf{BS}}(m, \mathsf{pk}), \mathcal{S}_{\mathsf{BS}}(\mathsf{sk}))$ it holds that $\mathsf{Verify}_{\mathsf{BS}}(m, \sigma, \mathsf{pk}) = 1$.

Definition 14 (Unforgeability). BS is *unforgeable* if for all PPT algorithms $\mathcal{A}$ having access to a signer oracle, there is a negligible function $\epsilon(\cdot)$ such that:

$$\Pr\left[\begin{array}{l} (\mathsf{sk}, \mathsf{pk}) \leftarrow \mathsf{KeyGen}_{\mathsf{BS}}(1^\kappa), \\ (m_i^*, \sigma_i^*)_{i=1}^{k+1} \leftarrow \mathcal{A}^{(\cdot, \mathcal{S}_{\mathsf{BS}}(\mathsf{sk}))}(\mathsf{pk}) \end{array} : \begin{array}{l} m_i^* \neq m_j^* \; \forall i, j \in [k+1], i \neq j \; \wedge \\ \mathsf{Verify}_{\mathsf{BS}}(m_i^*, \sigma_i^*, \mathsf{pk}) = 1 \; \forall i \in [k+1] \end{array}\right] \leq \epsilon(\kappa),$$

where k is the number of completed interactions with the oracle.

There are several flavors of blindness. The strongest definition is blindness in the *malicious signer* model [1,42], which allows the adversary to create pk, whereas in the *honest-signer* model the key pair is set up by the experiment. We prove our construction secure under the stronger notion, which was also considered by the recent round-optimal standard-model constructions [30,31].

Definition 15 (Blindness). BS is called *blind* if for all PPT algorithms $\mathcal{A}$ with one-time access to two user oracles, there is a negligible function $\epsilon(\cdot)$ such that:

$$\Pr\left[\begin{array}{l} b \xleftarrow{R} \{0,1\}, \; (\mathsf{pk}, m_0, m_1, \mathsf{st}) \leftarrow \mathcal{A}(1^\kappa), \\ \mathsf{st} \leftarrow \mathcal{A}^{(\mathcal{U}_{\mathsf{BS}}(m_b, \mathsf{pk}), \cdot)^{(1)}, (\mathcal{U}_{\mathsf{BS}}(m_{1-b}, \mathsf{pk}), \cdot)^{(1)}}(\mathsf{st}), \\ \text{Let } \sigma_b \text{ and } \sigma_{1-b} \text{ be the resp. outputs of } \mathcal{U}_{\mathsf{BS}}, \quad : \quad b^* = b \\ \text{If } \sigma_0 = \perp \text{ or } \sigma_1 = \perp \text{ then } (\sigma_0, \sigma_1) \leftarrow (\perp, \perp), \\ b^* \leftarrow \mathcal{A}(\mathsf{st}, \sigma_0, \sigma_1) \end{array}\right] - \frac{1}{2} \leq \epsilon(\kappa).$$

4.1 Construction

Our construction uses commitments to the messages and SPS-EQ to sign these commitments and to perform blinding and unblinding. Signing an equivalence class with an SPS-EQ scheme lets one derive a signature for arbitrary representatives of this class without knowing the private signing key. This concept provides an elegant way to realize a blind signing process as follows.

The signer's key contains an element Q under which the obtainer makes a Pedersen commitment $C = mP + rQ$ to the message m. (Since the commitment is perfectly hiding, the signer can be aware of q with $Q = qP$.) The obtainer then forms a vector (C, P), which can be seen as the canonical representative of equivalence class $[(C, P)]_\mathcal{R}$. Next, she picks $s \xleftarrow{R} \mathbb{Z}_p^*$ and moves (C, P) to a random representative (sC, sP), which hides C. She sends (sC, sP) to the signer and receives an SPS-EQ signature on it, from which she can derive a signature on the original message (C, P), which she can publish together with an opening of C. As verification will check validity of the SPS-EQ signature on a message ending with P, the unblinding is unambiguous.

Let us now discuss how the user opens the Pedersen commitment $C = mP + rQ$. Publishing (m, r) directly would break blindness of the scheme (a signer could link a pair $M = (D, S)$, received during signing, to a signature by checking whether $D = mS + rqS$). We therefore define a tweaked opening, for which we include $\hat{Q} = q\hat{P}$ in addition to $Q = qP$ in the signer's public key. We define the opening as (m, rP), which can be checked via the pairing equation $e(C - mP, \hat{P}) = e(rP, \hat{Q})$. This opening is still computationally binding under the co-DHI$_1^*$ assumption (in contrast to standard Pedersen commitments, which are binding under the discrete-log assumption). Hiding of the commitment still holds unconditionally, and we will prove the constructed blind-signature scheme secure in the malicious-signer model without requiring a trusted setup.

The scheme is presented as Scheme 3. (Note that for simplicity the blind signature contains $T = rQ$ instead of C.) Correctness follows by inspection.

4.2 Security

Theorem 2. *If the underlying SPS-EQ scheme is EUF-CMA secure and the co-DHI$_1^*$ assumption holds then Scheme 3 is unforgeable.*

The proof, which is given in the full version, follows the intuition that a forger must either forge an SPS-EQ signature on a new commitment or open a commitment in two different ways. The reduction has a natural security loss proportional to the number of signing queries.

Blindness. For the honest-signer model, blindness follows from the DDH assumption and perfect adaption of signatures (Definition 10) of the underlying SPS-EQ scheme. Let $Q \leftarrow qP$ and let q be part of the signing key, and let (P, rP, sP, tP) be a DDH instance. In the blindness game we compute M as $(m \cdot sP + q \cdot tP, sP)$. When the adversary returns a signature on M, we must adapt it to the unblinded message—which we cannot do as we do not know the

---

$\mathsf{KeyGen_{BS}}(1^\kappa)$: Compute $\mathsf{BG} \leftarrow \mathsf{BGGen_R}(1^\kappa)$, $(\mathsf{sk}, \mathsf{pk_R}) \xleftarrow{R} \mathsf{KeyGen_R}(\mathsf{BG}, \ell = 2)$, pick $q \xleftarrow{R} \mathbb{Z}_p^*$ and set $Q \leftarrow qP$, $\hat{Q} \leftarrow q\hat{P}$. Output $(\mathsf{sk}, \mathsf{pk} = (\mathsf{pk_R}, Q, \hat{Q}))$.

$\mathcal{U}_{\mathsf{BS}}^{(1)}(m, \mathsf{pk})$: Given $\mathsf{pk} = (\mathsf{pk_R}, Q, \hat{Q})$ and $m \in \mathbb{Z}_p$, compute $\mathsf{BG} \leftarrow \mathsf{BGGen_R}(1^\kappa)$. If $Q = 0_{\mathbb{G}_1}$ or $e(Q, \hat{P}) \neq e(P, \hat{Q})$ then return $\bot$; else choose $s \xleftarrow{R} \mathbb{Z}_p^*$ and $r \xleftarrow{R} \mathbb{Z}_p$ s.t. $mP + rQ \neq 0_{\mathbb{G}_1}$ and output

$$M \leftarrow (s(mP + rQ), sP) \qquad \mathsf{st} \leftarrow (\mathsf{BG}, \mathsf{pk_R}, Q, M, r, s)$$

$\mathcal{S}_{\mathsf{BS}}(M, \mathsf{sk})$: Given $M \in (\mathbb{G}_1^*)^2$ and secret key sk, output $\pi \leftarrow \mathsf{Sign_R}(M, \mathsf{sk})$.

$\mathcal{U}_{\mathsf{BS}}^{(2)}(\mathsf{st}, \pi)$: Parse st as $(\mathsf{BG}, \mathsf{pk_R}, Q, M, r, s)$. If $\mathsf{Verify_R}(M, \pi, \mathsf{pk_R}) = 0$, return $\bot$. Run $((mP + rQ, P), \sigma) \leftarrow \mathsf{ChgRep_R}(M, \pi, \frac{1}{s}, \mathsf{pk_R})$ and output $\tau \leftarrow (\sigma, rP, rQ)$.

$\mathsf{Verify_{BS}}(m, \tau, \mathsf{pk})$: Given $m \in \mathbb{Z}_p^*$, blind signature $\tau = (\sigma, R, T)$ and $\mathsf{pk} = (\mathsf{pk_R}, Q, \hat{Q})$, with $Q \neq 0_{\mathbb{G}_1}$ and $e(Q, \hat{P}) = e(P, \hat{Q})$, output 1 if the following holds and 0 otherwise.

$$\mathsf{Verify_R}((mP + T, P), \sigma, \mathsf{pk_R}) = 1 \qquad e(T, \hat{P}) = e(R, \hat{Q})$$

Scheme 3. Blind signature scheme from SPS-EQ

blinding factor s. By perfect adaption however, an adapted signature is distributed as a fresh signature on the unblinded message, so, knowing the secret key, we can compute a signature σ on $(m \cdot P + q \cdot rP, P)$ and return the blind signature $(\sigma, rP, q \cdot rP)$. If the DDH instance was *real*, i.e., $t = s \cdot r$, then we perfectly simulated the game; if t was random then the adversary's view during issuing was independent of m.

For blindness in the malicious-signer model, we have to deal with two obstacles. (1) We do not have access to the adversarially generated signing key, meaning we cannot recompute the signature on the unblinded message. (2) The adversarially generated public-key values $Q, \hat{Q}$ do not allow us to embed a DDH instance for blinding and unblinding.

We overcome (1) by using the adversary $\mathcal{A}$ itself as a signing oracle by rewinding it. We first run $\mathcal{A}$ to obtain a signature on $(s'(mP + rQ), s'P)$, which, knowing s', we can transform into a signature on $(mP + rQ, P)$. We then rewind $\mathcal{A}$ to the point after outputting its public key and run it again, this time embedding our challenge. In the second run we cannot transform the received signature, instead we use the signature from the first run, which is distributed identically, due to perfect adaption under malicious keys (Definition 11) of the SPS-EQ scheme.

To deal with the second obstacle, we use an interactive variant of the DDH assumption: Instead of being given P, rP, sP and having to distinguish rsP from random, the adversary, for some Q of its choice, is given rP, rQ, sP and must distinguish rsQ from random.

Definition 16 (Assumption 1). Let BGGen be a bilinear-group generator that outputs $\mathsf{BG} = (p, \mathbb{G}_1, \mathbb{G}_2, \mathbb{G}_T, e, P_1 = P, P_2 = \hat{P})$. We assume that for all PPT algorithms $\mathcal{A}$ there is a negligible function $\epsilon(\cdot)$ such that:

$$\Pr\left[\begin{array}{l} b \xleftarrow{R} \{0,1\}, \ \mathsf{BG} = \mathsf{BGGen}_\mathcal{R}(1^\kappa) \qquad\qquad e(Q,\hat{P}) = e(P,\hat{Q}) \\ (\mathsf{st}, Q, \hat{Q}) \leftarrow \mathcal{A}(\mathsf{BG}), \ r,s,t \xleftarrow{R} \mathbb{Z}_p \qquad : \qquad\quad b^* = b \\ b^* \leftarrow \mathcal{A}(\mathsf{st}, rP, rQ, sP, ((1-b)\cdot t + b\cdot rs)Q) \end{array}\right] - \frac{1}{2} \leq \epsilon(\kappa) \ .$$

Proposition 3. *The assumption in Definition 16 holds in generic groups and reaches the optimal, quadratic simulation-error bound.*

Theorem 3. *If the underlying SPS-EQ scheme has perfect adaption of signatures under malicious keys and Assumption 1 holds then Scheme 3 is blind.*

The proofs can be found in the full version.

4.3 Discussion

Basing Our Scheme on Non-interactive Assumptions. Fischlin and Schröder [27] show that the unforgeability of a blind-signature scheme cannot be based on non-interactive hardness assumptions if (1) the scheme has 3 moves or less, (2) its blindness holds statistically and (3) from a transcript one can efficiently decide whether the interaction yielded a valid blind signature. Our scheme satisfies (1) and (3), whereas blindness only holds computationally.

They extend their result in [27] to computationally blind schemes that meet the following conditions: (4) One can efficiently check whether a public key has a matching secret key; this is the case in our setting because of group-membership tests and pairings. (5) Blindness needs to hold relative to a forgery oracle. As written in [27], this does e.g. not hold for Abe's scheme [2], where unforgeability is based on the discrete-log problem and blindness on the DDH problem.

This is the case in our construction too (as one can forge signatures by solving discrete logarithms), hence the impossibility result does not apply to our scheme. Our blind signature construction is black-box from any SPS-EQ with perfect adaption under malicious keys (Definition 11). However, the only known such scheme is the one from [29], which is EUF-CMA secure in the generic-group model, that is, it is based on an interactive assumption. Plugging this scheme into Scheme 3 yields a round-optimal blind signature scheme with unforgeability under this interactive assumption and co-DHI$_1^*$, and blindness (under adversarially chosen keys) under Assumption 1 (Definition 16), which is also interactive.

To construct a scheme under non-interactive assumptions, we would thus have to base blindness on a non-interactive assumption; and find an SPS-EQ scheme satisfying Definition 11 whose unforgeability is proven under a non-interactive assumption.

Efficiency of the Construction. When instantiating our blind-signature construction with the SPS-EQ scheme from [29] (given as Scheme 1), which we showed optimal, this yields a public key size of $1 \ \mathbb{G}_1 + 3 \ \mathbb{G}_2$, a communication complexity of $4 \ \mathbb{G}_1 + 1 \ \mathbb{G}_2$ and a signature size of $4 \ \mathbb{G}_1 + 1 \ \mathbb{G}_2$ elements. For a 80-bit security setting, a blind signature has thus 120 Bytes.

The most efficient scheme from standard assumptions is based on DLIN [30]. Ignoring the increase of the security parameter due to complexity leveraging, their scheme has a public key size of 43 $\mathbb{G}_1$ elements, communication complexity $18 \log_2 q + 41$ $\mathbb{G}_1$ elements (where, e.g., we have $\log_2 q = 155$ when assuming that the adversary runs in $\leq 2^{80}$ steps) and a signature size of 183 $\mathbb{G}_1$ elements.

4.4 Round-Optimal Partially Blind Signatures

Partially blind signatures are an extension of blind signatures, where messages contain *common information* γ, which is agreed between the user and the signer. This requires slight modifications to the unforgeability and blindness notions: An adversary breaks unforgeability if after k signing queries it outputs $k + 1$ distinct valid message-signature pairs for the same common information γ^*. In the partial-blindness game m_0 and m_1 must have the same common information γ to prevent the adversary from trivially winning the game. (Formal definitions for partially blind signatures can be found in the full version.)

Construction. We construct a round-optimal partially blind signature scheme PBS $= (\mathsf{KeyGen_{PBS}}, (\mathcal{U}_{\mathsf{PBS}}, \mathcal{S}_{\mathsf{PBS}}), \mathsf{Verify_{PBS}})$ secure in the standard model from an SPS-EQ scheme SPS-EQ by modifying Scheme 3 as follows. To include common information $\gamma \in \mathbb{Z}_p^*$, SPS-EQ is set up for $\ell = 3$. On input $M \leftarrow (s(mP + rQ), sP)$, $\mathcal{S}_{\mathsf{PBS}}$ returns a signature for $M \leftarrow (s(mP + rQ), \gamma \cdot sP, sP)$ and $\mathcal{U}_{\mathsf{PBS}}^{(2)}$ additionally checks correctness of the included γ and returns $\perp$ if this is not the case. Otherwise, it runs $((mP + rQ, \gamma P, P), \sigma) \leftarrow \mathsf{ChgRep}_{\mathcal{R}}(M, \pi, \frac{1}{s}, \mathsf{pk})$ and outputs signature $\tau \leftarrow (\sigma, rP, rQ)$ for message m and common information γ. For this construction we obtain the following, whose proofs are analogous to those for Scheme 3.

Theorem 4. *If SPS-EQ is EUF-CMA secure and the co-DHI$_1^*$ assumption holds, then the resulting partially blind signature scheme is unforgeable.*

Theorem 5. *If SPS-EQ has perfect adaption under malicious keys and Assumption 1 holds, then the resulting partially blind signature scheme is partially blind.*

5   One-Show Anonymous Credentials from SPS-EQ

Baldimtsi and Lysyanskaya [8] introduced blind signatures with attributes and show that they directly yield a one-show anonymous credential system in the vein of Brands [15]. In contrast to Brands' original construction, their construction relies on a provably secure three-move blind signature scheme (in the ROM). In this section we show how to construct two-move blind signatures on message vectors, which straightforwardly yield anonymous one-show credentials that are secure in the standard model.

5.1 Blind Signatures on Message Vectors

Our construction BSV of round-optimal blind signatures on message vectors $\boldsymbol{m} \in \mathbb{Z}_p^n$ simply replaces the Pedersen commitment $mP + rQ$ in Scheme 3 with a generalized Pedersen commitment $\sum_{i \in [n]} m_i P_i + rQ$. Thus, $\mathsf{KeyGen_{BSV}}$, on input $1^\kappa, n$, additionally outputs generators $(P_i)_{i \in [n]}$ and $\mathsf{Verify_{BSV}}(\boldsymbol{m}, (\sigma, R, T), \mathsf{pk})$ checks $\mathsf{Verify_{\mathcal{R}}}((\sum_{i \in [n]} m_i P_i + T, P), \sigma, \mathsf{pk}_{\mathcal{R}}) = 1$ and $e(T, \hat{P}) = e(R, \hat{Q})$. Due to space constraints, the construction BSV is detailed in the full version, where we also show the following.

Theorem 6. *If the underlying SPS-EQ scheme is EUF-CMA secure and the co-DHI_1^* assumption holds then* BSV *is unforgeable.*

Theorem 7. *If the underlying SPS-EQ scheme has perfect adaption under malicious keys and Assumption 1 holds then* BSV *is blind.*

5.2   Anonymous Credentials Light

The intuition behind our construction is comparable to [8], which roughly works as follows. In the *registration phase*, a user registers (once) a generalized Pedersen commitment C to her attributes and gives a zero-knowledge (ZK) proof of the opening (some attributes may be opened and some may remain concealed). In the *preparation* and *validation phase*, the user engages in a blind-signature-with-attributes protocol for some message m (which is considered the credential serial number) and another commitment C'. C' is a so-called combined commitment obtained from C and a second credential-specific commitment provided by the user. Finally, the credential is the user output of a blind-signature-with-attributes protocol resulting in a signature on message m and a so-called blinded Pedersen commitment C''. The latter contains the same attributes as C, but is unlinkable to C' and C''. Showing a credential amounts to presenting C'' along with the blind signature and proving in ZK a desired relation about attributes within C''.

Our construction combines BSV with efficient ZK proofs and is conceptually simpler than the one in [8]. For issuing, the user sends the issuer a blinded version $M \leftarrow (sC, sP)$ of a commitment C to the user's attributes (M corresponds to the blinded generalized Pedersen commitment in [8]). In addition, the user engages in a ZK proof (denoted PoK) proving knowledge of an opening of C (potentially revealing some of the committed attributes). The user obtains a BSV-signature π on M and turns it into a blind signature σ for commitment C by running $((C, P), \sigma) \leftarrow \mathsf{ChgRep_{\mathcal{R}}}(M, \pi, \frac{1}{s}, \mathsf{pk})$. The credential consists of C, σ and the randomness r used to produce the commitment. It is showed by sending C and σ and proving in ZK a desired relation about attributes within C.

For ease of presentation, we only consider selective attribute disclosure below. We note that proofs for a rich class of relations [17,20,24] w.r.t. generalized Pederson commitments, as used by our scheme, could be used instead. Henceforth, we denote by S the index set of attributes to be shown and by U those to be withheld. During a showing, a ZK proof of knowledge for a commitment $C = \sum_{i \in [n]} m_i P_i + rQ$ to attributes $(m_i)_{i \in [n]}$ amounts to proving

$$\mathsf{PoK_P}\{((\alpha_j)_{j \in U}, \beta) : C = \sum_{i \in S} m_i P_i + \sum_{j \in U} \alpha_j P_j + \beta Q\}. \tag{1}$$

The proof for a *blinded* commitment $(A, B) = (sC, sP)$ during the obtain phase is done as follows.

$$\mathsf{PoK_{BP}} \left\{ ((\alpha_j)_{j \in U}, \beta, \gamma) : \quad \begin{array}{l} A = \sum_{i \in S} m_i H_i + \sum_{j \in U} \alpha_j H_j + \beta H_Q \ \wedge \\ \bigwedge_{i \in [n]} (H_i = \gamma P_i) \wedge H_Q = \gamma Q \wedge B = \gamma P \end{array} \right\}. \quad (2)$$

Here the representation is with respect to bases $H_i = sP_i$, $H_Q = sQ$, which are published and guaranteed to be correctly formed by $\mathsf{PoK_{BP}}$.[2]

Construction. As we combine scheme BSV with ZK proofs, we need the following conceptual modifications. The signature $\tau \leftarrow (\sigma, R, T)$ reduces to $\tau \leftarrow \sigma$, since the user provides a ZK-PoK proving knowledge of the randomness r in C. Moreover, verification takes C instead of m as verifiers have only access to the commitment. Consequently, $\mathsf{Verify_{BSV}}$ of scheme BSV only runs $\mathsf{Verify_{\mathcal{R}}}$.

Setup. The issuer runs $(\mathsf{sk}, \mathsf{pk}) \leftarrow \mathsf{KeyGen_{BSV}}(1^\kappa, n)$, where n is the number of attributes in the system, and publishes pk as her public key.

Issuing. A user with attribute values m runs $(M, \mathsf{st}) \leftarrow \mathcal{U}^{(1)}_{\mathsf{BSV}}(m, \mathsf{pk}; (s, r))$ (where (s, r) is the chosen randomness), sends the blinded commitment $M = (sC, sP)$ to the issuer and gives a proof $\mathsf{PoK_{BP}}$ from (2) that M commits to m (where the sets U and S depend on the application). The issuer returns $\pi \leftarrow \mathcal{S}_{\mathsf{BSV}}(M, \mathsf{sk})$ and after running $\sigma \leftarrow \mathcal{U}^{(2)}_{\mathsf{BSV}}(\mathsf{st}, \pi)$ (the outputs rP and rQ are not needed), the user holds a credential (C, σ, r).

Showing. Assume a user with credential (C, σ, r) to the attributes $m = (m_i)_{i \in [n]}$ wants to conduct a selective showing of attributes with a verifier who holds the issuer's public key pk. They engage in a proof $\mathsf{PoK_P}$ from (1) and the verifier additionally checks the signature for the credential by running $\mathsf{Verify_{BSV}}(C, \sigma, \mathsf{pk})$. If both verifications succeed, the verifier accepts the showing.

Let us finally note that there is no formal security model for one-show credentials. Theorem 2 in [8] informally states that a secure commitment scheme together with a blind signature scheme with attributes implies a one-show credential system. Using the same argumentation as [8], our construction yields a one-show credential system in the standard model.

Acknowledgements. We would like to thank the anonymous reviewers for their valuable comments.

References

1. Abdalla, M., Namprempre, C., Neven, G.: On the (im)possibility of blind message authentication codes. In: Pointcheval, D. (ed.) CT–RSA 2006. LNCS, vol. 3860, pp. 262–279. Springer, Heidelberg (2006)

---

[2] In the blindness game, given $B = sP$ from a DDH instance, these bases are simulated as $H_j \leftarrow p_j B$ and $H_Q \leftarrow qB$. We can even prove security in the malicious-signer model by extending the assumption from Definition 16: in addition to Q the adversary outputs $(P_i)_{i \in [n]}$ and receives $(sP_i)_{i \in [n]}$ and sQ.

2. Abe, M.: A secure three-move blind signature scheme for polynomially many signatures. In: Pfitzmann, B. (ed.) EUROCRYPT 2001. LNCS, vol. 2045, pp. 136–151. Springer, Heidelberg (2001)
3. Abe, M., Fuchsbauer, G., Groth, J., Haralambiev, K., Ohkubo, M.: Structure-preserving signatures and commitments to group elements. In: Rabin, T. (ed.) CRYPTO 2010. LNCS, vol. 6223, pp. 209–236. Springer, Heidelberg (2010)
4. Abe, M., Groth, J., Haralambiev, K., Ohkubo, M.: Optimal structure-preserving signatures in asymmetric bilinear groups. In: Rogaway, P. (ed.) CRYPTO 2011. LNCS, vol. 6841, pp. 649–666. Springer, Heidelberg (2011)
5. Abe, M., Groth, J., Ohkubo, M.: Separating short structure-preserving signatures from non-interactive assumptions. In: Lee, D.H., Wang, X. (eds.) ASIACRYPT 2011. LNCS, vol. 7073, pp. 628–646. Springer, Heidelberg (2011)
6. Abe, M., Groth, J., Ohkubo, M., Tibouchi, M.: Structure-preserving signatures from type II pairings. In: Garay, J.A., Gennaro, R. (eds.) CRYPTO 2014, Part I. LNCS, vol. 8616, pp. 390–407. Springer, Heidelberg (2014)
7. Abe, M., Okamoto, T.: Provably secure partially blind signatures. In: Bellare, M. (ed.) CRYPTO 2000. LNCS, vol. 1880, pp. 271–286. Springer, Heidelberg (2000)
8. Baldimtsi, F., Lysyanskaya, A.: Anonymous credentials light. In: CCS. ACM (2013)
9. Baldimtsi, F., Lysyanskaya, A.: On the security of one-witness blind signature schemes. In: Sako, K., Sarkar, P. (eds.) ASIACRYPT 2013, Part II. LNCS, vol. 8270, pp. 82–99. Springer, Heidelberg (2013)
10. Barreto, P.S.L.M., Naehrig, M.: Pairing-friendly elliptic curves of prime order. In: Preneel, B., Tavares, S. (eds.) SAC 2005. LNCS, vol. 3897, pp. 319–331. Springer, Heidelberg (2005)
11. Bellare, M., Namprempre, C., Pointcheval, D., Semanko, M.: The one-more-RSA-inversion problems and the security of chaum's blind signature scheme. J. Cryptology **16**(3), 185–215 (2003)
12. Blazy, O., Fuchsbauer, G., Pointcheval, D., Vergnaud, D.: Signatures on randomizable ciphertexts. In: Catalano, D., Fazio, N., Gennaro, R., Nicolosi, A. (eds.) PKC 2011. LNCS, vol. 6571, pp. 403–422. Springer, Heidelberg (2011)
13. Blazy, O., Pointcheval, D., Vergnaud, D.: Compact round-optimal partially-blind signatures. In: Visconti, I., De Prisco, R. (eds.) SCN 2012. LNCS, vol. 7485, pp. 95–112. Springer, Heidelberg (2012)
14. Boldyreva, A.: Threshold signatures, multisignatures and blind signatures based on the gap-diffie-hellman-group signature scheme. In: Desmedt, Y.G. (ed.) PKC 2003. LNCS, vol. 2567, pp. 31–46. Springer, Heidelberg (2003)
15. Brands, S.: Rethinking public-key infrastructures and digital certificates: building in privacy. MIT Press (2000)
16. Brands, S., Paquin, C.: U-Prove Cryptographic Specification v1 (2010)
17. Bresson, E., Stern, J.: Proofs of knowledge for non-monotone discrete-log formulae and applications. In: Chan, A.H., Gligor, V.D. (eds.) ISC 2002. LNCS, vol. 2433, pp. 272–288. Springer, Heidelberg (2002)
18. Camenisch, J., Dubovitskaya, M., Haralambiev, K.: Efficient structure-preserving signature scheme from standard assumptions. In: Visconti, I., De Prisco, R. (eds.) SCN 2012. LNCS, vol. 7485, pp. 76–94. Springer, Heidelberg (2012)
19. Camenisch, J.L., Koprowski, M., Warinschi, B.: Efficient blind signatures without random oracles. In: Blundo, C., Cimato, S. (eds.) SCN 2004. LNCS, vol. 3352, pp. 134–148. Springer, Heidelberg (2005)
20. Camenisch, J.L., Michels, M.: Proving in zero-knowledge that a number is the product of two safe primes. In: Stern, J. (ed.) EUROCRYPT 1999. LNCS, vol. 1592, pp. 107–122. Springer, Heidelberg (1999)

21. Chatterjee, S., Menezes, A.: On cryptographic protocols employing asymmetric pairings - the role of ψ revisited. Discrete Appl. Math. **159**(13), 1311–1322 (2011)
22. Chaum, D.: Blind signatures for untraceable payments. In: CRYPTO 1982, pp. 199–203. Plenum Press (1982)
23. Chaum, D.: Blind signature system. In: Chaum, D. (ed.) CRYPTO 1983, p. 153. Springer, New York (1983)
24. Cramer, R., Damgård, I., Schoenmakers, B.: Proofs of partial knowledge and simplified design of witness hiding protocols. In: Desmedt, Y.G. (ed.) CRYPTO 1994. LNCS, vol. 839, pp. 174–187. Springer, Heidelberg (1994)
25. Fischlin, M.: Round-optimal composable blind signatures in the common reference string model. In: Dwork, C. (ed.) CRYPTO 2006. LNCS, vol. 4117, pp. 60–77. Springer, Heidelberg (2006)
26. Fischlin, M., Schröder, D.: Security of blind signatures under aborts. In: Jarecki, S., Tsudik, G. (eds.) PKC 2009. LNCS, vol. 5443, pp. 297–316. Springer, Heidelberg (2009)
27. Fischlin, M., Schröder, D.: On the impossibility of three-move blind signature schemes. In: Gilbert, H. (ed.) EUROCRYPT 2010. LNCS, vol. 6110, pp. 197–215. Springer, Heidelberg (2010)
28. Fuchsbauer, G.: Breaking existential unforgeability of a signature scheme from asiacrypt 2014. Cryptology ePrint Archive, report 2014/892 (2014)
29. Fuchsbauer, G., Hanser, C., Slamanig, D.: EUF-CMA-secure structure-preserving signatures on equivalence classes. Cryptology ePrint Archive, report 2014/944 (2014)
30. Garg, S., Gupta, D.: Efficient round optimal blind signatures. In: Nguyen, P.Q., Oswald, E. (eds.) EUROCRYPT 2014. LNCS, vol. 8441, pp. 477–495. Springer, Heidelberg (2014)
31. Garg, S., Rao, V., Sahai, A., Schröder, D., Unruh, D.: Round optimal blind signatures. In: Rogaway, P. (ed.) CRYPTO 2011. LNCS, vol. 6841, pp. 630–648. Springer, Heidelberg (2011)
32. Ghadafi, E., Smart, N.P.: Efficient two-move blind signatures in the common reference string model. In: Gollmann, D., Freiling, F.C. (eds.) ISC 2012. LNCS, vol. 7483, pp. 274–289. Springer, Heidelberg (2012)
33. Groth, J., Sahai, A.: Efficient non-interactive proof systems for bilinear groups. In: Smart, N.P. (ed.) EUROCRYPT 2008. LNCS, vol. 4965, pp. 415–432. Springer, Heidelberg (2008)
34. Hanser, C., Slamanig, D.: Structure-preserving signatures on equivalence classes and their application to anonymous credentials. In: Sarkar, P., Iwata, T. (eds.) ASIACRYPT 2014. LNCS, vol. 8873, pp. 491–511. Springer, Heidelberg (2014)
35. Hazay, C., Katz, J., Koo, C.-Y., Lindell, Y.: Concurrently-secure blind signatures without random oracles or setup assumptions. In: Vadhan, S.P. (ed.) TCC 2007. LNCS, vol. 4392, pp. 323–341. Springer, Heidelberg (2007)
36. Juels, A., Luby, M., Ostrovsky, R.: Security of blind digital signatures. In: Kaliski Jr., B.S. (ed.) CRYPTO 1997. LNCS, vol. 1294, pp. 150–164. Springer, Heidelberg (1997)
37. Kiayias, A., Zhou, H.-S.: Concurrent blind signatures without random oracles. In: De Prisco, R., Yung, M. (eds.) SCN 2006. LNCS, vol. 4116, pp. 49–62. Springer, Heidelberg (2006)
38. Lindell, Y.: Bounded-concurrent secure two-party computation without setup assumptions. In: STOC, pp. 683–692. ACM (2003)

39. Lysyanskaya, A., Rivest, R.L., Sahai, A., Wolf, S.: Pseudonym systems. In: Heys, H., Adams, C. (eds.) SAC 2000. LNCS, vol. 1758, pp. 184–199. Springer, Heidelberg (2000)

40. Meiklejohn, S., Shacham, H., Freeman, D.M.: Limitations on transformations from composite-order to prime-order groups: the case of round-optimal blind signatures. In: Abe, M. (ed.) ASIACRYPT 2010. LNCS, vol. 6477, pp. 519–538. Springer, Heidelberg (2010)

41. Okamoto, T.: Provably secure and practical identification schemes and corresponding signature schemes. In: Brickell, E.F. (ed.) CRYPTO 1992. LNCS, vol. 740, pp. 31–53. Springer, Heidelberg (1993)

42. Okamoto, T.: Efficient blind and partially blind signatures without random oracles. In: Halevi, S., Rabin, T. (eds.) TCC 2006. LNCS, vol. 3876, pp. 80–99. Springer, Heidelberg (2006)

43. Pointcheval, D., Stern, J.: Security arguments for digital signatures and blind signatures. J. Cryptology 13(3), 361–396 (2000)

44. Rückert, M.: Lattice-based blind signatures. In: Abe, M. (ed.) ASIACRYPT 2010. LNCS, vol. 6477, pp. 413–430. Springer, Heidelberg (2010)

45. Schröder, D., Unruh, D.: Security of blind signatures revisited. In: Fischlin, M., Buchmann, J., Manulis, M. (eds.) PKC 2012. LNCS, vol. 7293, pp. 662–679. Springer, Heidelberg (2012)

46. Seo, J.H., Cheon, J.H.: Beyond the limitation of prime-order bilinear groups, and round optimal blind signatures. In: Cramer, R. (ed.) TCC 2012. LNCS, vol. 7194, pp. 133–150. Springer, Heidelberg (2012)

Programmable Hash Functions Go Private: Constructions and Applications to (Homomorphic) Signatures with Shorter Public Keys

Dario Catalano[1], Dario Fiore[2], and Luca Nizzardo[2][(✉)]

[1] Dipartimento di Matematica e Informatica,
Università di Catania, Catania, Italy
catalano@dmi.unict.it
[2] IMDEA Software Institute, Madrid, Spain
{dario.fiore,luca.nizzardo}@imdea.org

Abstract. We introduce the notion of asymmetric programmable hash functions (APHFs, for short), which adapts Programmable Hash Functions, introduced by Hofheinz and Kiltz at Crypto 2008, with two main differences. First, an APHF works over bilinear groups, and it is asymmetric in the sense that, while only *secretly* computable, it admits an isomorphic copy which is publicly computable. Second, in addition to the usual programmability, APHFs may have an alternative property that we call *programmable pseudorandomness*. In a nutshell, this property states that it is possible to embed a pseudorandom value as part of the function's output, akin to a random oracle. In spite of the apparent limitation of being only secretly computable, APHFs turn out to be surprisingly powerful objects. We show that they can be used to generically implement both regular and linearly-homomorphic signature schemes in a simple and elegant way. More importantly, when instantiating these generic constructions with our concrete realizations of APHFs, we obtain: (1) the *first* linearly-homomorphic signature (in the standard model) whose public key is *sub-linear* in both the dataset size and the dimension of the signed vectors; (2) short signatures (in the standard model) whose public key is shorter than those by Hofheinz-Jager-Kiltz from Asiacrypt 2011, and essentially the same as those by Yamada, Hannoka, Kunihiro, (CT-RSA 2012).

1 Introduction

PROGRAMMABLE HASH FUNCTIONS. Programmable Hash Functions (PHFs) were introduced by Hofheinz and Kiltz [26] as an information theoretic tool to "mimic" the behavior of a random oracle in finite groups. In a nutshell, a PHF H is an efficiently computable function that maps suitable inputs (e.g., binary strings) into a group $\mathbb{G}$, and can be generated in two different, indistinguishable, ways. In the standard modality, H hashes inputs X into group elements $H(X) \in \mathbb{G}$. When generated in trapdoor mode, a trapdoor allows one to express

R. Gennaro and M. Robshaw (Eds.): CRYPTO 2015, Part II, LNCS 9216, pp. 254–274, 2015.
DOI: 10.1007/978-3-662-48000-7_13

every output in terms of two (user-specified) elements $g, h \in \mathbb{G}$, i.e., one can compute two integers a_X, b_X such that $\mathsf{H}(X) = g^{a_X} h^{b_X}$. Finally, H is programmable in the sense that it is possible to program the behavior of H so that its outputs contain (or not) g with a certain probability. More precisely, H is said (m, n)-*programmable* if for all disjoint sets of inputs $\{X_1, \ldots, X_m\}$ and $\{Z_1, \ldots, Z_n\}$, the joint probability that $\forall i, a_{X_i} = 0$ and $\forall j, a_{Z_j} \neq 0$ is significant (e.g., $1/\mathsf{poly}(\lambda)$). Programmability turns out to be particularly useful in several security proofs. For instance, consider a security proof where a signature on $\mathsf{H}(X)$ can be simulated as long as $a_X = 0$ (i.e., g does not appear) while a forgery on $\mathsf{H}(Z)$ can be successfully used if $a_Z \neq 0$ (i.e., g does appear). Then one could rely on an $(m, 1)$-programmability of H to "hope" that all the queried messages $X_1, \ldots, X_m$ are simulatable, i.e., $\forall i, a_{X_i} = 0$, while the forgery message Z is not, i.e., $a_Z \neq 0$. PHFs essentially provide a nice abstraction of the so-called partitioning technique used in many cryptographic proofs.

1.1 Our Contribution

ASYMMETRIC PROGRAMMABLE HASH FUNCTIONS. We introduce the notion of *asymmetric programmable hash functions* (asymmetric PHFs) which modifies the original notion of PHFs [26] in two main ways. First, an asymmetric PHF H maps inputs into a *bilinear* group $\mathbb{G}$ and is only *secretly computable*. At the same time, an isomorphic copy of H can be *publicly computed* in the target group $\mathbb{G}_T$, i.e., anyone can compute $e(\mathsf{H}(X), g)$.[1] Second, when generated in trapdoor mode, for two given group elements $g, h \in \mathbb{G}$ such that $h = g^z$, the trapdoor allows one to write every $\mathsf{H}(X)$ as $g^{c_X(z)}$ for a degree-d polynomial $c_X(z)$.

We define two main programmability properties of asymmetric PHFs. The first one is an adaptation of the original programmability notion, and it says that H is (m, n, d)-programmable if it is (m, n)-programmable as before except that, instead of looking at the probability that $a_X = 0$, one now looks at whether $c_{X,0} = 0$, where $c_{X,0}$ is the coefficient of the degree-0 term of the polynomial $c_X(\cdot)$ obtained using the trapdoor.[2] The second programmability property is new and is called *programmable pseudo-randomness*. Roughly speaking, programmable pseudo-randomness says that one can program H so that the values $g^{c_{X,0}}$ look random to any polynomially-bounded adversary who observes the public hash key and the outputs of H on a set of adaptively chosen inputs. This functionality turns out to be useful in security proofs where one needs to cancel some random values for simulation purposes (we explain this in slightly more detail later in the introduction). In other words, programmable pseudo-randomness provides another random-oracle-like property for standard model hash functions, that is to "hide" a PRF inside the hash function. This is crucial in our security proofs, and we believe it can have further applications.

APPLICATIONS. In principle, secretly computable PHFs seem less versatile than regular PHFs. In this work, however, we show that, for applications such as digital signatures, asymmetric PHFs turn out to be *more* powerful than their

---

[1] Because of such asymmetric behavior we call these functions "asymmetric".

[2] For $d = 1$, this is basically the same programmability of [26].

publicly computable counterparts. Specifically, we show how to use asymmetric PHFs to realize both *regular* and *linearly-homomorphic* signatures secure in the standard model. Next, we show efficient realizations of asymmetric PHFs that, when plugged in our generic constructions, yield new and existing schemes that improve the state-of-the-art in the following way. First, we obtain the *first* linearly homomorphic signature scheme, secure in the standard model, achieving a public key which is *sub-linear* in both the dataset size and the dimension of the signed vectors. Second, we obtain regular signature schemes, matching the efficiency of the ones in [31], thus providing the shortest signatures in the standard model with a public key shorter than in [25].

In the following we elaborate more on these solutions.

Linearly-Homomorphic Signatures with Short Public Key in the Standard Model. Imagine a user Alice stores one or more datasets $D_1, D_2, \ldots, D_\ell$ on a cloud server. Imagine also that some other user, Bob, is allowed to perform queries over Alice's datasets, i.e., to compute one or more functions $F_1, \ldots, F_m$ over any D_i. The crucial requirement here is that Bob wants to be ensured about the correctness of the computation's results $F_j(D_i)$, even if the server is not trusted. An obvious way to do this (reliably) is to ask Alice to sign all her data $D_i = m_1^{(i)}, \ldots, m_N^{(i)}$. Later, Bob can check the validity of the computation by (1) downloading the full dataset locally, (2) checking all the signatures and (3) redoing the computation from scratch. Efficiency-wise, this solution is clearly undesirable in terms of bandwidth, storage (Bob has to download and store potentially large amount of data) and computation (Bob has to recompute everything on his own).

A much better solution comes from the notion of homomorphic signatures [9]. These allow to overcome the first issue (bandwidth) in a very elegant way. Using such a scheme, Alice can sign $m_1, \ldots, m_N$, thus producing signatures $\sigma_1, \ldots, \sigma_N$, which can be verified exactly as ordinary signatures. In addition, the homomorphic property provides the extra feature that, given $\sigma_1, \ldots, \sigma_N$ and some function $F : \mathcal{M}^N \to \mathcal{M}$, one can compute a signature $\sigma_{F,y}$ on the value $y = F(m_1, \ldots, m_N)$ *without* knowledge of the secret signing key sk. In other words, for a set of signed messages and any function F, it is possible to provide $y = F(m_1, \ldots, m_N)$ along with a signature $\sigma_{F,y}$ vouching for the correctness of y. The security of homomorphic signatures guarantees that creating a signature σ_{F,y^*} for a $y^* \neq F(m_1, \ldots, m_N)$ is computationally hard, unless one knows sk.

To solve the second issue and allow Bob to *verify efficiently* such signatures (i.e., by spending less time than that required to compute F), one can use *homomorphic signatures with efficient verification*, recently introduced in [15].

The notion of homomorphic signature was first introduced by Johnson *et al.* [28]. Since then several schemes have been proposed. The first schemes were homomorphic only for linear functions over vector spaces [1–3, 8, 10, 12–14, 17, 19, 30] and have nice applications to network coding and proofs of retrievability. More recent works proposed realizations that can support more expressive functionalities such as polynomials [9, 15] or general circuits of bounded polynomial depth [11, 21].

Despite the significant research work in the area, it is striking that *all* the existing homomorphic signature schemes that are proven secure in the standard model [1–3,11,13–15,17,21,30] suffer from a public key that is *at least linear* in the size N of the signed datasets. On one hand, the cost of storing such large public key can be, in principle, amortized since the key can be re-used for multiple datasets. On the other hand, this limitation still represents a challenging open question from both a theoretical and a practical point of view. From a practical perspective, a linear public key might be simply unaffordable by a user Bob who has limited storage capacity. From a theoretical point of view, considered the state-of-the-art, it seems unclear whether achieving a standard-model scheme with a key of length $o(N)$ is possible at all. Technically speaking, indeed, all these schemes in the standard model somehow rely on a public key as large as one dataset for simulation purposes. This essentially hints that any solution for this problem would require a novel proof strategy.

OUR CONTRIBUTION. We solve the above open problem by proposing the *first* standard-model homomorphic signature scheme that achieves a public key whose size is *sub-linear* in the maximal size N of the supported datasets. Slightly more in detail, we show how to use asymmetric PHFs in a generic fashion to construct a linearly-homomorphic signature scheme based on bilinear maps that can sign datasets, each consisting of up to N vectors of dimension T. The public key of our scheme mainly consists of the public hash keys of two asymmetric PHFs. By instantiating these using (one of) our concrete realizations we obtain a linearly-homomorphic signature with a public key of length $O(\sqrt{N} + \sqrt{T})$. We stress that ours is also the *first* linearly-homomorphic scheme where the public key is sub-linear in the dimension T of the signed vectors. Concretely, if one considers applications with datasets of 1 million of elements and a security parameter of 128bits, previous solutions (e.g., [2,14]) require a public key of at least 32 MB, whereas our solution simply works with a public key below 100 KB.

ON THE POWER OF SECRETLY-COMPUTABLE PHFS. The main technical idea nderlying this result is a new proof technique that builds on *asymmetric hash functions with programmable pseudo-randomness*. We illustrate the technique via a toy example inspired by our linearly-homomorphic signature scheme. The scheme works over asymmetric bilinear groups $\mathbb{G}_1, \mathbb{G}_2$, and with an asymmetric PHF $\mathsf{H} : [N] \to \mathbb{G}_1$ that has programmable pseudo-randomness w.r.t. $d = 1$. To sign a *random* message $M \in \mathbb{G}_1$ w.r.t. a label τ, one creates the signature

$$S = (\mathsf{H}(\tau) \cdot M)^{1/z}$$

where z is the secret key. The signature is linearly-homomorphic – $S_1 S_2 = (\mathsf{H}(\tau_1)\mathsf{H}(\tau_2)M)^{1/z}$, for $M = M_1 M_2$ – and it can be efficiently checked using a pairing – $e(S, g_2^z) = \prod_i e(\mathsf{H}(\tau_i), g_2)e(M, g_2)$ – and by relying on that $e(\mathsf{H}(\cdot), g_2)$ is publicly computable.

The first interesting thing to note is that having H *secretly* computable is necessary: if H is public the scheme could be easily broken, e.g., choose $M^* = \mathsf{H}(\tau)^{-1}$. Let us now show how to prove its security assuming that we want to do a reduction to the following assumption: given g_1, g_2, g_2^z, the challenge is to compute $W^{1/z} \in \mathbb{G}_1$ for $W \neq 1$ of adversarial choice. Missing g_1^z seems to make hard the simulation of signatures since $M, S \in \mathbb{G}_1$. However, we can use the trapdoor generation of H for $d = 1$ (that for asymmetric pairings takes $g_1, h_1 = g_1^{y_1}, g_2, h_2 = g_2^{y_2}$ and allows to express $\mathsf{H}(X) = g_1^{c_X(y_1, y_2)}$), by plugging $h_1 = 1, h_2 = g_2^z$. This allows to write every output as $\mathsf{H}(\tau) = g_1^{c_\tau(z)} = g_1^{c_{\tau,0} + c_{\tau,1}z}$. Every signing query with label τ is simulated by setting $M_\tau = g^{-c_{\tau,0}}$ and $S_\tau = (g_1^{c_{\tau,1}})$. The signature is correctly distributed since (1) $S_\tau = (\mathsf{H}(\tau) \cdot M_\tau)^{1/z}$, and (2) M_τ looks random thanks to the programmable pseudo-randomness of H. To conclude the proof, assume that the adversary comes up with a forgery M^*, S^* for label τ^* such that τ^* was already queried, and let $\hat{S}, \hat{M}$ be the values in the simulation of the signing query for τ^*. Now, $\hat{S} = (\mathsf{H}(\tau^*) \cdot \hat{M})^{1/z}$ holds by correctness, while $S^* = (\mathsf{H}(\tau^*) \cdot M^*)^{1/z}$ holds for $M^* \neq \hat{M}$ by definition of forgery. Then $(M^*/\hat{M}, S^*/\hat{S})$ is clearly a solution to the above assumption. This essentially shows that we can sign as many M's as the number of τ's, that is N. And by using our construction $\mathsf{H} = \mathsf{H}_{\mathsf{sqrt}}$ this is achievable with a key of length $O(\sqrt{N})$. Let us stress that the above one is an incomplete proof sketch, that we give only to illustrate the core ideas of using programmable pseudo-randomness. We defer the reader to Sect. 4 for a precise description of our signature scheme and its security proof.

Short Signatures from Bilinear Maps in the Standard Model. Hofheinz and Kiltz [26] proposed efficient realizations of PHFs, and showed how to use them to obtain black-box proofs of several cryptographic primitives. Among these applications, they use PHFs to build generic, standard-model, signature schemes from the Strong RSA problem and the Strong q-Diffie Hellman problem. Somewhat interestingly, these schemes (in particular the ones over bilinear groups) can enjoy very short signatures. The remarkable contribution of the generic construction in [26] is that signatures can be made short by reducing the size ρ of the randomness used (and included) in the signature so that ρ can go beyond the birthday bound. Precisely, by using an $(m, 1)$-programmable hash function, m can control the size of the randomness so that the larger is m, the smaller is the randomness. However, although this would call for $(m, 1)$-PHFs with a large m, the original work [26] described PHFs realizations that are only $(2, 1)$-programmable.[3]

Later, Hofheinz, Jager and Kiltz [25] showed constructions of $(m, 1)$-PHFs for any $m \geq 1$. By choosing a larger m, these new PHFs realizations yield the shortest known signatures in the standard model. On the negative side, however, this also induces much larger public keys. For instance, to obtain a signature of 302 bits from bilinear maps, they need a public key of more than 8MB. The reason of such inefficiency is that their realizations of (deterministic) $(m, 1)$-PHFs have keys of length $O(m^2 \ell)$, where ℓ is the bit size of the inputs. In a subsequent work, Yamada et al. [31] improved on this aspect by proposing a signature scheme with a public key of length $O(m\sqrt{\ell})$. Their solution followed a different approach: instead of

---

[3] [26] gives also a $(1, \mathsf{poly})$-programmable PHF which allows for different applications.

relying on $(m, 1)$-PHFs they obtained the signature by applying the Naor's transformation [7] to a new identity-based key encapsulation mechanism (IBKEM).

OUR RESULTS. Our results are mainly two. First, we revisit the generic signature constructions of [25,26] in order to work with asymmetric $(m, 1, d)$-PHFs. Our generic construction is very similar to that in [25,26], and, as such, it inherits the same property: the larger is m, the shorter can be the randomness.

Second we show the construction of an asymmetric PHF, $\mathsf{H_{acfs}}$, that is $(m, 1, 2)$-programmable and has a hash key consisting of $O(m\sqrt{\ell})$ group elements. By plugging $\mathsf{H_{acfs}}$ into our generic construction we immediately obtain standard-model signatures that achieve the same efficiency as the scheme of Yamada et al. [31]. Namely, they are the shortest standard model signature schemes with a public key of length $O(m\sqrt{\ell})$, that concretely allows for signatures of 302bits and a public key of 50KB. One of our two schemes recover the one in [31]. In this sense we provide a different conceptual approach to construct such signatures. While Yamada et al. obtained this result by going through an IBKEM, our solution revisits the original Hofheinz-Kiltz's idea of applying programmable functions.

Other Related Work. Hanaoka, Matsuda and Schuldt [23] show that there cannot be any black-box construction of a $(\mathsf{poly}, 1)$-PHF. The latter result has been overcome by the recent work of Freire et al. [18] who propose a $(\mathsf{poly}, 1)$-PHF based on multilinear maps. The latter result is obtained by slightly changing the definition of PHFs in order to work in the multilinear group setting. Their $(\mathsf{poly}, 1)$-PHF leads to several applications, notably standard-model versions (over multilinear groups) of BLS signatures, the Boneh-Franklin IBE, and identity-based non-interactive key-exchange. While the notion of PHFs in the multilinear setting of [18] is different from our asymmetric PHFs (with the main difference being that ours are secretly computable), it is worth noting that the two notions have some relation. As we discuss in the full version of our paper, our asymmetric PHFs indeed imply PHFs in the *bilinear* setting (though carrying the same degree of programmability).

The idea of using bilinear maps to reduce the size of public keys was used previously by Haralambiev et al. [24] in the context of public-key encryption, and by Yamada et al. [31] in the context of digital signatures. We note that our solutions use a similar approach in the construction of APHFs, which however also include the important novelty of programmable pseudorandomness, that turned out to be crucial in our proofs for the linearly-homomorphic signature.

2   Preliminaries

Bilinear Groups and Complexity Assumptions. Let $\lambda \in \mathbb{N}$ be a security parameter and let $\mathcal{G}(1^\lambda)$ be an algorithm which takes as input the security parameter and outputs the description of (asymmetric) bilinear groups $\mathsf{bgp} = (p, \mathbb{G}_1, \mathbb{G}_2, \mathbb{G}_T, e, g_1, g_2)$ where $\mathbb{G}_1$, $\mathbb{G}_2$ and $\mathbb{G}_T$ are groups of the same prime order $p > 2^\lambda$, $g_1 \in \mathbb{G}_1$ and $g_2 \in \mathbb{G}_2$ are two generators, and $e : \mathbb{G}_1 \times \mathbb{G}_2 \to \mathbb{G}_T$ is an efficiently computable, non-degenerate, bilinear map, and there is no efficiently computable isomorphism between $\mathbb{G}_1$ and $\mathbb{G}_2$. We call such an algorithm $\mathcal{G}$ a *bilinear group generator*. In the case $\mathbb{G}_1 = \mathbb{G}_2$, the groups are said *symmetric*, else they are said *asymmetric*.

In our work we rely on specific complexity assumptions in such bilinear groups: q-Strong Diffie-Hellman [6], q-Diffie-Hellman-Inversion [5], and External DDH in $\mathbb{G}_1$. For lack of space, we defer the interested reader to the corresponding references or the full version of our paper for their definition.

Finally, we introduce the following static assumption over asymmetric bilinear groups, that we call "Flexible Diffie-Hellman Inversion" (FDHI) for its similarity to Flexible Diffie-Hellman [22]. As we discuss in the full version of our paper, FDHI is hard in the generic bilinear group model.

Definition 1 (Flexible Diffie-Hellman Inversion Assumption). *Let $\mathcal{G}$ be a generator of asymmetric bilinear groups, and let* $\mathsf{bgp} = (p, \mathbb{G}_1, \mathbb{G}_2, \mathbb{G}_T, g_1, g_2, e)$ $\xleftarrow{\$} \mathcal{G}(1^\lambda)$. *We say that the Flexible Diffie-Hellman Inversion (FDHI) Assumption is ϵ-hard for $\mathcal{G}$ if for random $z, r, v \xleftarrow{\$} \mathbb{Z}_p$ and for every PPT adversary $\mathcal{A}$:*

$$\mathbf{Adv}_{\mathcal{A}}^{FDHI}(\lambda) = \Pr[W \in \mathbb{G}_1 \setminus \{1_{\mathbb{G}_1}\} : (W, W^{\frac{1}{z}}) \leftarrow \mathcal{A}(g_1, g_2, g_2^z, g_2^v, g_1^{\frac{z}{v}}, g_1^r, g_1^{\frac{r}{v}})] \leq \epsilon$$

3 Asymmetric Programmable Hash Functions

In this section we present our new notion of asymmetric programmable hash functions.

Let $\mathsf{bgp} = (p, \mathbb{G}_1, \mathbb{G}_2, \mathbb{G}_T, g_1, g_2, e)$ be a family of asymmetric bilinear groups induced by a bilinear group generator $\mathcal{G}(1^\lambda)$ for a security parameter $\lambda \in \mathbb{N}$.[4] An *asymmetric group hash function* $\mathsf{H} : \mathcal{X} \to \mathbb{G}_1$ consists of three PPT algorithms $(\mathsf{H.Gen}, \mathsf{H.PriEval}, \mathsf{H.PubEval})$ working as follows:

$\mathsf{H.Gen}(1^\lambda, \mathsf{bgp}) \to (\mathsf{sek}, \mathsf{pek})$: on input the security parameter $\lambda \in \mathbb{N}$ and a bilinear group description bgp, the PPT key generation algorithm outputs a (secret) evaluation key sek and a (public) evaluation key pek.

$\mathsf{H.PriEval}(\mathsf{sek}, X) \to Y \in \mathbb{G}_1$: given the secret evaluation key sek and an input $X \in \mathcal{X}$, the deterministic evaluation algorithm returns an output $Y = \mathsf{H}(X) \in \mathbb{G}_1$.

$\mathsf{H.PubEval}(\mathsf{pek}, X) \to \hat{Y} \in \mathbb{G}_T$: on input the public evaluation key pek and an input $X \in \mathcal{X}$, the public evaluation algorithm outputs a value $\hat{Y} \in \mathbb{G}_T$ such that $\hat{Y} = e(\mathsf{H}(X), g_2)$.

For asymmetric hash functions satisfying the syntax described above, we define two different properties that model their possible programmability.

The first property is a generalization of the notion of programmable hash functions of [26,27] to our asymmetric setting (i.e., where the function is only

---

[4] Our definition can be easily adapted to work in symmetric bilinear groups where $\mathbb{G}_1 = \mathbb{G}_2$.

secretly-computatble), and to the more specific setting of bilinear groups. The basic idea is that it is possible to generate the function in a trapdoor-mode that allows one to express every output of H in relation to some specified group elements. In particular, the most useful fact of programmability is that for two arbitrary disjoint sets of inputs $\bar{X}, \bar{Z} \subset \mathcal{X}$, the joint probability that some of these group elements appear in $\mathsf{H}(Z), \forall Z \in \bar{Z}$ and do not appear in $\mathsf{H}(X), \forall X \in \bar{X}$ is significant.

Definition 2 (Asymmetric Programmable Hash Functions). *An asymmetric group hash function* $\mathsf{H} = (\mathsf{H.Gen}, \mathsf{H.PriEval}, \mathsf{H.PubEval})$ *is* $(m, n, d, \gamma, \delta)$-*programmable if there exist an efficient trapdoor generation algorithm* $\mathsf{H.TrapGen}$ *and an efficient trapdoor evaluation algorithm* $\mathsf{H.TrapEval}$ *such that:*

Syntax: $\mathsf{H.TrapGen}(1^\lambda, \mathsf{bgp}, \hat{g}_1, \hat{h}_1, \hat{g}_2, \hat{h}_2) \rightarrow (\mathsf{td}, \mathsf{pek})$ *takes as input the security parameter* λ, *bilinear group description* bgp *and group elements* $\hat{g}_1, \hat{h}_1 \in \mathbb{G}_1, \hat{g}_2, \hat{h}_2 \in \mathbb{G}_2$, *and it generates a public hash key* pek *along with a trapdoor* td. $\mathsf{H.TrapEval}(\mathsf{td}, X) \rightarrow \mathbf{c}_X$ *takes as input the trapdoor information* td *and an input* $X \in \mathcal{X}$, *and outputs a vector of integer coefficients* $\mathbf{c}_X = (c_0, \ldots, c_{d'}) \in \mathbb{Z}^{d'}$ *of a 2-variate polynomial* $c_X(y_1, y_2)$ *of degree* $\leq d$.

Correctness: *For all group elements* $\hat{g}_1, \hat{h}_1 \in \mathbb{G}_1, \hat{g}_2, \hat{h}_2 \in \mathbb{G}_2$ *such that* $\hat{h}_1 = \hat{g}_1^{y_1}$ *and* $\hat{h}_2 = \hat{g}_2^{y_2}$ *for some* $y_1, y_2 \in \mathbb{Z}_p$, *for all trapdoor keys* $(\mathsf{td}, \mathsf{pek}) \xleftarrow{\$} \mathsf{H.TrapGen}(1^\lambda, \hat{g}_1, \hat{h}_1, \hat{g}_2, \hat{h}_2)$, *and for all inputs* $X \in \mathcal{X}$, *if* $\mathbf{c}_X \leftarrow \mathsf{H.TrapEval}(\mathsf{td}, X)$, *then*

$$\mathsf{H}(X) = \hat{g}_1^{c_X(y_1, y_2)}$$

Statistically-Close Trapdoor Keys: *For all generators* $\hat{g}_1, \hat{h}_1 \in \mathbb{G}_1, \hat{g}_2, \hat{h}_2 \in \mathbb{G}_2$ *and for all* $(\mathsf{sek}, \mathsf{pek}) \xleftarrow{\$} \mathsf{H.Gen}(1^\lambda)$, $(\mathsf{td}, \mathsf{pek}') \xleftarrow{\$} \mathsf{H.TrapGen}(1^\lambda, \hat{g}_1, \hat{h}_1, \hat{g}_2, \hat{h}_2)$, *the distribution of the public keys* pek *and* pek' *is within statistical distance* γ.

Well Distributed Logarithms: *For all* $\hat{g}_1, \hat{h}_1 \in \mathbb{G}_1, \hat{g}_2, \hat{h}_2 \in \mathbb{G}_2$, *all keys* $(\mathsf{td}, \mathsf{pek}) \xleftarrow{\$} \mathsf{H.TrapGen}(1^\lambda, \hat{g}_1, \hat{h}_1, \hat{g}_2, \hat{h}_2)$, *and all inputs* $X_1, \ldots, X_m \in \mathcal{X}$ *and* $Z_1, \ldots, Z_n \in \mathcal{X}$ *such that* $X_i \neq Z_j$ *for all* i, j, *we have*

$$\Pr[c_{X_1,0} = \cdots = c_{X_m,0} = 0 \ \wedge \ c_{Z_1,0}, \ldots, c_{Z_n,0} \neq 0] \geq \delta$$

where $\mathbf{c}_{X_i} \leftarrow \mathsf{H.TrapEval}(\mathsf{td}, X_i)$ *and* $\mathbf{c}_{Z_j} \leftarrow \mathsf{H.TrapEval}(\mathsf{td}, Z_j)$, *and* $c_{X_i,0}$ *(resp.* $c_{Z_j,0}$) *is the coefficient of the term of degree 0.*

If γ *is negligible and* δ *is noticeable we simply say that* H *is* (m, n, d)-*programmable. Furthermore, if* m *(resp.* n) *is an arbitrary polynomial in* λ, *then we say that* H *is* (poly, n, d)-*programmable (resp.* (m, poly, d)-*programmable). Finally, if* H *admits trapdoor algorithms that satisfy only the first three properties, then* H *is said simply* (d, γ)-*programmable. Note that any* H *that is* $(m, n, d, \gamma, \delta)$-*programmable is also* (d, γ)-*programmable.*

Programmable Pseudo-randomness. The second main programmability property that we define for asymmetric hash functions is quite different from the previous one. It is called *programmable pseudo-randomness*, and very intuitively it says that, when using the hash function in trapdoor mode, it is possible to "embed" a PRF into it. More precisely, the trapdoor algorithms satisfy programmable pseudo-randomness if they allow to generate keys such that even by observing pek and $H(X)$ for a bunch of inputs X, then the elements $g_1^{c_X,0}$ look random. The formal definition follows:

Definition 3 (Asymmetric Hash Functions with Programmable Pseudorandomness). *An asymmetric hash function* H = (H.Gen, H.PriEval, H.PubEval) *has* (d, γ, ϵ)-*programmable pseudorandomness if there exist efficient trapdoor algorithms* H.TrapGen, H.TrapEval *that satisfy the properties of syntax, correctness, and* γ-*statistically-close trapdoor keys as in Definition 2, and additionally satisfy the following property with parameter* ϵ:

Pseudorandomness: Let $b \in \{0, 1\}$ *and let* $\mathbf{Exp}_{\mathcal{A},H}^{PRH-b}(\lambda)$ *be the following experiment between an adversary* $\mathcal{A}$ *and a challenger.*

1. *Generate* bgp $\overset{\$}{\leftarrow} \mathcal{G}(1^\lambda)$, *and run* $\mathcal{A}(\text{bgp})$, *that outputs two generators* $h_1 \in \mathbb{G}_1, h_2 \in \mathbb{G}_2$.

2. *Compute* (td, pek) $\overset{\$}{\leftarrow}$ H.TrapGen$(1^\lambda, g_1, h_1, g_2, h_2)$ *and run* $\mathcal{A}(\text{pek})$ *with access to the following oracle:*
   - *If* $b = 0$, $\mathcal{A}$ *is given* $\mathcal{O}(\cdot)$ *that on input* $X \in \mathcal{X}$ *returns* $H(X) = g_1^{c_X(y_1, y_2)}$ *and* $g_1^{c_X,0}$, *where* $c_X \leftarrow$ H.TrapEval(td, X);
   - *If* $b = 1$, $\mathcal{A}$ *is given* $\mathcal{R}(\cdot)$ *that on input* $X \in \mathcal{X}$ *returns* $H(X) = g_1^{c_X(y_1, y_2)}$ *and* $g_1^{r_X}$, *for a randomly chosen* $r_X \overset{\$}{\leftarrow} \mathbb{Z}_p$ *(which is unique for every* $X \in \mathcal{X}$).

3. *At the end the adversary outputs a bit* b', *and* b' *is returned as the output of the experiment.*

   Then we say that H.TrapGen, H.TrapEval *satisfy pseudo-randomness for* ϵ, *if for all PPT* $\mathcal{A}$

$$\left| \Pr[\mathbf{Exp}_{\mathcal{A},H}^{PRH-0}(\lambda) = 1] - \Pr[\mathbf{Exp}_{\mathcal{A},H}^{PRH-1}(\lambda) = 1] \right| \leq \epsilon$$

   where the probabilities are taken over all the random choices of TrapGen, *the oracle* $\mathcal{R}$ *and the adversary* $\mathcal{A}$.

Other Variants of Programmability. Here we define two other variants of the programmability notion given in Definition 2. Formal definitions appear in the full version of our paper.

WEAK PROGRAMMABILITY. We consider a weak version of the above programmability property in which one fixes at key generation time the n inputs Z_j on which $c_{Z_j,0} \neq 0$.

Remark 1. We remark that for those (deterministic) functions H whose domain $\mathcal{X}$ has polynomial size any weak programmability property for an arbitrary $m =$ poly trivially holds with $\delta = 1$.

DEGREE-d PROGRAMMABILITY. In our work we also consider a variant of the above definition in which the property of well distributed logarithms is stated with respect to the *degree-d coefficients* of the polynomials generated by H.TrapEval. In this case, we say that H is $(m, n, d, \gamma, \delta)$-*degree-d-programmable*.

3.1 An Asymmetric PHF Based on Cover-Free Sets

In this section we present the construction of an asymmetric hash function, H_{acfs}, based on cover-free sets. Our construction uses ideas similar to the ones used by Hofheinz, Jager and Kiltz [25] to design a (regular) programmable hash function. Our construction extends these ideas with a technique that allows us to obtain a much shorter public key. Concretely, for binary inputs of size ℓ, the programmable hash function H_{cfs} in [25] is $(m, 1)$-programmable with a hash key of length $O(\ell m^2)$. In contrast, our new construction H_{acfs} is $(m, 1)$-programmable with a hash key of length $O(m\sqrt{\ell})$. While such improvement is obtained at the price of obtaining the function in the secret-key model, our results of Sect. 5 show that *asymmetric* programmable hash are still useful to build short bilinear-map signatures, whose efficiency, in terms of signature's and key's length matches that of state-of-the-art schemes [31].

Before proceeding with describing our function, below we recall the notion of cover-free sets.

COVER-FREE FAMILIES. If S, V are sets, we say that S does not cover V if $S \not\supseteq V$. Let T, m, s be positive integers, and let $F = \{F_i\}_{i \in [s]}$ be a family of subsets of $[T]$. A family F is said to be m-*cover-free* over $[T]$, if for any subset $I \subseteq [s]$ of cardinality at most m, then the union $\cup_{i \in I} F_i$ does not cover F_j for all $j \notin I$. More formally, for any $I \subseteq [s]$ such that $|I| \leq m$, and any $j \notin I$, $\cup_{i \in I} F_i \not\supseteq F_j$. Furthermore, we say that F is w-uniform if every subset F_i in the family have size w. In our construction, we use the following fact from [16,29]:

Lemma 1 ([16,29]). *There is a deterministic polynomial time algorithm that, on input integers $s = 2^\ell$ and m, returns w, T, F where $F = \{F_i\}_{i \in [s]}$ is a w-uniform, m-cover-free family over $[T]$, for $w = T/4m$ and $T \leq 16m^2\ell$.*

THE CONSTRUCTION OF H_{acfs}. Let $\mathcal{G}(1^\lambda)$ be a bilinear group generator, let $\mathsf{bgp} = (p, \mathbb{G}_1, \mathbb{G}_2, \mathbb{G}_T, g_1, g_2, e)$ be an instance of bilinear group parameters generated by $\mathcal{G}$. Let $\ell = \ell(\lambda)$ and $m = m(\lambda)$ be two polynomials in the security parameter. We set $s = 2^\ell$, $T = 16m^2\ell$, and $w = T/4m$ as for Lemma 1, and define $t = \lceil \sqrt{T} \rceil$. Note that every integer $k \in [T]$ can be written as a pair of integers $(i, j) \in [t] \times [t]$ using some canonical mapping. For the sake of simplicity, sometimes we abuse notation and write $(i, j) \in [T]$ where $i, j \in [t]$.

In the following we describe the asymmetric hash function $H_{acfs} = (H.\mathsf{Gen}, H.\mathsf{PriEval}, H.\mathsf{PubEval})$ that maps $H_{acfs} : \mathcal{X} \to \mathbb{G}_1$ where $\mathcal{X} = \{0, 1\}^\ell$. In particular, every input $X \in \{0, 1\}^\ell$ is associated to a set F_i, $i \in [2^\ell]$, by interpreting X as an integer in $\{0, \ldots, 2^\ell - 1\}$ and by setting $i = X + 1$. We call F_X such subset associated to X.

H.Gen(1^λ, bgp): for $i = 1$ to t, sample $\alpha_i, \beta_i \xleftarrow{\$} \mathbb{Z}_p$ and compute $A_i = g_1^{\alpha_i}, B_i = g_2^{\beta_i}$. Finally, set sek $= \{\alpha_i, \beta_i\}_{i=1}^t$, pek $= \{A_i, B_i\}_{i=1}^t$, and return (sek, pek).

H.PriEval(sek, X): first, compute the subset $F_X \subseteq [T]$ associated to $X \in \{0,1\}^\ell$, and then return

$$Y = g_1^{\sum_{(i,j) \in F_X} \alpha_i \beta_j} \in \mathbb{G}_1$$

H.PubEval(pek, X): let $F_X \subseteq [T]$ be the subset associated to X, and compute

$$\hat{Y} = \prod_{(i,j) \in F_X} e(A_i, B_j) = e(\mathsf{H}(X), g_2)$$

Theorem 1. *Let $\mathcal{G}$ be a bilinear group generator. The hash function $\mathsf{H}_{\mathsf{acfs}}$ described above is an asymmetric $(m, n, d, \gamma, \delta)$-programmable hash function with $n = 1$, $d = 2$, $\gamma = 0$ and $\delta = 1/T$.*

We show a proof sketch by giving the description of the trapdoor algorithms. A full proof showing that these algorithms satisfy the desired programmability property appears in the full version.

H.TrapGen(1^λ, bgp, $\hat{g}_1, \hat{h}_1, \hat{g}_2, \hat{h}_2$): first, sample $a_i, b_i \xleftarrow{\$} \mathbb{Z}_p$ for all $i \in [t]$, and pick a random index $\tau \xleftarrow{\$} [T]$. Parse $\tau = (i^*, j^*) \in [t] \times [t]$. Next, set $A_{i^*} = \hat{g}_1 \hat{h}_1^{a_{i^*}}$, $B_{j^*} = \hat{g}_2 \hat{h}_2^{b_{j^*}}$, $A_i = \hat{h}_1^{a_i}$, $\forall i \neq i^*$, and $B_j = \hat{h}_2^{b_j}$, $\forall j \neq j^*$. Finally, set td $= (\tau, \{a_i, b_i\}_{i=1}^t)$, pek $= \{A_i, B_i\}_{i=1}^t$, and output (td, pek).

H.TrapEval(td, X): first, compute the subset $F_X \subseteq [T]$ associated to $X \in \{0,1\}^\ell$, and then return the coefficients of the degree-2 polynomial $c_X(y_1, y_2) = \sum_{(i,j) \in F_X} \alpha_i(y_1) \cdot \beta_j(y_2)$, where every $\alpha_i(y_1)$ (resp. $\beta_j(y_2)$) is the discrete logarithm of A_i (resp. B_j) in base $\hat{g}_1$ (resp. $\hat{g}_2$), viewed as a degree-1 polynomial in the unknown y_1 (resp. y_2).

3.2   An Asymmetric PHF with Small Domain

In this section, we present the construction of an asymmetric hash function, $\mathsf{H}_{\mathsf{sqrt}}$, whose domain is of polynomial size T. $\mathsf{H}_{\mathsf{sqrt}}$ has a public key of length $O(\sqrt{T})$, and it turns out to be very important for obtaining our linearly-homomorphic signature scheme with short public key presented in Sect. 4. Somewhat interestingly, we show that this new function $\mathsf{H}_{\mathsf{sqrt}}$ satisfies several programmability properties, that make it useful in the context of various security proofs.

Let $\mathcal{G}(1^\lambda)$ be a bilinear group generator, let $T = \mathsf{poly}(\lambda)$ and $t = \lceil \sqrt{T} \rceil$. The hash function $\mathsf{H}_{\mathsf{sqrt}} = (\mathsf{H.Gen, H.PriEval, H.PubEval})$ that maps $\mathsf{H}_{\mathsf{sqrt}} : \mathcal{X} \to \mathbb{G}_1$ with $\mathcal{X} = [T]$ is defined as follows.

H.Gen(1^λ, bgp): for $i = 1$ to t, sample $\alpha_i, \beta_i \xleftarrow{\$} \mathbb{Z}_p$ and compute $A_i = g_1^{\alpha_i}, B_i = g_2^{\beta_i}$. Finally, set sek $= \{\alpha_i, \beta_i\}_{i=1}^t$, pek $= \{A_i, B_i\}_{i=1}^t$, and return(sek, pek).

H.PriEval(sek, X): first, write $X \in [T]$ as a pair of integer $(i, j) \in [t] \times [t]$, and then return

$$Y = g_1^{\alpha_i \beta_j} \in \mathbb{G}_1$$

H.PubEval(pek, X): let $X = (i, j)$. The public evaluation algorithm returns

$$\hat{Y} = e(A_i, B_j) = e(\mathsf{H}(X), g_2)$$

Here we show that $\mathsf{H}_{\mathsf{sqrt}}$ satisfies the programmable pseudo-randomness property of Definition 3.

Theorem 2 (Programmable Pseudorandomness of $\mathsf{H}_{\mathsf{sqrt}}$). *Let $\mathbb{G}_1$ be a bilinear group of order p over which the XDDH assumption is ϵ'-hard. Then the asymmetric hash function $\mathsf{H}_{\mathsf{sqrt}}$ described above satisfies $(2, 0, \epsilon)$-programmable pseudo-randomness with $\epsilon = T \cdot \epsilon'$. Furthermore, in the case when $h_1 = 1 \in \mathbb{G}_1$ or $h_1 = g_1$, $\mathsf{H}_{\mathsf{sqrt}}$ has $(1, 0, \epsilon)$-programmable pseudo-randomness.*

Proof. First, we describe the trapdoor algorithms:

H.TrapGen($1^\lambda, g_1, h_1, g_2, h_2$): first, sample $a_i, r_i, s_i, b_i \xleftarrow{\$} \mathbb{Z}_p$ for all $i \in [t]$ and
   then set $A_i = h_1^{r_i} g_1^{a_i}, B_i = h_2^{s_i} g_2^{b_i}$. Finally, set $\mathsf{td} = (\{a_i, r_i, s_i, b_i\}_{i=1}^t)$, $\mathsf{pek} = \{A_i, B_i\}_{i=1}^t$, and output $(\mathsf{td}, \mathsf{pek})$.
H.TrapEval(td, X): let $X = (i, j)$, and then return the coefficients of the degree-2
   polynomial

$$c_X(y_1, y_2) = (y_1 r_i + a_i)(y_2 s_j + b_j)$$

First, it is easy to see that the two algorithms satisfy the syntax and correctness properties. Also, in the case $h_1 = 1$ (i.e., $y_1 = 0$) or $h_1 = g_1$ (i.e., $y_1 = 1$), we obtain a degree-1 polynomial $c_X(y_2)$. Second, observe that each element A_i (resp. B_i) in pek is a uniformly distributed group element in $\mathbb{G}_1$ (resp. $\mathbb{G}_2$), as in H.Gen, hence $\gamma = 0$. Third, we show that the function satisfies the pseudo-randomness property under the assumption that XDDH holds in $\mathbb{G}_1$. The main observation is that for every $X = (i, j)$, we have $c_{X,0} = a_i b_j$ where all the values b_i are uniformly distributed and information-theoretically hidden to an adversary who only sees pek. In particular, this holds even if $h_1 = 1$.

To prove the pseudo-randomness we make use of Lemma 2 below, which shows that for a uniformly random choice of $a, b \xleftarrow{\$} \mathbb{Z}_p^t, c \xleftarrow{\$} \mathbb{Z}_p^{t \times t}$ the distributions $(g_1^a, g_1^{a \cdot b^\top}) \in \mathbb{G}_1^{t \times (t+1)}$ and $(g_1^a, g_1^c) \in \mathbb{G}_1^{t \times (t+1)}$ are computationally indistinguishable.

Lemma 2. *Let $a, b \xleftarrow{\$} \mathbb{Z}_p^t, c \xleftarrow{\$} \mathbb{Z}_p^{t \times t}$ be chosen uniformly at random. If the XDDH assumption is ϵ'-hard in $\mathbb{G}_1$, then for any PPT $\mathcal{B}$ it holds $|\Pr[\mathcal{B}(g_1^a, g_1^{a \cdot b^\top}) = 1] - \Pr[\mathcal{B}(g_1^a, g_1^c) = 1]| \leq T \cdot \epsilon'$.*

We first show how to use Lemma 2 to prove that $\mathsf{H}_{\mathsf{sqrt}}$ has programmable pseudo-randomness. The proof of Lemma 2 appears in the full version.

Let $\mathcal{A}$ be an adversary that breaks the ϵ-programmable pseudo-randomness of $\mathsf{H}_{\mathsf{sqrt}}$. We construct a simulator $\mathcal{B}$ that can distinguish the two distributions $(g_1^a, g_1^{a \cdot b^\top})$ and (g_1^a, g_1^c) described above with advantage greater than ϵ.

$\mathcal{B}$'s input is a tuple $(A', C) \in \mathbb{G}_1^t \times \mathbb{G}_1^{t \times t}$ and its goal is to decide about the distribution of C. First, $\mathcal{B}$ runs $\mathcal{A}(\mathsf{bgp})$ which outputs the generators h_1, h_2.

$\mathcal{B}$ then samples two random vectors $r, \beta \xleftarrow{\$} \mathbb{Z}_p^t$, computes $B = g_2^\beta \in \mathbb{G}_2^t$, $A = h_1^r \cdot A' \in \mathbb{G}_1^t$, sets $\mathsf{pek} = (A, B)$, and runs $\mathcal{A}(\mathsf{pek})$ Next, for every oracle query (i, j) made by $\mathcal{A}$, $\mathcal{B}$ simulates the answer by returning to $\mathcal{A}$: $\mathsf{H}(i, j) = A_i^{\beta_j}$ and $C_{i,j}$. It is easy to see that if $C = g_1^{a \cdot b^\top}$ then $\mathcal{B}$ is perfectly simulating $\mathbf{Exp}_{\mathcal{A}, \mathsf{H}_{\mathsf{sqrt}}}^{PRH\text{-}0}$, otherwise, if C is random and independent, then $\mathcal{B}$ is simulating $\mathbf{Exp}_{\mathcal{A}, \mathsf{H}_{\mathsf{sqrt}}}^{PRH\text{-}1}$. As a final note, we observe that the above proof works even in the case $h_1 = 1$. $\square$

In the following theorems (whose proofs appear in the full version of our paper) we show that $\mathsf{H}_{\mathsf{sqrt}}$ satisfies programmability with various parameters.

Theorem 3 ((poly, 0, 2)-programmability of $\mathsf{H}_{\mathsf{sqrt}}$). *The asymmetric hash function $\mathsf{H}_{\mathsf{sqrt}}$ described above is $(\mathsf{poly}, 0, d, \gamma, \delta)$-programmable with $d = 2$, $\gamma = 0$ and $\delta = 1$. Furthermore, in the case when either $\hat{h}_1 = \hat{g}_1$ or $\hat{h}_2 = \hat{g}_2$, $\mathsf{H}_{\mathsf{sqrt}}$ is $(\mathsf{poly}, 0, d, \gamma, \delta)$-programmable with $d = 1$, $\gamma = 0$ and $\delta = 1$.*

Theorem 4 (Weak $(\mathsf{poly}, 1, 2)$-programmability of $\mathsf{H}_{\mathsf{sqrt}}$). *The asymmetric hash function $\mathsf{H}_{\mathsf{sqrt}}$ described above is weakly $(\mathsf{poly}, 1, d, \gamma, \delta)$-programmable with $d = 2$, $\gamma = 0$ and $\delta = 1$.*

Theorem 5 (Weak $(\mathsf{poly}, 1, 2)$-degree-2-programmability of $\mathsf{H}_{\mathsf{sqrt}}$). *The asymmetric hash function $\mathsf{H}_{\mathsf{sqrt}}$ described above is weakly $(\mathsf{poly}, 1, d, \gamma, \delta)$-degree-2 programmable with $d = 2$, $\gamma = 0$ and $\delta = 1$.*

4 Linearly-Homomorphic Signatures with Short Public Keys

In this section, we show a new linearly-homomorphic signature scheme that uses asymmetric PHFs in a generic way. By instantiating the asymmetric PHFs with our construction $\mathsf{H}_{\mathsf{sqrt}}$ given in Sect. 3, we obtain the *first* linearly-homomorphic signature scheme that is secure in the standard model, and whose public key has a size that is *sub-linear* in both the dataset size and the dimension of the signed vectors. Precisely, if the signature scheme supports datasets of maximal size N and can sign vectors of dimension T, then the public key of our scheme is of size $O(\sqrt{N} + \sqrt{T})$. All previously existing constructions in the standard model achieved only public keys of length $O(N + T)$. Furthermore, our scheme is adaptive secure and achieves the interesting property of *efficient verification* that allows to use the scheme for verifiable delegation of computation in the preprocessing model [15].

4.1 Homomorphic Signatures for Multi-Labeled Programs

First we recall the definition of homomorphic signatures as presented in [15]. This definition extends the one by Freeman in [17] in order to work with the general notion of multi-labeled programs [4, 20].

Multi-Labeled Programs. A *labeled program* $\mathcal{P}$ is a tuple $(f, \tau_1, ..., \tau_n)$ such that $f : \mathcal{M}^n \to \mathcal{M}$ is a function of n variables (e.g., a circuit) and $\tau_i \in \{0, 1\}^*$

is a label of the i-th input of f. Labeled programs can be composed as follows: given $\mathcal{P}_1, \ldots, \mathcal{P}_t$ and a function $g : \mathcal{M}^t \to \mathcal{M}$, the composed program $\mathcal{P}^*$ is the one obtained by evaluating g on the outputs of $\mathcal{P}_1, \ldots, \mathcal{P}_t$, and it is denoted as $\mathcal{P}^* = g(\mathcal{P}_1, \ldots, \mathcal{P}_t)$. The labeled inputs of $\mathcal{P}^*$ are all the distinct labeled inputs of $\mathcal{P}_1, \cdots \mathcal{P}_t$ (all the inputs with the same label are grouped together and considered as a unique input of $\mathcal{P}^*$). Let $f_{id} : \mathcal{M} \to \mathcal{M}$ be the identity function and $\tau \in \{0,1\}^*$ be any label. We refer to $\mathcal{I}_\tau = (f_{id}, \tau)$ as the identity program with label τ. Note that a program $\mathcal{P} = (f, \tau_1, \cdots, \tau_n)$ can be expressed as the composition of n identity programs $\mathcal{P} = f(\mathcal{I}_{\tau_1}, \cdots, \mathcal{I}_{\tau_1})$.

A *multi-labeled program* $\mathcal{P}_\Delta$ is a pair $(\mathcal{P}, \Delta)$ in which $\mathcal{P} = (f, \tau_1, \cdots, \tau_n)$ is a labeled program while $\Delta \in \{0,1\}^*$ is a *data set identifier*. Given $(\mathcal{P}_1, \Delta), \ldots, (\mathcal{P}_t, \Delta)$ which has the same data set identifier Δ, and given a function $g : \mathcal{M}^t \to \mathcal{M}$, the composed multi-labeled program $\mathcal{P}_\Delta^*$ is the pair $(\mathcal{P}^*, \Delta)$ where $\mathcal{P}^* = g(\mathcal{P}_1, \cdots, \mathcal{P}_t)$, and Δ is the common data set identifier for all the $\mathcal{P}_i$. As for labeled programs, one can define the notion of a multi-labeled identity program as $\mathcal{I}_{(\Delta, \tau)} = ((f_{id}, \tau), \Delta)$.

Definition 4 (Homomorphic Signatures). *A homomorphic signature scheme* HSig *consists of a tuple of PPT algorithms* (KeyGen, Sign, Ver, Eval) *satisfying the following four properties:* authentication correctness, evaluation correctness, succinctness *and* security.

KeyGen($1^\lambda, \mathcal{L}$) *the key generation algorithm takes as input a security parameter λ, the description of the label space $\mathcal{L}$ (which fixes the maximum data set size N), and outputs a public key* vk *and a secret key* sk. *The public key* vk *defines implicitly a message space $\mathcal{M}$ and a set $\mathcal{F}$ of admissible functions.*

Sign(sk, Δ, τ, m) *the signing algorithm takes as input a secret key* sk, *a data set identifier Δ, a label $\tau \in \mathcal{L}$ a message $m \in \mathcal{M}$, and it outputs a signature σ.*

Ver(vk, $\mathcal{P}_\Delta, m, \sigma$) *the verification algorithm takes as input a public key* vk, *a multi-labeled program $\mathcal{P}_\Delta = ((f, \tau_1, \ldots, \tau_n), \Delta)$ with $f \in \mathcal{F}$, a message $m \in \mathcal{M}$, and a signature σ. It outputs either 0 (reject) or 1 (accept).*

Eval(vk, $f, \boldsymbol{\sigma}$) *the evaluation algorithm takes as input a public* vk, *a function $f \in \mathcal{F}$ and a tuple of signatures $\{\sigma_i\}_{i=1}^n$ (assuming that f takes n inputs). It outputs a new signature σ.*

AUTHENTICATION CORRECTNESS. The scheme HSig satisfies the authentication correctness property if for a given label space $\mathcal{L}$, all key pairs (sk, vk) $\leftarrow$ KeyGen($1^\lambda, \mathcal{L}$), any label $\tau \in \mathcal{L}$, data identifier $\Delta \in \{0,1\}^*$, and any signature $\sigma \leftarrow$ Sign(sk, Δ, τ, m), Ver(vk, $\mathcal{I}_{\Delta,\tau}, m, \sigma$) outputs 1 with all but negligible probability.

EVALUATION CORRECTNESS. Fix a key pair (vk, sk) $\xleftarrow{\$}$ KeyGen($1^\lambda, \mathcal{L}$), a function $g : \mathcal{M}^t \to \mathcal{M}$, and any set of program/message/signature triples $\{(\mathcal{P}_i, m_i, \sigma_i)\}_{i=1}^t$ such that Ver(vk, $\mathcal{P}_i, m_i, \sigma_i$) = 1. If $m^* = g(m_1, \ldots, m_t)$, $\mathcal{P}^* = g(\mathcal{P}_1, \cdots, \mathcal{P}_t)$, and $\sigma^* =$ Eval(vk, $g, (\sigma_1, \ldots, \sigma_t)$), then Ver(vk, $\mathcal{P}^*, m^*, \sigma^*$) = 1 holds with all but negligible probability.

SUCCINTNESS. A homomorphic signature scheme is said to be succint if, for a fixed security parameter λ, the size of signatures depends at most logarithmically on the data set size N.

SECURITY. To define the security notion of homomorphic signatures we define the following experiment HomUF-CMA$_{\mathcal{A},\text{HomSign}}(\lambda)$ between an adversary $\mathcal{A}$ and a challenger $\mathcal{C}$:

Key Generation $\mathcal{C}$ runs $(\text{vk}, \text{sk}) \xleftarrow{\$} \text{KeyGen}(1^\lambda, \mathcal{L})$ and gives vk to $\mathcal{A}$.

Signing Queries $\mathcal{A}$ can adaptively submit queries of the form (Δ, τ, m), where Δ is a data set identifier, $\tau \in \mathcal{L}$, and $m \in \mathcal{M}$. The challenger $\mathcal{C}$ proceeds as follows: if (Δ, τ, m) is the first query with the data set identifier Δ, the challenger initializes an empty list $T_\Delta = \emptyset$ for Δ. If T_Δ does not already contain a tuple $(\tau, \cdot)$, the challenger $\mathcal{C}$ computes $\sigma \xleftarrow{\$} \text{Sign}(\text{sk}, \Delta, \tau, m)$, returns σ to $\mathcal{A}$ and updates the list $T_\Delta \leftarrow T_\Delta \cup (\tau, m)$. If $(\tau, m) \in T_\Delta$ then $\mathcal{C}$ replies with the same signature generated before. If T_Δ contains a tuple (τ, m') for some message $m' \neq m$, then the challenger ignores the query.

Forgery At the end $\mathcal{A}$ outputs a tuple $(\mathcal{P}^*_{\Delta^*}, m^*, \sigma^*)$.

The experiment HomUF-CMA$_{\mathcal{A},\text{HomSign}}(\lambda)$ outputs 1 if the tuple returned by $\mathcal{A}$ is a forgery, and 0 otherwise. To define what is a forgery in such a game we recall the notion of well defined program with respect to a list T_Δ [15].

Definition 5. *A labeled program* $\mathcal{P}^* = (f^*, \tau_1^*, \ldots, \tau_n^*)$ *is well defined with respect to* T^*_Δ *if* $\exists \ m_1, \ldots, m_n$ *s.t.* $(\tau_i^*, m_i) \in T_{\Delta^*} \ \forall i = 1, \ldots, n,$ *or if.* $\exists \ i \in \{1, \cdots, n\}$ *s.t.* $(\tau_i, \cdot) \notin T_{\Delta^*}$ *and* $f^*(\{m_j\}_{(\tau_j, m_j) \in T_{\Delta^*}} \cup \{\tilde{m}_{(\tau_j, \cdot) \notin T_{\Delta^*}}\})$ *does not change for all possible choices of* $\tilde{m}_j \in \mathcal{M}$.

Using this notion, it is then possible to define the three different types of forgeries that can occur in the experiment HomUF-CMA:

Type 1: $\text{Ver}(\text{vk}, \mathcal{P}^*_{\Delta^*}, m^*, \sigma^*) = 1$ and T_{Δ^*} was not initialized in the game

Type 2: $\text{Ver}(\text{vk}, \mathcal{P}^*_{\Delta^*}, m^*, \sigma^*) = 1$, $\mathcal{P}^*$ is well defined w.r.t. T_{Δ^*} and $m^* \neq f^*(\{m_j\}_{(\tau_j, m_j) \in T_{\Delta^*}})$

Type 3: $\text{Ver}(\text{vk}, \mathcal{P}^*_{\Delta^*}, m^*, \sigma^*) = 1$ and $\mathcal{P}^*$ is *not* well defined w.r.t. T_{Δ^*}.

Then we say that HSig is a secure homomorphic signature if for any PPT adversary $\mathcal{A}$, we have that $\Pr[\text{HomUF-CMA}_{\mathcal{A},\text{HomSign}}(\lambda) = 1] \leq \epsilon(\lambda)$ where $\epsilon(\lambda)$ is a negligible function.

Finally, we recall that, as proved by Freeman in [17], in a linearly-homomorphic signatures scheme any adversary who outputs a Type 3 forgery can be converted into one that outputs a Type 2 forgery.

Homomorphic Signatures with Efficient Verification. We recall the notion of homomorphic signatures with efficient verification introduced in [15]. Informally, the property states that the verification algorithm can be split in two phases: an *offline* phase where, given the verification key vk and a labeled program $\mathcal{P}$, one precomputes a concise key $\text{vk}_\mathcal{P}$; an *online* phase in which $\text{vk}_\mathcal{P}$ can be used to verify signatures w.r.t. $\mathcal{P}$ and *any* dataset Δ. To achieve (amortized) efficiency, the idea is that $\text{vk}_\mathcal{P}$ can be reused an unbounded number of times, and the online verification is cheaper than running $\mathcal{P}$.

4.2   Our Construction

Let $\Sigma' = (\mathsf{KeyGen}', \mathsf{Sign}', \mathsf{Ver}')$ be a regular signature scheme, and $F : \mathcal{K} \times \{0,1\}^* \to \mathbb{Z}_p$ be a pseudorandom function with key space $\mathcal{K}$. Our linearly-homomorphic signature scheme signs T-dimensional vectors of messages in $\mathbb{Z}_p$, and supports datasets of size N, with both $N = \mathsf{poly}(\lambda)$ and $T = \mathsf{poly}(\lambda)$. Let $\mathsf{H} = (\mathsf{H.Gen}, \mathsf{H.PriEval}, \mathsf{H.PubEval})$ and $\mathsf{H}' = (\mathsf{H.Gen}', \mathsf{H.PriEval}', \mathsf{H.PubEval}')$ be two asymmetric programmable hash functions such that $\mathsf{H} : [N] \to \mathbb{G}_1$ and $\mathsf{H}' : [T] \to \mathbb{G}_1$. We construct a homomorphic signature $\mathsf{HSig} = (\mathsf{KeyGen}, \mathsf{Sign}, \mathsf{Ver}, \mathsf{Eval})$ as follows:

$\mathsf{KeyGen}(1^\lambda, \mathcal{L}, T)$. Let λ be the security parameter, $\mathcal{L}$ be a set of admissible labels where $\mathcal{L} = \{1, \dots, N\}$, and T be an integer representing the dimension of the vectors to be signed. The key generation algorithm works as follows.
- Generate a key pair $(\mathsf{vk}', \mathsf{sk}') \xleftarrow{\$} \mathsf{KeyGen}'(1^\lambda)$ for the regular scheme.
- Run $\mathsf{bgp} \xleftarrow{\$} \mathcal{G}(1^\lambda)$ to generate the bilinear groups parameters $\mathsf{bgp} = (p, \mathbb{G}_1, \mathbb{G}_2, \mathbb{G}_T, g_1, g_2, e)$.
- Choose a random seed $K \xleftarrow{\$} \mathcal{K}$ for the PRF $F_K : \{0,1\}^* \to \mathbb{Z}_p$.
- Run $(\mathsf{sek}, \mathsf{pek}) \xleftarrow{\$} \mathsf{H.Gen}(1^\lambda, \mathsf{bgp})$ and $(\mathsf{sek}', \mathsf{pek}') \xleftarrow{\$} \mathsf{H.Gen}'(1^\lambda, \mathsf{bgp})$ to generate the keys of the asymmetric hash functions.
- Return $\mathsf{vk} = (\mathsf{vk}', \mathsf{bgp}, \mathsf{pek}, \mathsf{pek}')$ and $\mathsf{sk} = (\mathsf{sk}', K, \mathsf{sek}, \mathsf{sek}')$.

$\mathsf{Sign}(\mathsf{sk}, \Delta, \tau, \boldsymbol{m})$. The signing algorithm takes as input the secret key sk, a data set identifier $\Delta \in \{0,1\}^*$, a label $\tau \in [N]$ and a message vector $\boldsymbol{m} \in \mathbb{Z}_p^T$, and proceeds as follows:
1. Derive the integer $z \leftarrow F_K(\Delta)$ using the PRF, and compute $Z = g_2^z$.
2. Compute $\sigma_\Delta \leftarrow \mathsf{Sign}'(\mathsf{sk}', \Delta | Z)$ to bind Z to the dataset identifier Δ.
3. Choose a random $R \xleftarrow{\$} \mathbb{G}_1$ and compute

$$S = \left(\mathsf{H.PriEval}(\mathsf{sek}, \tau) \cdot R \cdot \prod_{j=1}^{T} \mathsf{H.PriEval}'(\mathsf{sek}', j)^{m_j} \right)^{1/z}$$

4. Return a signature $\sigma = (\sigma_\Delta, Z, R, S)$.

Essentially, the algorithm consists of two main steps. First, it uses the PRF F_K to derive a common parameter z which is related to the data set Δ, and it signs the public part, $Z = g_2^z$, of this parameter using the signature scheme Σ'. Second, it uses z to create the homomorphic component R, S of the signature, such that S is now related to all $(\Delta, \tau, \boldsymbol{m})$.

$\mathsf{Eval}(\mathsf{vk}, f, \boldsymbol{\sigma})$. The public evaluation algorithm takes as input the public key vk, a linear function $f : \mathbb{Z}_p^\ell \to \mathbb{Z}_p$ described by its vector of coefficients $\boldsymbol{f} = (f_1, \dots, f_\ell)$, and a vector $\boldsymbol{\sigma}$ of ℓ signatures $\sigma_1, \dots, \sigma_\ell$ where $\sigma_i = (\sigma_{\Delta, i}, Z_i, R_i, S_i)$ for $i = 1, \dots, \ell$. Eval returns a signature $\sigma = (\sigma_\Delta, Z, R, S)$ that is obtained by setting $Z = Z_1$, $\sigma_\Delta = \sigma_{\Delta,1}$, and by computing

$$R = \prod_{i=1}^{\ell} R_i^{f_i}, \quad S = \prod_{i=1}^{\ell} S_i^{f_i}$$

$\mathsf{Ver}(\mathsf{vk}, \mathcal{P}_\Delta, \boldsymbol{m}, \sigma)$. Let $\mathcal{P}_\Delta = ((f, \tau_1, \ldots, \tau_\ell), \Delta)$ be a multi-labeled program such that $f : \mathbb{Z}_p^\ell \to \mathbb{Z}_p$ is a linear function described by coefficients $\boldsymbol{f} = (f_1, \ldots, f_\ell)$. Let $\boldsymbol{m} \in \mathbb{Z}_p^T$ be a message-vector and $\sigma = (\sigma_\Delta, Z, R, S)$ be a signature.

First, run $\mathsf{Ver}'(\mathsf{vk}', \Delta | Z, \sigma_\Delta)$ to check that σ_Δ is a valid signature for Z and the dataset identifier Δ taken as input by the verification algorithm. If σ_Δ is not valid, stop and return 0 (reject).

Otherwise, output 1 if and only if the following equation is satisfied

$$e(S, Z) = \left(\prod_{i=1}^{\ell} \mathsf{H.PubEval}(\mathsf{pek}, \tau_i)^{f_i} \right) \cdot e(R, g_2) \cdot \left(\prod_{j=1}^{T} \mathsf{H.PubEval}'(\mathsf{pek}', j)^{m_j} \right)$$

Finally, we describe the algorithms for efficient verification:

$\mathsf{VerPrep}(\mathsf{vk}, \mathcal{P})$. Let $\mathcal{P} = (f, \tau_1, \ldots, \tau_\ell)$ be a labeled program for a linear function $f : \mathbb{Z}_p^\ell \to \mathbb{Z}_p$. The algorithm computes $H = \prod_{i=1}^{\ell} \mathsf{H.PubEval}(\mathsf{pek}, \tau_i)^{f_i}$, and returns the concise verification key $\mathsf{vk}_\mathcal{P} = (\mathsf{vk}', \mathsf{bgp}, H, \mathsf{pek}')$.

$\mathsf{EffVer}(\mathsf{vk}_\mathcal{P}, \Delta, \boldsymbol{m}, \sigma)$. The online verification is the same as Ver except that in the verification equation the value H has been already computed in the off-line phase (and is included in $\mathsf{vk}_\mathcal{P}$).

Clearly, running the combination of $\mathsf{VerPrep}$ and EffVer gives the same result as running Ver, and EffVer's running time is independent of f's complexity ℓ.

The following theorem states the security of the scheme. Formal proofs of correctness and security appear in the full version of our paper.

Theorem 6. *Assume that Σ' is an unforgeable signature scheme, F is a pseudorandom function, and $\mathcal{G}$ is a bilinear group generator such that: H has $(1, \gamma, \epsilon)$-programmable pseudorandomness; H' is weakly $(\mathsf{poly}, 1, 2, \gamma', \delta')$-degree-2-programmable, weakly $(\mathsf{poly}, 1, 2, \gamma', \delta')$-programmable and $(\mathsf{poly}, 0, 1, \gamma', \delta')$-programmable; the 2-DHI and the FDHI assumptions hold. Then HSig is a secure linearly-homomorphic signature scheme.*

We note that our scheme HSig can be instantiated by instantiating both H and H' with two different instances of our programmable hash $\mathsf{H}_{\mathsf{sqrt}}$ described in Sect. 3.2. As one can check in Sect. 3.2, $\mathsf{H}_{\mathsf{sqrt}}$ allows for the multiple programmability modes required in our Theorem 6. Let us stress that requiring the same function to have multiple programmability modes is not contradictory, as such modes do not have to hold simultaneously. It simply means that for the same function there exist different pairs of trapdoor algorithms each satisfying programmability with different parameters.[5]

5   Short Signatures with Shorter Public Keys from Bilinear Maps

In this section we describe how to use asymmetric PHFs to construct in a generic fashion standard-model signature schemes over bilinear groups. We propose two

---

[5] We also stress that, by definition, the outputs of these trapdoor algorithms are statistically indistinguishable.

constructions that are provably-secure under the q-Strong Diffie-Hellman [6] and the q-Diffie-Hellman [5] assumptions. These constructions are the analogues of the schemes in [26] and [25] respectively. The basic idea behind the constructions is to replace a standard $(m, 1)$-PHF with an *asymmetric* $(m, 1, d)$-PHF. In fact, in this context, having a secretly-computable H does not raise any issue when using H in the signing procedure as the signer already uses a secret key. At the same time, for verification purposes, computing the (public) isomorphic copy of H in the target group is also sufficient. Our proof confirms that the $(m, 1, d)$-programmability can still be used to control the size of the randomness in the same way as in [25, 26]. One difference in the security proof is that the schemes in [25, 26] are based on the q-(S)DH assumption, where q is the number of signing queries made by the adversary, whereas ours have to rely on the $(q+d-1)$-(S)DH problem. Since our instantiations use $d = 2$, the difference (when considering concrete security) is very minor.

When plugging into these generic constructions our new asymmetric PHF, H_{acfs}, described in Sect. 3.1, which is $(m, 1, 2)$-programmable, we obtain schemes that, for signing ℓ-bits messages, allow for public keys of length $O(m\sqrt{\ell})$ as in [31].

Below we describe the scheme based on q-SDH. For lack of space, the one based on q-DH (which uses similar ideas) appears in the full version. As discussed in [25], the advantage of the scheme from q-DH compared to the one from q-SDH is to be based on a weaker assumption.

A q-Strong Diffie-Hellman Based Solution. Here we revisit the q-SDH based solution of [26]. The signature $\Sigma_{\mathsf{qSDH}} = (\mathsf{KeyGen}, \mathsf{Sign}, \mathsf{Ver})$ is as follows:

$\mathsf{KeyGen}(1^\lambda)$. Let λ be the security parameter, and let $\ell = \ell(\lambda)$ and $\rho = \rho(\lambda)$ be arbitrary polynomials. Our scheme can sign messages in $\{0, 1\}^\ell$ using randomness in $\{0, 1\}^\rho$. The key generation algorithm works as follows:
- Run $\mathsf{bgp} \xleftarrow{\$} \mathcal{G}(1^\lambda)$ to generate the bilinear groups parameters $\mathsf{bgp} = (p, \mathbb{G}_1, \mathbb{G}_2, \mathbb{G}_T, g_1, g_2, e)$.
- Run $(\mathsf{sek}, \mathsf{pek}) \xleftarrow{\$} \mathsf{H.Gen}(1^\lambda, \mathsf{bgp})$ to generate the keys of the asymmetric hash function.
- Choose a random $x \xleftarrow{\$} \mathbb{Z}_p^*$ and set $X \leftarrow g_2^x$. Return $\mathsf{vk} = (\mathsf{bgp}, \mathsf{pek}, X)$ and $\mathsf{sk} = (\mathsf{sek}, x)$.

$\mathsf{Sign}(\mathsf{sk}, M)$. The signing algorithm takes as input the secret key sk, and a message $M \in \{0, 1\}^\ell$. It starts by generating a random $r \xleftarrow{\$} \{0, 1\}^\rho$. Next, it computes $\sigma = \mathsf{H.PriEval}(\mathsf{sek}, M)^{\frac{1}{x+r}}$ and outputs (σ, r).

$\mathsf{Ver}(\mathsf{vk}, M, (\sigma, r))$. To check that (σ, r) is a valid signature, check that r is of length ρ and that $e(\sigma, X \cdot g_2^r) = \mathsf{H.PubEval}(\mathsf{pek}, M)$.

We state the security of the scheme in the following theorem (whose proof appears in the full version). We note that for simplicity our proof assumes an asymmetric $(m, 1, d)$-PHF for $d = 2$, which matches our realization. A generalization of the theorem for a generic d can be immediately obtained, in which case one would rely on the $(q + d - 1)$-SDH assumption.

Theorem 7. *Assume that $\mathcal{G}$ is a bilinear group generator such that the $(q+1)$-SDH assumption holds in $\mathbb{G}_1$ and H is $(m,1,2,\gamma,\delta)$-programmable, then Σ_{qSDH} is a secure signature scheme. More precisely, let $\mathcal{B}$ be an efficient (probabilistic) algorithm that runs in time t, asks (up to) q signing queries and produces a valid forgery with probability ϵ, then there exists an equally efficient algorithm $\mathcal{A}$ that confutes the $(q+1)$-SDH assumption with probability $\epsilon' \geq \frac{\delta}{q}\left(\epsilon - \gamma - \frac{q}{p} - \frac{q^{m+1}}{2^{\rho m}}\right)$.*

References

1. Attrapadung, N., Libert, B.: Homomorphic network coding signatures in the standard model. In: Catalano, D., Fazio, N., Gennaro, R., Nicolosi, A. (eds.) PKC 2011. LNCS, vol. 6571, pp. 17–34. Springer, Heidelberg (2011)
2. Attrapadung, N., Libert, B., Peters, T.: Computing on authenticated data: new privacy definitions and constructions. In: Wang, X., Sako, K. (eds.) ASIACRYPT 2012. LNCS, vol. 7658, pp. 367–385. Springer, Heidelberg (2012)
3. Attrapadung, N., Libert, B., Peters, T.: Efficient completely context-hiding quotable and linearly homomorphic signatures. In: Kurosawa, K., Hanaoka, G. (eds.) PKC 2013. LNCS, vol. 7778, pp. 386–404. Springer, Heidelberg (2013)
4. Backes, M., Fiore, D., Reischuk, R.M.: Verifiable delegation of computation on outsourced data. In: Sadeghi, A.-R., Gligor, V.D., Yung, M. (eds.) ACM CCS 13, pp. 863–874. ACM Press, New York (2013)
5. Boneh, D., Boyen, X.: Efficient selective-id secure identity-based encryption without random oracles. In: Cachin, C., Camenisch, J.L. (eds.) EUROCRYPT 2004. LNCS, vol. 3027, pp. 223–238. Springer, Heidelberg (2004)
6. Boneh, D., Boyen, X.: Short signatures without random oracles. In: Cachin, C., Camenisch, J.L. (eds.) EUROCRYPT 2004. LNCS, vol. 3027, pp. 56–73. Springer, Heidelberg (2004)
7. Boneh, D., Franklin, M.: Identity-based encryption from the weil pairing. In: Kilian, J. (ed.) CRYPTO 2001. LNCS, vol. 2139, pp. 213–239. Springer, Heidelberg (2001)
8. Boneh, D., Freeman, D., Katz, J., Waters, B.: Signing a linear subspace: signature schemes for network coding. In: Jarecki, S., Tsudik, G. (eds.) PKC 2009. LNCS, vol. 5443, pp. 68–87. Springer, Heidelberg (2009)
9. Boneh, D., Freeman, D.M.: Homomorphic signatures for polynomial functions. In: Paterson, K.G. (ed.) EUROCRYPT 2011. LNCS, vol. 6632, pp. 149–168. Springer, Heidelberg (2011)
10. Boneh, D., Freeman, D.M.: Linearly homomorphic signatures over binary fields and new tools for lattice-based signatures. In: Catalano, D., Fazio, N., Gennaro, R., Nicolosi, A. (eds.) PKC 2011. LNCS, vol. 6571, pp. 1–16. Springer, Heidelberg (2011)
11. Boyen, X., Fan, X., Shi, E.: Adaptively secure fully homomorphic signatures based on lattices. Cryptology ePrint Archive, report 2014/916 (2014). http://eprint.iacr.org/2014/916
12. Catalano, D., Fiore, D., Gennaro, R., Vamvourellis, K.: Algebraic (Trapdoor) one-way functions and their applications. In: Sahai, A. (ed.) TCC 2013. LNCS, vol. 7785, pp. 680–699. Springer, Heidelberg (2013)

13. Catalano, D., Fiore, D., Warinschi, B.: Adaptive pseudo-free groups and applications. In: Paterson, K.G. (ed.) EUROCRYPT 2011. LNCS, vol. 6632, pp. 207–223. Springer, Heidelberg (2011)

14. Catalano, D., Fiore, D., Warinschi, B.: Efficient network coding signatures in the standard model. In: Fischlin, M., Buchmann, J., Manulis, M. (eds.) PKC 2012. LNCS, vol. 7293, pp. 680–696. Springer, Heidelberg (2012)

15. Catalano, D., Fiore, D., Warinschi, B.: Homomorphic signatures with efficient verification for polynomial functions. In: Garay, J.A., Gennaro, R. (eds.) CRYPTO 2014, Part I. LNCS, vol. 8616, pp. 371–389. Springer, Heidelberg (2014)

16. Erdös, P., Frankel, P., Furedi, Z.: Families of finite sets in which no set is covered by the union of r others. Israeli J. Math. **51**, 79–89 (1985)

17. Freeman, D.M.: Improved security for linearly homomorphic signatures: a generic framework. In: Fischlin, M., Buchmann, J., Manulis, M. (eds.) PKC 2012. LNCS, vol. 7293, pp. 697–714. Springer, Heidelberg (2012)

18. Freire, E.S.V., Hofheinz, D., Paterson, K.G., Striecks, C.: Programmable hash functions in the multilinear setting. In: Canetti, R., Garay, J.A. (eds.) CRYPTO 2013, Part I. LNCS, vol. 8042, pp. 513–530. Springer, Heidelberg (2013)

19. Gennaro, R., Katz, J., Krawczyk, H., Rabin, T.: Secure network coding over the integers. In: Nguyen, P.Q., Pointcheval, D. (eds.) PKC 2010. LNCS, vol. 6056, pp. 142–160. Springer, Heidelberg (2010)

20. Gennaro, R., Wichs, D.: Fully homomorphic message authenticators. In: Sako, K., Sarkar, P. (eds.) ASIACRYPT 2013, Part II. LNCS, vol. 8270, pp. 301–320. Springer, Heidelberg (2013)

21. Gorbunov, S., Vaikuntanathan, V., Wichs, D.: Leveled fully homomorphic signatures from standard lattices. In: 47th ACM STOC. ACM Press (2015). To appear

22. Green, M., Hohenberger, S.: Practical adaptive oblivious transfer from simple assumptions. In: Ishai, Y. (ed.) TCC 2011. LNCS, vol. 6597, pp. 347–363. Springer, Heidelberg (2011)

23. Hanaoka, G., Matsuda, T., Schuldt, J.C.N.: On the impossibility of constructing efficient key encapsulation and programmable hash functions in prime order groups. In: Safavi-Naini, R., Canetti, R. (eds.) CRYPTO 2012. LNCS, vol. 7417, pp. 812–831. Springer, Heidelberg (2012)

24. Haralambiev, K., Jager, T., Kiltz, E., Shoup, V.: Simple and efficient public-key encryption from computational diffie-hellman in the standard model. In: Nguyen, P.Q., Pointcheval, D. (eds.) PKC 2010. LNCS, vol. 6056, pp. 1–18. Springer, Heidelberg (2010)

25. Hofheinz, D., Jager, T., Kiltz, E.: Short signatures from weaker assumptions. In: Lee, D.H., Wang, X. (eds.) ASIACRYPT 2011. LNCS, vol. 7073, pp. 647–666. Springer, Heidelberg (2011)

26. Hofheinz, D., Kiltz, E.: Programmable hash functions and their applications. In: Wagner, D. (ed.) CRYPTO 2008. LNCS, vol. 5157, pp. 21–38. Springer, Heidelberg (2008)

27. Hofheinz, D., Kiltz, E.: Programmable hash functions and their applications. J. Cryptology **25**(3), 484–527 (2012)

28. Johnson, R., Molnar, D., Song, D., Wagner, D.: Homomorphic signature schemes. In: Preneel, B. (ed.) CT-RSA 2002. LNCS, vol. 2271, pp. 244–262. Springer, Heidelberg (2002)

29. Kumar, R., Rajagopalan, S., Sahai, A.: Coding constructions for blacklisting problems without computational assumptions. In: Wiener, M. (ed.) CRYPTO 1999. LNCS, vol. 1666, pp. 609–623. Springer, Heidelberg (1999)
30. Libert, B., Peters, T., Joye, M., Yung, M.: Linearly homomorphic structure-preserving signatures and their applications. In: Canetti, R., Garay, J.A. (eds.) CRYPTO 2013, Part II. LNCS, vol. 8043, pp. 289–307. Springer, Heidelberg (2013)
31. Yamada, S., Hanaoka, G., Kunihiro, N.: Two-dimensional representation of cover free families and its applications: short signatures and more. In: Dunkelman, O. (ed.) CT-RSA 2012. LNCS, vol. 7178, pp. 260–277. Springer, Heidelberg (2012)

Structure-Preserving Signatures from Standard Assumptions, Revisited

Eike Kiltz[1], Jiaxin Pan[1(✉)], and Hoeteck Wee[2]

[1] Ruhr-Universität Bochum, Bochum, Germany
{eike.kiltz,jiaxin.pan}@rub.de
[2] ENS, Paris, France

Abstract. Structure-preserving signatures (SPS) are pairing-based signatures where all the messages, signatures and public keys are group elements, with numerous applications in public-key cryptography. We present new, simple and improved SPS constructions under standard assumptions via a conceptually different approach. Our constructions significantly narrow the gap between existing constructions from standard assumptions and optimal schemes in the generic group model.

1 Introduction

Structure-preserving signatures (SPS) [4] are pairing-based signatures where all the messages, signatures and public keys are group elements, verified by testing equality of products of pairings of group elements. They are useful building blocks in modular design of cryptographic protocols, in particular in combination with non-interactive zero-knowledge (NIZK) proofs for algebraic relations in a group [29]. Structure-preserving signatures have found numerous applications in public-key cryptography, such as blind signatures [4,25], group signatures [4,25,27,28,40], homomorphic signatures [38], delegatable anonymous credentials [11,24], compact verifiable shuffles [18], network encoding [9], oblivious transfer [26] and e-cash [13].

A systematic treatment of structure-preserving signatures was initiated by Abe et al. in 2010 [4], building upon previous constructions in [17,26,27]. In the past few years, substantial and rapid progress were made in our understanding of the construction of structure-preserving signatures, yielding both efficient schemes under standard assumptions [2–4,30] as well as "optimal" schemes in the generic group model with matching upper and lower bounds on the efficiency of the schemes [5–8,10]. The three important measures of efficiency in structure-preserving signatures are (i) signature size, (ii) public key size (also per-user public key size for applications like delegatable credentials where we need to

E. Kiltz—Supported by a Sofja Kovalevskaja Award of the Alexander von Humboldt Foundation, the German Israel Foundation, and ERC Project ERCC (FP7/615074).
H. Wee—CNRS, INRIA and Columbia University. Partially supported by the Alexander von Humboldt Foundation, NSF Award CNS-1445424 and ERC Project aSCEND (639554).

© International Association for Cryptologic Research 2015
R. Gennaro and M. Robshaw (Eds.): CRYPTO 2015, Part II, LNCS 9216, pp. 275–295, 2015.
DOI: 10.1007/978-3-662-48000-7_14

sign user public keys), and (iii) number of pairing equations during verification, which in turn affects the efficiency of the NIZK proofs.

One of the main advantages of designing cryptographic protocols starting from structure-preserving signatures is that we can obtain efficient protocols that are secure under standard cryptographic assumptions without the use of random oracles. Ideally, we want to build efficient SPS based on the well-understood k-Lin assumption, which can then be used in conjunction with Groth-Sahai proofs [29] to derive protocols based on the same assumption. In contrast, if we start with SPS that are only secure in the generic group model, then the ensuing protocols would also only be secure in the generic group model, which offer little theoretical or practical benefits over alternative – and typically more efficient and pairing-free – solutions in the random oracle model.

Unfortunately, there is still a big efficiency gap between existing constructions of structure-preserving signatures from the k-Lin assumption and the optimal schemes in the generic group model. For instance, to sign a single group element, the best construction under the SXDH (1-Lin) assumption contains 11 and 21 group elements in the signature and the public key [2], whereas the best construction in the generic group model contains 3 and 3 elements (moreover, this is "tight") [5]. The goal of this work is to bridge this gap.

1.1   Our Results

We present clean, simple, and improved constructions of structure-preserving signatures via a conceptually novel approach. Our constructions are secure under the k-Lin assumption; under the SXDH assumption (i.e., $k = 1$), we achieve 7 group elements in the signature.

Previous constructions use fairly distinct techniques, resulting in a large family of schemes with incomparable efficiency and security guarantees. We obtain a family of schemes that simultaneously match – and in many settings, improve upon – the efficiency, assumptions, and security guarantees of all of the previous constructions. Figure 1 summarizes the efficiency of our constructions. (The work of [41] is independent and concurrent.) Our schemes are fully explicit and simple to describe. Furthermore, our schemes have a natural derivation from a symmetric-key setting, and the derivation even extends to a modular and intuitive proof of security.

We highlight two results:

- For Type III asymmetric pairings, under the SXDH assumption, we can sign a vector of n elements in $\mathbb{G}_1$ with 7 group elements. This improves upon the prior SXDH-based scheme in [2] which requires 11 group elements, and matches the signature size of the scheme in [4] based on (non-standard) q-type assumptions;
- For Type I symmetric pairings, under the 2-Lin assumption, we can sign a vector of n elements with 10 group elements, improving upon that in [3] which requires 14 group elements.

In each of these cases, we also improve the size of the public key, as well as the number of equations used in verification. Finally, we extend our schemes to obtain

	Security Assumption		$|\mathbf{m}|$	$|\sigma|$	$|\mathsf{pk}|$	# equ.
AFGHO10 [4]	OT	2-KerLin ($\mathbb{G}_2$)	$(n_1,0)$	$(3,0)$	$2n_1+5$	2
$\mathsf{SPS}_{\mathsf{ot}}$ (Fig 2)	OT	$\mathcal{D}_k$-KerMDH ($\mathbb{G}_2$)	$(n_1,0)$	$(k+1,0)$	$(n_1+1)k+\mathsf{RE}(\mathcal{D}_k)$	k
AGHO11 [5]	full	Interactive (Generic)	(n_1,n_2)	$(2,1)$	n_1+n_2+2	2
AGHO11 [5]	full	Non-interactive (Generic)	(n_1,n_2)	$(3,3)$	n_1+n_2+2	2
AGHO11 [5]	full	Non-interactive (Generic)	$(n_1,0)$	$(3,1)$	n_1+2	2
ACDKNO12 [2]	full	SXDH, XDLIN	$(n_1,0)$	$(7,4)$	$20+n_1$	4
ACDKNO12 [2]	full	SXDH, XDLIN	(n_1,n_2)	$(8,6)$	$22+n_1+n_2$	5
ADKNO13 [3]	full	2-Lin ($\mathbb{G}_1=\mathbb{G}_2$)	n	14	$22+n$	7
AFGHO10 [4]	full	q-SFP	$(n_1,0)$	$(5,2)$	$13+n_1$	2
LPY15 [41]	full	SXDH, XDLIN	$(n_1,0)$	$(9,1)$	$2n_1+21$	5
$\mathsf{SPS}_{\mathsf{full}}$ (Fig 3)	full	$\mathcal{D}_k$-MDDH ($\mathbb{G}_1,\mathbb{G}_2$)	$(n_1,0)$	$(3k+3,1)$	$(n_1+2k+3)k+\mathsf{RE}(\mathcal{D}_k)$	$2k+1$
$\mathsf{BSPS}_{\mathsf{full}}$ (Fig 4)	full	$\mathcal{D}_k$-MDDH ($\mathbb{G}_1,\mathbb{G}_2$)	(n_1,n_2)	$(4k+3,k+2)$	$(n_1+n_2+3k+3)k+2\mathsf{RE}(\mathcal{D}_k)$	$3k+1$

Fig. 1. Structure-preserving signatures for message space $\mathcal{M} = \mathbb{G}_1^{n_1} \times \mathbb{G}_2^{n_2}$ or $\mathcal{M} = \mathbb{G}^n$ if $\mathbb{G} = \mathbb{G}_1 = \mathbb{G}_2$. Notation (x,y) means x elements in $\mathbb{G}_1$ and y elements in $\mathbb{G}_2$. $\mathsf{RE}(\mathcal{D}_k)$ denotes the number of group elements needed to represent $[\mathbf{A}]$. In case of k-Lin, we have $\mathsf{RE}(\mathcal{D}_k) = k$. Recall that k-Lin is a special case of $\mathcal{D}_k$-MDDH (decisional assumptions) and k-KerLin is a special case of $\mathcal{D}_k$-KerMDH (search assumptions), for $\mathcal{D}_k = \mathcal{L}_k$, the linear distribution. For $k=1$ (SXDH) and $n_1 = 1$, we obtain $(|\mathsf{pk}|,|\sigma|,\#\text{equations}) = (7,7,3)$ for $\mathcal{M} = \mathbb{G}_1^{n_1}$. For comparison, the known lower bound [5,6] is $(|\sigma|,\#\text{equations}) \geq (4,2)$.

efficient SPS for signing bilateral messages in $\mathbb{G}_1^{n_1} \times \mathbb{G}_2^{n_2}$ for Type III asymmetric pairings. Particularly, under the SXDH assumption, our scheme can sign messages in $\mathbb{G}_1^{n_1} \times \mathbb{G}_2^{n_2}$ with 10 group elements in the signature, 4 pairing product equations for verification, and $(n_1 + n_2 + 8)$ group elements in the public key. Prior SXDH-based schemes from [2] required 14 group elements in the signature, 5 pairing product equations, and $(n_1 + n_2 + 22)$ elements in the public key.

At a high level, our constructions and techniques borrow heavily from the recent work of Kiltz and Wee [36] which addresses a different problem of constructing pairing-based non-interactive zero-knowledge arguments [29,33]. We exploit recent developments in obtaining adaptively secure identity-based encryption (IBE) schemes, notably the use of pairing groups to "compile" a symmetric-key primitive into an asymmetric-key primitive [14,19,44], and the dual system encryption methodology for achieving adaptive security against unbounded collusions [37,43]. Along the way, we have to overcome a new technical hurdle which is specific to structure-preserving cryptography.

1.2   Our Approach: SPS from MACs

We provide an overview of our construction of structure-preserving signatures. Throughout this overview, we fix a pairing group $(\mathbb{G}_1, \mathbb{G}_2, \mathbb{G}_T)$ with $e : \mathbb{G}_1 \times \mathbb{G}_2 \to \mathbb{G}_T$, and rely on implicit representation notation for group elements, as explained in Sect. 2.1.[1] As a warm-up, we explain in some detail how to build a one-time structure-preserving signature scheme, following closely the exposition in [36]. While we do not obtain significant improvement in this setting (nonetheless, we do simplify and generalize prior one-time schemes [4]), we believe it already

---

[1] For fixed generators g_1 and g_2 of $\mathbb{G}_1$ and $\mathbb{G}_2$, respectively, and for a matrix $\mathbf{M} \in \mathbb{Z}_q^{n \times t}$, we define $[\mathbf{M}]_1 := g_1^{\mathbf{M}}$ and $[\mathbf{M}]_2 := g_2^{\mathbf{M}}$ (componentwise).

illustrates the conceptual simplicity and novelty of our approach over previous constructions of structure-preserving signatures.

Warm-Up: One-Time SPS. We want to build a one-time signature scheme for signing a vector $[\mathbf{m}]_1 \in \mathbb{G}_1^n$ of group elements. The starting point of our construction is a one-time "structure-preserving" information-theoretic MAC for vectors of group elements. We pick a secret MAC key $\mathbf{K} \leftarrow_R \mathbb{Z}_q^{(n+1) \times (k+1)}$ known to the verifier ($k \geq 1$ is a parameter of the security assumption), and the MAC on $[\mathbf{m}]_1$ is given by

$$\sigma := [(1, \mathbf{m}^\top)\mathbf{K}]_1 \in \mathbb{G}_1^{1 \times (k+1)}$$

Verification is straight-forward: check if

$$\sigma \stackrel{?}{=} (1, \mathbf{m}^\top)\mathbf{K} \tag{1}$$

Security follows readily from the fact that for any pair of distinct vectors $\mathbf{m}, \mathbf{m}^* \in \mathbb{Z}_q^n$, the vectors $(1, \mathbf{m}^\top)$ and $(1, \mathbf{m}^{*\top})$ are linearly independent, and therefore the quantities

$$(1, \mathbf{m}^\top)\mathbf{K}, (1, \mathbf{m}^{*\top})\mathbf{K} \in \mathbb{Z}_q^{(k+1)}$$

are two independently random values; this holds even if $\mathbf{m}^* \neq \mathbf{m}$ is chosen adaptively after seeing $(1, \mathbf{m}^\top)\mathbf{K}$.

To achieve public verifiability as is required for a signature scheme, we publish a "partial commitment" to $\mathbf{K}$ in $\mathbb{G}_2$ as given by $[\mathbf{A}]_2, [\mathbf{KA}]_2$, where the choice of $\mathbf{A} \in \mathbb{Z}_q^{(k+1) \times k}$ is defined by the security assumption. The signature on $[\mathbf{m}]_1$ is the same as the MAC value, and verification is the natural analogue of Eq. (1) with the pairing:

$$e(\sigma, [\mathbf{A}]_2) \stackrel{?}{=} e([(1, \mathbf{m}^\top)]_1, [\mathbf{KA}]_2)$$

As $[\mathbf{A}]_2, [\mathbf{KA}]_2$ leaks additional information about the secret MAC key $\mathbf{K}$, we can only prove computational adaptive soundness. In particular, we rely on the $\mathcal{D}_k$-KerMDH Assumption [42], which stipulates that given a random $[\mathbf{A}]_2$ drawn from a matrix distribution $\mathcal{D}_k$, it is hard to find a non-zero $[\mathbf{s}]_1 \in \mathbb{G}_1^{k+1}$ such that $\mathbf{s}^\top \mathbf{A} = \mathbf{0}$; this is implied by the $\mathcal{D}_k$-MDDH Assumption [22], a generalization of the k-Lin Assumption.[2] Therefore, for any $([\mathbf{m}^*]_1, [\sigma]_1)$ produced by an efficient adversary,

$$\sigma \mathbf{A} = (1, \mathbf{m}^{*\top})\mathbf{KA} \implies (\sigma - (1, \mathbf{m}^{*\top})\mathbf{K})\mathbf{A} = \mathbf{0}$$

$$\stackrel{\text{using assumption}}{\implies} \sigma - (1, \mathbf{m}^{*\top})\mathbf{K} = \mathbf{0} \implies \sigma = (1, \mathbf{m}^{*\top})\mathbf{K}.$$

That is, security of the signature reduces to the security for the MAC, with a little more work to account for the leakage from $\mathbf{KA}$. Moreover, adaptive security for the MAC (which is easy to analyze via a purely information-theoretic argument) carries over to adaptive security for the signature.

General SPS. To achieve unforgeability against multiple signature queries, we move from a one-time MAC to a randomized MAC that is secure against multiple

---

[2] We refer the reader to Sect. 2.2 for a more detailed treatment of the assumptions.

queries. As shown in [14, 36], we know that under the $\mathcal{D}_k$-MDDH assumption in $\mathbb{G}_1$, the following construction is a randomized PRF

$$\tau \mapsto \left([\mathbf{t}^\top (\mathbf{K}_0 + \tau \mathbf{K}_1)]_1, [\mathbf{t}^\top]_1 \right) \in (\mathbb{G}_1^{1 \times (k+1)})^2, \tag{2}$$

where $\mathbf{K}_0, \mathbf{K}_1$ is the seed and $\mathbf{t}$ is the randomness. We now use the randomized PRF to additively mask the one-time MAC value $[(1, \mathbf{m}^\top)\mathbf{K}]_1$. The new randomized MAC takes as input a vector of group elements $[\mathbf{m}]_1 \in \mathbb{G}_1^n$ as before, picks a random tag $\tau \in \mathbb{Z}_q$ and a fresh $\mathbf{t}$ and outputs

$$(\sigma_1, \sigma_2) := ([[(1, \mathbf{m}^\top)\mathbf{K}]_1 + \boxed{[\mathbf{t}^\top (\mathbf{K}_0 + \tau \mathbf{K}_1)]_1}, \boxed{[\mathbf{t}^\top]_1}) \in (\mathbb{G}_1^{1 \times (k+1)})^2 \tag{3}$$

where $\mathbf{K}$ and $\mathbf{K}_0, \mathbf{K}_1 \leftarrow_{\mathrm{R}} \mathbb{Z}_q^{(k+1) \times (k+1)}$ constitute the key. The boxed terms correspond to the additive mask from Eq. (2). We want to argue that an adversary upon obtaining MAC values on Q message vectors $[\mathbf{m}_1]_1, \ldots, [\mathbf{m}_Q]_1$, cannot compute the MAC value on a new message vector $[\mathbf{m}^*]_1$. First, we may assume that the MAC values on $[\mathbf{m}_1]_1, \ldots, [\mathbf{m}_Q]_1$ use distinct tags $\tau_1, \ldots, \tau_Q$. Then, we consider two cases:

- case 1: the adversary uses a fresh tag for $[\mathbf{m}^*]_1$. This immediately breaks the pseudorandomness of the security of the construction in Eq. (2);
- case 2: the adversary reuses tag τ_i. Again, we know from pseudorandomness that the MAC values on the remaining $Q-1$ tags do not leak any information $\mathbf{K}$; therefore, the only leakage about $\mathbf{K}$ in the Q queries comes from $(1, \mathbf{m}_i^\top)\mathbf{K}$. We may then rely on the security of the one-time MAC to argue that given only $(1, \mathbf{m}_i^\top)\mathbf{K}$, it is hard to compute $(1, \mathbf{m}^{*\top})\mathbf{K}$.

As before, to obtain a signature scheme, we then publish $[\mathbf{A}]_2, [\mathbf{KA}]_2, [\mathbf{K}_0\mathbf{A}]_2$, $[\mathbf{K}_1\mathbf{A}]_2$ for public verification:

$$e(\sigma_1, [\mathbf{A}]_2) \stackrel{?}{=} e([(1, \mathbf{m}^\top)]_1, [\mathbf{KA}]_2) \cdot e(\sigma_2, [\mathbf{K}_0\mathbf{A}]_2 \cdot [\tau \mathbf{K}_1 \mathbf{A}]_2)$$

Note that the above verification requires knowledge of $\tau \in \mathbb{Z}_q$ to compute $[\tau \mathbf{K}_1 \mathbf{A}]_2$.

To obtain a structure-preserving signature, we cannot publish $\tau \in \mathbb{Z}_q$ in the signature. The main technical challenge in this work is to find a way to embed τ as a group element that enables both verification and a security reduction. The natural work-around is to add $[\tau \mathbf{K}_1 \mathbf{A}]_2$ and $[\tau]_1$ to the signature, but the proof breaks down. Instead, we add $[\tau]_2$ and $[\tau \mathbf{t}^\top]_1$ to the signature to enable verification. This yields a signature with $3k + 4$ group elements.

An Alternative Interpretation. Linearly homomorphic signatures (LHS) [15, 21, 32] are signatures where the messages consist of vectors over group $\mathbb{G}_1$ such that from any set of signatures on $[\mathbf{m}_i]_1 \in \mathbb{G}_1^n$, one can efficiently derive a signature σ on any element message $[\mathbf{m}]_1 := [\sum \omega_i \mathbf{m}_i]_1$ in the span of $\mathbf{m}_1, \ldots, \mathbf{m}_Q$. For security, one requires that it is infeasible to produce a signature on a message outside of the span of all previously signed messages. Linearly homomorphic structure preserving signatures (LHSPS) [16,36,38] have the additional property that signatures and public keys are all elements of the groups $\mathbb{G}_1, \mathbb{G}_2, \mathbb{G}_T$, while allowing the use of a tag which is a scalar.

We can construct a SPS with message space $\mathbb{G}_1^n$ from a LHSPS with message space $\mathbb{G}_1^{n+1}$ as follows: to sign a message $[\mathbf{m}]_1$, we use a LHSPS to sign the $(n+1)$-dimensional vector $[1, \mathbf{m}]_1$ on a random tag. Suppose the SPS adversary forges a signature on $[\mathbf{m}^*]_1$. First, we may assume that all the signatures from the signing queries $[\mathbf{m}_1]_1, \ldots, [\mathbf{m}_Q]_1$ are on distinct tags $\tau_1, \ldots, \tau_Q$. Then, we consider two cases:

- case 1: the adversary uses a fresh tag. Then, security of LHSPS tells us that the adversary can only sign the vector $\mathbf{0} \in \mathbb{G}_1^{n+1}$, which does not correspond to a valid message in the SPS.
- case 2: the adversary reuses tag τ_i. Then, $(1, \mathbf{m}^{*\top})$ must lie in the span of $(1, \mathbf{m}_i^\top)$, which means $\mathbf{m}^* = \mathbf{m}_i$. Here, we crucially rely on the fact that $\tau_1, \ldots, \tau_Q$ are distinct, which ensures that the adversary has seen at most one signature corresponding to τ_i.

At this point, we can then embed $\tau \in \mathbb{Z}_q$ as a group element as described earlier. Our constructions may also be viewed as instantiating the above paradigm with the state-of-the-art LHSPS in [36].

1.3 Discussion

Optimality. The linearity in the verification equation of SPS poses severe restrictions on the efficiency of such constructions. In both Type I and III bilinear groups, it was proved in [5,8] that any fully secure SPS requires at least 2 verification equations, at least 3 group elements, the 3 elements not all the same group (for Type III asymmetric pairings).In fact, [5] shows the above lower bounds by giving attacks the weaker security model of unforgeability against two random message queries. Furthermore, one-time secure SPS against random message attack (RMA) in Type I bilinear groups require at least 2 group elements and 2 equations [8].Furthermore, SPSs in Type III bilinear groups require at least 4 group elements [6] for unforgeability against adaptive chosen message attack under *non-interactive assumptions* (such as k-Lin).

Interestingly, for one-time RMA-security, we can match the lower bounds. By combining our main result on the one-time CMA-secure SPS and the techniques used in [36] to obtain shorter QANIZK, we obtain an optimal RMA-secure one-time SPS (Sect. 5). In Type III asymmetric groups, under the SXDH assumption, signatures requires 1 group element and 1 verification equation which is clearly optimal; in Type I symmetric groups, under the 2-Lin assumption, our scheme requires 2 elements and 2 verification equations, matching the lower bound for one-time RMA-secure SPS from [8].

Comparison with Previous Approaches. The prior works of Abe, et al. [2,3] presented two generic approaches for constructing SPS from SXDH and

2-Lin assumptions: both constructions combine a structure-preserving one-time signature and random-message secure signatures ala [23], with slightly different syntax and security notions for the two underlying building blocks; the final signature is the concatenation of the two underlying signatures. Our construction has a similar flavor in that we combine a one-time MAC with a randomized PRF. However, we are able to exploit the common structure in both building blocks to compress the output; interestingly, working with the matrix Diffie-Hellman framework [22] makes it easier to identity such common structure. In particular, the output length of the randomized MAC with unbounded security is that of the PRF and not the sum of the output lengths of the one-time MAC and the PRF; this is akin to combining a one-time signature and a random-message secure signature in such a way that the combined signature size is that of the latter rather than the sum of the two.

Signatures from IBE. While our construction of signatures exploits techniques from the literature on IBE, it is quite different from the well-known Naor's derivation of a signature scheme from an IBE. There, the signature on a message $m \in \mathbb{Z}_q$ corresponds to an IBE secret key for the identity m. This approach seems to inherently fail for structure-preserving signatures as all known pairings-based IBE schemes need to treat the identity as a scalar. In our construction, a signature on $[\mathbf{m}]_1$ also corresponds to an IBE secret key: the message vector (specifically, a one-time MAC applied to the message vector) is embedded into the master secret key component of an IBE, and a fresh random tag $\tau \in \mathbb{Z}_q$ is chosen and used as the identity. The idea of embedding $[\mathbf{m}]_1$ into the master secret key component of an IBE also appeared in earlier constructions of linearly homomorphic structure-preserving schemes [36,38,39]; a crucial difference is that these prior constructions allow the use of a scalar tag in the signature.

Towards Shorter SPS? One promising approach to get even shorter SPS against adaptive chosen message attack by using our approach is to improve upon the underlying MAC in the computational core lemma (Lemma 3). Currently, the MAC achieves security against chosen message attacks, whereas it suffices to use one that is secure against random message attacks. Saving one group element in this MAC would likely yield a saving of two group elements in the SPS, which would in turn yield a SXDH-based signature with 5 group elements. Note that the state-of-the-art standard signature from SXDH contains 4 group elements [20]. Together with existing lower bounds for SPS, this indicates a barrier of 5 group elements for SXDH-based SPS; breaking this barrier would likely require improving upon the best standard signatures from SXDH.

Perspective. As noted at the beginning of the introduction, structure-preserving signatures have been a target of intense scrutiny in recent years. We presented a conceptually different yet very simple approach for building structure-preserving signatures. We are optimistic that our approach will yield further insights into structure-preserving signatures as well as concrete improvements to the numerous applications that rely on such signatures.

2   Definitions

Notation. If $\mathbf{x} \in \mathcal{B}^n$, then $|\mathbf{x}|$ denotes the length n of the vector. Further, $x \leftarrow_R \mathcal{B}$ denotes the process of sampling an element x from set $\mathcal{B}$ uniformly at random. If $\mathbf{A} \in \mathbb{Z}_q^{n \times k}$ is a matrix with $n > k$, then $\overline{\mathbf{A}} \in \mathbb{Z}_q^{k \times k}$ denotes the upper square matrix of $\mathbf{A}$ and then $\underline{\mathbf{A}} \in \mathbb{Z}_q^{(n-k) \times k}$ denotes the remaining $n - k$ rows of $\mathbf{A}$. We use $span()$ to denote the column span of a matrix.

2.1   Pairing Groups

Let GGen be a probabilistic polynomial time (PPT) algorithm that on input 1^λ returns a description $\mathcal{PG} = (\mathbb{G}_1, \mathbb{G}_2, \mathbb{G}_T, q, g_1, g_2, e)$ of asymmetric pairing groups where $\mathbb{G}_1, \mathbb{G}_2, \mathbb{G}_T$ are cyclic groups of order q for a λ-bit prime q, g_1 and g_2 are generators of $\mathbb{G}_1$ and $\mathbb{G}_2$, respectively, and $e : \mathbb{G}_1 \times \mathbb{G}_2$ is an efficiently computable (non-degenerate) bilinear map. Define $g_T := e(g_1, g_2)$, which is a generator in $\mathbb{G}_T$.

We use implicit representation of group elements as introduced in [22]. For $s \in \{1, 2, T\}$ and $a \in \mathbb{Z}_q$, define $[a]_s = g_s^a \in \mathbb{G}_s$ as the *implicit representation* of a in $\mathbb{G}_s$. More generally, for a matrix $\mathbf{A} = (a_{ij}) \in \mathbb{Z}_q^{n \times m}$ we define $[\mathbf{A}]_s$ as the implicit representation of $\mathbf{A}$ in $\mathbb{G}_s$:

$$[\mathbf{A}]_s := \begin{pmatrix} g_s^{a_{11}} \cdots g_s^{a_{1m}} \\ g_s^{a_{n1}} \cdots g_s^{a_{nm}} \end{pmatrix} \in \mathbb{G}_s^{n \times m}$$

We will always use this implicit notation of elements in $\mathbb{G}_s$, i.e., we let $[a]_s \in \mathbb{G}_s$ be an element in $\mathbb{G}_s$. Note that from $[a]_s \in \mathbb{G}_s$ it is generally hard to compute the value a (discrete logarithm problem in $\mathbb{G}_s$). Further, from $[b]_T \in \mathbb{G}_T$ it is hard to compute the value $[b]_1 \in \mathbb{G}_1$ and $[b]_2 \in \mathbb{G}_2$ (pairing inversion problem). Obviously, given $[a]_s \in \mathbb{G}_s$ and a scalar $x \in \mathbb{Z}_q$, one can efficiently compute $[ax]_s \in \mathbb{G}_s$. Further, given $[a]_1, [a]_2$ one can efficiently compute $[ab]_T$ using the pairing e. For two matrices $\mathbf{A}, \mathbf{B}$ with matching dimensions define $e([\mathbf{A}]_1, [\mathbf{B}]_2) := [\mathbf{AB}]_T \in \mathbb{G}_T$.

2.2   Matrix Diffie-Hellman Assumption

We recall the definitions of the Matrix Decision Diffie-Hellman (MDDH) and the Kernel Diffie-Hellman assumptions [22,42].

Definition 1 (Matrix Distribution). *Let $k \in \mathbb{N}$. We call $\mathcal{D}_k$ a matrix distribution if it outputs matrices in $\mathbb{Z}_q^{(k+1) \times k}$ of full rank k in polynomial time.*

Without loss of generality, we assume the first k rows of $\mathbf{A} \leftarrow_R \mathcal{D}_k$ form an invertible matrix. The $\mathcal{D}_k$-Matrix Diffie-Hellman problem is to distinguish the two distributions $([\mathbf{A}], [\mathbf{Aw}])$ and $([\mathbf{A}], [\mathbf{u}])$ where $\mathbf{A} \leftarrow_R \mathcal{D}_k$, $\mathbf{w} \leftarrow_R \mathbb{Z}_q^k$ and $\mathbf{u} \leftarrow_R \mathbb{Z}_q^{k+1}$.

Definition 2 ($\mathcal{D}_k$-Matrix Diffie-Hellman Assumption $\mathcal{D}_k$-MDDH). *Let $\mathcal{D}_k$ be a matrix distribution and $s \in \{1, 2, T\}$. We say that the $\mathcal{D}_k$-Matrix Diffie-Hellman ($\mathcal{D}_k$-MDDH) Assumption holds relative to* GGen *in group $\mathbb{G}_s$ if for all PPT adversaries $\mathcal{A}$,*

$$\mathbf{Adv}^{\mathrm{mddh}}_{\mathcal{D}_k, \mathsf{GGen}}(\mathcal{A}) := |\Pr[\mathcal{A}(\mathcal{G}, [\mathbf{A}]_s, [\mathbf{Aw}]_s) = 1] - \Pr[\mathcal{A}(\mathcal{G}, [\mathbf{A}]_s, [\mathbf{u}]_s) = 1]| = \mathrm{negl}(\lambda),$$

where the probability is taken over $\mathcal{G} \leftarrow_{\mathrm{R}} \mathsf{GGen}(1^\lambda)$, $\mathbf{A} \leftarrow_{\mathrm{R}} \mathcal{D}_k, \mathbf{w} \leftarrow_{\mathrm{R}} \mathbb{Z}_q^k, \mathbf{u} \leftarrow_{\mathrm{R}} \mathbb{Z}_q^{k+1}$.

The Kernel-Diffie-Hellman assumption $\mathcal{D}_k$-KerMDH [42] is a natural *computational* analogue of the $\mathcal{D}_k$-MDDH Assumption.

Definition 3 ($\mathcal{D}_k$-Kernel Diffie-Hellman Assumption $\mathcal{D}_k$-KerMDH). *Let $\mathcal{D}_k$ be a matrix distribution and $s \in \{1, 2\}$. We say that the $\mathcal{D}_k$-Kernel Diffie-Hellman ($\mathcal{D}_k$-KerMDH) Assumption holds relative to* GGen *in group $\mathbb{G}_s$ if for all PPT adversaries $\mathcal{A}$,*

$$\mathbf{Adv}^{\mathrm{kmdh}}_{\mathcal{D}_k, \mathsf{GGen}}(\mathcal{A}) := \Pr[\mathbf{c}^\top \mathbf{A} = \mathbf{0} \wedge \mathbf{c} \neq \mathbf{0} \mid [\mathbf{c}]_{3-s} \leftarrow_{\mathrm{R}} \mathcal{A}(\mathcal{G}, [\mathbf{A}]_s)] = \mathrm{negl}(\lambda),$$

where the probability is taken over $\mathcal{G} \leftarrow_{\mathrm{R}} \mathsf{GGen}(1^\lambda)$, $\mathbf{A} \leftarrow_{\mathrm{R}} \mathcal{D}_k$.

Note that we can use a non-zero vector in the kernel of $\mathbf{A}$ to test membership in the column space of $\mathbf{A}$. This means that the $\mathcal{D}_k$-KerMDH assumption is a relaxation of the $\mathcal{D}_k$-MDDH assumption, as captured in the following lemma from [42].

Lemma 1. *For any matrix distribution $\mathcal{D}_k$, $\mathcal{D}_k$-MDDH $\Rightarrow$ $\mathcal{D}_k$-KerMDH.*

For each $k \geq 1$, [22,42] specify distributions $\mathcal{L}_k, \mathcal{SC}_k, \mathcal{U}_k$ (and others) such that the corresponding $\mathcal{D}_k$-MDDH and $\mathcal{D}_k$-KerMDH assumptions are generically secure in bilinear groups and form a hierarchy of increasingly weaker assumptions.

$$\mathcal{SC}_k : \mathbf{A} = \begin{pmatrix} 1 & 0 & 0 & \cdots & 0 \\ a & 1 & 0 & \cdots & 0 \\ 0 & a & 1 & \cdots & 0 \\ 0 & 0 & a & & 0 \\ & & & \ddots & \\ 0 & 0 & 0 & \cdots & a \end{pmatrix}, \mathcal{L}_k : \mathbf{A} = \begin{pmatrix} 1 & 1 & 1 & \cdots & 1 \\ a_1 & 0 & 0 & \cdots & 0 \\ 0 & a_2 & 0 & \cdots & 0 \\ 0 & 0 & a_3 & & 0 \\ & & & \ddots & \\ 0 & 0 & 0 & \cdots & a_k \end{pmatrix}, \mathcal{U}_k : \mathbf{A} = \begin{pmatrix} a_{1,1} & \cdots & a_{1,k} \\ \vdots & \ddots & \vdots \\ a_{k+1,1} & \cdots & a_{k+1,k} \end{pmatrix},$$

where $a, a_i, a_{i,j} \leftarrow \mathbb{Z}_q$. We define the *representation size* $\mathsf{RE}(\mathcal{D}_k)$ of a given matrix distribution $\mathcal{D}_k$ as the minimal number of group elements needed to represent $[\mathbf{A}]_s$, where $\mathbf{A} \leftarrow_{\mathrm{R}} \mathcal{D}_k$. Then $\mathsf{RE}(\mathcal{SC}_k) = 1$, $\mathsf{RE}(\mathcal{L}_k) = k$ and $\mathsf{RE}(\mathcal{U}_k) = k(k+1)$. As shown in [22], $\mathcal{SC}_k$-MDDH offers the same security guarantees as $\mathcal{L}_k$-MDDH (k-Linear Assumption of [31]), while having the advantage of a more compact representation. We define k-Lin := $\mathcal{L}_k$-MDDH and k-KerLin := $\mathcal{L}_k$-KerMDH. Note that 2-KerLin = SDP (Simultaneous Double Pairing Assumption of [17]). The relations between the different assumptions for $\mathcal{D}_k = \mathcal{L}_k$ are as follows:

$$\begin{array}{ccccccc}
\mathsf{DDH} & \longrightarrow & \text{2-Lin} & \longrightarrow & \text{3-Lin} & \longrightarrow & \cdots \\
\Downarrow & & \Downarrow & & \Downarrow & & \\
\text{1-KerLin} & \Rightarrow & \text{2-KerLin} = \mathsf{SDP} & \Rightarrow & \text{3-KerLin} & \Rightarrow \cdots & \Rightarrow \mathsf{CDH}
\end{array}$$

2.3 Structure-Preserving Signatures

Let par be some parameters that contain a pairing group $\mathcal{PG}$. In a structure-preserving signature (SPS) [4], both the messages and signatures are group elements, verification proceeds via a pairing-product equation.

Definition 4 (Structure-preserving signature). *A structure-preserving signature scheme* SPS *is defined as a triple of probabilistic polynomial time (PPT) algorithms* SPS = (Gen, Sign, Verify):

- *The probabilistic key generation algorithm* Gen(par) *returns the public/secret key* (pk, sk), *where* pk $\in \mathbb{G}^{n_{pk}}$ *for some* $n_{pk} \in \mathrm{poly}(\lambda)$. *We assume that* pk *implicitly defines a message space* $\mathcal{M} := \mathbb{G}^n$ *for some* $n \in \mathrm{poly}(\lambda)$.
- *The probabilistic signing algorithm* Sign(sk, [m]) *returns a signature* $\sigma \in \mathbb{G}^{n_\sigma}$ *for* $n_\sigma \in \mathrm{poly}(\lambda)$.
- *The deterministic verification algorithm* Verify(pk, [m], σ) *only consists of pairing product equations and returns 1 (accept) or 0 (reject).*

(Perfect correctness.) for all (pk, sk) $\leftarrow_R$ Gen(par) *and all messages* [m] $\in \mathcal{M}$ *and all* $\sigma \leftarrow_R$ Sign(sk, [m]) *we have* Verify(pk, [m], σ) = 1.

Definition 5 (Unforgeablility against chosen message attack). *To an adversary* $\mathcal{A}$ *and* SPS *we associate the advantage function*

$$\mathbf{Adv}_{\mathsf{SPS}}^{\mathsf{cma}}(\mathcal{A}) := \Pr\left[[\mathbf{m}^*] \notin \mathcal{Q}_{\mathrm{msg}} \wedge \mathsf{Verify}(\mathsf{pk}, [\mathbf{m}^*], \sigma^*) = 1 \,\middle|\, \begin{matrix} (\mathsf{pk}, \mathsf{sk}) \leftarrow_R \mathsf{Gen}(\mathsf{par}) \\ ([\mathbf{m}^*], \sigma^*) \leftarrow_R \mathcal{A}^{\mathsf{SignO}(\cdot)}(\mathsf{pk}) \end{matrix} \right],$$

where SignO([m]) *runs* $\sigma \leftarrow_R$ Sign(sk, [m]), *adds the vector* [m] *to* $\mathcal{Q}_{\mathrm{msg}}$ *(initialized with $\emptyset$) and returns σ to $\mathcal{A}$.* SPS *is said to be (unbounded)* CMA-secure *if for all PPT adversaries* $\mathcal{A}$, $\mathbf{Adv}_{\mathsf{SPS}}^{\mathsf{cma}}(\mathcal{A})$ *is negligible.* SPS *is said to be* one-time CMA-secure *with corresponding advantage function* $\mathbf{Adv}_{\mathsf{SPS}}^{\mathsf{ot\text{-}cma}}(\mathcal{A})$, *if $\mathcal{A}$ is restricted to make at most one query to oracle* SignO.

3 One-Time CMA-Secure SPS

The scheme is given in Fig. 2 and its parameters are:

$$|\mathsf{pk}| = (n+1)k + \mathsf{RE}(\mathcal{D}_k), \qquad |\sigma| = k+1.$$

As defined in Sect. 2.2, $\mathsf{RE}(\mathcal{D}_k)$ denotes the number of group elements needed to represent $[\mathbf{A}]_s$, where $\mathbf{A} \leftarrow_R \mathcal{D}_k$. For k-Lin, we achieve 2 group elements in the signature for $k = 1$ and 3 group elements for $k = 2$. Moreover, we note that the verification needs k pairing product equations: for $e(\sigma, [\mathbf{A}]_2) = e([(1, \mathbf{m})]_1, [\mathbf{C}]_2)$ we need to pair the vector σ with every column of $[\mathbf{A}]_2$ and thus this check needs k pairing product equations.

We will exploit the following lemma in the analysis of our scheme. Informally, the lemma says that $\mathbf{m} \mapsto (1, \mathbf{m}^\top)\mathbf{K}$ is a secure information-theoretic one-time MAC even if the adversary first sees $(\mathbf{A}, \mathbf{KA})$.

Lemma 2 (Core lemma for adaptive soundness). *Let n, k be integers. For any* $\mathbf{A} \in \mathbb{Z}_q^{(k+1) \times k}$ *and any (possibly unbounded) adversary $\mathcal{A}$,*

$$\Pr\left[\mathbf{m}^* \neq \mathbf{m} \wedge \mathbf{z}^\top = (1, \mathbf{m}^{*\top})\mathbf{K} \,\middle|\, \begin{matrix} \mathbf{K} \leftarrow_R \mathbb{Z}_q^{(n+1) \times (k+1)} \\ (\mathbf{z}, \mathbf{m}^*) \leftarrow_R \mathcal{A}^{\mathcal{O}(\cdot)}(\mathbf{A}, \mathbf{KA}) \end{matrix} \right] \leq \frac{1}{q}, \qquad (4)$$

where $\mathcal{O}(\mathbf{m} \in \mathbb{Z}_q^n)$ *returns* $(1, \mathbf{m}^\top)\mathbf{K}$ *and $\mathcal{A}$ only gets a single call to $\mathcal{O}$.*

Gen(par):	Sign(sk, $[\mathbf{m}]_1$):
$\mathbf{A} \leftarrow_R \mathcal{D}_k; \mathbf{K} \leftarrow_R \mathbb{Z}_q^{(n+1)\times(k+1)}$	$\sigma := [(1, \mathbf{m}^\top)\mathbf{K}]_1$
$\mathbf{C} := \mathbf{KA} \in \mathbb{Z}_q^{(n+1)\times k}$	Return $\sigma \in \mathbb{G}_1^{1\times(k+1)}$
sk $:= \mathbf{K}$	
pk $:= ([\mathbf{C}]_2, [\mathbf{A}]_2)$	Verify(pk, $[\mathbf{m}]_1, \sigma$):
Return (pk, sk)	Check: $e(\sigma, [\mathbf{A}]_2) = e([(1, \mathbf{m}^\top)]_1, [\mathbf{C}]_2)$

Fig. 2. One-time CMA-secure structure-preserving signature SPS_{ot} with message-space $\mathcal{M} = \mathbb{G}_1^n$.

This lemma can be seen as an adaptive version of a special case of [36, Lemma 2] in that we fix $t = 1$, $\mathbf{M}$ to be the matrix $(1, \mathbf{m}^\top) \in \mathbb{Z}_q^{1\times(n+1)}$, and we use the fact that if $\mathbf{m}^* \neq \mathbf{m}$, then $(1, \mathbf{m}^\top) \notin span(\mathbf{M})$. In our adaptive version, $\mathbf{m}$ may depend on $\mathbf{KA}$ but the proof is essentially the same as in [36]. Lemma 2 implies the security of SPS_{ot}. Formal proofs of Lemma 2 and Theorem 1 are given in [35].

Theorem 1. *Under the $\mathcal{D}_k$-KerMDH Assumption in $\mathbb{G}_2$, SPS_{ot} from Fig. 2 is a one-time CMA-secure structure-preserving signature scheme.*

4   Unbounded CMA-Secure SPS

4.1   Computational Core Lemma

We present a variant of the computational core lemma from [36, Lemma 3].

Lemma 3 (Computational core lemma for unbounded CMA-security). *For all adversaries $\mathcal{A}$, there exists an adversary $\mathcal{B}$ with $\mathbf{T}(\mathcal{A}) \approx \mathbf{T}(\mathcal{B})$ and*

$$
\Pr\left[
\begin{array}{l|l}
\tau^* \notin \mathcal{Q}_{tag} & \mathbf{A}, \mathbf{B} \leftarrow_R \mathcal{D}_k \\
\wedge\, b' = b & \mathbf{K}_0, \mathbf{K}_1 \leftarrow_R \mathbb{Z}_q^{(k+1)\times(k+1)} \\
& (\mathbf{P}_0, \mathbf{P}_1) := (\mathbf{B}^\top \mathbf{K}_0, \mathbf{B}^\top \mathbf{K}_1) \\
& \mathsf{pk} := ([\mathbf{P}_0]_1, [\mathbf{P}_1]_1, [\mathbf{B}]_1, \mathbf{K}_0\mathbf{A}, \mathbf{K}_1\mathbf{A}, \mathbf{A}) \\
& b \leftarrow_R \{0,1\}; b' \leftarrow_R \mathcal{A}^{\mathcal{O}_b(\cdot), \mathcal{O}^*(\cdot)}(\mathsf{pk})
\end{array}
\right]
$$
$$
\leq \frac{1}{2} + 2Q \cdot \mathbf{Adv}_{\mathcal{D}_k, \mathsf{GGen}}^{mddh}(\mathcal{B}) + Q/q,
$$

where

- *$\mathcal{O}_b(\tau)$ returns $([b\mu\mathbf{a}^\perp + \mathbf{r}^\top(\mathbf{P}_0 + \tau\mathbf{P}_1)]_1, [\mathbf{r}^\top \mathbf{B}^\top]_1) \in (\mathbb{G}_1^{1\times(k+1)})^2$ with $\mu \leftarrow_R \mathbb{Z}_q$, $\mathbf{r} \leftarrow_R \mathbb{Z}_q^k$ and adds τ to $\mathcal{Q}_{msg}$. Here, $\mathbf{a}^\perp$ is non-zero vector in $\mathbb{Z}_q^{1\times(k+1)}$ that satisfies $\mathbf{a}^\perp \mathbf{A} = \mathbf{0}$.*
- *$\boxed{\mathcal{O}^*([\tau^*]_2) \text{ returns } [\mathbf{K}_0 + \tau^*\mathbf{K}_1]_2}$. $\mathcal{A}$ only gets a single call τ^* to $\mathcal{O}^*$.*
- *Q is the number of queries $\mathcal{A}$ makes to $\mathcal{O}_b$.*

Compared to [36, Lemma 3], oracle $\mathcal{O}^*$ is modified as follows. Instead of getting tag τ^* and returning $\mathbf{K}_0 + \tau^*\mathbf{K}_1$ in the clear, both the query and the output are encoded in $\mathbb{G}_2$. The change is boxed in the lemma. It is straight-forward to check that the proof goes through as in [36]:

- the security reduction knows $\mathbf{K}_0, \mathbf{K}_1$, and therefore it can compute $[\mathbf{K}_0 + \tau^* \mathbf{K}_1]_2$ given $[\tau^*]_2$;
- the quantity $[\mathbf{K}_0 + \tau^* \mathbf{K}_1]_2$ does not reveal any additional information about $\mathbf{K}_0, \mathbf{K}_1$ beyond $\mathbf{K}_0 + \tau^* \mathbf{K}_1$.

For completeness, a formal proof of the lemma is given in [35].

4.2 Our Scheme

The parameters are:

$$|\mathsf{pk}| = (n+1)k + 2(k+1)k + \mathsf{RE}(\mathcal{D}_k), \qquad |\sigma| = (3(k+1), 1),$$

where notation (x, y) represents x elements in $\mathbb{G}_1$ and y elements in $\mathbb{G}_2$. For k-Lin, this yields $(n+6, (6,1))$ for $k = 1$ and $(2n+16, (9,1))$ for $k = 2$. Moreover, we note that the verification needs $2k+1$ pairing product equations: for $e(\sigma_1, [\mathbf{A}]_2) = e([(1, \mathbf{m})]_1, [\mathbf{C}]_2) \cdot e(\sigma_2, [\mathbf{C}_0]_2) \cdot e(\sigma_3, [\mathbf{C}_1]_2)$ we need to pair the vector σ_1 with every column of $[\mathbf{A}]_2$ and thus this check needs k pairing product equations; and for $e(\sigma_2, [\tau]_2) = e(\sigma_3, [1]_2)$ we need to pair every element from σ_2 with $[\tau]_2 \in \mathbb{G}_2$ and thus this requires $k+1$ pairing product equations.

Gen(par):	Sign(sk, $[\mathbf{m}]_1$):
$\mathbf{A}, \mathbf{B} \leftarrow_{\mathrm{R}} \mathcal{D}_k; \mathbf{K} \leftarrow_{\mathrm{R}} \mathbb{Z}_q^{(n+1)\times(k+1)}$	$\mathbf{r} \leftarrow_{\mathrm{R}} \mathbb{Z}_q^k; \tau \leftarrow_{\mathrm{R}} \mathbb{Z}_q;$
$\mathbf{K}_0, \mathbf{K}_1 \leftarrow_{\mathrm{R}} \mathbb{Z}_q^{(k+1)\times(k+1)}$	$\sigma_1 := \left[(1, \mathbf{m}^\top)\mathbf{K} + \mathbf{r}^\top (\mathbf{P}_0 + \tau \mathbf{P}_1) \right]_1 \in \mathbb{G}_1^{1\times(k+1)}$
$\mathbf{C} := \mathbf{K}\mathbf{A} \in \mathbb{Z}_q^{(n+1)\times k}$	$\sigma_2 := \left[\mathbf{r}^\top \mathbf{B}^\top \right]_1 \in \mathbb{G}_1^{1\times(k+1)}$
$(\mathbf{C}_0, \mathbf{C}_1) := (\mathbf{K}_0\mathbf{A}, \mathbf{K}_1\mathbf{A})$	$\sigma_3 := \left[\mathbf{r}^\top \mathbf{B}^\top \tau \right]_1 \in \mathbb{G}_1^{1\times(k+1)}$
$\in (\mathbb{Z}_q^{(k+1)\times k})^2$	$\sigma_4 := [\tau]_2 \in \mathbb{G}_2$
$(\mathbf{P}_0, \mathbf{P}_1) := (\mathbf{B}^\top \mathbf{K}_0, \mathbf{B}^\top \mathbf{K}_1)$	Return $(\sigma_1, \sigma_2, \sigma_3, \sigma_4)$
$\in (\mathbb{Z}_q^{k\times(k+1)})^2$	
$\mathsf{sk} := (\mathbf{K}, [\mathbf{P}_0]_1, [\mathbf{P}_1]_1, [\mathbf{B}]_1)$	Verify(pk, $[\mathbf{m}]_1, \sigma$):
$\mathsf{pk} := ([\mathbf{C}_0]_2, [\mathbf{C}_1]_2, [\mathbf{C}]_2, [\mathbf{A}]_2)$	Parse $\sigma = (\sigma_1, \sigma_2, \sigma_3, \sigma_4 = [\tau]_2)$
Return $(\mathsf{pk}, \mathsf{sk})$	Check:
	$e(\sigma_1, [\mathbf{A}]_2) = e([(1, \mathbf{m})]_1, [\mathbf{C}]_2) \cdot e(\sigma_2, [\mathbf{C}_0]_2) \cdot$
	$e(\sigma_3, [\mathbf{C}_1]_2)$
	$\wedge \quad e(\sigma_2, [\tau]_2) = e(\sigma_3, [1]_2)$

Fig. 3. Structure-preserving signature $\mathsf{SPS}_{\mathsf{full}}$ with message-space $\mathcal{M} = \mathbb{G}_1^n$.

Theorem 2. *Under the $\mathcal{D}_k$-MDDH Assumption in $\mathbb{G}_1$ and $\mathcal{D}_k$-KerMDH Assumption in $\mathbb{G}_2$, $\mathsf{SPS}_{\mathsf{full}}$ from Fig. 3 is an unbounded CMA-secure structure-preserving signature scheme.*

Proof. Perfect correctness and the structure-preserving property are straight-forward. We proceed to establish the unbounded CMA-security. We will show that for any adversary $\mathcal{A}$ that makes at most Q signing queries, there exists adversaries $\mathcal{B}_0, \mathcal{B}_1$ with $\mathbf{T}(\mathcal{A}) \approx \mathbf{T}(\mathcal{B}_0) \approx \mathbf{T}(\mathcal{B}_1)$ and

$$\mathbf{Adv}^{\mathsf{cma}}_{\mathsf{SPS}_{\mathsf{full}}}(\mathcal{A}) \leq \mathbf{Adv}^{\mathsf{kmdh}}_{\mathcal{D}_k, \mathsf{GGen}}(\mathcal{B}_0) + 2Q(Q+1) \cdot \mathbf{Adv}^{\mathsf{mddh}}_{\mathcal{D}_k, \mathsf{GGen}}(\mathcal{B}_1) + (Q+1)^2/q + Q^2/2q. \quad (5)$$

We proceed via a series of games and we use $\mathbf{Adv}_i$ to denote the advantage of $\mathcal{A}$ in Game i.

Game 0. This is the CMA-security experiment from Definition 5.

$$\mathbf{Adv}_{\mathsf{SPS_{full}}}^{\mathrm{cma}}(\mathcal{A}) = \mathbf{Adv}_0$$

Game 1. Switch Verify to Verify*:

Verify*(pk, $[\mathbf{m}]_1, \sigma$):

Parse $\sigma = (\sigma_1, \sigma_2, \sigma_3, \sigma_4 = [\tau]_2)$

Check: $e(\sigma_1, [1]_2) = e([(1, \mathbf{m}^\top)\mathbf{K}]_1, [1]_2) \cdot e(\sigma_2, [\mathbf{K}_0 + \tau\mathbf{K}_1]_2)$

$\wedge \quad e(\sigma_2, [\tau]_2) = e(\sigma_3, [1]_2)$

Suppose $e(\sigma_2, [\tau]_2) = e(\sigma_3, [1]_2)$. We note that

$$e(\sigma_1, [\mathbf{A}]_2) = e([(1, \mathbf{m}^\top)]_1, [\mathbf{C}]_2) \cdot e(\sigma_2, [\mathbf{C}_0]_2) \cdot e(\sigma_3, [\mathbf{C}_1]_2)$$

$$\iff e(\sigma_1, [\mathbf{A}]_2) = e([(1, \mathbf{m}^\top)]_1, [\mathbf{K}\mathbf{A}]_2) \cdot e(\sigma_2, [\mathbf{K}_0\mathbf{A}]_2) \cdot e(\sigma_3, [\mathbf{K}_1\mathbf{A}]_2)$$

$$\Longleftarrow e(\sigma_1, [1]_2) = e([(1, \mathbf{m}^\top)]_1, [\mathbf{K}]_2) \cdot e(\sigma_2, [\mathbf{K}_0]_2) \cdot e(\sigma_3, [\mathbf{K}_1]_2)$$

$$\iff e(\sigma_1, [1]_2) = e([(1, \mathbf{m}^\top)]_1, [\mathbf{K}]_2) \cdot e(\sigma_2, [\mathbf{K}_0 + \tau\mathbf{K}_1]_2)$$

Hence, for any $([\mathbf{m}]_1, \sigma)$ that passes Verify but not Verify*, the value

$$\sigma_1 - ([(1, \mathbf{m}^\top)\mathbf{K}]_1 + \sigma_2\mathbf{K}_0 + \sigma_3\mathbf{K}_1) \in \mathbb{G}_1^{1\times(k+1)}$$

is a non-zero vector in the kernel of $\mathbf{A}$, which is hard to be computed under the $\mathcal{D}_k$-KerMDH assumption in $\mathbb{G}_2$. This means that

$$|\mathbf{Adv}_0 - \mathbf{Adv}_1| \leq \mathbf{Adv}_{\mathcal{D}_k, \mathsf{GGen}}^{\mathrm{kmdh}}(\mathcal{B}_0).$$

Game 2. Let $\tau_1, \ldots, \tau_Q$ denote the randomly chosen tags in the Q queries to SignO. We abort if $\tau_1, \ldots, \tau_Q$ are not all distinct.

$$\mathbf{Adv}_2 > \mathbf{Adv}_1 - Q^2/2q.$$

Game 3. We define $\tau_{Q+1} := \tau^*$. Now, pick $i^* \leftarrow_\mathrm{R} [Q+1]$ and abort if i^* is not the smallest index i for which $\tau^* = \tau_i$. In the rest of the proof, we focus on the case we do not abort, which means that $\tau^* = \tau_{i^*}$ and $\tau_1, \ldots, \tau_{i^*-1}$ are all different from τ^*. This means that given τ, SignO can check whether τ^* equals τ: for the rest $i^* - 1$ queries, answer NO, and starting from the i^*'th query, we know τ^*. It is easy to see that

$$\mathbf{Adv}_3 \geq \frac{1}{Q+1}\mathbf{Adv}_2.$$

Game 4. Switch SignO to SignO* where

SignO*($[\mathbf{m}]_1$):   // adds $\mu\mathbf{a}^\perp$ for $\tau \neq \tau^*$

$\mathbf{r} \leftarrow_\mathrm{R} \mathbb{Z}_q^k; \ \tau \leftarrow_\mathrm{R} \mathbb{Z}_q; \ \mu \leftarrow_\mathrm{R} \mathbb{Z}_q;$

if $\tau = \tau^*$ then $\mu := 0$

$\sigma_1 := \left[(1, \mathbf{m}^\top)\mathbf{K} + \mu\mathbf{a}^\perp + \mathbf{r}^\top(\mathbf{P}_0 + \tau\mathbf{P}_1)\right]_1$

$\sigma_2 := \left[\mathbf{r}^\top\mathbf{B}^\top\right]_1$

$\sigma_3 := \left[\mathbf{r}^\top\mathbf{B}^\top\tau\right]_1$

$\sigma_4 := [\tau]_2$

Return $(\sigma_1, \sigma_2, \sigma_3, \sigma_4) \in \mathbb{G}_1^{1\times(k+1)} \times \mathbb{G}_1^{1\times(k+1)} \times \mathbb{G}_1^{1\times(k+1)} \times \mathbb{G}_2$

We will use Lemma 3 to show that

$$|\mathbf{Adv}_3 - \mathbf{Adv}_4| \leq 2Q\mathbf{Adv}_{\mathcal{D}_k, \mathsf{GGen}}^{\mathrm{mddh}}(\mathcal{B}_1) + Q/q$$

Basically, we pick $\mathbf{K}$ ourselves and use $\mathcal{O}_b$ to simulate either SignO or SignO* and $\mathcal{O}^*$ to simulate Verify* as follows:

- For the i'th signing query $[\mathbf{m}]_1$ where $i \neq i^*$, we query $\mathcal{O}_b$ at $\tau \leftarrow_R \mathbb{Z}_q$ to obtain

$$(\sigma_1', \sigma_2) := \left(\left[b\mu\mathbf{a}^\perp + \mathbf{r}^\top(\mathbf{P}_0 + \tau\mathbf{P}_1)\right]_1, [\mathbf{r}^\top\mathbf{B}^\top]_1\right),$$

and we return

$$(\sigma_1 := [(1, \mathbf{m}^\top)\mathbf{K}]_1 \cdot \sigma_1', \ \sigma_2, \ \sigma_3 := \sigma_2\tau, \ \sigma_4 := [\tau]_2)$$

- For the i^*'th signing query $[\mathbf{m}]_1$ where $i^* \leq Q$, we run Sign honestly using our knowledge of $\mathbf{K}, [\mathbf{P}_0]_1, [\mathbf{P}_1], [\mathbf{B}]_1$.
- For Verify*, we will query $\mathcal{O}^*$ on $[\tau^*]_2$ to get $[\mathbf{K}_0 + \tau^*\mathbf{K}_1]_2$. The latter is sufficient to simulate the Verify* query by computing $e(\sigma_2, [\mathbf{K}_0 + \tau^*\mathbf{K}_1]_2)$.

This allows us to then build a distinguisher for Lemma 3.

Game 5. Switch $\mathbf{K} \leftarrow_R \mathbb{Z}_q^{(n+1)\times(k+1)}$ in Gen to $\mathbf{K} := \mathbf{K}' + \mathbf{u}\mathbf{a}^\perp$, where $\mathbf{K}' \leftarrow_R \mathbb{Z}_q^{(n+1)\times(k+1)}, \mathbf{u} \leftarrow_R \mathbb{Z}_q^{n+1}$.

Since $\mathbf{u}\mathbf{a}^\perp$ is masked by a uniform matrix $\mathbf{K}'$, $\mathbf{K}$ in Game 5 is still uniformly random and thus Game 4 and 5 are identical. We have

$$\mathbf{Adv}_5 = \mathbf{Adv}_4.$$

To conclude the proof, we bound the adversarial advantage in Game 5 via an information-theoretic argument. We first consider the information about $\mathbf{u}$ leaked from pk and signing queries:

- $\mathbf{C} = (\mathbf{K}' + \mathbf{u}\mathbf{a}^\perp)\mathbf{A} = \mathbf{K}'\mathbf{A}$ completely hides $\mathbf{u}$;
- the output of SignO* on $(\mathbf{m}, \tau)$ for $\tau \neq \tau^*$ completely hides $\mathbf{u}$, since $(1, \mathbf{m}^\top)(\mathbf{K}' + \mathbf{u}\mathbf{a}^\perp) + \mu\mathbf{a}^\perp$ is identically distributed to $(1, \mathbf{m}^\top)\mathbf{K}' + \mu\mathbf{a}^\perp$ (namely, $(1, \mathbf{m}^\top)\mathbf{u}$ is masked by $\mu \leftarrow_R \mathbb{Z}_q$).
- the output of SignO* on τ^* leaks $(1, \mathbf{m}^\top)(\mathbf{K}' + \mathbf{u}\mathbf{a}^\perp)$, which is captured by $(1, \mathbf{m}^\top)\mathbf{u}$.

To convince Verify* to accept a signature σ^* on $\mathbf{m}^*$, the adversary must correctly compute

$$(1, \mathbf{m}^{*\top})(\mathbf{K}' + \mathbf{u}\mathbf{a}^\perp)$$

and thus $(1, \mathbf{m}^{*\top})\mathbf{u} \in \mathbb{Z}_q$. Given $(1, \mathbf{m}^\top)\mathbf{u}$, for any adaptively chosen $\mathbf{m}^* \neq \mathbf{m}$, we have that $(1, \mathbf{m}^{*\top})\mathbf{u}$ is uniformly random over $\mathbb{Z}_q$ from the adversary's view-point. Therefore, $\mathbf{Adv}_5 \leq 1/q$. $\qquad\square$

4.3 Extension: SPS for Bilateral Message Spaces

Let $\mathcal{M} := \mathbb{G}_1^{n_1} \times \mathbb{G}_2^{n_2}$ be a message space. In Type III pairing groups, $\mathcal{M}$ is bilateral if both $n_1 \neq 0$ and $n_2 \neq 0$; otherwise, $\mathcal{M}$ is unilateral. We extend the construction from Sect. 4.2 to sign bilateral message spaces.

The main idea of our construction is to use the Even-Goldreich-Micali (EGM) framework [23] and a method of Abe et al. [2]: for $\mathbf{m} = ([\mathbf{m}_1]_1, [\mathbf{m}_2]_2) \in \mathbb{G}_1^{n_1} \times \mathbb{G}_2^{n_2}$ we sign $[\mathbf{m}_1]_1$ by using a one-time SPS with a fresh public key pk_{ot} over $\mathbb{G}_2$ and then sign message $([\mathbf{m}_2]_2, \mathsf{pk}_{ot})$ using an unbounded CMA-secure SPS; the signature on $([\mathbf{m}_1]_1, [\mathbf{m}_2]_2)$ is pk_{ot} together with the concatenation of both signatures. However, this yields long signatures as pk_{ot} contains $O(n_1 k)$ group element for the best known one-time SPS. Next, we observe that our one-time SPS is in fact a so-called "two-tier" signature scheme [12], i.e. opk can decomposed into a reusable long *primary key* plus a one-time short *secondary key* which contains only k group elements. For the transformation sketched above it is sufficient to put the short secondary key in the signature which leads to short signatures.

Details about our two-tier SPS and generic transformation are given in the full version [35]. The resulting unbounded CMA-secure SPS for bilateral message spaces is shown in Fig. 4. Its parameters are: $|\mathsf{pk}| = (n_1 + n_2)k + 3(k+1)k + 2\mathsf{RE}(\mathcal{D}_k)$, $|\sigma| = (4k+3, k+2)$, and #equations $= 3k+1$. Notation (x, y) represents x elements in $\mathbb{G}_1$ and y elements in $\mathbb{G}_2$. Under the SXDH assumption, our scheme achieves $(|\mathsf{pk}|, |\sigma|,$ #equations$) = (n_1 + n_2 + 8, (7, 3), 4)$. Compared with $(n_1 + n_2 + 22, (8, 6), 5)$ of [2], we obtain better efficiency under standard assumptions. The following theorem is proved in the full version [35].

Theorem 3. *Under the* $\mathcal{D}_k$*-MDDH Assumption in* $\mathbb{G}_1$ *and* $\mathcal{D}_k$ *KerMDH Assumption in both* $\mathbb{G}_1$ *and* $\mathbb{G}_2$, BSPS$_{\mathsf{full}}$ *from Fig. 4 is an unbounded* CMA*-secure structure-preserving signature scheme.*

Gen(par):	Sign(sk, $([\mathbf{m}_1]_1, [\mathbf{m}_2]_2))$:
$\mathbf{A}, \mathbf{B} \leftarrow_{\mathrm{R}} \mathcal{D}_k; \mathbf{K} \leftarrow_{\mathrm{R}} \mathbb{Z}_q^{(n_1+k+1)\times(k+1)}$	$\mathbf{x} \leftarrow_{\mathrm{R}} \mathbb{Z}_q^{k+1}; \mathbf{z} := \mathbf{x}^\top \mathbf{A}' \in \mathbb{Z}_q^{1\times k}$
$\mathbf{K}_0, \mathbf{K}_1 \leftarrow_{\mathrm{R}} \mathbb{Z}_q^{(k+1)\times(k+1)}$	$\mathbf{r} \leftarrow_{\mathrm{R}} \mathbb{Z}_q^k; \tau \leftarrow_{\mathrm{R}} \mathbb{Z}_q;$
$\mathbf{C} := \mathbf{KA} \in \mathbb{Z}_q^{(n_1+k+1)\times k}$	$\sigma_1 := \left[(1, \mathbf{m}_1^\top, \mathbf{z})\mathbf{K} + \mathbf{r}^\top(\mathbf{P}_0 + \tau\mathbf{P}_1)\right]_1$
$(\mathbf{C}_0, \mathbf{C}_1) := (\mathbf{K}_0\mathbf{A}, \mathbf{K}_1\mathbf{A})$	$\in \mathbb{G}_1^{1\times(k+1)}$
$\in (\mathbb{Z}_q^{(k+1)\times k})^2$	$\sigma_2 := \left[\mathbf{r}^\top\mathbf{B}^\top\right]_1 \in \mathbb{G}_1^{1\times(k+1)}$
$(\mathbf{P}_0, \mathbf{P}_1) := (\mathbf{B}^\top\mathbf{K}_0, \mathbf{B}^\top\mathbf{K}_1)$	$\sigma_3 := \left[\mathbf{r}^\top\mathbf{B}^\top\tau\right]_1 \in \mathbb{G}_1^{1\times(k+1)}$
$\in (\mathbb{Z}_q^{k\times(k+1)})^2$	$\sigma_4 := [\tau]_2 \in \mathbb{G}_2$
$\mathbf{A}' \leftarrow_{\mathrm{R}} \mathcal{D}_k; \mathbf{X} \leftarrow_{\mathrm{R}} \mathbb{Z}_q^{n_2\times(k+1)}$	$\sigma_5 := [\mathbf{x} + \mathbf{X}^\top\mathbf{m}_2]_2 \in \mathbb{G}_2^{k+1}$
$\mathbf{Z} := \mathbf{XA}' \in \mathbb{Z}_q^{n_2\times k}$	Return $([\mathbf{z}]_1, \sigma_1, \sigma_2, \sigma_3, \sigma_4, \sigma_5)$
$\mathsf{sk} := (\mathbf{K}, \mathbf{X}, [\mathbf{P}_0]_1, [\mathbf{P}_1]_1, [\mathbf{B}]_1)$	
$\mathsf{pk} := ([\mathbf{C}_0]_2, [\mathbf{C}_1]_2, [\mathbf{C}]_2, [\mathbf{Z}]_1, [\mathbf{A}]_2,$	Verify(pk, $([\mathbf{m}_1]_1, [\mathbf{m}_2]_2), \sigma$):
$[\mathbf{A}']_1)$	Parse $\sigma = ([\mathbf{z}]_1, \sigma_1, \sigma_2, \sigma_3, \sigma_4, \sigma_5)$
Return $(\mathsf{pk}, \mathsf{sk})$	Check:
	$e(\sigma_1, [\mathbf{A}]_2) = e([1, \mathbf{m}_1^\top, \mathbf{z}]_1, [\mathbf{C}]_2) \cdot e(\sigma_2, [\mathbf{C}_0]_2) \cdot$
	$e(\sigma_3, [\mathbf{C}_1]_2) \wedge e(\sigma_2, \sigma_4) = e(\sigma_3, [1]_2)$
	$\wedge e([\mathbf{A}']_1^\top, \sigma_5) = e([\mathbf{z}]_1^\top, [1]_2) \cdot e([\mathbf{Z}]_1^\top, \mathbf{m}_2)$

Fig. 4. Structure-preserving signature BSPS$_{\mathsf{full}}$ for bilateral message spaces $\mathcal{M} = \mathbb{G}_1^{n_1} \times \mathbb{G}_2^{n_2}$.

5   Security Against Random Message Attacks

In this section, we consider possible efficiency improvements on the structure-preserving signatures (SPS) from Sects. 3 and 4 for the weaker security notion of *unforgeability against random message attacks* (RMA). Precisely, we obtain a one-time RMA-secure SPS with signature size one less than that from Fig. 2 and an unbounded RMA-secure SPS with signature size $k+1$ less than that from Fig. 3. Figure 5 summarizes our results.

Our $\mathsf{rSPS_{ot}}$ is optimal for both the Type I and III settings: in the Type I setting, under the 2-Lin assumption, $\mathsf{rSPS_{ot}}$ requires 2 elements and 2 verification equations, matching the lower bound for one-time RMA-secure SPS from [8]; in the Type III setting, under the SXDH assumption, $\mathsf{rSPS_{ot}}$ requires 1 element and 1 verification equation, which is clearly optimal.

	Security	Assumption	$|\mathbf{m}|$	$|\sigma|$	$|\mathsf{pk}|$	# equ.
AGOT14 (Fig. 2) [8]	OT	Generic (Type I)	1	2	3	2
AGOT14 (Fig. 3) [8]	OT	Generic (Type III)	n	$(1,0)$	$n+3$	1
ACDKNO12 [2]	full	2-Lin	6	8	13	7
$\mathsf{rSPS_{ot}}$ (Fig 6)	OT	$\mathcal{D}_k$-KerMDH ($\mathbb{G}_2$)	n	$(k,0)$	$(n+1)k + \mathsf{RE}(\mathcal{D}_k)$	k
$\mathsf{rSPS_{full}}$ (Fig 7)	full	$\mathcal{D}_k$-MDDH ($\mathbb{G}_1, \mathbb{G}_2$)	n	$(2k+2, 1)$	$(n+2k+3)k + \mathsf{RE}(\mathcal{D}_k)$	$2k+1$

Fig. 5. Structure-preserving signatures secure against random message attacks for $\mathcal{M} = \mathbb{G}_1^n$ in the Type I and III setting. For the Type I setting we have $\mathbb{G} = \mathbb{G}_1 = \mathbb{G}_2$. Notation (x, y) represents x elements in $\mathbb{G}_1$ and y elements in $\mathbb{G}_2$.

5.1 Unforgeability Against Random Message Attacks

RMA-security states that it is hard for an adversary to forge a signature even if he sees many signatures on randomly chosen messages. The security is formally defined as follows:

Definition 6 (Unforgeability against random message attacks). *To an adversary $\mathcal{A}$ and SPS we associate the advantage function*

$$\mathbf{Adv}_{\mathsf{SPS}}^{\mathrm{rma}}(\mathcal{A}) := \Pr\left[[\mathbf{m}^*] \notin \mathcal{Q}_{\mathrm{msg}} \wedge \mathsf{Verify}(\mathsf{pk}, [\mathbf{m}^*], \sigma^*) = 1 \,\middle|\, \begin{array}{l} (\mathsf{pk}, \mathsf{sk}) \leftarrow_\mathrm{R} \mathsf{Gen}(\mathsf{par}) \\ ([\mathbf{m}^*], \sigma^*) \leftarrow_\mathrm{R} \mathcal{A}^{\mathsf{SignO}()}(\mathsf{pk}) \end{array} \right],$$

where $\mathsf{SignO}()$ chooses a random message $[\mathbf{m}] \leftarrow_\mathrm{R} \mathbb{G}^n$, runs $\sigma \leftarrow_\mathrm{R} \mathsf{Sign}(\mathsf{sk}, [\mathbf{m}])$, adds the vector $[\mathbf{m}]$ to $\mathcal{Q}_{\mathrm{msg}}$ (initialized with $\emptyset$) and returns $([\mathbf{m}], \sigma)$ to $\mathcal{A}$. SPS is said to be RMA-secure if for all PPT adversaries $\mathcal{A}$, $\mathbf{Adv}_{\mathsf{SPS}}^{\mathrm{rma}}(\mathcal{A})$ is negligible. SPS is said to be one-time RMA-secure with corresponding advantage function $\mathbf{Adv}_{\mathsf{SPS}}^{\mathrm{ot\text{-}rma}}(\mathcal{A})$, if $\mathcal{A}$ is restricted to make at most one query to oracle SignO.

5.2 One-Time RMA-Secure SPS

Motivated by the techniques used in [1,34,36] to obtain shorter QANIZK proofs for linear subspaces, we construct a one-time RMA-secure SPS in Fig. 6 with the following parameters:

$$|\mathsf{pk}| = (n+1)k + \mathsf{RE}(\mathcal{D}_k), \qquad |\sigma| = k.$$

For k-Lin, this yields $(n+2, 1)$ for $k = 1$ and $(2n+4, 2)$ for $k = 2$. Moreover, we note that verification needs k pairing product equations for $e(\sigma_1, [\mathbf{A}]_2) = e([(1, \mathbf{m})]_1, [\mathbf{C}]_2)$. Compared with $\mathsf{SPS_{ot}}$, we reduce the signature size by one element.

Theorem 4. *Under the $\mathcal{D}_k$-KerMDH Assumption in $\mathbb{G}_2$, $\mathsf{rSPS_{ot}}$ from Fig. 6 is a one-time RMA-secure structure-preserving signature scheme.*

Gen(par):	Sign(sk, $[\mathbf{m}]_1$):
$\mathbf{A} \leftarrow_R \mathcal{D}_k; \mathbf{K} \leftarrow_R \mathbb{Z}_q^{(n+1)\times k}$	$\sigma := \left[(1, \mathbf{m}^\top)\mathbf{K}\right]_1$
$\mathbf{C} := \mathbf{K}\overline{\mathbf{A}} \in \mathbb{Z}_q^{(n+1)\times k}$	Return $\sigma \in \mathbb{G}_1^{1\times k}$
sk := $\mathbf{K}$	
pk := $([\mathbf{C}]_2, [\overline{\mathbf{A}}]_2)$	Verify(pk, $[\mathbf{m}]_1, \sigma$):
Return (pk, sk)	Check: $e(\sigma, [\overline{\mathbf{A}}]_2) = e([(1, \mathbf{m}^\top)]_1, [\mathbf{C}]_2)$

Fig. 6. One-time RMA-secure structure-preserving signature $\mathsf{rSPS_{ot}}$ with message-space $\mathcal{M} = \mathbb{G}_1^n$. Recall that $\overline{\mathbf{A}}$ denotes the upper $k \times k$ submatrix of $\mathbf{A}$.

Our proof is similar to that in [36, Theorem 2]. As we choose $\mathbf{m} \in \mathbb{Z}_q^n$ in the security game ourselves, we can compute the kernel basis $\mathbf{M}^\perp \in \mathbb{Z}_q^{(n+1)\times n}$ of $(1, \mathbf{m}^\top)$ such that $(1, \mathbf{m}^\top) \cdot \mathbf{M}^\perp = \mathbf{0}$ and then we embed $\mathbf{M}^\perp$ in the secret key $\mathbf{K}$. This way we do not need to compute the kernel of $[\mathbf{A}]_2$ when answering the signing query. However, for the forgery $\mathbf{m}^* \neq \mathbf{m}$, since $(1, \mathbf{m}^{*\top})\mathbf{M}^\perp \neq \mathbf{0}$, the adversary has to compute an element from the kernel to break RMA-security, which is infeasible under the $\mathcal{D}_k$-KerMDH Assumption.

5.3 Unbounded RMA-Secure SPS

Consider the scheme $\mathsf{SPS_{full}}$ from Fig. 3 with the modification that in the signing algorithm, vector $\mathbf{Br}$ is chosen as a random vector as $\mathbf{t} \leftarrow_R \mathbb{Z}_q^{k+1}$. Clearly, under the $\mathcal{D}_k$-MDDH Assumption, this modified scheme is also a CMA-secure SPS. Suppose that the message space is $\mathbb{G}_1^n$ with $n = n' + k + 1 \geq k + 1$. Then we can view the random

Gen(par):	Sign(sk, $[\mathbf{m}]_1$):
$\mathbf{A} \leftarrow_R \mathcal{D}_k; \mathbf{K} \leftarrow_R \mathbb{Z}_q^{(n+1)\times(k+1)}$	Parse $[\mathbf{m}]_1 = ([\mathbf{s}]_1, [\mathbf{t}]_1) \in \mathbb{G}_1^{n'} \times \mathbb{G}_1^{k+1}$
$\mathbf{K}_0, \mathbf{K}_1 \leftarrow_R \mathbb{Z}_q^{(k+1)\times(k+1)}$	$\tau \leftarrow_R \mathbb{Z}_q$;
$\mathbf{C} := \mathbf{K}\overline{\mathbf{A}} \in \mathbb{Z}_q^{(n+1)\times k}$	$\sigma_1 := \left[(1, \mathbf{m}^\top)\mathbf{K} + \mathbf{t}^\top(\mathbf{K}_0 + \tau\mathbf{K}_1)\right]_1$
$(\mathbf{C}_0, \mathbf{C}_1) := (\mathbf{K}_0\overline{\mathbf{A}}, \mathbf{K}_1\overline{\mathbf{A}})$	$\sigma_2 := \left[\tau\mathbf{t}^\top\right]_1$
$\in (\mathbb{Z}_q^{(k+1)\times k})^2$	$\sigma_3 := [\tau]_2$
sk := $(\mathbf{K}, \mathbf{K}_0, \mathbf{K}_1)$	Return $(\sigma_1, \sigma_2, \sigma_3) \in \mathbb{G}_1^{1\times(k+1)} \times \mathbb{G}_1^{1\times(k+1)} \times \mathbb{G}_2$
pk := $([\mathbf{C}_0]_2, [\mathbf{C}_1]_2, [\mathbf{C}]_2, [\overline{\mathbf{A}}]_2)$	
Return (pk, sk)	Verify(pk, $[\mathbf{m}]_1, \sigma$):
	Parse $\sigma = (\sigma_1, \sigma_2, \sigma_3 = [\tau]_2)$
	Parse $[\mathbf{m}]_1 = ([\mathbf{s}]_1, [\mathbf{t}]_1)$
	Check:
	$e(\sigma_1, [\overline{\mathbf{A}}]_2) = e([(1, \mathbf{m}^\top)]_1, [\mathbf{C}]_2) \cdot e([\mathbf{t}^\top]_1, [\mathbf{C}_0]_2) \cdot$
	$e(\sigma_2, [\mathbf{C}_1]_2)$
	$\wedge \quad e(\sigma_2, [\mathbf{1}]_2) = e([\mathbf{t}^\top]_1, [\tau]_2)$

Fig. 7. An unbounded RMA-secure structure-preserving signature $\mathsf{rSPS_{full}}$ with message-space $\mathcal{M} = \mathbb{G}_1^n$ where $n = n' + k + 1 \geq k + 1$.

vector $[\mathbf{t}]_1 \in \mathbb{G}_1^{k+1}$ as part of the message space which reduces the signature size from $3k + 4$ elements to $2k + 3$. The modified scheme is presented in Fig. 7. Its parameters are:

$$|\mathsf{pk}| = (n+1)k + 2(k+1)k + \mathsf{RE}(\mathcal{D}_k), \qquad |\sigma| = (2(k+1), 1),$$

where notation (x, y) represents x elements in $\mathbb{G}_1$ and y elements in $\mathbb{G}_2$. For k-Lin, $(|\mathsf{pk}|, |\sigma|) = (n + 6, (4, 1))$ for $k = 1$ and $(2n + 16, (6, 1))$ for $k = 2$. Moreover, we note that the verification needs $2k + 1$ pairing product equations. Compared to the $\mathsf{SPS}_{\mathsf{full}}$ from Fig. 3, $\mathsf{rSPS}_{\mathsf{full}}$ requires $(k + 1)$ elements less in the signature.

Theorem 5. *Under the $\mathcal{D}_k$-MDDH Assumption in $\mathbb{G}_1$ and $\mathcal{D}_k$-KerMDH Assumption in $\mathbb{G}_2$, $\mathsf{rSPS}_{\mathsf{full}}$ from Fig. 7 is an unbounded RMA-secure structure-preserving signature scheme.*

The proof is given in [35].

Acknowledgments. We thank Olivier Blazy and Georg Fuchsbauer for helpful discussions.

References

1. Abdalla, M., Benhamouda, F., Pointcheval, D.: Disjunctions for hash proof systems: new constructions and applications. In: Oswald, E., Fischlin, M. (eds.) EUROCRYPT 2015. LNCS, vol. 9057, pp. 69–100. Springer, Heidelberg (2015)
2. Abe, M., Chase, M., David, B., Kohlweiss, M., Nishimaki, R., Ohkubo, M.: Constant-size structure-preserving signatures: generic constructions and simple assumptions. In: Wang, X., Sako, K. (eds.) ASIACRYPT 2012. LNCS, vol. 7658, pp. 4–24. Springer, Heidelberg (2012)
3. Abe, M., David, B., Kohlweiss, M., Nishimaki, R., Ohkubo, M.: Tagged one-time signatures: tight security and optimal tag size. In: Kurosawa, K., Hanaoka, G. (eds.) PKC 2013. LNCS, vol. 7778, pp. 312–331. Springer, Heidelberg (2013)
4. Abe, M., Fuchsbauer, G., Groth, J., Haralambiev, K., Ohkubo, M.: Structure-preserving signatures and commitments to group elements. In: Rabin, T. (ed.) CRYPTO 2010. LNCS, vol. 6223, pp. 209–236. Springer, Heidelberg (2010)
5. Abe, M., Groth, J., Haralambiev, K., Ohkubo, M.: Optimal structure-preserving signatures in asymmetric bilinear groups. In: Rogaway, P. (ed.) CRYPTO 2011. LNCS, vol. 6841, pp. 649–666. Springer, Heidelberg (2011)
6. Abe, M., Groth, J., Ohkubo, M.: Separating short structure-preserving signatures from non-interactive assumptions. In: Lee, D.H., Wang, X. (eds.) ASIACRYPT 2011. LNCS, vol. 7073, pp. 628–646. Springer, Heidelberg (2011)
7. Abe, M., Groth, J., Ohkubo, M., Tibouchi, M.: Structure-preserving signatures from type II pairings. In: Garay, J.A., Gennaro, R. (eds.) CRYPTO 2014, Part I. LNCS, vol. 8616, pp. 390–407. Springer, Heidelberg (2014)
8. Abe, M., Groth, J., Ohkubo, M., Tibouchi, M.: Unified, minimal and selectively randomizable structure-preserving signatures. In: Lindell, Y. (ed.) TCC 2014. LNCS, vol. 8349, pp. 688–712. Springer, Heidelberg (2014)
9. Attrapadung, N., Libert, B., Peters, T.: Efficient completely context-hiding quotable and linearly homomorphic signatures. In: Kurosawa, K., Hanaoka, G. (eds.) PKC 2013. LNCS, vol. 7778, pp. 386–404. Springer, Heidelberg (2013)

10. Barthe, G., Fagerholm, E., Fiore, D., Scedrov, A., Schmidt, B., Tibouchi, M.: Strongly-optimal structure preserving signatures from type II pairings: synthesis and lower bounds. In: Katz, J. (ed.) PKC 2015. LNCS, vol. 9020, pp. 355–376. Springer, Heidelberg (2015)
11. Belenkiy, M., Camenisch, J., Chase, M., Kohlweiss, M., Lysyanskaya, A., Shacham, H.: Randomizable proofs and delegatable anonymous credentials. In: Halevi, S. (ed.) CRYPTO 2009. LNCS, vol. 5677, pp. 108–125. Springer, Heidelberg (2009)
12. Bellare, M., Shoup, S.: Two-tier signatures, strongly unforgeable signatures, and fiat-shamir without random oracles. In: Okamoto, T., Wang, X. (eds.) PKC 2007. LNCS, vol. 4450, pp. 201–216. Springer, Heidelberg (2007)
13. Blazy, O., Canard, S., Fuchsbauer, G., Gouget, A., Sibert, H., Traoré, J.: Achieving optimal anonymity in transferable e-cash with a judge. In: Nitaj, A., Pointcheval, D. (eds.) AFRICACRYPT 2011. LNCS, vol. 6737, pp. 206–223. Springer, Heidelberg (2011)
14. Blazy, O., Kiltz, E., Pan, J.: (Hierarchical) Identity-based encryption from affine message authentication. In: Garay, J.A., Gennaro, R. (eds.) CRYPTO 2014, Part I. LNCS, vol. 8616, pp. 408–425. Springer, Heidelberg (2014)
15. Boneh, D., Freeman, D., Katz, J., Waters, B.: Signing a linear subspace: signature schemes for network coding. In: Jarecki, S., Tsudik, G. (eds.) PKC 2009. LNCS, vol. 5443, pp. 68–87. Springer, Heidelberg (2009)
16. Catalano, D., Marcedone, A., Puglisi, O.: Authenticating computation on groups: new homomorphic primitives and applications. In: Sarkar, P., Iwata, T. (eds.) ASIACRYPT 2014, Part II. LNCS, vol. 8874, pp. 193–212. Springer, Heidelberg (2014)
17. Cathalo, J., Libert, B., Yung, M.: Group encryption: non-interactive realization in the standard model. In: Matsui, M. (ed.) ASIACRYPT 2009. LNCS, vol. 5912, pp. 179–196. Springer, Heidelberg (2009)
18. Chase, M., Kohlweiss, M., Lysyanskaya, A., Meiklejohn, S.: Malleable proof systems and applications. In: Pointcheval, D., Johansson, T. (eds.) EUROCRYPT 2012. LNCS, vol. 7237, pp. 281–300. Springer, Heidelberg (2012)
19. Chen, J., Gay, R., Wee, H.: Improved dual system ABE in prime-order groups via predicate encodings. In: Oswald, E., Fischlin, M. (eds.) EUROCRYPT 2015. LNCS, vol. 9057, pp. 595–624. Springer, Heidelberg (2015)
20. Chen, J., Lim, H.W., Ling, S., Wang, H., Wee, H.: Shorter IBE and signatures via asymmetric pairings. In: Abdalla, M., Lange, T. (eds.) Pairing 2012. LNCS, vol. 7708, pp. 122–140. Springer, Heidelberg (2013)
21. Desmedt, Y.: Computer security by redefining what a computer is. In: New Security Paradigms Workshop (NSPW) (1993)
22. Escala, A., Herold, G., Kiltz, E., Ràfols, C., Villar, J.: An algebraic framework for diffie-hellman assumptions. In: Canetti, R., Garay, J.A. (eds.) CRYPTO 2013, Part II. LNCS, vol. 8043, pp. 129–147. Springer, Heidelberg (2013)
23. Even, S., Goldreich, O., Micali, S.: On-line/off-line digital signatures. J. Cryptology 9(1), 35–67 (1996)
24. Fuchsbauer, G.: Commuting signatures and verifiable encryption. In: Paterson, K.G. (ed.) EUROCRYPT 2011. LNCS, vol. 6632, pp. 224–245. Springer, Heidelberg (2011)
25. Fuchsbauer, G., Vergnaud, D.: Fair Blind Signatures without Random Oracles. In: Bernstein, D.J., Lange, T. (eds.) AFRICACRYPT 2010. LNCS, vol. 6055, pp. 16–33. Springer, Heidelberg (2010)

26. Green, M., Hohenberger, S.: Universally composable adaptive oblivious transfer. In: Pieprzyk, J. (ed.) ASIACRYPT 2008. LNCS, vol. 5350, pp. 179–197. Springer, Heidelberg (2008)

27. Groth, J.: Simulation-sound NIZK proofs for a practical language and constant size group signatures. In: Lai, X., Chen, K. (eds.) ASIACRYPT 2006. LNCS, vol. 4284, pp. 444–459. Springer, Heidelberg (2006)

28. Groth, J.: Fully anonymous group signatures without random oracles. In: Kurosawa, K. (ed.) ASIACRYPT 2007. LNCS, vol. 4833, pp. 164–180. Springer, Heidelberg (2007)

29. Groth, J., Sahai, A.: Efficient non-interactive proof systems for bilinear groups. In: Smart, N.P. (ed.) EUROCRYPT 2008. LNCS, vol. 4965, pp. 415–432. Springer, Heidelberg (2008)

30. Hofheinz, D., Jager, T.: Tightly secure signatures and public-key encryption. In: Safavi-Naini, R., Canetti, R. (eds.) CRYPTO 2012. LNCS, vol. 7417, pp. 590–607. Springer, Heidelberg (2012)

31. Hofheinz, D., Kiltz, E.: Secure hybrid encryption from weakened key encapsulation. In: Menezes, A. (ed.) CRYPTO 2007. LNCS, vol. 4622, pp. 553–571. Springer, Heidelberg (2007)

32. Johnson, R., Molnar, D., Song, D., Wagner, D.: Homomorphic signature schemes. In: Preneel, B. (ed.) CT-RSA 2002. LNCS, vol. 2271, pp. 244–262. Springer, Heidelberg (2002)

33. Jutla, C.S., Roy, A.: Shorter quasi-adaptive NIZK proofs for linear subspaces. In: Sako, K., Sarkar, P. (eds.) ASIACRYPT 2013, Part I. LNCS, vol. 8269, pp. 1–20. Springer, Heidelberg (2013)

34. Jutla, C.S., Roy, A.: Switching lemma for bilinear tests and constant-size NIZK proofs for linear subspaces. In: Garay, J.A., Gennaro, R. (eds.) CRYPTO 2014, Part II. LNCS, vol. 8617, pp. 295–312. Springer, Heidelberg (2014)

35. Kiltz, E., Pan, J., Wee, H.: Structure-preserving signatures from standard assumptions, revisited. Cryptology ePrint Archive, Full version of this paper (2015)

36. Kiltz, E., Wee, H.: Quasi-adaptive NIZK for linear subspaces revisited. In: Oswald, E., Fischlin, M. (eds.) EUROCRYPT 2015. LNCS, vol. 9057, pp. 101–128. Springer, Heidelberg (2015)

37. Lewko, A., Waters, B.: New techniques for dual system encryption and fully secure HIBE with short ciphertexts. In: Micciancio, D. (ed.) TCC 2010. LNCS, vol. 5978, pp. 455–479. Springer, Heidelberg (2010)

38. Libert, B., Peters, T., Joye, M., Yung, M.: Linearly homomorphic structure-preserving signatures and their applications. In: Canetti, R., Garay, J.A. (eds.) CRYPTO 2013, Part II. LNCS, vol. 8043, pp. 289–307. Springer, Heidelberg (2013)

39. Libert, B., Peters, T., Joye, M., Yung, M.: Non-malleability from malleability: simulation-sound quasi-adaptive nizk proofs and cca2-secure encryption from homomorphic signatures. In: Nguyen, P.Q., Oswald, E. (eds.) EUROCRYPT 2014. LNCS, vol. 8441, pp. 514–532. Springer, Heidelberg (2014)

40. Libert, B., Peters, T., Yung, M.: Group signatures with almost-for-free revocation. In: Safavi-Naini, R., Canetti, R. (eds.) CRYPTO 2012. LNCS, vol. 7417, pp. 571–589. Springer, Heidelberg (2012)

41. Boneh, D., Boyen, X., Shacham, H.: Short group signatures. In: Franklin, M. (ed.) CRYPTO 2004. LNCS, vol. 3152, pp. 41–55. Springer, Heidelberg (2004)

42. Morillo, P., Ràfols, C., Villar, J.L.: Matrix computational assumptions in multilinear groups. Cryptology ePrint Archive, Report 2015/353 (2015)

43. Waters, B.: Dual system encryption: realizing fully secure IBE and HIBE under simple assumptions. In: Halevi, S. (ed.) CRYPTO 2009. LNCS, vol. 5677, pp. 619–636. Springer, Heidelberg (2009)
44. Wee, H.: Dual system encryption via predicate encodings. In: Lindell, Y. (ed.) TCC 2014. LNCS, vol. 8349, pp. 616–637. Springer, Heidelberg (2014)

Short Group Signatures via Structure-Preserving Signatures: Standard Model Security from Simple Assumptions

Benoît Libert[1]([✉]), Thomas Peters[2], and Moti Yung[3]

[1] Ecole Normale Supérieure de Lyon, Lyon, France
benoit.libert@ens-lyon.fr
[2] Ecole Normale Supérieure, CNRS, INRIA, Paris, France
[3] Google Inc. and Columbia University, New York, USA

Abstract. Group signatures are a central cryptographic primitive which allows users to sign messages while hiding their identity within a crowd of group members. In the standard model (without the random oracle idealization), the most efficient constructions rely on the Groth-Sahai proof systems (Eurocrypt'08). The structure-preserving signatures of Abe *et al.* (Asiacrypt'12) make it possible to design group signatures based on well-established, constant-size number theoretic assumptions (a.k.a. "simple assumptions") like the Symmetric eXternal Diffie-Hellman or Decision Linear assumptions. While much more efficient than group signatures built on general assumptions, these constructions incur a significant overhead w.r.t. constructions secure in the idealized random oracle model. Indeed, the best known solution based on simple assumptions requires 2.8 kB per signature for currently recommended parameters. Reducing this size and presenting techniques for shorter signatures are thus natural questions. In this paper, our first contribution is to significantly reduce this overhead. Namely, we obtain the first fully anonymous group signatures based on simple assumptions with signatures shorter than 2 kB at the 128-bit security level. In dynamic (resp. static) groups, our signature length drops to 1.8 kB (resp. 1 kB). This improvement is enabled by two technical tools. As a result of independent interest, we first construct a new structure-preserving signature based on simple assumptions which shortens the best previous scheme by 25 %. Our second tool is a method for attaining anonymity in the strongest sense using a new CCA2-secure encryption scheme which is also a Groth-Sahai commitment.

Keywords: Group signatures · Standard model · Simple assumptions · Efficiency · Structure-preserving cryptography · QA-NIZK arguments

1 Introduction

As introduced by Chaum and van Heyst [27] in 1991, group signatures allow members of a group administered by some authority to anonymously sign messages on behalf of the group. In order to prevent abuses, an opening authority has the power to uncover a signer's identity if the need arises.

© International Association for Cryptologic Research 2015
R. Gennaro and M. Robshaw (Eds.): CRYPTO 2015, Part II, LNCS 9216, pp. 296–316, 2015.
DOI: 10.1007/978-3-662-48000-7_15

The usual approach for building a group signature consists in having the signer encrypt his group membership credential under the public key of the opening authority while appending a non-interactive zero-knowledge (NIZK) proof, which is associated with the message, claiming that things were done correctly. Until 2006, efficient instantiations of this primitive were only available under the random oracle idealization [14], which is limited to only provide heuristic arguments in terms of security [24]. This state of affairs changed in the last decade, with the emergence of solutions [20,21,35,36] enabled by breakthrough results in the design of relatively efficient non-interactive witness indistinguishable (NIWI) proofs [37]. While drastically more efficient than solutions based on general NIZK proofs [12,15], the constructions of [20,21,35,36] still incur a substantial overhead when compared with their random-oracle-based counterparts [10,18,30]. Moreover, their most efficient variants [21,36] tend to rely on parametrized assumptions — often referred to as "q-type" assumptions — where the number of input elements is determined by a parameter q which, in turn, depends on the number of users in the system or the number of adversarial queries (or both). Since the assumption becomes stronger as q increases, a different assumption is needed for every adversary (based on its number of queries) and every maximal number of users in the group. Not only does it limit the scalability of realizations, it also restricts the level of confidence in their security.

In this paper, we consider the problem of devising as short as possible group signatures based on simple assumptions. By "simple assumption", we mean a well-established assumption, like the Decision Diffie-Hellman assumption, which is simultaneously non-interactive and described using a constant number of elements, regardless of the number of users in the system or the number of adversarial queries. We remark that even in the random oracle model, this problem turns out to be highly non-trivial as non-simple assumptions (like the Strong RSA [10,42] or Strong Diffie-Hellman [18,30]) are frequently relied on. In the standard model, our main contribution is designing the first group signatures based on simple assumptions and whose size is less than 2 kB for the currently recommended 128-bit security level. In static groups, our most efficient scheme features signatures slightly longer than 1 kB. So far, the best standard-model group signature based on simple assumptions was obtained from the structure-preserving signatures (SPS) of Abe et al. [1,2] and required 2.875 kB per signature. Along the way and as a result of independent interest, we also build a new structure-preserving signature (SPS) with the shortest length among those based on simple assumptions. Concretely, the best previous SPS based on similar assumptions [1,2] is shortened by 25%.

RELATED WORK. Group signatures have a long history. Still, efficient and provably coalition-resistant constructions (in the random oracle model) remained elusive until the work of Ateniese, Camenisch, Joye and Tsudik [10] in 2000. At that time, however, there was no proper formalization of the security properties that can be naturally expected from group signatures. This gap was filled in 2003 by Bellare, Micciancio and Warinschi [12] (BMW) who captured all the requirements of group signatures in three properties. In (a variant of) this model, Boneh, Boyen and Shacham [18] obtained very short signatures using the random oracle methodology [14].

The BMW model assumes static groups where the set of members is frozen after the setup phase beyond which no new member can be added. The setting of dynamic groups was explored later on by Bellare-Shi-Zhang [15] and, independently, by Kiayias and Yung [42]. In these models [15,42], short signature lengths were obtained in [30]. A construction based on interactive assumptions in the standard model was also put forth by Ateniese *et al.* [9]. Using standard assumptions, Boyen and Waters gave a different solution [20] based on the Groth-Ostrovsky-Sahai NIZK proof system [34]. They subsequently managed to obtain $O(1)$-size signatures at the expense of appealing to a q-type assumption [21]. Their constructions [20,21] were both analyzed in (a relaxation of) the BMW model [12] where the adversary is not granted access to a signature opening oracle. In dynamic groups [15], Groth [35] obtained constant-size signatures in the standard model but, due to huge hidden constants, his result was mostly a proof of concept. By making the most of Groth-Sahai NIWI proofs [37], he subsequently reduced signatures to 48 group elements [36] with the caveat of resting on relatively *ad hoc* q-type assumptions. For the time being, the best group signatures based on standard assumptions are enabled by the structure-preserving signatures of Abe, Chase, David, Kohlweiss, Nishimaki, and Ohkubo [1]. In asymmetric pairings $e : \mathbb{G} \times \hat{\mathbb{G}} \to \mathbb{G}_T$ (where $\mathbb{G} \neq \hat{\mathbb{G}}$), anonymously signing messages requires at least 40 elements of $\mathbb{G}$ and 26 elements of $\hat{\mathbb{G}}$.

In 2010, Abe *et al.* [3,8] advocated the use of *structure-preserving* cryptography as a general tool for building privacy-preserving protocols in a modular fashion. In short, structure-preserving signatures (SPS) are signature schemes that smoothly interact with Groth-Sahai proofs [37] as messages, signatures public keys all live in the source groups $(\mathbb{G}, \hat{\mathbb{G}})$ of a bilinear map $e : \mathbb{G} \times \hat{\mathbb{G}} \to \mathbb{G}_T$. SPS schemes were initially introduced by Groth [35] and further studied in [25,31]. In the last three years, a large body of work was devoted to the feasibility and efficiency of structure-preserving signatures [1–4,8,23,25,26,31,35,38]. In Type III pairings (i.e., where $\mathbb{G} \neq \hat{\mathbb{G}}$ and no isomorphism is computable from $\hat{\mathbb{G}}$ to $\mathbb{G}$ or backwards), Abe *et al.* [4] showed that any SPS scheme must contain at least 3 group elements per signature. For a natural class of reductions, the security of optimally short signatures was also shown [5] *unprovable* under any non-interactive assumption. These impossibility results were recently found [7] not to carry over to Type II pairings (i.e., where $\mathbb{G} \neq \hat{\mathbb{G}}$ and an efficiently computable isomorphism $\psi : \hat{\mathbb{G}} \to \mathbb{G}$ is available).

To the best of our knowledge, the minimal length of structure-preserving signatures based on simple assumptions remains an unsettled open question. We believe it to be of primary importance considering the versatility of structure-preserving cryptography in the design of privacy-related protocols, including group signatures [8], group encryption [25] or adaptive oblivious transfer [33].

OUR RESULTS. The first contribution of this paper is to describe a new structure-preserving signature based on the standard Symmetric eXternal Diffie-Hellman (SXDH) assumption and an asymmetric variant of the Decision Linear assumption with only 10 group elements (more precisely, 9 elements of $\mathbb{G}$ and one element of $\hat{\mathbb{G}}$) per signature. So far, the best instantiation of [1,2] required 7 elements of $\mathbb{G}$ and 4 elements of $\hat{\mathbb{G}}$. Since the representation of $\hat{\mathbb{G}}$ elements is at least twice as long as that of $\mathbb{G}$ elements, our scheme thus saves 26 % in terms of signature length. Armed with our new SPS and other tools, we then construct dynamic group signatures using only 32 elements of $\mathbb{G}$ and 14 elements of $\hat{\mathbb{G}}$ in each signature, where Abe et al. [1,2] need at least 40 elements of $\mathbb{G}$ and 26 elements of $\hat{\mathbb{G}}$. For typical parameters, our signatures are thus 37 % shorter with a total length of only 1.8 kB at the 128-bit security level. In an independent work, Kiltz, Pan and Wee [45] managed to obtain even shorter structure-preserving signatures than ours under the SXDH assumption. If their construction is used in our dynamic group signature, it allows eliminating at least 4 more elements of $\mathbb{G}$ from signatures. In the static model of Bellare, Micciancio and Warinschi [12], we describe an even more efficient realization where the signature length decreases to almost 1 kB.

OUR TECHNIQUES. Our structure-preserving signature can be seen as a non-trivial optimization of a modular design, suggested by Abe et al. [1], which combines a weakly secure SPS scheme and a tagged one-time signature (TOTS). In a TOTS scheme, each signature contains a fresh tag and, without knowing the private key, it should be computationally infeasible to generate a signature on a new message for a previously used tag. The construction of [1] obtains a full-fledged SPS by combining a TOTS scheme with an SPS system that is only secure against extended random message attacks (XRMA). As defined in [1], XRMA security basically captures security against an adversary that only obtains signatures on random group elements even knowing some auxiliary information used to sample these elements (typically their discrete logarithms). While Abe et al. [1] make use of the discrete logs of signed messages in their proofs of XRMA security, their modular construction does not. Here, by explicitly using the discrete logarithms in the construction, we obtain significant efficiency improvements. Using Waters' dual system techniques [51], we construct an SXDH-based F-unforgeable signature scheme which, according to the terminology of Belenkiy et al. [11], is a signature scheme that remains verifiable and unforgeable even if the adversary only outputs an injective function of the forgery message. Our new SPS is the result of combining our F-unforgeable signature and the TOTS system of [2]. We stress that our scheme can no longer be seen as an instantiation of a generic construction. Still, at the natural expense of sacrificing modularity, it does provide shorter signatures.

In turn, our F-unforgeable signatures are obtained by taking advantage of the quasi-adaptive NIZK (QA-NIZK) arguments of linear subspace membership suggested by Jutla and Roy [40] and further studied in [41,47], where the CRS may depend on the language for which proofs have to be generated. In a nutshell, our starting point is a signature scheme suggested by Jutla and Roy (inspired by ideas due to Camenisch et al. [22]) where each signature is a CCA2-secure encryption of the private key (made verifiable via QA-NIZK proofs) and the message is included in the label [50]. We rely on the observation that QA-NIZK proofs for linear subspaces [40] (or their optimized variants [41,47]) make it possible to verify signatures even if the message is only available in the exponent.

In order to save the equivalent of 15 elements of the group $\mathbb{G}$ and make the group signature as short as possible, we also design a new CCA2-secure tag-based encryption (TBE) scheme [44,48] which incorporates a Groth-Sahai commitment. In fully anonymous group signatures, CCA2-anonymity is usually acquired by verifiably encrypting the signer's credential using a CCA2-secure cryptosystem while providing evidence that the plaintext coincides with a committed group element. Inspired by a lossy encryption scheme [13] suggested by Hemenway *et al.* [39], we depart from this approach and rather use a CCA2-secure encryption scheme which simultaneously plays the role of a Groth-Sahai commitment. That is, even when the Groth-Sahai CRS is a perfectly hiding CRS, we are able to extract committed group elements for any tag but a specific one, where the encryption scheme behaves like a perfectly hiding commitment and induces perfectly NIWI proofs. In order to make the validity of TBE ciphertexts publicly verifiable, we rely on the QA-NIZK proofs of Libert *et al.* [47] which are well-suited to the specific subspaces encountered[1] in this context. We believe this encryption scheme to be of interest in its own right since it allows shortening other group signatures based on Groth-Sahai proofs (e.g., [36]) in a similar way.

Our group signature in the static BMW model [12] does not build on structure-preserving signatures but rather follows the same design principle as the constructions of Boyen and Waters [20,21]. It is obtained by extending our F-unforgeable signature into a 2-level hierarchical signature [43] (or, equivalently, an identity-based signature [49]) where first-level messages are implicit in the exponent. In spirit and from an efficiency standpoint, our static group signature is thus similar to the second construction [21] of Boyen and Waters, with the benefit of providing full anonymity while relying on the sole SXDH assumption.

2 Background

2.1 Hardness Assumptions

We use bilinear maps $e : \mathbb{G} \times \hat{\mathbb{G}} \to \mathbb{G}_T$ over groups of prime order p where $e(g, \hat{h}) \neq 1_{\mathbb{G}_T}$ if and only if $g \neq 1_{\mathbb{G}}$ and $\hat{h} \neq 1_{\hat{\mathbb{G}}}$. We rely on hardness assumptions that are non-interactive and described using a constant number of elements.

Definition 1. *The* **Decision Diffie-Hellman** *(DDH) problem in $\mathbb{G}$, is to distinguish the distributions (g^a, g^b, g^{ab}) and (g^a, g^b, g^c), with $a, b, c \xleftarrow{R} \mathbb{Z}_p$. The DDH assumption is the intractability of the problem for any PPT distinguisher.*

In the following, we will rely on the Symmetric external Diffie-Hellman (SXDH) assumption which posits the hardness of DDH in $\mathbb{G}$ and $\hat{\mathbb{G}}$ in asymmetric pairing configurations. We also assume the hardness of the following problem, which generalizes the Decision Linear problem [18] to asymmetric pairings.

---

[1] Specifically, we have to prove membership of a $t \times n$ subspace of rank t described by a $2t \times n$ matrix and the security proofs of [46,47] still work in this case.

Definition 2 ([1]). *In bilinear groups* $(\mathbb{G}, \hat{\mathbb{G}}, \mathbb{G}_T)$ *of prime order* p, *the* **eXternal Decision Linear Problem** 2 *(XDLIN$_2$) is to distinguish the distribution*

$$D_1 = \{(g, g^a, g^b, g^{ac}, g^{bd}, \hat{g}, \hat{g}^a, \hat{g}^b, \hat{g}^{ac}, \hat{g}^{bd}, \hat{g}^{c+d}) \in \mathbb{G}^5 \times \hat{\mathbb{G}}^6 \mid a, b, c, d \xleftarrow{R} \mathbb{Z}_p\}$$

$$D_2 = \{(g, g^a, g^b, g^{ac}, g^{bd}, \hat{g}, \hat{g}^a, \hat{g}^b, \hat{g}^{ac}, \hat{g}^{bd}, \hat{g}^z) \in \mathbb{G}^5 \times \hat{\mathbb{G}}^6 \mid a, b, c, d, z \xleftarrow{R} \mathbb{Z}_p\}.$$

The XDLIN$_1$ assumption is defined analogously and posits the infeasibility of distinguishing g^{c+d} and g^z given $(g, g^a, g^b, g^{ac}, g^{bd}, \hat{g}, \hat{g}^a, \hat{g}^b, \hat{g}^{ac}, \hat{g}^{bd})$.

2.2 Linearly Homomorphic Structure-Preserving Signatures

Structure-preserving signatures [3,8] are signature schemes where messages and public keys all consist of elements of a group over which a bilinear map $e : \mathbb{G} \times \hat{\mathbb{G}} \to \mathbb{G}_T$ is efficiently computable.

Libert *et al.* [46] considered structure-preserving signatures with linear homomorphic properties. This section recalls the one-time linearly homomorphic structure-preserving signature (LHSPS) of [46]. In the description below, we assume that all algorithms take as input the description of common public parameters cp consisting of asymmetric bilinear groups $(\mathbb{G}, \hat{\mathbb{G}}, \mathbb{G}_T, p)$ of prime order $p > 2^\lambda$, where λ is the security parameter.

In [46], Libert *et al.* suggested the following construction which can be proved secure under the SXDH assumption.

Keygen(cp, n): Given common public parameters cp $= (\mathbb{G}, \hat{\mathbb{G}}, \mathbb{G}_T, p)$ and the dimension $n \in \mathbb{N}$ of the subspace to be signed. Then, choose $\hat{g}_z, \hat{g}_r \xleftarrow{R} \hat{\mathbb{G}}$. For $i = 1$ to n, pick $\chi_i, \gamma_i \xleftarrow{R} \mathbb{Z}_p$ and compute $\hat{g}_i = \hat{g}_z^{\chi_i} \hat{g}_r^{\gamma_i}$. The private key is sk $= \{(\chi_i, \gamma_i)\}_{i=1}^n$ while the public key is pk $= (\hat{g}_z, \hat{g}_r, \{\hat{g}_i\}_{i=1}^n) \in \hat{\mathbb{G}}^{n+2}$.

Sign(sk, $(M_1, \ldots, M_n)$): In order to sign a vector $(M_1, \ldots, M_n) \in \mathbb{G}^n$ using sk $= \{(\chi_i, \gamma_i)\}_{i=1}^n$, output $\sigma = (z, r) = \left(\prod_{i=1}^n M_i^{-\chi_i}, \prod_{i=1}^n M_i^{-\gamma_i}\right)$.

SignDerive(pk, $\{(\omega_i, \sigma^{(i)})\}_{i=1}^\ell$): given pk as well as ℓ tuples $(\omega_i, \sigma^{(i)})$, parse $\sigma^{(i)}$ as $\sigma^{(i)} = (z_i, r_i)$ for $i = 1$ to ℓ. Return $\sigma = (z, r) = \left(\prod_{i=1}^\ell z_i^{\omega_i}, \prod_{i=1}^\ell r_i^{\omega_i}\right)$.

Verify(pk, σ, $(M_1, \ldots, M_n)$): Given a signature $\sigma = (z, r) \in \mathbb{G}^2$ and a vector $(M_1, \ldots, M_n)$, return 1 if and only if $(M_1, \ldots, M_n) \neq (1_{\mathbb{G}}, \ldots, 1_{\mathbb{G}})$ and (z, r) satisfy $1_{\mathbb{G}_T} = e(z, \hat{g}_z) \cdot e(r, \hat{g}_r) \cdot \prod_{i=1}^n e(M_i, \hat{g}_i)$.

In [47], (a variant of) this scheme was used to construct constant-size QA-NIZK arguments [40] showing that a vector $v \in \mathbb{G}^n$ belongs to a linear subspace of rank t spanned by a matrix $\rho \in \mathbb{G}^{t \times n}$. Under the SXDH assumption, each argument is comprised of two elements of $\mathbb{G}$, independently of t or n.

3   An F-Unforgeable Signature

As a technical tool, our constructions rely on a signature scheme which we prove F-unforgeable under the SXDH assumption. As defined by Belenkiy *et al.* [11], F-unforgeability refers to the inability of the adversary to output a valid signature for a non-trivial message M without outputting the message itself. Instead, the adversary is only required to output $F(M)$, for an injective but not necessarily efficiently invertible function F.

The scheme extends ideas used in signature schemes suggested in [22,40], where each signature is a CCA2-secure encryption —using the message to be signed as a label—of the private key accompanied with a QA-NIZK proof that the encrypted value is the private key. In their most efficient variant, Jutla and Roy observed [40, Sect. 5] that it suffices to encrypt private keys g^ω with a projective hash value $(v^M \cdot w)^r$ [29] so as to obtain signatures of the form $(\sigma_1, \sigma_3, \sigma_3) = (g^\omega \cdot (v^M \cdot w)^r, g^r, h^r)$, which is reminiscent of selectively secure Boneh-Boyen signatures [16].

As in [32,51], the security proof proceeds with a sequence of games to gradually reach a game where the signing oracle never uses the private key, in which case it becomes easier to prove security. In the final game, signatures always encrypt a random value while QA-NIZK proofs are simulated. When transitioning from one hybrid game to the next one, the crucial step is to argue that, even if the signing oracle produces fewer and fewer signatures using the private key, the adversary's forgery will still encrypt the private key. This is achieved via an information theoretic argument borrowed from hash proof systems [28,29].

In order to obtain an F-unforgeable signature which is verifiable given only $F(M)$, our key observation is that QA-NIZK proofs make it possible to verify signatures even if M appears only implicitly in a tuple $(g^{s \cdot M}, g^s, h^{s \cdot M}, h^s) \in \mathbb{G}^4$.

Keygen(cp): Given common public parameters $\mathsf{cp} = (\mathbb{G}, \hat{\mathbb{G}}, \mathbb{G}_T, p)$ consisting of asymmetric bilinear groups of prime order $p > 2^\lambda$, do the following.

1. Choose $\omega, a \xleftarrow{R} \mathbb{Z}_p$, $g, v, w \xleftarrow{R} \mathbb{G}$, $\hat{g} \xleftarrow{R} \hat{\mathbb{G}}$ and set $h = g^a$, $\Omega = h^\omega$.
2. Define a matrix $\mathbf{M} = (M_{j,i})_{j,i}$ given by

$$\mathbf{M} = \left(\begin{array}{c|c|c|c|c|c} g & 1 & 1 & 1 & 1 & h \\ \hline v & g & 1 & h & 1 & 1 \\ \hline w & 1 & g & 1 & h & 1 \end{array} \right) \in \mathbb{G}^{3 \times 6}. \tag{1}$$

3. Generate a key pair $(\mathsf{sk}_{hsps}, \mathsf{pk}_{hsps})$ for the one-time linearly homomorphic signature of Sect. 2.2 in order to sign vectors of dimension $n = 6$. Let $\mathsf{sk}_{hsps} = \{(\chi_i, \gamma_i)\}_{i=1}^6$ be the private key, of which the corresponding public key is $\mathsf{pk}_{hsps} = (\hat{g}_z, \hat{g}_r, \{\hat{g}_i\}_{i=1}^6)$.
4. Using $\mathsf{sk}_{hsps} = \{\chi_i, \gamma_i\}_{i=1}^6$, generate one-time homomorphic signatures $\{(z_j, r_j)\}_{j=1}^3$ on the rows $\mathbf{M}_j = (M_{j,1}, \ldots, M_{j,6}) \in \mathbb{G}^6$ of $\mathbf{M}$. These are obtained as $(z_j, r_j) = \left(\prod_{i=1}^6 M_{j,i}^{-\chi_i}, \prod_{i=1}^6 M_{j,i}^{-\gamma_i} \right)$, for each $j \in \{1, 2, 3\}$ and, as part of the common reference string for the QA-NIZK proof system of [47], they will be included in the public key.

The private key is $\mathsf{sk} := \omega$ and the public key is defined as

$$\mathsf{pk} = \left((\mathbb{G}, \hat{\mathbb{G}}, \mathbb{G}_T), \ p, \ g, \ h, \ \hat{g}, \ (v, w), \ \Omega = h^\omega, \ \mathsf{pk}_{hsps}, \ \{(z_j, r_j)\}_{j=1}^3 \right).$$

Sign(sk, M): given sk $= \omega$ and a message $M \in \mathbb{Z}_p$, choose $s \xleftarrow{R} \mathbb{Z}_p$ to compute

$$\sigma_1 = g^\omega \cdot (v^M \cdot w)^s, \qquad\qquad \sigma_2 = g^{s \cdot M}, \qquad\qquad \sigma_3 = g^s$$
$$\sigma_4 = h^{s \cdot M} \qquad\qquad\qquad \sigma_5 = h^s$$

Then, generate a QA-NIZK proof that the vector $(\sigma_1, \sigma_2, \sigma_3, \sigma_4, \sigma_5, \Omega) \in \mathbb{G}^6$ is in the row space of **M**. This QA-NIZK proof $(z, r) \in \mathbb{G}^2$ is obtained as

$$z = z_1^\omega \cdot (z_2^M \cdot z_3)^s, \qquad\qquad r = r_1^\omega \cdot (r_2^M \cdot r_3)^s. \qquad (2)$$

Return the signature $\sigma = (\sigma_1, \sigma_2, \sigma_3, \sigma_4, \sigma_5, z, r)$.

Verify(pk, σ, M): parse σ as above and return 1 if and only if it holds that

$$e(z, \hat{g}_z) \cdot e(r, \hat{g}_r) = e(\sigma_1, \hat{g}_1)^{-1} \cdot e(\sigma_3, \hat{g}_3 \cdot \hat{g}_2{}^M)^{-1} \cdot e(\sigma_5, \hat{g}_5 \cdot \hat{g}_4{}^M)^{-1}$$
$$\cdot e(\Omega, \hat{g}_6)^{-1}$$

and $(\sigma_2, \sigma_4) = (\sigma_3^M, \sigma_5^M)$.

Note that a signature can be verified given only $F(M) = \hat{g}^M$ by testing the equalities $e(\sigma_2, \hat{g}) = e(\sigma_3, F(M))$, $e(\sigma_4, \hat{g}) = e(\sigma_5, F(M))$ and

$$e(z, \hat{g}_z) \cdot e(r, \hat{g}_r)$$
$$= e(\sigma_1, \hat{g}_1)^{-1} \cdot e(\sigma_2, \hat{g}_2)^{-1} \cdot e(\sigma_3, \hat{g}_3)^{-1} \cdot e(\sigma_4, \hat{g}_4)^{-1} \cdot e(\sigma_5, \hat{g}_5)^{-1} \cdot e(\Omega, \hat{g}_6)^{-1}.$$

In order to keep the description as simple as possible, the above description uses the QA-NIZK argument system of [47], which is based on linearly homomorphic signatures. However, the security proof goes through if we use the more efficient SXDH-based QA-NIZK argument of Jutla and Roy [41], as explained in the full version of the paper. The pair (z, r) can thus be replaced by a single $\mathbb{G}$-element.

Under the SXDH assumption, the scheme can be proved to be F-unforgeable for the injective function $F(M) = \hat{g}^M$. The proof of this result is implied by the security result of Sect. 4 where we describe a generalization of the scheme that will be used to build a group signature in the BMW model.

4   A Two-Level SXDH-based Hierarchical Signature

This section extends our F-unforgeable signature into a 2-level hierarchical signature with partially hidden messages. In a 2-level hierarchical signature [43] (a.k.a. identity-based signature), a signature on a message ID (called "identity") can be used as a delegated key for signing messages of the form (ID, M) for any M. In order to construct group signatures, Boyen and Waters [21] used hierarchical signatures that can be verified even when identities (i.e., first-level messages) are not explicitly given to the verifier, but only appear implicitly in the exponent. The syntax and security definition are given in [20, 21].

In their most efficient construction [21], Boyen and Waters used a nonstandard q-type assumption. This section gives a very efficient solution based on the standard SXDH assumption. It is obtained from our signature of Sect. 3 by having a signature $(g^\omega \cdot (v^{\mathsf{ID}} \cdot w)^s, g^s, h^s)$ on a given identity ID serve as a private key for this identity modulo the introduction of a delegation component t^s akin to those of the Boneh-Boyen-Goh hierarchical IBE [17]. For the security proof to go through, we need to make sure that pairs $(g^{s \cdot M}, g^s)$, $(h^{s \cdot M}, h^s)$ hide the same message M, which is not immediately verifiable in the SXDH setting. To enforce this condition, we thus include $\hat{g}^M$ in each signature.

Setup(cp): Given public parameters $\mathsf{cp} = (\mathbb{G}, \hat{\mathbb{G}}, \mathbb{G}_T, p)$, do the following.

1. Choose $\omega, a \xleftarrow{R} \mathbb{Z}_p$, $g, t, v, w \xleftarrow{R} \mathbb{G}$, $\hat{g} \xleftarrow{R} \hat{\mathbb{G}}$ and set $h = g^a$, $\Omega = h^\omega$.
2. Define a matrix $\mathbf{M} = (M_{j,i})_{j,i}$ given by

$$\mathbf{M} = \left(\begin{array}{c|c|c|c|c|c|c|c} g & 1 & 1 & 1 & 1 & 1 & 1 & h \\ \hline v & g & 1 & h & 1 & 1 & 1 & 1 \\ \hline w & 1 & g & 1 & h & 1 & 1 & 1 \\ \hline t & 1 & 1 & 1 & 1 & g & h & 1 \end{array} \right) \in \mathbb{G}^{4 \times 8}. \quad (3)$$

3. Generate a key pair $(\mathsf{sk}_{hsps}, \mathsf{pk}_{hsps})$ for the one-time linearly homomorphic signature of Sect. 2.2 in order to sign vectors of dimension $n = 8$. Let $\mathsf{sk}_{hsps} = \{(\chi_i, \gamma_i)\}_{i=1}^8$ be the private key, of which the corresponding public key is $\mathsf{pk}_{hsps} = (\hat{g}_z, \hat{g}_r, \{\hat{g}_i\}_{i=1}^8)$.
4. Using $\mathsf{sk}_{hsps} = \{\chi_i, \gamma_i\}_{i=1}^8$, generate one-time homomorphic signatures $\{(z_j, r_j)\}_{j=1}^4$ on the rows $\mathbf{M}_j = (M_{j,1}, \ldots, M_{j,8}) \in \mathbb{G}^8$ of $\mathbf{M}$. These are obtained as $(z_j, r_j) = \left(\prod_{i=1}^8 M_{j,i}^{-\chi_i}, \prod_{i=1}^8 M_{j,i}^{-\gamma_i} \right)$ each for $j \in \{1, \ldots, 4\}$ and, as part of the common reference string for the QA-NIZK proof system of [47], they will be included in the public key.

The master secret key is $\mathsf{msk} := \omega$ and the master public key is defined as

$$\mathsf{mpk} = \left((\mathbb{G}, \hat{\mathbb{G}}, \mathbb{G}_T), \ p, \ g, \ h, \ \hat{g}, \ (t, v, w), \ \Omega = h^\omega, \ \mathsf{pk}_{hsps}, \ \{(z_j, r_j)\}_{j=1}^4 \right).$$

Extract($\mathsf{msk}, \mathsf{ID}$): given $\mathsf{msk} = \omega$ and $\mathsf{ID} \in \mathbb{Z}_p$, choose $s \xleftarrow{R} \mathbb{Z}_p$ to compute

$$K_1 = g^\omega \cdot (v^{\mathsf{ID}} \cdot w)^s, \qquad K_2 = g^{s \cdot \mathsf{ID}}, \qquad K_3 = g^s$$
$$K_4 = h^{s \cdot \mathsf{ID}} \qquad\qquad\qquad K_5 = h^s \qquad\qquad K_6 = t^s$$

as well as $\hat{K}_7 = \hat{g}^{\mathsf{ID}}$. Looking ahead, K_6 will serve as a delegation component in the generation of level 2 signatures. Then, generate a QA-NIZK proof that the vector $(K_1, K_2, K_3, K_4, K_5, 1, 1, \Omega) \in \mathbb{G}^8$ is in the row space of the first 3 rows of $\mathbf{M}$. This QA-NIZK proof $(z, r) \in \mathbb{G}^2$ is obtained as

$$z = z_1^\omega \cdot (z_2^{\mathsf{ID}} \cdot z_3)^s, \qquad\qquad r = r_1^\omega \cdot (r_2^{\mathsf{ID}} \cdot r_3)^s. \quad (4)$$

Then, generate a QA-NIZK proof (z_d, r_d) that the delegation component K_6 is well-formed. This proof consists of $(z_d, r_d) = (z_4^s, r_4^s)$. The private key is

$$K_{\mathsf{ID}} = (K_1, K_2, K_3, K_4, K_5, K_6, \hat{K}_7, z, r, z_d, r_d). \quad (5)$$

Sign$(\mathsf{mpk}, K_{\mathsf{ID}}, M)$: to sign $M \in \mathbb{Z}_p$, parse K_{ID} as in (5) and do the following.

1. Choose $s' \xleftarrow{R} \mathbb{Z}_p$ and compute

$$\sigma_1 = K_1 \cdot K_6^M \cdot (v^{\mathsf{ID}} \cdot t^M \cdot w)^{s'} = g^\omega \cdot (v^{\mathsf{ID}} \cdot t^M \cdot w)^{\tilde{s}},$$

where $\tilde{s} = s + s'$, as well as

$$\sigma_2 = K_2 \cdot g^{s' \cdot \mathsf{ID}} = g^{\tilde{s} \cdot \mathsf{ID}}, \qquad \sigma_3 = K_3 \cdot g^{s'} = g^{\tilde{s}}, \qquad \hat{\sigma}_6 = \hat{K}_7 = \hat{g}^{\mathsf{ID}}$$
$$\sigma_4 = K_4 \cdot h^{s' \cdot \mathsf{ID}} = h^{\tilde{s} \cdot \mathsf{ID}}, \qquad \sigma_5 = K_5 \cdot h^{s'} = h^{\tilde{s}}.$$

2. Using (z, r) and (z_d, r_d), generate a QA-NIZK proof $(\tilde{z}, \tilde{r}) \in \mathbb{G}^2$ that the vector $(\sigma_1, \sigma_2, \sigma_3, \sigma_4, \sigma_5, \sigma_3^M, \sigma_5^M, \Omega) \in \mathbb{G}^8$ is in the row space of $\mathbf{M}$. Namely, compute $\tilde{z} = z \cdot z_d^M \cdot (z_2^{\mathsf{ID}} \cdot z_4^M \cdot z_3)^{s'}$ and $\tilde{r} = r \cdot r_d^M \cdot (r_2^{\mathsf{ID}} \cdot r_4^M \cdot r_3)^{s'}$.

Return the signature $\sigma = (\sigma_1, \sigma_2, \sigma_3, \sigma_4, \sigma_5, \tilde{z}, \tilde{r}, \hat{\sigma}_6) \in \mathbb{G}^7 \times \hat{\mathbb{G}}$.

Verify$(\mathsf{mpk}, \sigma, M)$: parse σ as above and return 1 if and only if it holds that

$$e(\tilde{z}, \hat{g}_z) \cdot e(\tilde{r}, \hat{g}_r) = e(\sigma_1, \hat{g}_1)^{-1} \cdot e(\sigma_2, \hat{g}_2)^{-1} \cdot e(\sigma_3, \hat{g}_3 \cdot \hat{g}_6^M)^{-1}$$
$$\cdot e(\sigma_4, \hat{g}_4)^{-1} \cdot e(\sigma_5, \hat{g}_5 \cdot \hat{g}_7^M)^{-1} \cdot e(\Omega, \hat{g}_8)^{-1}$$

as well as $e(\sigma_2, \hat{g}) = e(\sigma_3, \hat{\sigma}_6)$ and $e(\sigma_4, \hat{g}) = e(\sigma_5, \hat{\sigma}_6)$.

As in Sect. 3, the technique of [41] can be used to shorten the signature by one element of $\mathbb{G}$ as it allows replacing $(\tilde{z}, \tilde{r})$ by one element of $\mathbb{G}$.

We prove that, under the sole SXDH assumption, the scheme is secure in the sense of the natural security definition used by Boyen and Waters [20,21]. In short, this definition requires that the adversary be unable to forge a valid signature for a pair $(\mathsf{ID}^\star, M^\star)$ such that no private key query was made for $\mathsf{ID}^\star$ and no signing query was made for the pair $(\mathsf{ID}^\star, M^\star)$.

Theorem 1. *The above hierarchical signature is secure under chosen-message attacks if the SXDH assumption holds in $(\mathbb{G}, \hat{\mathbb{G}}, \mathbb{G}_T)$.* (The proof is available the full version of the paper).

A simple reduction shows that the signature scheme of Sect. 3 is F-unforgeable so long as the above scheme is a secure 2-level hierarchical signature.

Theorem 2. *The signature scheme of Sect. 3 is F-unforgeable under chosen-message attacks for the function $F(M) = \hat{g}^M$ if the SXDH assumption holds in $(\mathbb{G}, \hat{\mathbb{G}}, \mathbb{G}_T)$.* (The proof is available in the full version of the paper).

5 A Structure-Preserving Signature from the SXDH and XDLIN$_2$ Assumptions

Our F-unforgeable signature of Sect. 3 can be combined with the tagged one-time signature of Abe *et al.* [2] (or, more precisely, an adaption of [2] to asymmetric pairings) so as to obtain a new structure-preserving signature based on the SXDH and XDLIN$_2$ assumptions. Like [1], we obtain an SPS scheme based on simple assumptions with only 11 group elements per signature. However, only one of them has to be in $\hat{\mathbb{G}}$, instead of 4 in [1]. Considering that $\hat{\mathbb{G}}$ elements are at least twice as long to represent as those of $\mathbb{G}$, we thus shorten signatures by the equivalent of 3 elements of $\mathbb{G}$ (or 20%).

Our construction can be seen as an optimized instantiation of a general construction [1] that combines a tagged one-time signature and an SPS scheme which is only secure against extended random-message (XRMA) attacks. A tagged one-time signature (TOTS) is a signature scheme where each signature contains a single-use tag: namely, only one signature is generated w.r.t. each tag. The generic construction of [1] proceeds by certifying the tag of the TOTS scheme using an XRMA-secure SPS scheme. Specifically, our F-unforgeable signature assumes the role of the XRMA-secure signature and its shorter message space allows us to make the most of the optimal tag size of [2]. In [1], the proofs of XMRA security rely on the property that, when the reduction signs random groups elements of its choice, it is allowed to know their discrete logarithms. However, this property is only used in the security proof and not in the scheme itself. Here, we also use the discrete logarithm of the tag in the SPS construction itself, which allows our F-unforgeable signature to supersede the XRMA-secure signature. By exploiting the smaller message space of our F-unforgeable signature, we can leverage the optimal tag size of [2]. Unlike the SPS of [2], we do not need to expand the tag from one to three group elements before certifying it.

Keygen(cp, n): given the length n of messages to be signed and common parameters cp specifying the description of bilinear groups $(\mathbb{G}, \hat{\mathbb{G}}, \mathbb{G}_T)$ of prime order $p > 2^\lambda$, do the following.

a. Generate a key pair $(\mathsf{sk}_{fsig}, \mathsf{pk}_{fsig}) \leftarrow \mathsf{Setup}(\mathsf{cp})$ for the F-unforgeable signature of Sect. 3. Namely,

1. Choose $\omega, a \xleftarrow{R} \mathbb{Z}_p$, $g \xleftarrow{R} \mathbb{G}$, $\hat{g} \xleftarrow{R} \hat{\mathbb{G}}$ and set $h = g^a$, $\Omega = h^\omega$. Then, choose $v, w \xleftarrow{R} \mathbb{G}$.

2. Define a matrix $\mathbf{M} = (M_{j,i})_{j,i}$ given by

$$
\mathbf{M} = \left(
\begin{array}{c|c|c|c|c|c}
g & 1 & 1 & 1 & 1 & h \\
\hline
v & g & 1 & h & 1 & 1 \\
\hline
w & 1 & g & 1 & h & 1
\end{array}
\right) \in \mathbb{G}^{3 \times 6}. \tag{6}
$$

3. Generate a key pair $(\mathsf{sk}_{hsps}, \mathsf{pk}_{hsps})$ for the linearly homomorphic signature of Sect. 2.2 in order to sign vectors of dimension $n = 6$. Let $\mathsf{sk}_{hsps} = \{(\chi_{0,i}, \gamma_{0,i})\}_{i=1}^{6}$ be the private key, of which the corresponding public key is $\mathsf{pk}_{hsps} = (\hat{g}_z, \hat{g}_r, \{\hat{g}_i\}_{i=1}^{6})$.

4. Using $\mathsf{sk}_{hsps} = \{\chi_{0,i}, \gamma_{0,i}\}_{i=1}^{6}$, generate one-time homomorphic signatures $\{(z_j, r_j)\}_{j=1}^{3}$ on the rows $\boldsymbol{M}_j = (M_{j,1}, \ldots, M_{j,6}) \in \mathbb{G}^6$ of $\mathbf{M}$. These are obtained as $(z_j, r_j) = \left(\prod_{i=1}^{6} M_{j,i}^{-\chi_{0,i}}, \prod_{i=1}^{6} M_{j,i}^{-\gamma_{0,i}} \right)$, for $j \in \{1, 2, 3\}$ and, as part of the common reference string for the QA-NIZK proofs of [47], they will be included in the public key.

b. Generate a key pair $(\mathsf{pk}_{pots}, \mathsf{sk}_{pots})$ for the partial one-time SPS of Abe et al. [1]. Namely, choose $w_z, w_r, \mu_z, \mu_u, w_t \xleftarrow{R} \mathbb{Z}_p$ and set

$$\hat{G}_z = \hat{g}^{w_z}, \qquad \hat{G}_r = \hat{g}^{w_r}, \qquad \hat{G}_t = \hat{g}^{w_t}, \qquad \hat{H}_z = \hat{g}^{\mu_z}, \qquad \hat{H}_u = \hat{g}^{\mu_u}$$
$$G_z = g^{w_z}, \qquad G_r = g^{w_r}, \qquad G_t = g^{w_t}, \qquad H_z = g^{\mu_z}, \qquad H_u = g^{\mu_u}$$

Then, for $i = 1$ to n, choose $\chi_i, \gamma_i, \delta_i \xleftarrow{R} \mathbb{Z}_p$ and compute $\hat{G}_i = \hat{G}_z{}^{\chi_i} \cdot \hat{G}_r{}^{\gamma_i}$ and $\hat{H}_i = \hat{G}_z{}^{\chi_i} \cdot \hat{G}_r{}^{\delta_i}$. Define $\mathsf{sk}_{pots} := \{(\chi_i, \gamma_i, \delta_i)\}_{i=1}^{n}$ and

$$\mathsf{pk}_{pots} := \big(G_z, G_r, G_t, H_z, H_u, \hat{G}_z, \hat{G}_r, \hat{G}_t, \hat{H}_z, \hat{H}_u, \{\hat{G}_i, \hat{H}_i\}_{i=1}^{n}\big).$$

The private key is $SK = (\omega, w_r, \mu_u, \mathsf{sk}_{pots})$ and the public key consists of

$$PK = \Big(\; g, \; h, \; \hat{g}, \; (v, w), \; \Omega = h^{\omega}, \; \mathsf{pk}_{pots}, \; \mathsf{pk}_{hsps}, \; \{(z_j, r_j)\}_{j=1}^{3} \Big).$$

Sign(SK, M): given $SK = (\omega, w_r, \mu_u, \mathsf{sk}_{pots})$ and $M = (M_1, \ldots, M_n) \in \mathbb{G}^n$,
1. Choose $s, \tau \xleftarrow{R} \mathbb{Z}_p$ to compute

$$\sigma_1 = g^{\omega} \cdot (v^{\tau} \cdot w)^s, \qquad\qquad \sigma_2 = g^{s \cdot \tau}, \qquad\qquad \sigma_3 = g^s,$$
$$\sigma_4 = h^{s \cdot \tau} \qquad\qquad\qquad \sigma_5 = h^s, \qquad\qquad \tilde{\sigma}_6 = \hat{g}^{\tau}.$$

Then, generate a QA-NIZK proof that the vector $(\sigma_1, \sigma_2, \sigma_3, \sigma_4, \sigma_5, \Omega)$ is in the row space of $\mathbf{M}$. This proof $(z, r) \in \mathbb{G}^2$ is computed as

$$z = z_1^{\omega} \cdot (z_2^{\tau} \cdot z_3)^s, \qquad\qquad r = r_1^{\omega} \cdot (r_2^{\tau} \cdot r_3)^s. \qquad (7)$$

2. Choose $\zeta \xleftarrow{R} \mathbb{Z}_p$ and compute $Z = g^{\zeta} \cdot \prod_{i=1}^{n} M_i^{-\chi_i}$ as well as

$$R = (G_t^{\tau} \cdot G_z^{-\zeta})^{1/w_r} \cdot \prod_{i=1}^{n} M_i^{-\gamma_i}, \qquad\qquad U = (H_z^{-\zeta})^{1/\mu_u} \cdot \prod_{i=1}^{n} M_i^{-\delta_i}.$$

Return $\sigma = \big(\sigma_1, \sigma_2, \sigma_3, \sigma_4, \sigma_5, \hat{\sigma}_6, z, r, Z, R, U\big) \in \mathbb{G}^5 \times \hat{\mathbb{G}} \times \mathbb{G}^5$.

Verify(PK, σ, M): given $M = (M_1, \ldots, M_n) \in \mathbb{G}^n$, parse σ as above. Return 1 if and only if $e(\sigma_2, \hat{g}) = e(\sigma_3, \hat{\sigma}_6)$ and $e(\sigma_4, \hat{g}) = e(\sigma_5, \hat{\sigma}_6)$ as well as

$$e(z, \hat{g}_z) \cdot e(r, \hat{g}_r) = \prod_{i=1}^{5} e(\sigma_i, \hat{g}_i)^{-1} \cdot e(\Omega, \hat{g}_6)^{-1}$$

$$e(G_t, \hat{\sigma}_6) = e(Z, \hat{G}_z) \cdot e(R, \hat{G}_r) \cdot \prod_{i=1}^{n} e(M_i, \hat{G}_i) \qquad (8)$$

$$1_{\mathbb{G}_T} = e(Z, \hat{H}_z) \cdot e(U, \hat{H}_u) \cdot \prod_{i=1}^{n} e(M_i, \hat{H}_i).$$

Each signature requires 10 elements of $\mathbb{G}$ and one element of $\hat{\mathbb{G}}$. Using the optimized F-unforgeable signature based on the Jutla-Roy QA-NIZK proof [41], we can also save one more element of $\mathbb{G}$ and obtain signatures in $\mathbb{G}^9 \times \hat{\mathbb{G}}$, which shortens the signatures of Abe *et al.* [1] by 26 %. In the full version of the paper, we give more detailed comparisons among all SPS based on non-interactive assumptions.

In the application to group signatures, it is desirable to minimize the number of signature components that need to appear in committed form. To this end, signatures must be randomizable in such a way that (σ_3, σ_5) can appear in the clear modulo a re-randomization of $s \in \mathbb{Z}_p$. To enable this randomization, it is necessary to augment signatures (similarly to [6]) with a randomization token $(g^\tau, h^\tau, v^\tau, z_2^\tau, r_2^\tau)$. We will prove that the scheme remains unforgeable even when the signing oracle also outputs these randomization tokens at each invocation.[2] We call this notion *extended existential unforgeability* (or EUF-CMA* for short).

When the re-randomization tokens are used, proving the knowledge of a signature on a committed message $M \in \mathbb{G}^n$ requires $2n + 24$ elements of $\mathbb{G}$ and 12 elements of $\hat{\mathbb{G}}$. In comparison, the best previous solution of Abe *et al.* costs $2n + 26$ elements of $\mathbb{G}$ and 18 elements of $\hat{\mathbb{G}}$.

Theorem 3. *The scheme provides EUF-CMA* security if the SXDH and XDLIN$_2$ assumptions hold in* $(\mathbb{G}, \hat{\mathbb{G}}, \mathbb{G}_T)$. (The proof is given in the full version of the paper).

In short, the proof of Theorem 3 considers two kinds of forgeries. In Type I forgeries, the adversary's forgery contains an element $\hat{\sigma_6}^\star$ that did not appear in any signature obtained by the forger during the game. In contrast, Type II forgeries are those for which $\hat{\sigma_6}^\star$ is recycled from a response of the signing oracle. It is easy to see that a Type I forger allows breaking the security of the F-unforgeable signature. As for Type II forgeries, they are shown to contradict the XDLIN$_2$ assumption via a careful adaptation of the proof given by Abe *et al.* for their TOTS scheme [2]. While the latter was originally presented in symmetric pairings, it goes through in Type 3 pairings modulo natural changes that consist in making sure that most handled elements of $\hat{\mathbb{G}}$ have a counterpart in $\mathbb{G}$. One difficulty is that, at each query, the reduction must properly simulate the randomization tokens $(v^\tau, g^\tau, h^\tau, z_2^\tau, r_2^\tau)$ as well as an instance of the F-unforgeable signature without knowing the discrete logarithm $\log_{\hat{g}}(\hat{\sigma_6}) = \hat{g}^\tau$ or that of its shadow $\log_g(\sigma_6) = g^\tau$ in $\mathbb{G}$. Fortunately, this issue can be addressed by letting the reduction know $\log_g(v)$ and $\log_g(w)$.

In an independent work [45], Kiltz, Pan and Wee obtained even shorter signatures, which live in $\mathbb{G}^6 \times \hat{\mathbb{G}}$ under the SXDH assumption. On the other hand, their security reduction is looser than ours as the gap between the adversary's advantage and the reduction's probability to break the underlying assumption is quadratic (instead of linear in our case) in the number of signing queries.

---

[2] Note, however, that the adversary is not required to produce any randomization token as part of its forgery.

6 A Publicly Verifiable Tag-Based Encryption Scheme

As a tool for constructing a CCA2-anonymous group signature, we describe a new tag-based encryption scheme [44,48] which is inspired by the lossy encryption scheme [13] of [39]. In our group signature, we will exploit the fact that the DDH-based lossy encryption scheme of Bellare *et al.* [13] can also be seen as a Groth-Sahai commitment.

Keygen(cp): Given public parameters $\mathsf{cp} = (\mathbb{G}, \hat{\mathbb{G}}, \mathbb{G}_T, p)$ specifying asymmetric bilinear groups of prime order $p > 2^\lambda$, conduct the following steps.

1. Choose $g, h \xleftarrow{R} \hat{\mathbb{G}}$. Choose $x, \alpha, \beta \xleftarrow{R} \mathbb{Z}_p$ and set $X_1 = g^x$, $X_2 = h^x$, $S = g^\alpha$, $T = g^\beta$, $W = h^\alpha$ and $V = h^\beta$.
2. Generate a key pair $(\mathsf{pk}'_{hsig}, \mathsf{sk}'_{hsig})$ for the homomorphic signature of Sect. 2.2 in order to sign vectors in $\mathbb{G}^3$. Let $\mathsf{pk}'_{hsig} = (\hat{G}_z, \hat{G}_r, \{\hat{G}_i\}_{i=1}^3)$ be the public key and let $\mathsf{sk}'_{hsig} = \{(\varphi_i, \vartheta_i)\}_{i=1}^3$ be the private key.
3. Use sk'_{hsig} to generate linearly homomorphic signatures $\{(Z_i, R_i)\}_{i=1}^4$ on the rows of the matrix

$$
\mathbf{L} = \begin{pmatrix} g & 1 & T \\ h & 1 & V \\ 1 & g & S \\ 1 & h & W \end{pmatrix} \in \mathbb{G}^{4 \times 3}
$$

which form a subspace of rank 2. The key pair consists of $\mathsf{sk} = (x, \alpha, \beta)$ and $\mathsf{pk} := \left(g, h, X_1, X_2, S, W, T, V, \mathsf{pk}'_{hsig}, \{(Z_i, R_i)\}_{i=1}^4 \right)$.

Encrypt(pk, M, τ): To encrypt $M \in \mathbb{G}$ under the tag τ, choose $\theta_1, \theta_2 \xleftarrow{R} \mathbb{Z}_p$ and compute the ciphertext $C = (C_0, C_1, C_2, Z, R)$ as

$$
C = \left(M \cdot X_1^{\theta_1} \cdot X_2^{\theta_2}, \; g^{\theta_1} \cdot h^{\theta_2}, \; (S^\tau \cdot T)^{\theta_1} \cdot (W^\tau \cdot V)^{\theta_2}, \right.
$$
$$
\left. (Z_3^\tau \cdot Z_1)^{\theta_1} \cdot (Z_4^\tau \cdot Z_2)^{\theta_2}, (R_3^\tau \cdot R_1)^{\theta_1} \cdot (R_4^\tau \cdot R_2)^{\theta_2} \right).
$$

Here, (Z, R) serves as a proof that the vector (C_1, C_1^τ, C_2) is in the row space of $\mathbf{L}$ and satisfies

$$
e(Z, \hat{G}_z) \cdot e(R, \hat{G}_r) = e(C_1, \hat{G}_1^\tau \cdot \hat{G}_2)^{-1} \cdot e(C_2, \hat{G}_2)^{-1} \tag{9}
$$

Decrypt(sk, C, τ): Parse C as above. Return $\bot$ if (Z, R) does not satisfy (9). Otherwise, return $M = C_0 / C_1^x$.

We observe that (C_0, C_1) form a Groth-Sahai commitment based on the DDH assumption in $\mathbb{G}$. If $\log_g(X_1) = \log_h(X_2)$, the commitment is extractable. Otherwise, it is perfectly hiding. We will use this CCA2-secure scheme as a commitment that is extractable on all tags, except one τ^* where it behaves as a perfectly hiding commitment. The above system achieves this while only expanding the original Groth-Sahai commitment (C_0, C_1) by 3 elements of $\mathbb{G}$.

This scheme will save our group signatures from having to contain (beyond (C_0, C_1)) an additional CCA2-secure encryption and a NIZK proof that the plaintext coincides with the content of a Groth-Sahai commitment. The above technique allows saving the equivalent of 16 elements of $\mathbb{G}$. We thus believe this cryptosystem to be of interest in its own right since it can be used in a similar way to shorten other group signatures (e.g., [36]) based on Groth-Sahai proofs.

In the full paper, the scheme is proved secure in the sense of [44].

Theorem 4. *The above scheme is selective-tag weakly IND-CCA2-secure if the SXDH assumption holds.* (The proof is given in the full paper).

7   Short Group Signatures in the BMW Model

The TBE scheme of Sect. 6 allows us to achieve anonymity in the CCA2 sense by encrypting an encoding of the group member's identifier. In order to minimize the signature length, we let the TBE ciphertext live in $\mathbb{G}$ instead of $\hat{\mathbb{G}}$. To open signatures in constant time, however, the opening algorithm uses the extraction trapdoor of a Groth-Sahai commitment in $\hat{\mathbb{G}}^2$ rather than the private key sk_{tbe} of the TBE system. The latter key is only used in the proof of anonymity where the reduction uses a somewhat inefficient opening algorithm of complexity $O(N)$.

Keygen(λ, N): given a security parameter $\lambda \in \mathbb{N}$ and the number of users N, choose asymmetric bilinear groups $\mathsf{cp} = (\mathbb{G}, \hat{\mathbb{G}}, \mathbb{G}_T, p)$ of order $p > 2^\lambda$.

1. Generate a key pair $(\mathsf{msk}, \mathsf{mpk})$ for the two-level hierarchical signature of Sect. 4. Let

$$\mathsf{mpk} := \left((\mathbb{G}, \hat{\mathbb{G}}, \mathbb{G}_T), \ p, \ g, \ h, \ \hat{g}, \ (t, v, w), \ \Omega = h^\omega, \ \mathsf{pk}_{hsps}, \ \{(z_j, r_j)\}_{j=1}^4 \right)$$

be the master public key and $\mathsf{msk} := \omega \in \mathbb{Z}_p$ be the master secret key.

2. Generate a key pair $(\mathsf{sk}_{tbe}, \mathsf{pk}_{tbe})$ for the tag-based encryption scheme of Sect. 6. Let $\mathsf{pk}_{tbe} = \left(g, h, X_1, X_2, S, W, T, V, \mathsf{pk}'_{hsig}, \{(Z_i, R_i)\}_{i=1}^4 \right)$ be the public key and $\mathsf{sk}_{tbe} = (x, \alpha, \beta)$ be the underlying private key. For simplicity, the element g can be recycled from mpk.

3. Choose a vector $\hat{\boldsymbol{u}_1} = (\hat{u}_{11}, \hat{u}_{12}) \xleftarrow{R} \hat{\mathbb{G}}^2$ and set $\hat{\boldsymbol{u}_2} = \hat{\boldsymbol{u}_1}^\xi$, where $\xi \xleftarrow{R} \mathbb{Z}_p$. Also, define the vectors $\boldsymbol{u}_1 = (g, X_1)$ and $\boldsymbol{u}_2 = (h, X_2)$. These vectors will form Groth-Sahai CRSes $(\boldsymbol{u}_1, \boldsymbol{u}_2)$ and $(\hat{\boldsymbol{u}_1}, \hat{\boldsymbol{u}_2})$ in the perfectly binding setting. Although sk_{tbe} serves as an extraction trapdoor for commitments generated on the CRS $(\boldsymbol{u}_1, \boldsymbol{u}_2)$, the group manager will more efficiently use $\zeta = \log_{\hat{u}_{11}}(\hat{u}_{12})$ to open signatures.

4. Choose a chameleon hash function $\mathsf{CMH} = (\mathsf{CMKg}, \mathsf{CMhash}, \mathsf{CMswitch})$ with a key pair (hk, tk) and randomness space $\mathcal{R}_{hash}$.

5. For each group member i, choose an identifier $\mathsf{ID}_i \xleftarrow{R} \mathbb{Z}_p$ and use msk to compute $K_{\mathsf{ID}_i} = (K_1, K_2, K_3, K_4, K_5, K_6, \hat{K}_7, z, r, z_d, r_d)$, where

$$K_1 = g^\omega \cdot (v^{\mathsf{ID}_i} \cdot w)^{s\cdot}, \qquad K_2 = g^{s\cdot\mathsf{ID}_i}, \qquad\qquad K_3 = g^s$$
$$K_4 = h^{s\cdot\mathsf{ID}_i} \qquad\qquad\quad K_5 = h^s \qquad\qquad\qquad K_6 = t^s$$
$$z = z_1^\omega \cdot (z_2^{\mathsf{ID}_i} \cdot z_3)^s \qquad\quad r = r_1^\omega \cdot (r_2^{\mathsf{ID}_i} \cdot r_3)^s \qquad \hat{K}_7 = \hat{g}^{\mathsf{ID}_i}$$

and $(z_d, r_d) = (z_4^s, r_4^s)$. For each $i \in \{1, \ldots, N\}$, the i-th group member's private key is $\mathsf{gsk}[i] = (\mathsf{ID}_i, K_{\mathsf{ID}_i})$.

The group manager's secret key is $\mathsf{gsk} := \big(\mathsf{msk}, \zeta = \log_{\hat{u}_{11}}(\hat{u}_{12})\big)$ while the group public key consists of

$$\mathsf{gpk} := \Big((\mathbb{G}, \hat{\mathbb{G}}, \mathbb{G}_T), \ \mathsf{mpk}, \ \mathsf{pk}_{tbe}, \ (\boldsymbol{u_1}, \boldsymbol{u_2}), \ (\hat{\boldsymbol{u_1}}, \hat{\boldsymbol{u_2}}), \ \mathsf{CMH}, \ hk\Big).$$

Sign$(\mathsf{gpk}, \mathsf{gsk}[i], M)$: In order to sign a message $M \in \mathbb{Z}_p$ using the i-th group member's private key $\mathsf{gsk}[i] = (\mathsf{ID}_i, K_{\mathsf{ID}_i})$, conduct the following steps.

1. Using $K_{\mathsf{ID}_i} = (K_1, K_2, K_3, K_4, K_5, K_6, \hat{K}_7, z, r, z_d, r_d)$, derive a second-level hierarchical signature. Namely, choose $s' \xleftarrow{R} \mathbb{Z}_p$ and compute

$$\sigma_1 = K_1 \cdot K_6^M \cdot (v^{\mathsf{ID}_i} \cdot t^M \cdot w)^{s'} \qquad \sigma_2 = K_2 \cdot g^{s' \cdot \mathsf{ID}_i} = g^{\tilde{s} \cdot \mathsf{ID}_i}$$
$$= g^\omega \cdot (v^{\mathsf{ID}_i} \cdot t^M \cdot w)^{\tilde{s}} \qquad \sigma_3 = K_3 \cdot g^{s'} = g^{\tilde{s}}$$
$$\sigma_4 = K_4 \cdot h^{s' \cdot \mathsf{ID}_i} = h^{\tilde{s} \cdot \mathsf{ID}_i} \qquad \sigma_5 = K_5 \cdot h^{s'} = h^{\tilde{s}},$$

and $\hat{\sigma}_6 = \hat{K}_7$, where $\tilde{s} = s + s'$, as well as

$$\tilde{z} = z \cdot z_d^M \cdot (z_2^{\mathsf{ID}_i} \cdot z_4^M \cdot z_3)^{s'} \qquad \tilde{r} = r \cdot r_d^M \cdot (r_2^{\mathsf{ID}_i} \cdot r_4^M \cdot r_3)^{s'}$$
$$= z_1^\omega \cdot (z_2^{\mathsf{ID}_i} \cdot z_4^M \cdot z_3)^{\tilde{s}} \qquad = r_1^\omega \cdot (r_2^{\mathsf{ID}_i} \cdot r_4^M \cdot r_3)^{\tilde{s}}.$$

2. Choose $\theta_1, \ldots, \theta_{12} \xleftarrow{R} \mathbb{Z}_p$ and compute Groth-Sahai commitments

$$\boldsymbol{C}_{\sigma_1} = (1, \sigma_1) \cdot \boldsymbol{u_1}^{\theta_1} \cdot \boldsymbol{u_2}^{\theta_2}, \qquad \boldsymbol{C}_{\sigma_2} = (1, \sigma_2) \cdot \boldsymbol{u_1}^{\theta_3} \cdot \boldsymbol{u_2}^{\theta_4},$$
$$\boldsymbol{C}_{\sigma_4} = (1, \sigma_4) \cdot \boldsymbol{u_1}^{\theta_5} \cdot \boldsymbol{u_2}^{\theta_6}, \qquad \boldsymbol{C}_{\hat{\sigma}_6} = (1, \hat{\sigma}_6) \cdot \hat{\boldsymbol{u_1}}^{\theta_7} \cdot \hat{\boldsymbol{u_2}}^{\theta_8}.$$
$$\boldsymbol{C}_{\tilde{z}} = (1, \tilde{z}) \cdot \boldsymbol{u_1}^{\theta_9} \cdot \boldsymbol{u_2}^{\theta_{10}}, \qquad \boldsymbol{C}_{\tilde{r}} = (1, \tilde{r}) \cdot \boldsymbol{u_1}^{\theta_{11}} \cdot \boldsymbol{u_2}^{\theta_{12}}.$$

Note that $\boldsymbol{C}_{\sigma_2}$ can be written as $(C_1, C_0) = (g^{\theta_3} \cdot h^{\theta_4}, \sigma_2 \cdot X_1^{\theta_3} \cdot X_2^{\theta_4})$.

3. Generate Groth-Sahai NIWI proofs $\boldsymbol{\pi}_1 \in \hat{\mathbb{G}}^2$, $\boldsymbol{\pi}_2 \in \mathbb{G}^2 \times \hat{\mathbb{G}}^2$ and $\boldsymbol{\pi}_3 \in \mathbb{G}^2 \times \hat{\mathbb{G}}^2$ that committed variables $(\tilde{z}, \tilde{r}, \sigma_1, \sigma_2, \sigma_4, \hat{\sigma}_6)$ satisfy

$$e(\boxed{\tilde{z}}, \hat{g}_z) \cdot e(\boxed{\tilde{r}}, \hat{g}_r) = e(\boxed{\sigma_1}, \hat{g}_1)^{-1} \cdot e(\boxed{\sigma_2}, \hat{g}_2)^{-1} \cdot e(\sigma_3, \hat{g}_3 \cdot \hat{g}_6{}^M)^{-1} \quad (10)$$
$$\cdot e(\boxed{\sigma_4}, \hat{g}_4)^{-1} \cdot e(\sigma_5, \hat{g}_5 \cdot \hat{g}_7{}^M)^{-1} \cdot e(\Omega, \hat{g}_8)^{-1}$$

and

$$e(\boxed{\sigma_2}, \hat{g}) = e(\sigma_3, \boxed{\hat{\sigma}_6}), \qquad e(\boxed{\sigma_4}, \hat{g}) = e(\sigma_5, \boxed{\hat{\sigma}_6}). \quad (11)$$

4. Choose $r_{hash} \xleftarrow{R} \mathcal{R}_{hash}$ and compute a chameleon hash value

$$\tau = \mathsf{CMhash}(hk, (\boldsymbol{C}_{\sigma_1}, \boldsymbol{C}_{\sigma_2}, \sigma_3, \boldsymbol{C}_{\sigma_4}, \sigma_5, \boldsymbol{C}_{\hat{\sigma}_6}, \boldsymbol{C}_{\tilde{z}}, \boldsymbol{C}_{\tilde{r}}, \boldsymbol{\pi}_1, \boldsymbol{\pi}_2, \boldsymbol{\pi}_3), r_{hash}).$$

Then, using τ and $(\theta_3, \theta_4) \in \mathbb{Z}_p^2$, compute $C_2 = (S^\tau \cdot T)^{\theta_3} \cdot (W^\tau \cdot V)^{\theta_4}$. Using pk'_{hsig}, compute $(Z, R) = ((Z_3^\tau \cdot Z_1)^{\theta_3} \cdot (Z_4^\tau \cdot Z_2)^{\theta_4}, (R_3^\tau \cdot R_1)^{\theta_3} \cdot (R_4^\tau \cdot R_2)^{\theta_4})$ as a QA-NIZK argument that (C_1, C_1^τ, C_2) is in the row space of $\mathbf{L}$. This allows turning $\boldsymbol{C}_{\sigma_2} = (C_1, C_0)$ into a TBE ciphertext $\tilde{\boldsymbol{C}}_{\sigma_2} = (C_0, C_1, C_2, Z, R)$ as

$$\tilde{\boldsymbol{C}}_{\sigma_2} = \left(\sigma_2 \cdot X_1^{\theta_3} \cdot X_2^{\theta_4}, \ g^{\theta_3} \cdot h^{\theta_4}, \ (S^\tau \cdot T)^{\theta_3} \cdot (W^\tau \cdot V)^{\theta_4}, \right.$$
$$\left. (Z_3^\tau \cdot Z_1)^{\theta_3} \cdot (Z_4^\tau \cdot Z_2)^{\theta_4}, \ (R_3^\tau \cdot R_1)^{\theta_3} \cdot (R_4^\tau \cdot R_2)^{\theta_4} \right) \in \mathbb{G}^5$$

for the tag τ. Note that $\tilde{\boldsymbol{C}}_{\sigma_2}$ contains the original commitment $\boldsymbol{C}_{\sigma_2}$. Return $\sigma = \left(\boldsymbol{C}_{\sigma_1}, \tilde{\boldsymbol{C}}_{\sigma_2}, \sigma_3, \boldsymbol{C}_{\sigma_4}, \sigma_5, \boldsymbol{C}_{\hat{\sigma}_6}, \boldsymbol{C}_{\tilde{z}}, \boldsymbol{C}_{\tilde{r}}, \boldsymbol{\pi}_1, \boldsymbol{\pi}_2, \boldsymbol{\pi}_3, r_{hash} \right)$.

Verify(gpk, M, σ): Parse σ as above. Return 1 if and only if: (i) The proofs $\boldsymbol{\pi}_1, \boldsymbol{\pi}_2, \boldsymbol{\pi}_3$ verify; (ii) $\tilde{\boldsymbol{C}}_{\sigma_2}$ is a valid TBE ciphertext (i.e., (9) holds) for the tag $\tau = \mathsf{CMhash}(hk, (\boldsymbol{C}_{\sigma_1}, \boldsymbol{C}_{\sigma_2}, \sigma_3, \boldsymbol{C}_{\sigma_4}, \sigma_5, \boldsymbol{C}_{\hat{\sigma}_6}, \boldsymbol{C}_{\tilde{z}}, \boldsymbol{C}_{\tilde{r}}, \boldsymbol{\pi}_1, \boldsymbol{\pi}_2, \boldsymbol{\pi}_3), r_{hash})$.

Open(gpk, gmsk, M, σ): To open σ using gmsk $= (\mathsf{msk}, \zeta)$, parse σ as above and return $\perp$ if it is not a valid signature w.r.t. gpk and M. Otherwise, use $\zeta = \log_{\hat{u}_{11}}(\hat{u}_{12})$ to decrypt the Elgamal ciphertext $\boldsymbol{C}_{\hat{\sigma}_6} \in \hat{\mathbb{G}}^2$. Then, check if the resulting plaintext is $\hat{g}^{\mathsf{ID}}$ for some group member's identifier ID. If so, output ID. Otherwise, return $\perp$.

The signature consists of 19 elements of $\mathbb{G}$, 8 elements of $\hat{\mathbb{G}}$ and one element of $\mathbb{Z}_p$. If each element of $\mathbb{G}$ (resp. $\hat{\mathbb{G}}$) has a 256-bit (resp. 512-bit) representation, the entire signature fits within 9216 bits (or 1.125 kB). By using the technique of Jutla and Roy [41] to shorten the hierarchical signature, it is possible to shorten the latter by one group element (as explained in Sect. 4), which saves two elements of $\mathbb{G}$ in the group signature without modifying the underlying assumption. In this case, the signature length reduces to 8704 bits (or 1.062 kB). Using the technique of Boyen, Mei and Waters [19], it is also possible to eliminate the randomness r_{hash} and replace the chameleon hash function by an ordinary collision-resistant hash function, as explained in the full version of the paper. By doing so, at the expense of a group public key made of $\Theta(\lambda)$ elements of $\hat{\mathbb{G}}$, we can further compress signatures down to 8448 bits (or 1.031 kB).

To give a concrete comparison with earlier constructions, an implementation of the Boyen-Waters group signature [21] in asymmetric prime order groups requires 8 elements of $\mathbb{G}$ and 8 elements of $\hat{\mathbb{G}}$ for a total of 6400 bits per signature. However, besides the SXDH assumption, the resulting scheme relies on the non-standard q-Hidden Strong Diffie-Hellman assumption [21] and only provides anonymity in the CPA sense.

Theorem 5. *The scheme provides full traceability under the SXDH assumption.*

The proof of Theorem 5 relies on the unforgeability of the two-level hierarchical signature of Sect. 4. By preparing extractable Groth-Sahai CRSes $(\boldsymbol{u}_1, \boldsymbol{u}_2)$ and $(\hat{\boldsymbol{u}}_1, \hat{\boldsymbol{u}}_2)$, the reduction can always turn a full traceability adversary (see [12] for a definition) into a forger for the hierarchical signature. The proof is straightforward and the details are omitted.

Theorem 6. *The scheme provides full anonymity assuming that: (i) The SXDH assumption holds in* $(\mathbb{G}, \hat{\mathbb{G}}, \mathbb{G}_T)$*; (ii)* CMhash *is a collision-resistant chameleon hash function.* (The proof is given in the full version of the paper).

In the full version of the paper, we extend the above system to obtain dynamic group signatures based on the SXDH and XDLIN$_2$ assumption. The signature length is only 1.8 kB, which gives us the shortest dynamic group signatures based on constant-size assumptions to date. The construction builds on our structure-preserving signature and the encryption scheme of Sect. 6 in a modular manner. Detailed efficiency comparisons are given in the full paper.

Acknowledgements. The first author's work was supported by the "Programme Avenir Lyon Saint-Etienne de l'Université de Lyon" in the framework of the programme "Inverstissements d'Avenir" (ANR-11-IDEX-0007). The second author was supported by the European Research Council (FP7/2007-2013 Grant Agreement no. 339563 CryptoCloud). Part of this work of the third author was done while visiting the Simons Institute for Theory of Computing, U.C. Berkeley.

References

1. Abe, M., Chase, M., David, B., Kohlweiss, M., Nishimaki, R., Ohkubo, M.: Constant-size structure-preserving signatures: generic constructions and simple assumptions. In: Wang, X., Sako, K. (eds.) ASIACRYPT 2012. LNCS, vol. 7658, pp. 4–24. Springer, Heidelberg (2012)
2. Abe, M., David, B., Kohlweiss, M., Nishimaki, R., Ohkubo, M.: Tagged one-time signatures: tight security and optimal tag size. In: Kurosawa, K., Hanaoka, G. (eds.) PKC 2013. LNCS, vol. 7778, pp. 312–331. Springer, Heidelberg (2013)
3. Abe, M., Fuchsbauer, G., Groth, J., Haralambiev, K., Ohkubo, M.: Structure-preserving signatures and commitments to group elements. In: Rabin, T. (ed.) CRYPTO 2010. LNCS, vol. 6223, pp. 209–236. Springer, Heidelberg (2010)
4. Abe, M., Groth, J., Haralambiev, K., Ohkubo, M.: Optimal structure-preserving signatures in asymmetric bilinear groups. In: Rogaway, P. (ed.) CRYPTO 2011. LNCS, vol. 6841, pp. 649–666. Springer, Heidelberg (2011)
5. Abe, M., Groth, J., Ohkubo, M.: Separating short structure-preserving signatures from non-interactive assumptions. In: Lee, D.H., Wang, X. (eds.) ASIACRYPT 2011. LNCS, vol. 7073, pp. 628–646. Springer, Heidelberg (2011)
6. Abe, M., Groth, J., Ohkubo, M., Tibouchi, M.: Unified, minimal and selectively randomizable structure-preserving signatures. In: Lindell, Y. (ed.) TCC 2014. LNCS, vol. 8349, pp. 688–712. Springer, Heidelberg (2014)
7. Abe, M., Groth, J., Ohkubo, M., Tibouchi, M.: Structure-preserving signatures from type II pairings. In: Garay, J.A., Gennaro, R. (eds.) CRYPTO 2014, Part I. LNCS, vol. 8616, pp. 390–407. Springer, Heidelberg (2014)
8. Abe, M., Haralambiev, K., Ohkubom, M.: Signing on elements in bilinear groups for modular protocol design. Cryptology ePrint Archive: Report 2010/133 (2010)
9. Ateniese, G., Camenisch, J., Hohenberger, S., de Medeiros, B.: Practical group signatures without random oracles. Cryptology ePrint Archive: Report 2005/385 (2005)
10. Ateniese, G., Camenisch, J.L., Joye, M., Tsudik, G.: A practical and provably secure coalition-resistant group signature scheme. In: Bellare, M. (ed.) CRYPTO 2000. LNCS, vol. 1880, p. 255. Springer, Heidelberg (2000)
11. Belenkiy, M., Chase, M., Kohlweiss, M., Lysyanskaya, A.: P-signatures and noninteractive anonymous credentials. In: Canetti, R. (ed.) TCC 2008. LNCS, vol. 4948, pp. 356–374. Springer, Heidelberg (2008)

12. Bellare, M., Micciancio, D., Warinschi, B.: Foundations of group signatures: Formal definitions, simplified requirements, and a construction based on general assumptions. In: Biham, E. (ed.) EUROCRYPT 2003. LNCS, vol. 2656, pp. 614–629. Springer, Heidelberg (2003)

13. Bellare, M., Hofheinz, D., Yilek, S.: Possibility and impossibility results for encryption and commitment secure under selective opening. In: Joux, A. (ed.) EUROCRYPT 2009. LNCS, vol. 5479, pp. 1–35. Springer, Heidelberg (2009)

14. Bellare, M., Rogaway, P.: Random oracles are practical: a paradigm for designing efficient protocols. In: 1st ACM Conference on Computer and Communications Security, pp. 62–73. ACM Press (1993)

15. Bellare, M., Shi, H., Zhang, C.: Foundations of group signatures: the case of dynamic groups. In: Menezes, A. (ed.) CT-RSA 2005. LNCS, vol. 3376, pp. 136–153. Springer, Heidelberg (2005)

16. Boneh, D., Boyen, X.: Efficient selective-ID secure identity-based encryption without random oracles. In: Cachin, C., Camenisch, J.L. (eds.) EUROCRYPT 2004. LNCS, vol. 3027, pp. 223–238. Springer, Heidelberg (2004)

17. Boneh, D., Boyen, X., Goh, E.-J.: Hierarchical identity based encryption with constant size ciphertext. In: Cramer, R. (ed.) EUROCRYPT 2005. LNCS, vol. 3494, pp. 440–456. Springer, Heidelberg (2005)

18. Boneh, D., Boyen, X., Shacham, H.: Short group signatures. In: Franklin, M. (ed.) CRYPTO 2004. LNCS, vol. 3152, pp. 41–55. Springer, Heidelberg (2004)

19. Boyen, X., Mei, Q., Waters, B.: Direct chosen-ciphertext security from identity-based techniques. In: ACM-CCS 2005, pp. 320–329. ACM Press (2006)

20. Boyen, X., Waters, B.: Compact group signatures without random oracles. In: Vaudenay, S. (ed.) EUROCRYPT 2006. LNCS, vol. 4004, pp. 427–444. Springer, Heidelberg (2006)

21. Boyen, X., Waters, B.: Full-domain subgroup hiding and constant-size group signatures. In: Okamoto, T., Wang, X. (eds.) PKC 2007. LNCS, vol. 4450, pp. 1–15. Springer, Heidelberg (2007)

22. Camenisch, J., Chandran, N., Shoup, V.: A public key encryption scheme secure against key dependent chosen plaintext and adaptive chosen ciphertext attacks. In: Joux, A. (ed.) EUROCRYPT 2009. LNCS, vol. 5479, pp. 351–368. Springer, Heidelberg (2009)

23. Camenisch, J., Dubovitskaya, M., Haralambiev, K.: Efficient structure-preserving signature scheme from standard assumptions. In: Visconti, I., De Prisco, R. (eds.) SCN 2012. LNCS, vol. 7485, pp. 76–94. Springer, Heidelberg (2012)

24. Canetti, R., Goldreich, O., Halevi, S.: The random oracle methodology, revisited. J. ACM 51(4), 557–594 (2004)

25. Cathalo, J., Libert, B., Yung, M.: Group encryption: non-interactive realization in the standard model. In: Matsui, M. (ed.) ASIACRYPT 2009. LNCS, vol. 5912, pp. 179–196. Springer, Heidelberg (2009)

26. Chase, M., Kohlweiss, M.: A new hash-and-sign approach and structure-preserving signatures from DLIN. In: Visconti, I., De Prisco, R. (eds.) SCN 2012. LNCS, vol. 7485, pp. 131–148. Springer, Heidelberg (2012)

27. Chaum, D., van Heyst, E.: Group signatures. In: Davies, D.W. (ed.) EUROCRYPT 1991. LNCS, vol. 547, pp. 257–265. Springer, Heidelberg (1991)

28. Cramer, R., Shoup, V.: A practical public key cryptosystem provably secure against adaptive chosen ciphertext attack. In: Krawczyk, H. (ed.) CRYPTO 1998. LNCS, vol. 1462, p. 13. Springer, Heidelberg (1998)

29. Cramer, R., Shoup, V.: Universal hash proofs and a paradigm for adaptive chosen ciphertext secure public-key encryption. In: Knudsen, L.R. (ed.) EUROCRYPT 2002. LNCS, vol. 2332, p. 45. Springer, Heidelberg (2002)

30. Delerablée, C., Pointcheval, D.: Dynamic fully anonymous short group signatures. In: Nguyên, P.Q. (ed.) VIETCRYPT 2006. LNCS, vol. 4341, pp. 193–210. Springer, Heidelberg (2006)

31. Fuchsbauer, G.: Automorphic signatures in bilinear groups and an application to round-optimal blind signatures. Cryptology ePrint Archive: Report 2009/320 (2009)

32. Gerbush, M., Lewko, A., O'Neill, A., Waters, B.: Dual form signatures: an approach for proving security from static assumptions. In: Wang, X., Sako, K. (eds.) ASIACRYPT 2012. LNCS, vol. 7658, pp. 25–42. Springer, Heidelberg (2012)

33. Green, M., Hohenberger, S.: Universally composable adaptive oblivious transfer. In: Pieprzyk, J. (ed.) ASIACRYPT 2008. LNCS, vol. 5350, pp. 179–197. Springer, Heidelberg (2008)

34. Groth, J., Ostrovsky, R., Sahai, A.: Perfect non-interactive zero knowledge for NP. In: Vaudenay, S. (ed.) EUROCRYPT 2006. LNCS, vol. 4004, pp. 339–358. Springer, Heidelberg (2006)

35. Groth, J.: Simulation-sound NIZK proofs for a practical language and constant size group signatures. In: Lai, X., Chen, K. (eds.) ASIACRYPT 2006. LNCS, vol. 4284, pp. 444–459. Springer, Heidelberg (2006)

36. Groth, J.: Fully anonymous group signatures without random oracles. In: Kurosawa, K. (ed.) ASIACRYPT 2007. LNCS, vol. 4833, pp. 164–180. Springer, Heidelberg (2007)

37. Groth, J., Sahai, A.: Efficient non-interactive proof systems for bilinear groups. In: Smart, N.P. (ed.) EUROCRYPT 2008. LNCS, vol. 4965, pp. 415–432. Springer, Heidelberg (2008)

38. Hofheinz, D., Jager, T.: Tightly secure signatures and public-key encryption. In: Safavı-Naini, R., Canetti, R. (eds.) CRYPTO 2012. LNCS, vol. 7417, pp. 590–607. Springer, Heidelberg (2012)

39. Hemenway, B., Libert, B., Ostrovsky, R., Vergnaud, D.: Lossy encryption: constructions from general assumptions and efficient selective opening chosen ciphertext security. In: Lee, D.H., Wang, X. (eds.) ASIACRYPT 2011. LNCS, vol. 7073, pp. 70–88. Springer, Heidelberg (2011)

40. Jutla, C.S., Roy, A.: Shorter quasi-adaptive NIZK proofs for linear subspaces. In: Sako, K., Sarkar, P. (eds.) ASIACRYPT 2013, Part I. LNCS, vol. 8269, pp. 1–20. Springer, Heidelberg (2013)

41. Jutla, C.S., Roy, A.: Switching lemma for bilinear tests and constant-size NIZK proofs for linear subspaces. In: Garay, J.A., Gennaro, R. (eds.) CRYPTO 2014, Part II. LNCS, vol. 8617, pp. 295–312. Springer, Heidelberg (2014)

42. Kiayias, A., Yung, M.: Secure scalable group signature with dynamic joins and separable authorities. Int. J. Secur. Netw. (IJSN) 1(1/2), 24–45 (2006)

43. Kiltz, E., Mityagin, A., Panjwani, S., Raghavan, B.: Append-only signatures. In: Caires, L., Italiano, G.F., Monteiro, L., Palamidessi, C., Yung, M. (eds.) ICALP 2005. LNCS, vol. 3580, pp. 434–445. Springer, Heidelberg (2005)

44. Kiltz, E.: Chosen-ciphertext security from tag-based encryption. In: Halevi, S., Rabin, T. (eds.) TCC 2006. LNCS, vol. 3876, pp. 581–600. Springer, Heidelberg (2006)

45. Kiltz, E., Pan, J., Wee, H.: Structure-preserving signatures from standard assumptions, revisited. In: Gennaro, R., Robshaw, M. (eds.) CRYPTO 2015, Part II. LNCS, vol. 9216, pp. 275–295. Springer, Heidelberg (2015)

46. Libert, B., Peters, T., Joye, M., Yung, M.: Linearly homomorphic structure-preserving signatures and their applications. In: Canetti, R., Garay, J.A. (eds.) CRYPTO 2013, Part II. LNCS, vol. 8043, pp. 289–307. Springer, Heidelberg (2013)
47. Libert, B., Peters, T., Joye, M., Yung, M.: Non-malleability from malleability: simulation-sound quasi-adaptive NIZK proofs and CCA2-secure encryption from homomorphic signatures. In: Nguyen, P.Q., Oswald, E. (eds.) EUROCRYPT 2014. LNCS, vol. 8441, pp. 514–532. Springer, Heidelberg (2014)
48. MacKenzie, P.D., Reiter, M.K., Yang, K.: Alternatives to non-malleability: definitions, constructions, and applications. In: Naor, M. (ed.) TCC 2004. LNCS, vol. 2951, pp. 171–190. Springer, Heidelberg (2004)
49. Shamir, A.: Identity-based cryptosystems and signature schemes. In: Blakely, G.R., Chaum, D. (eds.) CRYPTO 1984. LNCS, vol. 196, pp. 47–53. Springer, Heidelberg (1985)
50. Shoup, V.: A proposal for an ISO standard for public key encryption. Manuscript, 20 December 2001
51. Waters, B.: Efficient identity-based encryption without random oracles. In: Cramer, R. (ed.) EUROCRYPT 2005. LNCS, vol. 3494, pp. 114–127. Springer, Heidelberg (2005)

Multiparty Computation II

Efficient Constant Round Multi-party Computation Combining BMR and SPDZ

Yehuda Lindell[1], Benny Pinkas[1][(✉)], Nigel P. Smart[2], and Avishay Yanai[1]

[1] Department of Computer Science, Bar-Ilan University, Ramat Gan, Israel
benny@pinkas.net
[2] Department of Computer Science, University of Bristol, Bristol, UK

Abstract. Recently, there has been huge progress in the field of concretely efficient secure computation, even while providing security in the presence of *malicious adversaries*. This is especially the case in the two-party setting, where constant-round protocols exist that remain fast even over slow networks. However, in the multi-party setting, all concretely efficient fully-secure protocols, such as SPDZ, require many rounds of communication.

In this paper, we present an MPC protocol that is fully-secure in the presence of malicious adversaries and for any number of corrupted parties. Our construction is based on the constant-round BMR protocol of Beaver et al., and is the first fully-secure version of that protocol that makes black-box usage of the underlying primitives, and is therefore concretely efficient.

Our protocol includes an online phase that is extremely fast and mainly consists of each party locally evaluating a garbled circuit. For the offline phase we present both a generic construction (using any underlying MPC protocol), and a highly efficient instantiation based on the SPDZ protocol. Our estimates show the protocol to be considerably more efficient than previous fully-secure multi-party protocols.

1 Introduction

Background: Protocols for secure multi-party computation (MPC) enable a set of mutually distrustful parties to securely compute a joint functionality of their inputs. Such a protocol must guarantee *privacy* (meaning that only the output is learned), *correctness* (meaning that the output is correctly computed from the inputs), and *independence of inputs* (meaning that each party must choose its input independently of the others). Formally, security is defined by comparing the distribution of the outputs of all parties in a real protocol to an ideal model where an incorruptible trusted party computes the functionality for the parties. The two main types of adversaries that have been considered are *semi-honest adversaries* who follow the protocol specification but try to learn more than allowed by inspecting the transcript, and *malicious adversaries* who can run any arbitrary strategy in an attempt to break the protocol. Secure MPC has been studied since the late 1980s, and powerful feasibility results were proven showing that *any*

© International Association for Cryptologic Research 2015
R. Gennaro and M. Robshaw (Eds.): CRYPTO 2015, Part II, LNCS 9216, pp. 319–338, 2015.
DOI: 10.1007/978-3-662-48000-7_16

two-party or multi-party functionality can be securely computed [10,22], even in the presence of malicious adversaries. When an honest majority (or 2/3 majority) is assumed, then security can even be obtained information theoretically [3,4,19]. In this paper, we focus on the problem of security in the presence of malicious adversaries, and a dishonest majority.

Recently, there has been much interest in the problem of concretely efficient secure MPC, where "concretely efficient" refers to protocols that are sufficiently efficient to be implemented in practice (in particular, these protocols should only make black-box usage of cryptographic primitives; they must not, say, use generic ZK proofs that operate on the circuit representation of these primitives). In the last few years there has been tremendous progress on this problem, and there now exist extremely fast protocols that can be used in practice; see [8,13,14,16,17] for just a few examples. In general, there are two approaches that have been followed; the first uses Yao's garbled circuits [22] and the second utilizes interaction for every gate like the GMW protocol [10].

There are extremely efficient variants of Yao's protocol for the two party case that are secure against malicious adversaries (e.g., [14,16]). These protocols run in a constant number of rounds and therefore remain fast over slow networks. The BMR protocol [1] is a variant of Yao's protocol that runs in a multi-party setting with more than two parties. This protocol works by the parties jointly constructing a garbled circuit (possibly in an offline phase), and then later computing it (possibly in an online phase). However, in the case of malicious adversaries this protocol suffers from two main drawbacks: (1) Security is only guaranteed if at most a *minority* of the parties are corrupt; (2) The protocol uses generic protocols secure against malicious adversaries (say, the GMW protocol) that evaluate the pseudorandom generator used in the BMR protocol. This non black-box construction results in an extremely high overhead.

The TinyOT and SPDZ protocols [8,17] follow the GMW paradigm, and have offline and online phases. Both of these protocols overcome the issues of the BMR protocol in that they are secure against any number of corrupt parties, make only black-box usage of cryptographic primitives, and have very fast online phases that require only very simple (information theoretic) operations. (A black-box constant-round MPC construction appears in [11]; however, it is not "concretely efficient".) In the case of multi-party computation with more than two parties, these protocols are currently the *only practical* approach known. However, since they follow the GMW paradigm, their online phase requires a communication round for every multiplication gate. This results in a large amount of interaction and high latency, especially over slow networks. To sum up, there is no known concretely efficient constant-round protocol for the multi-party case (with the exception of [5] that considers the specific three-party case only). Our work introduces the first protocol with these properties.

Our Contribution: In this paper, we provide the first *concretely efficient constant-round* protocol for the general *multi-party* case, with security in the presence of malicious adversaries. The basic idea behind the construction is to use an efficient non-constant round protocol – with security for malicious adversaries – to compute the gate tables of the BMR garbled circuit (and since the computation of these tables is of constant depth, this step is constant round). A crucial observation, resulting in a great performance improvement, shows that in the offline stage it is *not required to verify the correctness* of the computations of the different tables. Rather, validation of the correctness is an immediate by product of the online computation phase, and therefore does not add any overhead to the computation. Although our basic generic protocol can be instantiated with any non-constant round MPC protocol, we provide an optimized version that utilizes specific features of the SPDZ protocol [8].

In our general construction, the new constant-round MPC protocol consists of two phases. In the first (offline) phase, the parties securely compute *random shares* of the BMR garbled circuit. If this is done naively, then the result is highly inefficient since part of the computation involves computing a pseudorandom generator or pseudorandom function multiple times for every gate. By modifying the original BMR garbled circuit, we show that it is possible to actually compute the circuit very efficiently. Specifically, each party locally computes the pseudorandom function as needed for every gate (in our construction we use a pseudorandom function rather than a pseudorandom generator), and uses the results as input to the secure computation. Our proof of security shows that if a party cheats and inputs incorrect values then no harm is done, since it can only cause the honest parties to abort (which is anyway possible when there is no honest majority). Next, in the online phase, all that the parties need to do is reconstruct the single garbled circuit, exchange garbled values on the input wires and locally compute the garbled circuit. The online phase is therefore very fast.

In our concrete instantiation of the protocol using SPDZ [8], there are actually three separate phases, with each being faster than the previous. The first two phases can be run offline, and the last phase is run online after the inputs become known.

- The first (slow) phase depends only on an upper bound on the number of wires and the number of gates in the function to be evaluated. This phase uses Somewhat Homomorphic Encryption (SHE) and is equivalent to the offline phase of the SPDZ protocol.
- The second phase depends on the function to be evaluated but not the function inputs; in our proposed instantiation this mainly involves information theoretic primitives and is equivalent to the online phase of the SPDZ protocol.
- In the third phase the parties provide their input and evaluate the function; this phase just involves exchanging shares of the circuit and garbled values on the input wire and locally computing the BMR garbled circuit.

We stress that our protocol is constant round *in all phases* since the depth of the circuit required to compute the BMR garbled circuit is constant. In addition, the computational cost of preparing the BMR garbled circuit is not much more than the cost of using SPDZ itself to compute the functionality directly. However, the key advantage that we gain is that our online time is extraordinarily fast, requiring only two rounds and local computation of a single garbled circuit. *This is faster than all other existing circuit-based multi-party protocols.*

Finite Field Optimization of BMR: In order to efficiently compute the BMR garbled circuit, we define the garbling and evaluation operations over a finite field. A similar technique of using finite fields in the BMR protocol was introduced in [2] in the case of semi-honest security against an honest majority. In contrast to [2], our utilization of finite fields is carried out via *vectors* of field elements, and uses the underlying arithmetic of the field as opposed to using very large finite fields to simulate integer arithmetic. This makes our modification in this respect more efficient.

2   The General Protocol

2.1   The BMR Protocol

To aid the reader we provide here a high-level description of the BMR protocol of [1]. A detailed description of the protocol can be found in [1,2] or in the full version of our paper [15]. We describe here the version of the protocol that is secure against semi-honest adversaries. The protocol is comprised of an offline-phase, where the garbled circuit is created by the players, and an online-phase, where garbled inputs are exchanged between the players and the circuit is evaluated.

Seeds and Superseeds: Each player associates random 0-seed and 1-seed with each wire. Input wires of the circuit are treated differently, and there only the player which provides the corresponding input bit knows the seeds of the wire. The 0-superseed (resp. 1-superseed) of a wire is the concatenation of all 0-seeds (1-seeds) of this wire, and its *components* are the seeds.

Garbling: For each of the four combinations of input values to a gate, the garbling produces an encryption of the corresponding superseed of the output wire, with the keys being each of the component seeds of the corresponding superseeds of the input wires.

The Offline Phase: In the offline-phase, the players run (in parallel) a secure computation for each gate, which computes the garbled table of the gate as a function of the 0/1-seeds of each of the players for the input/output wires of the gate, and of the truth table of the gate. This computation runs in a constant number of rounds. The resulting garbled table enables to compute the superseed of the output wire of the gate, given the superseeds of its input wires.

The Online Phase: In the online-phase each player which is assigned an input wire, and which has an input value b on that wire, sends the b-superseed of the wire to all other players. Then, every player is able evaluate the circuit on its own, without any further interaction with the other players.

2.2   Modified BMR Garbling

In order to facilitate fast secure computation of the garbled circuit in the offline phase, we make some changes to the original BMR garbling above. First, instead of using XOR of bit strings, and hence a binary circuit to instantiate the garbled gate, we use additions of elements in a finite field, and hence an arithmetic circuit. This idea was used by [2] in the FairplayMP system, which used the BGW protocol [3] in order to compute the BMR circuit. Note that FairplayMP achieved semi-honest security with an honest majority, whereas our aim is *malicious security* for *any number of corrupted parties*.

Second, we observe that the external values[1] do not need to be explicitly encoded, since each party can learn them by looking at its own "part" of the garbled value. In the original BMR garbling, each superseed contains n seeds provided by the parties. Thus, if a party's zero-seed is in the decrypted superseed then it knows that the external value (denoted by Λ) is zero, and otherwise it knows that it is one.

Naively, it seems that independently computing each gate securely in the offline phase is insufficient, since the corrupted parties might use inconsistent inputs for the computations of different gates. For example, if the output wire of gate g is an input to gate g', the input provided for the computation of the table of g might not agree with the inputs used for the computation of the table of g'. It therefore seems that the offline computation must verify the consistency of the computations of different gates. This type of verification would greatly increase the cost since the evaluation of the pseudorandom functions (or pseudorandom generator in the original BMR) used in computing the tables needs to be checked inside the secure computation. This means that the pseudorandom function is not treated as a black box, and the circuit for the offline phase would be huge (as it would include multiple copies of a subcircuit for computing pseudorandom function computations for every wire). Instead, we prove that this type of corrupt behavior can only result in an abort in the online phase, which would not affect the security of the protocol. This observation enables us to compute each gate independently and model the pseudorandom function used in the computation as a black box, thus simplifying the protocol and optimizing its performance.

We also encrypt garbled values as *vectors*; this enables us to use a finite field that can encode $\{0,1\}^\kappa$ (for each vector coordinate), rather than a much larger finite field that can encode all of $\{0,1\}^{n\cdot\kappa}$. Due to this, the parties choose *keys* (for a pseudorandom function) rather than *seeds* for a pseudorandom generator. The keys that P_i chooses for wire w are denoted $k_{w,0}^i$ and $k_{w,1}^i$, which will be elements in a finite field $\mathbb{F}_p$ such that $2^\kappa < p < 2^{\kappa+1}$. In fact we pick p to be the smallest prime number larger than 2^κ, and set $p = 2^\kappa + \alpha$, where (by the prime number theorem) we expect $\alpha \approx \kappa$. We shall denote the pseudorandom function by $F_k(x)$, where the key and output will be interpreted as elements of $\mathbb{F}_p$ in much of our MPC protocol. In practice the function $F_k(x)$ we suggest will be implemented using CBC-MAC using a block cipher enc with key and block size κ bits, as $F_k(x) = \mathsf{CBC\text{-}MAC_{enc}}(k \pmod{2^\kappa}, x)$. Note that the inputs x to our pseudorandom function will all be of the same length and so using naive CBC-MAC will be secure.

---

[1] The external values (as denoted in [2]) are the *signals* (as denoted in [1]) observable by the parties when evaluating the circuit in the online phase.

We interpret the κ-bit output of $F_k(x)$ as an element in $\mathbb{F}_p$ where $p = 2^\kappa + \alpha$. Note that a mapping which sends an element $k \in \mathbb{F}_p$ to a κ-bit block cipher key by computing $k \pmod{2^\kappa}$ induces a distribution on the key space of the block cipher which has statistical distance from uniform of

$$\frac{1}{2}\left((2^\kappa - \alpha) \cdot \left(\frac{1}{2^\kappa} - \frac{1}{p}\right) + \alpha \cdot \left(\frac{2}{p} - \frac{1}{2^\kappa}\right)\right) \approx \frac{\alpha}{p} \approx \frac{\kappa}{2^\kappa}.$$

The output of the function $F_k(x)$ will also induce a distribution which is close to uniform on $\mathbb{F}_p$. In particular the statistical distance of the output in $\mathbb{F}_p$, for a block cipher with block size κ, from uniform is given by

$$\frac{1}{2}\left(2^\kappa \cdot \left(\frac{1}{2^\kappa} - \frac{1}{p}\right) + \alpha \cdot \left(\frac{1}{p} - 0\right)\right) = \frac{\alpha}{p} \approx \frac{\kappa}{2^\kappa}$$

(note that $1 - \frac{2^\kappa}{p} = \frac{\alpha}{p}$). In practice we set $\kappa = 128$, and use the AES cipher as the block cipher enc. The statistical difference is therefore negligible.

Functionality 1 (The SFE Functionality: $\mathcal{F}_{\mathsf{SFE}}$)

The functionality is parameterized by a function $f(x_1, \ldots, x_n)$ which is input as a binary circuit C_f. The protocol consists of 3 externally exposed commands **Initialize**, **InputData**, and **Output** and one internal subroutine **Wait**.

Initialize: On input $(init, C_f)$ from all parties, the functionality activates and stores C_f.

Wait: This waits on the adversary to return a $GO/NO\text{-}GO$ decision. If the adversary returns $NO\text{-}GO$ then the functionality aborts.

InputData: On input $(input, P_i, varid, x_i)$ from P_i and $(input, P_i, varid, ?)$ from all other parties, with $varid$ a fresh identifier, the functionality stores $(varid, x_i)$. The functionality then calls **Wait**.

Output: On input $(output)$ from all honest parties the functionality computes $y = f(x_1, \ldots, x_n)$ and outputs y to the adversary. The functionality then calls **Wait**. Only if **Wait** does not abort it outputs y to all parties.

The goal of this paper is to present a protocol Π_{SFE} which implements the Secure Function Evaluation (SFE) functionality of Functionality 1 in a constant number of rounds in the case of a malicious dishonest majority. Our constant round protocol Π_{SFE} implementing $\mathcal{F}_{\mathsf{SFE}}$ is built in the $\mathcal{F}_{\mathsf{MPC}}$-hybrid model, i.e. utilizing a sub-protocol Π_{MPC} which implements the functionality $\mathcal{F}_{\mathsf{MPC}}$ given in Functionality 2. The generic MPC functionality $\mathcal{F}_{\mathsf{MPC}}$ is *reactive*. We require a *reactive* MPC functionality because our protocol Π_{SFE} will make repeated sequences of calls to $\mathcal{F}_{\mathsf{MPC}}$ involving both output and computation commands. In terms of round complexity, all that we require of the sub-protocol Π_{MPC} is that each of the commands which it implements can be implemented in constant rounds. Given this requirement our larger protocol Π_{SFE} will be constant round.

Functionality 2 (The Generic Reactive MPC Functionality: $\mathcal{F}_{\mathsf{MPC}}$)

The functionality consists of five externally exposed commands **Initialize, InputData, Add, Multiply,** and **Output,** and one internal subroutine **Wait.**

Initialize: On input $(init, p)$ from all parties, the functionality activates and stores p. All additions and multiplications below will be mod p.

Wait: This waits on the adversary to return a $GO/NO\text{-}GO$ decision. If the adversary returns $NO\text{-}GO$ then the functionality aborts.

InputData: On input $(input, P_i, varid, x)$ from P_i and $(input, P_i, varid, ?)$ from all other parties, with $varid$ a fresh identifier, the functionality stores $(varid, x)$. The functionality then calls **Wait.**

Add: On command $(add, varid_1, varid_2, varid_3)$ from all parties (if $varid_1, varid_2$ are present in memory and $varid_3$ is not), the functionality retrieves $(varid_1, x)$, $(varid_2, y)$ and stores $(varid_3, x + y \bmod p)$. The functionality then calls **Wait.**

Multiply: On input $(multiply, varid_1, varid_2, varid_3)$ from all parties (if $varid_1, varid_2$ are present in memory and $varid_3$ is not), the functionality retrieves $(varid_1, x)$, $(varid_2, y)$ and stores $(varid_3, x \cdot y \bmod p)$. The functionality then calls **Wait.**

Output: On input $(output, varid, i)$ from all honest parties (if $varid$ is present in memory), the functionality retrieves $(varid, x)$ and outputs either $(varid, x)$ in the case of $i \neq 0$ or $(varid)$ if $i = 0$ to the adversary. The functionality then calls **Wait,** and only if **Wait** does not abort then it outputs x to all parties if $i = 0$, or it outputs x only to party i if $i \neq 0$.

In what follows we use the notation $[varid]$ to represent the result stored in the variable $varid$ by the $\mathcal{F}_{\mathsf{MPC}}$ or $\mathcal{F}_{\mathsf{SFE}}$ functionality. In particular we use the arithmetic shorthands $[z] = [x] + [y]$ and $[z] = [x] \cdot [y]$ to represent the result of calling the **Add** and **Multiply** commands on the $\mathcal{F}_{\mathsf{MPC}}$ functionality.

2.3 The Offline Functionality: preprocessing-I and preprocessing-II

Our protocol, Π_{SFE}, is comprised of an offline-phase and an online-phase, where the offline-phase, which implements the functionality $\mathcal{F}_{\mathsf{offline}}$, is divided into two subphases: preprocessing-I and preprocessing-II. To aid exposition we first present the functionality $\mathcal{F}_{\mathsf{offline}}$ in Functionality 3. In the next section, we present an efficient methodology to implement $\mathcal{F}_{\mathsf{offline}}$ which uses the SPDZ protocol as the underlying MPC protocol for securely computing functionality $\mathcal{F}_{\mathsf{MPC}}$; while in the full version of the paper [15] we present a generic implementation of $\mathcal{F}_{\mathsf{offline}}$ based on any underlying protocol Π_{MPC} implementing $\mathcal{F}_{\mathsf{MPC}}$.

In describing functionality $\mathcal{F}_{\mathsf{offline}}$ we distinguish between *attached* wires and *common* wires: the attached wires are the circuit-input-wires that are directly connected to the parties (i.e., these are inputs wires to the circuit). Thus, if every party has ℓ inputs to the functionality f then there are $n \cdot \ell$ attached wires. The rest of the wires are considered as *common* wires, i.e. they are directly connected to *none* of the parties.

Functionality 3 (The Offline Functionality – $\mathcal{F}_{\text{offline}}$)

This functionality runs the same **Initialize**, **Wait**, **InputData** and **Output** commands as $\mathcal{F}_{\text{MPC}}$ (Functionality 2). In addition, the functionality has two additional commands **preprocessing-I** and **preprocessing-II**, as follows.

preprocessing-I: On input (preprocessing-I, W, G), for all wires $w \in [1, \ldots, W]$:
- The functionality chooses and stores a random masking value $[\lambda_w]$ where $\lambda_w \in \{0, 1\}$.
- For $1 \le i \le n$ and $\beta \in \{0, 1\}$,
  - The functionality stores a key of user i for wire w and value β, $[k^i_{w,\beta}]$ where $k^i_{w,\beta} \in \mathbb{F}_p$
  - The functionality outputs $[k^i_{w,\beta}]$ to party i by running **Output** as in functionality $\mathcal{F}_{\text{MPC}}$.

preprocessing-II: On input of (preprocessing-II, C_f) for a circuit C_f with at most W wires and G gates.
- For all wires w which are attached to party P_i the functionality opens $[\lambda_w]$ to party P_i by running **Output** as in functionality $\mathcal{F}_{\text{MPC}}$.
- For all output wires w the functionality opens $[\lambda_w]$ to all parties by running **Output** as in functionality $\mathcal{F}_{\text{MPC}}$.
- For every gate g with input wires $1 \le a, b \le W$ and output wire $1 \le c \le W$.
  - Party P_i provides the following values for $x \in \{a, b\}$ by running **InputData** as in functionality $\mathcal{F}_{\text{MPC}}$:

$$F_{k^i_{x,0}}(0\|1\|g), \ldots, F_{k^i_{x,0}}(0\|n\|g) \qquad F_{k^i_{x,0}}(1\|1\|g), \ldots, F_{k^i_{x,0}}(1\|n\|g)$$
$$F_{k^i_{x,1}}(0\|1\|g), \ldots, F_{k^i_{x,1}}(0\|n\|g) \qquad F_{k^i_{x,1}}(1\|1\|g), \ldots, F_{k^i_{x,1}}(1\|n\|g)$$

  - Define the selector variables

$$\chi_1 = \begin{cases} 0 & \text{if } f_g(\lambda_a, \lambda_b) = \lambda_c \\ 1 & \text{otherwise} \end{cases} \qquad \chi_2 = \begin{cases} 0 & \text{if } f_g(\lambda_a, \overline{\lambda_b}) = \lambda_c \\ 1 & \text{otherwise} \end{cases}$$

$$\chi_3 = \begin{cases} 0 & \text{if } f_g(\overline{\lambda_a}, \lambda_b) = \lambda_c \\ 1 & \text{otherwise} \end{cases} \qquad \chi_4 = \begin{cases} 0 & \text{if } f_g(\overline{\lambda_a}, \overline{\lambda_b}) = \lambda_c \\ 1 & \text{otherwise} \end{cases}$$

  - Set $\mathbf{A}_g = (A^1_g, \ldots, A^n_g)$, $\mathbf{B}_g = (B^1_g, \ldots, B^n_g)$, $\mathbf{C}_g = (C^1_g, \ldots, C^n_g)$, and $\mathbf{D}_g = (D^1_g, \ldots, D^n_g)$ where for $1 \le j \le n$:

$$A^j_g = \left(\sum_{i=1}^{n} F_{k^i_{a,0}}(0\|j\|g) + F_{k^i_{b,0}}(0\|j\|g) \right) + k^j_{c,\chi_1}$$

$$B^j_g = \left(\sum_{i=1}^{n} F_{k^i_{a,0}}(1\|j\|g) + F_{k^i_{b,1}}(0\|j\|g) \right) + k^j_{c,\chi_2}$$

$$C^j_g = \left(\sum_{i=1}^{n} F_{k^i_{a,1}}(0\|j\|g) + F_{k^i_{b,0}}(1\|j\|g) \right) + k^j_{c,\chi_3}$$

$$D^j_g = \left(\sum_{i=1}^{n} F_{k^i_{a,1}}(1\|j\|g) + F_{k^i_{b,1}}(1\|j\|g) \right) + k^j_{c,\chi_4}$$

  - The functionality stores the values $[\mathbf{A}_g], [\mathbf{B}_g], [\mathbf{C}_g], [\mathbf{D}_g]$.

Our preprocessing-I takes as input an upper bound W on the number of wires in the circuit, and an upper bound G on the number of gates in the circuit.

The upper bound G is not strictly needed, but will be needed in any efficient instantiation based on the SPDZ protocol. In contrast preprocessing-II requires knowledge of the precise function f being computed, which we assume is encoded as a binary circuit C_f.

Protocol 1 (Π_{SFE}: Securely Computing $\mathcal{F}_{\mathsf{SFE}}$ in the $\mathcal{F}_{\mathsf{offline}}$-Hybrid Model)

On input of a circuit C_f representing the function f which consists of at most W wires and at most G gates the parties execute the following commands.

Pre-Processing: This procedure is performed as follows

1. Call **Initialize** on $\mathcal{F}_{\mathsf{offline}}$ with the smallest prime p in $\{2^\kappa, \ldots, 2^{\kappa+1}\}$.
2. Call **Preprocessing-I** on $\mathcal{F}_{\mathsf{offline}}$ with input W and G.
3. Call **Preprocessing-II** on $\mathcal{F}_{\mathsf{offline}}$ with input C_f.

Online Computation: This procedure is performed as follows

1. For all input wires w for party P_i the party takes his input bit ρ_w and computes $\Lambda_w = \rho_w \oplus \lambda_w$, where λ_w was obtained in the preprocessing stage. The value Λ_w is broadcast to all parties.
2. Party i calls **Output** on $\mathcal{F}_{\mathsf{offline}}$ to open $[k^i_{w,\Lambda_w}]$ for all his input wires w, we denote the resulting value by k^i_w.
3. The parties call **Output** on $\mathcal{F}_{\mathsf{offline}}$ to open $[\mathbf{A}_g]$, $[\mathbf{B}_g]$, $[\mathbf{C}_g]$ and $[\mathbf{D}_g]$ for every gate g.
4. Passing through the circuit topologically, the parties can now locally compute the following operations for each gate g
   - Let the gates input wires be labeled a and b, and the output wire be labeled c.
   - For $j = 1, \ldots, n$ compute k^j_c according to the following cases:
     - *Case 1 – $(\Lambda_a, \Lambda_b) = (0,0)$:* compute
       $$k^j_c = A^j_g - \left(\sum_{i=1}^n F_{k^i_a}(0\|j\|g) + F_{k^i_b}(0\|j\|g) \right).$$
     - *Case 2 – $(\Lambda_a, \Lambda_b) = (0,1)$:* compute
       $$k^j_c = B^j_g - \left(\sum_{i=1}^n F_{k^i_a}(1\|j\|g) + F_{k^i_b}(0\|j\|g) \right).$$
     - *Case 3 – $(\Lambda_a, \Lambda_b) = (1,0)$:* compute
       $$k^j_c = C^j_g - \left(\sum_{i=1}^n F_{k^i_a}(0\|j\|g) + F_{k^i_b}(1\|j\|g) \right).$$
     - *Case 4 – $(\Lambda_a, \Lambda_b) = (1,1)$:* compute
       $$k^j_c = D^j_g - \left(\sum_{i=1}^n F_{k^i_a}(1\|j\|g) + F_{k^i_b}(1\|j\|g) \right).$$
   - If $k^i_c \notin \{k^i_{c,0}, k^i_{c,1}\}$, then P_i outputs abort. Otherwise, it proceeds. If P_i aborts it notifies all other parties with that information. If P_i is notified that another party has aborted it aborts as well.
   - If $k^i_c = k^i_{c,0}$ then P_i sets $\Lambda_c = 0$; if $k^i_c = k^i_{c,1}$ then P_i sets $\Lambda_c = 1$.
   - The output of the gate is defined to be $(k^1_c, \ldots, k^n_c)$ and Λ_c.
5. Assuming party P_i does not abort it will obtain Λ_w for every circuit-output wire w. The party can then recover the actual output value from $\rho_w = \Lambda_w \oplus \lambda_w$, where λ_w was obtained in the preprocessing stage.

In order to optimize the performance of the preprocessing-II phase, the secure computation does not evaluate the pseudorandom function $F()$, but rather has the parties compute $F()$ and provide the results as an input to the protocol. Observe that corrupted parties may provide *incorrect* input values $F_{k_{x,j}^i}()$ and thus the resulting garbled circuit may not actually be a valid BMR garbled circuit. Nevertheless, we show that such behavior can only result in an abort. This is due to the fact that if a value is incorrect and honest parties see that their key (coordinate) is not present in the resulting vector then they will abort. In contrast, if their seed is present then they proceed and the incorrect value had no effect. Since the keys are secret, the adversary cannot give an incorrect value that will result in a correct *different* key, except with negligible probability. This is important since otherwise correctness would be harmed. Likewise, a corrupted party cannot influence the masking values λ, and thus they are consistent throughout (when a given wire is input into multiple gates), ensuring correctness.

2.4   Securely Computing $\mathcal{F}_{\mathsf{SFE}}$ in the $\mathcal{F}_{\mathsf{offline}}$-Hybrid Model

We now define our protocol Π_{SFE} for securely computing $\mathcal{F}_{\mathsf{SFE}}$ (using the BMR garbled circuit) in the $\mathcal{F}_{\mathsf{offline}}$-hybrid model, see Protocol 1. In the full version of this paper [15], we prove the following theorem:

Theorem 1. *If F is a pseudorandom function, then Protocol Π_{SFE} securely computes $\mathcal{F}_{\mathsf{MPC}}$ in the $\mathcal{F}_{\mathsf{offline}}$-hybrid model, in the presence of a static malicious adversary corrupting any number of parties.*

2.5   Implementing $\mathcal{F}_{\mathsf{offline}}$ in the $\mathcal{F}_{\mathsf{MPC}}$-Hybrid Model

At first sight, it may seem that in order to construct an entire garbled circuit (i.e. the output of $\mathcal{F}_{\mathsf{offline}}$), an ideal functionality that computes each garbled gate can be used separately for each gate of the circuit (that is, for each gate the parties provide their PRF results on the keys and shares of the masking values associated with that gate's wires). This is sufficient when considering semi-honest adversaries. However, in the setting of malicious adversaries, this can be problematic since parties may input inconsistent values. For example, the masking values λ_w that are common to a number of gates (which happens when any wire enters more than one gate) need to be identical in all of these gates. In addition, the pseudorandom function values may not be correctly computed from the pseudorandom function keys that are input. In order to make the computation of the garbled circuit efficient, we will not check that the pseudorandom function values are correct. However, it is necessary to ensure that the λ_w values are correct, and that they (and likewise the keys) are consistent between gates (e.g., as in the case where the same wire is input to multiple gates). We achieve this by computing the entire circuit at once, via a single functionality.

The cost of this computation is actually almost the same as separately computing each gate. The single functionality receives from party P_i the values

$k_{w,0}^i, k_{w,1}^i$ and the output of the pseudorandom function applied to the keys *only once*, regardless of the number of gates to which w is input. Thereby consistency is immediate throughout, and this potential attack is prevented. Moreover, the λ_w values are generated once and used consistently by the circuit, making it easy to ensure that the λ values are correct.

Another issue that arises is that the single garbled gate functionality expects to receive a single masking value for each wire. However, since this value is secret, it must be generated from shares that are input by the parties. In the full version of the paper [15] we describe the full protocol for securely computing $\mathcal{F}_{\text{offline}}$ in the $\mathcal{F}_{\text{MPC}}$-hybrid model (i.e., using *any* protocol that securely computes the $\mathcal{F}_{\text{MPC}}$ ideal functionality). In short, the parties input shares of λ_w to the functionality, the single masking value is computed from these shares, and then input to all the necessary gates.

In the semi-honest case, the parties could contribute a share which is random in $\{0,1\}$ (interpreted as an element in $\mathbb{F}_p$) and then compute the product of all the shares (using the underlying MPC) to obtain a random masking value in $\{0,1\}$. This is however not the case in the malicious case since parties might provide a share that is not from $\{0,1\}$ and thus the resulting masking value wouldn't likewise be from $\{0,1\}$.

This issue is solved in the following way. The computation is performed by having the parties input random masking values $\lambda_w^i \in \{1,-1\}$, instead of bits. This enables the computation of a value μ_w to be the *product* of $\lambda_w^1, \ldots, \lambda_w^n$ and to be random in $\{-1,1\}$ as long as one of them is random. The product is then mapped to $\{0,1\}$ in $\mathbb{F}_p$ by computing $\lambda_w = \frac{\mu_w+1}{2}$.

In order to prevent corrupted parties from inputting λ_w^i values that are not in $\{-1,+1\}$, the protocol for computing the circuit outputs $(\prod_{i=1}^n \lambda_w^i)^2 - 1$, for every wire w (where λ_w^i is the share contributed from party i for wire w), and the parties can simply check whether it is equal to zero or not. Thus, if any party cheats by causing some $\lambda_w \notin \{-1,+1\}$, then this will be discovered since the circuit outputs a non-zero value for $(\prod_{i=1}^n \lambda_w^i)^2 - 1$, and so the parties detect this and can abort. Since this occurs before any inputs are used, nothing is revealed by this. Furthermore, if $\prod_{i=1}^n \lambda_w^i \in \{-1,+1\}$, then the additional value output reveals nothing about λ_w itself.

In the next section we shall remove *all* of the complications by basing our implementation for $\mathcal{F}_{\text{MPC}}$ upon the specific SPDZ protocol. The reason why the SPDZ implementation is simpler – and more efficient – is that SPDZ provides generation of such shared values effectively for free.

3   The SPDZ Based Instantiation

3.1   Utilizing the SPDZ Protocol

As discussed in Sect. 2.2, in the offline-phase we use an underlying secure computation protocol, which, given a binary circuit and the matching inputs to its input wires, securely and distributively computes that binary circuit. In this section we simplify and optimize the implementation of the protocol Π_{offline} which implements the functionality $\mathcal{F}_{\text{offline}}$ by utilizing the specific SPDZ MPC protocol as the underlying implementation of $\mathcal{F}_{\text{MPC}}$. These optimizations are possible because the SPDZ MPC protocol provides a richer interface to the protocol designer than the naive generic MPC interface given in functionality $\mathcal{F}_{\text{MPC}}$. In particular, it provides the capability of directly generating shared random bits and strings. These are used for generating the masking values and pseudorandom function keys. Note that one of the most expensive steps in FairplayMP [2] was coin tossing to generate the masking values; by utilizing the specific properties of SPDZ this is achieved essentially for free.

Functionality 4 (The SPDZ Functionality: $\mathcal{F}_{\text{SPDZ}}$)

The functionality consists of seven externally exposed commands **Initialize**, **InputData**, **RandomBit**, **Random**, **Add**, **Multiply**, and **Output** and one internal subroutine **Wait**.

Initialize: On input $(init, p, M, B, R, I)$ from all parties, the functionality activates and stores p. Pre-processing is performed to generate data needed to respond to a maximum of M **Multiply**, B **RandomBit**, R **Random** commands, and I **InputData** commands per party.

Wait: This waits on the adversary to return a *GO/NO-GO* decision. If the adversary returns *NO-GO* then the functionality aborts.

InputData: On input $(input, P_i, varid, x)$ from P_i and $(input, P_i, varid, ?)$ from all other parties, with $varid$ a fresh identifier, the functionality stores $(varid, x)$. The functionality then calls **Wait**.

RandomBit: On command $(randombit, varid)$ from all parties, with $varid$ a fresh identifier, the functionality selects a random value $r \in \{0, 1\}$ and stores $(varid, r)$. The functionality then calls **Wait**.

Random: On command $(random, varid)$ from all parties, with $varid$ a fresh identifier, the functionality selects a random value $r \in \mathbb{F}_p$ and stores $(varid, r)$. The functionality then calls **Wait**.

Add: On command $(add, varid_1, varid_2, varid_3)$ from all parties (if $varid_1, varid_2$ are present in memory), the functionality retrieves $(varid_1, x)$, $(varid_2, y)$, stores $(varid_3, x + y)$ and then calls **Wait**.

Multiply: On input $(multiply, varid_1, varid_2, varid_3)$ from all parties (if $varid_1, varid_2$ are present in memory), the functionality retrieves $(varid_1, x)$, $(varid_2, y)$, stores $(varid_3, x \cdot y)$ and then calls **Wait**.

Output: On input $(output, varid, i)$ from all honest parties (if $varid$ is present in memory), the functionality retrieves $(varid, x)$ and outputs either $(varid, x)$ in the case of $i \neq 0$ or $(varid)$ if $i = 0$ to the adversary. The functionality then calls **Wait**, and only if **Wait** does not abort then it outputs x to all parties if $i = 0$, or it outputs x only to party i if $i \neq 0$.

In Sect. 3.2 we describe explicit operations that are to be carried out on the inputs in order to achieve the desired output; the circuit's complexity analysis appears in Sect. 3.3 and the expected results from an implementation of the circuit using the SPDZ protocol are in Sect. 3.4.

Throughout, we utilize $\mathcal{F}_{\text{SPDZ}}$ (Functionality 4), which represents an idealized representation of the SPDZ protocol, akin to the functionality $\mathcal{F}_{\text{MPC}}$ from

Sect. 2.2. Note that in the real protocol, $\mathcal{F}_{\text{SPDZ}}$ is implemented itself by an offline phase (essentially corresponding to our preprocessing-I) and an online phase (corresponding to our preprocessing-II). We fold the SPDZ offline phase into the **Initialize** command of $\mathcal{F}_{\text{SPDZ}}$. In the SPDZ offline phase we need to know the maximum number of multiplications, random values and random bits required in the online phase. In that phase the random shared bits and values are produced, as well as the "Beaver Triples" for use in the multiplication gates performed in the SPDZ online phase. In particular the consuming of shared random bits and values results in no cost during the SPDZ online phase, with all consumption costs being performed in the SPDZ offline phase. The protocol, which utilizes Somewhat Homomorphic Encryption to produce the shared random values/bits and the Beaver multiplication triples, is given in [7].

As before, we use the notation $[varid]$ to represent the result stored in the variable $varid$ by the functionality. In particular we use the arithmetic shorthands $[z] = [x] + [y]$ and $[z] = [x] \cdot [y]$ to represent the result of calling the **Add** and **Multiply** commands on the functionality $\mathcal{F}_{\text{SPDZ}}$.

3.2   The Π_{offline} SPDZ Based Protocol

As remarked earlier $\mathcal{F}_{\text{offline}}$ can be securely computed using *any* secure multi-party protocol. This is advantageous since it means that future efficiency improvements to concretely secure multi-party computation (with dishonest majority) will automatically make our protocol faster. However, currently the best option is SPDZ. Specifically, it utilizes the fact that SPDZ can very efficiently generate coin tosses. This means that it is not necessary for the parties to input the λ_w^i values, to multiply them together to obtain λ_w and to output the check values $(\lambda_w)^2 - 1$. Thus, this yields a significant efficiency improvement. We now describe the protocol which implements $\mathcal{F}_{\text{offline}}$ in the $\mathcal{F}_{\text{SPDZ}}$-hybrid model

preprocessing-I

1. **Initialize the MPC Engine:** Call **Initialize** on the functionality $\mathcal{F}_{\text{SPDZ}}$ with input p, a prime with $p > 2^k$ and with parameters

$$M = 13 \cdot G, \quad B = W, \quad R = 2 \cdot W \cdot n, \quad I = 2 \cdot G \cdot n + W,$$

   where G is the number of gates, n is the number of parties and W is the number of input wires per party. In practice the term W in the calculation of I needs only be an upper bound on the total number of input wires per party in the circuit which will eventually be evaluated.
2. **Generate Wire Masks:** For every circuit wire w we need to generate a sharing of the (secret) masking-values λ_w. Thus for *all* wires w the parties execute the command **RandomBit** on the functionality $\mathcal{F}_{\text{SPDZ}}$, the output is denoted by $[\lambda_w]$. The functionality $\mathcal{F}_{\text{SPDZ}}$ guarantees that $\lambda_w \in \{0, 1\}$.
3. **Generate Keys:** For every wire w, each party $i \in [1, \ldots, n]$ and for $j \in \{0, 1\}$, the parties call **Random** on the functionality $\mathcal{F}_{\text{SPDZ}}$ to obtain output $[k_{w,j}^i]$. The parties then call **Output** to open $[k_{w,j}^i]$ to party i for all j and w. The vector of shares $[k_{w,j}^i]_{i=1}^n$ we shall denote by $[\mathbf{k}_{w,j}]$.

preprocessing-II (This protocol implements the computation gate table as it is detailed in the BMR protocol. The correctness of this construction is explained in the full version of the paper.)

1. **Output Input Wire Values:** For all wires w which are attached to party P_i we execute the command **Output** on the functionality $\mathcal{F}_{\mathsf{SPDZ}}$ to open $[\lambda_w]$ to party i.

2. **Output Masks for Circuit-Output-Wires:** In order to reveal the real values of the circuit-output-wires it is required to reveal their masking values. That is, for every circuit-output-wire w, the parties execute the command **Output** on the functionality $\mathcal{F}_{\mathsf{SPDZ}}$ for the stored value $[\lambda_w]$.

3. **Calculate Garbled Gates:** This step is operated for each gate g in the circuit in parallel. Specifically, let g be a gate whose input wires are a, b and output wire is c. Do as follows:

   (a) **Calculate Output Indicators:** This step calculates four indicators $[x_a], [x_b], [x_c], [x_d]$ whose values will be in $\{0,1\}$. Each one of the garbled labels $\mathbf{A}_g, \mathbf{B}_g, \mathbf{C}_g, \mathbf{D}_g$ is a vector of n elements that hide either the vector $\mathbf{k}_{c,0} = k_{c,0}^1, \ldots, k_{c,0}^n$ or $\mathbf{k}_{c,1} = k_{c,1}^1, \ldots, k_{c,1}^n$; which one it hides depends on these indicators, i.e. if $x_a = 0$ then $\mathbf{A}_g$ hides $\mathbf{k}_{c,0}$ and if $x_a = 1$ then $\mathbf{A}_g$ hides $\mathbf{k}_{c,1}$. Similarly, $\mathbf{B}_g$ depends on x_b, $\mathbf{C}_g$ depends on x_c and $\mathbf{D}_c$ depends on x_d. Each indicator is determined by some function on $[\lambda_a], [\lambda_b], [\lambda_c]$ and the truth table of the gate f_g. Every indicator is calculated slightly different, as follows (concrete examples are given after the preprocessing specification):

   $$[x_a] = \left(f_g([\lambda_a], [\lambda_b]) \stackrel{?}{\neq} [\lambda_c] \right) = (f_g([\lambda_a], [\lambda_b]) - [\lambda_c])^2$$

   $$[x_b] = \left(f_g([\lambda_a], [\overline{\lambda_b}]) \stackrel{?}{\neq} [\lambda_c] \right) = (f_g([\lambda_a], (1 - [\lambda_b])) - [\lambda_c])^2$$

   $$[x_c] = \left(f_g([\overline{\lambda_a}], [\lambda_b]) \stackrel{?}{\neq} [\lambda_c] \right) = (f_g((1 - [\lambda_a]), [\lambda_b]) - [\lambda_c])^2$$

   $$[x_d] = \left(f_g([\overline{\lambda_a}], [\overline{\lambda_b}]) \stackrel{?}{\neq} [\lambda_c] \right) = (f_g((1 - [\lambda_a]), (1 - [\lambda_b])) - [\lambda_c])^2$$

   where the binary operator $\stackrel{?}{\neq}$ is defined as $[a] \stackrel{?}{\neq} [b]$ equals $[0]$ if $a = b$, and equals $[1]$ if $a \neq b$. For the XOR function on a and b, for example, the operator can be evaluated by computing $[a] + [b] - 2 \cdot [a] \cdot [b]$. Thus, these can be computed using **Add** and **Multiply**.

   (b) **Assign the Correct Vector:** As described above, we use the calculated indicators to choose for every garbled label either $\mathbf{k}_{c,0}$ or $\mathbf{k}_{c,1}$. Calculate:

   $$[\mathbf{v}_{c,x_a}] = (1 - [x_a]) \cdot [\mathbf{k}_{c,0}] + [x_a] \cdot [\mathbf{k}_{c,1}]$$
   $$[\mathbf{v}_{c,x_b}] = (1 - [x_b]) \cdot [\mathbf{k}_{c,0}] + [x_a] \cdot [\mathbf{k}_{c,1}]$$
   $$[\mathbf{v}_{c,x_c}] = (1 - [x_c]) \cdot [\mathbf{k}_{c,0}] + [x_a] \cdot [\mathbf{k}_{c,1}]$$
   $$[\mathbf{v}_{c,x_d}] = (1 - [x_d]) \cdot [\mathbf{k}_{c,0}] + [x_a] \cdot [\mathbf{k}_{c,1}]$$

In each equation either the value $\mathbf{k}_{c,0}$ or the value $\mathbf{k}_{c,1}$ is taken, depending on the corresponding indicator value. Once again, these can be computed using **Add** and **Multiply**.

(c) **Calculate Garbled Labels:** Party i knows the value of $k^i_{w,b}$ (for wire w that enters gate g) for $b \in \{0,1\}$, and so can compute the $2 \cdot n$ values $F_{k^i_{w,b}}(0\,\|\,1\,\|\,g), \ldots, F_{k^i_{w,b}}(0\,\|\,n\,\|\,g)$ and $F_{k^i_{w,b}}(1\,\|\,1\,\|\,g), \ldots, F_{k^i_{w,b}}(1\,\|\,n\,\|\,g)$. Party i inputs them by calling **InputData** on the functionality $\mathcal{F}_{\mathsf{SPDZ}}$. The resulting input pseudorandom vectors are denoted by

$$[F^0_{k^i_{w,b}}(g)] = [F_{k^i_{w,b}}(0\,\|\,1\,\|\,g), \ldots, F_{k^i_{w,b}}(0\,\|\,n\,\|\,g)]$$
$$[F^1_{k^i_{w,b}}(g)] = [F_{k^i_{w,b}}(1\,\|\,1\,\|\,g), \ldots, F_{k^i_{w,b}}(1\,\|\,n\,\|\,g)].$$

The parties now compute $[\mathbf{A}_g], [\mathbf{B}_g], [\mathbf{C}_g], [\mathbf{D}_g]$, using **Add**, via

$$[\mathbf{A}_g] = \sum_{i=1}^{n} \left([F^0_{k^i_{a,0}}(g)] + [F^0_{k^i_{b,0}}(g)] \right) + [\mathbf{v}_{c,x_a}]$$
$$[\mathbf{B}_g] = \sum_{i=1}^{n} \left([F^1_{k^i_{a,0}}(g)] + [F^0_{k^i_{b,1}}(g)] \right) + [\mathbf{v}_{c,x_b}]$$
$$[\mathbf{C}_g] = \sum_{i=1}^{n} \left([F^0_{k^i_{a,1}}(g)] + [F^1_{k^i_{b,0}}(g)] \right) + [\mathbf{v}_{c,x_c}]$$
$$[\mathbf{D}_g] = \sum_{i=1}^{n} \left([F^1_{k^i_{a,1}}(g)] + [F^1_{k^i_{b,1}}(g)] \right) + [\mathbf{v}_{c,x_d}]$$

where every $+$ operation is performed on vectors of n elements.

4. **Notify Parties:** Output construction-done.

The functions f_g in Step 3a above depend on the specific gate being evaluated. For example, on clear values we have,

- If $f_g = \wedge$ (i.e. the AND function), $\lambda_a = 1$, $\lambda_b = 1$ and $\lambda_c = 0$ then $x_a = ((1 \wedge 1) - 0)^2 = (1 - 0)^2 = 1$. Similarly $x_b = ((1 \wedge (1-1)) - 0)^2 = (0-0)^2 = 0$, $x_c = 0$ and $x_d = 0$. The parties can compute f_g on shared values $[x]$ and $[y]$ by computing $f_g([x], [y]) = [x] \cdot [y]$.
- If $f_g = \oplus$ (i.e. the XOR function), then $x_a = ((1 \oplus 1) - 0)^2 = (0 - 0)^2 = 0$, $x_b = ((1 \oplus (1-1)) - 0)^2 = (1-0)^2 = 1$, $x_c = 1$ and $x_d = 0$. The parties can compute f_g on shared values $[x]$ and $[y]$ by computing $f_g([x], [y]) = [x] + [y] - 2 \cdot [x] \cdot [y]$.

Below, we will show how $[x_a], [x_b], [x_c]$ and $[x_d]$ can be computed more efficiently.

3.3 Circuit Complexity

In this section we analyze the complexity of the above circuit in terms of the number of multiplication gates and its depth. We are highly concerned with multiplication gates since, given the SPDZ shares $[a]$ and $[b]$ of the secrets a, and b resp., an interaction between the parties is required to achieve a secret sharing of the secret $a \cdot b$. Achieving a secret sharing of a linear combination of a and b (i.e. $\alpha \cdot a + \beta \cdot b$ where α and β are constants), however, can be done locally and is thus considered negligible. We are interested in the depth of the circuit because it gives a lower bound on the number of rounds of interaction that our circuit requires (note that here, as before, we are concerned with the depth in terms of multiplication gates).

Multiplication Gates: We first analyze the number of multiplication operations that are carried out per gate (i.e. in step 3) and later analyze the entire circuit.

– **Multiplications Per Gate.** We will follow the calculation that is done per gate chronologically as it occurs in step 3 of preprocessing-II phase:

1. In order to calculate the indicators in step 3a it suffices to compute one multiplication and 4 squares. We can do this by altering the equations a little. For example, for $f_g = AND$, we calculate the indicators by first computing $[t] = [\lambda_a] \cdot [\lambda_b]$ (this is the only multiplication) and then $[x_a] = ([t] - [\lambda_c])^2$, $[x_b] = ([\lambda_a] - [t] - [\lambda_c])^2$, $[x_c] = ([\lambda_b] - [t] - [\lambda_c])^2$, and $[x_d] = (1 - [\lambda_a] - [\lambda_b] + [t] - [\lambda_c])^2$.

    As another example, for $f_g = XOR$, we first compute $[t] = [\lambda_a] \oplus [\lambda_b] = [\lambda_a] + [\lambda_b] - 2 \cdot [\lambda_a] \cdot [\lambda_b]$ (this is the only multiplication), and then $[x_a] = ([t] - [\lambda_c])^2$, $[x_b] = (1 - [\lambda_a] - [\lambda_b] + 2 \cdot [t] - [\lambda_c])^2$, $[x_c] = [x_b]$, and $[x_d] = [x_a]$. Observe that in XOR gates only two squaring operations are needed.

2. To obtain the correct vector (in step 3b) which is used in each garbled label, we carry out 8 multiplications. Note that in XOR gates only 4 multiplications are needed, because $\mathbf{k}_{c,x_c} = \mathbf{k}_{c,x_b}$ and $\mathbf{k}_{c,x_d} = \mathbf{k}_{c,x_a}$.

Summing up, we have 4 squaring operations in addition to 9 multiplication operations per AND gate and 2 squarings in addition to 5 multiplications per XOR gate.

– **Multiplications in the Entire Circuit.** Denote the number of multiplication operation per gate (i.e. 13 for AND and 7 for XOR) by c, we get $G \cdot c$ multiplications for garbling all gates (where G is the number of gates in the boolean circuit computing the functionality f). Besides garbling the gates we have no other multiplication operations in the circuit. Thus we require $c \cdot G$ multiplications in total.

Depth of the Circuit and Round Complexity: Each gate can be garbled by a circuit of depth 3 (two levels are required for step 3a and another one for step 3b). Recall that additions are local operations only and thus we measure depth in terms of multiplication gates only. Since all gates can be garbled in parallel this implies an overall depth of three. (Of course in practice it may be more efficient to garble a set of gates at a time so as to maximize the use of bandwidth and CPU resources.) Since the number of rounds of the SPDZ protocol is in the order of the depth of the circuit, it follows that $\mathcal{F}_{\text{offline}}$ can be securely computed in a constant number of rounds.

Other Considerations: The overall cost of the pre-processing does not just depend on the number of multiplications. Rather, the parties also need to produce the random data via calls to **Random** and **RandomBit** to the functionality $\mathcal{F}_{\text{SPDZ}}$.[2] It is clear all of these can be executed in parallel. If W is the number of wires in the circuit then the total number of calls to **RandomBit** is equal to W, whereas the total number of calls to **Random** is $2 \cdot n \cdot W$.

Arithmetic vs Boolean Circuits: Our protocol will perform favourably for functions which are reasonably represented as boolean circuit, but the low round complexity may be outweighed by other factors when the function can be expressed much more succinctly using an arithmetic circuit, or other programatic representation as in [12]. In such cases, the performance would need to be tested for the specific function.

3.4 Expected Runtimes

To estimate the running time of our protocol, we extrapolate from known public data [7,8]. The offline phase of our protocol runs both the offline and online phases of the SPDZ protocol. The numbers below refer to the SPDZ offline phase, as described in [7], with covert security and a 20 % probability of cheating, using finite fields of size 128-bits, to obtain the following generation times (in milliseconds). As described in [7], comparable times are obtainable for running in the fully malicious mode (but more memory is needed) (Table 1).

Table 1. SPDZ offline generation times in milliseconds per operation

No. parties	Beaver triple	RandomBit	Random	Input
2	0.4	0.4	0.3	0.3
3	0.6	0.5	0.4	0.4
4	0.9	1.2	0.9	0.9

The implementation of the SPDZ online phase, described in both [7] and [12], reports online throughputs of between 200,000 and 600,000 per second for multiplication, depending on the system configuration. As remarked earlier the online time of other operations is negligible and are therefore ignored.

To see what this would imply in practice consider the AES circuit described in [18]; which has become the standard benchmarking case for secure computation calculations. The basic AES circuit has around 33,000 gates and a similar number of wires, including the key expansion within the circuit.[3] Assuming the parties share a XOR sharing of the AES key, (which adds an additional $2 \cdot n \cdot 128$ gates and wires to the circuit), the parameters for the **Initialize** call to the $\mathcal{F}_{\text{SPDZ}}$ functionality in the preprocessing-I protocol will be

$$M \approx 429,000, \quad B \approx 33,000, \quad R \approx 66,000 \cdot n, \quad I \approx 66,000 \cdot n + 128.$$

---

[2] These **Random** calls are followed immediately with an **Open** to a party. However, in SPDZ **Random** followed by **Open** has roughly the same cost as **Random** alone.

[3] Note that unlike [18] and other Yao based techniques we cannot process XOR gates for free. On the other hand we are not restricted to only two parties.

Using the above execution times for the SPDZ protocol we can then estimate the time needed for the two parts of our processing step for the AES circuit. The expected execution times, in seconds, are given in the following table. These expected times, due to the methodology of our protocol, are likely to estimate both the latency and throughput amortized over many executions.

No. parties	preprocessing-I	preprocessing-II
2	264	0.7–2.0
3	432	0.7–2.0
4	901	0.7–2.0

The execution of the online phase of our protocol, when the parties are given their inputs and actually want to compute the function, is very efficient: all that is needed is the evaluation of a garbled circuit based on the data obtained in the offline stage. Specifically, for each gate each party needs to process two input wires, and for each wire it needs to expand n seeds to a length which is n times their original length (where n denotes the number of parties). Namely, for each gate each party needs to compute a pseudorandom function $2n^2$ times (more specifically, it needs to run $2n$ key schedulings, and use each key for n encryptions). We examined the cost of implementing these operations for an AES circuit of 33,000 gates when the pseudorandom function is computed using the AES-NI instruction set. The run times for $n = 2, 3, 4$ parties were 6.35 ms, 9.88 ms and 15 ms, respectively, for C code compiled using the gcc compiler on a 2.9 GHZ Xeon machine. The actual run time, including all non-cryptographic operations, should be higher, but of the same order.

Our run-times estimates compare favourably to several other results on implementing secure computation of AES in a multiparty setting:

- In [6] an actively secure computation of AES using SPDZ took an offline time of over five minutes per AES block, with an online time of around a quarter of a second; that computation used a security parameter of 64 as opposed to our estimates using a security parameter of 128.
- In [12] another experiment was shown which can achieve a latency of 50 ms in the online phase for AES (but no offline times are given).
- In [17] the authors report on a two-party MPC evaluation of the AES circuit using the Tiny-OT protocol; they obtain for 80 bits of security an amortized offline time of nearly three seconds per AES block, and an amortized online time of 30 ms; but the reported non-amortized latency is much worse. Furthermore, this implementation is limited to the case of *two parties*, whereas we obtain security for multiple parties.

Most importantly, all of the above experiments were carried out in a LAN setting where communication latency is very small. However, in other settings where parties are not connect by very fast connections, the effect of the number of rounds on the protocol will be extremely significant. For example, in [6], an arithmetic circuit for AES is constructed of depth 120, and this is then reduced to depth 50 using a bit decomposition technique. Note that if parties are in separate geographical locations, then this number of rounds will very quickly dominate the running time. For example, the latency on Amazon EC2 between Virginia and Ireland is 75ms. For a circuit depth of 50, and even assuming just a *single* round per level, the running-time cannot be less than 3750 ms (even if computation takes *zero time*). In contrast, our online phase has just 2 rounds of communication and so will take in the range of 150 ms. We stress that even on a much faster network with latency of just 10ms, protocols with 50 rounds of communication will still be slow.

Acknowledgments. The first and fourth authors were supported in part by the European Research Council under the European Union's Seventh Framework Programme (FP/2007-2013) / ERC consolidators grant agreement n. 615172 (HIPS). The second author was supported under the European Union's Seventh Framework Program (FP7/2007-2013) grant agreement n. 609611 (PRACTICE), and by a grant from the Israel Ministry of Science, Technology and Space (grant 3-10883). The third author was supported in part by ERC Advanced Grant ERC-2010-AdG-267188-CRIPTO and by EPSRC via grant EP/I03126X. The first and third authors were also supported by an award from EPSRC (grant EP/M012824), from the Ministry of Science, Technology and Space, Israel, and the UK Research Initiative in Cyber Security.

References

1. Beaver, D., Micali, S., Rogaway, P.: The round complexity of secure protocols. In: Ortiz, H. (ed.) 22nd STOC, pp. 503–513. ACM (1990)
2. Ben-David, A., Nisan, N., Pinkas, B.: FairplayMP: a system for secure multi-party computation. In: Ning, P., Syverson, P.F., Jha, S. (eds.) ACM CCS, pp. 257–266. ACM (2008)
3. Ben-Or, M., Goldwasser, S., Wigderson, A.: Completeness theorems for non-cryptographic fault-tolerant distributed computation. In: Simon [21], pp. 1–10
4. Chaum, D., Crépeau, C., Damgård, I.: Multiparty unconditionally secure protocols. In: Simon [21], pp. 11–19
5. Choi, S.G., Katz, J., Malozemoff, A.J., Zikas, V.: Efficient three-party computation from cut-and-choose. In: Garay, Gennaro [9], pp. 513–530
6. Damgård, I., Keller, M., Larraia, E., Miles, C., Smart, N.P.: Implementing AES via an actively/covertly secure dishonest-majority MPC protocol. In: Visconti, I., De Prisco, R. (eds.) SCN 2012. LNCS, vol. 7485, pp. 241–263. Springer, Heidelberg (2012)
7. Damgård, I., Keller, M., Larraia, E., Pastro, V., Scholl, P., Smart, N.P.: Practical covertly secure MPC for dishonest majority – or: breaking the SPDZ limits. In: Crampton, J., Jajodia, S., Mayes, K. (eds.) ESORICS 2013. LNCS, vol. 8134, pp. 1–18. Springer, Heidelberg (2013)
8. Damgård, I., Pastro, V., Smart, N.P., Zakarias, S.: Multiparty computation from somewhat homomorphic encryption. In: Safavi-Naini, Canetti [20], pp. 643–662
9. Garay, J.A., Gennaro, R. (eds.): CRYPTO 2014, Part II. LNCS, vol. 8617, pp. 458–475. Springer, Heidelberg (2014)
10. Goldreich, O., Micali, S., Wigderson, A.: How to play any mental game or A completeness theorem for protocols with honest majority. In: Aho, A.V. (ed.) Proceedings STOC 1987, pp. 218–229. ACM (1987)

11. Goyal, V.: Constant round non-malleable protocols using one way functions. In: Fortnow, L., Vadhan, S.P. (eds.) Proceedings STOC 2011, pp. 695–704. ACM (2011)

12. Keller, M., Scholl, P., Smart, N.P.: An architecture for practical actively secure MPC with dishonest majority. In: Sadeghi, A., Gligor, V.D., Yung, M. (eds.) 2013 ACM CCS 2013, pp. 549–560. ACM (2013)

13. Larraia, E., Orsini, E., Smart, N.P.: Dishonest majority multi-party computation for binary circuits. In: Garay, Gennaro [9], pp. 495–512

14. Lindell, Y.: Fast cut-and-choose based protocols for malicious and covert adversaries. In: Canetti, R., Garay, J.A. (eds.) CRYPTO 2013, Part II. LNCS, vol. 8043, pp. 1–17. Springer, Heidelberg (2013)

15. Lindell, Y., Pinkas, B., Smart, N.P., Yanai, A.: Efficient constant round multi-party computation combining BMR and SPDZ (Full Version). IACR Cryptology ePrint Archive 2015:523 (2015)

16. Lindell, Y., Riva, B.: Cut-and-choose yao-based secure computation in the online/offline and batch settings. In: Garay, Gennaro [9], pp. 476–494

17. Nielsen, J.B., Nordholt, P.S., Orlandi, C., Burra, S.S.: A new approach to practical active-secure two-party computation. In: Safavi-Naini, Canetti [20], pp. 681–700

18. Pinkas, B., Schneider, T., Smart, N.P., Williams, S.C.: Secure two-party computation is practical. In: Matsui, M. (ed.) ASIACRYPT 2009. LNCS, vol. 5912, pp. 250–267. Springer, Heidelberg (2009)

19. Rabin, T., Ben-Or, M.: Verifiable secret sharing and multiparty protocols with honest majority. In: Johnson, D.S. (ed.) Proceedings STOC 1989, pp. 73–85. ACM (1989)

20. Safavi-Naini, R., Canetti, R. (eds.): CRYPTO 2012. LNCS, vol. 7417. Springer, Heidelberg (2012)

21. Simon, J. (ed.) Proceedings STOC 1988. ACM (1988)

22. Yao, A.C.: Protocols for secure computations. In: Proceedings FOCS 1982, pp. 160–164. IEEE Computer Society (1982)

Round-Optimal Black-Box Two-Party Computation

Rafail Ostrovsky[1], Silas Richelson[1], and Alessandra Scafuro[2(✉)]

[1] UCLA, Los Angeles, USA
[2] Boston University and Northeastern University, Boston, USA
{rafail,sirichel}@ucla.edu, scafuro@bu.edu

Abstract. In [Eurocrypt 2004] Katz and Ostrovsky establish the *exact* round complexity of secure two-party computation with respect to black-box proofs of security. They prove that 5 rounds are necessary for secure two-party protocols (4-round are sufficient if only one party receives the output) and provide a protocol that matches such lower bound. The main challenge when designing such protocol is to parallelize the proofs of consistency provided by both parties – necessary when security against malicious adversaries is considered – in 4 rounds. Toward this goal they employ specific proofs in which the statement can be unspecified till the last round but that require non-black-box access to the underlying primitives.

A rich line of work [1,9,11,13,24] has shown that the non-black-box use of the cryptographic primitive in secure two-party computation is not necessary by providing black-box constructions matching basically all the feasibility results that were previously demonstrated only via non-black-box protocols.

All such constructions however are far from being round optimal. The reason is that they are based on cut-and-choose mechanisms where one party can safely take an action only *after* the other party has successfully completed the cut-and-choose phase, therefore requiring additional rounds.

A natural question is whether round-optimal constructions do inherently require non-black-box access to the primitives, and whether the lower bound shown by Katz and Ostrovsky can only be matched by a non-black-box protocol.

In this work we show that round-optimality is achievable even with only black-box access to the primitives. We provide the first 4-round black-box oblivious transfer based on any enhanced trapdoor permutation. Plugging a parallel version of our oblivious transfer into the black-box non-interactive secure computation protocol of [12] we obtain the first round-optimal black-box two-party protocol in the plain model for any functionality.

1 Introduction

Secure two-party computation allows two mutually distrustful parties to compute a function of their secret inputs without revealing any information except

© International Association for Cryptologic Research 2015
R. Gennaro and M. Robshaw (Eds.): CRYPTO 2015, Part II, LNCS 9216, pp. 339–358, 2015.
DOI: 10.1007/978-3-662-48000-7_17

what can be gathered from the output. It is known that achieving secure two-party computation information theoretically is impossible, and thus computation assumptions are required. In this work we are interested in construction for two-party computation based on general hardness assumptions in the plain model. Protocols based on general assumptions are flexible in that they allow the protocol to be implemented based on a variety concrete assumptions; even possibly ones which where not considered when the protocol was designed. Constructions based on general assumptions may use the cryptographic primitive based on the assumption in two ways: **black-box usage**, if the construction refers only to the input/output behavior of the underlying primitive; **non-black-box usage**, if the construction uses the code computing the functionality of the primitive. The advantage of black-box constructions is that their complexity is independent of the complexity of the implementation of the underlying primitive and are typically considered the first step towards practical constructions.

Secure Two-Party Computation Under General Assumptions. Yao [29] provided an elegant construction which securely realizes any two-party functionality, and which uses the underlying cryptographic primitives as black-box. This construction however guarantees security only against semi-honest adversaries, *i.e.*, adversaries that honestly follow the protocol. [5] show that semi-honest security is sufficient, as any protocol tolerating semi-honest adversaries can be compiled into one secure against malicious adversaries (*i.e.*, adversaries who can arbitrarily deviate from the protocol), by forcing the parties to prove, after each step, that they behaved honestly. Roughly, in this compiler, each party commits to his input at the very beginning and use a coin-flipping protocol to define the randomness that will be used in the semi-honest protocol. Then, for each protocol message, they add a zero-knowledge "proof of consistency" proving that the message was correctly computed according to the input committed and the randomness generated in the coin-flipping.

Unfortunately, this compiler is highly inefficient as the proofs of consistency require Karp reductions involving the circuits of the cryptographic primitives used. The exact complexity of these reductions grows more than linearly in the circuit complexity of the cryptographic primitive.[1] Researchers naturally began to wonder whether security against malicious adversaries could be achieved without relying on non-black-box use of cryptographic primitives.

Black-Box Secure Two-Party Computation. Ishai et al. [9,11] show that malicious security can be achieved without using expensive zero-knowledge proofs involving the code of the cryptographic primitives. Their work is based on the following observation: we can check that a party is honestly computing the protocol messages, by challenging the party to reveal the input and the randomness

---

[1] We note that a different approach altogether was taken by Kilian in [16], which does not require the use of cryptographic assumptions at all. However, his compiler works in the OT-hybrid model, thus we still need a protocol that implements the oblivious transfer functionality against malicious adversaries.

used in the computation. While this will certainly prove consistency[2], it is not zero-knowledge, as it leaks parties' entire inputs. Thus, the next idea is to have the parties engage in several parallel executions of the semi-honest protocol, where they run with random inputs. When a party is challenged on a random subset of protocol executions, she can safely reveal the randomness/inputs used in those executions which are independent of her actual inputs. If the party provides all convincing answers, then the challenger is guaranteed that the majority of the remaining executions are honestly computed as well. This step is repeated again in the opposite direction. Eventually after both parties have passed the tests, they run an additional step to "connect" the random inputs with their actual inputs and combine them across the remaining executions.

Following [9,11] subsequent work have shown black-box construction for adaptively secure two-party protocols [1], and constant-round black-box two-party protocols [13,24].

Round-Optimal Secure Two-Party Computation. In [15] Katz and Ostrovsky establish the *exact* round complexity of secure two-party computation from general assumptions. They show that 5 rounds are necessary and sufficient to compute any two-party functionality where both parties obtain the output, and 4 rounds are sufficient if only one party receives the output. To prove the upper bound they give a protocol that uses non-black-box proofs of consistency to enforce semi-honest behavior. The main technical difficulty they face is getting these proofs to complete in only 4 rounds — not an easy task as zero-knowledge in the standard model requires at least 4 rounds. Nevertheless, they manage to parallelize the proof and the computation into just 4 rounds using special properties of certain constructions of witness-indistinguishable (WI) proofs of knowledge. Namely, they crucially use the fact that in the WI proof of [19], the statement can be specified in the last round. Unfortunately, however, the statements to be proved concern values committed or computed in the protocol, and require the use of the circuits of the cryptographic primitives used in the protocol.

Previous results [1,11,24] have shown that essentially any feasibility result for two-party computation demonstrated using non-black-box techniques can also be obtained via black-box constructions. A natural question however, which so far has not been answered, is whether this is true for round optimal non-black-box constructions. This question is the focus of the current work. Namely,

> *Can we construct a round-optimal fully black-box protocol for two-party computation based on general assumptions?*

Black-Box Round-Optimal Two-Party Computation? When it comes to round optimality, the current state of the art suggests a negative answer. All known black-box protocols for secure computation achieve malicious security using a

---

[2] For sake of better clarity we are oversimplifying here. In [11] they introduce the definition of defensible adversaries and show how to use it in the cut-and-choose. We refer the reader to [11] for more details.

cut-and-choose mechanism that introduces additional rounds. The need for additional rounds seems inherent because in such mechanisms a party will take an action only *after* the other party has successfully completed the cut-and-choose phase. Additional rounds are used to combine and connect the random inputs used in the unopened sessions of the cut-and-choose with the real inputs.

An alternative to the traditional cut-and-choose approach for black-box construction was shown by Ishai et al. in [12], where they provide a black-box protocol for non-interactive secure two-party computation (NISC) based on the "MPC-in-the-head" paradigm of [13]. This approach however, following [14,16], works in the OT-hybrid model, and thus can only hope to achieve round optimality in the plain model if there exists a 4-round black-box oblivious transfer protocol in the plain model with parallel security.

One might hope that perhaps we can build a 4-round black-box oblivious transfer in the plain model starting from the 2-round OT protocol of Peikert et al. [25] − whose security is in the CRS model − by running a two-party coin flipping protocol to generate the CRS. We note that this approach seems doomed to fail because, as proved in [15], secure coin-flipping requires at least 5 rounds, *regardless of the use of the underlying cryptographic primitives*.

Our Contribution. In this paper we answer the above question positively by constructing a 4-round black-box oblivious transfer protocol based on the existence of (enhanced) trapdoor permutations. Our construction is easily extended to achieve parallel secure oblivious transfer which, using the compiler of [12], gives a round-optimal black-box protocol for two-party computation in the plain model.

1.1   Our Techniques

As mentioned above, it suffices to build a 4-round black-box oblivious transfer protocol based on general assumptions. We start with a high-level overview of the main ideas behind the construction.

Our starting point is the following basic 3-round protocol for OT based on black-box use of enhanced trapdoor permutations (TDP).

1. S chooses trapdoor permutation $(f, f^{-1}) \leftarrow \mathsf{Gen}(1^\kappa)$ and sends f to R.
2. R chooses $x \xleftarrow{\text{R}} \{0,1\}^\kappa$, and sends (z_0, z_1) to S where $z_b = f(x)$ and where $z_{1-b} \xleftarrow{\text{R}} \{0,1\}^\kappa$ is random.
3. S returns (w_0, w_1) where $w_a = s_a \oplus \mathsf{hc}\big(f^{-1}(z_a)\big)$, $a \in \{0,1\}$

where $\mathsf{hc}(\cdot)$ is a hardcore bit of f. If both parties follow the protocol then S can't learn anything about R's input bit b as both z_0 and z_1 are just random κ−bit strings. Similarly, the security of the TDP f ensures that R cannot distinguish w_{1-b} from random as long as z_{1-b} was truly chosen randomly. Unfortunately, there are two serious problems with this protocol. First, there is nothing to stop a malicious R from sending (z_0, z_1) such that he knows the pre-images of *both* values under f, thus allowing him to learn both s_0 and s_1. Indeed, the above protocol only offers security against a semi-honest receiver. Second, while the above protocol leaks no information to S about R's input bit, it is not simulateably secure. Input indistinguishably is often sufficient if a protocol is to be executed once in isolation, however we aim to use our OT as a building block for general 2PC, as such, stronger security is required.

Katz and Ostrovsky [15] solve the first problem by having the parties engage in a secure coin-flipping protocol to produce a random $r \in \{0,1\}^\kappa$ and forcing R to prove that either $z_0 = r$ or $z_1 = r$ using a witness-indistinguishable proof of knowledge. This denies R the freedom to generate both z_0 and z_1. Such WI proofs, however, require using the underlying commitment scheme, used for the coin-flipping, in a non-black-box way. Our solution to this problem can be seen as implementing the coin-flipping idea of [15] while making only black-box use of the commitment scheme. For this we use an adaptation of the black-box commit-and-prove protocol of Kilian [17].

We solve the second problem by having S commit the inputs already in the second round and prove that such committed inputs are the ones used for the OT. Doing this naïvely would require making non-black-box use of cryptographic primitives, so, in typical cut-and-choose style, we instead have S commit to *shares* of the inputs, and play the protocol many times in parallel where R opens mostly the shares corresponding to his input bit (to enable reconstruction of s_b) but enough shares of s_{1-b} to be convinced that S is playing fairly. This introduces several subtleties, that we discuss in the next paragraph.

We construct our OT protocol in two steps. First, we construct a 4-round OT protocol, $\Pi_{\mathsf{OT}}^{\mathsf{R}}$, that is simulatable only against a malicious receiver. Then, we use $\Pi_{\mathsf{OT}}^{\mathsf{R}}$ as a building block to build the final OT protocol that is simulatable for both parties. In the next two paragraphs we describe the ideas outlined above in greater details.

A 4-Round Black-Box OT Secure Against Malicious Receivers. We want to implement the coin-flipping that we mentioned above, without requiring R to give non-black-box proofs about the committed values. We do it by recasting the above problem in terms of equivocal and binding commitments, and having the output of the coin-flipping to be the pair of strings (z_0, z_1) (instead of a random r such that either $z_0 = r$ or $z_1 = r$). We provide a mechanism that allows R to compute one binding commitment and one equivocal commitment. In the coin-flipping, R first sends such commitments to S, then after seeing S's random strings, she opens both commitments. The crucial point is that R can control the output of one of the strings by equivocating one the commitments, while the other string will be truly random. With this tool we can directly obtain a black-box OT protocol that is simulatable for the receiver as follows.

1. R, on secret input b, chooses random strings r_0, r_1. Then sends commitments C_0, C_1 such that commitment C_b is equivocal. R proves that one of the commitments is binding.
2. S chooses trapdoor permutation $(f, f^{-1}) \leftarrow \mathsf{Gen}(1^\kappa)$ and sends f to R. Additionally S sends a random string r to R.

3. R chooses $x \overset{R}{\leftarrow} \{0,1\}^\kappa$ and computes $z_b = f(x)$. Then it equivocates commitment C_b so that it opens to $r_b = z_b \oplus r$, while it honestly opens value r_{1-b}.
4. S upon receiving r_0, r_1, computes $z_0 = r \oplus r_0$ and $z_1 = r \oplus r_1$ and sends (w_0, w_1) where $w_a = s_a \oplus \mathsf{hc}(f^{-1}(z_a))$.

If the proof in Step 1 is sound, R can only equivocate one string and thus knows the preimage of one value only. If the proofs and the commitments are hiding, the sender has no advantage in distinguishing which string is controlled by the receiver. Additionally, if we make the proof extractable, then the above protocol is simulatable against a malicious receiver.

Thus, what is left to do is to construct the tool that allows R to compute an equivocal commitments and a binding commitment and a WI proof of the binding of one of the two. This proof must be black-box and 3 rounds only. We implement this proof, by employing ideas from the black-box commit-and-prove protocol due to Kilian [17] which allows a party to commit to two bits x_0 and x_1 and prove the *equality* $x_0 = x_1$ without revealing the value of the committed bit. Kilian's protocol for proving equality of two committed bits goes as follows. (In the following matrix, think of each column of the matrix as the shares of one bit.)

1. R chooses $\mathbf{M} = \begin{pmatrix} x_{0,0} & x_{0,1} \\ x_{1,0} & x_{1,1} \end{pmatrix} \in \{0,1\}^{2 \times 2}$ randomly such that $x_{0,a} \oplus x_{1,a} = x_a$
   for $a \in \{0,1\}$. R then computes and sends $\mathsf{Com}(x_{a,a'})$ for $a, a' \in \{0,1\}$ over to S along with $v = x_{0,0} \oplus x_{0,1}$.
2. S sends a random $b \overset{R}{\leftarrow} \{0,1\}$ to R.
3. R sends to S the decommitments to $x_{b,0}$ and $x_{b,1}$. S verifies that $v = x_{b,0} \oplus x_{b,1}$.

Note that the sum of the columns of $\mathbf{M}$ are equal iff $x_0 = x_1$, in which case the sum of the rows of $\mathbf{M}$ are also equal, and so if R is honest the protocol will complete and S's verification will succeed. On the other hand, if $x_0 \neq x_1$ then the sum of the rows of $\mathbf{M}$ are different and so no matter which value v was sent by R in Step 1, there is only a $1/2$ chance that S will ask for the row which sums to v. To decommit, R decommits to one of the remaining two values that he has not yet revealed, $x_{1-b,0}$ and $x_{1-b,1}$. Revealing one is enough since either one can be used to reconstruct x_0. The interesting feature of this protocol, which was already used in [4], is that opening only one of the columns, instead of two, can enable equivocality: assume R can guess the row S will ask to open, then R could commit to a matrix where each column sums to a different bit, and compute v as the xor of the row that S will select. In this way S will be convinced and later R can adaptively choose whether to decommit to 0 or 1, by opening one column or another. This observation is particularly useful when combined with a standard trick for composing two Σ-protocols to compute the OR of two statements [2]. Recall that Σ-protocols satisfy the property that, if the challenge is known in advance, then one can simulate an accepting transcript without knowledge of the witness. The trick is to run two independent executions, say Σ_0, Σ_1, in parallel, but have the challenges c_0, c_1 derived in such a way that the prover can control exactly one of the challenge c_b while the other c_{1-b} will be totally random, and the verifier cannot tell the difference. In this way, the prover can successfully finish both protocols Σ_0, Σ_1 by simulating one of the transcripts and computing the other one honestly.

Putting the two ideas together, we can build a protocol where R commits to a bit x and a bit y, using two executions of the above protocol for equality proofs, and then using the trick for OR composition, R can cheat in one of the equality proofs. Thus one of the value between x and y is equivocal.

One can extend this idea to a string commitment having R commits to two strings X and Y by committing each single bit and then cheat in all the proof for bits belonging to one string, and being honest in all bits belonging to the other string, by using the OR trick as before. Note however, that in the string case we must show that a malicious committer, cannot gain advantage by committing equivocally only *some* of the bits of each string. We protect our protocol from such behavior by using error-correction: we expand each κ-bit string into a 3κ bit string, while having the committer being able to control in total only κ bits for both strings. We are able to prove that due to error-correcting property, corrupting only some bits for each string is not enough to control the final value of the string. We provide more details on how this mechanism is implemented in Sect. 3.

From One-Side Simulatable OT to Fully Simulatable OT. The protocol Π_{OT}^R is not simulatable against a malicious sender. Just as with the basic protocol, S is not committed to any value till the last round so any rewinding strategy will be ineffective. Therefore we have the sender commit to two secret keys in the second round via an extractable commitment[3] and then have him play Π_{OT}^R using the decommitments as inputs. In the last round the server encrypts the actual inputs using the committed keys. This gives the simulator some hope of extracting S's secret inputs by rewinding. This idea by itself doesn't exactly work; the simulator has no guarantee that S used valid decommitments as inputs in the OT played with R. This opens the door to input-dependent abort attacks. We fix this by having S first secret share his inputs and commit to the shares. Then R and S run many executions of the OT protocol Π_{OT}^R, where in the i-th execution, S uses as input the decommitments to the i-th shares. Intuitively, this helps solving the input-depended abort attack, because now R will also check some of the shares corresponding to s_{b-1}. This check, however, must be done in such a way that the probability of S passing the check is independent of the bit b. A bit more in details, obtaining a fully secure protocol requires dealing with two types of malicious behavior. First, we need a mechanism that allows S to prove that he committed to valid shares of a secret. For this we use t-out-of-κ Shamir secret sharing scheme and another variant of Kilian's commit-and-prove protocol. Our main observation is that Kilian's technique is actually quite general and can be used, not only to prove equality of committed values, but that the committed

---

[3] Note that extractable commitments can be built from black-box use of any one-way permutation [21]. In particular it does *not* require trapdoor permutation.

values satisfy *any* linear relation. In particular, it can be used to prove that a committed vector is a set of valid shares of some secret according to Shamir secret sharing scheme.

Secondly, we must give R a strategy to detect the case in which S is not using valid decommitment of the shares in some of the OT executions. Consider, for example, what happens if S were to give correct decommitment in all of the parallel executions of Π_{OT}^R except one, where he uses a wrong decommitment in correspondence of the bit 1. Then since R opens more of the Π_{OT}^R using input b he is noticeably more likely to notice the bad input $b = 1$. We fix this problem by having R performing first a test, which is independent on his secret bit b. R opens an equal number of execution of Π_{OT}^R, say $\kappa/4$, using inputs $b = 0$ and $b = 1$. This test is clearly independent on R's actual input and allows R to check that S is playing honestly in most of the OT executions for *both* inputs. If the test passes, then R is guaranteed that he will obtain at least $t - n/4$ more valid decommitments from the remaining OTs and will be able to reconstruct the secret.

1.2   Further Discussions

Following the OT protocol used in [15], our protocol is based only on *enhanced* trapdoor permutation. We do not require any additional assumption. Moreover, we stress that the lower bound of 4 rounds for secure two-party computation only applies in the plain model. Indeed, in the UC-setting we know how to construct 2-round OT [25] (although under different, standard, assumptions).

We also emphasize that aim of this paper is to match the upper bound of 4-round for two-party computation from general assumptions, that so far was achieved only with a non-black-box construction. As such, our result should be seen as a feasibility result rather than an attempt of building more efficient two-party protocols under general assumptions. It is an interesting direction to improve our techniques to achieve better efficiency.

1.3   Other Related Work on Black-Box Secure Computation

We mention additional related work that are less relevant for our result but that have contributed in the understanding of the power of black-box access to cryptographic primitives. In [3] Damgaard and Ishai show a constant round multi-party protocol where the party have only black-box access to a PRG. This work assumes honest majority. In [27], Wee shows the first black-box constructions with sub-linear round complexity for MPC, which Goyal [6] improves to obtain constant-round MPC constructions based on the black-box use of any OWF. In [7] black-box use of OWFs has been shown to be sufficient to construct constant-round concurrent non-malleable commitments. Other black-box constructions for commitment schemes have been considered w.r.t. selective opening attacks in [22,28]. In [20] Lin and Pass showed the first black-box construction for MPC in the standard model that satisfies a non-trivial form of concurrent security. Their construction requires a non-constant number of rounds. Very recently, Kiyoshima et al. in [18] improved on the round complexity providing a constant- round construction for the same result. Finally, another line of research has looked at achieving black-box construction for protocols that requires non-black-box simulation, such as black-box public coin ZK [8] and resettably-sound ZK from OWF [23].

2   Preliminaries

General Notation. We denote by κ the security parameter, and by PPT a machine running in probabilistic polynomial time. We denote vector using bold notation $\mathbf{v}$ and we denote the i-th coordinate of a vector $\mathbf{v}$ using notation $[v]_i$. We denote a matrix using capital and bold letters $\mathbf{M}$, and we denote the element in position i, j of $\mathbf{M}^{\mathsf{index}}$ by $x_{i,j}^{\mathsf{index}}$. Let $[n]$ be the set $\{1, \ldots, n\}$ and $\mathbb{Z}_q$ be the integers mod q. For a bit $b \in \{0, 1\}$ we write $\bar{b}$ as shorthand for $1 - b$. We write $\mathbf{negl}(\cdot)$ for an unspecified negligible function.

Trapdoor Permutations. Trapdoor permutations are permutations which are easy to compute and hard to invert *unless* you know the trapdoor, in which case they are easy to invert. The formal definition is as follows.

Definition 1 (Trapdoor Permutation). Let $\mathcal{F} = (\mathsf{Gen}, \mathsf{Eval}, \mathsf{Invert})$ be three PPT algorithms such that

- $\mathsf{Gen}(1^\kappa)$ outputs a pair (f, trap) where $f : \{0, 1\}^\kappa \rightarrow \{0, 1\}^\kappa$ is a permutation;
- $\mathsf{Eval}(f, \cdot) = f(\cdot)$ evaluates f; and
- $\mathsf{Invert}(f, \mathsf{trap}, \cdot) = f^{-1}(\cdot)$ evaluates f^{-1}.

We say that $\mathcal{F}$ is a *family of trapdoor permutations* (TDPs) if for any PPT algorithm R R

$$\Pr_{(f,\mathsf{trap}) \leftarrow \mathsf{Gen}(1^\kappa), y \leftarrow \{0,1\}^\kappa} \left(\mathrm{R}(f, y) = f^{-1}(x) \right) = \mathbf{negl}(\kappa).$$

Additionally, we assume that our TDP families have a weak form of certifiability. Namely, we assume that given some f output by $\mathsf{Gen}(1^\kappa)$ it is possible to tell in polynomial time whether f is a permutation on $\{0, 1\}^\kappa$ or not. It will be convenient for us to have trapdoor permutations which act on vector spaces over fields instead of just $\{0, 1\}^\kappa$. This can be arranged by identifying $\{0, 1\}^\kappa$ with $\mathbb{F}_2^\kappa$, or if we need a larger alphabet, we can identify $\{0, 1\}^\kappa$ with $\mathbb{F}_{2^k}^{\kappa/k}$. When we are using this point of view we will write $(f, \mathsf{trap}) \xleftarrow{\mathrm{R}} \mathsf{Gen}(\mathbb{F}_2^\kappa)$.

Hard-Core Bits. We assume the reader is familiar with the notion of a hard-core bit of a oneway permutation. Briefly, we say that a family of predicates $\mathcal{H} = \{h : \{0, 1\}^\kappa \rightarrow \{0, 1\}\}$ is *hard-core* for the TDP family $\mathcal{F}$ if for random $(f, \mathsf{trap}) \xleftarrow{\mathrm{R}} \mathsf{Gen}(1^\kappa)$, $h \xleftarrow{\mathrm{R}} \mathcal{H}$, and $x \xleftarrow{\mathrm{R}} \{0, 1\}^\kappa$, $h(x)$ is hard to predict given $f(x)$. A hardcore big can be extended to output a vector in a natural way: $\mathbf{h}(x) = h(x) \circ h\big(f(x)\big) \circ \cdots \circ h\big(f^{k-1}(x)\big)$, which is indistinguishable from random, given $f^k(x)$. When we identify the domain $\{0, 1\}^\kappa$ of f with a κ−dimensional vector space over $\mathbb{F}_2$, we will likewise identify the output of $\mathbf{h}(\cdot)$ with an $\mathbb{F}_2$−vector of the same dimension.

Oblivious Transfer. Oblivious Transfer (OT) is a two-party functionality $\mathcal{F}_{OT}$, in which a sender S holds a pair of strings (s_0, s_1), and a receiver R holds an a bit b, and wants to obtain the string s_b. The security requirement for the $\mathcal{F}_{OT}$ functionality is that any malicious receiver does not learn anything about the string s_{1-b} and any malicious sender does not learn which string has been transfered. This security requirement is formalized via the ideal/real world paradigm. In the ideal world, the functionality is implemented by a trusted party that takes the inputs from S and R and provides the output to R and is therefore secure by definition. A real world protocol Π securely realizes the ideal $\mathcal{F}_{OT}$ functionalities, if the following two conditions hold. (a) **Security against a malicious receiver.** The output of any malicious receiver R* running one execution of Π with an honest sender S can be simulated by a PPT simulator Sim that has only access to the ideal world functionality $\mathcal{F}_{OT}$ and oracle access to R*. (b) **Security against a malicious sender.** The joint view of output of any malicious sender S* running one execution of Π with R and the output of R can be simulated by a PPT simulator Sim that has only access to the ideal world functionality functionality $\mathcal{F}_{OT}$ and oracle access to S*. In this case the output of the malicious S* is combined with the output of R in the ideal world.

We also consider a weaker definition of $\mathcal{F}_{OT}$ that is called **one-sided** simulatable $\mathcal{F}_{OT}$, in which we do not demand the existence of a simulator against a malicious sender, but we only require that a malicious sender cannot distinguish whether the honest receiver is playing with bit 0 or 1. A bit more formally, we require that for any PPT malicious sender S* the view obtained from executing Π when the receiver R plays with bit 0 is computationally indistinguishable from the view obtained when R is playing with bit 1.

Finally, we consider the $\mathcal{F}_{OT}^m$ functionality where the sender S and the receiver R runs m execution of OT in parallel.

Secure Two-Party Computation. Let $\mathcal{F}(x_1, x_2)$ be a two-party functionality run between parties P_1 holding input x_1 and P_2 holding input x_2. In the ideal world, P_i (with $i \in \{1, 2\}$) sends its input x_i to the f and obtains only $y = \mathcal{F}(x_1, x_2)$. We say that a protocol Π securely realizes $\mathcal{F}(\cdot, \cdot)$ if the view of any malicious P_i^* executing Π with an honest P_j with $i \neq j$ combined with the output of P_j (if any) can be simulated by a PPT simulator that has only access to $\mathcal{F}$ and has oracle access to P_i^*.

Shamir Secret Sharing Scheme. A t-out-of-n secret sharing scheme gives a way to break a secret into shares in such a way so that any set of shares either reveals nothing about the secret, if the set has size less than t, or allows one to reconstruct the entire secret, if the set has size at least t. Shamir secret sharing [26] constructs such a scheme using polynomials. Fix a prime $q > n$. To share a secret field element $\alpha \in \mathbb{Z}_q$, the function Share chooses a random polynomial $f(x) \in \mathbb{Z}_q[x]$ of degree at most $t - 1$ and defines the vector $[\alpha] = ([\alpha]_1, \ldots, [\alpha]_n) \in \mathbb{Z}_q^n$, by setting $[\alpha]_i = f(i)$. That this is a t-out-of-n secret sharing scheme follows from basic properties of polynomials. To reconstruct a secret, the fuction Recon takes in input a set of $t + 1$ valid shares and uses

Lagrange interpolation to compute the unique t-degree polynomial f defined by such shares and output the free coefficient of f.

We briefly comment on another property of this scheme that we will use in our protocol. The map which sends a degree $t - 1$ polynomial to its vector of shares is linear over $\mathbb{Z}_q$. It follows that the set of vectors in $\mathbb{Z}_q^n$ which are valid sharings is a $t-$plane in $\mathbb{Z}_q^n$, or equivalently, that there exists a linear map $\psi : \mathbb{Z}_q^n \to \mathbb{Z}_q^{n-t}$ such that $\psi(\mathbf{v}) = 0$ iff $\mathbf{v}$ is a valid sharing.

Extractable Commitments. A commitment scheme scheme is a two-party functionality run between a sender with input a secret message m and a committer that has no input, and consists of two phase: commitment and decommitment phase. In the commitment phase the sender commits to its message m. A commitment scheme is hiding if any PPT malicious receiver cannot distinguish the secret message m in this phase. In the decommitment phase the message reveals m and the randomness used to compute the commitment. This phase is statistically binding if any malicious sender cannot successfully open to any message $m' \neq m$ in this phase.

We say that a commitment scheme is extractable if there exists an efficient extractor that having black-box access to any malicious sender that successfully performs the commitment phase, is able to efficiently extract the committed string. In the paper we employ the extractable commitment provided in [21]. The commitment phase consists of 3 rounds, that we denote by (ExtCom1, ExtCom2, ExtCom3). The decommitment phase is non-interactive.

3   Four-Round Black-Box Oblivious Transfer

In this section we describe our 4-round black-box OT protocol Π_{OT} in details. We present it in two steps. First we give an OT protocol that is simulatable against a malicious receiver and provides only indistinguishability security against a malicious sender. We denote this protocol by Π_{OT}^R. We then show how to use (black-box) extractable commitments and Shamir secret sharing, to compile Π_{OT}^R into a protocol that is fully simulatable.

3.1   Four-Round Black-Box OT Secure Against Malicious Receivers

The building block for Π_{OT}^R is a protocol that allows the receiver to compute two string commitments, C_0, C_1, such that one commitment is equivocal, and prove that at least one commitment is binding.

As a warm up for our construction we show how to implement such building block for the simpler case where the receiver commits to two bits, and then he is able to equivocate one bit. The soundness of the warm up protocol is $1/2$. The idea is to have two executions of Kilian's black-box commit-and-prove protocol (outlined in Sect. 1.1), and combine the two proofs using the OR trick of Σ-protocols. The details are shown in Protocol 1.

Protocol 1. Compute One Biding and One Equivocal Commitment.

Input to R: *A bit b indicating which commitment should be equivocal.*

1. R *chooses two matrices* M^0 *and* M^1 *where each* $M^a = \begin{pmatrix} x^a_{0,0} & x^a_{0,1} \\ x^a_{1,0} & x^a_{1,1} \end{pmatrix} \in \{0,1\}^{2 \times 2}$
   is random such that:
   - *Matrix* M^b: *this matrix represent two different bits, therefore the xor of the first column is 0, and the xor of the second column is 1. Namely,* $x^b_{0,0} \oplus x^b_{1,0} = 0$ *and* $x^b_{0,1} \oplus x^b_{1,1} = 1$;
   - *Matrix* M^{1-b}: *both columns are representing the same bit:* $x^{1-b}_{0,0} \oplus x^{1-b}_{1,0} = x^{1-b}_{0,1} \oplus x^{1-b}_{1,1} = r_{1-b}$ *for* $r_{1-b} \in \{0,1\}$.
   R *commits to all of the* $x^a_{a',a''}$ *in both matrixes and sends to* S *values* v^0 *and* v^1 *computed as follows:*
   - v^{1-b} *is honestly computed as the xor of the first row (as in Kilian's protocol), namely,* $v^{1-b} = x^{1-b}_{0,0} \oplus x^{1-b}_{0,1}$.
   - v^b *is a random bit.*
2. S *sends* R *a random* $r' \xleftarrow{R} \{0,1\}$ *and a challenge* $c \in \{0,1\}$.
3. R *computes challenges* (c_0, c_1) *such that* $c_0 \oplus c_1 = c$ *and the challenge* c_b *is pointing exactly to the row of* M^b *the xor of which is* v_b. *Namely,* c_b *is such that* $v^b = x^b_{c_b,0} \oplus x^b_{c_b,1}$. *Note this is always possible as* $x^b_{0,0} \oplus x^b_{0,1} \neq x^b_{1,0} \oplus x^b_{1,1}$. *Next,* R *decommits to* $x^0_{c_0,0}$, $x^0_{c_0,1}$ *from matrix* M^0 *as well as* $x^1_{c_1,0}$, $x^1_{c_1,1}$ *from matrix* M^1.
   Finally, for each matrix, R *decommits to one column. For* M^{1-b}, *which is honestly computed,* R *opens one column chosen at random (*R *will need to decommit one value of the column as the other one was already opened to answer the challenge). For* M^b, R *will decommit to the column* r_b *such that* $r_b \oplus r' = s_b$ *where* s_b *is the bit that* R *wants to obtain out of the coin-flipping of the bit in position b. Formally,* R *decommits to one of* $x^b_{\bar{c}_b,0}$ *and* $x^b_{\bar{c}_b,1}$ *at random (using the shorthand* $\bar{b} = 1 - b$*), completing a decommitment to the value* $s_{1-b} = r_{1-b}$. R *decommits to* $x^b_{\bar{c}_b,r_b}$ *(completing a decommitment to* $s_b = r_b \oplus r'$*).* R *also sends* (c_0, c_1).
4. **Verification:** S *checks that* $c_0 \oplus c_1 = c$ *and that* $x^a_{c_a,0} \oplus x^a_{c_a,1} = v^a$ *for* $a \in \{0,1\}$. *If not* S *aborts.*
5. **Output:** *Both parties set output to* (z_0, z_1) *where* $z_a = s_a \oplus r'$.

If R correctly follows the protocol then the output (z_0, z_1) satisfies $z_b = r_b$ while z_{1-b} is random. Furthermore, if R, in an attempt to cheat, chooses M^0 and M^1 both such that $x^a_{0,0} \oplus x^a_{1,0} \neq x^a_{1,0} \oplus x^a_{1,1}$, then S will abort whenever $c \neq d_0 \oplus d_1$ (which happens with probability $1/2$), where $d_a \in \{0,1\}$ is such that $v^a = x^a_{d_a,0} \oplus x^a_{d_a,1}$. This protocol can be seen as partial coin-flipping protocol that output two coins and guarantee that at least one coin is fair.

The ability for R to completely control one but not both of the output bits in the above protocol is essentially exactly what we need in order to compile the OT which is secure only against a semi-honest R into one which is maliciously secure. The basic idea is to extend the above coin-flipping to strings and enable

R to obtain two strings $z_0, z_1 \in \{0,1\}^\kappa$ such that $z_b = f^k(x)$ for some value x chosen by her. As mentioned, this forces z_{1-b} to be random, and so R cannot know a preimage without breaking the trapdoor permutation.

However, when extending the above warm-up protocol to a string via bit-wise commit-and-proofs we must enforce that a malicious receiver cannot cheat by controlling *some* of the bits of both z_0 and z_1 and wind up knowing preimages of both values. We protect our protocol from such behavior by letting $\mathbf{A} \in \mathbb{Z}_q^{3\kappa \times \kappa}$ be a matrix with good error correcting properties (such as a Vandermonde matrix) and working in the image of $\mathbf{A}$.

This introduces some complications. Specifically it requires moving to a non-binary base field as we need an error correcting code with (constant but) large distance. In our actual protocol we use a variant of the unfair coin flipping described above, adapted to work over over $\mathbb{Z}_q$ for some prime power $q = \emptyset(\kappa)$. The major difference is that instead of committing to every entry in $2-\text{by}-2$ matrices, R commits to every entry in $2-\text{by}-q$ matrices. For each matrix R proves that the sum of the elements in each column is the same. In order to commit equivocally, R chooses the q columns of the matrices corresponding to his bit b to have distinct sums. Namely, for every $\alpha \in \mathbb{Z}_q$ there is exactly one column whose entries add to α. The final protocol $\Pi_{\mathsf{OT}}^{\mathsf{R}}$ is formally described in Protocol 2.

Protocol 2 ($\Pi_{\mathsf{OT}}^{\mathsf{R}}$). Public Input: *A prime* $q = \mathcal{O}(\kappa)$, *a Vandermonde matrix* $\mathbf{A} \in \mathbb{Z}_q^{3\kappa \times \kappa}$ *and a statistically binding commitment scheme* Com.
Sender's Input: $\mathbf{s}_0, \mathbf{s}_1 \in \mathbb{Z}_q^\kappa$.   **Receiver's Input:** $b \in \{0,1\}$.

1. $(\mathsf{R} \longrightarrow \mathsf{S})$: R *chooses* $\mathbf{r}^{\overline{b}} \xleftarrow{R} \mathbb{Z}_q^\kappa$ *and sets* $\hat{\mathbf{r}}^{\overline{b}} = \mathbf{Ar}^{\overline{b}} \in \mathbb{Z}_q^{3\kappa}$. R *then chooses*
   6κ *matrices* $\left\{(\mathbf{M}^{0,i}, \mathbf{M}^{1,i})\right\}_{i=1,\ldots,3\kappa}$ *where* $\mathbf{M}^{a,i} = \begin{pmatrix} x_{0,0}^{a,i} & x_{0,1}^{a,i} & \cdots & x_{0,q-1}^{a,i} \\ x_{1,0}^{a,i} & x_{1,1}^{a,i} & \cdots & x_{1,q-1}^{a,i} \end{pmatrix} \in$
   $\mathbb{Z}_q^{2\times q}$ *is random such that:*
   - $x_{0,0}^{\overline{b},i} + x_{1,0}^{\overline{b},i} = \cdots = x_{0,q-1}^{\overline{b},i} + x_{1,q-1}^{\overline{b},i} = [\hat{\mathbf{r}}^{\overline{b}}]_i, \; \forall \; i.$
   - $x_{0,0}^{b,i} + x_{1,0}^{b,i} = \sigma_i(0), \ldots, x_{0,q-1}^{b,i} + x_{1,q-1}^{b,i} = \sigma_i(q-1) \; \forall \; i,$ *where the* σ_i *are random permutations of* $\mathbb{Z}_q$.
   R *commits to all of the* $x_{a',a''}^{a,i}$ *using* Com. *Let* $\mathbf{x}_0^{a,i} = (x_{0,0}^{a,i}, \ldots, x_{0,q-1}^{a,i}) \in \mathbb{Z}_q^q$
   be the top row vector of $\mathbf{M}^{a,i}$. *Similarly, let* $\mathbf{x}_1^{a,i}$ *be the bottom row of* $\mathbf{M}^{a,i}$.
   Also let $\psi : \mathbb{Z}_q^q \to \mathbb{Z}_q^{q-1}$ *be the linear map* $\psi : \mathbf{x} = (x_0, \ldots, x_{q-1}) \mapsto (x_1 - x_0, \ldots, x_{q-1} - x_0)$. R *sends vectors* $\{\mathbf{v}^{0,i}, \mathbf{v}^{1,i}\}_{i=1,\ldots,3\kappa}$ *where each* $\mathbf{v}^{a,i} \in \mathbb{Z}_q^{q-1}$
   is generated as follows:
   - $\mathbf{v}^{\overline{b},i} = \psi(\mathbf{x}_0^{\overline{b},i})$;
   - *draw* $c_b \xleftarrow{R} \{0,1\}^{3\kappa}$ *and set* $\mathbf{v}^{b,i} = \psi(\mathbf{x}_0^{b,i})$ *if* $c_{b,i} = 0$, $\mathbf{v}^{b,i} = -\psi(\mathbf{x}_1^{b,i})$ *if*
     $c_{b,i} = 1$.
2. $(\mathsf{S} \longrightarrow \mathsf{R})$: S *chooses random* $c \xleftarrow{R} \{0,1\}^{3\kappa}$, $\mathbf{r}' \xleftarrow{R} \mathbb{Z}_q^\kappa$, *and sends* c *and*
   $\mathbf{r}'$. *Additionally,* S *chooses a trapdoor permutation* $(f, f^{-1}) \xleftarrow{R} \mathsf{Gen}(\mathbb{Z}_q^\kappa)$ *and sends* f *to* R.

3. $(\text{R} \longrightarrow \text{S})$: R *parses* c *into* (c_0, c_1) *such that* $c_0 \oplus c_1 = c$ *where* c_b *is as in step 1. For both* $a \in \{0,1\}$, R *decommits to every coordinate of* $\mathbf{x}^{a,i}_{c_{a,i}}$ *as well as to one coordinate,* $[x^{a,i}_{\bar{c}_{a,i}}]_j$, *of* $\mathbf{x}^{a,i}_{\bar{c}_{a,i}}$. *When* $a = \bar{b}$, *this coordinate* j *is chosen randomly, completing a decommitment to* $\hat{\mathbf{r}}^{\bar{b}}$ *(defined in step 1). When* $a = b$, R *draws a random* $\mathbf{y} \xleftarrow{R} \mathbb{Z}^\kappa_q$ *and sets* $\hat{\mathbf{r}}^b = \mathbf{A}\big(f^\kappa(\mathbf{y}) - \mathbf{r}'\big) \in \mathbb{Z}^{3\kappa}_q$. *Finally,* R *decommits to* $x^{b,i}_{\bar{c}_{b,i},j}$ $\forall\ i$, *where* j *is such that* $[\hat{\mathbf{r}}^b]_i = x^{b,i}_{0,j} + x^{b,i}_{1,j}$. *Note that* R *has decommitted to* $(\hat{\mathbf{r}}^0, \hat{\mathbf{r}}^1)$.
4. $(\text{S} \longrightarrow \text{R})$: *For all* (a, i), S *has received decommitments to all of the coordinates of exactly one of* $\mathbf{x}^{a,i}_0$ *and* $\mathbf{x}^{a,i}_1$. S *checks either that* $\mathbf{v}^{i,a} = \psi(\mathbf{x}^{a,i}_0)$ *or that* $\mathbf{v}^{a,i} + \psi(\mathbf{x}^{a,i}_1) = 0$. *If any of these checks fails,* S *aborts. Otherwise,* S *computes vectors* $(\mathbf{z}_0, \mathbf{z}_1)$ *where* $\mathbf{z}_a \in \mathbb{Z}^\kappa_q$ *is the unique vector such that* $\mathbf{A}\mathbf{z}_a = \hat{\mathbf{r}}^a + \mathbf{A}\mathbf{r}'$ *(such a value exists by linearity). If no such* $\mathbf{z}_a$ *exists for some* a *then* S *aborts.* S *sends* $(\mathbf{w}_0, \mathbf{w}_1)$ *to* R *where* $\mathbf{w}_a = \mathbf{s}_a - \mathbf{h}\big(f^{-\kappa}(\mathbf{z}_a)\big)$.

Output: R *outputs* $\mathbf{s}_b = \mathbf{w}_b + \mathbf{h}(\mathbf{y})$.

3.2   Four-Round Fully Simulatable Oblivious Transfer from $\Pi^{\text{R}}_{\text{OT}}$

We transform the one-sided simulatable $\Pi^{\text{R}}_{\text{OT}}$ into an OT which is simulatable for both the sender and the receiver using the following ingredients. We use a $(\kappa + 1, 2\kappa)$-secure Shamir Secret sharing scheme. Let $A \in \mathbb{Z}^{2\kappa \times \kappa}_q$ be the Vandermonde matrix and let ϕ be a linear map such that $\phi(A) = 0$.

First, the sender picks two *random* keys x_0, x_1, and computes their correspondent vectors of 2κ shares $\mathbf{v}_0, \mathbf{v}_1$ according to Shamir secret sharing. Then, the sender commits to each coordinate of vectors $\mathbf{v}_0, \mathbf{v}_1$ and proves that they are valid shares, in a black-box way. We build this proof using the observation that $\mathbf{v}$ is a valid vector of shares for a $(\kappa + 1, 2\kappa)$-secure Shamir secret sharing, iff $\phi(\mathbf{v}) = 0$, and that for any pair of vectors $\mathbf{a}, \mathbf{b}$ it holds that if $\mathbf{a} + \mathbf{b} = \mathbf{v}$ then also $\phi(\mathbf{a}) + \phi(\mathbf{b}) = 0$. Thus, to prove that a vector $\mathbf{v}$ is a vector of valid shares, the sender will commit to κ pairs of vectors $\mathbf{a}_j, \mathbf{b}_j$ such that $\mathbf{v} = \mathbf{a}_j + \mathbf{b}_j$, and prove that there exists at least a j such that the predicate $\phi(\mathbf{a}) + \phi(\mathbf{b}) = 0$ holds. This proof is easily implemented by having the sender commit to $\mathbf{a}_j, \mathbf{b}_j$ and $\mathbf{z}_j = \phi(\mathbf{a}_j)$ and having the receiver ask to either open $\mathbf{a}_j$ and check that $\mathbf{z}_j = \phi(\mathbf{a}_j)$, or to open $\mathbf{b}_j$ and check that $\phi(\mathbf{b}_j) + \mathbf{z}_j = 0$. Note that this proof only guarantees that there exists at least one j for which the condition $\phi(\mathbf{a}) + \phi(\mathbf{b}) = 0$ is true. Summing up, S will commit to vectors $\mathbf{a}_{0,j} \mathbf{b}_{0,j}$ and $\mathbf{a}_{1,j} \mathbf{b}_{1,j}$ (for shares $\mathbf{v}_0, \mathbf{v}_1$) with an extractable commitment scheme, and run a proof of validity for each such pair. In the last round S will send the encryptions $x_0 + s_0, x_1 + s_1$ of his actual secret inputs. Now we need a way for R to retrieve the decommitments of the shares for the secret he is interested in, without the server knowing which decommitments are revealed. We accomplish this by using the OT protocol $\Pi^{\text{R}}_{\text{OT}}$ implemented above. Therefore, in parallel to such extractable commitments and proofs, the sender and the receiver will engage in 2κ parallel executions of $\Pi^{\text{R}}_{\text{OT}}$: in the i-th OT execution S plays with inputs the opening of the i-th coordinate of $(\mathbf{a}_{0,j}, \mathbf{b}_{0,j})$ and $(\mathbf{a}_{1,j}, \mathbf{b}_{1,j})$ for *all* j, and R plays with bit b_i. Note that, opening to the i-th coordinate of all j vectors allows the receiver to check that all j vectors agree on the *same* coordinate $[v_{b_i}]_i$. This check, together with the proof of consistency provided above, will guarantee that most of the shares received via OT (and extracted by the simulator via the extractable commitments) are valid.

Before attempting to reconstruct the secret, R will test the consistency of $\kappa/2$ coordinates for vector $\mathbf{v}_0$ and $\mathbf{v}_1$ by playing with bit 0 and 1 accordingly, in the correspondent OTs, while he plays with the his secret bit b for the remaining κ executions. R will attempt to reconstruct the vector $\mathbf{v}_b$ only if the consistency test passes. We provide a formal description of such steps in Protocol 3.

Protocol 3 (Π_{OT}). Sub-protocols. *Let* $\Pi_{\mathsf{OT}}^{\mathsf{R}} = \{\mathsf{OT1}, \mathsf{OT2}, \mathsf{OT3}, \mathsf{OT4}\}$ *denote the 4 messages exchanged in protocol* $\Pi_{\mathsf{OT}}^{\mathsf{R}}$ *(Prot. 2). Let* $\mathsf{OT}[i]$ *denote the i-th parallel execution of* $\Pi_{\mathsf{OT}}^{\mathsf{R}}$. *Let* $\mathsf{ExtCom} = (\mathsf{ExtCom1}, \mathsf{ExtCom2}, \mathsf{ExtCom3})$ *be a 3-round statistically binding extractable commitment scheme with non-interactive decommitment* ExtDec. *Let* $\mathsf{Share}, \mathsf{Recon}$ *be a* $(\kappa + 1)$-*out-of-*2κ *Shamir secret sharing scheme over* $\mathbb{Z}_p$, *together with a linear map* $\psi : \mathbb{Z}_p^{2\kappa} \to \mathbb{Z}_p^{\kappa-1}$ *such that* $\psi(\mathbf{v}) = 0$ *iff* $\mathbf{v}$ *is a valid sharing of some secret.*

Public Input: *A prime p and $\ell = \lfloor \log q \rfloor$ st $2^\ell/p = 1 - \mathsf{negl}(\kappa)$, a Vandermonde matrix $A \in \mathbb{Z}_p^{2\kappa \times \kappa}$, linear map ϕ.*

Sender's Input: $s_0, s_1 \in \mathbb{Z}_p$.   **Receiver's Input:** $b \in \{0, 1\}$.

1. $(\mathsf{R} \longrightarrow \mathsf{S})$: R *randomly chooses a set* $\mathsf{T}_{1-b} \in [2\kappa]$ *of* $\kappa/2$ *coordinates. R plays the i-th execution of* $\Pi_{\mathsf{OT}}^{\mathsf{R}}$ *with input* $b_i = (1 - b)$. *For the remaining* $i \notin \mathsf{T}_{1-b}$ *set* $b_i = b$ R *sends* $(\mathsf{OT1}[1], \dots, \mathsf{OT1}[2\kappa])$ *to* S, *where* $\mathsf{OT1}[i]$ *is computed on input* b_i.

2. $(\mathsf{S} \longrightarrow \mathsf{R})$: *Upon receiving a correct first message, S proceeds as follows.*
   - *Pick random strings* $x_0, x_1 \in \mathbb{Z}_p$ *and secret share each string: Compute shares* $\mathbf{v}_b = ([v_b]_1, \dots, [v_b]_{2\kappa}) \leftarrow \mathsf{Share}(x_b)$ *for* $b \in \{0, 1\}$.
   - *To commit to shares* $\mathbf{v}_0, \mathbf{v}_1$ *and prove that they are valid shares of a κ-degree polynomial S proceeds as follows.*
     - *For* $j = 1, \dots, \kappa$, *pick random* $\mathbf{a}_{0,j}, \mathbf{b}_{0,j} \in \mathbb{Z}_p^{2\kappa}$ *such that* $\mathbf{a}_{0,j} + \mathbf{b}_{0,j} = \mathbf{v}_0$ *and compute* $\mathbf{z}_{0,j} = \phi(\mathbf{a}_{0,j})$ *for all j. Resp., compute* $\mathbf{a}_{1,j}, \mathbf{b}_{1,j} = \mathbf{v}_1$
     - *Commit to each coordinate of* $\mathbf{a}_{b,j}$ *and* $\mathbf{b}_{b,j}$ *using* ExtCom, *namely send* $\mathsf{acom}_{b,j,i} = \mathsf{ExtCom1}([a_{b,j}]_i))$, $\mathsf{bcom}_{b,j,i} = \mathsf{ExtCom1}([b_{b,j}]_i)$.
   S *sends to R the messages* $(\mathsf{OT2}[1], \dots, \mathsf{OT2}[2\kappa])$, $\{\mathsf{ExtCom1}([a_{b,j}]_i)$, $\{\mathsf{ExtCom1}([b_{b,j}]_i)\}_{i \in 2\kappa}$, *and* $\mathbf{z}_{b,j}$ *for* $b = 0, 1$ *and* $j \in [\kappa]$.

3. $(\mathsf{R} \longrightarrow \mathsf{S})$: R *sends* $(\mathsf{OT3}[1], \dots, \mathsf{OT3}[2\kappa])$, *the second message* $\mathsf{ExtCom2}$ *for the extractable commitment, and a random challenge* $c_1, \dots, c_\kappa \in \{0, 1\}^\kappa$.

4. $(\mathsf{S} \longrightarrow \mathsf{R})$: S *computes OT message* $\mathsf{OT4}[i]$ *using as inputs the i-th coordinate of all j vectors committed before. Specifically, in the i-th OT it uses decommitment to values* $([a_{0,j}]_i, [b_{0,j}]_i) \, \forall j$; $([a_{1,j}]_i, [b_{1,j}]_i) \, \forall j$. *Additionally, for each j, S reveals vector* $\mathbf{a}_{0,j}, \mathbf{a}_{1,j}$ *if* $c_j = 0$; *or vectors* $\mathbf{b}_{0,j}, \mathbf{b}_{1,j}$ *if* $c_j = 1$; *and the messages for the third round of the extractable commitments, namely* $\{\mathsf{ExtCom3}([a_{b,j}]_i), \{\mathsf{ExtCom3}([b_{b,j}]_i)\}_{i \in 2\kappa, j \in \kappa}$. *Finally, S sends* $C_0 = s_0 \oplus x_0$ *and* $C_1 = s_1 \oplus x_1$.

Verification and Output: *If the extractable commitments are all successfully completed, proceeds as follows.*

- **Check Validity of Shares.** *For $j = 1, \ldots, \kappa$, if $c_j = 0$ check that $\mathbf{z}_{0,j} = \phi(\mathbf{a}_{0,j})$ and $\mathbf{z}_{1,j} = \phi(\mathbf{a}_{1,j})$. Else, if $c_j = 1$ check that $\phi(\mathbf{b}_{0,j}) + \mathbf{z}_{0,j} = 0$ and $\phi(\mathbf{b}_{1,j}) + \mathbf{z}_{0,j} = 1$.*
- **Test Phase.** R *randomly chooses a set T_b of $\kappa/2$ coordinates in $\{[2\kappa]/\mathsf{T}_{1-b}\}$. For each $i \in \mathsf{T}_\sigma$, with $\sigma \in \{0,1\}$; let $[a_{\sigma,j}]_i, [b_{\sigma,j}]_i$ be the coordinates obtained from the i-th OT. R checkes that, for all j, there exists a unique $[v_\sigma]_i$ such that $[a_{\sigma,j}]_i + [b_{\sigma,j}]_i = [v_\sigma]_i$. If so, $[v_\sigma]_i$ is then marked as* **consistent.** *If all shares obtained in this phase are consistent, R proceeds to the reconstruction phase. Else abort.*
- **Reconstruction Phase.** *For $i \in \{[2\kappa]/\mathsf{T}_{1-b}\}$, if there exists a unique $[v_b]_i$ such that $[a_{b,j}]_i + [b_{b,j}]_i = [v_b]_i$, mark share $[v_b]_i$ as* **consistent.** *If R obtains less than $\kappa + 1$ consistent shares, he aborts. Else, let $[v_b]_{j_1}, \ldots, [v_b]_{j_{\kappa+1}}$ be any set of $\kappa + 1$ consistent shares. R computes $x_b \leftarrow \mathsf{Recon}([v_b]_{j_1}, \ldots, [v_b]_{j_{\kappa+1}})$ and outputs $s_b = C_b \oplus x_b$.*

3.3   Proof of Security

In this section we provide the intuition behind the security of our constructions. The reader is referred to the full version for the complete proof.

Security of $\Pi_{\mathsf{OT}}^{\mathsf{R}}$. We start by proving that $\Pi_{\mathsf{OT}}^{\mathsf{R}}$ is one-sided simulatable.

Indistinguishability Against a Malicious Sender. It follows from the hiding of the commitment scheme used by R to commit to the secret vectors $\mathbf{r}^0, \mathbf{r}^1$. Indeed, the only difference between the transcript of a completed execution of $\Pi_{\mathsf{OT}}^{\mathsf{R}}$ when R uses input bit $b = 0$ and when R uses bit 1 is in the matrices that are computed equivocally. In turn, the equivocal matrix differs from a binding matrix in that the sum of the rows of an equivocal $\mathbf{M}^{i,b}$ leads to the vector of all permuted values in $\mathbb{Z}_q$ while in a binding matrix the sum of the row of the i-th matrix corresponds to the vector $\hat{\mathbf{r}}_i^b$.

Simulatability Against a Malicious Receiver. For the case of a malicious receiver, we build a simulator who rewinds S and extracts R's input bit from the coin-flipping protocol. Note that when R's input bit is b, R commits equivocally to the matrices $\mathbf{M}^{b,i}$, and therefore cannot commit equivocally to the $\mathbf{M}^{\bar{b},i}$. It follows that when R is rewound and asked a new query, his decommitment from the $\bar{b}$ matrices will be the same. In this way, our simulator can figure out R's input bit. It remains to show that a malicious R cannot gain some advantage by committing equivocally to some of the $\mathbf{M}^{0,i}$ and some of the $\mathbf{M}^{1,i}$. This follows from the error-correction property guaranteed by the choice of the matrix $\mathbf{A}$. A more detailed proof is provided in the full version.

Security of Π_{OT}. We now sketch the main ideas behind the security of Π_{OT}. Correctness follows from the correctness of the underlying $\Pi_{\text{OT}}^{\text{R}}$ protocol, the correctness of the statistically binding commitment scheme and the Shamir secret sharing scheme: the receiver will be able to retrieve more than $\kappa + 1$ shares and reconstruct the key x_b that allows to decrypt s_b. We now analyze the security of the protocol in case either of the parties is corrupted.

Simulatability Against Malicious Receiver. We show a PPT simulator that simulates the attack of the receiver in the ideal world as follows. Sim computes the messages of protocol Π_{OT} honestly till the third round, by committing to randomly selected x_0, x_1. In parallel, Sim extracts the bits played by R* in protocol $\Pi_{\text{OT}}^{\text{R}}$ by running the simulator $\text{Sim}_{\text{R}}^{\text{OT}}$ guaranteed by the one-sided simulatability property of $\Pi_{\text{OT}}^{\text{R}}$. $\text{Sim}_{\text{R}}^{\text{OT}}$ outputs the bits $b_1, \ldots, b_{2\kappa}$ which are the selections made by R* in the first 3 rounds of protocol $\Pi_{\text{OT}}^{\text{R}}$. (Note that in $\Pi_{\text{OT}}^{\text{R}}$ the server commits to its input only in the fourth round, when the selection has already beed committed. However, this will not be a problem because in Π_{OT} the sender is still using its secret input only in the last round). If there are more than $\kappa + 1$ bits pointing to the same bit b then Sim sends this bit to $\mathcal{F}_{\text{OT}}$ and receives the string s_b. Otherwise it will just send a random bit and continue the simulation of the protocol with random values. In the last round the simulator uses $\text{Sim}_{\text{R}}^{\text{OT}}$ to complete the OT using in input the shares that were dictated by the bits $b_1, \ldots, b_{2\kappa}$ and it obtains messages OT4[i] for $i \in [2\kappa]$. Finally, Sim completes the protocol by honestly computing message CPmsg3, but it prepares $c_b = x_b \oplus s_b, c_{1-b} = r$, where r is a randomly chosen string.

The indistinguishability of the simulation follows from the simulatability of the underlying OT, the security of Shamir secret sharing and the hiding of the underlying commitment scheme. We stress that in the proof we need to argue about the hiding of the unopened shares. Namely, we require to prove that the protocol satisfies a form of hiding in presence of selective opening attack. This is not a problem as our protocol is interactive and the positions that the receiver is choosing to open are fixed in advance before observing any commitment. This property allows us to prove indistinguishability by relying on standard hiding definition.

Simulatability Against Malicious Sender. We show a simulator that, having oracle access to the malicious sender S*, extracts both inputs s_0, s_1. Sim runs as receiver in the Π_{OT} protocol by choosing sets T_0 and T_1, and playing with a random bit in the remaining OT executions. Then, if the Test phase passes, Sim rewinds S* to extract the vectors $(\mathbf{a}_{0,j}, \mathbf{b}_{0,j})$ and $(\mathbf{a}_{01,j}, \mathbf{b}_{1,j})$ from the extractable commitments. Due to the indistinguishability property of the underlying $\Pi_{\text{OT}}^{\text{R}}$ we have that any malicious sender cannot detect on which coordinates he will be tested. Therefore, if the test phase passes, then it holds that, for each bit, at least $\kappa/2 + 1$ of the remaining OT were computed correctly for that bit. Due to the binding of the commitment scheme, to the correctness of Shamir's secret sharing, and the correctness of the proof of consistency of the shares, the values reconstructed from the shares extracted by the simulator in the extractable commitments correspond to the unique value that a honest receiver would have obtained from the shares retrieved via Π_{OT}^R.

3.4 Parallel OT

Protocol Π_{OT} can be used as a building block for constructing a protocol implementing the $\mathcal{F}_{OT}^m$ functionality. The idea is to have the Sender S and the receiver R compute m executions of Π_{OT} in parallel, and accepting a round of communication if and only if all the m executions are computed correctly.

3.5 Round-Optimal Secure Two-Party Computation

The non-interactive secure two-party protocol proposed in [12] it is based on Yao [29] garbled circuits and works in the OT-hybrid model. The main contribution of [12] is to show an (asymptotically) more efficient black-box cut-and-choose for proving that a garbled circuit is computed correctly. The cut-and-choose is non-interactive in the OT-hybrid model. We can cast their construction to the simpler setting of stand-alone two-party computation and replace the ideal calls to the OT with our parallel OT Π_{OT}^m.

Acknowledgments. We thank the anonymous reviewers for helpful comments. Work supported in part by NSF grants 09165174, 1065276, 1118126 and 1136174, US-Israel BSF grant 2008411, OKAWA Foundation Research Award, IBM Faculty Research Award, Xerox Faculty Research Award, B. John Garrick Foundation Award, Teradata Research Award, and Lockheed-Martin Corporation Research Award. This material is based upon work supported by the Defense Advanced Research Projects Agency through the U.S. Office of Naval Research under Contract N00014 -11 -1-0392. The views expressed are those of the author and do not reflect the official policy or position of the Department of Defense or the U.S. Government.

References

1. Choi, S.G., Dachman-Soled, D., Malkin, T., Wee, H.: Simple, Black-box constructions of adaptively secure protocols. In: Reingold, O. (ed.) TCC 2009. LNCS, vol. 5444, pp. 387–402. Springer, Heidelberg (2009)
2. Cramer, R., Damgård, I.B., Schoenmakers, B.: Proof of partial knowledge and simplified design of witness hiding protocols. In: Desmedt, Y.G. (ed.) CRYPTO 1994. LNCS, vol. 839, pp. 174–187. Springer, Heidelberg (1994). http://dx.doi.org/10.1007/3-540-48658-5_19
3. Damgård, I.B., Ishai, Y.: Constant-round multiparty computation using a black-box pseudorandom generator. In: Shoup, V. (ed.) CRYPTO 2005. LNCS, vol. 3621, pp. 378–394. Springer, Heidelberg (2005). http://dx.doi.org/10.1007/11535218_23
4. Damgård, I., Scafuro, A.: Unconditionally secure and universally composable commitments from physical assumptions. In: Sako, K., Sarkar, P. (eds.) ASIACRYPT 2013, Part II. LNCS, vol. 8270, pp. 100–119. Springer, Heidelberg (2013). http://dx.doi.org/10.1007/978-3-642-42045-0_6

5. Goldreich, O., Micali, S., Wigderson, A.: How to play any mental game or A completeness theorem for protocols with honest majority. In: Aho, A.V. (ed.) Proceedings of the 19th Annual ACM Symposium on Theory of Computing, pp. 218–229. ACM, New York (1987). http://doi.acm.org/10.1145/28395.28420
6. Goyal, V.: Constant round non-malleable protocols using one-way functions. In: Proceedings of the 43rd Annual ACM Symposium on Theory of Computing, STOC 2011, pp. 695–704. ACM (2011)
7. Goyal, V., Lee, C.K., Ostrovsky, R., Visconti, I.: Constructing non-malleable commitments: A black-box approach. In: FOCS, pp. 51–60. IEEE Computer Society (2012)
8. Goyal, V., Ostrovsky, R., Scafuro, A., Visconti, I.: Black-box non-black-box zero knowledge. In: Symposium on Theory of Computing, STOC 2014, pp. 515–524 (2014)
9. Haitner, I.: Semi-honest to malicious oblivious transfer—the black-box way. In: Canetti, R. (ed.) TCC 2008. LNCS, vol. 4948, pp. 412–426. Springer, Heidelberg (2008)
10. Hazay, C., Lindell, Y.: Efficient secure two-party protocols - techniques and constructions. In: Information Security and Cryptography. Springer (2010). http://dx.doi.org/10.1007/978-3-642-14303-8
11. Ishai, Y., Kushilevitz, E., Lindell, Y., Petrank, E.: Black-box constructions for secure computation. In: Proceedings of the 38th Annual ACM Symposium on Theory of Computing, STOC 2006, pp. 99–108 (2006)
12. Ishai, Y., Kushilevitz, E., Ostrovsky, R., Prabhakaran, M., Sahai, A.: Efficient non-interactive secure computation. In: Paterson, K.G. (ed.) EUROCRYPT 2011. LNCS, vol. 6632, pp. 406–425. Springer, Heidelberg (2011)
13. Ishai, Y., Kushilevitz, E., Ostrovsky, R., Sahai, A.: Zero-knowledge from secure multiparty computation. In: Proceedings of the 39th Annual ACM Symposium on Theory of Computing, STOC 2007, pp. 21–30 (2007)
14. Ishai, Y., Prabhakaran, M., Sahai, A.: Founding cryptography on oblivious transfer – efficiently. In: Wagner, D. (ed.) CRYPTO 2008. LNCS, vol. 5157, pp. 572–591. Springer, Heidelberg (2008). http://dx.doi.org/10.1007/978-3-540-85174-5_32
15. Katz, J., Ostrovsky, R.: Round-optimal secure two-party computation. In: Franklin, M. (ed.) CRYPTO 2004. LNCS, vol. 3152, pp. 335–354. Springer, Heidelberg (2004)
16. Kilian, J.: Founding cryptography on oblivious transfer. In: Proceedings of the 20th Annual ACM Symposium on Theory of Computing, May 2–4, 1988, Chicago, Illinois, USA, pp. 20–31. ACM (1988)
17. Kilian, J.: A note on efficient zero-knowledge proofs and arguments. In: STOC, pp. 723–732 (1992)
18. Kiyoshima, S., Manabe, Y., Okamoto, T.: Constant-round black-box construction of composable multi-party computation protocol. In: Lindell, Y. (ed.) TCC 2014. LNCS, vol. 8349, pp. 343–367. Springer, Heidelberg (2014)
19. Lapidot, D., Shamir, A.: Publicly verifiable non-interactive zero-knowledge proofs. In: Menezes, A., Vanstone, S.A. (eds.) CRYPTO 1990. LNCS, vol. 537, pp. 353–365. Springer, Heidelberg (1991)
20. Lin, H., Pass, R.: Black-box constructions of composable protocols without setup. In: Safavi-Naini, R., Canetti, R. (eds.) CRYPTO 2012. LNCS, vol. 7417, pp. 461–478. Springer, Heidelberg (2012)
21. Micciancio, D., Ong, S.J., Sahai, A., Vadhan, S.P.: Concurrent zero knowledge without complexity assumptions. In: Halevi, S., Rabin, T. (eds.) TCC 2006. LNCS, vol. 3876, pp. 1–20. Springer, Heidelberg (2006)

22. Ostrovsky, R., Rao, V., Scafuro, A., Visconti, I.: Revisiting lower and upper bounds for selective decommitments. In: Sahai, A. (ed.) TCC 2013. LNCS, vol. 7785, pp. 559–578. Springer, Heidelberg (2013)

23. Ostrovsky, R., Scafuro, A., Venkitasubramanian, M.: Resettably sound zero-knowledge arguments from OWFs - the (semi) black-box way. In: Dodis, Y., Nielsen, J.B. (eds.) TCC 2015, Part I. LNCS, vol. 9014, pp. 345–374. Springer, Heidelberg (2015)

24. Pass, R., Wee, H.: Black-box constructions of two-party protocols from one-way functions. In: Reingold, O. (ed.) TCC 2009. LNCS, vol. 5444, pp. 403–418. Springer, Heidelberg (2009)

25. Peikert, C., Vaikuntanathan, V., Waters, B.: A framework for efficient and composable oblivious transfer. In: Wagner, D. (ed.) CRYPTO 2008. LNCS, vol. 5157, pp. 554–571. Springer, Heidelberg (2008)

26. Shamir, A.: How to share a secret. Commun. ACM **22**(11), 612–613 (1979)

27. Wee, H.: Black-box, round-efficient secure computation via non-malleability amplification. In: Proceedings of the 51th Annual IEEE Symposium on Foundations of Computer Science, pp. 531–540 (2010)

28. Xiao, D.: (Nearly) round-optimal black-box constructions of commitments secure against selective opening attacks. In: Ishai, Y. (ed.) TCC 2011. LNCS, vol. 6597, pp. 541–558. Springer, Heidelberg (2011)

29. Yao, A.C.C.: How to generate and exchange secrets (extended abstract). In: FOCS, pp. 162–167 (1986)

Secure Computation with Minimal Interaction, Revisited

Yuval Ishai[1], Ranjit Kumaresan[2]($\boxtimes$), Eyal Kushilevitz[1],
and Anat Paskin-Cherniavsky[3]

[1] Department of Computer Science, Technion, Haifa, Israel
{yuvali,eyalk}@cs.technion.ac.il
[2] MIT CSAIL, Cambridge, USA
ranjit@csail.mit.edu
[3] Department of Computer Science, Ariel University, Melbourne, Australia
anps83@gmail.com

Abstract. Motivated by the goal of improving the concrete efficiency of secure multiparty computation (MPC), we revisit the question of MPC with only two rounds of interaction. We consider a minimal setting in which parties can communicate over secure point-to-point channels and where no broadcast channel or other form of setup is available.

Katz and Ostrovsky (Crypto 2004) obtained negative results for such protocols with $n = 2$ parties. Ishai et al. (Crypto 2010) showed that if only one party may be corrupted, then $n \geq 5$ parties can securely compute any function in this setting, with guaranteed output delivery, assuming one-way functions exist. In this work, we complement the above results by presenting positive and negative results for the cases where $n = 3$ or $n = 4$ and where there is a *single malicious party*.

When $n = 3$, we show a 2-round protocol which is secure with "selective abort" against a single malicious party. The protocol makes a black-box use of a pseudorandom generator or alternatively can offer unconditional security for functionalities in NC^1. The concrete efficiency of this protocol is comparable to the efficiency of secure two-party computation protocols for *semi-honest* parties based on garbled circuits.

When $n = 4$ in the setting described above, we show the following:

- A *statistical VSS* protocol that has a 1-round sharing phase and 1-round reconstruction phase. This improves over the state-of-the-art result of Patra et al. (Crypto 2009) whose VSS protocol required 2 rounds in the reconstruction phase.
- A 2-round statistically secure protocol for *linear functionalities* with guaranteed output delivery. This implies a 2-round 4-party fair coin tossing protocol. We complement this by a negative result, showing that there is a (nonlinear) function for which there is no 2-round statistically secure protocol.

Y. Ishai—Research supported by the European Union's Tenth Framework Programme (FP10/2010-2016) under grant agreement no. 259426 ERC-CaC, ISF grant 1709/14 and BSF grant 2012378.

R. Kumaresan—Supported by Qatar Computing Research Institute. Work done in part while at the Technion.

E. Kushilevitz—Research supported by ISF grant 1709/14 and BSF grant 2012378.

R. Gennaro and M. Robshaw (Eds.): CRYPTO 2015, Part II, LNCS 9216, pp. 359–378, 2015.
DOI: 10.1007/978-3-662-48000-7_18

- A 2-round computationally secure protocol for *general functionalities* with guaranteed output delivery, under the assumption that injective (one-to-one) one-way functions exist.
- A 2-round protocol for general functionalities with guaranteed output delivery in the *preprocessing model*, whose correlated randomness complexity is proportional to the length of the inputs. This protocol makes a black-box use of a pseudorandom generator or alternatively can offer unconditional security for functionalities in NC^1.

Prior to our work, the feasibility results implied by our positive results were not known to hold even in the stronger MPC model considered by Gennaro et al. (Crypto 2002), where a broadcast channel is available.

Keywords: Secure multiparty computation · Round complexity · Efficiency

1 Introduction

Suppose that two or more parties wish to compute some function on their sensitive inputs while hiding the inputs from each other to the extent possible. One solution would be to employ an external trusted server. Such a trust assumption gives rise to the following minimalist protocol: each party sends its input to the server, who computes the result and sends only the output back to the parties.

However, trusting an external server has several drawbacks, such as being susceptible to server breaches. To eliminate the single point of failure, the parties may employ a secure multiparty computation (MPC) protocol for distributing the trust between the parties. When replacing the external trusted server with an MPC protocol, a major practical disadvantage is that we lose the minimalist structure of the earlier protocol. Indeed, MPC protocols that offer security against malicious parties typically require a substantial amount of interaction. For instance,

- Implementing *broadcast* (a special case of MPC) over secure point-to-point channels generally requires more than two rounds [12].
- Even if broadcast is given for free, 3 or more rounds are necessary for general MPC protocols that tolerate $t \geq 2$ malicious parties and guarantee fairness [15].

Fortunately, neither of the above limitations rules out the possibility of obtaining 2-round MPC protocols secure against a *single malicious party*. This was exploited in the work of Ishai et al. [19], who showed that if only one party can be corrupted, then $n \geq 5$ parties can securely compute any function of their inputs, with guaranteed output delivery, by using only two rounds of interaction over secure point-to-point channels, and *without assuming broadcast or any additional setup*. Since a similar result can be ruled out in the case of $n = 2$ parties [21], the work of [19] leaves open the corresponding question for $n = 3$ and $n = 4$.

This question may be highly relevant to real world situations where the number of parties is small and the existence of two or more corrupted parties is unlikely. Indeed, the only real world deployment of MPC that we are aware of is for the case of $n = 3$ and $t = 1$ (cf. [5,6]). Furthermore, in settings where secure computation between multiple servers involves long-term secrets, such as cryptographic keys or sensitive databases, it may be preferable to employ three or more servers as opposed to two for the purpose of recovery from faults. Indeed, in secure 2-server solutions the long-term secrets are lost forever if one of the servers malfunctions. Finally, the existence of a strict honest majority allows for achieving stronger security goals, such as fairness and strong forms of composability, that are provably unrealizable in the two-party setting and, moreover, it gives hope for designing leaner protocols that use weaker cryptographic assumptions and have better concrete efficiency. Thus, positive results in this regime (i.e., 2-round protocols for $n = 3$ and $n = 4$) may have strong relevance to the goal of practically efficient secure computation.

Our interest in this problem is motivated not only by the quantitative goal of minimizing the amount of interaction, but also by qualitative advantages of 2-round protocols over protocols with more rounds. For instance, as pointed out in [19], the minimal interaction pattern of 2-round protocols makes it possible to divide the secure computation process into two non-interactive stages of input contribution and output delivery. These stages can be performed independently of each other in an asynchronous manner, allowing clients to go online only when their inputs change, and continue to (passively) receive periodic outputs while inputs of other parties may change.

Our Results. We obtain several results on the existence of 2-round MPC protocols over secure point-to-point channels, without broadcast or any additional setup, which tolerate a single malicious party out of $n = 3$ or $n = 4$ parties.

Three-Party Setting. In an information-theoretic setting without a broadcast channel, the broadcast functionality itself is unrealizable for $n = 3$ and $t = 1$ [22]. Therefore, if we wish to obtain secure computation protocols with perfect/statistical security, with *guaranteed output delivery*, then we have to assume a broadcast channel. In the computational setting, broadcast is realizable in two rounds using digital signatures (assuming a public key infrastructure setup). Further, assuming indistinguishability obfuscation and a CRS setup, there exist 2-round protocols which tolerate an arbitrary number of corruptions $t < n$ [2,13]. These protocols guarantee fairness when $t = 1$ and $n = 3$ (more generally, when $t < n/2$), and also have nearly optimal communication complexity. However, the above computationally secure protocols require a trusted setup and, perhaps more importantly, they rely on strong cryptographic assumptions and have poor concrete efficiency.

Fortunately, as we show, it turns out that a further relaxation of this notion, referred to as "security-with-selective-abort," allows us to obtain statistical security even without resorting to the use of a broadcast channel or a trusted setup. This notion of security, introduced in [17], differs from the standard notion of security-with-abort in that it allows the adversary (after learning its own outputs) to individually decide for each uncorrupted party whether this party will obtain its correct output or will abort with the special output "⊥". Our main result in this setting is the following:

- There exists a 2-round, 3-party general MPC protocol over secure point-to-point channels, that provides security-with-selective-abort in the presence of a single malicious party. The protocol provides statistical security for functionalities in NC^1 and computational security for general functionalities by making a black-box use of a PRG.[1]

The above protocol is very efficient in concrete terms. There is a large body of recent work on optimizing the efficiency of 2-party protocols based on garbled circuits. A recent work of Choi et al. [8] considered the 3-party setting, but required security against 2 malicious parties and thus did not offer better efficiency than that of 2-party protocols. Our work suggests that settling for security against a single party can lead to better overall efficiency while also minimizing round complexity. In particular, our 3-party protocol is roughly as efficient as 2-party *semi-honest* garbled circuit protocols. See discussion in Sect. 3.

Four-Party Setting. Gennaro et al. [14] show the impossibility of 2-round *perfectly secure* protocols for secure computation for $n = 4$ and $t = 1$, even assuming a broadcast channel. Ishai et al. [19] show a secure-with-selective-abort protocol in this setting over point-to-point channels. Their protocol does not guarantee output delivery. We complete the picture in several ways. We start by focusing on the simpler question of designing verifiable secret sharing (VSS) protocols. Prior to our work, for the case when $n = 4$ and $t = 1$, it was known that (1) there exists a 1-round sharing and 2-round reconstruction statistical VSS protocol [24], and (2) there exists a 2-round sharing and 1-round reconstruction statistical VSS protocol [1]. We improve the state-of-the-art by showing that:

- There exists a 4-party statistically secure VSS protocol over point-to-point channels that tolerates a single malicious party and requires one round in the sharing phase and one round in the reconstruction phase.

The above result is somewhat unexpected in light of the results from [1,24], and the corresponding protocol is significantly more involved than other 1-round VSS protocols. Our 1-round VSS protocol implies statistically secure 2-round protocols for fair coin-tossing and simultaneous broadcast over point-to-point channels. More generally, we show that:

- There exists a 2-round 4-party statistically secure MPC protocol for *linear functionalities* (that compute a linear mapping from inputs to outputs) over secure point-to-point channels, providing full security against a single malicious party.

---

[1] Our information-theoretic protocols are limited to NC^1 like all known constant-round protocols, even in the semi-honest model. However, settling for computational security, all our protocols apply to general circuits by using any PRG as a black box.

We complement the above positive result by proving the following negative result:

– There exists a nonlinear function which cannot be realized by a protocol as above.

Taken together, the two results above showcase a unique provable separation between the round complexity of linear functionalities (which capture coin-tossing and secure multicast as special cases) and that of higher degree functions. Next, we show that settling for computational security allows us to beat the previous negative result.

– Assuming the existence of injective (one-to-one) one-way functions, there exists a 2-round 4-party *computationally* secure MPC protocol for *general functionalities* over secure point-to-point channels, providing full security against a single malicious party.

None of our previous results require a setup assumption. A natural question is whether it is possible to obtain statistical security (at least for functionalities in NC^1) in the same setting by relying on some form of setup. Several prior works [4,7,9,10,18] obtain information-theoretic security in a so-called pre-processing model, where the parties are given access to a source of correlated randomness before the inputs are known. However, these protocols either have a higher round complexity, or alternatively make use of correlated randomness whose size grows exponentially with the input length [3,18]. We present a protocol in this setting where the size of correlated randomness is exactly the length of the inputs. In the full version, we show that:

– Assuming a correlated randomness setup, there exists a 2-round 4-party MPC protocol over secure point-to-point channels, providing full security against a single malicious party. The protocol provides statistical security for functionalities in NC^1 and computational security for general functionalities by making a black-box use of a PRG. The size of the correlated randomness is linear in the input size.

Prior to our work, our positive results in either the 3-party or 4-party settings were not known to hold even in the setting considered where a broadcast channel is available, which was studied in the line of work originating from [14,15]. Moreover, our protocols are secure against adaptive and rushing adversaries. Finally, while we analyze our protocols in the standalone setting, they are in fact composable (in particular, none of our simulators is rewinding).

Technical Overview. We now give a very brief and high level overview of some of our results. The main primitives that we use in our protocols are private simultaneous message (PSM) protocols [11] and 1-private secret sharing schemes (cf. Sect. 2). Our high level strategy is similar to the one used in [19]. The parties secret share their inputs among other parties in the first round. Then, in the second round, they make use of PSM subprotocols to reconstruct parties' inputs from the shares, and also to evaluate a function on the reconstructed inputs. Given the above, there are still two main issues that need to be resolved: (1) a malicious PSM client may supply inconsistent shares of honest parties inputs inside the PSM, and (2) a malicious party may supply inconsistent shares of its own input to honest parties. Thus, different PSM instances may reconstruct different inputs thereby generating different outputs all of which seem correct.

Ishai et al. [19] get around (1) and (2) by using $(n-2)$-client PSM. Note that for $n \geq 5$ there are at least two honest clients and these two clients hold *all* the shares of all parties. Thus, it is easy to detect inconsistent input shares *inside* the PSM, and it is possible to either apply a "correction" inside the PSM or easily ensure that incorrect PSM outputs are discarded. In our setting, i.e., $n \in \{3, 4\}$, we have to deal with 2-client PSMs. This is obviously necessary when $n = 3$. We can use 3-client PSM when $n = 4$, but this PSM cannot be expected to deliver output since a malicious client can simply abort this PSM. For these reasons, techniques from [19] do not work when $n \in \{3, 4\}$. We can no longer apply corrections inside the PSM or easily identify incorrect PSM outputs.

To get around (1), we use a novel "view reconstruction" technique (cf. Sect. 3). When $n = 3$, this technique suffices, together with some additional ideas, to get around both (1) and (2). To get around (2), when $n = 4$, we use information-theoretic MACs for secure linear function evaluation and non-interactive commitments for general secure function evaluation. Additional complications arise when using MACs inside the PSM and we overcome these by employing a cut-and-choose technique (cf. Sect. 4).

2   Preliminaries

In this section, we provide definitions of verifiable secret sharing (VSS) and private simultaneous message (PSM) protocols. We also describe the secret sharing schemes we use.

Verifiable Secret Sharing (VSS). In this work, we focus on the statistical variant of verifiable secret sharing. We give the general definition below, but will construct protocols for the specific case of $n = 4$ and $t = 1$.

Definition 1. *Let σ be a statistical security parameter. A two-phase protocol for parties $\mathcal{P} = \{P_1, \ldots, P_n\}$, where a distinguished dealer $D \in \mathcal{P}$ holds initial input $s \in \mathbb{F}$, is a statistical VSS protocol tolerating t malicious parties if the following conditions hold for any adversary controlling at most t parties:*

- **Privacy.** *If the dealer is honest at the end of the first phase (the sharing phase), then at the end of this phase the joint view of the malicious parties is independent of the dealer's input s.*
- **Correctness.** *Each honest party P_i outputs a value s_i at the end of the second phase (the reconstruction phase). If the dealer is honest, then except with probability negligible in σ, it holds that $s_i = s$.*

– **Commitment.** *Except with probability negligible in σ, the joint view of the honest parties at the end of the sharing phase defines a value s' such that $s_i = s'$ for every honest P_i.*   ◇

The PSM Model. A private simultaneous messages (PSM) protocol [11] is a non-interactive protocol involving m parties $P_1, \ldots, P_m$, who share a common random string $r = r^{\text{psm}}$, and an external referee who has no access to r. In such a protocol, each party P_i sends a single message to the referee based on its input x_i and r. These m messages should allow the referee to compute some function of the inputs without revealing any additional information about the inputs. Our definitions below are taken almost verbatim from [19].

Formally, a PSM protocol π for a function $f : \{0,1\}^{\ell \times m} \to \{0,1\}^*$ is defined by $R(\ell)$, a randomness length parameter, m message algorithms $A_1, \ldots, A_m$ and a reconstruction algorithm Rec, such that the following requirements hold.

– *Correctness:* for every input length ℓ, all $x_1, \ldots, x_m \in \{0,1\}^\ell$, and all $r \in \{0,1\}^{R(\ell)}$, we have $\mathsf{Rec}(A_1(x_1, r), \ldots, A_m(x_m, r)) = f(x_1, \ldots, x_m)$.
– *Privacy:* there is a simulator $\sim_\pi^{\text{trans}}$ such that, for all $x_1, \ldots, x_m$ of length ℓ, the distribution $\sim_\pi^{\text{trans}}(1^\ell, f(x_1, \ldots, x_m))$ is indistinguishable from $(A_1(x_1, r), \ldots, A_m(x_m, r))$.

We consider either perfect or computational privacy, depending on the notion of indistinguishability. (For simplicity, we use the input length ℓ also as security parameter, as in [16]; this is without loss of generality, by padding inputs to the required length.)

A *robust* PSM protocol π should additionally guarantee that even if a subset of the m parties is malicious, the protocol still satisfies a notion of "security with abort." That is, the effect of the messages sent by corrupted parties on the output can be simulated by either inputting to f a valid set of inputs (independently of the honest parties' inputs) or by making the referee abort. This is formalized as follows.

– *Statistical Robustness:* For any subset $T \subset [m]$, there is an efficient (black-box) simulator $\sim_\pi^{\text{ext}}$ which, given access to the common r and to the messages sent by (possibly malicious) parties P_i^*, $i \in T$, can generate a distribution x_T^* over x_i, $i \in T$, such that the output of Rec on inputs $A_T(x_T^*, r), A_{\overline{T}}(x_{\overline{T}}, r)$ is statistically close to the "real-world" output of Rec when receiving messages from the m parties on a randomly chosen r. The latter real-world output is defined by picking r at random, letting party P_i pick a message according to A_i, if $i \notin T$, and according to P_i^* for $i \in T$, and applying Rec to the m messages. We allow $\sim_\pi^{\text{ext}}$ to produce a special symbol $\bot$ (indicating abort) on behalf of some party P_i^*, in which case Rec outputs $\bot$ as well.

The following theorem summarizes some known facts about PSM protocols.

Theorem 1 ([11, 19, 23]). *(i) For any $f \in \mathrm{NC}^1$, there is a polynomial-time, perfectly private, and statistically robust PSM protocol. (ii) For any polynomial-time computable f, there is a polynomial-time, computationally private, and statistically robust PSM protocol which uses any PRG as a black box.*

Secret Sharing. In a t-private n-party secret sharing scheme every t parties learn nothing about the secret, and every $t+1$ parties can jointly reconstruct it. A secret sharing scheme is *efficiently extendable*, if for any subset $T \subseteq [n]$, it is possible to efficiently check whether the (purported) shares to T are consistent with a valid sharing of some secret s. Additionally, in case the shares are consistent, it is possible to efficiently sample a (full) sharing of some secret which is consistent with that partial sharing. In our protocols, we use 2-out-of-2 additive secret sharing and 1-private 3-party CNF secret sharing.

Additive Sharing. In 2-out-of-2 additive sharing over $\mathbb{F}_2$, given both shares r_1, r_2, we can reconstruct the secret as $s = r_1 \oplus r_2$. On the other hand, given the secret s and one of the shares r_1, we can determine the remaining share $r_2 = s \oplus r_1$.

CNF Sharing [20]. In 1-private 3-party CNF sharing over $\mathbb{F}_2$, we choose random $r_1, r_2 \in \mathbb{F}_2$, compute $r_3 = s \oplus r_1 \oplus r_2$, and set the CNF shares held by P_1, P_2, P_3 as $\langle r_2, r_3 \rangle, \langle r_3, r_1 \rangle, \langle r_1, r_2 \rangle$ respectively. Given two of the three CNF shares, say $\langle r_1, r_2 \rangle, \langle r_2, r_3 \rangle$ we can reconstruct the secret $s = r_1 \oplus r_2 \oplus r_3$. Also, given s and one of the shares say $\langle r_1, r_2 \rangle$, we can determine the remaining shares as $\langle r_2, s \oplus r_1 \oplus r_2 \rangle$ and $\langle s \oplus r_1 \oplus r_2, r_1 \rangle$. We say that P_1, P_2 hold "consistent" CNF shares if P_1, P_2 respectively hold $\langle r_2, r_3 \rangle, \langle r_3', r_1 \rangle$ with $r_3' = r_3$.

Notation. We let n denote the number of parties. In this paper $n \in \{3, 4\}$. We denote by T_i (resp. $T_{i,j}$) the set $[n] \setminus \{i\}$ (resp. $[n] \setminus \{i, j\}$), where the value of n is clear from the context. Throughout this paper, the number of corrupted parties $t = 1$. Since this is the case, we sometimes abuse notation and use t as a variable to denote parties' index (e.g., P_t). We let $r_{i,j}^{\mathrm{psm}} = r_{j,i}^{\mathrm{psm}}$ to denote the shared randomness for PSM executions involving clients P_i and P_j.

3   2-Round 3-Party Computation with Selective Abort Security

Recall that in security with selective abort, the adversary is able to deny output to an honest party (i.e., there is no guaranteed output delivery), and further it can choose to do so individually for each honest party. We wish to stress that the abort is dependent only on the inputs/outputs of the corrupt party and is otherwise (statistically) independent of the inputs/outputs of the honest parties.

A First Attempt. Consider the following protocol which makes use of additive sharing and PSM subprotocols. Each party P_i first additively shares its input x_i into $x_{i,j}$ and $x_{i,k}$ (i.e., $x_i = x_{i,j} \oplus x_{i,k}$) and sends $x_{i,j}$ to party P_j and $x_{i,k}$ to party P_k. In the second round, parties execute pairwise (robust) PSMs that first reconstruct each party's input from the additive shares possessed by the PSM clients, and then compute the output from the reconstructed inputs. It should be clear that the above yields a secure protocol in the semi-honest setting.

Predictably, things go wrong in the presence of a malicious adversary. Specifically, an adversary that corrupts, say, P_1 can carry out the following attack: Party P_1 can use input 0 in the PSM execution where P_1 and P_2 are the PSM clients and P_3 is the PSM referee. Then, P_1 uses a different input, say 1 in the PSM execution where P_1 and P_3 are the PSM clients and P_2 is the PSM referee. This results in the undesirable situation where P_2 and P_3 disagree on the output and, furthermore, are not even aware that there may be a disagreement. Note that this does not yield security with selective abort, since honest parties accept outputs that are computed using different values for the corrupt input. In other words, there is no single effective corrupt input (to be extracted by the 'simulator' in the ideal execution) that explains all honest outputs. To counter this attack, we employ the following "view reconstruction trick."

View Reconstruction Trick. Essentially this trick tries to reconstruct the (first round) view of the PSM referee using the views supplied by the PSM clients. Note that the "view" in the naïve protocol described above consists of additive shares supplied by the parties. Fortunately, the *efficient extendability* of linear secret sharing schemes such as the additive secret sharing and CNF secret sharing, enables us to reconstruct the unique share that must be held by the PSM referee. (For more details see Sect. 2 and [19].)

To see this trick in action, consider a concrete example. Suppose P_i and P_j are PSM clients and P_k is the PSM referee. Note that P_k's view consists of the shares $x_{i,k}$ sent by P_i and $x_{j,k}$ sent by P_j. Now in the PSM subprotocol (instantiated in the naïve protocol) suppose party P_i supplies input x_i' and party P_j supplies input x_j'. (If P_i (resp. P_j) is not honest then $x_i' = x_i$ (resp. $x_j' = x_j$) may not hold.) In the PSM protocol, we now ask P_i to supply in addition to its input $x_i' = x_i$ also the shares obtained in round 1, namely $x_{j,i}' = x_{j,i}$ obtained from P_j and $x_{k,i}' = x_{k,i}$ obtained from P_k. We ask P_j to do the same as well, i.e., P_j supplies $x_j' = x_j$, $x_{i,j}' = x_{i,j}$, $x_{k,j}' = x_{k,j}$. Of course, a malicious party, say P_i, may not supply the correct inputs or shares as it obtained from the honest parties (i.e., it may be the case that $x_i' \neq x_i$ or $x_{j,i}' \neq x_{j,i}$ or $x_{k,i}' \neq x_{k,i}$). Anyway, we can compute the values that *ought* to be held by P_k using the values supplied by P_i and P_j. For instance, the values $x_{k,i}, x_{k,j}$ can directly be obtained from P_i, P_j since they supplied $x_{k,i}', x_{k,j}'$ (respectively) to the PSM subprotocol. The values $x_{i,k}$ (resp. $x_{j,k}$) can be reconstructed as $x_i' \oplus x_{i,j}'$ where x_i' was supplied by P_i and $x_{i,j}'$ was supplied by P_j.

In our modified protocol, we let the PSM referee, say P_k to accept the final output only if the reconstructed view from the PSM protocol matches its first round view, i.e., only if $x_{k,i}' = x_{k,i}$, $x_{k,j}' = x_{k,j}$, $x_{i,k}' = x_{i,k}$, and $x_{j,k}' = x_{j,k}$ all hold. We prove the following theorem.

Theorem 2. *There exists a 2-round 3-party secure-with-selective-abort protocol for secure function evaluation over point-to-point channels that tolerates a single malicious party. The protocol provides statistical security for functionalities in* NC^1 *and computational security for general functionalities by making a black-box use of a pseudorandom generator.*

Proof. The formal protocol is described in Fig. 1. We provide a sketch of the simulation and the analysis below.

Simulation Sketch. Denote the corrupt party by P_ℓ. Let P_i, P_j be the remaining (honest) parties. The simulator begins by sending random additive shares to the corrupt party on behalf of the honest parties. It also sends and receives randomness to be used in the PSM executions in the next round. Note that the simulator also receives additive shares from the corrupt party. Using the additive shares, the simulator computes the effective input say $\hat{x}_\ell$ of the corrupt party (i.e., by simply xor-ing the additive shares). Then, the simulator sends $\hat{x}_\ell$ to the trusted party first, and obtains the output z_ℓ.

Next the simulator invokes the PSM simulator $\sim_{\pi_{i,j}}^{\text{trans}}$ (guaranteed by the privacy property) on inputs z_ℓ and the additive shares sent on behalf of the honest parties. Denote the output of the $\sim_{\pi_{i,j}}^{\text{trans}}$ by $\tau_{i,\ell}$ and $\tau_{j,\ell}$. Acting as the honest party P_i (resp. P_j), the simulator sends $\tau_{i,\ell}$ (resp. $\tau_{j,\ell}$) to the corrupt party. It remains to be shown how the simulator decides which uncorrupted parties learn the output and which receive $\perp$. To do this, the simulator does the following. First, acting as the honest party P_i the simulator receives the PSM message $\tau_{\ell,i}$ that P_ℓ sends to P_i as part of PSM execution $\pi_{\ell,j}$. Similarly, acting as P_j, the simulator also receives $\tau_{\ell,j}$. Next, the simulator invokes the PSM simulator $\sim_{\pi_{\ell,i}}^{\text{ext}}$ on the PSM message $\tau_{\ell,i}$ (and also the PSM randomness) to decide what effective input P_ℓ used in PSM subprotocol $\pi_{\ell,j}$. Depending on this input, the simulator then decides whether P_i will accept the output of $\pi_{\ell,j}$ or not. Specifically as in the real execution, the simulator checks if the shares input by P_ℓ are consistent with those held by P_i. If this is indeed the case, then the simulator asks the trusted party to deliver output to P_i, else it asks the trusted party to deliver $\perp$ to P_i. Whether P_j gets the output or not is also handled similarly by the simulator.

Analysis Sketch. We first consider a hybrid experiment which is exactly the same as the real execution except that the PSM messages sent by the honest parties to P_ℓ are replaced by the simulated PSM transcripts generated by $\sim_{\pi_{i,j}}^{\text{trans}}$. To generate these transcripts we first extract the input $\hat{x}_\ell$ by xor-ing the additive shares sent by P_ℓ, and then compute the output of $\pi_{i,j}$ using inputs provided by honest parties and $\hat{x}_\ell$. We then supply this output to $\sim_{\pi_{i,j}}^{\text{trans}}$ to generate the simulated PSM transcripts. The privacy property of the PSM protocol implies that the joint distribution of the view of the adversary and honest outputs in the real protocol is indistinguishable from the corresponding distribution in the hybrid execution.

Note that the distribution of the additive shares and the PSM randomness sent by the simulator in the ideal execution is identical to the distribution of the corresponding values in the hybrid execution. Thus, to prove indistinguishability of the hybrid execution and the ideal execution it suffices to focus on the distribution of honest outputs. Note that in the ideal execution the honest outputs are generated using the true honest inputs and extracted input $\hat{x}_\ell$.

We first show that honest party P_i (resp. P_j) that accepts a non-$\perp$ output in the hybrid execution is ensured that this output is computed using the true honest inputs and the corrupt input $\hat{x}_\ell$. It is here that we use the view reconstruction trick. Specifically now, (1) if P_ℓ supplied incorrect input, then the reconstructed

---

Round 1.
- For $i \in [3]$, each P_i additively shares its input x_i into $x_{i,j}$ and $x_{i,k}$, and sends $x_{i,j}$ to P_j, and $x_{i,k}$ to P_k for distinct $j, k \in [3] \setminus \{i\}$.
- Every pair of parties P_i, P_j, $i, j \in [3]$ and $i < j$, exchange randomness $r_{i,j}^{\mathrm{psm}}$. (For instance, by letting P_i pick $r_{i,j}^{\mathrm{psm}}$ and send $r_{i,j}^{\mathrm{psm}}$ to P_j.)

Round 2.
- Every pair of parties P_i and P_j, $i, j \in [3]$ and $i < j$, use shared randomness $r_{i,j}$ to execute a robust PSM protocol $\pi_{i,j}$, that
  - takes input $\tilde{x}_i = (x'_{k,i}, x'_i, x'_{j,i})$ from P_i where $x'_{k,i} = x_{k,i}, x'_i = x_i, x'_{j,i} = x_{j,i}$,
  - takes input $\tilde{x}_j = (x'_{k,j}, x'_j, x'_{i,j})$ from P_j where $x'_{k,j} = x_{k,j}, x'_j = x_j, x'_{i,j} = x_{i,j}$,
  - reconstructs $x'_k = x_{k,i} \oplus x_{k,j}$,
  - computes $z'_k = f_k(x'_1, x'_2, x'_3)$, $x'_{i,k} = x'_i \oplus x_{i,j}$, $x'_{j,k} = x'_j \oplus x_{j,i}$, and
  - delivers output $(z'_k, x'_{i,k}, x'_{j,k}, x'_{k,i}, x'_{k,j})$ to P_k for $k \in [3]$ and $k \notin \{i, j\}$.

Output. Each P_k outputs z'_k if $x'_{i,k} = x_{i,k}$, $x'_{j,k} = x_{j,k}$, $x'_{k,i} = x_{k,i}$, and $x'_{k,j} = x_{k,j}$ hold, else it outputs $\bot$.

Fig. 1. 2-round 3-party secure-with-selective-abort protocol.

share $x'_{\ell,i}$ (which is revealed as part of the output of $\pi_{\ell,j}$) does not equal $x_{\ell,i}$ possessed by P_i and thus the final output is rejected, and (2) if P_ℓ supplied inconsistent share $x'_{i,\ell} \neq x_{i,\ell}$ inside $\pi_{\ell,j}$, then since this value is revealed as part of the output of $\pi_{\ell,j}$, the final output will be rejected by P_i.

Given the above it remains to be shown that the set of honest parties that receive $\bot$ in the ideal execution equals the set of honest parties that output $\bot$ in the hybrid execution. To prove the above, we use the fact that for all $j \in T_\ell$, with all but negligible probability the PSM simulator $\sim_{\pi_{\ell,j}}^{\mathrm{ext}}$ extracts the input supplied by P_ℓ in the PSM execution $\pi_{\ell,j}$. It follows by simple inspection that the criterion used to add i to S_ℓ in the simulation is essentially the same as the criterion used by P_i to reject the final output of $\pi_{\ell,j}$ in the hybrid execution. □

Concrete Efficiency. Robust PSM subprotocols can be based on Yao garbled circuits [11,23]. The concrete cost of such a robust PSM protocol is essentially the same as a single Yao garbled circuit and incurs an additional cost proportional to the length of the inputs (and is otherwise independent of the complexity of f). Thus our 3-party protocol costs essentially the same as cost of transmitting and evaluating 3 garbled circuits, i.e., thrice the cost of semi-honest 2-party Yao. Contrast this with the concrete cost of realizing state-of-the-art maliciously secure *two-party* protocols which is essentially the cost of transmitting and evaluating roughly σ garbled circuits where σ denotes the statistical security parameter. We previously argued that 3-party protocols provide more redundancy and stability compared to 2-party protocols. Now by settling for just security-with-selective-abort, our three-party protocol provides a much better alternative from a cost perspective as well. All this is in addition to the fact that our 3-party protocol requires only two rounds over point-to-point channels. In contrast, current implementations of 3-party protocols [5,6] require rounds proportional to the depth of the circuit, provide only semi-honest security, or require use of broadcast.

4 4-Party Statistical VSS in a Total of 2 Rounds

Let the set of parties be $\{D, P_1, P_2, P_3\}$. First, let us look at a naïve protocol that assumes the existence of a broadcast channel. Here, the dealer CNF shares its input in the sharing phase. Then in the reconstruction phase, parties simply broadcast the CNF shares they obtained from the dealer. To decide on the output, parties construct an "inconsistency graph" G which tells which parties broadcasted consistent CNF shares.

Sharing Phase. The dealer CNF shares (according to a 1-private 3-party CNF scheme) its secret s among P_1, P_2, P_3. That is, it chooses random s_1, s_2, s_3 subject to $\bigoplus_{i=1,2,3} s_i = s$, and sends CNF share $\{s_j\}_{j \neq i}$ to party P_i for $i \in [3]$.

Reconstruction Phase. Each party P_i broadcasts its share $\{s_j^{(i)} = s_j\}_{j \neq i}$.

Local Computation. D outputs s and terminates the protocol. For every $j, k \in [3]$, define $\mathsf{rec}_{j,k} = s_j^{(k)} \oplus \bigoplus_{i \neq j} s_i^{(j)}$ (i.e., secret reconstructed from CNF shares possessed by P_j and P_k). Let G denote the 3-vertex inconsistency graph which contains an edge between vertices $i, j \in [3]$ iff $\exists k \in [3] \setminus \{i,j\}$ such that $s_k^{(i)} \neq s_k^{(j)}$. (That is, P_i and P_j disagree on the share s_k.)

- (Single-edge case) If G contains exactly one edge, output $\perp$.
- (Even-edge case) Else, if $\exists (j, k) \notin G$, then each party outputs $\mathsf{rec}_{j,k}$.
- (Triple-edge case) If there is no such j, k, then output default value say $\perp$.

It can be easily shown that the above protocol works as long as G does not contain exactly one edge. The difficulty in handling the single-edge case comes because parties do not know which of the inconsistent CNF shares to trust, i.e., which of $s_k^{(i)} \neq s_k^{(j)}$ when $(i,j) \in G$. In the computational setting, this is solved by a trivial use of signatures. In the information-theoretic setting, we can substitute signatures with information-theoretic MACs, but this is not sufficient since such MACs do not have public verification. Fortunately, a combination of MACs with a cut-and-choose technique helps us in this case.

Protocol Overview. The high level idea is to use MACs and then apply the cut-and-choose technique to ensure that (1) parties reveal their true share when D is honest, and (2) detect an inconsistent sharing by a dishonest D. In more detail, now we require D to send, in addition to the CNF shares, also authentication information in the form of information-theoretic MACs (such that a forgery is possible only with probability $\mathsf{negl}(\sigma)$). Specifically for each CNF share s_j, the dealer D sends s_j along with σ MAC values $\{M_{j,\ell}^{(i)}\}_{\ell \in [\sigma]}$ to each party P_i for each $j \neq i$, while each party P_j receives the corresponding keys $\{K_{j,\ell}^{(i)}\}_{\ell \in [\sigma]}$ for each $i \neq j$. Each share is authenticated multiple times to allow application of the cut-and-choose technique.

The reconstruction phase is modified to handle, in particular, the case when the inconsistency graph contains exactly one edge. (All other cases are handled exactly as in the naïve attempt described above.) Now we ask each P_i to broadcast its CNF share $\{s_j^{(i)}\}_{j \neq i}$ (as in the naïve construction), and in addition broadcast its MAC values $\{M_{j,\ell}^{(i)}\}_{j \neq i, \ell \in [\sigma]}$. Also we ask each party P_j to pick for every $i \neq j$, a random subset $S_{j,i} \subset [\sigma]$ (this corresponds to the check set for the cut-and-choose step), and send (1) keys $K_{j,\ell}^{(i)}$ for $\ell \in S_{j,i}$ to P_i, and (2) all keys (i.e., $K_{j,\ell}^{(i)}$ for all $\ell \in [\sigma]$) to P_k where $k \in [3] \setminus \{i, j\}$.

Now we explain in more detail how the cut-and-choose technique helps to resolve the single-edge case. Let $(i, j) \in G$ and let $k \notin \{i, j\}$. We consider two cases depending on whether D is honest or not. Note that in either case, we are assured that P_k is honest, and in fact, our protocol will use MAC keys held by P_k to anchor the parties' output towards the correct output. First consider the case when D is honest. Wlog assume P_i is dishonest, and that P_i disagrees with P_j on the value s_k that is supposed to be held by both of them. Note that while P_k does not hold s_k, it does hold the keys $\{K_{k,\ell}^{(i)}\}_{\ell \in [\sigma]}$ to verify the MACs that P_i possesses. Note that the protocol asks P_i to broadcast all its MACs on s_k, and P_k to send half its keys, say corresponding to some subset $S_{k,i} \subset [\sigma]$, to P_i and all its keys to P_j. While a rushing P_i can wait to receive (half) the keys from P_k to allow forging the corresponding MACs, note that it cannot forge the MACs for the remaining half (except with negligible probability) for which it simply does not know the keys. In other words, when P_i tries to reveal $s_k' \neq s_k$ along with MACs $\{\widetilde{M}_{k,\ell}^{(i)}\}_{\ell \in [\sigma]}$, then with high probability the MAC verification will fail for *all* keys that P_i does not know. Thus, by asking honest P_j and P_k to accept P_i's reveal only if MACs revealed by P_i is consistent with all keys in $\{K_{k,\ell}^{(i)}\}_{\ell \in S_{k,i}}$ (i.e., those that were sent to P_i) and at least one key in $\{K_{k,\ell}^{(i)}\}_{\ell \notin S_{k,i}}$ (i.e., those that were *not* sent to P_i), we are ensured (except with negligible probability) that P_i's reveal $s_k' \neq s_k$ will be rejected by P_j and P_k. Finally note that honest P_j's share s_k is always accepted by the honest parties.

Next, consider the case when D is dishonest. In this case, a single-edge in the inconsistency graph is induced by the inconsistent shares dealt to P_i, P_j. Therefore, the main challenge here is to ensure that all parties agree that D dealt inconsistent shares (as opposed to suspecting that one of the honest parties is deviating from the protocol). Once again, the keys held by P_k serve to anchor all honest parties' decisions on whether to accept or reject reveals made by P_i, P_j. The crux of the argument is the following: except with negligible probability, all parties P_i, P_j, P_k unanimously agree on their decision to accept/reject each of P_i, P_j's reveals. Before we show this, observe that this suffices to achieve resilience against a malicious D. For e.g., suppose both parties' reveals get accepted then if they revealed inconsistent values then all parties agree to output some default value. The case when both parties' reveals get rejected is handled similarly. Finally, when only one of P_i, P_j's reveal is accepted, then all parties can simply agree to output the value corresponding to the reveal that got accepted.

Now we argue that except with negligible probability, all parties will unanimously agree on whether to accept or reject reveals made by P_i, P_j. First observe that the reveals made by a party, say P_j, are either unanimously accepted or unanimously rejected by both P_i and P_k. This is because both P_i and P_k make decisions using the same algorithm on the same values. Next, in our protocol, P_j will accept or reject its own reveal by checking whether its reveal is consistent with the keys that P_k sent to it (i.e., those corresponding to the subset $S_{k,i}$). Thus, if P_j's reveal is rejected by P_j itself, then obviously it will also be rejected by P_i and P_k. Therefore, by way of contradiction, wlog assume that P_j's reveal is rejected by P_i, P_k while it is accepted by P_j. Clearly this happens only if P_k chooses its random subset $S_{k,j}$ such that *all* the MAC values held by P_j corresponding to $S_{k,j}$ are consistent with the keys held by P_k, while *all* the MAC values held by P_j corresponding to $[\sigma] \setminus S_{k,j}$ are *not* consistent with the keys held by P_k. Obviously such an event happens with probability $\binom{\sigma}{\sigma/2}^{-1} = \mathsf{negl}(\sigma)$. Hence we have that with all but negligible probability, all parties P_i, P_j, P_k unanimously agree whether to accept/reject reveals made by P_i and P_j. As explained before, this suffices to prove that agreement holds even when D is dishonest. Fortunately, we can remove the use of broadcast channel in the above protocol. In the full version, we prove the following theorem.

Theorem 3. *There exists a 4-party statistically secure protocol for VSS over point-to-point channels that tolerates a single malicious party and requires one round in the sharing phase and one round in the reconstruction phase.*

5   2-Round 4-Party Statistically Secure Computation for Linear Functions over Point-to-Point Channels

Overview. In the first round of the protocol parties verifiably secret share their inputs (using the protocol from the previous section), and also exchange randomness for running pairwise (robust) PSM executions. Loosely speaking, the PSM executions serve two purposes: (1) parties can evaluate the function on their inputs while preserving privacy, and (2) parties can learn the inconsistency graph corresponding to each VSS sharing. To do (1), the PSM protocol first attempts to reconstruct parties' inputs from the CNF shares held by the PSM clients, and if successful, evaluates the function on these inputs. To do (2), the PSM protocol makes use of the "view reconstruction trick." Note that in the case of VSS, learning the inconsistency graphs was trivial, since parties would broadcast their shares during the reconstruction phase. Unlike VSS, here it is important to protect privacy of these shares throughout the computation. The view reconstruction trick enables us to construct the inconsistency graphs while preserving privacy of the shares.

Recall that each party could potentially receive PSM outputs from three PSM executions. Computing the final output from these PSM outputs is not straightforward, and we will need the inconsistency graphs (generated using outputs of the PSM protocols) to help us. To explain how this is done, we will adopt the perspective of the simulation extraction procedure. Let $m \in [4]$ denote the index of the corrupt party. The extraction procedure constructs the inconsistency graph G' adding edges between vertices if the CNF shares held by corresponding parties are not consistent. If the graph contains all three edges, then the effective input used in this case is 0. We call this the *identifiable triple-edge* case since it is clear that P_m is corrupt. Next, if the graph contains two edges or no edges (i.e., an even number of edges), then we are now assured that there exists a pair of (honest) parties that hold consistent CNF shares of P_m's input. In this case, we can extract the effective input as the secret reconstructed from these consistent CNF shares. We call this case the *resolvable even-edge* case. As was the case in VSS, if G' contains a single-edge then the procedure performs a vote computation step using the MAC values and the corresponding keys. This is to find out which of the two parties is supported by P_m. If there is a unique party that is supported by P_m, then the inconsistency in CNF shares is resolved by using the CNF share possessed by this party. We call this the *resolvable single-edge* case. On the other hand if there is no unique party supported by P_m, then it is clear that P_m is corrupt. We call this the *identifiable single-edge* case. In this case, we extract the effective input used for P_m as the xor of all unique shares (including the inconsistent CNF shares) possessed by all remaining parties.

Observe that the extraction procedure is identical to the VSS extraction procedure except in the identifiable single-edge case. In VSS, it was possible to simply output 0 in the identifiable single-edge case. Here we are not able to replace the corrupt party's input by 0 and then evaluate the function while simultaneously preserving privacy of honest inputs. However, if we use the effective input extracted as described above, then we can exploit the linearity of f to force parties' outputs to be consistent with the extracted input.

Clearly we are done if we force honest parties' outputs in the real protocol to be consistent with the corrupt input extracted by the simulator while preserving privacy of honest parties' inputs. The main obstacle in the implementation is that different honest parties' may hold different inconsistency graphs. The challenge therefore is to design an output computation procedure that allows honest parties' to end up with the same correct output even though they may possess different inconsistency graphs. Also, unlike VSS, here we do not have the luxury of a reconstruction phase where parties can freely disclose their secret shares.

Our output computation procedure makes use of the view reconstruction trick to help each party compute its inconsistency graph, and adapts the cut-and-choose idea from our VSS protocol to help compute the votes (which we can ensure whp that parties agree on). In addition, our procedure exploits the linearity of f to compute the correct output in the identifiable single-edge case. To ensure parties' compute the same output in the resolvable cases, we make use of an "accusation graph" which parties use to determine a pair of honest parties that hold consistent shares of the corrupt input extracted by the simulation procedure described above. For a detailed step-by-step overview of the protocol, please see the full version where we prove:

Theorem 4. *There exists a 2-round 4-party statistically secure protocol for secure linear function evaluation over point-to-point channels that tolerates a single malicious party.*

5.1   Impossibility of 2-Round Statistically Secure 4-Party Computation

In this section, we prove the following:

Theorem 5. *There exists a function which cannot be information-theoretically realized by a 2-round 4-party protocol over point-to-point channels that tolerates a single corrupt party.*

Proof. Assume by way of contradiction that there exists a 2-round statistically secure 4-party protocol π for general secure computation. Let us further set up some notation related to protocol π. Let $A_{i,j}^{(r)}$ denote the algorithm specified by protocol π that is to be executed by (honest) party P_i to generate its r-th round message to P_j. We use the notation

$$m_{i,j}^{(r)} \leftarrow A_{i,j}^{(r)}(x_i, \{\{m_{k,i}^{(s)}\}_{k \in K_i^{(s)}}\}_s : 0 < s < r; \omega_i)$$

where x_i (resp. ω_i) represents P_i's input (resp. internal randomness), and $m_{i,j}^{(r)}$ represents P_i's message to P_j in round r, and $K_i^{(s)}$ represents the subset of parties from which P_i receives a message in round s. Wlog, we assume that algorithm $A_{i,i}^{(3)}$ computes the final output of honest P_i.

The function that we consider is a simple non-linear function and is inspired by the oblivious transfer functionality. Let f be such that $f(b, \perp, \perp, (y_0, y_1)) = (y_b, \perp, \perp, \perp)$. That is, f takes as input a bit $b \in \{0,1\}$ from P_1 and a pair of bits $y_0, y_1 \in \{0,1\}$ from P_4, and returns y_b to P_1. The parties P_2, P_3 supply no inputs, and parties P_2, P_3, P_4 receive no outputs.

The high level strategy is to launch an attack on the real protocol that cannot be simulated in the ideal execution. We let P_1 be the corrupt party, and show that it can obtain *both* y_0 and y_1 in the real protocol with non-negligible probability. Clearly, no ideal process adversary can do the same, and hence the negative result is establised. At a high level, the adversarial strategy of P_1 is to set things up such that the joint view of P_2 and P_4 would infer that P_1's input is 0, while the joint view of P_3 and P_4 would infer that P_1's input is 1. To do this, P_1 chooses internal randomness ω_1 and computes its first round messages $\tilde{m}_{1,2}^{(1)}, \tilde{m}_{1,4}^{(1)}$ to send to P_2 and P_4 assuming that its input equals 0. Then, it samples uniform randomness $\tilde{\omega}$ such that its first round message to P_4 computed assuming input 1 and randomness $\tilde{\omega}$ matches $\tilde{m}_{1,4}^{(1)}$. Since we are in the information-theoretic regime, note that we can allow P_1 to perform arbitrary computations. Then it will follow from the privacy property of π that P_1 will be able to sample $\tilde{\omega}$ with all but negligible probability. P_1 then computes its first round message to P_3 assuming input 1 and internal randomness $\tilde{\omega}$. It then sends its first round messages to the parties, and accepts messages from them. In the second round, it does not send any messages and only accepts messages from other parties. Next, P_1 computes a value y_0' by invoking its output computation algorithm on input 0, internal randomness ω_1, round 1 messages received from all parties, and round 2 messages received from P_2 and P_4. Similarly, P_1 computes y_1' by invoking its output computation algorithm on input 1, internal randomness $\tilde{\omega}$, round 1 messages from all parties, and round 2 messages from P_3 and P_4. Finally, P_1 outputs the values y_0', y_1' as part of its view. We will show that with all but negligible probability it will hold that $y_0' = y_0$ and $y_1' = y_1$. Since an ideal-process adversary has access to P_4's input only via the trusted party implementing f, it is clear that it can obtain either y_0 or y_1 but not both. Thus, this suffices to establish the theorem. This is the high level idea; we now proceed to the formal details. Formally, P_1 does the following:

- Choose randomness ω_1 and compute $\tilde{m}_{1,2}^{(1)} \leftarrow A_{1,2}^{(1)}(0, \perp, \omega_1)$, and $\tilde{m}_{1,4}^{(1)} \leftarrow A_{1,4}^{(1)}(0, \perp, \omega_1)$.
- Choose random $\tilde{\omega}$ such that $A_{1,4}^{(1)}(1, \perp, \tilde{\omega}) = \tilde{m}_{1,4}^{(1)}$. If no such $\tilde{\omega}$ exists, output fail_1 and terminate.
- Compute $\tilde{m}_{1,3}^{(1)} \leftarrow A_{1,3}^{(1)}(1, \perp, \tilde{\omega})$.
- For $j = 2, 3, 4$, send message $\tilde{m}_{1,j}^{(1)}$ to P_j in round 1.
- Receive round 1 messages $m_{2,1}^{(1)}, m_{3,1}^{(1)}, m_{4,1}^{(1)}$, from other parties. Do not send any round 2 messages to any party. Receive round 2 messages $m_{2,1}^{(2)}, m_{3,1}^{(2)}, m_{4,1}^{(2)}$, from other parties and terminate the protocol.
- Compute and output $y_0' \leftarrow A_{1,1}^{(3)}(0, \{\{m_{k,i}^{(1)}\}_{k \in T_1}, \{m_{k,i}^{(2)}\}_{k \in \{2,4\}}\}; \omega_1)$, $y_1' \leftarrow A_{1,1}^{(3)}(1, \{\{m_{k,i}^{(1)}\}_{k \in T_1}, \{m_{k,i}^{(2)}\}_{k \in \{3,4\}}\}; \tilde{\omega})$.

First, we claim that corrupt P_1 does not output fail_1 with all but negligible probability, i.e., P_1 will be able to successfully find $\tilde{\omega}$ satisfying the conditions above. To show this, we rely on the privacy property of π against an (all powerful) P_4. Clearly, if there exists no $\tilde{\omega}$ such that the output of $A_{1,4}^{(1)}$ on input 1 and internal randomness $\tilde{\omega}$, it is obvious to P_4 that P_1's input is 0, and thus privacy is violated. Therefore, it must hold with all but negligible probability (over the choice of ω) that such $\tilde{\omega}$ exists.

Next, we first assert that $y_0' = y_0$ holds with all but negligible probability. The key observation is that messages input to $A_{1,1}^{(3)}$ that are distributed identically to an execution where P_1 holds input 0 and a corrupt P_3 behaves honestly except it does not send its round 2 messages (i.e., aborts after round 1). Thus, it follows from the correctness of π that $y_0 = y_0'$ holds with all but negligible probability. Similarly, we assert that $y_1' = y_1$ holds with all but negligible probability. This is because the messages input to $A_{1,1}^{(3)}$ are distributed identically to an execution where P_1 holds input 1 and a corrupt party P_2 behaves honestly except it does not send its round 2 messages. Thus it follows from the correctness of π that $y_1' = y_1$ holds with all but negligible probability.

Finally we claim that no ideal-process adversary can generate a view with (y_0', y_1') such that these equal P_4's inputs with probability greater than $1/2$. The key observation is that an ideal-process adversary has access to P_4's input only via the trusted party implementing f, it is clear that it can obtain either y_0 or y_1 but not both. In such a case, the best strategy for the ideal process adversary is to obtain one of them, and then simply try and guess the value of the other (thereby succeeding with probability $1/2$). $\qquad\square$

It is instructive to note why the above impossibility does not apply to linear functions. Specifically for a linear function f, if the adversary P_1 can obtain an evaluation of f on input x_1 and honest inputs, then it can trivially obtain an evaluation of f on input $x_1' \neq x_1$ and the same honest inputs. Finally, we note that our negative result can be easily extended to hold in a setting with broadcast.

6   2-Round Computationally Secure 4-Party Computation

Protocol Overview. For simplicity let us assume the existence of a broadcast channel. Our protocol proceeds by letting each party to broadcast a commitment of its input, and then CNF share the corresponding decommitment among the remaining parties. In the second round, parties execute pairwise PSMs that first attempts to reconstruct the inputs of all parties, and then compute the output from the reconstructed inputs. Unfortunately the general framework described as-is does not suffice for secure computation. For one, it may not always be possible to reconstruct input from shares distributed by a malicious party. Further, it may be the case that one pair of honest parties may hold consistent CNF shares from the malicious party while a different pair of honest parties may not. This is exacerbated by the fact that an honest party is guaranteed to receive output from only one PSM instance. In other words, even guaranteeing agreement on output seems somewhat nontrivial.

To circumvent the problems mentioned above, our protocol first detects whether the joint view of honest parties suffices to reconstruct the input of all parties. We do this by enhancing the PSM functionality in a way that lets parties ascertain if for every broadcasted commitment, there exists some pair of parties that hold (consistent) shares of the corresponding decommitment. (Indeed, this is our strategy for extracting the adversary's input in the simulation.) If a pair of parties do not hold consistent shares of a valid decommitment for some party's commitment, then the pairwise PSM in which the parties act as clients delivers as outputs the first round *views* of the honest clients. This in turn lets the referee to determine if its own shares coupled with shares from one of the clients suffices to reconstruct valid decommitments for all commitments. If this is indeed the case, then the referee can reconstruct all inputs from the joint views and then evaluate the function from scratch. On the other hand if there is some party whose commitment cannot be decommitted using the joint views, then the referee simply substitutes that party's input with 0, and evaluates the function from scratch using this new set of inputs. Of course, care must be taken not to reveal honest inputs to a malicious referee. We achieve this by letting the PSM check if the referee's commitment can be decommitted using shares held by honest clients, and then revealing the client views only if this check passes.

The ideas described above still do not suffice to address the somewhat subtler issue of agreement on output. We describe this issue in more detail below. Note that a malicious party that distributed shares of an invalid decommitment can ensure that all inputs are reconstructed successfully in *exactly one* of the PSM instances where it participated as a client and supplied shares of a valid decommitment. Thus, in this PSM instance the function will be evaluated on the reconstructed inputs. Note that this strategy lets exactly one honest party (that acted as referee in the PSM instance described above) to obtain directly the output of the function, while all other honest parties evaluate the function from scratch after substituting the malicious party's input with 0. In other words, the adversary can succeed in forcing different honest parties to obtain evaluations of the function on different sets of inputs. We use a somewhat counterintuitive idea to counter this adversarial strategy. Namely, we force the honest referee in the PSM instance to disregard the output of the function, and instead evaluate the function from scratch (using honest clients' views output in a different PSM instance) after substituting the malicious party's input with 0. To do this, we design the PSM functionality in a way that allows an honest referee to infer whether the joint view of the honest parties indeed contains a valid decommitments to all broadcasted commitments. In more detail, the PSM functionality will attempt to reconstruct the first round view of the referee from the views of the participating clients. (Note that this is possible due to the *efficient extendability* property of CNF sharing schemes.) Upon receiving this reconstructed view, the referee outputs the PSM output only if its view agrees with the reconstructed views. For a formal description of the protocol, and how to remove the use of broadcast, please see the full version where we prove:

Theorem 6. *Assuming the existence of one-way permutations (alternatively, one-to-one one-way functions), there exists a 2-round 4-party computationally secure protocol over point-to-point channels for secure function evaluation that tolerates a single malicious party.*

References

1. Agrawal, S.: Verifiable secret sharing in a total of three rounds. Info. Process. Lett. **112**(22), 856–859 (2012)
2. Asharov, G., Jain, A., López-Alt, A., Tromer, E., Vaikuntanathan, V., Wichs, D.: Multiparty computation with low communication, computation and interaction via threshold FHE. In: Pointcheval, D., Johansson, T. (eds.) EUROCRYPT 2012. LNCS, vol. 7237, pp. 483–501. Springer, Heidelberg (2012)
3. Beimel, A., Ishai, Y., Kumaresan, R., Kushilevitz, E.: On the cryptographic complexity of the worst functions. In: Lindell, Y. (ed.) TCC 2014. LNCS, vol. 8349, pp. 317–342. Springer, Heidelberg (2014)
4. Bendlin, R., Damgård, I., Orlandi, C., Zakarias, S.: Semi-homomorphic encryption and multiparty computation. In: Paterson, K.G. (ed.) EUROCRYPT 2011. LNCS, vol. 6632, pp. 169–188. Springer, Heidelberg (2011)
5. Bogdanov, D., Laur, S., Willemson, J.: Sharemind: a framework for fast privacy-preserving computations. In: Jajodia, S., Lopez, J. (eds.) ESORICS 2008. LNCS, vol. 5283, pp. 192–206. Springer, Heidelberg (2008)
6. Bogetoft, P., Christensen, D.L., Damgård, I., Geisler, M., Jakobsen, T., Krøigaard, M., Nielsen, J.D., Nielsen, J.B., Nielsen, K., Pagter, J., Schwartzbach, M., Toft, T.: Secure multiparty computation goes live. In: Dingledine, R., Golle, P. (eds.) FC 2009. LNCS, vol. 5628, pp. 325–343. Springer, Heidelberg (2009)

7. Choi, S.G., Elbaz, A., Malkin, T., Yung, M.: Secure multi-party computation minimizing online rounds. In: Matsui, M. (ed.) ASIACRYPT 2009. LNCS, vol. 5912, pp. 268–286. Springer, Heidelberg (2009)

8. Choi, S.G., Katz, J., Malozemoff, A., Zikas, V.: Efficient three-party computation from cut-and-choose. Crypto **2**, 513–530 (2014)

9. Damgård, I., Pastro, V., Smart, N., Zakarias, S.: Multiparty computation from somewhat homomorphic encryption. In: Safavi-Naini, R., Canetti, R. (eds.) CRYPTO 2012. LNCS, vol. 7417, pp. 643–662. Springer, Heidelberg (2012)

10. Damgård, I., Zakarias, S.: Constant-overhead secure computation of boolean circuits using preprocessing. In: Sahai, A. (ed.) TCC 2013. LNCS, vol. 7785, pp. 621–641. Springer, Heidelberg (2013)

11. Feige, U., Kilian, J., Naor, M.: A minimal model for secure computation (extended abstract). In: 26th ACM STOC, Annual ACM Symposium on Theory of Computing (STOC), pp. 554–563. ACM Press, May 1994

12. Fischer, M.J., Lynch, N.A.: A lower bound for the time to assure interactive consiistency. Info. Process. Lett. **14**(4), 183–186 (1982)

13. Garg, S., Gentry, C., Halevi, S., Raykova, M.: Two-round secure MPC from indistinguishability obfuscation. In: Lindell, Y. (ed.) TCC 2014. LNCS, vol. 8349, pp. 74–94. Springer, Heidelberg (2014)

14. Gennaro, R., Ishai, Y., Kushilevitz, E., Rabin, T.: The round complexity of verifiable secret sharing and secure multicast. In: 33rd ACM STOC, Annual ACM Symposium on Theory of Computing (STOC), pp. 580–589. ACM Press, July 2001

15. Gennaro, R., Ishai, Y., Kushilevitz, E., Rabin, T.: On 2-round secure multiparty computation. In: Yung, M. (ed.) CRYPTO 2002. LNCS, vol. 2442, pp. 178–193. Springer, Heidelberg (2002)

16. Goldreich, O.: Foundations of Cryptography: Basic Applications, vol. 2. Cambridge University Press, Cambridge (2004)

17. Goldwasser, S., Lindell, Y.: Secure multi-party computation without agreement. J. Cryptol. **18**(3), 247–287 (2005)

18. Ishai, Y., Kushilevitz, E., Meldgaard, S., Orlandi, C., Paskin-Cherniavsky, A.: On the power of correlated randomness in secure computation. In: Sahai, A. (ed.) TCC 2013. LNCS, vol. 7785, pp. 600–620. Springer, Heidelberg (2013)

19. Ishai, Y., Kushilevitz, E., Paskin, A.: Secure multiparty computation with minimal interaction. In: Rabin, T. (ed.) CRYPTO 2010. LNCS, vol. 6223, pp. 577–594. Springer, Heidelberg (2010)

20. Ito, M., Saito, A., Nishizeki, T.: Secret sharing schemes realizing general access structure. In: GLOBECOM, pp. 99–102 (1987)

21. Katz, J., Ostrovsky, R.: Round-optimal secure two-party computation. In: Franklin, M. (ed.) CRYPTO 2004. LNCS, vol. 3152, pp. 335–354. Springer, Heidelberg (2004)

22. Katz, J., Koo, C.-Y.: Round-efficient secure computation in point-to-point networks. In: Naor, M. (ed.) EUROCRYPT 2007. LNCS, vol. 4515, pp. 311–328. Springer, Heidelberg (2007)

23. Paskin-Cherniavsky, A.: Secure computation with minimal interaction. Ph.D. Thesis, Technion (2012)

24. Patra, A., Choudhary, A., Rabin, T., Rangan, C.P.: The round complexity of verifiable secret sharing revisited. In: Halevi, S. (ed.) CRYPTO 2009. LNCS, vol. 5677, pp. 487–504. Springer, Heidelberg (2009)

PoW-Based Distributed Cryptography with No Trusted Setup

Marcin Andrychowicz and Stefan Dziembowski$^{(\boxtimes)}$

University of Warsaw, Warsaw, Poland
stefan@dziembowski.net

Abstract. Motivated by the recent success of Bitcoin we study the question of constructing distributed cryptographic protocols in a fully peer-to-peer scenario under the assumption that the adversary has limited computing power and there is *no* trusted setup (like PKI, or an unpredictable beacon). We propose a formal model for this scenario and then we construct a broadcast protocol in it. This protocol is secure under the assumption that the honest parties have computing power that is some non-negligible fraction of computing power of the adversary (this fraction can be small, in particular it can be much less than 1/2), and a (rough) total bound on the computing power in the system is known.

Using our broadcast protocol we construct a protocol for simulating any trusted functionality. A simple application of the broadcast protocol is also a scheme for generating an unpredictable beacon (that can later serve, e.g., as a genesis block for a new cryptocurrency).

Under a stronger assumption that the majority of computing power is controlled by the honest parties we construct a protocol for simulating any trusted functionality with guaranteed termination (i.e. that cannot be interrupted by the adversary). This could in principle be used as a provably-secure substitute of the blockchain technology used in the cryptocurrencies.

Our main tool for verifying the computing power of the parties are the Proofs of Work (Dwork and Naor, CRYPTO 92). Our broadcast protocol is built on top of the classical protocol of Dolev and Strong (SIAM J. on Comp. 1983).

1 Introduction

Distributed cryptography is a term that refers to cryptographic protocols executed by a number of mutually distrusting parties in order to achieve a common goal. One of the first primitives constructed in this area were the *broadcast protocols* [14,24] using which a party P can send a message over a point-to-point network in such a way that all the other parties will reach *consensus* about the value that was sent (even if P is malicious). Another standard example

This work was supported by the Foundation for Polish Science WELCOME/2010-4/2 grant founded within the framework of the EU Innovative Economy (National Cohesion Strategy) Operational Programme.

R. Gennaro and M. Robshaw (Eds.): CRYPTO 2015, Part II, LNCS 9216, pp. 379–399, 2015.
DOI: 10.1007/978-3-662-48000-7_19

are the secure multiparty computations (MPCs) [7,11,20,30], where the goal of the parties is to simulate a trusted functionality. The MPCs turned out to be a very exciting theoretical topic. They have also found some applications in practice (in particular they are used to perform the secure on-line auctions [8]). Despite of this, the MPCs unfortunately still remain out of scope of interest for most of the security practitioners, who are generally more focused on more basic cryptographic tools such as encryption, authentication or the digital signature schemes.

One of very few examples of distributed cryptography techniques that attracted attention from general public are the *cryptographic currencies* (also dubbed the *cryptocurrencies*), a fascinating recent concept whose popularity exploded in the past 1-2 years. Historically the first, and the most prominent of them is the *Bitcoin*, introduced in 2008 by an anonymous developer using a pseudonym "Satoshi Nakamoto" [26]. Bitcoin works as a peer-to-peer network in which the participants jointly emulate the central server that controls the correctness of transactions, in particular: it ensures that there was no "double spending", i.e., a given coin was not spent twice by the same party. Although the idea of multiple users jointly "emulating a digital currency" sounds like a special case of the MPCs, the creators of Bitcoin did not directly use the tools developed in this area, and it is not clear even to which extend they were familiar with this literature (in particular, Nakamoto [26] did not cite any of MPC papers in his work). Nevertheless, at the first sight, there are some resemblances between these areas. In particular: the Bitcoin system works under the assumption that the majority of computing power in the system is under control of the honest users, while the classical results from the MPC literature state that in general constructing MPC protocols is possible when the majority of the users is honest.

At a closer look, however, it becomes clear that there are some important differences between both areas. In particular the main reason why the MPCs cannot be used directly to construct the cryptocurrencies is that the scenarios in which these protocols are used are fundamentally different. The MPCs are supposed to be executed by a fixed (and known in advance) set of parties, out of which some may be honestly following the protocol, and some other ones may be corrupt (i.e. controlled by the adversary). In the most standard case the number of misbehaving parties is bounded by some threshold parameter t. This can be generalized in several ways. Up to our knowledge, however, until now all these generalizations use a notion of a "party" as a separate and well-defined entity that is either corrupt or honest.

The model for the cryptocurrencies is very different, as they are supposed to work in a purely peer-to-peer environment, and hence the notion of a "party" becomes less clear. This is because they are constructed with a minimal trusted setup (as we explain below the only "trusted setup" in Bitcoin was the generation of an unpredictable "genesis block"), and in particular they do not rely on any Public Key Infrastructure (PKI), or any type of a trusted authority that would, e.g., "register" the users. Therefore the adversary can always launch a so-called *Sybil attack* [15] by creating a large number k of "virtual" parties that remain under his control. In this way, even if in reality he is just a single entity, from the point of view of the other participants he will control a large number of parties. In some sense the cryptocurrencies lift the "lack of trust" assumption to a whole new level, by considering the situation when it is not even clear who is a "party". The Bitcoin system overcomes this problem in the following way: the honest majority is defined in terms of the "majority of computing power". This is achieved by having all the honest participants to constantly prove that they devote certain computing power to the system, via the so-called "Proofs of Work" (PoWs) [16, 17].

The high level goal for this work is to bridge the gap between these two areas. In particular, we propose a formal model for the peer-to-peer communication and the Proofs of Work concept used in Bitcoin. We also show how some standard primitives from the distributed computation, like broadcast and MPCs, can be implemented in this model. Our protocols do not require any trusted setup assumptions, unlike Bitcoin that assumes a trusted generation of an unpredictable "genesis block" (see below for more details). Besides of being of general interest, our work is motivated twofold.

Firstly, recently discovered weaknesses of Bitcoin [5, 19] come, in our opinion, partially from the lack of a formal framework for this system. Our work can be viewed as a step towards better understanding of this model. We also believe that the "PoW-based distributed cryptography" can find several other applications in the peer-to-peer networks (we describe some of them). In particular, as the Bitcoin example shows, the "lack of trusted setup" can be very attractive to users[1]. In fact, there are already some ongoing efforts to use the Bitcoin paradigm for purposes other than the cryptocurrencies (see full version of this paper [1] for more on this). We would like to stress however, that this is not the main purpose of our work, and that we do not provide a full description of a new currency. Our goal is also not the full analysis of the security of Bitcoin (which would be a very ambitious project that would also need to take into account the economical incentives of the participants).

Secondly, what may be considered unsatisfactory in Bitcoin is the fact that its security relies on the fact that the so-called *genesis block* B_0, announced by Satoshi Nakamoto on January 3, 2009, was generated using heuristic methods. More concretely, in order to prove that he did not know B_0 earlier, he included the text *The Times 03/Jan/2009 Chancellor on brink of second bailout for banks* in B_0 (taken from the front page of the London Times on that day). The *unpredictability* of B_0 is important for Bitcoin to work properly, as otherwise a "malicious Satoshi Nakamoto" $\mathcal{A}$ that knew B_0 beforehand could start the mining process much earlier, and publish an alternative block chain at some later point. Since he would have more time to work on his chain, it would be longer than the "official" chain, even if $\mathcal{A}$ controls only a small fraction of the total computing power. Admittedly, its now practically certain that no attack like this was performed, and that B_0 was generated honestly, as it is highly

---

[1] Actually, probably one of the reasons why the MPCs are not widely used in practice is that the typical users do not see a fundamental difference between assuming a trusted setup and delegating the whole computation to a trusted third party.

unlikely that any $\mathcal{A}$ invested more computing power in Bitcoin mining than all the other miners combined, even if $\mathcal{A}$ started the mining process long before January 3, 2009.

However, if we want to use the Bitcoin paradigm for some other purpose (including starting a new currency), it may be desirable to have an automatic and non-heuristic method of generating unpredictable strings of bits. The problem of generating such *random beacons* [27] has been studied in the literature for a long time. Informally: a random beacon scheme is a method (possibly involving a trusted party) of generating uniformly random (or indistinguishable from random) strings that are unknown before the moment of their generation. The beacons have found a number of applications in cryptography and information security, including the secure contract signing protocols [18,27], voting schemes [25], or zero-knowledge protocols [3,21]. Note that a random beacon is a stronger concept than the *common reference string* frequently used in cryptography, as it has to be unpredictable before it was generated (for every instance of the protocol using it). Notice also that for Bitcoin we actually need something weaker than uniformity of the B_0, namely it is enough that B_0 is hard to predict for the adversary.

Constructing random beacons is generally hard. Known practical solutions are usually based on a trusted third party (like the servers www.random.org and beacon.nist.gov). Since we do not want to base the security of our protocols on trusted third parties thus using such services is not an option for our applications. Another method is to use public data available on the Internet, e.g. the financial data [12] (the Bitcoin genesis block generation can also be viewed as an example of this method). Using publicly-available data makes more sense, but also this reduces the overall security of the constructed system. For example, in any automated solution the financial data would need to come from a trusted third party that would need to certify that the data was correct. The same problem applies to most of other data of this type (like using a sentence from a newspaper article). One could also consider using the Bitcoin blocks as such beacons (in fact recently some on-line lotteries started using them for this purpose). We discuss the problems with this approach in the full version of this paper [1].

Our Contribution. Motivated by the cryptocurrencies we initiate a formal study of the distributed peer-to-peer cryptography based on the Proofs of Work. From the theory perspective the first most natural questions in this field is what is the right model for communication and computation in this scenario? And then, is it possible to construct in this model some basic primitives from the distributed cryptography area, like: (a) broadcast, (b) unpredictable beacon generation, or (c) general secure multiparty computations? We propose such a model (in Sect. 2). Our model does not assume any trusted setup (in particular: we do not assume any trusted beacon generation). Then, in Sect. 4 we answer the questions (a)-(c) positively. To describe our results in more detail let n denote the number of honest parties, let π be the computing power of each honest party (for simplicity we assume that all the honest parties have the same computing power), let π_{max} be the maximal computing power of all the participants of the protocol (the honest parties and the adversary), and let $\pi_{\mathcal{A}} \leq \pi_{max} - n\pi$ be the actual computing power of the adversary. We allow the adversary to adaptively corrupt at most t parties, in which case he takes the full control over them (however, we do not allow him to use the computing power of the corrupt parties, or in other words: once he corrupts a party he is also responsible for computing the Proofs of Work for her). Of course in general it is better to have protocols depending on $\pi_{\mathcal{A}}$, not on π_{max}. On the other hand, sometimes the dependence from π_{max} is unavoidable, as the participants need to have some rough estimate on the power of the adversary (e.g. clearly it is hard to construct any protocol when π is negligible compared to π_{max}). Note that also Bitcoin started with some arbitrary assumption on the computing power of the participant (this was reflected by setting the initial "mining difficulty" to 2^{32} hash computations). Our contribution is as follows. First, we construct a broadcast protocol secure against any π_{max}, working in time linear in $\lceil \pi_{max}/\pi \rceil$. Then, using this broadcast protocol, we argue how to construct a protocol for executing any functionality in our model. In case the adversary controls the minority of the computing power (i.e. $n \geq \lceil \pi_{\mathcal{A}}/\pi \rceil + t)^2$ that were user ber our protocol cannot be aborted prematurely by her. This could in principle be used as a provably-secure substitute of the blockchain technology used in the cryptocurrencies. Using the broadcast protocol as a subroutine we later (in Sect. 5) construct a scheme for an unpredictable beacon generation.

One thing that needs to be stressed is that our protocols do not require an unpredictable trusted beacon to be executed (and actually, as described above, constructing a protocol that emulates such a beacon is one of our contributions). This poses a big technical challenge, since we have to prevent the adversary from launching a "pre-computation" attack, i.e., computing solutions to some puzzles before the execution of the protocol started.

The only thing that we assume is that the participating parties know a session identifier (sid), which can be known publicly long time before the protocol starts. Observe that some sort of mechanism of this type is always needed, as the parties need to know in advance, e.g., the time when the execution starts.

One technical problem that we need to address is that, since we work in a purely peer-to-peer model, an adversary can always launch a Denial of Service Attack, by "flooding" the honest parties with his messages, hence forcing them to work forever. Thus, in order for the protocols to terminate in a finite time we also need some mild upper bound θ on the number of messages that the adversary can send (much greater than what the honest parties will send).

---

[2] The reader might be confused we in this inequality t appears on the righ hand side, as it may look like contradicting the assumption that the adversary does not take the control of the computing power of the corrupt parties. The reason for having this term is the adaptivity: the adversary can corrupt a party at the very end of the protocol, hence, in some sense taking advantage of her computing resources before she was corrupted.

We write more on this in Sect. 2. Although our motivation is mostly theoretic, we believe that our ideas can lead to practical implementations (probably after some optimizations and simplifications). We discuss some possible applications of our protocols in Sect. 5.

Independent Work. Recently an interesting paper by Katz, Miller and Shi [23] with a motivation similar to ours was published on the Eprint archive. While their high-level goal is similar to ours, there are some important technical differences. First of all, their solution essentially assumes existence of a trusted unpredictable beacon (technically: they assume that the parties have access to a random oracle that was not available to the adversary before the execution started). This simplifies the design of the protocols significantly, as it removes the need for every party to ensure that "her" challenge was used to compute the Proof-of-Work (that in our work we need to address to deal with the precomputation attacks described above). Secondly, they assume that the proof verification takes zero time (we note that with such an assumption our protocols would be significantly simpler, and in particular we would not need an additional paramter θ that measures the number of messages sent by the adversary). Thirdly, unlike us, they assume that the number of parties executing the protocol is known from the beginning. On the other hand, their work covers also the "sequential puzzles" (see [23]), while in this work we focus on parallelizable puzzles.

2   Our Model

In this section we present our model for reasoning about computing power and the peer-to-peer protocols. We first do it informally, and then formalize it using the *universal composability framework* of Canetti [9].

Modeling Hashrate. Since in general proving lower bounds on the computational hardness is very difficult, we make some simplifying assumptions about our model. In particular, following a long line of previous works both in theory and in the systems community (see e.g. [4,17,26]), we establish the lower bounds on computational difficulty by counting the number of times a given algorithm calls some random oracle H [6]. In our protocols the size of the input of H will be linear in the security parameter κ (usually it will be 2κ at most). Hence it is realistic to assume that each invocation of such a function takes some fixed unit of time.

Our protocols are executed in real time by a number of devices and attacked by an adversary $\mathcal{A}$. The exact way in which time is measured is not important, but it is useful to fix a unit of time Δ (think of it as 1 minute, say). Each device D that participates in our protocols will be able to perform some fixed number π_D of queries to H in time Δ. The parameter π_D is called the *hashrate of D* *(per time Δ)*. The hashrate of the adversary is denoted by $\pi_{\mathcal{A}}$. The other steps of the algorithms do not count as far as the hashrate is considered (they will count, however, when we measure the efficiency of our protocols, see paragraph *Computational complexity* below). Moreover we assume that the parties have access to a "cheap" random oracle, calls to this oracle do not count as far as the hashrate is considered. This assumption is made to keep the model as simple as possible. It should be straightforward that in our protocols we do not abuse this assumption, and in on any reasonable architecture the time needed for computing H's would be the dominating factor during the Proofs of Work. In particular: any other random oracles will be invoked a much smaller number of times than H. Note that, even if these numbers were comparable, one could still make H evaluate much longer than any other hash function F, e.g., by defining H to be equal to multiple iterations of F.

In this paper we will assume that every party (except of the adversary) has the same hashrate per time Δ (denoted π). This is done only to make the exposition simpler. Our protocols easily generalize to the case when each party has a device with hashrate π_i and the π_i's are distinct. Note that if a party has a hashrate $t\pi$ (for natural t) then we can as well think about her as of t parties of hashrate π each. Making it formal would require changing the definition of the "honest majority" in the MPCs to include also "weights" of the parties.

The Communication Model. Unlike in the traditional MPC settings, in our case the number of parties executing the protocol is not known in advance to the parties executing it. Because of this it makes no sense to specify a protocol by a finite sequence $(M_1, \ldots, M_n)$ of Turing machines. Instead, we will simply assume that there is *one* Turing machine M whose code will be executed by each party participating in the protocol (think of it as many independent executions of the same program). This, of course, does no mean that these parties have identical behavior, since their actions depend also on their inputs, the party identifier (pid), and the random coins.

Since we do not assume any trusted set-up (like a PKI or shared private keys) modeling the communication between the parties is a bit tricky. We assume that the parties have access to a public channel which allows every party and the adversary to post a message on it. One can think of this channel as being implemented using some standard (cryptographically insecure) "network broadcast protocol" like the one in Bitcoin [29]. The contents of the communication channel is publicly available. The message m sent in time t by some P_i is guaranteed to arrive to P_j within time t' such that $t' - t \leq \Delta$. Note that some assumption of this type needs to be made, as if the messages can be delayed arbitrarily then there is little hope to measure the hashrate reliably. Also observe that we have to assume that the messages always reach their destinations, as otherwise an honest party could be "cut of" the network. Similar assumptions are made (implicitly) in Bitcoin. Obviously without assumptions like this, Bitcoin would be easy to attack (e.g. if the miners cannot send messages to each other reliably then it is easy to make a "fork" in the blockchain).

To keep the model simple we will assume that the parties have perfectly synchronized clocks. This assumption could be easily relaxed by assuming that clocks can differ by a small amount of time δ, and our model from Sect. 2.1 could be easily extended to cover also this case, using the techniques, e.g., from [22]. We decided not to do it to in order to keep the exposition as simple as possible.

We give to the adversary full access to the communication between the parties: he learns (without any delay) every message that is sent through the communication channel, and he can insert messages into it. The adversary may decide that the messages inserted into the channel by him arrive only to a certain subset of the parties (he also has a full control over the timing when they arrive). The only restriction is that he cannot erase or modify the messages that were sent by the other parties (but he can delay them for time at most Δ).

Resistance to the Denial of Service Attacks. As already mentioned in the introduction, in general a complete prevention of the denial of service attacks against fully distributed peer-to-peer protocols seems very hard. Since we do not assume any trusted set-up phase, hence from the theoretical point of view the adversary is indistinguishable from the honest users, and hence he can always initiate a connection with an honest user forcing it to perform some work. Even if this work can be done very efficiently, it still costs some effort (e.g. it requires the user to verify a PoW solution), and hence it allows a powerful (yet poly-time bounded) adversary to force each party to work for a very long amount of time, and in particular to exceed some given deadline for communicating with the other parties. Since any PoW-based protocol inherently needs to have such deadlines, thus we need to somehow restrict the power of adversary. We do it in the following way.

First of all, we assume that if a message m sent to P_i is longer than the protocols specifies then P_i can discard it without processing it.[3] Secondly, we assume that there is a total bound θ on the number of messages that all the participants can send during each interval Δ. Since this includes also the messages sent by the honest parties, thus the bound on the number of messages that the adversary $\mathcal{A}$ sends will be slightly more restrictive, but from practical point of view (since the honest parties send very few messages) it is approximately equal to θ. This bound can be very generous, and, moreover it will be much larger than the number of messages sent by the honest users[4]. In practice such a bound could be enforced using some ad-hoc methods. For example each party could limit the number of messages it can receive from a given IP address. Although from the theoretical perspective no heuristic method is fully satisfactory, in practice they seem to work. For example Bitcoin seems to resist pretty well the DoS attacks thanks to over 30 ad-hoc methods of mitigating them (see [28]). Hence, we believe that some bound on θ is reasonable to assume (and, as argued above, seems necessary). We will use this bound in a weak way, in

---

[3] Discarding incorrect messages is actually a standard assumption in the distributed cryptography. Here we want to state it explicitly to make it clear that the processing time of too long messages does not count into the computing steps of the users.

[4] This is important, since otherwise we could trivialize the problem by asking each user to prove that he is honest by sending a large number of messages.

particular the number of messages sent by the honest parties will not depend on it, and the communication complexity will (for any practical choice of parameters) be linear in θ for every party (in other words: by sending θ messages the adversary can force an honest party to send one long message of length $O(\theta)$). The real time of the execution of the protocol can depend on θ. Formally it is a linear dependence (again: this seems to be unavoidable, since every message that is sent to an honest party P_i forces P_i to do some non-trivial work). Fortunately, the constant of this linear function will be really small. For example, in the RankedKeys (Fig. 3, Page 16) the time each round takes (in the "key ranking phase") will be $\Delta + \theta \cdot \mathsf{time}_V / \pi$, where time_V is small. Observe that, e.q, $\theta/\pi = 1$ if the adversary can send the messages at the same speed as the honest party can compute the $\mathcal{H}^\kappa$ queries, hence it is reasonable to assume that $\theta/\pi < 1$.

Communication, Message and Computational Complexity. In the full version of this paper [1] we define and analyze the communication complexity of our protocols. We also analyze their computational complexity. We also extend our model to cover the case of non-authenticated bilateral channels.

2.1 Formal Definition

Formally, a *multiparty protocol (in the $(\pi, \pi_\mathcal{A}, \theta)$ -model)* is an ITM (Interactive Turing Machine) M. It is executed together with an ITM $\mathcal{A}$ representing the *adversary*, and and ITM $\mathcal{E}$ representing the *environment*. The *real execution* of the system essentially follows that scheme from [9]. Every ITM gets as input a security parameter 1^κ. Initially, the environment takes some input $z \in \{0,1\}^*$ and then it activates an adversary $\mathcal{A}$ and a set $\mathcal{P}$ of parties. The adversary may (actively and adaptively) corrupt some parties. The environment is notified about these corruptions.

The set $\mathcal{P}$ (or even its size) will *not* be given as input to the honest parties. In other words: the protocol should work in the same way for any $\mathcal{P}$. On the other hand: each $P \in \mathcal{P}$ will get as input her own hashrate π and the upper bound π_{max} on the total combined hashrate of all the parties and the adversary (this will be the paramters of the protocol). The running time of $P \in \mathcal{P}$ can depend on these parameters. Note that $|\mathcal{P}| \cdot \pi + \pi_\mathcal{A} \leq \pi_{\mathsf{max}}$, but this inequality may be sharp, and even $|\mathcal{P}| \cdot \pi + \pi_\mathcal{A} \ll \pi_{\mathsf{max}}$ is possible, as, e.g., the adversary can use much less hashrate than the maximal amount that he is allowed to[5].

Each party $P \in \mathcal{P}$ runs the code of M. It gets as input its party identifier (pid) and some random input. We assume that all the pid's are distinct, which can be easily obtained by choosing them at random from a large domain ($\{0,1\}^\kappa$, say). Moreover the environment sends to each P some input $x_P \in \{0,1\}^*$, and at the end of its execution it outputs to $\mathcal{E}$ some value $y_P \in \{0,1\}^*$. We assume that at a given moment only one machine is active. For a detailed description on

---

[5] In particular it is important to stress that the assumption that the majority of the computing power is honest means that $n \cdot \pi > \pi_\mathcal{A}$, and *not*, as one might think, $n \cdot \pi > \pi_{\mathsf{max}}/2$ (assuming the number t of corrupt parties is zero).

Functionality $\mathcal{F}_{\mathsf{syn}}^{\pi,\pi_{\mathcal{A}},\theta}$

$\mathcal{F}_{\mathsf{syn}}^{\pi,\pi_{\mathcal{A}},\theta}$ receives a session ID sid $\in \{0,1\}^*$. Moreover we assume that it obtains a list $\mathcal{P}$ of parties that were activated with sid, i.e., those parties among which synchronization is to be provided and that will issue the random oracle queries.

1. At the first activation, the functionality chooses at random a random oracle H. It then waits for queries from the adversary $\mathcal{A}$ of a form (Hash, w) (where w is from the domain of H). Each such a query is answered with $H(w)$. This phase ends when $\mathcal{A}$ sends a query Next or when it terminates its operation.
2. Initialize a round counter $r := 1$, for every party $P \in \mathcal{P}$ initialize variables $h_P := 0$ and $L_P^1 = \emptyset$. Initialize $h_{\mathcal{A}} := 0$. Send a public delayed output (Init, sid) to all parties in $\mathcal{P}$.
3. Upon receiving input (Send, sid, m) from a party $P \in \mathcal{P}$, for every $P' \in \mathcal{P}$ set $L_{P'}^r := L_{P'}^r \cup \{m\}$ and output (sid, P, m, r) to the adversary.
4. Upon receiving input (Send, sid, P', m) from $\mathcal{A}$ (where $P' \in \mathcal{P}$) set $L_{P'}^r := L_{P'}^r \cup \{m\}$.
5. Upon receiving (Hash, w) from $P' \in \mathcal{P} \cup \{\mathcal{A}\}$ (note that P' can either be a party or the adversary) do
   (a) if $P' \in \mathcal{P}$, where P' is not corrupt and $h_{P'} < \pi$ then reply with $H(w)$ and increment the counter: $h_{P'} := h_{P'} + 1$,
   (b) if $P' = \mathcal{A}$ and $h_{\mathcal{A}} < \pi_{\mathcal{A}}$ then reply with H and increment the counter: $h_{\mathcal{A}} := h_{\mathcal{A}} + 1$,
   (c) otherwise do nothing (since P' has already exceeded the number of allowed queries to $\mathcal{H}^{\kappa}$ in this round).
6. Upon receiving input (Receive, sid, r') from a party $P \in \mathcal{P}$, do:
   (a) If $r' = r$ (i.e., r' is the current round), and you have received the Send message from every non-corrupt party in this round then:
      i. Increment the round number: $r := r + 1$.
      ii. For every $P' \in \mathcal{P} \cup \{\mathcal{A}\}$ reset the variable $h_{P'} := 0$.
      iii. If the size of L_P^{r-1} is at most θ then output (Received, sid, L_P^{r-1}) to P, otherwise output $\perp$ to P.
   (b) If $r' < r$ and the size of $L_P^{r'}$ is at most θ then output (Received, sid, $L_P^{r'}$) to P, otherwise output $\perp$ to P.
   (c) Else (i.e., $r' > r$ or not all parties in $\mathcal{P}$ have sent their messages for round r), output Round Incomplete to P.

Fig. 1. Functionality $\mathcal{F}_{\mathsf{syn}}^{\pi,\pi_{\mathcal{A}},\theta}$.

how the control is passed between the machines see [9], but let us only say that it is done via sending messages (if one party sends a message to the other one then it "activates it"). The environment $\mathcal{E}$ can communicate with $\mathcal{A}$ and with the parties in $\mathcal{P}$. The adversary controls the network. However, we require that every message sent between two parties is always eventually delivered. Moreover, since the adversary is poly-time bounded, thus he always eventually terminates. If he does so without sending any message then the control passed to the environment (that chooses which party will be activated next).

We assume that all the parties have access to an ideal functionality $\mathcal{F}_{\mathsf{syn}}^{\pi,\pi_{\mathcal{A}},\theta}$ (depicted on Fig. 1) and possibly to some random oracles. The ideal functionality $\mathcal{F}_{\mathsf{syn}}^{\pi,\pi_{\mathcal{A}},\theta}$ is used to formally model the setting described informally above. Since we assumed that every message is delivered in time Δ we can think of the whole execution as divided into rounds (implicitly: of length Δ). This is essentially the "synchronous communication" model from [9] (see Sect. 6.5 of the Eprint version of that paper). As it is the case there, the notion of a "round" is controlled by a counter r, which is initially set to 1 and is increased each time all the honest parties send all their inputs for a given round to $\mathcal{F}_{\text{syn}}^{\pi,\pi_{\mathcal{A}},\theta}$. The messages that are sent to P in a given round r are "collected" in a buffer denote L_P^r and delivered to P at the end of the rounds (on P's request). The fact that every message sent by an honest party has to arrive to another honest party within a given round is reflected as follows: the round progresses only if every honest party sent her message for a given round to the functionality (and this happens only if all of them received messages from the previous round). Recall also that sending "delayed output x to a P" means that the x is first received by $\mathcal{A}$ who can decide when x is delivered to P.

Compared to [9] there are some important differences though. First of all, since in our model the set $\mathcal{P}$ of the parties participating in the execution is known to the honest participants, thus we cannot assume that $\mathcal{P}$ is a part of the session identifier. We therefore give it directly to the functionality (note that this set is anyway known to $\mathcal{E}$, which can share it with $\mathcal{A}$).

Secondly, we do not assume that the parties can send messages directly to each other. The only communication that is allowed is through the "public channel". This is reflected by the fact that the "Send" messages produced by the parties do not specify the sender and the receiver (cf. Step 3), and are delivered to everybody . In contrast, the adversary can send messages to concrete parties (cf. Step 4).

Thirdly, and probably most importantly, the functionality $\mathcal{F}_{\text{syn}}^{\pi,\pi_{\mathcal{A}},\theta}$ also keeps track on how much computational resources (in terms of access to the oracle H) were used by each participant in each round. To take into account the fact that the adversary may get access the oracle long before the honest parties started the execution we first allow him (in Step 1) to query this oracle adaptively (the number of these queries is bounded only by the running time of the adversary, and hence it has to be polynomial in the security parameter). Then, in each round every party $P \in \mathcal{P}$ can query H. The number of such queries is bounded by π.

We use a counter h_P (reset to 0 at the beginning of each new round) to track the number of times the user P queried H. The number of oracle queries that $\mathcal{A}$ can ask is bounded by $\pi_{\mathcal{A}}$ and controlled by the counter $h_{\mathcal{A}}$. Note that, once a party $P \in \mathcal{P}$ gets corrupted by $\mathcal{A}$ it looses access the oracle H. This reflects the fact that from this point the computing power of P does not count anymore as being controlled by the honest parties, and hence every call to H made by such a P has to be "performed" by the adversary (and consequently increase $h_{\mathcal{A}}$). The output of the environment on input z interacting with M, $\mathcal{A}$ and the ideal functionality $\mathcal{F}_{\text{syn}}^{\pi,\pi_{\mathcal{A}},\theta}$ will be denoted $\text{exec}_{M,\mathcal{A},\mathcal{E}}^{\pi,\pi_{\mathcal{A}},\theta}(z)$.

In order to define security of such execution we define an ideal functionality $\mathcal{F}$ that is also an ITM that can interact with the adversary. Its interaction with the parties is pretty simple: each party simply interacts with $\mathcal{F}$ directly (with no disturbance from the adversary). The adversary may corrupt some of the parties, in which case he learns their inputs and outputs. The functionality is notified about each corruption. At the end the environment outputs a value $\mathrm{exec}_{\mathcal{F},\mathcal{A},\mathcal{E}}(z)$.

Definition 1. *We say that a protocol M securely implements a functionality $\mathcal{F}$ in the $(\pi, \pi_{\mathcal{A}}, \theta)$-model if for every polynomial-time adversary $\mathcal{A}$ there exists a polynomial-time simulator S such that for every environment $\mathcal{Z}$ the distribution ensemble $\mathrm{exec}_{M,\mathcal{A},\mathcal{E}}^{\pi,\pi_{\mathcal{A}},\theta}$ and the distribution ensemble $\mathrm{exec}_{\mathcal{F},\mathcal{A},\mathcal{E}}$ are computationally indistinguishable (see [9] for the definition of the distribution ensembles and the computational indistinguishability).*

3   The Security Definition of Broadcast

In this section we present the security definitions of our main construction, i.e., the broadcast protocol. We first describe its informal properties and then specify it as an ideal functionality. Let $\mathcal{P}$ be the set of parties executing Π, each of them having a device with hashrate $\pi > 0$ per time Δ. Each $P \in \mathcal{P}$ takes as input $x_P \in \{0,1\}^\kappa$, and it produces as output a multiset $\mathcal{Y}_P \subset \{0,1\}^\kappa$. The protocol is called a π_{max}-*secure broadcast protocol* if it terminates is some finite time and for any poly-time adversary $\mathcal{A}$ whose device has hashrate $\pi_{\mathcal{A}} < \pi_{\mathsf{max}}$ and who attacks this protocol the following conditions hold except with probability

---

$\mathcal{F}_{\mathsf{syn}}^T$ receives a session ID sid $\in \{0,1\}^*$. Moreover it obtains a list $\mathcal{P}$ of parties that were activated with sid.

1. At the first activation initialize the variables $\mathcal{X} := \emptyset$ and $\mathcal{X}_S := \emptyset$, where $\mathcal{X}$ and $\mathcal{X}_S$ are multisets. Send a public delayed output (Init, sid) to all parties in $\mathcal{P}$.
2. Upon receiving input (Broadcast, sid, x) (where $x \in \{0,1\}^*$) from $P \in \mathcal{P}$ (with PID pid) do the following:
   (a) add x to $\mathcal{X}$, i.e., let $\mathcal{X} := \mathcal{X} \cup \{x\}$, moreover send (Broadcast, sid, pid, x) to S,
   (b) otherwise do nothing.
3. Upon receiving (Broadcast, sid, x) from S:
   (a) if $|\mathcal{X}_S| < T$ then let $\mathcal{X}_S := \mathcal{X}_S \cup \{x\}$,
   (b) otherwise do nothing.
4. Upon receiving (Remove, sid, pid) from S: if P with PID pid is not corrupt or such a message has already been received before then ignore it.
   Otherwise look for a string x that was added by a party with PID pid to $\mathcal{X}$ in Step 2. If no such string exists do nothing. Otherwise: remove x from the multiset $\mathcal{X}$.
5. Upon receiving (Receive, sid) from some $P \in \mathcal{P}$:
   (a) If there is some non-corrupt party $P \in \mathcal{P}$ from which no message (Broadcast, sid, x) has been received yet then ignore this message.
   (b) Otherwise:
        i. If it is the first message (Receive, sid) received then set $\mathcal{Y} := \mathcal{X} \cup \mathcal{X}_S$ and send $\mathcal{Y}$ to the adversary.
        ii. Output (Received, sid, $\mathcal{Y}$) to P.

Fig. 2. Functionality $\mathcal{F}_{\mathsf{bc}}^T$, where T is the bound on the number of "fake identities" that the adversary can create. Our security definition requires that $T = \lceil \pi_{\mathcal{A}} / \pi \rceil$.

negligible in κ (let $\mathcal{H}$ denote the set of parties in $\mathcal{P}$ that were not corrupted by the adversary): (1) *Consistency:* All the sets $\mathcal{Y}_P$ are equal for all non-corrupt P's, i.e.: there exists a set $\mathcal{Y}$ such that for every $P \in \mathcal{H}$ we have $\mathcal{Y}_P = \mathcal{Y}$, (2) *Validity:* For every $P \in \mathcal{H}$ we have $x_i \in \mathcal{Y}$, and (3) *Bounded creation of inputs:* The number of elements in $\mathcal{Y}$ that do not come from the honest parties (i.e.:$|\mathcal{Y} \setminus \{x_P\}_{P \in \mathcal{P}}|$) is at most $\lceil \pi_{\mathcal{A}}/\pi \rceil$. This is formally defined by specifying an ideal functionality $\mathcal{F}_{\mathrm{bc}}^{\lceil \pi_{\mathcal{A}}/\pi \rceil}$ see Fig. 2. The formal definition is given below.

Definition 2. *An ITM M is a $(\pi_{\max}, \theta)$-secure broadcast protocol if for any π and $\pi_{\mathcal{A}}$ it securely implements the functionality $\mathcal{F}_{\mathrm{bc}}^{\lceil \pi_{\mathcal{A}}/\pi \rceil}$ in the $(\pi, \pi_{\mathcal{A}}, \theta)$-model (see Definition 1 from Sect. 2.1), as long as the number $|\mathcal{P}|$ of parties running the protocol (i.e. invoked by the environment) is such that $|\mathcal{P}| \cdot \pi + \pi_{\mathcal{A}} \leq \pi_{\max}$.*

Note that we do not require any lower bound on π other than 0. In practice, however, running this protocol will make sense only for π being a noticeable fraction of $\pi_{\max}$, since the running time of our protocol is linear in $\pi_{\max}/\pi$. This protocol is implemented in the next section.

4 The Construction of the Broadcast Protocol

We are now ready to present the constructions of the protocols specified in Sect. 3. In our protocols the computational effort will be verified using so-called Proofs of Work. A *Proof-of-Work (PoW)* scheme [16], for a fixed security parameter κ is a pair of randomized algorithms: a *prover* P and a verifier V, having access to a random oracle H (in our constructions the typical input to H will be of size 2κ). The algorithm P takes as input a *challenge* $c \in \{0,1\}^{\kappa}$ and produces as output a *solution* $s \in \{0,1\}^*$. The algorithm V takes as input (c, s) and outputs true or false. We require that for every $c \in \{0,1\}^*$ it is that case that $V(c, P(c)) = \mathsf{true}$.

We say that a PoW (P, V) has *prover complexity t* if on every input $c \in \{0,1\}^*$ the prover P makes at most t queries to the oracle H. We say that (P, V) has *verifier complexity t'* if for every $c \in \{0,1\}^{\kappa}$ and $s \in \{0,1\}^*$ the verifier V makes at most t' queries to the oracle H. Defining security is a little bit tricky, since we need to consider also the malicious provers that can spend considerable amount of computational effort *before* they get the challenge c. We will therefore have two parameters: $\hat{t}_0, \hat{t}_1 \in \mathbf{N}$, where $\hat{t}_0$ will be the bound on the *total time* that a malicious prover has, and $\hat{t}_1 \leq \hat{t}_0$ will be the bound on the time that a malicious prover got after he learned c. Consider the following game between a malicious prover $\hat{P}$ and a verifier V: (1) $\hat{P}$ adaptively queries the oracles H on the inputs of his choice, (2) $\hat{P}$ receives $c \leftarrow \{0,1\}^{\kappa}$, (3) $\hat{P}$ again adaptively queries the oracles H on the inputs of his choice, (4) $\hat{P}$ sends a value $s \in \{0,1\}^*$ to V. We say that $\hat{P}$ *won* if $V(c, s) = \mathsf{true}$. We say that (P, V) is $(\hat{t}_0, \hat{t}_1)$ *-secure with ϵ -error (in the H-model)* if for a uniformly random $c \leftarrow \{0,1\}^*$ and every malicious prover $\hat{P}$ that makes in total at most $\hat{t}_0$ queries to H in the game above, and at most $\hat{t}_1$ queries after receiving c we have that $\mathbb{P}\left(\hat{P}(c) \text{ wins the game}\right) \leq \epsilon$. It will also be useful to use the asymptotic variant of this notion (where κ is the security parameter). Consider a family $\{(\mathsf{P}^\kappa, \mathsf{V}^\kappa)\}_{\kappa=1}^\infty$. We will say that it is $\hat{t}_1$-*secure* if for every polynomial $\hat{t}_0$ there exists a negligible ϵ such that $(\mathsf{P}^\kappa, \mathsf{V}^\kappa)$ is $(\hat{t}_0(\kappa), \hat{t}_1)$-secure with error $\epsilon(\kappa)$. Our protocols will be based on the PoW based on the Merkle trees combined with the Fiat-Shamir transform. The following lemma is proved in the full version of this paper [1].

Lemma 1. *For every function $t : \mathbf{N} \to \mathbf{N}$ s.t. $t(\kappa) \geq \kappa$ there exists a family of PoWs $(\mathsf{PTree}_{t(\kappa)}^\kappa, \mathsf{VTree}_{t(\kappa)}^\kappa)$ has prover complexity t and verifier complexity $\lceil \kappa \log^2 t \rceil$. Moreover the family $\{(\mathsf{PTree}_{t(\kappa)}^\kappa, \mathsf{VTree}_{t(\kappa)}^\kappa)\}_{\kappa=1}^\infty$ is ξt-secure for every constant $\xi \in [0, 1)$.*

One of the main challenges will be to prevent the adversary from precomputing the solutions to PoW, as given enough time every puzzle can be solved even by a device with a very small hashrate. Hence, each honest party P_i can accept a PoW proof only if it is computed on some string that contains a freshly generated challenge c. Since we work in a completely distributed scenario, and in particular we do not want to assume existence of a trusted beacon, thus the only way a P_i can be sure that a challenge c was fresh is that she generated it herself at some recent moment in the past (and, say, sent it to all the other parties).

This problem was already considered in [2], where the following solution was proposed. At the beginning of the protocol each party P_i creates a fresh (public key, secret key) pair $(\mathsf{pk}_i, \mathsf{sk}_i)$ (we will call the public keys *identities*) and sends to all other parties a random challenge c_i. Then, each party computes a Proof of Work on her public key and all received challenges. Finally, each party sends her public key with a Proof of Work to all other parties. Moreover, whenever a party receives a message with a given key for the first time, than it forwards it to all other parties. An honest party P_i accepts only these public keys which: (1) she received before some agreed deadline, and (2) are accompanied with a Proof of Work containing her challenge c_i. It is clear that each honest party accepts a public key of each other honest party and that after this process an adversary can not control a higher fraction of all identities that his fraction of the computational power. Hence, it may seem that the parties can later execute protocols assuming channels that are authenticated with the secret keys corresponding to these identities.

Unfortunately there is a problem with this solution. Namely it is easy to see that the adversary can cause a situation where some of his identities will be accepted by some honest parties and not accepted by some other honest parties[6]. We present a solution to this problem in the next sections.

4.1 Ranked Key Sets

The main idea behind our protocol is that parties assign *ranks* to the keys they have received. If a key was received before the deadline and the corresponding

---

[6] This discrepancy can come from two reasons: (1) some messages could be received by some honest parties before deadline and by some other after it, and (2) a Proof of Work can containing challenges of some of the honest parties, but not all.

proof contains the appropriate challenge, then the key is assigned a rank 0. In particular, keys belonging to honest parties are always assigned a rank 0. The rank bigger than 0 means that the key was received with some discrepancy from the protocol (e.g. it was received slightly after the deadline) and the bigger the rank is, the bigger this discrepancy was. More precisely each party P_i computes a function rank_i from the set of keys she knows $\mathcal{K}_i$ into the set $\{0, \ldots, \ell\}$ for some parameter ℓ. Note that this primitive bares some similarities with the "proxcast" protocol of Considine et al. [13]. Since we will use this protocol only as a subroutine for our broadcast protocol, to save space, we present its definition without using the "ideal functionality" paradigm.

Let $\Sigma = (\mathsf{Gen}, \mathsf{Sign}, \mathsf{Vrfy})$ be a signature scheme and let $\ell \in \mathbf{N}$ be an arbitrary parameter. Consider a multi-party protocol Π in the model from Sect. 2, i.e., having access to an ideal functionality $\mathcal{F}_{\mathsf{syn}}^{\pi, \pi_{\mathcal{A}}}$, where π is interpreted as the hashrate of each of the parties, and $\pi_{\mathcal{A}}$ as the hashrate of the adversary.

Each party P takes as input a security parameter 1^{κ}, and it produces as output a tuple $(\mathsf{sk}_P, \mathsf{pk}_P, \mathcal{K}_P, \mathsf{rank}_P)$, where $(\mathsf{sk}_P, \mathsf{pk}_P) \in \{0, 1\}^* \times \{0, 1\}^*$ is called a *(private key, public key) pair of* P, the finite set $\mathcal{K}_P \subset \{0, 1\}^*$ will be some set of public keys, and $\mathsf{rank}_P : \mathcal{K}_P \to \{0, \ldots, \ell\}$ will be called a *key-ranking function (of* P). We will say that an *identity* pk *was created during the execution* Π if $\mathsf{pk} \in \mathcal{K}_P$ for at least one honest P (regardless of the value of $\mathsf{rank}_P(\mathsf{pk})$). The protohancol Π is called a $\pi_{\mathcal{A}}$-*secure* ℓ-*ranked* Σ-*key generation protocol* if for any poly-time adversary $\mathcal{A}$ who attacks this protocol (in the model from Sect. 2) the following conditions hold: (1) *Key-generation:* Π is a key-generation algorithm for every P, by which we mean the following. First of all, for every $i = 1, \ldots, n$ and every $m \in \{0, 1\}^*$ we have that $\mathsf{Vrfy}(\mathsf{pk}_P, \mathsf{Sign}(\mathsf{sk}_P, m)) = \mathsf{true}$. Moreover sk_P can be securely used for signing messages in the following sense. Suppose the adversary $\mathcal{A}$ learns the entire information received by all the parties except of some P, and later $\mathcal{A}$ engages in the "chosen message attack" against an oracle that signs messages with key sk_P. Then any such $\mathcal{A}$ has negligible (in κ) probability of forging a valid (under key pk_P) signature on a fresh message. (2) *Bounded creation of identities:* We require that the number of created identities is at most $n + \lceil \pi_{\mathcal{A}}/\pi \rceil$ except with probability negligible in κ. (3) *Validity:* For every two honest parties P and P' we have that $\mathsf{rank}_P(P') = 0$. (4) *Consistency:* For every two honest parties P and P' and every key $\mathsf{pk} \in \mathcal{K}_P$ such that $\mathsf{rank}_P(\mathsf{pk}) < \ell$ we have that $\mathsf{pk} \in \mathcal{K}_{P'}$ and moreover $\mathsf{rank}_{P'}(\mathsf{pk}) \le \mathsf{rank}_P(\mathsf{pk}) + 1$.

Our construction of a ranked key generation protocol $\mathsf{RankedKeys}$ is presented on Fig. 3. The protocol $\mathsf{RankedKeys}$ uses a Proof of Work scheme (P, V) with prover time time_P and verifier time time_V. Note that the algorithms P and V query the oracle H. Technically this is done by sending Hash queries to the $\mathcal{F}_{\mathsf{syn}}^{\pi, \pi_{\mathcal{A}}}$ oracle, in the $\mathcal{H}^{\kappa}$-model (it also uses another hash function $F : \{0, 1\}^* \to \{0, 1\}^{\kappa}$ that is modeled as a random oracle, but its computation does not count into the hashrate). We can instantiate this PoW scheme with the scheme $(\mathsf{PTree}, \mathsf{VTree})$ described in the full version of this paper [1]. The parameter ℓ will be equal to $\lceil \pi_{\mathsf{max}}/\pi \rceil$. The notation $\prec$ is described below.

Let us present some intuitions behind our protocol. First, recall that the problem with the protocol from [2] (described at the beginning of this section)

The **challenges phase** consists of $\ell + 2$ rounds:

- *Round 0:* Each party P draws a random challenge $c_P \leftarrow \{0,1\}^\kappa$ and sends his *challenge message of level* 0 equal to $(\mathsf{Challenge}^0, c_P^0)$ to all parties (including herself).
- For $k = 1$ to $\ell + 1$ in *round k* each party P does the following. It waits for the messages of a form $(\mathsf{Challenge}^{k-1}, a)$ that were sent in the previous round (note that some of them might have already arrived earlier, but, by our assumptions they are all guaranteed to arrive before round k ends). Of course if the adversary does not perform any attack then there will be exactly n such messages (one from every party), but in general there can be much more of them. Let $(\mathsf{Challenge}^{k-1}, a_1), \ldots, (\mathsf{Challenge}^{k-1}, a_m)$ be all messages received by P. Denote $A_P^k = (a_1, \ldots, a_m)$. Then P computes her challenge in round k as $c_P^k = F(A_P^k)$ and sends $(\mathsf{Challenge}^k, c_P^k)$ to all parties (this is not needed in the last rounds, i.e., when $k = \ell + 1$).

In the **Proof of Work phase** each party P performs the following.

1. Generate a fresh key pair $(\mathsf{sk}_P, \mathsf{pk}_P) \leftarrow \mathsf{Gen}(1^k)$ and compute $\mathsf{Sol}_P = \mathsf{P}(F(\mathsf{pk}_P, A_P^{\ell+1}))$ (recall that $A_P^{\ell+1}$ contains all the challenges that P received in the last round of the "challenges phase"). Note that this phase takes $\lceil \mathsf{time}_P / (\pi \cdot \Delta) \rceil$ rounds.
2. Send to all the other parties a message $(\mathsf{Key}^0, \mathsf{pk}_P, A_P^{\ell+1}, \mathsf{Sol}_P)$. This message contains P's public key pk_P, the sequence $A_P^{\ell+1}$ of challenges that he received in the last round of the "challenges phase", and a Proof of Work Sol_P. The reason why she sends the entire $A_P^{\ell+1}$, instead of $F(\mathsf{pk}_P, A_P^{\ell+1})$, is that in this way every other party will be able check if her challenge was used as an input to F when $F(\mathsf{pk}_P, A_P^{\ell+1})$ was computed (this check will be performed in the next phase).

The **key ranking phase** consists of $\ell + 1$ steps, each lasting $1 + \lceil (\theta \cdot \mathsf{time}_V) / (\pi \cdot \Delta) \rceil$ rounds. During these steps each party P constructs a set $\mathcal{K}_P$ of ranked keys, together with a ranking function $\mathsf{rank}_P : \mathcal{K}_P \rightarrow \{0, \ldots, \ell\}$ (the later a key is added to $\mathcal{K}_P$ the higher will be its rank). Initially all $\mathcal{K}_P$'s are empty.

- *Step 0:* Each party P waits for one round for the messages of the form $(\mathsf{Key}^0, \mathsf{pk}, B^{\ell+1}, \mathsf{Sol})$ sent in the PoW phase. Then, for each such message she checks the following conditions:
  - Sol is a correct PoW solution for the challenge $F(\mathsf{pk}, B^{\ell+1})$, i.e., if $\mathsf{V}(F(\mathsf{pk}, B^{\ell+1}), \mathsf{Sol}) = \mathsf{true}$,
  - c_P^ℓ appears in $B^{\ell+1}$, i.e., $c_P^\ell \prec B^{\ell+1}$.

  If both of these conditions hold then P accepts the key pk with rank 0, i.e., P adds pk to the set $\mathcal{K}_P$ and sets $\mathsf{rank}_P(\mathsf{pk}) := 0$. Moreover P notifies all the other parties about this fact by sending to every other party a message $(\mathsf{Key}^1, \mathsf{pk}, A_P^\ell, B^{\ell+1}, \mathsf{Sol})$.
- For $k = 1$ to ℓ in *step k* each party P does the following. She waits for one round for the messages of a form $(\mathsf{Key}^k, \mathsf{pk}, B^{\ell+1-k}, \ldots, B^{\ell+1}, \mathsf{Sol})$. Then she stops listening and for each received message she checks the following conditions:
  - the key pk has not been yet added to $\mathcal{K}_P$, i.e.: $\mathsf{pk} \notin \mathcal{K}_P$,
  - Sol is a correct PoW solution for the challenge $F(\mathsf{pk}, B^{\ell+1})$, i.e., if $\mathsf{V}(F(\mathsf{pk}, B^{\ell+1}), \mathsf{Sol}) = \mathsf{true}$,
  - $c_P^{\ell-k} \prec B^{\ell+1-k}$ and for every $i = \ell + 1 - k$ to ℓ it holds that $F(B^i) \prec B^{i+1}$.

  If all of these conditions hold then P accepts the key pk with rank k, i.e., P adds pk to the set $\mathcal{K}_P$ and sets $\mathsf{rank}_P(\mathsf{pk}) := k$. Moreover if $k < \ell$ then P notifies all the other parties about this fact by sending at the end of the round to every other party a message $(\mathsf{Key}^{k+1}, \mathsf{pk}, A_P^{\ell-k}, B^{\ell+1-k}, \ldots, B^{\ell+1}, \mathsf{Sol})$ (recall that A_P^k is equal to the set of challenges received by P in the k-th round of the "challenges phase", and $F(A_P^k) = c_P^k$).

At the end of the protocol each party P outputs $(\mathsf{sk}_P, \mathsf{pk}_P, \mathcal{K}_P, \mathsf{rank}_P)$.

Fig. 3. The RankedKeys protocol.

was that some public keys could be recognized only by a subset of the honest parties. A key could be dropped because: (1) it was received too late; or (2) the corresponding proof did not contained the appropriate challenge. Informally, the idea behind the RankedKeys protocol is to make these conditions more granular. If we forget about the PoWs, and look only at the time constrains then our protocol could be described as follows: keys received in the first round are assigned rank 0, keys received in the second round are assigned rank 1, and so on. Since we instruct every honest party to forward to everybody all the keys that she receives, hence if a key receives rank k from some honest party, then it receives rank at most $k + 1$ from all the other honest parties.

If we also consider the PoWs then the description of the protocol becomes a bit more complicated. The RankedKeys protocol consists of 3 phases. We now sketch them informally. The "challenges phase" is divided into $\ell + 2$ rounds. At the beginning of the first round each P generates his challenge c_P^0 randomly and sends it to all the other parties. Then, in each k-th round each P collects the messages $a_1, \ldots, a_m$ sent in the previous round, concatenates then into $A_P^k = (a_1, \ldots, a_m)$, hashes them, and sends the result $c_P^k = F(A_P^k)$ to all the other parties.

Let $a \prec (b_1, \ldots, b_m)$ denote the fact that $a = b_i$ for some i. We say that the string b *dependents* on a if there exists a sequence $a = v_1, \ldots, v_m = b$, such that for every $1 \leq i < m$, it holds that $F(v_i) \prec v_{i+1}$. The idea behind this notion is that b could not have been predicted before a was revealed, because b is created using a series of concatenations and queries to the random oracle starting from the string a. Note that in particular c_P^k depends on $c_{P'}^{k-1}$ for any honest P, P'[7] and $1 \leq k \leq \ell$ and hence c_P^k depends on $c_{P'}^0$ for any honest P, P' and an arbitrary $1 \leq k \leq \ell + 1$.

Then, during the "Proof of Work" phase each honest party P draws a random key pair $(\mathsf{sk}_P, \mathsf{pk}_P)$ and creates a proof of work[8] $\mathsf{P}(F(\mathsf{pk}_P, A_P^{\ell+1}))$. Then, she sends her public key together with the proof to all the other parties.

Later, during the "key ranking phase" the parties receive the public keys of the other parties and assign them ranks. To assign the public key pk rank k the party P requires that she receives it in the k-th round in this phase and that it is accompanied with a proof $\mathsf{P}(F(\mathsf{pk}_P, s))$ for some string s, which depends on $c_P^{\ell-k}$. Such a proof could not have been precomputed, because $c_P^{\ell-k}$ depends on c_P^0, which was drawn randomly by P at the beginning of the protocol and hence could not been predicted before the execution of the protocol. If those conditions are met, then P forwards the message with the key to the other parties. This message will be accepted by other parties, because it will be received by them in the $(k+1)$-st round of this phase and because s depends on $c_P^{\ell-k}$, which depends on $c_{P'}^{\ell-(k+1)}$ for any honest P'. In the effect, all other honest parties, which have not yet assigned pk a rank will assign it a rank $k + 1$.

---

[7] This is because $c_{P'}^{k-1} \prec A_P^k$ and $F(A_P^k) = c_P^k$.

[8] The reason why we hash the input before computing a PoW is that the PoW definition requires that the challenges are random.

Let $\mathsf{RankedKeys}_{\mathsf{PTree}}$ denote the $\mathsf{RankedKeys}$ scheme instantiated with the PoW scheme $(\mathsf{PTree}^\kappa_{\mathsf{time_P}}, \mathsf{VTree}^\kappa_{\mathsf{time_P}})$ (from Lemma 1), where $\mathsf{time_P} := \kappa^2 \cdot (\ell + 2)\Delta \cdot \pi$ and $\mathsf{time_V} := \kappa \lceil \log_2 \mathsf{time_P} \rceil$. We have the following fact (its proof appears in the full version of this paper [1]).

Lemma 2. *Assume the total hashrate of all the participants is at most π_{max}, the hashrate of each honest party if π, and the adversary can not send more than $(\theta - \lceil \pi_{\mathsf{max}}/\pi \rceil)$ messages in every round. Then the $\mathsf{RankedKeys}_{\mathsf{PTree}}$ protocol is a $\pi_{\mathcal{A}}$-secure ℓ-ranked key generation protocol, for $\ell = \lceil \pi_{\mathsf{max}}/\pi \rceil$, whose total execution takes $(2\ell + 3) + \lceil \mathsf{time_P}/(\pi \cdot \Delta) \rceil + \lceil (\ell+1)(\theta \cdot \mathsf{time_V})/(\pi \cdot \Delta) \rceil$ rounds.*

The communication and message complexity of the $\mathsf{RankedKeys}$ protocol are analysed in the full version of this paper [1].

The Broadcast Protocol. The reason why ranked key sets are useful is that they allow to construct a reliable broadcast protocol, which is secure against an adversary that has an arbitrary hashrate. The only assumption that we need to make is that the total hashrate in the system is bounded by some π_{max} and the adversary cannot send more than $\theta - n$ messages in one interval (for some parameter θ). Our protocol, denoted $\mathsf{Broadcast}$, works in time that is linear in $\ell = \lceil \pi_{\mathsf{max}}/\pi \rceil$ plus the execution time of $\mathsf{RankedKeys}$. It is based on a classical authenticated Byzantine agreement by Dolev and Strong [14] (and is similar

---

1. Each party P takes as input $(\mathsf{Broadcast}, \mathsf{sid}, x_P)$.
2. The parties run the $\mathsf{RankedKeys}$ protocol (attaching sid to every message). Let $(\mathsf{sk}_P, \mathsf{pk}_P, \mathcal{K}_P, \mathsf{rank}_P)$ be the output of each $P \in \mathcal{P}$.
3. Each party P initializes, for every $\mathsf{pk} \in \mathcal{K}_P$, a variable $\mathcal{Z}^{\mathsf{pk}}_P = \emptyset$.
4. Each $D \in \mathcal{P}$ performs the following procedure that consists of $\ell + 1$ rounds (this can be executed in parallel for every D):
   - *Round 0:* D (we will call him the *Dealer*) sends to every other party a message $(\mathsf{sid}, x_D, \mathsf{pk}_D, \mathsf{Sign}_{\mathsf{pk}_D}(x_D, \mathsf{pk}_D))$.
   - *Round k, for $1 \leq k \leq \ell$:* Each party P except of the dealer D waits for the messages of the form $(\mathsf{sid}, v, \mathsf{pk}_D, \mathsf{Sign}_{\mathsf{sk}_{a_1}}(v, \mathsf{pk}_D), \ldots, \mathsf{Sign}_{\mathsf{sk}_{a_k}}(v, \mathsf{pk}_D))$. Such a message is accepted by P if:
     (1) all signatures are valid and are corresponding to different public keys,
     (2) $\mathsf{pk}_{a_1} = \mathsf{pk}_D$,
     (3) $\mathsf{pk}_{a_j} \in \mathcal{K}_P$ and $\mathsf{rank}_P(\mathsf{pk}_{a_j}) \leq k$ for $1 \leq j \leq k$, and
     (4) $v \notin \mathcal{Z}^{\mathsf{pk}_D}_P$ and $|\mathcal{Z}^{\mathsf{pk}_D}_P| < 2$.
     If a message is accepted then P adds v to her set $\mathcal{Z}^{\mathsf{pk}_D}_P$ and if moreover $k < \ell$, than she sends a message $(\mathsf{sid}, v, \mathsf{pk}_D, \mathsf{Sign}_{\mathsf{pk}_{a_1}}(v, \mathsf{pk}_D), \ldots, \mathsf{Sign}_{\mathsf{pk}_{a_k}}(v, \mathsf{pk}_D), \mathsf{Sign}_{\mathsf{pk}_P}(v, \mathsf{pk}_D))$ to all other parties.
5. Each party P determines the set $\mathcal{Y}_P$ as the union over all $\mathcal{Z}^{\mathsf{pk}}_P$'s that are of size 1, i.e.: $\mathcal{Y}_P = \bigcup_{\mathsf{pk} : |\mathcal{Z}^{\mathsf{pk}}_P| = 1} \mathcal{Z}^{\mathsf{pk}}_P$. It outputs $(\mathsf{Received}, \mathsf{sid}, \mathcal{Y}_P)$.

Fig. 4. The Broadcast protocol.

to the technique used to construct broadcast from a proxcast protocol [13]). The protocol is depicted on Fig. 4 and it works as follows. First the parties execute the RankedKeys protocol with parameters π, π_{max} and θ, built on top of a signature scheme (Gen, Sign, Vrfy) — Sign_{pk} denotes a signatures computed using a *private* key corresponding to a public key pk. For convenience assume that every signature σ contains information identifying the public key that was used to compute it. Let $(\text{sk}_P, \text{pk}_P, \mathcal{K}_P, \text{rank}_P)$ be the output of each P after this protocol ends (recall that $(\text{sk}_P, \text{pk}_P)$ is her key pair, $\mathcal{K}_P$ is the set of public keys that she accepted, rank_P is the key ranking function). Then, each party $D \in \mathcal{P}$ executes in parallel the procedure from Step 4. During the execution each party P maintains a set $\mathcal{Z}_P^{\text{pk}_D}$ initialized with $\emptyset$. The output of each party is equal to the only elements of this set (if $\mathcal{Z}_P^{\text{pk}_D}$ is a singleton) or $\perp$ otherwise. The following lemma is proven in the full version of this paper [1].

Lemma 3. *The* Broadcast *protocol is a* (π_{max}, θ)-*secure broadcast protocol.*

5   Applications

Multiparty Computations. As already mentioned before, the Broadcast protocol can be used to establish a group of parties that can later perform the MPC protocols. For the lack of space we only sketch this method here. The main idea is as follows. First, each party P generates its key pair $(\text{sk}_P, \text{pk}_P)$. Then, it uses the broadcast protocol to send to all the other parties the public key pk_P. Let $\pi_{\mathcal{A}}$ be the computing power of the adversary, and let t be the number of parties that he corrupted. From the properties of the broadcast protocol he can make the honest parties accept at most $\lceil \pi_{\mathcal{A}}/\pi \rceil$ keys pk_P chosen by the adversary. Additionally, the adversary knowns up to t secret keys of the parties that she corrupted. Therefore altogether there are at most $\lceil \pi_{\mathcal{A}}/\pi \rceil + t$ keys pk_P such that the adversary knows the corresponding secret keys sk_P. The total number of keys is $\lceil \pi_{\mathcal{A}}/\pi \rceil + n$ (where n is the number of the honest parties).

Given such a setup the parties can now simulate the secure channels, even if initially they did not know each others identities (i.e. in the model from Sect. 2), by treating the public keys as the identities. More precisely: whenever a party P wants to send a message to P' (known to P by her public key $\text{pk}_{P'}$) she would use the standard method of encryption (note that in the adaptive case this is secure only if the encryption scheme is non-committing) and digital signatures to establish a secure channel (via the insecure broadcast channel available in the model) with P'. Hence the situation is exactly as in the standard MPC settings with the private channels between $\lceil \pi_{\mathcal{A}}/\pi \rceil + n$ parties. We can now use well-known fact that simulating any functionality is possible in this case [10]. In case we require that the protocol has guaranteed termination we need an assumption that the majority of the participants is honest [20], i.e., that $\lceil \pi_{\mathcal{A}}/\pi \rceil + t < (n - t)$. Suppose we ignore the rounding up (observe that in practice we can make $\lceil \pi_{\mathcal{A}}/\pi \rceil$ arbitrarily close to $\pi_{\mathcal{A}}/\pi$ by making π small). Then we obtain the condition $\pi_{\mathcal{A}} + t\pi < (n - t)\pi$. The left hand side of this inequality can be interpreted as the "total computing power of the adversary" (including his own computing power and the one of corrupt parties), and the right hand side can be interpreted as the total computing power of the honest parties. Therefore we get that every functionality can he simulated (with guaranteed termination) as long as the majority of the computing power is controlled by the honest parties. This argument will be formalized in the full version of this paper.

Unpredictable Beacon Generation. The Broadcast protocols can also be used to produce unpredictable beacons even if there is no honest majority of computing power in the system by letting every party broadcast a random nonce and then hashing the result. This is described in more detail in the full version of this paper [1], where we also discuss also the possibility of creating provable secure currencies using our techniques.

References

1. Andrychowicz, M., Dziembowski, S.: Distributed cryptography based on the proofs of work. Cryptology ePrint Archive, report 2014/796 (2014)
2. Aspnes, J., Jackson, C., Krishnamurthy, A.: Exposing computationally-challenged byzantine impostors. Department of CS, Yale University, Technical report (2005)
3. Babai, L.: Trading group theory for randomness. In: STOC (1985)
4. Back, A.: Hashcash - a denial of service counter-measure, Technical report (2002)
5. Bahack, V.: Theoretical bitcoin attacks with less than half of the computational power (draft). arXiv preprint arXiv:1312.7013 (2013)
6. Bellare, M., Rogaway, P.: Random oracles are practical: a paradigm for designing efficient protocols. In: ACM CCS (1993)
7. Ben-Or, M., Goldwasser, S., Wigderson, A.: Completeness theorems for non-cryptographic fault-tolerant distributed computation. In: STOC (1988)
8. Bogetoft, P., Christensen, D.L., Damgård, I., Geisler, M., Jakobsen, T., Krøigaard, M., Nielsen, J.D., Nielsen, J.B., Nielsen, K., Pagter, J., Schwartzbach, M., Toft, T.: Secure multiparty computation goes live. In: Dingledine, R., Golle, P. (eds.) FC 2009. LNCS, vol. 5628, pp. 325–343. Springer, Heidelberg (2009)
9. Canetti, R.: Universally composable security: a new paradigm for cryptographic protocols. In: FOCS (2001)
10. Canetti, R., Lindell, Y., Ostrovsky, R., Sahai, A.: Universally composable two-party and multi-party secure computation. In: STOC (2002)
11. Chaum, D., Crépeau, C., Damgård, I.: Multiparty unconditionally secure protocols (extended abstract). In: STOC (1988)
12. Clark, J., Hengartner, U.: On the use of financial data as a random beacon. In: The International Conference on Electronic Voting Technology/Workshop on Trustworthy Elections (2010)
13. Considine, J., Fitzi, M., Franklin, M.K., Levin, L.A., Maurer, U.M., Metcalf, D.: Byzantine agreement given partial broadcast. J. Cryptol. **18**(3), 191–217 (2005)
14. Dolev, D., Strong, H.R.: Authenticated algorithms for byzantine agreement. SIAM J. Comput. **12**(4), 656–666 (1983)
15. Douceur, J.R.: The sybil attack. In: Druschel, P., Kaashoek, M.F., Rowstron, A. (eds.) IPTPS 2002. LNCS, vol. 2429, pp. 251–260. Springer, Heidelberg (2002)

16. Dwork, C., Naor, M.: Pricing via processing or combatting junk mail. In: Brickell, E.F. (ed.) CRYPTO 1992. LNCS, vol. 740, pp. 139–147. Springer, Heidelberg (1993)
17. Dwork, C., Naor, M., Wee, H.M.: Pebbling and proofs of work. In: Shoup, V. (ed.) CRYPTO 2005. LNCS, vol. 3621, pp. 37–54. Springer, Heidelberg (2005)
18. Even, S., Goldreich, O., Lempel, A.: A randomized protocol for signing contracts. In: CRYPTO (1982)
19. Eyal, I., Sirer, E.G.: Majority is not enough: Bitcoin mining is vulnerable. In: Financial Cryptography (2014)
20. Goldreich, O., Micali, S., Wigderson, A.: How to play any mental game. In: STOC (1987)
21. Goldwasser, S., Sipser, M.: Private coins versus public coins in interactive proof systems. In: STOC (1986)
22. Katz, J., Maurer, U., Tackmann, B., Zikas, V.: Universally composable synchronous computation. In: Sahai, A. (ed.) TCC 2013. LNCS, vol. 7785, pp. 477–498. Springer, Heidelberg (2013)
23. Katz, J., Miller, A., Shi, E.: Pseudonymous secure computation from time-lock puzzles. Cryptology ePrint Archive, report 2014/857, 2014
24. Lamport, L., Shostak, R., Pease, M.: The byzantine generals problem. ACM Trans. Program. Lang. Syst. 4(3), 382–401 (1982)
25. Moran, T., Naor, M.: Split-ballot voting: everlasting privacy with distributed trust. In: ACM CCS (2007)
26. Nakamoto, S.: Bitcoin: A peer-to-peer electronic cash system (2008). Accessed http:bitcoin.org/bitcoin.pdf
27. Rabin, M.O.: Transaction protection by beacons. J. Comput. Syst. Sci. 27(2), 256–267 (1983)
28. Wiki, B.: Denial of service (dos) attacks. en.bitcoin.it/wiki/ Weaknesses, Accessed on 26.09.2014
29. Bitcoin Wiki. Network. en.bitcoin.it/wiki/Network. Accessed on 26.09.2014
30. Yao, A.C-C.: Protocols for secure computations (extended abstract). In: FOCS (1982)

Non-Signaling and
Information-Theoretic Crypto

Multi-prover Commitments Against Non-signaling Attacks

Serge Fehr$^{(\boxtimes)}$ and Max Fillinger

Centrum Wiskunde and Informatica (CWI), Amsterdam, The Netherlands
{serge.fehr,M.J.Fillinger}@cwi.nl

Abstract. We reconsider the concept of two-prover (and more generally: multi-prover) commitments, as introduced in the late eighties in the seminal work by Ben-Or *et al.* As was recently shown by Crépeau *et al.*, the security of known two-prover commitment schemes not only relies on the explicit assumption that the two provers cannot communicate, but also depends on what their information processing capabilities are. For instance, there exist schemes that are secure against classical provers but insecure if the provers have *quantum* information processing capabilities, and there are schemes that resist such quantum attacks but become insecure when considering general so-called *non-signaling* provers, which are restricted *solely* by the requirement that no communication takes place.

This poses the natural question whether there exists a two-prover commitment scheme that is secure under the *sole* assumption that no communication takes place, and that does not rely on any further restriction of the information processing capabilities of the dishonest provers; no such scheme is known.

In this work, we give strong evidence for a negative answer: we show that any single-round two-prover commitment scheme can be broken by a non-signaling attack. Our negative result is as bad as it can get: for any candidate scheme that is (almost) perfectly hiding, there exists a strategy that allows the dishonest provers to open a commitment to an arbitrary bit (almost) as successfully as the honest provers can open an honestly prepared commitment, i.e., with probability (almost) 1 in case of a perfectly sound scheme. In the case of multi-round schemes, our impossibility result is restricted to perfectly hiding schemes.

On the positive side, we show that the impossibility result can be circumvented by considering *three* provers instead: there exists a three-prover commitment scheme that is secure against arbitrary non-signaling attacks.

Keywords: Non-signaling · Bit-commitment · Multi-prover

1 Introduction

Background. A commitment scheme is an important primitive in theoretical cryptography with various applications, for instance to zero-knowledge proofs

© International Association for Cryptologic Research 2015
R. Gennaro and M. Robshaw (Eds.): CRYPTO 2015, Part II, LNCS 9216, pp. 403–421, 2015.
DOI: 10.1007/978-3-662-48000-7_20

and multiparty computation, which themselves are fundamentally important concepts in modern cryptography. For a commitment scheme to be secure, it must be *hiding* and *binding*. The former means that after the commit phase, the committed value is still hidden from the verifier, and the latter means that the prover (also referred to as committer) can open a commitment only to one value. Unfortunately, a commitment scheme cannot be unconditionally hiding *and* unconditionally binding at the same time. This is easy to see in the classical setting, and holds as well when using quantum communication [9,10]. Thus, we have to put some limitation on the capabilities of the dishonest party. One common approach is to assume that the dishonest prover (or, alternatively, the dishonest verifier) has limited computing resources, so that he cannot solve certain computational problems (like factoring large integers). Another approach was suggested by Ben-Or, Goldwasser, Kilian and Wigderson in their seminal paper [2] in the late eighties. They assume that the prover consists of two (or more) agents that cannot communicate with each other, and they show the existence of a secure commitment scheme in this two-prover setting. Based on this two-prover commitment scheme, they then show that every language in NP has a two-prover perfect zero-knowledge interactive proof system (though there are some subtle issues in this latter result, as discussed in [15]).

A simple example of a two-prover commitment scheme, due to [4], is the following. The verifier chooses a uniformly random string $a \in \{0,1\}^n$ and sends it to the first prover, who sends back $x := r \oplus a \cdot b$ as the commitment for bit $b \in \{0,1\}$, where $r \in \{0,1\}^n$ is a uniformly random string known (only) to the two provers, and where "$\oplus$" is bit-wise XOR and "$\cdot$" scalar multiplication (of the scalar b with the vector a). In order to open the commitment (to b), the second prover sends back $y := r$, and the verifier checks the obvious: whether $y = x \oplus a \cdot b$. It is clear that this scheme is hiding: $x := r \oplus a \cdot b$ is uniformly random and independent of a no matter what b is, and the intuition behind the binding property is the following. In order to open the commitment to $b = 0$, the second prover needs to announce $y = x$; in order to open to $b = 1$, he needs to announce $y = x \oplus a$. Therefore, in order to open to *both*, he must know x *and* $x \oplus a$, which means he knows a, but this is a contradiction to the no-communication assumption, because a was sent only to the first prover.

In [4], Crépeau, Salvail, Simard and Tapp show that, as a matter of fact, the security of such two-prover commitment schemes not only relies on the explicit assumption that the two provers cannot communicate, but the security also crucially depends on the information processing capabilities of the dishonest provers. Indeed, they show that a slight variation of the above two-prover commitment scheme (where some slack is given to the verification $y = x \oplus a \cdot b$) is secure against classical provers, but is completely insecure if the provers have *quantum* information processing capabilities and can obtain x and y by means of doing local measurements on an entangled quantum state.[1] Furthermore, they

---

[1] The above intuition for the binding property of the scheme (which also applies to the variation considered in [4]) fails in the quantum setting where x and y are obtained by means of *destructive* measurements.

show that the above two-prover commitment scheme remains secure against such quantum attacks, but becomes insecure against so-called *non-signaling* provers. The notion of non-signaling was first introduced by Khalfin and Tsirelson [14] and by Rastall [12] in the context of Bell-inequalities, and later reintroduced by Popescu and Rohrlich [11]. Non-signaling provers are restricted *solely* by the requirement that no communication takes place — no additional restriction limits their information processing capabilities (not even the laws of quantum mechanics) — and thus considering non-signaling provers is the *minimal* assumption for the two-prover setting to make sense.

This gives rise to the following question. Does there exist a two-prover commitment scheme that is secure against arbitrary non-signaling provers? Such a scheme would *truly* be based on the sole assumption that the provers cannot communicate. No such scheme is known. Clearly, from a practical point of view, asking for such a scheme may be overkill; given our strong believe in quantum mechanics, relying on a scheme that resists quantum attacks seems to be a safe bet. But from a theoretical perspective, this question is certainly in line with the general goal of theoretical cryptography: to find the strongest possible security based on the weakest possible assumption.

Our Results. In this work, we give strong evidence for a negative answer: we show that there exists no single-round two-prover commitment scheme that is secure against general non-signaling attacks. Our impossibility result is as strong as it can get. We show that for any candidate single-round two-prover commitment scheme that is (almost) perfectly hiding, the binding property can be (almost) completely broken: there exists a non-signaling strategy that allows the dishonest provers to open a commitment to an arbitrary bit (almost) as successfully as the honest provers can open an honestly prepared commitment, i.e., with probability (almost) 1 in case of a perfectly sound scheme. Furthermore, for a restricted but natural class of schemes, namely for schemes that have the same communication pattern as the above example scheme, our impossibility result is tight: for every (rational) parameter $0 < \varepsilon \leq 1$ there exists a perfectly sound two-prover commitment scheme that is ε-hiding and as binding as allowed by our negative result (which is almost not binding if ε is small).

In the case of multi-round schemes, our impossibility result is limited and applies to perfectly hiding schemes only. Proving the impossibility of non-perfectly-hiding multi-round schemes remains open.

On the positive side, we show the existence of a secure *three*-prover commitment scheme against non-signaling attacks. Thus, our impossibility result can be circumvented by considering three instead of two provers.

Related Work. Two-prover commitments are closely related to *relativistic commitments*, as introduced by Kent in [8]. In a nutshell, a relativistic commitment scheme is a two-prover commitment scheme where the no-communication requirement is enforced by having the actions of the two provers separated by a space-like interval, i.e., the provers are placed far enough apart, and the scheme is executed quickly enough, so that no communication can take place by the laws of special relativity. As such, our impossibility result immediately implies impossibility of relativistic commitment schemes of the form we consider (e.g., we do not consider quantum schemes) against general non-signaling attacks.

Very generally speaking, and somewhat surprisingly, the (in)security of cryptographic primitives against non-signaling attacks may have an impact on more standard cryptographic settings, as was recently demonstrated by Kalai, Raz and Rothblum [7], who showed the (computational) security of a *delegation scheme* based on the security of an underlying multi-party interactive proof system against non-signaling (or statistically-close-to-non-signaling) adversaries.

2   Preliminaries

2.1   (Conditional) Distributions

For the purpose of this work, a *(probability) distribution* is a function $p : \mathcal{X} \to \mathbb{R}$, $x \mapsto p(x)$, where $\mathcal{X}$ is a finite non-empty set, with the properties that $p(x) \geq 0$ for every $x \in \mathcal{X}$ and $\sum_{x \in \mathcal{X}} p(x) = 1$. For any subset $\Lambda \subset \mathcal{X}$, $p(\Lambda)$ is naturally defined as $p(\Lambda) = \sum_{x \in \Lambda} p(x)$, and it holds that

$$p(\Lambda) + p(\Gamma) = p(\Lambda \cup \Gamma) - p(\Lambda \cap \Gamma) \leq 1 + p(\Lambda \cap \Gamma) \tag{1}$$

for all $\Lambda, \Gamma \subset \mathcal{X}$. A probability distribution is *bipartite* if it is of the form $p : \mathcal{X} \times \mathcal{Y} \to \mathbb{R}$. In case of such a bipartite distribution $p(x, y)$, probabilities like $p(x=y)$, $p(x=f(y))$, $p(x \neq y)$ etc. are naturally understood as

$$p(x=y) = p(\{(x, y) \in \mathcal{X} \times \mathcal{Y} \mid x = y\}) = \sum_{\substack{x \in \mathcal{X}, y \in \mathcal{Y} \\ \text{s.t. } x=y}} p(x, y)$$

etc. Also, for a bipartite distribution $p : \mathcal{X} \times \mathcal{Y} \to \mathbb{R}$, the *marginals* $p(x)$ and $p(y)$ are given by $p(x) = \sum_y p(x, y)$ and $p(y) = \sum_x p(x, y)$, respectively. We note that this notation may lead to an ambiguity when writing $p(w)$ for some $w \in \mathcal{X} \cap \mathcal{Y}$; we avoid this by writing $p(x = w)$ or $p(y = w)$ instead, which are naturally understood. The above obviously extends to arbitrary *multipartite* distributions $p(x, y, z)$ etc.

A *conditional (probability) distribution* is a function $p : \mathcal{X} \times \mathcal{A} \to \mathbb{R}$, $(x, a) \mapsto p(x|a)$, for finite non-empty sets $\mathcal{X}$ and $\mathcal{A}$, such that for every fixed $a^* \in \mathcal{A}$, the function $p(x|a^*)$ is a probability distribution in the above sense, which we also write as $p(x|a = a^*)$. As such, the above naturally extends to bi- and multipartite conditional probability distributions; e.g., if $p(x, y|a, b)$ is a conditional distribution then $p(x|a, b)$, $p(y|a, b)$, $p(x = y|a, b)$ etc. are all naturally defined. However, we emphasize that for instance $p(x|a)$ is in general *not* well defined—unless the corresponding conditional distribution $p(b|a)$ is given, or unless $p(x|a, b)$ does not depend on b.

Remark 1. By convention, we write $p(x|a, b) = p(x|a)$ to express that $p(x|a, b)$ does not depend on b, i.e., that $p(x|a, b_1) = p(x|a, b_2)$ for all b_1 and b_2, and as such $p(x|a)$ *is* well defined and equals $p(x|a, b)$.

A distribution $\delta(x)$ over $\mathcal{X}$ is called a *Dirac* distribution if there exists $x^* \in \mathcal{X}$ so that $\delta(x = x^*) = 1$, and a conditional distribution $\delta(x|a)$ over $\mathcal{X}$ is called a conditional *Dirac* distribution if $\delta(x|a = a^*)$ is a *Dirac* distribution for every $a^* \in \mathcal{A}$, i.e., for every $a^* \in \mathcal{A}$ there exists $x^* \in \mathcal{X}$ so that $\delta(x = x^*|a = a^*) = 1$.

Note that we often abuse notation slightly and simply write $p(x)$ instead of $p : \mathcal{X} \to \mathbb{R},\ x \mapsto p(x)$; furthermore, we may use p for different distributions and distinguish between them by using different names for the variable, like when we consider the two marginals $p(x)$ and $p(y)$ of a bipartite distribution $p(x, y)$. Finally, given two distributions $p(x_0)$ and $q(x_1)$ over the same set $\mathcal{X}$ (and similarly if we use the above convention and denote them by $p(x_0)$ and $p(x_1)$ instead), we write $p(x_0) = q(x_1)$ to denote that $p(x_0 = w) = q(x_1 = w)$ for all $w \in \mathcal{X}$. In a corresponding way, equalities like $p(x_0, x_0', y) = q(x_1, x_1', y)$ should be understood; in situations where we feel it is helpful, we may clarify that "x_0 is associated with x_1, and x_0' with x_1'"; similarly for conditional distributions.

2.2 Gluing Together Distributions

We recall the definition of the statistical distance.

Definition 1. *Let $p(x_0)$ and $p(x_1)$ be two distributions over the same set $\mathcal{X}$.[2] Then, their statistical distance is defined as*

$$d\big(p(x_0), p(x_1)\big) = \frac{1}{2} \cdot \sum_{x \in \mathcal{X}} \big| p(x_0 = x) - p(x_1 = x) \big|.$$

The following property of the statistical distance is well known (see e.g. [13]).

Proposition 1. *Let $p(x_0)$ and $p(x_1)$ be two distributions over the same set $\mathcal{X}$ with $d\big(p(x_0), p(x_1)\big) = \varepsilon$. Then, there exists a distribution $p'(x_0, x_1)$ over $\mathcal{X} \times \mathcal{X}$ with marginals $p'(x_0) = p(x_0)$ and $p'(x_1) = p(x_1)$, and such that $p'(x_0 \neq x_1) = \varepsilon$.*

The following is an immediate consequence.

Lemma 1. *Let $p(x_0, y_0)$ and $p(x_1, y_1)$ be distributions with $d\big(p(x_0), p(x_1)\big) = \varepsilon$. Then, there exists a distribution $p'(x_0, x_1, y_0, y_1)$ with marginals $p'(x_0, y_0) = p(x_0, y_0)$ and $p'(x_1, y_1) = p(x_1, y_1)$, and such that $p'(x_0 \neq x_1) = \varepsilon$ and, as a consequence, $d\big(p'(x_0, y_1), p'(x_1, y_1)\big) \leq \varepsilon$.*

Proof. We first apply Proposition 1 to $p(x_0)$ and $p(x_1)$ to obtain $p'(x_0, x_1)$, and then we set

$$p'(x_0, x_1, y_0, y_1) = p'(x_0, x_1) \cdot p(y_0|x_0) \cdot p(y_1|x_1).$$

The claims on the marginals and on $p'(x_0 \neq x_1)$ follow immediately, and for the last claim we note that

$$p'(x_0, y_1) = p'(x_0 = x_1) \cdot p'(x_0, y_1|x_0 = x_1) + p'(x_0 \neq x_1) \cdot p'(x_0, y_1|x_0 \neq x_1)$$
$$= p'(x_0 = x_1) \cdot p'(x_1, y_1|x_0 = x_1) + p'(x_0 \neq x_1) \cdot p'(x_0, y_1|x_0 \neq x_1)$$

---

[2] This is without loss of generality: the domain can always be extended by including zero-probability elements.

and

$$p'(x_1, y_1) = p'(x_0 = x_1) \cdot p'(x_1, y_1 | x_0 = x_1) + p'(x_1 \neq x_1) \cdot p'(x_1, y_1 | x_0 \neq x_1)$$

and the claim follows because $p'(x_1 \neq x_1) = \varepsilon$.   □

Remark 2. Note that due to the consistency of the marginals, it makes sense to write $p(x_0, x_1, y_0, y_1)$ instead of $p'(x_0, x_1, y_0, y_1)$. We say that we "glue together" $p(x_0, y_0)$ and $p(x_1, y_1)$ along x_0 and x_1.

Remark 3. In the special case where $p(x_0)$ and $p(x_1)$ are identically distributed, i.e., $d\big(p(x_0), p(x_1)\big) = 0$, we obviously have $p(x_0, y_1) = p(x_1, y_1)$.

Remark 4. It is easy to see from the proof of Lemma 1 that the following natural property holds. If $p(x_0, x_1, y_0, y_1, y_0', y_1')$ is obtained by gluing together $p(x_0, y_0, y_0')$ and $p(x_1, y_1, y_1')$ along x_0 and x_1, then the marginal $p(x_0, x_1, y_0, y_1)$ coincides with the distribution obtained by gluing together the marginals $p(x_0, y_0)$ and $p(x_1, y_1)$ along x_0 and x_1.

3   Bipartite Systems and Two-Prover Commitments

3.1   One-Round Bipartite Systems

Informally, a *bipartite system* consists of two subsystem, which we refer to as the left and the right subsystem. Upon input a to the left and input a' to the right subsystem, the left subsystem outputs x and the right subsystem outputs x' (see Fig. 1, left). Formally, the behavior of such a system is given by a conditional distribution $q(x, x'|a, a')$, with the interpretation that given input (a, a'), the system outputs a specific pair (x, x') with probability $q(x, x'|a, a')$. Note that we leave the sets $\mathcal{A}, \mathcal{A}', \mathcal{X}$ and $\mathcal{X}'$, from which a, a', x and x' are respectively sampled, implicit.

If we do not put any restriction upon the system, then *any* conditional distribution $q(x, x'|a, a')$ is eligible, i.e., describes a bipartite system. However, we are interested in systems where the two subsystems cannot communicate with each other. How exactly this requirement restricts $q(x, x'|a, a')$ depends on the available "resources". For instance, if the two subsystems are deterministic, i.e., compute x and x' as *deterministic* functions of a and a' respectively, then this restricts $q(x, x'|a, a')$ to be of the form $q(x, x'|a, a') = \delta(x|a) \cdot \delta(x'|a')$ for conditional Dirac distributions $\delta(x|a)$ and $\delta(x'|a')$. If in addition to allowing them to compute deterministic functions, we give the two subsystem *shared randomness*, then $q(x, x'|a, a')$ may be of the form

$$q(x, x'|a, a') = \sum_r p(r) \cdot \delta(x|a, r) \cdot \delta(x'|a', r)$$

for a distribution $p(r)$ and conditional Dirac distributions $\delta(x|a, r)$ and $\delta(x'|a', r)$. Such a system is called *classical* or *local*. Interestingly, this is not the end of the story. By the laws of *quantum mechanics*, if the two subsystems share an entangled quantum state and obtain x and x' without communication as the result of local measurements that may depend on a and a', respectively, then this gives rise to conditional distributions $q(x, x'|a, a')$ of the form

$$q(x, x'|a, a') = \langle \psi | (E_x^a \otimes F_{x'}^{a'}) | \psi \rangle,$$

where $|\psi\rangle$ is a quantum state and $\{E_x^a\}_x$ and $\{F_{x'}^{a'}\}_{x'}$ are so-called POVMs. What this exactly means is not important for us; what *is* important is that this leads to a *strictly larger* class of bipartite systems. This is typically referred to as a *violation of Bell inequalities* [1], and is nicely captured by the notion of *nonlocal games*. A famous example is the so-called CHSH-game [3], which is closely connected to the example two-prover commitment scheme from the introduction, and which shows that the variant considered in [4] is insecure against quantum attacks.

The largest possible class of bipartite systems that is compatible with the requirement that the two subsystem do not communicate, but otherwise does not assume anything on the available resources and/or the underlying physical theory, are the so-called *non-signaling* systems, defined as follows.

Definition 2. *A conditional distribution* $q(x, x'|a, a')$ *is called a* non-signaling (one-round) *bipartite system if it satisfies*

$$q(x|a, a') = q(x|a) \qquad \text{(NS)}$$

as well as with the roles of the primed and unprimed variables exchanged, i.e.,

$$q(x'|a, a') = q(x'|a') \qquad \text{(NS')}$$

Recall that, by the convention in Remark 1, the equality (NS) is to be understood in the sense that $q(x|a, a')$ does not depend on a', i.e., that $q(x|a, a_1') = q(x|a, a_2')$ for all a_1', a_2', and correspondingly for (NS').

We emphasize that this is the *minimal* necessary condition for the requirement that the two subsystems do not communicate. Indeed, if e.g. $q(x|a, a_1') \neq q(x|a, a_2')$, i.e., if the input-output behavior of the left subsystem depends on the input to the right subsystem, then the system can be used to communicate by giving input a_1' or a_2' to the right subsystem, and observing the input-output behavior of the left subsystem. Thus, in such a system, communication does take place.

The non-signaling requirement for a bipartite system is — conceptually and formally — equivalent to requiring that the two subsystems can (in principle) be queried *in any order*. Conceptually, it holds because the left subsystem should be able to deliver its outputs *before* the right subsystem has received any input if and only if the output does not depend on the right subsystem's input (which means that no information is communicated from right to left), and similarly the other way round. And, formally, we see that the non-signaling requirement from Definition 2 is equivalent to asking that $q(x, x'|a, a')$ can be written as

$$q(x, x'|a, a') = q(x|a) \cdot q(x'|x, a, a') \quad \text{and} \quad q(x, x'|a, a') = q(x'|a') \cdot q(x|x', a, a')$$

for some respective conditional distributions $q(x|a)$ and $q(x'|a')$. This characterization is a convenient way to "test" whether a given bipartite system is non-signaling without doing the maths.

Clearly, all classical systems are non-signaling. Also, any quantum system is non-signaling.[3] But there are non-signaling systems that are not quantum (and thus in particular not classical). The typical example is the *NL-box* (non-local box; also known as *PR-box*) [11], which, upon input bits a and a' outputs *random* output bits x and x' subject to

$$x \oplus x' = a \cdot a'.$$

This system is indeed non-signaling, as it can be queried in any order: submit a to the left subsystem to obtain a uniformly random x, and then submit a' to the right subsystem to obtain $x' := x \oplus a \cdot b$, and correspondingly the other way round.

3.2   Two-Round Systems

We now consider bipartite systems as discussed above, but where one can interact with the two subsystems multiple times. We restrict to two rounds: after having input a to the left subsystem and obtained x as output, one can now input b into the left subsystem and obtain output y, and similarly with the right subsystem (see Fig. 1, right). In such a two-round setting, the non-signaling condition needs to be paired with *causality*, which captures that the output of the first round does not depend on the input that will be given in the second round.

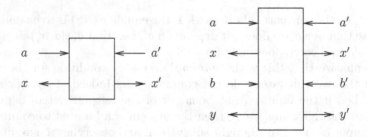

Fig. 1. A one-round (left) and two-round (right) bipartite system.

Definition 3. *A conditional distribution* $q(x, x', y, y'|a, a', b, b')$ *is called a* non-signaling *two-round bipartite system if it satisfies the following two causality constraints*

$$q(x, x'|a, a', b, b') = q(x, x'|a, a') \tag{C1}$$

---

[3] Indeed, the two parts of an entangled quantum state can be measured in any order, and the outcome of the first measurement does not depend on how the other part is going to be measured.

$$and \quad q(x'|x, y, a, a', b, b') = q(x'|x, y, a, a', b) \tag{C2}$$

and the following two non-signaling constraints

$$q(x, y|a, a', b, b') = q(x, y|a, b) \tag{NS1}$$

$$and \quad q(y|x, x', a, a', b, b') = q(y|x, x', a, a', b) \tag{NS2}$$

as well as with the roles of the primed and unprimed variables exchanged.

(C1) captures causality of the overall system, i.e., when considering the left and the right system as one "big" multi-round system. (C2) captures that no matter what interaction there is with the left system, the right system still satisfies causality. Similarly, (NS1) captures that the left and the right system are non-signaling over both rounds, and (NS2) captures that no matter what interaction there was in the first round, the left and the right system remain non-signaling in the second round.

It is rather clear that these are *necessary* conditions; we argue that they are *sufficient* to capture a non-signaling two-round system in the full version [6].

3.3   Two-Prover Commitments

We consider two-prover commitments of the following form. To commit to bit b, the two provers P and Q receive respective "questions" a and a' from the verifier V, and they compute, without communicating with each other, respective replies x and x' and send them to V. To open the commitment, P and Q send respectively y and y'. Finally, V performs some check to decide whether to accept or not.

In case of *classical* provers P and Q, restricting the opening phase to one round with one-way communication is without loss of generality: one may always assume that in the opening phase P and Q simply reveal the shared randomness, and V checks whether x and x' had been correctly computed, consistent with the claimed bit b. Restricting the commit phase to one round is, as far as we can see, *not* without loss of generality; we discuss the multi-round case later.

Formally, this can be captured as follows.

Definition 4. *A (single-round) two-prover commitment scheme* Com *consists of a probability distribution* $p(a, a')$, *two conditional distributions* $p_0(x, x', y, y'|a, a')$ *and* $p_1(x, x', y, y'|a, a')$, *and an acceptance predicate* $\mathsf{Acc}(x, x', y, y'|a, a', b)$.

We say that Com *is classical/quantum/non-signaling if* $p_0(x, x', y, y'|a, a')$ *and* $p_1(x, x', y, y'|a, a')$ *are both classical/quantum/non-signaling when parsed as bipartite one-round systems* $p_b((x, y), (x', y')|a, a')$. *By default, any two-prover commitment scheme* Com *is assumed to be non-signaling.*

The distribution $p(a, a')$ captures how V samples the "questions" a and a', $p_b(x, x', y, y'|a, a')$ describes the choices of x and x' and of y and y', given that the bit to commit to is b, and $\mathsf{Acc}(x, x', y, y'|a, a', b)$ determines whether V accepts the opening or not. Whether a scheme is classical, quantum or non-signaling captures the restrictions of the honest provers.

Given a two-prover commitment scheme Com, we define

$$\mathrm{Prob}[\mathsf{Acc}|b] := \sum_{a,a',x,x',y,y'} p(a,a') \cdot p_b(x,x',y,y'|a,a') \cdot \mathsf{Acc}(x,x',y,y'|a,a',b),$$

which is the probability that a correctly formed commitment to bit b is successfully opened.

Definition 5. *A commitment scheme* Com *is* θ-*sound if* $\mathrm{Prob}_p[\mathsf{Acc}|b] \geq \theta$ *for* $b \in \{0,1\}$. *We say that it is* perfectly sound *if it is* 1-*sound.*

It will be convenient to write $p(x_0, x_0', y_0, y_0'|a, a')$ instead of $p_0(x, x', y, y'|a, a')$ and $p(x_1, x_1', y_1, y_1'|a, a')$ instead of $p_1(x, x', y, y'|a, a')$. Switching to this notation, the hiding property is expressed as follows.

Definition 6. Com *is called* ε-*hiding if* $d\big(p(x_0, x_0'|a, a'), p(x_1, x_1'|a, a')\big) \leq \varepsilon$ *for all* a, a'. *If* Com *is* 0-*hiding, we also say it is* perfectly hiding.

Capturing the binding property is more subtle. From the classical approach of defining the binding property for a commitment scheme, one is tempted to require that once the commit phase is over and a, a', x and x' are fixed, adversarial provers $\hat{P}$ and $\hat{Q}$ cannot come up with an opening to $b = 0$ and *simultaneously* with an opening to $b = 1$, i.e., with y_0, y_0' and y_1, y_1' such that $\mathsf{Acc}(x, x', y_0, y_0'|a, a', b = 0)$ and $\mathsf{Acc}(x, x', y_1, y_1'|a, a', b = 1)$ are both satisfied (except with small probability). However, as pointed out by Dumais, Mayers and Salvail [5], in the context of a general physical theory where y and y' may possibly be obtained as respective outcomes of *destructive* measurements (as is the case in quantum mechanics), such a definition is too weak. It does not exclude that $\hat{P}$ and $\hat{Q}$ can freely choose to open the commitment to $b = 0$ or to $b = 1$, whatever they want, but they cannot do both *simultaneously*; once they have produced one opening, their respective states got disturbed and the other opening can then not be obtained anymore.

Our definition for the binding property is based on the following game between the (honest) verifier V and the adversarial provers $\hat{P}, \hat{Q}$.

1. The commit phase is executed: V samples a and a' according to $p(a, a')$, and sends a to $\hat{P}$ and a' to $\hat{Q}$, upon which $\hat{P}$ and $\hat{Q}$ send x and x' back to V, respectively.
2. V sends a bit $b \in \{0,1\}$ to $\hat{P}$ and $\hat{Q}$.
3. $\hat{P}$ and $\hat{Q}$ try to open the commitment to b: they prepare y and y' and send them to V.
4. V checks if the verification predicate $\mathsf{Acc}(x, x', y, y'|a, a', b)$ is satisfied.

We emphasize that even though in the actual binding game above, *the same* bit b is given to the two provers, we require that the response of the provers is well determined by their strategy even in the case that $b \neq b'$. Of course, if the provers are allowed to communicate, they are able to detect when $b \neq b'$ and could reply with, e.g., $y = y' = \perp$ in that case. However, if we restrict to non-signaling provers, we assume that it is *physically* impossible for them to communicate with each other and distinguish the case of $b = b'$ from $b \neq b'$.

As such, a non-signaling attack strategy against the binding property of a two-prover commitment scheme Com is given by a non-signaling two-round bipartite system $q(x, x', y, y'|a, a', b, b')$, as specified in Definition 3. For any such bipartite system, representing a strategy for $\hat{P}$ and $\hat{Q}$ in the above game, the probability that $\hat{P}$ and $\hat{Q}$ win the game, in that $\mathsf{Acc}(x, x', y, y'|a, a', b)$ is satisfied when they have to open to the bit b, is given by

$$\mathrm{Prob}^*_q[\mathsf{Acc}|b] := \sum_{a,a',x,x',y,y'} p(a, a') \cdot q(x, x', y, y'|a, a', b, b) \cdot \mathsf{Acc}(x, x', y, y'|a, a', b) \,.$$

We are now ready to define the binding property.

Definition 7. *A two-prover commitment scheme* Com *is* δ-binding *(against non-signaling attacks) if it holds for any non-signaling two-round bipartite system* $q(x, x', y, y'|a, a', b, b')$ *that*

$$\mathrm{Prob}^*_q[\mathsf{Acc}|0] + \mathrm{Prob}^*_q[\mathsf{Acc}|1] \leq 1 + \delta \,.$$

In other words, a scheme is δ-binding if in the above game the dishonest provers win with probability at most $(1 + \delta)/2$ when $b \in \{0, 1\}$ is chosen uniformly at random. If a commitment scheme is binding (for a small δ) in the sense of Definition 7, then for any strategy q for $\hat{P}$ and $\hat{Q}$, they can just as well *honestly* commit to a bit $\hat{b}$, where $\hat{b}$ is set to 0 with probability $p_0 = \mathrm{Prob}^*_q[\mathsf{Acc}|0]$ and to 1 with probability $p_1 = 1 - p_0 \approx \mathrm{Prob}^*_q[\mathsf{Acc}|1]$, and they will have essentially the same respective success probabilities in opening the commitment to $b = 0$ and to $b = 1$.

4   Impossibility of Two-Prover Commitments

In this section, we show impossibility of secure single-round two-prover commitments against arbitrary non-signaling attacks. We start with the analysis of a restricted class of schemes which are easier to understand and for which we obtained stronger results.

4.1   Simple Schemes

We first consider a special, yet natural, class of schemes. We call a two-prover commitment scheme Com *simple* if it has the same communication pattern as the scheme described in the introduction. More formally, it is called simple if a', x' and y are "empty" (or fixed), i.e., if Com is given by $p(a)$, $p_0(x, y'|a)$, $p_1(x, y'|a)$ and $\mathsf{Acc}(x, y'|a, b)$; to simplify notation, we then write y instead of y'. In other words, P is only involved in the commit phase, where, in order to commit to bit b, he outputs x upon input a, and Q is only involved in the opening phase, where he outputs y. The non-signaling requirement for Com then simplifies to

$p_b(y|a) = p_b(y)$. Recall that by our convention, we may write $p(x_0, y_0|a)$ instead of $p_0(x, y|a)$ and $p(x_1, y_1|a)$ instead of $p_1(x, y|a)$.

In case of such a simple two-prover commitment scheme Com, a non-signaling two-prover strategy reduces to a non-signaling *one-round* bipartite system as specified in Definition 2 (see Fig. 2).

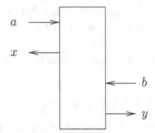

Fig. 2. The adversaries' strategy $q(x, y|a, b)$ in case of a *simple* commitment scheme.

As a warm-up exercise, we first consider a simple two-prover commitment scheme that is *perfectly hiding* and *perfectly sound*. Recall that formally, a simple scheme is given by $p(a)$, $p_0(x, y|b)$, $p_1(x, y|a)$ and $\mathsf{Acc}(x, y|a, b)$, and the perfect hiding property means that $p_0(x|a) = p_1(x|a)$ for any a. To show that such a scheme cannot be binding, we have to show that there exists a non-signaling one-round bipartite system $q(x, y|a, b)$ such that $\mathrm{Prob}_q^*[\mathsf{Acc}|0] + \mathrm{Prob}_q^*[\mathsf{Acc}|1]$ is significantly larger than 1. But this is actually trivial: we can simply set $q(x, y|a, b) := p_b(x, y|a)$. It then holds trivially that

$$\mathrm{Prob}_q^*[\mathsf{Acc}|b] = \sum_{a,x,y} p(a)\, q(x, y|a, b)\, \mathsf{Acc}(x, y|a, b)$$

$$= \sum_{a,x,y} p(a)\, p_b(x, y|a)\, \mathsf{Acc}(x, y|a, b)$$

$$= \mathrm{Prob}_p[\mathsf{Acc}|b]$$

and thus that the dishonest provers are as successful in opening the commitment as are the honest provers in opening an honestly prepared commitment. Thus, the binding property is broken as badly as it can get. The only thing that needs to be verified is that $q(x, y|a, b)$ is non-signaling, i.e., that $q(x|a, b) = q(x|a)$ and $q(y|a, b) = q(y|b)$. To see that the latter holds, note that $q(y|a, b) = p_b(y|a)$, and because Com is non-signaling we have that $p_b(y|a) = p_b(y)$, i.e., does not depend on a. Thus, the same holds for $q(y|a, b)$ and we have $q(y|a, b) = q(y|b)$. The former condition follows from the (perfect) hiding property: $q(x|a, b) = p_b(x|a) = p_{b'}(x|a) = q(x|a, b')$ for arbitrary $b, b' \in \{0, 1\}$, and thus $q(x|a, b) = q(x|a)$.

Below, we show how to extend this result to non-perfectly-binding simple schemes. In this case, we cannot simply set $q(x, y|a, b) := p_b(x, y|a)$, because such a q would not be non-signaling anymore—it would merely be "almost non-signaling". Instead, we have to find a strategy $q(x, y|a, b)$ that is (perfectly) non-signaling and close to $p_b(x, y|a)$; we will find such a strategy with the help of Lemma 1. In Sect. 4.2, we will then consider general schemes where *both* provers interact with the verifier in *both* phases. In this general case, further complications arise.

Theorem 1. *Consider a simple two-prover commitment scheme* Com *that is ε-hiding. Then, there exists a non-signaling strategy $q(x, y|a, b)$ such that*

$$\mathrm{Prob}_q^*[\mathrm{Acc}|0] = \mathrm{Prob}_p[\mathrm{Acc}|0] \quad and \quad \mathrm{Prob}_q^*[\mathrm{Acc}|1] \geq \mathrm{Prob}_p[\mathrm{Acc}|1] - \varepsilon.$$

If Com *is perfectly sound, it follows that*

$$\mathrm{Prob}_q^*[\mathrm{Acc}|0] + \mathrm{Prob}_q^*[\mathrm{Acc}|1] \geq 1 + (1 - \varepsilon)$$

and thus it cannot be δ-binding for $\delta < 1 - \varepsilon$.

Proof. Recall that Com is given by $p(a)$, $p_b(x, y|a)$ and $\mathrm{Acc}(x, y|a, b)$, and we write $p(x_b, y_b|a)$ instead of $p_b(x, y|a)$. Because Com is ε-hiding, it holds that $d\big(p(x_0|a), p(x_1|a)\big) \leq \varepsilon$ for any fixed a. Thus, using Lemma 1 for every a, we can glue together $p(x_0, y_0|a)$ and $p(x_1, y_1|a)$ along x_0 and x_1 to obtain a distribution $p(x_0, x_1, y_0, y_1|a)$ such that $p(x_0 \neq x_1|a) \leq \varepsilon$, and in particular $d\big(p(x_0, y_1|a), p(x_1, y_1|a)\big) \leq \varepsilon$.

We define a strategy q for the dishonest provers by setting $q(x, y|a, b) := p(x_0, y_b|a)$ (see Fig. 3). First, we show that q is non-signaling. Indeed, we have $q(x|a, b) = p(x_0|a)$ for any b, so $q(x|a, b) = q(x|a)$, and we have $q(y|a, b) = p(y_b|a) = p(y_b)$ for any a, and thus $q(y|a, b) = q(y|b)$.

As for the acceptance probability, for $b = 0$ we have $q(x, y|a, 0) = p(x_0, y_0|a)$ and as such $\mathrm{Prob}_q^*[\mathrm{Acc}|0]$ equals $\mathrm{Prob}_p[\mathrm{Acc}|0]$. For $b = 1$, we have

$$d\big(q(x, y|a, 1), p(x_1, y_1|a)\big) = d\big(p(x_0, y_1|a), p(x_1, y_1|a)\big) \leq \varepsilon$$

and since the statistical distance does not increase under data processing, it follows that $\mathrm{Prob}_p[\mathrm{Acc}|1]$ and $\mathrm{Prob}_q^*[\mathrm{Acc}|1]$ are ε-close; this proves the claim. $\square$

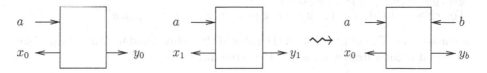

Fig. 3. Defining the strategy q by gluing together $p(x_0, y_0|a)$ and $p(x_1, y_1|a)$.

The bound on the binding property in Theorem 1 is tight, as the following theorem shows. The proof is given in the full version [6].

Theorem 2. *For all $\varepsilon \in \mathbb{Q}$ such that $0 < \varepsilon \leq 1$ there exists a classical simple two-prover commitment scheme that is perfectly sound, ε-hiding and $(1 - \varepsilon)$-binding against non-signaling adversaries.*

4.2 Arbitrary Schemes

We now remove the restriction on the scheme to be simple. As before, we first consider the case of a perfectly hiding scheme.

Theorem 3. *Let* Com *be a single-round two-prover commitment scheme. If* Com *is perfectly hiding, then there exists a non-signaling two-prover strategy* $q(x, x', y, y'|a, a', b, b')$ *such that*

$$\text{Prob}_q^*[\text{Acc}|b] = \text{Prob}_p[\text{Acc}|b]$$

for $b \in \{0, 1\}$.

Proof. Com being perfectly hiding means that $d(p(x_0, x_0'|a, a'), p(x_1, x_1'|a, a')) = 0$ for all a and a'. Gluing together the distributions $p(x_0, x_0', y_0, y_0'|a, a')$ and $p(x_1, x_1', y_1, y_1'|a, a')$ along (x_0, x_0') and (x_1, x_1') for every (a, a'), we obtain a distribution $p(x_0, x_0', x_1, x_1', y_0, y_0', y_1, y_1'|a, a')$ with the correct marginals and $p((x_0, x_0') \neq (x_1, x_1')|a, a') = 0$. That is, we have $x_0 = x_1$ and $x_0' = x_1'$ with certainty. We now define a strategy for dishonest provers as (Fig. 4).

$$q(x, x', y, y'|a, a', b, b') := p(x_0, x_0', y_b, y_{b'}'|a, a').$$

Since $p(x_0, x_0', y_b, y_b'|a, a') = p(x_b, x_b', y_b, y_b'|a, a')$, it holds that $\text{Prob}_q^*[\text{Acc}|b] = \text{Prob}_p[\text{Acc}|b]$. It remains to show that this distribution satisfies the non-signaling and causality constraints (C1) up to (NS2) of Definition 2. This is done below.

- For (C1), note that summing up over y and y' yields $q(x, x'|a, a', b, b') = p(x_0, x_0'|a, a')$, which indeed does not depend on b and b'.
- For (NS1), note that $q(x, y|a, a', b, b') = p(x_0, y_b|a, a') = p(x_b, y_b|a, a') = p(x_b, y_b|a)$, where the last equality holds by the non-signaling property of $p(x_b, y_b|a, a')$.
- For (C2), first note that

$$q(x, x', y|a, a', b, b') = p(x_0, x_0', y_b|a, a') \tag{2}$$

which does not depend on b'. We then see that (C2) holds by dividing by $q(x, y|a, a', b, b') = p(x_0, y_b|a, a')$.
- For (NS2), divide Eq. (2) by $q(x, x'|a, a', b, b') = p(x_0, x_0'|a, a')$

The properties (C1) to (NS2) with the roles of the primed and unprimed variables exchanged follows from symmetry. This concludes the proof.   □

The case of non-perfectly hiding schemes is more involved. At first glance, one might expect that by proceeding analogously to the proof of Theorem 3 — i.e., gluing together $p(x_0, x_0', y_0, y_0'|a, a')$ and $p(x_1, x_1', y_1, y_1'|a, a')$ along (x_0, x_0') and (x_1, x_1') and defining q the same way — one can obtain a strategy q that succeeds with probability $1 - \varepsilon$ if the scheme is ε-hiding. Unfortunately, this approach fails because in order to show (NS1) we use that $p(x_0, y_1|a, a') = p(x_1, y_1|a, a')$ which in general does not hold for commitment schemes that are not perfectly hiding. As a consequence, our proof is more involved, and we have a constant-factor loss in the parameter.

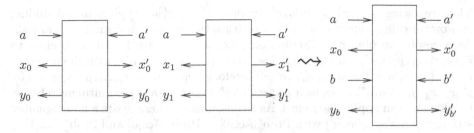

Fig. 4. Defining q from $p(x_0, x_0', y_0, y_0'|a, a')$ and $p(x_1, x_1', y_1, y_1'|a, a')$ glued together.

Theorem 4. *Let* Com *be a single-round two-prover commitment scheme and suppose that it is ε-hiding. Then there exists a non-signaling two-prover strategy $q(x, x', y, y'|a, a', b, b')$ such that*

$$\mathrm{Prob}_q^*[\mathrm{Acc}|0] = \mathrm{Prob}_p[\mathrm{Acc}|0] \quad and \quad \mathrm{Prob}_q^*[\mathrm{Acc}|1] \geq \mathrm{Prob}_p[\mathrm{Acc}|1] - 5\varepsilon.$$

Thus, if Com *is perfectly sound, it is at best $(1 - 5\varepsilon)$-binding.*

To prove this result, we use two lemmas. In the first one, we add the additional assumptions that $p(x_0|a, a') = p(x_1|a, a')$ and $p(x_0'|a, a') = p(x_1'|a, a')$. The second one shows that we can tweak an arbitrary scheme in such a way that these additional conditions hold. The proofs are given in the full version [6].

Lemma 2. *Let* Com *be a ε-hiding two-prover commitment scheme with the additional property that $p(x_0|a, a') = p(x_1|a, a')$ and $p(x_0'|a, a') = p(x_1'|a, a')$. Then, there exists a non-signaling $p'(x_1, x_1', y_1, y_1'|a, a')$ such that*

$$d\big(p'(x_1, x_1', y_1, y_1'|a, a'), p(x_1, x_1', y_1, y_1'|a, a')\big) \leq \varepsilon$$

and $p'(x_1, x_1'|a, a') = p(x_0, x_0'|a, a')$.

As usual, the non-signaling requirement on $p'(x_1, x_1', y_1, y_1'|a, a')$ is to be understood as $p'(x_1, y_1|a, a') = p'(x_1, y_1|a)$ and $p'(x_1', y_1'|a, a') = p'(x_1', y_1'|a')$.

Lemma 3. *Let* Com *be a ε-hiding two-prover commitment scheme. Then, there exists a non-signaling $\tilde{p}(x_1, x_1', y_1, y_1'|a, a')$ such that*

$$d\big(\tilde{p}(x_1, x_1', y_1, y_1'|a, a'), p(x_1, x_1', y_1, y_1'|a, a')\big) \leq 2\varepsilon$$

which has the property that $\tilde{p}(x_1|a, a') = p(x_0|a, a')$ and $\tilde{p}(x_1'|a, a') = p(x_0'|a, a')$.

With these two lemmas, Theorem 4 is easy to prove.

Proof (Theorem 4). We start with a ε-hiding non-signaling bit-commitment scheme Com. We apply Lemma 3 and obtain $\tilde{p}(x_1, x_1', y_1, y_1'|a, a')$ that is 2ε-close to $p(x_1, x_1', y_1, y_1'|a, a')$ and satisfies $\tilde{p}(x_1|a, a') = p(x_0|a, a')$ and $\tilde{p}(x_1'|a, a') = p(x_0'|a, a')$. Furthermore, by triangle inequality

$$d\big(\tilde{p}(x_1, x_1'|a, a'), p(x_0, x_0'|a, a')\big) \leq 3\varepsilon.$$

Thus, replacing $p(x_1, x_1', y_1, y_1 | a, a')$ by $\tilde{p}(x_1, x_1', y_1, y_1' | a, a')$ gives us a 3ε-hiding two-prover commitment scheme that satisfies the extra assumption in Lemma 2. As a result, we obtain a distribution $p'(x_1, x_1', y_1, y_1' | a, a')$ that is 3ε-close to $\tilde{p}(x_1, x_1', y_1, y_1' | a, a')$, and thus 5ε-close to $p(x_1, x_1', y_1, y_1' | a, a')$, with the property that $p'(x_1, x_1' | a, a') = p(x_0, x_0' | a, a')$. Therefore, replacing $\tilde{p}(x_1, x_1', y_1, y_1' | a, a')$ by $p'(x_1, x_1', y_1, y_1' | a, a')$ gives us a *perfectly-hiding* two-prover commitment scheme, to which we can apply Theorem 3. As a consequence, there exists a non-signaling strategy $q(x, x', y, y' | a, a')$ with $\mathrm{Prob}_q^*[\mathsf{Acc}|0] = \mathrm{Prob}_p[\mathsf{Acc}|0]$ and $\mathrm{Prob}_q^*[\mathsf{Acc}|1] \geq \mathrm{Prob}_p[\mathsf{Acc}|1] - 5\varepsilon$, as claimed.

Remark 5. If Com already satisfies $p(x_0 | a, a') = p(x_1 | a, a')$ and $p(x_0' | a, a') = p(x_1' | a, a')$, we can apply Lemma 2 right away and thus get a strategy q with $\mathrm{Prob}_q^*[\mathsf{Acc}|0] = \mathrm{Prob}_p[\mathsf{Acc}|0]$ and $\mathrm{Prob}_q^*[\mathsf{Acc}|1] \geq \mathrm{Prob}_p[\mathsf{Acc}|1] - \varepsilon$. Thus, with this additional condition, we still obtain a tight bound as in Theorem 1.

4.3   Multi-round Schemes

We briefly discuss a limited extension of our impossibility results for single-round schemes to schemes where during the commit phase, there is multi-round interaction between the verifier V and the two provers P and Q. We still assume the opening phase to be one-round; this is without loss of generality in case of *classical* two-prover commitment schemes (where the honest provers are restricted to be classical). In this setting, we have the following impossibility result, which is restricted to perfectly-hiding schemes.

Theorem 5. *Let* Com *be a multi-round two-prover commitment scheme. If* Com *is perfectly hiding, then there exists a non-signaling two-prover strategy that completely breaks the binding property, in the sense of Theorem 3.*

A formal proof of this statement requires a definition of n-round non-signaling bipartite systems for arbitrary n. Such a definition can be based on the intuition that it must be possible to query the left and right subsystem in any order. With this definition, the proof is a straightforward extension of the proof of Theorem 3: the non-signaling strategy is obtained by gluing together $p(\mathbf{x}_0, \mathbf{x}_0' | \mathbf{a}, \mathbf{a}')$ and $p(\mathbf{x}_1, \mathbf{x}_1' | \mathbf{a}, \mathbf{a}')$ along $(\mathbf{x}_0, \mathbf{x}_0')$ and $(\mathbf{x}_1, \mathbf{x}_1')$, and setting $q(\mathbf{x}, \mathbf{x}', y, y' | \mathbf{a}, \mathbf{a}', b, b') := p(\mathbf{x}_0, \mathbf{x}_0', y_b, y_{b'}' | \mathbf{a}, \mathbf{a}')$, where we use bold-face notation for the vectors that collect the messages sent during the multi-round commit phase: $\mathbf{a}$ collects all the messages sent by the verifier to the prover P, etc.

As far as we see, the proof of the non-perfect case, i.e. Theorem 4, does not generalize immediately to the multi-round case. As such, proving the impossibility of *non-perfectly-hiding multi-round* two-prover commitment schemes remains an open problem.

5   Possibility of Three-Prover Commitments

It turns out that we can overcome the impossibility results by adding a third prover. We will describe a scheme that is perfectly sound, perfectly hiding and

2^{-n}-binding with communication complexity $O(n)$. We now define what it means for three provers to be non-signaling; since our scheme is similar to a simple scheme, we can simplify this somewhat. We consider distributions $q(x, y, z|a, b, c)$ where a and x are input and output of the first prover P, b and y are input and output of the second prover Q and c and z are input and output of the third prover R.

Definition 8. *A conditional distribution* $q(x, y, z|a, b, c)$ *is called a non-signaling (one-round) tripartite system if it satisfies*

$$q(x|a, b, c) = q(x|a) , \quad q(y|a, b, c) = q(y|b) , \quad q(z|a, b, c) = q(z|c) ,$$

and

$$q(x, y|a, b, c) = q(x, y|a, b), \, q(x, z|a, b, c) = q(x, z|a, c), \, q(y, z|a, b, c) = q(y, z|b, c).$$

In other words, for any way of viewing q as a bipartite system by dividing in- and outputs consistently into two groups, we get a non-signaling bipartite system.

We restrict to *simple* schemes, where during the commit phase, only P is active, sending x upon receiving a from the verifier, and during the opening phase, only Q and R are active, sending y and z to the verifier, respectively.

Definition 9. *A simple three-prover commitment scheme* Com *consists of a probability distribution* $p(a)$, *two distributions* $p_0(x, y, z|a)$ *and* $p_1(x, y, z|a)$, *and an acceptance predicate* $\mathsf{Acc}(x, y, z|a, b)$.

It is called classical/quantum/non-signaling if $p_b(x, y, z|a)$ *is, when understood as a tripartite system* $p_b(x, y, z|a, \emptyset, \emptyset)$ *with two "empty" inputs.*

Soundness and the hiding-property are defined in the obvious way. As for the binding property, for a simple three-prover commitment scheme Com and a non-signaling strategy $q(x, y, z|a, b, c)$, let

$$\mathrm{Prob}_q^*[\mathsf{Acc}|b] = \sum_{a, x, y, z} p(a) \cdot q(x, y, z|a, b, b) \cdot \mathsf{Acc}(x, y, z|a, b) .$$

We say that Com is δ-binding if

$$\mathrm{Prob}_q^*[\mathsf{Acc}|0] + \mathrm{Prob}_q^*[\mathsf{Acc}|1] \leq 1 + \delta.$$

Theorem 6. *For every positive integer n, there exists a classical simple three-prover commitment scheme that is perfectly sound, perfectly hiding and 2^{-n}-binding. The verifier communicates n bits to the first prover and receives n bits from each prover.*

The scheme that achieves this is essentially the same as the example two-prover scheme described in the introduction, except that we add a third prover that imitates the actions of the second. To be more precise: the provers P, Q and R have as shared randomness a uniformly random $r \in \{0, 1\}^n$. The verifier V chooses a uniformly random $a \in \{0, 1\}^n$ and sends it to P. As commitment, P

returns $x := r \oplus a \cdot b$. To open the commitment to b, Q and R send $y := r$ and $z := r$ to V who accepts if and only if $y = z$ and $x = y \oplus a \cdot b$.

Before beginning with the formal proof that this scheme has the properties stated in our theorem, we give some intuition. Let a and x be the input and output of the dishonest first prover, P. To succeed, the second prover Q has to produce output $x \oplus a \cdot b$ where b is the second prover's input and the third prover R has to produce $x \oplus a \cdot c$ where c is the third prover's input. Our theorem implies that a strategy which always produces these outputs must be signaling. Why is that the case?

In the game that defines the binding-property, we always have $b = c$, but the dishonest provers must obey the non-signaling constraint even in the "impossible" case that $b \neq c$. Let us consider the XOR of Q's output and R's output in the case that $b \neq c$: we get $(x \oplus a \cdot b) \oplus (x \oplus a \cdot c) = a \cdot b \oplus a \cdot c = a$. But in the non-signaling setting, the joint distribution of Q's and R's output may not depend on a. Thus, the strategy we suggested does not satisfy the non-signaling constraint. Let us now prove the theorem.

Proof (Theorem 6). It is easy to see that the scheme is sound. Furthermore, for every fixed a and b, $p_b(x|a)$ is uniform, so the scheme is perfectly hiding. Now consider a non-signaling strategy q for dishonest provers. The provers succeed if and only if $y = z = x \oplus a \cdot b$. Define $q(a, x, y, z|b, c) = p(a) \cdot q(x, y, z|a, b, c)$. The non-signaling property implies that

$$q(y = x \oplus a \cdot b|a, b, c = 0) = q(y = x \oplus a \cdot b|a, b, c = 1) \quad \text{and} \quad (3)$$

$$q(z = x \oplus a \cdot c|a, b = 0, c) = q(z = x \oplus a \cdot c|a, b = 1, c). \quad (4)$$

It follows that

$$
\begin{aligned}
\text{Prob}_q^*[\text{Acc}|0] &+ \text{Prob}_q^*[\text{Acc}|1] \\
&= q(y = x \oplus a \cdot b, z = x \oplus a \cdot c|b = 0, c = 0) \\
&\quad + q(y = x \oplus a \cdot b, z = x \oplus a \cdot c|b = 1, c = 1) \\
&\leq q(y = x \oplus a \cdot b|b = 0, c = 0) + q(z = x \oplus a \cdot c|b = 1, c = 1) \\
&= q(y = x \oplus a \cdot b|b = 0, c = 1) + q(z = x \oplus a \cdot c|b = 0, c = 1) \\
&\quad \text{by Eqs. (3) and (4)} \\
&\leq 1 + q(y = x \oplus a \cdot b, z = x \oplus a \cdot c|b = 0, c = 1) \text{ by Eq. (1)}
\end{aligned}
$$

It now remains to upper-bound $q(y = x \oplus a \cdot b, z = x \oplus a \cdot c|b = 0, c = 1)$. Since $p(a)$ is uniform and $q(y, z|a, b, c)$ is independent of a, we have

$$q(y = x \oplus a \cdot b, z = x \oplus a \cdot c|b = 0, c = 1) \leq q(y \oplus z = a|b = 0, c = 1) = \frac{1}{2^n}$$

and thus our scheme is 2^{-n}-binding.   □

Remark 6. The three-prover scheme above has the drawback that *two* provers are involved in the opening phase; as such, there needs to be *agreement* on whether to open the commitment or not; if there is disagreement then this may be problematic in certain applications. However, P and Q are not allowed to communicate. One possible solution is to have V forward an *authenticated* "open" or "not open" message from P to Q and R. This allows for some communication from P to Q and R, but if the size of the authentication tag is small enough compared to the security parameter of the scheme, i.e., n, then security is still ensured.

Acknowledgements. We would like to thank Claude Crépeau for pointing out the issue addressed in Remark 6 and the solution sketched there, and Jed Kaniewski for helpful discussions regarding relativistic commitments.

References

1. John, S.B.: On the Einstein-Podolsky-Rosen paradox. Physics **1**, 195–200 (1964)
2. Ben-Or, M., Goldwasser, S., Kilian, J., Wigderson, A.: Multi-Prover Interactive Proofs: how to Remove Intractability Assumptions. In: Simon, J. (ed.) STOC, pp. 113–131. ACM (1988)
3. Clauser, J.F., Horne, M.A., Shimony, A., Holt, R.A.: Proposed experiment to test local hidden-variable theories. Phys. Rev. Lett. **23**, 880–884 (1969)
4. Crépeau, C., Salvail, L., Simard, J.-R., Tapp, A.: Two provers in isolation. In: Lee, D.H., Wang, X. (eds.) ASIACRYPT 2011. LNCS, vol. 7073, pp. 407–430. Springer, Heidelberg (2011)
5. Dumais, P., Mayers, D., Salvail, L.: Perfectly concealing quantum bit commitment from any quantum one-way permutation. In: Preneel, B. (ed.) EUROCRYPT 2000. LNCS, vol. 1807, pp. 300–315. Springer, Heidelberg (2000)
6. Fehr, S., Fillinger, M.: Multi-Prover Commitments Against Non-Signaling Attacks. ArXiv e-prints (2015). http://arxiv.org/abs/1505.03040
7. Kalai, Y.T., Raz, R., Rothblum, R.D.: How to delegate computations: the power of no-signaling proofs. In: Shmoys, D.B. (ed.) STOC, pp. 485–494. ACM (2014)
8. Kent, A.: Unconditionally secure bit commitment. Phys. Rev. Lett. **83**, 1447–1450 (1999)
9. Lo, H.-K., Chau, H.F.: Is quantum bit commitment really possible? Phys. Rev. Lett. **78**, 3410–3413 (1997)
10. Mayers, D.: Unconditionally secure quantum bit commitment is impossible. Phys. Rev. Lett. **18**, 3414–3417 (1997)
11. Popescu, S., Rohrlich, D.: Quantum nonlocality as an axiom. Found. Phys. **24**(3), 379–385 (1994)
12. Rastall, P.: Locality, bell's theorem, and quantum mechanics. Found. Phys. **15**(9), 963–972 (1985)
13. Renner, R.S., König, R.: Universally composable privacy amplification against quantum adversaries. In: Kilian, J. (ed.) TCC 2005. LNCS, vol. 3378, pp. 407–425. Springer, Heidelberg (2005)
14. Tsirelson, B.S., Khalfin, L.A.: Quantum and quasi-classical analogs of Bell inequalities. In: Symposium on the Foundations of Modern Physics, pp. 441–460 (1985)
15. Yang, N.: Zero-Knowledge Multi-Prover Interactive Proofs. Master's thesis, Concordia University Montreal (2013)

Arguments of Proximity
[Extended Abstract]

Yael Tauman Kalai[1]([✉]) and Ron D. Rothblum[2]

[1] Microsoft Research, Cambridge, USA
yael@microsoft.com
[2] Weizmann Institute of Science, Rehovot, Israel
ron.rothblum@weizmann.ac.il

Abstract. An interactive proof of proximity (IPP) is an interactive protocol in which a prover tries to convince a *sublinear-time* verifier that $x \in \mathcal{L}$. Since the verifier runs in sublinear-time, following the property testing literature, the verifier is only required to reject inputs that are *far* from $\mathcal{L}$. In a recent work, Rothblum *et. al* (STOC, 2013) constructed an IPP for every language computable by a low depth circuit.

In this work, we study the computational analogue, where soundness is required to hold only against a *computationally bounded* cheating prover. We call such protocols *interactive arguments of proximity*.

Assuming the existence of a sub-exponentially secure **FHE** scheme, we construct a *one-round* argument of proximity for *every language* computable in time t, where the running time of the verifier is $o(n)+$polylog(t) and the running time of the prover is poly(t).

As our second result, assuming sufficiently hard cryptographic PRGs, we give a lower bound, showing that the parameters obtained both in the IPPs of Rothblum *et al.*, and in our arguments of proximity, are close to optimal.

Finally, we observe that any one-round argument of proximity immediately yields a one-round delegation scheme (without proximity) where the verifier runs in *linear* time.

1 Introduction

With the prominent use of computers, tremendous amounts of data are available. For example, hospitals have massive amounts of medical data. This data is very precious as it can be used, for example, to learn important statistics about various diseases. This data is often too large to store locally, and thus is often stored on cloud platforms (or external servers). As a result, if a hospital (which has bounded storage and bounded computational power), wishes to perform some computation on its medical data, it would need to delegate this computation to the cloud. Since the cloud's computation may be faulty, the party delegating the computation (say, the hospital), may want a proof that the computation was done correctly. It is important that this proof can be verified very efficiently,

© International Association for Cryptologic Research 2015
R. Gennaro and M. Robshaw (Eds.): CRYPTO 2015, Part II, LNCS 9216, pp. 422–442, 2015.
DOI: 10.1007/978-3-662-48000-7_21

and that the prover's running time is not much larger than the time it takes to perform the computation, since otherwise, the solution will not be practical.

This problem is closely related to the problem of computation delegation, where a weak client delegates a computation to a powerful server, and the server needs to provide the client with a proof that the computation was done correctly. In contrast to the current setting, in the setting of computation delegation, the input is thought of as being small and the computation is thought of as being large. The client (verifier) is required to run in time that is proportional to the input size (but much smaller than the time it takes to do the computation), and the powerful server (prover) runs in time polynomially related to the time it takes to do the computation. Indeed the problem of computation delegation is extremely important, and received a lot of attention (e.g., [GKR08, Mic94, Gro10, GGP10, CKV10, AIK10, GLR11, Lip12, BCCT12a, DFH12, BCCT12b, GGPR12, PRV12, KRR13a, KRR13b]).

In reality, however, the input (data) is often very large, and the client cannot even store the data. Hence, we seek a solution in which the client runs in time that is *sub-linear* in the input size. The question is:

If the client cannot read the data, how can he verify the correctness of a computation on the data?

The work of [CKLR11], on memory delegation, considers this setting where the input (thought of as the client's memory) is large, and the client cannot store it locally. However, in memory delegation, it is assumed that the client (verifier) stores a short "commitment" of the input, and then can verify computations in sub-linear time. However, computing such a commitment takes time at least linear in the input length, which is infeasible in many settings.

Recently, Rothblum, Vadhan and Wigderson [RVW13], in their work on interactive proofs of proximity (IPP, a notion first studied by Ergün, Kumar and Rubinfeld [EKR04]), provide a solution where the verifier does not need to know such a commitment. Without such a commitment, the verifier cannot be sure that the computation is correct (since he cannot read the entire input), however they guarantee that the input is "close" to being correct. More specifically, they construct an interactive proof system for every language computable by a (log-space uniform) low depth circuit, where the verifier is given *oracle access* to the input (the data), and the verifier can check whether the input is *close* to being in the language in *sub-linear* time in the input (and linear time in the depth of the computation). We note that in many settings where the data is large (such as medical data) and the goal is to compute some statistics on this data, an approximate solution is acceptable. The work of [RVW13] is the starting point of our work.

1.1   Our Results in a Nutshell

We depart from the interactive proof of proximity setting, and consider *arguments of proximity*. In contrast to proofs of proximity, in an argument of proximity, soundness is required to hold only against *computationally bounded* cheating provers. Namely, the soundness guarantee is that any bounded cheating prover can convince the verifier to accept an input that is far from the language (in Hamming distance) only with small probability. By relaxing the power of the prover we obtain stronger results.

We construct *one-round* arguments of proximity for every deterministic language (without a dependency on the depth). Namely, fix any $t = t(n)$ and any language $\mathcal{L} \in \mathsf{DTIME}(t(n))$, we construct a one-round argument of proximity for $\mathcal{L}$ where the verifier runs in time $o(n) + \mathsf{polylog}(t)$, assuming the existence of a sub-exponentially secure fully homomorphic encryption (FHE) scheme.

Our one-round argument of proximity is constructed in two steps, and follows the outline of the recent works of Kalai *et al.* [KRR13a, KRR13b]. These works first show how to construct an MIP for all deterministic languages, that is sound against *no-signaling strategies*. Such no-signaling soundness is stronger than the typical notion of soundness, and is inspired by quantum physics and by the principal that information cannot travel faster than light (see Sect. 3.2 for the definition, and [KRR13a, KRR13b] for more background on this notion). They then show how to convert these no-signaling MIPs into one-round arguments.

As our first step, we combine the IPPs of [RVW13], and the no-signaling MIP construction of [KRR13b], to obtain a no-signaling *multi-prover interactive proof of proximity* (MIPP). This construction combines techniques and results of [RVW13] and [KRR13b], and may be of independent interest.

Then, similarly to [KRR13a], we show how to convert any no-signaling MIPP to a one-round argument of proximity. This transformation relies on a heuristic developed by Aiello *et al.* [ABOR00], which uses a (computational) PIR scheme (or a fully homomorphic encryption scheme) to convert any MIP into a one-round argument. This heuristic was proven to be secure in [KRR13a] if the underlying MIP is secure against no-signaling strategies. We extend the result of [KRR13a] to the proximity setting.

Finally, we provide a negative result, which shows that the parameters we obtain for MIPP and the parameters obtained in [RVW13], are somewhat tight. Proving such a lower bound was left as an open problem in [RVW13]. This part contains several new ideas, and is the main technical contribution of this work.

We also show that the parameters in our one-round argument of proximity are somewhat optimal, for arguments with adaptive soundness and are proven to be (adaptively) sound via a black-box reduction to a falsifiable assumption. See the full version for further details.

Linear-Time Delegation. We observe that both proofs and arguments of proximity, aside from being natural notions, can also be used as tools to obtain new results for delegating computation in the standard setting (i.e., where soundness is guaranteed for *every* $x \notin \mathcal{L}$). More specifically, using our results on arguments of proximity and the [RVW13] results on interactive proofs of proximity for low-depth circuits, we can construct (standard) one-round argument-systems for any deterministic computation, and interactive proof systems for low-depth circuits, where the verifier truly runs in *linear-time*. In contrast, the results of [GKR08] and [KRR13b] only give a *quasi-linear* time verifier.[1]

1.2   Our Results in More Detail

Our main result is a construction of a one-round argument of proximity for any deterministic language. Here, and throughout this work, we use n to denote the input length. Let $t = t(n)$, let $\mathcal{L} \in \mathsf{DTIME}(t)$ be a language. For a proximity parameter $\varepsilon = \varepsilon(n) \in (0, 1)$, we denote by ε-IPP an interactive proof for testing ε-proximity to $\mathcal{L}$.[2] Similarly we denote by ε-MIPP a multi-prover interactive proof for testing ε-proximity to $\mathcal{L}$.

Theorem 1 (Informal). *Suppose that there exists a sub-exponentially secure* FHE. *Fix a proximity parameter $\varepsilon \stackrel{\text{def}}{=} n^{-(1-\beta)}$, for some sufficiently small $\beta > 0$, and a security parameter τ (polynomially related to n).*

There exists a 1-round argument of ε-proximity for $\mathcal{L}$, where the verifier runs in time $n^{1-\gamma} + \mathsf{polylog}(t) + \mathsf{poly}_{\mathsf{FHE}}(\tau)$, where $\gamma > 0$ is a constant and $\mathsf{poly}_{\mathsf{FHE}}$ is a polynomial that depends only on the FHE *scheme, and makes $n^{1-\gamma} + \mathsf{polylog}(t)$ oracle queries to the main input. The prover runs in time $\mathsf{poly}(t)$. The total communication is of length $\mathsf{poly}_{\mathsf{FHE}}(\tau)$.*

Note that for languages in $\mathsf{DTIME}(2^{n^\alpha})$ for sufficiently small $\alpha > 0$ (and in particular for languages in P), the verifier in Theorem 1 runs in *sub-linear* time.

As mentioned previously, this result is obtained in two steps. We first construct an MIPP that is sound against no-signaling strategies, and then show how to convert any such MIPP into a one-round argument of proximity.

Theorem 2 (Informal). *Fix a proximity parameter $\varepsilon = \varepsilon(n) \in (0, 1)$. There exists an ε-MIPP that is secure against no-signaling strategies, where the verifier makes $q = (1/\varepsilon)^{1+o(1)}$ oracle queries to the input, the communication complexity $c = (\varepsilon n)^2 \cdot n^{o(1)} \cdot \mathsf{polylog}(t)$ and the running time of the verifier is $(\varepsilon n)^2 \cdot \mathsf{polylog}(t) + \left(\frac{1}{\varepsilon} + \varepsilon n\right)^{1+o(1)}$.*

We then show how to convert any no-signaling ε-MIPP to a one-round argument of ε-proximity. In the following we say that a fully homomorphic encryption scheme (FHE) is (T, δ) secure if every family of circuits of size T can break the semantic security of the FHE with probability at most δ.

Theorem 3 (Informal). *Fix a proximity parameter $\varepsilon = \varepsilon(n) \in (0, 1)$. Suppose that the language $\mathcal{L}$ has an ℓ-prover ε-MIPP that is sound against δ-no-signaling strategies, with communication complexity c. Suppose that there exists a $(T, \delta/\ell)$-secure* FHE, *where $T \geq 2^c$. Then $\mathcal{L}$ has a 1-round argument of ε-proximity where*

---

[1] Actually, by an observation of Vu *et al.* [VSBW13] (see also [Tha13, Lemma 3]), the verifier in the [GKR08] protocol can be directly implemented in linear-time. However the latter implementation would only guarantee *constant* soundness error.

[2] A string $x \in \{0, 1\}^n$ is ε-close to $\mathcal{L}$ if there exists $x' \in \{0, 1\}^n \cap \mathcal{L}$ such that $\triangle(x, x') \leq \varepsilon n$, where $\triangle$ denotes the Hamming distance between the two strings.

the running time of the prover and verifier and the communication complexity of the argument system, are proportional to those of the underlying MIPP *scheme.*

We note that the parameters in Theorem 2 are somewhat similar to the parameters of the interactive proof of proximity (IPP) in [RVW13]. In particular, in both constructions it holds that $c \cdot q = \Omega(n)$. The work of [RVW13] shows that this lower bound of $c \cdot q = \Omega(n)$ is inherent for IPPs with 2-messages (and that a weaker bound holds for IPPs with a constant number of rounds), and left open the question of whether this lower bound is inherent for general (multi-round) IPPs.

We resolve this question by showing that for every ε-IPP, and every ε-MIPP that is sound against no-signaling strategies, it must be the case that $c \cdot q = \Omega(n)$. For this result we assume the existence of exponentially hard pseudorandom generators.

Theorem 4 (Informal). *Assume the existence of exponentially hard pseudorandom generators. There exists a constant $\varepsilon > 0$ such that for every $q = q(n) \leq n$, there exists a language $\mathcal{L} \in$ P such that for every ε-IPP for $\mathcal{L}$, and for every ε-MIPP for $\mathcal{L}$ that sound against no-signaling adversaries, it holds that $q \cdot c = \Omega(n)$, where q is the query complexity and c is the communication complexity.*

In fact, assuming a slightly stronger cryptographic assumption, we can replace $\mathcal{L} \in$ P with $\mathcal{L} \in$ NC$_1$ (which shows that the [RVW13] upper bound for log-space uniform NC is essentially tight). See Sect. 4 for details.

We note that the [RVW13] lower bound for 2-message IPPs is unconditional (and in particular they do not assume that the verifier is *computationally bounded*). It remains an interesting open problem to obtain an *unconditional* lower bound for multi-message IPPs.

The parameters we obtain for the one-round argument also satisfy $q \cdot c = \Omega(n)$. We show that these parameters are close to optimal for arguments with adaptive soundness, that are proven sound via a black-box reduction to falsifiable assumptions. We refer the reader to the full version for details.

Finally, using the [RVW13] protocol or the protocol of Theorem 1 we construct delegation schemes in which the verifier runs in *linear-time*.

Theorem 5 (Informal). *For every language in (logspace-uniform) NC there exists an interactive proof system in which the verifier runs in time $O(n)$ and the prover runs in time* poly(n).

Theorem 6 (Informal). *Assume that there exists a sub-exponentially secure* FHE. *Then, for every language in P there exists a 1-round argument-system in which the verifier runs in time $O(n)$ and the prover runs in time* poly(n).

1.3   Related Work

As mentioned above, the work of [RVW13] and [KRR13a, KRR13b] are most related to ours. Both our work, and the work of [RVW13], lie in the intersection of property-testing and computation delegation. As opposed to property testing, where an algorithm is required to decide whether an input is close to the language *on its own* in sub-linear time, in our work the algorithm receives a proof, and only needs to verify correctness of the proof in sub-linear time. Thus, our task is significantly easier than the task in property testing. Indeed we get much stronger results. In particular, the works on property testing typically get sub-linear algorithms for specific languages, whereas our result holds *for all deterministic languages.*[3]

Another very related problem is that of constructing a *probabilistically checkable proof of proximity* (PCPP) [BSGH+06] (also known as *assignment testers* [DR06]). A PCPP consists of a prover who publishes a long proof, and a verifier, who gets oracle access to this proof and to the instance x, and needs to decide whether x is close to the language in sub-linear time. The significant difference between PCPP and proofs/argument of proximity is that in the PCPP setting the proof is a fixed string (and cannot be modified adaptively based on the verifier's messages).

The fundamental works of Kilian and Micali [Kil92, Mic94] show how to convert any probabilistically checkable proof (PCP) into a 2-round (4-message) argument. As pointed out by [RVW13], their transformation can be also used to convert any PCPP into a 2-round argument of proximity. Thus, obtaining a 2-round argument of proximity follows immediately by applying the transformation of [Kil92, Mic94] to any PCPP construction. Moreover, the parameters of the resulting 2-round argument are optimal (up to logarithmic factors); i.e., the query complexity, the communication complexity and the runtime of the verifier is $\mathrm{poly}(\log(t), \tau)$ where t is the time it takes to compute if x is in the language, and where τ is the security parameter.

The focus of this work is on constructing *one-round* arguments of proximity. Unfortunately, our parameters do not match those of the two-round arguments of proximity outlined above. However, we show that using our techniques (i.e., of constructing one-round arguments of proximity from no-signaling MIPPs), our parameters are almost optimal.

Other works that are related to ours are the work of Gur and Rothblum [GR13] on non-interactive proofs of proximity, and of Fischer *et al.* [FGL14] on partial testing. The former studies an NP version of property testing (which can be thought of as a 1-message variant of IPP), whereas the latter studies a model of property testing in which the tester needs to only accept a sub-property (we note that the two notions, which were developed independently, are tightly related, see [GR13, FGL14] for details).

Organization. In this extended abstract we give an overview of our techniques and only prove some of our results. In Sect. 2 we give a high level view of our techniques. In Sect. 3 we formally define arguments of proximity and the other central definitions that are used throughout this work. In Sect. 4 we show our

---

[3] Indeed, as shown by Goldwasser, Goldreich and Ron [GGR98], there are properties in very low complexity classes that require $\Omega(n)$ queries and running-time in order to test (without the help of a prover).

lower bound for no-signaling MIPPs. See the full version for the missing proofs and formal theorem statements.

2   Our Techniques

2.1   Our Positive Results

To construct arguments of proximity for languages in $\mathsf{DTIME}(t)$, we adapt the technique of [KRR13a] to the "proximity" setting. That is, we first construct an MIPP that has soundness against no-signaling strategies and then employ the technique of Aiello *et al.* [ABOR00] to obtain an argument of proximity. We elaborate on these two steps below. In what follows, we focus for simplicity on languages in P, though everything extends to languages in $\mathsf{DTIME}(t)$.

No-Signaling MIPPs for P. Our first step (which is technically more involved) is a construction of MIPPs that are sound against no-signaling strategies for any language $\mathcal{L} \in \mathsf{P}$. This construction is inspired by (and reminiscent of) the IPP construction of [RVW13]. The starting point for the [RVW13] IPP is the "Muggles" protocol of Goldwasser *et al.* [GKR08], whereas our starting point is the no-signaling MIP of [KRR13b].

The main technical difficulty in using both the [GKR08] and [KRR13b] protocols by a sublinear time verifier is that in both protocols, the verifier needs to compute an error corrected encoding of the input x. More specifically, the verifier needs to compute the low degree extension of x, denoted LDE_x. Since error-correcting codes are very sensitive to changes in the input, a sub-linear algorithm has no hope to compute LDE_x.

The key point is that in both the [GKR08] and the [KRR13b] protocols, it suffices for the verifier to check the value of LDE_x at relatively few *randomly* selected points (this property was also used by [CKLR11] in their work on memory delegation). Hence, it will be useful for us to view both the [GKR08] and [KRR13b] protocols as protocols for producing a sequence of points J in the low degree extension of x and a sequence of corresponding values v with the following properties:

- If $x \in \mathcal{L}$ and the prover(s) honestly follow the protocol then $\mathsf{LDE}_x(J) = v$.
- If $x \notin \mathcal{L}$ then no matter what the cheating prover does (resp., no-signaling cheating prover do), with high probability the verifier outputs J, v such that $\mathsf{LDE}_x(J) \neq v$.

Hence, the verifiers in both protocols first run this subroutine to produce J and v and then accept if and only if $\mathsf{LDE}_x(J) = v$. Remarkably, in both cases, in the protocol that produces J and v, the verifier does not need to access x.

The next step in [RVW13] is a parallel repetition of the foregoing protocol in order to reduce the soundness error. Once the soundness error is sufficiently small, [RVW13] argue that for every x that is ε-far from $\mathcal{L}$, no matter what the cheating prover does (in the parallel repetition of the base protocol), the verifier will output J, v such that not only $\mathsf{LDE}_x(J) \neq v$, but furthermore, x is far from *any* x' such that $\mathsf{LDE}_{x'}(J) = v$. This steps simply follows by taking a union bound over all x' that are close to x.

We borrow this step almost as-is from [RVW13] except for the following technical difficulty - it is not known whether parallel repetition decreases the soundness error of no-signaling MIP protocols.[4] However, we observe that the [KRR13b] protocol already allows for sufficient flexibility in choosing its soundness error so that the parallel repetition step can be avoided.

The last step of [RVW13] is designing an IPP protocol for a language that they call $\mathsf{PVAL}_{J,v}$ (for "polynomial evaluation"). This language, parameterized by J and v, consists of all strings x such that $\mathsf{LDE}_x(J) = v$. Using this IPP for PVAL, the IPP verifier for a language $\mathcal{L}$ first runs the (parallel repetition of the) [GKR08] protocol, to produce J, v as above. Then, the IPP verifier runs the $\mathsf{PVAL}_{J,v}$ protocol and accepts if and only if the PVAL-verifier accepts. If $x \in \mathcal{L}$ then we know that $\mathsf{LDE}_x(J) = v$ and therefore the PVAL-verifier will accept, whereas if x is far from $\mathcal{L}$ then x is far from $\mathsf{PVAL}_{J,v}$ and therefore the PVAL-verifier will reject. Hence the (parallel repetition of the) [GKR08] protocol is sequentially composed with the IPP for PVAL.

For the no-signaling case, we also use the [RVW13] IPP protocol for PVAL. A technical difficulty that arises is that in contrast to the IPP setting in which sequential composition (of two interactive proofs) is trivial, here we need to compose a 1-round no-signaling MIP with an IPP protocol, to produce a no-signalling MIPP. We indeed prove that such a composition holds thereby constructing a no-signaling MIPP as we desire.

From No-Signaling MIPP to Arguments of Proximity. The transformation from a no-signaling MIPP to an argument of proximity is based on the assumption that there exists a fully homomorphic encryption scheme (or alternatively, a computational private information retrieval scheme) and is practically identical to that in [KRR13a]. More specifically, the argument's verifier uses the MIPP verifier to generate a sequence of queries $q_1, \ldots, q_\ell$ to the ℓ provers. It encrypts each query using a fresh encryption key as follows: $\hat{q}_i \leftarrow Enc_{k_i}(q_i)$. The argument's verifier sends all the encrypted queries to the prover. Given $\hat{q}_1, \ldots, \hat{q}_\ell$, the prover uses the homomorphic evaluation algorithm to compute the MIPP answers "underneath" the encryption. It sends these answers back to the verifier, which can decrypt the encrypted answers and decide. As in [KRR13a] we show that if the MIPP is sound against no-signaling strategies then, assuming the semantic security of the FHE, the resulting protocol is sound against computationally bounded adversaries.

Linear-Time Delegation. We show that using the foregoing one-round argument of proximity for every language $\mathcal{L} \in \mathsf{P}$ and good error-correcting codes, one can easily construct a one-round delegation protocol where the verifier runs in *linear* time (in contrast, the verifier in [KRR13b] runs in *quasi*-linear time). A similar observation, in the context of PCPs, was previously pointed out by [EKR04].

---

[4] Holenstein [Hol09] showed a parallel repetition theorem for no-signaling 2-prover MIPs. It is not known whether this result can be extended to 3 or more provers.

Let $\mathcal{L} \in \mathsf{P}$ and consider $\mathcal{L}' = \{\mathsf{ECC}(x) \ : \ x \in \mathcal{L}\}$ where ECC is an error correcting code with constant rate, constant relative distance, linear-time encoding
and polynomial-time decoding[5]. Then, $\mathcal{L}' \in \mathsf{P}$ and so it has an argument of
proximity with a sublinear-time verifier. We construct a delegation scheme for
$\mathcal{L}$ by having both the verifier and the prover compute $x' = \mathsf{ECC}(x)$ and run
the argument of proximity protocol with respect to x'. Since the argument of
proximity verifier runs in sublinear time, and $\mathsf{ECC}(x)$ can be computed in linear-
time, the resulting delegation verifier runs in linear-time. Soundness follows from
the fact that a cheating prover that convinces the argument-system verifier to
accept $x \notin L$ can be used to convince the argument-of-proximity verifier to
accept $\mathsf{ECC}(x)$ which is indeed far from $\mathcal{L}'$.

A similar result can be obtained for interactive proofs for low-depth computation based on the results of [RVW13] by using an error-correcting code that
can be decoded in logarithmic-depth (such a code was constructed by Spielman [Spi96]).

2.2   Our Negative Results

We prove that assuming the existence of exponentially hard pseudorandom generators, there exists a constant $\varepsilon > 0$ for which there does not exist a no-signaling
ε-MIPP for all of P with query complexity q and communication complexity c
such that $q \cdot c = o(n)$ (where n is the input length). We also show a similar result
for ε-IPP.

We start by focusing on our lower bound for MIPP. The high-level idea is
the following: Suppose (towards contradiction) that every language in P has a
no-signaling MIPP with query complexity q and communication complexity c
where $q \cdot c = o(n)$. The fact that $q = o(n)$ implies that (for every language in P),
there is some set of coordinates $S \subseteq [n]$ of size $O(n/q)$ that with high (constant)
probability the verifier does not query.

As a first step, suppose for the sake of simplicity that there is a fixed (universal) set of coordinates $S \subseteq [n]$ such that with high probability the verifier never
queries the coordinates in S, for every language in P (for example, if the verifier's queries are non-adaptive and are generated before it communicates with
the prover, then such a set S must exist). We derive a contradiction by showing that one can use the no-signaling MIPP to construct a no-signaling MIP for
languages in NP\P with communication $c = o(n)$. The latter was shown to be
impossible, assuming that $\mathsf{NP} \nsubseteq \mathsf{DTIME}(2^{o(n)})$ [DLN+04] (see also [Ito10]).

The basic idea is the following: Take any language $\mathcal{L} \in \mathsf{NP} \backslash \mathsf{P}$ that is assumed
to be hard to compute in time $2^{o(n)}$, and convert it into the language $\mathcal{L}' \in \mathsf{P}$,
defined as follows: $x' \in \mathcal{L}'$ if and only if x'_S is a valid witness of $x'_{[n]\backslash S}$ in the
underlying NP language $\mathcal{L}$. The no-signaling MIP for $\mathcal{L}$ will simply be the no-
signaling ε-MIPP for $\mathcal{L}'$, where the MIP verifier simulates the ε-MIPP verifier
with oracle access to x' where $x'_{[n]\backslash S} = x$, and $x'_S = 0^{|S|}$. Note that the MIP
verifier, which takes as input x (supposedly in $\mathcal{L}$), cannot (efficiently) generate a

---

[5] Such codes are known to exist, see, e.g., [Spi96].

corresponding witness w and set $x'_S = w$. But the point is that it does not need to, since S was chosen so that with high probability the MIPP verifier for $\mathcal{L}'$ will not query x' on coordinates in S.

There are several problems with this approach. First, the witness can be very long compared to x, and the set S may be very small compared to n. In this case we will not be able to fit the entire witness in the coordinate set S. Second, after running the MIPP, the verifier is convinced that x' is close to an instance in $\mathcal{L}'$. However, this does not imply that x is in $\mathcal{L}$ (and can only imply that x is close to $\mathcal{L}$).

One can fix these two problems with a single solution: Instead of setting $x'_{[n] \setminus S} = x$ we set $x'_{[n] \setminus S} = \mathsf{ECC}(x)$, where ECC is a error-correcting code with efficient encoding, that is resilient to 2ε-fraction of errors. Now, we can take $\mathsf{ECC}(x)$ so that $|\mathsf{ECC}(x)|$ is very large compared to $|w|$, so that we can fit all of the witness in the coordinate set S. Moreover, if $|\mathsf{ECC}(x)| > |w|$ then if x' is ε-close to $\mathcal{L}'$ then $x'_{[n] \setminus S}$ is 2ε-close to $\mathcal{L}$. This, together with the fact that $\mathsf{ECC}(x)$ is resilient to 2ε-fraction of errors implies that the encoded element is indeed in $\mathcal{L}$.

The foregoing idea indeed seems to work if there was a fixed (universal) set S that the MIPP verifier does not query (with high probability). However, this is not necessarily the case, and this set S may be different for different languages in P. In particular, we cannot claim that for the language $\mathcal{L}'$ the set S is exactly where the witness lies. Namely, it may be that the verifier in the underlying MIPP always queries some coordinates in S.

We solve this problem by using repetitions. Namely, every element $x' \in \mathcal{L}'$ will consist of many instances (encoded using an error-correcting code) along with many witnesses; i.e., $x' = (\mathsf{ECC}(x_1, \ldots, x_m), w_1, \ldots, w_m)$, where each w_j is a witness for the NP statement $x_j \in \mathcal{L}$. Now, suppose that the verifier makes q queries to x' (where $q = o(n)$). Then if we take $m = 4q$ then we know that $3/4$ of the (x_j, w_j)'s are not queried.

As above, we derive a contradiction by showing that one can use the no-signaling MIPP to construct a no-signaling MIP for languages in $\mathsf{NP} \setminus \mathsf{P}$ with $o(n)$ communication, (which is known to be impossible for languages that cannot be computed in time $2^{o(n)}$ [DLN+04,Ito10]). However, now the no-signaling MIP construction will be different: Given an instance x (supposedly in $\mathcal{L}$), the MIP verifier will choose a random $i^* \in_R [m]$, along with m random instance and witness pairs $(x_1, w_1), \ldots, (x_m, w_m)$, where $x_{i^*} = x$ and w_{i^*} can be arbitrary (assumed not to be queried).

We need to argue that with probability at least $3/4$ the verifier will not query the coordinates of w_{i^*}, and thus with probability at least $3/4$ the MIP verifier will successfully simulate the MIPP verifier. If the queries of the MIPP verifier were chosen before interacting with the prover then this would follow immediately from the fact that $i^* \in [m]$ is chosen at random. However, the MIPP verifier may choose its oracle queries after interacting with the MIPP provers, and therefore we need to argue that the MIPP provers also do not know i^*. Note that the MIPP provers see all of $x_1, \ldots, x_m$. Hence, in order to claim that the provers cannot guess i^* it needs to be the case that x is distributed identically to the other $x_1, \ldots, x_m$.

Hence, we seek a language $\mathcal{L} \in \mathsf{NP}\backslash\mathsf{P}$ for which there exists a distribution $\mathcal{D}$ (distributed over $\mathcal{L}$) such that:

1. It is computationally hard to distinguish between $x \in_R \mathcal{D}$ and $x \notin \mathcal{L}$ (i.e., $\mathcal{L}$ is hard on the average); and
2. $x \in_R \mathcal{D}$ can be sampled together with a corresponding NP witness.

We note that the first requirement is needed to obtain a contradiction (and replaces the weaker assumption that $\mathcal{L} \in \mathsf{NP}\backslash\mathsf{P}$) whereas the second assumption is required so that we can sample $x_1, \ldots, x_m$ (together with the corresponding witnesses) so that MIPP protocol cannot distinguish between x and any of the x_j's (thereby hiding i^*). In can be easily verified that both requirement are met by considering $\mathcal{D}$ which is the output of a cryptographic pseudorandom generator (PRG). Hence the language $\mathcal{L}$ that we use is precisely the output of such a PRG.

Indeed, we can only argue that our no-signaling MIP has *average-case* completeness (with respect to the distribution $\mathcal{D}$), since if $x \in \mathcal{L}$ is distributed differently from $(x_1, \ldots, x_m)$ then the verifier of the MIPP may always query the coordinates where the witness of x is embedded, in which case the MIP verifier will fail to simulate. However, for random $x \in_R \mathcal{L}$ the provers (and verifier) in the MIPP cannot guess i^* with any non-negligible advantage, and therefore the verifier will not query the coordinates of w_{i^*} with probability at least $3/4$, in which case the MIP verifier will succeed in simulating the underlying ε-MIPP verifier. We refer the reader to Sect. 4 for further details.

A Lower Bound for IPP. To obtain a multiplicative lower bound for IPP, we follow the same paradigm outlined above for MIPP's with no-signaling soundness. More specifically, we consider a language $\mathcal{L} \in \mathsf{NP}$ and the corresponding language

$$\mathcal{L}' = \left\{ (\mathsf{ECC}(x_1, \ldots, x_m), w_1, \ldots, w_m) : w_j \text{ is an NP-witness for } x_j \right\}$$

as above. We show that an IPP protocol for $\mathcal{L}'$ implies a (standard) interactive-proof for $\mathcal{L}$ with similar communication complexity. Here we obtain a contradiction by arguing that (assuming exponential hardness) there are languages in $\mathsf{NP}\backslash\mathsf{P}$ for which every interactive proof require $\Omega(n)$ communication. The latter is based on the proof that $\mathsf{IP} \subseteq \mathsf{PSPACE}$ (i.e., the "easy" direction in the $\mathsf{IP} = \mathsf{PSPACE}$ theorem).

Given the [RVW13] positive result of IPP for low depth computations, we would like to show that our lower bound is not just for languages in P but even for languages, say, in NC_1 (thereby showing that the [RVW13] result is tight). To do so we observe that if (1) the error correcting code that we use has an encoding procedure that can be computed by an NC_1 circuit and (2) the cryptographic PRG can be computed in NC_1, then indeed $\mathcal{L}' \in \mathsf{NC}_1$.

A Lower Bound for One-Round Arguments of Proximity. For one-round arguments of proximity, we show a similar lower-bound of $q \cdot c = \Omega(n)$, assuming the argument has *adaptive* soundness, and the proof of (adaptive) soundness is via a *black-box reduction* to some *falsifiable* cryptographic assumption.

Loosely speaking, a cryptographic assumption is falsifiable (a notion due to Naor [Nao03]) if there is an *efficient* way to "falsify it", i.e., to demonstrate that it is false. We note that most standard cryptographic assumptions (e.g., one-way functions, public-key encryption, LWE etc.) are falsifiable. A black-box reduction of one cryptographic primitive to another, is a reduction that, using black-box access to any (possibly inefficient) adversary for the first primitive, breaks the security of the second primitive.

Similarly to the MIPP and IPP lower bounds, we consider the languages $\mathcal{L}$ and $\mathcal{L}'$, as above, where $\mathcal{L} \in$ NP is exponentially hard on average and $\mathcal{L} \in$ P. We prove that if there exists an adaptively sound one-round argument of proximity for $\mathcal{L}'$ with $q \cdot c = o(n)$ then there exists an adaptively sound one-round argument for $\mathcal{L}$ with $o(n)$ communication (in the crs model).

We then rely on a result of Gentry and Wichs [GW11], which shows that there does not exist a one-round argument for exponentially hard (on average) NP languages, with adaptive soundness and black-box reduction to a falsifiable assumption.

We conclude that P does not have an adaptively sound one-round argument of proximity with $q \cdot c = o(n)$, and a black-box reduction to a falsifiable assumption. We refer the reader to the full version for details.

3   Definitions

In this section we define arguments of proximity and MIPs of proximity (with soundness against no-signaling strategies). See the full version for additional standard definitions.

Notation. For $x, y \in \{0,1\}^n$, we denote the Hamming distance of x and y by $\Delta(x,y) \overset{\text{def}}{=} |\{i \in [n] : x_i \neq y_i\}|$. We say that x is ε-close to y if $\Delta(x,y) \leq \delta$. We say that x is ε-close to a set $S \subseteq \{0,1\}^n$ if there exists $y \in S$ such that x is ε-close to y.

If A is an oracle machine, we denote by $A^x(z)$ the output of A when given oracle access to x and explicit access to z.

For a vector $a = (a_1, \ldots, a_\ell)$ and a subset $S \subseteq [\ell]$, we denote by a_S the sequence of elements of a that are indexed by indices in S, that is, $a_S = (a_i)_{i \in S}$.

For a distribution $\mathcal{A}$, we denote by $a \in_R \mathcal{A}$ a random variable distributed according to $\mathcal{A}$ (independently of all other random variables). We will measure the distance between two distributions by their *statistical distance*, defined as half the l_1-distance between the distributions. We will say that two distributions are δ-*close* if their statistical distance is at most δ.

3.1   Arguments of Proximity

An interactive argument of proximity for a language $\mathcal{L}$ consists of a polynomial-time verifier that wishes to verify that x is close (in Hamming distance) to some

x' such that $x' \in \mathcal{L}$, and a prover that helps the verifier to decide. The verifier is given as input $n \in \mathbb{N}$, a proximity parameter $\varepsilon = \varepsilon(n) > 0$ and oracle access to $x \in \{0,1\}^n$ (and its oracle queries are counted). The prover gets as input ε and x. The two parties interact and at the end of the interaction the verifier either accepts or rejects. We require that if $x \in \mathcal{L}$ then the verifier accepts with high probability but if x is ε-far from $\mathcal{L}$, then no *computationally bounded* prover can convince the verifier to accept with non-negligible (in n) probability.

We focus on 1-round arguments of proximity systems. Such an argument-system consists of a single message sent from the verifier V to the prover P, followed by a single message sent from the prover to the verifier.

Let $\varepsilon = \varepsilon(n) \in (0,1)$ be a proximity parameter. Let $T : \mathbb{N} \to \mathbb{N}$ and $s : \mathbb{N} \to [0,1]$ be parameters. We say that (V, P) is a one-round argument of ε-proximity for $\mathcal{L}$, with soundness (T, s), if the following two properties are satisfied:

1. **Completeness:** For every $x \in \mathcal{L}$, the verifier $V^x(|x|, \varepsilon)$ accepts with overwhelming probability, after interacting with $P(\varepsilon, x)$.
2. **Soundness:** For every family of circuits $\{P_n^*\}_{n \in \mathbb{N}}$ of size $\mathsf{poly}(T(n))$ and for all sufficiently large $x \notin \mathcal{L}$, the verifier $V^x(|x|, \varepsilon)$ rejects with probability $\geq 1 - s(|x|)$, after interacting with $P_{|x|}^*(\varepsilon, x)$.

3.2   Multi-prover Interactive Proofs (MIP)

Let $\mathcal{L}$ be a language and let x be an input of length n. In a one-round ℓ-prover interactive proof, ℓ computationally unbounded provers, $P_1, \ldots, P_\ell$, try to convince a (probabilistic) $\mathsf{poly}(n)$-time verifier, V, that $x \in \mathcal{L}$. The input x is known to all parties.

The proof consists of only one round. Given x and its random string, the verifier generates ℓ queries, $q_1, \ldots, q_\ell$, one for each prover, and sends them to the ℓ provers. Each prover responds with an answer that depends only on its own individual query. That is, the provers respond with answers $a_1, \ldots, a_\ell$, where for every i we have $a_i = P_i(q_i)$. Finally, the verifier decides wether to accept or reject based on the answers that it receives (as well as the input x and its random string).

We say that $(V, P_1, \ldots, P_\ell)$ is a one-round multi-prover interactive proof system (MIP) for $\mathcal{L}$, with completeness $c \in [0,1]$ and soundness $s \in [0,1]$ (think of $s < c$) if the following two properties are satisfied:

1. **Completeness:** For every $x \in \mathcal{L}$, the verifier V accepts with probability c, over the random coins of $V, P_1, \ldots, P_\ell$, after interacting with $P_1, \ldots, P_\ell$, where c is a parameter referred to as the *completeness* of the proof system.
2. **Soundness:** For every $x \notin \mathcal{L}$, and any (computationally unbounded, possibly cheating) provers $P_1^*, \ldots, P_\ell^*$, the verifier V rejects with probability $\geq 1 - s$, over the random coins of V, after interacting with $P_1^*, \ldots, P_\ell^*$, where s is a parameter referred to as the *error* or *soundness* of the proof system.

Important parameters of an MIP are the number of provers, the length of queries, the length of answers, and the error. We say that the proof-system has *perfect completeness* If completeness hold with probability 1 (i.e. $c = 1$).

No-Signaling MIP. We will consider a variant of the MIP model, where the cheating provers are more powerful. In the MIP model, each prover answers its own query locally, without knowing the queries that were sent to the other provers. The no-signaling model allows each answer to depend on all the queries, as long as for any subset $S \subset [\ell]$, and any queries q_S for the provers in S, the distribution of the answers a_S, conditioned on the queries q_S, is independent of all the other queries.

Intuitively, this means that the answers a_S do not give the provers in S information about the queries of the provers outside S, except for information that they already have by seeing the queries q_S.

Formally, denote by D the alphabet of the queries and denote by Σ the alphabet of the answers. For every $q = (q_1, \ldots, q_\ell) \in D^\ell$, let $\mathcal{A}_q$ be a distribution over Σ^ℓ. We think of $\mathcal{A}_q$ as the distribution of the answers for queries q.

We say that the family of distributions $\{\mathcal{A}_q\}_{q \in D^\ell}$ is *no-signaling* if for every subset $S \subset [\ell]$ and every two sequences of queries $q, q' \in D^\ell$, such that $q_S = q'_S$, the following two random variables are identically distributed:

- a_S, where $a \in_R \mathcal{A}_q$
- a'_S where $a' \in_R \mathcal{A}_{q'}$

If the two distributions are δ-close, rather than identical, we say that the family of distributions $\{\mathcal{A}_q\}_{q \in D^\ell}$ is δ-*no-signaling*.

An MIP $(V, P_1, \ldots, P_\ell)$ for a language $\mathcal{L}$ is said to have soundness s against no-signaling strategies (or provers) if the following (more general) soundness property is satisfied:

2. **Soundness:** For every $x \notin \mathcal{L}$, and any no-signaling family of distributions $\{\mathcal{A}_q\}_{q \in D^\ell}$, the verifier V rejects with probability $\geq 1 - s$, where on queries $q = (q_1, \ldots, q_\ell)$ the answers are given by $(a_1, \ldots, a_\ell) \in_R \mathcal{A}_q$, and s is the soundness parameter.

If the property is satisfied for any δ-no-signaling family of distributions $\{\mathcal{A}_q\}_{q \in D^\ell}$, we say that the MIP has soundness s against δ-no-signaling strategies (or provers).

MIP of Proximity (MIPP). Let $\mathcal{L}$ be a language, let x be an input of length n (which we refer to as the main input) and let $\varepsilon = \varepsilon(n) \in (0, 1)$ be a proximity parameter. In a one-round ℓ-prover interactive proof of proximity, ℓ computationally unbounded provers, $P_1, \ldots, P_\ell$, try to convince a (probabilistic) polynomial-time verifier, V, that the input x is ε-close (in relative Hamming distance) to some $x' \in \mathcal{L}$. The provers have free access to n, ε and x. The verifier has free access to n and ε and oracle access to x (and the number of oracle queries is counted).

We say that $(V, P_1, \ldots, P_\ell)$ is a one-round multi-prover interactive proof system of ε-proximity (ε-MIPP) for $\mathcal{L}$, with completeness $c \in [0, 1]$ and soundness $s \in [0, 1]$, if the following properties are satisfied:

1. **Running Time:** The verifier runs in polynomial time, i.e., time polynomial in the communication complexity and the number of oracle queries.

2. **Completeness:** For every $x \in \mathcal{L}$ the verifier V accepts with probability c, after interacting with $P_1, \ldots, P_\ell$.
3. **Soundness:** For every x that is ε-far from $\mathcal{L}$, and any (computationally unbounded, possibly cheating) provers $P_1^*, \ldots, P_\ell^*$, the verifier V rejects with probability $\geq 1 - s$, after interacting with $P_1^*, \ldots, P_\ell^*$.

We denote such a proof system by ε-MIPP (and omit the soundness and completeness parameters from the notation). We say that the proof-system has *perfect completeness* if completeness hold with probability 1 (i.e. $c = 1$). The parameters we are mainly interested in are the query complexity and the communication complexity.

No-Signaling MIPP. An ε-MIPP, $(V, P_1, \ldots, P_\ell)$ for a language $\mathcal{L}$ is said to have soundness s against no-signaling strategies (or provers) if the following (more general) soundness property is satisfied:

2. **Soundness:** For every x that is ε-far from $\mathcal{L}$, and any no-signaling family of distributions $\{\mathcal{A}_q\}_{q \in D^\ell}$, the verifier V rejects with probability $\geq 1 - s$, where on queries $q = (q_1, \ldots, q_\ell)$ the answers are given by $(a_1, \ldots, a_\ell) \in_R \mathcal{A}_q$, and s is the error parameter.

If the property is satisfied for any δ-no-signaling family of distributions $\{\mathcal{A}_q\}_{q \in D^\ell}$, we say that the MIP has soundness s against δ-no-signaling strategies (or provers).

4   Lower Bound for No-Signaling MIPP

In this section we prove a lower bound, showing that there does not exist a no-signaling MIPP for all of P with query complexity q and communication complexity c such that $q \cdot c = o(n)$ (where n is the input length). More specifically, for every q we construct a language $\mathcal{L}$ in P and prove that if exponentially hard pseudo-random generators exist then for any no-signaling ε-MIPP for $\mathcal{L}$ with query complexity q and communication complexity c, it must be the case that $q \cdot c = \Omega(n)$. In the full version we show how to extend the result to IPPs and to arguments of proximity.

In what follows we denote by τ the security parameter.

Definition 1. *A pseudo-random generator* $G : \{0,1\}^n \to \{0,1\}^{\ell(n)}$ *(with stretch* $\ell(n) > n$*) is said to be* exponentially hard *if for every circuit family* $\{\mathcal{A}_\tau\}_\tau$ *of size* $2^{o(\tau)}$,

$$\left| \Pr_{s \in_R \{0,1\}^\tau} [\mathcal{A}_\tau(1^\tau, G(s)) = 1] - \Pr_{y \in_R \{0,1\}^{\ell(\tau)}} [\mathcal{A}_\tau(1^\tau, y) = 1] \right| = \mathrm{negl}(\tau).$$

Theorem 7. *Assume the existence of exponentially hard pseudo-random generators. There exists a constant* $\varepsilon > 0$ *such that for every* $q = q(n) \leq n$*, there exists a language* $\mathcal{L} \in \mathsf{P}$ *such that every MIPP for testing* ε-proximity to $\mathcal{L}$ *with completeness* $2/3$*, soundness* $1/3$*, query complexity* q *and communication complexity* c *it holds that* $q \cdot c = \Omega(n)$.

Remark 1. The above theorem holds with respect to any constant completeness parameter $c > 0$ and constant soundness parameter s such that $s < c$, and we chose $c = 2/3$ and $s = 1/3$ only for the sake of concreteness.

Remark 2. The assumption in Theorem 7 can be reduced to sub-exponentially hard pseudo-random generators (i.e., it is infeasible for circuits of size 2^{τ^δ} to distinguish the output of the generator from uniform, for some $\delta > 0$), rather than exponential hardness, at the cost of a weaker implication (i.e., $q \cdot c = \Omega(n^\delta)$).

Proof of Theorem 7. We start by defining the notion of average-case no-signaling MIP (in the crs model), which is used in the proof of Theorem 7. We note that this average-case completeness seems too weak for applications and we define this weak notion only for the sake of the proof of Theorem 7.

Definition 2. *An average-case no-signaling MIP in the common random string (crs) model, for a language $\mathcal{L}$, with completeness c and soundness s, consists of $(V, P_1, \ldots, P_\ell, crs)$, where as before V is the verifier, $P_1, \ldots, P_\ell$ are the provers, and crs is a common random string of length $\mathsf{poly}(n)$, chosen uniformly at random and given to all parties. In particular, V's queries and decision may depend on the crs, and the answers generated by both honest and cheating provers may depend on the crs. The following completeness and soundness conditions are required:*

- **Average-Case Completeness.** *For all sufficiently large $n \in \mathbb{N}$,*

$$\Pr\left[(V, P_1, \ldots, P_\ell)(x, crs) = 1\right] \geq c,$$

  where the probability is over uniformly distributed $x \in_R \mathcal{L} \cap \{0, 1\}^n$, over uniformly generated $crs \in_R \{0, 1\}^{\mathsf{poly}(n)}$, and over the random coin tosses of the verifier V.
- **Soundness Against No-Signaling Provers.** *For every $x \notin \mathcal{L}$, and every family of distributions $\{\mathcal{A}_{q,crs}\}_{q \in D^\ell, crs \in \{0,1\}^{\mathsf{poly}(n)}}$ such that for every $crs \in \{0, 1\}^{\mathsf{poly}(n)}$ the family of distributions $\{\mathcal{A}_{q,crs}\}_{q \in D^\ell}$ is no-signaling, the verifier V rejects with probability $\geq 1 - s$, where the answers corresponding to (q, crs) are given by $(a_1, \ldots, a_\ell) \in_R \mathcal{A}_{q,crs}$.*

The following proposition, which we use in the proof of Theorem 7, follows from [DLN+04] (see also [Ito10]).

Proposition 1. *Suppose that a language $\mathcal{L}$ has an average-case no-signaling MIP in the crs model, communication complexity $c = c(n)$ (where n is the instance length), and with constant completeness and soundness (where the soundness parameter is smaller than the completeness parameter). Then, there exists a randomized algorithm D that runs in time $\mathsf{poly}(n, 2^c)$ such that:*

- *For every $n \in \mathbb{N}$,*

$$\Pr_{x \in_R \mathcal{L} \cap \{0,1\}^n} [D(x) = 1] \geq 2/3$$

  where the probability is also over the coin tosses of D.

– For every $x \notin \mathcal{L}$ it holds that

$$\Pr[D(x) = 1] \leq 1/3$$

where the probability is over the coins tosses of D.

We note that [DLN+04,Ito10] did not consider the crs model nor average-case completeness, but the claim extends readily to this setting as well.

We are now ready to prove Theorem 7.

Proof of Theorem 7. Assume that there exists a pseudo-random generator (PRG), denoted by $G : \{0,1\}^\tau \to \{0,1\}^{2\tau}$, that is exponentially secure. Namely, every adversary of size $2^{o(\tau)}$ cannot distinguish between uniformly distributed $r \in_R \{0,1\}^{2\tau}$ and $G(s)$ for uniformly distributed $s \in_R \{0,1\}^\tau$, with non-negligible advantage. For sake of simplicity, we assume that G is injective[6].

Let $\varepsilon > 0$ be a constant for which there exists a (good) error-correcting-code, denoted by ECC, with constant rate and efficient encoding that is resilient to (2ε)-fraction of adversarially chosen errors.

Fix any query complexity $q = o(n)$.[7] We show that there exists a language $\mathcal{L} \in \mathsf{P}$ such that for every no-signaling ε-MIPP for $\mathcal{L}$ with query complexity q and communication complexity c (and completeness $\frac{2}{3}$ and soundness $\frac{1}{3}$) it must be the case that $q \cdot c = \Omega(n)$.

Consider the following language:

$$\mathcal{L} = \{(\mathsf{ECC}(r_1, \ldots, r_m), s_1, \ldots, s_m) : \forall i \in [m], G(s_i) = r_i\},$$

where $m = 4q$ and $\tau = |s_i| = \Theta(n/q)$, where $n = |(\mathsf{ECC}(r_1, \ldots, r_m), s_1, \ldots, s_m)|$. The fact that $|s_i| = \Theta(n/q)$ follows from the fact that ECC has constant rate (i.e., $|\mathsf{ECC}(z)| = O(|z|)$).

The fact that ECC is efficiently decodable and G is efficiently computable implies that $\mathcal{L} \in \mathsf{P}$. Suppose for contradiction that there exists a no-signaling ε-MIPP for $\mathcal{L}$, denoted by $(V, P_1, \ldots, P_\ell)$, with communication complexity c such that $c = o(n/q)$.

Consider the following NP language

$$\mathcal{L}_G = \{r : \exists s \text{ s.t. } G(s) = r\}.$$

Proposition 1, together with the fact that G is exponentially secure, implies that $\mathcal{L}_G$ does not have an average-case MIP in the crs model with soundness against no-signaling strategies, with communication complexity $o(\tau)$ for instances of length τ.

We obtain a contradiction by constructing an average-case MIP in the crs model with soundness against no-signaling strategies, with communication complexity $o(\tau)$. To this end, consider the following MIP in the crs model for $\mathcal{L}_G$, denoted by $(V', P'_1, \ldots, P'_\ell, \mathsf{crs})$.

---

[6] We note that this assumption can be easily removed by replacing the use of the uniform distribution over the language $\mathcal{L}'$ (defined below) with the distribution $G(s)$ for $s \in_R \{0,1\}^\tau$.

[7] Note that for $q = \Omega(n)$ the theorem is trivially true.

- The crs consists of m uniformly distributed seeds $s_1, \ldots, s_m \in_R \{0,1\}^\tau$, and a random coordinate $i \in_R [m]$.
- The verifier V', on input $r \in \{0,1\}^{2\tau}$, does the following:
    1. Let $r_i = r$, and for every $j \in [m] \setminus \{i\}$, let $r_j = G(s_j)$.
    2. Emulate V with oracle access to $(\mathsf{ECC}(r_1, \ldots, r_m), s_1, \ldots, s_m)$.
    (Note that with overwhelming probability $r \neq G(s_i)$, and thus $r_i \neq G(s_i)$. However V will not notice this unless it queries coordinates that belong to s_i.)
- The provers $P_1', \ldots, P_\ell'$, emulate $P_1, \ldots, P_\ell$ on input $(\mathsf{ECC}(r_1, \ldots, r_m), s_1, \ldots, s_m)$, while setting $r_i = r$ and setting $s_i = s$ where $r = G(s)$ (assuming that such s exists).[8] If such s does not exist then the provers $P_1', \ldots, P_\ell'$ send a reject message, and abort.

Note that the communication complexity of $(V', P_1', \ldots, P_\ell', \mathsf{crs})$ is equal to the communication complexity of $(V, P_1, \ldots, P_\ell, \mathsf{crs})$, denoted by c. By our assumption, $c = o(n/q) = o(\tau)$, as desired.

Average-Case Completeness. We need to prove that $\Pr[(V', P_1', \ldots, P_\ell') (r, \mathsf{crs}) = 1] \geq \frac{1}{2}$, where the probability is over *uniformly distributed* $r \in_R (\mathcal{L}_G)_\tau$, over uniformly generated $\mathsf{crs} = (s_1, \ldots, s_m, i)$ where each $s_j \in_R \{0,1\}^\tau$, $i \in_R [m]$, and over the random coin tosses of the verifier V.

Let GOOD denote the event that V' does not query any of the coordinates that belong to s_i, where $i \in [m]$ is the random coordinate chosen by V'. Notice that for every $r \in \mathcal{L}_G$,

$$\Pr\left[(V', P_1', \ldots, P_\ell')(r, \mathsf{crs}) = 1 \mid \mathsf{GOOD}\right] =$$

$$\Pr\left[(V, P_1, \ldots, P_\ell)(\mathsf{ECC}(r_1, \ldots, r_m), s_1, \ldots, s_m) = 1 \mid s_i \text{ is not queried}\right] \geq \frac{2}{3}$$

where the probabilities are over a uniformly distributed crs and the random coin tosses of V' and V, and where in the second equation $r_i = r$ and $s_i = s$, where $r = G(s)$. Recall that the fact that $r \in \mathcal{L}_G$ implies that such s exists.

The fact that

$$\Pr[(V', P_1', \ldots, P_\ell')(r, \mathsf{crs}) = 1] \geq \Pr[(V', P_1', \ldots, P_\ell')(r, \mathsf{crs}) = 1 \mid \mathsf{GOOD}] \cdot \Pr[\mathsf{GOOD}]$$

implies that it suffices to prove that $\Pr[\mathsf{GOOD}] \geq \frac{3}{4}$, where the probability is over uniformly distributed $r \in_R \mathcal{L}_G$, uniformly distributed crs, and over the random coin tosses of V'.

Note that $r_1, \ldots, r_m$ are all distributed identically to r, and thus $V, P_1, \ldots, P_\ell$, which all receive as input $(\mathsf{ECC}(r_1, \ldots, r_m), s_1, \ldots, s_m)$, where $r_i = r$, do not have any advantage in guessing i (here we crucially use the fact that the MIPP provers are not given access to the crs). Therefore, since V makes at most q queries,

---

[8] This step can be done by a brute force search (since the honest provers are also computationally unbounded). Nevertheless, we note that typically in proof-systems for languages in NP the prover is given the NP witness and so this step can also be done efficiently.

and since $m = 4q$, it follows from the union bound that V queries any location of s_i with probability at most $\frac{q}{m} = \frac{1}{4}$. Hence, $\Pr[\mathsf{GOOD}] \geq \frac{3}{4}$ and (average-case) completeness follows.

Soundness Against No-Signaling Strategies. We prove that for every $r \notin \mathcal{L}_G$, every $\mathsf{crs} = (s_1, \ldots, s_m, i)$, and every no-signaling cheating strategy $P^{\mathsf{NS}} = (P_1^*, \ldots, P_\ell^*)$, it holds that $\Pr[(V', P^{\mathsf{NS}})(r, \mathsf{crs}) = 1] \leq \frac{1}{3}$, where the probability is over the random coin tosses of V' and P^{NS}.

To this end, fix any $r \notin \mathcal{L}_G$ and any $\mathsf{crs} = (s_1, \ldots, s_m, i)$ where each $s_j \in \{0, 1\}^\tau$ and $i \in [m]$. Suppose for the sake of contradiction that there exists a no-signaling cheating strategy $P^{\mathsf{NS}} = (P_1^*, \ldots, P_\ell^*)$ such that $\Pr[(V', P^{\mathsf{NS}})(r, \mathsf{crs}) = 1] > \frac{1}{3}$, where the probability is over the random coin tosses of V' and P^{NS}.

Recall that V' runs V on input $(\mathsf{ECC}(r_1, \ldots, r_m), s_1, \ldots, s_m)$, where $r_i = r$ and where $r_j = G(s_j)$ for every $j \in [m] \setminus \{i\}$. We prove that there exists a no-signaling cheating strategy, denoted by $\hat{P}^{\mathsf{NS}}$, such that

$$\Pr\left[\left(V, \hat{P}^{\mathsf{NS}}\right)(\mathsf{ECC}(r_1, \ldots, r_m), s_1, \ldots, s_m) = 1\right] > \frac{1}{3}, \tag{1}$$

where the probability is over the random coin tosses of V and $\hat{P}^{\mathsf{NS}}$.

The cheating strategy $\hat{P}^{\mathsf{NS}}$ simply emulates P^{NS}. Namely, $\hat{P}^{\mathsf{NS}}$, upon receiving queries $(q_1, \ldots, q_\ell)$, will emulate $P^{\mathsf{NS}}(r, \mathsf{crs})$ upon receiving $(q_1, \ldots, q_\ell)$, where $r = r_i$ and $\mathsf{crs} = (s_1, \ldots, s_m, i)$. Note that $\hat{P}^{\mathsf{NS}}$ simulates P^{NS} perfectly, and therefore indeed Equation (1) holds. Also note that the fact that P^{NS} is a no-signaling strategy immediately implies that $\hat{P}^{\mathsf{NS}}$ is also a no-signaling strategy.

To get a contradiction, it thus remains to show that $(\mathsf{ECC}(r_1, \ldots, r_m), s_1, \ldots, s_m)$ is ε-far from $\mathcal{L}$. Indeed, the fact that ECC is an error correcting code resilient to 2ε-fraction of adversarial errors, together with the fact that $r \notin \mathcal{L}_G$ implies that $(\mathsf{ECC}(r_1, \ldots, r_m), s_1, \ldots, s_m)$ is ε-far from $\mathcal{L}$, as desired. $\square$

Acknowledgments.. We thank Guy Rothblum for pointing out to us the question about arguments of proximity for P - the question that initiated this work. The second author was supported by the Israel Science Foundation (grant No. 671/13).

References

[ABOR00] Aiello, W., Bhatt, S., Ostrovsky, R., Rajagopalan, S.R.: Fast verification of any remote procedure call: short witness-indistinguishable one-round proofs for NP. In: Welzl, E., Montanari, U., Rolim, J.D.P. (eds.) ICALP 2000. LNCS, vol. 1853, pp. 463–474. Springer, Heidelberg (2000)

[AIK10] Applebaum, B., Ishai, Y., Kushilevitz, E.: From secrecy to soundness: efficient verification via secure computation. In: Abramsky, S., Gavoille, C., Kirchner, C., Meyer auf der Heide, F., Spirakis, P.G. (eds.) ICALP 2010. LNCS, vol. 6198, pp. 152–163. Springer, Heidelberg (2010)

[BCCT12a] Bitansky, N., Canetti, R., Chiesa, A., Tromer, E.: From extractable collision resistance to succinct non-interactive arguments of knowledge, and back again. In: ITCS, pp. 326–349 (2012)

[BCCT12b] Bitansky, N., Canetti, R., Chiesa, A., Tromer, E.: Recursive composition and bootstrapping for snarks and proof-carrying data. IACR Cryptology ePrint Archive 2012:95 (2012)

[BSGH+06] Ben-Sasson, E., Goldreich, O., Harsha, P., Sudan, M., Vadhan, S.P.: Robust PCPs of proximity, shorter PCPs, and applications to coding. SIAM J. Comput. **36**(4), 889–974 (2006)

[CKLR11] Chung, K.-M., Kalai, Y.T., Liu, F.-H., Raz, R.: Memory delegation. In: Rogaway, P. (ed.) CRYPTO 2011. LNCS, vol. 6841, pp. 151–168. Springer, Heidelberg (2011)

[CKV10] Chung, K.-M., Kalai, Y., Vadhan, S.: Improved delegation of computation using fully homomorphic encryption. In: Rabin, T. (ed.) CRYPTO 2010. LNCS, vol. 6223, pp. 483–501. Springer, Heidelberg (2010)

[DFH12] Damgård, I., Faust, S., Hazay, C.: Secure two-party computation with low communication. In: Cramer, R. (ed.) TCC 2012. LNCS, vol. 7194, pp. 54–74. Springer, Heidelberg (2012)

[DLN+04] Dwork, C., Langberg, M., Naor, M., Nissim, K., Reingold, O.: Succinct proofs for NP and spooky interactions. Unpublished manuscript (2004). http://www.cs.bgu.ac.il/kobbi/papers/spooky_sub_crypto.pdf

[DR06] Dinur, I., Reingold, O.: Assignment testers: Towards a combinatorial proof of the PCP theorem. SIAM J. Comput. **36**(4), 975–1024 (2006)

[EKR04] Funda Ergün, Ravi Kumar, and Ronitt Rubinfeld: Fast approximate probabilistically checkable proofs. Inf. Comput. **189**(2), 135–159 (2004)

[FGL14] Fischer, E., Goldhirsh, Y., Lachish, O.: Partial tests, universal tests and decomposability. In: ITCS, pp. 483–500 (2014)

[GGP10] Gennaro, R., Gentry, C., Parno, B.: Non-interactive verifiable computing: outsourcing computation to untrusted workers. In: Rabin, T. (ed.) CRYPTO 2010. LNCS, vol. 6223, pp. 465–482. Springer, Heidelberg (2010)

[GGPR12] Gennaro, R., Gentry, C., Parno, B., Raykova, M.: Quadratic span programs and succinct NIZKs without PCPs. IACR Cryptology ePrint Archive 2012:215 (2012)

[GGR98] Goldreich, O., Goldwasser, S., Ron, D.: Property testing and its connection to learning and approximation. J. ACM (JACM) **45**(4), 653–750 (1998)

[GKR08] Goldwasser, S., Kalai, Y.T., Rothblum, G.N.: Delegating computation: interactive proofs for muggles. In: STOC, pp. 113–122 (2008)

[GLR11] Goldwasser, S., Lin, H., Rubinstein, A.: Delegation of computation without rejection problem from designated verifier cs-proofs. IACR Cryptology ePrint Archive 2011:456 (2011)

[GR13] Gur, T., Rothblum, R.D.: Non-interactive proofs of proximity. Electronic Colloquium on Computational Complexity (ECCC), 20:78 (2013)

[Gro10] Groth, J.: Short pairing-based non-interactive zero-knowledge arguments. In: Abe, M. (ed.) ASIACRYPT 2010. LNCS, vol. 6477, pp. 321–340. Springer, Heidelberg (2010)

[GW11] Gentry, C., Wichs, D.: Separating succinct non-interactive arguments from all falsifiable assumptions. In: STOC, pp. 99–108 (2011)

[Hol09] Holenstein, T.: Parallel repetition: simplification and the no-signaling case. Theory Comput. **5**(1), 141–172 (2009)

[Ito10] Ito, T.: Polynomial-space approximation of no-signaling provers. In: Abramsky, S., Gavoille, C., Kirchner, C., Meyer auf der Heide, F., Spirakis, P.G. (eds.) ICALP 2010. LNCS, vol. 6198, pp. 140–151. Springer, Heidelberg (2010)

[Kil92] Kilian, J.: A note on efficient zero-knowledge proofs and arguments (extended abstract). In: STOC, pp. 723–732 (1992)

[KRR13a] Kalai, Y.T., Raz, R., Rothblum, R.D.: Delegation for bounded space. In: STOC, pp. 565–574 (2013)

[KRR13b] Kalai, Y.T., Raz, R., Rothblum, R.D.: How to delegate computations: The power of no-signaling proofs. Electronic Colloquium on Computational Complexity (ECCC), 20:183 (2013)

[Lip12] Lipmaa, H.: Progression-free sets and sublinear pairing-based non-interactive zero-knowledge arguments. In: Cramer, R. (ed.) TCC 2012. LNCS, vol. 7194, pp. 169–189. Springer, Heidelberg (2012)

[Mic94] Micali, S.: CS proofs (extended abstracts). In: FOCS, pp. 436–453 (1994)

[Nao03] Naor, M.: On cryptographic assumptions and challenges. In: Boneh, D. (ed.) CRYPTO 2003. LNCS, vol. 2729, pp. 96–109. Springer, Heidelberg (2003)

[PRV12] Parno, B., Raykova, M., Vaikuntanathan, V.: How to delegate and verify in public: verifiable computation from attribute-based encryption. In: Cramer, R. (ed.) TCC 2012. LNCS, vol. 7194, pp. 422–439. Springer, Heidelberg (2012)

[RVW13] Rothblum, G.N., Vadhan, S.P., Wigderson, A.: Interactive proofs of proximity: delegating computation in sublinear time. In: STOC, pp. 793–802 (2013)

[Spi96] Spielman, D.A.: Linear-time encodable and decodable error-correcting codes. IEEE Trans. Inf. Theory 42(6), 1723–1731 (1996)

[Tha13] Thaler, J.: Time-optimal interactive proofs for circuit evaluation. In: Canetti, R., Garay, J.A. (eds.) CRYPTO 2013, Part II. LNCS, vol. 8043, pp. 71–89. Springer, Heidelberg (2013)

[VSBW13] Vu, V., Setty, S.T.V., Blumberg, A.J., Walfish, M.: A hybrid architecture for interactive verifiable computation. In: IEEE Symposium on Security and Privacy, pp. 223–237 (2013)

Distributions Attaining Secret Key at a Rate of the Conditional Mutual Information

Eric Chitambar[1]([⊠]), Benjamin Fortescue[1], and Min-Hsiu Hsieh[2]

[1] Department of Physics and Astronomy, Southern Illinois University,
Carbondale, IL 62901, USA
echitamb@siu.edu

[2] Faculty of Engineering and Information Technology (FEIT), Centre
for Quantum Computation & Intelligent Systems (QCIS),
University of Technology Sydney (UTS), Sydney, NSW 2007, Australia

Abstract. In this paper we consider the problem of extracting secret key from an eavesdropped source p_{XYZ} at a rate given by the conditional mutual information. We investigate this question under three different scenarios: (i) Alice (X) and Bob (Y) are unable to communicate but share common randomness with the eavesdropper Eve (Z), (ii) Alice and Bob are allowed one-way public communication, and (iii) Alice and Bob are allowed two-way public communication. Distributions having a key rate of the conditional mutual information are precisely those in which a "helping" Eve offers Alice and Bob no greater advantage for obtaining secret key than a fully adversarial one. For each of the above scenarios, strong necessary conditions are derived on the structure of distributions attaining a secret key rate of $I(X : Y|Z)$. In obtaining our results, we completely solve the problem of secret key distillation under scenario (i) and identify $H(S|Z)$ to be the optimal key rate using shared randomness, where S is the Gács-Körner Common Information. We thus provide an operational interpretation of the conditional Gács-Körner Common Information. Additionally, we introduce simple example distributions in which the rate $I(X : Y|Z)$ is achievable if and only if two-way communication is allowed.

Keywords: Information-theoretic security · Public key agreement · Gács-Körner Common Information

1 Introduction

A basic information-processing task involves the exchange of secret information between Alice (X) and Bob (Y) in the presence of an eavesdropper, Eve (E). If Alice and Bob have some pre-established key that is secret from Eve, then any future message M can be transmitted using the key as a one-time pad. Thus, the problem of private communication can be reduced to the problem of *secret key distillation*, which studies the extraction of secret key $\Phi_{XY} \cdot q_Z$ from some initial tripartite correlation p_{XYZ}. Here, Φ_{XY} is a perfectly correlated bit and q_Z is an arbitrary distribution. Often, the correlations p_{XYZ} are presented as a

© International Association for Cryptologic Research 2015
R. Gennaro and M. Robshaw (Eds.): CRYPTO 2015, Part II, LNCS 9216, pp. 443–462, 2015.
DOI: 10.1007/978-3-662-48000-7_22

many-copy source p_{XYZ}^n, and Alice and Bob wish to know the optimal rate of secret bits per copy that they can distill from this source.

It turns out that Alice and Bob can often enhance their distillation capabilities by openly disclosing some information about X and Y through public communication [1,8]. In general, Alice and Bob's communication schemes can be interactive with one round of communication depending on what particular messages were broadcasted in previous rounds. Such interactive protocols are known to generate higher key rates than non-interactive protocols, at least in the absence of "noisy" local processing by Alice and Bob [8]. Thus, for a given distribution p_{XYZ}, one obtains a hierarchy of key rates pertaining to the respective scenarios of no communication, one-way communication, and two-way (interactive) communication. It is also possible to consider no-communication scenarios in which Alice and Bob have access to some publically shared randomness that is uncorrelated with their primary source p_{XYZ}. Clearly publically shared randomness is a weaker resource than public communication since the latter is able to generate the former. However, below we will prove even stronger that publically shared randomness offers no advantage whatsoever for secret key distillation.

For the one-way communication scenario, a single-letter characterization of the key rate has been proven by Ahlswede and Csiszár [1]. When the unidirectional communication is from Alice to Bob, we denote the key rate by $\overrightarrow{K}(X : Y|Z)$, while $\overleftarrow{K}(X : Y|Z)$ denotes the rate when communication is from Bob to Alice only. No formula is known for the two-way key rate of a given distribution, which we denote by $K(X : Y|Z)$, and the complexity of protocols utilizing interactive communication makes computing this a highly challenging open problem.

In the special case of an uncorrelated Eve in p_{XYZ}, the key rate is given by the mutual information $I(X : Y)$, and this can be achieved using one-way communication. For more general distributions in which Eve possesses some side information of XY, the conditional mutual information $I(X : Y|Z)$ is a known upper bound for the key rate under two-way communication [1,8]. In general this bound is not tight [9]. Rather, the conditional mutual information quantifies the key rate when Eve helps Alice and Bob by broadcasting her variable Z. Key obtained by a helping Eve is also known as *private key* [4], and private key is still secret from Eve even though she helps Alice and Bob obtain it. The relevance of private key naturally arises in situations where Eve functions as a central server who helps establish secret correlations between Alice and Bob. Thus, distributions with a secret key rate equaling the private key rate of $I(X : Y|Z)$ are precisely those in which nothing is gained by a helping Eve.

The objective of this paper is to investigate the types of distributions for which $I(X : Y|Z)$ is indeed an achievable secret key rate. This will be considered under the scenarios of (i) publically shared randomness but no communication, (ii) one-way communication, and (iii) two-way communication. A full solution to the problem would involve a structural characterization of the distributions p_{XYZ} whose key rates are $I(X : Y|Z)$. We are able to fully achieve this only for the no-communication setting, but we nevertheless derive strong necessary conditions for both the one-way and the two-way scenarios. In the case of one-way communication, our condition makes use of the key-rate formula derived by Ahlswede and Csiszár. For the statement of this formula, recall that three variables A, B, and C satisfy the Markov chain $A - B - C$ if C is conditionally independent of A given B; i.e. $p(c|b,a) = p(c|b)$ for letters in the range of A, B, and C. Then,

Lemma 1 ([1]). *For distribution p_{XYZ},*

$$\overrightarrow{K}(X:Y|Z) = \max_{KU|X} I(K:Y|U) - I(K:Z|U), \tag{1}$$

where the maximization is taken over all auxiliary variables K and U satisfying the Markov chain $KU - X - YZ$, with K and U ranging over sets of size no greater than $|\mathcal{X}| + 1$. In particular,

$$\overrightarrow{K}(X:Y|Z) \geqslant I(X:Y) - I(X:Z). \tag{2}$$

In this paper, we consider when variables KU can be found that satisfy both $KU - X - YZ$ and $I(K;Y|U) - I(K;Z|U) = I(X:Y|Z)$. Theorem 2 below offers a necessary condition on the structure of distributions for which this is possible. Turning to the scenario of two-way communication, we utilize the well-known intrinsic information upper bound on $K(X:Y|Z)$. For distribution p_{XYZ}, its intrinsic information is given by

$$I(X:Y \downarrow Z) := \min_{\overline{Z}|Z} I(X:Y|\overline{Z}) \tag{3}$$

where the minimization is taken over over all auxiliary variables $\overline{Z}$ satisfying $XY - Z - \overline{Z}$, with $\overline{Z}$ having the same range as Z [3]. Thus, the intrinsic information is the smallest conditional mutual information achievable after Eve processes her variable Z. The intrinsic information satisfies $K(X:Y|Z) \leqslant I(X:Y \downarrow Z)$. In Theorem 3 below, we identify a large class of distributions for which a channel $\overline{Z}|Z$ can be found satisfying $I(X:Y|\overline{Z}) < I(X:Y|Z)$. This allows us to derive a necessary condition on distributions having $K(X:Y|Z) = I(X:Y|Z)$.

A brief summary of our results is the following:

- For publically shared randomness with no communication, we identify $H(J_{XY}|Z)$ as the secret key rate, where J_{XY} is the Gács-Körner Common Information of Alice and Bob's marginal distribution p_{XY}. Moreover, this rate is achievable without using shared randomness. Using this result, the structure of distributions attaining $I(X:Y|Z)$ can easily be characterized.
- When one-way communication is permitted between Alice and Bob, we show that the distribution p_{XYZ} must satisfy a certain "block-like" structure in order to obtain the key rate $I(X:Y|Z)$. Specifically, given some outcome z of Eve, if there exists collections of events $\mathcal{X}_0$ and $\mathcal{Y}_0$ for Alice and Bob respectively that satisfy $p(\mathcal{Y}_0|\mathcal{X}_0, z) = p(\mathcal{X}_0|\mathcal{Y}_0, z) = 1$, then $p(\mathcal{Y}_0|\mathcal{X}_0) = p(\mathcal{X}_0|\mathcal{Y}_0) = 1$; i.e. conclusive determination of whether an event belongs to $\mathcal{X}_0 \times \mathcal{Y}_0$ can be done by each party, regardless of Eve's outcome.

- For key distillation with two-way communication, we show that distributions attaining a key rate of $I(X : Y|Z)$ must also satisfy a certain type of uniformity similar to the one-way case. One special class of distributions our necessary condition applies to are those obtained by mixing a perfectly correlated distribution p_{XY} with an uncorrelated one such that the marginals have the same range and such that Eve's variable Z specifies which one of the distributions Alice and Bob hold. We show that unless either Alice or Bob can likewise identify the distribution from his or her variable, a key rate of $I(X : Y|Z)$ is unattainable.
- We construct distributions in which a distillation rate of $I(X : Y|Z)$ is unachievable when the communication is restricted from Alice to Bob, and yet it becomes achievable if the communication direction is from Bob to Alice. We further provide an example when $I(X : Y|Z)$ is achievable only if two-way communication is used.

Before presenting these results in greater detail, we begin in Sect. 2 with a more precise overview of the key rates studied in this paper. In Sect. 3, we then present the Gács-Körner Common Information and prove some basic properties. Section 4 contains our main results, with longer proofs postponed to the appendix. Finally, Sect. 5 offers some concluding remarks.

2   Definitions

Let us review the relevant definitions of secret key rate under various communication scenarios. We consider random variables X, Y and Z ranging over finite alphabets $\mathcal{X}$, $\mathcal{Y}$, and $\mathcal{Z}$ respectively. For a general distribution q, we say its support (denoted by $supp[q]$) is the collection of x such that $q(x) > 0$. In all distillation tasks, we assume that Alice and Bob each have access to one part of an i.i.d. (identical and independently distributed) source XYZ whose distribution is p_{XYZ}. Hence, after n realizations of the source, X^n, Y^n and Z^n belong to Alice, Bob, and Eve respectively. In addition, Alice and Bob each possess a local random variable, Q_A and Q_B respectively, which are mutually independent from each other and from $X^nY^nZ^n$. This allows them to introduce local randomness into their processing of X^nY^n.

We first turn to the most restrictive scenario, which is key distillation using publicly shared randomness. The *common randomness (c.r.) key rate* of X, Y, and Z, denoted by $K^{c.r.}(X : Y|Z)$, is defined to be the largest R such that for every $\epsilon > 0$, there is an integer N such that $n \geqslant N$ implies the existence of (a) a random variable W independent of $X^nY^nZ^n$ and ranging over some set $\mathcal{W}$, (b) a random variable K ranging over some set $\mathcal{K}$, and (c) a pair of mappings $f(X^n, Q_A, W)$ and $g(Y^n, Q_B, W)$ for which

(i) $Pr[f = g = K] > 1 - \epsilon$;
(ii) $\log|\mathcal{K}| - H(K|Z^nW) < \epsilon$;
(iii) $\frac{1}{n}\log|\mathcal{K}| \geqslant R$.

We next move to the more general scenario of when Alice and Bob are allowed to engage in public communication. A *local operations and public communication* (LOPC) protocol consists of a sequence of public communication exchanges between Alice and Bob. The ith message exchanged between them is described by the variable M_i. If Alice (resp. Bob) is the broadcasting party in round i, then M_i is a function of X^n and Q_A (resp. Y^n and Q_B) as well as the previous messages $(M_1, M_2, \cdots, M_{i-1})$. The protocol is one-way if there is only one round of a message exchange.

For distribution p_{XYZ}, the *Alice-to-Bob secret key rate* $\overrightarrow{K}(X : Y|Z)$ is the largest R that satisfies the above three conditions except with W being replaced by some message M that is generated by Alice and therefore a function of (X^n, Q_A). We can likewise define the Bob-to-Alice key rate $\overleftarrow{K}(X : Y|Z)$. The *(two-way) secret key rate* of X and Y given Z, denoted by $K(X : Y|Z)$, is defined analogously except with $M = (M_1, M_2, \cdots, M_r)$ being any random variable generated by an LOPC protocol [1,8]. The key rates satisfy the obvious relationship:

$$K^{c.r.}(X : Y|Z) \leqslant \begin{cases} \overrightarrow{K}(X : Y|Z) \\ \overleftarrow{K}(X : Y|Z) \end{cases} \leqslant K(X : Y|Z). \tag{4}$$

3   The Gács-Körner Common Information

In this section, we introduce the Gács-Körner Common Information. For every pair of random variables XY, there exists a *maximal* common variable J_{XY} in the sense that J_{XY} is a function of both X and Y, and any other such common function of both X and Y is itself a function of J_{XY}. Hence, up to relabeling, the variable J_{XY} is unique for each distribution p_{XY}. In terms of its structure, a distribution p_{XY} can always be decomposed as

$$p(x, y) = \sum_{J_{XY}=j} p(x, y|j)p(j), \tag{5}$$

where for any $x, x' \in \mathcal{X}$ and $y, y' \in \mathcal{Y}$, the conditional distributions satisfy $p(x, y|j)p(x, y'|j') = 0$ and $p(x, y|j)p(x', y|j') = 0$ if $j \neq j'$. Gács and Körner identify $H(J_{XY})$ as the common information of XY [6].

It is instructive to rigorously prove the statements of the preceding paragraph. A *common partitioning of length t* for XY are pairs of subsets $(\mathcal{X}_i, \mathcal{Y}_i)_{i=1}^t$ such that

(i) $\mathcal{X}_i \cap \mathcal{X}_j = \mathcal{Y}_i \cap \mathcal{Y}_j = \emptyset$ for $i \neq j$,
(ii) $p(\mathcal{X}_i|\mathcal{Y}_j) = p(\mathcal{Y}_i|\mathcal{X}_j) = \delta_{ij}$, and
(iii) if $(x, y) \in \mathcal{X}_i \times \mathcal{Y}_i$ for some i, then $p_X(x)p_Y(y) > 0$.

For a given common partitioning, we refer to the subsets $\mathcal{X}_i \times \mathcal{Y}_i$ as the "blocks" of the partitioning. The subscript i merely serves to label the different blocks, and for any fixed labeling, we associate a random variable $C(X, Y)$ such that $C(x, y) = i$ if $(x, y) \in \mathcal{X}_i \times \mathcal{Y}_i$. Note that each party can determine the value of

J from their local information, and it is therefore called a *common function* of X and Y. A *maximal common partitioning* is a common partitioning of greatest length. The following proposition is proven in the appendix.

Proposition 1

(a) *Every pair of finite random variables XY has a unique maximal common partitioning, which we denote by J_{XY}.*

(b) *Variable J_{XY} satisfies*

$$H(J_{XY}) = \max_K \{H(K) : 0 = H(K|X) = H(K|Y)\}$$

iff J_{XY} is a common function for the maximal common partitioning of XY.

(c) *If $f(X) = g(Y) = C$ is any other common function of X and Y, then $C(J_{XY})$.*

With property (a), we can speak unambiguously of *the* maximal common partitioning of a distribution p_{XY}. Consequently the variable J_{XY} is unique up to a relabeling of its range. The following proposition from [6] provides a useful characterization of values x and x' that belong to the same block in a maximal common partitioning.

Proposition 2 ([6]). *If $J_{XY}(x) = J_{XY}(x')$ for $x, x' \in J_{XY}$, then there exists a sequence of values*

$$xy_1x_1y_2x_2 \cdots y_nx'$$

such that $p(x, y_1)p(y_1, x_1)p(x_1, y_2) \cdots p(y_n, x') > 0$.

4 Results

4.1 Key Distillation Using Auxiliary Public Randomness

The Gács and Körner Common Information plays a central role in the problem of key distillation with no communication. To see a preliminary connection, we recall an operational interpretation of $H(J_{XY})$ that Gács and Körner prove in Ref. [6]. The task involves Alice and Bob constructing faithful encodings of their respective sources X and Y, and $H(J_{XY})$ quantifies the asymptotic average sequence-length of codewords per copy such that both Alice and Bob's encodings output matching codewords with high probability over this sequence [6].

For the task of key distillation, Alice and Bob are likewise trying to convert their sources into matching sequences of optimal length. However, the key distillation problem is different in two ways. On the one hand there is the additional constraint that the common sequence should be nearly uncorrelated from Eve. On the other hand, unlike the Gács-Körner problem, it is not required that these sequences belong to faithful encodings of the sources X and Y. Nevertheless, we find that $H(J_{XY}|Z)$ quantifies the distillable key when Alice and Bob are unable to communicate with one another. This is also the rate even if Alice and Bob have access to auxillary public randomness which is uncorrelated with their primary distribution.

Theorem 1. $K^{c.r.}(X : Y|Z) = H(J_{XY}|Z)$. *Moreover,* $H(J_{XY}|Z)$ *is achievable with no additional common randomness.*

Proof. See the appendix. Many parts of the converse proof follow analogously to the converse proof of Theorem 2.6 in Ref. [4] (see also [5]).

One can also consider a related quantity known as the *maximal conditional common function* $J_{XY|Z}$, which is the collection of variables $\{J_{XY|Z=z} : z \in \mathcal{Z}\}$ with $J_{XY|Z=z}$ being a maximal common function of the conditional distribution $p_{XY|Z=z}$. The variable $J_{XY|Z}$ is again unique for every distribution p_{XYZ} up to relabeling. Since $J_{XY|Z=z}$ is computed from both X and Y with the additional information that $Z = z$, maximality of $J_{XY|Z=z}$ ensures that J_{XY} is a function of $J_{XY|Z=z}$ for each $z \in \mathcal{Z}$. In other words, a labeling of J_{XY} and $J_{XY|Z}$ can be chosen so that J_{XY} is a coarse-graining of $J_{XY|Z}$. Therefore, $H(J_{XY}|Z) \leqslant H(J_{XY|Z}|Z)$ with equality iff $H(J_{XY|Z}|ZJ_{XY}) = 0$. When the equality condition holds, it means that for each $z \in \mathcal{Z}$, the value of $J_{XY|Z=z}$ can be determined from J_{XY} alone. Hence, the variables J_{XY} and $J_{XY|Z}$ must be equivalent up to relabeling. From this it follows that a distribution satisfies $H(J_{XY|Z}|ZJ_{XY}) = 0$ iff it admits a decomposition of

$$p(x, y, z) = \sum_{J_{XY}=j} p(x, y|z, j)p(j|z)p(z), \tag{6}$$

where for any $x, x' \in \mathcal{X}$, $y, y' \in \mathcal{Y}$ and $z, z' \in \mathcal{Z}$ the conditional distributions satisfy

$$p(x, y|z, j)p(x, y'|z', j') = 0, \qquad p(x, y|j)p(x', y|z', j') = 0 \quad \text{if} \quad j \neq j'.$$

The class of distributions of this form we shall call *uniform block* (UB) (see Fig. 1).

The quantity $H(J_{XY|Z}|Z)$ is the private key rate when Eve is helping yet Alice and Bob are still prohibited from communicating with one another. Thus, the difference $H(J_{XY|Z}|Z) - H(J_{XY}|Z)$ quantifies how much Eve can assist Alice and Bob in distilling key when no communication is exchanged between the two. From the previous paragraph, it follows that Eve offers no assistance (i.e. the private key rate equals the secret key rate) in the no-communication scenario iff the distribution is UB.

Returning to Theorem 1, we can now answer the underlying question of this paper for no-communication distillation. By using the chain rule of conditional mutual information and the fact that J_{XY} is both a function of X and Y, we readily compute

$$I(X : Y|Z) = I(J_{XY}X : Y|Z) = I(J_{XY} : Y|Z) - I(X : Y|ZJ_{XY})$$
$$= H(J_{XY}|Z) - I(X : Y|ZJ_{XY}). \tag{7}$$

The conditional mutual information is thus an achievable rate whenever $I(X : Y|ZJ_{XY}) = 0$. Distributions satisfying this equality are uniform block with

(a) Not Uniform Block

Z = 0	0	1	2		Z = 1	0	1	2
0	1/2	.	.		0	.	1/3	.
1	.	1/2	.		1	.	1/3	.
2	.	.	.		2	.	.	1/3

(b) Uniform Block

Z = 0	0	1	2		Z = 1	0	1	2
0	1/2	.	.		0	.	.	.
1	.	1/2	.		1	.	.	1/3
2	.	.	.		2	.	1/3	1/3

Fig. 1. Example of a distribution that is not uniform block (a) and one that is (b). Each entry corresponds to a conditional probability value $p(x, y|z)$. UB distribution (b) is not uniform block independent (UBI) since the block in the $Z = 1$ plane contains correlations between Alice and Bob.

the extra condition that $p(x, y|z, j) = p(x|z, j)p(y|z, j)$ in Eq. (6). We shall call distributions having this form *uniform block independent* (UBI). Putting everything together, we find that

Corollary 1. *A distribution p_{XYZ} satisfies $K^{c.r.}(X : Y|Z) = I(X : Y|Z)$ if and only if it is uniformly block independent.*

Remark 1. The no-communication results discussed above and proven in the appendix are already implicit in the work of Csiszár and Narayan. In Ref. [4], they study various key distillation scenarios with Eve functioning as a helper and limited communication between Alice and Bob. Included in this is the no-communication scenario with and without helper. However, being very general in nature, Csiszár and Narayan's results involve optimizations over auxiliary random variables, and it is therefore still a non-trivial matter to discern Theorem 1 and Corollary 1 directly from their work. Additionally, they do not consider the scenario of just shared public randomness.

4.2 Obtaining $I(X : Y|Z)$ with One-Way Communication

In this section we want to identify the type of tripartite distributions from which secret key can be distilled at the rate $I(X : Y|Z)$ using one-way communication. Since $K(X : Y|Z) \leqslant I(X : Y|Z)$, our analysis deals with distributions for which one-way communication suffices to optimally distill secret key. Manipulating Eq. (1) of Lemma 1 allows us to determine when $\overrightarrow{K}(X : Y|Z) = I(X : Y|Z)$. We have that

$$I(K : Y|U) - I(K : Z|U) = I(K : Y|ZU) - I(K : Z|YU)$$
$$= I(KU : Y|Z) - I(U : Y|Z) - I(K : Z|YU)$$
$$= I(X : Y|Z) - I(X : Y|KUZ) - I(U : Y|Z) - I(K : Z|YU),$$

where K and U satisfy $KU - X - YZ$. From this and Lemma 1, we conclude the following.

Lemma 2. *Distribution p_{XYZ} has $\overrightarrow{K}(X : Y|Z) = I(X : Y|Z)$ iff there exists variables $KUXYZ$ with K and U ranging over sets of size no greater than $|\mathcal{X}|+1$ such that*

$$
\begin{array}{lll}
(1) & KU - X - YZ, \qquad & (2) & X - KUZ - Y, \\
(3) & U - Z - Y, & (4) & K - YU - Z.
\end{array} \tag{8}
$$

The conditions of Lemma 2 allow for the follow rough interpretation. (1) says that Alice is able to generate variables K and U from knowledge of her variable X. We think of K as containing the key that Alice and Bob will share and U as the public message sent from Alice to Bob. (2) says that from Eve's perspective, Alice and Bob share no more correlations given U and K. Likewise, (3) says that from Eve's perspective, the public message is uncorrelated with Bob. Finally, (4) says that after learning U, Bob can generate the key K that is independent from Eve.

Unfortunately, Lemma 2 does not provide a transparent characterization of the distributions for which $\overrightarrow{K}(X : Y|Z) = I(X : Y|Z)$. We next proceed to obtain a better picture of these distributions by exploring additional consequences of the Markov chains in Eq. (8). The following places a necessary condition on the distributions. We will see in Sect. 4.4, however, that it fails to be sufficient.

Theorem 2. *If distribution p_{XYZ} has either $\overrightarrow{K}(X : Y|Z) = I(X : Y|Z)$ or $\overleftarrow{K}(X : Y|Z) = I(X : Y|Z)$, then p_{XYZ} must have the following property: For any $z \subset \mathcal{Z}$, if $\mathcal{X}_i \times \mathcal{Y}_i$ and $\mathcal{X}_j \times \mathcal{Y}_j$ are two distinct blocks in the maximal common partitioning of $p_{XY|Z=z}$, then*

$$
p_{XY}(\mathcal{X}_i, \mathcal{Y}_j) = 0.
$$

Proof. Without loss of generality, assume that $\overrightarrow{K}(X : Y|Z) = I(X : Y|Z)$. For distribution $p_{XY|Z=z}$ with maximal common partition $(\mathcal{X}_\lambda, \mathcal{Y}_\lambda)_{\lambda=1}^t$, consider arbitrary $(x_i, y_i) \in \mathcal{X}_i \times \mathcal{Y}_i$ and $(x_j, y_j) \in \mathcal{X}_j \times \mathcal{Y}_j$. Note that from the definition of a maximal common partitioning, we have that $p(x_i, z)p(y_i, z) > 0$, but we need not have that $p(x_i, y_i, z) > 0$.

We will prove that $p(x_i, y_j, z') = 0$ for all $z' \in \mathcal{Z}$ (clearly this already holds when $z' = z$). Suppose on the contrary that $p(x_i, y_j, z') > 0$. Since $p(x_i, z) > 0$, there will exist some $y_i' \in \mathcal{Y}_i$ such that $p(x_i, y_i', z) > 0$. Then the Markov chain condition $KU - X - YZ$ implies that for some $(k, u) \in \mathcal{K} \times \mathcal{U}$ such that $p(k, u|x_i) > 0$, we have

$$
p(k, u|x_i) = p(k, u|x_i, y_i', z) = p(k, u|x_i, y_j, z') > 0. \tag{9}
$$

Equation (9) implies that both $p(k, u|y_i', z) > 0$ and $p(k, u|y_j, z') > 0$. From $p(u|y_i', z) > 0$ and the Markov chain $U - Z - Y$, we have that $p(u|y_j, z) > 0$.

(a)

$Z=0$	X → 0	1	2		$Z=1$	0	1	2		$Z=2$	0	1	2
Y↓ 0	1/2	.	.		0	.	.	1/3		0	.	.	.
1	.	1/2	.		1	.	.	1/3		1	.	.	1/3
2	.	.	.		2	.	1/3	.		2	.	1/3	1/3

(b)

$Z=0$	X → 0	1	2		$Z=1$	0	1	2		$Z=2$	0	1	2
Y↓ 0	1/2	.	.		0	.	.	1/3		0	.	.	.
1	.	1/2	.		1	.	1/3	1/3		1	.	.	1/3
2	.	.	.		2	.	.	.		2	.	1/3	1/3

Fig. 2. (a) The conditions of Theorem 2 are violated for this distribution. To see this, note that the events $(X = 1, Y = 2)$ and $(X = 2, Y = 1)$ are both possible when $Z = 1$. Hence, Theorem 2 necessitates $p(1,1) = 0$, which is not the case because of the plane $Z = 0$. Distribution (b) lacks this characteristic and therefore it satisfies the conditions of Theorem 2.

Then we can further derive

$$0 < p(k, u|y_j, z') = p(u|y_j, z')p(k|u, y_j, z') = p(u|y_j, z')p(k|u, y_j, z)$$
$$\Rightarrow \quad p(k|u, y_j, z) > 0,$$
$$\Rightarrow \quad p(k, u|y_j, z) = p(k|u, y_j, z)p(u|y_j, z) > 0,$$

where we have used the Markov chain $K - YU - Z$. From the last line, we must be able to find some $x'_j \in \mathcal{X}_j$ such that $p(x'_j, y_j, z) > 0$ and $p(k, u|x'_j, y_j, z) > 0$. Inverting probabilities gives that both $p(x'_j, y_j|k, u, z) > 0$ and $p(x_i, y'_i|k, u, z) > 0$. Hence,

$$I(X : Y|KUZ) = I(J_{XY|Z}X : Y|KUZ)$$
$$= I(X : Y|J_{XY|Z}KUZ) + \sum_{k,u,z} H(J_{XY|Z=z}|k, u, z)p(k, u, z) > 0,$$

since $H(J_{XY|Z=z}|k, u, z) > 0$ because $(x_i, y'_i) \in \mathcal{X}_i \times \mathcal{Y}_i$ and $(x'_j, y_j) \in \mathcal{X}_j \times \mathcal{Y}_j$. However, this strict inequality contradicts the Markov chain condition $X - KUZ - Y$. ∎

Figure 2 (a) provides an example distribution which does not satisfy the necessary conditions of Theorem 2 for $I(X : Y|Z)$ to be an achievable one-way key rate. On the other hand, Fig. 2 (b) depicts an distribution for which the conditions of the theorem are met. However, Theorem 3 in the next section will show that both distributions (a) and (b) have $K(X : Y|Z) < I(X : Y|Z)$.

4.3   Obtaining $I(X : Y|Z)$ with Two-Way Communication

We now turn to the general scenario of interactive two-way communication. Our main result is the necessary structural condition of Theorem 3. Its statement requires some new terminology.

For two distributions p_{XY} and q_{XY} over $\mathcal{X} \times \mathcal{Y}$, we say that $q_{XY} \blacktriangleleft p_{XY}$ if, up to a permutation between X and Y, the distributions satisfy $supp[q_X] \subset supp[p_X]$ and one of the three additional conditions: (i) q_{XY} is uncorrelated, (ii) $supp[q_Y] \subset supp[p_Y]$, or (iii) $y \in supp[q_Y] \setminus supp[p_Y]$ implies that $H(X|Y = y) = 0$.

Theorem 3. *Let p_{XYZ} be a distribution over $\mathcal{X} \times \mathcal{Y} \times \mathcal{Z}$ such that $p_{XY|Z=z_1} \blacktriangleleft p_{XY|Z=z_0}$ for some $z_0, z_1 \in \mathcal{Z}$. If there exists some pair $(x, y) \in supp[p_{X|Z=0}] \times supp[p_{Y|Z=0}]$ for which $p(x, y|z_1) > 0$ but $p(x, y|z_0) = 0$, then $K(X : Y|Z) < I(X : Y|Z)$.*

Proof. The proof will involve showing that there exists a channel $\overline{Z}|Z$ such that $I(X : Y|\overline{Z}) < I(X : Y|Z)$. The channel will involve mixing z_0 and z_1 but leaving all other elements unchanged. Define the function

$$f(t) = I(X : Y)_{(1-t)p_{XY|Z=z_0}+tp_{XY|Z=z_1}} \qquad t \in [0, 1], \qquad (10)$$

which gives the mutual information of the mixed distribution $(1 - t)p_{XY|Z=z_0} + tp_{XY|Z=z_1}$. The function f is continuous and twice differentiable in the open interval $(0, 1)$. To prove the theorem, we will need a simple general fact about functions of this sort.

Proposition 3. *Suppose that f is a continuous function on the closed interval $[0, 1]$ and twice differentiable in the open interval $(0, 1)$. Suppose there exists some $0 < \delta < 1$ such that f is strictly convex in the interval $\mathcal{I} = (0, \delta]$ and $f(1) - f(0) > f'(t)$ for all $t \in \mathcal{I}$. Then $f(t) < (1 - t)f(0) + tf(1)$ for all $t \in \mathcal{I}$.*

Continuing with the proof of Theorem 3, it will suffice to show that the function given by Eq. (10) satisfies the conditions of Proposition 3. For if this is true, then we can argue as follows. Choose ϵ sufficiently small so that $\frac{\epsilon p(z_1)}{p(z_0)+\epsilon p(z_1)} \in (0, \delta]$, where δ is described by the proposition. Define the channel $\overline{Z}|Z$ by $p(\overline{z}_0|z_1) = \epsilon$, $p(\overline{z}_1|z_1) = 1 - \epsilon$, and $p(\overline{z}|z) = 1$ for all $z \neq z_1 \in \mathcal{Z}$. This means that $p(\overline{z}_0) = p(z_0) + \epsilon p(z_1)$ and $p(\overline{z}_1) = (1 - \epsilon)p(z_1)$, and inverting the probabilities gives $p(z_1|\overline{z}_1) = 1$, $p(z_1|\overline{z}_0) = \frac{\epsilon p(z_1)}{p(z_0)+\epsilon p(z_1)}$, and $p(z_0|\overline{z}_0) = \frac{p(z_0)}{p(z_0)+\epsilon p(z_1)}$. Since $p(x, y|\overline{Z} = \overline{z}) = \sum_z p(x, y|Z = z)p(Z = z|\overline{Z} = \overline{z})$, the average conditional mutual information is

$$\sum_{z \neq z_0, z_1 \in \mathcal{Z}} I(X : Y|\overline{Z} = \overline{z})p(\overline{z}) + f(\tfrac{\epsilon p(z_1)}{p(z_0)+\epsilon p(z_1)})p(\overline{z}_0) + f(1)p(\overline{z}_1)$$

$$< \sum_{z \neq z_0, z_1 \in \mathcal{Z}} I(X : Y|Z - z)p(z)$$

$$+ \left(\tfrac{p(z_0)}{p(z_0)+\epsilon p(z_1)}f(0) + \tfrac{\epsilon p(z_1)}{p(z_0)+\epsilon p(z_1)}f(1) \right) p(\overline{z}_0) + f(1)(1 - \epsilon)p(z_1)$$

$$= I(X : Y|Z), \qquad (11)$$

where Proposition 3 at $x = \frac{\epsilon p(z_1)}{p(z_0)+\epsilon p(z_1)}$ has been invoked.

Let us then show that the conditions of Proposition 3 hold true for the function given by Eq. (10) whenever $p_{XY|Z=z_1} \blacktriangleleft p_{XY|Z=z_0}$; i.e. that there exists some interval $(0, \delta]$ for which f is strictly convex and $f(1) - f(0) > f'(t)$. We have

$$f(t) = -\sum_{x \in \mathcal{X}} [(1-t)p(x|z_0) + tp(x|z_1)] \log[(1-t)p(x|z_0) + tp(x|z_1)]$$

$$- \sum_{y \in \mathcal{Y}} [(1-t)p(y|z_0) + tp(y|z_1)] \log[(1-t)p(y|z_0) + tp(y|z_1)]$$

$$+ \sum_{x \in \mathcal{X}} \sum_{y \in \mathcal{Y}} [(1-t)p(x,y|z_0) + tp(x,y|z_1)] \log[(1-t)p(x,y|z_0) + tp(x,y|z_1)]. \quad (12)$$

We are interested in $\lim_{t \to 0} f'(t)$ and $\lim_{t \to 0} f''(t)$. To compute these, we use the fact that the function $g(t) = (r + st) \log(r + st)$ satisfies $g'(t) = s(1 + \log(r + st))$ and $g''(t) = \frac{s^2}{r+st}$. We separate the analysis into three cases. Without loss of generality, we will assume $supp[p_{X|Z=z_1}] \subset supp[p_{X|Z=z_0}]$.

Case (i): $p_{XY|Z=z_1}$ is Uncorrelated

Since $supp[p_{X|Z=z_1}] \subset supp[p_{X|Z=z_0}]$, we can assume that $p(x|z_0) \neq 0$ for all x; otherwise there is no term involving x in Eq. (12). Now suppose that $p(y|z_0) = 0$. Then for this fixed y, the summation over x in the third term of Eq. (12) becomes

$$\sum_{x \in \mathcal{X}} [(1-t)p(x,y|z_0) + tp(x,y|z_1)] \log[(1-t)p(x,y|z_0) + tp(x,y|z_1)]$$

$$= t \sum_{x \in \mathcal{X}} p(x|z_1)p(y|z_1) \log[tp(x|z_1)p(y|z_1)]$$

$$= tp(y|z_1) \log[tp(y|z_1)] + tp(y|z_1) \sum_{x \in \mathcal{X}} p(x|z_1) \log[p(x|z_1)]. \quad (13)$$

Hence, by letting $\mathcal{B}_I = \{y : p(y|z_I) > 0\}$ for $I \in \{0,1\}$, we can equivalently write Eq. (12) as

$$f(t) = -\sum_{x \in \mathcal{X}} [(1-t)p(x|z_0) + tp(x|z_1)] \log[(1-t)p(x|z_0) + tp(x|z_1)]$$

$$- \sum_{y \in \mathcal{B}_0} [(1-t)p(y|z_0) + tp(y|z_1)] \log[(1-t)p(y|z_0) + tp(y|z_1)]$$

$$+ \sum_{y \in \mathcal{B}_0} \sum_{x \in \mathcal{X}} [(1-t)p(x,y|z_0) + tp(x,y|z_1)] \log[(1-t)p(x,y|z_0) + tp(x,y|z_1)]$$

$$+ t \sum_{y \in \mathcal{B}_1 \setminus \mathcal{B}_0} p(y|z_1) \sum_{x \in \mathcal{X}} p(x|z_1) \log[p(x|z_1)]. \quad (14)$$

If $p(x,y|z_0) = 0$ for some $(x,y) \in \mathcal{X} \times \mathcal{B}_0$, then the first derivative of (14) will diverge to $-\infty$ as $t \to 0$ while its second derivative will diverge to $+\infty$ whenever $p(x,y|z_1) > 0$. But by assumption, there is at least one pair of (x,y) for which this latter case holds. Hence, an interval $(0, \delta]$ can always be found for which Proposition 3 can be applied to f.

Case (ii): $\mathcal{B}_1 \setminus \mathcal{B}_0 = \varnothing$

This is covered in case (iii).

Case (iii): $y \in \mathcal{B}_1 \setminus \mathcal{B}_0 \Rightarrow p(y|z_1) = p(x_y, y|z_1)$ for some particular $x_y \in \mathcal{X}$

The condition $p(y|z_1) = p(x_y, y|z_1)$ implies that $p(x, y|z_1) = 0$ for all $x \neq x_y$. Then similar to the previous case, when $y \in \mathcal{B}_1 \setminus \mathcal{B}_0$, the summation over x in the third term of Eq. (12) is

$$\sum_{x \in \mathcal{X}} tp(x, y|z_1) \log[tp(x, y|z_1)] = tp(x_y, y|z_1) \log[tp(x_y, y|z_1)]$$

$$= tp(y|z_1) \log[tp(y|z_1)]. \tag{15}$$

Hence each term with $y \in \mathcal{B}_1 \setminus \mathcal{B}_0$ becomes canceled in Eq. (12). Then Eq. (12) reduces to

$$
\begin{aligned}
f(t) = &-\sum_{x \in \mathcal{X}} [(1-t)p(x|z_0) + tp(x|z_1)] \log[(1-t)p(x|z_0) + tp(x|z_1)] \\
&- \sum_{y \in \mathcal{B}_0} [(1-t)p(y|z_0) + tp(y|z_1)] \log[(1-t)p(y|z_0) + tp(y|z_1)] \\
&+ \sum_{x \in \mathcal{X}} \sum_{y \in \mathcal{B}_0} [(1-t)p(x, y|z_0) + tp(x, y|z_1)] \log[(1-t)p(x, y|z_0) + tp(x, y|z_1)].
\end{aligned}
\tag{16}
$$

As in the previous case, the first derivative of this function will diverge to $-\infty$ while its second derivative will diverge to $+\infty$ whenever $p(x, y|z_1) > 0$ and $p(x, y|z_0) = 0$. By assumption, such a pair (x, y) exists, and so again, an interval $(0, \delta]$ can always be found for which Proposition 3 can be applied to f. Note that when $\mathcal{B}_1 \setminus \mathcal{B}_0 = \varnothing$, as in case (ii), Eq. (16) is equivalent to (12). The derivative argument can thus be applied directly to (12). ∎

Theorem 3 is quite useful in that it allows us to quickly eliminate many distributions from achieving the rate $I(X : Y|Z)$. For example, consider when $p_{XY|Z=z}$ is uncorrelated for some $z \in \mathcal{Z}$, but $p_{XY|Z=z'}$ is perfectly correlated for some other $z' \in \mathcal{Z}$ with either $supp[p_{X|Z=z}] \subset supp[p_{X|Z=z'}]$ or $supp[p_{Y|Z=z}] \subset supp[p_{Y|Z=z'}]$. Here, perfectly correlated means that $p(x, y|z') = p(x|z')\delta_{x,y}$ up to relabeling. Then from Theorem 3, it follows that $I(X : Y|Z)$ is an achievable rate only if

$$p(x, y|z) > 0 \quad \Rightarrow \quad p(x|z')p(y|z') = 0.$$

In other words, it is always possible for either Alice or Bob to identify when $Z \neq z'$.

Finally, we close this section by comparing Theorems 2 and 3. In short, neither one supersedes the other. As noted above, distribution (b) in Fig. 2 satisfies the necessary condition of Theorem 2 for $\overrightarrow{K}(X : Y|Z) = I(X : Y|Z)$. However, Theorem 3 can be used to show that $K(X : Y|Z) < I(X : Y|Z)$. This is because $p_{XY|Z=1} \blacktriangleleft p_{XY|Z=2}$ yet $p(1, 1|2) = 0$ while $p(1, 1|1) = 1/3$. Therefore its key rate is strictly less than $I(X : Y|Z)$. Figure 3 depicts a distribution for which Theorem 3 cannot be applied but Theorem 2 shows that $\overrightarrow{K}(X : Y|Z) < I(X : Y|Z)$. The two-way key rate for this distribution is still unknown.

	X →		
Z = 0	0	1	2
Y ↓ 0	1/7	1/7	·
1	1/7	1/7	1/7
2	·	1/7	1/7

Z = 1	0	1
0	1/2	·
1	·	1/2

Fig. 3. The event $(x, y) = (0, 1)$ has conditional probabilities $p(0, 1 | Z = 0) > 0$ and $p(0, 1 | Z = 1) = 0$. However, we cannot use these facts in conjunction with Theorem 3 to conclude that $K(X : Y | Z) < I(X : Y | Z)$ since the distribution does not satisfy $p_{XY|Z=0} \blacktriangleleft p_{XY|Z=1}$ (neither $supp[p_{X|Z=0}] \subset supp[p_{X|Z=1}]$ nor $supp[p_{Y|Z=0}] \subset supp[p_{Y|Z=1}]$). On the other hand, since $p(0, 1 | Z = 0) > 0$, Theorem 2 can be applied to conclude that the one-way rate is less than $I(X : Y | Z)$.

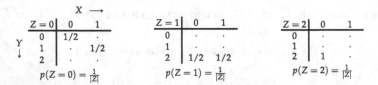

Fig. 4. A distribution requiring communication from Bob to Alice to achieve a key rate of $I(X : Y | Z)$.

4.4　Communication Dependency in Optimal Distillation

We next consider some general features of the public communication when performing optimal key distillation. Our main observations will be that (i) attaining a key rate of $I(X : Y | Z)$ by one-way communication may depend on the direction of the communication, and (ii) two-way communication may be necessary in order to achieve the key rate $I(X : Y | Z)$.

Example 1 (Optimal one-way distillation depends on communication direction). Consider the distribution depicted in Fig. 4 with $I(X : Y | Z) = 1/3$. When Bob is the communicating party, a protocol attaining this as a key rate is obvious: he simply announces whether or not $y \in \{0, 1\}$. If it is, they share one bit, otherwise they fail. Hence, $I(X : Y | Z) = 1/3$ is an achievable key rate.

However, the interesting question is whether or not the key rate $I(X : Y | Z)$ is achievable by one-way communication from Alice to Bob. We will now show that this is not possible. By Lemma 2, in order to obtain the rate $I(X : Y | Z)$, there must exist random variables U and V satisfying Eq. (8). Assume that such variables exist. If $U - Z - Y$, then $p(u | X = 0)p(u | X = 1) > 0$ for all $U = u$; otherwise, U and Y couldn't be independent. But then $X - KUZ - Y$ applied to $Z = 0$ means there must exist a pair $(k, u) \in \mathcal{K} \times \mathcal{U}$ such that

$$p(k, u | X = 0) = 0 \quad \& \quad p(k, u | X = 1) > 0.$$

Hence, $0 = p(k | Y = 2, U = u, Z = 2) < p(k | Y = 2, U = u, Z = 1)$, which contradicts $K - YU - Z$. Thus $\overrightarrow{K}(X : Y | Z) < I(X : Y | Z) = \overleftarrow{K}(X : Y | Z)$.

$$X \longrightarrow$$

$Z = 3$	0	1	2
Y 0	·	·	1/2
↓ 1	·	·	1/2

$$p(Z = 3) = \tfrac{1}{|Z|}$$

$Z = 4$	0	1	2
0	·	·	1
1	·	·	·

$$p(Z = 4) = \tfrac{1}{|Z|}$$

Fig. 5. Additional outcomes augmented to the distribution of Fig. 4. The enlarged distribution can no longer attain a key rate of $I(X : Y|Z)$ unless both parties communicate.

In this example, notice that if we restricted Eve's distribution to $\mathcal{Z} = \{0,1\}$ (i.e. $p(Z = 2) = 0$), then the rate $I(X : Y|Z)$ would indeed be achievable using one-way communication from Alice to Bob. This is because without the $z = 2$ outcome, the Markov Chain $X - Y - Z$ holds. Such a result is counterintuitive since Alice and Bob share no correlations when $z \in \{1,2\}$. And yet the distribution becomes one-way reversible from Alice to Bob when $p(Z = 2) = 0$, but otherwise it is not.

Example 2 (Optimal distillation requires two-way communication). The previous example can be generalized by adding two more outcomes for Eve so that $|Z| = 5$. The additional outcomes are shown in Fig. 5 and this is combined with Fig. 4 to give the full distribution. Notice that the distribution $p_{XY|Z=3}$ is obtained from $p_{XY|Z=1}$ simply by swapping Alice and Bob's variables, and likewise for $p_{XY|Z=4}$ and $p_{XY|Z=2}$. Hence by the argument of the previous example, if Eve were to reveal whether or not $z \in \{0,3,4\}$, then the average Bob-to-Alice distillable key c onditioned on this information would be less than $I(X : Y|Z)$. Likewise, if Eve were to reveal whether or not $z \in \{0,1,2\}$, then the Alice-to-Bob distillable key conditioned on this information would be less than $I(X : Y|Z)$. Thus since the average conditional key rate cannot exceed the key rate with no side information, we conclude that $I(X : Y|Z)$ is unattainable using one one-way communication in either direction. On the other hand, the distribution is easily seen to admit a key rate of $I(X : Y|Z)$ when the parties simply announce whether or not their variable belongs to the set $\{0,1\}$.

5   Conclusion

In this paper, we have considered when a secret key rate of $I(X : Y|Z)$ can be attained by Alice and Bob when working with a variety of auxiliary resources. The conditional mutual information quantifies the private key rate of p_{XYZ}, which is the rate of key private from Eve that is attainable when Eve helps Alice and Bob by announcing her variable. Therefore, distributions for which $K(X : Y|Z) = I(X : Y|Z)$ are those for which nothing is gained when Eve functions as a helper rather than a full adversary.

We have found that with no additional communication, the key rate is $I(X : Y|Z)$ if and only if the distribution is uniform block independent. Furthermore, supplying Alice and Bob with additional public randomness does not increase the distillable key rate. While this may not be overly surprising since the considered common randomness is uncorrelated with the source, it is nevertheless a nontrivial result because in general, randomness can serve a resource in distillation tasks [1, 10].

Turning to the one and two-way communication scenarios, we have presented in Theorems 2 and 3 necessary conditions for a distribution to attain the key rate $I(X:Y|Z)$. The conditions we have derived are all single-letter structural characterizations, and they are thus computationally easy to apply. We leave open the question of whether Theorem 3 is also sufficient for attaining $I(X:Y|Z)$, although we have no strong reason to believe this is true. Further improvements to the results of this paper can possibly be obtained by studying tighter bounds on $K(X:Y|Z)$ than the intrinsic information such as those presented in Refs. [11] and [7]. Nevertheless, we hope this paper has shed new light on the problem of secret key distillation under various communication settings.

6 Appendix

6.1 Proof of Propositions 1

Proof. (a) Trivially $\mathcal{X} \times \mathcal{Y}$ gives a common partitioning of length one, and any common partitioning cannot have length exceeding $\min\{|\mathcal{X}|, |\mathcal{Y}|\}$; hence a maximal common partitioning exists. To prove uniqueness, suppose that $(\mathcal{X}_i, \mathcal{Y}_i)_{i=1}^t$ and $(\mathcal{X}_i', \mathcal{Y}_i')_{i=1}^t$ are two maximal common partitionings. If they are not equivalent, then there must exist some subset, say $\mathcal{X}_{i_0}$ such that $\mathcal{X}_{i_0} \subset \cup_{\lambda=1}^K \mathcal{X}_\lambda'$ in which $\mathcal{X}_{i_0} \cap \mathcal{X}_\lambda' \neq \emptyset$ for $\lambda = 1, \cdots, K$, $K \geqslant 2$. Choose any such $\mathcal{X}_{\lambda_0}'$ from this collection and define the new sets $R_{i_0} = \mathcal{X}_{i_0} \cap \mathcal{X}_{\lambda_0}'$ and $\tilde{R}_{i_0} = \mathcal{X}_{i_0} \setminus \mathcal{X}_{\lambda_0}'$, which are both nonempty since $k \geqslant 2$ and the $\mathcal{X}_\lambda$ are disjoint. However, we also have the properties

$$x \in \mathcal{X}_{i_0} \Rightarrow p(\mathcal{Y}_{i_0}|x) = 1; \qquad x \in \mathcal{X}_{\lambda_0}' \Rightarrow p(\mathcal{Y}_{\lambda_0}'|x) = 1;$$
$$x \notin \mathcal{X}_{i_0} \Rightarrow p(\mathcal{Y}_{i_0}|x) = 0; \qquad x \notin \mathcal{X}_{\lambda_0}' \Rightarrow p(\mathcal{Y}_{\lambda_0}'|x) = 0.$$

(Here we are implicitly using condition (iii) in the above definition by assuming that $p(x) > 0$ thereby defining conditional distributions). Therefore, $p(S_{i_0}|R_{i_0}) = p(\tilde{S}_{i_0}|\tilde{R}_{i_0}) = 1$ and $p(S_{i_0}|\tilde{R}_{i_0}) = p(\tilde{S}_{i_0}|R_{i_0}) = 0$, where $S_{i_0} = \mathcal{Y}_{i_0} \cap \mathcal{Y}_{\lambda_0}'$ and $\tilde{S}_{i_0} = \mathcal{Y}_{i_0} \setminus \mathcal{Y}_{\lambda_0}'$. A similar argument shows that $p(R_{i_0}|S_{i_0}) = p(\tilde{R}_{i_0}|\tilde{S}_{i_0}) = 1$ and $p(R_{i_0}|\tilde{S}_{i_0}) = p(\tilde{R}_{i_0}|S_{i_0}) = 0$. Hence, $(\mathcal{X}_i, \mathcal{Y}_i)_{i \neq i_0}^t \bigcup (S_{i_0}, R_{i_0}) \bigcup (\tilde{S}_{i_0}, \tilde{R}_{i_0})$ is a common partitioning of length $t+1$. But this is a contradiction since $(\mathcal{X}_i, \mathcal{Y}_i)_{i=1}^t$ is a maximal common decomposition.

(b) Suppose that K satisfies $0 = H(K|X) = H(K|Y)$ so that $K = f(X) = g(Y)$ for some functions f and g. It is clear that f and g must be constant-valued for any pair of values taken from same block $\mathcal{X}_i \times \mathcal{Y}_i$ in the maximal common partitioning of XY. Hence the maximum possible entropy of K is then attained iff f and g take on a different value for each block in this partitioning.

(c) Suppose that C is not a function of J_{XY}. Then $H(CJ_{XY}) > H(J_{XY})$, which contradicts the maximality of J_{XY}. ∎

6.2   Proof of Theorem 1

Proof. **Achievability:** We will prove that $H(J_{XY}|Z)$ is an achievable rate without any auxiliary shared public randomness (i.e. W is constant). For n copies of p_{XYZ}, Alice and Bob extract their common information from each copy of p_{XYZ}. This will generate a sequence of J_{XY}^n, with Alice and Bob having identical copies of this sequence. It is now a matter of performing privacy amplification on this sequence to remove Eve's information [2]. The main construction is guaranteed to exist by the following lemma.

Lemma 3 (See Corollary 17.5 in [5]). *For an i.i.d. source of two random variables J_{XY} and Z with J_{XY} ranging over set $\mathcal{J}$, for any $\delta > 0$ and $k < 2^{n[H(J_{XY}|Z)-\delta]}$, there exists an $\epsilon > 0$ and a mapping $\kappa : \mathcal{J}^n \to \mathcal{K} = \{1, 2, \cdots, k\}$ such that*

$$\log |\mathcal{K}| - H(\kappa(J_{XY}^n)|Z^n) < 2^{-n\epsilon}.$$

From this lemma, it follows that $H(J_{XY}|Z)$ is an achievable key rate.

Converse: The converse proof follows analogously to the converse proof of Theorem 2.6 in Ref. [4] (see also [5]). We will first prove the converse under the assumption of no local randomness (i.e. Q_A and Q_B are constant). We will then show that adding local randomness does not change the result. Suppose that $K^{c.r.}(X : Y|Z) = R$. We consider a slightly weaker security condition than the one presented in Sect. 2. This is done by replacing (ii) with (ii'): $\frac{1}{n}(\log |\mathcal{K}| - H(K|Z^nW)) < \epsilon$. Under this weaker assumption, we can assume without loss of generality that K is a function of (X^n, Q_A, W); i.e. $K = f(X^n, Q_A, W)$ [5]. Then, for every $\delta, \epsilon > 0$ and n sufficiently large, there exists a random variable W independent of $X^nY^nZ^n$ along with functions $f(X^n, W)$ and $g(Y^n, W)$ satisfying (i) $Pr[f = g = K] > 1 - \epsilon$, (ii') $\frac{1}{n}(\log |\mathcal{K}| - H(K|Z^nW)) < \epsilon$ and (iii) $\frac{1}{n}\log |\mathcal{K}| \geqslant R$.

Note that from (i) in the security condition, Fano's Inequality together with data processing gives

$$H(K|Y^nW) < h(\epsilon) + \epsilon(\log |\mathcal{K}| - 1). \tag{17}$$

Combining this with (ii') gives

$$\frac{1}{n}(1 - \epsilon)\log |\mathcal{K}| < \frac{1}{n}[H(K|Z^nW) - H(K|Y^nW) + h(\epsilon) - \epsilon],$$

and so

$$R \leqslant \frac{1}{n}\log |\mathcal{K}| + \delta < \frac{1}{1-\epsilon} \cdot \frac{1}{n}[H(K|Z^nW) - H(K|Y^nW)] + \frac{h(\epsilon) - \epsilon}{1-\epsilon} \cdot \frac{1}{n} + \delta. \tag{18}$$

To analyze the quantity $H(K|Z^nW) - H(K|Y^nW)$, we will use a standard trick.

Lemma 4. *Let J be uniformly distributed over the set $\{1, \cdots, n\}$ and let $A^{(i)}$ denote the ith instance of A in A^n. Likewise, let $A^{(<i)} = A^{(1)} \cdots A^{(i-1)}$ and*

$A^{(>i)} = A^{(i+1)} \cdots A^{(n)}$ with $A^{(<1)} := \varnothing$ and $A^{(n+1)} := \varnothing$. Then for random variables P and Q and sequences of random variables A^n, B^n,

$$H(P|A^nQ) - H(P|B^nQ) = n[I(P : B^{(J)}|TQ) - I(P : A^{(J)}|TQ)], \qquad (19)$$

where $T = JA^{(>J)}B^{(<J)}$

Proof. See, e.g., proof of Lemma 17.12 in [5].

Then we can use Lemma 4 to obtain

$$H(K|Z^nW) - H(K|Y^nW) = n[I(K : Y^{(J)}|UW) - I(K : Z^{(J)}|UW)], \qquad (20)$$

where $U := JY^{(<J)}Z^{(>J)}$. Notice that for any $i \in \{1, \cdots, n\}$ we have

$$X^{(<i)}X^{(>i)}Y^{(<i)}Z^{(>i)} - X^{(i)} - Y^{(i)}Z^{(i)}, \qquad (21)$$

since the sampling is i.i.d.. Therefore, because K is a function of (X^n, W), we have $KU - X^{(J)}W - Y^{(J)}Z^{(J)}$. Removing the superscript "J" and taking $\epsilon, \delta \to 0$, we have the bound

$$R \leqslant I(K : Y|UW) - I(K : Z|UW) \qquad (22)$$

such that $KU - XW - YZ$.

Next, Eq. (17) gives

$$h(\epsilon) + \epsilon(\log |\mathcal{K}| - 1) > H(K|Y^nW) - H(K|X^nW)$$
$$= n[I(K : X^{(J)}|JY^{(<J)}X^{(>J)}W) - I(K : Y^{(J)}|JY^{(<J)}X^{(>J)}W)],$$

where the first inequality follows because $H(K|X^nW)$ is nonnegative and the equality follows from Lemma 4. We want to put this in terms of U. To do this, note that

$$I(K : X^{(J)}|JY^{(<J)}X^{(>J)}W)$$
$$= I(KY^{(<J)}X^{(>J)} : X^{(J)}|JW)$$
$$= I(KY^{(<J)}X^{(>J)}Z^{(>J)} : X^{(J)}|JW) - I(Z^{(>J)} : X^{(J)}|JKY^{(<J)}X^{(>J)}W)$$
$$= I(KUX^{(>J)} : X^{(J)}|JW) = I(KU : X^{(J)}|JW) + I(X^{(>J)} : X^{(J)}|KUW),$$

where the first equality follows from the chain rule and $I(Y^{(<J)}X^{(>J)} : X^{(J)}JW) = 0$, and in the second equality

$$I(Z^{(>J)} : X^{(J)}|JKY^{(<J)}X^{(>J)}W) \leqslant I(Z^{(>J)} : KX^{(J)}|JY^{(<J)}X^{(>J)}W)$$
$$= I(Z^{(>J)} : X^{(J)}|JY^{(<J)}X^{(>J)}W) = 0.$$

Here we use $I(Z^{(>J)} : K|JY^{(<J)}X^{(\geqslant J)}W) = 0$ since $K - JY^{(<J)}X^{(\geqslant J)}W - Z^{(>J)}$ is a Markov chain. Again this follows from the basic Markov condition

$K - WX^n - Y^nZ^n$ and the sampling is i.i.d.. The second equality follows from i.i.d. sampling and W independence of X^n, Y^n, Z^n.

A similar analysis likewise gives

$$I(K : Y^{(J)}|JY^{(<J)}X^{(>J)}W) = I(KU : Y^{(J)}|JW) + I(X^{(>J)} : Y^{(J)}|KUW)$$
$$\leqslant I(KU : Y^{(J)}|JW) + I(X^{(>J)} : X^{(J)}|KUW),$$

where the inequality follows from the Markov condition

$$X^{(>J)} - KUX^{(J)}W - Y^{(J)},$$

a consequence of the more obvious condition $KUX^n - JX^{(J)}W - Y^{(J)}$. Putting everything together yields

$$h(\epsilon) + \epsilon(\log |\mathcal{K}| - 1)$$
$$> I(KU : X^{(J)}|JW) - I(KU : Y^{(J)}|JW)$$
$$= I(KU : X^{(J)}Y^{(J)}|JW) - I(KU : Y^{(J)}|JX^{(J)}W) - I(KU : Y^{(J)}|JW) \quad (23)$$
$$= I(KU : X^{(J)}|JY^{(J)}W) + I(KU : Z^{(J)}|JY^{(J)}X^{(J)}W) \quad (24)$$
$$= I(KU : X^{(J)}Z^{(J)}|JY^{(J)}W),$$

where the second term in Eq. (23) is zero from the already proven Markov chain $KU - XW - YZ$, and in Eq. (24) we use the fact that $I(KU : Z^{(J)}|JY^{(J)}X^{(J)}W) = 0$. Removing the superscript "J" and taking $\epsilon \to 0$ necessitates the Markov chain $KU - YW - XZ$.

It is easy to verify that the double Markov chain $K - XW - Y$ and $K - YW - X$ implies that $I(K : XY|J_{XY}W) = 0$ (see Exercise 16.25 in [5]). Since K is a function of (X, W), we have that $H(K|J_{XY}W) = 0$. Thus, K must also be a function of (Y, W). Continuing Eq. (22) gives the bound

$$R \leqslant I(K : Y|UW) - I(K : Z|UW) = H(K|UW) - I(K : Z|UW)$$
$$= H(K|ZUW) \leqslant H(K|ZW). \quad (25)$$

We have therefore obtained the following:

$$R \leqslant \max H(K|ZW), \quad (26)$$

where the maximization is taken over all variables K such that $H(K|XW) = H(K|YW) = 0$.

This can be further bounded by using the following proposition.

Proposition 4. *If W is independent of XY and $H(K|XW) = H(K|YW) = 0$, then K is a function of (J_{XY}, W).*

Proof. The fact that $H(K|XW) = H(K|YW) = 0$ implies the existence of two functions $f(X, W)$ and $g(Y, W)$ such that $Pr[f(X, W) = g(Y, W)] = 1$. Consequently, if $p(x_1, y_1)p(x_1, y_2) > 0$, then $f(x_1, w) = g(y_1, w) = g(y_2, w)$ for all

$w \in \mathcal{W}$ with $p(w) > 0$. Indeed, if, say, $f(x_1, w) \neq g(y_1, w)$, then $Pr[f(X, W) \neq g(Y, W)] \geqslant p(x_1, y_1, w) = p(x_1, y_2)p(w) > 0$, where we have used the independence between XY and W. By the same reasoning, $p(x_1, y_1)p(y_1, x_2) > 0$ implies that $f(x_1, w) = f(x_2, w) = g(y_1, w)$ for all $w \in \mathcal{W}$. Turning to Proposition 2, if $J_{XY}(x) = J_{XY}(x')$, then there exists a sequence $xy_1x_1y_2x_2 \cdots y_nx'$ such that $p(xy_1)p(y_1x_1)p(x_1y_2) \cdots p(y_nx') > 0$. Therefore, as just argued, we must have that $f(x, w) = f(x', w)$ for all $w \in \mathcal{W}$. Hence K must be a function of (J_{XY}, W).

We now apply Proposition 4 to Eq. (26). Suppose that K obtains the maximization in Eq. (26). Then, since K is a function of (J_{XY}, W), we have that

$$H(K|ZW) \leqslant H(J_{XY}W|ZW) = H(J_{XY}|ZW) \leqslant H(J_{XY}|Z). \qquad (27)$$

This proves the desired upper bound under no local randomness.

To consider the case when Alice and Bob have local randomness Q_A and Q_B, respectively, define $\hat{X} := (X, Q_A)$ and $\hat{Y} := (Y, Q_B)$. Then repeating the above argument shows that $R \leqslant H(J_{\hat{X}\hat{Y}}|Z)$. It is straightforward to show that with Q_A and Q_B pairwise independent and independent of XY, we have $J_{\overline{X},\overline{Y}} = J_{XY}$. ∎

References

1. Ahlswede, R., Csiszár, I.: Common randomness in information theory and cryptography. i. secret sharing. IEEE Trans. Inf. Theory 39(4), 1121–1132 (1993)
2. Bennett, C., Brassard, G., Crepeau, C., Maurer, U.: Generalized privacy amplification. IEEE Trans. Inf. Theory **41**(6), 1915–1923 (1995)
3. Christandl, M., Renner, R., Wolf, S.: A property of the intrinsic mutual information. In: Proceedings of the IEEE International Symposium on Information Theory 2003, pp. 258–258, June 2003
4. Csiszár, I., Narayan, P.: Common randomness and secret key generation with a helper. IEEE Trans. Inf. Theory **46**(2), 344–366 (2000)
5. Csiszár, I., Körner, J.: Information Theory: Coding Theorems for Discrete Memoryless Systems. Cambridge University Press, Cambridge (2011)
6. Gács, P., Körner, J.: Common information is far less than mutual information. Probl. Control Inf. Theory **2**(2), 149 (1973)
7. Gohari, A., Anantharam, V.: Information-theoretic key agreement of multiple terminals; part i. IEEE Trans. Inf. Theory **56**(8), 3973–3996 (2010)
8. Maurer, U.: Secret key agreement by public discussion from common information. IEEE Trans. Inf. Theory **39**(3), 733–742 (1993)
9. Maurer, U., Wolf, S.: Unconditionally secure key agreement and the intrinsic conditional information. IEEE Trans. Inf. Theory **45**(2), 499–514 (1999)
10. Ozols, M., Smith, G., Smolin, J.A.: Bound entangled states with a private key and their classical counterpart. Phys. Rev. Lett. **112**, 110502 (2014)
11. Renner, R., Wolf, S.: New bounds in secret-key agreement: the gap between formation and secrecy extraction. In: Biham, E. (ed.) EUROCRYPT 2003. LNCS, vol. 2656, pp. 562–577. Springer, Heidelberg (2003)

Privacy with Imperfect Randomness

Yevgeniy Dodis[1]([✉]) and Yanqing Yao[2]

[1] Department of Computer Science, New York University, New York, USA
dodis@cs.nyu.edu
[2] School of Computer Science and Engineering, Beihang University, Beijing, China
yaoyanqing1984@gmail.com

Abstract. We revisit the impossibility of a variety of cryptographic tasks including privacy and differential privacy with imperfect randomness. For traditional notions of privacy, such as security of encryption, commitment or secret sharing schemes, dramatic impossibility results are known [MP90, DOPS04] for several concrete sources $\mathcal{R}$, including a (seemingly) very "nice and friendly" Santha-Vazirani (SV) source. Somewhat surprisingly, Dodis et al. [DLMV12] showed that non-trivial *differential* privacy is possible with the SV sources. This suggested a qualitative gap between traditional and differential privacy, and left open the question of whether differential privacy is possible with more realistic (i.e., less structured) sources than the SV sources.

Motivated by this question, we introduce a new, modular framework for showing strong impossibility results for (both traditional and differential) privacy under a *general* imperfect source $\mathcal{R}$. As direct corollaries of our framework, we get the following new results:

(1) Existing, but *quantitatively improved*, impossibility results for traditional privacy, but under a wider variety of sources $\mathcal{R}$.
(2) First impossibility results for *differential* privacy for a variety of realistic sources $\mathcal{R}$ (including most "block sources", but not the SV source).
(3) Any imperfect source allowing (either traditional or differential) privacy under $\mathcal{R}$ admits a certain type of deterministic bit extraction from $\mathcal{R}$.

1 Introduction

Traditional cryptographic tasks take for granted the availability of perfect random sources, i.e., sources that output unbiased and independent random bits. However, in many situations it seems unrealistic to expect a source to be perfectly random, and one must deal with various imperfect sources of randomness. Some well known examples of such imperfect random sources are physical sources [BST03, BH05], biometric data [BDK+05, DORS08], secrets with partial leakage, and group elements from Diffie-Hellman key exchange [GKR04, Kra10].

IMPERFECT SOURCES. To abstract this concept, several formal models of imperfect sources have been described (e.g., [vN51, CFG+85, Blu86, SV86, CG88,

---

Y. Yao—Most of this work was done while the author visited New York University.

R. Gennaro and M. Robshaw (Eds.): CRYPTO 2015, Part II, LNCS 9216, pp. 463–482, 2015.
DOI: 10.1007/978-3-662-48000-7_23

LLS89, Zuc96, ACRT99, Dod01]). Roughly, they can be divided into extractable and non-extractable. Extractable sources (e.g., [vN51, CFG+85, Blu86, LLS89]) allow for deterministic extraction of nearly perfect randomness. And, while the question of optimizing the extraction rate and efficiency has been very interesting, from the qualitative perspective such sources are good for any application where perfect randomness is sufficient. Unfortunately, it was quickly realized many imperfect sources are non-extractable [SV86, CG88, Dod01]. The simplest example is the Santha-Vazirani (SV) source [SV86], which produces an infinite sequence of bits $r_1, r_2, \ldots$, with the property that $\Pr[r_i = 0 \mid r_1 \ldots r_{i-1}] \in [\frac{1}{2}(1 - \gamma), \frac{1}{2}(1 + \gamma)]$, for any setting of the prior bits $r_1, \ldots, r_{i-1}$. Namely, each bit has almost one bit of fresh entropy, but can have a small bias $\gamma < 1$. Santha and Vazirani [SV86] showed that there exists no deterministic extractor Enc : $\{0, 1\}^n \to \{0, 1\}$ capable of extracting even a *single* bit of bias *strictly* less than γ from the γ-SV source, irrespective of how many SV bits $r_1, \ldots, r_n$ it is willing to wait for.

Despite this pessimistic result, ruling out the "black-box compiler" from imperfect (e.g., SV) to perfect randomness for *all* applications, one may still hope that specific "non-extractable" sources, such as SV-sources, might be sufficient for *concrete* applications, such as simulating probabilistic algorithms or cryptography. Indeed, a series of results [VV85, SV86, CG88, Zuc96, ACRT99] showed that very "weak" sources (including SV-sources and even much more realistic "weak" and "block" sources) are sufficient for simulating probabilistic polynomial-time algorithms; namely, for problems which do not inherently need randomness, but which could potentially be sped up using randomization. Moreover, even in the area of cryptography — where randomness is *essential* (e.g., for key generation) — it turns out that many "non-extractable" sources (again, including SV sources and more) are sufficient for *authentication* applications, such as the designs of MACs [MW97, DKRS06] and even signature schemes [DOPS04, ACM+14] (under appropriate hardness assumptions). Intuitively, the reason for the latter "success story" is that authentication applications only require that it is hard for the attacker to completely guess (i.e., "forge") some long string, so having min-entropy in our source should be sufficient to achieve this goal.

NEGATIVE RESULTS FOR PRIVACY WITH IMPERFECT RANDOMNESS. In contrast, the situation appears to be much less bright when dealing with *privacy* applications, such as encryption, commitment, zero-knowledge, and a few others. First, McInnes and Pinkas [MP90] showed that unconditionally secure symmetric encryption cannot be based on SV sources, even if one is restricted to encrypting a single bit. This result was subsequently strengthened by Dodis et al. [DOPS04], who showed that SV sources are not sufficient for building even computationally secure encryption (again, even of a single bit), and, in fact, essentially any other cryptographic task involving "privacy" (e.g., commitment, zero-knowledge, secret sharing and others). This was again strengthened by Austrin et al. [ACM+14], who showed that the negative results still hold even if the SV source is efficiently samplable. Finally, Bosley and Dodis [BD07] showed an even more negative result: if a source of randomness $\mathcal{R}$ is "good enough" to generate a secret key capable of encrypting k bits, then one can deterministically extract nearly k almost uniform bits from $\mathcal{R}$, suggesting that traditional privacy *requires* an "extractable" source of randomness.[1]

WHAT ABOUT DIFFERENTIAL PRIVACY? While the above series of negative results seem to strongly point in the direction that privacy inherently requires extractable randomness, a recent work of Dodis et al. [DLMV12] put a slight dent into this consensus, by showing that SV sources are provably sufficient for achieving a more recent notion of privacy, called *differential privacy* [DMNS06]. Intuitively, a differentially private mechanism $M(D, \mathbf{r})$ uses its randomness $\mathbf{r}$ to add some "noise" to the true answer $q(D)$, where D is some sensitive database of users, and q is some useful aggregate information (query) about the users of D. This noise is added in a way as to satisfy the following two conflicting properties (see Definitions 6 and 7 for formalism):

(a) *ε-differential privacy* (ε-DP): up to "advantage" ε, the returned value $z = M(D, \mathbf{r})$ does not tell any information about the value $D(i)$ of any individual user i, which was not already known to the attacker before z was returned;

(b) *ρ-utility*: on average (over $\mathbf{r}$), $|z - q(D)|$ is upper bounded by ρ, meaning that perturbed answer is not too far from the true answer.

Since we will be mainly talking about negative results, for the rest of this work we will restrict our attention to the simplest concrete example of differential privacy, where a "record" $D(i)$ is a single bit, and q is the Hamming weight $wt(D)$ of the corresponding bit-vector D (i.e., $wt(D) = \sum D(i)$). In this case, a very simple ε-DP mechanism [DMNS06] $M(D, \mathbf{r})$ would simply return $wt(D) + e(\mathbf{r})$ (possibly truncated to always be between 0 and $|D|$), where $e(\mathbf{r})$ is an appropriate noise[2] with $\rho = \mathbb{E}[|q(\mathbf{r})|] \approx 1/\varepsilon$. Intuitively, this setting ensures that when $D(i)$ changes from 0 to 1, the answer distribution $M(D, \mathbf{r})$ does not "change" by more than ε.

Coming back to Dodis et al. [DLMV12], the authors show that although no "additive noise" mechanism of the form $M(D, \mathbf{r}) = wt(D) + e(\mathbf{r})$ can simultaneously withstand all γ-SV-distributions $\mathbf{r} \leftarrow R$, a better designed mechanism (that they also constructed) is capable of working with all such distributions, provided that the utility ρ is now relaxed to be polynomial in $1/\varepsilon$, whose degree and coefficients depend on γ, but *not* on the size of the database D. Moreover, *the value ε can be made an arbitrarily small constant* (e.g., $\varepsilon \ll \gamma$). This should be contrasted with the impossibility results for the traditional privacy [MP90, DOPS04] with SV sources, where it was shown that $\varepsilon = \Omega(\gamma)$, meaning that even a fixed *constant* (let alone "negligible") security is impossible. Hence, the result of [DLMV12] suggested a *qualitative gap between traditional*

---

[1] On the positive side, [DS02, BD07] showed that extractable sources are not strictly necessary for encrypting a "very small" number of bits. Still, for natural "non-extractable" sources, such as SV sources, it is known that encrypting even a single bit is impossible [SV86, DOPS04, ACM+14].

[2] So called Laplacian distribution, but the details do not matter here.

and differential privacy, but left open the question of whether differential privacy is possible with more realistic (i.e., less structured) sources than the SV sources. Indeed, the SV sources seem to be primarily interesting from the perspective of negative results, since real-world distributions are unlikely to produce a sequence of bits, each of which has almost a full unit of fresh entropy.

OUR RESULTS IN BRIEF. In part motivated by solving this question, we abstract and generalize prior techniques for showing impossibility results for achieving privacy with various imperfect sources of randomness. Unlike prior work (with the exception of [BD07]), which focused on specific imperfect sources $\mathcal{R}$ (e.g., SV sources), we obtain most of our results for *general* sources $\mathcal{R}$, but then use various natural sources (namely, SV sources [SV86], weak/block sources [CG88], and Bias-Control Limited sources [Dod01]) as specific examples to illustrate our technique. In particular, we introduce the concepts of *expressiveness* and *separability* of a given imperfect source $\mathcal{R}$ as a measure of its "imperfectness", and show the following results:

- Low levels of expressiveness generically imply strong impossibility results for *differential* as well as traditional privacy.
- We reduce expressiveness to separability and prove the equivalence between "weak bit extraction" and NON-separability.
- Though the separability of some concrete (e.g., SV) sources $\mathcal{R}$ was implicitly known, we show new separability results for several important sources, including general "block sources".

We stress that the first two results are completely generic, and reduce the question of feasibility of privacy under $\mathcal{R}$ to a much easier and self-contained question of separability of $\mathcal{R}$. And establishing the latter is the only "source-specific" technical work which remains. In particular, after explicitly stating known separability results for weak and SV sources, and establishing our new separability results for block and Bias-Control Limited (BCL) sources, we obtain the following direct corollaries:

- Existing, but *quantitatively improved*, impossibility results for traditional privacy, but under a wider variety of sources $\mathcal{R}$ (i.e., weak, block, SV, BCL).
- First impossibility results for *differential* privacy. Although, unsurprisingly, these results (barely) miss the highly structured SV sources, they come back *extremely quickly* once the source becomes slightly more realistic (e.g., a very "constrained" weak/block/BCL source).
- Any imperfect source allowing (either traditional or differential) privacy admits a certain type of deterministic bit extraction. (This result is incomparable to the result of [BD07].)

We briefly expand on these results below, but conclude that, despite the result of [DLMV12], our results seem to unify and strengthen the belief that, for the most part, privacy with imperfect randomness is impossible, unless the source is (almost) deterministically extractable. More importantly, they provide an intuitive, modular and unified picture elucidating the (im)possibility of privacy with *general* imperfect sources.

1.1   Our Results in More Detail

At a high level, our results follow the blueprint of [DOPS04] (who concentrated exclusively on the SV sources), but in significantly more modular and quantitatively optimized way (making our proofs somewhat more illuminating, in our opinion). In essence, they establish an impossibility of a given privacy task P under a source $\mathcal{R}$ using three steps:

STEP 1: IMPOSSIBILITY OF TASK P UNDER $\mathcal{R}$ $\longrightarrow$ EXPRESSIVENESS OF $\mathcal{R}$.
Intuitively, *expressiveness* of $\mathcal{R}$ means that $\mathcal{R}$ is rich enough to "distinguish" any functions f and g which are not point-wise equal almost everywhere (see Definition 1): there exists $R \in \mathcal{R}$ s.t. $\mathsf{SD}(f(R), g(R))$ is "noticeable", where SD is the statistical distance between distributions.[3] With this clean abstraction, we almost trivially show (see Theorem 1) that most traditional privacy tasks P (extraction, encryption, secret sharing, commitment) imply the existence of sufficiently-distinct functions f and g that violate the expressiveness of $\mathcal{R}$. For example, such $f(\mathbf{r})$ and $g(\mathbf{r})$ are simply the encryptions of two different plaintexts under key $\mathbf{r}$ when P is encryption, and similar arguments hold for commitment, extraction and secret sharing schemes.

More interestingly, we show expressiveness is again sufficient to rule out even *differential* privacy (Theorem 2). The proof follows the same high-level intuition as for the traditional privacy, but is somewhat more involved. This is because DP only gives us security for "close" databases, while the utility guarantees are only meaningful for "far" databases. In particular, for this reason it will turn out that the expressiveness requirement on $\mathcal{R}$ for ruling out differential privacy will be slightly higher than that for traditional privacy (Theorem 2 vs. Theorem 1).[4] Still, aside from this quantitative difference, there is no qualitative difference between our arguments for traditional and differential privacy.

Overall, the deceptive simplicity of our "privacy-to-expressiveness" arguments is actually a *feature* of our framework, as these arguments are the *only place when the specific details of P matter*, as the rest of the framework — described below — will only concentrate on the expressiveness of $\mathcal{R}$!

STEP 2: EXPRESSIVENESS OF $\mathcal{R}$ $\longrightarrow$ SEPARABILITY OF $\mathcal{R}$.
Intuitively, *separability* of $\mathcal{R}$ means that $\mathcal{R}$ is rich enough to "separate" any sufficiently large disjoint sets G and B (see Definition 8; wlog, assume that $|G| \geq |B|$): there exists $R \in \mathcal{R}$ s.t. $(\Pr[R \in G] - \Pr[R \in B])$ is "noticeable".[5] A moment reflection shows that separability is closely related to expressiveness, but restricted to *boolean* functions f and g of disjoint support (i.e., the

---

[3] Like in [DOPS04] and unlike [MP90], our distinguishers between $f(R)$ and $g(R)$ will be very efficient, but we will not require this in order not to clutter the notation.

[4] Jumping ahead, this will be the reason although our new impossibility results for DP will (barely) miss the SV sources, they will come back very quickly once the source becomes more realistic.

[5] For example, if $\mathcal{R}$ only consists of the uniform distribution U_n, the latter is impossible when $|G| = |B|$. In contrast, we will see that natural "non-extractable" sources (i.e., weak, block, SV, and BCL sources) are separable.

characteristic functions of G and B), which makes it noticeably easier to work with (as we will see).

Nevertheless, we show that *separability generically implies expressiveness*, with nearly identical parameters (see Theorem 3). This is where we differ and quantitatively improve the argument implicit in [DOPS04]: while [DOPS04] used a bit-by-bit hybrid argument to show expressiveness (for the SV source), our proof of Theorem 3 used a more clever "universal hashing trick",[6] allowing us to obtain results which are independent of the ranges of f and g (which, in turn, will later correspond to bit sizes of ciphertexts, commitments, secret shares, etc.)

Of independent interest, we also show that NON-separability of $\mathcal{R}$ is equivalent to some type of "weak bit extraction" from $\mathcal{R}$ (see Theorem 4): (a) when produced, the extracted bit is guaranteed to be almost unbiased, (b) although the extractor is allowed to fail, it will typically succeed at least on the uniform distribution.[7]

Coupled with Step 1, we get the following two implications. First, we reduce the impossibility of many privacy tasks P under $\mathcal{R}$ to a much easier question of separability of $\mathcal{R}$ (which is independent of P). Second, we generically show that the feasibility of P under $\mathcal{R}$ implies deterministic weak bit extraction from $\mathcal{R}$, incomparably complementing the prior result of [BD07]. Namely, [BD07] showed that several traditional privacy primitives, including (only multi-bit) encryption and commitment (but not secret sharing) imply the existence of multi-bit deterministic extraction schemes capable of extracting almost the same number of bits as the plaintext. On the positive, our result applies to a much wider set of primitives P (e.g., secret-sharing, as well as even *single-bit* encryption and commitment). On the negative, we can only argue a rather weak kind of single-bit extraction, where the extractor is allowed to fail, while [BD07] showed traditional, and possibly multi-bit, extraction.

STEP 3: SEPARABILITY OF VARIOUS SOURCES $\mathcal{R}$.
Unlike the prior results in [MP90, DOPS04, ACM+14], all the above results are true for any imperfect source $\mathcal{R}$. To get concrete impossibility results for natural sources, though, we finally must establish good separability bounds for specific $\mathcal{R}$. Such bounds were already implicitly known [DOPS04] (or trivial to see) for the SV and general weak sources, but we show how they can also be demonstrated for other natural sources: block sources [CG88] and Bias-Control Limited sources [Dod01]. In particular, our separability bounds for block sources turned out to be quite non-trivial, and form one of the more technical contributions of this work. See the proof of Lemma 2(b).

Aside from being natural and interesting in their own right, the new separability results for block/BCL sources are especially interesting from the perspective of differential privacy (see below). Indeed, both of them can be viewed as realistic relaxations of highly-structured (and unrealistic!) SV sources, but yet not

---

[6] Similar trick with randomness extractors was used, in a slightly different context, by [ACM+14].

[7] Unfortunately, we demonstrate that the limitation of part (b) holding only for the uniform distribution is somewhat inherent *in this great level of generality*.

as general/unstructured as weak sources. And since we already know that DP *is* possible with SV sources [DLMV12], it is interesting to know how soon it will take for the impossibility results to come back, once the source slowly becomes more realistic/unstructured, but before going "all the way" to being weak.

PUTTING THEM ALL TOGETHER: NEW AND OLD IMPOSSIBILITY RESULTS. Applying Steps 1–3 to specific sources of interest (i.e., weak, block, SV, and BCL sources), we immediately derive a variety of impossibility results for traditional privacy (see Table 1). Although these results were derived mainly as a "warm-up" to our (completely new) impossibility results for differentially privacy, they offer quantitative improvements to the results of [DOPS04] (due to stronger expressiveness-to-separability reduction). For example, they rule out even constant (as opposed to negligible) security for encryption/commitment/secret sharing, irrespective of the sizes of ciphertexts/commitments/shares. Relatedly, we unsurprisingly get stronger impossibility results for block/BCL sources than the more structured SV sources.

More interestingly, we obtain first impossibility results for differential privacy with imperfect randomness. In light of the positive result of [DLMV12], our separability result for SV sources is (barely) not strong enough to rule out differential privacy under SV sources. As we explained, this failure happened not because our framework was too weak to apply to SV sources or differential privacy, but rather due to a "local-vs-global gap" between the privacy and utility requirements for differential privacy.

However, once we consider general weak sources, or even much more structured BCL/block sources, the impossibility results come back *extremely quickly!* For example, when studying ε-DP with utility ρ, n-bit weak sources of min-entropy k are ruled out the moment $k = n - \log(\varepsilon\rho) - O(1)$ (Theorem 6(a)),[8] while BCL sources are ruled out the moment the number of "SV bits" b the attacker can fix completely (instead of only bias by γ) is just $b = \Omega(\log(\varepsilon\rho)/\gamma)$ (Theorem 6(c)). As $\varepsilon\rho$ is typically desired to be a constant, $\log(\varepsilon\rho)$ is an even smaller constant, which means we even rule out *constant* entropy deficiency $(n - k)$ (or $m - k$ for block source) or number of "interventions" b, respectively. We also compare impossibility results for traditional and differential privacy in Table 2, and observe that the latter are only marginally weaker than the former. This leads us to the conclusion that differential privacy is still rather demanding to achieve with realistic imperfect sources of randomness.

Due to space limitations, most proofs are deferred to the full version [DY14].

2   Preliminaries

Let U_S be the uniform distribution over a set S. For simplicity, $U_n \stackrel{def}{=} U_{\{0,1\}^n}$. For a distribution or random variable R, let $\mathbf{r} \leftarrow R$ denote the operation of

---

[8] More generally, even n-bit block sources with block length m and fresh min-entropy k per block are ruled out when $k = m - \log(\varepsilon\rho) - O(1)$, irrespective of the number of blocks n/m. See Theorem 6(b).

sampling a random $\mathbf{r}$ according to R, and $\mathbf{H}_\infty(R) \overset{def}{=} \min_{\mathbf{r} \in \mathrm{supp}(R)} \log \frac{1}{\Pr[R=\mathbf{r}]}$ denote the min-entropy of R. We call a family of distributions over $\{0,1\}^n$ a source, denoted as $\mathcal{R}_n$. All logarithms are to the base 2.

For two random variables R and R' over $\{0,1\}^n$, the statistical distance between R and R' is defined as $\mathsf{SD}(R,R') \overset{def}{=} \frac{1}{2} \sum_{\mathbf{r} \in \{0,1\}^n} |\Pr[R = \mathbf{r}] - \Pr[R' = \mathbf{r}]|$. One can observe that $\mathsf{SD}(R,R') = \max_{\mathsf{Eve}} |\Pr[\mathsf{Eve}(R) = 1] - \Pr[\mathsf{Eve}(R') = 1]|$, where Eve is a distinguisher. We say that the relative distance between R and R' is ε, denoted as $\mathsf{RD}(R,R') = \varepsilon$, if ε is the smallest number such that $e^{-\varepsilon} \cdot \Pr[R' = \mathbf{r}] \leq \Pr[R = \mathbf{r}] \leq e^{\varepsilon} \cdot \Pr[R' = \mathbf{r}]$ for all $\mathbf{r} \in \{0,1\}^n$. It's easy to see that $\mathsf{RD}(R,R') \leq \varepsilon$ implies $\mathsf{SD}(R,R') \leq e^{\varepsilon} - 1$.

3   Expressiveness and Its Implications to Privacy

In this section, we introduce the concept of expressiveness of a source. Then we study its implications to both traditional and differential privacy.

Informally, an expressive source $\mathcal{R}_n$ can separate two distributions $f(R)$ and $g(R)$, unless the functions f and g are point-wise equal almost everywhere.

Definition 1. *We say that a source $\mathcal{R}_n$ is $(t, \delta)-$expressive if for any functions $f, g : \{0,1\}^n \to \mathcal{C}$, where $\mathcal{C}$ is any universe, such that $\Pr_{\mathbf{r} \leftarrow U_n} [f(\mathbf{r}) \neq g(\mathbf{r})] \geq \frac{1}{2^t}$ for some $t \geq 0$, there exists a distribution $R \in \mathcal{R}_n$ such that $\mathsf{SD}(f(R), g(R)) \geq \delta$.*

3.1   Implications to Traditional Privacy

We recall (or define) some cryptographic primitives related to traditional privacy: bit extractor, bit encryption scheme, weak bit commitment, and bit T-secret sharing as follows.

Definition 2. *We say that $\mathsf{Ext} : \{0,1\}^n \to \{0,1\}$ is $(\mathcal{R}_n, \delta)$-secure bit extractor if for every distribution $R \in \mathcal{R}_n$, $|\Pr_{\mathbf{r} \leftarrow R}[\mathsf{Ext}(\mathbf{r}) = 1] - \Pr_{\mathbf{r} \leftarrow R}[\mathsf{Ext}(\mathbf{r}) = 0]| < \delta$ (equivalently, $\mathsf{SD}(\mathsf{Ext}(R), U_1) < \delta/2$).*

In the following, we consider the simplest encryption scheme, where the plaintext is composed of a single bit x.

Definition 3. *A $(\mathcal{R}_n, \delta)-$secure bit encryption scheme is a tuple of functions $\mathsf{Enc} : \{0,1\}^n \times \{0,1\} \to \{0,1\}^\lambda$ and $\mathsf{Dec} : \{0,1\}^n \times \{0,1\}^\lambda \to \{0,1\}$, where, for convenience, $\mathsf{Enc}(\mathbf{r}, x)$ (resp. $\mathsf{Dec}(\mathbf{r}, \mathbf{c})$) is denoted as $\mathsf{Enc}_{\mathbf{r}}(x)$ (resp. $\mathsf{Dec}_{\mathbf{r}}(\mathbf{c})$), satisfying the following two properties:*

(a) Correctness: for all $\mathbf{r} \in \{0,1\}^n$ and $x \in \{0,1\}$, $\mathsf{Dec}_{\mathbf{r}}(\mathsf{Enc}_{\mathbf{r}}(x)) = x$;
(b) Statistical Hiding: $\mathsf{SD}(\mathsf{Enc}_R(0), \mathsf{Enc}_R(1)) < \delta$, for every distribution $R \in \mathcal{R}_n$.

Commitment schemes allow the sender Alice to commit a chosen value (or statement) while keeping it secret from the receiver Bob, with the ability to reveal the committed value in a later stage. Binding and hiding properties are essential to any commitment scheme. Informally, "binding" means that it's "hard" for Alice to alter her commitment after she has made it; "hiding" means that it's "hard" for Bob to find out the committed value without Alice revealing it.

Each of them can be computational or information theoretical. However, we can't achieve information theoretically binding and information theoretically hiding properties at the same time. Instead of defining computational notions, we relax binding to some very weak property, so that hiding and this new (very weak) binding properties both can be information theoretical. Since we aim to show an impossibility result, such relaxation is justified.

Definition 4. A $(\mathcal{R}_n, \delta)-$secure weak bit commitment is a function Com : $\{0,1\}^n \times \{0,1\} \to \{0,1\}^\lambda$ satisfying that: for any distribution $R \in \mathcal{R}_n$,

(a) Weak Binding: $\Pr_{\mathbf{r} \leftarrow U_n} [\mathsf{Com}(0; \mathbf{r}) \neq \mathsf{Com}(1; \mathbf{r})] \geq \frac{1}{2}$;

(b) Statistical Hiding: $SD(\mathsf{Com}(0; R), \mathsf{Com}(1; R)) < \delta$.

Note that in the traditional notion of commitment, the binding property holds if it is "hard" to find $\mathbf{r}_1$ and $\mathbf{r}_2$ such that $\mathsf{Com}(0; \mathbf{r}_1) = \mathsf{Com}(1; \mathbf{r}_2)$. Here we give a much weaker binding notion. We only require that the attacker can not win with probability $\geq \frac{1}{2}$ by choosing $\mathbf{r}_1 = \mathbf{r}_2$ uniformly at random. For example, $\mathsf{Com}(x; r) = x \oplus r$, where $x, r \in \{0,1\}$ can be easily verified to be a weak bit commitment for any $\delta > 0$ (despite not being a standard commitment).

In the notion of T-party Secret Sharing, two thresholds T_1 and T_2, where $1 \leq T_1 < T_2 \leq T$, are involved such that (a) any T_1 partics have "no information" about the secret, (b) any T_2 parties enable to recover the secret. Because our purpose is to show an impossibility result, we restrict to $T_1 = 1$ and $T_2 = T$, and only consider one bit secret x.

Definition 5. A $(\mathcal{R}_n, \delta)-$secure bit $T-$Secret Sharing scheme is a tuple $(\mathsf{Share}_1, \mathsf{Share}_2, \ldots, \mathsf{Share}_T, \mathsf{Rec})$ satisfying the following two properties:

(a) Correctness: $\mathsf{Rec}(\mathsf{Share}_1(x, \mathbf{r}), \ldots, \mathsf{Share}_T(x, \mathbf{r})) = x$ for all $\mathbf{r} \in \{0,1\}^n$ and each $x \in \{0,1\}$;

(b) Statistical Hiding: $SD(\mathsf{Share}_j(0; R), \mathsf{Share}_j(1; R)) < \delta$, for every index $j \in [T]$ and any distribution $R \in \mathcal{R}_n$.

Now we abstract and generalize the results of [MP90, DOPS04] to show that expressiveness implies the impossibility of security involving traditional privacy. See [DY14] for the proof.

Theorem 1. (a) When $\mathcal{R}_n$ is $(0, \delta)-$expressive, no $(\mathcal{R}_n, \delta)$-secure bit extractor exists.

(b) When $\mathcal{R}_n$ is $(0, \delta)-$expressive, no $(\mathcal{R}_n, \delta)$-secure bit encryption scheme exists.

(c) When $\mathcal{R}_n$ is $(1, \delta)-$expressive, no $(\mathcal{R}_n, \delta)$-secure weak bit commitment exists.

(d) When $\mathcal{R}_n$ is $(\log T, \delta)-$expressive, no $(\mathcal{R}_n, \delta)$-secure bit T-secret sharing exists.

3.2   Implications to Differential Privacy

Dodis et al. [DLMV12] have shown how to do differential privacy with respect to the γ-SV source for all "queries of low sensitivity". Since we aim to show impossibility results, henceforth we only consider the simplest case: let $\mathcal{D} = \{0,1\}^N$ be the space of all databases and for $D \in \mathcal{D}$, the query function q is the Hamming weight function $wt(D) = |\{i \mid D(i) = 1\}|$, where $D(i)$ means the i-th bit ("record") of D. If the source $\mathcal{R}_n$ has only one distribution U_n, $\mathcal{R}_n$ is denoted by U_n for simplicity. For any $D, D' \in \mathcal{D}$, the discrete distance function between them is defined by $\Delta(D, D') \stackrel{def}{=} wt(D \oplus D')$, where $\oplus$ is the bitwise exclusive OR operator. We say that D and D' are neighboring if $\Delta(D, D') = 1$. A mechanism M is an algorithm that takes as input a database $D \in \mathcal{D}$ and a distribution $R \in \mathcal{R}_n$, and outputs a random value z. Informally, we wish $z = M(D, R)$ to approximate the true value $wt(D)$ without revealing too much information about any individual $D(i)$. More formally, a mechanism is differentially private for the Hamming weight queries if replacing an entry in the database with one containing fake information only changes the output distribution of the mechanism by a small amount. In other words, evaluating the mechanism on two neighboring databases, does not change the outcome distribution by much. On the other hand, we define its utility to be the expected difference between the true answer $wt(D)$ and the output of the mechanism. More formally,

Definition 6. *Let $\varepsilon \geq 0$ and $\mathcal{R}_n$ be a source. A mechanism M (for the Hamming weight queries) is $(\mathcal{R}_n, \varepsilon)$-differentially private if for all neighboring databases $D_1, D_2 \in \mathcal{D}$, and all distributions $R \in \mathcal{R}_n$, we have $\mathsf{RD}(M(D_1, R), M(D_2, R)) \leq \varepsilon$. Equivalently, for any possible output z:*

$$\frac{\Pr_{\mathbf{r} \leftarrow R}[M(D_1, \mathbf{r}) = z]}{\Pr_{\mathbf{r} \leftarrow R}[M(D_2, \mathbf{r}) = z]} \leq e^\varepsilon.$$

Note that for $\varepsilon < 1$, we can rather accurately approximate e^ε by $1 + \varepsilon$.

Definition 7. *Let $0 < \rho \leq N/4$ and $\mathcal{R}_n$ be a source. A mechanism M has $(\mathcal{R}_n, \rho)$-utility for the Hamming weight queries, if for all databases $D \in \mathcal{D}$ and all distributions $R \in \mathcal{R}_n$, we have $\mathbb{E}_{\mathbf{r} \leftarrow R}[|M(D, \mathbf{r}) - wt(D)|] \leq \rho$.*

We show that, much like with traditional privacy, expressiveness implies impossibility of differential privacy with imperfect randomness, albeit with slightly more demanding parameters.

Theorem 2. *Assume $1/(8\rho) \leq \varepsilon \leq 1/4$ and the source $\mathcal{R}_n$ is $(\log(\frac{\rho\varepsilon}{\delta}) + 4, \delta)$-expressive, for some $2\varepsilon \leq \delta \leq 1$. Then no $(\mathcal{R}_n, \varepsilon)$-differentially private and (U_n, ρ)-accurate mechanism for the Hamming weight queries exists. In particular, plugging $\delta = 2\varepsilon$ and $\delta = \frac{1}{2}$, respectively, this holds if either*

(a) $\mathcal{R}_n$ is $(3 + \log(\rho), 2\varepsilon)$-expressive; or (b) $\mathcal{R}_n$ is $(5 + \log(\rho\varepsilon), \frac{1}{2})$-expressive.

The high-level idea is as follows. For two databases D and D', define two functions $f(\mathbf{r}) \overset{def}{=} M(D, \mathbf{r})$ and $g(\mathbf{r}) \overset{def}{=} M(D', \mathbf{r})$. Intuitively, for all $R \in \mathcal{R}_n$, since $\mathsf{RD}(f(R), g(R)) \le \varepsilon \cdot \Delta(D, D')$ implies $\mathsf{SD}(f(R), g(R)) \le e^{\varepsilon \cdot \Delta(D,D')} - 1$, we could use expressiveness to argue that $f(\mathbf{r}) = g(\mathbf{r})$ almost everywhere, which must eventually contradict utility (even for uniform distribution). However, we can't use this technique directly, because if $\varepsilon \cdot \Delta(D, D')$ is large enough, then $e^{\varepsilon \cdot \Delta(D,D')} - 1 > 1$, which is greater than the general upper bound 1 of the statistical distance. Instead, we simply use this trick on close-enough databases D and D', and then use a few "jumps" from D_0 to D_1, etc., until eventually we must violate the ρ-utility.

Proof. Assume for contradiction that there exists such a mechanism M. Let $\mathcal{D}' \overset{def}{=} \{D \mid wt(D) \le 4\rho\}$. Denote

$$\mathsf{Trunc}(x) \overset{def}{=} \begin{cases} 0, & if\ x < 0; \\ x, & if\ x \in \{0, 1, \ldots, 4\rho\}; \\ 4\rho, & otherwise. \end{cases}$$

For any $D \in \mathcal{D}'$, define the truncated mechanism $M' \overset{def}{=} \mathsf{Trunc}(M)$ by $M'(D, \mathbf{r}) \overset{def}{=} \mathsf{Trunc}(M(D, \mathbf{r}))$. Since for every $D \in \mathcal{D}'$, we have $wt(D) \in \{0, 1, \ldots, 4\rho\}$, M' still has (U_n, ρ)-utility on $\mathcal{D}'$. Additionally, from Definition 6, it's straightforward that M' is $(\mathcal{R}_n, \varepsilon)$-differentially private on $\mathcal{D}'$. In the following, we only consider the truncated mechanism M' on $\mathcal{D}'$.

Let $t = \log(\frac{\rho \varepsilon}{\delta}) + 4$ and $s = \frac{\delta}{2\varepsilon}$. Notice, $1 \le s \le 1/(2\varepsilon) \le 4\rho$, $e^{\varepsilon s} - 1 < \delta$, and $2^t = 8\rho/s$.

We start with the following claim:

Claim. Consider any databases $D, D' \in \mathcal{D}'$, s.t. $\Delta(D, D') \le s$, and denote $f(\mathbf{r}) \overset{def}{=} M'(D, \mathbf{r})$ and $g(\mathbf{r}) \overset{def}{=} M'(D', \mathbf{r})$. Then $\Pr_{\mathbf{r} \leftarrow U_n}[f(\mathbf{r}) \ne g(\mathbf{r})] < \frac{1}{2^t}$.

Proof. Since M' is $(\mathcal{R}_n, \varepsilon)$-differentially private, then for all $R \in \mathcal{R}_n$, we have $\mathsf{RD}(f(R), g(R)) \le \varepsilon \cdot \Delta(D, D') \le \varepsilon \cdot s$. Hence, $\mathsf{SD}(f(R), g(R)) \le e^{\varepsilon \cdot s} - 1 < \delta$, by our choice of s. Since this holds for all $R \in \mathcal{R}_n$ and $\mathcal{R}_n$ is (t, δ)-expressive, we conclude that it must be the case that $\Pr_{\mathbf{r} \leftarrow U_n}[f(\mathbf{r}) \ne g(\mathbf{r})] < \frac{1}{2^t}$. $\qquad\square$

Coming back to the main proof, consider a sequence of databases $D_0, D_1, \cdots$, $D_{4\rho/s}$ such that $wt(D_i) = i \cdot s$ and $\Delta(D_i, D_{i+1}) = s$. Denote $f_i(R) \overset{def}{=} M'(D_i, R)$ for all $i \in \{0, 1, \ldots, 4\rho/s\}$. From the above Claim, we get that $\Pr_{\mathbf{r} \leftarrow U_n}[f_i(\mathbf{r}) \ne f_{i+1}(\mathbf{r})] < \frac{1}{2^t}$. By the union bound and our choice of s and t,

$$\Pr_{\mathbf{r} \leftarrow U_n}[f_0(\mathbf{r}) \ne f_{4\rho/s}(\mathbf{r})] < \frac{4\rho}{2^t \cdot s} \le \frac{1}{2} \tag{1}$$

Let $\alpha \overset{def}{=} \mathbb{E}_{\mathbf{r} \leftarrow U_n}[\, f_{4\rho/s}(\mathbf{r}) - f_0(\mathbf{r}) \,]$. From (U_n, ρ)-utility, we get that

$$\alpha \ge (wt(D_{4\rho/s}) - \rho) - (wt(D_0) + \rho) = (4\rho - \rho) - (0 + \rho) = 2\rho.$$

On the other hand, from Inequation (1),

$$\alpha \le \Pr_{\mathbf{r} \leftarrow U_n} [f_0(\mathbf{r}) \ne f_{4\rho/s}(\mathbf{r})] \cdot \max_{\mathbf{r}} |f_{4\rho/s}(\mathbf{r}) - f_0(\mathbf{r})| < \frac{1}{2} \cdot 4\rho = 2\rho,$$

which is a contradiction.   □

4   Separability and Its Implications

Expressiveness is a powerful tool, but it's hard for us to use it directly. In this section, we introduce the concept of separability and show that it implies expressiveness, and also has its own applications to (weak) coin flipping. Several typical examples can been seen in Sect. 5.

Intuitively, separable sources $\mathcal{R}_n$ allow one to choose a distribution $R \in \mathcal{R}_n$ capable of "separating" any sufficiently large, disjoint sets G and B: increasing a relative weight of one set w.r.t. R without doing the same for the counterpart of the other one.

Definition 8. *We say that a source $\mathcal{R}_n$ is $(t, \delta)-$separable if for all $G, B \subseteq \{0,1\}^n$, where $G \cap B = \emptyset$ and $|G \cup B| \ge 2^{n-t}$, there exists a distribution $R \in \mathcal{R}_n$ such that $| \Pr_{\mathbf{r} \leftarrow R}[\mathbf{r} \in G] - \Pr_{\mathbf{r} \leftarrow R}[\mathbf{r} \in B] | \ge \delta$.*

4.1   Separability Implies Expressiveness

We investigate the relationship between separability and expressiveness. We show that separable sources must be expressive. The high-level idea of the proof comes from the work of [DOPS04] (who only applied it to SV sources), but we quantitatively improve the technique of [DOPS04], by making the gap between expressiveness and separability independent of the range $\mathcal{C}$ of the functions f and g. See [DY14] for the proof.

Theorem 3. *If a source $\mathcal{R}_n$ is $(t + 1, \delta)-$separable, then it's (t, δ)-expressive.*

Remark 1. Note that if the universe $\mathcal{C}$ is a subset of $\{0,1\}^{poly(n)}$, then the universal hash function family in the proof of Theorem 3 can be made efficient (in n). Hence, the distinguisher Eve can be made efficient as well. Therefore, there exists an efficient distinguisher Eve such that $| \Pr_{\mathbf{r} \leftarrow R}[\mathsf{Eve}(f(\mathbf{r})) = 1] - \Pr_{\mathbf{r} \leftarrow R}[\mathsf{Eve}(g(\mathbf{r})) = 1]| \ge \delta$. Namely, $f(R)$ is "δ- computationally distinguishable" from $g(R)$.

Combining Theorem 3 with Theorems 1 and 2, we get

Corollary 1. *(a) If $\mathcal{R}_n$ is $(1, \delta)-$separable, then no $(\mathcal{R}_n, \delta)$-secure bit extractor exists.*
(b) If $\mathcal{R}_n$ is $(1, \delta)-$separable, then no $(\mathcal{R}_n, \delta)$-secure bit encryption exists.
(c) If $\mathcal{R}_n$ is $(2, \delta)-$separable, then no $(\mathcal{R}_n, \delta)$-secure weak bit commitment exists.
(d) If $\mathcal{R}_n$ is $(\log T + 1, \delta)-$separable, then no $(\mathcal{R}_n, \delta)$-secure bit T-secret sharing exists.

(e) Assume $1/(8\rho) \leq \varepsilon \leq 1/4$ and $\mathcal{R}_n$ is $(\log(\frac{\rho\varepsilon}{\delta}) + 5, \delta)-separable$, for some $2\varepsilon \leq \delta \leq 1$. Then no $(\mathcal{R}_n, \varepsilon)-differentially$ private and (U_n, ρ)-accurate mechanism for the Hamming weight queries exists. In particular, plugging $\delta = 2\varepsilon$ and $\delta = \frac{1}{2}$, respectively, this holds if either (e.1) $\mathcal{R}_n$ is $(4 + \log(\rho), 2\varepsilon)-separable$; or (e.2) $\mathcal{R}_n$ is $(6 + \log(\rho\varepsilon), \frac{1}{2})-separable$.

The above results are illustrated by several typical sources in Sect. 5.

4.2   Separability and Weak Bit Extraction

In this section, we define weak bit extraction and show that weak bit extraction is equivalent to NON-separability. Then we propose its implications to privacy.

Recall, Bosley and Dodis [BD07] initiated the study of the general question: *does privacy inherently require "extractable" source of randomness?* A bit more formally, if a primitive P admits $(\mathcal{R}_n, \delta)$-secure implementation, does it mean one can construct a (deterministic, single- or multi-) bit extractor from $\mathcal{R}_n$?

They also obtained very strong affirmative answers to this question for several traditional privacy primitives, including (only multi-bit) encryption and commitment (but not secret sharing, for example). Here we make the observation that our impossibility results give an incomparable (to [BD07]) set of affirmative answers to this question. On the positive, our results apply to a much wider set of primitives P (e.g., secret-sharing, as well as even single-bit encryption and commitment). On the negative, we can only argue a rather weak kind of single-bit extraction (as opposed to [BD07], who showed traditional, and possibly multi-bit extraction). Our weak notion of extraction is defined below.

Definition 9. *We say that* $\mathsf{Ext} : \{0,1\}^n \to \{0, 1, \bot\}$ *is* $(\mathcal{R}_n, \delta, \tau)$-secure weak bit extractor if

(a) for every distribution $R \in \mathcal{R}_n$, $\left| \Pr_{r \leftarrow R}[\mathsf{Ext}(r) = 1] - \Pr_{r \leftarrow R}[\mathsf{Ext}(r) = 0] \right| < \delta$;

(b) $\Pr_{r \leftarrow U_n}[\mathsf{Ext}(r) \neq \bot] \geq \tau$.

We briefly discuss this notion, before showing our results. First, we notice that setting $\tau = 1$ recovers the notion of traditional bit-extractor given in Definition 2. And, even for general $\tau < 1$, the odds of outputting 0 or 1 are roughly the same, for any distribution R in the source. However, now the extractor is also allowed to output a failure symbol $\bot$, which means that each of the above two probabilities can occur with probabilities noticeably smaller than $1/2$. Hence, to make it interesting, we also add the requirement that Ext does not output $\bot$ all the time. This is governed by the second parameter τ requiring that $\Pr_{r \leftarrow R}[\mathsf{Ext}(r) \neq \bot] \geq \tau$. Ideally, we would like this to be true for any distribution R in the source. Unfortunately, such a desirable guarantee will not be achievable in our setting (see Remark 2). Thus, to salvage a meaningful and realizable notion, we will only require that this non-triviality guarantee at least holds for $R \equiv U_n$. Namely, while we do not rule out the possibility that some particular distributions R might force Ext to fail the extraction with high probability, we still ensure that:

(a) when the extraction succeeds, the extracted bit is unbiased for *any* R in the source; (b) the extraction succeeds with noticeable probability at least when R is ("close to") the uniform distribution U_n.

We now observe (and prove in [DY14]) that the notion of weak bit-extraction is simply a different way to express (the negation of) our notion of separability!

Lemma 1. $\mathcal{R}_n$ *has a* $(\mathcal{R}_n, \delta, 2^{-t})$-*secure weak bit extractor if and only if* $\mathcal{R}_n$ *is* <u>*not*</u> (t, δ)-*separable.*

Combining Lemma 1 with the counter-positive of Corollary 1, we get

Theorem 4. *(a) If* $(\mathcal{R}_n, \delta)$-*secure bit encryption scheme exists, then* $(\mathcal{R}_n, \delta, \frac{1}{2})$-*secure weak bit-extraction exists.*

(b) If $(\mathcal{R}_n, \delta)$-*secure weak bit commitment exists, then* $(\mathcal{R}_n, \delta, \frac{1}{4})$-*secure weak bit extraction exists.*

(c) If $(\mathcal{R}_n, \delta)$-*secure bit* T-*secret-sharing exists, then* $(\mathcal{R}_n, \delta, \frac{1}{2T})$-*secure weak bit extraction exists.*

(d) If $(\mathcal{R}_n, \varepsilon)$−*differentially private and* (U_n, ρ)-*accurate mechanism for the Hamming weight queries exists, then* $(\mathcal{R}_n, 2\varepsilon, \frac{1}{16\rho})$-*secure weak bit extraction exists.*

It is also instructive to see the explicit form of our weak bit extractor. For example, in the case of bit encryption (part (a), other examples similar), we get

$$\mathsf{Ext}(\mathbf{r}) \stackrel{def}{=} \begin{cases} 1, & \text{if } h^*(\mathsf{Enc_r}(1)) = 1 \text{ and } h^*(\mathsf{Enc_r}(0)) = 0, \\ 0, & \text{if } h^*(\mathsf{Enc_r}(1)) = 0 \text{ and } h^*(\mathsf{Enc_r}(0)) = 1, \\ \bot, & \text{otherwise (i.e., if } h^*(\mathsf{Enc_r}(1)) = h^*(\mathsf{Enc_r}(0))), \end{cases}$$

where h^* is the boolean universal hash function from the proof of Theorem 3, chosen as to ensure $\Pr_{\mathbf{r} \leftarrow U_n}[\mathsf{Ext}(\mathbf{r}) \neq \bot] = \Pr_{\mathbf{r} \leftarrow U_n}[h^*(\mathsf{Enc_r}(0)) \neq h^*(\mathsf{Enc_r}(1))] \geq \frac{1}{2}$. When the bit encryption (resp. commitment, secret sharing, DP mechanism) is computationally efficient (in n), our bit extractor is efficient too. This means that even computationally secure analogs of encryption (commitment, secret sharing, DP mechanism) imply efficient, statistically secure weak bit extraction.

Remark 2. As we mentioned, the major weakness of our weak bit extraction definition comes from the fact that the non-triviality condition $\Pr_{\mathbf{r} \leftarrow R}[\mathsf{Ext}(\mathbf{r}) \neq \bot] \geq \tau$ is only required for $R \equiv U_n$. Unfortunately, we observe that the analog of Theorem 4.(a)-(c) is no longer true if we require the extraction non-triviality to hold for all $R \in \mathcal{R}_n$. Indeed, this stronger notion of $(\mathcal{R}_n, \delta, \tau)$-secure weak bit extraction clearly implies traditional $(\mathcal{R}_n, 1 + \delta - \tau)$-secure bit extraction (by mapping $\bot$ to 1). On the other hand, Dodis and Spencer [DS02] gave an example of a source $\mathcal{R}_n$ for which, for any $\varepsilon > 0$, there exists $(\mathcal{R}_n, \varepsilon)$-secure bit encryption (and hence, weak commitment and 2-secret sharing) scheme, but no $(\mathcal{R}_n, 1 - 2^{1-n/2})$-secure bit-extraction. Thus, the only analogs of Theorem 4.(a)-(c) we could hope to prove using the strengthened notion of weak bit extraction would have to satisfy $\tau \leq \delta + 2^{1-n/2}$, which is not a very interesting weak bit extraction scheme (e.g., if δ is "negligible", then the extraction succeeds with "negligible" probability as well).[9]

5 Privacy with Several Typical Imperfect Sources

Now we define several imperfect sources $\mathcal{R}_n$: the (k,n)–source [CG88], n-bit (k,m)-block source [CG88], n-bit γ-Santha-Vazirani (SV) source [SV86], and (γ, b, n)-Bias-Control Limited (BCL) source [Dod01] below. Then we prove all these sources are separable. Based on this result, we show they are all expressive. Afterwards, we study the impossibility of traditional and differential privacy with weak, block and BCL sources, and explain why the SV source does not work. Finally, we compare the impossibility of traditional and differential privacy.

Definition 10. *The (k,n)-source (or n-bit weak source with min-entropy at least k) is defined by $\mathcal{W}eak(k,n) \overset{def}{=} \{R \mid \mathbf{H}_\infty(R) \geq k, \text{where } R \text{ is over } \{0,1\}^n\}$.*

Block sources are generalizations of weak sources, allowing n/m blocks $R_1, \ldots, R_{n/m}$ each having k fresh bits of entropy.[10]

Definition 11. *Let m divide n, and $R_1, \ldots, R_{n/m}$ be a sequence of Boolean random variables over $\{0,1\}^m$. A probability distribution $R = (R_1, \ldots, R_{n/m})$ over $\{0,1\}^n$ is an n-bit (k,m)-block distribution, denoted by $Block(k,m,n)$, if for all $i \in [n/m]$ and for every $s_1, \ldots, s_{i-1} \in \{0,1\}^m$, we have*

$$\mathbf{H}_\infty(R_i \mid R_1 \ldots R_{i-1} = s_1 \ldots s_{i-1}) \geq k.$$

We define the n-bit (k,m)-block source $Block(k,m,n)$ to be the set of all n-bit (k,m)-block distributions.

Hence, weak sources correspond to $m = n$ (i.e., one block). From the other extreme, SV sources as shown in Definition 12 correspond to 1-bit blocks (i.e., $m = 1$). In this case, it is customary to express the imperfectness of the source as the function of its "bias" γ instead of min-entropy k. Of course, for 1-bit random variables bias and min-entropy are related by $2^{-k} = (1 + \gamma)/2$.

Definition 12. *Let $r_1, \ldots, r_n$ be a sequence of Boolean random variables and $0 \leq \gamma < 1$. A probability distribution $R = (r_1, \ldots, r_n)$ over $\{0,1\}^n$ is an n-bit γ-Santha-Vazirani distribution, denoted by $SV(\gamma,n)$, if for all $i \in \{1, \ldots, n\}$ and every string $s \in \{0,1\}^{i-1}$, $\frac{1-\gamma}{2} \leq \Pr[r_i = 1 \mid r_1 \ldots r_{i-1} = s] \leq \frac{1+\gamma}{2}$ holds. We define the n-bit γ-SV source $\mathcal{SV}(\gamma,n)$ to be the set of all n-bit γ-SV distributions.*

---

[9] For differential privacy (part (d)), we do not have an analog of the counter-example in [DS02], and anyway the value $\tau = O(1/\rho) \ll \delta = O(\varepsilon)$ (so no contradiction). Of course, this does not imply that a stronger bit extraction result should be true; only that it is not definitely false.

[10] For consistency with prior work, we only assume that R_i has k fresh bits conditioned on the prior blocks, but our impossibility results easily extend to the case when we condition on both the past and the future blocks..

Finally, we define BCL sources [Dod01].

Definition 13. *Assume that* $0 \leq \gamma < 1$. *The* (γ, b, n)-*Bias-Control Limited (BCL) source* $\mathcal{BCL}(\gamma, b, n)$ *generates* n *bits* $r_1, \ldots, r_n$, *where for all* $i \in \{1, \ldots, n\}$, *the value of* r_i *can depend on* $r_1, \ldots, r_{i-1}$ *in one of the following two ways:*

(a) *r_i is determined by $r_1, \ldots, r_{i-1}$, but this can happen for at most b bits. This rule of determining a bit is called an intervention.*

(b) $\frac{1-\gamma}{2} \leq \Pr[r_i = 1 \mid r_1 r_2 \ldots r_{i-1}] \leq \frac{1+\gamma}{2}$.

Every distribution over $\{0, 1\}^n$ *generated from* $\mathcal{BCL}(\gamma, b, n)$ *is called a* (γ, b, n)-*BCL distribution* $BCL(\gamma, b, n)$.

In particular, if $b = 0$, $\mathcal{BCL}(\gamma, b, n)$ degenerates into $\mathcal{SV}(\gamma, n)$ [SV86]; if $\gamma = 0$, it yields the sequential-bit-fixing source of Lichtenstein, Linial, and Saks [LLS89].

5.1 Separability Results

In the following, we propose that the above sources are separable. It should be noted that: (a) The results for the weak and SV sources are implicitly known; (b) The BCL source was not considered before, but it is not hard to prove its separability given careful application of prior work; (c) The separability of the block source is new. It was not considered before because the SV source is a block source with each block of length 1, and [MP90, DOPS04] showed traditional privacy impossible even with the SV source (hence with the block source). But in light of [DLMV12], where differential privacy is possible with the SV source, we find it important to precisely figure out the separability of the block source. A naive approach would be to employ the so called γ-biased half-space source (see [DY14]), introduced by [RVW04] and [DOPS04], which is both γ-SV and $(m - \log \frac{1+\gamma}{1-\gamma}, m)$-block sources. We can easily conclude that (1) $\mathcal{SV}(\gamma, n)$ is $(t, \frac{\gamma}{2^{t+1}})$-separable, and (2) $\mathcal{Block}(k, m, n)$ is $(t, \frac{2^{m-k}-1}{2^{t+1} \cdot (2^{m-k}+1)})$-separable. However, these results are somewhat sub-optimal. Instead, we introduce a new separability bound for block sources in Lemma 2 (b), and use it to get an improved result about the SV sources as well (see [DY14] for the proof).

Lemma 2. (a) *Assume that $k \leq n - 1$. Then $\mathcal{Weak}(k, n)$ is $(t, 1)$-separable when $k \leq n - t - 1$, and $(t, 2^{n-t-k-1})$-separable when $n - t - 1 < k \leq n - 1$. In particular, it's $(t, \frac{1}{2})$-separable when $k \leq n - t$.*

(b) *$\mathcal{Block}(k, m, n)$ is $\left(t, \frac{1}{1 + 2^{t+1} \cdot \left(\frac{2^k - 1}{2^m - 2^k}\right)}\right)$-separable. In particular, it is $(t, 1/(1 + 2^{2+t+k-m}))$-separable when $k \leq m - 1$ (and, hence, $(t, \frac{1}{2})$-separable when $k \leq m - t - 2$).*

(c) *$\mathcal{SV}(\gamma, n)$ is $(t, \frac{\gamma}{2^t})$-separable.*

(d) *$\mathcal{BCL}(\gamma, b, n)$ is $(t, 1 - \frac{2^{t+2}}{(1+\gamma)^b})$-separable. In particular, it is $(t, \frac{1}{2})$-separable for $b \geq \frac{t+3}{\log(1+\gamma)} = \Theta(\frac{t+1}{\gamma})$.*

5.2   Implications to Traditional and Differential Privacy

IMPOSSIBILITY OF TRADITIONAL PRIVACY. From Lemma 2 and Corollary 1 (a)-(d), we conclude:

Theorem 5. *For the following values of δ, shown in Table 1, no $(\mathcal{R}_n, \delta)$−secure cryptographic primitive P exists, where $\mathcal{R}_n \in \{\mathcal{W}eak(k,n), \mathcal{B}lock(m - 1, m, n), \mathcal{SV}(\gamma, n), \mathcal{BCL}(\gamma, b, n)\}$ and $P \in \{bit\ extractor,\ bit\ encryption\ scheme, weak\ bit\ commitment,\ bit\ T\text{-}secret\ sharing\}$.*

Table 1. Values of δ for which no $(\mathcal{R}_n, \delta)$−secure cryptographic primitive P exists

$\mathcal{R}_n$ P	bit extractor	bit encryption scheme	weak bit commitment	bit T-secret sharing
$\mathcal{W}eak(k,n)$	1, if $k \leq n - 2$	1, if $k \leq n - 2$	1, if $k \leq n - 3$	1, if $k \leq n - \log T - 2$
$\mathcal{W}eak(n-1,n)$	$\frac{1}{2}$	$\frac{1}{2}$	$\frac{1}{4}$	$\frac{1}{2T}$
$\mathcal{B}lock(m-1,m,n)$	$\frac{1}{5}$	$\frac{1}{5}$	$\frac{1}{9}$	$\frac{1}{4T+1}$
$\mathcal{SV}(\gamma,n)$	$\frac{\gamma}{2}$	$\frac{\gamma}{2}$	$\frac{\gamma}{4}$	$\frac{\gamma}{2T}$
$\mathcal{BCL}(\gamma,b,n)$	$\frac{1}{2}$, if $b \geq \frac{4}{\log(1+\gamma)}$	$\frac{1}{2}$, if $b \geq \frac{4}{\log(1+\gamma)}$	$\frac{1}{2}$, if $b \geq \frac{5}{\log(1+\gamma)}$	$\frac{1}{2}$, if $b \geq \frac{\log T + 4}{\log(1+\gamma)}$

We notice that, while the impossibility results for the block and BCL sources are new, the prior work of [MP90, DOPS04] already obtained similar results for the weak and SV sources. However, our results still offer some improvements over the works of [MP90, DOPS04]. First, unlike the work of [MP90], our distinguisher is efficient (see Remark 1), ruling out even computationally secure encryption, commitment, and secret sharing schemes. Second, unlike the work of [DOPS04], our lower bound on δ does not depend on the sizes of ciphertext/commitment/shares. In particular, while [DOPS04] used a bit-by-bit hybrid argument to show their impossibility results, our proof of Theorem 3 used a more clever "universal hashing trick". More importantly, instead of focusing the entire proof on some specific weak/block/SV sources [MP90, DOPS04], our impossibility results for such sources were obtained in a more modular manner, making these proofs somewhat more illuminating.

IMPOSSIBILITY OF DIFFERENTIAL PRIVACY WITH THE WEAK, BLOCK AND BCL SOURCES. Now we apply the impossibility results of differential privacy to the sources $\mathcal{W}eak(k, n)$, $\mathcal{B}lock(k, m, n)$, and $\mathcal{BCL}(\gamma, b, n)$. In particular, by combining Corollary 1 (e.2) with Lemma 2 (a), (b), and (d), respectively, we get

Theorem 6. *For the following sources $\mathcal{R}_n$, no $(\mathcal{R}_n, \varepsilon)$−differentially private and $(\mathcal{U}_n, \rho)$-accurate mechanisms for the Hamming weight queries exist:*
(a) $\mathcal{W}eak(k, n)$ where $k \leq n - \log(\varepsilon\rho) - 6$;
(b) $\mathcal{B}lock(k, m, n)$ where $k \leq m - \log(\varepsilon\rho) - 8$;
(c) $\mathcal{BCL}(\gamma, b, n)$ where $b \geq \frac{\log(\varepsilon\rho)+9}{\log(1+\gamma)} = \Omega(\frac{\log(\varepsilon\rho)+1}{\gamma})$.

Table 2. Comparison about the Impossibility of Traditional Privacy and Differential Privacy.

Source	Traditional Privacy δ	Differential Privacy ε & Utility ρ
$\mathcal{B}lock(k, m, n)$	Impossible if $\delta \leq \frac{1}{9}$, even if $k = m - 1$	Impossible if $k \leq m - \log(\varepsilon\rho) - O(1)$
$\mathcal{SV}(\gamma, n)$	Impossible if $\delta = O(\gamma)$	Impossible if $\rho = O(\frac{1}{\varepsilon})$, even for U_n
		(Possible if $\rho = poly_{1/(1-\gamma)}(\frac{1}{\varepsilon}) \gg \frac{1}{\varepsilon}$)
$\mathcal{BCL}(\gamma, b, n)$	Impossible if $\delta = O(\gamma)$, even if $b = 0$;	Impossible if $b = \Omega(\frac{\log(\varepsilon\rho)+1}{\gamma})$
	Impossible if $\delta \leq \frac{1}{2}$ and $b = \Omega(\frac{1}{\gamma})$	

We discuss the (non-)implications to the SV source below, but notice the strength of these negative results the moment the source becomes a little bit more "adversarial" as compared to the SV source. In particular, useful mechanisms in differential privacy (called "non-trivial" by [DLMV12]) aim to achieve utility ρ (with respect to the uniform distribution) which only depends on the differential privacy ε, and not on the size N of the database D. This means that the value $\log(\varepsilon\rho)$ is typically upper bounded by some constant $c = O(1)$. For such "non-trivial" mechanisms, our negative results say that differential privacy is impossible with (1) weak sources even when the min-entropy $k = n - O(1)$; (2) block sources even when the min-entropy $k = m - O(1)$; (3) BCL sources even when the number of interventions $b = \Omega(1)$. So what prevented us from strong impossibility for the SV sources, as is expected given the feasibility results of [DLMV12]? The short answer is that the separability of the SV sources given by Lemma 2 (c) is just not good enough to yield very strong results. We explain it in more detail in [DY14].

5.3   Comparing Impossibility Results for Traditional and Differential Privacy

In this section, we compare the impossibility of traditional privacy and differential privacy (see Table 2). For traditional privacy, we consider bit extractor, bit encryption scheme, weak bit commitment, and bit T-secret sharing (i.e., set $T = 2$ for concreteness). We observe that the impossibility results for differential privacy are only marginally weaker than those for traditional privacy.

In particular, while a very "structured" (and, hence, rather unrealistic) SV source is sufficient to guarantee loose, but non-trivial differential privacy, without guaranteeing (strong-enough) traditional privacy, once the source becomes more realistic (e.g., number of interventions b becomes super-constant, or one removes the conditional entropy guarantee within different blocks), both notions of privacy become impossible *extremely quickly*. In other words, despite the surprising feasibility result of [DLMV12] regarding differential privacy with SV sources, the prevalent opinion that "privacy is impossible with realistic weak randomness" appears to be rather accurate.

Acknowledgments. The authors would like to thank Benjamin Fuller, Sasha Golovnev, Hamidreza Jahanjou, Zhoujun Li, Umut Orhan, and Abhishek Samanta. In particular, the authors thank Prof. Zhoujun Li very much for his great help about this paper. The authors also thank the anonymous reviewers for their helpful comments. Yevgeniy Dodis was partially supported by gifts from VMware Labs and Google, and NSF grants 1319051, 1314568, 1065288, and 1017471. Yanqing Yao was supported by NSFC grants 61170189 and 61370126, the Fund for the Doctoral Program of Higher Education of China 20111102130003, the Scholarship Award for Excellent Doctoral Student granted by Ministry of Education 400618, and CSC grant 201206020063.

References

[ACM+14] Austrin, P., Chung, K.-M., Mahmoody, M., Pass, R., Seth, K.: On the impossibility of cryptography with tamperable randomness. In: Garay, J.A., Gennaro, R. (eds.) CRYPTO 2014, Part I. LNCS, vol. 8616, pp. 462–479. Springer, Heidelberg (2014)

[ACRT99] Andreev, A.E., Clementi, A.E.F., Rolim, J.D.P., Trevisan, L.: Weak random sources, hitting sets, and BPP simulations. SIAM J. Comput. 28(6), 2103–2116 (1999)

[Blu86] Blum, M.: Independent unbiased coin-flips from a correlated biased source-a finite state Markov chain. Combinatorica 6(2), 97–108 (1986)

[BD07] Bosley, C., Dodis, Y.: Does privacy require true randomness? In: Vadhan, S.P. (ed.) TCC 2007. LNCS, vol. 4392, pp. 1–20. Springer, Heidelberg (2007)

[BDK+05] Boyen, X., Dodis, Y., Katz, J., Ostrovsky, R., Smith, A.: Secure remote authentication using biometric data. In: Cramer, R. (ed.) EUROCRYPT 2005. LNCS, vol. 3494, pp. 147–163. Springer, Heidelberg (2005)

[BH05] Barak, B., Halevi, S.: A model and architecture for pseudo-random generation with applications to /dev/random. In: Proceedings of the 12th ACM Conference on Computer and Communications Security, pp. 203–212 (2005)

[BST03] Barak, B., Shaltiel, R., Tromer, E.: True random number generators secure in a changing environment. In: Walter, C.D., Koç, Ç.K., Paar, C. (eds.) CHES 2003. LNCS, vol. 2779, pp. 166–180. Springer, Heidelberg (2003)

[CFG+85] Chor, B., Friedman, J., Goldreich, O., Håstad, J., Rudich, S., Smolensky, R.: The bit extraction problem or t-resilient functions. In: The 26th Annual Symposium on Foundations of Computer Science, pp. 396–407 (1985)

[CG88] Chor, B., Goldreich, O.: Unbiased bits from sources of weak randomness and probabilistic communication complexity. SIAM J. Comput. 17(2), 230–261 (1988)

[CW79] Carter, J.L., Wegman, M.N.: Universal classes of hash functions. J. Comput. Syst. Sci. 18(2), 143–154 (1979)

[Dod01] Dodis, Y.: New imperfect random source with applications to coin-flipping. In: Orejas, F., Spirakis, P.G., van Leeuwen, J. (eds.) ICALP 2001. LNCS, vol. 2076, p. 297. Springer, Heidelberg (2001)

[DKRS06] Dodis, Y., Katz, J., Reyzin, L., Smith, A.: Robust fuzzy extractors and authenticated key agreement from close secrets. In: Dwork, C. (ed.) CRYPTO 2006. LNCS, vol. 4117, pp. 232–250. Springer, Heidelberg (2006)

[DLMV12] Dodis, Y., López-Alt, A., Mironov, I., Vadhan, S.: Differential privacy with imperfect randomness. In: Safavi-Naini, R., Canetti, R. (eds.) CRYPTO 2012. LNCS, vol. 7417, pp. 497–516. Springer, Heidelberg (2012)

[DMNS06] Dwork, C., McSherry, F., Nissim, K., Smith, A.: Calibrating noise to sensitivity in private data analysis. In: Halevi, S., Rabin, T. (eds.) TCC 2006. LNCS, vol. 3876, pp. 265–284. Springer, Heidelberg (2006)

[DOPS04] Dodis, Y., Ong, S.J., Prabhakaran, M., Sahai, A.: On the (im)possibility of cryptography with imperfect randomness. In: 45th Symposium on Foundations of Computer Science, pp. 196–205 (2004)

[DORS08] Dodis, Y., Ostrovsky, R., Reyzin, L., Smith, A.: Fuzzy extractors: how to generate strong keys from biometrics and other noisy data. SIAM J. Comput. **38**(1), 97–139 (2008)

[DS02] Dodis, Y., Spencer, J.: On the (non)universality of the one-time pad. In: 43rd Symposium on Foundations of Computer Science, pp. 376–385 (2002)

[DY14] Dodis, Y., Yao, Y.Q.: Privacy with imperfect randomness. IACR Cryptology ePrint Archive, 2014/623 (2014)

[GKR04] Gennaro, R., Krawczyk, H., Rabin, T.: Secure hashed Diffie-Hellman over Non-DDH groups. In: Cachin, C., Camenisch, J.L. (eds.) EUROCRYPT 2004. LNCS, vol. 3027, pp. 361–381. Springer, Heidelberg (2004)

[Kra10] Krawczyk, H.: Cryptographic extraction and key derivation: The HKDF scheme. In: Rabin, T. (ed.) CRYPTO 2010. LNCS, vol. 6223, pp. 631–648. Springer, Heidelberg (2010)

[LLS89] Lichtenstein, D., Linial, N., Saks, M.E.: Some extremal problems arising form discrete control processes. Combinatorica **9**(3), 269–287 (1989)

[MP90] McInnes, J.L., Pinkas, B.: On the Impossibility of private key cryptography with weakly random keys. In: Menezes, A., Vanstone, S.A. (eds.) CRYPTO 1990. LNCS, vol. 537, pp. 421–435. Springer, Heidelberg (1991)

[MW97] Maurer, U.M., Wolf, S.: Privacy amplification secure against active adversaries. In: Kaliski Jr., B.S. (ed.) CRYPTO 1997. LNCS, vol. 1294, pp. 307–321. Springer, Heidelberg (1997)

[RVW04] Reingold, O., Vadhan, S., Widgerson, A.: No deterministic extraction from Santha-Vazirani sources: a simple proof (2004). http://windowsontheory. org/2012/02/21/nodeterministic-extraction-from-santha-vazirani-sources-a-simple-proof/

[SV86] Santha, M., Vazirani, U.V.: Generating quasi-random sequences from semi-random sources. J. Comput. Syst. Sci. **33**(1), 75–87 (1986)

[vN51] Neumann, J.V.: Various techniques used in connection with random digits. Nat. Bur. Stand. Appl. Math. Ser. **12**, 36–38 (1951)

[VV85] Vazirani, U.V., Vazirani, V.V.: Random polynomial time is equal to slightly random polynomial time. In: 26th Annual Symposium on Foundations of Computer Science, pp. 417–428 (1985)

[Zuc96] Zuckerman, D.: Simulating BPP using a general weak random source. Algorithmica **16**(4/5), 367–391 (1996)

Attribute-Based Encryption

Communication Complexity of Conditional Disclosure of Secrets and Attribute-Based Encryption

Romain Gay[1](✉), Iordanis Kerenidis[2], and Hoeteck Wee[3,4]

[1] ENS, Paris, France
rgay@di.ens.fr
[2] CNRS, LIAFA, University Paris Diderot, Paris, France
jkeren@liafa.univ-paris-diderot.fr
[3] ENS, CNRS, INRIA, Paris, France
[4] Columbia University, New York, USA
wee@di.ens.fr

Abstract. We initiate a systematic treatment of the communication complexity of conditional disclosure of secrets (CDS), where two parties want to disclose a secret to a third party if and only if their respective inputs satisfy some predicate. We present a general upper bound and the first non-trivial lower bounds for conditional disclosure of secrets. Moreover, we achieve tight lower bounds for many interesting setting of parameters for CDS with linear reconstruction, the latter being a requirement in the application to attribute-based encryption. In particular, our lower bounds explain the trade-off between ciphertext and secret key sizes of several existing attribute-based encryption schemes based on the dual system methodology.

1 Introduction

We revisit a fundamental question in the foundations of cryptography: what is the communication overhead of privacy in computation? This question has been considered in several different models and settings [2,12,14,41]. In this work, we focus on a very simple and natural model where non-private computation requires very little communication (just a single bit), whereas the best upper bound for private computation is exponential.

Namely, we consider two-party conditional disclosure of secrets (CDS) [19] (c.f. Fig. 2), a generalization of secret sharing [23,44]: two parties want to disclose a secret to a third party if and only if their respective inputs satisfy some fixed predicate P. Concretely, Alice holds x, Bob holds y and they both share a secret

---

R. Gay—Partially supported by ANR Project EnBiD (ANR-14-CE28-0003).

I. Kerenidis—Partially supported by the ERC project QCC and the ANR project RDAM.

H. Wee—Partially supported by ANR Project EnBiD (ANR-14-CE28-0003), NSF Award CNS-1445424 and the Alexander von Humboldt Foundation.

R. Gennaro and M. Robshaw (Eds.): CRYPTO 2015, Part II, LNCS 9216, pp. 485–502, 2015.
DOI: 10.1007/978-3-662-48000-7_24

$\alpha \in \{0, 1\}$ (along with some additional private randomness), whereas Carol knows x, y but not α. Alice and Bob want to disclose α to Carol iff $P(x, y) = 1$. How many bits do Alice and Bob need to communicate to Carol? In the non-private setting, Alice or Bob can send α to Carol, upon which Carol computes $P(x, y)$ and decides whether to output α or $\perp$. This trivial protocol with one-bit communication is not private because Carol learns α even when the predicate is false; in fact, the best upper bound we have for CDS for general predicates requires that Alice and Bob each transmits $2^{\Omega(|x|+|y|)}$ bits [7]. Here, we are interested not only in the total communication from Alice and Bob to Carol, but also in trade-offs between the length of Alice's message ℓ_A and that of Bob's message ℓ_B.

Connection to Attribute-Based Encryption. Attribute-based encryption (ABE) [20,43] is a new paradigm for public-key encryption that enables fine-grained access control for encrypted data. In attribute-based encryption, ciphertexts are associated with descriptive values x in addition to a plaintext, secret keys are associated with values y, and a secret key decrypts the ciphertext if and only if $P(x, y) = 1$ for some boolean predicate P. Note that x and y are public given the respective ciphertext and secret key. Here, y together with P may express an arbitrarily complex access policy, which is in stark contrast to traditional public-key encryption, where access is all or nothing. The simplest example of ABE is that of identity-based encryption (IBE) [8,13,45] where P corresponds to equality. The security requirement for attribute-based encryption enforces resilience to collusion attacks, namely any group of users holding secret keys for different values learns nothing about the plaintext if none of them is individually authorized to decrypt the ciphertext. This should hold even if the adversary *adaptively* decides which secret keys to ask for.

In [47], Waters introduced the powerful *dual system encryption* methodology for building adaptively secure IBE in bilinear groups; this has since been extended to obtain adaptively secure ABE for a large class of predicates [30,31,33,35,38,40]. In recent works [3,48], Attrapadung and Wee presented a unifying framework for the design and analysis of dual system ABE schemes, which decouples the predicate P from the security proof. Specifically, the latter work puts forth the notion of *predicate encoding*, a private-key, one-time, information-theoretic primitive similar to conditional disclosure of secrets, and provides a compiler from predicate encoding for a predicate P into an ABE for the same predicate using the dual system encryption methodology. Moreover, the parameters in the predicate encoding scheme and in CDS correspond naturally to ciphertext and key sizes in the ABE. In particular, Alice's message corresponds to the ciphertext, and Bob's message to the secret key. For these applications, we require that Alice's and Bob's messages are linear functions of the shared randomness, and also that Carol computes a linear function of the messages to reconstruct the secret α. These applications consider linear functions over $\mathbb{Z}_p$ where p is the order of the underlying bilinear group; in this work, we focus on lower bounds for the case $p = 2$ although our techniques do hold for general p. Note that while the parameters for ABE schemes coming from predicate encodings are not necessarily the best known parameters, they do match the state-of-the-art in terms of ciphertext and secret key sizes for many predicates such as inner product, index, and read-once formula.

CDS Parameters. Unlike in traditional communication complexity where the primary measure is the total communication from Alice and from Bob, we make a more fine-grained distinction between the lengths of Alice's and Bob's messages ℓ_A and ℓ_B. For instance, in the application to ABE, ℓ_A and ℓ_B correspond to ciphertext and secret key sizes respectively. Note that for ABE ciphertext and key sizes, we ignore the contributions from the descriptive values x, y as well as multiplicative factors in the security parameter.[1] We are particularly interested in three regimes of parameters for (ℓ_A, ℓ_B):

- How small can ℓ_B be when ℓ_A is constant? This corresponding to minimizing key sizes for schemes with constant-size ciphertexts;
- How small can ℓ_A be when ℓ_B is constant? This corresponding to minimizing ciphertext sizes for schemes with constant-size keys;
- How small can $\max(\ell_A, \ell_B)$ be? This corresponds to minimizing the overall parameter sizes of the scheme.

We also care about the complexity of the reconstruction function as computed by Carol, as a function of the messages from Alice and Bob; as noted earlier, for ABE, we will require linear reconstruction.

Prior Works. There have been several works studying CDS protocols (and strengthenings thereof) for a large class of predicates [3,19,22,48]: the best general upper bound achieves both linear reconstruction and communication that is linear in the size of the smallest (arithmetic) branching program computing the predicate [19,22]. However, we basically do not have any techniques for proving lower bounds on the communication complexity of CDS protocols. Here, even the probabilistic method or a counting argument does not seem to yield meaningful lower bounds for a random function (in contrast, these techniques do yield meaningful lower bounds for circuit complexity of a random function).

1.1   Our Results

We initiate a systematic treatment of the communication complexity of conditional disclosure of secrets (CDS). We present a general upper bound and the first non-trivial lower bounds for conditional disclosure of secrets, summarized in Fig. 1. Moreover, we achieve tight lower bounds for many interesting setting of parameters for CDS with linear reconstruction, the latter being a requirement in the application to attribute-based encryption; this addresses an open problem posed in [48]. Very informally, for CDS with linear reconstruction, we obtain lower bounds of the form:

$$\ell_A \cdot \ell_B \geq \text{"communication complexity of P"}$$

---

[1] The latter suppresses the distinction between counting bits and group elements, and also between working over $\mathbb{Z}_2$ vs $\mathbb{Z}_p$, where p is the order of the underlying bilinear group.

Predicate	ℓ_B, constant ℓ_A		ℓ_A, constant ℓ_B		$\max(\ell_A, \ell_B)$	
	upper	lower	upper	lower	upper	lower
index, prefix	$O(n)$	$\Omega(n)^*$	$O(n)$	$\Omega(\sqrt{n})$	$O(\sqrt{n})$	$\Omega(\sqrt{n})^*$
disjointness, inner product	$O(n)$	$\Omega(n)^*$	$O(n)$	$\Omega(n)^*$	$O(n)$	$\Omega(\sqrt{n})$
read-once span programs	$O(2^n)$	$\Omega(n)$	$O(2^n)$	$\Omega(n^2)$	$O(n)$	$\Omega(n)^*$

Fig. 1. Summary of our upper and lower bounds for linear CDS, where ℓ_A and ℓ_B denote the length of the messages from Alice and Bob respectively. We marked the tight lower bounds with an asterisk *.

For example, for inner product on n-bit vectors, we have $\ell_A \cdot \ell_B = \Omega(n)$. Our lower bounds partially explain the trade-off between ciphertext and secret key sizes of several existing attribute-based encryption schemes based on the dual system methodology, c.f. [3,10,31,35,39,48].

Proof Techniques. Since we want to argue about the lengths of the messages of Alice and Bob to Carol, the first idea would be to look at the communication complexity of the predicate P [29,49]. Informally, communication complexity measures how many bits of information about x and y we need to transmit in order to compute $P(x,y)$ (c.f. Fig. 2). Namely, Alice holds x and Bob holds y and each of them sends a message to a third party Carol who wants to compute $P(x,y)$. We also allow all three parties to share public randomness w. The goal is to minimize the communication from Alice and Bob to Carol, and there is no privacy requirement. There is now a large body of works in communication complexity giving tight upper and lower bounds for a large class of predicates. For instance, a classic result from communication complexity tells us that to compute the inner product of two vectors $\mathbf{x}, \mathbf{y} \in \{0,1\}^n$, each of Alice and Bob

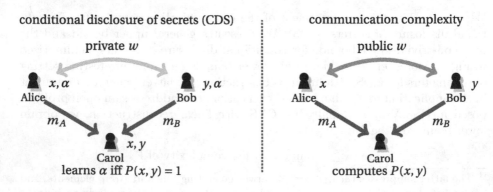

Fig. 2. Pictorial representation of CDS and communication complexity.

must send $n - \Omega(1)$ bits [11]. That is, we need to know essentially all of **x** and all of **y** in order to compute their inner product.

Our goal is to leverage the rich literature on lower bounds for communication complexity to obtain lower bounds for CDS. Namely, we want to transform any CDS Π_{cds} for a predicate P into a communication complexity protocol Π_{cc} for P with only a small blow-up in communication complexity. The crucial distinction between CDS and communication complexity is that Carol knows x, y in Π_{cds} but not in Π_{cc} (as shown in Fig. 2).

The first attempt would be to show that a Π_{cds} for a predicate P is also a Π_{cc} for P. Fix x, y to denote the inputs to Π_{cc}. That is, we would like to argue that Alice's message together with Bob's message in a CDS (even without x, y) must completely determine $P(x, y)$. Intuitively, this ought to be the case because if the CDS messages are consistent with both values of $P(x, y)$, then they must simultaneously uniquely determine α (via correctness) and hide α (via privacy), a contradiction. Indeed, if this worked out, we would have a lower bound of the form

$$\ell_A + \ell_B \geq \text{"communication complexity for P"}$$

Unfortunately, the above statement is false for inner product. The above statement implies a lower bound of $2n - \Omega(1)$ bits for inner product, but we have a CDS for inner product with $n + 1$ bits! It is instructive to understand why the above attempt fails. The issue arises in using correctness of CDS to argue that Alice's and Bob's message must determine α: specifically, it is necessary for Carol to specify inputs x', y' in order to reconstruct α from Alice's and Bob's messages. In fact, different inputs (x', y') could yield different values for α. We need to fix this issue.

– The first idea is to have Alice in Π_{cc} also send the secret α; Carol then tries all possible (x', y') for which $P(x', y') = 1$ and output 1 iff for some x, y the reconstructed secret indeed equals α. By the correctness of CDS, Carol will output 1 when $P(x, y) = 1$. However, there could be false positives, since even when $P(x, y) = 0$, there could be inputs (x', y') for which $P(x', y') = 1$ and the reconstructed secret matches α, upon which Carol will incorrectly output 1. In fact, privacy tells us that Carol will recover a random value for the secret for each choice of (x', y'), and with pretty good probability, at least one of them will match α.

– The second idea is to avoid false positives by having Alice and Bob run the CDS protocol Π_{cds} N times, with fresh independent private randomness and secrets across the repetitions. As before, Carol will try all possible (x', y') for which $P(x', y') = 1$ and output 1 iff for some x', y' the reconstructed secret equals α in all repetitions of the protocol. By the correctness of CDS, Carol will always output 1 when $P(x, y) = 1$. On the other hand, if $P(x, y) = 0$, a straight-forward union bound over $(x', y') \in P^{-1}(1)$ tells us Carol outputs 1 with probability at most $P^{-1}(1) \cdot 2^{-N}$, since Carol recovers a random value in each repetition. For inner product, we need to take a union bound over 2^{2n-1} possible pairs, which requires running $N = 2n - 1$ copies of the CDS protocol Π_{cds}; the communication complexity of Π_{cc} is then $2n - 1$ times that of Π_{cds}.

This does not yield any non-trivial lower bound for Π_{cds} since we have an upper bound of $2n$ for communication complexity.

Here comes our key observation: we can substantially reduce the number of repetitions needed if the CDS protocol Π_{cds} has small communication complexity! Suppose Π_{cds} has total communication $\ell_A + \ell_B \ll n$ bits. Observe that the reconstruction function computed by Carol in Π_{cds} is a function from $\{0,1\}^{\ell_A+\ell_B}$ to $\{0,1\}$. Now, instead of having Carol in Π_{cc} enumerate over all possible (x, y), she will instead enumerate over all functions from $\{0,1\}^{\ell_A+\ell_B}$ to $\{0,1\}$, and output 1 iff for some function the reconstructed secret equals α in all N repetitions. By the correctness of CDS, Carol will always output 1 when $P(x, y) = 1$. Moreover, there are $2^{2^{\ell_A+\ell_B}}$ possible functions, which means we will need to run $2^{\ell_A+\ell_B}$ copies of Π_{cds} in Π_{cc}; this already implies a $\Omega(\log n)$ lower bound for inner product! Moreover, if the CDS Π_{cds} admits linear reconstruction, then Carol in Π_{cc} will also need to enumerate over all $2^{\ell_A+\ell_B}$ linear functions from $\{0,1\}^{\ell_A+\ell_B}$ to $\{0,1\}$, which means we only need to run $\ell_A + \ell_B$ copies of Π_{cds} in Π_{cc}; this in turn yields a $\Omega(\sqrt{n})$ lower bound for inner product.

We obtain our lower bounds on CDS for concrete predicates by instantiating the above argument with existing lower bounds in communication complexity [4, 11, 24, 28, 36, 42] (c.f. Sect. 5).

Implications for Dual System ABE. As observed in [3, 10, 48], underlying most "information-theoretic" dual system ABE schemes for a predicate P is a CDS for the same predicate, and our lower bounds apply to ciphertext and secret key sizes for these dual system ABE schemes. On the other hand, we do have ABE schemes based on a "computational" dual system argument, such as those in [3, 9, 27, 32, 34], many of which are more efficient and do avoid the lower bounds in this work. Informally, underlying the "computational" dual system argument is a computational analogue of CDS, where the privacy requirement is computational rather than information-theoretic. As it turns out, formalizing the right notion of computational privacy in CDS is quite tricky.

Recall that CDS guarantees privacy of the secret α whenever $P(x, y) = 0$, and in the application to ABE, we require that privacy holds even if x, y are chosen adaptively, namely Alice's input x may be chosen depending on Bob's input y and Bob's message, and vice versa. Now, if the privacy guarantee is information-theoretic and perfect, then privacy for non-adaptive choices of x, y implies privacy for adaptive choices[2]; this equivalence dissipates as soon as we relax the privacy requirement to be statistical or computational. The "right" notion of computational privacy for use in ABE schemes is that of "doubly selective" security [3, 34], where "doubly" refers to the two possibilities depending on

---

[2] The easiest way to see this is via complexity leveraging: an adaptive distinguisher with advantage ε can be converted into a non-adaptive distinguisher with an exponential loss in ε via random guessing. Since any non-adaptive distinguisher has advantage 0, we must have $\varepsilon = 0$ to begin with.

whether x or y is chosen first. Unsurprisingly, proving[3] and using doubly selective security require substantially more delicate security reductions, and in most cases, stronger and less desirable q-type assumptions. This raises the natural question of whether the increased complexity in these proofs and assumptions are inherent, or simply a failure to find more clever and efficient CDS with information-theoretic privacy. Our work rules out the latter option.

1.2 Discussion

Perspective. Note that our set-up is quite different from previous lower bounds for private computation in the literature; to the best of our knowledge, this is the first super-constant lower bound in a setting where the price of privacy in computation is always bounded. For instance, in interactive secure two-party computation, some functions are impossible to compute securely [12], so the cost of privacy is infinite for these functions (whereas ours is bounded for all predicates). For secure computation in the FKN model [14,15], we do not have any techniques for super-constant gaps. For locally decodable codes, there is no gap for privacy in some ranges of parameters, for instance, when we want to minimize one-way communication from the client and communication from the server is essentially "free"; here, the server needs to send the entire database, whether or not we care about client privacy.

Additional Related Work. There is a large body of work on lower bounds on share sizes in secret-sharing (c.f. [5, Sect. 5]). Most of these works rely on Shannon-type inequalities on entropy of random variables, which do not seem applicable to our setting. Roughly speaking, in secret sharing, Carol either gets a share or not, whereas Alice and Bob in CDS can do more complex computations than simply computing shares and then deciding whether to send each share to Carol. The recent work of Data, Prabhakaran and Prabhakaran [14] draws upon tools from information theory to obtain new communication complexity lower bounds for secure computation in three-party setting. In their model which allows multiple rounds of interaction, the problem we consider admits a secure protocol with a single bit of communication, and their techniques do not yield better bounds in the non-interactive setting.

Open Problems. We conclude with a number of open problems:

- explore the power of non-linear reconstruction in CDS (that is, positive results, c.f. [6,46]);
- tight lower bounds for inner product with linear reconstruction (which we conjecture to be $\Omega(n)$);

---

[3] Typically, this entails two separate reductions, one for x being chosen first and the other for y. In [34], these correspond to selectively secure key-policy and ciphertext-policy ABE schemes; in [3], these correspond to so-called selective and co-selective security.

– obtain better lower bounds for multi-bit secrets (which is related to lower bounds for secret sharing for multi-bit secrets), or obtain upper bounds that are better than the naive "direct product" construction;
– improve the upper or lower bounds in CDS for read-once span programs for constant ℓ_A or constant ℓ_B. A related problem is to prove stronger communication complexity lower bounds for general span programs (which may not be read-once).

2   Preliminaries

Notations. We denote by $s \leftarrow_R S$ the fact that s is picked uniformly at random from a finite set S or from a distribution. Throughout this paper, we denote by log the logarithm of base 2.

2.1   Conditional Disclosure of Secrets

We recall the notion of conditional disclosure of secrets (CDS), c.f. Fig. 2. The definition we give here is for two parties Alice and Bob and a referee Carol, where Alice and Bob share randomness w and want to conditionally disclose a secret α to Carol. The general notion of conditional disclosure of secrets has first been investigated in [19]. Two-party CDS is closely related to the notions of predicate encoding [10,48] and pairing encoding [3]; in particular, the latter two notions imply two-party CDS with linear reconstruction.

Definition 1 (Conditional Disclosure of Secrets (CDS) [19,48]). *Fix a predicate* $P : \mathcal{X} \times \mathcal{Y} \to \{0,1\}$. *A* (ℓ_A, ℓ_B)-*conditional disclosure of secrets (CDS) for* P *is a triplet of deterministic functions* (A, B, C)

$$A : \mathcal{X} \times \mathcal{W} \times \mathcal{D} \to \{0,1\}^{\ell_A}, \quad B : \mathcal{Y} \times \mathcal{W} \times \mathcal{D} \to \{0,1\}^{\ell_B}, \quad C : \mathcal{X} \times \mathcal{Y} \times \{0,1\}^{\ell_A} \times \{0,1\}^{\ell_B} \to \mathcal{D}$$

satisfying the following properties:

(reconstruction.) *For all* $(x,y) \in \mathcal{X} \times \mathcal{Y}$ *such that* $P(x,y) = 1$, *for all* $w \in \mathcal{W}$, *and for all* $\alpha \in \mathcal{D}$:

$$C(x, y, A(x, w, \alpha), B(y, w, \alpha)) = \alpha$$

(privacy.) *For all* $(x,y) \in \mathcal{X} \times \mathcal{Y}$ *such that* $P(x,y) = 0$, *and for all* $C^* : \{0,1\}^{\ell_A} \times \{0,1\}^{\ell_B} \to \mathcal{D}$,

$$\Pr_{w \leftarrow \mathcal{W}, \alpha \leftarrow_R \mathcal{D}} \left[C^*\big(A(x, w, \alpha), B(y, w, \alpha)\big) = \alpha \right] \leq \frac{1}{|\mathcal{D}|}$$

Note that the formulation of privacy above with uniformly random secrets is equivalent to standard indistinguishability-based formulations

A useful measure for the complexity of a CDS is the complexity of reconstruction as a function of the outputs of A, B, as captured by the function C, with (x, y) hard-wired.

Definition 2 ($\mathcal{C}$-Reconstruction). *Given a set $\mathcal{C}$ of functions from $\{0,1\}^{\ell_A} \times \{0,1\}^{\ell_B} \to \mathcal{D}$, we say that a CDS $(\mathsf{A}, \mathsf{B}, \mathsf{C})$ admits $\mathcal{C}$-reconstruction if for all (x, y) such that $P(x, y) = 1$, $\mathsf{C}(x, y, \cdot, \cdot) \in \mathcal{C}$.*

Two examples of $\mathcal{C}$ of interest are:

- $\mathcal{C}_{\mathrm{all}}$ is the set of all functions from $\{0,1\}^{\ell_A} \times \{0,1\}^{\ell_B} \to \mathcal{D}$; that is, we do not place any restriction on the complexity of reconstruction. Note that $|\mathcal{C}_{\mathrm{all}}| = |\mathcal{D}|^{2^{\ell_A + \ell_B}}$.
- $\mathcal{C}_{\mathrm{lin}}$ is the set of all *linear* functions over $\mathbb{Z}_2$ from $\{0,1\}^{\ell_A} \times \{0,1\}^{\ell_B} \to \mathcal{D}$; that is, we require the reconstruction to be linear as a function of the outputs of A and B as bit strings (but may depend arbitrarily on x, y). This is the analogue of linear reconstruction in linear secret sharing schemes and is a requirement for the applications to attribute-based encryption [3,10,48]. Note that $|\mathcal{C}_{\mathrm{linear}}| \leq |\mathcal{D}|^{\ell_A + \ell_B}$ for $|\mathcal{D}| \geq 2$.

Remark 1. Note that while looking at $\mathcal{C}$, we consider $\mathsf{C}(x, y, \cdot, \cdot)$, which has (x, y) hard-wired, and takes an input of total length $\ell_A + \ell_B$. In particular, it could be that C runs in time linear in $|x| = |y| = n$, and yet $\ell_A = \ell_B = O(\log n)$ so C has "exponential" complexity w.r.t. $\ell_A + \ell_B$.

Definition 3 (Linear CDS). *We say that a CDS $(\mathsf{A}, \mathsf{B}, \mathsf{C})$ is linear if it admits $\mathcal{C}_{\mathrm{lin}}$-reconstruction.*

2.2 Communication Complexity

The description of communication complexity in Fig. 2 actually refers to the "simultaneous message" model, where A and B each sends a message to C. For our actual proof, it suffices to consider one way communication complexity, where there is no C, but either A sends a single message to B or B sends a single message to A. We now proceed to recall the basic definitions for communication complexity [29,49], specifically one-way communication complexity with one-sided error [1,28,37].

Definition 4 ([28,49]). *A one-way $(\mathsf{A} \to \mathsf{B})$ communication protocol for a predicate $P : \mathcal{X} \times \mathcal{Y} \to \{0,1\}$ it is a pair of deterministic functions (A, B) where*

$$\mathsf{A} : \mathcal{X} \times \mathcal{W} \times \{0,1\}^\ell \to \{0,1\}, \quad \mathsf{B} : \mathcal{Y} \times \mathcal{W} \times \{0,1\}^\ell \to \{0,1\},$$

and the following properties are satisfied for every $(x, y) \in \mathcal{X} \times \mathcal{Y}$:

- *If $P(x, y) = 1$, then $\Pr_{w \leftarrow_R \mathcal{W}}[\mathsf{B}(y, w, \mathsf{A}(x, w)) = 1] = 1$*
- *If $P(x, y) = 0$, then $\Pr_{w \leftarrow_R \mathcal{W}}[\mathsf{B}(y, w, \mathsf{A}(x, w)) = 0] \geq 1/2$.*

The one-way communication complexity of P, denoted by $\mathsf{R}^{\mathsf{A} \to \mathsf{B}}(P)$, is the minimum ℓ over all one-way communication protocols (A, B) for P.

We also denote by $\mathsf{R}^{\mathsf{B} \to \mathsf{A}}(P)$ the minimum ℓ over all one-way $(\mathsf{B} \to \mathsf{A})$ communication protocols (A, B), where

$$\mathsf{A} : \mathcal{X} \times \mathcal{W} \times \{0,1\}^\ell \to \{0,1\}, \quad \mathsf{B} : \mathcal{Y} \times \mathcal{W} \times \{0,1\}^\ell \to \{0,1\},$$

and the following properties are satisfied for every $(x, y) \in \mathcal{X} \times \mathcal{Y}$:

- If $P(x, y) = 1$, then $\Pr_{w \leftarrow_R W}[A(x, w, B(y, w)) = 1] = 1$
- If $P(x, y) = 0$, then $\Pr_{w \leftarrow_R W}[A(x, w, B(y, w)) = 0] \geq 1/2$.

3   CDS for General Predicates

We present a general upper bound for linear CDS for any predicate:

Theorem 1 (Generic Upper Bounds for Linear CDS). *Given any predi-cate* $P : \{0, 1\}^n \times \{0, 1\}^n \to \{0, 1\}$, *for any* $t \leq 2^n$, *there exists a linear* $(t, 2^n/t)$-*CDS for* P *with* $\mathcal{D} = \{0, 1\}$. *In particular, there exists a* $(1, 2^n)$-*CDS, a* $(2^n, 1)$-*CDS, a* $(2^{n/2}, 2^{n/2})$-*CDS for* P, *all three of which are linear.*

The result improves upon the $(2^{n/2}, 2^{n/2})$-CDS (but not linear) given in [7]; our construction is also considerably simpler.

Proof (sketch.) The construction follows from a standard reduction of any gen-eral predicate to the INDEX predicate on 2^n-dimensional vectors: Alice treats the truth table $P(x, \cdot)$ as a vector of length 2^n and Bob treats $y \in \{0, 1\}^n$ as an index, so that the INDEX predicate returns $P(x, y)$. Then, we can use the $(t, 2^n/t)$-linear CDS for the INDEX predicate on 2^n-dimensional vectors in [10,17]   □

More generally, for any predicate $P : \mathcal{X} \times \mathcal{Y} \to \{0, 1\}$, we have a $(t, \min(|\mathcal{X}|, |\mathcal{Y}|)/t)$-linear CDS, by treating either x or y as an index depending on whether $|\mathcal{X}| \leq |\mathcal{Y}|$ or not. This is essentially optimal for linear reconstruction, since we prove a tight lower bound for INDEX: $\{0, 1\}^n \times [n] \to \{0, 1\}$ in Sect. 5.

4   Lower Bounds for CDS

In this section, we present our lower bounds on the communication complexity of CDS.

Theorem 2 (Lower Bounds for Linear CDS). *Let* $P : \mathcal{X} \times \mathcal{Y} \to \{0, 1\}$ *be a predicate. For all linear* (ℓ_A, ℓ_B)-*CDS of* P *with* $|\mathcal{D}| \geq 2$, *we have*

$$\ell_A \cdot (\ell_A + \ell_B + 1) \geq R^{A \to B}(P) \quad and \quad \ell_B \cdot (\ell_A + \ell_B + 1) \geq R^{B \to A}(P).$$

We then derive our lower bounds for linear CDS by using existing lower bounds on one-way communication complexity; see Sect. 5. In fact, our techniques are fairly general and also yield lower bounds on non-linear CDS.

Theorem 3 (Lower Bounds for General CDS). *Let* $P : \mathcal{X} \times \mathcal{Y} \to \{0, 1\}$ *be a predicate. For all* (ℓ_A, ℓ_B)-*predicate CDS of* P *with* $|\mathcal{D}| \geq 2$, *we have*

$$\ell_A + \ell_B \geq \frac{1}{2} \log\left(R^{A \to B}(P) + R^{B \to A}(P)\right).$$

While the lower bounds for general CDS are exponentially smaller than those for linear CDS, we still do obtain non-trivial logarithmic lower bounds for many concrete predicates.

4.1 Main Lemma

We obtain both lower bounds via a general reduction from CDS for a predicate P to one-way communication protocols for the same predicate; the communication cost of the reduction depends crucially on the complexity of reconstruction (c.f. Definition 2):

Lemma 1 (Main Technical Lemma). *Let* $P : \mathcal{X} \times \mathcal{Y} \to \{0,1\}$ *be a predicate. Then, any* (ℓ_A, ℓ_B)-*CDS for P with* $|\mathcal{D}| \geq 2$ *and which admits* $\mathcal{C}$-*reconstruction satisfies*

$$(\log |\mathcal{C}| + 1) \cdot \ell_A \geq \mathsf{R}^{A \to B}(P) \cdot \log |\mathcal{D}| \quad and \quad (\log |\mathcal{C}| + 1) \cdot \ell_B \geq \mathsf{R}^{B \to A}(P) \cdot \log |\mathcal{D}|$$

Theorem 2 then follows from instantiating the lemma with $\mathcal{C} := \mathcal{C}_{\mathrm{lin}}$, where $\log |\mathcal{C}_{\mathrm{lin}}| = (\ell_A + \ell_B) \cdot \log |\mathcal{D}|$. Similarly, Theorem 3 uses $\mathcal{C} := \mathcal{C}_{\mathrm{all}}$ where $\log |\mathcal{C}_{\mathrm{all}}| = 2^{\ell_A + \ell_B} \cdot \log |\mathcal{D}|$.

Proof (of Lemma 1). Let $N := \frac{\log |\mathcal{C}| + 1}{\log |\mathcal{D}|}$. We build a one-way communication protocol $(\widetilde{A}, \widetilde{B})$ for the predicate P as follows:

- Sample $w_i \leftarrow_R \mathcal{W}, \alpha_i \leftarrow_R \mathcal{D}$ for $i = 1, \ldots, N$ and set

$$w := (w_1, \alpha_1, \ldots, w_N, \alpha_N)$$

- Alice computes

$$\widetilde{A}(x, w) := (A(x, w_1, \alpha_1), \ldots, A(x, w_N, \alpha_N))$$

- Bob outputs 1 iff there exists a function $C^* \in \mathcal{C}$ such that

$$C^*\big(A(x, w_i, \alpha_i), B(y, w_i, \alpha_i)\big) = \alpha_i, \quad \forall i = 1, \ldots, N$$

We proceed to analyze the protocol $(\widetilde{A}, \widetilde{B})$.

- **Completeness.** Suppose $P(x, y) = 1$. Then, by the reconstruction property, the function $C^*(\cdot) := C(x, y, \cdot) \in \mathcal{C}$ satisfies

$$C^*\big(A(x, w_i, \alpha_i), B(y, w_i, \alpha_i)\big) = \alpha_i, \quad \forall i = 1, \ldots, N$$

  for all $(w_1, \alpha_1, \ldots, w_N, \alpha_N)$. Therefore, $\widetilde{B}$ outputs 1 with probability 1.

- **Soundness.** Suppose $P(x, y) = 0$. Fix $C^* \in \mathcal{C}$. For each $i = 1, \ldots, N$, α-privacy implies that

$$\Pr_{w_i, \alpha_i}\Big[C^*\big(A(x, w_i, \alpha_i), B(y, w_i, \alpha_i)\big) = \alpha_i \Big] \leq \tfrac{1}{|\mathcal{D}|}$$

  Since the (w_i, α_i) are chosen independently at random, we have

$$\Pr_{w_1, \alpha_1, \ldots, w_N, \alpha_N}\Big[C^*\big(A(x, w_i, \alpha_i), B(y, w_i, \alpha_i)\big) = \alpha_i, \quad \forall i = 1, \ldots, N \Big] \leq \tfrac{1}{|\mathcal{D}|^N}$$

  By a union bound over all $|\mathcal{C}|$ functions $C^* \in \mathcal{C}$, we have

$$\Pr\Big[\widetilde{B} \text{ outputs } 1 \Big] \leq |\mathcal{C}| \cdot |\mathcal{D}|^{-N} \leq 1/2$$

  by our choice of N.

It is straightforward to check that $\widetilde{A}$ sends $\frac{\log|\mathcal{C}|+1}{\log|\mathcal{D}|} \cdot \ell_A$ bits to $\widetilde{B}$. Similarly, we can build a $(\widetilde{B}, \widetilde{A})$ protocol for P, where $\widetilde{B}$ sends $\frac{\log|\mathcal{C}|+1}{\log|\mathcal{D}|} \cdot \ell_B$ bits to $\widetilde{A}$. This completes the proof.   □

Remark 2 (Extensions). It is easy to see that the reduction also works for CDS with imperfect reconstruction and weak privacy. If the gap between the probability of reconstructing α when $P(x, y) = 1$ and the probability of recovering α when $P(x, y) = 0$ is δ, then it suffices to take $N := O\left(\frac{1}{\delta} \log |\mathcal{C}|\right)$ via a straightforward application of the Chernoff bound. The ensuing randomized protocol for communication complexity will then have a two-sided error.

Remark 3 (Beyond Linear CDS). Note that the bounds of Theorem 2 are much more general than just for linear CDS. For instance, if we require that reconstruction be carried out by circuits of size ℓ^c for some constant c (where $\ell := \ell_A + \ell_B$), or by polynomials of degree c, then we get lower bounds of the form

$$\ell_A + \ell_B = \Omega\left((R^{A \to B}(P) + R^{B \to A}(P))^{1/(c+1)}\right)$$

4.2 Lower Bounds for Multi-bit Secrets

We now look at CDS where the secret α is a multi-bit string; that is, $\mathcal{D}$ is of the form $\{0, 1\}^d$, for $d \geq 1$. There is a trivial upper bound for d-bit secrets obtained by running d times a CDS for single-bit secrets. Note, of course, that hiding a secret of size $d = 1$ is the easiest case, since we can simply embed this secret to a larger d-bit string by randomly adding $d - 1$ bits and use the CDS for the secret of size d. Hence, the lower bounds on the message lengths of the CDS for a secret of size $d = 1$ still hold for the CDS of secret of size $d \geq 1$. We would like a lower bound that grows with d.

Here, we prove that for any non-trivial predicate P, for any (ℓ_A, ℓ_B)-CDS of P, both ℓ_A and ℓ_B need to be at least d. A trivial predicate is one whose output is completely determined by either x or y (e.g. the output of the predicate is the first bit of x), for which there is a protocol with $\ell_A + \ell_B = d$. The intuition is that in any non-trivial predicate, Alice's message essentially serves as the secret key for a one-time pad, which is needed to "unlock" $\alpha \in \{0, 1\}^d$ from Bob's message. This means that Alice's message must itself be at least d bits.

It is easy to see that the lower bound is tight for the equality predicate. For all other non-trivial predicates, it remains an open problem to close the gap between lower and upper bounds for CDS of multi-bit secrets.

Theorem 4. *Let $\mathcal{D} := \{0, 1\}^d$, and let $P : \mathcal{X} \times \mathcal{Y} \to \{0, 1\}$ be a non-trivial predicate that depends on both inputs x and y; that is, there exists $x^* \in \mathcal{X}$, such that $P(x^*, \cdot)$ is not constant on $\mathcal{Y}$, and there exists $y^* \in \mathcal{Y}$ such that $P(\cdot, y^*)$ is not constant on $\mathcal{X}$. Then, for any (ℓ_A, ℓ_B)-CDS of P, we have*

$$\ell_A \geq d \quad and \quad \ell_B \geq d.$$

Proof. We begin with the lower bound on ℓ_A. Let $x_0, x_1 \in \mathcal{X}$ be such that

$$P(x_0, y^*) = 0 \quad and \quad P(x_1, y^*) = 1$$

Let $\mathsf{C}^* : \{0,1\}^{\ell_A + \ell_B} \to \{0,1\}^d$ be a randomized function defined as follows: on input $m_A \in \{0,1\}^{\ell_A}$ and $m_B \in \{0,1\}^{\ell_B}$,

- picks a message $m \leftarrow_R \{0,1\}^{\ell_A}$ at random (and ignores m_A);
- outputs $\mathsf{C}(x_1, y^*, m, m_B)$.

By α-reconstruction for $P(x_1, y^*) = 1$, for all $\alpha \in \mathcal{D}$, $w \in \mathcal{W}$, we have

$$\mathsf{C}(x_1, y^*, \mathsf{A}(x_1, w, \alpha), \mathsf{B}(y^*, w, \alpha)) = \alpha.$$

Therefore, for all $\alpha \in \mathcal{D}$, $w \in \mathcal{W}$, we have

$$\Pr_{m \leftarrow_R \{0,1\}^{\ell_A}} \left[\mathsf{C}(x_1, y^*, \mathsf{A}(x_1, w, \alpha), \mathsf{B}(y^*, w, \alpha)) = \alpha \text{ and } m = \mathsf{A}(x_1, w, \alpha) \right] = 1/2^{\ell_A}$$

Thus,

$$\Pr_{w \leftarrow \mathcal{W}, \alpha \leftarrow_R \mathcal{D}, \text{ coins of } \mathsf{C}^*} \left[\mathsf{C}^* (\mathsf{A}(x_1, w, \alpha), \mathsf{B}(y^*, w, \alpha)) = \alpha \right]$$

$$\geq \Pr_{w \leftarrow \mathcal{W}, \alpha \leftarrow_R \mathcal{D}, m \leftarrow_R \{0,1\}^{\ell_A}} \left[\mathsf{C}(x_1, y^*, \mathsf{A}(x_1, w, \alpha), \mathsf{B}(y^*, w, \alpha)) = \alpha \text{ and } m = \mathsf{A}(x_1, w, \alpha) \right]$$

$$= 1/2^{\ell_A}$$

Since C^* ignores m_A, this means that for all m_A, and in particular for $m_A = \mathsf{A}(x_0, w, \alpha)$, we have

$$\Pr_{w \leftarrow \mathcal{W}, \alpha \leftarrow_R \mathcal{D}, \text{ coins of } \mathsf{C}^*} \left[\mathsf{C}^* (\mathsf{A}(x_0, w, \alpha), \mathsf{B}(y^*, w, \alpha)) = \alpha \right] \geq 1/2^{\ell_A}$$

On the other hand, by α-privacy for $P(x_0, y^*) = 0$, we have

$$\Pr_{w \leftarrow \mathcal{W}, \alpha \leftarrow_R \mathcal{D}, \text{ coins of } \mathsf{C}^*} \left[\mathsf{C}^* (\mathsf{A}(x_0, w, \alpha), \mathsf{B}(y^*, w, \alpha)) = \alpha \right] \leq 1/2^d$$

Combining the two preceding inequalities, we have $1/2^{\ell_A} \leq 1/2^d$ and thus,

$$\ell_A \geq d.$$

For the same reason,

$$\ell_B \geq d. \qquad \square$$

5   Concrete Predicates

In this section, we describe how we can combine the results in the previous section with lower bounds in one-way communication complexity to obtain the results in Fig. 1. Each of these predicates has been studied in prior works on attribute-based encryption. For each of these predicates, we obtain non-trivial lower bounds for general (ℓ_A, ℓ_B)-CDS of the form:

$$\ell_A + \ell_B = \Omega(\log n).$$

We focus hence-forth on lower bounds for linear (ℓ_A, ℓ_B)-CDS, where linearity is over $\mathbb{Z}_2$. In the applications to ABE, we will typically work with linear functions over $\mathcal{D} = \mathbb{Z}_p$ (where $\log p$ is linear in the security parameter), in which case we lose a multiplicative $\log p$ factor in the lower bounds.

Index, Prefix. We consider the following predicates:

- Index: $\mathcal{X} := \{0, 1\}^n, \mathcal{Y} := [n]$ and

$$\mathsf{P}_{\mathrm{index}}(\mathbf{x}, i) = 1 \text{ iff } x_i = 1$$

  That is, $\mathbf{x}$ is the characteristic vector of a subset of $[n]$. In the context of ABE, this corresponds to broadcast encryption [16].
- Prefix: $\mathcal{X} := \{0, 1\}^n, \mathcal{Y} := \{0, 1\}^{\leq n}$ and

$$\mathsf{P}_{\mathrm{prefix}}(\mathbf{x}, \mathbf{y}) = 1 \text{ iff } \mathbf{y} \text{ is a prefix of } \mathbf{x}$$

  In the context of ABE, this corresponds to hierarchical identity-based encryption [18,21].

For both predicates, we have tight bounds for one-way communication complexity:

$$\mathsf{R}^{A \to B}(\mathsf{P}) = \Theta(n) \quad and \quad \mathsf{R}^{B \to A}(\mathsf{P}) = \Theta(\log n)$$

By Theorem 2, this means that any linear (ℓ_A, ℓ_B)-CDS for any of the two predicates must satisfy

$$\ell_A(\ell_A + \ell_B + 1) = \Omega(n)$$

This immediately yields

- $\ell_B = \Omega(n)$ if $\ell_A = O(1)$ and more generally, $\ell_B = \Omega(n/\ell_A)$ for any $\ell_A = o(\sqrt{n})$;
- $\ell_A = \Omega(\sqrt{n})$ if $\ell_B = O(1)$;
- $\max(\ell_A, \ell_B) = \Omega(\sqrt{n})$.

The first and third lower bounds are tight, as we have matching upper bounds in [3, 10, 48] exhibiting a linear $(t, n/t)$-CDS for both predicates and any $t \in [n]$.

Disjointness, Inner Product. We consider the following predicates:

- Disjointness: $\mathcal{X} = \mathcal{Y} := \{S \subseteq [n]\}$ and

$$P_{\mathrm{disj}}(X, Y) = 1 \text{ iff } X \cap Y = \emptyset$$

In the context of ABE, this is related to a special case of fuzzy IBE [43].
- Inner Product [26]: $\mathcal{X} = \mathcal{Y} := \mathbb{Z}_p^n$ and

$$P_{\mathrm{IP}}(\mathbf{x}, \mathbf{y}) = 1 \text{ iff } \mathbf{x}^\top \mathbf{y} = 0$$

For both predicates, we have tight bounds for one-way communication complexity:

$$R^{A \to B}(P) = \Theta(n) \quad and \quad R^{B \to A}(P) = \Theta(n)$$

given in [4,24,42] for disjointness, in [11] for inner product. By Theorem 2, this means that any linear (ℓ_A, ℓ_B)-CDS for any of the two predicates must satisfy

$$\ell_A(\ell_A + \ell_B + 1) = \Omega(n) \quad and \quad \ell_B(\ell_A + \ell_B + 1) = \Omega(n)$$

This immediately yields

- $\ell_B = \Omega(n)$ if $\ell_A = O(1)$;
- $\ell_A = \Omega(n)$ if $\ell_B = O(1)$;
- $\max(\ell_A, \ell_B) = \Omega(\sqrt{n})$.

The first and second lower bounds are tight, as we have matching upper bounds in [3,10,48] exhibiting a linear $(t, n - t + O(1))$-CDS for these predicates and any $t \in [n]$. It is open whether a CDS with overall parameter size of $O(\sqrt{n})$ is possible.

Read-Once Monotone Span Programs. We consider the following predicate:

- Read-once monotone span program: $\mathcal{X} := \{0, 1\}^n$, $\mathcal{Y} := \mathbb{Z}_p^{n \times n}$ is a collection of read-once monotone span programs [25] specified by a matrix $\mathbf{M}$ of height n and

$$P_{\mathrm{MSP}}(\mathbf{x}, \mathbf{M}) = 1 \text{ iff } \mathbf{x} \text{ satisfies } \mathbf{M}$$

Here, $\mathbf{x}$ satisfies $\mathbf{M}$ iff $(1, 0, \dots, 0)$ lies in the row span of $\{\mathbf{M}_j : x_j = 1\}$ where $\mathbf{M}_j$ is the j'th row of $\mathbf{M}$. In the context of ABE, this corresponds to key-policy ABE for access structures [20].

$$R^{A \to B}(P) = \Theta(n) \quad and \quad R^{B \to A}(P) = \Theta(n^2)$$

By Theorem 2, this means that any linear (ℓ_A, ℓ_B)-CDS for both predicates must satisfy

$$\ell_A(\ell_A + \ell_B + 1) = \Omega(n) \quad and \quad \ell_B(\ell_A + \ell_B + 1) = \Omega(n^2)$$

This immediately yields

- $\ell_B = \Omega(n)$ if $\ell_A = O(1)$;
- $\ell_A = \Omega(n^2)$ if $\ell_B = O(1)$;
- $\max(\ell_A, \ell_B) = \Omega(n)$.

The third lower bound is tight, as we have matching upper bounds in [3, 10, 48] exhibiting a linear (n, n)-CDS for the predicate. It is open what the optimal parameters are when we keep either the key or the ciphertext size constant.

Acknowledgments. We would like to thank Amos Beimel, Yuval Ishai and Sophie Laplante for insightful and inspiring discussions.

References

1. Ablayev, F.M.: Lower bounds for one-way probabilistic communication complexity and their application to space complexity. Theor. Comput. Sci. **157**(2), 139–159 (1996)
2. Ada, A., Chattopadhyay, A., Cook, S.A., Fontes, L., Koucký, M., Pitassi, T.: The hardness of being private. TOCT **6**(1), 1 (2014)
3. Attrapadung, N.: Dual system encryption via doubly selective security: framework, fully secure functional encryption for regular languages, and more. In: Nguyen, P.Q., Oswald, E. (eds.) EUROCRYPT 2014. LNCS, vol. 8441, pp. 557–577. Springer, Heidelberg (2014)
4. Bar-Yossef, Z., Jayram, T.S., Kumar, R., Sivakumar, D.: Information theory methods in communication complexity. In: IEEE Conference on Computational Complexity, pp. 93–102 (2002)
5. Beimel, A.: Secret-sharing schemes: a survey. In: Chee, Y.M., Guo, Z., Ling, S., Shao, F., Tang, Y., Wang, H., Xing, C. (eds.) IWCC 2011. LNCS, vol. 6639, pp. 11–46. Springer, Heidelberg (2011)
6. Beimel, A., Ishai, Y.: On the power of nonlinear secrect-sharing. In: Conference on Computational Complexity, pp. 188–202 (2001)
7. Beimel, A., Ishai, Y., Kumaresan, R., Kushilevitz, E.: On the cryptographic complexity of the worst functions. In: Lindell, Y. (ed.) TCC 2014. LNCS, vol. 8349, pp. 317–342. Springer, Heidelberg (2014)
8. Boneh, D., Franklin, M.K.: Identity-based encryption from the Weil pairing. SIAM J. Comput. **32**(3), 586–615 (2003)
9. Chen, J., Wee, H.: Fully, (almost) tightly secure IBE and dual system groups. In: Canetti, R., Garay, J.A. (eds.) CRYPTO 2013, Part II. LNCS, vol. 8043, pp. 435–460. Springer, Heidelberg (2013)
10. Chen, J., Gay, R., Wee, H.: Improved dual system ABE in prime-order groups via predicate encodings. In: Oswald, E., Fischlin, M. (eds.) EUROCRYPT 2015. LNCS, vol. 9057, pp. 595–624. Springer, Heidelberg (2015)
11. Chor, B., Goldreich, O.: Unbiased bits from sources of weak randomness and probabilistic communication complexity. SIAM J. Comput. **17**(2), 230–261 (1988)
12. Chor, B., Kushilevitz, E.: A zero-one law for boolean privacy. SIAM J. Discrete Math. **4**(1), 36–47 (1991)
13. Cocks, C., An identity based encryption scheme based on quadratic residues. In: IMA International Conference, pp. 360–363 (2001)

14. Data, D., Prabhakaran, M.M., Prabhakaran, V.M.: On the communication complexity of secure computation. In: Garay, J.A., Gennaro, R. (eds.) CRYPTO 2014, Part II. LNCS, vol. 8617, pp. 199–216. Springer, Heidelberg (2014)
15. Feige, U., Kilian, J., Naor, M.: A minimal model for secure computation. In: STOC, pp. 554–563 (1994)
16. Fiat, A., Naor, M.: Broadcast encryption. In: Stinson, D.R. (ed.) CRYPTO 1993. LNCS, vol. 773, pp. 480–491. Springer, Heidelberg (1994)
17. Garg, S., Kumarasubramanian, A., Sahai, A., Waters, B.: Building efficient fully collusion-resilient traitor tracing and revocation schemes. In: ACM Conference on Computer and Communications Security, pp. 121–130 (2010)
18. Gentry, C., Silverberg, A.: Hierarchical ID-based cryptography. In: Zheng, Y. (ed.) ASIACRYPT 2002. LNCS, vol. 2501, pp. 548–566. Springer, Heidelberg (2002)
19. Gertner, Y., Ishai, Y., Kushilevitz, E., Malkin, T.: Protecting data privacy in private information retrieval schemes. J. Comput. Syst. Sci. 60(3), 592–629 (2000)
20. Goyal, V., Pandey, O., Sahai, A., Waters, B.: Attribute-based encryption for fine-grained access control of encrypted data. In: ACM Conference on Computer and Communications Security, pp. 89–98 (2006)
21. Horwitz, J., Lynn, B.: Toward hierarchical identity-based encryption. In: Knudsen, L.R. (ed.) EUROCRYPT 2002. LNCS, vol. 2332, p. 466. Springer, Heidelberg (2002)
22. Ishai, Y., Wee, H.: Partial garbling schemes and their applications. In: Esparza, J., Fraigniaud, P., Husfeldt, T., Koutsoupias, E. (eds.) ICALP 2014. LNCS, vol. 8572, pp. 650–662. Springer, Heidelberg (2014)
23. Ito, M., Saito, A., Nishizeki, T.: Secret sharing schemes realizing general access structure. In: GLOBECOM, pp. 99–102 (1987)
24. Kalyanasundaram, B., Schnitger, G.: The probabilistic communication complexity of set intersection. SIAM J. Discrete Math. 5(4), 545–557 (1992)
25. Karchmer, M., Wigderson, A.: On span programs. In: Structure in Complexity Theory Conference, pp. 102–111 (1993)
26. Katz, J., Sahai, A., Waters, B.: Predicate encryption supporting disjunctions, polynomial equations, and inner products. In: Smart, N.P. (ed.) EUROCRYPT 2008. LNCS, vol. 4965, pp. 146–162. Springer, Heidelberg (2008)
27. Kowalczyk, L., Lewko, A.B.: Bilinear entropy expansion from the decisional linear assumption. In: Gennaro, R., Robshaw, M. (eds.) CRYPTO 2015, Part II, LNCS 9216, pp. 524–541 (2015)
28. Kremer, I., Nisan, N., Ron, D.: On randomized one-round communication complexity. Comput. Complex. 8(1), 21–49 (1999)
29. Kushilevitz, E., Nisan, N.: Commun. Compl. Cambridge University Press, Cambridge (1997). ISBN 978-0-521-56067-2
30. Lewko, A.: Tools for simulating features of composite order bilinear groups in the prime order setting. In: Pointcheval, D., Johansson, T. (eds.) EUROCRYPT 2012. LNCS, vol. 7237, pp. 318–335. Springer, Heidelberg (2012)
31. Lewko, A., Waters, B.: New techniques for dual system encryption and fully secure HIBE with short ciphertexts. In: Micciancio, D. (ed.) TCC 2010. LNCS, vol. 5978, pp. 455–479. Springer, Heidelberg (2010)
32. Lewko, A., Waters, B.: Unbounded HIBE and attribute-based encryption. In: Paterson, K.G. (ed.) EUROCRYPT 2011. LNCS, vol. 6632, pp. 547–567. Springer, Heidelberg (2011)
33. Lewko, A., Waters, B.: Decentralizing attribute-based encryption. In: Paterson, K.G. (ed.) EUROCRYPT 2011. LNCS, vol. 6632, pp. 568–588. Springer, Heidelberg (2011)

34. Lewko, A., Waters, B.: New proof methods for attribute-based encryption: achieving full security through selective techniques. In: Safavi-Naini, R., Canetti, R. (eds.) CRYPTO 2012. LNCS, vol. 7417, pp. 180–198. Springer, Heidelberg (2012)

35. Lewko, A., Okamoto, T., Sahai, A., Takashima, K., Waters, B.: Fully secure functional encryption: attribute-based encryption and (hierarchical) inner product encryption. In: Gilbert, H. (ed.) EUROCRYPT 2010. LNCS, vol. 6110, pp. 62–91. Springer, Heidelberg (2010)

36. Nayak, A.: Optimal lower bounds for quantum automata and random access codes. In: FOCS, pp. 369–377 (1999). CoRR quant-ph/9904093

37. Newman, I., Szegedy, M.: Public vs. private coin flips in one round communication games. In: STOC, pp. 561–570 (1996)

38. Okamoto, T., Takashima, K.: Fully secure functional encryption with general relations from the decisional linear assumption. In: Rabin, T. (ed.) CRYPTO 2010. LNCS, vol. 6223, pp. 191–208. Springer, Heidelberg (2010)

39. Okamoto, T., Takashima, K.: Achieving short ciphertexts or short secret-keys for adaptively secure general inner-product encryption. In: Lin, D., Tsudik, G., Wang, X. (eds.) CANS 2011. LNCS, vol. 7092, pp. 138–159. Springer, Heidelberg (2011)

40. Okamoto, T., Takashima, K.: Adaptively attribute-hiding (hierarchical) inner product encryption. In: Pointcheval, D., Johansson, T. (eds.) EUROCRYPT 2012. LNCS, vol. 7237, pp. 591–608. Springer, Heidelberg (2012)

41. Ostrovsky, R., Skeith III, W.E.: Communication complexity in algebraic two-party protocols. In: Wagner, D. (ed.) CRYPTO 2008. LNCS, vol. 5157, pp. 379–396. Springer, Heidelberg (2008)

42. Razborov, A.A.: On the distributional complexity of disjointness. Theor. Comput. Sci. **106**(2), 385–390 (1992)

43. Sahai, A., Waters, B.: Fuzzy identity-based encryption. In: Cramer, R. (ed.) EUROCRYPT 2005. LNCS, vol. 3494, pp. 457–473. Springer, Heidelberg (2005)

44. Shamir, A.: Factoring numbers in $O(\log n)$ arithmetic steps. Inf. Process. Lett. **8**(1), 28–31 (1979)

45. Shamir, A.: Identity-based cryptosystems and signature schemes. In: Blakely, G.R., Chaum, D. (eds.) CRYPTO 1984. LNCS, vol. 196, pp. 47–53. Springer, Heidelberg (1985)

46. Vaikuntanathan, V., Vasudevan, P.N.: From statistical zero knowledge to secret sharing. Cryptology ePrint Archive, Report 2015/281 (2015)

47. Waters, B.: Dual system encryption: realizing fully secure IBE and HIBE under simple assumptions. In: Halevi, S. (ed.) CRYPTO 2009. LNCS, vol. 5677, pp. 619–636. Springer, Heidelberg (2009)

48. Wee, H.: Dual system encryption via predicate encodings. In: Lindell, Y. (ed.) TCC 2014. LNCS, vol. 8349, pp. 616–637. Springer, Heidelberg (2014)

49. Yao, A.C.: Some complexity questions related to distributive computing. In: STOC, pp. 209–213 (1979)

Predicate Encryption for Circuits from LWE

Sergey Gorbunov[1]([✉]), Vinod Vaikuntanathan[1], and Hoeteck Wee[2]

[1] MIT, Boston, USA
sergeyg@mit.edu, vinodv@csail.mit.edu
[2] ENS, Paris, France
wee@di.ens.fr

Abstract. In predicate encryption, a ciphertext is associated with descriptive attribute values x in addition to a plaintext μ, and a secret key is associated with a predicate f. Decryption returns plaintext μ if and only if $f(x) = 1$. Moreover, security of predicate encryption guarantees that an adversary learns nothing about the attribute x or the plaintext μ from a ciphertext, given arbitrary many secret keys that are not authorized to decrypt the ciphertext individually.

We construct a leveled predicate encryption scheme for all circuits, assuming the hardness of the subexponential learning with errors (LWE) problem. That is, for any polynomial function $d = d(\lambda)$, we construct a predicate encryption scheme for the class of all circuits with depth bounded by $d(\lambda)$, where λ is the security parameter.

1 Introduction

Predicate encryption [BW07, SBC+07, KSW08] is a new paradigm for public-key encryption that supports searching on encrypted data. In predicate encryption, ciphertexts are associated with descriptive attribute values x in addition to plaintexts μ, secret keys are associated with a predicate f, and a secret key decrypts the ciphertext to recover μ if and only if $f(x) = 1$. The security requirement for predicate encryption enforces privacy of x and the plaintext even amidst multiple secret key queries: an adversary holding secret keys for different query predicates learns nothing about the attribute x and the plaintext (apart from the fact that x does not satisfy any of the query predicates) if none of them is individually authorized to decrypt the ciphertext.

Sergey Gorbunov — Supported in part by the Northrop Grumman Cybersecurity Research Consortium (CRC) and by a Microsoft PhD Fellowship.

Vinod Vaikuntanathan — Research supported in part by DARPA Grant number FA8750-11-2-0225, NSF Awards CNS-1350619 and CNS-1413920, an Alfred P. Sloan Research Fellowship, the Northrop Grumman Cybersecurity Research Consortium (CRC), Microsoft Faculty Fellowship, and a Steven and Renee Finn Career Development Chair from MIT.

Hoeteck Wee — CNRS, INRIA and Columbia University.Supported in part by NSF Award CNS-1445424, ERC project aSCEND (639554),the Alexander von Humboldt Foundation and a Google Faculty Research Award.

R. Gennaro and M. Robshaw (Eds.): CRYPTO 2015, Part II, LNCS 9216, pp. 503–523, 2015.
DOI: 10.1007/978-3-662-48000-7_25

Motivating Applications. We begin with several motivating applications for predicate encryption [BW07, SBC+07]:

- For inspecting recorded log files for network intrusions, we would encrypt network flows labeled with a set of attributes from the network header, such as the source and destination addresses, port numbers, time-stamp, and protocol numbers. We could then issue auditors with restricted secret keys that can only decrypt the network flows that fall within a particular range of IP addresses and some specific time period.
- For credit card fraud investigation, we would encrypt credit card transactions labeled with a set of attributes such as time, costs and zipcodes. We could then issue investigators with restricted secret keys that decrypt transactions over $1,000 which took place in the last month and originated from a particular range of zipcodes.
- For anti-terrorism investigation, we would encrypt travel records labeled with a set of attributes such as travel destination and basic traveller data. We could then issue investigators with restricted secret keys that match certain suspicious travel patterns.
- For online dating, we would encrypt personal profiles labeled with dating preferences pertaining to age, height, weight, salary and hobbies. Secret keys are associated with specific attributes and can only decrypt profiles for which the attributes match the dating preferences.

In all of these examples, it is important that unauthorized parties do not learn the contents of the ciphertexts, nor of the meta-data associated with the ciphertexts, such as the network header or dating preferences. On the other hand, it is often okay to leak the meta-data to authorized parties. We stress that privacy of the meta-data is an additional security requirement provided by predicate encryption but not by the related and weaker notion of attribute-based encryption (ABE) [SW05, GPSW06]; the latter only guarantees the privacy of the plaintext μ and not the attribute x.

Utility and Expressiveness. The utility of predicate encryption is intimately related to the class of predicates for which we could create secret keys. Ideally, we would like to support the class of all circuits. Over the past decade, substantial advances were made for the weaker primitive of ABE, culminating most recently in schemes supporting any policy computable by general circuits [GVW13, BGG+14] under the standard LWE assumption [Reg09]. However, the state-of-the-art for predicate encryption is largely limited to very simple functionalities related to computing an inner product [BW07, SBC+07, KSW08, AFV11, GMW15].

1.1   Our Contributions

In this work, we substantially advance the state of the art to obtain predicate encryption for all circuits (c.f. Fig. 1):

Theorem (Informal). Under the LWE assumption, there exists a predicate encryption scheme for all circuits, with succint ciphertexts and secret keys independent of the size of the circuit.

As with prior LWE-based ABE for circuits [GVW13, BGG+14], to support circuits of depth d, the parameters of the scheme grow with poly(d), and we require sub-exponential $n^{\Omega(d)}$ hardness of the LWE assumption. In addition, the security guarantee is selective, but can be extended to adaptive security via complexity leveraging [BB04].

Privacy Guarantees. The privacy notion we achieve is a *simulation-based* variant of "attribute-hiding" from the literature [SBC+07, OT10, AFV11]. That is, we guarantee privacy of the attribute x and the plaintext μ against collusions holding secret keys for functions f such that $f(x) = 0$. An even stronger requirement would be to require privacy of x even against authorized keys corresponding to functions f where $f(x) = 1$; in the literature, this stronger notion is referred to as "full attribute-hiding" [BW07, KSW08]. This stronger requirement is equivalent to "full-fledged" functional encryption [BSW11], for which we cannot hope to achieve simulation-based security for all circuits as achieved in this work [BSW11, AGVW13].

Relation to Prior Works. Our result subsumes all prior works on predicate encryption under standard cryptographic assumptions, apart from a few exceptions pertaining to the inner product predicate [BW07, KSW08, OT12]. These results achieve a stronger security notion for predicate encryption, known as full (or strong) security (please refer to Sect. 3.1, and the full version for definitions).

In a recent break-through work, Garg et al. [GGH+13b] gave a beautiful candidate construction of functional encryption (more general primitive than predicate encryption) for arbitrary circuits. However, the construction relies on "multi-linear maps" [GGH13a, CLT13, GGH15], for which we have few candidates and which rely on complex intractability assumptions that are presently poorly understood and not extensively studied in the literature. It remains an intriguing open problem to construct a functional encryption scheme from a standard assumption, such as LWE.

In contrast, if we consider functional encryption with *a-priori bounded collusions size* (that is, the number of secret keys any collusion of adversaries may obtain is fixed by the scheme at the setup phase), then it is possible to obtain functional encryption for general circuits under a large class of standard assumptions [SS10, GVW12, GKP+13]. This notion is *weaker* than standard notion of functional encryption, yet remains very meaningful for many applications.

1.2 Overview of Our Construction

Our starting point is the work of Goldwasser, Kalai, Popa, Vaikuntanathan and Zeldovich [GKP+13] who show how to convert an attribute-based encryption (ABE) scheme into a *single key secure* functional encryption (FE) scheme.

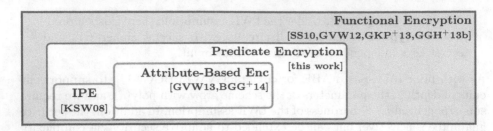

Fig. 1. State of the art in functional encryption. The white region refers to function-alities for which we have constructions under standard cryptographic assumptions like LWE or decisional problems in bilinear groups: these functionalities include inner prod-uct encryption (IPE), attribute-based encryption for general circuits (ABE) and pred-icate encryption for general circuits. The grey region refers to functionalities beyond predicate encryption for which we only have constructions for weaker security notions like bounded collusions, or under non-standard cryptographic assumptions like obfus-cation or multilinear maps.

	Interface	Security Guarantee given sk_f
ABE	$\mathsf{Enc}(x, \mu)$	μ is secret iff $f(x) = 0$
		x is always public
PE	$\mathsf{Enc}(x, \mu)$	(x, μ) is secret iff $f(x) = 0$
FE	$\mathsf{Enc}(x)$	user learns only $f(x)$

Fig. 2. Comparison of the security guarantees provided by attribute-based (ABE), predicate (PE) and functional encryption (FE), where secret keys are associated with a Boolean function f; the main distinction lies in how much information about x is potentially leaked to the adversary. The main distinction between ABE and PE is that x is always public in ABE, but remains secret in PE when the user is not authorized to decrypt. The main distinction between PE and FE is that x always remains hidden (even when $f(x) = 1$) and hence the user only learns the output of the computation of f on x.

Recall that in an attribute-based encryption scheme [GPSW06], a ciphertext is associated with a descriptive value (a public "attribute") x and plaintext μ, and it hides μ, but not x. The observation of Goldwasser et al. [GKP+13] is to hide x by encrypting it using a fully homomorphic encryption (FHE) scheme [Gen09,BV11b], and then using the resulting FHE ciphertext as the pub-lic "attribute" in an ABE scheme for general circuits [GVW13,BGG+14]. This has the dual benefit of guaranteeing privacy of x, while at the same time allowing homomorphic computation of predicates f on the encryption of x (Fig 2).

This initial idea quickly runs into trouble. The decryptor who is given the predicate secret key for f and a predicate encryption of (x, μ) can indeed com-pute an *FHE encryption* of $f(x)$. However, the decryption process is confronted with a decision, namely whether to release the message μ or not, and this decision depends on whether the *plaintext* $f(x)$ is 0 or 1.[1] Clearly, resolving this conundrum requires obtaining $f(x)$, which requires knowledge of the FHE secret key. Goldwasser et al. [GKP+13] solved this by employing a (single use) Yao garbling of the FHE decryption circuit, however this limited them to obtaining *single key secure* predicate/functional encryption schemes.[2]

Our first key idea is to embed the FHE secret key as part of the attributes in the ABE ciphertext. That is, in order to encrypt a plaintext μ with attributes x in the predicate encryption scheme, we first choose a symmetric key fhe.sk for the FHE scheme, encrypt x into a FHE ciphertext $\hat{x}$, and encrypt μ using the ABE scheme with (fhe.sk, $\hat{x}$) as the attributes to obtain an ABE ciphertext ct. Our predicate encryption ciphertext is then given by

$$(\hat{x}, \mathsf{ct})$$

To generate the predicate secret key for a function f, one simply generates the ABE secret key for the function g that takes as input (fhe.sk, $\hat{x}$) and computes

$$g(\mathsf{fhe.sk}, \hat{x}) = \mathsf{FHE.Dec}(\mathsf{fhe.sk}; \mathsf{FHE.Eval}(f, \hat{x}))$$

That is, g first homomorphically computes a FHE encryption of $f(x)$, and then decrypts it using the FHE secret key to output $f(x)$.

At first glance, this idea evokes strong and conflicting emotions as it raises two problems. The first pertains to correctness: we can no longer decrypt the ciphertext since the ABE decryption algorithm needs to know all of the attributes ($\hat{x}$ and fhe.sk), but fhe.sk is missing. The second pertains to security: the ABE ciphertext ct is not guaranteed to protect the privacy of the attributes, and could leak all of fhe.sk which together with $\hat{x}$ would leak all of x. Solving both of these problems seems to require designing a predicate encryption scheme from scratch!

Our next key observation is that the bulk of the computation in g, namely the homomorphic evaluation of the function f, is performed on the *public* attribute $\hat{x}$. The only computation performed on the secret value fhe.sk is FHE decryption which is a fairly lightweight computation. In particular, with all known FHE

---

[1] In fact, there is a syntactic mismatch since $\hat{f}(\cdot)$ is not a predicate, as it outputs an FHE ciphertext.

[2] A reader familiar with [GKP+13] might wonder whether replacing single-use garbled circuits in their construction with reusable garbled circuits (also from [GKP+13].) might remove this limitation. We remark that this does not seem possible, essentially because the construction in [GKP+13] relies crucially on the simplicity of computing garbled inputs from the "garbling key". In particular, in Yao's garbled circuit scheme, the garbling key is (many) pairs of "strings" L_0 and L_1, and a garbling of an input bit b is simply L_b. This fits perfectly with the semantics of ABE (rather, a variant termed two-input ABE in [GKP+13]) that releases one of two possible "messages" L_0 or L_1 depending on the outcome of a computation. In contrast, computing a garbled input in the reusable garbling scheme is a more complex and randomized function of the garbling key, and does not seem to align well with the semantics of ABE.

schemes [Gen09, BV11b, BV11a, BGV12, GSW13, BV14, AP14], decryption corresponds to computing an inner product followed by a threshold function. Furthermore, we do know how to construct lattice-based predicate encryption schemes for threshold of inner product [AFV11, GMW15]. We stress that the latter do not correspond to FHE decryption since the inner product is computed over a vector in the ciphertext and one in the key, whereas FHE decryption requires computing an inner product over two vectors in the ciphertext; nonetheless, we will build upon the proof techniques in achieving attribute-hiding in [AFV11, GMW15] in the proof of security.

In other words, if we could enhance ABE with a modicum of secrecy so that it can perform a heavyweight computation on public attributes followed by a lightweight privacy-preserving computation on *secret* attributes, we are back in business. Our first contribution is to define such an object, that we call *partially hiding predicate encryption*.

Partially Hiding Predicate Encryption. We introduce the notion of partially hiding predicate encryption (PHPE), an object that interpolates between attribute-based encryption and predicate encryption (analogously to partial garbling in [IW14]). In PHPE, the ciphertext, encrypting message μ, is associated with an attribute (x, y) where x is private but y is always public. The secret key is associated with a function f, and decryption succeeds iff $f(x, y) = 1$. On the one extreme, considering a dummy x or functions f that ignore x and compute on y, we recover attribute-based encryption. On the other end, considering a dummy y or functions f that ignore y and compute on x, we recover predicate encryption.

We will be interested in realizing PHPE for functions ϕ of the form $\phi(x, y) = g(x, h(y))$ for some functions g and h where h may perform arbitrary heavyweight computation on the public y and g only performs light-weight computation on the private x. Mapping back to our discussion, we would like to achieve PHPE for the "evaluate-then-decrypt" class of functions, namely where g is the FHE decryption function, h is the FHE evaluation function, x is the FHE secret key, and y is the FHE ciphertext. In general, we would like g to be simple and will allow h to be complex. It turns out that we can formalize the observation above, namely that PHPE for this class of functions gives us a predicate encryption scheme. The question now becomes: can we construct PHPE schemes for the "evaluate-then-decrypt" class of functions?

Assuming the subexponential hardness of learning with errors (LWE), we show how to construct a partially hiding predicate encryption for the class of functions $f : \mathbb{Z}_q^t \times \{0, 1\}^\ell \to \{0, 1\}$ of the form

$$f_\gamma(\mathbf{x}, \mathbf{y}) = \mathsf{IP}_\gamma(\mathbf{x}, h(\mathbf{y})),$$

where $h : \{0, 1\}^\ell \to \{0, 1\}^t$, $\gamma \in \mathbb{Z}_q$, and $\mathsf{IP}_\gamma(\mathbf{x}, \mathbf{z}) = 1$ iff $\langle \mathbf{x}, \mathbf{z} \rangle = \left(\sum_{i \in [t]} \mathbf{x}[i] \cdot \mathbf{z}[i] \right) = \gamma \mod q$.

This is almost what we want, but not quite. Recall that FHE decryption in many recent schemes [BV11b, BGV12, GSW13, BV14, AP14] is a function that checks whether an inner product of two vectors in $\mathbb{Z}_q^t$ (one of which could be over $\{0,1\}^t$) lies in a certain range. Indeed, if $\mathbf{z} \in \{0,1\}^t$ is an encryption of 1 and $\mathbf{x} \in \mathbb{Z}_q^t$ is the secret key, we know that $\langle \mathbf{x}, \mathbf{z} \rangle \in [q/2 - B, q/2 + B] \pmod{q}$, where B is the noise range. Applying the so-called "modulus reduction" [BV11b] transformation to all these schemes, we can assume that this range is polynomial in size.

In other words, we will manage to construct a partially hiding PE scheme for the function

$$f_\gamma(\mathbf{x}, \mathbf{y}) : \langle \mathbf{x}, h(\mathbf{y}) \rangle \overset{?}{=} \gamma \pmod{q}$$

whereas we need a partially hiding PE scheme for the FHE decryption function which is

$$f'_R(\mathbf{x}, \mathbf{y}) : \langle \mathbf{x}, h(\mathbf{y}) \rangle \overset{?}{\in} R \pmod{q}$$

where R is the polynomial size range $[q/2 - B, q/2 + B]$ from above. How do we reconcile this disparity?

The "Lazy OR" Trick. The solution, called the "lazy OR trick" [SBC+07, GMW15] is to publish secret keys for all functions f_γ for $\gamma \in R := [q/2 - B, q/2 + B]$. This will indeed allow us to test if the FHE decryption of the evaluated ciphertext is 1 (and reveal the message μ if it is), but it is also worrying. Publishing these predicate secret keys for the predicates f_γ reveals more information than whether $\langle \mathbf{x}, h(\mathbf{y}) \rangle \overset{?}{\in} R$. In particular, it reveals what $\langle \mathbf{x}, h(\mathbf{y}) \rangle$ is. This means that an authorized key would leak partial information about the attribute, which we do allow for predicate encryption. On the other hand, for an unauthorized key where the FHE decryption is 0, each of these $f_\gamma, \gamma \in R$ is also an unauthorized key in the PHPE and therefore leaks no information about the attribute. This extends to the collection of keys in R since the PHPE is secure against collusions. For simplicity, we assume in the rest of this overview that FHE decryption corresponds to exactly to inner product.

Asymmetry to the Rescue: Constructing Partially Hiding PE. Our final contribution is the construction of a partially hiding PE for the function class $f_\gamma(\mathbf{x}, \mathbf{y})$ above. We will crucially exploit the fact that f_γ computes an inner product on the private attribute $\mathbf{x}$. There are two challenges here: first, we need to design a decryption algorithm that knows f_γ and $\mathbf{y}$ but not $\mathbf{x}$ (this is different from decryption in ABE where the algorithm also knows $\mathbf{x}$); second, show that the ciphertext does not leak too much information about $\mathbf{x}$. We use the fully key-homomorphic encryption techniques developed by Boneh et al. [BGG+14] in the context of constructing an "arithmetic" ABE scheme. The crucial observation about the ABE scheme of [BGG+14] is that while it was not designed to hide the attributes, it can be made to partially hide them in exactly the way we want. In particular, the scheme allows us to carry out an inner product of a public attribute vector (corresponding to the evaluated FHE ciphertext) and a private attribute vector (corresponding to the FHE secret key fhe.sk), thanks to an inherent asymmetry in homomorphic evaluation of a multiplication gate on

ABE ciphertexts. More concretely, in the homomorphic evaluation of a ciphertext for a multiplication gate in [BGG+14], the decryption algorithm works even if one of the attribute remains private, and for addition gates, the decryption algorithms works even if both attributes remain private. This addresses the first challenge of a decryption algorithm that is oblivious to $\mathbf{x}$. For the second challenge of security, we rely on techniques from inner product predicate encryption [AFV11] to prove the privacy of $\mathbf{x}$ Note that in the latter, the inner product is computed over a vector in the ciphertext and one in the key, whereas in our scheme, the inner product is computed over two vectors in the ciphertext. Interestingly, the proof still goes through since the ciphertext in the ABE [BGG+14] has the same structure as the ciphertext in [AFV11]. We refer the reader to Sect. 3.2 for a detailed overview of the partial hiding PE, and to Sect. 4 for an overview of how we combine the partial hiding PE with FHE to obtain our main result.

Finally, we remark that exploiting asymmetry in multiplication has been used in fairly different contexts in both FHE [GSW13, BV14] and in ABE [GVW13, GV14]. In [GSW13] and in this work, the use of asymmetry was crucial for realizing the underlying cryptographic primitive; whereas in [GVW13, BV14, GV14], asymmetry was used to reduce the noise growth during homomorphic evaluation, thereby leading to quantitative improvements in the underlying assumptions and hence improved efficiency.

1.3   Discussion

Comparison with Other Approaches. The two main alternative approaches for realizing predicate and functional encryption both rely on multi-linear maps either implicitly, or explicitly. The first is to use indistinguishability obfuscation as in [GGH+13b], and the second is to extend the dual system encryption framework to multi-linear maps [Wat09, GGHZ14]. A crucial theoretical limitation of these approaches is that they all rely on non-standard assumptions; we have few candidates for multi-linear maps [GGH13a, CLT13, GGH15] and the corresponding assumptions are presently poorly understood and not extensively studied in cryptanalysis, and in some cases, broken [CHL+15]. In particular, the latest attack in [CHL+15] highlight the importance of obtaining constructions and developing techniques that work under standard cryptographic assumptions, as is the focus of this work.

Barriers to Functional Encryption from LWE. We note the two main barriers to achieving full-fledged functional encryption from LWE using our framework. First, the lazy conjunction approach to handle threshold inner product for FHE decryption leaks the exact inner product and therefore cannot be used to achieve full attribute-hiding. Second, we do not currently know of a fully attribute-hiding inner product encryption scheme under the LWE assumption, although we do know how to obtain such schemes under standard assumptions in bilinear groups [OT12, KSW08].

2 Preliminaries

We refer the reader to the full version for the background on lattices.

2.1 Fully-Homomorphic Encryption

We present a fairly minimal definition of fully homomorphic encryption (FHE) which is sufficient for our constructions. A leveled homomorphic encryption scheme is a tuple of polynomial-time algorithms (HE.KeyGen, HE.Enc, HE.Eval, HE.Dec):

- **Key generation.** HE.KeyGen($1^\lambda, 1^d, 1^k$) is a probablistic algorithm that takes as input the security parameter λ, a depth bound d and message length k and outputs a secret key sk.
- **Encryption.** HE.Enc(sk, μ) is a probabilistic algorithm that takes as input sk and a message $\mu \in \{0,1\}^k$ and outputs a ciphertext ct.
- **Homomorphic evaluation.** HE.Eval(f, ct) is a deterministic algorithm that takes as input a boolean circuit $C : \{0,1\}^k \to \{0,1\}$ of depth at most d and a ciphertext ct and outputs another ciphertext ct$'$.
- **Decryption.** HE.Dec(sk, ct$'$) is a deterministic algorithm that takes as input sk and ciphertext ct$'$ and outputs a bit.

Correctness. We require perfect decryption correctness with respect to homomorphically evaluated ciphertexts: namely for all λ, d, k and all sk $\leftarrow$ HE.KeyGen($1^\lambda, 1^d, 1^k$), all $\mu \in \{0,1\}^k$ and for all boolean circuits $C : \{0,1\}^k \to \{0,1\}$ of depth at most d:

$$\Pr\Big[\mathsf{HE.Dec}(\mathsf{sk}, \mathsf{HE.Eval}(C, \mathsf{HE.Enc}(\mathsf{sk}, \mu))) = C(\mu)\Big] = 1$$

where the probablity is taken over HE.Enc and HE.KeyGen.

Security. We require semantic security for a single ciphertext: namely for every stateful p.p.t. adversary $\mathcal{A}$ and for all $d, k = \mathrm{poly}(\lambda)$, the following quantity

$$\Pr\left[b = b' : \begin{array}{l} \mathsf{sk} \leftarrow \mathsf{Setup}(1^\lambda, 1^d, 1^k); \\ (\mu_0, \mu_1) \leftarrow \mathcal{A}(1^\lambda, 1^d, 1^k); \\ b \xleftarrow{\$} \{0,1\}; \\ \mathsf{ct} \leftarrow \mathsf{Enc}(\mathsf{sk}, \mu_b); \\ b' \leftarrow \mathcal{A}(\mathsf{ct}) \end{array} \right] - \frac{1}{2}$$

is negligible in λ.

FHE from LWE We will rely on an instantiation of FHE from the LWE assumption:

Theorem 2.1. (FHE from LWE [BV11b, BGV12, GSW13, BV14, AP14]**).**
There is a FHE scheme HE.KeyGen, HE.Enc, HE.Eval, HE.Dec *that works for any*
q *with* $q \geq O(\lambda^2)$ *with the following properties:*

- HE.KeyGen *outputs a secret key* $\mathsf{sk} \in \mathbb{Z}_q^t$ *where* $t = \mathrm{poly}(\lambda)$;
- HE.Enc *outputs a ciphertext* $\mathsf{ct} \in \{0,1\}^\ell$ *where* $\ell = \mathrm{poly}(k, d, \lambda, \log q)$;
- HE.Eval *outputs a ciphertext* $\mathsf{ct}' \in \{0,1\}^t$;
- *for any boolean circuit of depth* d, HE.Eval$(C, \cdot)$ *is computed by a boolean circuit of depth* $\mathrm{poly}(d, \lambda, \log q)$.
- HE.Dec *on input* $\mathsf{sk}, \mathsf{ct}'$ *outputs a bit* $b \in \{0,1\}$. *If* ct' *is an encryption of* 1 *then*

$$\sum_{i=1}^{t} \mathsf{sk}[i] \cdot \mathsf{ct}'[i] \in [\lfloor q/2 \rfloor - B, \lfloor q/2 \rfloor + B]$$

for some fixed $B = \mathrm{poly}(\lambda)$. *Otherwise, if* ct' *is an encryption of* 0, *then*

$$\sum_{i=1}^{t} \mathsf{sk}[i] \cdot \mathsf{ct}'[i] \notin [\lfloor q/2 \rfloor - B, \lfloor q/2 \rfloor + B];$$

- *security relies on* dLWE$_{\Theta(t), q, \chi}$.

We highlight several properties of the above scheme: (1) the ciphertext is a bit-string, (2) the bound B is a polynomial independent of q (here, we crucially exploit the new results in [BV14] together with the use of leveled bootstrapping)[3], (3) the size of normal ciphertexts is independent of the size of the circuit (this is the typical compactness requirement).

3   Partially Hiding Predicate Encryption

3.1   Definitions

We introduce the notation of partially hiding predicate encryption (PHPE), which interpolates attribute-based encryption and predicate encryption (analogously to partial garbling in [IW14]). In PHPE, the ciphertext, encrypting message μ, is associated with an attribute (x, y) where x is private but y is always public. The secret key is associated with a predicate C, and decryption succeeds iff $C(x, y) = 1$. The requirement is that a collusion learns nothing about (x, μ) if none of them is individually authorized to decrypt the ciphertext. Attribute-based encryption corresponds to the setting where x is empty, and predicate encryption corresponds to the setting where y is empty. We refer the reader to the full version for the standard notion of predicate encryption.

Looking ahead to our construction, we show how to:

- construct PHPE for a restricted class of circuits that is "low complexity" with respect to x and allows arbitrarily polynomial-time computation on y;
- bootstrap this PHPE using FHE to obtain PE for all circuits.

---

[3] Recall that no circular security assumption needs to be made for leveled bootstrapping.

Syntax. A Partially-Hiding Predicate Encryption scheme $\mathcal{PHPE}$ for a pair of input-universes $\mathcal{X}, \mathcal{Y}$, a predicate universe $\mathcal{C}$, a message space $\mathcal{M}$, consists of four algorithms (PH.Setup, PH.Enc, PH.Keygen, PH.Dec):

PH.Setup$(1^\lambda, \mathcal{X}, \mathcal{Y}, \mathcal{C}, \mathcal{M}) \rightarrow$ (ph.mpk, ph.msk). The setup algorithm gets as input the security parameter λ and a description of $(\mathcal{X}, \mathcal{Y}, \mathcal{C}, \mathcal{M})$ and outputs the public parameter ph.mpk, and the master key ph.msk.

PH.Enc(ph.mpk, $(x, y), \mu) \rightarrow \mathsf{ct}_y$. The encryption algorithm gets as input ph.mpk, an attribute $(x, y) \in \mathcal{X} \times \mathcal{Y}$ and a message $\mu \in \mathcal{M}$. It outputs a ciphertext ct_y.

PH.Keygen(ph.msk, $C) \rightarrow \mathsf{sk}_C$. The key generation algorithm gets as input ph.msk and a predicate $C \in \mathcal{C}$. It outputs a secret key sk_C.

PH.Dec$((\mathsf{sk}_C, C), (\mathsf{ct}_y, y)) \rightarrow \mu$. The decryption algorithm gets as input the secret key sk_C, a predicate C, and a ciphertext ct_y and the public part of the attribute y. It outputs a message $\mu \in \mathcal{M}$ or $\bot$.

Correctness. We require that for all PH.Setup$(1^\lambda, \mathcal{X}, \mathcal{Y}, \mathcal{C}, \mathcal{M}) \rightarrow$ (ph.mpk, ph.msk), for all $(x, y, C) \in \mathcal{X} \times \mathcal{Y} \times \mathcal{C}$, for all $\mu \in \mathcal{M}$,

- if $C(x, y) = 1$, $\Pr\left[\mathsf{PH.Dec}((\mathsf{sk}_C, C), (\mathsf{ct}_y, y)) = \mu\right] \geq 1 - \mathsf{negl}(\lambda)$,
- if $C(x, y) = 0$, $\Pr\left[\mathsf{PH.Dec}((\mathsf{sk}_C, C), (\mathsf{ct}_y, y)) = \bot\right] \geq 1 - \mathsf{negl}(\lambda)$,

where the probabilities are taken over $\mathsf{sk}_C \leftarrow \mathsf{PH.Keygen}(\mathsf{ph.msk}, C)$, $\mathsf{ct}_y \leftarrow \mathsf{PH.Enc}(\mathsf{ph.mpk}, (x, y), \mu)$ and coins of PH.Setup.

Definition 3.1 (PHPE   Attribute-Hiding). *Fix* (PH.Setup, PH.Enc, PH.Keygen, PH.Dec). *For every stateful p.p.t. adversary* Adv, *and a p.p.t. simulator* Sim, *consider the following two experiments:*

$\exp^{\mathsf{real}}_{\mathcal{PHPE}, \mathrm{Adv}}(1^\lambda)$:	$\exp^{\mathsf{ideal}}_{\mathcal{PHPE}, \mathrm{Sim}}(1^\lambda)$:
1: $(x, y) \leftarrow \mathrm{Adv}(1^\lambda, \mathcal{X}, \mathcal{Y}, \mathcal{C}, \mathcal{M})$	1: $(x, y) \leftarrow \mathrm{Adv}(1^\lambda, \mathcal{X}, \mathcal{Y}, \mathcal{C}, \mathcal{M})$
2: (ph.mpk, ph.msk) $\leftarrow$ PH.Setup$(1^\lambda, \mathcal{X}, \mathcal{Y}, \mathcal{C}, \mathcal{M})$	2: (ph.mpk, ph.msk) $\leftarrow$ PH.Setup$(1^\lambda, \mathcal{X}, \mathcal{Y}, \mathcal{C}, \mathcal{M})$
3: $\mu \leftarrow \mathrm{Adv}^{\mathsf{PH.Keygen}(\mathsf{msk}, \cdot)}(\mathsf{ph.mpk})$	3: $\mu \leftarrow \mathrm{Adv}^{\mathsf{PH.Keygen}(\mathsf{ph.msk}, \cdot)}(\mathsf{ph.mpk})$
4: $\mathsf{ct}_y \leftarrow \mathsf{PH.Enc}(\mathsf{ph.mpk}, (x, y), \mu)$	4: $\mathsf{ct}_y \leftarrow \mathrm{Sim}(\mathsf{mpk}, y, 1^{\lvert x \rvert}, 1^{\lvert \mu \rvert})$
5: $\alpha \leftarrow \mathrm{Adv}^{\mathsf{PH.Keygen}(\mathsf{ph.msk}, \cdot)}(\mathsf{ct}_y)$	5: $\alpha \leftarrow \mathrm{Adv}^{\mathsf{PH.Keygen}(\mathsf{msk}, \cdot)}(\mathsf{ct}_y)$
6: Output (x, y, μ, α)	6: Output (x, y, μ, α)

We say an adversary Adv *is admissible if all oracle queries that it makes* $C \in \mathcal{C}$ *satisfy* $C(x, y) = 0$. *The Partially-Hiding Predicate Encryption scheme* $\mathcal{PHPE}$ *is then said to be* attribute-hiding *if there is a p.p.t. simulator* Sim *such that for every stateful p.p.t. adversary* Adv, *the following two distributions are computationally indistinguishable:*

$$\left\{ \exp^{\mathsf{real}}_{\mathcal{PHPE}, \mathrm{Adv}}(1^\lambda) \right\}_{\lambda \in \mathbb{N}} \overset{c}{\approx} \left\{ \exp^{\mathsf{ideal}}_{\mathcal{PHPE}, \mathrm{Sim}}(1^\lambda) \right\}_{\lambda \in \mathbb{N}}$$

Remarks. We point out some remarks of our definition (SIM-AH) when treated as a regular predicate encryption (i.e. the setting where y is empty; see the full version for completeness) and how it compares to other definitions in the literature.

- We note the simulator for the challenge ciphertext gets y but not x; this captures the fact that y is public whereas x is private. In addition, the simulator is not allowed to program the public parameters or the secret keys. In the ideal experiment, the simulator does not explicitly learn any information about x (apart from its length); nonetheless, there is implicit leakage about x from the key queries made by an admissible adversary. Finally, we note that we can efficiently check whether an adversary is admissible.
- Our security notion is "selective", in that the adversary "commits" to (x, y) before it sees ph.mpk. It is possible to bootstrap selectively-secure scheme to full security using standard complexity leveraging arguments [BB04, GVW13], at the price of a $2^{|x|}$ loss in the security reduction.
- Our definition refers to a single challenge message, but the definition extends readily to a setting with multiple challenge messages. Moreover, our definition composes in that security for a single message implies security with multiple messages (see the full version). The following remarks refer to many messages setting.
- We distinguish between two notions of indistinguishability-based (IND) definitions used in the literature: attribute-hiding (IND-AH)[4] and strong attribute-hiding (IND-SAH)[5] [BW07, SBC+07, KSW08, AFV11]. In the IND-AH, the adversary should not be able to distinguish between two pairs of attributes/messages given that it is restricted to queries which do not decrypt the challenge ciphertext (See the full version for details). It is easy to see that our SIM-AH definition is stronger than IND-AH. Furthermore, IND-SAH also ensures that adversary cannot distinguish between the attributes even when it is allowed to ask for queries that decrypt the messages (in this case, it must output $\mu_0 = \mu_1$). Our SIM-AH definition is weaker than IND-SAH, since we explicitly restrict the adversary to queries that do not decrypt the challenge ciphertext.
- In the context of arbitrary predicates, *strong* variants of definitions (that is, IND-SAH and SIM-SAH) are equivalent to security notions for functional encryption (the simulation definition must be adjusted to give the simulated the outputs of the queries). However, the strong variant of notion (SIM-SAH) is impossible to realize for many messages [BSW11, AGVW13]. We refer the reader to the full version for a sketch of the impossibility. Hence, SIM-AH is the best-possible simulation security for predicate encryption which we realize in this work. The only problem which we leave open is to realize IND-SAH from standard LWE.

---

[4] Sometimes also referred as weak attribute-hiding.
[5] Sometimes also referred as full attribute-hiding.

3.2 Our Construction

We refer the reader to the full version for the complete description of our construction. Below, we provide an overview.

Overview. We construct a partially hiding predicate encryption for the class of predicate circuits $C : \mathbb{Z}_q^t \times \{0,1\}^\ell \to \{0,1\}$ of the form $\widehat{C} \circ \mathsf{IP}_\gamma$ where $\widehat{C} : \{0,1\}^\ell \to \{0,1\}^t$ is a boolean circuit of depth d, $\gamma \in \mathbb{Z}_q$, and

$$(\widehat{C} \circ \mathsf{IP}_\gamma)(\mathbf{x}, \mathbf{y}) = \mathsf{IP}_\gamma(\mathbf{x}, \widehat{C}(\mathbf{y})),$$

where $\mathsf{IP}_\gamma(\mathbf{x}, \mathbf{z}) = 1$ iff $\langle \mathbf{x}, \mathbf{z} \rangle = \left(\sum_{i \in [t]} \mathbf{x}[i] \cdot \mathbf{z}[i] \right) = \gamma \mod q$. We refer to circuit IP as the generic inner-product circuit of two vectors.

Looking ahead, $\widehat{C}$ corresponds to FHE evaluation of an arbitrary circuit C, whereas IP_γ corresponds to roughly to FHE decryption; in the language of the introduction in Sect. 1, $\widehat{C}$ corresponds to heavy-weight computation h, whereas IP_γ corresponds to light-weight computation g.

The scheme. The public parameters are matrices

$$\left(\mathbf{A}, \mathbf{A}_1, \ldots, \mathbf{A}_\ell, \mathbf{B}_1, \ldots, \mathbf{B}_t \right)$$

An encryption corresponding to the attribute $(\mathbf{x}, \mathbf{y}) \in \mathbb{Z}_q^t \times \{0,1\}^\ell$ is a GPV ciphertext (an LWE sample) corresponding to the matrix

$$\left[\mathbf{A} \mid \mathbf{A}_1 + \mathbf{y}[1] \cdot \mathbf{G} \mid \cdots \mid \mathbf{A}_\ell + \mathbf{y}[\ell] \cdot \mathbf{G} \mid \mathbf{B}_1 + \mathbf{x}[1] \cdot \mathbf{G} \mid \cdots \mid \mathbf{B}_t + \mathbf{x}[t] \cdot \mathbf{G} \right]$$

To decrypt the ciphertext given $\mathbf{y}$ and a key for $\widehat{C} \circ \mathsf{IP}_\gamma$, we apply the BGGHNSVV algorithm to first transform the first part of the ciphertext into a GPV ciphertext corresponding to the matrix

$$\left[\mathbf{A} \mid \mathbf{A}_{\widehat{C}_1} + \mathbf{z}[1] \cdot \mathbf{G} \mid \cdots \mid \mathbf{A}_{\widehat{C}_t} + \mathbf{z}[t] \cdot \mathbf{G} \right]$$

where $\widehat{C}_i$ is the circuit computing the i'th bit of $\widehat{C}$ and $\mathbf{z} = \widehat{C}(\mathbf{y}) \in \{0,1\}^t$. Next, observe that

$$-\left((\mathbf{A}_{\widehat{C}_i} + \mathbf{z}[i] \cdot \mathbf{G}) \cdot \mathbf{G}^{-1}(\mathbf{B}_i) \right) + \mathbf{z}[i] \cdot \left(\mathbf{B}_i + \mathbf{x}[i] \cdot \mathbf{G} \right) = -\mathbf{A}_{\widehat{C}_i} \mathbf{G}^{-1}(\mathbf{B}_i) + \mathbf{x}[i] \cdot \mathbf{z}[i] \cdot \mathbf{G}.$$

Summing over i, we have

$$\sum_{i=1}^{\ell} -\left((\mathbf{A}_{\widehat{C}_i} + \mathbf{z}[i] \cdot \mathbf{G}) \cdot \mathbf{G}^{-1}(\mathbf{B}_i) \right) + \mathbf{z}[i] \cdot \left(\mathbf{B}_i + \mathbf{x}[i] \cdot \mathbf{G} \right) = \mathbf{A}_{\widehat{C} \circ \mathsf{IP}} + \langle \mathbf{x}, \mathbf{z} \rangle \cdot \mathbf{G}$$

where

$$\mathbf{A}_{\widehat{C} \circ \mathsf{IP}} := -\left(\mathbf{A}_{\widehat{C}_1} \mathbf{G}^{-1}(\mathbf{B}_1) + \cdots + \mathbf{A}_{\widehat{C}_t} \mathbf{G}^{-1}(\mathbf{B}_t) \right).$$

Therefore, given only the public matrices and $\mathbf{y}$ (but not $\mathbf{x}$), we may transform the ciphertext into a GPV ciphertext corresponding to the matrix

$$\left[\, \mathbf{A} \mid \mathbf{A}_{\widehat{C} \circ \mathsf{IP}} + \langle \mathbf{x}, \mathbf{z} \rangle \cdot \mathbf{G} \,\right].$$

The secret key corresponding to $\widehat{C} \circ \mathsf{IP}_\gamma$ is essentially a "short basis" for the matrix

$$\left[\, \mathbf{A} \mid \mathbf{A}_{\widehat{C} \circ \mathsf{IP}} + \gamma \cdot \mathbf{G} \,\right]$$

which can be sampled using a short trapdoor $\mathbf{T}$ of the matrix $\mathbf{A}$.

Proof Strategy. There are two main components to the proof. Fix the selective challenge attribute $\mathbf{x}, \mathbf{y}$. First, we will simulate the secret keys without knowing the trapdoor for the matrix $\mathbf{A}$: here, we rely on the simulated key generation for the ABE [BGG+14]. Roughly speaking, we will need to generate a short basis for the matrix

$$\left[\, \mathbf{A} \mid \mathbf{A}\mathbf{R}_{\widehat{C} \circ \mathsf{IP}} + (\gamma - \widehat{C} \circ \mathsf{IP}(\mathbf{x}, \mathbf{y})) \cdot \mathbf{G} \,\right]$$

where $\mathbf{R}_{\widehat{C} \circ \mathsf{IP}}$ is a small-norm matrix known to the simulator. Now, whenever $\widehat{C} \circ \mathsf{IP}(\mathbf{x}, \mathbf{y}) \neq \gamma$ as is the case for admissible adversaries, we will be able to simultae secret keys using the puncturing techniques in [ABB10, AFV11, MP12].

Next, we will show that the attribute $\mathbf{x}$ is hidden in the challenge ciphertext. Here, we adopt the proof strategy for attribute-hiding inner product encryption in [AFV11, GMW15]. In the proof, we simulate the matrices $\mathbf{A}, \mathbf{B}_1, \ldots, \mathbf{B}_t$ using

$$\mathbf{A}, \mathbf{A}\mathbf{R}_1' - \mathbf{x}[1]\mathbf{G}, \ldots, \mathbf{A}\mathbf{R}_t' - \mathbf{x}[t]\mathbf{G}$$

where $\mathbf{R}_1', \ldots, \mathbf{R}_t' \xleftarrow{\$} \{\pm 1\}^{m \times m}$. In addition, we simulate the corresponding terms in the challenge ciphertext by $\mathbf{c}, \mathbf{c}^\mathsf{T}\mathbf{R}_1', \ldots, \mathbf{c}^\mathsf{T}\mathbf{R}_t'$, where $\mathbf{c}$ is a uniformly random vector, which we switched from $\mathbf{A}^\mathsf{T}\mathbf{s} + \mathbf{e}$ using the LWE assumption. Here we crucially rely on the fact that switched to simulation of secret keys without knowing the trapdoor of $\mathbf{A}$. Going further, once $\mathbf{c}$ is random, we can switch back to simulating secret keys using the trapdoor $\mathbf{T}$. Hence, the secret keys now do not leak any information about $\mathbf{R}_1', \ldots, \mathbf{R}_t'$. Therefore, we may then invoke the left-over hash lemma to argue that $\mathbf{x}$ is information-theoretically hidden.

4   Predicate Encryption for Circuits

In this section, we present our main construction of predicate encryption for circuits by bootstrapping on top of the partially-hiding predicate encryption. That is,

- We construct a Predicate Encryption scheme $\mathcal{PE} = (\mathsf{Setup}, \mathsf{Keygen}, \mathsf{Enc}, \mathsf{Dec})$ for boolean predicate family $\mathcal{C}$ bounded by depth d over k bit inputs.

starting from

- an FHE scheme $\mathcal{FHE} = (\mathsf{HE.KeyGen}, \mathsf{HE.Enc}, \mathsf{HE.Dec}, \mathsf{HE.Eval})$ with properties as described in Sect. 2.1. Define ℓ as the size of the initial ciphertext encrypting k bit messages, and t as the size of the FHE secret key and evaluated ciphertext vectors;

– a partially-hiding predicate encryption scheme $\mathcal{PHPE}$ = (PH.Setup, PH.Keygen, PH.Enc, PH.Dec) for the class $\mathcal{C}_{\mathsf{PHPE}}$ of predicates bounded by some depth parameter $d' = \mathrm{poly}(d, \lambda, \log q)$. Recall that

$$(\widehat{C} \circ \mathsf{IP}_\gamma)(\mathbf{x} \in \mathbb{Z}_q^t, \mathbf{y} \in \{0,1\}^t) = 1 \text{ iff } \left(\sum_{i \in [t]} \mathbf{x}[i] \cdot \widehat{C}(\mathbf{y})[i] \right) = \gamma \mod q$$

where $\widehat{C} : \{0,1\}^\ell \to \{0,1\}^t$ is a circuit of depth at most d'.

Overview. At a high level, the construction proceeds as follows:

– the $\mathcal{PE}$ ciphertext corresponding to an attribute $\mathbf{a} \in \{0,1\}^k$ is a $\mathcal{PHPE}$ ciphertext corresponding to an attribute $(\mathsf{fhe.sk}, \mathsf{fhe.ct})$ where $\mathsf{fhe.sk} \xleftarrow{\$} \mathbb{Z}_q^t$ is private and $\mathsf{fhe.ct} := \mathsf{HE.Enc}(\mathbf{a}) \in \{0,1\}^\ell$ is public;
– the $\mathcal{PE}$ secret key for a predicate $C : \{0,1\}^k \to \{0,1\} \in \mathcal{C}$ is a collection of $2B + 1$ $\mathcal{PHPE}$ secret keys for the predicates $\{\widehat{C} \circ \mathsf{IP}_\gamma : \mathbb{Z}_q^t \times \{0,1\}^\ell \to \{0,1\}\}_{\gamma = \lfloor q/2 \rfloor - B, \ldots, \lfloor q/2 \rfloor + B}$ where $\widehat{C} : \{0,1\}^\ell \to \{0,1\}$ is the circuit:

$$\widehat{C}(\mathsf{fhe.ct}) := \mathsf{HE.Eval}(\mathsf{fhe.ct}, C),$$

so $\widehat{C}$ is a circuit of depth at most $d' = \mathrm{poly}(d, \lambda, \log q)$;
– decryption works by trying all possible $2B + 1$ secret keys.

Note that the construction relies crucially on the fact that B (the bound on the noise in the FHE evaluated ciphertexts) is polynomial. For correctness, observe that for all $C, \mathbf{a}$:

$$C(\mathbf{a}) = 1$$
$$\Leftrightarrow \mathsf{HE.Dec}(\mathsf{fhe.sk}, \mathsf{HE.Eval}(C, \mathsf{fhe.ct})) = 1$$
$$\Leftrightarrow \exists \gamma \in [\lfloor q/2 \rfloor - B, \lfloor q/2 \rfloor + B] \text{ such that } \left(\sum_{i \in [t]} \mathsf{fhe.sk}[i] \cdot \mathsf{fhe.ct}[i] \right) = \gamma \mod q$$
$$\Leftrightarrow \exists \gamma \in [\lfloor q/2 \rfloor - B, \lfloor q/2 \rfloor + B] \text{ such that } (\widehat{C} \circ \mathsf{IP}_\gamma)(\mathsf{fhe.sk}, \mathsf{fhe.ct}) = 1$$

where $\mathsf{fhe.sk}, \mathsf{fhe.ct}, \widehat{C}$ are derived from $C, \mathbf{a}$ as in our construction.

4.1 Our Predicate Encryption Scheme

Our construction proceeds as follows:

– $\mathsf{Setup}(1^\lambda, 1^k, 1^d)$: The setup algorithm takes the security parameter λ, the attribute length k and the predicate depth bound d.
  1. Run the partially-hiding PE scheme for family $\mathcal{C}_{\mathsf{PHPE}}$ to obtain a pair of master public and secret keys:

$$(\mathsf{ph.mpk}, \mathsf{ph.msk}) \leftarrow \mathsf{PH.Setup}(1^\lambda, 1^t, 1^\ell, 1^{d'})$$

where for k-bit messages and depth d circuits: t is the length of FHE secret key, ℓ is the bit-length of the initial FHE ciphertext and d' is the bound on FHE evaluation circuit (as described at the beginning of this section).

2. Output $(\mathsf{mpk} \doteq \mathsf{ph.mpk}, \mathsf{msk} \doteq \mathsf{ph.msk})$.

– Keygen(msk, C): The key-generation algorithms takes as input the master secret key msk and a predicate C. It outputs a secret key sk_C computed as follows.

1. Let $\widehat{C}(\cdot) := \mathsf{HE.Eval}(\cdot, C)$ and let $(\widehat{C} \circ \mathsf{IP}_\gamma)$ be the predicates for $\gamma = \lfloor q/2 \rfloor - B, \ldots, \lfloor q/2 \rfloor + B$.

2. For all $\gamma = \lfloor q/2 \rfloor - B, \ldots, \lfloor q/2 \rfloor + B$, compute

$$\mathsf{sk}_{\widehat{C} \circ \mathsf{IP}_\gamma} \leftarrow \mathsf{PH.Keygen}(\mathsf{ph.msk}, \widehat{C} \circ \mathsf{IP}_\gamma)$$

3. Output the secret key as $\mathsf{sk}_C \doteq \left(\{\mathsf{sk}_{\widehat{C} \circ \mathsf{IP}}\}_{\gamma = \lfloor q/2 \rfloor - B, \ldots, \lfloor q/2 \rfloor + B}\right)$.

– Enc$(\mathsf{mpk}, \mathbf{a}, \mu)$: The encryption algorithm takes as input the public key mpk, the input attribute vector $\mathbf{a} \in \{0, 1\}^k$ and message $\mu \in \{0, 1\}$. It proceeds as follow.

1. Samples a fresh FHE secret key $\mathsf{fhe.sk} \in \mathbb{Z}_q^t$ by running $\mathsf{HE.KeyGen}(1^\lambda, 1^{d'}, 1^k)$.

2. Encrypt the input to obtain

$$\mathsf{fhe.ct} \leftarrow \mathsf{HE.Enc}(\mathsf{fhe.sk}, \mathbf{a}) \in \{0, 1\}^\ell$$

3. Compute
$$\mathsf{ct}_{\mathsf{fhe.ct}} \leftarrow \mathsf{PH.Enc}\big(\mathsf{mpk}, (\mathsf{fhe.sk}, \mathsf{fhe.ct}), \mu\big)$$

Note that the $\mathsf{fhe.sk}$ corresponds to the hidden attribute and $\mathsf{fhe.ct}$ corresponds to the public attribute.

4. Output the ciphertext $\mathsf{ct} = (\mathsf{ct}_{\mathsf{fhe.ct}}, \mathsf{fhe.ct})$.

– Dec$((\mathsf{sk}_C, C), \mathsf{ct})$: The decryption algorithm takes as input the secret key sk_C with corresponding predicate C and the ciphertext ct. If there exists $\gamma = \lfloor q/2 \rfloor - B, \ldots, \lfloor q/2 \rfloor + B$ such that

$$\mathsf{PH.Dec}((\mathsf{sk}_{\widehat{C} \circ \mathsf{IP}_\gamma}, \widehat{C} \circ \mathsf{IP}_\gamma), (\mathsf{ct}_{\mathsf{fhe.ct}}, \mathsf{fhe.ct})) = \mu \neq \perp$$

then output μ. Otherwise, output $\perp$.

4.2   Correctness

Lemma 4.1. *Let $\mathcal{C}$ be a family of predicates bounded by depth d and let $\mathcal{PHPE}$ be the partially-hiding PE and $\mathcal{FHE}$ be a fully-homomorphic encryption as per scheme description. Then, our predicate encryption scheme $\mathcal{PE}$ is correct. Moreover, the size of each secret key is $\mathrm{poly}(d, \lambda)$ and the size of each ciphertext is $\mathrm{poly}(d, \lambda, k)$.*

We refer the reader to the full version for the proof.

4.3   Security

Theorem 4.2. *Let $\mathcal{C}$ be a family of predicates bounded by depth d and let $\mathcal{PHPE}$ be the secure partially-hiding PE and $\mathcal{FHE}$ be the secure fully-homomorphic encryption as per scheme description. Then, our predicate encryption scheme $\mathcal{PE}$ is secure.*

Proof. We define p.p.t. simulator algorithms $\mathsf{Enc}_{\mathsf{Sim}}$ and argue that its output is indistinguishable from the output of the real experiment. Let $\mathsf{PH.Enc}_{\mathsf{Sim}}$ be the p.p.t. simulator for partially-hiding predicate encryption scheme.

- $\mathsf{Enc}_{\mathsf{Sim}}(\mathsf{mpk}, 1^{|\mathbf{a}|}, 1^{|\mu|})$: To compute the encryption, the simulator does the following. It samples FHE secret key $\mathsf{fhe.sk}$ by running $\mathsf{HE.KeyGen}(1^\lambda, 1^{d'}, 1^k)$. It encrypts a zero-string $\mathsf{fhe.ct} \leftarrow \mathsf{HE.Enc}(\mathsf{fhe.sk}, \mathbf{0})$. It obtains the ciphertext as $\mathsf{ct}_{\mathsf{fhe.ct}} \leftarrow \mathsf{PH.Enc}_{\mathsf{Sim}}(\mathsf{mpk}, \mathsf{fhe.ct}, 1^{|\mathsf{fhe.sk}|}, 1^{|\mu|})$.

We now argue via a series of hybrids that the output of the ideal experiment.

- **Hybrid 0**: The real experiment.
- **Hybrid 1**: The real encryption algorithm is replaced with Enc^*, where Enc^* is an auxiliary algorithm defined below. On the high level, Enc^* computes the FHE ciphertext honestly by sampling a secret key and using the knowledge of $\mathbf{a}$. It then invokes $\mathsf{PH.Enc}_{\mathsf{Sim}}$ on the honestly generated ciphertext.
- **Hybrid 2**: The simulated experiment.

Auxiliary Algorithms. We define the auxiliary algorithm Enc^* used in Hybrid 1.

- $\mathsf{Enc}^*(\mathbf{a}, 1^{|\mu|})$: The auxiliary encryption algorithm takes as input the attribute vector $\mathbf{a}$ and message length.
  1. Sample a fresh FHE secret key $\mathsf{fhe.sk}$ by running $\mathsf{HE.KeyGen}(1^\lambda, 1^{d'}, 1^k)$.
  2. Encrypt the input attribute vector to obtain a ciphertext

$$\mathsf{fhe.ct} \leftarrow \mathsf{HE.Enc}(\mathsf{fhe.sk}, \mathbf{a}) \in \{0,1\}^\ell$$

  3. Run $\mathsf{PH.Enc}_{\mathsf{Sim}}$ on input $(\mathsf{mpk}, \mathsf{fhe.ct}, 1^{|\mathsf{fhe.sk}|}, 1^{|\mu|})$ to obtain the ciphertext $\mathsf{ct}_{\mathsf{fhe.ct}}$.

Lemma 4.3. *The output of Hybrid 0 is computationally indistinguishable from the Hybrid 1, assuming security of Partially-Hiding Predicate Encryption.*

Proof. Assume there is an adversary Adv and a distinguisher $\mathcal{D}$ that distinguishes the output $(\mathbf{a}, \mu, \alpha)$ produced in either of the two hybrids. We construct an adversary Adv' and a distinguisher $\mathcal{D}'$ that break the security of the Partially-Hiding Predicate Encryption. The adversary Adv' does the following.

1. Invoke the adversary Adv to obtain an attribute vector $\mathbf{a}$.
2. Sample a fresh FHE secret key $\mathsf{fhe.sk}$ using $\mathsf{HE.KeyGen}(1^\lambda, 1^{d'}, 1^k)$. Encrypt the attribute vector

$$\mathsf{fhe.ct} \leftarrow \mathsf{HE.Enc}(\mathsf{fhe.sk}, \mathbf{a})$$

and output the pair $(\mathsf{fhe.sk}, \mathsf{fhe.ct})$ as the "selective" challenge attribute.

3. Upon receiving mpk, it forwards it to Adv.
4. For each oracle query C that Adv makes which satisfies $C(\mathbf{a}) \neq 0$, Adv$'$ uses its oracle to obtain secret keys $\mathsf{sk}_{\widehat{C} \circ \mathsf{IP}_\gamma}$ for $\gamma = \lfloor q/2 \rfloor - B, \ldots, \lfloor q/2 \rfloor + B$. It outputs $\mathsf{sk}_C = \left(\{\mathsf{sk}_{\widehat{C} \circ \mathsf{IP}_\gamma}\}_{\gamma = \lfloor q/2 \rfloor - B, \ldots, \lfloor q/2 \rfloor + B}\right)$.
5. It outputs message μ that Adv produces, obtains a ciphertext $\mathsf{ct}_{\mathsf{fhe.ct}}$ and sends $\mathsf{ct} = (\mathsf{ct}_{\mathsf{fhe.ct}}, \mathsf{fhe.ct})$ back to Adv to obtain α.

We note that given Adv that is admissible, Adv$'$ is also admissible. That is, for all queries $\widehat{C} \circ \mathsf{IP}_\gamma$ that Adv$'$ makes satisfies $(\widehat{C} \circ \mathsf{IP}_\gamma)(\mathsf{fhe.sk}, \mathsf{fhe.ct}) = 0$ since $\langle \mathsf{fhe.sk}, \widehat{C}(\mathsf{fhe.ct}) \rangle \neq \gamma$ for $\gamma = \lfloor q/2 \rfloor - B, \ldots, \lfloor q/2 \rfloor + B$ by the correctness of FHE in Sect. 2.1 and the fact that $C(\mathbf{a}) \neq 0$. Finally, the distinguisher $\mathcal{D}'$ on input $(\mathsf{fhe.sk}, \mathsf{fhe.ct}, \mu, \alpha)$ invokes $\mathcal{D}$ and outputs whatever it outputs. Now, in Hybrid 0 the algorithms used as PH.Setup, PH.Keygen, PH.Enc which corresponds exactly to the real security game of PHPE. However, in Hybrid 1 the algorithms correspond exactly to the simulated security game. Hence, we can distinguish between the real and simulated experiments contradicting the security of PHPE scheme.

Lemma 4.4. *The output of Hybrid 1 and Hybrid 2 are computationally indistinguishable, assuming semantic security of Fully-Homomorphic Encryption Scheme.*

Proof. The only difference in Hybrids 1 and 2 is how the FHE ciphertext is produced. In one experiment, it is computed honestly by encrypting the attribute vector $\mathbf{a}$, while in the other experiment it is always an encryption of $\mathbf{0}$. Hence, we can readily construct an FHE adversary that given $\mathbf{a}$, distinguishes encryption of $\mathbf{a}$ from encryption of $\mathbf{0}$ as follows:

1. Invoke the admissible PE adversary Adv to obtain an attribute vector $\mathbf{a}$.
2. Run the honest PH.Setup and forwards mpk to Adv.
3. For each oracle query C that Adv makes which satisfies $C(\mathbf{a}) \neq 0$, return $\mathsf{sk}_C = \left(\{\mathsf{sk}_{\widehat{C} \circ \mathsf{IP}_\gamma}\}_{\gamma = \lfloor q/2 \rfloor - B, \ldots, \lfloor q/2 \rfloor + B}\right)$ as computed using the honest PH.Keygen algorithm.
4. To simulate the ciphertext, first forward the pair $(\mathbf{a}, \mathbf{0})$ to the FHE challenger to obtain a ciphertext fhe.ct. Then, run $\mathsf{PH.Enc}_{\mathsf{Sim}}(\mathsf{mpk}, \mathsf{fhe.ct}, 1^{|\mathsf{fhe.sk}|}, 1^\mu)$ to obtain a ciphertext $\mathsf{ct}_{\mathsf{fhe.ct}}$ and forward it to Adv
5. Finally, it runs the PE distinguisher on input $(\mathbf{a}, \mu, \alpha)$ and outputs its guess.

The lemma then follows from semantic security of the FHE completing the security proof. We also refer the reader to the full version for the summary of parameters selection.

References

[ABB10] Agrawal, S., Boneh, D., Boyen, X.: Efficient lattice (H)IBE in the standard model. In: Gilbert, H. (ed.) EUROCRYPT 2010. LNCS, vol. 6110, pp. 553–572. Springer, Heidelberg (2010)

[AFV11] Agrawal, S., Freeman, D.M., Vaikuntanathan, V.: Functional encryption for inner product predicates from learning with errors. In: Lee, D.H., Wang, X. (eds.) ASIACRYPT 2011. LNCS, vol. 7073, pp. 21–40. Springer, Heidelberg (2011)

[AGVW13] Agrawal, S., Gorbunov, S., Vaikuntanathan, V., Wee, H.: Functional encryption: new perspectives and lower bounds. In: Canetti, R., Garay, J.A. (eds.) CRYPTO 2013, Part II. LNCS, vol. 8043, pp. 500–518. Springer, Heidelberg (2013)

[AP14] Alperin-Sheriff, J., Peikert, C.: Faster bootstrapping with polynomial error. In: Garay, J.A., Gennaro, R. (eds.) CRYPTO 2014, Part I. LNCS, vol. 8616, pp. 297–314. Springer, Heidelberg (2014)

[BB04] Boneh, D., Boyen, X.: Efficient selective-ID secure identity-based encryption without random oracles. In: Cachin, C., Camenisch, J.L. (eds.) EUROCRYPT 2004. LNCS, vol. 3027, pp. 223–238. Springer, Heidelberg (2004)

[BGG+14] Boneh, D., Gentry, C., Gorbunov, S., Halevi, S., Nikolaenko, V., Segev, G., Vaikuntanathan, V., Vinayagamurthy, D.: Fully key-homomorphic encryption, arithmetic circuit ABE and compact garbled circuits. In: Nguyen, P.Q., Oswald, E. (eds.) EUROCRYPT 2014. LNCS, vol. 8441, pp. 533–556. Springer, Heidelberg (2014)

[BGV12] Brakerski, Z., Gentry, C., Vaikuntanathan, V.: (Leveled) fully homomorphic encryption without bootstrapping. In: ITCS, pp. 309–325 (2012)

[BSW11] Boneh, D., Sahai, A., Waters, B.: Functional encryption: definitions and challenges. In: Ishai, Y. (ed.) TCC 2011. LNCS, vol. 6597, pp. 253–273. Springer, Heidelberg (2011)

[BV11a] Brakerski, Z., Vaikuntanathan, V.: Fully homomorphic encryption from ring-LWE and security for key dependent messages. In: Rogaway, P. (ed.) CRYPTO 2011. LNCS, vol. 6841, pp. 505–524. Springer, Heidelberg (2011)

[BV11b] Brakerski, Z., Vaikuntanathan, V.: Efficient fully homomorphic encryption from (standard) LWE. In: FOCS, pp. 97–106 (2011)

[BV14] Brakerski, Z., Vaikuntanathan, V.: Lattice-based FHE as secure as PKE. In: ITCS, pp. 1–12 (2014)

[BW07] Boneh, D., Waters, B.: Conjunctive, subset, and range queries on encrypted data. In: Vadhan, S.P. (ed.) TCC 2007. LNCS, vol. 4392, pp. 535–554. Springer, Heidelberg (2007)

[CHL+15] Cheon, J.H., Han, K., Lee, C., Ryu, H., Stehlé, D.: Cryptanalysis of the multilinear map over the integers. In: Oswald, E., Fischlin, M. (eds.) EUROCRYPT 2015. LNCS, vol. 9056, pp. 3–12. Springer, Heidelberg (2015)

[CLT13] Coron, J.-S., Lepoint, T., Tibouchi, M.: Practical multilinear maps over the integers. In: Canetti, R., Garay, J.A. (eds.) CRYPTO 2013, Part I. LNCS, vol. 8042, pp. 476–493. Springer, Heidelberg (2013)

[Gen09] Gentry, C.: Fully homomorphic encryption using ideal lattices. In: STOC, pp. 169–178 (2009)

[GGH13a] Garg, S., Gentry, C., Halevi, S.: Candidate multilinear maps from ideal lattices. In: Johansson, T., Nguyen, P.Q. (eds.) EUROCRYPT 2013. LNCS, vol. 7881, pp. 1–17. Springer, Heidelberg (2013)

[GGH+13b] Garg, S., Gentry, C., Halevi, S., Raykova, M., Sahai, A., Waters, B.: Candidate indistinguishability obfuscation and functional encryption for all circuits. In: FOCS, pp. 40–49, (2013). Also, Cryptology ePrint Archive, report 2013/451

[GGH15] Gentry, C., Gorbunov, S., Halevi, S.: Graph-induced multilinear maps from lattices. In: Dodis, Y., Nielsen, J.B. (eds.) TCC 2015, Part II. LNCS, vol. 9015, pp. 498–527. Springer, Heidelberg (2015)

[GGHZ14] Garg, S., Gentry, C., Halevi, S., Zhandry, M.: Fully secure functional encryption without obfuscation. Cryptology ePrint Archive, report 2014/666 (2014)

[GKP+13] Goldwasser, S., Kalai, Y.T., Popa, R.A., Vaikuntanathan, V., Zeldovich, N.: Reusable garbled circuits and succinct functional encryption. In: STOC, pp. 555–564 (2013)

[GMW15] Gay, R., Méaux, P., Wee, H.: Predicate encryption for multi-dimensional range queries from lattices. In: Public Key Cryptography (2015). Also, Cryptology ePrint Archive, report 2014/965

[GPSW06] Goyal, V., Pandey, O., Sahai, A., Waters, B.: Attribute-based encryption for fine-grained access control of encrypted data. In: ACM CCS, pp. 89–98 (2006)

[GSW13] Gentry, C., Sahai, A., Waters, B.: Homomorphic encryption from learning with errors: conceptually-simpler, asymptotically-faster, attribute-based. In: Canetti, R., Garay, J.A. (eds.) CRYPTO 2013, Part I. LNCS, vol. 8042, pp. 75–92. Springer, Heidelberg (2013)

[GV14] Gorbunov, S., Vinayagamurthy, D.: Riding on asymmetry: Efficient ABE for branching programs. Cryptology ePrint Archive, report 2014/819 (2014). http://eprint.iacr.org/

[GVW12] Gorbunov, S., Vaikuntanathan, V., Wee, H.: Functional encryption with bounded collusions via multi-party computation. In: Safavi-Naini, R., Canetti, R. (eds.) CRYPTO 2012. LNCS, vol. 7417, pp. 162–179. Springer, Heidelberg (2012)

[GVW13] Gorbunov, S., Vaikuntanathan, V., Wee, H.: Attribute-based encryption for circuits. In: STOC, pp. 545–554 (2013). Also, Cryptology ePrint Archive, report 2013/337

[IW14] Ishai, Y., Wee, H.: Partial garbling schemes and their applications. In: Esparza, J., Fraigniaud, P., Husfeldt, T., Koutsoupias, E. (eds.) ICALP 2014. LNCS, vol. 8572, pp. 650–662. Springer, Heidelberg (2014)

[KSW08] Katz, J., Sahai, A., Waters, B.: Predicate encryption supporting disjunctions, polynomial equations, and inner products. In: Smart, N.P. (ed.) EUROCRYPT 2008. LNCS, vol. 4965, pp. 146–162. Springer, Heidelberg (2008)

[MP12] Micciancio, D., Peikert, C.: Trapdoors for lattices: simpler, tighter, faster, smaller. In: Pointcheval, D., Johansson, T. (eds.) EUROCRYPT 2012. LNCS, vol. 7237, pp. 700–718. Springer, Heidelberg (2012)

[OT10] Okamoto, T., Takashima, K.: Fully Secure functional encryption with general relations from the decisional linear assumption. In: Rabin, T. (ed.) CRYPTO 2010. LNCS, vol. 6223, pp. 191–208. Springer, Heidelberg (2010)

[OT12] Okamoto, T., Takashima, K.: Adaptively attribute-hiding (hierarchical) inner product encryption. In: Pointcheval, D., Johansson, T. (eds.) EUROCRYPT 2012. LNCS, vol. 7237, pp. 591–608. Springer, Heidelberg (2012)

[Reg09] Regev, O.: On lattices, learning with errors, random linear codes, and cryptography. J. ACM **56**(6), 1–40 (2009)

[SBC+07] Shi, E., Bethencourt, J., Chan, H.T.-H., Song, D.X., Perrig, A.: Multidimensional range query over encrypted data. In: IEEE Symposium on Security and Privacy, pp. 350–364 (2007)

[SS10] Sahai, A., Seyalioglu, H.: Worry-free encryption: functional encryption with public keys. In: ACM Conference on Computer and Communications Security, pp. 463–472 (2010)

[SW05] Sahai, A., Waters, B.: Fuzzy identity-based encryption. In: Cramer, R. (ed.) EUROCRYPT 2005. LNCS, vol. 3494, pp. 457–473. Springer, Heidelberg (2005)

[Wat09] Waters, B.: Dual system encryption: realizing fully secure IBE and HIBE under simple assumptions. In: Halevi, S. (ed.) CRYPTO 2009. LNCS, vol. 5677, pp. 619–636. Springer, Heidelberg (2009)

Bilinear Entropy Expansion from the Decisional Linear Assumption

Lucas Kowalczyk(✉) and Allison Bishop Lewko

Columbia University, New York, USA
{luke,allison}@cs.columbia.edu

Abstract. We develop a technique inspired by pseudorandom functions
that allows us to increase the entropy available for proving the security
of dual system encryption schemes under the Decisional Linear Assumption. We show an application of the tool to Attribute-Based Encryption
by presenting a Key-Policy ABE scheme that is fully-secure under DLIN
with short public parameters.

1 Introduction

Since its conception in [31], attribute-based encryption (ABE) has served as a
demonstrably fertile ground for exploring the possible tradeoffs between expressibility, security, and efficiency in cryptographically enforced access control. In
addition to the potential applications it has in its own right, the primitive
of attribute-based encryption has been a catalyst for the definitions and constructions of further cryptographic primitives, such as functional encryption
for general circuits. The rich structure of secret keys demanded by expressive
attribute-based encryption has promoted a continuing evolution of proof techniques designed to meet the challenges inherent in balancing large and complex
structures on the pinhead of simple computational hardness assumptions.

The origins of attribute-based encryption can be traced back to identity-
based encryption [5,10], where users have identities that serve as public keys
and secret keys are generated on demand by a master authority. A desirable
notion of security for such schemes ensures resilience against arbitrary collusions
among users by allowing an attacker to demand many secret keys for individual
users and attack a ciphertext encrypted to any user not represented in the set
of obtained keys. Proving this kind of security requires a reduction design that
can satisfy the attacker's demands without fully knowing the master secret key.
This challenge is exacerbated in the (key-policy) attribute-based setting, where
user keys correspond to access policies expressed over attributes and ciphertexts
are associated with subsets of these attributes. Decryption is allowed precisely
when a single user's policy is satisfied by a ciphertext's attribute set. Thus, the

Lucas Kowalczyk is supported in part by NSF CNS 1445424 and the Office of Naval
Research (ONR) contract number N00014-14-C-0113.
Allison Lewko is supported in part by NSF CNS 1413971 and NSF CCF 1423306.

R. Gennaro and M. Robshaw (Eds.): CRYPTO 2015, Part II, LNCS 9216, pp. 524–541, 2015.
DOI: 10.1007/978-3-662-48000-7_26

structure of allowable keys that the attacker can request grows more complex as the scheme is equipped to express more complex policies.

As a consequence of this, the intuitive and elegant constructions of attribute-based encryption in bilinear groups in [17,33] were only proven secure in the selective security model: a weakened model of security that requires the attacker to declare the target of attack in advance, before seeing the public parameters of the system. This limitation of the model allows the security reduction to embed the computational challenge into its view of the public parameters of the scheme in a way that partitions the space of secret keys. Keys that do not satisfy the targeted ciphertext are able to be generated under the embedding, while keys that do satisfy the ciphertext cannot be generated. This approach does not extend well to the full security model, where this artificial limitation on the attacker is lifted.

The first fully secure ABE schemes appeared in [18], using the dual system encryption methodology [32] for designing the security reduction. In a dual system approach, there are typically multiple (computationally indistinguishable) forms of keys and ciphertexts. There are "norma" keys and ciphertexts that are employed in the real system, and then are various forms of "semi-functional" keys and ciphertexts. The core idea is to prove security via a hybrid argument, where the ciphertext is changed to semi-functional and keys are changed to semi-functional types one by one, until all the keys are of a semi-functional type incapable of decrypting the semi-functional ciphertext (it is important that they still decrypt normal ciphertexts, otherwise the hybrid transitions could be detected by the attacker who can create normal ciphertexts for itself using the public parameters). Once we reach a state where the key and ciphertexts distributions provided to the attacker are no longer bound by correct decrypt behavior, it is easier for the reduction to produce these without knowing the master secret key.

The most critical step of these dual system arguments occurs when a particular key changes from a type that can decrypt the challenge ciphertext to a type that cannot - the fact that this change is not detected by the attacker is where the reduction must use the criterion that the access policy is not satisfied. The security reductions in [18] and many subsequent works (e.g. [21,27]) used an information-theoretic argument for this step. However, this argument requires a great deal of entropy (specifically, fresh randomness for each attribute-use in a policy). This entropy was supplied by parameters in the semi-functional space that paralleled the published parameters of the normal space. This necessitated a blowup in public parameter and ciphertext sizes, specifically a multiplicative factor of the number of attribute-uses allowed for access policies.

In [25], it was observed that the initial steps of a typical dual system encryption hybrid argument could be re-interpreted as providing a "shadow copy" of the system parameters in the semi-functional space that does not have to be committed to when the public parameters for the normal space are provided. This perspective suggests that one can embed a computational challenge into these semi-functional space parameters as semi-functional objects are produced. For instance, when a portion of these parameters affect a single semi-functional key that is queried after the semi-functional ciphertext, one can essentially embed the challenge in the same way as the original selective security arguments in [17]. In the reverse case, where the semi-functional key is queried before the challenge ciphertext, the embedding can be similar to a selective security proof for a ciphertext-policy ABE scheme, where keys are associated with attributes and ciphertexts are associated with access policies. In [25], state of the art selective techniques for KP-ABE and CP-ABE systems were combined into a full security proof, avoiding the blowup in parameters incurred by the information-theoretic dual system techniques.

However, even selective security for CP-ABE systems remains a rather challenging task, and the state of the art technique in [33] introduces an undesirable q-type assumption into the fully secure ABE scheme. In the CP-ABE setting, selectivity means that the attacker declares a target access policy up front. This can then be leveraged by the security reduction to design public parameters so that it can create keys precisely for sets of attributes that do not satisfy this target policy. The q-type assumption in [33] was a consequence of the need to encode a potentially large access policy into small public parameters. This leaves us still searching for an ideal KP-ABE scheme in the bilinear setting that has parameter sizes comparable to the selectively secure scheme in [17] and a full security proof from a simple assumption such as the decisional linear assumption (DLIN). A security reduction for such a scheme must seemingly break outside the mold of using either a purely information-theoretic or purely computational argument for leveraging the fact that a requested key policy cannot be satisfied by the challenge ciphertext.

Our Results. To demonstrate our approach, we present a KP-ABE constructions in the composite-order bilinear setting which is proven fully secure from simple assumptions, and supports LSSS/MSP access policies (like its bilinear predecessors). Security is proven using a few specific instances of subgroup-decision assumptions and DLIN. Our scheme greatly reduces the size of the public parameters as compared to [18,27], as the number of group elements we need to include in the public parameters grows only logarithmically rather than linearly in the bound on the number of attribute-uses in an access policy.

Our Techniques. We intermix the computational and information-theoretic dual system encryption approaches, using computational steps to "boost" the entropy of a small set of (unpublished) semi-functional parameters to a level that suffices to make the prior information-theoretic argument work. Essentially, we use the fact that the semi-functional space parameters are never published to not only "delay" their definition as exploited in [25], but further to argue that they can (computationally) appear to provide more entropy than their size would information-theoretically allow. The gadget that allows us do this computational pre-processing before the running information-theoretic argument is presented as our "bilinear entropy expansion lemma."

The inspiration for the gadget construction comes from pseudorandom generators/pseudorandom functions. Naturally, if we want a small set of semi-functional generators to seemingly produce a large amount of entropy, we may want to view these parameters as the seed for a PRF, for example. Out-of-the-box PRF constructions like Naor-Reingold [26] and its DLIN-based extension [24] however are unsuitable in the bilinear setting (even though the DLIN version would remain secure) because they would require direct access to the seed for computation, and a secure bilinear construction will only provide indirect access to the seed as exponents of group elements.

To circumvent this difficulty, we use a subset-sum based construction that can be computed in a bilinear group with the seed elements in the exponents. Of course, using a naked linear structure would be detectable, but we are able to use a rather minimal amount of additional random exponents to push the linear sub-structure out of reach of detection by regular group or pairing operations.

We build our construction in two steps. First, we present a construction for a *one-use* KP-ABE system which only supports access policies where each attribute is used at most once. This scheme achieves ciphertext and key sizes which rival those of selectively secure schemes (up to constants), while significantly reducing public parameter size. Then, we apply a standard transformation to get from a one-use system to a system which allows multiple uses of attributes in policies (the number of uses allowed per attribute is constant and fixed at setup). The overhead of this transformation is drastically mitigated by our scheme's small public parameters. The effect on ciphertext and key sizes compared to previous applications of this transformation remain the same up to constants.

Further Discussion of Related Work. Additional work on ABE in the bilinear setting includes various constructions of KP-ABE and CP-ABE schemes (e.g. [4,16,30]), schemes supporting multiple authorities (e.g. [6,7,21,29]), and schemes supporting large attribute universes (e.g. [22,28]). Some of the structure for randomization in our schemes is inspired by [22]. The large universe scheme in [28] also achieves full-security with short public parameters using conceptually different techniques. We view the main contribution of this paper to be the entropy expansion lemma, which we believe is modular and potentially useful in other settings. Our approach lends a clear understanding of the roles of information-theoretic and computational techniques in dual-system encryption proofs.

There are also recent constructions of ABE schemes in the lattice setting. The construction of [15] allows access policies to be expressed as circuits, which makes it more expressive than any known bilinear scheme. It was proven selectively secure under the standard LWE assumption. Circuit policies are also supported by the construction in [12] based on multilinear maps. This scheme is also proven selectively secure, under a particular computational hardness assumption for multilinear groups. The very recent multilinear scheme in [13] achieves full security, relying on computational hardness assumptions in multilinear groups. The fully secure general functional encryption scheme in [34], which relies on indistinguishability obfuscation, can also be specialized to the ABE setting.

Some relationships between ABE and other cryptographic primitives have also been explored. The work of [2] derives schemes for verifiable computation from attribute-based encryption schemes, while [14] use attribute-based encryption as a tool in designing more general functional encryption and reusable garbling schemes. Dual system encryption proof techniques have also been further studied in the works of [1,9,20,34], applied to achieve leakage resilience in [11,19,23], and applied directly to computational assumptions in [8].

2   Preliminaries

Our construction uses composite order bilinear groups. Background on these groups and the (static) subgroup decision assumptions on which our composite order construction's security is based can be found in the full version of this paper. We now give required background material on Linear Secret Sharing Schemes. The formal definition of a KP-ABE scheme, and the security definition we will use can be found in the full version.

Linear Secret Sharing Schemes. Our construction uses linear secret-sharing schemes (LSSS). We use the following definition (adapted from [3]). In the context of ABE, attributes will play the role of parties and will be represented as nonempty subsets $K \subseteq [k]$ for a fixed k.

Definition 1 *(Linear Secret-Sharing Schemes (LSSS)). A secret sharing scheme Π over a set of attributes is called linear (over $\mathbb{Z}_p$) if the shares belonging to all attributes form a vector over $\mathbb{Z}_p$ and there exists an $\ell \times n$ matrix Λ called the share-generating matrix for Π. The matrix Λ has ℓ rows and n columns. For all $j = 1, \ldots, \ell$, the jth row of Λ is labeled by an attribute K. When we consider the column vector $v = (s, r_2, \ldots, r_n)$, where $s \in \mathbb{Z}_p$ is the secret to be shared and $r_2, \ldots, r_n \in \mathbb{Z}_p$ are randomly chosen, then Λv is the vector of ℓ shares of the secret s according to Π. The share $(\Lambda v)_j = \lambda_K$ belongs to attribute K.*

We note the *linear reconstruction* property: we suppose that Π is an LSSS. We let S denote an authorized set. Then there is a subset $S^* \subseteq S$ such that the vector $(1, 0, \ldots, 0)$ is in the span of rows of Λ indexed by S^*, and there exist constants $\{\omega_K \in \mathbb{Z}_p\}_{K \in S^*}$ such that, for any valid shares $\{\lambda_K\}$ of a secret s according to Π, we have: $\sum_{K \in S^*} \omega_K \lambda_K = s$. These constants $\{\omega_K\}$ can be found in time polynomial in the size of the share-generating matrix Λ [3]. For unauthorized sets, no such S^*, $\{\omega_K\}$ exist.

For our construction, we will employ LSSS matrices over $\mathbb{Z}_N$, where N is a product of three distinct primes p, q, w. As in the definition above over the prime order $\mathbb{Z}_p$, we say a set of attributes S is authorized if is a subset $S^* \subseteq S$ such that the rows of the access matrix A labeled by elements of S have the vector $(1, 0, \ldots, 0)$ in their span modulo N. In our security proof for our system, we will further assume that for an unauthorized set, the corresponding rows of A do not include the vector $(1, 0, \ldots, 0)$ in their span modulo q. We may assume this because if an adversary can produce an access matrix A over $\mathbb{Z}_N$ and an unauthorized set over $\mathbb{Z}_N$ that is authorized over $\mathbb{Z}_q$, then this can be used to produce a non-trivial factor of the group order N, which would violate our subgroup decision assumptions.

Transformation from One-Use to Multiple Use KP-ABE. Given a KP-ABE scheme which is fully-secure when attributes are used at most once in access policies, we can obtain a KP-ABE scheme which is fully-secure when each attribute is used at most some constant number of times in access policies using a standard transformation. Essentially, multiple uses of an attribute are treated as new "attributes" in the one-use system. For example, if we want an attribute x to be able to be used up to k_x times in access policies, we will instantiate our one-use system with k_x "attributes" $x : 1, ..., x : k_x$. Each time we want to label a row of an access matrix Λ with x, we label it with $x : i$ for a new value of i. Each time we want to associate a subset S of attributes to a ciphertext, we instead use the set $S' = \{x : 1, ..., x : k_x \mid x \in S\}$. We can then employ the one-use KP-ABE scheme on this new larger set of "attributes" and retain its full security and functionality.

Clearly, this transformation comes at a cost. Typically, the ciphertext and public parameter size of the KP-ABE scheme resulting from the transformation now scale linearly with the number of attribute-*uses* allowed in access policies, not just the number of attributes. This presents a problem if one desires policies which have high reuse of attributes. Our one-use KP-ABE scheme mitigates the problem with public parameter size by featuring public parameters that scale only logarithmically with the number of attributes supported by the system, compared to the linear scaling of the fully secure KP-ABE schemes based on static assumptions in [18, 21]. Note that [28] also achieves full-security from static assumptions with short parameters, using conceptually different techniques.

3   KP-ABE Construction

Our single-use KP-ABE construction assumes a polynomially sized attribute universe $\mathcal{U}$ where attributes are non-empty subsets $K \subseteq [k]$ for some fixed k. The prior fully secure single-use KP-ABE scheme in [18] required a fresh group element to appear in the public parameters for each attribute in the universe. After using the generic transformation discussed in Sect. 2, this results in the scheme requiring a fresh group element for each attribute-*use* allowed in access policies. As a concrete example, if one wanted to allow 9 attributes to be used up to 7 times each, one needed to have $9 \times 7 = 63$ group elements in the public parameters corresponding to this attribute. In our composite order scheme, to allow the same $63 = 2^6 - 1$ attribute-uses, we only need 2×6 group elements in the public parameters corresponding to the attribute. The way we accomplish this dramatic "compression" of public parameters is to note that the encryptor can produce 63 group elements from 6 by taking products of all non-empty subsets (these correspond to subset-sums in the exponent). More generally, given k group elements $g^{\alpha_1}, \ldots, g^{\alpha_k}$, we can produce $2^k - 1$ group elements by enumerating over all non-empty subsets $K \subseteq [k]$ and computing $g^{\sum\limits_{j \in K} a_j}$. We name the resulting collection of elements g^{A_K}, where $A_K := \sum\limits_{j \in K} a_j$. Our composite order scheme uses two parallel such subset constructions (causing the factor of 2).

These $2^k - 1$ group elements no longer look random - they have linear relationships in their exponents by construction. However, since we are assuming the decisional linear assumption is hard, if we choose $2^k - 1$ additional random exponents $\{t_K\}$, then the $2(2^k - 1)$ group elements formed as $\{g^{t_K}, g^{t_K A_K}\}$ are computationally indistinguishable from $2(2^k - 1)$ uniformly random group elements (which lack any hidden linear structures in their exponents). The proof of this is the core of bilinear entropy expansion lemma, though the full statement of the lemma includes some additional structure that is useful for linking into a KP-ABE construction. The dual system encryption framework allows us to apply this argument to the parameters in the semi-functional space, where we do not need to publish the values $\{g^{a_j}\}$. (Note that publishing these would make the structure of $\{g^{t_K}, g^{t_K A_K}\}$ detectable through applications of the bilinear map).

$Setup(\lambda, \mathcal{U}, k) \to PP, MSK$. The setup algorithm chooses a bilinear group G of order $N = pqw$ where p, q, w are primes. We let G_p, G_q, G_w represent the subgroup of order $p, q,$ and w respectively in G. It then draws $\alpha \leftarrow \mathbb{Z}_N$ and random group element $g_p \in G_p$. For each $j \in [k]$, it chooses values $a_j, b_j \leftarrow \mathbb{Z}_N$. The public parameters are $N, g_p, e(g_p, g_p)^\alpha, \{g_p^{a_j}, g_p^{b_j} : j \in [k]\}$. The MSK is α and a generator g_w of G_w. Such a construction is equipped to create keys for access policies which include attributes $K \subseteq [k]$ where K is not empty.

$KeyGen(MSK, \Lambda, PP) \to SK$. The key generation algorithm takes in the public parameters, master secret key, and LSSS access matrix Λ. First, the key generation algorithm generates $\{\lambda_K\}$: a linear sharing of α according to policy matrix Λ (the reader is referred to Sect. 2 for details). For each attribute K corresponding to a row in the policy matrix Λ, it then raises generator g_w to random exponents to create $g_w^{z_K}, g_w^{z'_K}, g_w^{z''_K} \in G_w$, chooses exponent $y_K \leftarrow \mathbb{Z}_N$ and computes $g_p^{A_K} = \prod\limits_{j \in K} g_p^{a_j}$ and $g_p^{B_K} = \prod\limits_{j \in K} g_p^{b_j}$. Note that here and throughout the rest of the description of our construction and its proof of security we will use the notation $A_K = \sum\limits_{j \in K} a_j$ and $B_K = \sum\limits_{j \in K} b_j$. It then outputs the secret key:

$$SK_\Lambda = \{g_p^{\lambda_K} g_p^{y_K A_K} g_w^{z_K}, \quad g_p^{y_K} g_w^{z'_K}, \quad g_p^{y_K B_K} g_w^{z''_K} : (\forall K \text{ labels} \in \Lambda)\}$$

$Encrypt(M, S, PP) \to CT$. The encryption algorithm first draws $s \leftarrow \mathbb{Z}_N$. For each $K \in S$, the encryption algorithm draws $t_K \leftarrow \mathbb{Z}_N$ and computes $g_p^{A_K} = \prod\limits_{j \in K} g_p^{a_j}$ and $g_p^{B_K} = \prod\limits_{j \in K} g_p^{b_j}$. It then outputs the ciphertext:

$$CT = Me(g_p, g_p)^{\alpha s}, \quad \{g_p^s, \quad g_p^{s A_K} g_p^{t_K B_K}, \quad g_p^{t_K} : (\forall K \in S)\}$$

$Decrypt(CT, SK, PP) \rightarrow M$. We let S correspond to the set of attributes associated to ciphertext CT, and Λ be the policy matrix. If S satisfies Λ, the decryption algorithm computes suitable constants ω_K such that $\sum_{K \in S^*} \omega_K \lambda_K = \alpha$ (recall Sect. 2). It then computes:

$$\prod_{K \in S^*} \left(e(g_p^s, \ g_p^{\lambda_K} g_p^{y_K A_K} g_w^{z_K}) \left(\frac{e(g_p^{y_K} g_w^{z'_K}, \ g_p^{s A_K} g_p^{t_K B_K})}{e(g_p^{t_K}, \ g_p^{y_K B_K} g_w^{z''_K})} \right)^{-1} \right)^{\omega_K}$$

$$= \prod_{K \in S^*} \left(\frac{e(g_p, g_p)^{s \lambda_K} e(g_p, g_p)^{s y_K A_K}}{e(g_p, g_p)^{s y_K A_K}} \right)^{\omega_K}$$

$$= \prod_{K \in S^*} \left(e(y_p, y_p)^{s \lambda_K} \right)^{\omega_K} = e(g_p, g_p)^{\sum_{K \in S^*} s \omega_K \lambda_K} = e(g_p, g_p)^{\alpha s}$$

The message can then be recovered by computing: $M e(g_p, g_p)^{\alpha s} / e(g_p, g_p)^{\alpha s} = M$. This demonstrates correctness of the scheme.

3.1 Security Proof Overview

Our security proof uses a hybrid argument over a sequence of games. We let $Game_{real}$ denote the real security game. The rest of the games use semi-functional keys and ciphertexts, which we describe below. We let g_q denote a fixed generator of the subgroup G_q, which will serve as the "semi-functional space."

Like a typical dual system encryption proof, we will begin by transitioning from a normal ciphertext to a semi-functional ciphertext with semi-functional components that mimic the structure of their normal counterparts. This kind of transition can be done with a basic subgroup decision assumption. We will then perform a hybrid over keys, gradually changing each one to a semi-functional form that does not properly decrypt the semi-functional ciphertext. To start, we can bring in semi-functional components for a particular key that mimic the structure of normal components, up to the constraint that the shared valued in the semi-functional space will be 0 (modulo q). Technically, this constraint arises because we will be taking a challenge term from a subgroup decision assumption that has an unknown exponent in the normal space and raising it to a share - so we have to make this a share of 0 and separately share the α value in the normal space so that the unknown exponent does not affect the correctness of the sharing in the normal space. At a higher level, this constraint explains why the simulator at this stage of the hybrid cannot solve the challenge problem for itself by test decrypting against a semi-functional ciphertext. Since the structure in the semi-functional space parallels the normal structure and the shared value here is zero, the semi-functional components will cancel out upon decryption.

So we can arrive at a stage where a key and ciphertext have semi-functional components structured just like the normal space, but with fresh parameters modulo q that are independent of the published parameters modulo p. This is a consequence of the Chinese Remainder Theorem, that ensures when we sample an exponent uniformly at random modulo N, its modulo p and modulo q reductions are independent and uniformly random in $\mathbb{Z}_p, \mathbb{Z}_q$ respectively. Since these implicit parameters in the semi-functional space are never published, we can use our bilinear entropy expansion lemma to argue that their subset-sum structure is hidden under the decisional linear assumption. This allows us to replace them with higher entropy parameters (lacking the subset-sum structure of the normal space), and then argue that the shared value in the semi-functional space is information-theoretically hidden (this is where we use that the access policy is not satisfied and that attributes are used at most once in the policy). This enables us to switch the semi-functional shares in the key to shares of a random value, now destroying correct decryption of a semi-functional ciphertext. We then remove some of the other (now unnecessary) semi-functional components of the key, to reclaim the entropy of those parameters to use in processing the next key in the hybrid. Finally, once we have reached a game where all keys are semi-functional with shares of a random secret modulo q, we can use Subgroup Decision Assumption 3 to create such keys without knowing the master secret and can hence complete the proof.

We now formally present our definitions of semi-functional ciphertexts and keys used in our hybrid proof:

Semi-functional Ciphertext. We will use 3 types of semi-functional ciphertexts. To produce a semi-functional ciphertext for an attribute set S, one first calls the normal encryption algorithm to produce a normal ciphertext consisting of:

$$Me(g_p, g_p)^{\alpha s}, \quad \{g_p^s, \quad g_p^{sA_K} g_p^{t_K B_K}, \quad g_p^{t_K} : (\forall K \in S)\}$$

One then draws $\tilde{s} \leftarrow \mathbb{Z}_N$. For each $K \in S$, an exponent $\tilde{t}_K \leftarrow \mathbb{Z}_N$ is chosen. The remaining composition of the semifunctional ciphertext depends on the type of ciphertext desired:

Type 1. The semi-functional ciphertext of Type 1 is formed as:

$$Me(g_p, g_p)^{\alpha s}, \quad \{g_p^s g_q^{\tilde{s}}, \quad g_p^{sA_K} g_p^{t_K B_K} g_q^{\tilde{s}A_K} g_q^{\tilde{t}_K B_K}, \quad g_p^{t_K} g_q^{\tilde{t}_K} : (\forall K \in S)\}$$

(again, here $A_K = \sum_{j \in K} a_j$ and $B_K = \sum_{j \in K} b_j$).

Type 2. The semi-functional ciphertext of Type 2 is formed as:

$$Me(g_p, g_p)^{\alpha s}, \quad \{g_p^s g_q^{\tilde{s}}, \quad g_p^{sA_K} g_p^{t_K B_K} g_q^{\tilde{s}A_K} g_q^{\tilde{t}_K \tilde{b}_K}, \quad g_p^{t_K} g_q^{\tilde{t}_K} : (\forall K \in S)\}$$

for fixed $\tilde{b}_K \in \mathbb{Z}_N$ which are chosen uniformly at random and fixed if they do not already exist (in a semi-functional key, for instance).

Type 3. The semi-functional ciphertext of Type 3 is formed as:

$$Me(g_p, g_p)^{\alpha s}, \quad \{g_p^s g_q^{\tilde{s}}, \quad g_p^{sA_K} g_p^{t_K B_K} g_q^{\tilde{s}\tilde{a}_K} g_q^{\tilde{t}_K \tilde{b}_K}, \quad g_p^{t_K} g_q^{\tilde{t}_K} : (\forall K \in S)\}$$

for fixed $\tilde{a}_K, \tilde{b}_K \in \mathbb{Z}_N$ which are chosen uniformly at random and fixed if they do not already exist.

Semi-functional Keys. We will use 7 types of semi-functional keys. To produce a semi-functional key for an access policy Λ, one first calls the normal key generation algorithm to produce a normal key consisting of:

$$\{g_p^{\lambda_K} g_p^{y_K A_K} g_w^{z_K}, \quad g_p^{y_K} g_w^{z_K'}, \quad g_p^{y_K B_K} g_w^{z_K''} : (\forall K \text{ labels} \in \Lambda)\}$$

The first 6 types of keys fall under 3 classes which have two variants each: a "Z" variant and an "R" variant. For Z-type keys one computes a linear sharing of 0 under access policy Λ, creating shares $\tilde{\lambda}_K$. For R-type keys one computes a linear sharing of *a random element* u of $\mathbb{Z}_q$ which is fixed once it is created and used for all R-type keys. u is shared under access policy Λ, to create shares $\tilde{\lambda}_K$. The next steps depend on the class of the key:

Class 1. First compute $g_q^{A_K}$ and $g_q^{B_K}$ (where, again, A_K and B_K represent the subset-sums of a_j and b_j). For each K label in the honest key, one then draws $\tilde{y}_K \leftarrow \mathbb{Z}_N$ and forms the semi-functional key of type 1Z or 1R (depending on the sharing $\tilde{\lambda}_K$) as:

$$\{g_p^{\lambda_K} g_p^{y_K A_K} g_q^{\tilde{\lambda}_K} g_q^{\tilde{y}_K A_K} g_w^{z_K}, \quad g_p^{y_K} g_q^{\tilde{y}_K} g_w^{z_K'}, \quad g_p^{y_K B_K} g_q^{\tilde{y}_K B_K} g_w^{z_K''} : (\forall K \text{ labels} \in \Lambda)\}$$

Class 2. First compute $g_q^{A_K}$. *Random values* $\tilde{b}_K \in \mathbb{Z}_N$ are chosen if they do not already exist (in a semi-functional ciphertext, for instance) and fixed. For each K label in the honest key, one then draws $\tilde{y}_K \leftarrow \mathbb{Z}_N$ and forms the semi-functional key of type 2Z or 2R as:

$$\{g_p^{\lambda_K} g_p^{y_K A_K} g_q^{\tilde{\lambda}_K} g_q^{\tilde{y}_K A_K} g_w^{z_K}, \quad g_p^{y_K} g_q^{\tilde{y}_K} g_w^{z_K'}, \quad g_p^{y_K B_K} g_q^{\tilde{y}_K \tilde{b}_K} g_w^{z_K''} : (\forall K \text{ labels} \in \Lambda)\}$$

Class 3. *Random values* $\tilde{a}_K, \tilde{b}_K \in \mathbb{Z}_N$ are chosen if they do not already exist and fixed. For each K label in the honest key, one then draws $\tilde{y}_K \leftarrow \mathbb{Z}_N$ and forms the semi-functional key of type 3Z or 3R as:

$$\{g_p^{\lambda_K} g_p^{y_K A_K} g_q^{\tilde{\lambda}_K} g_q^{\tilde{y}_K \tilde{a}_K} g_w^{z_K}, \quad g_p^{y_K} g_q^{\tilde{y}_K} g_w^{z_K'}, \quad g_p^{y_K B_K} g_q^{\tilde{y}_K \tilde{b}_K} g_w^{z_K''} : (\forall K \text{ labels} \in \Lambda)\}$$

Note we now have defined 6 types of keys: 1Z, 1R, 2Z, 2R, 3Z, and 3R, where the letter (Z/R) describes whether the $\tilde{\lambda}_K$ share zero or a random element of $\mathbb{Z}_q$ respectively, and the number (1/2/3) describes whether the semi-functional analogues of the $g_p^{A_K}$ and $g_p^{B_K}$ in the G_q group are structured as subset-sums or as random elements of G_q (Class 1 keys have both $g_q^{A_K}$ and $g_q^{B_K}$. Class 2 keys have just $g_q^{A_K}$ structured, with a random element $g_q^{\tilde{b}_K}$. Class 3 keys have both replaced by random elements $g_q^{\tilde{a}_K}, g_q^{\tilde{b}_K}$ of G_q). There is one final type of key, type 4R, which does not contain any of these elements:

Type 4R. Using shares $\tilde{\lambda}_K$ of u (which is randomly chosen from $\mathbb{Z}_p$ and fixed if it has not already been fixed), one forms the semi-functional key of type 4R as:

$$\{g_p^{\lambda_K} g_p^{y_K A_K} g_q^{\tilde{\lambda}_K} g_w^{z_K}, \quad g_p^{y_K} g_w^{z'_K}, \quad g_p^{y_K B_K} g_w^{z''_K} : (\forall K \text{ labels} \in \Lambda)\}$$

Proof Structure. Our hybrid proof takes place over a series of games defined as follows: Letting Q denote the total number of key queries that the attacker makes, we define Game_{ℓ_1}, Game_{ℓ_2}, Game_{ℓ_3}, Game_{ℓ_4}, Game_{ℓ_5}, Game_{ℓ_6}, and Game_{ℓ_7} for $\ell = 1, ..., Q$. In each game, the first $\ell - 1$ keys are semi-functional of type 4R, and all keys after the ℓth request are normal. They differ in the construction of the ℓth key and the ciphertext as follows:

Game ℓ_1. In this game, the ℓth key is type 1Z and the ciphertext is type 1.
Game ℓ_2. In this game, the ℓth key is type 2Z and the ciphertext is type 2.
Game ℓ_3. In this game, the ℓth key is type 3Z and the ciphertext is type 3.
Game ℓ_4. In this game, the ℓth key is type 3R and the ciphertext is type 3.
Game ℓ_5. In this game, the ℓth key is type 2R and the ciphertext is type 2.
Game ℓ_6. In this game, the ℓth key is type 1R and the ciphertext is type 1.
Game ℓ_7. In this game, the ℓth key is type 4R and the ciphertext is type 1.

Note that under this definition, we have that in Game_{0_7}, the ciphertext given to the attacker is type 1 and the keys are all normal.

The outer structure of our hybrid argument will progress as follows. First, we transition from Game_{real} to Game_{0_7}, then to Game_{1_1}, next to Game_{1_2}, next to Game_{1_3}, next to Game_{1_4}, next to Game_{1_5}, next to Game_{1_6}, next to Game_{1_7} and then to Game_{2_1} and so on. We then arrive at Game_{Q_7}, where the ciphertext is semifunctional of type 1 and *all* of the keys given to the attacker are type 4R. We then transition to one last game named Game_{final} which will complete our proof. Game_{final} uses a semi-functional ciphertext of a new type: type X:

Type X. The semi-functional ciphertext of Type X is formed as:

$$MX, \quad \{g_p^s g_q^{\tilde{s}}, \quad g_p^{sA_K} g_p^{t_K B_K} g_q^{\tilde{s}A_K} g_q^{\tilde{t}_K B_K}, \quad g_p^{t_K} g_q^{\tilde{t}_K} : (\forall K \in S)\} \text{ for } X \leftarrow G_T$$

Game$_{final}$. In this game, all keys are semi-functional of type 4R and the ciphertext is semi-functional of type X.

Note that a ciphertext of type X information-theoretically hides its message M because the message is multiplied by the uniform random X which is unused anywhere else. So, in Game_{final}, no polynomial time adversary will be able to achieve advantage in the security game, completing our proof. This hybrid argument is accomplished in the full version of this paper.

4   Bilinear Entropy Expansion Lemma

The main technical lemma used in our security argument is used to transition between hybrid games where semi-functional keys and ciphertexts have subset-sum structured $g_q^{B_K}$ components and games where they have random $g_q^{b_K}$ components. The relevant quantities in the following lemma are r_i where either r_i is a random exponent or is structured as a subset sum of c_i (which are analogous to the a_j, b_j in different applications of the lemma).

Definition 2. *Given G, a group of prime order q, and g a generator of that group, let $\mathcal{D}_1(m)$ be the distribution of:*

$$g^{\tilde{s}}, g^{\tilde{y}_1}, ..., g^{\tilde{y}_{M-1}},$$
$$g^{\tilde{y}_1 r_1}, ..., g^{\tilde{y}_{M-1} r_{M-1}},$$
$$g^{\tilde{y}_1 b_1}, ..., g^{\tilde{y}_{M-1} b_{M-1}},$$
$$g^{\tilde{t}_1}, ..., g^{\tilde{t}_{M-1}},$$
$$g^{\tilde{s} r_1 + \tilde{t}_1 b_1}, ..., g^{\tilde{s} r_{M-1} + \tilde{t}_{M-1} b_{M-1}}$$

where the $\tilde{y}_i, \tilde{t}_i, b_i, r_i, \tilde{s} \leftarrow \mathbb{Z}_q$ and $M = 2^m$.

Definition 3. *Given G, a group of prime order q, g a generator of that group, and $C = \{c_1, ..., c_m\}$ a set of m elements drawn uniformly at random from $\mathbb{Z}_q$, let $\mathcal{D}_2(m)$ be the distribution of the same elements (where the $\tilde{y}_i, \tilde{t}_i, b_i, \tilde{s} \leftarrow \mathbb{Z}_q$ and $M = 2^m$) EXCEPT that each $r_i = \sum_{j \in C_i} c_j$ where C_i denotes the ith indexed nonempty subset of C ($|C| = m$ and there are $M - 1 = 2^m - 1$ nonempty subsets).*

We show that the distributions $\mathcal{D}_1(m)$ and $\mathcal{D}_2(m)$ are computationally indistinguishable if $m = O(\lg poly(\lambda))$ through an inductive proof, beginning with the base case of $m = 2$, where a distinguisher for $\mathcal{D}_1(2)$ and $\mathcal{D}_2(2)$ ($C = \{c_1, c_2\}$) can be used to achieve the same advantage in the 2-Linear Problem.

Lemma 1. *If there exists a polynomial-time algorithm able to achieve advantage $2^2 \delta$ in distinguishing between the distributions $\mathcal{D}_1(2)$ and $\mathcal{D}_2(2)$, then there exists a polynomial-time algorithm able to achieve advantage δ in the 2-Linear Problem.*

Proof. If there exists a polynomial time algorithm $\mathcal{A}$ which distinguishes between $\mathcal{D}_1(2)$ and $\mathcal{D}_2(2)$ with advantage $2^2 \delta$, we can construct a distinguisher for the 2-Linear problem: $\mathcal{B}$. $\mathcal{B}$, upon receiving $g, g^{y_1}, g^{y_2}, g^{y_1 c_1}, g^{y_2 c_2}, g^{c_1 + c_2 + r}$, draws uniform random $\tilde{s}, b_3, \tilde{y}_3, \tilde{t}_1, \tilde{t}_2, \tilde{t}_3, \gamma_1, \gamma_2 \leftarrow \mathbb{Z}_q$, then creates the set:

$$g^{\tilde{s}}, g^{y_1}, \quad g^{y_2}, \quad g^{\tilde{y}_3},$$
$$g^{y_1 c_1}, \quad g^{y_2 c_2}, \quad (g^{c_1 + c_2 + r})^{\tilde{y}_3},$$
$$(g^{y_1 c_1})^{-\frac{\tilde{s}}{\tilde{t}_1}} (g^{y_1})^{\gamma_1}, \quad (g^{y_2 c})^{-\frac{\tilde{s}}{\tilde{t}_2}} (g^{y_2})^{\gamma_2}, \quad g^{\tilde{y}_3 b_3},$$
$$g^{\tilde{t}_1}, \quad g^{\tilde{t}_2}, \quad g^{\tilde{t}_3},$$
$$g^{\tilde{t}_1 \gamma_1}, \quad g^{\tilde{t}_2 \gamma_2}, \quad (g^{c_1 + c_2 + r})^{\tilde{s}} g^{\tilde{t}_3 b_3}$$

then runs $\mathcal{A}$ on this input and returns the output of $\mathcal{A}$.

Notice that if $r = 0$, this distribution is exactly $\mathcal{D}_2(2)$ (with $C = \{c_1, c_2\}$, $\tilde{y}_1 = y_1, \tilde{y}_2 = y_2, b_1 = -\frac{c_1 s}{\tilde{t}_1} + \gamma_1$, and $b_2 = -\frac{c_2 s}{\tilde{t}_2} + \gamma_2$). If r is instead random, this distribution is exactly $\mathcal{D}_1(2)$. Therefore, $\mathcal{B}$ will achieve the same advantage $2^2 \delta$ as $\mathcal{A}$ (which is greater than δ) in deciding the 2-Linear problem.

Lemma 2. *For all integers $m \geq 2$, if there exists a polynomial-time algorithm able to achieve an advantage of $2^{m+1}\delta$ deciding between distributions $\mathcal{D}_1(m+1)$ and $\mathcal{D}_2(m+1)$, then either there exists a polynomial-time algorithm able to achieve an advantage of $2^m\delta$ in deciding between distributions $\mathcal{D}_1(m)$ and $\mathcal{D}_2(m)$ or there exists a polynomial time algorithm able to achieve an advantage of δ in the 2-Linear Problem.*

Proof. If there exists a polynomial time algorithm $\mathcal{A}$ which distinguishes between $\mathcal{D}_1(m+1)$ and $\mathcal{D}_2(m+1)$ with non-negligible advantage $2^{m+1}\delta$, we construct $\mathcal{B}$: a distinguisher for $\mathcal{D}_1(m)$ and $\mathcal{D}_2(m)$.

$\mathcal{B}$, upon receiving:

$g^{\tilde{s}}, g^{\tilde{y}_1}, ..., g^{\tilde{y}_{M-1}}, g^{\tilde{y}_1 r_1}, ..., g^{\tilde{y}_{M-1} r_{M-1}}, g^{\tilde{y}_1 b_1}, ..., g^{\tilde{y}_{M-1} b_{M-1}}, g^{\tilde{t}_1}, ..., g^{\tilde{t}_{M-1}},$
$g^{\tilde{s} r_1 + \tilde{t}_1 b_1}, ..., g^{\tilde{s} r_{M-1} + \tilde{t}_{M-1} b_{M-1}}$ where $M = 2^m$, first draws:

$y_1^*, ..., y_{M-1}^*, \sigma_1, ..., \sigma_{M-1}, \gamma_1, ..., \gamma_{M-1}, \tilde{y}_M, \tilde{t}_M, b_M, c_{m+1} \leftarrow \mathbb{Z}_q$, and constructs:

$$g^{\tilde{s}}, g^{\tilde{y}_1}, ..., g^{\tilde{y}_{M-1}}, g^{\tilde{y}_M},$$
$$(g^{\tilde{y}_1})^{y_1^*}, ..., (g^{\tilde{y}_{M-1}})^{y_{M-1}^*},$$
$$g^{\tilde{y}_1 r_1}, ..., g^{\tilde{y}_{M-1} r_{M-1}}, g^{\tilde{y}_M c_{m+1}},$$
$$(g^{\tilde{y}_1})^{y_1^* c_{m+1}} (g^{\tilde{y}_1 r_1})^{y_1^*}, ..., (g^{\tilde{y}_{M-1}})^{y_{M-1}^* c_{m+1}} (g^{\tilde{y}_{M-1} r_{M-1}})^{y_{M-1}^*},$$
$$g^{\tilde{y}_1 b_1}, ..., g^{\tilde{y}_{M-1} b_{M-1}}, g^{\tilde{y}_M b_M},$$
$$(g^{\tilde{y}_1 b_1})^{y_1^*} (g^{\tilde{y}_1})^{\sigma_1 y_1^*}, ..., (g^{\tilde{y}_{M-1} b_{M-1}})^{y_{M-1}^*} (g^{\tilde{y}_{M-1}})^{\sigma_{M-1} y_{M-1}^*},$$
$$g^{\tilde{t}_1}, ..., g^{\tilde{t}_{M-1}}, g^{\tilde{t}_M},$$
$$g^{\tilde{t}_1} (g^{\tilde{y}_1})^{\gamma_1}, ..., g^{\tilde{t}_{M-1}} (g^{\tilde{y}_{M-1}})^{\gamma_{M-1}},$$
$$g^{\tilde{s} r_1 + \tilde{t}_1 b_1}, ..., g^{\tilde{s} r_{M-1} + \tilde{t}_{M-1} b_{M-1}}, (g^{\tilde{s}})^{c_{m+1}} g^{\tilde{t}_M b_M},$$
$$(g^{\tilde{s}})^{c_{m+1}} g^{\tilde{s} r_1 + \tilde{t}_1 b_1} (g^{\tilde{t}_1})^{\sigma_1} (g^{b_1 \tilde{y}_1})^{\gamma_1} (g^{\tilde{y}_1})^{\sigma_1 \gamma_1}, ...,$$
$$(g^{\tilde{s}})^{c_{m+1}} g^{\tilde{s} r_{M-1} + \tilde{t}_{M-1} b_{M-1}} (g^{\tilde{t}_{M-1}})^{\sigma_{M-1}} (g^{b_{M-1} \tilde{y}_{M-1}})^{\gamma_{M-1}} (g^{\tilde{y}_{M-1}})^{\sigma_{M-1} \gamma_{M-1}}$$

which is equal to:

$$g^{\tilde{s}}, g^{\tilde{y}_1}, ..., g^{\tilde{y}_{M-1}}, g^{\tilde{y}_M},$$
$$g^{\tilde{y}_1 y_1^*}, ..., g^{\tilde{y}_{M-1} y_{M-1}^*},$$
$$g^{\tilde{y}_1 r_1}, ..., g^{\tilde{y}_{M-1} r_{M-1}}, g^{\tilde{y}_M c_{m+1}},$$
$$g^{\tilde{y}_1 y_1^* (r_1 + c_{m+1})}, ..., g^{\tilde{y}_{M-1} y_{M-1}^* (r_{M-1} + c_{m+1})},$$
$$g^{\tilde{y}_1 b_1}, ..., g^{\tilde{y}_{M-1} b_{M-1}}, g^{\tilde{y}_M b_M},$$
$$g^{\tilde{y}_1 y_1^* (b_1 + \sigma_1)}, ..., g^{\tilde{y}_{M-1} y_{M-1}^* (b_{M-1} + \sigma_{M-1})},$$
$$g^{\tilde{t}_1}, ..., g^{\tilde{t}_{M-1}}, g^{\tilde{t}_M},$$
$$g^{\tilde{t}_1 + \tilde{y}_1 \gamma_1}, ..., g^{\tilde{t}_{M-1} + \tilde{y}_{M-1} \gamma_{M-1}},$$
$$g^{\tilde{s} r_1 + \tilde{t}_1 b_1}, ..., g^{\tilde{s} r_{M-1} + \tilde{t}_{M-1} b_{M-1}}, g^{\tilde{s} c_{m+1} + \tilde{t}_M b_M},$$
$$g^{\tilde{s}(r_1 + c_{m+1}) + (\tilde{t}_1 + \tilde{y}_1 \gamma_1)(b_1 + \sigma_1)}, ..., g^{\tilde{s}(r_{M-1} + c_{m+1}) + (\tilde{t}_{M-1} + \tilde{y}_{M-1} \gamma_{M-1})(b_{M-1} + \sigma_{M-1})}$$

Notice that if $\mathcal{B}$'s input is $\mathcal{D}_2(m)$, then the distribution of sets constructed by $\mathcal{B}$ is exactly $\mathcal{D}_2(m+1)$, where a new c_{m+1} element is drawn and added to form the subsets of the new augmented set C, $\tilde{y}_{M+i} = \tilde{y}_i y_i^*$, $b_{M+i} = b_i + \sigma_i$, and $\tilde{t}_{M+i} = \tilde{t}_i + \tilde{y}_i \gamma_i$ for $i = 1, ..., M-1$ which are all uniformly distributed at random. However, if $\mathcal{B}$'s input is $\mathcal{D}_1(m)$, then the distribution of sets constructed by $\mathcal{B}$ is not exactly $\mathcal{D}_1(m+1)$.

Definition 4. *Let $\mathcal{D}_1'(m+1)$ be the distribution of sets created by $\mathcal{B}$ given input sets from $\mathcal{D}_1(m)$.*

We have therefore only proved that if an algorithm is able to achieve advantage in distinguishing $\mathcal{D}_1'(m+1)$ and $\mathcal{D}_2(m+1)$, then it can be used to achieve that same advantage in deciding between $\mathcal{D}_1(m)$ and $\mathcal{D}_2(m)$. Fortunately, we can transition between $\mathcal{D}_1'(m+1)$ and $\mathcal{D}_1(m+1)$ using a hybrid proof. First we define $M = 2^m$ hybrid distributions indexed by (j):

Definition 5. *Let $\mathcal{D}_1'^{(j)}(m+1)$ be the distribution of:*

$$g^{\tilde{s}}, g^{\tilde{y}_1}, ..., g^{\tilde{y}_{M-1}}, g^{\tilde{y}_M},$$

$$g^{\tilde{y}_{M+1}}, ..., g^{\tilde{y}_{2M-1}},$$

$$g^{\tilde{y}_1 r_1}, ..., g^{\tilde{y}_{M-1} r_{M-1}}, g^{\tilde{y}_M c_{m+1}},$$

$$g^{\tilde{y}_{M+1}(r_1 + c_{m+1})}, ..., g^{\tilde{y}_{M+j}(r_j + c_{m+1})}, g^{\tilde{y}_{M+j+1} r_{M+j+1}}, ..., g^{\tilde{y}_{2M-1} r_{2M-1}},$$

$$g^{\tilde{y}_1 b_1}, ..., g^{\tilde{y}_{M-1} b_{M-1}}, g^{\tilde{y}_M b_M},$$

$$g^{\tilde{y}_{M+1} b_{M+1}}, ..., g^{\tilde{y}_{2M-1} b_{2M-1}},$$

$$g^{\tilde{t}_1}, ..., g^{\tilde{t}_{M-1}}, g^{\tilde{t}_M},$$

$$g^{\tilde{t}_{M+1}}, ..., g^{\tilde{t}_{2M-1}},$$

$$g^{\tilde{s} r_1 + \tilde{t}_1 b_1}, ..., g^{\tilde{s} r_{M-1} + \tilde{t}_{M-1} b_{M-1}}, g^{\tilde{s} c_{m+1} + \tilde{t}_M b_M},$$

$$g^{\tilde{s}(r_1 + c_{m+1}) + \tilde{t}_{M+1} b_{M+1}}, ..., g^{\tilde{s}(r_j + c_{m+1}) + \tilde{t}_{M+j} b_{M+j}}, g^{\tilde{s} r_{M+j+1} + \tilde{t}_{M+j+1} b_{M+j+1}}, ...,$$

$$g^{\tilde{s} r_{2M-1} + \tilde{t}_{2M-1} b_{2M-1}}$$

for $j = 0, ..., M-1$ where the r_{M+j+i} are distributed uniformly at random in $\mathbb{Z}_p$ for $i = 1, ... M - j - 1$.

Notice that $\mathcal{D}_1'^{(0)}(m+1) = \mathcal{D}_1(m+1)$ and $\mathcal{D}_1'^{(M-1)}(m+1) = \mathcal{D}_1'(m+1)$. So, if some adversary $\mathcal{A}$ could distinguish between $\mathcal{D}_1(m+1)$ and $\mathcal{D}_1'(m+1)$ with non-negligible advantage $2^m \delta$, then by the triangle inequality, there must exists some j such that: $\left| \Pr[\mathcal{A} = 1 | \mathcal{D}_1'^{(j+1)}(m+1)] - \Pr[\mathcal{A} = 1 | \mathcal{D}_1'^{(j)}(m+1)] \right| \geq \frac{2^m \delta}{M} = \delta$. Such an $\mathcal{A}$ can be used to construct a distinguisher for the 2-Linear Problem: $\mathcal{B}$ that achieves advantage δ:

$\mathcal{B}$, upon receiving $g, g^{y_1}, g^{y_2}, g^{y_1 c_1}, g^{y_2 c_2}, g^{c_1 + c_2 + r}$, relabels the elements as: $g, g^{y_1}, g^{y_2}, g^{y_1 r^*}, g^{y_2 c_{m+1}}, g^x$ (defining $y_1 = y_1, y_2 = y_2, r^* = c_1, c_{m+1} = c_2$, and $x = r^* + c_{m+1} + r$). $\mathcal{B}$ then draws $\tilde{s}, \gamma_{j+1}, \tilde{y}_1, ..., \tilde{y}_j, \tilde{y}_{j+2}, ..., \tilde{y}_{M-1}, y_M^*, ..., y_{M+j}^*,$ $\tilde{y}_{M+j+1}, ..., \tilde{y}_{2M-1}, \tilde{t}_1, ..., \tilde{t}_{2M-1}, \gamma_M, ..., \gamma_{M+j}, b_{j+2}, ..., b_{2M-1}, r_1, ..., r_j, r_{j+2}, ...,$ r_{2M-1} uniformly at random from $\mathbb{Z}_q$ and constructs:

$g^{\tilde{s}}, g^{\tilde{y}_1}, ..., g^{\tilde{y}_j}, g^{y_1}, g^{\tilde{y}_{j+2}}, ..., g^{\tilde{y}_{M-1}}, (g^{y_2})^{y_M^*},$

$(g^{y_2})^{y_{M+1}^*}, ..., (g^{y_2})^{y_{M+j}^*}, g^{\tilde{y}_{M+j+1}}, ..., g^{\tilde{y}_{2M-1}},$

$g^{\tilde{y}_1 r_1}, ..., g^{\tilde{y}_j r_j}, g^{y_1 r^*}, g^{\tilde{y}_{j+2} r_{j+2}}, ..., g^{\tilde{y}_{M-1} r_{M-1}}, (g^{y_2 c_{m+1}})^{y_M^*},$

$((g^{y_2})^{y_{M+1}^*})^{r_1}(g^{y_2 c_{m+1}})^{y_{M+1}^*}, ..., ((g^{y_2})^{y_{M+j}^*})^{r_j}(g^{y_2 c_{m+1}})^{y_{M+j}^*}, (g^x)^{\tilde{y}_{M+j+1}},$

$g^{\tilde{y}_{M+j+2} r_{M+j+2}}, ..., g^{\tilde{y}_{2M-1} r_{2M-1}},$

$g^{\tilde{y}_1 b_1}, ..., g^{\tilde{y}_j b_j}, (g^{y_1 r^*})^{-\frac{\tilde{s}}{\tilde{t}_{j+1}}}(g^{y_1})^{\gamma_{j+1}}, g^{\tilde{y}_{j+2} b_{j+2}}, ..., g^{\tilde{y}_{M-1} b_{M-1}}, (g^{y_2})^{y_{M-1}^* b_{M-1}},$

$(g^{y_2})^{y_M^* \gamma_M}(g^{y_2 c_{m+1}})^{-y_M^* \frac{\tilde{s}}{t_M}}$

$(g^{y_2})^{-y_{M+1}^*(\frac{\tilde{s} r_1}{t_{M+1}} - \gamma_{M+1})}(g^{y_2 c_{m+1}})^{-y_{M+1}^* \frac{\tilde{s}}{t_{M+1}}}, ...,$

$(g^{y_2})^{-y_{M+j}^*(\frac{\tilde{s} r_j}{t_{M+j}} - \gamma_{M+j})}(g^{y_2 c_{m+1}})^{-y_{M+j}^* \frac{\tilde{s}}{t_{M+j}}},$

$g^{\tilde{y}_{M+j+1} b_{M+j+1}}, ..., g^{\tilde{y}_{2M-1} b_{2M-1}},$

$g^{\tilde{t}_1}, ..., g^{\tilde{t}_M},$

$g^{\tilde{t}_{M+1}}, ..., g^{\tilde{t}_{2M-1}},$

$g^{\tilde{s} r_1 + \tilde{t}_1 b_1}, ..., g^{\tilde{s} r_j + \tilde{t}_j b_j}, g^{\tilde{t}_{j+1} \gamma_{j+1}}, g^{\tilde{s} r_{j+1} + \tilde{t}_{j+1} b_{j+1}}, ..., g^{\tilde{s} r_{M-1} + \tilde{t}_{M-1} b_{M-1}}, g^{\tilde{t}_M \gamma_M}$

$g^{\tilde{t}_{M+1} \gamma_{M+1}}, ..., g^{\tilde{t}_{M+j} \gamma_{M+j}}, (g^x)^{\tilde{s}} g^{\tilde{t}_{M+j+1} b_{M+j+1}}, g^{\tilde{s} r_{M+j+2} + \tilde{t}_{M+j+2} b_{M+j+2}}, ...,$

$g^{\tilde{s} r_{2M-1} + \tilde{t}_{2M-1} b_{2M-1}}$

where $\tilde{y}_{j+1} = y_1$, $r_{j+1} = r^*$, $b_{j+1} = -\frac{\tilde{s} r^*}{\tilde{t}_{j+1}} + \gamma_{j+1}$, $b_M = -\frac{\tilde{s} c_{m+1}}{t_M} + \gamma_M$, and the $b_{M+i} = -\frac{\tilde{s}(r_i + c_{m+1})}{\tilde{t}_{M+i}} + \gamma_{M+i}$ for $i = 1, ..., j$ and $\tilde{y}_{M+i} = y_2 \tilde{y}_{M+i}^*$ for $i = 0, ..., j$ are all distributed uniformly at random in $\mathbb{Z}_p$.

$\mathcal{B}$ then runs $\mathcal{A}$ on this input and outputs the same.

Note that if $x = r^* + c_{m+1} + 0$, then $\mathcal{B}$ has sampled an instance of $\mathcal{D}_1'^{(j+1)}(m+1)$. Otherwise, if $x = r^* + c_{m+1} + r$ for a uniform random r it has sampled an instance of $\mathcal{D}_1'^{(j)}(m + 1)$. So, $\mathcal{B}$ will enjoy the same advantage δ of $\mathcal{A}$ but in deciding the 2-Linear Problem.

We assumed there is a polynomial time algorithm $\mathcal{A}$ which distinguishes between $\mathcal{D}_1(m + 1)$ and $\mathcal{D}_2(m + 1)$ with advantage $2^{m+1}\delta$. By the triangle inequality, then $\mathcal{A}$ must be able to be used to either achieve advantage $2^m \delta$ in distinguishing between instances of $\mathcal{D}_1(m + 1)$ and $\mathcal{D}_1'(m + 1)$ or achieve advantage $2^m \delta$ in distinguishing between instances of between $\mathcal{D}_1'(m + 1)$ and $\mathcal{D}_2(m + 1)$.

In the first case, if $\mathcal{A}$ can be used to achieve advantage $2^m \delta$ in distinguishing between instances of $\mathcal{D}_1(m+1)$ and $\mathcal{D}_1'(m+1)$, then we showed in the first proof how such an algorithm could be used to distinguish between $\mathcal{D}_1(m)$ and $\mathcal{D}_2(m)$ with the same advantage $(2^m \delta)$.

In the second case, if $\mathcal{A}$ can be used to achieve advantage $2^m \delta$ in distinguishing between instances of $\mathcal{D}_1'(m+1)$ and $\mathcal{D}_2(m+1)$, then we showed in the second proof how such an algorithm could be used to break the 2-Linear problem with advantage $\frac{2^m \delta}{M} = \delta$.

Therefore, if there is a polynomial time algorithm $\mathcal{A}$ which distinguishes between $\mathcal{D}_1(m + 1)$ and $\mathcal{D}_2(m + 1)$ with advantage $2^{m+1}\delta$, then either there exists a polynomial-time algorithm able to achieve an advantage of $2^m\delta$ in deciding between distributions $\mathcal{D}_1(m)$ and $\mathcal{D}_2(m)$ or there exists a polynomial time algorithm able to achieve an advantage of δ in the 2-Linear Problem.

Lemma 3. *The distributions $\mathcal{D}_1(k)$ and $\mathcal{D}_2(k)$ are computationally indistinguishable under the 2-Linear computational hardness assumption if $k = O$ $(\lg poly(\lambda))$.*

Proof. We have shown that for all integers $m \geq 2$, if there exists a polynomial-time algorithm able to achieve an advantage of $2^{m+1}\delta$ deciding between distributions $\mathcal{D}_1(m + 1)$ and $\mathcal{D}_2(m + 1)$, then either there exists a polynomial-time algorithm able to achieve an advantage of $2^m\delta$ in deciding between distributions $\mathcal{D}_1(m)$ and $\mathcal{D}_2(m)$ or there exists a polynomial time algorithm able to achieve an advantage of δ in the 2-Linear Problem. We have also shown that if there exists a polynomial-time algorithm able to achieve advantage $2^2\delta$ in distinguishing between the distributions $\mathcal{D}_1(2)$ and $\mathcal{D}_2(2)$, then there exists a polynomial-time algorithm able to achieve advantage δ in the 2-Linear Problem. By induction, it follows that for all m, if an algorithm is able to achieve an advantage of $2^m\delta$ in distinguishing between distributions $\mathcal{D}_1(m)$ and $\mathcal{D}_2(m)$, then that algorithm can be used to achieve advantage δ in the 2-Linear problem.

If $k = O(\lg poly(\lambda))$, then any algorithm $\mathcal{A}$ able to achieve non-negligible advantage δ in distinguishing between $\mathcal{D}_1(k)$ and $\mathcal{D}_2(k)$ can be used to achieve non-negligible advantage $\Omega(\frac{\delta}{poly(\lambda)})$ in the 2-Linear problem. This violates our 2-Linear Assumption, so no such algorithm $\mathcal{A}$ can exist.

5   Concluding Remarks

We have presented a composite order KP-ABE scheme proven fully secure under the DLIN assumption and additional subgroup decision type assumptions. The scheme allows a bound of $2^k - 1$ attribute-uses in an access policy, where the number of group elements required in the public parameters per attribute-use grows polynomially with k. An interesting question for future work is whether the ciphertext sizes can be significantly reduced (our scheme has ciphertexts still growing linearly in size with $2^k - 1$). We have chosen to demonstrate our techniques on a KP-ABE scheme, though we note that they are equally applicable to the CP-ABE setting. The core of CP-ABE schemes often mirror the structure of KP-ABE schemes, and would benefit similarly from the reduced public parameter size our lemma enables. Finally, our bilinear entropy expansion lemma is not restricted to the ABE setting, and we suspect it may have applications to other cryptographic primitives. Primitive structure can be built around the lemma's core components of $\{g^{t_K}, g^{t_K A_K}\}$, which can be plugged in to replace a need for independent random group elements. Our composite order KP-ABE scheme demonstrates this usage.

References

1. Attrapadung, N.: Dual system encryption via doubly selective security: framework, fully secure functional encryption for regular languages, and more. In: Nguyen, P.Q., Oswald, E. (eds.) EUROCRYPT 2014. LNCS, vol. 8441, pp. 557–577. Springer, Heidelberg (2014)
2. Parno, B., Raykova, M., Vaikuntanathan, V.: How to delegate and verify in public: verifiable computation from attribute-based encryption. In: Cramer, R. (ed.) TCC 2012. LNCS, vol. 7194, pp. 422–439. Springer, Heidelberg (2012)
3. Beimel, A.: Secure schemes for secret sharing and key distribution. Ph.D. thesis, Israel Institute of Technology, Technion, Haifa, Israel (1996)
4. Bethencourt, J., Sahai, A., Waters, B.: Ciphertext-policy attribute-based encryption. In: Proceedings of the IEEE Symposium on Security and Privacy, pp. 321–334
5. Boneh, D., Franklin, M.: Identity-based encryption from the weil pairing. In: Kilian, J. (ed.) CRYPTO 2001. LNCS, vol. 2139, p. 213. Springer, Heidelberg (2001)
6. Chase, M.: Multi-authority attribute based encryption. In: Vadhan, S.P. (ed.) TCC 2007. LNCS, vol. 4392, pp. 515–534. Springer, Heidelberg (2007)
7. Chase, M., Chow, S.S.M.: Improving privacy and security in multi-authority attribute-based encryption. In: Proceedings of the 2009 ACM Conference on Computer and Communications Security, pp. 121–130 (2009)
8. Chase, M., Meiklejohn, S.: Déjà Q: using dual systems to revisit q-type assumptions. In: Nguyen, P.Q., Oswald, E. (eds.) EUROCRYPT 2014. LNCS, vol. 8441, pp. 622–639. Springer, Heidelberg (2014)
9. Chen, J., Wee, H.: Fully, (almost) tightly secure ibe and dual system groups. In: Canetti, R., Garay, J.A. (eds.) CRYPTO 2013, Part II. LNCS, vol. 8043, pp. 435–460. Springer, Heidelberg (2013)
10. Cocks, C.: An identity based encryption scheme based on quadratic residues. In: Honary, B. (ed.) Cryptography and Coding 2001. LNCS, vol. 2260, p. 360. Springer, Heidelberg (2001)
11. Dodis, Y., Lewko, A.B., Waters, B., Wichs, D.: Storing secrets on continually leaky devices. In: FOCS, pp. 688–697 (2011)
12. Garg, S., Gentry, C., Halevi, S., Sahai, A., Waters, B.: Attribute-based encryption for circuits from multilinear maps. In: Canetti, R., Garay, J.A. (eds.) CRYPTO 2013, Part II. LNCS, vol. 8043, pp. 479–499. Springer, Heidelberg (2013)
13. Garg, S., Gentry, C., Halevi, S., Zhandry, M.: Fully secure attribute based encryption from multilinear maps. IACR Cryptology ePrint Arch. **2014**, 622 (2014)
14. Goldwasser, S., Kalai, Y.T., Popa, R.A., Vaikuntanathan, V., Zeldovich, N.: How to run turing machines on encrypted data. In: Canetti, R., Garay, J.A. (eds.) CRYPTO 2013, Part II. LNCS, vol. 8043, pp. 536–553. Springer, Heidelberg (2013)
15. Gorbunov, S., Vaikuntanathan, V., Wee, H.: Attribute-based encryption for circuits. In: STOC, pp. 545–554 (2013)
16. Goyal, V., Jain, A., Pandey, O., Sahai, A.: Bounded ciphertext policy attribute based encryption. In: Aceto, L., Damgård, I., Goldberg, L.A., Halldórsson, M.M., Ingólfsdóttir, A., Walukiewicz, I. (eds.) ICALP 2008, Part II. LNCS, vol. 5126, pp. 579–591. Springer, Heidelberg (2008)
17. Goyal, V., Pandey, O., Sahai, A.B.: Waters. Attribute based encryption for fine-grained access control of encrypted data. In: ACM Conference on Computer and Communications Security, pp. 89–98 (2006)

18. Lewko, A., Okamoto, T., Sahai, A., Takashima, K., Waters, B.: Fully secure functional encryption: attribute-based encryption and (hierarchical) inner product encryption. In: Gilbert, H. (ed.) EUROCRYPT 2010. LNCS, vol. 6110, pp. 62–91. Springer, Heidelberg (2010)

19. Lewko, A., Rouselakis, Y., Waters, B.: Achieving leakage resilience through dual system encryption. In: Ishai, Y. (ed.) TCC 2011. LNCS, vol. 6597, pp. 70–88. Springer, Heidelberg (2011)

20. Lewko, A., Waters, B.: New techniques for dual system encryption and fully secure hibe with short ciphertexts. In: Micciancio, D. (ed.) TCC 2010. LNCS, vol. 5978, pp. 455–479. Springer, Heidelberg (2010)

21. Lewko, A., Waters, B.: Decentralizing attribute-based encryption. In: Paterson, K.G. (ed.) EUROCRYPT 2011. LNCS, vol. 6632, pp. 568–588. Springer, Heidelberg (2011)

22. Lewko, A., Waters, B.: Unbounded HIBE and attribute-based encryption. In: Paterson, K.G. (ed.) EUROCRYPT 2011. LNCS, vol. 6632, pp. 547–567. Springer, Heidelberg (2011)

23. Lewko, A.B., Lewko, M., Waters, B.: How to leak on key updates. In: STOC, pp. 725–734 (2011)

24. Lewko, A.B., Waters, B.: Efficient pseudorandom functions from the decisional linear assumption and weaker variants. In: Proceedings of the 2009 ACM Conference on Computer and Communications Security, pp. 112–120 (2009)

25. Lewko, A., Waters, B.: New proof methods for attribute-based encryption: achieving full security through selective techniques. In: Safavi-Naini, R., Canetti, R. (eds.) CRYPTO 2012. LNCS, vol. 7417, pp. 180–198. Springer, Heidelberg (2012)

26. Naor, M., Reingold, O.: Number-theoretic constructions of efficient pseudo-random functions. In: FOCS, pp. 458–467 (1997)

27. Okamoto, T., Takashima, K.: Fully secure functional encryption with general relations from the decisional linear assumption. In: Rabin, T. (ed.) CRYPTO 2010. LNCS, vol. 6223, pp. 191–208. Springer, Heidelberg (2010)

28. Okamoto, T., Takashima, K.: Fully secure unbounded inner-product and attribute-based encryption. In: Wang, X., Sako, K. (eds.) ASIACRYPT 2012. LNCS, vol. 7658, pp. 349–366. Springer, Heidelberg (2012)

29. Okamoto, T., Takashima, K.: Decentralized attribute-based signatures. In: PKC, pp. 125–142 (2013)

30. Ostrovksy, R., Sahai, A., Waters, B.: Attribute based encryption with non-monotonic access structures. In: ACM conference on Computer and Communications Security, pp. 195–203 (2007)

31. Sahai, A., Waters, B.: Fuzzy identity-based encryption. In: Cramer, R. (ed.) EUROCRYPT 2005. LNCS, vol. 3494, pp. 457–473. Springer, Heidelberg (2005)

32. Waters, B.: Dual system encryption: realizing fully secure IBE and HIBE under simple assumptions. In: Halevi, S. (ed.) CRYPTO 2009. LNCS, vol. 5677, pp. 619–636. Springer, Heidelberg (2009)

33. Waters, B.: Ciphertext-policy attribute-based encryption: an expressive, efficient, and provably secure realization. In: PKC, pp. 53–70 (2011)

34. Wee, H.: Dual system encryption via predicate encodings. In: Lindell, Y. (ed.) TCC 2014. LNCS, vol. 8349, pp. 616–637. Springer, Heidelberg (2014)

New Primitives

Data Is a Stream: Security of Stream-Based Channels

Marc Fischlin[1]([✉]), Felix Günther[1], Giorgia Azzurra Marson[1],
and Kenneth G. Paterson[2]

[1] Cryptoplexity, Technische Universität Darmstadt, Darmstadt, Germany
marc.fischlin@cryptoplexity.de, guenther@cs.tu-darmstadt.de,
giorgia.marson@cased.de
[2] Information Security Group, Royal Holloway, University of London,
London, UK
kenny.paterson@rhul.ac.uk

Abstract. The common approach to defining secure channels in the literature is to consider transportation of discrete messages provided via atomic encryption and decryption interfaces. This, however, ignores that many practical protocols (including TLS, SSH, and QUIC) offer streaming interfaces instead, moreover with the complexity that the network (possibly under adversarial control) may deliver arbitrary fragments of ciphertexts to the receiver. To address this deficiency, we initiate the study of stream-based channels and their security. We present notions of confidentiality and integrity for such channels, akin to the notions for atomic channels, but taking the peculiarities of streams into account. We provide a composition result for our setting, saying that combining chosen-plaintext confidentiality with integrity of the transmitted ciphertext stream lifts confidentiality of the channel to chosen-ciphertext security. Notably, for our proof of this theorem in the streaming setting we need an additional property, called error predictability. We finally give an AEAD-based construction that achieves our notion of a secure stream-based channel. The construction matches rather well the one used in TLS, providing validation of that protocol's design.

Keywords: Secure channel · Data stream · AEAD · Confidentiality · Integrity

1 Introduction

The most widely-used application for cryptography today is still secure communications—providing a 'secure channel' for the transmission of data between two parties. Secure channel protocols are numerous and diverse in their features, operating at different network layers and offering different security services. Prominent examples can be found in GSM, UMTS and LTE [1] mobile telecommunications systems, in WEP, WPA and WPA2 [19] (which secure wireless LAN communications), IPsec [22] (which provides security at the IP layer), TLS [15] and DTLS [31] (which run over TCP [30] and UDP [29], respectively), Google's QUIC protocol [33], and SSH [36] (an 'application layer' secure protocol).

© International Association for Cryptologic Research 2015
R. Gennaro and M. Robshaw (Eds.): CRYPTO 2015, Part II, LNCS 9216, pp. 545–564, 2015.
DOI: 10.1007/978-3-662-48000-7_27

AEAD and Secure Channels in the Literature. Authenticated Encryption with Associated Data (AEAD) [32] has emerged as being the right cryptographic tool for building secure channels. AEAD provides both confidentiality and integrity guarantees for data. However, on its own, AEAD is insufficient for constructing secure channels. For example, in most practical situations, a secure channel should provide more than simple encryption of messages, but also guarantee detection of (and possibly recovery from) out-of-order delivery and replays of messages. Furthermore, a secure channel should deal with error handling, with errors potentially arising from both cryptographic and non-cryptographic processing —whether or not to tear-down a secure channel session when an error is encountered, and how (and indeed whether) to signal errors to the other side. As another difference, some secure channel designs (such as IPsec and to a limited extent TLS) have additional features that can be used to provide protection against traffic analysis. A secure channel may accept messages of arbitrary length and need to fragment these before encryption, and may reassemble these fragments again after decryption; alternatively, it may present to applications a maximum message size that is well-matched to the underlying network infrastructure. Finally, and most importantly in the context of the paper here, a secure channel may be designed to protect a *stream* of data rather than the series of discrete messages that is usually found in cryptographic abstractions.

There is, then, a substantial gap between what the AEAD primitive can reasonably provide and the needs of secure channels. We are not the first to recognize this gap, of course. For example, Bellare et al. [5] extended the standard security notions of confidentiality and integrity for symmetric encryption to the stateful setting, enabling the treatment of security of the ordering of discrete messages in a secure channel, with application to the analysis of SSH being their principle motivation. Their notions were later extended by Black et al. [23] to include a richer variety of features, suitable for handling channels that permit (or deny) replays, message drops, and reordering. Additional literature concerning the formalization of secure channels includes [3,12,13,20,24–27,34].

Stream-Based Channels. Characteristic of all the above-mentioned prior works is that they treat secure channels as providing an *atomic* interface for messages, meaning that the channel is designed only for sending and receiving sequences of discrete messages. However, this only captures a fraction of secure channel designs that are actually used in the real world. In particular, TLS, SSH, and QUIC all provide a *streaming* interface for the applications that use them: applications submit segments (or fragments) of message (or plaintext) streams to an application programming interface (API), and similarly receive fragments of message streams from the API. The sending side may arbitrarily buffer and/or fragment the message stream before encapsulating it for sending. Moreover, in some cases, even under normal operations, it is not guaranteed by the network that the resulting stream of ciphertext fragments (which we refer to as *ciphertexts* henceforth treating them as opaque bit strings) that is sent will arrive at the receiver with the same pattern of fragmentation, even if the reconstructed message streams are in the end identical. Under adversarial conditions, such guarantees certainly do not hold: for example, TLS runs over TCP and an active man-in-the-middle adversary can tinker with the TCP segments, adding, removing and reordering TLS data at will. Thus practical secure channels need to securely process arbitrarily fragmented ciphertexts. Finally, to make things even more complex, and coming full circle, applications (like HTTP [17]) often attempt to use stream-oriented secure channels (like TLS) to perform secure, atomic message delivery.

This discussion points to a mismatch between atomic descriptions of secure channels in the cryptography literature and the reality of the operation of secure channels. As one may expect, such mismatches can have negative consequences for security. The starkest example of this comes from the plaintext recovery attack against SSH given by Albrecht et al. [2]. Their attack specifically exploits the adversary's ability to deliver arbitrary sequences of SSH packet fragments to the receiver (over TCP) and observe the receiver's behavior in response. The attack is possible despite the analysis of [5] which proved that the SSH secure channel satisfies suitable *atomic* stateful security notions. Related attacks against certain IPsec configurations (and exploiting IPsec's need to handle IP fragmentation) were presented in [14]. Attacks highlighting a disjunction between what applications expect and what secure channels provide, in the specific context of HTTP and TLS, can be found in [7,35]. All these attacks show the incompleteness of previous approaches to modeling and analyzing secure channels.

Boldyreva et al. [9] extended the classical, atomic secure channel notions to cover the case of SSH-like stream-based secure channels, broadening the SSH-specific work of [28]. However, while they allow for fragmented delivery of ciphertexts to the receiver, their work still assumes that the encryption process on the sender's side is atomic, meaning that there is a one-to-one correspondence between message and ciphertexts. This may be the case for SSH when used in interactive sessions, but it is not the case for the tunneling mode of SSH, and never the case for other secure channels protocols. For example, even though the TLS specification [15] does not include a formal API definition, it is clear that the design intention is to provide a secure channel for data streams (and the application programmer is in practice offered a TCP-like socket interface), and, as noted above, the sending side can arbitrarily buffer and fragment the message stream when preparing ciphertexts for sending.

Our Contributions. In this paper we develop formal functional specifications, security notions, and a construction (using AEAD as a building block) for *stream-based channels*. Our models are in the game-based tradition, and extend those of [5,9] to handle the streaming nature of the channels that we consider.

While our methodology and modeling closely resemble those of [9], and indeed build upon them, a crucial difference comes in our treatment of the sending (or encrypting) function of a stream-based channel: in [9], this is still atomic (while decryption is not), whereas in our stream-based channel setting, both the sending and receiving function support streams of data, with potentially arbitrary buffering and fragmentation on the sending and receiving side. This requires careful modification of the confidentiality definitions of [9]. In addition, we develop suitable integrity notions for the streaming setting, whereas [9] does not consider this aspect. This is important because the (informal) security properties that applications expect a secure channel to provide include confidentiality as well as integrity, while security in the most powerful 'chosen fragment attack' setting of [9] does not provide *any* integrity guarantees.

Bringing integrity into the picture for stream-based channels also enables us to prove a composition result analogous to the classical result of [6] for symmetric encryption schemes, which states that $\mathsf{IND} - \mathsf{CPA}$ security in combination with integrity of ciphertexts ($\mathsf{INT} - \mathsf{CTXT}$ security) guarantees $\mathsf{IND} - \mathsf{CCA}$ security. This provides an easy route to proving that a given stream-based channel construction provides appropriate confidentiality (indistinguishability under chosen ciphertext-fragment attacks, or $\mathsf{IND} - \mathsf{CCFA}$ security) and integrity (integrity of plaintext streams, $\mathsf{INT} - \mathsf{PST}$ security).

The composition theorem brings an interesting technical challenge to surmount: as was already recognized in [10] for the classical (atomic) setting, the possibility that realistic models of encryption schemes may involve *multiple* error messages means that the original composition proof of [6] does not go through. In [10], this was overcome by assuming the scheme is such that only one of the possible error messages has a non-negligible chance of being produced during operation of the scheme. Here we take a different tack, introducing the concept of *error predictability*, which guarantees the existence of an efficient algorithm that can predict which errors should be output during decryption of a ciphertext stream.

We demonstrate the feasibility of our security notions by providing a generic construction for a stream-based channel that uses AEAD as a component and achieves our strongest confidentiality and integrity notions. The resulting stream-based channel closely mimics the TLS Record Protocol. So our security results provide validation for this important real-world protocol design, whilst fully taking its streaming behavior into account. In the full version of this paper we moreover propose a generic construction of a stream-based channel from symmetric encryption supporting fragmentation as per [9].

Also in the full version, we return to the starting point of our discussion and analyze how applications can use stream-based channels to safely transport atomic messages by encoding distinguished end-of-message symbols into the sent message stream to identify the atomic messages' boundaries. Establishing the security of this simple and natural approach however requires the introduction of an additional technical property orthogonal to integrity and confidentiality. Our analysis sheds a new formal light on the truncation [35] and 'cookie-cutter' [7] attacks on HTTP running over TLS, showing how they can be seen as arising from a misunderstanding of the security guarantees that can be provided by a stream-based channel to applications expecting an atomic-message channel.

Further Related Work. Bhargavan et al. [8] have developed notions of security for stream-based channels as part of their detailed analysis of the TLS Record Protocol. Their approach involves expressing channel security properties as types in a programming language, and then formally proving that the type definitions are respected in an adversarial setting (where the adversary is modeled as another program interacting with the code for the send and receive functions of the channel).

A seemingly similar line of work to ours concerns blockwise-adaptive security and on-line symmetric encryption schemes, as developed in [4,11,18,21]. There, the schemes operate in an on-the-fly manner, processing one fixed-size block of plaintext or ciphertext at a time; meanwhile the adversary is given access to blockwise encryption (and possibly decryption) oracles. However, in these papers messages and ciphertexts are ultimately regarded as discrete entities, rather than as streams of message and ciphertext fragments as in our treatment.

Paper Organization. After introducing some basic notation and terminology in Sect. 2, we present in Sect. 3 our formal definition for stream-based channels. Section 4 contains our security notions for confidentiality and integrity of stream-based channels as well as our composition theorem. Finally, in Sect. 5 we show feasibility of our notions by providing a generic construction of a stream-based channel. We conclude with open questions arising from this work in Sect. 6.

2   Preliminaries

Notation. Let Σ be an alphabet and $s \in \Sigma^*$. We indicate by $|s|$ the length of s, by $s[i]$ its i-th character, and by $s[i, \ldots, j]$ the substring $s[i]|| \ldots ||s[j]$, where $||$ denotes the string concatenation. Let $s, t \in \Sigma^*$. We say that s is a *prefix* of t and write $s \prec t$ if there exists $r \in \Sigma^*$ such that $s||r = t$; in this case we write $r = t \% s$. We denote the longest common prefix of s and t by $[s, t] = [t, s]$. Note that $s \prec t$ if and only if $[s, t] = s$. Using the above notation we will often consider $s \% [s, t]$, i.e., the suffix of s with the longest common prefix of s and t stripped off. Let $\boldsymbol{s} = (s_1, \ldots, s_\ell) \in (\Sigma^*)^\ell$ be a vector of strings for some integer ℓ; if $\boldsymbol{s}$ is empty, i.e., $\ell = 0$, we denote this by $\boldsymbol{s} = ()$. For every $0 \le i \le j \le \ell$ we denote $\boldsymbol{s}[i] = s_i$ and $\boldsymbol{s}[i, \ldots, j] = (s_i, \ldots, s_j)$; we use the shortcut $||\boldsymbol{s}$ for the concatenation $s_1|| \ldots ||s_\ell$, and conventionally define $||() = \varepsilon$. We say that two vectors $\boldsymbol{s} = (s_1, \ldots, s_\ell)$ and $\boldsymbol{t} = (t_1, \ldots, t_{\ell'})$ are equal and write $\boldsymbol{s} = \boldsymbol{t}$ if and only if $\ell = \ell'$ and $s[i] = t[i]$ for all $1 < i < \ell$. Slightly overloading notation, we denote the merge of two vectors $\boldsymbol{s}$ and $\boldsymbol{t}$ as $\boldsymbol{s}||\boldsymbol{t} = (s_1, \ldots, s_\ell, t_1, \ldots, t_{\ell'})$.

Channel Terminology. Our syntax for channels is intentionally independent of the targeted security properties as these may vary from one specific application to another. To reflect the generic functionality of channels and maintain a higher level of abstraction than, e.g., in the case of authenticated encryption, we define sending (Send) and receiving (Recv) rather than encryption and decryption algorithms.

3   Stream-Based Channels

We capture the functionality of channel protocols that offer a reliable transmission of *streams* like the Transmission Control Protocol (TCP) [30] and, in a second step, we define confidentiality and integrity properties expected from (stream-based) secure channel protocols like the Transport Layer Security (TLS) Record Protocol [15] or the Secure Shell (SSH) Binary Packet Protocol [37].[1] To do so we first need to define the syntax of stream-based channels that, in constrast to previous models for channel, send fragments of a message (or plaintext) stream rather than atomic messages. In order to remain close to real-world implementations we restrict both the message space and the ciphertext space to the set of bit strings, where we understand 'messages' and 'ciphertexts' not as atomic units, but as fragments (i.e., substrings) of a message stream and a ciphertext stream.

Additionally, we do not stipulate a particular input/output behavior on the sender side, but instead allow the sending algorithm Send to process input data at its discretion, e.g., implementing some form of buffering. We enforce sending out particular chunks of the message stream by employing the established concept of 'flushing a stream' known from network socket programming, and provide the Send algorithm with an additional *flush* flag $f \in \{0,1\}$ which, if set to $f = 1$, ensures that all the message fragments fed so far are sent out instantaneously. Jumping ahead, in our security model this choice conservatively also allows the adversary to control fragmentation. If the flush flag is set to zero, Send may internally decide to keep accepting more message fragments or to send out a ciphertext fragment, depending on its implementation and resources. In our definition below we demand that each message fragment m_i processed by Send results in a ciphertext fragment c_i. Since a ciphertext fragments can be empty ($c_i = \varepsilon$), this implicitly enables Send to wait for more data by outputting empty ciphertext fragments. Figure 1 illustrates the behavior of the sending and receiving algorithms of a stream-based channel.

We proceed with defining syntax and correctness of stream-based channels.

Definition 1 (Syntax of stream-based channels). *A stream-based channel* Ch = (Init, Send, Recv) *with associated sending and receiving state space* $\mathcal{S}_S$ *resp.* $\mathcal{S}_R$ *and error space* $\mathcal{E}$ *consists of three efficient probabilistic algorithms:*

– Init. *On input of a security parameter* 1^λ, *this algorithm outputs initial states* $\mathsf{st}_{S,0} \in \mathcal{S}_S$, $\mathsf{st}_{R,0} \in \mathcal{S}_R$ *for the sender and the receiver, respectively. We write* $(\mathsf{st}_{S,0}, \mathsf{st}_{R,0}) \leftarrow_{\$} \mathsf{Init}(1^\lambda)$.

---

[1] Our model inherently assumes that, in a benign scenario, ciphertext fragments are delivered reliably and in order (i.e., in a TCP-like manner). While we recognize that efficient and secure transmission protocols can be designed also on top of unreliable protocols like the User Datagram Protocol (UDP) [29] as done, e.g., in Google's Quick UDP Internet Connections (QUIC) protocol [33], we deem these approaches orthogonal or unrelated to our work. In such cases, a reliable and ordered stream transmission can be implemented *non-cryptographically* either by TCP-like preprocessing of the UDP datagrams before handing them over to a stream-based channel according to our definition or by postprocessing UDP datagrams which are encrypted and authenticated in an isolated manner (e.g., using an AEAD scheme).

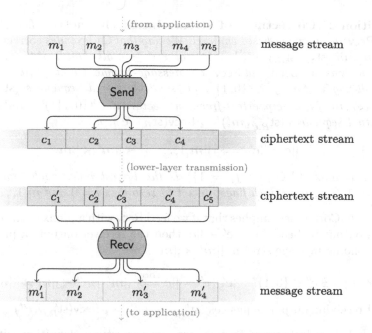

Fig. 1. Illustration of the behavior of the Send and Recv algorithms of a stream-based channel, indicating the message and ciphertext fragments being sent (m_i resp. c_i) and received (m'_i resp. c'_i).

- Send. *On input of a state* $\mathsf{st}_S \in \mathcal{S}_S$, *a fragment* $m \in \{0,1\}^*$, *and a flush flag* $f \in \{0,1\}$, *this algorithm outputs an updated state* $\mathsf{st}'_S \in \mathcal{S}_s$ *and a ciphertext fragment* $c \in \{0,1\}^*$. *We write* $(\mathsf{st}'_S \in \mathcal{S}_s, c) \leftarrow_\$ \mathsf{Send}(\mathsf{st}_S, m, f)$.
- Recv. *On input of a state* $\mathsf{st}_R \in \mathcal{S}_R$ *and a ciphertext fragment* $c \in \{0,1\}^*$, *this algorithm outputs an updated state* $\mathsf{st}'_R \in \mathcal{S}_R$ *and a message fragment* $m \in \{0,1\}^* \cup \mathcal{E}$. *We write* $(\mathsf{st}'_R, m) \leftarrow_\$ \mathsf{Recv}(\mathsf{st}_R, c)$.

Given a state pair $(\mathsf{st}_{S,0}, \mathsf{st}_{R,0})$, an integer $\ell \geq 0$, and tuples of message fragments $\boldsymbol{m} = (m_1, \ldots, m_\ell) \in (\{0,1\}^*)^\ell$ and of flush flags $\boldsymbol{f} = (f_1, \ldots, f_\ell) \in \{0,1\}^\ell$, let $(\mathsf{st}_S, \boldsymbol{c}) \leftarrow_\$ \mathsf{Send}(\mathsf{st}_{S,0}, \boldsymbol{m}, \boldsymbol{f})$ be shorthand for the sequential execution $(\mathsf{st}_{S,1}, c_1) \leftarrow_\$ \mathsf{Send}(\mathsf{st}_{S,0}, m_1, f_1), \ldots, (\mathsf{st}_{S,\ell}, c_\ell) \leftarrow_\$ \mathsf{Send}(\mathsf{st}_{S,\ell-1}, m_\ell, f_\ell)$ with $\boldsymbol{c} = (c_1, \ldots, c_\ell)$ and $\mathsf{st}_S = \mathsf{st}_{S,\ell}$. For $\ell = 0$ we define $\boldsymbol{c}$ to be the empty vector and $\mathsf{st}_{S,\ell} = \mathsf{st}_S$ to be the initial state. We use an analogous notation for the receiver's algorithm.

Intuitively, correctness of stream-based channels guarantees that for every message fragments input to Send, if the corresponding ciphertext stream is processed by Recv, then no matter how the ciphertext stream is (re)fragmented at the receiver side the returned message stream is a prefix of the initial message stream. Moreover, when Recv consumes a ciphertext fragment generated by a call to Send with the flush flag set to 1, its output stream contains all the message fragments input to Send up to that call. We next formalize this intuition.

Definition 2 (Correctness of stream-based channels). *Let* $\mathsf{Ch} = (\mathsf{Init},$ $\mathsf{Send}, \mathsf{Recv})$ *be a stream-based channel. We say that* Ch *provides correctness if for all state pair* $(\mathsf{st}_{S,0}, \mathsf{st}_{R,0}) \leftarrow_\$ \mathsf{Init}(1^\lambda)$, *all* $\ell, \ell' \geq 0$, *all choices of the randomness for algorithms* $\mathsf{Init}, \mathsf{Send}$ *and* Recv, *all message-fragment vectors* $\boldsymbol{m} \in (\{0,1\}^*)^\ell$, *all flush-flag vectors* $\boldsymbol{f} \in \{0,1\}^\ell$, *all sending output sequences* $(\mathsf{st}_{S,\ell}, \boldsymbol{c}) \leftarrow_\$$ $\mathsf{Send}(\mathsf{st}_{S,0}, \boldsymbol{m}, \boldsymbol{f})$, *all ciphertext-fragment vectors* $\boldsymbol{c}' \in (\{0,1\}^*)^{\ell'}$, *and all receiving output sequences* $(\mathsf{st}'_{R,\ell'}, \boldsymbol{m}') \leftarrow_\$ \mathsf{Recv}(\mathsf{st}_{R,0}, \boldsymbol{c}')$, *we have*

$$||\boldsymbol{c} = ||\boldsymbol{c}' \implies ||\boldsymbol{m}[1, \ldots, i] \prec ||\boldsymbol{m}' \prec ||\boldsymbol{m},$$

where $i = \max(\{0\} \cup \{j : f_j = 1\})$ *is the largest index such that the flush flag* $f_i = 1$ *(i.e., if all flush flags are set to zero then* $i = 0$ *and* $\boldsymbol{m}[1, \ldots, i] = \varepsilon$).

Remark 1. Correctness implies that if we feed Recv with a prefix of the ciphertext stream output by Send, i.e., $||\boldsymbol{c}' \prec ||\boldsymbol{c}$, then the receiver outputs a prefix of the corresponding message stream, $||\boldsymbol{m}' \prec ||\boldsymbol{m}$, since

$$||\boldsymbol{c}' \prec ||\boldsymbol{c} \Rightarrow \exists\, c'' \in \{0,1\}^* : ||\boldsymbol{c}'\,||\,c'' = ||\boldsymbol{c} \overset{\text{(corr.)}}{\implies} ||\boldsymbol{m}'\,||\,m'' \prec ||\boldsymbol{m} \Rightarrow ||\boldsymbol{m}' \prec ||\boldsymbol{m}$$

for all receiving output sequences $(\mathsf{st}'_{R,\ell'+1}, m'') \leftarrow_\$ \mathsf{Recv}(\mathsf{st}'_{R,\ell'}, c'')$.

Remark 2. It is instructive to compare our correctness definition with that of Boldyreva et al. [9]. There, correctness requires that if a sequence $\boldsymbol{m}$ of discrete messages is encrypted, and the resulting ciphertext stream $||\boldsymbol{c}$ is then decrypted (possibly in a fragmented manner), then the obtained message sequence (when message separators ¶ are removed) is identical to the original sequence $\boldsymbol{m}$. In the special case of a single message, this implies that encryption 'always flushes' in the setting of [9], and is in turn the reason why encryption is necessarily an atomic operation. By contrast, in our setting the Send algorithm is equipped with a flush flag and, when the latter is set to zero, potentially the entire message fragment is buffered for later sending. This is, then, an essential difference between the setting of Boldyreva et al. [9] and the streaming one. An additional difference is that the correctness condition in [9] is stronger than ours as it incorporates a certain amount of robustness. More specifically, the sequence of ciphertext fragments $\boldsymbol{c}'$ submitted for decryption in the correctness definition of [9] may extend the sequence produced by encryption (in other words, $||\boldsymbol{c}$ is only required to be a prefix of $||\boldsymbol{c}'$ for decryption to still work correctly up to $||\boldsymbol{c}$).

4   Security for Stream-Based Channels

In the following we introduce both confidentiality and integrity notions attuned to the stream-based setting and analyze their composition. We provide corresponding notions in terms of asymptotic security; analogous notions in the concrete setting are easy to infer.[2]

---

[2] It is straightforward to define a concrete notion of security by considering the advantage of the adversary as a concrete function of its running time, the numbers of oracle queries, and bounds on the size of the input streams for oracle queries.

4.1   Confidentiality

As in the ciphertext fragmentation setting introduced by Boldyreva et al. [9], whose confidentiality notion in turn is inspired by the $\mathsf{IND} - \mathsf{sfCCA}$ notion by Bellare et al. [5], our security notions have to deal with the fact that stream-based channels support processing of arbitrary fragments of the message resp. ciphertext stream. While Boldyreva et al. [9] considered only fragmented decryption (but atomic encryption) and therefore focused their attention on the CCA-like setting, the fragmented message processing of stream-based channels in our case also affects the adversarial capabilities in the CPA-like setting. We hence define security notions both for the case of *chosen plaintext-fragment* attacks ($\mathsf{IND} - \mathsf{CPFA}$) as well as *chosen ciphertext-fragment* attacks ($\mathsf{IND} - \mathsf{CCFA}$).

Adapting the chosen-plaintext capabilities of an adversary to the stream-based settings is relatively straightforward (incorporating the standard left-or-right oracle). However, deriving a sound security notion for an adversary controlling the fragmentation on the received ciphertext stream turns out to be more delicate. In general, chosen-ciphertext-like oracles strive to allow decryption of as much of the input as possible without enabling trivial attacks. We follow the approach of Bellare et al. [5] to model stateful (decryption) security notions by considering the receiving oracle $\mathcal{O}_{\mathsf{Recv}}$ to be in-sync and not returning a response to the adversary $\mathcal{A}$ as long as $\mathcal{A}$ supplies (parts of) the original ciphertext stream output by the left-or-right sending oracle $\mathcal{O}_{\mathsf{LoR}}$ in correct sequential order. When $\mathcal{A}$ deviates from the original ciphertext stream, the $\mathcal{O}_{\mathsf{Recv}}$ oracle is considered out of sync and, from that point on, the output of the Recv algorithm is given to the adversary.

For a sound definition we are faced with the question: At which point *exactly* shall $\mathcal{O}_{\mathsf{Recv}}$ be considered out-of-sync? Boldyreva et al. decided to stay close to the original definitions of Bellare et al. and conservatively defined synchronization to be lost at ciphertext boundaries (i.e., their notion reveals the decryption of the full ciphertext as output by Send whenever any part of it is modified). However this option is inappropriate in our stream-based setting where the output of Send is not necessarily an atomic unit.

As an example to illustrate this, consider the case of TLS and the Send algorithm being called on a $(2^{14} + 1)$-byte input message with the flush flag set to 1—mimicking the behavior of many TLS implementations that keep no send buffer. Obeying the limit of at most 2^{14} bytes payload in a single TLS record, Send is forced to output a ciphertext fragment which contains (at least) two TLS records. An adversary which now forwards this fragment to the decryption oracle in the $\mathsf{IND} - \mathsf{sfCFA}$ definition of Boldyreva et al. [9, Definition 4] with the second record modified but the first record untouched will be provided with the decryption of *both* records, thereby trivially revealing parts of the challenge message string.

Mindful of this example and taking into account that the output of Send in our case is a bit stream without any further structure in general, the natural choice appears to consider $\mathcal{O}_{\mathsf{Recv}}$ to become out-of-sync exactly when the first bit of its ciphertext stream input deviates from the genuine output of Send.

$\mathsf{Expt}_{\mathsf{Ch},\mathcal{A}}^{\mathsf{IND\text{-}atk},b}(1^\lambda)$:

1 $(\mathsf{st}_S, \mathsf{st}_R) \leftarrow_\$ \mathsf{Init}(1^\lambda)$
2 $\mathsf{sync} \leftarrow 1$
3 $C_S \leftarrow \varepsilon,\ C_R \leftarrow \varepsilon$
4 $b' \leftarrow_\$ \mathcal{A}(1^\lambda)^{\mathcal{O}_{\mathsf{LoR}}(\cdot,\cdot),\mathcal{O}_{\mathsf{Recv}}(\cdot)}$
5 return b'

If $\mathcal{A}$ queries $\mathcal{O}_{\mathsf{LoR}}(m_0, m_1, f)$:

1 if $|m_0| \neq |m_1|$ then
2   return ε to $\mathcal{A}$
3 $(\mathsf{st}_S, c) \leftarrow_\$ \mathsf{Send}(\mathsf{st}_S, m_b, f)$
4 $C_S \leftarrow C_S \| c$
5 return c to $\mathcal{A}$

If $\mathcal{A}$ queries $\mathcal{O}_{\mathsf{Recv}}(c)$:

1 if $\mathsf{sync} = 0$ then
2    $(\mathsf{st}_R, m) \leftarrow_\$ \mathsf{Recv}(\mathsf{st}_R, c)$
3   return m to $\mathcal{A}$
4 else if $C_R \| c \prec C_S$ then
5    $C_R \leftarrow C_R \| c$
6    $(\mathsf{st}_R, m) \leftarrow_\$ \mathsf{Recv}(\mathsf{st}_R, c)$
7   return ε to $\mathcal{A}$
8 else
9    $\mathsf{sync} \leftarrow 0$
10    $\widetilde{c} \leftarrow [C_R \| c, C_S] \% C_R$
11    $\widetilde{\mathsf{st}_R} \leftarrow \mathsf{st}_R$
12    $(\widetilde{\mathsf{st}_R}, \widetilde{m}) \leftarrow_\$ \mathsf{Recv}(\widetilde{\mathsf{st}_R}, \widetilde{c})$
13    $(\mathsf{st}_R, m) \leftarrow_\$ \mathsf{Recv}(\mathsf{st}_R, c)$
14    $m' \leftarrow m \% [m, \widetilde{m}]$
15   return m' to $\mathcal{A}$

Fig. 2. Security experiment for *confidentiality* (IND – atk) of stream-based channels. A CPFA-attacker only has access to the oracle $\mathcal{O}_{\mathsf{LoR}}$.

In more detail, we define our stream-based confidentiality notions IND – CPFA (indistinguishability under chosen plaintext-fragment attack) and IND – CCFA (indistinguishability under chosen ciphertext-fragment attack) through the experiment $\mathsf{Expt}_{\mathsf{Ch},\mathcal{A}}^{\mathsf{IND\text{-}atk},b}$ (where atk is a placeholder for either CPFA or CCFA), depicted in Fig. 2. The adversary's goal in the experiment $\mathsf{Expt}_{\mathsf{Ch},\mathcal{A}}^{\mathsf{IND\text{-}atk},b}$ is to guess the bit b. In the experiment the $\mathcal{O}_{\mathsf{LoR}}$ oracle provides the adversary with the response of Send to the (left or right) message fragment input. The oracle first checks if the input message fragments m_0 and m_1 have the same bit length (i.e., $|m_0| = |m_1|$). If this is the case, it invokes Send on m_b, adds its response c to the internal ciphertext stream variable C_S and provides $\mathcal{A}$ with c.

The $\mathcal{O}_{\mathsf{Recv}}$ oracle in the experiment processes the ciphertext fragment input (thereby updating the receiving state $\mathsf{st}_R,$), but artificially suppresses the output of Recv as long as the fragments are in sync. In case synchronization has been already lost (i.e., $\mathsf{sync} = 0$), $\mathcal{O}_{\mathsf{Recv}}$ simply passes the output of Recv to $\mathcal{A}$. Otherwise, it checks whether the concatenation C_R of ciphertext fragments seen so far together with the current fragment c is still a prefix of the ciphertext stream C_S output by $\mathcal{O}_{\mathsf{LoR}}$: if this is the case, Recv is invoked on c but its output is suppressed. Otherwise $\mathcal{O}_{\mathsf{Recv}}$ is now considered out-of-sync and there are two definitional options available, both following the paradigm of giving as much information to the adversary as possible without enabling trivial attacks: The first option is to split the call to the receiver into two, one for the longest common prefix $\widetilde{c}$ of the received ciphertext c which still matches the ciphertext stream C_S output by $\mathcal{O}_{\mathsf{LoR}}$, and one for the remaining ciphertext part where they diverge. The second option, and this is the one we use here and which turns out to be more appropriate than the first one (as we discuss in the full version), is to run the receiver on the full ciphertext c and later suppress parts of the message stream which the receiver *would* have obtained when run on $\widetilde{c}$.

More formally, our suppression strategy on the level of the message stream first simulates a Recv call on a copy of the current state st_R and $\tilde{c}$ and registers its output $\tilde{m}$. Second, Recv is regularly invoked (again for the original state st_R,) on the full ciphertext fragment c provided by the adversary, resulting in a message m being output. Finally, the common prefix of m and $\tilde{m}$ (i.e., any potential challenge message stream bits in m) is suppressed and the remaining part of m is passed to $\mathcal{A}$.

Definition 3 (IND − CPFA and IND − CCFA Security). *Let* Ch = (Init, Send, Recv) *be a stream-based channel and experiment* $\mathsf{Expt}_{\mathsf{Ch},\mathcal{A}}^{\mathsf{IND-atk},b}(1^\lambda)$ *for an adversary* $\mathcal{A}$ *and a bit* b *be defined as in Fig. 2, where* atk *is a placeholder for either* CPFA *or* CCFA. *Within the experiment the adversary* $\mathcal{A}$ *is given access to a (stateful) left-or-right sending oracle* $\mathcal{O}_{\mathsf{LoR}}$ *and, in the case of* IND − CCFA *security, a (stateful) receiving oracle* $\mathcal{O}_{\mathsf{Recv}}$. *We say that* Ch *provides indistinguishability under chosen plaintext-fragment (resp. ciphertext-fragment) attacks (*IND − CPFA *resp.* IND − CCFA*) if for all PPT adversaries* $\mathcal{A}$ *the following advantage function is negligible in the security parameter:*

$$\mathsf{Adv}_{\mathsf{Ch},\mathcal{A}}^{\mathsf{IND-atk},b}(\lambda) := \left| \Pr\left[\mathsf{Expt}_{\mathsf{Ch},\mathcal{A}}^{\mathsf{IND-atk},1}(1^\lambda) = 1 \right] - \Pr\left[\mathsf{Expt}_{\mathsf{Ch},\mathcal{A}}^{\mathsf{IND-atk},0}(1^\lambda) = 1 \right] \right|.$$

For the sake of completeness we comment on the alternative, intuitively appealing way for defining the receiving oracle by splitting the ciphertext in our setting in in the full version, which however leads to a confidentiality notion that only covers a smaller class of channels.

4.2 Integrity

In this section we formalize integrity notions for stream-based channels. We highlight that, while integrity properties for atomic messages (and atomic ciphertexts) are well-understood, no previous work considered integrity in the non-atomic setting. In particular Boldyreva et al. [9] only addressed confidentiality in the presence of ciphertext fragmentation. We define integrity notions for stream-based channels as refinements of standard (stateful) properties of plaintext integrity (INT − sfPTXT), resp., ciphertext integrity (INT − sfCTXT) from [5] and refer to the new properties as *plaintext-stream integrity*, resp., *ciphertext-stream integrity* (INT − PST, resp., INT − CST).

Similarly to the setting with atomic messages, INT − PST ensures that no adversarial query to the receiving oracle causes the message stream output by Recv to deviate from the message stream input to Send. Formalizing the stronger INT − CST property demands more care. Intuitively, from ciphertext integrity we expect that when processing any 'out-of-sync' ciphertext, the algorithm Recv should return an error message. However, when considering a stream-based interface it may happen that Recv processes an out-of-sync ciphertext which does not yet contain 'enough information' to be recognized as being invalid; in this case the receiving algorithm would buffer (part of) the ciphertext and wait for further fragments until a sufficiently long ciphertext string is available to be processed and deemed as valid or invalid. In such a scenario, a naive adaptation of the INT − sfCTXT definition of [5] would allow trivial attacks by declaring successful any adversary that makes the Recv buffer (part of) an out-of-sync ciphertext. Our notion of ciphertext-stream integrity carefully identifies the case just described and, by letting the receiving oracle wait for further ciphertext fragments, declares the adversary successful only if Recv outputs a non-emtpy message fragment resulting from an out-of-sync portion of the ciphertext stream.

$\mathsf{Expt}_{\mathsf{Ch},\mathcal{A}}^{\mathsf{INT-atk}}(1^\lambda)$:
1. $(st_S, st_R) \leftarrow_\$ \mathsf{Init}(1^\lambda)$
2. $\mathsf{sync} \leftarrow 1, \mathsf{win} \leftarrow 0$
3. $M_S, C_S \leftarrow \varepsilon, M_R, C_R \leftarrow \varepsilon$
4. $\mathcal{A}(1^\lambda)^{\mathcal{O}_{\mathsf{Send}}(\cdot,\cdot), \mathcal{O}_{\mathsf{Recv}}(\cdot)}$
5. return win

If $\mathcal{A}$ queries $\mathcal{O}_{\mathsf{Send}}(m, f)$:
1. $(st_S, c) \leftarrow_\$ \mathsf{Send}(st_S, m, f)$
2. $M_S \leftarrow M_S \| m$
3. $C_S \leftarrow C_S \| c$
4. return c to $\mathcal{A}$

INT-PST
If $\mathcal{A}$ queries $\mathcal{O}_{\mathsf{Recv}}(c)$:
1. $(st_R, m) \leftarrow_\$ \mathsf{Recv}(st_R, c)$
2. $M_R \leftarrow M_R \| m$
3. if $M_R \not\prec M_S$ and
   $M_R \% [M_R, M_S] \notin \mathcal{E}^*$ then
4.   win $\leftarrow 1$
5. return m to $\mathcal{A}$

INT-CST
If $\mathcal{A}$ queries $\mathcal{O}_{\mathsf{Recv}}(c)$:
1. if $\mathsf{sync} = 0$ then
2. $(st_R, m) \leftarrow_\$ \mathsf{Recv}(st_R, c)$
3.   if $m \notin \mathcal{E}^*$ then win $\leftarrow 1$
4. else if $C_R \| c \prec C_S$ then
5. $(st_R, m) \leftarrow_\$ \mathsf{Recv}(st_R, c)$
6. $C_R \leftarrow C_R \| c$
7. else
8. $\mathsf{sync} \leftarrow 0$
9. $\widetilde{c} \leftarrow [C_R \| c, C_S] \% C_R$
10. $\widetilde{st_R} \leftarrow st_R$
11. $(\widetilde{st_R}, \widetilde{m}) \leftarrow_\$ \mathsf{Recv}(\widetilde{st_R}, \widetilde{c})$
12. $(st_R, m) \leftarrow_\$ \mathsf{Recv}(st_R, c)$
13. $m' \leftarrow m \% [m, \widetilde{m}]$
14.   if $m' \notin \mathcal{E}^*$ then win $\leftarrow 1$
15. return m to $\mathcal{A}$

Fig. 3. Security experiment for *integrity* (INT − atk) of stream-based channels. An PST-attacker is provided with access to the middle $\mathcal{O}_{\mathsf{Recv}}$ oracle (INT − PST), whereas a CST-attacker is instead granted access to the oracle on the right-hand side (INT − CST).

We formalize integrity of plaintext and ciphertext streams through the security experiment $\mathsf{Expt}_{\mathsf{Ch},\mathcal{A}}^{\mathsf{INT-atk}}$ depicted in Fig. 3. The experiment provides the adversary with oracles $\mathcal{O}_{\mathsf{Send}}$ and $\mathcal{O}_{\mathsf{Recv}}$, where the former grants $\mathcal{A}$ access to algorithm Send under arbitrarily chosen message fragments and the latter gives $\mathcal{A}$ an interface with algorithm Recv. We highlight that, while the sending oracle $\mathcal{O}_{\mathsf{Send}}$ is common for both experiments INT − PST and INT − CST, the receiving oracle $\mathcal{O}_{\mathsf{Recv}}$ follows different procedures in the two cases, as we further explain below.

In the execution of the INT − PST experiment, $\mathcal{O}_{\mathsf{Send}}$ maintains in string M_S the stream of all sent message fragments and, analogously, $\mathcal{O}_{\mathsf{Recv}}$ maintains in M_R

the stream of all received message fragments (and/or error symbols). The adversary wins the game if it causes M_S and M_R to deviate in such a way that their difference contains more than error symbols. Formally, we demand that the string M_R output by the receiver is not a prefix of the sender's string M_S, but such that this prefix-freeness is not only due to error symbols from $\mathcal{E}$.

In the INT − CST experiment oracles $\mathcal{O}_{\mathsf{Send}}$ and $\mathcal{O}_{\mathsf{Recv}}$ maintain strings C_S and C_R to record the streams of sent ciphertexts resp. received ciphertext fragments. Furthermore, $\mathcal{O}_{\mathsf{Recv}}$ decides when the adversary wins by inspecting sent and received *ciphertext* streams, an inherently more complex task than looking for deviations in the underlying sequences of sent/received message fragments. Indeed, in a stream-based channel the algorithm Recv may need to buffer several ciphertexts before being able to recover the underlying message stream or detecting that an error occurred; such a behavior is reflected in our experiment. When processing in-sync ciphertexts $\mathcal{O}_{\mathsf{Recv}}$ simply appends each new fragment to C_R. In the moment when an out-of-sync ciphertext arrives, the oracle compares the outputs of algorithm Recv when processing (i) the current input ciphertext c and (ii) its longest in-sync prefix $\tilde{c}$. The adversary wins if $\mathcal{O}_{\mathsf{Recv}}$ outputs more in case (i) than it would in case (ii) and if the difference between the two outputs is a non-empty, valid message. It also wins if it is able to make Recv output a non-empty, valid message with a subsequent out-of-sync ciphertext.

Definition 4 (INT − PST and INT − CST Security). *Let* Ch = (Init, Send, Recv) *be a stream-based channel and experiment* $\mathsf{Expt}_{\mathsf{Ch},\mathcal{A}}^{\mathsf{INT-atk}}(1^\lambda)$ *for an adversary* $\mathcal{A}$ *be defined as in Fig. 2, where* atk *is a placeholder for either* PST *or* CST. *Within the experiment, the adversary* $\mathcal{A}$ *is given access to a sending oracle* $\mathcal{O}_{\mathsf{Send}}$ *and a receiving oracle* $\mathcal{O}_{\mathsf{Recv}}$. *We say that* Ch *provides integrity of plaintext streams (resp. ciphertext streams) (INT − PST resp.* INT − CST*) if for all PPT adversaries* $\mathcal{A}$ *the following advantage function is negligible in the security parameter:*

$$\mathsf{Adv}_{\mathsf{Ch},\mathcal{A}}^{\mathsf{INT-atk}}(\lambda) := \Pr\left[\mathsf{Expt}_{\mathsf{Ch},\mathcal{A}}^{\mathsf{INT-atk}}(1^\lambda) = 1\right].$$

Remark 3. Our definitions of integrity do not preclude from being secure those channels in which message bits can be output as a result of the adversary delivering *partial* ciphertexts to the Recv oracle. This is because in the streaming setting we care about the adversary's ability to force the receiver to accept message fragments corresponding to a part of the ciphertext stream that has gone out-of-sync, without attaching importance to ciphertext boundaries. Hence, this is quite distinct from the usual 'atomic' setting. In particular, applications that use a streaming channel to transmit atomic messages must take extra care to ensure no partially retrieved message fragment from the streaming channel is processed as if it was a complete (atomic) message, as such misinterpretation can lead—and in the past has led—to attacks [7,35].

We further note that stream-based integrity providing weaker guarantees than atomic-message integrity seems to be an intrinsic consequence of the nature of stream-based channels. In particular, apparent avenues of strengthening the given integrity definition lead to notions which are clearly inappropriate in the streaming setting. On the one hand, requiring a channel to output an error immediately after processing the first bit deviating from the sent ciphertext stream is, for most constructions, an unattainable goal as it is in general impossible to decide if an initial bit received is genuine or not. On the other hand, requiring that a channel does not output any message bit until a full ciphertext output by Send is received inappropriately enforces an atomic structure on the channel, i.e., basically the one of [9] which, as already discussed, is too strong for channels that, like TLS, might output ciphertexts which contain multiple, independent parts.

4.3   Relations Amongst Notions and Generic Composition Theorem

Due to space restrictions we comprehensively discuss the relations among the introduced security notions for the streaming setting only in the full version. In short, we show that, for both confidentiality and integrity, the stronger notion implies the weaker one, i.e., $\mathsf{IND} - \mathsf{CCFA} \Rightarrow \mathsf{IND} - \mathsf{CPFA}$ and $\mathsf{INT} - \mathsf{CST} \Rightarrow \mathsf{INT} - \mathsf{PST}$, as one might expect. Further, we extend the composition result from [6]—that (stateful) $\mathsf{IND} - \mathsf{CPA}$ and $\mathsf{INT} - \mathsf{CTXT}$ together imply (stateful) $\mathsf{IND} - \mathsf{CCA}$— to our streaming setting. Interestingly, the analogous prerequisites $\mathsf{IND} - \mathsf{CPFA}$ and $\mathsf{INT} - \mathsf{CST}$ alone are not sufficient to establish the composition result in our case: we additionally require the channel to be *error predictable* ($\mathsf{ERR} - \mathsf{PRE}$). The latter notion, defined only in the full version due to space restrictions, formalizes the ability to efficiently predict the error messages that should be obtained when the receiving algorithm fails.

Error predictability assists the security proof for our composition theorem in two ways. First, it allows us to deal with the problem of having multiple decryption errors [10]. This problem also appears in the atomic setting and has been surmounted there by considering only single error messages [6] or by restricting the likelihood of different error messages to appear [10]. Our notion of error predictability gives a more general approach which is also applicable in the atomic setting. Secondly, error predictability directly supports the reduction to the integrity property $\mathsf{INT} - \mathsf{CST}$ in our proof. In our stream-based scenario we basically must be able to tell if the receiver is still buffering ciphertext fragments, or if it can already produce an error message. Error predictability gives us exactly this.

We stress, and will expand in Sect. 5, that error predictability can be met by natural constructions. The composition result for stream-based channels is summarized in the theorem below. We provide a formal proof of this result in the full version.

Theorem 1 ($\mathsf{INT} - \mathsf{CST} \wedge \mathsf{IND} - \mathsf{CPFA} \wedge \mathsf{ERR} - \mathsf{PRE} \Rightarrow \mathsf{IND} - \mathsf{CCFA}$). *Let* $\mathsf{Ch} = (\mathsf{Init}, \mathsf{Send}, \mathsf{Recv})$ *be a (correct) stream-based channel with associated error space* $\mathcal{E}$. *If* Ch *provides integrity of ciphertext streams, error predictability, and indistinguishability under chosen plaintext-fragment attacks then it also provides indistinguishability under chosen ciphertext-fragment attacks. Formally, for* *every efficient* IND − CCFA *adversary* $\mathcal{A}$ *there exist efficient* INT − CST *adversary* $\mathcal{B}$, ERR − PRE *adversary* $\mathcal{C}$, *and* IND − CPFA *adversary* $\mathcal{D}$ *such that*

$$\mathsf{Adv}_{\mathsf{Ch},\mathcal{A}}^{\mathsf{IND-CCFA}} \leq 2 \cdot \mathsf{Adv}_{\mathsf{Ch},\mathcal{B}}^{\mathsf{INT-CST}} + 2 \cdot \mathsf{Adv}_{\mathsf{Ch},\mathcal{C}}^{\mathsf{ERR-PRE}} + \mathsf{Adv}_{\mathsf{Ch},\mathcal{D}}^{\mathsf{IND-CPFA}}.$$

5 Construction of Stream-Based Channels

In this section we demonstrate the feasibility of our security notions by providing a generic construction of stream-based channels which directly bases on the well-established primitive of authenticated encryption with associated data and provides strong security in terms of confidentiality as well as integrity. Although it is rather illustrative than definitive, we remark that our construction is quite close to the TLS Record Protocol.

We define the generic construction of a stream-based channel $\mathsf{Ch}_{\mathsf{AEAD}} = (\mathsf{Init}, \mathsf{Send}, \mathsf{Recv})$ based on an authenticated encryption with associated data (AEAD) scheme $\mathsf{AEAD} = (\mathsf{Enc}, \mathsf{Dec})$ with key space $\mathcal{K}$ and distinguished error symbol $\perp$ as introduced by Rogaway [32].[3] The encryption algorithm $\mathsf{Enc}\colon \mathcal{K} \times \{0,1\}^* \times \{0,1\}^* \to \{0,1\}^*$ on input a key, an associated data string, and a message, outputs a ciphertext. The decryption algorithm $\mathsf{Dec}\colon \mathcal{K} \times \{0,1\}^* \times \{0,1\}^* \to (\{0,1\}^* \cup \{\perp\})$ on input a key, an associated data string, and a ciphertext, outputs either a message or the distinguished error symbol. We assume that the AEAD scheme allows the encryption of variable-length messages of up to il bits and that the ciphertext output for such messages has length at most $2^{\mathsf{ol}} - 1$ bits. This enables us to encode the length of ciphertexts with a fixed-size string of ol bits.

Our channel construction $\mathsf{Ch}_{\mathsf{AEAD}}$ is displayed in Fig. 4 and has sending state space $\mathcal{S}_S = \mathcal{K} \times \mathbb{N} \times \{0,1\}^*$, receiving state space $\mathcal{S}_R = \mathcal{K} \times \mathbb{N} \times \{0,1\}^* \times \{0,1\}$, and error space $\mathcal{E} = \{\perp\}$. The channel works as follows.

- The Init algorithm first draws uniformly at random a key K for the AEAD scheme. It then initializes the sending and receiving state respectively as tuples containing key K, a sequence number set to 0, and a message-fragment resp. ciphertext-fragment buffer initially empty; the receiving state also contains a failure flag, initially set to 0.
- The Send algorithm keeps on buffering input message strings until it has collected at least il bits. If sufficiently many bits have been collected, then Send encrypts message chunks m' of length il bits using the AEAD scheme on input message m' and associated data a running sequence number seqno.[4] The ciphertext generated is then prepended with the binary encoding of its size

---

[3] Although our construction does not incorporate nonces it can easily be extended to the nonce-based setting as originally defined by Rogaway [32].

[4] A more natural construction in the nonce-based setting would use seqno as the encryption nonce and have empty associated data input. We have chosen the current construction because of its closeness to TLS, which treats its sequence number as associated data.

(with the fixed number of ol bits) and the result appended to the ciphertext string c to be output. Note that the size encoding is not authenticated. In case the Send algorithm was called with the flush flag set to 1, in a final step it also encrypts any remaining buffered message in the same way, in order to empty the message buffer (this message will potentially be of length smaller than il).

– The Recv algorithm outputs an error (without any further state modification) once a first error has emerged from the AEAD decryption algorithm in some previous call; otherwise, it appends the incoming ciphertext fragment to its buffer. In case enough bits to parse the length field of ol bits were received it does so. Next, it checks whether the buffer contains the complete AEAD ciphertext of the indicated length and, if so, strips it from the buffer, decrypts it (incrementing the sequence number used in the associated data), and appends the result to the message to be output. This process is repeated until there is no completely parsable ciphertext left. However, in case the AEAD decryption algorithms outputs an error, after appending this error symbol to the output message, the Recv algorithm sets the failure flag fail to 1 and stops parsing further input.

Correctness of Ch_{AEAD} follows from the correctness of the AEAD scheme.

Security Analysis. Our generic stream-based channel construction Ch_{AEAD} from Fig. 4 provides indistinguishability under chosen plaintext-fragment attacks (IND − CPFA), integrity of ciphertext streams (INT − CST), and error predictability (ERR − PRE), given that the underlying authenticated encryption with associated data scheme AEAD provides indistinguishability under chosen plaintext attacks (IND − CPA) and authenticity (AUTH) as defined by Rogaway [32].[5] Using Theorem 1 we can moreover infer that it also provides indistinguishability under chosen ciphertext-fragment attacks (IND − CCFA). We provide the detailed security analysis in the full version of this paper.

5.1   A Note on the TLS Record Protocol

As discussed earlier, the Transport Layer Security (TLS) Record Protocol implements a stream-based channel whose complete analysis as such lies outside of the scope of this work. However we do pause to note that our construction of a stream-based channel based on authenticated encryption with associated data is actually very close to the TLS Record Protocol when using an AEAD scheme as specified for TLS version 1.2 [15, Section 6.2.3.3] and in the current draft for TLS version 1.3 [16, Section 6.2.2]: the Record Protocol also incorporates a sequence number which is authenticated but not sent on the wire and a length field which is sent and authenticated in TLS 1.2 (and which is sent but *not* authenticated

---

[5] Note that Rogaway [32] actually defines the stronger IND$- CPA notion which implies IND − CPA security based on a standard left-or-right encryption oracle. We only require IND − CPA though as it is sufficient for our security proof.

```
Init(1^λ):                              Recv(st_R, c):
 1  K ←$ K                               1  parse st_R as (K, seqno, buf, fail)
 2  st_{S,0} = (K, 0, ε)                 2  if fail = 1 then
 3  st_{R,0} = (K, 0, ε, 0)              3    return (st_R, ⊥)
 4  return (st_{S,0}, st_{R,0})          4  buf ← buf||c
                                         5  m ← ε
                                         6  while |buf| ≥ ol do
Send(st_S, m, f):                        7    parse buf[1, ..., ol] as integer ℓ
 1  parse st_S as (K, seqno, buf)        8    if |buf| ≥ ol + ℓ then
 2  buf ← buf||m                         9      len ← buf[1, ..., ol]
 3  c ← ε                               10      c' ← buf[ol + 1, ..., ol + ℓ]
 4  while |buf| ≥ il do                 11      buf ← buf % len||c'
 5    m' ← buf[1, ..., il]              12      m' ← Dec_K(seqno, c')
 6    buf ← buf % m'                     13      seqno ← seqno + 1
 7    c' ← Enc_K(seqno, m')              14      m ← m||m'
 8    seqno ← seqno + 1                  15      if m' = ⊥ then
 9    c ← c || |c'| || c' for |c'| ∈ {0,1}^ol   16        fail ← 1
10  if f = 1 and buf ≠ ε then           17        break
11    c' ← Enc_K(seqno, buf)            18    else
12    seqno ← seqno + 1                 19      break
13    c ← c || |c'| || c' for |c'| ∈ {0,1}^ol   20  st_R ← (K, seqno, buf, fail)
14    buf ← ε                           21  return (st_R, m)
15  st_S ← (K, seqno, buf)
16  return (st_S, c)
```

Fig. 4. A generic construction of a stream-based channel $\mathsf{Ch_{AEAD}} = (\mathsf{Init}, \mathsf{Send}, \mathsf{Recv})$ from any authenticated encryption with associated data (AEAD) scheme $\mathsf{AEAD} = (\mathsf{Enc}, \mathsf{Dec})$ with key space $\mathcal{K}$ and distinguished error symbol $\perp$ which allows to encrypt variable-length messages of up to il bits and for which the ciphertext output has length at most $2^{ol} - 1$ bits.

in TLS 1.3).[6] However, the TLS Record Protocol additionally includes a 2-byte version number and a 1-byte content type; these are both sent and authenticated in the associated data. Moreover, the AEAD schemes used are considered to be nonce-based, though the exact nonce generation is left to be specified by the particular cipher suite in use.

The content type field in particular allows TLS to multiplex data streams for different purposes within a single connection stream, as TLS does for the Handshake Protocol, the Alert Protocol, the ChangeCipherSpec protocol, and the Application protocol. While our model does not capture multiplexing several message streams into one ciphertext stream, it can be augmented to do so. This brings additional complexity and is an avenue for future work.

---

[6] That is, our approach of using a length field which is sent on the wire but not part of the authenticated associated data conforms with the approach adopted in TLS 1.3.

6 Conclusion

In this work we approached the security of channels designed to (securely) convey a stream of data from one party to another, narrowing the gap between real-world transport layer security protocols (like TLS or SSH) and our theoretical understanding of them. For this purpose, we formalized the syntax of such stream-based channels, explored strong security notions, and demonstrated their feasibility by providing a natural and secure construction which closely mimics the operation of the TLS Record Protocol.

Our approach sheds a formal light on recent attacks, in particular concerning the use of HTTP over TLS, confirming a disjunction between applications' expectations on the one hand and the guarantees that secure streaming channels provide on the other. This highlights that there is a need for detailed specifications of APIs and security guarantees for such protocols.

Our work also raises new research questions. Naturally, exploring the exact relation between stream-based and atomic-message channels is an avenue that should be pursued, with the development of detailed relations between security notions in our work and those in [9] as a specific task. Considering established techniques, the open question remains whether the well-accepted concept of length-hiding encryption can be incorporated in the stream-based setting despite being intrinsically connected to atomic messages. It also seems worthwhile to extend our stream-based model to encompass channel protocol designs (such as TLS and QUIC) that allow multiplexing of several data streams within a single channel.

Acknowledgments. The authors thank the anonymous reviewers for their valuable comments. Marc Fischlin is supported by the Heisenberg grant Fi 940/3-2 of the German Research Foundation (DFG). Kenneth Paterson is supported by EPSRC Leadership Fellowship EP/H005455/1 and by EPSRC grant EP/M013472/1. This work has been co-funded by the DFG as part of projects P2 and S4 within the CRC 1119 CROSSING and by the EU COST Action IC 1306.

References

1. 3rd Generation Partnership Project (3GPP): GSM, UMTS, and LTE standards. http://www.3g.pp.org
2. Albrecht, M.R., Paterson, K.G., Watson, G.J.: Plaintext recovery attacks against SSH. In: 2009 IEEE Symposium on Security and Privacy, pp. 16–26. IEEE Computer Society Press, May 2009
3. Badertscher, C., Matt, C., Maurer, U., Rogaway, P., Tackmann, B.: Augmented secure channels and the goal of the TLS 1.3 record layer. Cryptology ePrint Archive, Report 2015/394 (2015). http://eprint.iacr.org/
4. Bellare, M., Boldyreva, A., Knudsen, L.R., Namprempre, C.: Online ciphers and the Hash-CBC construction. In: Kilian, J. (ed.) CRYPTO 2001. LNCS, vol. 2139, pp. 292–309. Springer, Heidelberg (2001)

5. Bellare, M., Kohno, T., Namprempre, C.: Breaking and provably repairing the SSH authenticated encryption scheme: A case study of the encode-then-encrypt-and-MAC paradigm. ACM Trans. Inf. Syst. Secur. **7**(2), 206–241 (2004)

6. Bellare, M., Namprempre, C.: Authenticated encryption: relations among notions and analysis of the generic composition paradigm. In: Okamoto, T. (ed.) ASI-ACRYPT 2000. LNCS, vol. 1976, pp. 531–545. Springer, Heidelberg (2000)

7. Bhargavan, K., Delignat-Lavaud, A., Fournet, C., Pironti, A., Strub, P.Y.: Triple handshakes and cookie cutters: breaking and fixing authentication over TLS. In: 2014 IEEE Symposium on Security and Privacy, pp. 98–113. IEEE Computer Society Press, May 2014

8. Bhargavan, K., Fournet, C., Kohlweiss, M., Pironti, A., Strub, P.Y.: Implementing TLS with verified cryptographic security. In: 2013 IEEE Symposium on Security and Privacy, pp. 445–459. IEEE Computer Society Press, May 2013

9. Boldyreva, A., Degabriele, J.P., Paterson, K.G., Stam, M.: Security of symmetric encryption in the presence of ciphertext fragmentation. In: Pointcheval, D., Johansson, T. (eds.) EUROCRYPT 2012. LNCS, vol. 7237, pp. 682–699. Springer, Heidelberg (2012)

10. Boldyreva, A., Degabriele, J.P., Paterson, K.G., Stam, M.: On symmetric encryption with distinguishable decryption failures. In: Moriai, S. (ed.) FSE 2013. LNCS, vol. 8424, pp. 367–390. Springer, Heidelberg (2014)

11. Boldyreva, A., Taesombut, N.: Online encryption schemes: new security notions and constructions. In: Okamoto, T. (ed.) CT-RSA 2004. LNCS, vol. 2964, pp. 1–14. Springer, Heidelberg (2004)

12. Canetti, R.: Universally composable security: A new paradigm for cryptographic protocols. Cryptology ePrint Archive, Report 2000/067 (2000). http://eprint.iacr.org/2000/067

13. Canetti, R., Krawczyk, H.: Analysis of key-exchange protocols and their use for building secure channels. In: Pfitzmann, B. (ed.) EUROCRYPT 2001. LNCS, vol. 2045, pp. 453–474. Springer, Heidelberg (2001)

14. Degabriele, J.P., Paterson, K.G.: On the (in)security of IPsec in MAC-then-encrypt configurations. In: Al-Shaer, E., Keromytis, A.D., Shmatikov, V. (eds.) ACM CCS 2010, pp. 493–504. ACM Press, October 2010

15. Dierks, T., Rescorla, E.: The Transport Layer Security (TLS) Protocol Version 1.2. RFC 5246 (Proposed Standard), August 2008. http://www.ietf.org/rfc/rfc5246.txt updated by RFCs 5746, 5878, 6176

16. Dierks, T., Rescorla, E.: The Transport Layer Security (TLS) Protocol Version 1.3. Internet-Draft (work in progress), January 2015. https://tools.ietf.org/id/draft-ietf-tls-tls13-04.txt (Expires: 7 July, 2015)

17. Fielding, R., Reschke, J.: Hypertext Transfer Protocol (HTTP/1.1): Message Syntax and Routing. RFC 7230 (Proposed Standard), June 2014. http://www.ietf.org/rfc/rfc7230.txt

18. Fouque, P.A., Joux, A., Martinet, G., Valette, F.: Authenticated on-line encryption. In: Matsui, M., Zuccherato, R.J. (eds.) SAC 2003. LNCS, vol. 3006, pp. 145–159. Springer, Heidelberg (2004)

19. Institute of Electrical and Electronics Engineers Inc: IEEE Standard 801.11: Wireless LAN Medium Access Control (MAC) and Physical Layer (PHY) Specifications. http://standards.ieee.org/about/get/802/802.11.html

20. Jager, T., Kohlar, F., Schäge, S., Schwenk, J.: On the security of TLS-DHE in the standard model. In: Safavi-Naini, R., Canetti, R. (eds.) CRYPTO 2012. LNCS, vol. 7417, pp. 273–293. Springer, Heidelberg (2012)

21. Joux, A., Martinet, G., Valette, F.: Blockwise-adaptive attackers. In: Yung, M. (ed.) CRYPTO 2002. LNCS, vol. 2442, pp. 17–30. Springer, Heidelberg (2002)

22. Kent, S., Seo, K.: Security Architecture for the Internet Protocol. RFC 4301 (Proposed Standard), December 2005. http://www.ietf.org/rfc/rfc4301.txt (updated by RFC 6040)

23. Kohno, T., Palacio, A., Black, J.: Building secure cryptographic transforms, or how to encrypt and MAC. Cryptology ePrint Archive, Report 2003/177 (2003). http://eprint.iacr.org/2003/177

24. Krawczyk, H., Paterson, K.G., Wee, H.: On the security of the TLS protocol: a systematic analysis. In: Canetti, R., Garay, J.A. (eds.) CRYPTO 2013, Part I. LNCS, vol. 8042, pp. 429–448. Springer, Heidelberg (2013)

25. Maurer, U., Tackmann, B.: On the soundness of authenticate-then-encrypt: formalizing the malleability of symmetric encryption. In: Al-Shaer, E., Keromytis, A.D., Shmatikov, V. (eds.) ACM CCS 2010. pp. 505–515. ACM Press, October 2010

26. Namprempre, C.: Secure channels based on authenticated encryption schemes: a simple characterization. In: Zheng, Y. (ed.) ASIACRYPT 2002. LNCS, vol. 2501, pp. 515–532. Springer, Heidelberg (2002)

27. Paterson, K.G., Ristenpart, T., Shrimpton, T.: Tag size *Does* matter: attacks and proofs for the TLS record protocol. In: Lee, D.H., Wang, X. (eds.) ASIACRYPT 2011. LNCS, vol. 7073, pp. 372–389. Springer, Heidelberg (2011)

28. Paterson, K.G., Watson, G.J.: Plaintext-dependent decryption: a formal security treatment of SSH-CTR. In: Gilbert, H. (ed.) EUROCRYPT 2010. LNCS, vol. 6110, pp. 345–361. Springer, Heidelberg (2010)

29. Postel, J.: User Datagram Protocol. RFC 768 (INTERNET STANDARD), August 1980. http://www.ietf.org/rfc/rfc768.txt

30. Postel, J.: Transmission Control Protocol. RFC 793 (INTERNET STANDARD), September 1981. http://www.ietf.org/rfc/rfc793.txt (updated by RFCs 1122, 3168, 6093, 6528)

31. Rescorla, E., Modadugu, N.: Datagram Transport Layer Security Version 1.2. RFC 6347 (Proposed Standard), January 2012. http://www.ietf.org/rfc/rfc6347.txt

32. Rogaway, P.: Authenticated-encryption with associated-data. In: Atluri, V. (ed.) ACM CCS 2002, pp. 98–107. ACM Press, November 2002

33. Roskind, J.: QUIC (Quick UDP Internet Connections): Multiplexed Stream Transport Over UDP, December 2013. https://docs.google.com/document/d/1RNHkx_VvKWyWg6Lr8SZ-saqsQx7rFV-ev2jRFUoVD34/ (retrieved on 23 Jan 2015)

34. Shoup, V.: On formal models for secure key exchange. Cryptology ePrint Archive, Report 1999/012 (1999). http://eprint.iacr.org/1999/012

35. Smyth, B., Pironti, A.: Truncating TLS connections to violate beliefs in web applications. In: WOOT 2013: 7th USENIX Workshop on Offensive Technologies. USENIX Association (2013) (first appeared at Black Hat USA 2013)

36. Ylonen, T., Lonvick, C.: The Secure Shell (SSH) Protocol Architecture. RFC 4251 (Proposed Standard), January 2006. http://www.ietf.org/rfc/rfc4251.txt

37. Ylonen, T., Lonvick, C.: The Secure Shell (SSH) Transport Layer Protocol. RFC 4253 (Proposed Standard), January 2006. http://www.ietf.org/rfc/rfc4253.txt, updated by RFC 6668

Bloom Filters in Adversarial Environments

Moni Naor$^{(\boxtimes)}$ and Eylon Yogev

Weizmann Institute of Science, Rehovot, Israel
{moni.naor,eylon.yogev}@weizmann.ac.il

Abstract. Many efficient data structures use randomness, allowing them to improve upon deterministic ones. Usually, their efficiency and/or correctness are analyzed using probabilistic tools under the assumption that the inputs and queries are *independent* of the internal randomness of the data structure. In this work, we consider data structures in a more robust model, which we call the *adversarial model*. Roughly speaking, this model allows an adversary to choose inputs and queries *adaptively* according to previous responses. Specifically, we consider a data structure known as "Bloom filter" and prove a tight connection between Bloom filters in this model and cryptography.

A Bloom filter represents a set S of elements approximately, by using fewer bits than a precise representation. The price for succinctness is allowing some errors: for any $x \in S$ it should always answer 'Yes', and for any $x \notin S$ it should answer 'Yes' only with small probability.

In the adversarial model, we consider both efficient adversaries (that run in polynomial time) and computationally unbounded adversaries that are only bounded in the amount of queries they can make. For computationally bounded adversaries, we show that non-trivial (memory-wise) Bloom filters exist if and only if one-way functions exist. For unbounded adversaries we show that there exists a Bloom filter for sets of size n and error ε, that is secure against t queries and uses only $O(n \log \frac{1}{\varepsilon} + t)$ bits of memory. In comparison, $n \log \frac{1}{\varepsilon}$ is the best possible under a non-adaptive adversary.

1 Introduction

Data structures are one of the most basic objects in Computer Science. They provide means to organize a large amount of data such that it can be queried efficiently. In general, constructing efficient data structures is key to designing efficient algorithms. Many efficient data structures use randomness, a resource that allows them to bypass lower bounds on deterministic ones. In these cases, their efficiency and/or correctness are analyzed in expectation or with high probability.

M. Naor—Incumbent of the Judith Kleeman Professorial Chair.

E. Yogev—Supported in part by a grant from the I-CORE Program of the Planning and Budgeting Committee, the Israel Science Foundation, BSF and the Israeli Ministry of Science and Technology.

R. Gennaro and M. Robshaw (Eds.): CRYPTO 2015, Part II, LNCS 9216, pp. 565–584, 2015.
DOI: 10.1007/978-3-662-48000-7_28

To analyze randomized data structures one must first define the underlying model of the analysis. Usually, the model assumes that the inputs and queries are *independent* of the internal randomness of the data structure. That is, the analysis is of the form: For any sequence of inputs, with high probability (or expectation) over its internal randomness, the data structure will yield a correct answer. This model is reasonable in a situation where the adversary picking the inputs gets no information about the randomness of the data structure (in particular, the adversary does not get the responses on previous inputs).

In this work, we consider data structures in a more robust model, which we call the *adversarial model*. Roughly speaking, this model allows an adversary to choose inputs and queries *adaptively* according to previous responses. That is, the analysis is of the form: With high probability over the internal randomness of the data structure, for any adversary adaptively choosing a sequence of inputs, the output of the data structure will be correct. Specifically, we consider a data structure known as "Bloom filter" and prove a tight connection between Bloom filters in this model and cryptography: We show that Bloom filters in an adversarial model exist if and only if one-way functions exist.

Bloom Filters in Adversarial Environments. The approximate set membership problem deals with succinct representations of a set S of elements from a large universe U, where the price for succinctness is allowing some errors. A data structure solving this problem is required to answer queries in the following manner: for any $x \in S$ it should always answer 'Yes', and for any $x \notin S$ it should answer 'Yes' only with small probability. The latter are called *false positive* errors.

The study of the approximate set membership problem began with Bloom's 1970 paper [4], introducing the so called "Bloom filter", which provided a simple and elegant solution to the problem. (The term "Bloom filter" may refer to Bloom's original construction, but we use it to denote any construction solving the problem.) The two major advantages of Bloom filters are: (i) they use significantly less memory (as opposed to storing S precisely) and (ii) they have very fast query time (even constant query time). Over the years, Bloom filters have been found to be extremely useful and practical in various areas. Some main examples are distributed systems [32], networking [10], databases [19], spam filtering [30], web caching [13], streaming algorithms [9,21] and security [17,31]. For a survey about Bloom filters and their applications see [6] and a more recent one [28].

Following Bloom's original construction many generalizations and variants have been proposed and extensively analyzed, proving better memory consumption and running time, see e.g. [1,8,24,27]. However, as discussed, all known constructions of Bloom filters work under the assumption that the input query x is fixed, and then the probability of an error occurs over the randomness of the construction. Consider the case where the query results are made public. What happens if an adversary chooses the next query according to the responses of previous ones? Does the bound on the error probability still hold? The traditional analysis of Bloom filters is no longer sufficient, and stronger techniques are required.

Let us demonstrate this need with a concrete scenario. Consider a system where a Bloom filter representing a *white list* of email addresses is used to filter spam mail. When an email message is received, the sender's address is checked against the Bloom filter, and if the result is negative it is marked as spam. Addresses not on the white list have only a small probability of being a false positive and thus not marked as spam. In this case, the results of the queries are public, as an attacker might check whether his emails are marked as spam[1]. The attacker (after a sequence of queries) might be able to find a bulk of email addresses that are not marked as spam although they are not in the white list, and thus, bypass the security of the system and flood users with spam mail.

Alternatively, Bloom filters are often used for holding the contents of a cache. For instance, a web proxy holds on a (slow) disk, a cache of locally available webpages. To improve performance, it maintains in (fast) memory a Bloom filter representing all addresses in the cache. When a user queries for a webpage, the proxy first checks the Bloom filter to see if the page is available in the cache, and only then does it search for the webpage on the disk. A false positive is translated to a cache miss, that is, an unnecessary (slow) disk lookup. In the standard analysis, one would set the error to be small such that cache misses happen very rarely (e.g., one in a thousand requests). However, by timing the results of the proxy, an adversary might learn the responses of the Bloom filter, enabling her to cause a cache miss for almost every query and, eventually, causing a Denial of Service (DoS) attack.

Under the adversarial model, we construct Bloom filters that are resilient to the above attacks. We consider both efficient adversaries (that run in polynomial time) and computationally unbounded adversaries that are only bounded in the amount of queries they can make. We define a Bloom filter that maintains its error probability in this setting and say it is *adversarial resilient* (or just resilient for shorthand).

The security of an adversarial resilient Bloom filter is defined as a game with an adversary. The adversary is allowed to make a sequence of t adaptive queries to the Bloom filter and get their responses. Note that the adversary has only oracle access to the Bloom filter and cannot see its internal memory representation. Finally, the adversary must output an element x^* (that was not queried before) which she believes is a false positive. We say that a Bloom filter is (n, t, ε)-adversarial resilient if when initialized over sets of size n then after t queries the probability of x^* being a false positive is at most ε. If a Bloom filter is resilient for any polynomially many queries we say it is *strongly resilient*.

A simple construction of a strongly resilient Bloom filter (even against computationally unbounded adversaries) can be achieved by storing S precisely. Then, there are no false positives at all and no adversary can find one. The drawback of this solution is that it requires a large amount of memory, whereas Bloom filters aim to reduce the memory usage. We are interested in Bloom filters that use a small amount of memory but remain nevertheless, resilient.

---

[1] For example, the attacker can spam his personal email account and see if the messages are being filtered.

1.1 Our Results

We introduce the notion of *adversarial-resilient Bloom filter* and show several possibility results (constructions of resilient Bloom filters) and impossibility results (attacks against any Bloom filter) in this context.

Our first result is that adversarial-resilient Bloom filters against computationally bounded adversaries that are non-trivial (i.e., they require less space than the amount of space it takes to store the elements explicitly) must use one-way functions. That is, we show that if one-way functions do not exist then any Bloom filter can be 'attacked' with high probability.

Theorem 1 (Informal). *Let* **B** *be a* non-trivial *Bloom filter. If* **B** *is strongly resilient against computationally bounded adversaries then one-way functions exist.*

Actually, we show a trade-off between the amount of memory used by the Bloom filter and the number of queries performed by the adversary. Carter et al. [7] proved a lower bound on the amount of memory required by a Bloom filter. To construct a Bloom filter for sets of size n and error rate ε one must use (roughly) $n \log \frac{1}{\varepsilon}$ bits of memory (and this is tight). Given a Bloom filter that uses m bits of memory we get a lower bound for its error rate ε and thus a lower bound for the (expected) number of false positives. As m is smaller the number of false positives is larger and we prove that adversary can perform fewer queries.

In the other direction, we show that using one-way functions one can construct a strongly resilient Bloom filter. Actually, we show that you can transform any Bloom filter to be strongly resilient with almost exactly the same memory requirements and at a cost of a single evaluation of a pseudorandom permutation (which can be constructed using one-way functions). Specifically, we show:

Theorem 2. *Let* **B** *be an* (n, ε)-*Bloom filter using* m *bits of memory. If pseudorandom permutations exist, then for large enough security parameter* λ *there exists an* $(n, \varepsilon + \mathsf{neg}(\lambda))$-*strongly resilient Bloom filter that uses* $m' = m + \lambda$ *bits of memory.*

Bloom filters consist of two algorithms: an initialization algorithm that gets a set and outputs a compressed representation of the set, and a membership query algorithm that gets a representation and an input. Usually, Bloom filters have a *randomized* initialization algorithm but a *deterministic* query algorithm that does not change the representation. We say that such Bloom filters have a "steady representation". We consider also Bloom filters with "unsteady representation" where the query algorithm is randomized and can change the underlying representation on each query. A randomized query algorithm may be more sophisticated and, for example, incorporate differentially private [12] algorithms in order to protect the internal memory from leaking. Differentially private algorithms are designed to protect a private database against adversarial and also adaptive queries from a data analyst. One might hope that such techniques can eliminate the need of one-way functions in order to construct resilient Bloom filters. However, we extend our results and show that they hold even for Bloom filter with unsteady representations, which proves that this approach cannot gain additional security.

In the context of unbounded adversaries, we show a positive result. For a set of size n and an error probability of ε most constructions use about $O(n \log \frac{1}{\varepsilon})$ bits of memory. We construct a resilient Bloom filter that does not use one-way functions, is resilient against t queries, uses $O(n \log \frac{1}{\varepsilon} + t)$ bits of memory, and has query time $O(\log \frac{1}{\varepsilon})$.

Theorem 3. *For any $n, t \in \mathbb{N}$, and $\varepsilon > 0$ there exists an (n, t, ε)-resilient Bloom filter (against unbounded adversaries) that uses $O(n \log \frac{1}{\varepsilon} + t)$ bits of memory.*

1.2   Related Work

One of the first works to consider an adaptive adversary that chooses queries based on the response of the data structure is by Lipton and Naughton [16], where adversaries that can measure the *time* of specific operations in a dictionary were addressed. They showed how such adversaries can be used to attack hash tables. Hash tables have some method for dealing with collisions. An adversary that can measure the time of an insert query, can determine whether there was a collision and might figure out the precise hash function used. She can then choose the next elements to insert accordingly, increasing the probability of a collision and hurting the overall performance.

Mironov et al. [18] considered the model of *sketching* in an adversarial environment. The model consists of several honest parties that are interested in computing a joint function in the presence of an adversary. The adversary chooses the inputs of the honest parties based on the common randomness shared among them. These inputs are provided to the parties in an on-line manner, and each party incrementally updates a compressed sketch of its input. The parties are not allowed to communicate, they do not share any secret information, and any public information they share is known to the adversary in advance. Then, the parties engage in a protocol in order to evaluate the function on their current inputs using only the compressed sketches. Mironov et al. construct explicit and efficient (optimal) protocols for two fundamental problems: testing equality of two data sets, and approximating the size of their symmetric difference.

In a more recent work, Hardt and Woodruff [14] considered linear sketch algorithms in a similar setting. They consider an adversary that can adaptively choose the inputs according to previous evaluations of the sketch. They ask whether linear sketches can be robust to adaptively chosen inputs. Their results are negative: They show that no linear sketch approximates the Euclidean norm of its input to within an arbitrary multiplicative approximation factor on a polynomial number of adaptively chosen inputs.

One may consider adversarial resilient Bloom filters in the framework of computational learning theory. The task of the adversary is to learn the private memory of the Bloom filter in the sense that it is able to predict on which elements the Bloom filter outputs a false positive. The connection between learning and cryptographic assumptions has been explored before (already in his 1984 paper introducing the PAC model Valiant's observed that the nascent pseudorandom random functions imply hardness of learning [29]). In particular Blum et al. [5] showed how to construct several cryptographic primitives (pseudorandom bit generators, one-way functions and private-key cryptosystems) based on certain assumptions on the difficulty of learning. The necessity of one-way functions for several cryptographic primitives has been shown in [15].

2   Model and Problem Definitions

Our model considers a universe U of elements, and a subset $S \subset U$. We denote the size of U by u, and the size of S by n. For the security parameter we use λ (sometimes we omit the explicit use of the security parameter and assume it is polynomial in n). We consider mostly the static problem, where the set is fixed throughout the lifetime of the data structure. We note that the lower bounds imply the same bounds for the dynamic case and the cryptographic upper bound (Theorem 4) can be adapted to the dynamic case.

A Bloom filter is a data structure that is composed of a setup algorithm and a query algorithm $\mathbf{B} = (\mathbf{B}_1, \mathbf{B}_2)$. The setup algorithm $\mathbf{B}_1$ is randomized, gets as input a set S, and outputs a compressed representation of it $\mathbf{B}_1(S) = M$. To denote the representation M on a set S with random string r we write $\mathbf{B}_1(S; r) = M_r^S$ and its size in bits is denoted as $|M_r^S|$.

The query algorithm answers membership queries to S given the compressed representation M. Usually in the literature, the query algorithm is deterministic and cannot change the representation. In this case we say $\mathbf{B}$ has a *steady representation*. However, we also consider Bloom filters where their query algorithm is *randomized* and can change the representation M after each query. In this case we say that $\mathbf{B}$ has an *unsteady representation*. We define both variants.

Definition 1 (Steady-representation Bloom filter). *Let $\mathbf{B} = (\mathbf{B}_1, \mathbf{B}_2)$ be a pair of polynomial-time algorithms where $\mathbf{B}_1$ is a randomized algorithm that gets as input a set S and outputs a representation, and $\mathbf{B}_2$ is a deterministic algorithm that gets as input a representation and a query element $x \in U$. We say that $\mathbf{B}$ is an (n, ε)-Bloom filter (with a steady representation) if for any set $S \subset U$ of size n it holds that:*

1. *Completeness: For any $x \in S$: $\Pr[\mathbf{B}_2(\mathbf{B}_1(S), x) = 1] = 1$*
2. *Soundness: For any $x \notin S$: $\Pr[\mathbf{B}_2(\mathbf{B}_1(S), x) = 1] \le \varepsilon$,*

where the probabilities are over the setup algorithm $\mathbf{B}_1$.

False Positive and Error Rate. Given a representation M of S, if $x \notin S$ and $\mathbf{B}_2(M, x) = 1$ we say that x is a *false positive*. Moreover, we say that ε is the *error rate* of $\mathbf{B}$.

Definition 1 considers only a single fixed input x and the probability is taken over the randomness of $\mathbf{B}$. We want to give a stronger soundness requirement that considers a sequence of inputs $x_1, x_2, \ldots, x_t$ that is not fixed but chosen by an adversary, where the adversary gets the responses of previous queries and can adaptively choose the next query accordingly. If the adversary's probability of finding a false positive x^* that was not queried before is bounded by ε, then we say that $\mathbf{B}$ is an (n, t, ε)-resilient Bloom filter (this notion is defined in the challenge $\mathsf{Challenge}_{A,t}$ which is described below). Note that in this case, the setup phase of the Bloom filter and the adversary get the security parameter 1^λ as an additional input (however, we usually omit it when clear from context). For a steady representation Bloom filter we define:

Definition 2 (Adversarial-resilient Bloom filter with a steady representation). *Let $\mathbf{B} = (\mathbf{B}_1, \mathbf{B}_2)$ be an (n, ε)-Bloom filter with a steady representation (see Definition 1). We say that $\mathbf{B}$ is an (n, t, ε)-adversarial resilient Bloom filter (with a steady representation) if for any set S of size n, for all sufficiently large $\lambda \in \mathbb{N}$ and for any probabilistic polynomial-time adversary A we have that the advantage of A in the following challenge is at most ε:*

1. Adversarial Resilient: $\Pr[\mathsf{Challenge}_{A,t}(\lambda) = 1] \leq \varepsilon,$

where the probabilities are taken over the internal randomness of $\mathbf{B}_1$ and A and where the random variable $\mathsf{Challenge}_{A,t}(\lambda)$ is the outcome of the following game:

$\mathsf{Challenge}_{A,t}(\lambda)$:

1. $M \leftarrow \mathbf{B}_1(S, 1^\lambda)$.
2. $x^* \leftarrow A^{\mathbf{B}_2(M, \cdot)}(1^\lambda, S)$ where A performs at most t queries $x_1, \ldots, x_t$ to the query oracle $\mathbf{B}_2(M, \cdot)$.
3. If $x^* \notin S \cup \{x_1, \ldots, x_t\}$ and $\mathbf{B}_2(M, x^*) = 1$ output 1, otherwise output 0.

Unsteady representations. When the Bloom filter has an unsteady representation, then the algorithm $\mathbf{B}_2$ is randomized and moreover can change the representation M. That is, $\mathbf{B}_2$ is a query algorithm that outputs the response to the query as well as a new representation. Thus, the user or the adversary do not interact directly with the $\mathbf{B}_2(M, \cdot)$ but with an interface $Q(\cdot)$ (initialized with M) to a process that on query x updates its representation M and outputs only the response to the query (i.e. it cannot issue successive queries to the same memory representation but to one that keeps changing). Formally, $Q(\cdot)$ initialized with M on input x acts as follows:

The interface $Q(x)$ (initialized with M):

1. $(M', y) \leftarrow \mathbf{B}_2(M, x)$.
2. $M \leftarrow M'$.
3. Output y.

We define an analogue of the original Bloom filter for unsteady representations and then define an adversarial resilient one.

Definition 3 (Bloom filter with an unsteady representation). *Let $S \subset U$ be a set of size n. Let $\mathbf{B} = (\mathbf{B}_1, \mathbf{B}_2)$ be a pair of probabilistic polynomial-time algorithms such that $\mathbf{B}_1$ gets as input the set S and outputs a representation M_0, and $\mathbf{B}_2$ gets as input a representation and query x and outputs a new representation and a response to the query. Let $Q(\cdot)$ be the process initialized with M_0. We say that $\mathbf{B}$ is an (n, ε)-Bloom filter (with an unsteady representation) if for any such set S the following two conditions hold:*

1. *Completeness: After any sequence of queries $x_1, x_2, \ldots$ performed to $Q(\cdot)$ we have that for any $x \in S$: $\Pr[Q(x) = 1] = 1$.*
2. *Soundness: After any sequence of queries $x_1, x_2, \ldots$ performed to $Q(\cdot)$ we have that for any $x \notin S$: $\Pr[Q(x) = 1] \leq \varepsilon$,*

where the probabilities are taken over the internal randomness of $\mathbf{B}_1$ and $\mathbf{B}_2$.

Definition 4 (Adversarial-resilient Bloom filter with an unsteady representation). *Let $\mathbf{B} = (\mathbf{B}_1, \mathbf{B}_2)$ be an (n, ε)-Bloom filter with an unsteady representation (see Definition 3). We say that $\mathbf{B}$ is an (n, t, ε)-adversarial resilient Bloom filter (with an unsteady representation) if for any set $S \subset U$ of size n, for all sufficiently large $\lambda \in \mathbb{N}$ and for any probabilistic polynomial-time adversary A it holds that:*

1. *Adversarial Resilient: $\Pr[\mathsf{Challenge}_{A,t}(\lambda) = 1] \leq \varepsilon$,*

where the probabilities are taken over the internal randomness of $\mathbf{B}_1, \mathbf{B}_2$ and A and where the random variable $\mathsf{Challenge}_{A,t}(\lambda)$ is the outcome of the following process:

$\underline{\mathsf{Challenge}_{A,t}(\lambda):}$

1. *$M_0 \leftarrow \mathbf{B}_1(S, 1^\lambda)$.*
2. *Initialize $Q(\cdot)$ with M_0.*
3. *$x^* \leftarrow A^{Q(\cdot)}(1^\lambda, S)$ where A performs at most t (adaptive) queries $x_1, \ldots, x_t$ to the interface $Q(\cdot)$.*
4. *If $x^* \notin S \cup \{x_1, \ldots, x_t\}$ and $Q(x^*) = 1$ output 1, otherwise output 0.*

If $\mathbf{B}$ is not (n, t, ε)-resilient then we say there exists an adversary A that can (n, t, ε)-attack $\mathbf{B}$.

If $\mathbf{B}$ is resilient for any polynomial number of queries we say it is *strongly resilient*.

Definition 5 (Strongly resilient). *We say that $\mathbf{B}$ is an (n, ε)-strongly resilient Bloom filter, if for large enough security parameter λ and any polynomial $t = t(\lambda)$ we have that $\mathbf{B}$ is an (n, t, ε)-adversarial resilient Bloom filter.*

Remark 1. Notice that in Definitions 2 and 4 the adversary gets the set S as an additional input. This strengthens the definition of the resilient Bloom filter such that even given the set S it is hard to find false positives. An alternative definition might be to not give the adversary the set and also not require that

$x^* \notin S$. However, our results of Theorem 1 hold even if the adversary does not get the set. That is, the algorithm that predicts a false positive makes no use of the set S, either then checking that $x^* \notin S$. Moreover, the construction in Theorem 2 holds in both cases, even against adversaries that do get the set.

An important parameter is the memory use of a Bloom filter $\mathbf{B}$. We say $\mathbf{B}$ uses $m = m(n, \lambda, \varepsilon)$ bits of memory if for any set S of size n the largest representation is of size at most m. The desired properties of Bloom filters is to have m as small as possible and to answer membership queries as fast as possible. Let $\mathbf{B}$ be a (n, ε)-Bloom filter that uses m bits of memory. Carter et al. [7] proved a lower bound on the memory use of any Bloom filter showing that $m \geq n \log \frac{1}{\varepsilon}$ (or written equivalently as $\varepsilon \geq 2^{-\frac{m}{n}}$). This leads us to defining the minimal error of $\mathbf{B}$.

Definition 6 (Minimal error). *Let $\mathbf{B}$ be an (n, ε)-Bloom filter that uses m bits of memory. We say that $\varepsilon_0 = 2^{-\frac{m}{n}}$ is the minimal error of $\mathbf{B}$.*

Note that using Carter's lower bound we get that for any (n, ε)-Bloom filter its minimal error ε_0 always satisfies $\varepsilon_0 \leq \varepsilon$. Also, a trivial Bloom filter can always store the set S precisely using $m = \log \binom{u}{n} \approx n \log \left(\frac{u}{n}\right)$ bits. Using the $m \geq n \log \frac{1}{\varepsilon}$ lower bound we get that a Bloom filter is trivial if $\varepsilon > \frac{n}{u}$. Moreover, if u is super-polynomial in n, and ε is negligible in n then any polynomial-time adversary has only negligible chance in finding any false positive, and again we say that the Bloom filter is trivial.

Definition 7 (Non-trivial Bloom filter). *Let $\mathbf{B}$ be an (n, ε)-Bloom filter that uses m bits of memory and let ε_0 be the minimal error of $\mathbf{B}$ (see Definition 6). We say that $\mathbf{B}$ is non-trivial if there exists a constant $c \geq 1$ such that $\varepsilon_0 > \max \left\{ \frac{n}{u}, \frac{1}{n^c} \right\}$.*

3 Our Techniques

3.1 One-Way Functions and Adversarial Resilient Bloom Filters

We present the main ideas and techniques of the equivalence of adversarial resilient Bloom filters and one-way functions (i.e., the proof of Theorems 1 and 2). The simpler direction is showing that the existence of one-way functions implies the existence of adversarial resilient Bloom filters. Actually, we show that any Bloom filter can be efficiently transformed to be adversarial resilient with essentially the same amount of memory. The idea is simple and works in general for other data structures as well: apply a pseudo-random permutation of the input and then send it to the original Bloom filter. The point is that an adversary has almost no advantage in choosing the inputs adaptively, as they are all randomized by the permutation, while the correctness properties remain under the permutation.

The other direction is more challenging. We show that if one-way functions do not exist then any non-trivial Bloom filter can be 'attacked' by an efficient adversary. That is, the adversary performs a sequence of queries and then outputs an element x^* (that was not queried before) which is a false positive with high probability. We give two proofs: One for the case where the Bloom filter has a steady representation and one for an unsteady representation.

The main idea is that although we are given only oracle access to the Bloom filter, we are able to construct an (approximate) simulation of it. We use techniques from machine learning to (efficiently) 'learn' the internal memory of the Bloom filter, and construct the simulation. The learning task for steady and unsteady Bloom filters is quite different and each yield a simulation with different guarantees. Then we show how to exploit each simulation to find false positives without querying the real Bloom filter.

In the steady case, we state the learning process as a 'PAC learning' [29] problem. We use what's known as 'Occam's Razor' which states that any hypothesis consistent on a large enough random training set will have a small error. Finally, we show that since we assume that one-way functions do not exist then we are able to find a consistent hypothesis in polynomial-time. Since the error is small, the set of false positive elements defined by the real Bloom filter is approximately the same set of false positive elements defined by the simulator.

Handling Bloom filters with an unsteady representation is more challenging. Recall that such Bloom filters are allowed to randomly change their internal representation after each query. In this case, we are trying to learn a distribution that might change after each sample. We describe two examples of Bloom filters with unsteady representations which seem to capture the main difficulties of the unsteady case.

The first example considers any ordinary Bloom filter with error rate $\varepsilon/2$, where we modify the query algorithm to first answer 'Yes' with probability $\varepsilon/2$ and otherwise continue with its original behavior. The resulting Bloom filter has an error rate of ε. However, its behaviour is tricky: When observing its responses, elements can alternate between being false positive and negatives, which makes the learning task much harder.

The second example consists of two ordinary Bloom filters with error rate ε, both initialized with the set S. At the beginning only the first Bloom filter is used, and after a number of queries (which may be chosen randomly) only the second one is used. Thus, when switching to the second Bloom filter the set of false positives changes completely. Notice that while first Bloom filter was used exclusively, no information was leaked about the second. This example proves that any algorithm trying to 'learn' the memory of the Bloom filter cannot perform a fixed number of samples (as does our learning algorithm for the steady representation case).

To handle these examples we apply the framework of adaptively changing distributions (ACDs) presented by Naor and Rothblum [20], which models the task of learning distributions that can adaptively change after each sample was studied. Their main result is that if one-way functions do not exist then there exists an efficient learning algorithm that can approximate the next activation of the ACD, that is, produce a distribution that is statistically close to the distribution of the next activation of the ACD. We show how to facilitate (a slightly modified version of) this algorithm to learn the unsteady Bloom filter and construct a simulation. One of the main difficulties is that since we get only a statistical distance guarantee, then a false positive for the simulation need not be a false positive for the real Bloom filter. Nevertheless, we show how to estimate whether an element is a false positive in the real Bloom filter.

3.2 Computationally Unbounded Adversaries

In Theorem 3 we construct a Bloom Filter that is resilient against any unbounded adversary for a given number (t) of queries. One immediate solution would be to imitate the construction of the computationally bounded case while replacing the pseudo-random permutation with a $k = (t + n)$-wise independent hash function. Then, any set of t queries along with the n elements of the set would behave as truly random under the hash function. The problem with this approach is that the representation of the hash function is too large: It is $O(k \log |U|)$ which is more than the number of bits needed for a precise representation of the set S. Turning to almost k-wise independence does not help either. First, the memory will still be too large (it can be reduced to $O(n \log n \log \frac{1}{\varepsilon} + t \log n \log \frac{1}{\varepsilon})$ bits) and second, almost k-wise guarantees works only for sets chosen in advance, where the point of a resilient Bloom filter is to handle adaptively chosen sets.

Carter et al. [7] presented a general transformation from any exact dictionary to a Bloom filter. The idea was simple: storing x in the Bloom filter translates to storing $g(x)$ in a dictionary for some (universal) hash function $g : U \to V$, where $|V| = \frac{n}{\varepsilon}$. The choice of the hash function and underlying dictionary are important as they determine the performance and memory size of the Bloom filter. Notice that, at this point replacing g with a $k = (t + n)$-wise independent hash function (or an almost k-independent hash function) yields the same problems discussed above. Nevertheless, this is our starting point where the final construction is quite different. Specifically, we combine two main ingredients: Cuckoo hashing and a highly independent hash function tailored for this construction.

For the underlying dictionary in the transformation we use the Cuckoo hashing construction [25,26]. Using cuckoo hashing as the underlying dictionary was already shown to yield good constructions for Bloom filters by Pagh et al. [24] and Arbitman et al. [1]. Among the many advantages of Cuckoo hashing (e.g., succinct memory representation, constant lookup time) is the simplicity of its structure. It consists of two tables T_1 and T_2 and two hash functions h_1 and h_2 and each element x in the Cuckoo dictionary resides in either $T_1[h_1(x)]$ or $T_2[h_2(x)]$. However, we use this structure a bit differently. Instead of storing $g(x)$ in the dictionary directly (as the reduction of Carter et al. suggests) which would resolve to storing $g(x)$ at either $T_1[h_1(g(x))]$ or $T_2[h_2(g(x))]$ we store $g(x)$ at either $T_1[h_1(x)]$ or $T_2[h_2(x)]$. That is, we use the full description of x to decide where x is stored but eventually store only a hash of x (namely, $g(x)$). Since each element is compared only with two cells, this lets us improve the analysis of the reduction which reduce the size of V to $O\left(\frac{1}{\varepsilon}\right)$ (instead of $\frac{n}{\varepsilon}$).

To initialize the hash function g, instead of using a universal hash function we use a very high independence function (which in turn is also constructed based on cuckoo hashing) based on the work of Pagh and Pagh [23] and Dietzfelbinger and Woelfel [11]. They show how to construct a family G of hash functions such that on any given set of k inputs it behaves like a truly random function with high probability. Furthermore, a function in G can be evaluated in constant time (in the RAM model), and its description can be stored using roughly $O(k \log |V|)$ bits (where V is the range of the function).

Note that the guarantee of the function acting random holds only for sets S of size k that are *chosen in advance*. In our case the set is not chosen in advance but rather chosen adaptively and adversarially. However, Berman et al. [3] showed that the same construction of Pagh and Pagh actually holds *even when the set of queries is chosen adaptively*.

At this point, one solution would be to use the family of functions G setting $k = t + n$, with the analysis of Berman et al. as the hash function g and the structure of the Cuckoo hashing dictionary. To get an error of ε, we set $|V| = O\left(\log \frac{1}{\varepsilon}\right)$ and get an adversarial resilient Bloom filter that is resilient for t queries and uses $O\left(n \log \frac{1}{\varepsilon} + t \log \frac{1}{\varepsilon}\right)$ bits of memory. However, our goal is to get a memory size of $O\left(n \log \frac{1}{\varepsilon} + t\right)$.

To reduce the memory of the Bloom filter even further, we use the family G a bit differently. Let $\ell = O\left(\log \frac{1}{\varepsilon}\right)$, and set $k = O\left(t/\ell\right)$. We define the function g to be a concatenation of ℓ independent instances g_i of functions from G, each outputting a single bit ($V = \{0, 1\}$). Using the analysis of Berman et al. we get that each of them behaves like a truly random function for any sequence of k adaptively chosen elements. Consider an adversary performing t queries. To see how this composition of hash functions helps reduce the independence needed, consider the comparisons performed in a query between $g(x)$ and some value y being performed bit by bit. Only if the first pair of bits are equal we continue to compare the next pair. The next query continues from the last pair compared, in a cyclic order. For any set of k elements, the probability of the two bits to be equal is $1/2$. Thus, with high probability, only a constant number of bits will be compared during a single query. That is, in each query only a constant number of functions g_i will be involved and "pay" in their independence, where the rest remain untouched. Altogether, we get that although there are t queries performed, we have ℓ different functions and each function g_i is involved in at most $O(t/\ell) = k$ queries (with high probability). Thus, the view of each function remains random on these elements. This results in an adversarial resilient Bloom filter that is resilient for t queries and uses only $O(n \log \frac{1}{\varepsilon} + k \log \frac{1}{\varepsilon}) = O(n \log \frac{1}{\varepsilon} + t)$ bits of memory.

4   Preliminaries

We start with some general notation. We denote by $[n]$ the set of numbers $\{1, 2, \ldots, n\}$. We denote by $\mathsf{neg} : \mathbb{N} \to \mathbb{R}$ a function such that for every positive integer c there exists an integer N_c such that for all $n > N_c$, $\mathsf{neg}(n) < 1/n^c$. Finally, throughout this paper we denote by log the base 2 logarithm.

Definition 8 (One-Way Functions). *A function f is said to be one-way if:*

1. *There exists a polynomial-time algorithm A such that $A(x) = f(x)$ for every $x \in \{0,1\}^*$.*
2. *For every probabilistic polynomial-time algorithm A' and large enough n,*

$$\Pr[A'(1^n, f(x)) \in f^{-1}(f(x))] < \mathsf{neg}(n),$$

where the probability is taken uniformly over $x \in \{0,1\}^n$ and the internal randomness of A'.

Definition 9 (Universal Hash Family). *A family of functions $\mathcal{H} = \{h : U \to [m]\}$ is called universal if for any $x_1 \neq x_2$: $\Pr_{h \in \mathcal{H}}[h(x_1) = h(x_2)] \leq \frac{1}{m}$.*

5   Adversarial Resilient Bloom Filters and One-Way Functions

In this section we show that adversarial resilient Bloom filters are (existentially) equivalent to one-way functions (see Definition 8). We begin by showing that if one-way functions do not exist, then any Bloom filter can be 'attacked' by an efficient algorithm in a strong sense:

Theorem 4. *Let $\mathbf{B} = (\mathbf{B}_1, \mathbf{B}_2)$ be any non-trivial Bloom filter of n elements that uses m bits of memory and let ε_0 be the minimal error of $\mathbf{B}$. If one-way function do not exist, then for any constant $\varepsilon < 1$, $\mathbf{B}$ is not (n, t, ε)-adversarial resilient for $t = O\left(m/\varepsilon_0^2\right)$.*

We give two different proofs; The first is self contained (e.g. we do not even have to use the Impagliazzo-Luby [15] technique of finding a random inverse), but, deals only with Bloom filters with steady representations. The second handles Bloom filters with unsteady representations, and uses the framework of adaptively changing distributions of [20].

5.1   A Proof for Bloom Filters with Steady Representations

Overview: We prove Theorem 4 for the case of steady representation (see Definition 1). Actually, for the steady case the theorem holds even for $t = O(m/\varepsilon_0)$.

Assume that there are no one-way functions. We want to construct an adversary that can attack the Bloom filter. We define a function f to be a function that gets a set S, random bits r, and elements $x_1, \ldots, x_t$, computes $M = \mathbf{B}_1(S; r)$ and outputs these elements along with their evaluation on $\mathbf{B}_2(M, \cdot)$ (i.e. for each element x_i the value $\mathbf{B}_2(M, x_i)$). Since f is not one-way, there is an efficient algorithm that can invert it with high probability[2]. That is, the algorithm is

---

[2] The algorithm can invert the function for infinitely many input sizes. Thus, the adversary we construct will succeed in its attack on the same (infinitely many) input sizes.

given a random set of elements labeled whether they are (false) positives or not and it outputs a set S' and bits r'. For $M' = \mathbf{B}_1(S'; r')$ the function $\mathbf{B}_2(M', \cdot)$ is consistent with $\mathbf{B}_2(M, \cdot)$ for all the elements $x_1, \ldots, x_t$. For a large enough set of queries we show that $\mathbf{B}_2(M', \cdot)$ is actually a good approximation of $\mathbf{B}_2(M, \cdot)$ as a boolean function. We use $\mathbf{B}_2(M', \cdot)$ to find an input x^* such that $\mathbf{B}_2(M', x^*) = 1$ and show that $\mathbf{B}_2(M, x^*) = 1$ as well with high probability. This contradicts $\mathbf{B}$ being adversarial-resilient and proves that f is a (weak) one-way function. See the full paper for more details [22].

5.2 Handling Unsteady Bloom Filters

We describe the proof of the general statement of Theorem 4, i.e., handle Bloom filters with an unsteady-representation as well. A Bloom filter with an unsteady representation (see Definition 3) has a *randomized* query algorithm and may change the underlying representation after each query. We want to show that if one-way functions do not exist then we can construct an adversary, `Attack`, that 'attacks' this Bloom filter. The proof of this case is more involved and we show a simpler version that has an additional assumption (for the full proof see [22]).

Hard-core Positives. Let $\mathbf{B} = (\mathbf{B}_1, \mathbf{B}_2)$ be an (n, ε)-Bloom filter with an unsteady representation that uses m bits of memory (see Definition 3). Let M and M' be two representations of a set S generated by $\mathbf{B}_1$. In the previous proof in Sect. 5.1, given a representation M we considered $\mathbf{B}_2(M, \cdot)$ as a boolean function. We defined the function $\mu(M)$ to measure the number of positives in $\mathbf{B}_2(M, \cdot)$ and we defined the error between two representations $\mathsf{err}(M, M')$ to measure the fraction of inputs that the two boolean functions agree on. These definitions make sense only when $\mathbf{B}_2$ is deterministic and does not change the representation. However, in the case of Bloom filters with unsteady representations we need to modify the definitions to have new meanings.

Given a representation M consider the query interface $Q(\cdot)$ initialized with M. For an element x, the probability of x being a false positive is $\Pr[Q(x) = 1] = \Pr[\mathbf{B}_2(M, x) = 1]$. Recall that after querying $Q(\cdot)$, the interface updates its representation and the probability of x being a false positive might change (it could be higher or lower). We say that x is a 'hard-core positive' if after any arbitrary sequence of queries we have that $\Pr[Q(x) = 1] = 1$. That is, the query interface will always response with a 'Yes' on x even after any sequence of queries. Then, we define $\mu(M)$ to be the set of hard-core positive elements in U. Note that over the time, the size of $\mu(M)$ might grow, but it can never become smaller. The following claim proves that for almost all sets S the number of *hard-core* positives is large (see [22] for the proof).

Claim. For any Bloom filter with minimal error ε_0 it holds that:

$$\Pr_S\left[\exists r : \mu\left(M_r^S\right) \leq \frac{\varepsilon_0}{8}\right] \leq 2^{-n}$$

where the probability is taken a random set S of size n from the universe U.

The distribution D_M. As we can not talk about the *function* $\mathbf{B}_2(M, \cdot)$ (as in the steady case) we use terms of *distributions*. For any representation M define the distribution D_M: Sample k elements at random $x_1, \ldots, x_k$ (k will be determined later), and output $(x_1, \ldots, x_k, Q(x_1), \ldots, Q(x_k))$. Note that the underlying representation M changes after each query. Formally, the algorithm for D_M is:

1. Sample $x_1, \ldots, x_k \in U$ uniformly at random.
2. For $i = 1, \ldots, k$: compute $y_i = Q(x_i)$.
3. Output $(x_1, \ldots, x_k, y_1, \ldots, y_k)$.

Let M_0 be a representation of a random set S generated by $\mathbf{B}_1$, and let ε_0 be the minimal error of $\mathbf{B}$. Assume that one-way functions do not exists. Our goal is to construct an algorithm Attack that will 'attack' $\mathbf{B}$, that is, it is given access to $Q(\cdot)$ initialized with M_0 (M_0 is secret and not known to Attack) it must find an non-set element x^* such that $\Pr[Q(x) = 1] \geq 2/3$.

Consider the distribution D_{M_0}, and notice that given access to $Q(\cdot)$ we can perform a single sample from D_{M_0}. Let M_1 be the random variable of the resulting representation after the sample. Then, we can sample from the distribution D_{M_1}, and then D_{M_2} and so on. We describe a simplified version of the proof where we assume that M_0 is known to the adversary. This version seems to captures the main ideas.

Attacking when M_0 *is known.* Suppose that after activating D_{M_0} for r rounds we are given the initial state M_0 (of course, in the actual execution M_0 is secret and later we show how to overcome this assumption). Let $p_1, \ldots, p_r$ be the outputs of the rounds (that is, $p_i = (x_1, \ldots, x_k, y_1, \ldots, y_k)$). For a specific output p_i we say that x_j was labeled '1' if $y_j = 1$.

Denote by $D_{M_0}(p_0, \ldots, p_r)$ the distribution over the $(r + 1)^{\text{th}}$ activation of D_{M_0} conditioned on the first r activations resulting in the states $p_0, \ldots, p_r$. Computational issues aside, the distribution $D_{M_0}(p_0, \ldots, p_r)$ can be sampled by enumerating all random strings such that when applied to D_{M_0} yield the output $p_0, \ldots, p_r$, sampling one of them, and outputting the representations generated by the random string chosen. Moreover, define $D_{M_0}(p_0, \ldots, p_r; x_1, \ldots, x_k)$ to be the distribution $D_{M_0}(p_0, \ldots, p_r)$ conditioned on that the elements chosen in the sample are $x_1, \ldots, x_k$. We also define $D(p_0, \ldots, p_r)$ to be the same distribution as $D_{M_0}(p_0, \ldots, p_r)$ only where the representation M_0 is also chosen at random (according to $\mathbf{B}_1(S)$).

We define an (inefficient) adversary Attack (see Fig. 1) that (given M_0) can attack the Bloom filter, that is, find an element x^* that was not queried before and is a false positive with high probability.

Set $k = 160/\varepsilon_0$ and $\ell = 100k$. Then we get the following claims.

Claim. There is a common x_j: With probability $99/100$ there exist a $1 \leq j \leq k$ such that for all $i \in [\ell]$ it holds that $y_{ij} = 1$, where the probability is over the random choice of S and $x_1, \ldots, x_k$.

The Algorithm Attack

Given: The representation M_0.

Input: 1^λ.

1. Sample $x_1, \ldots, x_k \in U$ at random.
2. For $i \in [\ell]$ sample $D_{M_0}(p_0, \ldots, p_r; x_1, \ldots, x_k)$ to get $y_{i1}, \ldots, y_{ik}$.
3. If there exists an index $j \in [k]$ such that for all $i \in [\ell]$ it holds that $y_{ij} = 1$:
   (a) Set $x^* = x_j$.
   (b) Query $Q(x_1), \ldots, Q(x_{j-1})$.
4. Otherwise set x^* to be an arbitrary element in U.
5. Output x^*.

Fig. 1. The description of the algorithm Attack.

Proof. Let M_r be the resulting representation of the rth activation of $D_{M_0}(p_0, \ldots, p_r; x_1, \ldots, x_k)$. We have seen that with probability $1 - 2^{-n}$ over the choice of S for any M_0 we have that the set of hard-core positives satisfy $|\mu(M_0)| \geq \varepsilon_0/16$. By the definition of the hard-core positives, the set $\mu(M_0)$ may only grow after each query. Thus, for each sample from $D_{M_0}(p_0, \ldots, p_r; x_1, \ldots, x_k)$ we have that $\mu(M_0) \subseteq \mu(M_r)$. If $x_j \in \mu(M_0)$ then $x_j \in \mu(M_r)$ and thus $y_{ij} = 1$ for all $i \in [\ell]$. The probability that all elements $x_1, \ldots, x_k$ are sampled outside the set $\mu(M_0)$ is at most $(1 - \varepsilon_0/16)^k \leq e^{-10}$ (over the random choices of the elements). All together we get that probability of choosing a 'good' S and a 'good' sequence $x_1, \ldots, x_t$ is at least $1 - 2^{-n} + e^{-10} \geq 99/100$.

Claim. Let M_r be the underlying representation of the interface $Q(\cdot)$ at the time right after sampling $p_0, \ldots, p_r$. Then, with probability at least $98/100$ the algorithm Attack outputs an element x^* such that $Q(x^*) = 1$, where the probability is taken over the randomness of Attack, the sampling of $p_0, \ldots, p_r$, and $\mathbf{B}$.

Proof. Consider the distribution $D_{M_0}(p_0, \ldots, p_r; x_1, \ldots, x_k)$ to work as follows: First a representation M is sampled conditioned on starting from M_0 and outputting the states $p_0, \ldots, p_r$ and then we compute $y_j = \mathbf{B}_2(M, x_j)$. Let $M_1', \ldots, M_\ell'$ be the representations chosen during the run of Attack. Note that M_r is chosen from the same distribution that $M_1', \ldots, M_\ell'$ are sampled from. Thus, we can think of M_r of being picked after the choice of $x_1, \ldots, x_k$. That is, we sample $M_1', \ldots, M_{\ell+1}'$, and choose one of them at random to be M_r, and the rest are relabeled as $M_1', \ldots, M_\ell'$. Now, for any x_j, the probability that for all i, M_i' will answer '1' on x_j but M_r will answer '0' on x_j is at most $1/\ell$. Thus, the probability that there exist any such x_j is at most $\frac{k}{\ell} = \frac{k}{100k} = 1/100$. Altogether, the probability that A find such an x_j that is always labeled '1' and that M_r answers '1' on it, is at least $99/100 - 1/100 = 98/100$.

We are left to show how to construct the algorithm Attack so that it will run in polynomial-time and perform the same tasks *without* knowing M_0. One difficulty (which was discussed in Sect. 3), is that the number of samples r must be chosen as a function of the samples and cannot be fixed in advance. Algorithms for such tasks were studied in the framework Naor and Rothblum [20] on adaptively changing distributions. The full proof is given at [22].

5.3 A Construction Using Pseudorandom Permutations

We have seen that Bloom filters that are adversarial resilient require using one-way functions. To complete the equivalence, we show that pseudorandom permutations and functions can be used to construct adversarial resilient Bloom filters. Actually, we show that any Bloom filter can be efficiently transformed to be adversarial resilient with essentially the same amount of memory. The idea is simple and can work in general for other data structures as well: On any input x we compute a pseudo-random permutation of x and send it to the original Bloom filter. The full proof is given at [22].

Theorem 5. *Let* **B** *be an* (n, ε)-*Bloom filter using* m *bits of memory. If pseudorandom permutations exist, then for any security parameter* λ *there exists an* $(n, \varepsilon + \mathsf{neg}(\lambda))$-*strongly resilient Bloom filter with memory* $m' = m + \lambda$.

6 Computationally Unbounded Adversary

In this section, we extend the discussion of adversarial resilient Bloom filters to ones against computationally *unbounded* adversaries. First, notice that the attack of Theorem 4 holds in this case as well, since an unbounded adversary can invert any function (with probability 1). Formally, we get the following:

Corollary 1. *Let* $\mathbf{B} = (\mathbf{B}_1, \mathbf{B}_2)$ *be any non-trivial Bloom filter of* n *elements that uses* m *bits of memory and let* ε_0 *be the minimal error of* **B**. *Then for any constant* $\varepsilon < 1$, **B** *is not* (n, t, ε)-*adversarial resilient against unbounded adversaries for* $t = O\left(\frac{m}{\varepsilon_0^2}\right)$.

As we saw, any (n, ε)-Bloom filter must use at least $n \log \frac{1}{\varepsilon}$ bits of memory. We show how to construct Bloom Filters that are resilient against *unbounded* adversaries for t of queries while using only $O\left(n \log \frac{1}{\varepsilon} + t\right)$ bits of memory (for a discussion on the optimality of the number of queries t see the full paper [22]).

Theorem 6. *For any* $n, t \in \mathbb{N}$, *and* $\varepsilon > 0$ *there exists an* (n, t, ε)-*resilient Bloom filter (against unbounded adversaries) that uses* $O(n \log \frac{1}{\varepsilon} + t)$ *bits of memory.*

Our construction uses two main ingredients: Cuckoo hashing and a very high independence hash family G. We begin by describing these ingredients.

The Hash Function Family G. Pagh and Pagh [23] and Dietzfelbinger and Woelfel [11] (see also Aumuller et al. [2]) showed how to construct a family G of hash functions $g : U \to V$ so that on any set of k inputs it behaves like a truly random function with high probability $(1 - 1/\mathsf{poly}(k))$. Furthermore, g can be evaluated in constant time (in the RAM model), and its description can be stored using $(1 + \alpha)k \log |V| + O(k)$ bits (where here α is an arbitrarily small constant).

Note that the guarantee of g acting as a random function holds for any set S that is *chosen in advance*. In our case the set is not chosen in advance but chosen adaptively and adversarially. However, Berman et al. [3] showed that the same line of constructions, starting with Pagh and Pagh, actually holds *even when the set of queries is chosen adaptively.* That is, for any distinguisher that can adaptively choose k inputs, the advantage of distinguishing a function $g \in_R G$ from a truly random function is polynomially small[3].

Set $\ell = 4 \log \frac{1}{\varepsilon}$. Our function g will be composed of the concatenation of ℓ one bit functions $g_1, g_2, \ldots g_\ell$ where each g_i is selected independently from a family G where $V = \{0, 1\}$ and $k = 2t/\log \frac{1}{\varepsilon}$. For a random $g_i \in_R G$:

- There is a constant c (which we can choose) so that for any adaptive distinguisher that issues a sequence of k adaptive queries g_i the advantage of distinguishing between g_i and an exact k-wise independent function $U \to V$ is bounded by $\frac{1}{k^c}$.
- g_i can be represented using $(1 + \alpha)k\ell = O(t)$ bits.
- g_i can be evaluated in constant time.

Thus, the representation of g requires $O(t)$ bits. The evaluation of g at a given point x takes $O(\ell) = O\left(\log \frac{1}{\varepsilon}\right)$ time.

Cuckoo Hashing. Cuckoo hashing is a data structure for dictionaries introduced by Pagh and Rodler [26]. It consists of two tables T_1 and T_2, each containing r cells where r is slightly larger than n (that is, $r = (1 + \alpha)n$ for some small constant α) and two hash functions $h_1, h_2 : U \to [r]$. The elements are stored in the two tables so that an element x resides at either $T_1[h_1(x)]$ or $T_2[h_2(x)]$. Thus, the lookup procedure consists of one memory accesses to each table plus computing the hash functions. (This description ignores insertions.)

We assume that $n > \log u$ (we can actually let n go as low as $O(\log \log u)$ using almost pair-wise independent hashing). Our construction of an adversarial resilient Bloom filter is:

Setup. The input is a set S of size n. Sample a function g by sampling ℓ functions $g_i \in_R G$ and initialize a Cuckoo hashing dictionary D of size n (with $\alpha = 0.1$) as described above. That is, D has two tables T_1 and T_2 each of size $1.1n$, two hash functions h_1 and h_2, and each element x will reside at either $T_1[h_1(x)]$ or $T_2[h_2(x)]$. Insert the elements of S into D. Then, go over the two tables T_1 and

---

[3] Any exactly k-wise independent function is also good against k adaptive queries, but this is not necessarily the case for *almost* k-wise.

T_2 and at each cell replace each x with $g(x)$. That is, now for each $x \in S$ we have that $g(x)$ resides at either $T_1[h_1(x)]$ or $T_2[h_2(x)]$. Put $\perp$ in the empty locations. The final memory of the Bloom Filter is the memory of D and the representation of g. The dictionary D consists of $O(n)$ cells, each of size $|g(x)| = O(\log \frac{1}{\varepsilon})$ bits and therefore D and g together can be represented by $O(n \log \frac{1}{\varepsilon} + t)$ bits.

Lookup. On input x we answer whether 'Yes' if either $T_1[h_1(x)] = g(x)$ or $T_2[h_2(x)] = g(x)$. The full proof of this construction is given at [22].

References

1. Arbitman, Y., Naor, M., Segev, G.: Backyard cuckoo hashing: constant worst-case operations with a succinct representation. In: FOCS, pp. 787–796. IEEE Computer Society (2010)
2. Aumüller, M., Dietzfelbinger, M., Woelfel, P.: Explicit and efficient hash families suffice for cuckoo hashing with a stash. Algorithmica **70**(3), 428–456 (2014). http://dx.doi.org/10.1007/s00453-013-9840-x
3. Berman, I., Haitner, I., Komargodski, I., Naor, M.: Hardness preserving reductions via cuckoo hashing. In: Sahai, A. (ed.) TCC 2013. LNCS, vol. 7785, pp. 40–59. Springer, Heidelberg (2013)
4. Bloom, B.H.: Space/time trade-offs in hash coding with allowable errors. Commun. ACM **13**, 422–426 (1970)
5. Blum, A., Furst, M.L., Kearns, M., Lipton, R.J.: Cryptographic primitives based on hard learning problems. In: Stinson, D.R. (ed.) CRYPTO 1993. LNCS, vol. 773, pp. 278–291. Springer, Heidelberg (1994)
6. Broder, A., Mitzenmacher, M.: Network applications of bloom filters: a survey. In: Internet Mathematics, pp. 636–646 (2002)
7. Carter, L., Floyd, R., Gill, J., Markowsky, G., Wegman, M.: Exact and approximate membership testers. In: Proceedings of the tenth annual ACM symposium on Theory of computing, STOC 1978, pp. 59–65. ACM, New York, NY, USA (1978). http://doi.acm.org/10.1145/800133.804332
8. Chazelle, B., Kilian, J., Rubinfeld, R., Tal, A.: The bloomier filter: an efficient data structure for static support lookup tables. In: SODA, pp. 30–39. SIAM (2004)
9. Deng, F., Rafiei, D.: Approximately detecting duplicates for streaming data using stable Bloom filters. In: Proceedings of the ACM SIGMOD International Conference on Management of Data, pp. 25–36. ACM (2006)
10. Dharmapurikar, S., Krishnamurthy, P., Sproull, T.S., Lockwood, J.W.: Deep packet inspection using parallel Bloom filters. IEEE Micro **24**(1), 52–61 (2004)
11. Dietzfelbinger, M., Woelfel, P.: Almost random graphs with simple hash functions. In: Proceedings of the 35th Annual ACM Symposium on Theory of Computing, June 9–11, 2003, San Diego, CA, USA, pp. 629–638. ACM (2003)
12. Dwork, C., McSherry, F., Nissim, K., Smith, A.: Calibrating noise to sensitivity in private data analysis. In: Halevi, S., Rabin, T. (eds.) TCC 2006. LNCS, vol. 3876, pp. 265–284. Springer, Heidelberg (2006)
13. Fan, L., Cao, P., Almeida, J.M., Broder, A.Z.: Summary cache: a scalable wide-area web cache sharing protocol. IEEE/ACM Trans. Netw. **8**(3), 281–293 (2000)
14. Hardt, M., Woodruff, D.P.: How robust are linear sketches to adaptive inputs? In: STOC, pp. 121–130. ACM (2013)

15. Impagliazzo, R., Luby, M.: One-way functions are essential for complexity based cryptography (extended abstract). In: FOCS, pp. 230–235. IEEE Computer Society (1989)

16. Lipton, R.J., Naughton, J.F.: Clocked adversaries for hashing. Algorithmica **9**(3), 239–252 (1993)

17. Manber, U., Wu, S.: An algorithm for approximate membership checking with application to password security. Inf. Process. Lett. **50**(4), 191–197 (1994)

18. Mironov, I., Naor, M., Segev, G.: Sketching in adversarial environments. SIAM J. Comput. **40**(6), 1845–1870 (2011). http://dx.doi.org/10.1137/080733772

19. Mullin, J.K.: Optimal semijoins for distributed database systems. IEEE Trans. Softw. Eng. **16**(5), 558–560 (1990)

20. Naor, M., Rothblum, G.N.: Learning to impersonate. In: ICML, ACM International Conference Proceeding Series, vol. 148, pp. 649–656. ACM (2006)

21. Naor, M., Yogev, E.: Sliding Bloom filters. In: Cai, L., Cheng, S.-W., Lam, T.-W. (eds.) ISAAC 2013. LNCS, vol. 8283, pp. 513–523. Springer, Heidelberg (2013)

22. Naor, M., Yogev, E.: Bloom filters in adversarial environments. CoRR abs/1412.8356 (2014). http://arxiv.org/abs/1412.8356

23. Pagh, A., Pagh, R.: Uniform hashing in constant time and optimal space. SIAM J. Comput. **38**(1), 85–96 (2008)

24. Pagh, A., Pagh, R., Rao, S.S.: An optimal Bloom filter replacement. In: SODA, pp. 823–829. SIAM (2005)

25. Pagh, R.: Cuckoo hashing. In: Kao, M.-Y. (ed.) Encyclopedia of Algorithms, pp. 1–99. Springer, New York (2008)

26. Pagh, R., Rodler, F.F.: Cuckoo hashing. J. Algorithms **51**(2), 122–144 (2004)

27. Putze, F., Sanders, P., Singler, J.: Cache-, hash-, and space-efficient bloom filters. ACM J. Exp. Algorithmics 14 (2009)

28. Tarkoma, S., Rothenberg, C.E., Lagerspetz, E.: Theory and practice of bloom filters for distributed systems. IEEE Commun. Surv. Tutorials **14**(1), 131–155 (2012)

29. Valiant, L.G.: A theory of the learnable. Commun. ACM **27**(11), 1134–1142 (1984)

30. Yan, J., Cho, P.L.: Enhancing collaborative spam detection with bloom filters. In: ACSAC, pp. 414–428. IEEE Computer Society (2006)

31. Zhang, L., Guan, Y.: Detecting click fraud in pay-per-click streams of online advertising networks. In: ICDCS, pp. 77–84. IEEE Computer Society (2008)

32. Zhu, Y., Jiang, H., Wang, J.: Hierarchical Bloom filter arrays (hba): a novel, scalable metadata management system for large cluster-based storage. In: CLUSTER, pp. 165–174. IEEE Computer Society (2004)

Proofs of Space

Stefan Dziembowski[1]([✉]), Sebastian Faust[2], Vladimir Kolmogorov[3],
and Krzysztof Pietrzak[3]

[1] University of Warsaw, Warszawa, Poland
stefan@dziembowski.net
[2] Ruhr-University Bochum, Bochum, Germany
[3] IST Austria, Klosterneuburg, Austria

Abstract. Proofs of work (PoW) have been suggested by Dwork and
Naor (Crypto'92) as protection to a shared resource. The basic idea is to
ask the service requestor to dedicate some non-trivial amount of computational work to every request. The original applications included prevention of spam and protection against denial of service attacks. More
recently, PoWs have been used to prevent double spending in the Bitcoin
digital currency system.

In this work, we put forward an alternative concept for PoWs –
so-called *proofs of space* (PoS), where a service requestor must dedicate a significant amount of disk space as opposed to computation. We
construct secure PoS schemes in the random oracle model (with one
additional mild assumption required for the proof to go through), using
graphs with high "pebbling complexity" and Merkle hash-trees. We discuss some applications, including follow-up work where a decentralized
digital currency scheme called Spacecoin is constructed that uses PoS
(instead of wasteful PoW like in Bitcoin) to prevent double spending.

The main technical contribution of this work is the construction of
(directed, loop-free) graphs on N vertices with in-degree $O(\log \log N)$
such that even if one places $\Theta(N)$ pebbles on the nodes of the graph,
there's a constant fraction of nodes that needs $\Theta(N)$ steps to be pebbled
(where in every step one can put a pebble on a node if all its parents
have a pebble).

1 Introduction

Proofs of Work (PoW). Dwork and Naor [16] suggested *proofs of work* (PoW)
to address the problem of junk emails (aka. Spam). The basic idea is to require
that an email be accompanied with some value related to that email that is

S. Dziembowski—Supported by the Foundation for Polish Science WELCOME/2010-
4/2 grant founded within the framework of the EU Innovative Economy (National
Cohesion Strategy) Operational Programme.

V. Kolmogorov—VK is supported by European Research Council under the European Unions Seventh Framework Programme (FP7/2007-2013)/ERC grant agreement no 616160.

K. Pietrzak—Research supported by ERC starting grant (259668-PSPC).

R. Gennaro and M. Robshaw (Eds.): CRYPTO 2015, Part II, LNCS 9216, pp. 585–605, 2015.
DOI: 10.1007/978-3-662-48000-7_29

moderately hard to compute but which can be verified very efficiently. Such a proof could for example be a value σ such that the hash value $\mathcal{H}(\mathsf{Email}, \sigma)$ starts with t zeros. If we model the hash function $\mathcal{H}$ as a random oracle [8], then the sender must compute an expected 2^t hashes until she finds such a σ.[1] A useful property of this PoW is that there is no speedup when one has to find many proofs, i.e., finding s proofs requires $s2^t$ evaluations. The value t should be chosen such that it is not much of a burden for a party sending out a few emails per day (say, it takes 10 s to compute), but is expensive for a Spammer trying to send millions of messages. Verification on the other hand is extremely efficient, the receiver will accept σ as a PoW for Email, if the hash $\mathcal{H}(\mathsf{Email}, \sigma)$ starts with t zeros, i.e., it requires only one evaluation of the hash funciton. PoWs have many applications, and are in particular used to prevent double spending in the Bitcoin digital currency system [38] which has become widely popular by now.

Despite many great applications, PoWs suffer from certain drawbacks. Firstly, running PoW costs energy – especially if they are used on a massive scale, like in the Bitcoin system. For this reason Bitcoin has even been labelled an "environmental disaster" [3]. Secondly, by using dedicated hardware instead of a general purpose processor, one can solve a PoW at a tiny fraction of the hardware and energy cost, this asymmetry is problematic for several reasons.

Proofs of Space (PoS). From a more abstract point of view, a proof of work is simply a means of showing that one invested a non-trivial amount of effort related to some statement. This general principle also works with resources other than computation like real money in micropayment systems [37] or human attention in CAPTCHAs [12,46]. In this paper we put forward the concept of *proofs of space* where the resource in question is disk space.

PoS are partially motivated by the observation that users often have a significant amount of free disk space available, and in this case using a PoS is essentially for free. This is in contrast to a PoW: even if one only contributes computation by processors that would otherwise be idle, this will still waste energy which usually does not come for free.

A PoS is a protocol between a prover P and a verifier V which has two distinct phases. After an initialisation phase, the prover P is supposed to store some data $\mathcal{F}$ of size N, whereas V only stores some small piece of information. At any later time point V can initialise a proof execution phase, at the end of which V outputs either reject or accept. We require that V is highly efficient in both phases, whereas P is highly efficient in the execution phase providing he stored and has random access to the data $\mathcal{F}$.

As an illustrative application for a PoS, suppose that the verifier V is an organization that offers a free email service. To prevent that someone registers a huge number of fake-addresses for spamming, V might require users to dedicate some nontrivial amount of disk space, say 100 GB, for every address registered. Occasionally, V will run a PoS to verify that the user really dedicates this space.

---

[1] The hashed Email should also contain the receiver of the email, and maybe also a timestamp, so that the sender has to search for a fresh σ for each receiver, and also when resending the email at a later point in time.

The simplest solution to prove that one really dedicates the requested space would be a scheme where the verifier V sends a pseudorandom file $\mathcal{F}$ of size 100 GB to the prover P during the initialization phase. Later, V can ask P to send back some bits of $\mathcal{F}$ at random positions, making sure V stores (at least a large fraction of) $\mathcal{F}$. Unfortunately, with this solution, V has to send a huge 100 GB file to P, which makes this approach pretty much useless in practice.

We require from a PoS that the computation, storage requirement and communication complexity of the verifier V during initialization and execution of the PoS is very small, in particular, at most polylogarithmic in the storage requirement N of the prover P and polynomial in some security parameter γ. In order to achieve small communication complexity, we must let the prover P generate a large file $\mathcal{F}$ locally during an initialization phase, which takes some time I. Note that I must be at least linear in N, our constructions will basically[2] achieve this lower bound. Later, P and V can run executions of the PoS which will be very cheap for V, and also for P, assuming it has stored $\mathcal{F}$.

Unfortunately, unlike in the trivial solution (where P sends $\mathcal{F}$ to V), now there is no way we can force a potentially cheating prover $\tilde{\mathsf{P}}$ to store $\mathcal{F}$ in-between the initialization and the execution of the PoS: $\tilde{\mathsf{P}}$ can delete $\mathcal{F}$ after initialization, and instead only store the (short) communication with V during the initialization phase. Later, before an execution of the PoS, P reconstructs $\mathcal{F}$ (in time I), runs the PoS, and deletes $\mathcal{F}$ again once the proof is over.

We will thus consider a security definition where one requires that a cheating prover $\tilde{\mathsf{P}}$ can only make V accept with non-negligible probability if $\tilde{\mathsf{P}}$ *either* uses N_0 bits of storage in-between executions of the PoS *or* if $\tilde{\mathsf{P}}$ invests time T for every execution. Here $N_0 \leq N$ and $T \leq I$ are parameters, and ideally we want them to be not much smaller than N and I, respectively. Our actual security definition in Sect. 2 is more fine-grained, and besides the storage N_0 that $\tilde{\mathsf{P}}$ uses in-between initialization and execution, we also consider a bound N_1 on the total storage used by $\tilde{\mathsf{P}}$ during execution (including N_0, so $N_1 \geq N_0$).

High Level Description of Our Scheme. We described above why the simple idea of having V send a large pseudorandom file $\mathcal{F}$ to P does not give a PoS as the communication complexity is too large. Another simple idea that comes to mind is to let V send a short description of a "randomly behaving"permutation $\pi : \{0,1\}^n \rightarrow \{0,1\}^n$ to P, who then stores a table of $N = n2^n$ bits where the entry at position i is $\pi^{-1}(i)$. During the execution phase, V asks for the preimage of a random value y, which P can efficiently provide as the value $\pi^{-1}(y)$ is stored at position y in the table. Unfortunately, this scheme is no a good PoS because of time-memory trade-offs [27] which imply that one can invert a random permutation over N values using only $\sqrt{N}$ time and space.[3] For random functions (as opposed to permutations), it's still possible to invert in time and space $N^{2/3}$. The actual PoS scheme we propose in this paper is based

---

[2] One of our constructions will achieve the optimal $I = \Theta(N)$ bound, our second construction achieves $I = O(N \log \log N)$.

[3] And initialising this space requires $O(N \log(N))$ time.

on hard to pebble graphs. During the initalisation phase, V sends the description of a hash function to P, who then labels the nodes of a hard to pebble graph using this function. Here the label of a node is computed as the hash of the labels of its children. V then computes a Merkle hash of all the labels, and sends this value to P. In the proof execution phase, V simply asks P to open labels corresponding to some randomly chosen nodes.

Outline and Our Contribution. In this paper we introduce the concept of a PoS, which we formally define in Sect. 2. In Sect. 3 we discuss and motivate the model in which we prove our constructions secure (It is basically the random oracle model, but with an additional assumption). In Sect. 4 we explain how to reduce the security of a simple PoS (with an inefficient verifier) to a graph pebbling game. In Sect. 5 we show how to use hash-trees to make the verifier in the PoS from Sect. 4 efficient. In Sect. 6 we define our final construction and prove its security in Sects. 6.1 and 6.2.

Our proof uses a standard technique for proving lower bounds on the space complexity of computational problems, called *pebbling*. Typically, the lower bounds shown using this method are obtained via the *pebbling games* played on a directed graph. During the game a player can place pebbles on some vertices. The game starts with some pebbles already on the graph. Informally, placing a pebble on a vertex v corresponds to the fact that an algorithm keeps the label of a vertex v in his memory. Removing a pebble from a vertex corresponds therefore to deleting the vertex label from the memory. A pebble can be placed on a vertex v only if the vertices in-going to v have pebbles, which corresponds to the fact that computing v's label is possible only if the algorithm keeps in his memory the labels of the in-going vertices (in our case this will be achieved by defining the label of v to be a hash of the labels of its in-going vertices). The goal of the player is to pebble a certain vertex of the graph. This technique was used in cryptography already before [17,19,20]. For an introduction to the graph pebbling see, e.g., [44].

In Sect. 6.1 we consider two different (infinite families of) graphs with different (and incomparable) pebbling complexities. These graphs then also give PoS schemes with different parameters (cf. Theorem 3). Informally, the construction given in Theorem 1 proves a $\Omega(N/\log N)$ bound on the storage required by a malicious prover. Moreover, no matter how much time he is willing to spend during the execution of the protocol, he is forced to use at least $\Omega(N/\log N)$ storage when executing the protocol. Our second construction from Theorem 2 gives a stronger bound on the storage. In particular, a successful malicious prover either has to dedicate $\Theta(N)$ storage (i.e., almost as much as the N stored by the honest prover) or otherwise it has to use $\Theta(N)$ time with every execution of the PoS (after the initialization is completed). The second construction, whose proof can be found in the full version of this paper [18], is based on superconcentrators, random bipartite expander graphs and on the graphs of Erdös, Graham and Szemerédi [21] is quite involved and is the main technical contribution of our paper.

More Related Work and Applications. Dwork and Naor [16] pioneered the concept of proofs of work as easy-to-check proofs of computational efforts. More concretely, they proposed to use the CPU running time that is required to carry out the proof as a measure of computational effort. In [1] Abadi, Burrows, Manasse and Wobber observed that CPU speeds may differ significantly between different devices and proposed as an alternative measure the number of times the memory is accessed (i.e., the number of cache misses) in order to compute the proof. This approach was formalized and further improved in [2,15,17,47], which use pebbling based techniques. Such memory-hard functions cannot be used as PoS as the memory required to compute and verify the function is the same for provers and verifiers. This is not a problem for memory-hard functions as the here the memory just has to be larger than the cache of a potential prover, whereas in a PoS the storage is the main resource, and will typically be in the range of terabytes.

Originally put forward to counteract spamming, PoWs have a vast number of different applications such as metering web-site access [22], countering denial-of-service attacks [6,30] and many more [29]. An important application for PoWs are digital currencies, like the recent Bitcoin system [38], or earlier schemes like the Micromint system of Rivest and Shamir [42]. The concept of using bounded resources such as computing power or storage to counteract the so-called "Sybil Attack", i.e., misuse of services by fake identities, has already mentioned in the work of Douceur [14].

PoW are used in Bitcoin to prevent double spending: honest *miners* must constantly devote more computational power towards solving PoWs than a potential adversary who tries to double spend. This results in a gigantic waste of energy [3] devoted towards keeping Bitcoin secure, and thus also requires some strong form of incentive for the miners to provide this computational power.[4] Recently a decentralized cryptocurrency called Spacecoin [39] was proposed which uses PoS instead of PoW to prevent double spending. In particular, a miner in Spacecoin who wants to dedicate N bits of disk space towards mining must just run the PoS initialisation phase once, and after that mining is extremely cheap: the miner just runs the PoS execution phase, which means accessing the stored space at a few positions, every few minutes.

A large body of work investigates the concepts of *proofs of storage* and *proofs of retrievability* (cf. [5,9,13,24,25,31] and many more). These are proof systems where a verifier sends a file $\mathcal{F}$ to a prover, and later the prover can convince the verifier that it really stored or received the file. As discussed above, proving that one stores a (random) file certainly shows that one dedicates space, but these proof systems are not good PoS because the verifier has to send at least $|\mathcal{F}|$ bits to the verifier, and hence does not satisfy our polylogarithmic bound on the communication complexity.

---

[4] There are two mechanisms to incentivise mining: miners who solve a PoW get some fixed reward, this is currently the main incentive, but Bitcoin specifies that this reward will decrease over time. A second mechanism are transactions fees.

Proof of Secure Erasure (PoSE) are related to PoS. Informally, a PoSE allows a space restricted prover to convince a verifier that he has erased its memory of size N. PoSE were suggested by Perito and Tsudik [41], who also proposed a scheme where the verifier sends a random file of size N to the prover, who then answers with a hash of this file. Dziembowski, Kazana and Wichs used graph pebbling to give a scheme with small communication complexity (which moreover is independent of N), but large $\Omega(N^2)$ computation complexity (for prover and verifier). Concurrently, and independently of our work, Karvelas and Kiayias [32], and also Ateniese et al. [4] construct PoSE using graphs with high pebbling complexity. Interestingly, their schemes are basically the scheme one gets when running the initialisation and execution phase of our PoS (as in Eq. (7) in Theorem 3).[5] References [32] and [4] give a security proof of their construction in the random oracle model, and do not make any additional assumptions as we do. The reason is that to prove that our "collapsed PoS" (as described above) is a PoSE it is sufficient to prove that a prover uses much space *either* during initialisation or during execution. This follows from a (by now fairly standard) "ex post facto" argument as originally used in [17]. We have to prove something much stronger, namely, that the prover needs much space (or at least time) in the execution phase, even if he makes an unbounded amount of queries in the initialisation phase (we will discuss this in more detail in Sect. 3.1). As described above, a PoS (to be precise, a PoS where the execution phase requires large space, not just time) implies a PoSE, but a PoSE does not imply a PoS, nor can it be used for any of the applications mentioned in this paper. The main use-case for PoSE we know of is the one put forward by Perito and Tsudik [41], namely, to verify that some device has erased its memory. A bit unfortunately, Ateniese et al. [4] chose to call the type of protocols they construct also "proofs of space" which led to some confusion in the past.

Finally, let us mention a recent beautiful paper [10] which introduces the concept of "catalytic space". They prove a surprising result showing that *using* and *erasing* space is not the same relative to some additional space that is filled with random bits and must be in its original state at the end of the computation (i.e., it's only used as a "catalyst"). Thus, relative to such catalytic space, proving that one has access to some space as in a PoS, and proving that one has erased it, like in PoSE, really can be different things.

2   Defining Proofs of Space

We denote with $(\mathsf{out_V}, \mathsf{out_P}) \leftarrow \langle \mathsf{V}(\mathsf{in_V}), \mathsf{P}(\mathsf{in_P}) \rangle (\mathsf{in})$ the execution of an interactive protocol between two parties P and V on shared input in, local inputs[6] $\mathsf{in_P}$

---

[5] There are some differences, the bounds in [4] are somewhat worse as they use hard-to-pebble graphs with worse parameters, and [32] do not use a Merkle hash-tree to make the computation of the verifier independent of N.

[6] We use the expression "local input/output" instead the usual "private input/output", because in our protocols no values will actually be secret. The reason to distinguish between the parties' inputs is only due to the fact that P's input will be very large, whereas we want V to use only small storage.

and in$_V$, and with local outputs out$_V$ and out$_P$, respectively. A proof of space (PoS) is given by a pair of interactive random access machines,[7] a prover P and a verifier V. These parties run the PoS protocol in two phases: a PoS initialization and a PoS execution as defined below. The protocols are executed with respect to some statement id, given as common input (e.g., an email address in the example from the previous section). The identifier id is only required to make sure that P cannot reuse the same space to execute PoS for different statements.

Initialization is an interactive protocol with shared inputs an identifier id, storage bound $N \in \mathbb{N}$ and potentially some other parameters, which we denote with prm $= (\text{id}, N, \ldots)$. The execution of the initialization is denoted by $(\Phi, S) \leftarrow \langle V, P \rangle(\text{prm})$, where Φ is short and S is of size N. V can output the special symbol $\Phi = \bot$, which means that it aborts (this can only be the case if V interacts with a cheating prover).

Execution is an interactive protocol during which P and V have access to the values stored during the initialization phase. The prover P has no output, the verifier V either accepts or rejects.

$$(\{\text{accept}, \text{reject}\}, \emptyset) \leftarrow \langle V(\Phi), P(S) \rangle(\text{prm})$$

In an honest execution the initialization is done once at the setup of the system, e.g., when the user registers with the email service, while the execution can be repeated very efficiently many times without requiring a large amount of computation.

To formally define a proof of space, we introduce the notion of a (N_0, N_1, T) (dishonest) prover $\tilde{P}$. $\tilde{P}$'s storage after the initiation phase is bounded by at most N_0, while during the execution phase its storage is bounded to N_1 and its running time is at most T (here $N_1 \geq N_0$ as the storage during execution contains at least the storage after initialization). We remark that $\tilde{P}$'s storage and running time is unbounded during the the initialization phase (but, as just mentioned, only N_0 storage is available in-between the initialization and execution phase).

A protocol (P, V) as defined above is a (N_0, N_1, T)-*proof of space*, if it satisfies the properties of completeness, soundness and efficiency defined below.

Completeness: We will require that for any honest prover P:

$$\Pr[\text{out} = \text{accept} : (\Phi, S) \leftarrow \langle V, P \rangle(\text{prm}), (\text{out}, \emptyset) \leftarrow \langle V(\Phi), P(S) \rangle(\text{prm})] = 1.$$

Note that the probability above is exactly 1, and hence the completeness is perfect.

Soundness: For any (N_0, N_1, T)-adversarial prover $\tilde{P}$ the probability that V accepts is negligible in some statistical security parameter γ. More precisely, we have

$$\Pr[\text{out} = \text{accept} : (\Phi, S) \leftarrow \langle V, \tilde{P} \rangle(\text{prm}), (\text{out}, \emptyset) \leftarrow \langle V(\Phi), \tilde{P}(S) \rangle(\text{prm})] \leq 2^{-\Theta(\gamma)} \quad (1)$$

---

[7] In a PoS, we want the prover P to run in time much less than its storage size. For this reason, we must model our parties as random access machines (and not, say Turing machines), where accessing a storage location is assumed to take constant (or at most polylogarithmic) time.

The probability above is taken over the random choice of the public parameters prm and the coins of $\tilde{\mathsf{P}}$ and V.[8]

Efficiency: We require the verifier V to be efficient, by which (here and below)
we mean at most polylogarithmic in N and polynomial in some security
parameter γ. Prover P must be efficient during execution, but can run in
time $\mathrm{poly}(N)$ during initialization.[9]

In the soundness definition above, we only consider the case where the PoS is
executed only once. This is without loss of generality for PoS where V is stateless
(apart from Φ) and holds no secret values, and moreover the honest prover P
uses only read access to the storage of size N holding S. The protocols in this
paper are of this form. We will sometimes say that a PoS is (N_0, N_1, T)-secure
if it is a (N_0, N_1, T)-*proof of space*.

It is instructive to observe what level of security trivially cannot be achieved
by a PoS. Below we use small letters n, t, c to denote values that are small, i.e.,
polylogarithmic in N and polynomial in a security parameter γ. If the honest
prover P is an $(N, N + n, t)$ prover, where t, n denote the time and storage
requirements of P during execution, then there exists *no*

1. $(N, N + n, t)$-PoS, as the honest prover "breaks" the scheme by definition,
   and
2. $(c, I + t + c, I + t)$-PoS, where c is the number of bits sent by V to P during initialization. To see this, consider a malicious prover $\tilde{\mathsf{P}}$ that runs the
   initialization like the honest P, but then only stores the messages sent by V
   during initialization instead of the entire large S. Later, during execution, $\tilde{\mathsf{P}}$
   can simply emulate the initialization process (in time I) to get back S, and
   run the normal execution protocol (in time t).

3   The Model

We analyze the security and efficiency of our PoS in the random oracle (RO)
model [8], making an additional assumption on the behavior of adversaries,
which we define and motivate below. Recall that in the RO model, we assume
that all parties (including adversaries) have access to the same random function
$\mathcal{H} : \{0, 1\}^* \rightarrow \{0, 1\}^L$. In practice, one must instantiate $\mathcal{H}$ with a real hash function like SHA3. Although security proofs in the RO model are just a heuristic
argument for real-world security, and there exist artificial schemes where this
heuristic fails [11, 23, 34], the model has proven surprisingly robust in practice.

---

[8] Our construction is based on a hash-function $\mathcal{H}$, which will be part of prm and we
require to be collision resistant. As assuming collision resistance for a fixed function
is not meaningful [43], we must either assume that the probability of Eq. (1) is over
some distribution of identities id (which can then be used as a hash key), or, if we
model $\mathcal{H}$ as a random oracle, over the choice of the random oracle.

[9] As explained in the introduction, P's running time I during initialization must be
at least linear in the size N of the storage. Our construction basically match this
$I = \Omega(N)$ lower bound as mentioned in Footnote 2.

Throughout, we fix the output length of our random oracle $\mathcal{H} : \{0,1\}^* \to \{0,1\}^L$ to some $L \in \mathbb{N}$, which should be chosen large enough, so it is infeasible to find collisions. As finding a collision requires roughly $2^{L/2}$ queries, setting $L = 512$ and assuming that the total number of oracle queries during the entire experiment is upper bounded by, say $2^{L/3}$, would be a conservative choice.

3.1 Modeling the Malicious Prover

In this paper, we want to make statements about adversaries (malicious provers) $\tilde{\mathsf{P}}$ with access to a random oracle $\mathcal{H} : \{0,1\}^* \to \{0,1\}^L$ and bounded by three parameters N_0, N_1, T. They run in two phases:

1. In a first (initialization) phase, $\tilde{\mathsf{P}}$ makes queries[10] $\mathcal{A} = (a_1, \ldots, a_q)$ to $\mathcal{H}$ (adaptively, i.e., a_i can be a function of $\mathcal{H}(a_1), \ldots, \mathcal{H}(a_{i-1})$). At the end of this phase, $\tilde{\mathsf{P}}$ stores a file S of size $N_0 L$ bits, and moreover he must commit to a subset of the queries $\mathcal{B} \subseteq \mathcal{A}$ of size N (technically, we'll do this by a Merkle hash-tree).
2. In a second phase, $\tilde{\mathsf{P}}(S)$ is asked to output $\mathcal{H}(b)$ for some random $b \in \mathcal{B}$. The malicious prover $\tilde{\mathsf{P}}(S)$ is allowed a total number T of oracle queries in this phase, and can use up to $N_1 L$ bits of storage (including the $N_0 L$ bits for S).

As $\mathcal{H}$ is a random oracle, one cannot compress its uniformly random outputs. In particular, as S is of size $N_0 L$, it cannot encode more than N_0 outputs of $\mathcal{H}$. We will make the simplifying assumption that we can explicitly state which outputs these are by letting $S_{\mathcal{H}} \subset \{0,1\}^L, |S_{\mathcal{H}}| \leq N_0$ denote the set of all possible outputs $\mathcal{H}(a), a \in \mathcal{A}$ that $\tilde{\mathsf{P}}(S)$ can write down during the second phase without explicitly querying $\mathcal{H}$ on input a in the 2nd phase.[11] Similarly, the storage bound $N_1 L$ during execution implies that $\tilde{\mathsf{P}}$ cannot store more than N_1 outputs of $\mathcal{H}$ at any particular time point, and we assume that this set of $\leq N_1$ inputs is well defined at any time-point. The above assumption will allow us to bound the advantage of a malicious prover in terms of a pebbling game.

The fact that we need the additional assumption outlined above and cannot reduce the security of our scheme to the plain random oracle model is a bit unsatisfactory, but unfortunately the standard tools (in particular, the elegant "ex post facto" argument from [17]), typically used to reduce pebbling complexity

---

[10] The number q of queries in this phase is unbounded, except for the huge exponential $2^{L/3}$ bound on the total number of oracle queries made during the entire experiment by all parties mentioned above.

[11] Let us stress that we do not claim that such an $S_{\mathcal{H}}$ exists for every $\tilde{\mathsf{P}}$, one can easily come up with a prover where this is not the case (as we will show below). All we need is that for every (N_0, N_1, T) prover $\tilde{\mathsf{P}}$, there exists another prover $\tilde{\mathsf{P}}'$ with (almost) the same parameters and advantage, that obeys our assumption.

An adversary with $N_0 = N_1 = T = 1$ not obeying our assumption is, e.g., a $\tilde{\mathsf{P}}$ that makes queries 0 and 1 and stores $S = \mathcal{H}(0) \oplus \mathcal{H}(1)$ in the first phase. In the second phase, $\tilde{\mathsf{P}}(S)$ picks a random $b \leftarrow \{0,1\}$, makes the query b, and can write down $\mathcal{H}(b), \mathcal{H}(1-b) = S \oplus \mathcal{H}(b)$. Thus, $\tilde{\mathsf{P}}(S)$ can write $2 > N_0 = 1$ values $\mathcal{H}(0)$ or $\mathcal{H}(1)$ without quering them in the 2nd phase.

to the number of random oracle queries, cannot be applied in our setting due to the auxiliary information about the random oracle the adversary can store. We believe that a proof exploiting the fact that random oracles are incompressible using techniques developed in [26,45] can be used to avoid this additional assumption, and we leave this question as interesting future work.

3.2 Storage and Time Complexity

Time Complexity. Throughout, we let the *running time* of honest and adversarial parties be the *number of oracle queries* they make. We also take into account that hashing long messages is more expensive by "charging" k queries for a single query on an input of bit-length $L(k-1)+1$ to Lk. Just counting oracle queries is justified by the fact that almost all computation done by honest parties consists of invocations of the random-oracle, thus we do not ignore any computation here. Moreover, ignoring any computation done by adversaries only makes the security proof stronger.

Storage Complexity. Unless mentioned otherwise, the storage of honest and adversarial parties is measured by the number of outputs $y = \mathcal{H}(x)$ stored. The honest prover P will only store such values by construction; for malicious provers $\tilde{P}$ this number is well defined under the assumption from Sect. 3.1.

4   PoS from Graphs with High Pebbling Complexity

The first ingredient of our proof uses graph pebbling. We consider a directed, acyclic graph $G = (V, E)$. The graph has $|V| = N$ vertices, which we label with numbers from the set $[N] = \{1, \ldots, N\}$. With every vertex $v \in V$ we associate a value $w(v) \in \{0,1\}^L$, and extend the domain of w to include also ordered tuples of elements from V in the following way: for $V' = (v_1, \ldots, v_n)$ (where $v_i \in V$) we define $w(V') = (w(v_1), \ldots, w(v_n))$. Let $\pi(v) = \{v' : (v', v) \in E\}$ denote v's predecessors (in some arbitrary, but fixed order). The value $w(v)$ of v is computed by applying the random oracle to the index v and the values of its predecessors

$$w(v) = \mathcal{H}(v, w(\pi(v))) . \tag{2}$$

Note that if v is a source, i.e., $\pi(v) = \emptyset$, then $w(v)$ is simply $\mathcal{H}(v)$. Our PoS will be an extension of the simple basic PoS $(P_0, V_0)[G, \Lambda]$ from Fig. 1, where Λ is an efficiently samplable distribution that outputs a subset of the vertices V of $G = (V, E)$. This PoS does not yet satisfy the efficiency requirement from Sect. 2, as the complexity of the verifier needs to be as high as the one of the prover. This is because, in order to perform the check in Step 3 of the execution phase, the verifier needs to compute $w(C)$ himself. In our model, as discussed in Sect. 3.1, the only way a malicious prover $\tilde{P}_0(S)$ can determine $w(v)$ is if $w(v) \in S_{\mathcal{H}}$ is in the encoded set of size at most N_0, or otherwise by explicitly making the oracle query $\mathcal{H}(v, w(\pi(v)))$ during execution. Note that if $w(i) \notin S_{\mathcal{H}}$ for some

---

Parameters prm = (id, N, $G = (V, E)$, Λ), where G is a graph on $|V| = N$ vertices and Λ is an efficiently samplable distribution over V^β (we postpone specifying β as well as the function of id to Sect. 6).

Initialization $(S, \emptyset) \leftarrow \langle \mathsf{P}_0, \mathsf{V}_0 \rangle(\mathrm{prm})$ where $S = w(V)$.

Execution $(\mathsf{accept/reject}, \emptyset) \leftarrow \langle \mathsf{V}(\emptyset), \mathsf{P}(S) \rangle(\mathrm{prm})$

1. $\mathsf{V}_0(\emptyset)$ samples $C \leftarrow \Lambda$ and sends C to P_0.
2. $\mathsf{P}_0(S)$ answers with $A = w(C) \subset S$.
3. $\mathsf{V}_0(\emptyset)$ outputs accept if $A = w(C)$ and reject otherwise.

Fig. 1. The basic PoS $(\mathsf{P}_0, \mathsf{V}_0)[G, \Lambda]$ (with inefficient verifier V_0).

$i \in \pi(v)$, then $\tilde{\mathsf{P}}_0(S)$ will have to make even more queries recursively to learn $w(v)$. Hence, in order to prove (N_0, N_1, T)-security of the PoS $(\mathsf{P}_0, \mathsf{V}_0)[G, \Lambda]$ in our idealized model, it suffices to upper bound the advantage of Player 1 in the following pebbling game on $G = (V, E)$:

1. Player 1 puts up to N_0 initial pebbles on the vertices of V.
2. Player 2 samples a subset $C \leftarrow \Lambda$ of size α of challenge vertices.
3. Player 1 applies a sequence of up to T steps according to the following rules:
   (i) it can place a pebble on a vertex v if (1) all its predecessors $u \in \pi(v)$ are pebbled and (2) there are currently less than N_1 vertices pebbled.
   (ii) it can remove a pebble from any vertex.
4. Player 1 wins if it places pebbles on all vertices of C.

In the pebbling game above, Step 1 corresponds to a malicious prover $\tilde{\mathsf{P}}_0$ choosing the set $S_{\mathcal{H}}$. Step 3 corresponds to $\tilde{\mathsf{P}}_0$ computing values according to the rules in Eq. (2), while obeying the N_1 total storage bound. Putting a pebble corresponds to invoking $y = \mathcal{H}(x)$ and storing the value y. Removing a pebble corresponds to deleting some previously computed y.

5   Efficient Verifiers Using Hash Trees

The PoS described in the previous section does not yet meet our Definition from Sect. 2 as V_0 is not efficient. In this section we describe how to make the verifier efficient, using hash-trees, a standard cryptographic technique introduced by Ralph Merkle [35].

Using Hash Trees for Committing. A hash-tree allows a party P to compute a commitment $\phi \in \{0,1\}^L$ to N data items $x_1, \ldots, x_N \in \{0,1\}^L$ using $N - 1$ invocations of a hash function $\mathcal{H} : \{0,1\}^* \rightarrow \{0,1\}^L$. Later, P can prove to a party holding ϕ what the value of any x_i is, by sending only $L \log N$ bits. For example, for $N = 8$, P commits to $x_1, \ldots, x_N$ by hashing the x_i's in a tree like structure as

$$\phi = \mathcal{H}(\ \mathcal{H}(\ \mathcal{H}(x_1, x_2), \mathcal{H}(x_3, x_4)\ ), \mathcal{H}(\ \mathcal{H}(x_5, x_6), \mathcal{H}(x_7, x_8)\ )\ )$$

We will denote with $\mathcal{T}^{\mathcal{H}}(x_1, \ldots, x_N)$ the $2N-1$ values of all the nodes (including the N leaves x_i and the root ϕ) of the hash-tree, e.g., for $N = 8$, where we define $x_{ab} = \mathcal{H}(x_a, x_b)$

$$\mathcal{T}^{\mathcal{H}}(x_1, \ldots, x_8) = \{x_1, \ldots, x_8, x_{12}, x_{34}, x_{56}, x_{78}, x_{1234}, x_{5678}, \phi = x_{12345678}\}$$

The prover P, in order to later efficiently open any x_i, will store all $2N-1$ values $\mathcal{T} = \mathcal{T}^{\mathcal{H}}(x_1, \ldots, x_N)$, but only send the single root element ϕ to a verifier V. Later P can "open" any value x_i to V by sending x_i and the $\log N$ values, which correspond to the siblings of the nodes that lie on the path from x_i to ϕ, e.g., to open x_3 P sends x_3 and $\mathsf{open}(\mathcal{T}, 3) = (x_{12}, x_4, x_{5678})$ and the prover checks if

$$\mathsf{vrfy}(\phi, 3, x_3, (x_{12}, x_4, x_{5678})) = \left(\mathcal{H}(x_{12}, \mathcal{H}(x_3, x_4)), x_{56789} \right) \stackrel{?}{=} \phi$$

As indicated above, we denote with $\mathsf{open}(\mathcal{T}, i) \subset \mathcal{T}$ the $\log N$ values P must send to V in order to open x_i, and denote with $\mathsf{vrfy}(\phi, i, x_i, o) \rightarrow \{\mathsf{accept}, \mathsf{reject}\}$ the above verification procedure. This scheme is correct, i.e., for $\phi, \mathcal{T}$ computed as above and any $i \in [N]$, $\mathsf{vrfy}(\phi, i, x_i, \mathsf{open}(\mathcal{T}, i)) = \mathsf{accept}$.

The security property provided by a hash-tree states that it is hard to open any committed value in more than one possible way. This "binding" property can be reduced to the collision resistance of $\mathcal{H}$: from any $\phi, i, (x, o), (x', o'), x \neq x'$ where $\mathsf{vrfy}(\phi, i, x, o) = \mathsf{vrfy}(\phi, i, x', o') = \mathsf{accept}$, one can efficiently extract a collision $z \neq z', \mathcal{H}(z) = \mathcal{H}(z')$ for $\mathcal{H}$.

We add an initialization phase to the graph based PoS from Fig. 1, where the prover P(prm) commits to $x_1 = w(v_1), \ldots, x_N = w(v_N)$ by computing a hash tree $\mathcal{T} = \mathcal{T}^{\mathcal{H}}(x_1, \ldots, x_N)$ and sending its root ϕ to V. In the execution phase, the prover must then answer a challenge c not only with the value $x_c = w(c)$, but also open c by sending $(x_c, \mathsf{open}(\mathcal{T}, c))$ which P can do without any queries to $\mathcal{H}$ as it stored $\mathcal{T}$.

If a cheating prover $\tilde{\mathsf{P}}(\mathsf{prm})$ sends a correctly computed ϕ during the initialization phase, then during execution $\tilde{\mathsf{P}}(\mathsf{prm}, S)$ can only make $\mathsf{V}(\mathsf{prm}, \phi)$ accept by either answering each challenge c with the correct value $w(c)$, or by breaking the binding property of the hash-tree (and thus the collision resistance of the underlying hash-function).

We are left with the challenge to deal with a prover who might cheat and send a wrongly computed $\tilde{\Phi} \neq \phi$ during initialization. Some simple solutions are

- Have V compute ϕ herself. This is not possible as we want V's complexity to be only polylog in N.
- Let P prove, using a proof system like computationally sound (CS) proofs [36] or universal arguments [7], that ϕ was computed correctly. Although these proof systems do have polylogarithmic complexity for the verifier, and thus formally would meet our efficiency requirement, they rely on the PCP theorem and thus are not really practical.

Dealing with Wrong Commitments. Unless $\tilde{\mathsf{P}}$ breaks the collision resistance of $\mathcal{H}$, no matter what commitment $\tilde{\Phi}$ the prover P sends to V, he can later only open it to some fixed N values which we will denote $\tilde{x}_1, \ldots, \tilde{x}_N$.[12] We say that $\tilde{x}_i$ is consistent if

$$\tilde{x}_i = \mathcal{H}(i, \tilde{x}_{i_1}, \ldots, \tilde{x}_{i_d}) \text{ where } \pi(i) = \{i_1, \ldots, i_d\} \tag{3}$$

Note that if *all* $\tilde{x}_i$ are consistent, then $\tilde{\Phi} = \phi$. We add a second initialization phase to the PoS, where V will check the consistency of α random $\tilde{x}_i$'s. This can be done by having $\tilde{\mathsf{P}}$ open $\tilde{x}_i$ and $\tilde{x}_j$ for all $j \in \pi(i)$. If $\tilde{\mathsf{P}}$ passes this check, we can be sure that with high probability a large fraction of the $\tilde{x}_i$'s is consistent. More concretely, if the number of challenge vertices is $\alpha = \varepsilon t$ for some $\varepsilon > 0$, then $\tilde{\mathsf{P}}$ will fail the check with probability $1 - 2^{-\Theta(t)}$ if more than an ε-fraction of the $\tilde{x}_i$'s are inconsistent.

A cheating $\tilde{\mathsf{P}}$ might still pass this phase with high probability with an $\tilde{\Phi}$ where only $1 - \varepsilon$ fraction of the $\tilde{x}_i$ are consistent for some sufficiently small $\varepsilon > 0$. As the inconsistent $\tilde{x}_i$ are not outputs of $\mathcal{H}$, $\tilde{\mathsf{P}}$ can chose their value arbitrarily, e.g., all being 0^L. Now $\tilde{\mathsf{P}}$ does not have to store this εN inconsistent values $\tilde{x}_j$ while still knowing them.

In our idealized model as discussed in Sect. 3.1, one can show that this is already all the advantage $\tilde{\mathsf{P}}$ gets. We can model an εN fraction of inconsistent $\tilde{x}_i$'s by slightly augmenting the pebbling game from Sect. 4. Let the pebbles from the original game be *white* pebbles. We now additionally allow player 1 to put εN *red* pebbles (apart from the N_0 white pebbles) on V during step 1. These red pebbles correspond to inconsistent values. The remaining game remains the same, except that player 1 is never allowed to remove red pebbles.

We observe that being allowed to initially put an additional εN red pebbles is no more useful than getting an additional εN white pebbles (as white pebbles are strictly more useful because, unlike red pebbles, they later can be removed.) Translated back to our PoS, in order prove (N_0, N_1, T)-security of our PoS allowing up to εN inconsistent values, it suffices to prove $(N_0 - \varepsilon N, N_1 - \varepsilon N, T)$-security of the PoS, assuming that the initial commitment is computed honestly, and there are no inconsistent values (and thus no red pebbles in the corresponding game).

6 Our Main Construction

Below we formally define our PoS (P, V). The common input to P, V are the parameters $\mathsf{prm} = (\mathsf{id}, 2N, \gamma, G, \Lambda)$, which contain the identifier $\mathsf{id} \in \{0, 1\}^*$, a storage bound $2N \in \mathbb{N}$ (i.e., $2NL$ bits),[13] a statistical security parameter γ, the description of a graph $G(V, E)$ on $|V| = N$ vertices and an efficiently samplable distribution Λ which outputs some "challenge" set $C \subset V$ of size $\alpha = \alpha(\gamma, N)$.

Below $\mathcal{H}$ denotes a hash function, that depends on id: given a hash function $\mathcal{H}'(.)$ (which we will model as a random oracle in the security proof), throughout

---

[12] Potentially, $\tilde{\mathsf{P}}$ cannot open some values at all, but wlog. we assume that it can open every value in exactly one way.

[13] We set the bound to $2N$, so if we let N denote the number of vertices in the underlying graph, we must store $2N - 1$ values of the hash-tree.

we let $\mathcal{H}(.)$ denote $\mathcal{H}'(\mathrm{id},.)$. The reason for this is simply so we can assume that the random oracles $\mathcal{H}'(\mathrm{id},.)$ and $\mathcal{H}'(\mathrm{id}',.)$ used in PoS with different identifiers $\mathrm{id} \neq \mathrm{id}'$ are independent, and thus anything stored for the PoS with identifier id is useless to answer challenges in a PoS with different identifier id'.

Initialization $(\varPhi, S) \leftarrow \langle \mathsf{V}, \mathsf{P} \rangle (\mathrm{prm})$:

1. **P sends V a commitment** ϕ **to** $w(V)$
   - P computes the values $x_i = w(i)$ for all $i \in V$ as in Eq. (2).
   - P's output is a hash-tree $S = \mathcal{T}^{\mathcal{H}}(x_1, \ldots, x_N)$, which requires $|S| = (2N - 1)L$ bits) as described in Sect. 5.
   - P sends the root $\phi \in S$ to V.
2. **P proves consistency of** ϕ **for** $\alpha = \alpha(\gamma, N)$ **random values**
   - V picks a set of challenges $C \leftarrow \varLambda$ where the size of C is α and sends C to P.
   - For all $c \in C$, P opens the value corresponding to c and all its predecessors to V by sending, for all $c \in C$

   $$\{(x_i, \mathsf{open}(S, i)) \; : \; i \in \{c, \pi(c)\}\}$$

   - V verifies that P sends all the required openings, and they are consistent, i.e., for all $c \in C$ the opened values $\tilde{x}_c$ and $\tilde{x}_i, i \in \pi(c) = (i_1, \ldots, i_d)$ must satisfy $\tilde{x}_c = \mathcal{H}(c, \tilde{x}_{i_1}, \ldots, \tilde{x}_{i_d})$, and the verification of the opened commitments passes. If either check fails, V outputs $\varPhi = \bot$ and aborts. Otherwise, V outputs $\varPhi = \phi$, and the initialization phase is over.

Execution $(\mathsf{accept}/\mathsf{reject}, \emptyset) \leftarrow \langle \mathsf{V}(\varPhi), \mathsf{P}(S) \rangle (\mathrm{prm})$:

P proves it stores the committed values by opening a random $\beta = \varTheta(\gamma)$ **subset of them**
   - V picks a challenge set $C \subset V$ of size $|C| = \beta$ at random, and sends C to P.
   - P answers with $\{o_c = (x_c, \mathsf{open}(S, c)) \; : \; c \in C\}$.
   - V checks for every $c \in C$ if $\mathsf{vrfy}(\varPhi, c, o_c) \overset{?}{=} \mathsf{accept}$. V outputs accept if this is the case and reject otherwise.

6.1 Constructions of the Graphs

We consider the following pebbling game, between a player and a challenger, for a directed acyclic graph $G = (V, E)$ and a distribution λ over V.

1. Player puts initial pebbles on some subset $U \subseteq V$ of vertices.
2. Challenger samples a "challenge vertex" $c \in V$ according to λ.
3. Player applies a sequence of steps according to the following rules:
   (i) it can place a pebble on a vertex v if all its predecessors $u \in \pi(v)$ are pebbled.

(ii) it can remove a pebble from any vertex.
4. Player wins if it places a pebble on c.

Let $S_0 = |U|$ be the number of initial pebbles, S_1 be the total number of used pebbles (or equivalently, the maximum number of pebbles that are present in the graph at any time instance, including initialization), and let T be the number of pebbling steps given in 3i). The definition implies that $S_1 \geq S_0$ and $T \geq S_1 - S_0$. Note, with $S_0 = |V|$ pebbles the player can always achieve time $T = 0$: it can just place initial pebbles on V.

Definition 1. Consider functions $f = f(N, S_0)$ and $g = g(N, S_0, S_1)$. A family of graphs $\{G_N = (V_N, E_N) \mid |V_N| = N \in \mathbb{N}\}$ is said to have *pebbling complexity* $\Omega(f, g)$ if there exist constants $c_1, c_2, \delta > 0$ and distributions λ_N over V_N such that for any player that wins the pebbling game on (G_N, λ_N) (as described above) with probability 1 it holds that

$$\Pr[\, S_1 \geq c_1 f(N, S_0) \,\wedge\, T \geq c_2 g(N, S_0, S_1) \,] \geq \delta \tag{4}$$

Let $\mathcal{G}(N, d)$ be the set of directed acyclic graphs $G = (V, E)$ with $|V| = N$ vertices and the maximum in-degree at most d. We now state our two main pebbling theorems:

Theorem 1. *There exists an explicit family of graphs $G_N \in \mathcal{G}(N, 2)$ with pebbling complexity*

$$\Omega(N/\log N, 0) \tag{5}$$

In the next theorem we use the *Iverson bracket* notation: $[\phi] = 1$ if statement ϕ is true, and $[\phi] = 0$ otherwise.

Theorem 2. *There exists a family of graphs $G_N \in \mathcal{G}(N, O(\log \log N))$ with pebbling complexity*

$$\Omega(0, [S_0 < \tau N] \cdot \max\{N, N^2/S_1\}) \tag{6}$$

for some constant $\tau \in (0, 1)$. It can be constructed by a randomized algorithm with a polynomial expected running time that produces the desired graph with probability at least $1 - 2^{-\Theta(N/\log N)}$.

Complete proofs of these theorems are given in the full version of this paper [18]; here we give a brief summary of our techniques. For Theorem 1 we use the construction of Paul, Tarjan and Celoni [40], and derive the theorem as a corollary of their Lemma 2. For Theorem 2 we use a new construction which relies on three building blocks: (i) random bipartite graphs $R_{(m)}^d \in \mathcal{G}(2m, d)$ with m inputs and m outputs; (ii) superconcentrator graphs $C_{(m)}$ with m inputs and m outputs; (iii) graphs $D_t = ([t], E_t)$ of Erdös, Graham and Szemerédi [21] with *dense long paths*. These are directed acyclic graphs with t vertices and $\Theta(t \log t)$ edges (of the form (i, j) with $i < j$) that satisfy the following for some constant $\eta \in (0, 1)$ and a sufficiently large t: for any subset $X \subseteq [t]$ of size at most ηt graph D_t contains a path of length at least ηt that avoids X. We show that family D_t can be chosen so that the maximum in-degree is $\Theta(\log t)$. The main component of our construction is graph $\tilde{G}_{(m,t)}^d$ defined as follows:

- Add mt nodes $\tilde{V} = V_1 \cup \ldots \cup V_t$ to $\tilde{G}^d_{(m,t)}$ where $|V_1| = \ldots = |V_t| = m$. This will be the set of challenges.
- For each edge (i,j) of graph D_t add a copy of graph $R^d_{(m)}$ from V_i to V_j, i.e. identify the inputs of $R^d_{(m)}$ with nodes in V_i (using an arbitrary permutation) and the outputs of $R^d_{(m)}$ with nodes in V_j (again, using an arbitrary permutation).

We set $d = \Theta(1)$, $t = \Theta(\log N)$ and $m = \Theta(N/t)$ (with specific constants), then $\tilde{G}^d_{(m,t)} \in \mathcal{G}(mt, O(\log \log N))$.

Note that a somewhat similar graph was used by Dwork, Naor and Wee [17]. They connect bipartite graphs $R^d_{(m)}$ consecutively, i.e. instead of graph D_t they use a chain graph with t nodes. Dwork et al. give an intuitive argument that removing at most τm nodes from each layer $V_1, \ldots, V_t$ (for some constant $\tau < 1$) always leaves a graph which is "well-connected": informally speaking, many nodes of V_1 are still connected to many nodes of V_t. (We give a formal treatment of their argument in the full version of this paper [18].) However, this does not hold if more than $m = \Theta(N/\log N)$ nodes are allowed to be removed: by placing initial pebbles on, say, the middle layer $V_{t/2}$ one can completely disconnect V_1 from V_t.

In contrast, in our construction removing any $\tau' N$ nodes still leaves a graph which is "well-connected". Our argument is as follows. If constant τ' is sufficiently small then there can be at most ηt layers with more than τm initial pebbles (for a given constant $\tau < 1$). By the property of D_t, there exists a sufficiently long path P in D_t that avoids those layers. We can thus use the argument above for the subgraph corresponding to P. We split P into three parts of equal size, and show that many nodes in the first part are connected to many nodes in the third part.

In this way we prove that graphs $\tilde{G}^d_{(m,t)}$ have pebbling complexity $\Omega(0, [S_0 < \tau N] \cdot N)$. To get complexity $\Omega(0, [S_0 < \tau N] \cdot \max\{N, N^2/S_1\})$, we add mt extra nodes V_0 and a copy of superconcentrator $C_{(mt)}$ from V_0 to $\tilde{V}$. We then use a standard "basic lower bound argument" for superconcentrators [33].

Remark 1. As shown in [28], any graph $G \in \mathcal{G}(N, O(1))$ can be entirely pebbled using $S_1 = O(N/\log N)$ pebbles (without any initial pebbles). This implies that expression $N/\log N$ in Theorem 1 cannot be improved upon. Note, this still leaves the possibility of a graph that can be pebbled using $O(N/\log N)$ pebbles only with a large time T (e.g. superpolynomial in N). Examples of such graph for a non-interactive version of the pebble game can be found in [33]. Results stated in [33], however, do not immediately imply a similar bound for our interactive game.

6.2   Putting Things Together

Combining the results and definitions from the previous sections, we can now state our main theorem.

Theorem 3. *In the model from Sect. 3.1, for constants $c_i > 0$, the PoS from Sect. 6 instantiated with the graphs from Theorem 1 is a*

$$(c_1(N/\log N), c_2(N/\log N), \infty)\text{-secure PoS.} \tag{7}$$

Instantiated with the graphs from Theorem 2 it is a

$$(c_3 N, \infty, c_4 N)\text{-secure PoS.} \tag{8}$$

Efficiency, measured as outlined in Sect. 3.2, is summarized in the table below where γ is the statistical security parameter

	Communication	Computation P	Computation V
PoS Eq. (7) Initialization	$O(\gamma \log^2 N)$	$4N$	$O(\gamma \log^2 N)$
PoS Eq. (7) Execution	$O(\gamma \log N)$	0	$O(\gamma \log N)$
PoS Eq. (8) Initialization	$O(\gamma \log N \log \log N)$	$O(N \log \log N)$	$O(\gamma \log N \log \log N)$
PoS Eq. (8) Execution	$O(\gamma \log N)$	0	$O(\gamma \log N)$

Equation (8) means that a successful cheating prover must either store a file of size $\Omega(N)$ (in L bit blocks) after initialization, or make $\Omega(N)$ invocations to the RO. Equation (7) gives a weaker $\Omega(N/\log N)$ bound, but forces a potential adversary not storing that much after initialization, to use at least $\Omega(N/\log N)$ storage during the execution phase, no matter how much time he is willing to invest. This PoS could be interesting in contexts where one wants to be sure that one talks with a prover who has access to significant memory during execution.

Below we explain how security and efficiency claims in the theorem were derived. We start by analyzing the basic (inefficient verifier) PoS $(P_0, V_0)[G, \Lambda]$ from Fig. 1 if instantiated with the graphs from Theorems 1 and 2.

Proposition 1. *For some constants $c_i > 0$, if G_N has pebbling complexity $\Omega(f(N), 0)$ according to Definition 1, then the basic PoS $(P_0, V_0)[G_N, \Lambda_N]$ as illustrated in Fig. 1, where the distribution Λ_N samples $\Theta(\gamma)$ (for a statistical security parameter γ) vertices according to the distribution λ_N from Definition 1, is*

$$(S_0, c_1 f(N), \infty)\text{-secure (for any } S_0 \leq c_1 f(N)) \tag{9}$$

If G_N has pebbling complexity $(0, g(N, S_0, S_1))$, then for any S_0, S_1 the PoS $(P_0, V_0)[G_N, \Lambda_N]$ is

$$(S_0, S_1, c_2 g(N, S_0, S_1))\text{-secure.} \tag{10}$$

Above, secure means secure in the model from Sect. 3.1.

(The proof of appears in the full version [18].) Instantiating the above proposition with the graphs G_N from Theorems 1 and 2, we can conclude that the simple (inefficient verifier) PoS $(P_0, V_0)[G_N, \Lambda_N]$ is

$$(c_1 N/\log N, c_2 N/\log N, \infty) \quad \text{and} \quad (S_0, S_1, c_3 \cdot [S_0 \leq \tau N] \cdot \max\{N, N^2/S_1\}) \tag{11}$$

secure, respectively (for constants $c_i > 0$, $0 < \tau < 1$ and $[S_0 < \tau N] = 1$ if $S_0 \leq \tau N$ and 0 otherwise). If we set $S_0 = \lfloor \tau N \rfloor = c_4 N$, the right side of Eq. (11) becomes $(c_4 N, S_1, c_3 \cdot \max\{N, N^2/S_1\})$ and further setting $S_1 = \infty$ $(c_4 N, \infty, c_3 N)$ As explained in Sect. 5, we can make the verifier V_0 efficient during initialization, by giving up on εN in the storage bound. We can choose ε ourselves, but must check $\Theta(\gamma/\varepsilon)$ values for consistency during initialization (for a statistical security parameter γ). For our first PoS, we set $\varepsilon = \frac{c_1}{2\log N}$ and get with $c_5 = c_1/2$ using $c_2 \geq c_1$

$$(\underbrace{c_1 \cdot N/\log N - \varepsilon \cdot N}_{=c_5 N/\log N}, \underbrace{c_2 \cdot N/\log N - \varepsilon \cdot N}_{\geq c_5 N/\log N}, \infty)$$

security as claimed in Eq. (7). For the second PoS, we set $\varepsilon = \frac{c_4}{2}$ which gives with $c_6 = c_4/2$

$$(\underbrace{c_4 N - \varepsilon N}_{\geq c_6 N}, \infty - \varepsilon N, c_3 N)$$

security, as claimed in Eq. (8). Also, note that the PoS described above are PoS as defined in Sect. 6 if instantiated with the graphs from Theorems 1 and 2, respectively.

Efficiency of the PoS Eq. (7). We analyze the efficiency of our PoS, measuring time and storage complexity as outlined in Sect. 3.2. Consider the $(c_1 N/\log N, c_2 N/\log N, \infty)$-secure construction from Eq. (7). In the first phase of the initialization, P needs roughly $4N = \Theta(N)$ computation: using that the underlying graph has max in-degree 2, computing $w(V)$ according to Eq. (2) requires N hashes on inputs of length at most $2L + \log N \leq 3L$, and P makes an additional $N - 1$ hashes on inputs of length $2L$ to compute the hash-tree. The communication and V's computation in the first phase of initialization is $\Theta(1)$ (as V just receives the root $\phi \in \{0,1\}^L$).

During the 2nd phase of the initialization, V will challenge P on α (to be determined) vertices to make sure that with probability $1 - 2^{-\Theta(\gamma)}$, at most an $\varepsilon = \Theta(1/\log N)$ fraction of the $\hat{x}_i$ are inconsistent. As discussed above, for this we have to set $\alpha = \Theta(\gamma \log N)$. Because this PoS is based on a graph with degree 2 (cf. Theorem 1), to check consistency of a $\hat{x}_i$ one just has to open 3 values. Opening the values requires to send $\log N$ values (and the verifier to compute that many hashes). This adds up to an $O(\gamma \log^2 N)$ communication complexity during initialization, V's computation is of the same order.

During execution, P opens ϕ on $\Theta(\gamma)$ positions, which requires $\Theta(\gamma \log N)$ communication (in L bit blocks), and $\Theta(\gamma \log N)$ computation by V.

Efficiency of the PoS Eq. (8). Analyzing the efficiency of the second PoS is analogous to the first. The main difference is that now the underlying graph has larger degree $O(\log \log N)$ (cf. Theorem 2), and we only need to set $\varepsilon = \Theta(1)$.

References

1. Abadi, M., Burrows, M., Wobber, T.: Moderately hard and memory-bound functions. In: NDSS 2003. The Internet Society, February 2003
2. Alwen, J., Serbinenko, V.: High parallel complexity graphs and memory-hard functions. In: Symposium on Theory of Computing, STOC 2015 (2015)
3. Anderson, N.: Mining Bitcoins takes power, but is it an "environmental disaster"? April 2013. http://tinyurl.com/cdh95at
4. Ateniese, G., Bonacina, I., Faonio, A., Galesi, N.: Proofs of space: when space is of the essence. In: Abdalla, M., De Prisco, R. (eds.) SCN 2014. LNCS, vol. 8642, pp. 538–557. Springer, Heidelberg (2014)
5. Ateniese, G., Burns, R.C., Curtmola, R., Herring, J., Kissner, L., Peterson, Z.N.J., Song, D.: Provable data possession at untrusted stores. In: Ning, P., De Capitani di Vimercati, S., Syverson, P.F. (eds.) ACM CCS 2007, pp. 598–609. ACM Press, October 2007
6. Back, A.: Hashcash - a denial of service counter-measure (2002). http://www.hashcash.org/papers/hashcash.pdf
7. Barak, B., Goldreich, O.: Universal arguments and their applications. SIAM J. Comput. 38(5), 1661–1694 (2008)
8. Bellare, M., Rogaway, P.: Random oracles are practical: a paradigm for designing efficient protocols. In: Ashby, V. (ed.) ACM CCS 1993, pp. 62–73. ACM Press, November 1993
9. Bowers, K.D., Juels, A., Oprea, A.: Proofs of retrievability: theory and implementation. In: CCSW, pp. 43–54 (2009)
10. Buhrman, H., Cleve, R., Koucký, M., Loff, B., Speelman, P.: Computing with a full memory: catalytic space. In: Symposium on Theory of Computing, STOC 2014, May 31 - June 03 2014, pp. 857–866, New York, NY, USA (2014)
11. Canetti, R., Goldreich, O., Halevi, S.: The random oracle methodology, revisited (preliminary version). In: 30th ACM STOC, pp. 209–218. ACM Press, May 1998
12. Canetti, R., Halevi, S., Steiner, M.: Mitigating dictionary attacks on password-protected local storage. In: Dwork, C. (ed.) CRYPTO 2006. LNCS, vol. 4117, pp. 160–179. Springer, Heidelberg (2006)
13. Di Pietro, R., Mancini, L.V., Law, Y.W., Etalle, S., Havinga, P.: Lkhw: a directed diffusion-based secure multicast scheme for wireless sensor networks. In: 2003 Proceedings of the International Conference on Parallel Processing Workshops, pp. 397–406 (2003)
14. Douceur, J.R.: The sybil attack. In: Druschel, P., Kaashoek, M.F., Rowstron, A. (eds.) IPTPS 2002. LNCS, vol. 2429, pp. 251–260. Springer, Heidelberg (2002)
15. Dwork, C., Goldberg, A.V., Naor, M.: On memory-bound functions for fighting spam. In: Boneh, D. (ed.) CRYPTO 2003. LNCS, vol. 2729, pp. 426–444. Springer, Heidelberg (2003)
16. Dwork, C., Naor, M.: Pricing via processing or combatting junk mail. In: Brickell, E.F. (ed.) CRYPTO 1992. LNCS, vol. 740, pp. 139–147. Springer, Heidelberg (1993)
17. Dwork, C., Naor, M., Wee, H.M.: Pebbling and proofs of work. In: Shoup, V. (ed.) CRYPTO 2005. LNCS, vol. 3621, pp. 37–54. Springer, Heidelberg (2005)
18. Dziembowski, S., Faust, S., Kolmogorov, V., Pietrzak, K.: Proofs of space. Cryptology ePrint Archive, Report 2013/796 (2013). http://eprint.iacr.org/2013/796
19. Dziembowski, S., Kazana, T., Wichs, D.: Key-evolution schemes resilient to space-bounded leakage. In: Rogaway, P. (ed.) CRYPTO 2011. LNCS, vol. 6841, pp. 335–353. Springer, Heidelberg (2011)

20. Dziembowski, S., Kazana, T., Wichs, D.: One-time computable self-erasing functions. In: Ishai, Y. (ed.) TCC 2011. LNCS, vol. 6597, pp. 125–143. Springer, Heidelberg (2011)

21. Erdös, P., Graham, R.L., Szemerédi, E.: On sparse graphs with dense long paths. Technical report STAN-CS-75-504, Stanford University, Computer Science Department (1975)

22. Franklin, K.M., Malkhi, D.: Auditable metering with lightweight security. In: Luby, M., Rolim, J.D.P., Serna, M. (eds.) FC 1997. LNCS, vol. 1318, pp. 151–160. Springer, Heidelberg (1997)

23. Goldwasser, S., Kalai, Y.T.: On the (in)security of the Fiat-Shamir paradigm. In: 44th FOCS, pp. 102–115. IEEE Computer Society Press, October 2003

24. Golle, P., Jarecki, S., Mironov, I.: Cryptographic primitives enforcing communication and storage complexity. In: Blaze, M. (ed.) FC 2002. LNCS, vol. 2357, pp. 120–135. Springer, Heidelberg (2003)

25. Gratzer, V., Naccache, D.: Alien vs. quine. IEEE Secur. Priv. 5(2), 26–31 (2007)

26. Haitner, I., Hoch, J.J., Reingold, O., Segev, G.: Finding collisions in interactive protocols - a tight lower bound on the round complexity of statistically-hiding commitments. In: 48th FOCS, pp.669–679. IEEE Computer Society Press, October 2007

27. Hellman, M.E.: A cryptanalytic time-memory trade-off. IEEE Trans. Inf. Theory 26(4), 401–406 (1980)

28. Hopcroft, J., Paul, W., Valiant, L.: On time versus space. J. ACM 24(2), 332–337 (1977)

29. Jakobsson, M., Juels, A.: Proofs of work and bread pudding protocols. In: Preneel, B. (ed.) Proceedings of the IFIP Conference on Communications and Multimedia Security, vol. 152, pp. 258–272. Kluwer (1999)

30. Juels, A., Brainard, J.G.: Client puzzles: a cryptographic countermeasure against connection depletion attacks. In: NDSS 1999. The Internet Society, February 1999

31. Juels, A., Kaliski Jr., B.S.: Pors: proofs of retrievability for large files. In: Ning, P., De Capitani di Vimercati, S., Syverson, P.F. (eds.) ACM CCS 07, pp. 584–597. ACM Press, October 2007

32. Karvelas, N.P., Kiayias, A.: Efficient proofs of secure erasure. In: Abdalla, M., De Prisco, R. (eds.) SCN 2014. LNCS, vol. 8642, pp. 520–537. Springer, Heidelberg (2014)

33. Lengauer, T., Tarjan, R.E.: Asymptotically tight bounds on time-space trade-offs in a pebble game. J. ACM 29(4), 1087–1130 (1982)

34. Maurer, U.M., Renner, R.S., Holenstein, C.: Indifferentiability, impossibility results on reductions, and applications to the random oracle methodology. In: Naor, M. (ed.) TCC 2004. LNCS, vol. 2951, pp. 21–39. Springer, Heidelberg (2004)

35. Merkle, R.C.: Method of providing digital signatures. US Patent 4309569, 5 January 1982

36. Micali, S.: Computationally sound proofs. SIAM J. Comput. 30(4), 1253–1298 (2000)

37. Micali, S., Rivest, R.L.: Micropayments revisited. In: Preneel, B. (ed.) CT-RSA 2002. LNCS, vol. 2271, pp. 149–163. Springer, Heidelberg (2002)

38. Nakamoto, S.: Bitcoin: a peer-to-peer electronic cash system (2009). http://bitcoin.org/bitcoin.pdf

39. Park, S., Pietrzak, K., Alwen, J., Fuchsbauer, G., Gazi, P.: Spacecoin: a cryptocurrency based on proofs of space. Cryptology ePrint Archive, Report 2015/528 (2015). http://eprint.iacr.org/2015/528

40. Paul, W.J., Tarjan, R.E., Celoni, J.R.: Space bounds for a game on graphs. Math. Syst. Theory 10(1), 239–251 (1976–1977)
41. Perito, D., Tsudik, G.: Secure code update for embedded devices via proofs of secure erasure. In: Gritzalis, D., Preneel, B., Theoharidou, M. (eds.) ESORICS 2010. LNCS, vol. 6345, pp. 643–662. Springer, Heidelberg (2010)
42. Rivest, R.L., Shamir, A.: Payword and micromint: two simple micropayment schemes. In: CryptoBytes, pp. 69–87 (1996)
43. Rogaway, P.: Formalizing human ignorance. In: Nguyên, P.Q. (ed.) VIETCRYPT 2006. LNCS, vol. 4341, pp. 211–228. Springer, Heidelberg (2006)
44. Savage, J.E.: Models of Computation: Exploring the Power of Computing, 1st edn. Addison-Wesley Longman Publishing Co. Inc., Boston (1997)
45. Simon, D.R.: Findings collisions on a one-way street: can secure hash functions be based on general assumptions? In: Nyberg, K. (ed.) EUROCRYPT 1998. LNCS, vol. 1403, pp. 334–345. Springer, Heidelberg (1998)
46. Von Ahn, L., Blum, M., Hopper, N.J., Langford, J.: CAPTCHA: Using hard AI problems for security. In: Biham, E. (ed.) EUROCRYPT 2003. LNCS, vol. 2656, pp. 246–256. Springer, Heidelberg (2003)
47. Waters, B., Juels, A., Halderman, J.A., Felten, E.W.: New client puzzle outsourcing techniques for dos resistance. In: Proceedings of the 11th ACM Conference on Computer and Communications Security, CCS 2004, pp. 246–256. ACM, New York (2004)

Fully Homomorphic/Functional Encryption

Quantum Homomorphic Encryption for Circuits of Low T-gate Complexity

Anne Broadbent[1]([✉]) and Stacey Jeffery[2]

[1] Department of Mathematics and Statistics,
University of Ottawa, Ottawa, ON, Canada
abroadbe@uottawa.ca
[2] Institute for Quantum Information and Matter,
California Institute of Technology, Pasadena, CA, USA
sjeffery@caltech.edu

Abstract. Fully homomorphic encryption is an encryption method with the property that any computation on the plaintext can be performed by a party having access to the ciphertext only. Here, we formally define and give schemes for *quantum* homomorphic encryption, which is the encryption of *quantum* information such that *quantum* computations can be performed given the ciphertext only. Our schemes allow for arbitrary Clifford group gates, but become inefficient for circuits with large complexity, measured in terms of the non-Clifford portion of the circuit (we use the "$\pi/8$" non-Clifford group gate, also known as the T-gate).

More specifically, two schemes are proposed: the first scheme has a decryption procedure whose complexity scales with the square of the *number* of T-gates (compared with a trivial scheme in which the complexity scales with the total number of gates); the second scheme uses a quantum evaluation key of length given by a polynomial of degree exponential in the circuit's T-gate depth, yielding a homomorphic scheme for quantum circuits with constant T-depth. Both schemes build on a classical fully homomorphic encryption scheme.

A further contribution of ours is to formally define the security of encryption schemes for quantum messages: we define *quantum indistinguishability under chosen plaintext attacks* in both the public- and private-key settings. In this context, we show the equivalence of several definitions. Our schemes are the first of their kind that are secure under modern cryptographic definitions, and can be seen as a quantum analogue of classical results establishing homomorphic encryption for circuits with a limited number of *multiplication* gates. Historically, such results appeared as precursors to the breakthrough result establishing classical fully homomorphic encryption.

1   Introduction

An encryption scheme is *homomorphic* over some set of circuits $\mathscr{S}$ if any circuit in $\mathscr{S}$ can be evaluated on an encrypted input. That is, given an encryption of the message m, it is possible to produce a ciphertext that decrypts to the output

© International Association for Cryptologic Research 2015
R. Gennaro and M. Robshaw (Eds.): CRYPTO 2015, Part II, LNCS 9216, pp. 609–629, 2015.
DOI: 10.1007/978-3-662-48000-7_30

of the circuit C on input m, for any $C \in \mathscr{S}$. In *fully homomorphic encryption (FHE)*, $\mathscr{S}$ is the set of all classical circuits. FHE was introduced in 1978 [26], but the existence of such a scheme was an open problem for over 30 years. Some early public-key encryption schemes were homomorphic over the set of circuits consisting of only additions [18, 23] or over the set of circuits consisting of only multiplications [12]. Several steps were made towards FHE, with schemes that were homomorphic over increasingly large circuit classes, such as circuits containing additions and a single multiplication [4], or of logarithmic depth [29], until finally in 2009, Gentry established a breakthrough result by giving the first fully homomorphic encryption scheme [15]. Follow-up work showed that FHE could be simplified [11], and based on standard assumptions, such as *learning with errors* [5]. The advent of FHE has unleashed a series of far-reaching consequences, such as delegating computations, and functional encryption [17]. For a survey on FHE, see [32].

A number of works have studied the secure delegation of quantum computation [1, 6–8, 10, 13, 33]. None directly address the question of quantum homomorphic encryption, since they are interactive schemes, and the work of the client is proportional to the size of the circuit being evaluated (and thus, they do not satisfy the *compactness* requirement of FHE, even if we allow interaction). Non-interactive approaches are given by [3, 27] and [31]. However, none of these approaches are applicable to universal circuit families. Furthermore, in the case of [3], security is given only in terms of cheat sensitivity, while both [27] and [31] only bound the leakage of their encoding schemes.

Recent work [36] examines the question of perfect security and correctness for quantum fully homomorphic encryption (QFHE), concluding that the trivial scheme is optimal in this context. In light of this result, it is natural to consider computational assumptions in achieving QFHE. Indeed, the question of computationally secure QFHE remains an open problem; our contribution makes progress in this direction by presenting the first schemes that are homomorphic for a large class of quantum circuits.

1.1 Summary of Contributions and Techniques

We introduce schemes for *quantum homomorphic encryption (QHE)*, the quantum version of homomorphic encryption; we are interested in the evaluation of *quantum* circuits on encrypted *quantum* data. In terms of definitions, we contribute by giving the first definition of quantum homomorphic encryption (QHE) in the computational setting, in the case of both public-key and symmetric-key cryptosystems. As a consequence, we give the first formal definition (and scheme) for the public-key encryption of quantum information, where security is given in terms of *quantum indistinguishability under chosen plaintext attacks*—for which we show the equivalence of a number of definitions, including security for multiple messages. Prior work considered the computational setting for quantum encryption of classical plaintexts only [20, 22, 35].

In terms of QHE schemes, we start by using straightforward techniques to construct a scheme that is homomorphic for Clifford circuits. This can be seen as an analogue to a classical scheme that is homomorphic for linear circuits (circuits performing only additions). While Clifford circuits are not universal for quantum computation, this already yields a range of applications for quantum information processing, including encoding and decoding into stabilizer codes. Our quantum public-key encryption scheme is a hybrid of a classical public-key fully homomorphic encryption scheme and the quantum one-time pad [2]. Intuitively, the scheme works by encrypting the quantum register with a quantum one-time pad, and then encrypting the one-time pad encryption keys with a classical public-key FHE scheme. Since Clifford circuits conjugate Pauli operators to Pauli operators, any Clifford circuit can be directly applied to the encrypted quantum register; the homomorphic property of the classical encryption scheme is used to update the encryption key. Of course, we specify that the classical FHE scheme should be secure against quantum adversaries. By using, *e.g.*, the scheme from [5], we get security based on the *learning with errors* (LWE) assumption [24,25]; this has been equated with worst-case hardness of "short vector problems"on arbitrary lattices [21], which is widely believed to be a quantum-safe (or "post-quantum") assumption.

For universal quantum computations, we must evaluate a non-Clifford gate, for which we choose the "T" gate (also known as "R" or "$\pi/8$"). Applying the above principle we run into trouble, since $TX^aZ^b = X^aZ^{a\oplus b}P^aT$. That is, conditioned on the quantum one-time pad encryption key $a, b \in \{0, 1\}$, the output picks up an undesirable non-Pauli error. Our main contribution is to present two schemes, EPR and AUX, that deal with this situation in two different ways:

EPR: The main idea of EPR is to use entangled quantum registers to enable corrections *within the circuit* at the time of decryption. This scheme is efficient for any quantum circuit, however, it fails to meet a requirement for fully homomorphic encryption called *compactness*, which requires that the complexity of the decryption procedure be independent of the evaluated circuit. More specifically, the complexity of the decryption procedure for EPR scales with the square of the number of T-gates. This gives an advantage over the trivial scheme whenever the number of T-gates in the evaluated circuit is less than the squareroot of the number of gates. (The *trivial* scheme consists of appending to the ciphertext a description of the circuit to be evaluated, and specifying that it should be applied as part of the decryption procedure.)

AUX: Compared to EPR, the scheme AUX takes a more proactive approach to performing the correction required for a T-gate: to do this, it uses a number of auxiliary qubits that are given as part of the evaluation key. Intuitively, these auxiliary qubits encode the required corrections. In order to ensure universality, a large number of possible corrections must be available — the length of the evaluation key is thus given by a polynomial of degree exponential in the circuit's T-gate *depth*, yielding a homomorphic scheme that is efficient for quantum circuits with constant T-depth.

The two main schemes are incomparable. The scheme EPR becomes less *compact* (and therefore less interesting, since it approaches the trivial scheme), as the *number* of T-gates increases, while the scheme AUX becomes inefficient (*extremely* rapidly) as the *depth* of T-gates increases.

Our results can be viewed as a quantum analogue of precursory results to classical fully homomorphic encryption, which established the homomorphic property of encryption schemes that tolerate a limited amount of operations. One difference is that, while these schemes started with the modest goal of just a *single* multiplication (the addition operation being "easy"), we have already allowed for at the very least a *constant* number, and, depending on the circuit, up to a polynomial number of "hard" operations, namely of T-gates.

Our schemes use the existence of classical FHE, although at the expense of a slightly more complicated exposition, a classical scheme that is homomorphic only for linear circuits would actually suffice. We see the relationship between our schemes and classical FHE as a strength of our result, via the following interpretation: classical FHE is sufficient to enable QHE for a large family of circuits, and perhaps by taking greater advantage of the *fully* homomorphic property of the classical scheme in some as yet unknown way, our ideas might be extended to larger classes of quantum circuits. With this in mind, and for ease of exposition, we use a classical fully homomorphic encryption scheme for all of our quantum homomorphic encryption schemes.

Some preliminaries and notation are given in Sect. 2. We give formal definitions of quantum homomorphic encryption and related concepts, including security definitions, in Sect. 3; this allows us to formally state our results in Sect. 4. Section 5 contains a basic quantum homomorphic encryption scheme, CL, for Clifford circuits that is used as a basis for EPR (Sect. 6), and AUX (Sect. 7). Further details, including proofs of our main theorems, can be found in the full version [9].

2   Preliminaries and Notation

A negligible function, denoted $\eta(\cdot)$, is a function such that for every polynomial $p(\cdot)$, there exists an N such that for all integers $n > N$ it holds that $\eta(n) < \frac{1}{p(n)}$. As a convention, if a is a classical plaintext, we denote its encryption by $\tilde{a}$. Throughout this work we use κ to indicate the security parameter.

A *quantum register* is a quantum system, which we view as a physical object that stores quantum information. The contents of a quantum register are mathematically modelled as the set of trace-1, positive semidefinite operators, called *density operators*, on $\mathcal{X}$, where $\mathcal{X}$ is a complex Euclidean space. We denote the set of density operators on any space $\mathcal{X}$ by $D(\mathcal{X})$.

Quantum registers are denoted with calligraphic typeset. Two quantum systems, $\mathcal{X}$ and $\mathcal{Y}$, form a composite system by the tensor product, $\mathcal{X} \otimes \mathcal{Y}$. If $\rho \in D(\mathcal{X} \otimes \mathcal{Y})$ is a state on the joint system, we write $\rho^{\mathcal{X}}$ to denote $Tr_{\mathcal{Y}}(\rho)$. If $\mathcal{X}$ and $\mathcal{Y}$ have the same dimension, we denote this by $\mathcal{X} \equiv \mathcal{Y}$. The *trace distance* between two states, ρ and σ, is defined $\Delta(\rho, \sigma) := Tr\left(\sqrt{(\rho - \sigma)^{\dagger}(\rho - \sigma)}\right)$.

A density matrix that is diagonal in the computational basis corresponds to a classical random variable. For a random variable X on some set Σ_X, we define $\rho(X) := \sum_{x \in \Sigma_X} \Pr[X = x]|x\rangle\langle x|$, the density matrix corresponding to X. A *classical-quantum* state is a state of the form $\rho^{\mathcal{MA}} = \sum_x \Pr[X = x]|x\rangle\langle x|^{\mathcal{M}} \otimes \rho_x^{\mathcal{A}}$.

One special quantum state on any system $\mathcal{X}$ is the *completely mixed state*, $\frac{1}{\dim \mathcal{X}} \mathbb{I}_{\mathcal{X}}$, which we will sometimes denote by \$ (where $\mathcal{X}$ should be implicit from the context). When $\mathcal{X}$ is interpreted as $\mathbb{C}^S$ for some finite set S, then \$ corresponds to the uniform distribution on S.

A *quantum channel* $\Phi : D(\mathcal{A}) \to D(\mathcal{B})$ refers to any physically-realizable mapping on quantum registers. The identity channel on register $\mathcal{R}$ is denoted $\mathbb{I}_{\mathcal{R}}$. Let Φ be a quantum channel acting on register $\mathcal{A}$, and $\rho^{\mathcal{AE}}$ a quantum system held in the joint registers $\mathcal{A} \otimes \mathcal{E}$. Then to simplify notation, when it is clear from the context, we write $\Phi(\rho^{\mathcal{AE}})$ to mean $(\Phi \otimes \mathbb{I})(\rho^{\mathcal{AE}})$.

We work with the gate set $\{X, Z, P, CNOT, H\}$. This gate set applied to arbitrary wires (redundantly) generates the Clifford group, and adding any non-Clifford gate, such as T, gives a generating set for all quantum circuits.

For a single-qubit register $\mathcal{R}$, and $a, b \in \{0, 1\}$, we denote by $\mathsf{QEnc}_{a,b} : \mathcal{R} \to \mathcal{R}$ the quantum one-time pad encryption and by $\mathsf{QDec}_{a,b} : \mathcal{R} \to \mathcal{R}$ the quantum one-time pad decryption [2], $\mathsf{QEnc}_{a,b} : \rho \mapsto X^a Z^b \rho Z^b X^a$ and $\mathsf{QDec}_{a,b} = \mathsf{QEnc}_{a,b}$. It is easy to see that $\mathsf{QDec}_{a,b} \circ \mathsf{QEnc}_{a,b} = \mathbb{I}_{\mathcal{R}}$. By specifying that (a, b) be chosen uniformly at random, we get that the encryption maps any input to the completely mixed state (from the point of view of the adversary), since for all ρ, $\frac{1}{4} \sum_{a,b} X^a Z^b \rho Z^b X^a = \frac{\mathbb{I}_2}{2}$.

3 Definitions

We now formally define QHE schemes and their properties. In Sect. 3.1, we define QHE in the public-key setting. Section 3.2 carefully defines the security of QHE, giving two definitions for security under chosen plaintext attacks, shown in the full version [9] to be equivalent. Section 3.3 defines correctness and compactness for QHE, culminating in a complete definition of quantum fully homomorphic encryption. Section 3.4 deals with an important subtlety that arises in the quantum case: due to the no-cloning theorem, when a large system is encrypted with some auxiliary quantum information needed for decryption, that auxiliary information cannot be copied and given to every subsystem, but rather, the system must now be decrypted as a whole, rather than subsystem-by-subsystem. We also define compactness and quasi-compactness in this context. Finally, one of our schemes (AUX) must be used in the symmetric-key setting, defined in Sect. 3.5. We do not address the issue of *circuit privacy* [16], leaving this question for future work.

3.1 Classical and Quantum Homomorphic Encryption

Our schemes rely on a classical fully homomorphic encryption scheme. Since our adversaries are modelled as being *quantum* polynomial-time, we need a further security guarantee on the classical scheme, namely that it is secure against *quantum* adversaries (see Definition 1). Fortunately, much of classical fully homomorphic encryption uses lattice-based cryptography, which exploits one of the few conjectured "quantum-safe" assumptions [21]. Among all known solutions, the scheme of [5] appears to be the best for our purposes, as it bases its security on the *learning with errors* (LWE) assumption [24,25], which has been equated with worst-case hardness of "short vector problems" on arbitrary lattices.

Definition 1 (q-IND-CPA). *A classical homomorphic encryption scheme* HE *is q-IND-CPA secure if for any* quantum *polynomial-time adversary* $\mathscr{A}$, *there exists a negligible function* η *such that for* $(pk, evk, sk) \leftarrow$ HE.Keygen(1^κ):

$$|\Pr[\mathscr{A}(pk, evk, \mathsf{HE.Enc}_{pk}(0)) = 1] - \Pr[\mathscr{A}(pk, evk, \mathsf{HE.Enc}_{pk}(1)) = 1]| \le \eta(\kappa).$$

Although a classical scheme that is q-IND-CPA is also IND-CPA, the converse may not be true. Note, however, that any proof that a scheme is IND-CPA can potentially be turned into a proof for q-IND-CPA if all statements still hold when "probabilistic polynomial-time adversary" is replaced by "quantum polynomial-time adversary" (see [30]).

We now give our new definitions for quantum homomorphic encryption. In our definitions, both pk, the public encryption key, and sk, the secret decryption key, are classical, whereas the evaluation key is allowed to be a quantum state.

Definition 2 (QHE). *A quantum homomorphic encryption scheme is a 4-tuple of quantum algorithms* (QHE.KeyGen, QHE.Enc, QHE.Eval, QHE.Dec):

Key Generation. QHE.KeyGen : $1^\kappa \rightarrow (pk, sk, \rho_{evk})$. *This algorithm takes a unary representation of the security parameter as input and outputs a classical public encryption key* pk, *a classical secret decryption key* sk *and a quantum evaluation key* $\rho_{evk} \in D(\mathcal{R}_{evk})$.

Encryption. QHE.Enc$_{pk}$: $D(\mathcal{M}) \rightarrow D(\mathcal{C})$. *For every possible* pk, *the quantum channel* Enc$_{pk}$ *maps a state in the message space* $\mathcal{M}$ *to a state (the cipherstate) in the cipherspace* $\mathcal{C}$.

Homomorphic Evaluation. QHE.Eval$^\mathsf{C}$: $D(\mathcal{R}_{evk} \otimes \mathcal{C}^{\otimes n}) \rightarrow D(\mathcal{C}'^{\otimes m})$. *For every quantum circuit* C, *with induced channel* $\Phi_\mathsf{C} : D(\mathcal{M}^{\otimes n}) \rightarrow D(\mathcal{M}^{\otimes m})$, *we define a channel* Eval$^\mathsf{C}$ *that maps an* n-fold cipherstate to an m-fold cipherstate, consuming the evaluation key in the process.

Decryption. QHE.Dec$_{sk}$: $D(\mathcal{C}') \rightarrow D(\mathcal{M})$. *For every possible* sk, Dec$_{sk}$ *is a quantum channel that maps the state in* $D(\mathcal{C}')$ *to a quantum state in* $D(\mathcal{M})$.

3.2   Security of Quantum Homomorphic Encryption

We now define a notion of security for QHE analogous to the classical notion of indistinguishability under chosen plaintext attack. We note that, by taking the evaluation key to be empty, our definitions are trivially applicable to the scenario of quantum public-key encryption (*i.e.* without a homomorphic property).

The CPA indistinguishability experiment is given below and illustrated in Fig. 1. The experiment interacts with an adversary $\mathscr{A}$, which is a pair of polynomial-time quantum algorithms ($\mathscr{A}_1, \mathscr{A}_2$) (which we also call adversaries).

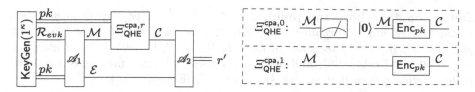

Fig. 1. The quantum CPA indistinguishability experiment

The quantum CPA indistinguishability experiment $\mathsf{PubK}^{\mathsf{cpa}}_{\mathscr{A},\mathsf{QHE}}(\kappa)$

1. $\mathsf{KeyGen}(1^\kappa)$ is run to obtain keys (pk, sk, ρ_{evk}).
2. Adversary $\mathscr{A}_1$ is given (pk, ρ_{evk}) and outputs a quantum state on $\mathcal{M} \otimes \mathcal{E}$.
3. For $r \in \{0,1\}$, let $\Xi^{\mathsf{cpa},r}_{\mathsf{QHE}} : D(\mathcal{M}) \to D(\mathcal{C})$ be: $\Xi^{\mathsf{cpa},0}_{\mathsf{QHE}}(\rho) = \mathsf{QHE.Enc}_{pk}(|0\rangle\langle 0|)$
   and $\Xi^{\mathsf{cpa},1}_{\mathsf{QHE}}(\rho) = \mathsf{QHE.Enc}_{pk}(\rho)$. A random bit $r \in \{0,1\}$ is chosen and $\Xi^{\mathsf{cpa},r}_{\mathsf{QHE}}$
   is applied to the state in $\mathcal{M}$ (the output being a state in $\mathcal{C}$).
4. Adversary $\mathscr{A}_2$ obtains the system in $\mathcal{C} \otimes \mathcal{E}$ and outputs a bit r'.
5. The output of the experiment is defined to be 1 if $r' = r$ and 0 otherwise. In
   case $r = r'$, we say that $\mathscr{A}$ *wins* the experiment.

Definition 3 (Quantum Indistinguishability under Chosen Plaintext Attack (q-IND-CPA)). *A quantum homomorphic encryption scheme* QHE *is* q-IND-CPA *secure if for any quantum polynomial-time adversary* $\mathscr{A} = (\mathscr{A}_1, \mathscr{A}_2)$ *there exists a negligible function* η *such that* $\Pr[\mathsf{PubK}^{\mathsf{cpa}}_{\mathscr{A},\mathsf{QHE}}(\kappa) = 1] \leq \frac{1}{2} + \eta(\kappa)$.

In the case of classical cryptosystems, it is known that IND-CPA security, the classical analogue of Definition 1, implies a seemingly stronger security against an adversary who can send multiple messages to a challenger. In the quantum case, we can analogously define an experiment similar to $\mathsf{PubK}^{\mathsf{cpa}}_{\mathscr{A},\mathsf{QHE}}$, but where the adversary prepares a state in $\mathcal{M}^{\otimes t} \otimes \mathcal{M}^{\otimes t}$ and sends it to the challenger, who traces out either the first half or the second half of the system, before applying an encryption map to each of the remaining subspaces. The adversary must then decide which system was traced out. In the full version [9], we give a formal definition of this notion of security, which we call q-IND-CPA-mult, and prove the equivalence of q-IND-CPA and q-IND-CPA-mult. This strengthens our results since security in the most general case (q-IND-CPA-mult) follows from security for the simplest definition (q-IND-CPA).

3.3 Correctness and Compactness of QHE

Next, we give a notion that encapsulates correctness of both encryption and evaluation, with respect to a class $\mathscr{S}$ of quantum circuits. In the classical context, it is common to restrict attention to circuits that output a single bit, since any deterministic string can be computed bit-by-bit. We cannot do this quantumly, as a quantum state cannot be described qubit-by-qubit. We therefore consider correctness as a global property of the output. Furthermore, as quantum data can be entangled, we require that a correct scheme preserve this entanglement and thus explicitly include an auxiliary space in the definition below.

Definition 4 *($\mathscr{S}$-homomorphic).* *Let $\mathscr{S} = \{\mathscr{S}_\kappa\}_{\kappa \in \mathbb{N}}$ be a class of quantum circuits. A quantum encryption scheme QHE is $\mathscr{S}$-homomorphic (or homomorphic for $\mathscr{S}$) if for any sequence of circuits $\{C_\kappa \in \mathscr{S}_\kappa\}_\kappa$ with induced channels $\Phi_{C_\kappa} : \mathcal{M}^{\otimes n(\kappa)} \to \mathcal{M}^{\otimes m(\kappa)}$, and input $\rho \in D(\mathcal{M}^{\otimes n(\kappa)} \otimes \mathcal{E})$, there exists a negligible function η such that for $(pk, sk, \rho_{evk}) \leftarrow$ QHE.Keygen(1^κ):*

$$\Delta\left(\text{QHE.Dec}_{sk}^{\otimes m(\kappa)}\left(\text{QHE.Eval}^{C_\kappa}\left(\rho_{evk}, \text{QHE.Enc}_{pk}^{\otimes n}(\rho) \right) \right), \Phi_{C_\kappa}(\rho) \right) = \eta(\kappa). \quad (1)$$

We point out two properties of the above definition. First, we do not require that ciphertexts be decryptable themselves, only that they become decryptable after homomorphic evaluation, however, as long as QHE is homomorphic for the class of identity circuits, we can effectively decrypt a ciphertext by first homomorphically evaluating the identity. Second, we do not require that the output of QHE.Eval be able to undergo additional homomorphic evaluations; indeed, if the evaluation key ρ_{evk} is quantum, it will in general be "consumed" by the QHE.Eval process, rendering any future applications of QHE.Eval impossible.

Analogously to the classical case, we define compactness, which requires that the complexity of QHE.Dec be independent of the evaluated circuit, ruling out schemes where applying the circuit is delayed until after decryption.

Definition 5 *($\mathscr{S}$-compactness).* *Let $\mathscr{S} = \{\mathscr{S}_\kappa\}_{\kappa \in \mathbb{N}}$ be a class of quantum circuits. A quantum encryption scheme QHE is $\mathscr{S}$-compact if there exists a polynomial p such that for any sequence of circuits $\{C_\kappa \in \mathscr{S}_\kappa\}_\kappa$, the circuit complexity of applying QHE.Dec to the output of QHE.Eval$^{C_\kappa}$ is at most $p(\kappa)$.*

If QHE is $\mathscr{S}$-compact for $\mathscr{S}$ the class of all quantum circuits over some universal gate set, then we simply say that QHE is compact.

Although this work leaves open the question of quantum fully homomorphic encryption, we have established all the machinery relevant for a formal definition:

Definition 6 (Quantum Fully Homomorphic Encryption). *A scheme is a quantum fully homomorphic encryption scheme if it is both compact and homomorphic for the class of all quantum circuits over some universal gate set.*

3.4 Indivisible Schemes

In general, a quantum system is not equal to the sum of its parts. Because of this, for one of our schemes (as given in Sect. 6), it is convenient (if not necessary, by the no-cloning theorem [34]) to define the output of QHE.Eval as containing, in addition to a series of cipherstates corresponding to each qubit, some auxiliary quantum register, possibly entangled with each cipherstate. Then the decryption operation, QHE.Dec must operate on the entire quantum system, rather than qubit-by-qubit. This is in contrast to a classical scheme, in which we could make a copy of the auxiliary register for each encrypted bit, enabling the decryption of individual bits, without decrypting the entire system.

Definition 7. *An* indivisible *quantum homomorphic encryption scheme is a QHE scheme with* QHE.Eval *and* QHE.Dec *re-defined as:*

Homomorphic Evaluation. QHE.Eval^C : $D(\mathcal{R}_{evk} \otimes \mathcal{C}^{\otimes n}) \to D(\mathcal{R}_{aux} \otimes \mathcal{C}'^{\otimes m})$. *Compared to* QHE.Eval *in a standard QHE, this algorithm outputs an additional auxiliary quantum register* $\mathcal{R}_{aux}$. *This extra information is used in the decryption phase. Since the state of* $\mathcal{R}_{aux}$ *may be entangled with the state of each* $\mathcal{C}'$, *the system in* $\mathcal{R}_{aux} \otimes \mathcal{C}'^{\otimes m}$ *can no longer be considered subsystem-by-subsystem.*

Decryption. QHE.Dec$_{sk}$: $D(\mathcal{R}_{aux} \otimes \mathcal{C}'^{\otimes m}) \to D(\mathcal{M}^{\otimes m})$. *For every possible value of* sk, Dec$_{sk}$ *is a quantum channel that maps an auxiliary register, together with an m-fold cipherstate, to an m-fold message in* $D(\mathcal{M}^{\otimes m})$.

We need to define compactness for an indivisible scheme.

Definition 8 *($\mathscr{S}$-compactness for an indivisible scheme). Fix a class of quantum circuits,* $\mathscr{S} = \{\mathscr{S}_\kappa\}_{\kappa \in \mathbb{N}}$. *An indivisible QHE scheme* QHE *is $\mathscr{S}$-compact if there exists a polynomial p such that for any sequence of circuits* $\{C_\kappa \in \mathscr{S}_\kappa\}_\kappa$ *with channels* $\Phi_{C_\kappa} : \mathcal{M}^{\otimes n(\kappa)} \to \mathcal{M}^{\otimes m(\kappa)}$, *the circuit complexity of applying* QHE.Dec$^{\otimes m(\kappa)}$ *to the output of* QHE.Eval$^{C_\kappa}$ *is at most* $p(\kappa, m(\kappa))$.

The trivial quantum fully homomorphic encryption scheme, TRIV, is easily phrased as an indivisible scheme. Informally, TRIV is defined by taking TRIV.KeyGen and TRIV.Enc from any public-key encryption scheme, letting TRIV.Eval^C append a description of C to the cipherstate, and TRIV.Dec decode the cipherstate, and then apply C. Clearly, TRIV is homomorphic, but it is not compact, since TRIV.Dec must evaluate the quantum circuit C, and so its complexity scales with $G(C)$, the number of gates in C.

Although a decryption procedure with any dependence on G, or any other property of C, is not compact, it is still interesting to consider schemes whose decryption procedure has complexity that scales sublinearly in G (such schemes are called *quasi-compact* schemes [14]). We give a formal definition that quantifies this notion for indivisible quantum homomorphic encryption schemes.

Definition 9 (quasi-compactness). *Let* $\mathscr{S} = \{\mathscr{S}_\kappa\}_\kappa$ *be the set of all quantum circuits over some fixed universal gate set. For any* $f : \mathscr{S} \to \mathbb{R}_{\geq 0}$, *an indivisible QHE scheme* QHE *is f-quasi-compact if there exists a polynomial p such that for any sequence of circuits* $\{C_\kappa \in \mathscr{S}_\kappa\}_\kappa$ *with induced channels* $\Phi_{C_\kappa} : \mathcal{M}^{\otimes n(\kappa)} \to \mathcal{M}^{\otimes m(\kappa)}$, *the circuit complexity of decrypting the output of* QHE.Eval$^{C_\kappa}$ *is at most* $f(C_\kappa)p(\kappa, m(\kappa))$.

This definition allows us to consider schemes whose decryption complexity scales with some property of the evaluated circuit. We consider such a scaling non-trivial when it is smaller than $G(C)$, the number of gates in C.

3.5   Symmetric-Key Quantum Homomorphic Encryption

We have defined quantum homomorphic encryption as a *public-key* encryption scheme. For technical reasons, our final scheme, AUX is given in the symmetric-key setting, so in this section we define *symmetric-key* quantum homomorphic encryption. In the case of classical FHE, symmetric-key encryption is known to be *equivalent* to public-key encryption [28]. In the quantum case, this is not known. This section also contains the definition of a *bounded* QHE scheme, which we again require for technical reasons in our symmetric-key scheme, AUX.

Definition 10. *A* symmetric-key *QHE scheme is a quantum homomorphic encryption scheme with* QHE.KeyGen *and* QHE.Enc *re-defined as:*

Key Generation. QHE.KeyGen : $1^\kappa \to (sk, \rho_{evk})$. *This algorithm takes a unary representation of the security parameter as input and outputs a secret encryption/decryption key sk and a quantum evaluation key* $\rho_{evk} \in D(\mathcal{R}_{evk})$.
Encryption. QHE.Enc$_{sk}$: $D(\mathcal{M}) \to D(\mathcal{C})$. *For every possible value of sk, the quantum channel* Dec$_{sk}$ *maps a state in the message space* $\mathcal{M}$ *to a state (the* cipherstate*) in the cipherspace* $\mathcal{C}$.

Next, we define a quantum homomorphic encryption scheme that is *bounded* by n, which forces the number of ciphertexts encrypted by sk to be at most n. Furthermore, the scheme maintains a counter, d, of the number of previous encryptions, which can be thought of as allowing the scheme to avoid key reuse.

Definition 11. *A* bounded *symmetric-key QHE scheme is a symmetric-key QHE scheme with* QHE.KeyGen, QHE.Enc, *and* QHE.Dec *re-defined as:*

Key Generation. QHE.KeyGen : $(1^\kappa, 1^n) \to (sk, \rho_{evk})$.
Encryption. QHE.Enc$_{sk,d}$: $D(\mathcal{M}) \to D(\mathcal{C})$. *Every time* QHE.Enc$_{sk,d}$ *is called, the register containing d is incremented:* $d \leftarrow d + 1$. *If* $d > n$, QHE.Enc$_{sk,d}$ *outputs* $\bot$, *indicating an error.*
Decryption. QHE.Dec$_{sk,d}$: $D(\mathcal{C}') \to D(\mathcal{M})$.

We can define q-IND-CPA security for the symmetric-key setting by allowing the adversary access to an encryption oracle Enc$_{sk}(\cdot)$. We give details in [9].

4   Main Contributions

We now formally state our main results. Our first theorem, Theorem 1, establishes quantum homomorphic encryption for Clifford circuits.

Theorem 1. *(Clifford scheme,* CL*). Let* $\mathscr{S}$ *be the class of Clifford circuits. Then assuming the existence of a classical fully homomorphic encryption scheme that is q-IND-CPA secure, there exists a quantum homomorphic encryption scheme that is q-IND-CPA, compact and* $\mathscr{S}$-*homomorphic.*

Next, we consider two variants of the scheme given by Theorem 1. Each variant deals with non-Clifford T-gates in a different way. The first scheme, described in Theorem 2 and formally defined in Sect. 6, uses entanglement to implement T-gates, resulting in a QHE scheme in which the complexity of decryption scales with the number of T-gates in the homomorphically evaluated circuit.

Theorem 2. *(entanglement-based scheme, EPR). Let $\mathscr{S}$ be the set of all quantum circuits over the universal gate set $\{X, Z, P, H, CNOT, T\}$. Then assuming the existence of a classical fully homomorphic encryption scheme that is q-IND-CPA secure, there exists an indivisible quantum homomorphic encryption scheme that is q-IND-CPA, $\mathscr{S}$-homomorphic and R^2-quasi-compact, where $R(C)$ is the number of T-gates in a circuit C.*

The compactness of the scheme EPR is nontrival for all circuits in which $R^2 \ll G$, where G is the number of gates.

Our second scheme, formally defined in Sect. 7, is based on the use of auxiliary qubits to implement T-gates, resulting in a QHE scheme that is homomorphic for circuits with constant T-depth, as described in the following theorem:

Theorem 3. *(auxiliary-qubit scheme, AUX). Fix a constant L. Let $\mathscr{S}$ be the set of quantum circuits over the universal gate set $\{X, Z, P, H, CNOT, T\}$ with T-depth at most L. Then assuming the existence of a classical fully homomorphic encryption scheme that is q-IND-CPA secure, there exists a bounded symmetric-key quantum homomorphic encryption scheme that is q-IND-CPA, $\mathscr{S}$-homomorphic and compact.*

The QHE scheme in Theorem 3 can be seen as somewhat analogous to an important building block in classical fully homomorphic encryption: a *levelled* fully homomorphic scheme, which is a scheme that takes a parameter L, which is an a-priori bound on the *depth* of the circuit that can be evaluated. However, we note that in contrast to a levelled fully homomorphic scheme, in which operations are polynomial in L, the complexity of our scheme is a polynomial of degree exponential in L, so we really require L to be constant.

As previously noted, Theorems 2 and 3 are complementary: the scheme EPR becomes less compact as the *number* of T-gates increases, while the scheme AUX becomes inefficient as the *depth* of T-gates increases.

5   Homomorphic Encryption for Clifford Circuits: CL

In this section, we present CL, a compact quantum homomorphic encryption scheme for Clifford circuits. This is a building block for the schemes that follow in Sects. 6 and 7. In the full version [9], we prove that CL is q-IND-CPA secure, and homomorphic for Clifford circuits, hence proving Theorem 1.

By definition, Clifford circuits conjugate Pauli operators to Pauli operators [19]. In other words, for any Clifford C, and any Pauli, Q, there exists a

Pauli Q' such that $CQ = Q'C$. Furthermore, applying a random Pauli operator is a perfectly secure symmetric-key quantum encryption scheme: the quantum one-time pad. Thus, it is possible to perform any Clifford circuit on quantum data that is encrypted using the quantum one-time pad. We can apply the desired Clifford, C, to the encrypted state $Q|\psi\rangle$ to get $Q'(C|\psi\rangle)$. Now decrypting the state requires applying the Pauli Q'. If Q can be described by the encryption key $(a_1, \ldots, a_n, b_1, \ldots, b_n)$ — that is, $Q = X^{a_1}Z^{b_1} \otimes \cdots \otimes X^{a_n}Z^{b_n}$ — then Q' can be described by some key $(a_1', \ldots, a_n', b_1', \ldots, b_n')$ depending on C and $(a_1, \ldots, a_n, b_1, \ldots, b_n)$. We describe this dependence by a function $f^C : \mathbb{F}_2^{2n} \to \mathbb{F}_2^{2n}$, which we call a *key update rule*. We need only consider key update rules for each gate in our gate set, which consists of the one- and two-qubit gates in $\{X, Z, P, CNOT, H\}$. For a single-qubit gate C, since the only keys that are affected are those corresponding to the wire to which C is applied, an update rule can be more succinctly described by a pair of functions $f_a^C, f_b^C : \mathbb{F}_2^2 \to \mathbb{F}_2$ such that when C is applied to the ith wire, $a_i' = f_a^C(a_i, b_i)$ and $b_i' = f_b^C(a_i, b_i)$:

$$X^{a_i}Z^{b_i}|\psi\rangle \!-\!\boxed{C}\!-\! X^{a_i'}Z^{b_i'}C|\psi\rangle \qquad a_i \leftarrow a_i' = f_a^C(a_i, b_i), \;\; b_i \leftarrow b_i' = f_b^C(a_i, b_i)$$

For the CNOT-gate, the update rule is described by a 4-tuple of functions, since CNOT acts on two wires. We give the key update rules for all gates in the full version [9, App. C] (We also give key update rules for single-qubit measurement and qubit preparation, so that our scheme is actually homomorphic for stabilizer circuits.) By applying these rules after each gate, we can update the key so that the output is correctly decrypted (since we are actually carrying out computations on encrypted quantum data—in contrast to merely simulating a quantum computation—we note that all gates except the Pauli gates require quantum operations). Such a technique was already used, *e.g.* in [6, 10, 13].

This solution, however, requires that the key updates be executed by the party holding the encryption keys: an "easy" classical computation, but nevertheless a computation that is polynomial in the *size* of the circuit. In the context of quantum homomorphic encryption, the challenge is therefore to allow the execution of *arbitrary* Clifford circuits, while maintaining the compactness condition. Here, we present a quantum public-key encryption scheme which is a hybrid of the quantum one-time pad and of a classical fully homomorphic encryption scheme. This encryption scheme is used to perform key updates on encrypted quantum one-time pad keys, enabling the computation of arbitrary Clifford group circuits on the encrypted quantum states, while maintaining the compactness condition. More precisely, to homomorphically evaluate a Clifford circuit consisting of a sequence of gates $c_1, \ldots, c_G$, we apply the gates to the quantum one-time pad encrypted message, and homomorphically evaluate the function $f^{c_1} \circ \cdots \circ f^{c_G}$ on the encrypted one-time pad keys $a_1, \ldots, a_n, b_1, \ldots, b_n$, where $\circ$ denotes function composition. To accomplish this, we keep track of functions for each bit of the quantum one-time pad encryption key, $\{f_{a,i}, f_{b,i}\}_{i=1}^n$. Since each of the key update rules (see [9]) is linear, each $f_{a,i}$ and $f_{b,i}$ is a linear polynomial in $\mathbb{F}_2[a_1, \ldots, a_n, b_1, \ldots, b_n]$ (from the perspective of the evaluation procedure, $a_1, \ldots, a_n, b_1, \ldots, b_n$ are unknowns), so we refer to them as *key-polynomials.* Before we begin to evaluate the circuit, the key polynomials are the monomials $f_{a,i} = a_i$ and $f_{b,i} = b_i$. As we evaluate each gate c_j, we update the key-polynomials corresponding to the affected wires by composing them with the key update rules. To compute the new encrypted one-time pad keys once the circuit is complete, we homomorphically evaluate each key-polynomial on the old encrypted one-time pad keys. We note that since the key update rules (see [9]) are all linear, for the scheme CL, the underlying classical fully homomorphic scheme only needs to be additively homomorphic.

We define our scheme CL as a QHE scheme. Here and throughout, we assume HE to be a classical FHE scheme that is q-IND-CPA secure (see Definition 1). As noted, such a scheme could be derived from [5]. All of our schemes operate on qubit circuits, and encrypt qubit-by-qubit. Thus we fix $\mathcal{M} = \mathbb{C}^{\{0,1\}}$. Ciphertexts consist of quantum states in $\mathbb{C}^{\{0,1\}}$, combined with classical strings. Specifically, if C is the output space of HE.Enc, and C' is the output space of HE.Eval, then we define $\mathcal{C} = \mathbb{C}^{C \times C} \otimes \mathcal{X}$, where $\mathcal{X} \equiv \mathbb{C}^{\{0,1\}}$, and $\mathcal{C}' = \mathbb{C}^{C' \times C'} \otimes \mathcal{X}$.

Key Generation. CL.KeyGen(1^κ). For key generation, execute $(pk, sk, evk) \leftarrow$ HE.Keygen(1^κ). Output the obtained secret key, sk, and public key, pk. The evaluation key ρ_{evk} takes the value of the classical state $\rho(evk)$.

Encryption. CL.Enc$_{pk} : D(\mathcal{M}) \to D(\mathcal{C})$. Encryption is defined as

$$\mathsf{CL.Enc}_{pk}(\rho^{\mathcal{M}}) = \sum_{a,b \in \{0,1\}} \frac{1}{4} \rho(\mathsf{HE.Enc}_{pk}(a), \mathsf{HE.Enc}_{pk}(b)) \otimes \mathsf{QEnc}_{a,b}(\rho^{\mathcal{M}}).$$

Homomorphic Evaluation. CL.Eval$^C : D(\mathcal{R}_{evk} \otimes \mathcal{C}^{\otimes n}) \to D(\mathcal{C}'^{\otimes m})$.

Suppose $\mathsf{C} = c_1, \ldots, c_G$ is a Clifford circuit.

1. For all $i \in [n]$, set $f_{a,i} \leftarrow a_i$, $f_{b,i} \leftarrow b_i$.
2. For $j = 1, \ldots, G$ such that c_j is a gate or a measurement:
   (a) Apply the gate c_j to the state: $\rho \leftarrow c_j \rho c_j^{-1}$.
   (b) Compose the key update rules with the key-polynomials of the affected wires: if c_j is a single qubit gate or measurement acting on the ith wire, update as $(f_{a,i}, f_{b,i}) \leftarrow (f_{a,i} \circ f_a^{c_j}, f_{b,i} \circ f_b^{c_j})$. If c_j is a CNOT-gate acting on wires i and i', update $(f_{a,i}, f_{a,i'}, f_{b,i}, f_{b,i'})$.
3. Update the classical encryptions by computing

$$c_i = (\mathsf{HE.Eval}_{evk}^{f_{a,i}}(\tilde{a}_i), \mathsf{HE.Eval}_{evk}^{f_{b,i}}(\tilde{b}_i)).$$

4. Output $(c_1, \ldots, c_m, \rho)$.

Decryption. CL.Dec$_{sk} : D(\mathcal{C}') \to D(\mathcal{M})$. For $\tilde{a}, \tilde{b} \in C'$, decryption is defined:

$$\mathsf{CL.Dec}_{sk} : |\tilde{a}\rangle\langle\tilde{a}| \otimes |\tilde{b}\rangle\langle\tilde{b}| \otimes \rho^{\mathcal{X}} \mapsto \mathsf{QDec}_{\mathsf{HE.Dec}_{sk}(\tilde{a}), \mathsf{HE.Dec}_{sk}(\tilde{b})}(\rho^{\mathcal{X}}),$$

We prove the homomorphic and security properties of CL in [9].

6 T-gate Computation Using Entanglement: EPR

In order to achieve universality for quantum circuits, we need to add a non-Clifford group gate, such as the T-gate. As noted in Sect. 1.1, if we apply the same technique as in Sect. 5 (*i.e.* to apply the T-gate on the encrypted quantum data) we run into a problem, since $\mathsf{T}\mathsf{X}^a\mathsf{Z}^b = \mathsf{X}^a\mathsf{Z}^{a\oplus b}\mathsf{P}^a\mathsf{T}$ That is, conditioned on a, the output picks up an undesirable P error, which cannot be corrected by applying Pauli corrections. In [10], Childs arrives at the same conclusion, and makes the observation that, in the case where $a = 1$, the evaluation algorithm could be made to *correct* this erroneous P-gate. As long as the evaluation algorithm does not find out if this correction is being executed or not, security holds. The solution in [10] involves quantum interaction; this was recently improved to a single auxiliary qubit, coupled with classical interaction [6,13]. As a proof technique (for establishing security), [6,13] considers an equivalent, entanglement-based protocol. Here, we use the idea of exploiting entanglement in order to *delay* the correction required for the evaluation of the T-gate on encrypted data. The protocol is illustrated in Fig. 2. Correctness of Fig. 2 is proven in the full version [9].

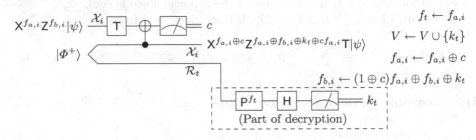

Fig. 2. Evaluation protocol for the tth T-gate, on the ith wire. The key-polynomials $f_{a,i}$ and $f_{b,i}$ are in $\mathbb{F}_2[V]$. After the protocol, V gains a new variable corresponding to the unknown measurement result k_t. The dashed box shows part of the decryption, which happens at some point in the future, after the complete evaluation is finished.

Figure 2 shows that, using the state $|\Phi^+\rangle = \frac{1}{\sqrt{2}}(|00\rangle + |11\rangle)$, the conditional P correction can be delayed. The cost of this is that the value of the measurement result, k_t, on auxiliary register $\mathcal{R}_t$, is undetermined until later, when it is measured as part of the decryption. Thus we view the key updates as a symbolic computation: each time a T-gate is applied, an extra variable, k_t, is introduced.

For the first T-gate evaluation ($t = 1$), the evaluation procedure does not have the knowledge to evaluate $f_1 = f_{a,i}$, where i is the wire upon which the gate is performed, in order to perform the correction. It is possible (using the classical scheme HE), to compute a classical ciphertext $\widetilde{f}_1$ that decrypts to $f_1(a_1, b_1, \ldots, a_n, b_n)$. Thus, for this T-gate, the output part of the auxiliary system contains both $\widetilde{f}_1$ and the register $\mathcal{R}_1$. As part of the decryption operation, compute $f_1 \leftarrow \mathsf{HE.Dec}(\widetilde{f}_1)$, and apply P^{f_1} on $\mathcal{R}_1$ before measuring in the Hadamard basis and obtaining k_1. From the point of view of the evaluation procedure, k_1 is unknown and so it becomes an *unknown* part of the encryption key (in contrast with the previous keys, which are also "unknown", but to a lesser degree, since we have access to the classical encrypted values of these keys). The algorithm Eval continues in this fashion for values of t up to R; each time, the set of unknown variables increasing by one. Note that, according to Fig. 2, as well as the linearity of the key update rules, for all t, $f_t \in \mathbb{F}_2[a_1, \ldots, a_n, b_1, \ldots, b_n, k_1, \ldots, k_{t-1}]$ is linear (since c is a known constant), so we can write $f_t = f_t^k + f_t^{ab}$ for $f_t^k \in \mathbb{F}_2[k_1, \ldots, k_{t-1}]$ and $f_t^{ab} \in \mathbb{F}_2[a_1, \ldots, a_n, b_1, \ldots, b_n]$.

The cost of this construction is that each T-gate adds to the complexity of the decryption procedure, since, in particular, for each T-gate, we must perform a possible P-correction and a measurement on an auxiliary qubit. In addition, we cannot evaluate the key-polynomials, nor the f_t, until the variables k_t have been measured, so this evaluation must take place in the decryption phase, increasing the dependence on R, the number of T-gates, to $O(R^2)$ (see full version [9]).

We now formally define the indivisible QHE scheme, EPR. As in CL, we have message space $\mathcal{M} = \mathbb{C}^{\{0,1\}}$ and cipherspace $\mathcal{C} = \mathbb{C}^{C \times C} \otimes \mathcal{X}$, where C is the output space of HE.Enc and $\mathcal{X} \equiv \mathbb{C}^{\{0,1\}}$. Since EPR is indivisible, the output space of EPR.Eval^C has the form $\mathcal{R}_{aux} \otimes \mathcal{C}'^{\otimes m}$. In our case, we have $\mathcal{R}_{aux} = \mathcal{R}_1 \otimes \cdots \otimes \mathcal{R}_R \otimes (\mathbb{C}^{\{0,1\}^{R+1}})^{\otimes R} \otimes (\mathbb{C}^{C'})^{\otimes R}$, where R is the number of T-gates, C' is the output space of HE.Eval, and $\mathcal{R}_t \equiv \mathbb{C}^{\{0,1\}}$. The classical parts of the auxiliary space allow us to output R linear polynomials in $\mathbb{F}_2[k_1, \ldots, k_R]$ corresponding to $\{f_t^k\}_{t=1}^R$, each of which can be represented with $R+1$ bits; as well as R HE.Eval outputs, corresponding to encryptions of $\{f_t^{ab}(a_1, \ldots, a_n, b_1, \ldots, b_n)\}_{t=1}^R$. Similarly, we have $\mathcal{C}' = (\mathbb{C}^{\{0,1\}^{R+1}})^{\otimes 2} \otimes \mathbb{C}^{C' \times C'} \otimes \mathcal{X}$.

The key generation, EPR.KeyGen, and encryption, EPR.Enc, are defined exactly as CL.KeyGen and CL.Enc. We now define EPR.Eval and EPR.Dec.

Evaluation. EPR.Eval$_{evk}$. As in CL, apply gates in $\{X, Z, P, H, CNOT\}$ directly on the encrypted quantum registers. For the T-gate, use the gadget defined in Fig. 2. This gadget differs from previous gadgets in that it uses an auxiliary Bell state, $|\Phi^+\rangle$. After the system of the ith wire, $\mathcal{X}_i$, is measured, relabel half of the Bell state as $\mathcal{X}_i$, and the other half as $\mathcal{R}_t$, which is returned as part of $\mathcal{R}_{aux}$. The full evaluation procedure is as follows.

1. Set $V \leftarrow \{a_i, b_i\}_{i \in [n]}$, and $\forall i \in [n]$, $f_{a,i} \leftarrow a_i$, $f_{b,i} \leftarrow b_i$.
2. Let $g_1, \ldots, g_G$ be a topological ordering of the gates in C. For $j = 1, \ldots, G$, evaluate g_j using the appropriate gadget.
3. Let S be the set of output wires. Let $\mathcal{L}$ be the set of labels $\mathcal{L} = \{(a, i), (b, i) : i \in S\} \cup \{1, \ldots, R\}$. For each $\alpha \in \mathcal{L}$, we want to homomorphically evaluate f_α to obtain the actual (encrypted) key, but we can only actually evaluate the part of f_α that is in the variables $\{a_i, b_i\}_i$ — the $\{k_t\}_t$ are still unknown. Recall that we can write $f_\alpha = f_\alpha^k + f_\alpha^{ab}$ for $f_\alpha^k \in \mathbb{F}[k_1, \ldots, k_R]$ and $f_\alpha^{ab} \in \mathbb{F}_2[a_1, \ldots, a_n, b_1, \ldots, b_n]$. Compute $\widetilde{f_\alpha^{ab}} \leftarrow \mathsf{HE.Eval}_{evk}^{f_\alpha^{ab}}(\tilde{a}_1, \ldots, \tilde{a}_n, \tilde{b}_1, \ldots, \tilde{b}_n)$.
4. Output: the $m = |S|$ qubit registers $\{\mathcal{X}_i : i \in S\}$ corresponding to the encrypted output of the circuit; the R qubit registers $\mathcal{R}_1, \ldots, \mathcal{R}_R$ correspond-

ing to auxiliary states created by T-gadgets; the polynomials $\{f_\alpha^k\}_{\alpha \in \mathcal{L}} \subset \mathbb{F}_2[k_1, \ldots, k_R]$ and the homomorphically evaluated polynomials $\{\widetilde{f_\alpha^{ab}}\}_{\alpha \in \mathcal{L}}$.

Decryption. $\mathsf{EPR.Dec}_{sk}$. In order to decrypt, measure the $\mathcal{R}_t$ in order from 1 to R, computing $f_t(k_1, \ldots, k_{t-1})$ as required. Formally:

1. For $t = 1, \ldots, R$:
   (a) Decrypt $f_t^{ab} \leftarrow \mathsf{HE.Dec}_{sk}(\widetilde{f_t^{ab}})$.
   (b) Compute $a \leftarrow f_t^k(k_1, \ldots, k_{t-1}) \oplus f_t^{ab}$ and apply HP^a to $\mathcal{R}_t$.
   (c) Measure $\mathcal{R}_t$ to get k_t.
2. Let S be the set of indices of the output qubit registers. For $i \in S$:
   (a) Decrypt $f_{a,i}^{ab} \leftarrow \mathsf{HE.Dec}_{sk}(\widetilde{f_{a,i}^{ab}})$ and $f_{b,i}^{ab} \leftarrow \mathsf{HE.Dec}_{sk}(\widetilde{f_{b,i}^{ab}})$.
   (b) Compute $a_i \leftarrow f_{a,i}^k(k_1, \ldots, k_t) \oplus f_{a,i}^{ab}$ and $b_i \leftarrow f_{b,i}^k(k_1, \ldots, k_t) \oplus f_{b,i}^{ab}$.
3. To each register $\mathcal{X}_i$, apply the map QDec_{a_i, b_i}. Output registers $\mathcal{X}_1, \ldots, \mathcal{X}_m$.

We prove that EPR is homomorphic for all quantum circuits in the universal gate set $\{\mathsf{X}, \mathsf{Z}, \mathsf{P}, \mathsf{CNOT}, \mathsf{H}, \mathsf{T}\}$, R^2-quasi-compact, and q-IND-CPA, in [9].

7   T-gate Computation Using Auxiliary States: AUX

In the previous QHE scheme, we solved the problem of performing the P correction by *delaying* the correction via entanglement. In this section, we present a quantum homomorphic encryption scheme, AUX, that takes a more proactive approach to dealing with the P correction. At a high level, AUX can be understood as the following: as part of the evaluation key, AUX.Keygen outputs a number of auxiliary states. These states "encode" parts of the original encryption key, and are used to correct for the errors induced by the straightforward application of the T-gate on the cipherstates. In more details, the auxiliary states encode hidden versions of P corrections, such as $|+_{a,k}\rangle := \mathsf{Z}^k \mathsf{P}^a |+\rangle$ (where k is a random bit and a is an encryption key) that are useful for the evaluation of the T-gate (see Fig. 3). In general (after having applied prior gates), the exact auxiliary state will not be available; instead, the Eval procedure combines a number of auxiliary states in order to create a single copy of a state that is useful for performing the correction. This combination operation, however, is expensive as it introduces new unknowns (in terms of new variables as well as "cross-terms"), that need to be corrected in any future T-gate. Thus the size of the evaluation key grows rapidly, as a polynomial whose degree is exponential in the T-depth. We can thus tolerate only a *constant* T-gate depth for this scheme to be efficient.

We further specify that AUX is a symmetric-key encryption scheme. This is because AUX.KeyGen generates auxiliary qubits that depend on the quantum one-time pad encryption keys. Also, KeyGen takes an extra parameter 1^n, where n is an upper bound on the total number of qubits that can be encrypted (AUX acts much like a classical one-time pad scheme that picks a fixed-length encryption key ahead of time). After this bound on the number of encryptions has been attained, no further qubits can be encrypted. We will suppose without loss of

$$\mathsf{X}^{f_{a,i}}\mathsf{Z}^{f_{b,i}} \underset{\mathcal{X}_i}{\boxed{\mathsf{T}}} \oplus \boxed{\diagup\!\!\!\!\diagup} = c$$

$$|+_{f_{a,i},k}\rangle \underset{\mathcal{X}_i}{\bullet} \mathsf{X}^{f_{a,i}\oplus c}\mathsf{Z}^{f_{a,i}\oplus f_{b,i}\oplus k\oplus cf_{a,i}}\mathsf{T}$$

$$f_{a,i} \leftarrow f_{a,i} \oplus c$$
$$f_{b,i} \leftarrow f_{a,i} \oplus f_{b,i} \oplus k \oplus cf_{a,i}$$
$$V \leftarrow V \cup \mathrm{var}(k)$$

Fig. 3. A T-gadget for the scheme AUX consists of the above circuit and key-update rules. We use $\mathrm{var}(k)$ to denote the set of variables in the polynomial k, which depends on the construction of the auxiliary state $|+_{f_{a,i},k}\rangle$, described below.

generality that a circuit being homomorphically evaluated is on n wires. Furthermore, the number and type of auxiliary qubits will depend on the T-depth of the circuit to be evaluated, L. The scheme will not be able to homomorphically evaluate circuits with T-depth greater than L. Fix a constant L. We will now define a scheme $\mathsf{AUX} = \mathsf{AUX}_L$ that is homomorphic for all circuits with T-depth at most L.

Providing the necessary auxiliary states for each T-gate would require advance knowledge of the key $f_{a,i}$ at the time a T-gate is applied to the ith wire. Since this depends on both the circuit being applied and prior measurement results, we appear to be at an impasse. The key observation that allows us to continue with this approach is that, given auxiliary states $|+_{f_1,k_1}\rangle$ and $|+_{f_2,k_2}\rangle$, we can combine them to get $|+_{f_1\oplus f_2,k}\rangle$, for some k, using the following circuit:

$$|+_{f_1,k_1}\rangle \underset{}{\bullet}\qquad\qquad |+_{f_1\oplus f_2,k_1\oplus k_2\oplus(f_1\oplus c)f_2}\rangle$$
$$|+_{f_2,k_2}\rangle \oplus \boxed{\diagup\!\!\!\!\diagup} = c$$

By iterating this procedure, given auxiliary states $|+_{f_1,k_1}\rangle, \ldots, |+_{f_r,k_r}\rangle$, we can construct $|+_{f_1\oplus\cdots\oplus f_r,k}\rangle$, where $k = \bigoplus_{i=1}^m k_i \oplus \bigoplus_{i=2}^r c_i f_i \oplus \bigoplus_{i=1}^r \bigoplus_{j=1}^{i-1} f_i f_j$ for known values c_i. Thus, if we give many initial auxiliary states of the form $\{|+_{a_i,k_{a,i}}\rangle, |+_{b_i,k_{b,i}}\rangle\}_i$ (with different keys for different copies), we can construct $|+_{f,k}\rangle$ for f a linear function of $\{a_i, b_i\}_{i\in[n]}$. However, using an auxiliary state $|+_{f_{a,i},k}\rangle$ to facilitate a T-gate on the ith wire introduces the unknown k into $f_{b,i}$. In particular, suppose $f_{a,i} = \bigoplus_{j=1}^r t_j$ for some monomial terms $t_j \in \mathbb{F}_2[V]$. Then we will need to construct it from auxiliary states $|+_{t_1,k_1}\rangle, \ldots, |+_{t_r,k_r}\rangle$, to get $|+_{f_{a,i},k}\rangle$ for $k = \bigoplus_{i=1}^m k_i \oplus \bigoplus_{i=2}^r c_i t_i \oplus \bigoplus_{i=1}^r \bigoplus_{j=1}^{i-1} t_i t_j$. Thus, after the T-gadget, the new keys $f'_{a,i}, f'_{b,i}$ are in unknowns $V \cup \{k_1, \ldots, k_r\}$. Furthermore, because of the cross terms $t_i t_j$, the degree of the key-polynomials increases, so we can no longer assume they are linear. Since we can't produce $|+_{f_1 f_2,k}\rangle$ from $|+_{f_1,k_1}\rangle$ and $|+_{f_2,k_2}\rangle$, we need to provide additional auxiliary states for every possible term. We discuss this more formally below and in the full version [9].

As in CL and EPR, we work with qubits: $\mathcal{M} \equiv \mathbb{C}^{\{0,1\}}$. In contrast to our previous schemes, the classical encryptions of quantum one-time pad keys is part of the evaluation key (for convenience only), so we have $\mathcal{C} \equiv \mathbb{C}^{\{0,1\}}$. However, after evaluation, the classical encryption of the new one-time pad keys is needed for decryption, so as in CL, we have $\mathcal{C}' \equiv \mathbb{C}^{C'\times C'} \otimes \mathcal{X}$, where C' is the output space of HE.Eval, and $\mathcal{X} \equiv \mathbb{C}^{\{0,1\}}$.

Key Generation. AUX.Keygen($1^\kappa, 1^n$). The evaluation key contains auxiliary states that allow each of L layers of T-gates to be implemented. Thus, for each layer, since every wire must have the possibility to implement a T-gate, for each wire, we need to be able to construct an auxiliary state $|+_{f_{a,i,k}}\rangle$ for some k. Since we can add auxiliary states, we can construct this auxiliary state if we have an auxiliary state for each term in $f_{a,i}$. Since $f_{a,i}$ depends on the circuit, which we do not know in advance, we need to provide an auxiliary state for every term that could possibly be in $f_{a,i}$ at the ℓth layer of T-gates, for $\ell = 1, \ldots, L$.

We now define sets of monomials $T_1, \ldots, T_L$ such that the keys in the ℓth layer consist of sums of terms from T_ℓ. Let $V_1 := \{a_i, b_i\}_{i \in [n]}$, and define $T_1 \subset \mathbb{F}_2[V_1]$ by $T_1 := \{a_i, b_i\}_{i \in [n]}$. The monomials in T_1 represent the possible terms in the key-polynomials before the first layer of T-gates. Each of the up to n T-gates in the first layer requires a copy of each of $\{|+_{t, k_t^{(1)}}\rangle\}_{t \in T_1}$, with independent random keys for each, for a total of $n|T_1|$ auxiliary states. More generally, for the ℓth layer of T-gates, we let T_ℓ be the set of possible terms in the key-polynomials before applying the ℓth layer of T-gates. We can see from the T-gadget, as well as the construction for adding auxiliary states that the keys from the previous layer's auxiliary states, $\{k_{1,i}^{(\ell-1)}, \ldots, k_{|T_{\ell-1}|,i}^{(\ell-1)}\}_{i=1}^n$, may now be variables in the key-polynomials, and that products of terms from the previous layer may now be terms in the key-polynomials of the current layer. (This is caused by auxiliary state addition. See [9] for details). Thus, for $\ell > 1$, we can define $T_\ell \subset \mathbb{F}_2[V_\ell]$, where $V_\ell := V_{\ell-1} \cup \{k_{1,i}^{(\ell-1)}, \ldots, k_{|T_{\ell-1}|,i}^{(\ell-1)}\}_{i=1}^n$, by

$$T_\ell := T_{\ell-1} \cup \{tt' : t, t' \in T_{\ell-1}, t \neq t'\} \cup \left\{k_{1,i}^{(\ell-1)}, \ldots, k_{|T_{\ell-1}|,i}^{(\ell-1)}\right\}_{i=1}^n.$$

We then provide each of the n wires with an auxiliary state for each term in T_ℓ, for $\ell = 1, \ldots, L$. We now make this more precise.

To each T_ℓ, we associate a family of strings $\{s^{(\ell)}(x)\}_{x \in \{0,1\}^{V_\ell}}$ in $\{0,1\}^{T_\ell}$, defined so that for every $f \in T_\ell$, the f-entry of $s^{(\ell)}(x)$ is $s_f^{(\ell)}(x) = f(x)$. That is, $s^{(\ell)}(x)$ represents evaluating every monomial in T_ℓ at x. For instance, we have, for any strings $a, b \in \{0,1\}^n$, $s^{(1)}(a,b) = (a_1, \ldots, a_n, b_1, \ldots, b_n)$.

For any strings $s, k \in \{0,1\}^n$, define $\sigma(s, k) := \bigotimes_{i=1}^n |+_{s_i, k_i}\rangle\langle+_{s_i, k_i}|$.

For any string s, let s^{*n} denote the concatenation of n copies of s. For any $a, b \in \{0,1\}^n$ and $k = (k^{(1)}, \ldots, k^{(L)}) \in \{0,1\}^{n|T_1|} \times \cdots \times \{0,1\}^{n|T_L|}$, define

$$\sigma_{aux}^{a,b,k} := \sigma(s^{(1)}(a,b)^{*n}, k^{(1)}) \otimes \cdots \otimes \sigma(s^{(L)}(a, b, k^{(1)}, \ldots, k^{(L-1)})^{*n}, k^{(L)}).$$

We can now define the procedure AUX.KeyGen($1^\kappa, 1^n$):

1. Execute $(pk, sk, evk) \leftarrow$ HE.KeyGen($1^{\kappa+n}$).
2. Choose uniform random $a, b \in \{0,1\}^n$ and $k = (k^{(1)}, \ldots, k^{(L)}) \in \{0,1\}^{n|T_1|} \times \cdots \times \{0,1\}^{n|T_L|}$.
3. Output secret key (sk, a, b, k).

4. Output evaluation key: pk, evk, $\tilde{a}_1 = \mathsf{HE.Enc}_{pk}(a_1), \ldots, \tilde{a}_n = \mathsf{HE.Enc}_{pk}(a_n)$,
$\tilde{b}_1 = \mathsf{HE.Enc}_{pk}(b_1), \ldots, \tilde{b}_n = \mathsf{HE.Enc}_{pk}(b_n)$, $\left(\tilde{k}_i^{(\ell)} = \mathsf{HE.Enc}_{pk}\left(k_{j,i}^{(\ell)} \right) \right)_{\substack{\ell \in [L] \\ i \in [n] \\ j \in [|T_\ell|]}}$,
and $\sigma_{aux}^{a,b,k}$.

Encryption. $\mathsf{AUX.Enc}_{(sk,a,b,k),d} : D(\mathcal{M}) \to D(\mathcal{C})$. The encryption procedure takes an extra parameter d that keeps track of the number of qubits already encrypted (we assume d is initially 1 and not modified outside of $\mathsf{AUX.Enc}$). If $d \leq n$, it applies the quantum one-time pad channel $\mathsf{QEnc}_{a_d,b_d} : D(\mathcal{M}) \to D(\mathcal{C})$. The output is the cipherstate in register $\mathcal{C}$; the parameter d is updated as $d \leftarrow d + 1$. If $d > n$, then output $\perp$ to indicate an error.

Decryption. $\mathsf{AUX.Dec}_{(sk,a,b,k),d} : D(\mathcal{C}') \to D(\mathcal{M})$. The decryption is defined the same as $\mathsf{CL.Dec}_{sk}$.

Homomorphic Evaluation. $\mathsf{AUX.Eval}^\mathsf{C} : D(\mathcal{R}_{evk} \otimes \mathcal{C}^{\otimes n}) \to D(\mathcal{C}'^{\otimes m})$. For Clifford group gates, we apply the gadgets as in $\mathsf{CL.Eval}$. For T-gates, we apply the gadget in Fig. 3. The full evaluation procedure is as follows:

1. Set $V \leftarrow \{a_i, b_i\}_{i \in [n]}$, and $\forall i \in [n]$, $f_{a,i} \leftarrow a_i$, $f_{b,i} \leftarrow b_i$.
2. Let $\mathsf{g}_1, \ldots, \mathsf{g}_G$ be a topological ordering of the gates in C. For $i = 1, \ldots, G$, evaluate g_i using the appropriate gadget.
3. Let S be the set of output wire labels. For each $i \in S$:
   (a) Homomorphically evaluate $f_{a,i}$ and $f_{b,i}$ to obtain updated (encrypted) keys: $\tilde{a}_i \leftarrow \mathsf{HE.Eval}_{evk}^{f_{a,i}}(\tilde{v} : v \in V)$ and $\tilde{b}_i \leftarrow \mathsf{HE.Eval}_{evk}^{f_{b,i}}(\tilde{v} : v \in V)$.
4. Output in $\mathcal{C}'_i$ the classical-quantum system given by:
   – The encrypted keys $\{\tilde{a}_i, \tilde{b}_i\}_{i \in S}$.
   – The output corresponding to the encrypted output qubit i of the circuit.

The correctness of this scheme depends on two facts, which we prove in [9]. First, for every unknown $v \in V$, we have an encrypted copy of $\tilde{v}$, encrypted using HE.Enc. We need these to compute the final keys $\{\tilde{a}_i, \tilde{b}_i\}$ using $f_{a,i}, f_{b,i} \in \mathbb{F}_2[V]$. Finally, for each level ℓ, for each wire label i, we need an auxiliary state $|+_{t,k}\rangle$ for every term that may appear in the key $f_{a,i}$ going into the ℓth level. This allows us to construct the auxiliary qubit required to execute each T-gadget. In the full version [9], we prove that AUX requires $O(n^{2^{L-1}+1})$ auxiliary qubits, from which it follows that AUX is homomorphic for quantum circuits with T-depth L. We further show that AUX is q-IND-CPA and compact.

We remark that if we only had a classical encryption scheme that was homomorphic over linear circuits, and not fully homomorphic, then we could get the same functionality from a slightly modified version of this scheme, in which we include with every auxiliary qubit $|+_{s,k}\rangle\langle+_{s,k}|$, $\mathsf{HE.Enc}_{pk}(s)$ — at the moment we only include some of these, but not those auxiliary states arising from *products* of terms, since we can compute products homomorphically. Since we have classical fully homomorphic encryption, we use this to slightly simplify the scheme, however the observation that the fully homomorphic property is not fully taken advantage of strengthens the idea that Clifford circuits are analogous to classical linear circuits in the context of QIIE.

References

1. Aharonov, D., Ben-Or, M., Eban, E.: Interactive proofs for quantum computations. In: Proceeding of Innovations in Computer Science 2010 (ICS 2010), pp. 453–469 (2010)
2. Ambainis, A., Mosca, M., Tapp, A., De Wolf, R.: Private quantum channels. In: Proceedings of the 41st Annual IEEE Symposium on Foundations of Computer Science (FOCS 2000), pp. 547–553 (2000)
3. Arrighi, P., Salvail, L.: Blind quantum computation. Int. J. Quantum Inf. **4**, 883–898 (2006)
4. Boneh, D., Goh, E.-J., Nissim, K.: Evaluating 2-DNF formulas on ciphertexts. In: Proceedings of the Second Theory of Cryptography Conference (TCC 2005), pp. 325–341 (2005)
5. Brakerski, Z., Vaikuntanathan, V.: Efficient fully homomorphic encryption from (standard) LWE. In: Proceedings of the 52nd Annual IEEE Symposium on Foundations of Computer Science (FOCS 2011), pp. 97–106 (2011). Full version available at Cryptology ePrint Archive, Report 2011/344
6. Broadbent, A.: Delegating private quantum computations. Canadian Journal of Physics (2015). arXiv:1506.01328 [quant-ph]
7. Broadbent, A., Fitzsimons, J., Kashefi, E.: Universal blind quantum computation. In: Proceedings of the 50th Annual IEEE Symposium on Foundations of Computer Science (FOCS 2009), pp. 517–526 (2009)
8. Broadbent, A., Gutoski, G., Stebila, D.: Quantum one-time programs. In: Canetti, R., Garay, J.A. (eds.) CRYPTO 2013, Part II. LNCS, vol. 8043, pp. 344–360. Springer, Heidelberg (2013)
9. Broadbent, A., Jeffery, S.: Quantum homomorphic encryption for circuits of low T-gate complexity (2014). arXiv:1412.8766 [quant-ph]
10. Childs, A.: Secure assisted quantum computation. Quantum Inf. Comput. **5**, 456–466 (2005)
11. van Dijk, M., Gentry, C., Halevi, S., Vaikuntanathan, V.: Fully homomorphic encryption over the integers. In: Gilbert, H. (ed.) EUROCRYPT 2010. LNCS, vol. 6110, pp. 24–43. Springer, Heidelberg (2010)
12. El Gamal, T.: A public key cryptosystem and a signature scheme based on discrete logarithms. In: Blakely, G.R., Chaum, D. (eds.) CRYPTO 1984. LNCS, vol. 196, pp. 10–18. Springer, Heidelberg (1985)
13. Fisher, K.A.G., Broadbent, A., Shalm, L.K., Yan, Z., Lavoie, J., Prevedel, R., Jennewein, T., Resch, K.J.: Quantum computing on encrypted data. Nat. Commun. **5**, 1–7 (2014)
14. Gentry, C.: A fully homomorphic encryption scheme. Ph.D. thesis, Stanford University (2009). http://crypto.stanford.edu/craig
15. Gentry, C.: Fully homomorphic encryption using ideal lattices. In: Proceedings of the 41st annual ACM symposium on Theory of Computing (STOC 2009), pp. 169–178 (2009)
16. Gentry, C., Halevi, S., Vaikuntanathan, V.: i-hop homomorphic encryption and rerandomizable Yao circuits. In: Rabin, T. (ed.) CRYPTO 2010. LNCS, vol. 6223, pp. 155–172. Springer, Heidelberg (2010)
17. Goldwasser, S., Kalai, Y., Popa, R.A., Vaikuntanathan, V., Zeldovich, N.: Reusable garbled circuits and succinct functional encryption. In: Proceedings of the 45th Annual ACM Symposium on Theory of Computing (STOC 2013), pp. 555–564 (2013)

18. Goldwasser, S., Micali, S.: Probabilistic encryption. J. Comput. Syst. Sci. **28**(2), 270–299 (1984)
19. Gottesman, D.: The Heisenberg representation of quantum computers. In: Group 22: Proceedings of the XXII International Colloquium on Group Theoretical Methods in Physics, pp. 32–43 (1998)
20. Koshiba, T.: Security notions for quantum public-key cryptography. IEICE Trans. Fundam. Electron. Commun. Comput. Sci. (Jpn. Ed.) **J90**-A(5), 367–375 (2007). arXiv:quant-ph/0702183
21. Micciancio, D., Regev, O.: Lattice-based cryptography. In: Bernstein, D.J., Buchmann, J., Dahmen, E. (eds.) Post-Quantum Cryptography, pp. 147–191. Springer, Berlin (2009)
22. Okamoto, T., Tanaka, K., Uchiyama, S.: Quantum public-key cryptosystems. In: Bellare, M. (ed.) CRYPTO 2000. LNCS, vol. 1880, pp. 147–165. Springer, Heidelberg (2000)
23. Paillier, P.: Public-key cryptosystems based on composite degree residuosity classes. In: Stern, J. (ed.) EUROCRYPT 1999. LNCS, vol. 1592, pp. 223–238. Springer, Heidelberg (1999)
24. Regev, O.: On lattices, learning with errors, random linear codes, and cryptography. In: Proceedings of the 37th annual ACM symposium on Theory of computing, (STOC 2005), pp. 84–93 (2005)
25. Regev, O.: On lattices, learning with errors, random linear codes, and cryptography. J. ACM **56**(6), 34:1–34:40 (2009)
26. Rivest, R., Adleman, L., Dertouzos, M.: On data banks and privacy homomorphisms. Found. Secure Comput. **4**(11), 169–177 (1978)
27. Rohde, P.P., Fitzsimons, J.F., Gilchrist, A.: Quantum walks with encrypted data. Phys. Rev. Lett. **109**, 150501 (2012)
28. Rothblum, R.: Homomorphic encryption: from private-key to public-key. In: Ishai, Y. (ed.) TCC 2011. LNCS, vol. 6597, pp. 219–234. Springer, Heidelberg (2011)
29. Sander, T., Young, A., Yung, M.: Non-interactive cryptocomputing for NC^1. In: Proceedings of the 40th Annual Symposium on the Foundations of Computer Science (FOCS 1999), pp. 554–566 (1999)
30. Song, F.: A note on quantum security for post-quantum cryptography. In: Mosca, M. (ed.) PQCrypto 2014. LNCS, vol. 8772, pp. 246–265. Springer, Heidelberg (2014)
31. Tan, S.-H., Kettlewell, J.A., Ouyang, Y., Chen, L., Fitzsimons, J.F.: A quantum approach to homomorphic encryption (2014). http://arxiv.org/abs/1411.5254
32. Vaikuntanathan, V.: Computing blindfolded: new developments in fully homomorphic encryption. In: Proceedings of the 52nd Annual IEEE Symposium on Foundations of Computer Science, (FOCS 2011), pp. 5–16 (2011)
33. Dunjko, V., Fitzsimons, J.F., Portmann, C., Renner, R.: Composable security of delegated quantum computation. In: Sarkar, P., Iwata, T. (eds.) ASIACRYPT 2014, Part II. LNCS, vol. 8874, pp. 406–425. Springer, Heidelberg (2014)
34. Wootters, W.K., Zurek, W.H.: A single quantum cannot be cloned. Nature **299**(5886), 802–803 (1982)
35. Xiang, C., Yang, L.: Indistinguishability and semantic security for quantum encryption scheme. In: Proceedings of the SPIE 8554, Quantum and Nonlinear Optics II, p. 85540G (2012)
36. Yu, L., Perez-Delgado, C.A., Fitzsimons, J.F.: Limitations on information-theoretically-secure quantum homomorphic encryption. Phys. Rev. A **90**(5), 050303 (2014)

Multi-identity and Multi-key Leveled FHE from Learning with Errors

Michael Clear and Ciarán McGoldrick[✉]

School of Computer Science and Statistics, Trinity College,
Dublin, Republic of Ireland
Ciaran.McGoldrick@scss.tcd.ie

Abstract. Gentry, Sahai and Waters recently presented the first (leveled) identity-based fully homomorphic (IBFHE) encryption scheme (CRYPTO 2013). Their scheme however only works in the single-identity setting; that is, homomorphic evaluation can only be performed on ciphertexts created with the same identity. In this work, we extend their results to the multi-identity setting and obtain a multi-identity IBFHE scheme that is selectively secure in the random oracle model under the hardness of Learning with Errors (LWE). We also obtain a multi-key fully-homomorphic encryption (FHE) scheme that is secure under LWE in the standard model. This is the first multi-key FHE based on a well-established assumption such as standard LWE. The multi-key FHE of López-Alt, Tromer and Vaikuntanathan (STOC 2012) relied on a non-standard assumption, referred to as the Decisional Small Polynomial Ratio assumption.

1 Introduction

Fully homomorphic encryption (FHE) is a cryptographic primitive that facilitates arbitrary computation on encrypted data. Since Gentry's breakthrough realization of FHE in 2009 [1], many improved variants have appeared in the literature [2–6].

A leveled FHE scheme allows an evaluator to evaluate a circuit of limited depth L. The parameter L must be specified in advance when generating the public parameters of the scheme, whose size may depend on L. Furthermore, a leveled homomorphic scheme allows L to be polynomial in the security parameter. A "pure" fully homomorphic encryption scheme allows circuits of unlimited depth to be evaluated. However, for many applications in practice, a leveled scheme is adequate.

Identity-Based Encryption (IBE) is centered around the notion that a user's public key can be efficiently derived from an identity string and system-wide public parameters / master public key. The public parameters are chosen by a trusted authority (TA) along with a secret trapdoor (master secret key), which is used to extract secret keys for user identities. The first secure IBE schemes were presented in 2001 by Boneh and Franklin [7] (based on bilinear pairings), and Cocks [8] (based on the quadratic residuosity problem).

© International Association for Cryptologic Research 2015
R. Gennaro and M. Robshaw (Eds.): CRYPTO 2015, Part II, LNCS 9216, pp. 630–656, 2015.
DOI: 10.1007/978-3-662-48000-7_31

At Crypto 2013, Gentry, Sahai and Waters presented the first (leveled) identity-based fully homomorphic encryption (IBFHE) scheme [6]. Their scheme is secure under the hardness of the Learning with Errors (LWE) problem, a problem introduced by Regev [9] that has received considerable attention in cryptography due to a known worst-case reduction to a hard lattice problem.

Gentry, Sahai and Waters described a compiler [6], which we call the GSW compiler, to transform an LWE-based IBE satisfying certain properties into a leveled IBFHE. They showed that all known LWE-based IBE schemes are compatible with their compiler. However, the GSW compiler only works in the *single-identity* setting. In other words, the resulting IBFHE can only evaluate on ciphertexts created with the same identity. Recently, a multi-identity IBFHE was described in [10], but that construction relies heavily on indistinguishability obfuscation [11], and is therefore highly inefficient at the present time. Furthermore, security cannot be based on a well-established computational problem. Our construction does not require indistinguishability obfuscation and is the first multi-identity IBFHE, to the best of our knowledge, whose security can be based on well-established problem.

Remark 1. Like [6], we omit the qualifier "leveled" for the rest of this paper since we focus only on leveled (IB)FHE in this work.

Note that our multi-identity and multi-key leveled IBFHE are 1-hop homomorphic insofar as after evaluation is complete, no further homomorphic evaluation can be carried out.

1.1 Multi-identity Setting

Consider the following simplified scenario. Alice and Bob work in an organization C that avails of a semi-trusted cloud server E. Let a and b denote the identity strings of Alice and Bob respectively. Their organization C serves as a trusted authority and issues them secret keys for their respective identity strings. Public users can send confidential data to Alice and Bob by encrypting it with their identity string and the master public key (public parameters) published by C. Suppose this encrypted data is sent by external users to the cloud server E. Furthermore, suppose some entity would like to perform some computation on E using encrypted data intended for Alice and encrypted data intended for Bob. The result should only be decryptable (assuming C is honest) by a collaborative effort made by Alice and Bob; they can run a multi-party computation protocol to collaboratively decrypt the result without leaking their secret keys to each other.

Let c_a and c_b be ciphertexts created with identities a and b respectively. The goal is to allow computation on c_a and c_b together. Assuming this could be achieved, let c' denote the ciphertext that encrypts the result of the computation. Intuitively, we expect the size of c' to depend on the number of distinct identities (2 in our example above i.e. a and b) because information about each identity must be "encoded" in c'. But like the single-identity setting, the size of c' should be independent of the size of the circuit evaluated. Of course we can naturally extend this notion to ciphertexts created under k distinct identities.

In the syntax of multi-identity IBFHE, a parameter $\mathcal{D}$ representing the number of distinct identities tolerated in an evaluation is specified in advance of generating the public parameters. Like the parameter L (the circuit depth supported), the size of the public parameters may depend on $\mathcal{D}$. A multi-identity and multi-attribute IBFHE and ABFHE that rely on indistinguishability obfuscation were described in [12].

Disjunctive Policies. There is another way of viewing multi-identity IBFHE, which might be more useful in some settings. It was mentioned in [13][1] that access policies consisting of disjunctions can be achieved with IBE. In this case, to issue a secret key for a policy $\hat{f}(X) \triangleq X =$ "MATH" OR $X =$ "CS", the TA issues a secret key for identity string "MATH" and a secret key for identity string "CS". In this case, we view the "identities" as attributes.

Suppose the TA issues a secret key $\mathsf{SK}_{\hat{f}} = \{\mathsf{sk}_{\text{"MATH"}}, \mathsf{sk}_{\text{"CS"}}\}$ for $\hat{f}$ to a professor working in both the Mathematics and Computer Science departments in a university; this secret key comprises an IBE secret key for identity string "MATH" and an IBE secret key for identity string "CS". The professor can decrypt the result of computation performed on ciphertexts with both attributes. This matches our intuition because her policy $\hat{f}$ permits her access to both attributes.

1.2   Our Results

Multi-identity IBFHE. Our central result in this paper is informally summarized in the following theorem statement. The theorem is formally stated and proven later in Appendix A.1.

Theorem 1 (Informal). *There exists a multi-identity IBFHE scheme that is selectively secure under the Learning With Errors problem in the random oracle model.*

Multi-key FHE. Our compiler for multi-identity IBFHE also works in the public-key setting. As a result, we can obtain a multi-key FHE [14] from LWE in the standard model. In fact, multi-identity IBFHE can be seen as an identity-based analog to multi-key FHE. The syntax of multi-key FHE from [14] entails a parameter M, which specifies the maximum number of independent keys tolerated in an evaluation. The size of the parameters and ciphertexts are allowed to depend polynomially on M. Note that M is fixed and specified in advance of generating the scheme's parameters. To the best of our knowledge, our multi-key FHE scheme is the first such scheme (for a non-constant number of keys) that is based on a well-established problem such as LWE; the construction from [14]

---

[1] The paper [13] attributes this observation to Brent Waters.

relies on a non-standard computational assumption referred to therein as the Decisional Small Polynomial Ratio (DSPR) assumption. Our scheme positively answers the question raised in [14] as to whether other multi-key FHE schemes exist supporting polynomially-sized M.

1.3 Our Approach: Intuition

We now give an informal sketch of our approach to achieving multi-identity IBFHE. This section is intended to provide an intuition and many of the details are deferred to later in the paper. We remind the reader that a matrix $\mathbf{M}$ is denoted by an uppercase symbol written in boldface, and a vector $\boldsymbol{v}$ is denoted by a lowercase symbol written in boldface. The i-th element of $\boldsymbol{v}$ is denoted by v_i. The inner product of two vectors $\boldsymbol{a}, \boldsymbol{b} \in \mathbb{Z}_q^n$ for some dimension n is written as $\langle a, b \rangle$.

GSW Single-Identity IBFHE. We start by briefly discussing the homomorphic properties of the GSW IBFHE schemes from [6]. This discussion applies to *any* IBFHE constructed with their compiler. A ciphertext in their scheme is an $N \times N$ matrix $\mathbf{C}$ over $\mathbb{Z}_q$ whose entries are "small" with respect to q. Note that N is a parameter that will be discussed later. A secret key for an identity id is an N-dimensional vector $\boldsymbol{v_{id}} \in \mathbb{Z}_q^N$ with at least one "large" coefficient; let this coefficient (say the i-th one) be $v_{id,i} \in \mathbb{Z}_q$. The scheme can encrypt "small" messages μ; an example to keep in mind is a message in $\{0, 1\}$. We say the matrix $\mathbf{C}$ *encrypts* μ under identity id if $\mathbf{C} \cdot \boldsymbol{v_{id}} = \mu \cdot \boldsymbol{v_{id}} + \boldsymbol{e} \in \mathbb{Z}_q^N$ where $\boldsymbol{e}$ is a "small" noise vector (i.e. roughly speaking, each of its coefficients is much less than q). As such, $\boldsymbol{v_{id}}$ is an *approximate* eigenvector for the matrix $\mathbf{C}$ with eigenvalue μ.

Homomorphic Operations

Suppose $\mathbf{C_1}$ and $\mathbf{C_2}$ encrypt μ_1 and μ_2 respectively; that is, $\mathbf{C_j} \cdot \boldsymbol{v_{id}} = \mu_j \cdot \boldsymbol{v_{id}} + \boldsymbol{e_j}$ for $j \in \{1, 2\}$. An additive homomorphism is supported. Let $\mathbf{C^+} = \mathbf{C_1} + \mathbf{C_2}$. Then we have $\mathbf{C^+} \cdot \boldsymbol{v_{id}} = (\mu_1 + \mu_2) \cdot \boldsymbol{v_{id}} + (\boldsymbol{e_1} + \boldsymbol{e_2})$. The error only grows slightly here, and as long as it remains "small", we can recover the sum $(\mu_1 + \mu_2)$. A multiplicative homomorphism is also supported. Let $\mathbf{C^\times} = \mathbf{C_1} \cdot \mathbf{C_2}$. Then we have

$$\begin{aligned}
\mathbf{C^\times} \cdot \boldsymbol{v_{id}} &= \mathbf{C_1} \cdot (\mu_2 \cdot \boldsymbol{v_{id}} + \boldsymbol{e_2}) \\
&= \mu_2 \cdot (\mu_1 \cdot \boldsymbol{v_{id}} + \boldsymbol{e_1}) + \mathbf{C_1} \cdot \boldsymbol{e_2} \\
&= \mu_1 \cdot \mu_2 \cdot \boldsymbol{v_{id}} + \mu_2 \cdot \boldsymbol{e_1} + \mathbf{C_1} \cdot \boldsymbol{e_2} \\
&= \mu_1 \cdot \mu_2 \cdot \boldsymbol{v_{id}} + \text{"small"}.
\end{aligned}$$

Different Identities. Now we give a flavor of how our multi-identity scheme operates. Suppose $\mathbf{C_1}$ encrypts μ_1 under identity id_1 and $\mathbf{C_2}$ encrypts μ_2 under identity id_2. Let $\boldsymbol{v_1}$ and $\boldsymbol{v_2}$ be the secret key vectors for id_1 and id_2 respectively. It holds that $\mathbf{C_1} \cdot \boldsymbol{v_1} = \mu_1 \cdot \boldsymbol{v_1} + \boldsymbol{e_1}$ and $\mathbf{C_2} \cdot \boldsymbol{v_2} = \mu_2 \cdot \boldsymbol{v_2} + \boldsymbol{e_2}$ where $\boldsymbol{e_1}, \boldsymbol{e_2} \in \mathbb{Z}_q^N$ are short vectors.

We would like to be able to perform homomorphic computation on both $\mathbf{C_1}$ and $\mathbf{C_2}$ together; that is, use them both as inputs to the same circuit. Here we denote the circuit by $C \in \mathbb{C}$. Suppose we could produce a resulting $2N \times 2N$ ciphertext matrix $\hat{\mathbf{C}}' \in \mathbb{Z}_q^{2N \times 2N}$ that encrypts $\mu' = C(\mu_1, \mu_2)$. More precisely, suppose that

$$\hat{\mathbf{C}}' \cdot \begin{bmatrix} v_1 \\ v_2 \end{bmatrix} = \mu' \cdot \begin{bmatrix} v_1 \\ v_2 \end{bmatrix} + e'$$

where e' is "short". Note that the size of $\hat{\mathbf{C}}'$ just depends (polynomially) on the number of distinct identities (2 in this example).

Let $v \in \mathbb{Z}_q^{2N}$ be the vertical concatenation of the two vectors v_1 and v_2. We could exploit the homomorphic properties described above to obtain $\hat{\mathbf{C}}'$ if we could somehow transform $\mathbf{C_1}$ and $\mathbf{C_2}$ into $2N \times 2N$ matrices $\hat{\mathbf{C}}_1$ and $\hat{\mathbf{C}}_2$ respectively such that $\hat{\mathbf{C}}_j \cdot v = \mu_j \cdot v +$ "small" for $j \in \{1,2\}$. Technically this transformation turns out to be difficult; we show how to abstractly accomplish it in Sect. 3 and concretely in Sect. 4.

2 Preliminaries

2.1 Notation

A quantity is said to be negligible with respect to some parameter λ, written $\mathsf{negl}(\lambda)$, if it is asymptotically bounded from above by the reciprocal of all polynomials in λ. We use the notation $[k]$ for an integer k to denote the set $\{1, \ldots, k\}$.

Distributions. For a probability distribution $\mathcal{D}$, we denote by $x \xleftarrow{\$} \mathcal{D}$ the fact that x is sampled according to $\mathcal{D}$. We overload the notation for a set S i.e. $y \xleftarrow{\$} S$ denotes that y is sampled uniformly from S. Let $\mathcal{D}_0$ and $\mathcal{D}_1$ be distributions. We denote by $\mathcal{D}_0 \underset{C}{\approx} \mathcal{D}_1$ and the $\mathcal{D}_0 \underset{S}{\approx} \mathcal{D}_1$ the facts that $\mathcal{D}_0$ and $\mathcal{D}_1$ are computationally indistinguishable and statistically indistinguishable respectively.

Definition 1 (B-Bounded Distributions (Definition 2 [6])). *A distribution ensemble $\{D_n\}_{n \in \mathbb{N}}$, supported over the integers, is called B-bounded if*

$$\Pr_{e \xleftarrow{\$} D_n}[|e| > B] = \mathsf{negl}(n).$$

Matrices and Vectors. A matrix $\mathbf{M}$ is denoted by an uppercase symbol written in boldface, and a vector v is denoted by a lowercase symbol written in boldface. The i-th element of v is denoted by v_i. The inner product of two vectors $a, b \in \mathbb{Z}_q^n$ for some dimension n is written as $\langle a, b \rangle$.

2.2   Multi-identity IBFHE

Definition 2. *A Multi-Identity (Leveled) IBFHE scheme is defined with respect to a message space $\mathcal{M}$, an identity space $\mathcal{I}$, a class of circuits $\mathbb{C} \subseteq \mathcal{M}^* \to \mathcal{M}$ and ciphertext space $\mathcal{C}$. A Multi-Identity IBHE scheme is a tuple of PPT algorithms* (Setup, KeyGen, Encrypt, Decrypt, Eval) *defined as follows:*

- Setup$(1^\lambda, L, \mathcal{D})$:
  On input (in unary) a security parameter λ, a number of levels L (circuit depth to support) and the number of distinct identities $\mathcal{D}$ that can be tolerated in an evaluation, generate public parameters PP and a master secret key MSK. Output (PP, MSK).
- KeyGen(MSK, id):
  On input master secret key MSK and an identity id: derive and output a secret key $\mathsf{sk_{id}}$ for identity id.
- Encrypt(PP, id, m):
  On input public parameters PP, an identity id, and a message $m \in \mathcal{M}$, output a ciphertext $c \in \mathcal{C}$ that encrypts m under identity id.
- Decrypt$(\mathsf{sk_{id_1}}, \dots, \mathsf{sk_{id_d}}, c)$:
  On input $d \leq \mathcal{D}$ secret keys $\mathsf{sk_{id_1}}, \dots, \mathsf{sk_{id_d}}$ for (resp.) identities $\mathsf{id_1}, \dots, \mathsf{id_d}$ and a ciphertext $c \in \mathcal{C}$, output $m' \in \mathcal{M}$ if c is a valid encryption under identities $\mathsf{id_1}, \dots, \mathsf{id_d}$; output a failure symbol $\perp$ otherwise.
- Eval(PP, $C, c_1, \dots, c_\ell$): *On input public parameters PP, a circuit $C \in \mathbb{C}$ and ciphertexts $c_1, \dots, c_\ell \in \mathcal{C}$, output an evaluated ciphertext $c' \in \mathcal{C}$.*

More precisely, the scheme is required to satisfy the following properties:

- *Over all choices of* (PP, MSK) $\leftarrow$ Setup(1^λ), $d \leq \mathcal{D}$, $\mathsf{id_1}, \dots, \mathsf{id_d} \in \mathcal{I}$, C : $\mathcal{M}^\ell \to \mathcal{M} \in \{C \in \mathbb{C} : \mathsf{depth}(C) \leq L\}$, $j_1, \dots, j_\ell \subset [d]$, $\mu_1, \dots, \mu_\ell \in \mathcal{M}$, $c_i \leftarrow$ Encrypt(PP, id_{j_i}, μ_i) *for $i \in [\ell]$, and $c' \leftarrow$ Eval(PP, $C, c_1, \dots, c_\ell$):*
  - **Correctness**

$$\mathsf{Decrypt}(\mathsf{sk_1}, \dots, \mathsf{sk_d}, c') = C(\mu_1, \dots, \mu_\ell) \qquad (2.1)$$

  for any $\mathsf{sk}_i \leftarrow$ KeyGen(MSK, id_i) for $i \in [k]$
  - **Compactness**

$$|c'| \leq \mathsf{poly}(\lambda, L, d) \qquad (2.2)$$

where $d \leq \mathcal{D}$ is the number of distinct identities; that is, $d = |\{j_1, \dots, i_\ell\}|$.

The size of evaluated ciphertexts in our construction grows with $d \leq \mathcal{D}$.

The security definition for multi-identity IBFHE is the same as that for single-identity IBFHE. In this work, we focus on IND-sID-CPA security whose definition remains the same for the multi-identity setting.

2.3   Learning with Errors

The Learning with Errors (LWE) problem was introduced by Regev [9]. The goal of the computational form of the LWE problem is to determine an n-dimensional secret vector $s \in \mathbb{Z}_q^n$ given a polynomial number of samples $(\boldsymbol{a_i}, b_i) \in \mathbb{Z}_q^{n+1}$ where $\boldsymbol{a_i}$ is uniform over $\mathbb{Z}_q^n$ and $b_i \leftarrow \langle \boldsymbol{a_i}, \boldsymbol{s} \rangle + e_i \in \mathbb{Z}_q$ is the inner product of $\boldsymbol{a_i}$ and $\boldsymbol{s_i}$ perturbed by a small *error* $e_i \in \mathbb{Z}$ that is sampled from a distribution χ over $\mathbb{Z}$. We call the distribution χ an error distribution (or noise distribution). The decision variant of the problem is to distinguish such samples $(\boldsymbol{a_i}, b_i) \in \mathbb{Z}_q^{n+1}$ from uniform vectors over $\mathbb{Z}_q^{n+1}$. The decisional variant is more commonly used in cryptography, and is most relevant to our own work. As a result, without further qualification, when we refer to LWE throughout this thesis we are referring to the decisional variant.

Definition 3 ((Decisional) Learning with Errors (LWE) Problem [9]**).** *Let λ be a security parameter. For parameters $n = n(\lambda)$, $q = q(\lambda) \geq 2$, and a distribution $\chi = \chi(\lambda)$ over $\mathbb{Z}$, the $LWE_{n,q,\chi}$ problem is to distinguish the following distributions:*

- ***Distribution 0:*** *The i-th sample $(\boldsymbol{a_i}, b_i) \in \mathbb{Z}_q^{n+1}$ is computed by uniformly sampling $\boldsymbol{a_i} \xleftarrow{\$} \mathbb{Z}_q^n$ and $b_i \xleftarrow{\$} \mathbb{Z}_q$.*

- ***Distribution 1:*** *Generate uniform vector $\boldsymbol{s} \xleftarrow{\$} \mathbb{Z}_q^n$. The i-th sample $(\boldsymbol{a_i}, b_i) \in \mathbb{Z}_q^{n+1}$ is computed by uniformly sampling $\boldsymbol{a_i} \xleftarrow{\$} \mathbb{Z}_q^n$, sampling an error value $e_i \xleftarrow{\$} \chi$ and computing $b_i \leftarrow \langle \boldsymbol{a_i}, \boldsymbol{s} \rangle + e_i$.*

Definition 4 (B-Bounded Distributions (Definition 2 [6]**)).** *A distribution ensemble $\{D_n\}_{n \in \mathbb{N}}$, supported over the integers, is called B-bounded if*

$$\Pr_{e \xleftarrow{\$} D_n}[|e| > B] = \mathsf{negl}(n).$$

Definition 5 (GapSVP$_\gamma$). *Let n be a lattice dimension, and let d be a real number. Then GapSVP$_\gamma$ is the problem of deciding whether an n-dimensional lattice has a nonzero vector shorter than d (an algorithm should accept in this case) or no nonzero vector shorter than $\gamma(n) \cdot d$ (an algorithm should reject in this case); an algorithm is allowed to error otherwise.*

Theorem 2 (Theorem 1 [6]**).** *Let $q = q(n) \in \mathbb{N}$ be either a prime power or a product of small (poly(n)) distinct primes, and let $B \geq \omega(\log n) \cdot \sqrt{n}$. Then there exists an efficient sampleable B-bounded distribution χ such that if there is an efficient algorithm that solves the average-case $LWE_{n,q,\chi}$ problem, then:*

- *There is an efficient quantum algorithm that solves GapSVP$_{\tilde{O}(nq/B)}$ on any n-dimensional lattice.*
- *If $q > \tilde{O}(2^{n/2})$, then there is an efficient classical algorithm for GapSVP$_{\tilde{O}(nq/B)}$ on any n-dimensional lattice.*

2.4   GSW Approximate Eigenvector Cryptosystem

Recall our brief overview of the GSW IBFHE construction earlier from Sect. 1.3. The following exposition describes this construction in more detail. Note that the public-key GSW scheme is similar to the identity-based variant. As such, to simplify the notation, the following discussion deals with the public-key setting, but the ideas apply to both.

Definition 6 (Sect. 1.3.2 from [6]). *B-boundedness: Let $B < q$ be an integer. Let $\mathbf{C}$ be a ciphertext matrix that encrypts μ. Let v be a secret key vector such that $\mathbf{C} \cdot v = \mu \cdot v + e$. Then $\mathbf{C}$ is said to be B-bounded (with respect to v) if the magnitude of μ is at most B, the magnitude of all the entries of $\mathbf{C}$ is at most B, and $\|\|e\|\|_\infty \leq B$.*

Let $\mathbf{C_1}$ and $\mathbf{C_2}$ be two B-bounded ciphertext matrices. Then $\mathbf{C}^+ = \mathbf{C_1} + \mathbf{C_2}$ is $2B$-bounded. Furthermore, $\mathbf{C}^\times = \mathbf{C_1} \cdot \mathbf{C_2}$ is $(N+1)^{B^2}$-bounded. As the authors of [6] point out, the error grows worse than B^{2^L}, where L is the multiplicative depth of a circuit being evaluated. The modulus q can be chosen to exceed this bound, but we must be careful to ensure that the ratio q/B is at most subexponential in N to guarantee security (see Theorem 2). Hence, only circuits of logarithmic multiplicative depth can be evaluated. This gives us a somewhat-homomorphic scheme.

   To evaluate deeper circuits, namely those with polynomial multiplicative depth, we must keep the entries of the ciphertext matrices "small". To achieve this, Gentry, Sahai and Waters propose a technique called *flattening*. Consider the following definition.

Definition 7 (Sect. 1.3.3 from [6]). *B-strong-boundedness: Let $B < q$ be an integer. Let $\mathbf{C}$ be a ciphertext matrix that encrypts μ. Let v be a secret key vector such that $\mathbf{C} \cdot v = \mu \cdot v + e$. Then $\mathbf{C}$ is said to be B-strongly-bounded (with respect to v) if the magnitude of μ is at most 1, the magnitude of all the entries of $\mathbf{C}$ is at most 1, and $\|\|e\|\|_\infty \leq B$.*

An example of a B-strongly-bounded ciphertext is a matrix $\mathbf{C}$ with binary entries that encrypts a plaintext bit $\mu \in \{0, 1\}$, provided the coefficients of its corresponding e vector have magnitude at most B. Let $\mathbf{C_1}$ and $\mathbf{C_2}$ be ciphertext matrices that encrypt $\mu_1 \in \{0, 1\}$ and $\mu_2 \in \{0, 1\}$ respectively. A NAND gate can be evaluated on two ciphertexts $\mathbf{C_1}$ and $\mathbf{C_2}$ as follows:

$$\mathbf{C_3} = \mathbf{I_N} - \mathbf{C_1} \cdot \mathbf{C_2},$$

where $\mathbf{I_N}$ is the $N \times N$ identity matrix. The matrix $\mathbf{C_3}$ encrypts $\mu_1 \text{NAND} \mu_2 \in \{0, 1\}$. Now if $\mathbf{C_1}$ and $\mathbf{C_2}$ are B-strongly-bounded, then the coefficients of $\mathbf{C_3}$'s error vector have magnitude at most $(N+1)B$, which is in contrast to $(N+1)B^2$ above where $\mathbf{C_1}$ and $\mathbf{C_2}$ were just B-bounded. Suppose there were some way to preserve strong-boundedness in $\mathbf{C_3}$ (i.e. to ensure the magnitude of its entries remained at most 1). Then it would be the case that $\mathbf{C_3}$ is $(N+1)B$-strongly-bounded. As a result, the error level would grow to at most $(N+1)^L B$ when evaluating a circuit of NAND gates of depth L. Therefore it would be possible to evaluate circuits of polynomial depth by letting q/B be subexponential. However, how can we preserve strong-boundedness? It is necessary to introduce some basic operations to help describe how strong boundedness is preserved. These operations serve as useful tools for our own constructions later.

Basic Operations. Let $\ell_q = \lfloor \lg q \rfloor + 1$. Let $v \in \mathbb{Z}_q^{m'}$ be a vector of some dimension m' over $\mathbb{Z}_q$. Let $N = m' \cdot \ell_q$.

- **BitDecomp(v)**: We define an algorithm BitDecomp that takes as input a vector $v \in \mathbb{Z}_q^{m'}$ and outputs an N-dimensional vector $(v_{1,0}, \ldots, v_{1,\ell_q-1}, \ldots, v_{k,0}, \ldots, v_{k,\ell_q-1})$ where $v_{i,j}$ is the j-th bit in v_i's binary representation (ordered from least significant to most significant).
- **BitDecomp$^{-1}(v')$**: We define an "inverse" algorithm BitDecomp^{-1} that takes an N-dimensional vector $v' = (v'_{1,0}, \ldots, v'_{1,\ell_q-1}, \ldots, v'_{k,0}, \ldots, v'_{k,\ell_q-1})$, and outputs a m'-dimensional vector $(\sum_{j=0}^{\ell_q-1} 2^j \cdot v'_{1,j}, \ldots, \sum_{j=0}^{\ell_q-1} 2^j \cdot v'_{k,j})$. Note that the input vector v' need not be binary, the algorithm is well-defined for any input vector in $\mathbb{Z}_q^N$.
- **Flatten(v')**: The algorithm Flatten takes as input an N-dimensional vector $v' \in \mathbb{Z}_q^N$ and outputs an N-dimensional binary vector BitDecomp(BitDecomp^{-1} $(v')) \in \{0,1\}^N$.
- **Powersof2(v)**: The algorithm Powersof2 takes a m'-dimensional vector $v \in \mathbb{Z}_q^{m'}$ and outputs an N-dimensional vector $(v_1, 2v_1, \ldots, 2^{\ell_q-1}v_1, \ldots, v_k, 2v_k, \ldots, 2^{\ell_q-1}v_k)$.

We also define BitDecomp, BitDecomp^{-1} and Flatten for matrix inputs; in this case, the respective algorithm is applied to each row independently.

We restate the following straightforward facts from [6] (Sect. 1.3.3): Let $a, b \in \mathbb{Z}_q^{m'}$ be m'-dimensional vectors, and let $a' \in \mathbb{Z}_q^N$ be an N-dimensional vector:

- $\langle \text{BitDecomp}(a), \text{Powersof2}(b) \rangle = \langle a, b \rangle$.
- $\langle a', \text{Powersof2}(b) \rangle = \langle \text{BitDecomp}^{-1}(a'), b \rangle = \langle \text{Flatten}(a'), \text{Powersof2}(b) \rangle$.

Flattening. With the help of BitDecomp, BitDecomp^{-1}, Powersof2 and Flatten, we can tackle the problem of preserving strong boundedness after a NAND operation. In order to make the coefficients of $\mathbf{C_3}$ above have magnitude at most 1, Gentry, Sahai and Waters propose to apply Flatten to the matrix $\mathbf{C_3}$. Thus, we compute $\mathbf{C}^{\mathsf{NAND}} \leftarrow \text{Flatten}(\mathbf{C_3})$ to produce the output ciphertext of the NAND gate. Now for this to work, the vector v must have a special form. More precisely, v is computed as Powersof2$(s) \in \mathbb{Z}_q^N$ for some secret key vector $s \in \mathbb{Z}_q^{m'}$ for some m'. Furthermore, the parameter N is defined as $N = m' \cdot \ell_q$, where $\ell_q = \lfloor \lg q \rfloor + 1$. With this form of secret key vector v, it holds that Flatten$(\mathbf{C}) \cdot v = \mathbf{C} \cdot v$ for any $N \times N$ matrix $\mathbf{C}$. So $\mathbf{C}^{\mathsf{NAND}}$ will have entries in $\{0,1\}$ and thus be strongly-bounded.

2.5   GSW Compiler for IBE in the Single-Identity Setting

The Gentry, Sahai and Waters (GSW) compiler from Crypto 2013 [6] (Sect. 4) allows transformation of an IBE scheme based on the Learning with Errors (LWE) problem into a related IBFHE scheme, provided the IBE scheme satisfies the following properties:

1. **Property 1 (Ciphertext and Secret Key Vectors):** The secret key for identity id and a ciphertext created under id are vectors $s_{id}, c_{id} \in \mathbb{Z}_q^{m'}$ for some m'. The first coefficient of s_{id} is 1.
2. **Property 2 (Small Dot Product):** If c_{id} encrypts 0, then $\langle c_{id}, s_{id} \rangle$ is "small".
3. **Property 3 (Security):** Encryptions of 0 are indistinguishable from uniform vectors over $\mathbb{Z}_q$ under the hardness of LWE.

As noted in [6] all known LWE-based IBE schemes satisfy the above properties e.g.: [15–18].

Let $\mathcal{E}$ be an IBE satisfying the Properties 1-3 above. Then $\mathcal{E}$ can be transformed into a single-identity IBFHE scheme $\mathcal{E}'$.

The public parameters PP generated by $\mathcal{E}$.Setup includes a modulus q and an integer m' representing the length of both secret key and ciphertext vectors in $\mathcal{E}$. Let $\ell_q = \lfloor \lg q \rfloor + 1$ and $N = m' \times \ell_q$.

To encrypt a message $\mu \in \{0,1\}$ under identity id $\in \mathcal{I}$, the encryptor generates N encryptions of 0 using $\mathcal{E}$. More precisely, she computes $e_i \leftarrow \mathcal{E}$.Encrypt$(PP, id, 0) \in \mathbb{Z}_q^{m'}$ for every $i \in [N]$. The set of N vectors $e_1, \ldots, e_N$ form the rows of an $N \times m'$ matrix $E \in \mathbb{Z}_q^{N \times m'}$. Finally the encryptor computes the $N \times N$ ciphertext matrix $\mathbf{C} \in \{0,1\}^{N \times N}$ as follows

$$\mathbf{C} \leftarrow \mathsf{Flatten}(\mu \cdot \mathbf{I_N} + \mathsf{BitDecomp}(\mathbf{E}))$$

where $\mathbf{I_N}$ denotes the $N \times N$ identity matrix.

A secret key in $\mathcal{E}'$ for identity id is an N-dimensional vector v_{id} derived from a secret key s_{id} for identity id in $\mathcal{E}$. This is computed as $v_{id} \leftarrow \mathsf{Powersof2}(s_{id})$. Decryption of a ciphertext $\mathbf{C}$ with v_{id} is as follows. By construction of v_{id}, it has at least one "large" coefficient; denote this by $v_{id,i}$, To perform decryption, we take the i-th row c_i of matrix $\mathbf{C}$, compute the inner product $x \leftarrow \langle c_i, v_{id} \rangle = \mu \cdot v_{id,i} + e_i$ and output the plaintext $\mu \leftarrow \lfloor x/v_{id,i} \rceil$. This is correct because

$$\mathbf{C} \cdot v_{id} = \mu \cdot v_{id} + \mathbf{E} \cdot s_{id} = \mu \cdot v_{id} + \text{"small"}$$

where $\mathbf{E} \cdot s_{id}$ is "small" as a consequence of Property 2. It is also easy to see that semantic security for $\mathcal{E}'$ follows immediately from the fact that $\mathcal{E}$ satisfies Property 3.

3   A Compiler for Multi-identity Leveled IBFHE

In this section, we present a new compiler that can transform an LWE-based IBE into a *multi-identity* IBFHE. As we will see, achieving multi-identity IBFHE is far more difficult than single-identity IBFHE.

3.1   Intuition

Suppose $\mathcal{E}$ is an LWE-based IBE that satisfies properties 1 - 3 above. We can apply the GSW compiler to yield an IBFHE scheme $\mathcal{E}'$ in the single-identity setting. Our goal is to construct a compiler for the multi-identity setting. Consider two ciphertexts $\mathbf{C_1}$ and $\mathbf{C_2}$ that encrypt μ_1 and μ_2 under identities id_1 and id_2 respectively. Let s_1 and s_2 be secret keys in the scheme $\mathcal{E}$ for identities id_1 and id_2 respectively. Accordingly, a decryptor can compute $v_1 \leftarrow \mathsf{Powersof2}(s_1)$ and $v_2 \leftarrow \mathsf{Powersof2}(s_2)$. It holds that $\mathbf{C_1} \cdot v_1 = \mu_1 \cdot v_1 + e_1$ and $\mathbf{C_2} \cdot v_2 = \mu_2 \cdot v_2 + e_2$ where $e_1, e_2 \in \mathbb{Z}_q^N$ are short vectors.

We would like to be able to perform homomorphic computation on both $\mathbf{C_1}$ and $\mathbf{C_2}$ together; that is, use them both as inputs in the same circuit. Here we denote the circuit by $C \in \mathbb{C}$. We expect the size of the resulting ciphertext to grow if $\mathsf{id}_1 \neq \mathsf{id}_2$. This is intuitive because the resulting ciphertext must *encode* information about *both* identities. Assume that $\mathsf{id}_1 \neq \mathsf{id}_2$. The compactness condition of multi-identity IBFHE allows the size of the resulting ciphertext to depend polynomially on the number of *distinct* identities d (in this case $d = 2$). Suppose we could produce a resulting $2N \times 2N$ ciphertext matrix $\mathbf{C}' \in \mathbb{Z}_q^{2N \times 2N}$ that encrypts $\mu' = C(\mu_1, \mu_2)$. More precisely, suppose that

$$\mathbf{C}' \cdot \begin{bmatrix} v_1 \\ v_2 \end{bmatrix} = \mu' \cdot \begin{bmatrix} v_1 \\ v_2 \end{bmatrix} + e'$$

where e' is "short". The size of the ciphertext matrix is quadratic in the number of distinct identities, and thus satisfies the compactness condition. How can such a matrix $\mathbf{C}'$ be computed?

The main idea behind our approach is to transform each input ciphertext matrix (i.e. $\mathbf{C_1}$ and $\mathbf{C_2}$ in this example) into a corresponding $dN \times dN$ "expanded matrix" where d is the number of distinct identities (i.e. $d = 2$ in our example).

Consider any input ciphertext matrix $\mathbf{C} \in \mathbb{Z}_q^{N \times N}$ that encrypts a plaintext μ under identity id_1. We denote by $\hat{\mathbf{C}} \in \mathbb{Z}_q^{2N \times 2N}$ its corresponding "expanded matrix". We require this expanded matrix to satisfy

$$\hat{\mathbf{C}} \cdot \begin{bmatrix} v_1 \\ v_2 \end{bmatrix} = \mu \cdot \begin{bmatrix} v_1 \\ v_2 \end{bmatrix} + \text{"small"}.$$

Now $\hat{\mathbf{C}}$ can be viewed as consisting of 2×2 submatrices in $\mathbb{Z}_q^{N \times N}$. We denote the submatrix on row i and column j as $\hat{\mathbf{C}}_{i,j} \in \mathbb{Z}_q^{N \times N}$. To satisfy the "top" part of the above equation, it is sufficient to set $\hat{\mathbf{C}}_{1,1} \leftarrow \mathbf{C}$ and $\hat{\mathbf{C}}_{1,2} \leftarrow \mathbf{0}$. To satisfy the "bottom" part of the equation, we need to find matrices $\mathbf{X}, \mathbf{Y} \in \{0,1\}^{N \times N}$ such that

$$\mathbf{X} \cdot v_1 + \mathbf{Y} \cdot v_2 = \mu \cdot v_2 + \text{"small"}.$$

We refer to a pair of solution matrices $(\mathbf{X}, \mathbf{Y})$ as a "mask" because of the fact that they hide the plaintext μ from a party that does not have a secret key for the recipient identity. In this section, we will abstract over the process of finding solution matrices $\mathbf{X}$ and $\mathbf{Y}$ with respect to arbitrary identities. Towards this goal, we introduce an abstraction called a *masking system*. In short, a masking system allows an encryptor to produce information $U \in \{0,1\}^*$ that allows an evaluator to derive matrices $\mathbf{X}$ and $\mathbf{Y}$ that solve the above equation with respect to any arbitrary identity. Informally, an adversary without a secret key for the *recipient identity* (id_1 in the above example) learns nothing about μ given U, but can still efficiently derive solution matrices $\mathbf{X}$ and $\mathbf{Y}$ with respect to any chosen identity. This notion is formalized in the next section, where we present our compiler. A concrete construction of a masking system is presented in Sect. 4.2.

3.2 Abstract Compiler

We start by describing an abstract framework for multi-identity IBFHE from Learning with Errors (LWE). Our compiler uses the aforementioned abstraction which we call a *masking system*. An additional prerequisite for an IBE scheme $\mathcal{E}$ (beyond Properties 1-3) to work with our compiler is that there exists a masking system $\mathsf{MS}_{\mathcal{E}}$ for $\mathcal{E}$. First we provide a formal definition of a masking system.

Definition 8. *Let $\mathcal{E}$ be an IBE scheme satisfying Properties 1-3. A masking system for $\mathcal{E}$ is a pair of PPT algorithms* (GenUnivMask, DeriveMask) *defined as follows:*

- GenUnivMask(PP, id, μ) *takes as input public parameters* PP *for $\mathcal{E}$, an identity* id $\in \mathcal{I}$ *and a message* $\mu \in \{0,1\}$, *and outputs* $U \in \{0,1\}^*$ *(referred to as a universal mask).*
- DeriveMask(PP, U, id') *takes as input public parameters* PP *for $\mathcal{E}$, a universal mask* $U \in \{0,1\}^*$ *and an identity* id' $\in \mathcal{I}$, *and outputs a pair of matrices* $(\mathbf{X}, \mathbf{Y}) \in (\mathbb{Z}_q^{N \times N})^2$.

A masking system (GenUnivMask, DeriveMask) *must satisfy the following properties:*

- **Correctness:** *Let $w(\cdot)$ be a polynomial associated with the masking system. Let $w = w(\lambda)$. We refer to w as the error expansion factor. For correctness, it is required that for any* (PP, MSK) $\leftarrow \mathcal{E}.\mathsf{Setup}(1^\lambda)$, *any identities* id, id' $\in \mathcal{I}$, *any secret keys* $\mathbf{v}_{\mathsf{id}} \leftarrow$ Powersof2($\mathcal{E}.\mathsf{KeyGen}(\mathsf{MSK}, \mathsf{id})) \in \mathbb{Z}_q^N$ *and* $\mathbf{v}_{\mathsf{id}'} \leftarrow$ Powersof2($\mathcal{E}.\mathsf{KeyGen}(\mathsf{MSK}, \mathsf{id}')) \in \mathbb{Z}_q^N$, *and any* $\mu \in \{0,1\}$, *and over all*
  - $U \leftarrow$ GenUnivMask(PP, id, μ),
  - $(\mathbf{X}, \mathbf{Y}) \leftarrow$ DeriveMask(PP, U, id')
  it holds that
  $$\mathbf{X}\mathbf{v}_{\mathsf{id}} + \mathbf{Y}\mathbf{v}_{\mathsf{id}'} = \mu \cdot \mathbf{v}_{\mathsf{id}'} + e \qquad (3.1)$$
  where $\||e|\|_\infty \leq w \cdot B$.
- **Security:** *The masking system is said to be secure if all PPT adversaries have a negligible advantage in the following modified* IND-X-CPA *game for $\mathcal{E}$ where $X \in \{\mathsf{sID}, \mathsf{ID}\}$. The only change in the security game is that the adversary is given* $U^* \leftarrow$ GenUnivMask(PP, id*, μ_b) *in place of the challenge ciphertext in the original game, where $b \xleftarrow{\$} \{0,1\}$ is the challenger's random bit, id* is the adversary's target identity, and μ_0 and μ_1 are the challenge messages chosen by the adversary.*

Our compiler can compile an IBE scheme $\mathcal{E}$ into a IBFHE scheme $\mathcal{E}'$ if the following conditions are met (for completeness, we restate Properties 1-3 above):

CP.1: (Ciphertext and secret key vectors): The secret key for identity id and a ciphertext created under id are vectors $s_{id}, c_{id} \in \mathbb{Z}_q^{m'}$ for some m'. The first coefficient of s_{id} is 1.

CP.2: (Small Dot Product): If c_{id} encrypts 0 under identity id, then $e = \langle c_{id}, s_{id} \rangle$ is "small" where s_{id} is generated as in CP.1. Formally, e is B-bounded; that is, $\||e\||_\infty \le B$.

CP.3: (Security): Encryptions of 0 are indistinguishable from uniform vectors over $\mathbb{Z}_q$ under the hardness of LWE.

CP.4: (Masking System): There exists a masking system (GenUnivMask, DeriveMask) for $\mathcal{E}$ meeting the correctness and security conditions of Definition 8.

Let $\mathsf{MS}_\mathcal{E} = (\mathsf{MS}_\mathcal{E}\mathsf{GenUnivMask}, \mathsf{MS}_\mathcal{E}\mathsf{DeriveMask})$ be a *masking system* for $\mathcal{E}$ that satisfies CP.4. A formal description is now given of a generic scheme, which we call mIBFHE, that uses $\mathcal{E}$ and $\mathsf{MS}_\mathcal{E}$. We have mIBFHE.Setup = $\mathcal{E}$.Setup and mIBFHE.KeyGen = $\mathcal{E}$.KeyGen. The remaining algorithms are described as follows.

Encryption. To encrypt a message μ under identity id $\in \mathcal{I}$, an encryptor performs the following steps. The encryptor computes the universal mask

$$U \leftarrow \mathsf{MS}_\mathcal{E}.\mathsf{GenUnivMask}(\mathsf{PP}, \mathsf{id}, \mu)$$

and outputs the ciphertext CT := (id, type := 0, enc := U). Setting the type component of CT to 0 indicates a "fresh" ciphertext.

Evaluation. The evaluator is given as input a circuit $C \in \mathcal{C}$ and a collection of ℓ ciphertexts $\mathsf{CT}_1 := (\mathsf{id}_1, \mathsf{type} := 0, \mathsf{enc} := U_1), \ldots, \mathsf{CT}_\ell := (\mathsf{id}_\ell, \mathsf{type} := 0, \mathsf{enc} := U_\ell)$.

Consider the set of *distinct* identities $I = \{\mathsf{id}_1, \ldots, \mathsf{id}_\ell\}$. Suppose that $|I| = d \le \ell$ is the number of distinct identities. If $d > \mathcal{D}$ (i.e. the maximum supported number of distinct identities is exceeded), the evaluator aborts the evaluation. For simplicity we re-label the distinct identities as $\mathsf{id}_1, \ldots, \mathsf{id}_d$. Thus, each distinct identity in the collection is associated with a unique index in $[d]$. Before evaluation can be performed, each ciphertext must be "transformed" into a $dN \times dN$ matrix, which we call an *expanded matrix*. This is achieved as follows.

Let $(\mathsf{id}_r, \mathsf{type} := 0, \mathsf{enc} := U)$ be a ciphertext whose associated identity has been assigned the index $r \in [d]$. A matrix $\hat{\mathbf{C}} \in \mathbb{Z}_q^{dN \times dN}$ is formed as follows. Start by setting $\hat{\mathbf{C}}$ to the zero matrix. Now $\hat{\mathbf{C}}$ can be viewed as consisting of $d \times d$ submatrices in $\mathbb{Z}_q^{N \times N}$. We denote the submatrix on row i and column j as $\hat{\mathbf{C}}_{i,j} \in \mathbb{Z}_q^{N \times N}$.

For $i \in [d]$:

1. Run $(\mathbf{X_i}, \mathbf{Y_i}) \leftarrow \mathsf{MS}_{\mathcal{E}}.\mathsf{DeriveMask}(\mathsf{PP}, U, \mathsf{id}_i)$.
2. Set $\hat{\mathbf{C}}_{\mathbf{i},\mathbf{i}} \leftarrow \mathbf{Y_i}$.
3. Set $\hat{\mathbf{C}}_{\mathbf{i},\mathbf{r}} \leftarrow \mathsf{Flatten}(\hat{\mathbf{C}}_{\mathbf{i},\mathbf{r}} + \mathbf{X_i})$. (The reason for addition here is to handle the special case of $i = r$).

This completes the process for computing the expanded matrix $\hat{\mathbf{C}}$. Consider an example where $r = 1$ and $d > 2$. The expanded matrix looks like the following:

$$\hat{\mathbf{C}} = \begin{pmatrix} (\mathsf{Flatten}(\mathbf{X_1} + \mathbf{Y_1}) & & \\ \mathbf{X_2} & \mathbf{Y_2} & \\ \vdots & & \ddots \\ \mathbf{X}_d & & \mathbf{Y}_d \end{pmatrix}$$

Perform the steps above to produce the expanded matrix $\hat{\mathbf{C}}^{(i)}$ for every input ciphertext CT_i. Then the circuit $C \in \mathcal{C}$ is evaluated gate-by-gate (NAND gates) on the expanded matrices to yield a $dN \times dN$ matrix $\hat{\mathbf{C}}'$. Suppose each $\hat{\mathbf{C}}^{(i)}$ encrypts $\mu_i \in \{0, 1\}$. Then $\hat{\mathbf{C}}'$ encrypts $C(\mu_1, \ldots, \mu_\ell)$. Finally, the evaluation algorithm outputs the tuple $\mathsf{CT}' := (\mathsf{id}_1, \ldots, \mathsf{id}_d, \mathsf{type} := 1, \mathsf{enc} := \hat{\mathbf{C}}')$. Setting the type component to 1 indicates an evaluated ciphertext. Note that the scheme is 1-hop homomorphic.

Decryption. On input a ciphertext $\mathsf{CT} := (\mathsf{id}_1, \ldots, \mathsf{id}_d, \mathsf{type}, \mathsf{enc})$ and a sequence of secret keys $v_{\mathsf{id}_1}, \ldots, v_{\mathsf{id}_d} \in \mathbb{Z}_q^N$ where v_{id_i} is a secret key for id_i for $i \in [d]$, the decryptor performs the following steps. Form the column vector v as the vertical concatenation of the column vectors $v_{\mathsf{id}_1}, \ldots, v_{\mathsf{id}_d}$. If $\mathsf{type} = 0$, parse enc as the universal mask U, compute $(\mathbf{X}, \mathbf{Y}) \leftarrow \mathsf{MS}_{\mathcal{E}}.\mathsf{DeriveMask}(\mathsf{PP}, U, \mathsf{id}_1)$ and set $\mathbf{C} \leftarrow \mathbf{X} + \mathbf{Y}$. Else if $\mathsf{type} = 1$, parse enc as $\hat{\mathbf{C}}$ and set $\mathbf{C} \leftarrow \hat{\mathbf{C}}$.

Let i be an index such that $v_i = 2^i \in (q/4, q/2]$. Compute $d_i \leftarrow \langle c_i, v \rangle$ where c_i is the i-th row of $\mathbf{C}$ and output $\mu' \leftarrow \lfloor d_i / v_i \rceil \in \{0, 1\}$. This works to recover the message because as a result of Eq. 3.1 (in Definition 8), we have

$$\mathbf{C}v = \mu \cdot v + e$$

with $\|\!|e|\!\|_\infty \le w \cdot B$, where w is the error expansion factor associated with the masking system $\mathsf{MS}_{\mathcal{E}}$.

Lemma 1. *Let B be a bound such that all freshly encrypted ciphertexts are B-strongly-bounded. Let $\mathcal{D}$ and L be positive integers. If $q > 8 \cdot w \cdot B(\mathcal{D}N+1)^{L2}$, then the scheme mIBFHE is correct and can evaluate NAND-based Boolean circuits of depth L with any number of distinct identities $d \le \mathcal{D}$.*

See Appendix B for the proof of Lemma 1.

Theorem 3. *Let $\mathcal{E}$ be an IBE scheme satisfying CP.1 - CP.4. Then $\mathcal{E}$ can be transformed into a multi-identity IBFHE scheme $\mathcal{E}'$.*

---

[2] Note that N (which depends on n) is itself dependent on $\lg q$. For security, it is required that $q/B = 2^{n^\epsilon}$ for some $\epsilon \in (0, 1)$. A discussion on parameters is provided in Appendix C.

Proof. The proof of the theorem is constructive. By CP.4, there exists a masking system $MS_{\mathcal{E}}$ for $\mathcal{E}$. The multi-identity IBFHE scheme $\mathcal{E}'$ that we obtain is mIBFHE instantiated with $\mathcal{E}$ and $MS_{\mathcal{E}}$. By Lemma 1, the scheme is correct. CP.4 implies that $\mathcal{E}'$ is IND-X-CPA secure for some $X \in \{\text{sID}, \text{ID}\}$.

4 Concrete Construction of Multi-identity Leveled IBFHE

To exploit our compiler from the last section to obtain a multi-identity IBFHE, we need to find an LWE-based IBE scheme $\mathcal{E}$ that satisfies CP.1 - CP.4. The major obstacle is finding a scheme for which a secure masking system can be constructed. A natural starting point is the IBE of Cash, Hofheinz, Kiltz and Peikert (CHKP) [18], which is IND-ID-CPA secure in the standard model. This IBE was adapted by Gentry, Sahai and Waters ([6] Appendix A.1) to work with their compiler. There are difficulties however in developing a secure masking system for this IBE. Instead, we consider the IBE of Gentry, Peikert and Vaikuntanathan (GPV) [15]. Unfortunately this scheme is only secure under LWE in the random oracle model. On the plus side, we show that it enjoys the distinction of admitting a secure masking system, and as a consequence of Theorem 3 can be compiled into a multi-identity IBFHE scheme.

4.1 The Gentry, Peikert and Vaikuntanthan (GPV) IBE

In the GPV scheme, the TA needs to use a lookup table[3] to store secret keys that are issued to users in order to ensure that only a single unique secret key is ever issued for a given identity. This is required for the security proof in the random oracle model.

A hash function $H : \{0,1\}^* \rightarrow \mathbb{Z}_q^n$ (modeled as a random oracle in the security proof) is used to map an identity string $\text{id} \in \{0,1\}^*$ to a vector $z_{\text{id}} \in \mathbb{Z}_q^n$. Due to space constraints a formal description of the GPV scheme is deferred to Appendix A. It is easy to see that GPV fulfills CP.1 and CP.2. Furthermore, GPV can be shown to be IND-sID-CPA secure in the random oracle model [15] under LWE, and CP.3 follows from the security proof. It remains to construct a masking system for GPV.

4.2 A Masking System for GPV

Relaxation: Support for a Single Identity. As a warm up, we consider a relaxation of a masking system. In this relaxation, it is sufficient to find $\mathbf{X}$ and $\mathbf{Y}$ for only *one* identity id', specified by the encryptor. More precisely, let id be the recipient's identity and let $\text{id}' \neq \text{id}$ be another identity known to the encryptor.

---

[3] Alternatively with the additional assumption of a PRF, a lookup table could be avoided by deterministically deriving secret keys (i.e. obtaining random coins from the PRF).

Furthermore, let v be a secret key for id and let v' be a secret key for id$'$. Then the goal is to allow the evaluator to find matrices $\mathbf{X}$ and $\mathbf{Y}$ satisfying

$$\mathbf{X} \cdot v + \mathbf{Y} \cdot v' = \mu \cdot v' + \text{``small''},$$

where μ is the plaintext. For every $i \in N$, we need to find row vectors x_i and y_i with $\langle x_i, v \rangle + \langle y_i, v' \rangle = \mu \cdot v' + \text{``small''}$.

A trivial way to do this is for the encryptor to set $x_i \leftarrow \mathbf{0}$ and $y_i \leftarrow$ $\mathsf{Flatten}((\underbrace{0}_{1,\dots,i-1}, \mu, \underbrace{0}_{i+1,\dots,N})) + \mathsf{BitDecomp}(\mathcal{E}.\mathsf{Encrypt}(PP, \mathsf{id}', 0)) \in \{0,1\}^N$ where the latter is a GSW row encryption of μ under identity id$'$. Observe that such an x_i and y_i serve as a solution to the above equation. However, it is easy to see that such a trivial solution violates semantic security, since a decryptor with a secret key v' for id$'$ (and no secret key for id) can still recover the plaintext μ.

One strategy for remedying the above approach is to prevent a key holder for identity id$'$ from recovering μ from y_i by appropriately hiding some components of y_i. Let us take a look at the structure of y_i when $\mathcal{E}$ is GPV. It is of the form

$$\mathsf{Flatten}((\underbrace{0}_{1,\dots,i-1}, \mu, \underbrace{0}_{i+1,\dots,N})) + \mathsf{BitDecomp}((\langle z_{\mathsf{id}'}, r \rangle + e, r \cdot \mathbf{A} + f) \in \mathbb{Z}_q^{m'})$$

where $e \xleftarrow{\$} \chi$, $f \xleftarrow{\$} \chi^m$, $r \xleftarrow{\$} \mathbb{Z}_q^n$ and $z_{\mathsf{id}'} = H(\mathsf{id}') \in \mathbb{Z}_q^n$. Suppose we instead generate y_i as

$$y_i \leftarrow \mathsf{Flatten}((\underbrace{0}_{1,\dots,i-1}, \mu, \underbrace{0}_{i+1,\dots,N})) + \mathsf{BitDecomp}((0, r \cdot \mathbf{A} + f)).$$

Now what we have done here is effectively set the first ℓ_q components of y_i to 0 with the exception of the special case $i \in [\ell_q]$ which we will handle separately later. As a result of this modification, we will have $\langle y_i, v' \rangle \approx -\langle z_{\mathsf{id}'}, r \rangle + \mu \cdot 2^i \bmod \ell_q$ (the symbol $\approx$ denotes equality up to "small" differences). Therefore, to cancel out the term $-\langle z_{\mathsf{id}'}, \rangle$, we need to ensure that we set x_i such that $\langle x_i, v \rangle \approx \langle z_{\mathsf{id}'}, r \rangle$.

The approach we take to achieve this is to *blind* the element $\langle z_{\mathsf{id}'}, r \rangle$ with a GPV encryption of zero under identity id such that it can only be *unblinded* with a secret key for identity id (note that the value cannot be recovered outright; instead a noisy approximation is obtained). For simplicity we define the algorithm Blind which takes an identity id and a value $v \in \mathbb{Z}_q$ and outputs a vector $\mathsf{Flatten}((c_1 + v, c_2, \dots, c_{m'}))$ where $c \leftarrow \mathcal{E}.\mathsf{Encrypt}(PP, \mathsf{id}, 0)$. So to provide an x_i counterpart to the vector y_i we generated above, we set $x_i \leftarrow \mathsf{Blind}(\mathsf{id}, (\langle z_{\mathsf{id}'}, r \rangle))$ where r is the vector used in the generation of y_i above. It follows that $\langle x_i, v \rangle + \langle y_i, v' \rangle = \mu \cdot v' + \text{``small''}$.

There are subtleties that we have overlooked. For security reasons, we need to change how we generate x_i and y_i for $i \in [\ell_q]$. This is because for the first ℓ_q components of y_i as generated above, the plaintext μ is not hidden; it is effectively sent in the clear. However we can resolve this issue by setting $x_i \leftarrow$ $\mathsf{Blind}(\mathsf{id}, \mu \cdot 2^{i-1}$ $y_i \leftarrow \mathbf{0}$ and simply setting $y_i \leftarrow \mathbf{0}$.

However there is still a major weakness in this approach. Suppose a decryptor has access to two decryption vectors $u', v' \in \mathbb{Z}_q^N$ that decrypt ciphertexts with identity id'. For example, the TA might have generated distinct secret key vectors when issuing keys to different parties, and the parties may have shared that information.

It is easy to see that

$$\mathbf{Y} \cdot u' - \mathbf{Y} \cdot v' = \mu \cdot (u' - v') + \text{"small"},$$

which allows the decryptor to easily determine $\mu \in \{0,1\}$. Hence a necessary condition for the approach to work is that there be a unique secret key vector for every identity. In fact, this is the primary reason our techniques do not work for ABE. Technically, this restriction means that the system can only support simple classes of access policies, namely classes of predicates with disjoint support sets, which includes the special case of IBE. Fortunately, in the GPV scheme, only a single secret key is ever issued for a given identity.

Support for All Identities. The algorithm above allows an encryptor to create a secure "mask" for a specific identity that he knows. But how can we create a succinct "universal mask" from which "masks" for arbitrary identities can be derived? To achieve this, we need to take a look at the structure of vector x_i in our masking system, which is constructed as $x_i \leftarrow \text{Blind}(\text{id}, \langle z_{\text{id}'}, r \rangle)$ where id' is known to the encryptor. But what if id' is an arbitrary identity (i.e. not simply one that is known beforehand by the encryptor but one that is chosen by the evaluator at evaluation time)? In this case, we need to obtain an x_i that blinds $\langle z_{\text{id}'}, r \rangle$. Our goal is to include information in the universal mask that we derive so that for any identity id' one can derive an x_i that blinds $\langle z_{\text{id}'}, r \rangle$ where $z_{\text{id}'} = H(\text{id}')$.

Recall the following property of BitDecomp from Sect. 2.4:

$$\langle z_{\text{id}'}, r \rangle = \langle \text{BitDecomp}(z_{\text{id}'}), \text{Powersof2}(r) \rangle.$$

Our approach is to *blind* each coefficient of $\text{Powersof2}(r)$, whose length is $\ell_q \cdot n$. We produce a matrix $\mathbf{B}^{(i)} \in \mathbb{Z}_q^{(\ell_q \cdot n) \times m'}$ by letting $b_j^{(i)} \leftarrow \text{BitDecomp}^{-1}(\text{Blind}(\text{id}, p_j))$ where p_j be the j-th coefficient of $\text{Powerof2}(r)$. Then to generate x_i, one computes $x_i \leftarrow \text{Flatten}(\text{BitDecomp}(z_{\text{id}'}) \cdot \mathbf{B}^{(i)})$. Note that y_i is generated as before.

More precisely what we have is shown is how to generate $\mathbf{B}^{(i)}$ and y_i for $i \in [\ell_q]$. Recall that in our previous masking system we generated x_i and y_i differently for $i \in [\ell_q]$. This will also apply here. Instead of computing $\mathbf{B}^{(i)}$ for $i \in [\ell_q]$, we instead merely compute $x_i \leftarrow \text{Blind}(\text{id}, \mu \cdot 2^{i-1})$ and $y_i \leftarrow \mathbf{0}$. This completes the description of our masking system.

We now formally present our masking system for GPV. (which we call MS_{GPV}). Let $\eta = \ell_q \cdot n$.

$\text{MS}_{\text{GPV}}.\text{GenUnivMask}(\text{PP}, \text{id}, \mu):$

1. For $i \in [\ell_q]$:
   (a) Set $\boldsymbol{x_i} \leftarrow \mathsf{Blind}(\mathsf{id}, \mu \cdot 2^{i-1})$
   (b) Set $\boldsymbol{y_i} \leftarrow \boldsymbol{0}$
2. For $\ell_q < i \leq N$:
   (a) Generate $\boldsymbol{r} \xleftarrow{\$} \mathbb{Z}_q^n$ and sample a short error vector $\boldsymbol{e} \xleftarrow{\$} \chi^{m'}$.
   (b) For $j \in [\eta]$:
      (i) Set $\boldsymbol{b}_j^{(i)} \leftarrow \mathsf{BitDecomp}^{-1}(\mathsf{Blind}(\mathsf{id}, p_j)) \in \mathbb{Z}_q^{m'}$ where p_j be the j-th coefficient of $\mathsf{Powerof2}(\boldsymbol{r})$
   (c) Form matrix $\mathbf{B}^{(i)}$ from rows $\boldsymbol{b}_1^{(i)}, \ldots, \boldsymbol{b}_\eta^{(i)}$.
   (d) Set $\boldsymbol{y_i} \leftarrow \mathsf{Flatten}((\underbrace{\ \ \mathbf{0}\ \ }_{1,\ldots,i-1}, \mu, \underbrace{\ \ \mathbf{0}\ \ }_{i+1,\ldots,N}) + \mathsf{BitDecomp}((0, \boldsymbol{r} \cdot \mathbf{A} + \boldsymbol{f})))$
3. Form matrix $\mathbf{Y}$ from rows $\boldsymbol{y_1}, \ldots, \boldsymbol{y_N}$.
4. Output $U := (\boldsymbol{x_1}, \ldots, \boldsymbol{x_{\ell_q}}, \mathbf{Y}, \mathbf{B}^{(\ell_q+1)}, \ldots, \mathbf{B}^{(N)})$.

$\mathsf{MS_{GPV}}.\mathsf{DeriveMask}(\mathsf{PP}, U, \mathsf{id}')$:

1. Parse U as $(\boldsymbol{x_1}, \ldots, \boldsymbol{x_{\ell_q}}, \mathbf{Y}, \mathbf{B}^{(\ell_q+1)}, \ldots, \mathbf{B}^{(N)})$.
2. Compute $\boldsymbol{z}_{\mathsf{id}'} \leftarrow H(\mathsf{id}')$.
3. For $\ell_q < i \leq N$:
   (a) Set $\boldsymbol{x_i} \leftarrow \mathsf{Flatten}(\mathsf{BitDecomp}(\boldsymbol{z}_{\mathsf{id}'}) \cdot \mathbf{B}^{(i)})$
4. Form $\mathbf{X} \in \{0,1\}^{N \times N}$ from $\boldsymbol{x_1}, \ldots, \boldsymbol{x_N}$.
5. Output $(\mathbf{X}, \mathbf{Y})$.

It is easy to see from the definition of $\mathsf{MS_{GPV}}.\mathsf{DeriveMask}$ that the error expansion factor is $w = \eta + 1$. This is because each row in an expanded matrix is formed from a row of $\mathbf{X}$ and a row of $\mathbf{Y}$. But the former decomposes into a sum of η ciphertexts (and hence error terms).

Theorem 4. *[Informal] The masking system* $\mathsf{MS_{GPV}}$ *is selectively secure in the random oracle model (i.e.* $\mathsf{MS_{GPV}}$ *meets the security condition of Definition 8).*

A formal statement of Theorem 4 along with the proof is given in Appendix E. See Appendix A.1 on how to apply the compiler.

5   Multi-key FHE

If we replace the GPV IBE with the Dual-Regev public-key encryption scheme from [15], then we can obtain a multi-key FHE. The only change in the masking system is that identity vectors (i.e. $\boldsymbol{z}_{\mathsf{id}} = H(\mathsf{id}) \in \mathbb{Z}_q^n$) are replaced with public-key vectors in $\mathbb{Z}_q^n$. As a result, the random oracle H is no longer needed, and security holds in the standard model. Our multi-key scheme is the first to the best of our knowledge that is based on well-established problem such as LWE in the standard model (recall that the scheme from [14] requires the non-standard Decisional Small Polynomial Ratio (DSPR) problem). See the full version [19] of this work for a description of an adaptation of our masking systm to the RLWE setting.

Acknowledgments. We would like to thank the anonymous reviewers of for their helpful comments. The authors would like to thank Fuqun Wang for pointing out errors in an earlier version of this paper.

A   The Gentry, Peikert and Vaikuntanthan (GPV) IBE

Note that this variant has been adapted in the same manner as CHKP in [6] for compatibility with the GSW compiler.

Let $\mathbf{A} \in \mathbb{Z}_q^{n \times m}$ be a matrix. We define the lattice $\Lambda^{\perp}(\mathbf{A}) = \{\boldsymbol{x} \in \mathbb{Z}^m : \mathbf{A} \cdot \boldsymbol{x} = \mathbf{0} \mod q\}$ as the space of vectors orthogonal to the rows of $\mathbf{A}$ modulo q. GPV depends on two efficient probabilistic algorithms, which are informally presented as follows:

- **TrapGen**(n, m, q): [21, 22] Generate a statistically uniform matrix $\mathbf{A} \in \mathbb{Z}_q^{n \times m}$ together with a short basis $\mathbf{S} \in \mathbb{Z}^{m \times m}$ for $\Lambda^{\perp}(\mathbf{A})$. Output $(\mathbf{A}, \mathbf{S})$.
- **SamplePre**$(\mathbf{S}, \mathbf{A}, \boldsymbol{u})$: [15] Generate a "short" solution $\boldsymbol{x} \in \mathbb{Z}_q^m$ to the equation $\mathbf{A} \cdot \boldsymbol{x} = \boldsymbol{u} \in \mathbb{Z}_q^n$.

See Appendix C.1 for more background on these algorithms. Furthermore, see Appendix C for a discussion on suitable parameter settings.

GPV.Setup(1^{λ}): Choose parameters $n = n(\lambda)$, $m = m(\lambda)$, $q = q(\lambda)$, a noise distribution $\chi : \mathbb{Z}$. Let $m' = m + 1$. These parameters are implicit in the public parameters PP below. Generate statistically uniform $\mathbf{A} \in \mathbb{Z}_q^{n \times m}$ together with a short basis $\mathbf{S} \in \mathbb{Z}^{m \times m}$ of $\Lambda^{\perp}(\mathbf{A})$ by running $(\mathbf{A}, \mathbf{S}) \leftarrow \mathsf{TrapGen}(n, m, q)$. Choose a collision-resistant hash function $H : \{0, 1\}^t \to \mathbb{Z}_q^n$. Output PP $:= (\mathbf{A}, H)$ and MSK $:= \mathbf{S}$.

GPV.KeyGen$(\mathsf{MSK}, \mathsf{id} \in \{0, 1\}^*)$: If $(\mathsf{id}, s_{\mathsf{id}}) \in$ store, output s_{id} and abort.

Compute $z_{\mathsf{id}} \leftarrow H(\mathsf{id}) \in \mathbb{Z}_q^n$. Compute $w_{\mathsf{id}} \leftarrow \mathsf{SamplePre}(\mathbf{S}, \mathbf{A}, z_{\mathsf{id}}) \in \mathbb{Z}_q^m$. Set $s_{\mathsf{id}} \leftarrow (1, -w_{\mathsf{id}}) \in \mathbb{Z}_q^{m'}$. Add $(\mathsf{id}, s_{\mathsf{id}})$ to store. Output s_{id}.

Let $\mathbf{A}'_{\mathsf{id}} = z_{\mathsf{id}} \parallel \mathbf{A} \in \mathbb{Z}_q^{m'}$. Observe that $\mathbf{A}'_{\mathsf{id}} \cdot s_{\mathsf{id}} = \mathbf{0} \in \mathbb{Z}_q^n$.

GPV.Encrypt$(\mathsf{PP}, \mathsf{id} \in \{0, 1\}^*, \mu \in \{0, 1\})$: Compute $z_{\mathsf{id}} \leftarrow H(\mathsf{id}) \in \mathbb{Z}_q^n$. Let $\mathbf{A}'_{\mathsf{id}} = z_{\mathsf{id}} \parallel \mathbf{A} \in \mathbb{Z}_q^{m'}$. Let $\boldsymbol{\mu} \in \mathbb{Z}_q^{m'}$ be the vector of 0's except with $\mu \cdot \lfloor q/2 \rfloor$ in the first coefficient. Choose random $\boldsymbol{r} \xleftarrow{\$} \mathbb{Z}_q^n$ and small error vector $\boldsymbol{e} \xleftarrow{\$} \chi^{m'}$. Output $c_{\mathsf{id}} \leftarrow \boldsymbol{r} \cdot \mathbf{A}'_{\mathsf{id}} + \boldsymbol{e} + \boldsymbol{\mu} \in \mathbb{Z}_q^{m'}$.

GPV.Decrypt$(s_{\mathsf{id}}, c_{\mathsf{id}})$: Set $\delta \leftarrow \langle c_{\mathsf{id}}, s_{\mathsf{id}} \rangle \in \mathbb{Z}_q$. If δ is small, output 0; if $\delta - q/2 \mod q$ is small, output 1; otherwise, output $\perp$.

A.1   Proof of Theorem 1

It is now possible to put all the pieces together. In more detail, we can now apply our compiler to the IBE scheme GPV with the masking system $\mathsf{MS_{GPV}}$ to yield an IND-sID-CPA secure multi-identity IBFHE in the random oracle model.

Theorem 1. *There exists a multi-identity leveled IBFHE scheme that is* IND-sID-CPA *secure in the random oracle model under the hardness of LWE.*

Proof. Let $\mathcal{D}$ be a maximum degree of composition to support, and let L be a desired number of levels. Let λ be the security parameter. We show there exists a leveled IBFHE scheme with maximum degree of composition $\mathcal{D}$, maximum circuit depth L and security parameter λ.

Choose dimension parameter $n = n(\lambda, L)$ and bound $B = B(n)$. Lemma 1 requires

$$q > 8 \cdot w \cdot B(\mathcal{D}N + 1)^L \tag{A.1}$$

to ensure correctness. Note that w is the expansion factor of the masking system. Now the error expansion factor of $\mathsf{MS_{GPV}}$ is $w = \eta + 1$. But this can be simplified to N^4. Theorem 4 requires $m \geq 2n \lg q$, and we have $N = (m + 1) \lg q$. We need to set q first before setting these parameters (m and N) because of their dependence on q. To do so, q must be expressed without dependence on N. It can be straightforwardly derived from the inequality A.1 that a suitable q is given by

$$q = B \cdot 2^{O(L \lg n\mathcal{D})}$$

with additional care taken to ensure q/B is subexponential in n.

Our parameter settings ensure that the GPV scheme meets CP.1, CP.2 and CP.3, three of the prerequisites for our compiler in Sect. 3. Furthermore, the masking system $\mathsf{MS_{GPV}}$ is secure (via Theorem 4). As a result, CP.4 is additionally satisfied. Therefore, Theorem 3 ensures there exists a secure leveled IBFHE scheme, which by virtue of our parameter settings above (which meet Lemma 1), can correctly evaluate L-depth circuits over ciphertexts with at most $\mathcal{D}$ distinct identities.

B   Proof of Lemma 1

Lemma 1. *Let B be a bound such that all freshly encrypted ciphertexts are B-strongly-bounded. Let $\mathcal{D}$ and L be positive integers. If $q > 8 \cdot w \cdot B(\mathcal{D}N + 1)^{L^5}$, then the scheme mIBFHE is correct and can evaluate NAND-based Boolean circuits of depth L with any number of distinct identities $d \leq \mathcal{D}$.*

Proof. Let the $d \leq \mathcal{D}$ distinct identities involved in an evaluation be $\mathsf{id}_1, \ldots, \mathsf{id}_d$. Consider an expanded matrix derived from a "fresh" ciphertext $\mathsf{CT} = (\mathsf{id}_i, \mathsf{type} := 0, \mathsf{enc} := U)$ associated with identity id_i for some $i \in [d]$. Let v_j be a secret key that decrypts ciphertexts with identity id_j for $j \in [d]$. Let $\hat{v}$ be the column vector consisting of the concatenation of $v_1, \ldots, v_d$. Let $\hat{\mathbf{C}}$ be the expanded matrix for CT computed with respect to identities $\mathsf{id}_1, \ldots, \mathsf{id}_d$ and $(\mathbf{X}_j, \mathbf{Y}_j) \leftarrow \mathsf{MS_{\mathcal{E}}}.\mathsf{DeriveMask}(\mathsf{PP}, U, \mathsf{id}_j)$ for $j \in [d]$. Now by construction, $\hat{\mathbf{C}}$ consists of $d \times d$ submatrices in $\mathbb{Z}_q^{N \times N}$. There are 2 non-zero submatrices on $N - 1$ rows when $\hat{\mathbf{C}}$

---

[4] $w = \eta + 1 = \ell_q \cdot n + 1 \leq \ell_q \cdot m < N$.

[5] Note that N (which depends on n) is itself dependent on $\lg q$. For security, it is required that $q/B = 2^{n^\epsilon}$ for some $\epsilon \in (0, 1)$. A discussion on parameters is provided in Appendix C.

is viewed as $d \times d$ matrix over $\mathbb{Z}_q^{N \times N}$, and one non-zero submatrix on the i-th row. The correctness condition for the masking system $\mathsf{MS}_{\mathcal{E}}$ gives us

$$
\begin{pmatrix} \mathbf{Y}_1 & & \mathbf{X}_1 \\ & \ddots & \vdots \\ & \text{Flatten}(\mathbf{X}_i + \mathbf{Y}_i) & \\ & \vdots & \ddots \\ \mathbf{X}_d & & \mathbf{Y}_d \end{pmatrix} \cdot \begin{bmatrix} v_1 \\ \vdots \\ v_i \\ \vdots \\ v_d \end{bmatrix} = \begin{bmatrix} \mathbf{X}_1 v_1 + \mathbf{Y}_1 v_1 \\ \vdots \\ \mathbf{X}_i v_i + \mathbf{Y}_i v_i \\ \vdots \\ \mathbf{X}_d v_d + \mathbf{Y}_d v_d \end{bmatrix} = \mu \cdot \begin{bmatrix} v_1 \\ \vdots \\ v_i \\ \vdots \\ v_d \end{bmatrix} + \text{'small'}.
$$

Since each of these submatrices is B-strongly-bounded, it follows that $\hat{\mathbf{C}} \cdot \hat{v} = \mu \cdot \hat{v} + \hat{e}$ where the coefficients of the error vector $\hat{e}$ are bounded by $w \cdot B$. Therefore, $\hat{C}$ is $w \cdot B$-strongly-bounded. Multiplying two $dN \times dN$ expanded matrices in a NAND operation produces a matrix that is $w \cdot B(dN + 1)$-strongly-bounded. After L successive levels, the bound on the error is $w \cdot B(dN+1)^L$. For correctness of decryption we need $w \cdot B(dN + 1)^L < q/8$. Since we have $d \leq \mathcal{D}$, it follows that

$$
w \cdot B(dN + 1)^L \leq w \cdot B(\mathcal{D}N + 1)^L \leq \frac{8 \cdot w \cdot B(\mathcal{D}N + 1)^L}{8} < \frac{q}{8}.
$$

$\square$

C   Parameters for Our Scheme

Before discussing how parameters are chosen for our scheme, more background is needed on preimage sampling.

C.1   Background on Preimage Sampling

Let $\mathbf{A} \in \mathbb{Z}_q^{n \times m}$ be a matrix. We define the lattice $\Lambda^{\perp}(\mathbf{A}) = \{x \in \mathbb{Z}^m : \mathbf{A} \cdot x = 0 \bmod q\}$ as the space of vectors orthogonal to the rows of $\mathbf{A}$ modulo q. There exist efficient algorithms to generate a statistically uniform matrix $\mathbf{A} \in \mathbb{Z}_q^{n \times m}$ together with a short basis $\mathbf{S} \in \mathbb{Z}^{m \times m}$ for $\Lambda^{\perp}(\mathbf{A})$ [21,22]. Such an algorithm will be simply called TrapGen here; that is, we will write $(\mathbf{A}, \mathbf{S}) \leftarrow \mathsf{TrapGen}(n, m, q)$. We denote by $\tilde{\mathbf{S}}$ the Gram-Schmidt orthonormalization of a basis $\mathbf{S}$. Let $\mathfrak{L} = \|\tilde{\mathbf{S}}\|$ be the norm of $\mathbf{S}$. There are instances of TrapGen that achieve $\mathfrak{L} = m^{1+\epsilon}$ for any $\epsilon > 0$ [15], although this has been improved upon in other works [23]. Hence, our setting of $\mathfrak{L}$ later will be a conservative choice.

Let d and t be positive integers with $d \leq t$. Let $\mathbf{B} \in \mathbb{R}^{d \times t}$ be a basis for a d-dimensional lattice $\Lambda(\mathbf{B}) \subset \mathbb{R}^t$. Then the discrete Gaussian distribution on $\Lambda(\mathbf{B})$ with center $c \in \mathbb{R}^t$ and standard deviation $\sigma \in \mathbb{R}$ is denoted by $D_{\Lambda(\mathbf{B}),s,c}$. When c is understood to be zero, the center parameter is omitted.

Gentry, Peikert and Vaikuntanthan [15] describe an algorithm to sample from a discrete Gaussian distribution on an arbitrary lattice. They describe an efficient probabilistic algorithm $\mathsf{SampleD}(\mathbf{B}, \sigma, c)$ that samples from a distribution that is statistically close to $D_{\Lambda(\mathbf{B}),\sigma,c}$, provided $\sigma \geq \|\tilde{\mathbf{B}}\| \cdot \omega(\sqrt{\log d})$.

Consider the function $f_A : \mathbb{Z}_q^m \to \mathbb{Z}_q^n$ defined by $f(x) = \mathbf{A} \cdot x \in \mathbb{Z}_q^n$. Given any vector $u \in \mathbb{Z}_q^n$, a *preimage* of u under f_A is any $x \in \mathbb{Z}_q^m$ with $f_A(x) = u$.

It turns out SampleD can be used to efficiently to find *short preimages* $x \in \mathbb{Z}_q^m$ such that $\mathbf{A} \cdot x = u \in \mathbb{Z}_q^n$ for an arbitrary vector $u \in \mathbb{Z}_q^n$. Consider the following algorithm SamplePre from [15]. Note that s is a parameter for which possible settings are given in the next section.

- **SamplePre($\mathbf{S}, \mathbf{A}, u$):** Find an arbitrary solution $t \in \mathbb{Z}_q^m$ (via linear algebra) such that $\mathbf{A} \cdot t = u \mod q$. Sample a vector $e \xleftarrow{\$} D_{\Lambda^\perp(\mathbf{A}), s, -t}$ by running $e \leftarrow \mathsf{SampleD}(\mathbf{S}, s, -t)$, and output the vector $x \leftarrow e + t$.

We remind the reader that there are improved variants of SamplePre in the literature [23].

C.2 Preimage Distribution

We need $s \geq \mathfrak{L} \cdot \omega(\sqrt{\log m})$ to satisfy Theorem 5.9 of [15]. Let $B_{\mathsf{preimage}} \geq \sqrt{n} \cdot s$. Then the probability of the magnitude of any coefficient of a preimage vector exceeding B_{preimage} is exponentially small in n via a standard tail inequality for a normal distribution[6]. One possible setting is $s = \mathfrak{L} \cdot \log m$, and $B_{\mathsf{preimage}} = \sqrt{n} \cdot s$.

C.3 Noise Distribution

To satisfy Theorem 2, we need the noise distribution χ to be B_χ-bounded for some B_χ (to satisfy Theorem 2, we require q/B_χ to be at most subexponential). Setting $\chi \leftarrow D_{\mathbb{Z},r}$ with $r = \log m$ and $B_\chi \geq \sqrt{n} \cdot r$ ensures that χ is B_χ-bounded, since by the aforementioned tail inequality, we have that $\Pr[x \xleftarrow{\$} D_{\mathbb{Z},r}, |x| > B_\chi]$ is exponential in n.

C.4 Parameter B (B-Strong-Boundedness)

"Fresh" ciphertexts in our scheme are B-strongly-bounded. The parameter B is derived from the product of B_{preimage} and B_χ, since when the ciphertext matrix is multiplied by a secret key vector, the resulting error vector is formed from the inner product of the noise vector in the ciphertext (drawn from χ) and the secret key (a sampled preimage). Concretely, with the suggested parameter setting, we have $B = \mathfrak{L} \cdot n \cdot \log^2 m$. It is necessary that q/B_1 is at most subexponential in N. However, our analysis simplifies this by taking q/B to be subexponential; however, since B_{preimage} is polynomial in N, it also holds that q/B_χ is subexponential.

---

[6] A normal variable with standard deviation σ is within $t \cdot \sigma$ standard deviations of its mean, except with probability at most $\frac{1}{t} \cdot \frac{1}{e^{t^2/2}}$ [15].

C.5   Sample Parameters and Ciphertext Size

Gentry, Sahai and Waters simplify their analysis by taking n to be a fixed parameter. This is a simplification because q/B must be subexponential in n, and q depends on L; therefore in actuality n depends on L.

Let L be the desired number of levels and let $\mathcal{D}$ be the desired maximum number of distinct identities to support in an evaluation. According to Lemma 1, correctness requires that

$$q > 8 \cdot w \cdot B(\mathcal{D}N + 1)^L. \tag{C.1}$$

In Appendix C.1, it was mentioned that $\mathfrak{L} \approx m$. Putting this together with the derivation of B above in Appendix C.4 gives $B = mn \cdot \log^2 m$, where $m \geq 2n \lg q$ from Theorem 4. Choosing B in this way means that it is not too large and allows us to derive $\lg q$ from the inequality C.1 above as follows: $\lg q = O(L(\lg \mathcal{D} + \lg n))$.

Consider the following concrete parameters. Suppose we require a circuit depth of $L = 40$ and a number of distinct identities up to $\mathcal{D} = 100$. We can satisfy the correctness constraint given by C.1 by setting $\lg q = \lceil c \cdot L(\lg \mathcal{D} + \lg L) = 4 \cdot 40(\lg 100 + \lg 40) \rceil = 1915$ (the constant $c = 4$ was chosen to meet the condition) and choosing the dimension to be $n = 2000$. However the size of freshly encrypted ciphertexts in our leveled IBFHE scheme with these parameters is greater than one exabyte (i.e. $> 2^{30}$ gigabytes) per bit of plaintext, which is extremely impractical. This illustrates the impracticality of our scheme, but it also highlights the impracticality at the present time of the GSW leveled IBFHE and ABFHE schemes.

D   Size of Evaluated Ciphertexts

As mentioned in the previous section, n is not a fixed parameter that depends solely on the security level λ. Instead n grows with both L and $\mathcal{D}$ because q/B must be subexponential in n to guarantee security. There is an optimization that applies to both our construction and the GSW constructions in terms of the size of evaluated ciphertexts. Decryption only requires a single row of a ciphertext matrix (see Sect. 3.2), so an evaluated ciphertext can have size $d \cdot N$ where d is the number of distinct identities in the evaluation. Let this vector be denoted by $\hat{c} \in \{0,1\}^{d \cdot N}$. Applying $\mathsf{BitDecomp}^{-1}$, the vector $c \leftarrow \mathsf{BitDecomp}^{-1}(\hat{c}) \in \mathbb{Z}_q^{m'}$ is obtained. As explained in [6], if we include additional information in the public parameters, the technique of modulus reduction [5] can be employed to each coefficient in c so that the size of each coefficient can be made independent of $\mathcal{D}$ and L; their size must still depend on d to ensure correctness, but this is allowed for by the compactness condition. However, while every coefficient can be reduced, the dimension cannot be reduced. This is because the technique of dimension reduction [5] appears to be only compatible with the public key setting since it relies on publishing encryptions of the secret key. We defer the details to [5]. So the length of the ciphertext vector is the length of c, namely

m', which in turn depends on both L and $\mathcal{D}$. Therefore, technically speaking, our multi-identity IBFHE in addition to both the IBFHE and ABFHE constructions of Gentry, Sahai and Waters are not *leveled* in the strict sense of the size of an evaluated ciphertext being independent of L.

E   Proof of Theorem 4

Corollary 1 (Corollary 5.4 [15]). *Let n be a positive integer, and let q be a prime. Let $m \geq 2n \lg q$. Then for all but a $2q^{-n}$ fraction of all $\mathbf{A} \in \mathbb{Z}_q^{n \times m}$ and for any $s \geq \omega(\sqrt{\log m})$, the distribution of the syndrome $\boldsymbol{u} = \mathbf{A}\boldsymbol{e} \bmod q$ is statistically close to uniform over $\mathbb{Z}_q^n$, where $\boldsymbol{e} \sim D_{\mathbb{Z}^m, s}$.*

Theorem 4. *Let n, m, q be chosen to meet Corollary 1. Let χ be a B_χ-bounded distribution where B_χ satisfies Theorem 2. Let TrapGen be an algorithm that generates a statistically uniform matrix $\mathbf{A} \in \mathbb{Z}_q^{n \times m}$ together with a basis $\mathbf{S} \in \mathbb{Z}^{m \times m}$ such that $\|\tilde{\mathbf{S}}\| \leq \mathcal{L}$ except with negligible probability. Let $s \geq \mathcal{L} \cdot \omega(\sqrt{\log m})$. Let the scheme GPV be instantiated with TrapGen and the SamplePre algorithm (with parameter s) described in Appendix C.1.*

Then the masking system $\mathsf{MS}_{\mathsf{GPV}}$ is selectively secure in the random oracle model (i.e. $\mathsf{MS}_{\mathsf{GPV}}$ meets the security condition of Definition 8) under the hardness of $LWE_{n,q,\chi}$.

Proof. We prove the theorem by means of a hybrid argument.
Game 0: This is the standard selective security game described in Definition 8.

Game 1: The following changes are made in this game. Let $\mathsf{id}^* \in \mathcal{I}$ be the adversary's target identity.

1. The matrix $\mathbf{A} \xleftarrow{\$} \mathbb{Z}_q^{n \times m}$ is generated as uniformly random.
2. The vector $z_{\mathsf{id}^*} \xleftarrow{\$} \mathbb{Z}_q^n$ is generated as uniformly random.
3. The random oracle H is simulated as follows: if the adversary $\mathcal{A}$ queries H on identity $\mathsf{id} \in \mathcal{I}$, run:
   (a) If $\mathsf{id} = \mathsf{id}^*$, then return z_{id^*}.
   (b) Else if $(\mathsf{id}, s_{\mathsf{id}}, z_{\mathsf{id}}) \in$ store, return z_{id}.
   (c) Else sample $t_{\mathsf{id}} \xleftarrow{\$} D_{\mathbb{Z}^{m'-1}, s}$, compute $z_{\mathsf{id}} \leftarrow \mathbf{A} \cdot t_{\mathsf{id}} \bmod q$, set $s_{\mathsf{id}} \leftarrow (1, -t_{\mathsf{id}}) \in \mathbb{Z}_q^{m'}$, add $(\mathsf{id}, s_{\mathsf{id}}, z_{\mathsf{id}})$ to store and return z_{id}.
   (d) Secret key queries are answered as follows. Suppose $\mathcal{A}$ queries a secret key for identity $\mathsf{id} \neq \mathsf{id}^*$. We assume w.l.o.g. that $\mathcal{A}$ has first queried H on id. In response to the query, s_{id} is returned where $(\mathsf{id}, s_{\mathsf{id}}, z_{\mathsf{id}}) \in$ store.

We claim that $\mathcal{A}$'s view in Game 0 is statistically close to $\mathcal{A}$'s view in Game 1. The first two changes above follow immediately from the definition of GPV (in particular, the trapdoor basis generation algorithm employed guarantees that a near uniform $\mathbf{A}$ can be generated). In regard to the simulation of H, Corollary 1 implies that the vector $H(\mathsf{id})$ when $\mathsf{id} \neq \mathsf{id}^*$ is statistically close to uniform. Finally, with regard to the distribution of secret keys, Lemma 5.2 from [15] states that a preimage t_{id} sampled with SamplePre (with parameter s) in GPV.KeyGen is identically distributed to $t_{id} \sim D_{\mathbb{Z}^{m'-1},s}$ conditioned on $\mathbf{A}_{id} \cdot t_{id} = z_{id} \bmod q$. It follows that the secret keys s_{id} in Game 1 have the same distribution as Game 0.

For $i \in [\ell_q]$:

Game $i + 1$: This game is the same as the previous game except that Step 1a of $\mathsf{MS_{GPV}.GenUnivMask}$ for iteration i (*only*) is replaced with

$$x_i \leftarrow \mathsf{BitDecomp}(t).$$

where $t \xleftarrow{\$} \mathbb{Z}_q^{m'}$.

Given an LWE instance $x^* \in \mathbb{Z}_q^{m'}$, one can easily generate x_i according to Game i or Game $i+1$. Suppose a distinguisher $\mathcal{D}$ has a non-negligible advantage distinguishing between Game i and Game $i + 1$. We can use $\mathcal{D}$ to construct an algorithm $\mathcal{B}$ that can solve an LWE instance. Given an appropriate number of samples from either the distribution $D_0 := \{\{(u_j, \langle u_j, s \rangle + e_j) : u_j \xleftarrow{\$} \mathbb{Z}_q^n, e)_j \xleftarrow{\$} \chi\} : s \xleftarrow{\$} \mathbb{Z}_q^n\}$ or the distribution $D_1 := \{\{(u_j, v_j) : u_j, v_j \xleftarrow{\$} \mathbb{Z}_q^n\}\}$, the u_j are used to construct $\mathbf{A} \in \mathbb{Z}_q^{n \times m}$ and $z_{id^*} \in \mathbb{Z}_q^n$. The algorithm $\mathcal{B}$ simulates the random oracle H as explained above, and answers secret key queries in the manner described above. Note that the distribution of $\mathbf{A}$ and z_{id^*} remain unchanged.

The algorithm $\mathcal{B}$ runs the same variant of $\mathsf{MS_{GPV}.GenUnivMask}$ as the previous game. The only difference is that on the i-th iteration, it replaces Step 1a with

$$x_i \leftarrow \mathsf{BitDecomp}(x^* + (\mu \cdot 2^i, 0, \dots, 0))$$

where $x^* \in \mathbb{Z}_q^{m'}$ is an LWE challenge vector that is either $s \cdot z_{id^*} \parallel \mathbf{A} + e \in \mathbb{Z}_q^{m'}$ or a uniformly random $t^* \in \mathbb{Z}_q^{m'}$. In the former case, the view is statistically close to Game i whereas the view in the latter case is statistically close to Game $i + 1$. It follows that $\mathcal{B}$ can output $\mathcal{D}$'s guess to solve an LWE instance. The games are thus indistinguishable by the hypothesized hardness of LWE.

As a shorthand for Game $(\ell_q + 1) + (i - \ell_q - 1) \cdot (\eta + 1) + j$, we use the notation Game (i, j) for $\ell_q < i \le N$ and $j \in [\eta + 1]$.

For $\ell_q < i \le N$:

    For $j \in [\eta]$:

- **Game (i,j):** This game is the same as the previous game except that we change the way that the j-th row of $\mathbf{B}^{(i)}$ is generated in $\mathsf{MS_{GPV}.GenUnivMask}$. More precisely, Step 2(b)i of algorithm $\mathsf{MS_{GPV}.GenUnivMask}$ is replaced with

$$b_j^{(i)} \leftarrow \mathsf{BitDecomp}(t)$$

    with $t \xleftarrow{\$} \mathbb{Z}_q^{m'}$. for the *specific case* of the i-th iteration of the outer loop and the j-th iteration of the inner loop.

An analogous argument to the argument made above concerning the indistinguishability of Game i and $i + 1$ for $i \in [\ell_q]$ can be made here to show that a non-negligible advantage distinguishing between the games implies a non-negligible advantage against LWE.

Remark 2. At this stage, note that $\mathbf{B}^{(i)}$ from $\mathsf{MS_{GPV}.GenUnivMask}$ is uniform over $\mathbb{Z}_q^{\eta \times m'}$; in particular it does not rely on any r associated with a y_i nor does it rely on μ.

Game $(i, \eta + 1)$: The modification in this game is as follows. Step 2d of $\mathsf{MS_{GPV}.GenUnivMask}$ for the i-th iteration is replaced with

$$y_i \leftarrow \mathsf{Flatten}((\mathsf{BitDecomp}((0, t)).$$

with $t \overset{\$}{\leftarrow} \mathbb{Z}_q^{m'}$.

Once again an analogous LWE-based argument to that above shows that one can embed an LWE challenge when generating y_i such that indistinguishability between the games implies a non-negligible advantage against LWE.

We conclude the proof by observing that in Game $(N, \eta + 1)$, the plaintext bit μ has been eliminated entirely from the generation of the universal mask U. It follows that an adversary has a zero advantage guessing the challenger's bit b, since no information about b is incorporated in the universal mask U given to the adversary. □

References

1. Gentry, C.: Fully homomorphic encryption using ideal lattices. In: Proceedings of the 41st Annual ACM Symposium on Theory of Computing, STOC 2009, p. 169 (2009)
2. Smart, N.P., Vercauteren, F.: Fully homomorphic encryption with relatively small key and ciphertext sizes. In: Nguyen, P.Q., Pointcheval, D. (eds.) PKC 2010. LNCS, vol. 6056, pp. 420–443. Springer, Heidelberg (2010)
3. van Dijk, M., Gentry, C., Halevi, S., Vaikuntanathan, V.: Fully homomorphic encryption over the integers. In: Gilbert, H. (ed.) EUROCRYPT 2010. LNCS, vol. 6110, pp. 24–43. Springer, Heidelberg (2010)
4. Brakerski, Z., Vaikuntanathan, V.: Fully homomorphic encryption from ring-lwe and security for key dependent messages. In: Rogaway, P. (ed.) CRYPTO 2011. LNCS, vol. 6841, pp. 505–524. Springer, Heidelberg (2011)
5. Brakerski, Z., Vaikuntanathan, V.: Efficient fully homomorphic encryption from (standard) LWE. In: Ostrovsky, R. (ed.) FOCS 2011, pp. 97–106. IEEE, Olympia (2011)
6. Gentry, C., Sahai, A., Waters, B.: Homomorphic encryption from learning with errors: conceptually-simpler, asymptotically-faster, attribute-based. In: Canetti, R., Garay, J.A. (eds.) CRYPTO 2013, Part I. LNCS, vol. 8042, pp. 75–92. Springer, Heidelberg (2013)
7. Boneh, D., Franklin, M.: Identity-based encryption from the weil pairing. In: Kilian, J. (ed.) CRYPTO 2001. LNCS, vol. 2139, pp. 213–229. Springer, Heidelberg (2001)

8. Cocks, C.: An identity based encryption scheme based on quadratic residues. In: Honary, B. (ed.) Cryptography and Coding 2001. LNCS, vol. 2260, pp. 360–363. Springer, Heidelberg (2001)

9. Regev, O.: On lattices, learning with errors, random linear codes, and cryptography. In: Proceedings of the Thirty-Seventh Annual ACM symposium on Theory of Computing, STOC 2005, pp. 84–93. ACM, New York, NY, USA (2005)

10. Clear, M., McGoldrick, C.: Bootstrappable identity-based fully homomorphic encryption. Cryptology ePrint Archive, report 2014/491 (2014). http://eprint.iacr.org/

11. Garg, S., Gentry, C., Halevi, S., Raykova, M., Sahai, A., Waters, B.: Candidate indistinguishability obfuscation and functional encryption for all circuits. In: FOCS, pp. 40–49. IEEE Computer Society (2013)

12. Clear, M., McGoldrick, C.: Bootstrappable identity-based fully homomorphic encryption. In: Gritzalis, D., Kiayias, A., Askoxylakis, I. (eds.) CANS 2014. LNCS, vol. 8813, pp. 1–19. Springer, Heidelberg (2014)

13. Agrawal, S., Freeman, D.M., Vaikuntanathan, V.: Functional encryption for inner product predicates from learning with errors. In: Lee, D.H., Wang, X. (eds.) ASIACRYPT 2011. LNCS, vol. 7073, pp. 21–40. Springer, Heidelberg (2011)

14. López-Alt, A., Tromer, E., Vaikuntanathan, V.: On-the-fly multiparty computation on the cloud via multikey fully homomorphic encryption. In: Proceedings of the 44th Symposium on Theory of Computing, STOC 2012, pp. 1219–1234. ACM, New York, NY, USA (2012)

15. Gentry, C., Peikert, C., Vaikuntanathan, V.: Trapdoors for hard lattices and new cryptographic constructions. In: STOC 2008: Proceedings of the 40th Annual ACM Symposium on Theory of Computing, pp. 197–206. ACM, New York, (2008)

16. Agrawal, S., Boneh, D., Boyen, X.: Efficient lattice (H)IBE in the standard model. In: Gilbert, H. (ed.) EUROCRYPT 2010. LNCS, vol. 6110, pp. 553–572. Springer, Heidelberg (2010)

17. Agrawal, S., Boneh, D., Boyen, X.: Lattice basis delegation in fixed dimension and shorter-ciphertext hierarchical IBE. In: Rabin, T. (ed.) CRYPTO 2010. LNCS, vol. 6223, pp. 98–115. Springer, Heidelberg (2010)

18. Cash, D., Hofheinz, D., Kiltz, E., Peikert, C.: Bonsai trees, or how to delegate a lattice basis. In: Gilbert, H. (ed.) EUROCRYPT 2010. LNCS, vol. 6110, pp. 523–552. Springer, Heidelberg (2010)

19. Clear, M., McGoldrick, C.: Multi-identity and multi-key leveled FHE from learning with errors. IACR Cryptology ePrint Archive 2014, p. 798 (2014). http://eprint.iacr.org/2014/798

20. Gilbert, H. (ed.): Advances in Cryptology - EUROCRYPT 2010. LNCS, vol. 6110. Springer, Heidelberg (2010)

21. Ajtai, M.: Generating hard instances of the short basis problem. In: Wiedermann, J., Van Emde Boas, P., Nielsen, M. (eds.) ICALP 1999. LNCS, vol. 1644, pp. 1–9. Springer, Heidelberg (1999)

22. Alwen, J., Peikert, C.: Generating shorter bases for hard random lattices. Cryptology ePrint Archive, report 2008/521 (2008). http://eprint.iacr.org/2008/521

23. Micciancio, D., Peikert, C.: Trapdoors for lattices: simpler, tighter, faster, smaller. In: Pointcheval, D., Johansson, T. (eds.) EUROCRYPT 2012. LNCS, vol. 7237, pp. 700–718. Springer, Heidelberg (2012)

From Selective to Adaptive Security in Functional Encryption

Prabhanjan Ananth[1], Zvika Brakerski[2], Gil Segev[3]([✉]),
and Vinod Vaikuntanathan[4]

[1] University of California, Los Angeles, USA
prabhanjan@cs.ucla.edu
[2] Weizmann Institute of Science, Rehovot, Israel
zvika.brakerski@weizmann.ac.il
[3] Hebrew University of Jerusalem, Jerusalem, Israel
segev@cs.huji.ac.il
[4] Massachusetts Institute of Technology, Cambridge, USA
vinodv@mit.edu

Abstract. In a functional encryption (FE) scheme, the owner of the secret key can generate restricted decryption keys that allow users to learn specific functions of the encrypted messages and nothing else. In many known constructions of FE schemes, security is guaranteed only for messages that are fixed ahead of time (i.e., before the adversary even interacts with the system). This so-called *selective security* is too restrictive for many realistic applications. Achieving *adaptive security* (also called *full security*), where security is guaranteed even for messages that are adaptively chosen at any point in time, seems significantly more challenging. The handful of known adaptively-secure schemes are based on specifically tailored techniques that rely on strong assumptions (such as obfuscation or multilinear maps assumptions).

We show that any sufficiently-expressive *selectively-secure* FE scheme

P. Ananth–This work was done in part while visiting MIT, and was supported in part by the Northrop Grumman Cybersecurity Consortium. Research supported in part from a DARPA/ONR PROCEED award, NSF Frontier Award 1413955, NSF grants 1228984, 1136174, 1118096, and 1065276. This material is based upon work supported by the Defense Advanced Research Projects Agency through the U.S. Office of Naval Research under Contract N00014-11- 1-0389. The views expressed are those of the author and do not reflect the official policy or position of the Department of Defense, the National Science Foundation, or the U.S. Government.

Z. Brakerski–Supported by the Israel Science Foundation (Grant No. 468/14) and by the Alon Young Faculty Fellowship.

G. Segev–Supported by the European Union's Seventh Framework Programme (FP7) via a Marie Curie Career Integration Grant, by the Israel Science Foundation (Grant No. 483/13), and by the Israeli Centers of Research Excellence (I-CORE) Program (Center No. 4/11).

V. Vaikuntanathan–Research supported in part by DARPA Grant number FA8750-11-2-0225, an Alfred P. Sloan Research Fellowship, the Northrop Grumman Cybersecurity Research Consortium (CRC), Microsoft Faculty Fellowship, and a Steven and Renee Finn Career Development Chair from MIT.

R. Gennaro and M. Robshaw (Eds.): CRYPTO 2015, Part II, LNCS 9216, pp. 657–677, 2015.
DOI: 10.1007/978-3-662-48000-7_32

can be transformed into an *adaptively-secure* one without introducing any additional assumptions. We present a black-box transformation, for both public-key and private-key schemes, making novel use of *hybrid encryption*, a classical technique that was originally introduced for improving the efficiency of encryption schemes. We adapt the hybrid encryption approach to the setting of functional encryption via a technique for embedding a "hidden execution thread" in the decryption keys of the underlying scheme, which will only be activated within the proof of security of the resulting scheme. As an additional application of this technique, we show how to construct functional encryption schemes for arbitrary circuits starting from ones for shallow circuits (NC1 or even TC0).

Keywords: Functional encryption · Adaptive security · Generic constructions

1 Introduction

Traditional notions of public-key encryption provide all-or-nothing access to data: owners of the secret key can recover the entire message from a ciphertext, whereas those who do not know the secret key learn nothing at all. Functional encryption, a revolutionary notion originating from the work of Sahai and Waters [SW05], is a modern type of encryption scheme where the owner of the (master) secret key can release function-specific secret keys sk_f, referred to as *functional keys*, which enable a user holding an encryption of a message x to compute $f(x)$ but nothing else (see [KSW08, LOS+10, BSW11, O'N10] and many others). Intuitively, in terms of indistinguishability-based security, encryptions of any two messages, x_0 and x_1, should be computationally indistinguishable given access to functional keys for any function f such that $f(x_0) = f(x_1)$.

While initial constructions of functional encryption schemes [BF03, BCO+04, KSW08, LOS+10] were limited to restricted function classes such as point functions and inner products, recent developments have dramatically improved the state of the art. In particular, the works of Sahai and Seyalioglu [SS10] and Gorbunov, Vaikuntanathan and Wee [GVW12] showed that a scheme supporting a single functional key can be based on any semantically-secure encryption scheme. This result can be extended to the case where the number of functional keys is polynomial and known a-priori [GVW12]. Goldwasser, Kalai, Popa, Vaikuntanathan and Zeldovich [GKP+13] constructed a scheme with succinct ciphertexts based on a specific hardness assumption (Learning with Errors).

The first functional encryption scheme that supports a-priori unbounded number of functional keys was constructed by Garg, Gentry, Halevi, Raykova, Sahai and Waters [GGH+13], based on the existence of a general-purpose indistinguishability obfuscator (for which a heuristic construction is presented in the same paper). Garg et al. showed that given any such obfuscator, their functional encryption scheme is *selectively secure*. At a high level, selective security guarantees security only for messages that are fixed ahead of time (i.e., before the adversary even interacts with the system). Whereas security only for such messages may be justified in some cases, it is typically too restrictive for realistic applications. A more realistic notion is that of *adaptive security* (often called *full security*), which guarantees security even for messages that can be adaptively chosen at any point in time.

Historically, the first functional encryption schemes were only proven selectively secure [BB04, GPS+06, KSW08, GVW13, GKP+13]. The problem of constructing adaptively secure schemes seems significantly more challenging and only few approaches are known. A simple observation is that if a selectively-secure scheme's message space is not too large, e.g., $\{0, 1\}^n$ for a relatively small n, then any adaptively-chosen message x can be guessed ahead of time with probability 2^{-n}. Starting with a *sub-exponential* hardness assumption, and taking the security parameter to be polynomial in n allows us to argue that the selectively-secure scheme is in fact also adaptively secure. This observation is known as "complexity leveraging" and is clearly not satisfactory in general.

The powerful "dual system" approach, put forward by Waters [Wat09], has been used to construct adaptively-secure attribute-based encryption scheme (a restricted notion of functional encryption) for formulas, as well as an adaptively-secure functional encryption scheme for linear functions [LOS+10]. However, this method is a general outline, and each construction was so far required to tailor the solution based on its specialized assumption. In some cases, such as attribute-based encryption for circuits, it is still not known how to implement dual system encryption to achieve adaptive security (although Garg, Gentry, Halevi and Zhandry [GGH+14a] show how to do this with custom-built methods and hardness assumptions).

Starting with [GGH+13], there has been significant effort in the research community to construct an adaptively-secure general-purpose functional encryption scheme with an unbounded number of functional keys. Boyle, Chung and Pass [BCP14] constructed an adaptively secure scheme, under the assumption that differing-input obfuscators exist (these are stronger primitives than the indistinguishability obfuscators used by [GGH+13]). Following their work, Waters [Wat14] and Garg, Gentry, Halevi and Zhandry [GGH+14b] constructed specific adaptively-secure schemes assuming indistinguishability obfuscation and assuming non-standard assumptions on multilinear maps, respectively. Despite this significant progress, each of these constructions relies on somewhat tailored methods and techniques.

1.1   Our Results: From Selective to Adaptive Security

We show that any selectively-secure functional encryption scheme implies an adaptively-secure one, without relying on any additional assumptions. Our transformation applies equally to public-key schemes and to private-key ones, where the resulting adaptive scheme inherits the public-key or private-key flavor of the underlying scheme. The following theorem informally summarizes our main contribution.

Theorem 1.1 (Informal). *Given any public-key (resp. private-key) selectively-secure functional encryption scheme for the class of all polynomial size circuits, there exists an adaptively-secure public-key (resp. private-key) functional encryption scheme with similar properties.*

Specifically, the adaptive scheme supports slightly smaller circuits than those supported by the selective scheme we started with.

Our transformation can be applied, in particular, to the selectively-secure schemes of Garg et al. [GGH+13] and Waters [Wat14], resulting in adaptively-secure schemes based on indistinguishability obfuscation (and one-way functions).[1]

We view the significance of our result in a number of dimensions. First of all, it answers the basic call of cryptographic research to substantiate the existence of rather complex primitives on that of somewhat simpler ones. We feel that this is of special interest in the case of adaptive security, where it seemed that ad-hoc methods were required. Secondly, our construction, being of fairly low overhead, will allow to focus the attention of the research community in studying selectively-secure functional encryption schemes, rather than investing unwarranted efforts in obtaining adaptively-secure ones. Lastly, we hope that our methods will be extended towards weaker forms of functional encryption schemes for which adaptive security is yet unattained generically, such as attribute-based encryption for all polynomial-size circuits.

1.2   Our Techniques

Our result is achieved by incorporating a number of techniques which will be explained in this section. In a nutshell, our main observation is that *hybrid encryption* (a.k.a key encapsulation) can be employed in the context of functional encryption, and has great potential in going from selective to adaptive security of encryption schemes. At a first glance, *hybrid functional encryption* should lead to a selective-to-adaptive transformation, given an additional weak component: A *symmetric* FE which is adaptively secure when only a single message query is allowed. We show that the latter can be constructed from any one-way function as a corollary of [GVW12,BS15]. However, the intuitive reasoning fails to translate into a proof of security. To resolve this issue, we use a technique we call *The Trojan Method*, which originates from De Caro et al.'s "trapdoor circuits" [CIJ+13] (similar ideas had been since used by Gentry et al. [GHR+14] and Brakerski and Segev [BS15]).

We conclude this section with a short comparison of our technique with the aforementioned "dual system encryption" technique that had been used to achieve adaptively secure attribute based encryption.

Hybrid Functional Encryption. Hybrid encryption is a veteran technique in cryptography and has been used in a variety of settings. We show that in the context of functional encryption it is especially powerful.

---

[1] Waters [Wat14] also constructed an adaptively-secure scheme, but using specific ad-hoc techniques and in a significantly more complicated manner.

The idea in hybrid encryption is to combine two encryption schemes: An "external" scheme (sometimes called KEM – Key Encapsulation Mechanism) and an "internal" scheme (sometimes called DEM – Data Encapsulation Mechanism). In order to encrypt a message in the hybrid scheme, a fresh key is generated for the internal scheme, and is used to encrypt the message. Then the key itself is encrypted using the external scheme. The final hybrid ciphertext contains the two ciphertexts: $(\mathsf{Enc}_{\mathsf{ext}}(k), \mathsf{Enc}_{\mathsf{int},k}(m))$ (all external ciphertexts use the same key). To decrypt, one first decrypts the external ciphertext, retrieves k and applies it to the internal ciphertext. Note that if, for example, the external scheme is public-key and the internal is symmetric key, then the resulting scheme will also be public key. Hybrid encryption is often used in cases where the external scheme is less efficient (e.g. in encrypting long messages) and thus there is an advantage in using it to encrypt only a short key, and encrypt the long message using the more efficient internal scheme. Lastly, note that the internal scheme only needs to be able to securely encrypt a single message.

The intuition as to why hybrid encryption may be good for achieving adaptive security is that the external scheme only encrypts keys for the internal scheme. Namely, it only encrypts messages from a predetermined and known distribution, so selective security should be enough for the external scheme. The hardness of adaptive security is "pushed" to the internal scheme, but there the task is easier since the internal scheme only needs to be able to encrypt a single message, and it can be private-key rather than public-key.

Let us see how to employ this idea in the case where both the internal and external schemes are FE schemes. To encrypt, we will generate a fresh master secret key for the internal scheme, and encrypt it under the external scheme. To generate a key for the function f, the idea is to generate a key for the function $G_f(\mathsf{msk}_{\mathsf{int}})$ which takes a master key for the internal scheme, and outputs a secret key for function f under the internal scheme, using $\mathsf{msk}_{\mathsf{int}}$ (randomness is handled using a PRF). This will allow to decrypt in a two-step process as above. First apply the external secret-key for G_f to the external ciphertext, this will give you an internal secret key for f, which is in turn applied to the internal ciphertext to produce $f(x)$.

For the external scheme, we will use a selectively secure FE scheme (for the sake of concreteness, let us say public-key FE). As explained above, selective security is sufficient here since all the messages encrypted using the external scheme can be generated ahead of time (i.e. they do not depend on the actual x's that the user wishes to encrypt).

For the internal scheme, we require an FE scheme that is *adaptively secure*, but only supports the encryption of a single message. Fortunately, such a primitive can be derived from the works of [GVW12,BS15]. In [GVW12], the authors present an adaptively secure one-time bounded FE scheme. This scheme allows to only generate a key for one function, and to encrypt as many messages as the user wishes. This construction is based on the existence of semantically secure encryption, so the public-key version needs public-key encryption and the symmetric version needs symmetric encryption. While this primitive seems dual to what we need for our purposes, [BS15] shows how to transform private-key FE schemes into *function private* FE. In function-private FE, messages and functions enjoy the same level of privacy, in the sense that a user that produces x_0, x_1, f_0, f_1 such that $f_0(x_0) = f_1(x_1)$ cannot distinguish between $(\mathsf{Enc}(x_0), \mathsf{sk}_{f_0})$ and $(\mathsf{Enc}(x_1), \mathsf{sk}_{f_1})$. Therefore, after applying the [BS15] transformation, we can switch the roles of the functions and messages, and obtain a symmetric FE scheme which is adaptively secure for a *single message and many functions*. (We note that the symmetric version of the [GVW12] scheme can be shown to be function private even without the [BS15] transformation, however since this claim is not made explicitly in the paper we choose not to rely on it.)

Whereas intuitively this should solve the problem, it is not clear how to prove security of the new construction. Standard security proofs for hybrid encryption follow by first relying on the security of the external scheme and removing the encapsulated key, and then relying on the security of the internal scheme and removing the message. However, in our case, removing the encapsulated key is easily distinguishable, since the adversary is allowed to obtain functional keys and apply them to the ciphertext (so long as $f(x_0) = f(x_1)$). Without the internal key, the decryption process no longer works. To resolve this difficulty, we use the Trojan method.

Before we describe the Trojan method, we pause to note that our idea so far can be thought of as "boosting" a single-message, many-key, adaptive symmetric-key FE into a many-message, many-key, adaptive public-key FE (using a selective public-key FE as a "catalyst"). The recent work of Waters [Wat14] proceeds along a similar train of thought, and indeed, motivated our approach. However, while our transformation is simple and general, Waters has to rely on a powerful catalyst, namely an indistinguishability obfuscator.

The Trojan Method. The Trojan Method, which is a generalization of techniques used in [CIJ+13] and later in [GHR+14, BS15], is a way to embed a hidden functionality thread in an FE secret-key that can only be invoked by special ciphertexts generated using special (secret) back-door information. This thread remains completely unused in the normal operation of the scheme (and can be instantiated with meaningless functionality). In the proof, however, the secret thread will be activated by the challenge ciphertext in such a way that is indistinguishable to the user (= attacker). Namely, the user will not be able to tell that it is executing the secret thread and not the main thread. This will be extremely beneficial to prove security. We wish to argue that in the view of the user, the execution of the main thread does not allow to distinguish between the encryption of two messages x_0, x_1. The problem is that for functionality purposes, the main thread has to know which input it is working on. This is where the hidden thread comes into the play. We will design the hidden thread so that in the eyes of the user, it is computationally indistinguishable from the main thread on the special messages x_0, x_1. However, in the hidden thread, the output can be computed in a way that does not distinguish between x_0 and x_1 (either by a statistical or a computational argument), which will allow us to conclude that encryptions of x_0, x_1 are indistinguishable.

In particular, this method will resolve the aforementioned conundrum in our proof outline above. In the proof, we will use the Trojan method to embed a hidden thread in which msk_{int} is not used at all, but rather G_f produces a precomputed internal sk_f. This will allow us to remove msk_{int} from the challenge ciphertext and use the security properties of the internal scheme to argue that a internal encryption of x_0, x_1 are identical so long as $f(x_0) = f(x_1)$.

We note that an important special case of the above outline is when the trojan thread is a constant function. This had been the case in [CIJ+13, GHR+14], and this is the case in this work as well. However, we emphasize that our description here allows for greater generality since we allow the trojan thread to implement functionality that depends on the input x. We feel that this additional power may be useful for future applications.

Technically, the hidden thread is implemented using (standard) symmetric-key encryption, which in turn can be constructed starting with any one-way function. In the functional secret-key generation process for a function f, the secret-key generation process will produce a symmetric-key ciphertext c (which can just be encryption of 0 or another fixed message, since it only needs to have meaningful content in the security proof). It will then consider the function $G_{f,c}$ that takes as input a pair (x, s), and first checks whether it can decrypt c using s as a symmetric key. If it cannot, then it just runs f on x and returns the output. If s actually decrypts c, we consider $f^* = \mathsf{Dec}_s(c)$ (i.e. c encrypts a description of a function), and the output is the execution of $f^*(x)$. The value c is therefore used as a Trojan Horse: Its contents are hidden from the users of the scheme, however given a hidden command (in the form of the symmetric s) it can embed functionality that "takes over" the functional secret-key.

We note that in order to support the Trojan method, the decryption keys of our FE scheme need to perform symmetric decryption, branch operations, and execution of the function f^*. Thus we need to start with an FE scheme which allows for the generation of sufficiently expressive keys.

Our Trojan method can be seen as a weak form of function privacy in FE, but one that can be applied even in the context of public-key FE. In essence, we cannot hide the main thread of the evaluated function (this is unavoidable in public-key FE). However, we can hide the secret thread and thus allow the function to operate in a designated way for specially generated ciphertexts. (This interpretation is not valid for previous variants of this method such as "trapdoor circuits" [CIJ+13].)

A simple application of the Trojan method is our reduction in Sect. 4, showing that FE that only supports secret-keys for functions with shallow circuits (e.g. logarithmic depth) implies a scheme that works for circuits of arbitrary depth (although with a size bound). Essentially, instead of producing a secret key for the desired functionality, we output a key for the function that computes a *randomized encoding* of that functionality. A *(computational) randomized encoding* [IK00, AIK05] of an input-function pair $\mathsf{RE}(f, x)$ is, in a nutshell, a representation of $f(x)$ that reveals no information except $f(x)$ on one hand, but can be computed with less resources on the other (in our case, lower depth).

To make the proof work, the Trojan thread will contain a precomputed $RE(f, x_0)$ value, which will allow us to use the security property of the encoding scheme and switch it to $RE(f, x_1)$. See Sect. 4 for details. We note that a similar approach is used in [GHR+14] to achieve FE that works for RAM machines.

Relation to Dual-System Encryption. Our approach takes some resemblance to the "Dual-System Encryption" method of Waters [Wat09] and followup works [LW10,LW12]. This method had been used to prove adaptive security for Identity Based Encryption and Attribute Based Encryption, based on the hardness of some problems on groups with bilinear-maps. In broad terms, in their proof the distribution of the ciphertext is changed into "semi-functional" mode in a way that is indiscoverable by an observer. A semi-functional ciphertext is still decryptable by normal secret keys. Then, the secret-keys are modified into semi-functional form, which is useless in decrypting semi-functional ciphertexts. This is useful since in IBE and ABE, the challenge ciphertext is not supposed to be decryptable by those keys given to the adversary. Still, a host of algebraic techniques are used to justify the adversary's inability to produce other semi-functional ciphertexts in addition to the challenge, which would foil the reduction.

Our proof technique also requires changing the distributions of the keys and challenge ciphertext. However, there are also major differences. Our modified ciphertext is not allowed to interact with properly generated secret keys, and therefore the distinction between "normal" and "semi-functional" does not fit here. Furthermore, in Identity Based and Attribute Based Encryption, the attacker in the security game is not allowed to receive keys that reveal any information on the message, which allows to generate semi-functional ciphertexts that do not contain any information, whereas in our case, there is a structured and well-defined output for any ciphertext and any key. This means that the information required for decryption (which can be a-priori unbounded) needs to be embedded in the keys. Lastly, our proof is completely generic and does not rely on the algebraic structure of the underlying hardness assumption as in previous implementations of this method.

2   Preliminaries

In this section we present the notation and basic definitions that are used in this work. For a distribution X we denote by $x \leftarrow X$ the process of sampling a value x from the distribution X. Similarly, for a set $\mathcal{X}$ we denote by $x \leftarrow \mathcal{X}$ the process of sampling a value x from the uniform distribution over $\mathcal{X}$. For a randomized function f and an input $x \in \mathcal{X}$, we denote by $y \leftarrow f(x)$ the process of sampling a value y from the distribution $f(x)$. A function $\mathsf{negl} : \mathbb{N} \to \mathbb{R}$ is *negligible* if for any polynomial $p(\lambda)$ it holds that $\mathsf{negl}(\lambda) < 1/p(\lambda)$ for all sufficiently large $\lambda \in \mathbb{N}$.

2.1   Pseudorandom Functions and Symmetric Encryption

Pseudorandom Functions. We rely on the following standard notion of a pseudorandom function family [GGM86], asking that a pseudorandom function be computationally indistinguishable from a truly random function via oracle access.

Definition 2.1. *A family* $\mathcal{F} = \{\mathsf{PRF_K} : \{0,1\}^n \to \{0,1\}^m : \mathsf{K} \in \mathcal{K}\}$ *of efficiently-computable functions is* pseudorandom *if for every PPT adversary $\mathcal{A}$ there exists a negligible function* negl$(\cdot)$ *such that*

$$\left| \Pr_{\mathsf{K} \leftarrow \mathcal{K}} \left[\mathcal{A}^{\mathsf{PRF_K}(\cdot)}(1^\lambda) = 1 \right] - \Pr_{\mathsf{R} \leftarrow \mathcal{U}} \left[\mathcal{A}^{\mathsf{R}(\cdot)}(1^\lambda) = 1 \right] \right| \leq \mathsf{negl}(\lambda),$$

for all sufficiently large $\lambda \in \mathbb{N}$, where $\mathcal{U}$ is the set of all functions from $\{0,1\}^n$ to $\{0,1\}^m$.

We say that a pseudorandom function family $\mathcal{F}$ is implementable in NC^1 if every function in $\mathcal{F}$ can be implemented by a circuit of depth $c \cdot \log(n)$, for some constant c. We also consider the notion of a *weak* pseudorandom function family, asking that the above definition holds for adversaries that may access the functions on random inputs (that is, the oracles $\mathsf{PRF_K}(\cdot)$ and $\mathsf{R}(\cdot)$ take no input, and on each query they sample a uniform input r and output $\mathsf{PRF_K}(r)$ and $\mathsf{R}(r)$, respectively).

Symmetric Encryption with Pseudorandom Ciphertexts. A symmetric encryption scheme consists of a tuple of PPT algorithms (Sym.Setup, Sym.Enc, Sym.Dec). The algorithm Sym.Setup takes as input a security parameter λ in unary and outputs a key K_E. The encryption algorithm Sym.Enc takes as input a symmetric key K_E and a message m and outputs a ciphertext CT. The decryption algorithm Sym.Dec takes as input a symmetric key K_E and a ciphertext CT and outputs the message m.

In this work, we require a symmetric encryption scheme Π where the ciphertexts produced by Sym.Enc are pseudorandom strings. Let $\mathsf{OEnc}_K(\cdot)$ denote the (randomized) oracle that takes as input a message m, chooses a random string r and outputs Sym.Enc(Sym.K, $m; r$). Let $R_{\ell(\lambda)}(\cdot)$ denote the (randomized) oracle that takes as input a message m and outputs a uniformly random string of length $\ell(\lambda)$ where $\ell(\lambda)$ is the length of the ciphertexts. More formally, we require that for every PPT adversary $\mathcal{A}$ the following advantage is negligible in λ:

$$\mathsf{Adv}_{\Pi,\mathcal{A}}^{\mathsf{symPR}}(\lambda) = \left| \Pr\left[\mathcal{A}^{\mathsf{OEnc_{Sym.K}}(\cdot)}(1^\lambda) = 1 \right] - \Pr\left[\mathcal{A}^{\mathsf{R}_{\ell(\lambda)}(\cdot)}(1^\lambda) = 1 \right] \right|$$

where the probability is taken over the choice of Sym.K $\leftarrow$ Sym.Setup(1^λ), and over the internal randomness of $\mathcal{A}$, OEnc and $R_{\ell(\lambda)}$.

We note that such a symmetric encryption scheme with pseudorandom ciphertexts can be constructed from one-way functions, e.g. using weak pseudorandom functions by defining Sym.Enc(K, $m; r$) = $(r, \mathsf{PRF_K}(r) \oplus m)$ (see [Gol04] for more details).

2.2   Public-Key Functional Encryption

A public-key functional encryption (FE) scheme Π_{Pub} over a message space $\mathcal{M} = \{\mathcal{M}_\lambda\}_{\lambda \in \mathbb{N}}$ and a function space $\mathcal{F} = \{\mathcal{F}_\lambda\}_{\lambda \in \mathbb{N}}$ is a tuple (Pub.Setup, Pub.KeyGen, Pub.Enc, Pub.Dec) of PPT algorithms with the following properties:

- Pub.Setup(1^λ): The setup algorithm takes as input the unary representation of the security parameter, and outputs a public key MPK and a secret key MSK.
- Pub.KeyGen(MSK, f): The key-generation algorithm takes as input a secret key MSK and a function $f \in \mathcal{F}_\lambda$, and outputs a functional key sk_f.
- Pub.Enc(MPK, m): The encryption algorithm takes as input a public key MPK and a message $m \in \mathcal{M}_\lambda$, and outputs a ciphertext CT.
- Pub.Dec(sk_f, CT): The decryption algorithm takes as input a functional key sk_f and a ciphertext CT, and outputs $m' \in \mathcal{M}_\lambda \cup \{\bot\}$.

We say that such a scheme is defined for a complexity class $\mathcal{C}$ if it supports all the functions that can be implemented in $\mathcal{C}$. In terms of correctness, we require that there exists a negligible function $\mathsf{negl}(\cdot)$ such that for all sufficiently large $\lambda \in \mathbb{N}$, for every message $m \in \mathcal{M}_\lambda$, and for every function $f \in \mathcal{F}_\lambda$ it holds that $\Pr[\mathsf{Pub.Dec}(\mathsf{Pub.KeyGen}(\mathsf{MSK}, f), \mathsf{Pub.Enc}(\mathsf{MPK}, m)) = f(m)] \geq 1 - \mathsf{negl}(\lambda)$, where $(\mathsf{MPK}, \mathsf{MSK}) \leftarrow \mathsf{Pub.Setup}(1^\lambda)$, and the probability is taken over the random choices of all algorithms.

We consider the standard selective and adaptive indistinguishability-based notions for functional encryption (see, for example, [BSW11,O'N10]). Intuitively, these notions ask that encryptions of any two messages, m_0 and m_1, should be computationally indistinguishable given access to functional keys for any function f such that $f(m_0) = f(m_1)$. In the case of selective security, adversaries are required to specify the two messages in advance (i.e., before interacting with the system). In the case of adaptive security, adversaries are allowed to specify the two messages even after obtaining the public key and functional keys.[2]

Definition 2.2. (Selective Security). *A public-key functional encryption scheme* $\Pi = (\mathsf{Sel.Setup}, \mathsf{Sel.KeyGen}, \mathsf{Sel.Enc}, \mathsf{Sel.Dec})$ *over a function space* $\mathcal{F} = \{\mathcal{F}_\lambda\}_{\lambda \in \mathbb{N}}$ *and a message space* $\mathcal{M} = \{\mathcal{M}_\lambda\}_{\lambda \in \mathbb{N}}$ *is selectively secure if for any PPT adversary* $\mathcal{A}$ *there exists a negligible function* $\mathsf{negl}(\cdot)$ *such that*

$$\mathsf{Adv}^{\mathsf{Sel}}_{\Pi, \mathcal{A}}(\lambda) = \left| \Pr[\mathsf{Expt}^{\mathsf{Sel}}_{\Pi, \mathcal{A}}(\lambda, 0) = 1] - \Pr[\mathsf{Expt}^{\mathsf{Sel}}_{\Pi, \mathcal{A}}(\lambda, 1) = 1] \right| \leq \mathsf{negl}(\lambda)$$

for all sufficiently large $\lambda \in \mathbb{N}$, *where for each* $b \in \{0, 1\}$ *and* $\lambda \in \mathbb{N}$ *the experiment* $\mathsf{Expt}^{\mathsf{Sel}}_{\Pi, \mathcal{A}}(\lambda, b)$, *modeled as a game between the adversary* $\mathcal{A}$ *and a challenger, is defined as follows:*

1. **Setup phase:** *The challenger samples* $(\mathsf{Sel.MPK}, \mathsf{Sel.MSK}) \leftarrow \mathsf{Sel.Setup}(1^\lambda)$.

---

[2] Our notions of security consider a single challenge, and in the public-key setting these are known to be equivalent to their multi-challenge variants via a standard hybrid argument.

2. **Challenge phase:** *On input 1^λ the adversary submits (m_0, m_1), and the challenger replies with* Sel.MPK *and* CT $\leftarrow$ Sel.Enc(Sel.MPK, m_b).
3. **Query phase:** *The adversary adaptively queries the challenger with any function $f \in \mathcal{F}_\lambda$ such that $f(m_0) = f(m_1)$. For each such query, the challenger replies with* Sel.sk_f $\leftarrow$ Sel.KeyGen(Sel.MSK, f).
4. **Output phase:** *The adversary outputs a bit b' which is defined as the output of the experiment.*

Definition 2.3. (Adaptive Security). *A public-key functional encryption scheme* $\Pi = ($Ad.Setup, Ad.KeyGen, Ad.Enc, Ad.Dec$)$ *over a function space* $\mathcal{F} = \{\mathcal{F}_\lambda\}_{\lambda \in \mathbb{N}}$ *and a message space* $\mathcal{M} = \{\mathcal{M}_\lambda\}_{\lambda \in \mathbb{N}}$ *is adaptively secure if for any PPT adversary $\mathcal{A}$ there exists a negligible function* negl($\cdot$) *such that*

$$\mathsf{Adv}_{\Pi,\mathcal{A}}^{\mathsf{Ad}}(\lambda) = \left| \Pr[\mathsf{Expt}_{\Pi,\mathcal{A}}^{\mathsf{Ad}}(\lambda, 0) = 1] - \Pr[\mathsf{Expt}_{\Pi,\mathcal{A}}^{\mathsf{Ad}}(\lambda, 1) = 1] \right| \leq \mathsf{negl}(\lambda)$$

for all sufficiently large $\lambda \in \mathbb{N}$, where for each $b \in \{0, 1\}$ and $\lambda \in \mathbb{N}$ the experiment $\mathsf{Expt}_{\Pi,\mathcal{A}}^{\mathsf{Ad}}(1^\lambda, b)$, *modeled as a game between the adversary $\mathcal{A}$ and a challenger, is defined as follows:*

1. **Setup phase:** *The challenger samples* (Ad.MPK, Ad.MSK) $\leftarrow$ Ad.Setup(1^λ), *and sends* Ad.MPK *to the adversary.*
2. **Query phase I:** *The adversary adaptively queries the challenger with any function $f \in \mathcal{F}_\lambda$. For each such query, the challenger replies with* Ad.sk_f $\leftarrow$ Ad.KeyGen(Ad.MSK, f).
3. **Challenge Phase:** *The adversary submits (m_0, m_1) such that $f(m_0) = f(m_1)$ for all function queries f made so far, and the challenger replies with* CT $\leftarrow$ Ad.Enc(Ad.MSK, m_b).
4. **Query phase II:** *The adversary adaptively queries the challenger with any function $f \in \mathcal{F}_\lambda$ such that $f(m_0) = f(m_1)$. For each such query, the challenger replies with* Ad.sk_f $\leftarrow$ Ad.KeyGen(Ad.MSK, f).
5. **Output phase:** *The adversary outputs a bit b' which is defined as the output of the experiment.*

3 Our Transformation in the Public-Key Setting

In this section we present our transformation from selective security to adaptive security for public-key functional encryption schemes. In addition to any selectively-secure public-key functional encryption scheme (see Definition 2.2), our transformation requires a *private-key* functional encryption scheme that is adaptively-secure for a single message query and many function queries. Based on [GVW12, BS15], such a scheme can be based on any one-way function[3].

---

[3] Gorbunov et al. [GVW12] constructed a private-key functional encryption scheme that is adaptively secure for a single function query and many message queries based on any private-key encryption scheme (and thus based on any one-way function). Any such scheme can be turned into a function private one using the generic transformation of Brakerski and Segev [BS15], and then one can simply switch the roles of functions and messages [AAB+13, BS15]. This results in a private-key scheme that is adaptively secure for a single message query and many function queries.

More specifically, we rely on the following building blocks (all of which are implied by any selectively-secure public-key functional encryption scheme):

1. A selectively-secure public-key functional encryption scheme Sel = (Sel.Setup, Sel.KeyGen, Sel.Enc, Sel.Dec).
2. An adaptively-secure single-ciphertext private-key functional encryption scheme[4] OneCT = (OneCT.Setup, OneCT.KeyGen, OneCT.Enc, OneCT.Dec).
3. A symmetric encryption scheme with pseudorandom ciphertexts SYM = (Sym.Setup, Sym.Enc, Sym.Dec).
4. A pseudorandom function family $\mathcal{F}$ with a key space $\mathcal{K}$.

Our adaptively-secure scheme Ad = (Ad.Setup, Ad.KeyGen, Ad.Enc, Ad.Dec) is defined as follows.

- **The setup algorithm:** On input 1^λ the setup algorithm Ad.Setup samples (Sel.MPK, Sel.MSK) $\leftarrow$ Sel.Setup(1^λ), and outputs Ad.MPK = Sel.MPK and Ad.MSK = Sel.MSK.
- **The key-generation algorithm:** On input the secret key Ad.MSK = Sel.MSK and a function f, the key-generation algorithm Ad.KeyGen first samples $C_E \leftarrow \{0,1\}^{\ell_1(\lambda)}$ and $\tau \leftarrow \{0,1\}^{\ell_2(\lambda)}$ uniformly and independently. Then, it computes and outputs Ad.sk_f = Sel.$sk_G \leftarrow$ Sel.KeyGen(Sel.MSK, $G_{f,C_E,\tau}$), where the function $G_{f,C_E,\tau}$ is defined in Fig. 1.
- **The encryption algorithm:** On input the public key Ad.MPK = Sel.MPK and a message m, the encryption algorithm Ad.Enc first samples K $\leftarrow \mathcal{K}_\lambda$ and OneCT.MSK $\leftarrow$ OneCT.Setup(1^λ). Then, it outputs CT = (CT$_0$, CT$_1$), where

$$CT_0 \leftarrow \text{OneCT.Enc(OneCT.MSK}, m) \text{ and}$$

$$CT_1 \leftarrow \text{Sel.Enc(Sel.MPK}, (\text{OneCT.MSK}, \text{K}, 0^\lambda, 0)).$$

- **The decryption algorithm:** On input a functional key Ad.sk_f = Sel.sk_G and a ciphertext CT = (CT$_0$, CT$_1$), the decryption algorithm Ad.Dec first computes OneCT.$sk_f \leftarrow$ Sel.Dec(Sel.sk_G, CT$_1$). Then, it computes $m \leftarrow$ OneCT.Dec(OneCT.sk_f, CT$_0$) and outputs m.

$G_{f,C_E,\tau}$(**OneCT.MSK, K, Sym.K,** β):

1. If $\beta = 1$ output OneCT.$sk_f \leftarrow$ Sym.Dec(Sym.K, C_E).
2. Otherwise, output OneCT.$sk_f \leftarrow$ OneCT.KeyGen(OneCT.MSK, f; PRF$_K(\tau)$).

Fig. 1. The function $G_{f,C_E,\tau}$.

The correctness of the above scheme easily follows from that of its underlying building blocks, and in the remainder of this section we prove the following theorem:

---

[4] That is, a private-key functional encryption scheme that is adaptively-secure for a single message query and many function queries (as discussed above).

Theorem 3.1. *Assuming that: (1)* Sel *is a selectively-secure public-key functional encryption scheme, (2)* OneCT *is an adaptively-secure single-ciphertext private-key functional encryption scheme, (3)* SYM *is a symmetric encryption scheme with pseudorandom ciphertexts, and (4)* $\mathcal{F}$ *is a pseudorandom function family, then* Ad *is an adaptively-secure public-key functional encryption scheme.*

Proof. We show that any PPT adversary $\mathcal{A}$ succeeds in the adaptive security game (see Definition 2.3) with only negligible probability. We will show this in a sequence of hybrids. The advantage of the adversary in Hybrid$_{i.b}$ is defined to be probability that the adversary outputs 1 in Hybrid$_{i.b}$ and this quantity is denoted by Adv$_{i.b}^{\mathcal{A}}$. For $b \in \{0, 1\}$, we define the following hybrids.

Hybrid$_{1.b}$: This corresponds to the real experiment when the challenger encrypts the message m_b. More precisely, the challenger produces an encryption CT $=$ (CT$_0$, CT$_1$) where

$$CT_0 \leftarrow OneCT.Enc(OneCT.MSK, m) \text{ and}$$

$$CT_1 \leftarrow Sel.Enc(Sel.MPK, (OneCT.MSK, K, 0^\lambda, 0)).$$

Hybrid$_{2.b}$: The challenger replaces the hard-coded ciphertext C_E in every functional key corresponding to a query f made by the adversary, with a symmetric key encryption of OneCT.sk_f (note that each key has its own different C_E). Here, OneCT.sk_f is the output of OneCT.KeyGen(OneCT.MSK*, f; PRF$_{K^*}(\tau)$) and K^* is a PRF key drawn from the key space $\mathcal{K}$. Further, the symmetric encryption is computed with respect to Sym.K^*, where Sym.K^* is the output of Sym.Setup(1^λ) and τ is the tag associated to the functional key of f. The same Sym.K^* and K^* are used while generating all the functional keys, and K^* is used for generating the challenge ciphertext CT$^* =$ (CT$_0^*$, CT$_1^*$) (that is, CT$_0^* \leftarrow$ OneCT.Enc(OneCT.MSK*, m_b) and CT$_1^* \leftarrow$ Sel.Enc(Sel.MSK, (OneCT.MSK*, K*, 0$^\lambda$, 0))). The rest of the hybrid is the same as the previous hybrid, Hybrid$_{1.b}$.

Note that the symmetric key Sym.K^* is not used for any purpose other than generating the values C_E. Therefore, the pseudorandom ciphertexts property of the symmetric scheme implies that Hybrid$_{2.b}$ and Hybrid$_{1.b}$ are indistinguishable.

Claim 3.2. *Assuming the pseudorandom ciphertexts property of* SYM, *for each* $b \in \{0, 1\}$ *we have* $|Adv_{1.b}^{\mathcal{A}} - Adv_{2.b}^{\mathcal{A}}| \leq negl(\lambda)$.

Proof. Suppose there exists an adversary such that the difference in the advantages is non-negligible, then we construct a reduction that can break the security of SYM. The reduction internally executes the adversary by simulating the role of the challenger in the adaptive public-key FE game. It answers both the message and the functional queries made by the adversary as follows. The reduction first executes OneCT.Setup(1^λ) to obtain OneCT.MSK*. It then samples K^* from $\mathcal{K}$. Further, the reduction generates Sel.MSK, which is the output of

Sel.Setup(1^λ) and Sym.K*, which is the output of Sym.Setup(1^λ). When the adversary submits a functional query f, the reduction first picks τ at random. The reduction executes OneCT.KeyGen(OneCT.MSK*, f; PRF(K*(τ))) to obtain OneCT.sk_f. It then sends OneCT.sk_f to the challenger of the symmetric encryption scheme. The challenger returns back with C_E, where C_E is either a uniformly random string or it is an encryption of OneCT.sk_f. The reduction then generates a selectively-secure FE functional key of $G_{f,C_E,\tau}$ and denote the result by Sel.sk_G which is sent to the adversary. The message queries made by the adversary are handled as in Hybrid$_1$. That is, the adversary submits the message-pair query (m_0, m_1) and the reduction sends CT* $= (\text{CT}_0^*, \text{CT}_1^*)$ back to the adversary, where $\text{CT}_0^* = $ OneCT.Enc(OneCT.MSK*, m_b) and $\text{CT}_1^* = $ Sel.Enc(Sel.MSK, (OneCT.MSK*, K*, 0^λ, 0)).

If the challenger of the symmetric key encryption scheme sends a uniformly random string back to the reduction every time the reduction makes a query to the challenger then we are in Hybrid$_{1.b}$, otherwise we are in Hybrid$_{2.b}$. Since the adversary can distinguish both the hybrids with non-negligible probability, we have that the reduction breaks the security of the symmetric key encryption scheme with non-negligible probability. From our hypothesis, we have that the reduction breaks the security of the symmetric key encryption scheme with non-negligible probability. This proves the claim.   □

Hybrid$_{3.b}$: The challenger modifies the challenge ciphertext CT* $= (\text{CT}_0^*, \text{CT}_1^*)$ so that CT_1^* is an encryption of $(0^\lambda, 0^\lambda, \text{Sym.K*}, 1)$. The ciphertext component CT_0^* is not modified (i.e., $\text{CT}_0^* = $ OneCT.Enc(OneCT.MSK*, m_b)). The rest of the hybrid is the same as the previous hybrid, Hybrid$_{2.b}$.

Note that the functionality of the functional keys generated using the underlying selectively-secure scheme is unchanged with the modified CT_1^*. Therefore, its selective security implies that Hybrid$_{3.b}$ and Hybrid$_{2.b}$ are indistinguishable.

Claim 3.3. *Assuming the selective security of* Sel, *for each* $b \in \{0, 1\}$ *we have* $|\text{Adv}_{2.b}^{\mathcal{A}} - \text{Adv}_{3.b}^{\mathcal{A}}| \leq \text{negl}(\lambda)$.

Proof. Suppose the claim is not true for some adversary $\mathcal{A}$, we construct a reduction that breaks the security of Sel. Our reduction will internally execute $\mathcal{A}$ by simulating the role of the challenger of the adaptive FE game.

Our reduction first executes OneCT.Setup(1^λ) to obtain OneCT.MSK*. It then samples K* from $\mathcal{K}$. It also executes Sym.Setup(1^λ) to obtain Sym.K*. The reduction then sends the message pair ((OneCT.MSK*, K*, 0^λ, 0), $(0^\lambda, 0^\lambda, \text{Sym.K*}, 1)$) to the challenger of the selective game. The challenger replies back with the public key Sel.MPK and the challenge ciphertext CT_1^*. The reduction is now ready to interact with the adversary $\mathcal{A}$. If $\mathcal{A}$ makes a functional query f then the reduction constructs the circuit $G_{f,C_E,\tau}$ as in Hybrid$_{2.b}$. It then queries the challenger of the selective game with the function G and in return it gets the key Sel.sk_G. The reduction then sets Ad.sk_f to be Sel.sk_G which it then sends back to $\mathcal{A}$. If $\mathcal{A}$ submits a message pair (m_0, m_1), the reduction executes OneCT.Enc(OneCT.MSK*, m_0) to obtain CT_0^*. It then sends the ciphertext

$CT^* = (CT_0^*, CT_1^*)$ to the adversary. The output of the reduction is the output of $\mathcal{A}$.

We claim that the reduction is a legal adversary in the selective security game of Sel, i.e., for challenge message query $(M_0 = (\text{OneCT.MSK}^*, \text{K}^*, 0^\lambda, 0)$, $M_1 = (0^\lambda, 0^\lambda, \text{Sym.K}^*, 1))$ and every functional query of the form $G_{f,C_E,\tau}$ made by the reduction, we have that $G_{f,C_E,\tau}(M_0) = G_{f,C_E,\tau}(M_1)$: By definition, $G_{f,C_E,\tau}(M_0)$ is the functional key of f, with respect to key OneCT.MSK^* and randomness $\text{PRF}_{\text{K}^*}(\tau)$. Further, $G_{f,C_E,\tau}(M_1)$ is the decryption of C_E which is nothing but the functional key of f, with respect to key OneCT.MSK^* and randomness $\text{PRF}_{\text{K}^*}(\tau)$. This proves that the reduction is a legal adversary in the selective security game.

If the challenger of the selective game sends back an encryption of $(\text{OneCT.MSK}^*, \text{K}^*, 0^\lambda, 0)$ then we are in $\text{Hybrid}_{2.b}$ else if the challenger encrypts $(0^\lambda, 0^\lambda, \text{Sym.K}^*, 1)$ then we are in $\text{Hybrid}_{3.b}$. By our hypothesis, this means the reduction breaks the security of the selective game with non-negligible probability that contradicts the security of Sel. This completes the proof of the claim.

Hybrid$_{4.b}$: For every function query f made by the adversary, the challenger generates C_E by executing $\text{Sym.Enc}(\text{Sym.K}^*, \text{OneCT}.sk_f)$, with $\text{OneCT}.sk_f$ being the output of $\text{OneCT.KeyGen}(\text{OneCT.MSK}^*, f; R)$, where R is picked at random. The rest of the hybrid is the same as the previous hybrid.

Note that the PRF key K^* is not explicitly needed in the previous hybrid, and therefore the pseudorandomness of $\mathcal{F}$ implies that $\text{Hybrid}_{4.b}$ and $\text{Hybrid}_{3.b}$ are indistinguishable.

Claim 3.4. *Assuming that $\mathcal{F}$ is a pseudorandom function family, for each $b \in \{0,1\}$ we have $|\text{Adv}_{3.b}^{\mathcal{A}} - \text{Adv}_{4.b}^{\mathcal{A}}| \leq \text{negl}(\lambda)$.*

Proof. Suppose the claim is false for some PPT adversary $\mathcal{A}$, we construct a reduction that internally executes $\mathcal{A}$ and breaks the security of the pseudorandom function family $\mathcal{F}$. The reduction simulates the role of the challenger of the adaptive game when interacting with $\mathcal{A}$. The reduction answers the functional queries, made by the adversary as follows; the message queries are answered as in $\text{Hybrid}_{3.b}$ (or $\text{Hybrid}_{4.b}$). For every functional query f made by the adversary, the reduction picks τ at random which is then forwarded to the challenger of the PRF security game. In response it receives R^*. The reduction then computes C_E to be $\text{Sym.Enc}(\text{Sym.K}^*, \text{OneCT}.sk_f)$, where $\text{OneCT}.sk_f = \text{OneCT.KeyGen}(\text{OneCT.MSK}^*, f; R^*)$. The reduction then proceeds as in the previous hybrids to compute the functional key $\text{Ad}.sk_f$ which it then sends to $\mathcal{A}$.

If the challenger of the PRF game sent $R^* = \text{PRF}_{\text{K}^*}(\tau)$ back to the reduction then we are in $\text{Hybrid}_{3.b}$ else if R^* is generated at random by the challenger then we are in $\text{Hybrid}_{4.b}$. From our hypothesis this means that the probability that the reduction distinguishes the pseudorandom value from random (at the point τ) is non-negligible, contradicting the security of the pseudorandom function family.□

We now conclude the proof of the theorem by showing that $\text{Hybrid}_{4.0}$ is computationally indistinguishable from $\text{Hybrid}_{4.1}$ based on the adaptive security of the underlying single-ciphertext scheme.

Claim 3.5. *Assuming the adaptive security of the scheme* OneCT, *we have* $|\mathsf{Adv}_{4.0}^{\mathcal{A}} - \mathsf{Adv}_{4.1}^{\mathcal{A}}| \leq \mathsf{negl}(\lambda)$.

Proof. Suppose there exists a PPT adversary $\mathcal{A}$, such that the claim is false. We design a reduction $\mathcal{B}$ that internally executes $\mathcal{A}$ to break the adaptive security of OneCT.

The reduction simulates the role of the challenger of the adaptive public-key FE game. It answers both the functional as well as message queries made by the adversary as follows. If $\mathcal{A}$ makes a functional query f then it forwards it to the challenger of the adaptively-secure single-ciphertext FE scheme. In return it receives OneCT.sk_f. It then encrypts it using the symmetric encryption scheme, where the symmetric key is picked by the reduction itself, and denote the resulting ciphertext to be C_E. The reduction then constructs the circuit $G_{f,C_E,\tau}$, with τ being picked at random, as in the previous hybrids. Finally, the reduction computes the selective public-key functional key of $G_{f,C_E,\tau}$, where the reduction itself picks the master secret key of selective public-key FE scheme. The resulting functional key is then sent to $\mathcal{A}$. If $\mathcal{A}$ makes a message-pair query (m_0, m_1), the reduction forwards this message pair to the challenger of the adaptive game. In response it receives CT_0^*. The reduction then generates CT_1^* on its own where CT_1^* is the selective FE encryption of $(0^\lambda, 0^\lambda, \mathsf{Sym.K}^*, 1)$. The reduction then sends $\mathsf{CT}^* = (\mathsf{CT}_0^*, \mathsf{CT}_1^*)$ to $\mathcal{A}$. The output of the reduction is the output of $\mathcal{A}$.

We note that the reduction is a legal adversary in the adaptive game of OneCT, i.e., for every challenge message query (m_0, m_1), functional query f, we have that $f(m_0) = f(m_1)$: this follows from the fact that (i) the functional queries (resp., challenge message query) made by the adversary (of Ad) is the same as the functional queries (resp., challenge message query) made by the reduction, and (ii) the adversary (of Ad) is a legal adversary. This proves that the reduction is a legal adversary in the adaptive game.

If the challenger sends an encryption of m_0 then we are in $\mathsf{Hybrid}_{4.0}$ and if the challenger sends an encryption of m_1 then we are in $\mathsf{Hybrid}_{4.1}$. From our hypothesis, this means that the reduction breaks the security of OneCT. This proves the claim.   □

4   From Shallow Circuits to All Circuits

In this section we show that a functional encryption scheme that supports functions computable by shallow circuits can be transformed into one that supports functions computable by arbitrarily deep circuits. In particular, the shallow class can be any class in which weak pseudorandom functions can be computed and has some composition properties.[5] For concreteness we consider here the class NC^1, which can compute weak pseudorandom functions under standard cryptographic assumptions such as DDH or LWE (a lower complexity class such as TC^0 is also sufficient under standard assumptions). We focus here on private-key

---

[5] Similarly to the class WEAK defined in [App14].

functional encryption schemes, and note that an essentially identical transformation applies for public-key scheme.

While we present a direct reduction below, we notice that this property can be derived from the transformation in Sect. 3, by recalling some properties of Gorbunov et al.'s [GVW12] single-key functional encryption scheme. One can verify that their setup algorithm can be implemented in NC^1 (under the assumption that it can evaluate weak pseudorandom functions), regardless of the depth of the function being implemented. This property carries through even after applying the function privacy transformation of Brakerski and Segev [BS15]. Lastly, to implement our approach we need a symmetric encryption scheme with decryption in NC^1, which again translates to the evaluation of a weak pseudorandom function [NR04, BPR12].

(Computational) Randomized Encodings [IK00, AIK05]**.** A (computational) randomized encoding scheme for a function class $\mathcal{F}$ consists of two PPT algorithms (RE.Encode, RE.Decode). The PPT algorithm RE.Encode takes as input $(1^\lambda, F, x, r)$, where λ is the security parameter, $F : \{0,1\}^\lambda \to \{0,1\}$ is a function in $\mathcal{F}$, instance $x \in \{0,1\}^\lambda$ and randomness r. The output is denoted by $\hat{F}(x; r)$. The PPT algorithm RE.Decode takes as input $\hat{F}(x; r)$ and outputs $y = F(x)$.

The security property states that there exists a PPT algorithm RE.Sim that takes as input $(1^\lambda, F(x))$ and outputs $\mathsf{SimOut}_{F(x)}$ such that any PPT adversary cannot distinguish the distribution $\{\hat{F}(x; r)\}$ from the distribution $\{\mathsf{SimOut}_{F(x)}\}$. The following corollary is derived from applying Yao's garbled circuit technique using a weak PRF based encryption algorithm.

Corollary 4.1. *Assuming a family of weak pseudorandom functions that can be evaluated in NC^1, there exists a randomized encoding scheme (RE.Encode, RE.Decode) for the class of polynomial size circuits, such that RE.Encode is computable in NC^1.*

Our Transformation. Let $\mathcal{NCFE} = $ (NCFE.Setup, NCFE.KeyGen, NCFE.Enc, NCFE.Dec) be a private-key functional encryption scheme for the class NC^1. We assume that $\mathcal{NCFE}$ supports functions with multi-bit outputs, as otherwise it is always possible to produce a functional key for each output bit separately. We also use a pseudorandom function family denoted by $\mathcal{F} = \{\mathsf{PRF}_K(\cdot)\}_{K \in \mathcal{K}}$ and a symmetric encryption scheme SYM = (Sym.Setup, Sym.Enc, Sym.Dec). We construct a private-key functional encryption scheme $\mathcal{PFE} = $ (PFE.Setup, PFE.KeyGen, PFE.Enc, PFE.Dec) as follows.

- **The setup algorithm:** On input 1^λ the algorithm PFE.Setup samples and outputs $MSK \leftarrow$ NCFE.Setup(1^λ).
- **The key-generation algorithm:** On input the secret key MSK and a circuit F, the algorithm PFE.KeyGen first samples $C_E \leftarrow \{0,1\}^{\ell_1(\lambda)}$ and $\tau \leftarrow \{0,1\}^\lambda$ uniformly and independently. Then, it computes a functional key $SK_G \leftarrow$ NCFE.KeyGen$(MSK, G_{F,C_E,\tau})$, where the function $G_{F,C_E,\tau}$ is defined in Fig. 2, and outputs SK_G.

- **The encryption algorithm:** On input the secret key MSK and a message x, the algorithm PFE.Enc first samples $K_P \leftarrow \{0,1\}^\lambda$, and then computes and outputs $C \leftarrow \mathsf{NCFE.Enc}(MSK, (x, K_P, 0^\lambda, 0))$.
- **The decryption algorithm:** On input a functional key $SK_F = SK_G$, and a ciphertext C, the decryption algorithm PFE.Dec computes $\widehat{F}(x) \leftarrow \mathsf{NCFE.Dec}(SK_G, C)$ and then outputs $\mathsf{RE.Decode}(\widehat{F}(x))$.

$G_{F,C_E,\tau}(x, K_P, K_E, \beta):$

1. If $\beta = 1$ output $\mathsf{Sym.Dec}_{K_E}(C_E)$.
2. Otherwise, output $\widehat{F}(x; \mathsf{PRF}_{K_P}(\tau)) = \mathsf{RE.Encode}(F, x; \mathsf{PRF}_{K_P}(\tau))$.

Fig. 2. The function $G_{F,C_E,\tau}$.

The correctness of the above scheme easily follows from that of its underlying building blocks, and in the remainder of this section we provide a sketch for proving the following theorem:

Theorem 4.2. *Assuming that: (1) $\mathcal{NCFE}$ is a selectively-secure private-key functional encryption scheme for* NC^1, *(2)* SYM *is a symmetric encryption scheme with pseudorandom ciphertexts whose decryption circuit is in* NC^1, *(3)* PRF *is a weak pseudorandom function family which can be evaluated in* NC^1, *and (4)* $(\mathsf{RE.Encode}, \mathsf{RE.Decode})$ *is a randomized encoding scheme with encoding in* NC^1, *then* $\mathcal{PFE}$ *is a selectively-secure private-key functional encryption scheme for P.*

Proof Sketch. The proof proceeds by a sequence of hybrids. For simplicity, we consider the case when the adversary submits a single challenge pair (m_0, m_1), and the argument can be easily generalized to the case of multiple challenges.

Hybrid$_0$: This corresponds to the real experiment where the challenger sends an encryption of m_0 to the adversary.

Hybrid$_1$: For every functional query F, the challenger replaces C_E with a symmetric encryption $\mathsf{Sym.Enc}(K_E, \widehat{F}(m_0; \mathsf{PRF}_{K_P}(t)))$ in the functional key for F. By a sequence of intermediate hybrids (as many as the number of function queries), Hybrid$_1$ can be shown to be computationally indistinguishable from Hybrid$_0$ based on the pseudorandom ciphertexts property of the symmetric encryption scheme.

Hybrid$_2$: The challenge ciphertext will consist of an encryption of $(m_0, 0, K_E, 1)$ instead of $(m_0, K_P, 0^\lambda, 0)$. This hybrid is computationally indistinguishable from Hybrid$_1$ by the security of the underlying functional encryption scheme.

Hybrid$_3$: For every function query F, the challenger replaces C_E in all the functional keys with $\mathsf{Sym.Enc}(K_E, \widehat{F}(m_0; r))$ for a uniform r. By a sequence of intermediate hybrids (as many as the number of function queries), Hybrid$_3$ can be shown to be computationally indistinguishable from Hybrid$_2$ based on the security of PRF.

Hybrid$_4$: Finally, for every function query F, the challenger replaces $\widehat{F}(m_0; r)$ in the ciphertext hardwired in the functional key for F by the simulated randomized encoding $\mathsf{RE.Sim}(1^\lambda, F(m_0))$. By a sequence of intermediate hybrids (as many as the number of function queries), Hybrid$_4$ can be shown to be computationally indistinguishable from Hybrid$_3$ based on the security of randomized encodings. Note that the this hybrid does not depend on whether m_0 or m_1 was encrypted since for all function queries F it holds that $F(m_0) = F(m_1)$, and this proves the security of $\mathcal{PFE}$.

References

[AAB+13] Agrawal, S., Agrawal, S., Badrinarayanan, S., Kumarasubramanian, A., Prabhakaran, M., Sahai, A.: Function private functional encryption and property preserving encryption: New definitions and positive results. Cryptology ePrint Archive, report 2013/744 (2013)

[AIK05] Applebaum, B., Ishai, Y., Kushilevitz, E.: Computationally private randomizing polynomials and their applications. In: CCC, pp. 260–274. IEEE Computer Society (2005)

[App14] Applebaum, B.: Bootstrapping obfuscators via fast pseudorandom functions. In: Sarkar, P., Iwata, T. (eds.) ASIACRYPT 2014, Part II. LNCS, vol. 8874, pp. 162–172. Springer, Heidelberg (2014)

[BB04] Boneh, D., Boyen, X.: Efficient selective-ID secure identity-based encryption without random oracles. In: Cachin, C., Camenisch, J.L. (eds.) EUROCRYPT 2004. LNCS, vol. 3027, pp. 223–238. Springer, Heidelberg (2004)

[BCO+04] Boneh, D., Di Crescenzo, G., Ostrovsky, R., Persiano, G.: Public key encryption with keyword search. In: Cachin, C., Camenisch, J.L. (eds.) EUROCRYPT 2004. LNCS, vol. 3027, pp. 506–522. Springer, Heidelberg (2004)

[BCP14] Boyle, E., Chung, K.-M., Pass, R.: On extractability obfuscation. In: Lindell, Y. (ed.) TCC 2014. LNCS, vol. 8349, pp. 52–73. Springer, Heidelberg (2014)

[BF03] Boneh, D., Franklin, M.K.: Identity-based encryption from the Weil pairing. SIAM J. Comput. **32**(3), 586–615 (2003)

[BPR12] Banerjee, A., Peikert, C., Rosen, A.: Pseudorandom functions and lattices. In: Pointcheval, D., Johansson, T. (eds.) EUROCRYPT 2012. LNCS, vol. 7237, pp. 719–737. Springer, Heidelberg (2012)

[BS15] Brakerski, Z., Segev, G.: Function-private functional encryption in the private-key setting. In: Dodis, Y., Nielsen, J.B. (eds.) TCC 2015, Part II. LNCS, vol. 9015, pp. 306–324. Springer, Heidelberg (2015)

[BSW11] Boneh, D., Sahai, A., Waters, B.: Functional encryption: definitions and challenges. In: Ishai, Y. (ed.) TCC 2011. LNCS, vol. 6597, pp. 253–273. Springer, Heidelberg (2011)

[CIJ+13] De Caro, A., Iovino, V., Jain, A., O'Neill, A., Paneth, O., Persiano, G.: On the achievability of simulation-based security for functional encryption. In: Canetti, R., Garay, J.A. (eds.) CRYPTO 2013, Part II. LNCS, vol. 8043, pp. 519–535. Springer, Heidelberg (2013)

[GGH+13] Garg, S., Gentry, C., Halevi, S., Raykova, M., Sahai, A., Waters, B.: Candidate indistinguishability obfuscation and functional encryption for all circuits. In: FOCS, pp. 40–49 (2013)

[GGH+14a] Garg, S., Gentry, C., Halevi, S., Zhandry, M.: Fully secure attribute based encryption from multilinear maps. IACR Cryptol. ePrint Arch. 2014, 622 (2014)

[GGH+14b] Garg, S., Gentry, C., Halevi, S., Zhandry, M.: Fully secure functional encryption without obfuscation. Cryptology ePrint Archive, report 2014/666 (2014)

[GGM86] Goldreich, O., Goldwasser, S., Micali, S.: How to construct random functions. J. ACM 33(4), 792–807 (1986)

[GHR+14] Gentry, C., Halevi, S., Raykova, M., Wichs, D.: Outsourcing private RAM computation. In: FOCS, pp. 404–413. IEEE Computer Society (2014)

[GKP+13] Goldwasser, S., Kalai, Y., Popa, R.A., Vaikuntanathan, V., Zeldovich, N.: Reusable garbled circuits and succinct functional encryption. In: ACM STOC, pp. 555–564 (2013)

[Gol04] Goldreich, O.: Foundations of Cryptography - Volume 2: Basic Applications. Cambridge University Press, Cambridge (2004)

[GPS+06] Goyal, V., Pandey, O., Sahai, A., Waters, B.: Attribute-based encryption for fine-grained access control of encrypted data. In: ACM CCS, pp. 89–98 (2006)

[GVW12] Gorbunov, S., Vaikuntanathan, V., Wee, H.: Functional encryption with bounded collusions via multi-party computation. In: Safavi-Naini, R., Canetti, R. (eds.) CRYPTO 2012. LNCS, vol. 7417, pp. 162–179. Springer, Heidelberg (2012)

[GVW13] Gorbunov, S., Vaikuntanathan, V., Wee, H.: Attribute-based encryption for circuits. In : ACM STOC, pp. 545–554 (2013)

[IK00] Ishai, Y., Kushilevitz, E.: Randomizing polynomials: a new representation with applications to round-efficient secure computation. In: FOCS, pp. 294–304 (2000)

[KSW08] Katz, J., Sahai, A., Waters, B.: Predicate encryption supporting disjunctions, polynomial equations, and inner products. In: Smart, N.P. (ed.) EUROCRYPT 2008. LNCS, vol. 4965, pp. 146–162. Springer, Heidelberg (2008)

[LOS+10] Lewko, A., Okamoto, T., Sahai, A., Takashima, K., Waters, B.: Fully Secure functional encryption: attribute-based encryption and (hierarchical) inner product encryption. In: Gilbert, H. (ed.) EUROCRYPT 2010. LNCS, vol. 6110, pp. 62–91. Springer, Heidelberg (2010)

[LW10] Lewko, A., Waters, B.: New techniques for dual system encryption and fully secure HIBE with short ciphertexts. In: Micciancio, D. (ed.) TCC 2010. LNCS, vol. 5978, pp. 455–479. Springer, Heidelberg (2010)

[LW12] Lewko, A., Waters, B.: New proof methods for attribute-based encryption: achieving full security through selective techniques. In: Safavi-Naini, R., Canetti, R. (eds.) CRYPTO 2012. LNCS, vol. 7417, pp. 180–198. Springer, Heidelberg (2012)

[NR04]   Naor, M., Reingold, O.: Number-theoretic constructions of efficient pseudo-random functions. J. ACM **51**(2), 231–262 (2004)

[O'N10] O'Neill, A.: Definitional issues in functional encryption. Cryptology ePrint Archive, report 2010/556 (2010)

[SS10]   Sahai, A., Seyalioglu, H.: Worry-free encryption: functional encryption with public keys. In: ACM CCS, pp. 463–472 (2010)

[SW05]   Sahai, A., Waters, B.: Fuzzy identity-based encryption. In: Cramer, R. (ed.) EUROCRYPT 2005. LNCS, vol. 3494, pp. 457–473. Springer, Heidelberg (2005)

[Wat09] Waters, B.: Dual system encryption: realizing fully secure IBE and HIBE under simple assumptions. In: Halevi, S. (ed.) CRYPTO 2009. LNCS, vol. 5677, pp. 619–636. Springer, Heidelberg (2009)

[Wat14] Waters, B.: A punctured programming approach to adaptively secure functional encryption. Cryptology ePrint Archive, report 2014/588 (2014)

A Punctured Programming Approach to Adaptively Secure Functional Encryption

Brent Waters[✉]

University of Texas, Austin, USA
bwaters@cs.utexas.edu

Abstract. We propose the first construction for achieving adaptively secure functional encryption (FE) for poly-sized circuits (without complexity leveraging) from indistinguishability obfuscation ($i\mathcal{O}$). Our reduction has polynomial loss to the underlying primitives. We develop a "punctured programming" approach to constructing and proving systems where outside of obfuscation we rely only on primitives realizable from pseudo random generators.

Our work consists of two constructions. Our first FE construction is provably secure against any attacker that is limited to making all of its private key queries *after* it sees the challenge ciphertext. (This notion implies selective security.) Our construction makes use of an we introduce called puncturable deterministic encryption (PDE) which may be of independent interest. With this primitive in place we show a simple FE construction.

We then provide a second construction that achieves adaptive security from indistinguishability obfuscation. Our central idea is to achieve an adaptively secure functional encryption by bootstrapping from a one-bounded FE scheme that is adaptively secure. By using bootstrapping we can use "selective-ish" techniques at the outer level obfuscation level and push down the challenge of dealing with adaptive security to the one-bounded FE scheme, where it has been already been solved. We combine our bootstrapping framework with a new "key signaling" technique to achieve our construction and proof. Altogether, we achieve the first construction and proof for adaptive security for functional encryption.

1 Introduction

In traditional encryption systems a message, m, is encrypted with a particular user's public key PK. Later a user that holds the corresponding secret key will be able to decrypt the ciphertext and learn the contents of the message. At the same time any computationally bounded attacker will be unable to get any additional information on the message.

Brent Waters is supported by NSF CNS-0915361 and CNS-0952692, CNS-1228599 DARPA through the U.S. Office of Naval Research under Contract N00014-11-1-0382, DARPA N11AP20006, Google Faculty Research award, the Alfred P. Sloan Fellowship, Microsoft Faculty Fellowship, and Packard Foundation Fellowship.

R. Gennaro and M. Robshaw (Eds.): CRYPTO 2015, Part II, LNCS 9216, pp. 678–697, 2015.
DOI: 10.1007/978-3-662-48000-7_33

While this communication paradigm is appropriate for many scenarios such as targeted sharing between users, there exist many applications that demand a more nuanced approach to sharing encrypted data. For example, suppose that an organization encrypts video surveillance images and stores these ciphertexts in a large online database. Later, we would like to give an analyst the ability to view all images that match a particular pattern such as ones that include a facial image that pattern matches with a particular individual. In a traditional encryptions system we would be forced to either give the analyst the secret key enabling them to view everything or give them nothing and no help at all.

The concept of functional encryption (FE) was proposed to move beyond this all or nothing view of decryption. In a functional encryption system a secret key SK_f is associated with a function f. When a user attempts to decrypt a ciphertext CT encrypted for message m with secret key SK_f, he will learn $f(m)$. The security of functional encryption states that an attacker that receives keys for any polynomial number of functions $f_1, \ldots, f_Q$ should not be able to distinguish between an encryption of m_0, m_1 as long as $\forall i\ f_i(m_0) = f_i(m_1)$.

The concept of functional encryption first appeared under the guise of predicate encryption [BW07,KSW08] with the nomenclature later being updated [SW08,BSW11] to functional encryption. In addition, functional encryption has early roots in Attribute-Based Encryption [SW05] and searching on encrypted data [BCOP04].

A central challenge is to achieve functional encryption for as expressive functionality classes as possible — ideally one would like to achieve it for any poly-time computable function. Until recently, the best available was roughly limited to the inner product functionality proposed by Katz, Sahai, and Waters [KSW08]. This state of affairs changed dramatically with the introduction of a candidate indistinguishability obfuscation [BGI+12] system for all poly-size circuits by Garg, Gentry, Halevi, Raykova, Sahai, and Waters [GGH+13] (GGHRSW). The authors showed that a functional encryption system for any poly-sized circuits can be built from an indistinguishability obfuscator plus public key encryption and statistically simulation sound non-interactive zero knowledge proofs.

Thinking of Adaptive Security. While the jump from inner product functionality to any poly-size circuit is quite significant, one limitation of the GGHRSW functional encryption system is that it only offers a *selective* proof of security where the attacker must declare the challenge messages before seeing the parameters of the FE system. Subsequently, Boyle, Chung and Pass [BCP14] proposed an FE construction based on an obfuscator that is differing inputs secure. We briefly recall that an obfuscator $\mathcal{O}$ is indistinguishability secure if it is computationally difficult for an attacker to distinguish between obfuscations $\mathcal{O}(C_0)$ and $\mathcal{O}(C_1)$ for any two (similar sized) circuits that are functionally equivalent (i.e. $\forall x\ C_0(x) = C_1(x)$). Recall, that differing inputs [BCP14,ABG+13] security allows for an attacker to use circuits C_0 and C_1 that are not functionally equivalent, but requires that for any PPT attacker that distinguishes between obfuscations of the two circuits there must a PPT extraction algorithm that finds some x such that $C_0(x) \neq C_1(x)$. Thus, differing inputs obfuscation is in a qualitatively different class of "knowledge definitions". Furthermore, there is significant evidence [GGHW14] that there exist certain functionalities with auxiliary input that are impossible to build obfuscate under the differing inputs definition.

Our goal is to build adaptively secure functional encryption systems from indistinguishability obfuscation. We require that our reductions have polynomial loss of security relative to the underlying primitives. (In particular, we want to avoid the folklore complexity leveraging transformation of simply guessing the challenge messages with an exponential loss.) In addition, we want to take a minimalist approach to the primitives we utilize outside of obfuscation. In particular, we wish to avoid the use of additional "strong tools" such as non-interactive zero knowledge proofs or additional assumptions over algebraic groups. We note that our focus is on indistinguishability notions of functional encryption as opposed to simulation definitions [BSW11, O'N10].

Our Results. In this work we propose two new constructions for achieving secure functional encryption (for poly-sized circuits) from indistinguishability obfuscation. We develop a "punctured programming" approach [SW14] to constructing and proving systems where our main tools in addition to obfuscation are a selectively secure puncturable pseudo random functions. We emphasize puncturable PRFs are themselves constructible from pseudo random generators [GGM84, BW13, BGI13, KPTZ13].

We start toward our FE construction which is provably secure against any attacker that is limited to making all of its private key queries *after* it sees the challenge ciphertext.[1] While this is attacker is still restricted relative to a fully adaptive attacker, we observe that such a definition is already stronger than the commonly used selective restriction.

To build our system we first introduce an abstraction that we call puncturable deterministic encryption (PDE). The main purpose of this abstraction is to serve in some places as a slightly higher level and more convenient abstraction to work with than puncturable PRFs. A PDE system is a symmetric key and deterministic encryption scheme and consists of four algorithms: $\mathsf{Setup}_{\mathsf{PDE}}(1^\lambda), \mathsf{Encrypt}_{\mathsf{PDE}}(K, m), \quad \mathsf{Decrypt}_{\mathsf{PDE}}(K, \mathrm{CT}), \quad$ and $\quad \mathsf{Puncture}_{\mathsf{PDE}}$ (K, m_0, m_1). The first three algorithms have the usual correctness semantics. The fourth puncture algorithm takes as input a master key and two messages (m_0, m_1) and outputs a punctured key that can decrypt all ciphertexts except for those encrypted for either of the two messages — recall encryption is deterministic so there are only two such ciphertexts. The security property of PDE is stated as a game where the attacker gives two messages (m_0, m_1) to the attacker and then returns back a punctured key as well as two ciphertexts, one encrypted under each message. In a secure system no PPT attacker will be able to distinguish which ciphertext is associated with which message.

---

[1] This model has been called semi-adaptive in other contexts [CW14].

Our PDE encryption mechanism is rather simple and is derived from the hidden trigger mechanism from the Sahai-Waters [SW14] deniable encryption scheme. PDE Ciphertexts are of the form:

$$CT = (A = F_1(K_1, m), \quad B = F_2(K_2, A) \oplus m).$$

where F_1 and F_2 are puncturable pseudo random functions, with F_1 being an injective function. Decryption requires first computing $m' = B \oplus F_2(K_2, A)$ and then checking that $F_1(K_1, m') = A$.[2]

With this tool in place we are now ready to describe our first construction. The setup algorithm will first choose a puncturable PRF key K for function F. Next, it will create the public parameters PP as an obfuscation of a program called INITIALENCRYPT. The INITIALENCRYPT program will take in randomness r and compute a tag $t = \mathrm{PRG}(r)$. Then it will output t and a PDE key k that is derived from $F(K, t)$. The encryption algorithm can use this obfuscated program to encrypt as follows. It will simply choose a random value $r \in \{0, 1\}^\lambda$, where λ is the security parameter. It then runs the obfuscated program on r to receive (t, k) and then creates the ciphertext CT as $(t, c = \mathsf{Encrypt}_{\mathsf{PDE}}(k, m))$.

The secret key SK_f for a function f will be created as an obfuscated program. This program will take as input a ciphertext $CT = (t, c)$. The program first computes k from $F(K, t)$, then uses k to decrypt c to a message m and outputs $f(m)$. The decryption algorithm is simply to run the obfuscated program on the ciphertext.

The proof of security of our first system follows what we can a "key-programming" approach. The high level idea is that for each key we will hardwire in the decryption response into each secret key obfuscated program for when the input is the challenge ciphertext. For all other inputs the key computes decryption normally. Our key-programming approach is enabled by two important factors. First, in the security game there is a single challenge ciphertext so only one hardwiring needs to be done per key. Second, since all queries come after the challenge messages (m_0, m_1) are declared we will know where we need to puncture to create our hardwiring.

Intuitively, our proof can be broken down into two high level steps. First, we will perform a set of steps that allow us to hardwire the decryption answers to all of the secret keys for the challenge ciphertext. Next, we use PDE security to move from encrypting m_b for challenge bit $b \in \{0, 1\}$ to always encrypting m_0— independent of the bit b. (The actual proof contains multiple hybrids and is more intricate.)

Handling Full Security. We now move to dealing with full security where we need to handle private key queries on both sides of the challenge ciphertext. At this point it is clear that relying only on key-programming will not suffice. First, a pre-challenge ciphertext key for function f will need to be created before the

---

[2] Despite sharing the term deterministic, our security definition of PDEs does not have much in common with deterministic encryption [BFO08, BFOR08] which has a central goal of hiding information among message distributions of high entropy.

challenge messages (m_0, m_1) are declared, so it will not even be known at key creation time what $f(m_0) = f(m_1)$ will be.

Our central idea is to achieve an adaptively secure functional encryption by bootstrapping from a one-bounded FE scheme that is adaptively secure. At a high level a ciphertext is associated with a tag t and a private key with a tag y. From the pair of tags (t, y) one can (with the proper key material) pseudorandomly derive a master secret key k for a one bounded FE system. The ciphertext will be equipped with an obfuscated program, C, which on input of a key tag y will generate the one bounded key k (associated with the pair (t, y)) and then uses this to create an encryption of the message m under the one-bounded scheme with key k. Likewise, the private key for functionality f comes equipped with an obfuscated program P_f which on input of a ciphertext tag t derives the one bounded secret key k and uses this to create a one-bounded secret key.

The decryption algorithm will pass the key tag y to the ciphertext program to get a one bounded ciphertext $\mathrm{CT_{OB}}$ and the ciphertext tag t to the key program to get a one bound key $\mathrm{SK_{OB}}$. Finally, it will apply the one bounded decryption algorithm as $\mathsf{DecryptOB}(\mathrm{CT_{OB}}, \mathrm{SK_{OB}})$ to learn the message m. The one bounded key and ciphertext are compatible since they are both derived psuedorandomly from the pair (t, y) to get *same* one-bounded key k. (Note a different pair $(t', y') \neq (t, y)$ corresponds to a different one bounded FE key k' with high probability.)

Our bootstrapping proof structure allows us to develop "selective-ish" techniques at the outer level since in our reductions the ciphertext and private key tags can be chosen randomly ahead of time before the challenge message or any private key queries are known. Then the challenge of dealing with adaptive security is then "pushed down" to the one bounded FE scheme, where it has been solved in previous work [GVW12].

In the description above we have so far omitted one critical ingredient. In addition to generating a one bounded secret key on input t, the program P_f on input t will also generate an encrypted signal a that is passed along with the tag y to the ciphertext program C on decryption to let it know that it is "okay" to generate the one-bounded ciphertext for the pair (t, y). In the actual use of the system, this is the only functionality of the signal. However, looking ahead to our proof we will change the signal encrypted to tell the program C to switch the message for which it generates one bounded encryption encryptions of.

Our proof replaces key programming with a method we call "key-signaling". In a key-signaling system a normal ciphertext will be associated with a single message m which we refer to as an α-message. The decryption algorithm will use the secret key to prepare an α-signal for the ciphertext which will enable normal decryption. However, the ciphertext can also have a second form in which it is associated with two messages m_α and m_β. The underlying semantics are that if it receives an α-signal it uses m_α and if it receives a β-signal it uses m_β.

These added semantics open up new strategy for proving security. In the initial security game the challenge ciphertext encrypts m_b for challenge bit b. It will only receive α-signals from keys. Next we (indistinguishably) move the challenge ciphertext to encrypt m_b as the α-message and m_0 as the β-message. All keys still send only α-signals. Now one by one we change each key to send an β-signal to the challenge ciphertext as opposed to an α-signal. This step is feasible since for any queried function f we must have that $f(m_b) = f(m_0)$. Finally, we are able to erase the message m_b since no key is signaling for it.

Stepping back we can see that instead of storing the response of decryption for the challenge ciphertext at each key, we are storing the fact that it is using the second message in decryption.

We note that we can instantiate the one-bounded system using the construction of Gorbunov, Vaikuntanathan and Wee [GVW12] (GVW) who proved adaptive security of a public key FE 1-bounded scheme from IND-CPA secure public key encryption. Since we actually only need master key encryption, we observe that this can be achieved from IND-CPA symmetric key encryption. Thus, we maintain our goal of not using heavy weight primitives outside of obfuscation. One important fact is that the GVW scheme is proven to be 1-bounded adaptively secure regardless of whether the private key query comes before or after the challenge ciphertext. We note that the GVW system actually allows for a single key, but many ciphertexts; however, we only require security for a single ciphertext. The actual proof of security requires several hybrid steps and we defer further details to Sect. 5.

Recent Work. Recently, Garg, Gentry, Halevi, and Zhandry [GGHZ14a] showed how to realize adaptively secure Attribute-Based Encryption from multilinear graded encodings. It is based on $\mathcal{U}$-graded encodings.

Subsequent to both of these works, the same authors [GGHZ14b] gave a construction of Functional Encryption from multilinear encodings. This construction required a new multilinear encoding functionality of allowing the "encoding grades" to be dynamically extended by any party using just the public parameters. Their scheme crucially leverages this capability and is also reflected in the assumption.

There are different tradeoffs between and pure indistinguishability obfuscation approach and that used in [GGHZ14b]. On one hand the approach of [GGHZ14b] allows one to directly get to mutlilinear encodings. On the other hand the novel use of extensions of grades both gives a novel technical idea, but possibly presents new risks. For example, there has been a flurry of recent activity consisting of attacks and responses to certain candidate constructions and assumptions of multilinear enocdings [CHL+14, BWZ14, GHMS14, CLT14].

If one reduces to indistinguishability obfuscation, it can potentially be realized from different types of assumptions, including different forms of multilinear encodings or potentially entirely different number theory. An interesting open question is whether indistinguishability obfuscation or some close variant of it can be reduced to a basic number theoretic assumption that does not rely on sub exponential hardness. One interesting variant of this direction is to consider different variations of $i\mathcal{O}$ that are more amenable to such proofs, but can be leveraged in similar ways.

Bootstrapping with a Flipped One-time FE Scheme. More recently, Ananth, Brakerski, Segev and Vaikuntanathan [ABSV14] showed an eloquent adaptation of our technique of bootstrapping from an adaptive 1-bounded scheme. Instead of starting with the 1-bounded FE scheme of GVW, they use a simple transformation on GVW due Brakerski and Segev [BS14] and applying universal circuits to create a flipped version of it. While the GVW scheme we used can handle a single key and many ciphertexts, the flipped version does the opposite. It can handle multiple keys, but only generating one ciphertext (this is done with secret key encryption).

They go on to show that using the flipped version of one-bounded FE for bootstrapping enables simplifications in the construction and proof. Instead of having attaching an obfuscated program to the ciphertext to generate one-bounded ciphertexts, the composite ciphertext contains a single 1-bounded ciphertext. In addition, it has a separate ("trojan") component that allows for transmitting the 1-bounded secret key used create a ciphertext to a program on the key side. Taken together the flipping and the trojan transmission allow for the private key to consist of a selectively secure functional encryption system.

2   Functional Encryption

Definition 1 (Functional Encryption). *A functional encryption scheme for a class of functions $\mathcal{F} = \mathcal{F}(\lambda)$ over message space $\mathcal{M} = \mathcal{M}(\lambda)$ consists of four algorithms $\mathcal{FE} = \{\mathsf{Setup}, \mathsf{KeyGen}, \mathsf{Encrypt}, \mathsf{Decrypt}\}$:*

$\mathsf{Setup}(1^\lambda)$ – *a polynomial time algorithm that takes the unary representation of the security parameter λ and outputs public parameters PP and a master secret key MSK.*

$\mathsf{KeyGen}(\mathrm{MSK}, f)$ – *a polynomial time algorithm that takes as input the master secret key MSK and a description of function $f \in \mathcal{F}$ and outputs a corresponding secret key SK_f.*

$\mathsf{Encrypt}(\mathrm{PP}, x)$ – *a polynomial time algorithm that takes the public parameters PP and a string x and outputs a ciphertext CT.*

$\mathsf{Decrypt}(\mathrm{SK}_f, \mathrm{CT})$ – *a polynomial time algorithm that takes a secret key SK_f and ciphertext encrypting message $m \in \mathcal{M}$ and outputs $f(m)$.*

A functional encryption scheme is correct for $\mathcal{F}$ if for all $f \in \mathcal{F}$ and all messages $m \in \mathcal{M}$:

$$\Pr[\,(\mathrm{PP}, \mathrm{MSK}) \leftarrow \mathsf{Setup}(1^\lambda);$$

$$\mathsf{Decrypt}(\mathsf{KeyGen}(\mathrm{MSK}, f), \mathsf{Encrypt}(\mathrm{PP}, m)) \neq f(m)\,] = negl(\lambda).$$

Indistinguishability Security for Functional Encryption. We describe indistinguishability security as a multi-phased game between an attacker $\mathcal{A}$ and a challenger.

Setup: The challengers runs $(\mathrm{PP}, \mathrm{MSK}) \leftarrow \mathsf{Setup}(1^\lambda)$ and gives PP to $\mathcal{A}$.

Query Phase 1: $\mathcal{A}$ adaptively submits queries f in $\mathcal{F}$ and is given $\mathrm{SK}_f \leftarrow$ KeyGen(MSK, f). This step can be repeated any polynomial number of times by the attacker.

Challenge: $\mathcal{A}$ submits two messages $m_0, m_1 \in \mathcal{M}$ such that $f(m_0) = f(m_1)$ for all functions f queried in the key query phase. The challenger then samples $\mathrm{CT}^* \leftarrow$ Encrypt(PP, m_b) for the attacker.

Query Phase 2: $\mathcal{A}$ continues to issue key queries as before subject to the restriction that any f queried must satisfy $f(m_0) = f(m_1)$.

Guess: $\mathcal{A}$ eventually outputs a bit b' in $\{0, 1\}$.

The advantage of an algorithm $\mathcal{A}$ in this game is $\mathsf{Adv}_\mathcal{A} = \Pr[b' = b] - \frac{1}{2}$.

Definition 2. *A functional encryption scheme is indistinguishability secure if for all poly-time $\mathcal{A}$ the function $\mathsf{Adv}_\mathcal{A}(\lambda)$ is negligible.*

Definition 3. *In the above security game we define a* post challenge ciphertext *attacker as one that does not make any key queries in Phase 1. We define a functional encryption scheme to be* post challenge ciphertext *indistinguishability secure if for any poly-time algorithm $\mathcal{A}$ that is a post challenge ciphertext attacker the advantage of $\mathcal{A}$ is negligible in the indistinguishability security game.*[3]

3   Puncturable Deterministic Encryption

In this section we define a primitive of puncturable deterministic encryption and show how to build it from (injective) puncturable PRFs. The main purpose of this abstraction is to give a slightly higher level tool (relative to puncturable PRFs) to work with in our punctured programming construction and proofs.

Definition 4 (Puncturable Deterministic Encryption). *A puncturable deterministic encryption (PDE) scheme is defined over a message space $\mathcal{M} = \mathcal{M}(\lambda)$ and consists of four algorithms: (possibly) randomized algorithms Setup$_{\mathsf{PDE}}$, and Puncture$_{\mathsf{PDE}}$ along with deterministic algorithms Encrypt$_{\mathsf{PDE}}$ and Decrypt$_{\mathsf{PDE}}$. All algorithms will be poly-time in the security parameter.*

Setup$_{\mathsf{PDE}}(1^\lambda)$ *The setup algorithm takes a security parameter and uses its random coins to generate a key K from a keyspace $\mathcal{K}$.*

Encrypt$_{\mathsf{PDE}}(K, m)$ *The encrypt algorithm takes as input a key K and a message m. It outputs a ciphertext CT. The algorithm is deterministic.*

Decrypt$_{\mathsf{PDE}}(K, \mathrm{CT})$ *The decrypt algorithm takes as input a key K and ciphertext CT. It outputs either a message $m \in \mathcal{M}$ or a special reject symbol $\bot$.*

Puncture$_{\mathsf{PDE}}(K, m_0, m_1)$ *The puncture algorithm takes as input a key $K \in \mathcal{K}$ as well as two messages m_0, m_1. It creates and outputs a new key $K(m_0, m_1) \in \mathcal{K}$. The parentheses are used to syntactically indicate what is punctured.*

---

[3] We remark that any system that is *post challenge ciphertext* secure must also be selectively secure.

Correctness. A punctured deterministic encryption scheme is correct if there exists a negligible function negl such that the following holds for all λ and all pairs of messages $m_0, m_1 \in \mathcal{M}(\lambda)$.

Let $K = \mathsf{Setup}_{\mathsf{PDE}}(1^\lambda)$ and $K(m_0, m_1) \leftarrow \mathsf{Puncture}_{\mathsf{PDE}}(K, m_0, m_1)$. Then for all $m \neq m_0, m_1$

$$\Pr[\mathsf{Decrypt}_{\mathsf{PDE}}(K(m_0, m_1), \mathsf{Encrypt}_{\mathsf{PDE}}(K, m)) \neq m] = \mathrm{negl}(\lambda).$$

In addition, we have that for all m (including m_0, m_1)

$$\Pr[\mathsf{Decrypt}_{\mathsf{PDE}}(K, \mathsf{Encrypt}_{\mathsf{PDE}}(K, m)) \neq m] = \mathrm{negl}(\lambda).$$

Definition 5. *We say that a correct scheme is perfectly correct if the above probability is 0 and otherwise say that it is statistically correct.*

(Selective) Indistinguishability Security for Punctured Deterministic Encryption. We describe indistinguishability security as a multi-phased game between an attacker $\mathcal{A}$ and a challenger.

Setup: The attacker selects two messages $m_0, m_1 \in \mathcal{M}$ and sends these to the challenger. The challenger runs $K = \mathsf{Setup}_{\mathsf{PDE}}(1^\lambda)$ and $K(m_0, m_1) = \mathsf{Puncture}_{\mathsf{PDE}}(K, m_0, m_1)$. It then chooses a random bit $b \in \{0, 1\}$ and computes
$$T_0 = \mathsf{Encrypt}_{\mathsf{PDE}}(K, m_b), \ T_1 = \mathsf{Encrypt}_{\mathsf{PDE}}(K, m_{1-b}).$$

It gives the punctured key $K(m_0, m_1)$ as well as T_0, T_1 to the attacker.
Guess: $\mathcal{A}$ outputs a bit b' in $\{0, 1\}$.

The advantage of an algorithm $\mathcal{A}$ in this game is $\mathsf{Adv}_\mathcal{A} = \Pr[b' = b] - \frac{1}{2}$.

Definition 6. *A puncturable deterministic encryption scheme is indistinguishability secure if for all poly-time $\mathcal{A}$ the function $\mathsf{Adv}_\mathcal{A}(\lambda)$ is negligible.*

Sampling Master Keys. At times instead of running the $\mathsf{Setup}_{\mathsf{PDE}}(1^\lambda)$ algorithm to generate the master key for a PDE scheme we will generate the master key by simply sampling a uniformly random string $k \in \{0, 1\}^\lambda$ where λ is the security parameter. We can also do this without loss of generality.

In our full version [Wat14] we give a construction of puncturable deterministic encryption puncturable PRFs. This follows the hidden triggers construction from [SW14].

4   A Post Challenge Ciphertext Secure Construction

We now describe our construction for a functional encryption (FE) scheme that is post challenge ciphertext secure. We let the message space $\mathcal{M} = \mathcal{M}(\lambda) = \{0, 1\}^{\ell(\lambda)}$ for some polynomial function ℓ and the function class be $\mathcal{F} = \mathcal{F}(\lambda)$.

We will use a puncturable PRF $F(\cdot, \cdot)$ such that when we fix the key K we have that $F(K, \cdot)$ takes in a 2λ bit input and outputs λ bits. In addition, we use a puncturable deterministic encryption scheme (PDE) where the message space $\mathcal{M}$ is the same as that of the (FE) system. In our PDE systems master (non-punctured) keys are sampled uniformly at random from $\{0,1\}^\lambda$. Finally, we use an indistinguishability secure obfuscator and a length doubling pseudo random generator $\mathrm{PRG} : \{0,1\}^\lambda \to \{0,1\}^{2\lambda}$.

Our Construction. In our system the setup algorithm will produce an obfuscated program P that serves as the public parameters. Encryption proceeds in two steps. First the encryptor will choose a random string r and run $P(r)$. The obfuscated program will first use r to generate a tag t. Next the program will apply a (puncturable) psuedorandom function on t with global key K to generate a PDE key k. The program outputs both the tag t and PDE key k to the encryptor. Finally, the encryptor will use k to perform an encryption of the actual message m getting PDE ciphertext c. The (total) ciphertext CT consists of the tag t and c. Intuitively, the ciphertext component c is the "core encryption" of the message and the tag t tells how one can derive the PDE key k (if one knows the system's puncturable PRF key).

The authority generates a private key for function f as an obfuscated program P_f. To decrypt a ciphertext CT $= (t, c)$ the decrypt or simply runs $P_f(t, c)$. The obfuscated program will first generate *the same* PDE key k that was used to encrypt the ciphertext.

We make two intuitive remarks about security. First, we note that the system's puncturable PRF key K only appears in obfuscated programs and not in the clear. Second, it is not necessarily a problem perform the core encryption of the message under a *deterministic* scheme. The reason is that the encryption procedure implicitly chooses a fresh k so with high probability any single PDE key should only be used once. (Clearly, performing a deterministic encryption step more than once with the same key would be problematic.)

We now give our construction in detail.

Setup(1^λ)

The setup algorithm first chooses a random punctured PRF key $K \leftarrow \mathrm{Key}_F(1^\lambda)$ and sets this as the master secret key MSK. Next it creates an obfuscation of the program Initial-Encrypt as $P \leftarrow i\mathcal{O}(1^\lambda, \textsc{Initial-Encrypt}{:}1[K])$.[4] (See Fig. 1) This obfuscated program, P, serves as the public parameters PP.

Encrypt(PP $= P(\cdot), m \in \mathcal{M}$)

The encryption algorithm chooses random $r \in \{0,1\}^\lambda$. It then runs the obfuscated program P on r to get:

$$(t, k) \leftarrow P(r).$$

It then computes $\mathsf{Encrypt}_{\mathsf{PDE}}(k, m) = c$. The output ciphertext is CT $= (t, c)$.

---

[4] The program Initial-Encrypt:1 is padded to be the same size as Initial-Encrypt:2.

KeyGen(MSK, $f \in \mathcal{F}(\lambda)$) The KeyGen algorithm produces an obfuscated program P_f by obfuscating[5]

$$P_f \leftarrow i\mathcal{O}(\text{KEY-EVAL:}1[K, f]).$$

Decrypt(CT = (t, c), SK = P_f) The decryption algorithm takes as input a ciphertext CT and a secret key SK which is an obfuscated program P_f. It runs $P_f(t, c)$ and outputs the response.

Correctness. Correctness follows in a rather straightforward manner from the correctness of the underlying primitives. We briefly sketch the correctness argument. Suppose we call the encryption algorithm for message m with randomness r. The obfuscated program generates $(t, k) = (\text{PRG}(r), F(K, t))$. Then it creates the ciphertext CT = $(t, c = \text{Encrypt}_{\text{PDE}}(k, m))$. Now let's examine what occurs when Decrypt(CT = (t, c), SK$_f$ = P_f) is called where P_f was a secret key created from function f. The decryption algorithm calls $P_f(t, c)$. The (obfuscated) program will compute the same PDE key $k = F(K, t)$ as used to create the ciphertext. Then it will use the PDE decryption algorithm and obtain m. This follows via the correctness of the PDE scheme. Finally, it outputs $f(m)$ which is the correct output.

Initial-Encrypt:1

Constants: Puncturable PRF key K.
Input: Randomness $r \in \{0, 1\}^\lambda$.

1. Let $t = \text{PRG}(r)$.
2. Compute: $k = F(K, t)$.
3. Output: (t, k).

Fig. 1. Program Initial-Encrypt:1

4.1 Proof of Security

Before delving into our formal security proof we give a brief overview with some intuition. In our system a challenge ciphetext CT* will be a pair (t^*, c^*) of a tag and PDE ciphertext. The first step of our proof is to use pseudorandom generator security to (indetectably) move t^* out of the set of tags $\mathcal{T}$ that might be generated from the program P. (Note the set $\mathcal{T}$ corresponds to the possible outputs of the pseudorandom generator.) This then enables us to perform multiple puncturing and hardwiring steps detailed below. Eventually, instead deriving the PDE key k^* as $F(K, t^*)$, it will be chosen uniformly at random. (Here k^* is the PDE key used in creating the challenge ciphertext.)

---

[5] The program KEY-EVAL:1 (of Fig. 2) is padded to be the same size as KEY-EVAL:2.

Key-Eval:1

Constants: PRF key K, function description $f \in \mathcal{F}$.
Input: (t, c).

1. Compute: $k = F(K, t)$.
2. Output $f(\mathsf{Decrypt}_{\mathsf{PDE}}(k, c))$. (If $\mathsf{Decrypt}_{\mathsf{PDE}}(k, c)$ evaluates to $\bot$ the program outputs $\bot$.)

Fig. 2. Program Key-Eval:1

Furthermore, instead of putting the PDE key k^* into the obfuscated programs given out as keys we will put a punctured version k'. This punctured version is can decrypt all ciphertexts *except* It cannot tell the difference between a PDE encryption of the challenge message m_0 from m_1. However, by the rules of the security game it must be the case that the bit $d_f = f(m_0) = f(m_1)$ for any queried private key function f. Therefore, an obfuscated program for private key f can output d_f when either of the two PDE ciphertexts arises without knowing which one is which. We note that the reduction knows which messages (m_0, m_1) to puncture the PDE key k at since in this security game all keys are given out after the challenge ciphertext is generated.

Finally, at this stage we can simply apply the PDE security game to argue that the message is hidden. We note that the first steps of the proof have similarities to prior programming puncturing proofs [SW14], but we believe the introduction of and the way we utilize puncturable deterministic encryption are novel to this construction. Details of our formal proof are in our full version [Wat14].

5   An Adaptively Secure Construction

We now describe our construction of a functional encryption (FE) scheme that is adaptively secure. We let the message space $\mathcal{M} = \{0, 1\}^{\ell(\lambda)}$ for some polynomial function ℓ and the function class be $\mathcal{F}(\lambda) = \mathcal{F}$.

We will use two puncturable PRFs F_1, F_2 such that when we fix the keys K we have that $F_1(K, \cdot)$ takes in a 2λ bit input and outputs two bit strings of length λ and $F_2(K, \cdot)$ takes λ bits to five bitstrings of length λ. In addition, we use a puncturable deterministic encryption scheme where the message space is $\{0, 1\}^\lambda$. In our Puncturable PRF and PDE systems master keys are sampled uniformly at random from $\{0, 1\}^\lambda$. Finally, we use an indistinguishability secure obfuscator and an *injective* length doubling pseudo random generator PRG : $\{0, 1\}^\lambda \rightarrow \{0, 1\}^{2\lambda}$.

Finally, we use a one-bounded secure functional encryption system with master key encryption consisting of algorithms: KeyGenOB, EncryptOB, DecryptOB. We assume without loss of generality that the master key is chosen uniformly from $\{0,1\}^\lambda$. The message space $\mathcal{M}$ and key description space $f \in \mathcal{F}$ of the one bounded scheme is the same as the scheme we are constructing.

Our Construction. Our construction achieves an adaptively secure functional encryption by bootstrapping from a one-bounded FE scheme that is adaptively secure. At a high level a ciphertext is associated with a tag t and a private key with a tag y. From the pair of tags (t, y) one can (with the proper key material) pseudorandomly derive a master secret key k for a one bounded FE system. The ciphertext will be equipped with an obfuscated program, C, which on input of a key tag y will generate the one bounded key k (associated with the pair (t, y)) and then uses this to create an encryption of the message m under the one-bounded scheme with key k. Likewise, the private key for functionality f comes equipped with an obfuscated program P_f which on input of a ciphertext tag t derives the one bounded secret key k and uses this to create a one-bounded secret key.

The decryption algorithm will pass the key tag y to the ciphertext program to get a one bounded ciphertext CT_{OB} and the ciphertext tag t to the key program to get a one bound key SK_{OB}. Finally, it will apply the one bounded decryption algorithm as $\mathsf{DecryptOB}(CT_{OB}, SK_{OB})$ to learn the message m. The one bounded key and ciphertext are compatible since they are both derived psuedorandomly from the pair (t, y) to get *same* one-bounded key k. (Note a different pair $(t', y') \neq (t, y)$ corresponds to a different one bounded FE key k' with high probability.)

Our bootstrapping proof structure allows us to develop "selective-ish" techniques at the outer level since in our reductions the ciphertext and private key tags can be chosen randomly ahead of time before the challenge message or any private key queries are known. Then the challenge of dealing with adaptive security is then "pushed down" to the one bounded FE scheme, where it has been solved in previous work [GVW12].

In the description above we have so far omitted one critical ingredient. In addition to generating a one bounded secret key on input t, the program P_f on input t will also generate an encrypted signal a that is passed along with the tag y to the ciphertext program C on decryption to let it know that it is "okay" to generate the one-bounded ciphertext for the pair (t, y). In the actual use of the system, this is the only functionality of the signal. However, looking ahead to our proof we will change the signal encrypted to tell the program C to switch the message for which it generates one bounded encryption encryptions of.

Setup(1^λ)
The algorithm first chooses a random punctured PRF key $K \leftarrow \mathsf{Key}_{F_1}(1^\lambda)$ which is set as the master secret key MSK. Next it creates an obfuscation of the program Initial-Encrypt as $P \leftarrow i\mathcal{O}(1^\lambda, \text{INITIAL-ENCRYPT:}1[K])$.[6]

Encrypt$(\text{PP} = P(\cdot), m \in \mathcal{M})$
The encryption algorithm performs the following steps in sequence.

---

[6] The program INITIAL-ENCRYPT:1 is padded to be the same size as INITIAL-ENCRYPT:2.) This obfuscated program, P serves as the public parameters PP.

1. Chooses random $r \in \{0,1\}^\lambda$.
2. Sets $(t, K_t, \alpha) \leftarrow P(r)$.
3. Sets $\tilde{\alpha} = \text{PRG}(\alpha)$.
4. Creates the program $C \leftarrow i\mathcal{O}(1^\lambda, \text{CT-EVAL:}1[K_t, \tilde{\alpha}, m])$.[7]
5. The output ciphertext is $\text{CT} = (t, C)$.

$KeyGen(\text{MSK}, f \in \mathcal{F}(\lambda))$
The KeyGen algorithm first chooses a random $y \in \{0,1\}^\lambda$. It next produces an obfuscated program P_f by obfuscating $P_f \leftarrow i\mathcal{O}(\text{KEY-SIGNAL:}1[K, f, y])$.[8]
    The secret key is $\text{SK} = (y, P_f)$.

$Decrypt(\text{CT} = (t, C), \text{SK} = (y, P_f))$
The decryption algorithm takes as input a ciphertext $\text{CT} = (t, C)$ and a secret key $\text{SK} = (y, P_f)$. It first computes $(a, \text{SK}_{\text{OB}}) = P_f(t)$. Next it computes $\text{CT}_{\text{OB}} = C(a, y)$. Finally, it will use the produced secret key to decrypt the produced ciphertext as $\text{DecryptOB}(\text{CT}_{\text{OB}}, \text{SK}_{\text{OB}})$ and outputs the result.

Correctness. We briefly sketch a correctness argument. Consider a ciphertext $\text{CT} = (t, C)$ created for message m that is associated with tag t and a key for function f that is associated with tag y. On decryption the algorithm first calls $(a, \text{SK}_{\text{OB}}) = P_f(t)$. Here the obfuscated program computes: $(K_t, \alpha) = F_1(K, t)$, $(d, k, s_1, s_2, s_3) = F_2(K_t, y)$, and $a = \text{Encrypt}_{\text{PDE}}(d, \alpha)$ and $\text{SK}_{\text{OB}} = \text{KeyGenOB}(k, f; s_2)$.

Next, it calls $\text{CT}_{\text{OB}} = C(a, y)$, where C was generated as an obfuscation of program $\text{CT-EVAL:}1[K_t, \tilde{\alpha}, m]$ where $\tilde{\alpha} = \text{PRG}(\alpha)$. This obfuscated program will compute the same values of $(d, k, s_1, s_2, s_3) = F_2(K_t, y)$ as the key signal program. By correctness of the PDE system we will have that $\text{Decrypt}_{\text{PDE}}(d, a) = \alpha$ and thus the program will output $\text{EncryptOB}(k, m; s_1)$. At this point the decryption algorithm has a one bounded private key for function f and a one bounded ciphertext for message m both created under the same master key k. Therefore, running the one-bounded decryption algorithm will produce $f(m)$ (Figs. 3, 4 and 5).

5.1   Proof of Security

Before delving into our formal security proof we will give a brief intuitive overview of its structure and sequence of games steps. In the first steps of our sequence of games proof we will use pseudorandom generator security to (inde-tectably) move t^* out of the set of tags $\mathcal{T}$ that might be generated from the program P.[9] Then we use use puncturing techniques to remove the key material, $K_{t^*}^*$, associated with t^* from the obfuscated program given in the public parameters. In addition, the proof will hardwire in the response of all private

---

[7] The program CT-EVAL:1 is padded to be the same size as the maximum of CT-EVAL:2 and CT-EVAL:3.

[8] The program KEY-SIGNAL:1 is padded to be the same size as KEY-SIGNAL:2.

[9] Note the set $\mathcal{T}$ corresponds to the possible outputs of the pseudorandom generator.

Initial-Encrypt:1

Constants: Puncturable PRF key K.
Input: Randomness $r \in \{0,1\}^\lambda$.

1. Let $t = \mathrm{PRG}(r)$.
2. Compute $(K_t, \alpha) = F_1(K, t)$.
3. Output: (t, K_t, α).

Fig. 3. Program Initial-Encrypt:1

CT-Eval:1

Constants: PRF key K_t, $\tilde{\alpha} \in \{0,1\}^{2\lambda}$, message $m \in \mathcal{M}$.
Input: PDE ciphertext a and value $y \in \{0,1\}^\lambda$.

1. Compute $(d, k, s_1, s_2, s_3) = F_2(K_t, y)$.
2. Compute $e = \mathsf{Decrypt}_{\mathsf{PDE}}(d, a)$.
3. If $\mathrm{PRG}(e) = \tilde{\alpha}$ output $\mathsf{EncryptOB}(k, m; s_1)$.
4. Else output a rejecting $\perp$.

Fig. 4. Program CT-Eval:1

Key-Signal:1

Constants: PRF key K, function description $f \in \mathcal{F}$, tag $y \in \{0,1\}^\lambda$.
Input: $t \in \{0,1\}^{2\lambda}$.

1. Compute $(K_t, \alpha) = F_1(K, t)$.
2. Compute $(d, k, s_1, s_2, s_3) = F_2(K_t, y)$.
3. Compute and output $a = \mathsf{Encrypt}_{\mathsf{PDE}}(d, \alpha)$ and $\mathrm{SK}_{\mathrm{OB}} = \mathsf{KeyGenOB}(k, f; s_2)$.

Fig. 5. Program Key-Signal:1

keys $P_{f_1}, \ldots, P_{f_Q}$ to the input of t^*, where Q is the number of queries issued. These actions are covered in moving from Game 1 to Game 5.

In the next grouping of steps we will introduce a *second alternative message* m_0 into the challenge ciphertext program C^* to go along with the message m_b for $b \in \{0,1\}$. The behavior of the obfuscated program is now (by Game 7) such that if C^* receives an an "α-signal" as input it will output a one-bounded FE encryption of m_b and if it receives a "β-signal" it will output a one-bounded FE encryption of m_0. However, the private key programs P_{f_i} are only set to generate α signals. Before this grouping of steps was executed only α-signals existed.

Subsequently, each private key program P_f is transformed one by one such that they are programmed to send out β-signals upon receiving the tag t^*.

When used in decryption this will cause the challenge ciphertext to output one time encryptions of m_0 instead of m_1. Intuitively, this is undetectable because $f(m_b) = f(m_0)$ for all private key functions f that can legally be requested. Executing this transformation requires multiple sub steps and is the most complex piece of the proof. It is also where the security one bounded FE scheme is invoked.

Finally, after the above transformations are made we are able to execute two final cleanup steps that remove the message m_b from the ciphertext program C^*. At this point all information about the bit b is removed from the challenge ciphertext and the advantage of any attacker is 0.

Theorem 1. *The above functional encryption scheme is adaptively secure if instantiated with a secure punctured PRF, puncturable deterministic encryption scheme, pseudo random generator, an adaptively secure one-bounded functional encryption scheme and indistinguishability secure obfuscator.*

To prove the above theorem, we first define a sequence of games where the first game is the original FE security game. We begin by with describing Game 1 in detail, which is the adaptive FE security game instantiated with our construction. From there we describe the sequence of games, where each game is described by its modification from the previous game.

In the main body we describe the proof hybrid structure. In our full version [Wat14] we provide the lemmas showing that any poly-time attacker's advantage in each game must be negligibly close to that of the previous game (based on the security of different primitives).

Game 1 The first game is the original security game instantiated for our construction.

1. Challenger computes keys $K \leftarrow \mathrm{Key}_{F_1}(1^\lambda)$ and randomly chooses the challenge bit $b \in \{0,1\}$.
2. Challenger chooses random $r^* \in \{0,1\}^\lambda$ and computes $t^* = \mathrm{PRG}(r^*)$.
3. Challenger computes $K_{t^*}^*, \alpha^* = F_1(K, t^*)$.
4. Challenger sets $\tilde{\alpha}^* = \mathrm{PRG}(\alpha^*)$.
5. Challenger creates $P \leftarrow i\mathcal{O}(1^\lambda, \text{INITIAL-ENCRYPT:}1[K])$ and passes P to attacker.
6. Phase 1 Queries: Let f_j be the function of associated with the j-th query. Choose random $y_j \in \{0,1\}^\lambda$. Generate the j-th private key by computing $P_{f_j} \leftarrow i\mathcal{O}(\text{KEY-SIGNAL:}1[K, f_j, y_j])$. Output the key as (y_j, P_{f_j}).
7. Attacker gives messages $m_0, m_1 \in \mathcal{M}$ to challenger.
8. Challenger sets the program $C^* \leftarrow i\mathcal{O}(1^\lambda, \text{CT-EVAL:}1[K_{t^*}^*, \tilde{\alpha}^*, m_b])$.
9. The output ciphertext is $\mathrm{CT} = (t^*, C^*)$.
10. Phase 2 Queries: Same as Phase 1 in step 6.
11. The attacker gives a bit b' and wins if $b' = b$.

Game 2
2. Challenger chooses random $t^* \in \{0,1\}^{2\lambda}$.

Game 3
2. Challenger chooses random $t^* \in \{0,1\}^{2\lambda}$ and sets $K(t^*) = \text{Puncture}_F(K, t^*)$.
5. Challenger creates $P \leftarrow i\mathcal{O}(1^\lambda, \text{INITIAL-ENCRYPT:2}[K(t^*)])$ and passes P to attacker.

Game 4
6. Phase 1 Queries: Let f_j be the function of associated with the j-th query.
   (a) Choose random $y_j \in \{0,1\}^\lambda$.
   (b) Compute $(d_j^*, k_j^*, s_{1,j}^*, s_{2,j}^*, s_{3,j}^*) = F_2(K_{t^*}, y_j)$.
   (c) Compute $a_j^* = \text{Encrypt}_{\text{PDE}}(d_j^*, \alpha^*)$ and $\text{SK}_{\text{OB},j}^* = \text{KeyGenOB}(k_j^*, f_j; s_{2,j}^*)$.
   (d) Compute $P_{f_j} \leftarrow i\mathcal{O}(\text{KEY-SIGNAL:2}[K(t^*), t^*, a_j^*, \text{SK}_{\text{OB},j}^*, f_j, y_j])$.
   (e) Output the key as (y_j, P_{f_j}).
10. Phase 2 Queries: Same as Phase 1 in step 6. (These are also changed as described above.)

Game 5
3. Challenger chooses random $K_{t^*}^*, \alpha^*$.

Game 6
4. Challenger sets $\tilde{\alpha}^* = \text{PRG}(\alpha^*)$ and chooses random $\tilde{\beta}^* \in \{0,1\}^{2\lambda}$.
8. Challenger sets the program $C^* \leftarrow i\mathcal{O}(1^\lambda, \text{CT-EVAL:2}[K_{t^*}^*, \tilde{\alpha}^*, \tilde{\beta}^*, m_b, m_0])$.

Game 7
4. Challenger sets $\tilde{\alpha}^* = \text{PRG}(\alpha^*)$, chooses $\beta^* \in \{0,1\}^\lambda$ at random and sets $\tilde{\beta}^* = \text{PRG}(\beta^*)$.

Game 8, i Defined for $i = 0$ to Q. (Q is number of key queries.)
6. Phase 1 Queries: Let f_j be the function of associated with the j-th query.
   (a) Choose random $y_j \in \{0,1\}^\lambda$.
   (b) Compute $(d_j^*, k_j^*, s_{1,j}^*, s_{2,j}^*, s_{3,j}^*) = F_2(K_{t^*}, y_j)$.
   (c) If $j > i$ then set $a_j^* = \text{Encrypt}_{\text{PDE}}(d_j^*, \alpha^*)$; otherwise if $j \le i$ set $a_j^* = \text{Encrypt}_{\text{PDE}}(d_j^*, \beta^*)$.
       Let $\text{SK}_{\text{OB},j}^* = \text{KeyGenOB}(k_j^*, f_j; s_{2,j}^*)$.
   (d) Compute $P_{f_j} \leftarrow i\mathcal{O}(\text{KEY-SIGNAL:2}[K(t^*), t^*, a_j^*, \text{SK}_{\text{OB},j}^*, f_j, y_j])$.
   (e) Output the key as (y_j, P_{f_j}).

Game 9
4. Challenger chooses $\tilde{\alpha}^* \in \{0,1\}^{2\lambda}$ at random, chooses $\beta^* \in \{0,1\}^\lambda$ at random and sets $\tilde{\beta}^* = \text{PRG}(\beta^*)$.
6. Phase 1 Queries: Let f_j be the function of associated with the j-th query.
   (a) Choose random $y_j \in \{0,1\}^\lambda$.
   (b) Compute $(d_j^*, k_j^*, s_{1,j}^*, s_{2,j}^*, s_{3,j}^*) = F_2(K_{t^*}, y_j)$.
   (c) Set $a_j^* = \text{Encrypt}_{\text{PDE}}(d_j^*, \beta^*)$. Let $\text{SK}_{\text{OB},j}^* = \text{KeyGenOB}(k_j^*, f_j; s_{2,j}^*)$.
   (d) Compute $P_{f_j} \leftarrow i\mathcal{O}(\text{KEY-SIGNAL:2}[K(t^*), t^*, a_j^*, \text{SK}_{\text{OB},j}^*, f_j, y_j])$.
   (e) Output the key as (y_j, P_{f_j}).

Initial-Encrypt:2

Constants: Puncturable PRF key $K(t^*)$.
Input: Randomness r.

1. Let $t = \mathrm{PRG}(r)$.
2. Compute $(K_t, \alpha) = F_1(K(t^*), t)$.
3. Output: (t, K_t, α).

Fig. 6. Program Initial-Encrypt:2

Key-Signal:2

Constants: PRF key $K(t^*)$, t^*, a^*, $\mathrm{SK}^*_{\mathrm{OB}}$, function description f, tag $y \in \{0,1\}^\lambda$.
Input: $t \in \{0,1\}^{2\lambda}$.

1. If $t = t^*$ output $a^*, \mathrm{SK}^*_{\mathrm{OB}}$.
2. Compute $(K_t, \alpha) = F_1(K(t^*), t)$.
3. Compute $(d, k, s_1, s_2, s_3) = F_2(K_t, y)$.
4. Compute and output $a = \mathsf{Encrypt}_{\mathsf{PDE}}(d, \alpha)$ and $\mathrm{SK}_{\mathrm{OB}} = \mathsf{KeyGenOB}(k, f; s_2)$

Fig. 7. Program Key-Signal:2

CT-Eval:2

Constants: PRF key K_t, $\breve{\alpha}, \breve{\beta} \in \{0,1\}^{2\lambda}$, messages $m, m_{\mathrm{fixed}} \in \mathcal{M}$.
Input: (a, y).

1. Compute $(d, k, s_1, s_2, s_3) = F_2(K_t, y)$.
2. Compute $e = \mathsf{Decrypt}_{\mathsf{PDE}}(d, a)$.
3. If $\mathrm{PRG}(e) = \tilde{\alpha}$ output $\mathsf{EncryptOB}(k, m; s_1)$.
4. If $\mathrm{PRG}(e) = \tilde{\beta}$ output $\mathsf{EncryptOB}(k, m_{\mathrm{fixed}}; s_3)$.
5. Else output a rejecting $\perp$.

Fig. 8. Program CT-Eval:2

Game 10
8. Challenger sets the program $C^* \leftarrow i\mathcal{O}(1^\lambda, \text{CT-EVAL:}1[K^*_{t^*}, \tilde{\beta}^*, m_0])$.

We observe at this stage the interaction with the challenger is completely independent of b — note the message m_0 is encrypted regardless of b — and thus the attacker's advantage is 0 in this final game (Figs. 6, 7 and 8).

References

[ABG+13] Ananth, P., Boneh, D., Garg, S., Sahai, A., Zhandry, M.: Differing-inputs obfuscation and applications. Cryptology ePrint Archive, Report 2013/689 (2013). http://eprint.iacr.org/

[ABSV14] Ananth, P., Brakerski, Z., Segev, G., Vaikuntanathan, V.: The trojan method in functional encryption: from selective to adaptive security, generically. Cryptology ePrint Archive, Report 2014/917 (2014). http://eprint.iacr.org/

[BCOP04] Boneh, D., Di Crescenzo, G., Ostrovsky, R., Persiano, G.: Public key encryption with keyword search. In: Cachin, C., Camenisch, J.L. (eds.) EUROCRYPT 2004. LNCS, vol. 3027, pp. 506–522. Springer, Heidelberg (2004)

[BCP14] Boyle, E., Chung, K.-M., Pass, R.: On extractability obfuscation. In: Lindell, Y. (ed.) TCC 2014. LNCS, vol. 8349, pp. 52–73. Springer, Heidelberg (2014)

[BFO08] Boldyreva, A., Fehr, S., O'Neill, A.: On notions of security for deterministic encryption, and efficient constructions without random oracles. In: Wagner, D. (ed.) CRYPTO 2008. LNCS, vol. 5157, pp. 335–359. Springer, Heidelberg (2008)

[BFOR08] Bellare, M., Fischlin, M., O'Neill, A., Ristenpart, T.: Deterministic encryption: definitional equivalences and constructions without random oracles. In: Wagner, D. (ed.) CRYPTO 2008. LNCS, vol. 5157, pp. 360–378. Springer, Heidelberg (2008)

[BGI+12] Barak, B., Goldreich, O., Impagliazzo, R., Rudich, S., Sahai, A., Vadhan, S.P., Yang, K.: On the (im)possibility of obfuscating programs. J. ACM 59(2), 6 (2012)

[BGI13] Boyle, E., Goldwasser, S., Ivan, I.: Functional signatures and pseudorandom functions. IACR Cryptology ePrint Archive, 2013/401 (2013)

[BS14] Brakerski, Z., Segev, G.: Function-private functional encryption in the private-key setting. Cryptology ePrint Archive, Report 2014/550 (2014). http://eprint.iacr.org/

[BSW11] Boneh, D., Sahai, A., Waters, B.: Functional encryption: definitions and challenges. In: Ishai, Y. (ed.) TCC 2011. LNCS, vol. 6597, pp. 253–273. Springer, Heidelberg (2011)

[BW07] Boneh, D., Waters, B.: Conjunctive, subset, and range queries on encrypted data. In: Vadhan, S.P. (ed.) TCC 2007. LNCS, vol. 4392, pp. 535–554. Springer, Heidelberg (2007)

[BW13] Boneh, D., Waters, B.: Constrained pseudorandom functions and their applications. IACR Cryptology ePrint Archive, 2013/352 (2013)

[BWZ14] Boneh, D., Wu, D.J., Zimmerman, J.: Immunizing multilinear maps against zeroizing attacks. Cryptology ePrint Archive, Report 2014/930 (2014). http://eprint.iacr.org/

[CHL+14] Cheon, J.H., Han, K., Lee, C., Ryu, H., Stehle, D.: Cryptanalysis of the multilinear map over the integers. Cryptology ePrint Archive, Report 2014/906 (2014). http://eprint.iacr.org/

[CLT14] Coron, J.-S., Lepoint, T., Tibouchi, M.: Cryptanalysis of two candidate fixes of multilinear maps over the integers. Cryptology ePrint Archive, Report 2014/975 (2014). http://eprint.iacr.org/

[CW14]   Chen, J., Wee, H.: Semi-adaptive attribute-based encryption and improved delegation for Boolean formula. In: Abdalla, M., De Prisco, R. (eds.) SCN 2014. LNCS, vol. 8642, pp. 277–297. Springer, Heidelberg (2014)

[GGH+13]   Garg, S., Gentry, C., Halevi, S., Raykova, M., Sahai, A., Waters, B.: Candidate indistinguishability obfuscation and functional encryption for all circuits. In: FOCS (2013)

[GGHW14]   Garg, S., Gentry, C., Halevi, S., Wichs, D.: On the implausibility of differing-inputs obfuscation and extractable witness encryption with auxiliary input. In: Garay, J.A., Gennaro, R. (eds.) CRYPTO 2014, Part I. LNCS, vol. 8616, pp. 518–535. Springer, Heidelberg (2014)

[GGHZ14a] Garg, S., Gentry, C., Halevi, S., Zhandry, M.: Fully secure attribute based encryption from multilinear maps. Cryptology ePrint Archive, Report 2014/622 (2014). http://eprint.iacr.org/

[GGHZ14b] Garg, S., Gentry, C., Halevi, S., Zhandry, M.: Fully secure functional encryption without obfuscation. Cryptology ePrint Archive, Report 2014/666 (2014). http://eprint.iacr.org/

[GGM84]   Goldreich, O., Goldwasser, S., Micali, S.: How to construct random functions (extended abstract). In: FOCS, pp. 464–479 (1984)

[GHMS14]   Gentry, C., Halevi, S., Maji, H.K., Sahai, A.: Zeroizing without zeroes: cryptanalyzing multilinear maps without encodings of zero. Cryptology ePrint Archive, Report 2014/929 (2014). http://eprint.iacr.org/

[GVW12]   Gorbunov, S., Vaikuntanathan, V., Wee, H.: Functional encryption with bounded collusions via multi-party computation. In: Safavi-Naini, R., Canetti, R. (eds.) CRYPTO 2012. LNCS, vol. 7417, pp. 162–179. Springer, Heidelberg (2012)

[KPTZ13]   Kiayias, A., Papadopoulos, S., Triandopoulos, N., Zacharias, T.: Delegatable pseudorandom functions and applications. IACR Cryptology ePrint Archive, 2013/379 (2013)

[KSW08]   Katz, J., Sahai, A., Waters, B.: Predicate encryption supporting disjunctions, polynomial equations, and inner products. In: Smart, N.P. (ed.) EUROCRYPT 2008. LNCS, vol. 4965, pp. 146–162. Springer, Heidelberg (2008)

[O'N10]   O'Neill, A.: Definitional issues in functional encryption. Cryptology ePrint Archive, Report 2010/556 (2010)

[SW05]   Sahai, A., Waters, B.: Fuzzy identity-based encryption. In: Cramer, R. (ed.) EUROCRYPT 2005. LNCS, vol. 3494, pp. 457–473. Springer, Heidelberg (2005)

[SW08]   Sahai, A., Waters, B.: Slides on functional encryption. PowerPoint presentation (2008). http://www.cs.utexas.edu/~bwaters/presentations/files/functional.ppt

[SW14]   Sahai, A., Waters, B.: How to use indistinguishability obfuscation: deniable encryption, and more. In: STOC, pp. 475–484 (2014)

[Wat14]   Waters, B.: A punctured programming approach to adaptively secure functional encryption. Cryptology ePrint Archive, Report 2014/588 (2014). http://eprint.iacr.org/

Multiparty Computation III

Secure Computation from Leaky Correlated Randomness

Divya Gupta[1], Yuval Ishai[2], Hemanta K. Maji[1,3]($\boxtimes$), and Amit Sahai[1]

[1] University of California, Los Angeles and Center for Encrypted Functionalities,
Los Angeles, USA
{divyag,sahai}@cs.ucla.edu
[2] Technion, Haifa, Israel
yuvali@cs.technion.ac.il
[3] Purdue University, West Lafayette, USA
hemanta.maji@gmail.com

Abstract. Correlated secret randomness is an essential resource for information-theoretic cryptography. In the context of secure two-party computation, the high level of efficiency achieved by information-theoretic protocols has motivated a paradigm of starting with correlated randomness, specifically random oblivious transfer (OT) correlations. This correlated randomness can be generated and stored during an offline preprocessing phase, long before the inputs are known. But what if some information about the correlated randomness is leaked to an adversary or to the other party? Can we still recover "fresh" correlated randomness after such leakage has occurred?

This question is a direct analog of the classical question of privacy amplification, which addresses the case of a *shared* random secret key, in the setting of *correlated* random secrets. Remarkably, despite decades of study of OT-based secure computation, very little is known about this question. In particular, the question of how much leakage is tolerable when recovering OT correlations has remained wide open. In our work, we resolve this question.

Prior to our work, the work of Ishai, Kushilevitz, Ostrovsky, and Sahai (FOCS 2009) obtained an initial feasibility result, tolerating only a tiny constant leakage rate. In our work, we show that starting with n

D. Gupta, H.K. Maji, A. Sahai–Research supported in part from a DARPA/ONR PROCEED award, a DARPA/ARL SAFEWARE award, NSF Frontier Award 1413955, NSF grants 1228984, 1136174, 1118096, and 1065276, a Xerox Faculty Research Award, a Google Faculty Research Award, an equipment grant from Intel, and an Okawa Foundation Research Grant. This material is based upon work supported by the Defense Advanced Research Projects Agency through the U.S. Office of Naval Research under Contract N00014-11-1-0389. The views expressed are those of the author and do not reflect the official policy or position of the Department of Defense, the National Science Foundation, or the U.S. Government.
Y. Ishai–Research supported by the European Union's Tenth Framework Programme (FP10/2010-2016) under grant agreement no. 259426 ERC-CaC, ISF grant 1709/14 and BSF grant 2012378.

© International Association for Cryptologic Research 2015
R. Gennaro and M. Robshaw (Eds.): CRYPTO 2015, Part II, LNCS 9216, pp. 701–720, 2015.
DOI: 10.1007/978-3-662-48000-7_34

random OT correlations, where each party holds $2n$ bits, up to $(1 - \epsilon)\frac{n}{2}$ bits of leakage are tolerable. This result is optimal, by known negative results on OT combiners.

We then ask the same question for other correlations: is there a correlation that is more leakage-resilient than OT correlations, and also supports secure computation? We answer in the affirmative, by showing that there exists a correlation that can tolerate up to $1/2 - \epsilon$ fractional leakage, for any $\epsilon > 0$ (compared to the optimal $1/4$ fractional leakage for OT correlations).

1   Introduction

Secure two-party computation [17,39] allows two mutually distrusting parties to perform secure computation using their private inputs without revealing any extra information to each other. It is known that even against semi-honest adversaries, i.e. adversaries who follow the prescribed protocol but are curious to find additional information, achieving information theoretic security in the plain model is impossible for most tasks [2,3,27,29,30]. For example, even the seemingly simple task of securely computing the AND of two bits is not possible. On the other hand, if suitable correlated randomness is provided as setup to the parties, then general secure two-party computation becomes possible [8,26,28]. A particularly useful type of correlated randomness is the random oblivious transfer (OT) correlation, where the sender gets two random bits (s_0, s_1) and the receiver gets (c, s_c), where c is a random bit.

One reason for the usefulness of OT correlations is the existence of highly efficient OT-based secure computation protocols both in theory and in practice. Indeed, protocols such as TinyOT [35] have popularized the approach of starting with random bit OT correlations for obtaining practically efficient secure computation protocols. Random OT correlations can be distributed or securely generated in an offline phase, long before the inputs are known, and later used in an online phase to perform a desired secure computation. But what if some information about the correlated randomness is leaked to an adversary? Can we still extract "fresh" correlated randomness after such leakage has occurred?

This question is a direct analog of the classical question of privacy amplification [4,5] that arose in the context of secure *communication*. Privacy amplification asks the following question: given shared secret randomness which has been partially leaked to an eavesdropper, can parties agree upon a common key which remains hidden from the eavesdropper? In our setting, we ask the same question for correlated randomness, which is useful for secure *computation*. Note, however, that participants in a privacy amplification protocol protect their secret only from an outsider. Instead, in our setting, each party must protect its secrets against the other party. For example, a fresh oblivious transfer correlation ensures that the bit c is hidden from the sender and the bit s_{1-c} is hidden from the receiver.

Quite surprisingly, very little is known about our question. This is in sharp contrast to the problem of privacy amplification, and despite decades of study of

OT-based secure computation. In particular, the question of how much leakage can be tolerated when recovering OT correlations has remained wide open. In our work, we resolve this question.

Prior to our work, Ishai, Kushilevitz, Ostrovsky and Sahai [24] studied this question, introducing the notion of correlation extractors. Concretely, they consider the setting of extracting fresh OT correlations from n independent copies of random OT correlations that have been subject to leakage. (One can either consider a deterministic leakage, captured by an arbitrary function f with t bits of output, or general probabilistic leakage, subject to the constraint that the secret has expected min-entropy of t bits conditioned on the leakage.) The main result of [24] is an interactive protocol for extracting OT correlations that remains secure even when some constant fraction of the $2n$ secret bits of each party can be leaked to the other party. Unfortunately, the concrete fractional leakage tolerated by this protocol is extremely small, approximately 10^{-7}. So, at best, this result serves as a proof of concept.

Since their work in 2009, there has not been any progress on this problem. In our work, we show that given n OT correlations as setup, one can tolerate $(1 - \varepsilon)n/2$ bits of leakage, for an arbitrarily small constant $\varepsilon > 0$, with negligible error. This leakage rate is near-optimal [25]. Moreover, in contrast to the previous protocol of [24], our protocol uses a minimal amount of interaction, requiring only two messages as opposed to the 4 messages that are inherently required by the technique from [24]. Finally, our protocol is conceptually simpler, completely avoiding the use of Algebraic-Geometric codes [16,19] needed in [24] and replacing them with simple families of binary linear codes.

Having settled the question of leakage-resilience for OT correlations, we then step back, and consider the question more broadly. While OT correlations are extremely useful and have a long history of applicability, perhaps there are other correlations that are better with respect to leakage-resilience, and still allow for secure computation. More precisely, we ask if there are correlations (X, Y) such that both parties receive $2n$ bits but where even after greater than $n/2$ bits of leakage, it is still possible to produce fresh secure OT correlations. We answer this question in the affirmative. We show that the so-called inner product correlation, where parties receive random binary vectors and additive shares of their inner product, can tolerate a significantly higher fractional leakage. Concretely, we show how to extract a fresh OT from such an inner product correlation while tolerating up to $1/2-\epsilon$ fractional leakage, for any $\epsilon > 0$ (compared to the optimal leakage rate of $1/4$ for OT correlations). This opens up a new set of questions to explore in future work (for more discussion refer to the full version of the paper [20]).

Finally, we note that while the primary focus of this work is on the information-theoretic setting for secure computation, the problem we consider is well motivated even in the setting of computational security. The reason is twofold. First, the fresh OTs produced by our extraction procedures can be used by computationally secure OT-based protocols such as those based on garbled circuits [39]. Second, these extraction procedures can be applied even when a computationally secure protocol is used for realizing the offline generation of correlated randomness. Suppose that a computationally secure two-party protocol Π is used for this purpose. If the leakage occurs after the execution of Π terminates (and the two parties erase everything but the output), then our protocols are guaranteed to produce clean OTs that can be consumed by subsequent (computational or information-theoretic) protocols. Moreover, if Π is so-called "leakage-tolerant" [6,7,21], then the same holds even if leakage can occur *during* the execution of Π. Such leakage-tolerant protocols can be constructed under standard intractability assumptions.

1.1   Our Contribution

In this section we give a more detailed overview of our main results.

Oblivious Transfer Correlation Extractor. We present our results in the terminology of "random oblivious transfer extractors." A random oblivious transfer (ROT) is a two party primitive where client S receives random bits (s_0, s_1); and the client R receives random bit c and s_c. Random oblivious transfer correlations can be easily converted into standard oblivious transfers, where a receiver R selects one of two bits held by a sender S. The latter can serve as a basis for general secure multiparty computation.

More concretely, Oblivious Transfer (OT) is a two-party functionality where client S (sender) has inputs (s_0, s_1), client R (receiver) has input c, and client R obtains output s_c. A Random OT correlation, referred to as ROT, provides (s_0, s_1) to one party and (c, s_c) to the other party, where s_0, s_1, c are uniform random bits. We work in the ROT^n-hybrid model, that is, there are n copies of ROT correlation provided to the two parties. A semi-honest client S can leak t_S bits from the correlation and a semi-honest client R can leak t_R bits from the correlation, where by default we define t bits of leakage as the output of an adversarially chosen function with t output bits. (However, all of our results extend to leakage measured in terms of average conditional min-entropy.) An $(n, t_S, t_R, \varepsilon)$ OT extractor is a two-party protocol between client S and client R such that it produces a $(1 - \varepsilon)$-secure copy of oblivious transfer despite prior leakage obtained by the clients.

Our first result shows the following feasibility result:

Theorem 1 (OT Extractor). *For any $n, t_S, t_R \in \mathbb{N}$, there exists a 2-message $(n, t_S, t_R, \varepsilon)$ OT Extractor protocol which produces a secure OT, such that $\varepsilon \leq 2^{-(g/4)+1}$ where $g := n - (t_S + t_R)$.*

Note that our result shows that if there is sufficient *gap* between n and the total leakage $(t_S + t_R)$, then we can securely extract one oblivious transfer. Further, the simulation error decreases exponentially in the gap. For example, $t_S = t_R = 0.49n$ leakage tolerant extractors exist by our result. Contrast this to the result of [24] who can tolerate leakage up to cn bits of leakage where c is a minuscule small constant. Thus, ours is the first feasibility result in the regime of high leakage tolerance. Moreover, our leakage resilience is (near) optimal due to the negative result of [25]. The negative result states that there does not exist any OT combiner (let alone an extractor) which can tolerate up to $n/2 - O(1)$ bits of leakage. Our protocol also improves upon the round complexity of [24] from 4 messages to 2 messages, which are clearly necessary.

We show that if the gap $g = n - (t_S + t_R)$ is at least cn, for some constant $c \in (0, 1)$, then we can trade off simulation error and increase the production rate of our extractor. That is, in the leaky ROT^n hybrid, we can produce large number of secure independent copies of oblivious transfer. Our result is summarized in the following theorem:

Theorem 2 (High Production). *For every $n, t_S, t_R \in \mathbb{N}$, such that $g = n - (t_S + t_R) = \Theta(n)$, and $\rho = \omega(\log n)$, there exists a 2-message (n, t_S, t_R) OT Extractor with production $p = n/\rho$ and $\varepsilon \leq \mathsf{negl}(n)$.*

Intuitively, this theorem states that if the gap is linear in n then we can obtain slightly sub-linear number of secure oblivious transfers while incurring negligible security error. Although our production rate is sub-constant, we show that it is possible to extract large number of secure oblivious transfers even if parties are permitted to perform $t_S = t_R = 0.49n$ bits of leakage. Contrasting this with the result of [24], for practical and typical n the number of oblivious transfers produced in our scheme surpasses the number of oblivious transfers produced in their protocol. Because their production rate, although linear, is a very small constant; even a generous estimate of the rate of production puts it below $1.2 \cdot 10^{-7}$. The hidden constant in our asymptotic production rate is small, say upper bounded by 10^{-1}. So, in concrete terms, our production rate is $\sim (g/n)/10 \log^2 n$, which is higher than the rate achieved by [24] for a practical range of parameters (we use $\rho - \log^2 n$ to derive this bound). An obvious open problem is to explore whether our approach can be extended to achieve the ideal goal of producing a linear number of secure OTs even if the gap is an arbitrarily small linear function of n.

Overall, our construction significantly simplifies the prior construction of [24] at a conceptual level by forgoing usage of Algebraic Geometric [16,19] codes and instead relying on binary linear codes generated by generator matrices whose parity check matrices are random Toeplitz matrices.

Unlike [24], we do not achieve constant (multiplicative) communication overhead per instance of oblivious transfer produced. Our communication complexity overhead per oblivious transfer produced is linear in n. We also do not consider the problem of error tolerance, another important area of exploration in future work.

Restriction to Combiners. Combiners are special types of extractors where parties's leakage functions are restricted. Parties are allowed to only indicate $T \subseteq [n]$ as their leakage function. The client S can send $|T| \leq t_S$ and client R can send $|T| \leq t_R$. The leakage provided is (s_0, s_1, c, s_c) of all ROT correlations indexed by T. Note that the actual information learned by the clients is one-bit per index (because all bits can be reconstructed from input and one bit leakage).

We show that our construction yields slightly better simulation error than the general analysis of Theorem 1.

Theorem 3 (OT Combiner). *For any $n, t_S, t_R \in \mathbb{N}$, there exists a 2-message $(n, t_S, t_R, \varepsilon)$ OT-Combiner which produces one secure OT using $O(n)$ bits of communication, where $\varepsilon \leq 2^{-g/2}$ and $g := n - (t_S + t_R)$.*

Note that the construction presented in [25] achieves similar bounds but the communication complexity in their construction is quadratic in n; while ours is linear in n. We emphasize that the higher production result of Theorem 2 also applies to the setting of combiners.

Large Correlations. We show that, in fact, there are correlations that can tolerate a fractional leakage close to $1/2$.

Theorem 4 (High Tolerance). *For any $s, t \in \mathbb{N}$, there exists a correlation (X, Y) over a pair of $(s + 1)$-bit strings such that, even after any party leaks t bits on the correlation (X, Y), they can securely realize OT using 2-message communication with simulation error $\varepsilon \leq 2^{-(g/2+1)}$, where $g := s/2 - t$.*

The correlation (X, Y) used to prove the above theorem is the so-called *inner-product correlation*, where each party receives a random s-bit vector and the mod-2 inner product of the two vectors is secret-shared between the parties. Moreover, it is not hard to show that our protocol cannot tolerate leakage rate bigger than $1/2$. We leave open the question whether the $1/2$ leakage rate is optimal for arbitrary correlations.

1.2   Prior Related Works

Most relevant work to our work is the work of Ishai, Kushilevitz, Ostrovsky, and Sahai [24], where the notion of correlation extractors was proposed. They showed that if the parties are allowed to leak a small linear amount of leakage, then a small linear number of correlations can be extracted. Both the leakage and production rates are a minuscule fraction of the initial number of correlations.

A closely related concept is the notion of OT combiners, which are a restricted variant of OT extractors in which leakage is limited to local information about individual OT correlations and there is no global leakage. The study of OT combiners was initiated by Harnik et al. [23]. Since then, there has been work on several variants and extensions of OT combiners [22,26,32,33,37].

Recently, [25] constructed OT combiners with nearly optimal leakage parameters. Our protocols were inspired by the OT combiners from [25], but the results we achieve are stronger in several imporant ways. First, whereas [25] only considers t *physical* bits of leakage, we tolerate a arbitrary bits of leakage (similarly to [24], though with a much better leakage rate). Second, even in the case of physical leakage, our solutions improve over [25] by reducing the communication and randomness complexity from quadratic to linear. Finally, our protocols can be used to produce a near-linear number of OTs without significantly compromising the leakage rate, whereas [25] only considers the case of producing a single OT.

Another related work is that of Dziembowski and Faust [14] which (similarly to our Theorem 4) obtains some form of leakage-resilient secure computation from the inner product correlation. However, the construction from [14] requires multiple independent instances of an inner product correlation even for producing just a single OT, and moreover the model considered in [14] assumes that the leakage applies *individually* to each instance. Even if the analysis of [14] could somehow be strengthened to tolerate some amount of global leakage, the tolerable leakage rate must inevitably be small (since even with a leakage rate of $1/4$, one can entirely compromise one of the inner product instances in [14]). Thus, the approach of [14] does not seem relevant to our goal of maximizing the leakage rate.

1.3   Technical Overview

We provide a short overview of our construction which proves Theorem 1. Our construction is inspired by the Massey secret sharing scheme [31]. Our construction is closely related to the constructions of [24,25]. The central novelty in our construction approach is that we choose a different class of matrices (thus, reducing communication complexity of our algorithm), but the primary technical contribution of our work is our new analysis in the context of leakage. We consider general leakage (unlike the setting of [25] which considers physical bits of leakage) and, hence, lose a small quadratic factor in simulation error. But the same construction when used in the setting of combiners yields identical simulation error as [25].

For $i \in [n]$, suppose the client S receives random pair of bits (a_i, b_i) and client R receives (x_i, z_i), such that x_i is a random bit and $z_i = a_i x_i \oplus b_i$, from the setup. Client S picks a random codeword $(u_0, u_1, \ldots, u_n)$ in a binary linear code $\mathcal{C}$ of length $(n+1)$. Client R picks a random codeword $(r_0, r_1, \ldots, r_n)$ in the binary linear code $\mathcal{C}^\perp$ of length $(n+1)$. Note that the set of all component-wise product of such codewords has non-trivial distance. Hence, they can correct one erasure. In particular, $u_0 r_0 = \sum_{i \in [n]} u_i r_i$. Hence, the clients need not explicitly compute $u_0 r_0$; but, instead, it suffices to compute $u_i r_i$ for all $i \in [n]$ and recovering one erasure thereafter.

For this section, we shall only consider privacy of client R against a semi-honest client S. Consider the following protocol: For each $i \in [n]$,

1. Client R sends $m_i = x_i \oplus r_i$.
2. Client S sends $\alpha_i = a_i \oplus u_i$. Client S sends $\beta_i = a_i m_i \oplus b_i$.

Note that client R can compute $\beta_i \oplus \alpha_i r_i \oplus z_i = u_i r_i$. To argue the privacy of client R, we need to show that r_0 remains hidden from the view of client S. Let H be the generator matrix of $\mathcal{C}^\perp$ and H is interpreted as $[H_0 | H']$, where H_0 is the first column of H and H' is the remaining n columns. Note that the ability of client S to predict r_0 can be abstracted out as follows: For λ uniform random vector, given $(\lambda H' \oplus x_{[n]}, H)$, client S needs to predict λH_0.

Note that since client S is permitted to perform t_S bits of leakage on $x_{[n]}$, we have the guarantee that $x_{[n]}$ has high min-entropy on average. Now, the experiment is reminiscent of min-entropy extraction from high min-entropy sources via masking with small bias distributions. But, the uniform distribution over codes of a fixed binary linear codespace $\mathcal{C}^\perp$ is not a small-bias source (projection on every dual codewords has full bias). So, we consider a set of codes $(\mathcal{C}_I, \mathcal{C}_I^\perp)$, where I is the index, such that on average these codewords have small bias. Such a distribution suffices in our setting, because leakage is performed in an offline phase and the random linear code or $\mathcal{C}_I$ is chosen only in the online phase. The class of matrices chosen are binary matrices in systematic form whose parity check matrices are uniformly chosen Toeplitz matrices. This, intuitively, is the basic argument which all our proofs reduce to.

Theorem 2 is obtained by sampling $\{S_1, \ldots, S_m\}$ such that they are all disjoint and each S_i indexes a set of servers. One OT is extracted by applying Theorem 1 on each index set S_i.

2 Preliminaries

Notations. We represent random variables by capital letters, for example X, and the values they take by small letters, for example $\Pr[X = x]$. The set $\{1, \ldots, n\}$ is represented by $[n]$, for $n \in \mathbb{N}$. Given a vector $v = (v_1, \ldots, v_n)$ and $T = \{i_1, \ldots, i_{|T|}\} \subseteq [n]$, we represent $(v_{i_1}, \ldots, v_{i_{|T|}})$ by v_T. Similarly, given a $k \times n$ matrix G, we represent by G_T the sub-matrix of G formed by columns indexed by T. For brevity, we use G_i instead of $G_{\{i\}}$, where $i \in [n]$.

Probability Basics. The support of a probability distribution X, represented as $\mathsf{Supp}(X)$ is the set of elements in the sample space which are assigned non-zero probability by X. A uniform distribution over a set S is represented by $\mathbf{U}_S$. A probability distribution X over a universe U is a flat source if there exists a constant $c \in (0, 1]$ such that $\Pr[X = x]$ is either 0 or c, for all $x \in U$. Further, we say that X is a flat-source of size $1/c$. Given a joint distribution (X, Y) over sample space $U \times V$, the conditional distribution $(X|y)$ represents the distribution over sample space U such that the probability of $x \in U$ is $\Pr[X = x | Y = y]$.

The statistical distance between two distributions X and Y over a finite sample space U is defined to be: $\frac{1}{2} \sum_{u \in U} |\Pr[X = u] - \Pr[Y = u]|$.

Entropy Definitions. For a probability distribution X over a sample space U, we define entropy of x as $H_X(x) := -\lg \Pr[X = x]$, for every $x \in U$. The entropy of X, represented by $\mathbf{H}(X)$, is defined to be $\mathbb{E}[H_X(x)]$. The min-entropy of X, represented by $\mathbf{H}_\infty(X)$, is defined to be $\min_{x \in \mathsf{Supp}(X)} H_X(x)$. If $\mathbf{H}_\infty(X) \geq n$, then X can be written as convex linear combination of distributions, each of which are flat sources of size $\geq 2^n$. The average min-entropy [10], represented by $\widetilde{\mathbf{H}}_\infty(X|Y)$, is defined to be $-\lg \mathbb{E}_{y \sim Y} \left[2^{-\mathbf{H}_\infty(X|y)}\right]$. Following lemma is useful for lower bounding average min-entropy after leakage on a high min-entropy source.

Lemma 1 (Chain Rule [10]). *If* $\mathbf{H}_\infty(X) \geq n$ *and* L *be arbitrary* ℓ-*bit leakage on* X, *then* $\widetilde{\mathbf{H}}_\infty(X|L) \geq n - \ell$.

2.1 Elementary Fourier Analysis

We define character $\chi_S(x) = (-1)^{\sum_{i \in S} x_i}$, where $S \subseteq [n]$ and $x \in \{0,1\}^n$. The inner product of two functions $f \colon \{0,1\}^n \to \mathbb{R}$ and $g \colon \{0,1\}^n \to \mathbb{R}$ is defined by $\mathbb{E}_{x \xleftarrow{\$} \{0,1\}^n} [f(x)g(x)]$. Given a probability distribution M over the sample space $\{0,1\}^n$, the function $f = M$ represents the function $f(x) = \Pr[M = x]$.

Definition 1 (Bias of a Distribution). *Let* $f : \{0,1\}^n \to \mathbb{R}$ *be a probability function. The bias of* f *with respect to subset* $S \subseteq [n]$ *is defined to be:*

$$\mathsf{Bias}_S(f) := \left| \Pr_{x \sim f}[\chi_S(x) = 1] - \Pr_{x \sim f}[\chi_S(x) = -1] \right|$$

Definition 2 (Small-bias Distribution Family [11]). *Let* $\mathcal{F} = \{F_1, \ldots, F_k\}$ *be a family of distributions over sample space* $\{0,1\}^n$ *such that for every* $\emptyset \neq S \subseteq [n]$, *we have:*

$$\mathbb{E}_{i \xleftarrow{\$} [k]} \left[\mathsf{Bias}_S(F_i)^2 \right] \leq \delta^2$$

Then the distribution family $\mathcal{F}$ *is called an* δ^2-*biased family.*

Lemma 2 (Min-entropy Extraction [1,11,18,34]). *Let* $\mathcal{F} - \{F_1, \ldots, F_\mu\}$ *be* δ^2-*biased family of distributions over the sample space* $\{0,1\}^n$. *Let* (M, L) *be a joint distribution such that the marginal distribution* M *is over* $\{0,1\}^n$ *and* $\widetilde{\mathbf{II}}_\infty(M|L) \geq m$. *Then, the following holds:*

$$\mathsf{SD}\left((F_I \oplus M, L, I), (U_{\{0,1\}^n}, L, I) \right) \leq \frac{\delta}{2} \left(\frac{2^n}{2^m} \right)^{1/2},$$

where I *is a uniform distribution over* $[\mu]$.

2.2 Functionalities

We introduce some useful functionalities in this section.

Oblivious Transfer. A *2-choose-1 bit Oblivious Transfer* (referred to as OT) is a two party functionality which takes input $(s_0 s_1) \in \{0,1\}^2$ from the *sender* and input $c \in \{0,1\}$ from the *receiver* and outputs s_c to the receiver.

Random Oblivious Transfer. A *random 2-choose-1 bit Oblivious Transfer* (referred to as ROT) is an input-less two party functionality which samples uniformly random bits s_0, s_1, c and outputs (s_0, s_1) to the sender and (c, s_c) to the receiver. The joint distribution of sender-receiver outputs is called an ROT-correlation.

Oblivious Linear-Function Evaluation. Let $(\mathbb{F}, +, \cdot)$ be an arbitrary field. An *Oblivious Linear-function Evaluation* over $\mathbb{F}$ is a two party functionality which takes inputs $(u, v) \in \mathbb{F}^2$ from the sender and $x \in \mathbb{F}$ from the receiver and outputs $u \cdot x + v$ to the receiver. This functionality is referred to as $\mathrm{OLE}(\mathbb{F})$. A random oblivious linear-function evaluation (ROLE) can be defined analogous to ROT.

The special case when $\mathbb{F} = \mathrm{GF}(2)$, is simply referred to as OLE and is equivalent to OT.

Random Inner Product Correlation. This is an input-less two party functionality which samples $x_{[n]}, y_{[n]} \xleftarrow{\$} \{0,1\}^n$, $a \xleftarrow{\$} \{0,1\}$ and $b = a + \langle x_{[n]}, y_{[n]} \rangle$. It outputs $(x_{[n]}, a)$ to party A and $(y_{[n]}, b)$ to party B. Note that for $n = 1$, this is equivalent to random oblivious transfer correlation and oblivious linear function evaluation.

2.3 Combiners and Extractors

In this section, we define oblivious transfer combiners and extractors.

Definition 3 $((n, p, t_S, t_R, \varepsilon)$ (Single Use) OT-Combiner). *An $(n, p, t_S, t_R, \varepsilon)$ (single use) OT-Combiner is an interactive protocol in the clients-servers setting. There are two clients S and R; and n servers. Each server implements one instance of oblivious transfer on inputs from S and R. We consider a semi-honest adversary who can either corrupt the client S and t_S servers or client R and t_R servers. The protocol implements p independent copies of secure oblivious transfer instances with correctness and simulation error at most ε.*

The correctness conditions for the protocol says that the receiver's output is correct in all p-instances of OT with probability at least $(1 - \varepsilon)$.

The privacy requirement says that the adversary should not learn more than it should. Let $(s_0^{(i)}, s_1^{(i)})$ and $c^{(i)}$ be the inputs of the sender and the receiver, respectively, in i^{th} copy of OT produced. Then a corrupt sender (resp., corrupt receiver) cannot output $c^{(i)}$ (resp., $s_{1-c}^{(i)}$) with probability more than $\frac{1}{2} + \varepsilon$ for any instance of OT produced.

Leakage Model and Correlation Extractors. Here we begin by describing our leakage model for ROLE correlations formally followed by defining correlation extractors for OLE. Recall that OT and OLE are just local renaming of each other. Our leakage model is as follows:

1. **n-Random OLE Correlation Generation phase:** For $i \in [n]$, the sender S gets random $(a_i, b_i) \in \{0,1\}^2$ and receiver R gets (x_i, z_i), where $x_i \in \{0,1\}$ is chosen uniformly at random and $z_i = a_i x_i + b_i$.
2. **Corruption and leakage phase.** A semi-honest adversary corrupts either the sender and sends a leakage function $L : \{0,1\}^n \to \{0,1\}^{t_S}$. It receives $L(\{x_i\}_{i \in [n]})$. Or, it corrupts the receiver and send a leakage function $L : \{0,1\}^n \to \{0,1\}^{t_R}$. It receives $L(\{a_i\}_{i \in [n]})$.

   Note that without loss of generality any leakage on sender (resp., receiver) can be seen as a leakage on $\{a_i\}$ (resp., $\{x_i\}$).

Let (X, Y) be the random OT correlation. We denote (t_S, t_R)-leaky version of $(X, Y)^n$ described above as $((X, Y)^n)^{[t_S, t_R]}$.

Definition 4 $((n, p, t_S, t_R, \varepsilon)$ **OT-Extractor).** *An $(n, p, t_S, t_R, \varepsilon)$ OT-Extractor is an interactive protocol between two parties S and R in the $((X, Y)^n)^{[t_S, t_R]}$ hybrid described above. The protocol implements p independent copies of secure oblivious transfer instances with simulation error ε.*

The correctness and privacy requirements are same as those defined above for $(n, p, t_S, t_R, \varepsilon)$ (Single Use) OT-Combiner.

Note that in our setting, in $(X, Y)^n$ hybrid, parties only get one sample from this correlation; unlike the typical setting where parties can invoke the trusted functionality of the hybrid multiple times. The maximum *fractional leakage resilience* is defined by the ordered tuple $(t_S/n, t_R/n)$; and the *production rate* is defined by p/n.

Remark: An $(n, p, t_S, t_R, \varepsilon)$ OT extractor is also an $(n, p, t_S, t_R, \varepsilon)$ OT combiner.

Noisy Leakage Model. The leakage model described above is referred to as the "bounded leakage" model since we restrict the number of bits output by the leakage function. But this model is sometimes too restrictive and does not capture many side channel attacks, which are the main cause of leakage in real world applications. A more realistic assumption one can make is to assume that leakages are sufficiently noisy. It is observed via experiments that the real-world physical leakages are inherently noisy. There have been many works trying to model noisy leakage and present solutions in this setting [9,12,13,15,36]. At a high level the noisy feature of a leakage function f is captured by assuming that an observation of $f(x)$ only implies a bounded bias in the probability distribution of x. More formally, f is said to be δ-*noisy* if

$$\delta = \mathrm{SD}\left((X), (X|f(X))\right).$$

Note that if $\mathbf{H}_\infty(X) \geq n$ then for any $k < n$, we can choose appropriate δ, such that $\widetilde{\mathbf{H}}_\infty(X|f(X)) \geq k$, where f is a δ-noisy channel.

We emphasize that all our protocols only rely on the fact that the initial correlation given to any party has high average min-entropy ($\widetilde{\mathbf{H}}_\infty$) after the leakage. Hence, all our protocols directly work even in the general setting of noisy leakage.

2.4   Distribution over Matrices

An $k \times n$ matrix M with $\{0, 1\}$ entries is in systematic form if $M = [I_{k \times k} \| P]$, where $I_{k \times k}$ is the identity matrix of dimension k and P is the parity check matrix of dimension $k \times (n - k)$. The matrix P is a Toeplitz matrix if $P_{i,j} = P_{i-1, j-1}$, for all $i \in (1, k]$ and $j \in (1, n - k]$. So, a Toeplitz matrix is uniquely defined by its first row and first column. We shall consider uniform distribution over $k \times n$ binary matrices in systematic form such that their parity check matrices are uniformly chosen Toeplitz matrices. A salient feature of family of such matrices is proved in Lemma 3.

Let $\mathbb{T}_{(k,n)}$ is a uniform distribution over matrices M of the following form. Let $M \equiv [I_{k\times k}|P_{k\times(n-k)}]$, where P is a binary Toeplitz matrix of dimension $k \times (n-k)$.

Define $\mathbb{T}_{\perp,(k,n)}$ is a uniform distribution over matrices M of the following form. Let $M \equiv [P_{k\times(n-k)}|I_{k\times k}]$, where P is a binary Toeplitz matrix of dimension $k \times (n-k)$.

Note that there exists an bijection between the matrices in $\mathbb{T}_{(k,n)}$ and $\mathbb{T}_{\perp,(n-k,n)}$ established by the function which maps dual matrices to each other.

For a given $G \in \mathbb{T}_{(k,n)}$, the distribution F_G corresponds to a uniform distribution over the codewords generated by G. We have the following lemma, which will be used to prove the main unpredictability lemma in the next section.

Lemma 3. *For the distribution of matrices $\mathbb{T}_{(k,n)}$, the following holds. For any $\emptyset \neq T \subseteq [n]$,*

$$\mathbb{E}_{G \xleftarrow{s} \mathbb{T}_{(k,n)}} \left[\mathsf{Bias}_T(F_G)^2\right] \leq 2^{-k}$$

Proof. Since F_G is a uniform distribution of codewords over a linear code, for any G, either $\mathsf{Bias}_T(F_G)^2$ is either 0 or 1. Moreover, $\mathsf{Bias}_T(F_G)^2 = 1$ if and only if $\sum_{i \in T} G_i = 0^k$. Hence, it suffices to show the following: For any fixed column $c \in \{0,1\}^k$ and non-empty set $T \subseteq [n]$, $\Pr[\sum_{i \in T} G_i = c] \leq 2^{-k}$. We prove this using a sequence of observations.

Note that: $G_i = c$, for $i > k$, happens with probability 2^{-k}.

Next, we claim that: $G_i + G_j = c$, for $i > j > k$, happens with probability 2^{-k}. This is so because the probability that the $G_{i,k} + G_{j,k} = c_k$ happens with probability $1/2$. Fixing the values of $G_{i,k}$ and $G_{j,k}$, the probability that we have $G_{i,k-1} + G_{j,k-1} = c_{k-1}$ is $1/2$; because the random variable $G_{j,k-1}$ is not fixed (columns $\{k+1, \ldots, n\}$ form a Toeplitz matrix). Extending this argument, we get for any $T' \subseteq \{k+1, \ldots, n\}$, $\Pr[\sum_{i \in T'} G_i = c] \leq 2^{-k}$.

To prove full claim, note that $\Pr[\sum_{i \in T} G_i = c] = \Pr[\sum_{i \in T:i>k} G_i = c + \sum_{i \in T':i\leq k} G_i] \leq 2^{-k}$ using the above conclusion.

3 Unpredictability Lemma

In this section we present the main unpredictability lemma.

Lemma 4 (Unpredictability Lemma). *Let $\mathbb{G} \in \{\mathbb{T}_{(k,n+1)}, \mathbb{T}_{\perp,(k,n+1)}\}$. Consider the following game between a honest challenger $\mathcal{H}$ and an adversary $\mathcal{A}$:*

---
1. $\mathcal{H}$ samples $m_{[n]} \sim U_{\{0,1\}^n}$.
2. $\mathcal{A}$ sends a leakage function $\mathcal{L}: \{0,1\}^n \to \{0,1\}^t$.
3. $\mathcal{H}$ sends $\mathcal{L}(m_{[n]})$ to $\mathcal{A}$. $\mathcal{H}$ samples $x_{[k]} \sim U_{\{0,1\}^k}$, $G \sim \mathbb{G}$; and computes $y_{\{0\}\cup[n]} = x \cdot G \oplus (0, m_{[n]})$. $\mathcal{H}$ sends $(y_{[n]}, G)$ to $\mathcal{A}$.
4. $\mathcal{A}$ outputs a bit $\tilde{y}$.

The adversary $\mathcal{A}$ wins the game if $y_0 = \tilde{y}$. For any $\mathcal{A}$, the advantage of the adversary, i.e. $\mathsf{Adv}(\mathcal{A}) = \Pr(y_0 = \tilde{y}) - 1/2 \leq \frac{1}{2}\sqrt{\frac{2}{2^{k-t}}}$.

Proof. Let $\mathbb{G}$ be the distribution $\mathbb{T}_{(k,n+1)}$. The proof for the other case will work similarly.

Given a $G \in \mathbb{G}$, the distribution F_G corresponds to a uniform distribution over the codewords generated by G. Note that over choice of G, they form a $\delta^2 = 2^{-k}$ biased family of distributions (by Lemma 3).

By Lemma 1, $\widetilde{\mathbf{H}}_\infty(M_{[n]}|\mathcal{L}(M_{[n]})) \geq n - t$. Let $M = (0, M_{[n]})$, then putting these in Lemma 2, we get

$$\mathsf{SD}\left((F_G \oplus M, L, G), \left(U_{\{0,1\}^{n+1}}, L, G\right)\right) \leq \frac{1}{2}\sqrt{\frac{2^{n+1}}{2^{k+n-t}}}$$

The lemma follows by noting that $\mathsf{Adv}(\mathcal{A}) \leq \mathsf{SD}\left((F_G \oplus M, L, G), \left(U_{\{0,1\}^{n+1}}, L, G\right)\right)$.

All our security proofs will directly reduce to this unpredictability lemma, i.e. Lemma 4.

4 Oblivious Transfer Extractor

4.1 Extracting One Oblivious Transfer

In this section, we shall prove Theorem 1 by presenting our $(n, t_S, t_R, \varepsilon)$ OT extractor which extracts one copy of secure OT. For ease of presentation, we provide our construction in the random oblivious linear evaluation (ROLE) correlation hybrid; and also produce one secure copy of oblivious linear evaluation. Recall that a ROLE correlation provides $(a, b) \xleftarrow{\$} \{0,1\}^2$ to the sender and $(x, z = ax \oplus b)$, where $x \xleftarrow{\$} \{0,1\}$, to the receiver. The security requirement insists that the sender cannot predict x and the receiver cannot predict a. Note that $(s_0 \oplus s_1)c \oplus s_0$ is identical to oblivious transfer. So, oblivious transfer and OLE are equivalent to each other; consequently, it suffices to construct a OLE extractor in ROLE^n hybrid.

The construction provided here is similar to the construction provided in [25]. But we deal with general leakage, instead of restricted leakage of physical bits in the combiner setting, using more sophisticated analysis tools. We also achieve lower communication complexity. In particular, we improve the communication complexity from $\Theta(n^2)$ in [25] to $\Theta(n)$ in the current work. When analyzed appropriately for the combiner setting, our current protocol achieves identical simulation error as in that paper (but reduces the communication complexity to linear from quadratic).

Note that after the correlation generation step, the protocol is only two rounds, i.e. client R sends one message (by combining steps 1 and 2.c) and client S replies with one message (step 2.d).

Extract-One (n, t_S, t_R):

Define $g := n - (t_S + t_R)$.

Private Inputs: The clients S and R have private inputs $(s_0, s_1) \in \{0,1\}^2$ and $c \in \{0,1\}$, respectively.

Hybrid (Random Correlations): For $i \in [n]$, client S gets random $(a_i, b_i) \in \{0,1\}^2$ and client R gets (x_i, z_i), such that $x_i \in \{0,1\}$ is chosen uniformly at random and $z_i = a_i x_i \oplus b_i$.

1. *Random Code Generation.* Client R picks a binary matrix $G = [I_{k \times k} \| P_{k \times (n+1-k)}]$ of dimension $k \times (n+1)$, where $k = \lceil t_R + g/2 \rceil$ and $P_{k \times (n+1-k)}$ is a uniformly random Toeplitz matrix. Let $\mathcal{C}$ be the code generated by the generator matrix G; and H be a generator matrix for the dual code $\mathcal{C}^\perp$. If the first column of H is all-zero column then **abort**; otherwise continue.

2. *Random OLE Extraction.*
   (a) Client S picks a random $(u_0, \ldots, u_n) \in \mathcal{C}$. Let $\mathcal{C}_{\text{parity}} \subseteq \{0,1\}^{n+1}$ be the (linear) code consisting of every length $(n+1)$ string of even parity. Client S picks a random $(v_0, \ldots, v_n) \in \mathcal{C}_{\text{parity}}$.
   (b) Client R picks a random $(r_0, \ldots, r_n) \in \mathcal{C}^\perp$.
   (c) For each $i \in [n]$, client R sets $m_i = x_i \oplus r_i$. Client R also sets $m = r_0 \oplus c$. Client R sends $(\{m_i\}_{i \in [n]}, m)$ to client S.
   (d) For each $i \in [n]$, client S sets $\alpha_i = a_i \oplus u_i$ and $\beta_i = a_i m_i \oplus b_i \oplus v_i$. Client S also sets $\alpha = u_0 \oplus s_0$ and $\beta = u_0 m \oplus v_0 \oplus s_1$. Client S sends $(\{(\alpha_i, \beta_i)\}_{i \in [n]}, \alpha, \beta)$ to client R.
   (e) Client R computes $t_i = \beta_i \oplus \alpha_i r_i \oplus z_i$ and $z = \oplus_{i \in [n]} t_i$. Finally, client R outputs $y = \beta \oplus \alpha c \oplus z$.

Fig. 1. Round optimal correlation extractor protocol which extracts one copy of Oblivious Linear Function Evaluation from n copies of Random Oblivious Linear Functions Evaluations.

No Corruption Case. We will first prove the correctness of the protocol presented in Fig. 1 for the case when all clients and servers are honest and there is no leakage.

The construction does not output **abort** with probability $1 - 2^{-(n+1-k)}$, because the algorithm aborts if and only if the first row of the parity check matrix of G is all 0s. Conditioned on not aborting, we show that the protocol is perfectly correct. Following lemma proves correctness.

Lemma 5. *In the protocol in Fig. 1 the client R outputs $y = s_0 c \oplus s_1$.*

Proof. We first show that $t_i = u_i r_i \oplus v_i$.

$$t_i = \beta_i \oplus \alpha_i r_i \oplus z_i = (a_i m_i \oplus b_i \oplus v_i) \oplus (a_i r_i \oplus u_i r_i) \oplus z_i$$
$$= a_i x_i \oplus a_i r_i \oplus b_i \oplus v_i \oplus a_i r_i \oplus u_i r_i \oplus a_i x_i \oplus b_i = u_i r_i \oplus v_i$$

This shows that $z = \oplus_{i \in [n]} t_i = u_0 r_0 \oplus v_0$. This follows from $\oplus_{i=0}^{n} u_i \cdot r_i = 0$ and $\oplus_{i=0}^{n} v_i = 0$. Now for y we have the following:

$$y = \beta \oplus \alpha c + z = (u_0 m \oplus v_0 \oplus s_1) \oplus (u_0 c \oplus s_0 c) \oplus z$$
$$= u_0 r_0 \oplus u_0 c \oplus v_0 \oplus s_1 \oplus u_0 c \oplus s_0 c \oplus u_0 r_0 \oplus v_0 = s_0 c \oplus s_1.$$

Sender Privacy and Receiver Privacy. In order to give a modular analysis, we consider a simpler protocol of 4-rounds which is equivalent to the protocol presented in Fig. 1. In the simpler protocol, the first two rounds correspond to ROLE extraction, where the receiver sends the messages $\{m_i\}_{i \in [n]}$ and receives $\{(\alpha_i, \beta_i)\}_{i \in [n]}$ and computes ROLE $z = u_0 r_0 \oplus v_0$. In the following, we will refer to this as *ROLE extraction phase*. In the next two rounds, it uses this ROLE to compute the OLE on inputs s_0, s_1, c as follows: Receiver sends message m and gets back α, β and computes y. Note that since we only consider semi-honest adversaries and leakage only occurs before the start of the protocol, these two protocols are equivalent in correctness and security guarantees.

Below, in order to prove the sender and receiver privacy we analyze this protocol. For security of both sides, it is sufficient to prove that extracted ROLE is secure in first phase.

Receiver Privacy. In order to prove receiver privacy, we need to show that the choice bit c is hidden from the semi-honest sender who can obtain t_S bits of leakage. We note that it suffices to show that at the end of the ROLE extraction phase (described above), the choice bit r_0 is hidden.

Let L denote the random variable for leakage obtained by the semi-honest sender. We will denote the random variable for the choice bit vector $x_{[n]}$ for the receiver in the correlation generation phase by $X_{[n]}$. Note that $X_{[n]}$ is identical to uniform distribution over $\{0,1\}^n$. Note that L has at most t_S bits of leakage on $X_{[n]}$.

The view of client S at the end of the random correlation extraction phase is:

$$\vartheta = (a_{[n]}, b_{[n]}, G, (u_0, \ldots, u_n), (v_0, \ldots, v_n), m_{[n]}, L = \ell)$$

Below we show that for any semi-honest client S, we have $\Pr(S(\vartheta) = r_0)$ is at most $1/2 + 2^{-g/4-1}$.

Note that $\Pr(S(\vartheta) = r_0) = \Pr(S(H, m_{[n]}, L) = r_0)$, where H is the generator matrix for $\mathcal{C}^\perp$. Recall that $H \in \mathbb{T}_{\perp,(n+1-k,n+1)}$, where $k = t_R + g/2$. In Fig. 1, the client R picks a random codeword $(r_0, \ldots, r_n) \in \mathcal{C}^\perp$. Alternatively, this can be done by picking $w \xleftarrow{\$} \{0,1\}^{n+1-k}$ and $(r_0, \ldots, r_n) = w \cdot H$, where H is the generator matrix for $\mathcal{C}^\perp$. Note that $m_{[n]} = (w \cdot H)_{[n]} \oplus x_{[n]}$ and $r_0 = \langle H_0, w \rangle$.

Since, the sender can leak t_S bits on $x_{[n]}$, we have: $\widetilde{\mathbf{H}}_\infty(X_{[n]} | L) \geq m = (n - t_S)$. By Lemma 4, the advantage of predicting $\langle H_0, w \rangle$ is at most: $2^{-g/4-1}$.

Sender Privacy. In order to prove sender privacy for Fig. 1, we need to show that the bit s_0 is hidden from the receiver after the protocol. Note that it suffices to show that at the end of the ROLE extraction phase (for the simpler protocol described above) bit u_0 is hidden.

Let L denote the random variable for leakage on vector $a_{[n]}$ obtained by the semi-honest adversary who corrupts the receiver after the random correlation generation phase. We will denote the random variable for the bit vector $a_{[n]}$ for the sender in the correlation generation phase by $A_{[n]}$. Note that $A_{[n]}$ is identical to uniform distribution over $\{0,1\}^n$ and L has at most t_R bits of leakage on $A_{[n]}$. So, we get $\widetilde{\mathbf{H}}_\infty(A_{[n]}|L) \geq m = n - t_R$.

The view of client R at the end of the random correlation extraction phase is:

$$\vartheta = (x_{[n]}, z_{[n]}, G, (r_0, \ldots, r_n), m_{[n]}, \alpha_{\{0,1\},[n]}, L = \ell)$$

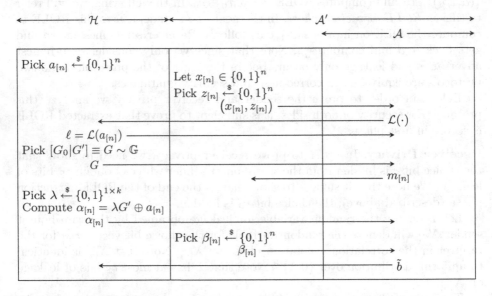

Fig. 2. Simulator for Sender Privacy. The distribution $\mathbb{G}$ is uniform distribution over $k \times (n+1)$ binary matrices in systematic form whose parity check matrices are uniform Toeplitz matrices.

Let U_0 denote the random variable for u_0. We are interested in the conditional distribution $(U_0|\vartheta)$. Below we will show that for any semi-honest client R, $\Pr(R(\vartheta) = u_0)$ is at most $1/2 + 2^{-(g/4)}$.

We show this via a reduction to Lemma 4 in Fig. 2. Given any adversary $\mathcal{A}$ who can predict u_0, we convert it into an adversary $\mathcal{A}'$ against the honest challenger $\mathcal{H}$ of Lemma 4 with identical advantage. It is easy to see that this reduction is perfect. Note that the only difference in the simulator from the actual protocol is that the generator matrix G is being generated by the honest party $\mathcal{H}$ instead of being obtained from $\mathcal{A}$. This does not cause any issues, because we are only dealing with semi-honest adversaries. At the end of random correlation extraction phase, the advantage in predicting U_0 is at most: $2^{-(g/4)}$.

Note that our simulation works even for arbitrary choice of $x_{[n]}$ and $m_{[n]}$. In particular, it works when these vectors are chosen uniformly at random.

4.2 Trading Off Simulation Error with Production Rate

In this section we use sub-sampling techniques to trade-off simulation-error to get improved production rate. The main idea is to sample small subsets of disjoint correlations and, subsequently, run the protocol in Fig. 1 on those subsets independently. This increases the simulation error (due to smaller number of OTs used to output each fresh OT i.e. smaller value of n), but yields higher production rates.

In our case, we use the trivial sub-sampling technique of picking indices at random with suitable probability; in case of a sample repeating itself, we discard it and re-sample. This technique yields distinct samples and has identical properties as the naïve subsampling technique (see [38]). The sophisticated techniques of [38] are also relevant to our setting; but they do not yield any reduction in "simulation error increase." They are useful only to reduce the communication complexity of the protocols.

We only work in the setting where $g = n - (t_S + t_R)$ is at least cn, for some constant $c \leq 1$. In general c could have been a function of n, but we forgo those cases. The main technical lemma is the following:

Lemma 6 (Sub-sampling [38]). *Let $(A_{[n]}, L)$ be a joint distribution such that, there exists a constant $\mu \in (0, 1)$ such that, $\widetilde{\mathbf{H}}_\infty(A_{[n]}|L) \geq \mu n$. For every constant $\varepsilon \in (0, \mu)$ and $\rho = \omega(\log n)$, there exists an efficient algorithm which outputs $(S_1, \ldots, S_m) \in \left(2^{[n]}\right)^m$ such that $m = n/\rho$ and with probability $1 - \mathsf{negl}(n)$, the following holds:*

1. *Large and Distinct: There exists a constant $\lambda \in (0, 1)$ such that $|S_i| = \lambda \rho$. We have $S_i \cap S_j = \emptyset$, for all $i, j \in [m]$ and $i \neq j$.*
2. *High Entropy: $\widetilde{\mathbf{H}}_\infty(S_{i+1}|S_{[i]}, L) \geq (\mu - \varepsilon)|S_{i+1}|$.*

Obtaining the Result of Theorem 2. We obtain this theorem as a direct application of Lemma 6. Recall that we will be working in the setting when $g = n - (t_S + t_R) \geq cn$ for some constant $c \in (0, 1]$. Now we apply Lemma 6 to obtain the disjoint sets $S_1, \ldots, S_m$ for $m = n/\rho$ where $\rho = \omega(\log n)$. Next, we apply the protocol in Fig. 1 to each of the sets independently for the following choice of parameters: $n' = |S_i|$, $t'_S = (\frac{t_S}{n} + \varepsilon)|S_i|$, and $t'_R = (\frac{t_R}{n} + \varepsilon)|S_i|$. Note that new gap $g' = (\frac{g}{n} - 2\varepsilon)|S_i|$. The simulation error obtained for any OT produced will be bounded by $2^{-\Theta(g')} = \mathsf{negl}(n)$.

We observe that the approach of subsampling to obtain "disjoint subsets" while preserving min-entropy is unlikely to yield constant production rate extractors.

5 Inner Product Correlation

In this section we prove Theorem 4. Our protocol is provided in Fig. 3.

When both parties are honest, we need to prove the correctness of the protocol which trivially follows.

Sender Corrupt. Suppose a semi-honest client A can leak t bits on information from $(y_{[n]}, b)$. In this case, we have $\widetilde{\mathbf{H}}_\infty(Y_{[n]}|L) \geq m = n - t$. For security, we

---

Extract-IP (n):

Hybrid (Random Correlations): Client A gets random $(x_{[n]}, a) \in \{0,1\}^{n+1}$ and client B gets random $(y_{[n]}, b) \in \{0,1\}^{n+1}$, such that $a + b = \langle x_{[n]}, y_{[n]} \rangle$.

1. *Random Code Generation.* Client R picks a binary matrix $G = [I_{k \times k} \| P_{k \times (n+1-k)}]$ of dimension $k \times (n+1)$, where $k = n/2$ and $P_{k \times (n+1-k)}$ is a uniformly chosen random Toeplitz matrix. Let C be the code generate by the generator matrix G; and H be a generator matrix for the dual code $C^{\perp}$. If the first column of H is all-zero column then **abort**; otherwise continue.

2. *Random ROLE Extraction.*
   (a) Client A picks a random $(u_0, \ldots, u_n) \in C$ and a random $v_0 \in \{0,1\}$.
   (b) Client B picks a random $(r_0, \ldots, r_n) \in C^{\perp}$.
   (c) Client B sends $m_{[n]} = y_{[n]} \oplus r_{[n]}$ to client A.
   (d) Client A sends $\alpha_{[n]} = x_{[n]} \oplus u_{[n]}$ and $\beta = \langle x_{[n]}, m_{[n]} \rangle \oplus a \oplus v_0$ to client B.
   (e) Client B computes $z = \beta \oplus b \oplus \langle \alpha_{[n]}, r_{[n]} \rangle$.
   (f) Client A outputs (u_0, v_0) and client B outputs (r_0, z).
      Note that $z = u_0 r_0 \oplus v_0$, because $\langle u_{[n]}, r_{[n]} \rangle = u_0 r_0$.

Fig. 3. Random oblivious function evaluation extractor from one inner product correlation over n-bits.

need to prove the hiding of the bit r_0 given $r_{[n]} \oplus y_{[n]}$, where $r_{[n]}$ is a uniformly chosen codeword from the image of "H with its first column punctured." Now, we can directly invoke Lemma 4 and get that the distribution $(R_0|\vartheta)$ is $= 2^{-(g/2+1)}$ close to the uniform distribution over $\{0,1\}$, where ϑ is the view of client A at the end of the protocol and $g = n/2 - t$.

Receiver Corrupt. For this case, we construct a reduction similar to the reduction provided in Fig. 2. Again, in this case we assume that client A sends the matrix G instead of client B (which is acceptable because the adversaries are semi-honest). Suppose there exists an adversary $\mathcal{A}$ which can distinguish U_0 from a uniformly random bit with certain advantage. We shall construct an adversary $\mathcal{A}'$ which uses $\mathcal{A}$ to break the unpredictability experiment of Lemma 4 with identical advantage using a simulation similar to Fig. 2.

Note that as before this will be a perfect simulation of the view of $\mathcal{A}$ because the bit v_0 is uniformly random in the actual protocol. Thus, if $\mathcal{A}$ can predict $u_0 = \lambda G_0$ then the adversary $\mathcal{A}'$ can also predict λG_0 with identical advantage. By Lemma 4, the distribution $(U_0|\vartheta)$ is at most $2^{-(g/2+1)}$ far from the uniform distribution over $\{0,1\}$.

References

1. Alon, N., Roichman, Y.: Random cayley graphs and expanders. Random Struct. Algorithms **5**(2), 271–285 (1994). http://dx.doi.org/10.1002/rsa.3240050203
2. Beaver, D.: Perfect privacy for two-party protocols. In: Feigenbaum, J., Merritt, M. (eds.) Proceedings of DIMACS Workshop on Distributed Computing and Cryptography. vol. 2, pp. 65–77. American Mathematical Society (1989)

3. Ben-Or, M., Goldwasser, S., Wigderson, A.: Completeness theorems for non-cryptographic fault-tolerant distributed computation (extended abstract). In: STOC. pp. 1–10 (1988)
4. Bennett, C.H., Brassard, G., Crépeau, C., Maurer, U.M.: Generalized privacy amplification. IEEE Trans. Inf. Theor. **41**(6), 1915–1923 (1995). http://dx.doi.org/10.1109/18.476316
5. Bennett, C.H., Brassard, G., Robert, J.: Privacy amplification by public discussion. SIAM J. Comput. **17**(2), 210–229 (1988). http://dx.doi.org/10.1137/0217014
6. Bitansky, N., Canetti, R., Halevi, S.: Leakage-tolerant interactive protocols. In: Cramer, R. (ed.) TCC 2012. LNCS, vol. 7194, pp. 266–284. Springer, Heidelberg (2012)
7. Boyle, E., Garg, S., Jain, A., Kalai, Y.T., Sahai, A.: Secure computation against adaptive auxiliary information. In: Canetti, R., Garay, J.A. (eds.) CRYPTO 2013, Part I. LNCS, vol. 8042, pp. 316–334. Springer, Heidelberg (2013)
8. Canetti, R., Lindell, Y., Ostrovsky, R., Sahai, A.: Universally composable two-party and multi-party secure computation. In: STOC. pp. 494–503 (2002)
9. Chari, S., Jutla, C.S., Rao, J.R., Rohatgi, P.: Towards sound approaches to counteract power-analysis attacks. In: Wiener, M. (ed.) CRYPTO 1999. LNCS, vol. 1666, p. 398. Springer, Heidelberg (1999)
10. Dodis, Y., Ostrovsky, R., Reyzin, L., Smith, A.: Fuzzy extractors: how to generate strong keys from biometrics and other noisy data. SIAM J. Comput. **38**(1), 97–139 (2008). http://dx.doi.org/10.1137/060651380
11. Dodis, Y., Smith, A.: Correcting errors without leaking partial information. In: STOC. pp. 654–663 (2005)
12. Duc, A., Dziembowski, S., Faust, S.: Unifying leakage models: from probing attacks to noisy leakage. In: Nguyen, P.Q., Oswald, E. (eds.) EUROCRYPT 2014. LNCS, vol. 8441, pp. 423–440. Springer, Heidelberg (2014)
13. Duc, A., Faust, S., Standaert, F.-X.: Making masking security proofs concrete. In: Oswald, E., Fischlin, M. (eds.) EUROCRYPT 2015. LNCS, vol. 9056, pp. 401–429. Springer, Heidelberg (2015)
14. Dziembowski, S., Faust, S.: Leakage resilient circuits without computational assumptions. In: Cramer, R. (ed.) TCC 2012. LNCS, vol. 7194, pp. 230–247. Springer, Heidelberg (2012)
15. Dziembowski, S., Faust, S., Skorski, M.: Noisy leakage revisited. In: Oswald, E., Fischlin, M. (eds.) EUROCRYPT 2015. LNCS, vol. 9057, pp. 159–188. Springer, Heidelberg (2015)
16. Garcia, A., Stichtenoth, H.: On the asymptotic behaviour of some towers of function fields over finite fields. J. Number Theor. **61**(2), 248–273 (1996)
17. Goldreich, O., Micali, S., Wigderson, A.: How to play any mental game or a completeness theorem for protocols with honest majority. In: STOC. pp. 218–229 (1987)
18. Goldreich, O., Wigderson, A.: Tiny families of functions with random properties: a quality-size trade-off for hashing. Random Struct. Algorithms **11**(4), 315–343 (1997). http://dx.doi.org/10.1002/(SICI)1098-2418(199712)11:4h(315:AID-RSA3) 3.0.CO;2-1
19. Goppa, V.D.: Codes on algebraic curves. Soviet Math. Dokl **24**(1), 170–172 (1981)
20. Gupta, D., Ishai, Y., Maji, H.K., Sahai, A.: Secure computation from leaky correlated randomness (2014). (full version: www.cs.ucla.edu/hmaji/papers/GuptaIsMaSa14.pdf)
21. Halevi, S., Lin, H.: After-the-fact leakage in public-key encryption. In: Ishai, Y. (ed.) TCC 2011. LNCS, vol. 6597, pp. 107–124. Springer, Heidelberg (2011)

22. Harnik, D., Ishai, Y., Kushilevitz, E., Nielsen, J.B.: OT-combiners via secure computation. In: Canetti, R. (ed.) TCC 2008. LNCS, vol. 4948, pp. 393–411. Springer, Heidelberg (2008)

23. Harnik, D., Kilian, J., Naor, M., Reingold, O., Rosen, A.: On robust combiners for oblivious transfer and other primitives. In: Cramer, R. (ed.) EUROCRYPT 2005. LNCS, vol. 3494, pp. 96–113. Springer, Heidelberg (2005)

24. Ishai, Y., Kushilevitz, E., Ostrovsky, R., Sahai, A.: Extracting correlations. In: FOCS, pp. 261–270 (2009)

25. Ishai, Y., Maji, H.K., Sahai, A., Wullschleger, J.: Single-use ot combiners with near-optimal resilience. In: ISIT, IEEE (2014)

26. Ishai, Y., Prabhakaran, M., Sahai, A.: Founding cryptography on oblivious transfer – efficiently. In: Wagner, D. (ed.) CRYPTO 2008. LNCS, vol. 5157, pp. 572–591. Springer, Heidelberg (2008)

27. Kilian, J.: Founding cryptography on oblivious transfer. In: STOC, pp. 20–31 (1988)

28. Kilian, J.: More general completeness theorems for secure two-party computation. In: STOC, pp. 316–324 (2000)

29. Kushilevitz, E.: Privacy and communication complexity. In: FOCS, pp. 416–421. IEEE (1989)

30. Maji, H.K., Prabhakaran, M., Rosulek, M.: Complexity of multi-party computation problems: the case of 2-party symmetric secure function evaluation. In: Reingold, O. (ed.) TCC 2009. LNCS, vol. 5444, pp. 256–273. Springer, Heidelberg (2009)

31. Massey, J.L.: Some applications of coding theory in cryptography. In: Codes and Ciphers: Cryptography and Coding IV. pp. 33–47 (1995)

32. Meier, R., Przydatek, B.: On robust combiners for private information retrieval and other primitives. In: Dwork, C. (ed.) CRYPTO 2006. LNCS, vol. 4117, pp. 555–569. Springer, Heidelberg (2006)

33. Meier, R., Przydatek, B., Wullschleger, J.: Robuster Combiners for Oblivious Transfer. In: Vadhan, S.P. (ed.) TCC 2007. LNCS, vol. 4392, pp. 404–418. Springer, Heidelberg (2007)

34. Naor, J., Naor, M.: Small-bias probability spaces: efficient constructions and applications. In: STOC, pp. 213–223 (1990)

35. Nielsen, J.B., Nordholt, P.S., Orlandi, C., Burra, S.S.: A new approach to practical active-secure two-party computation. In: Safavi-Naini, R., Canetti, R. (eds.) CRYPTO 2012. LNCS, vol. 7417, pp. 681–700. Springer, Heidelberg (2012)

36. Prouff, E., Rivain, M.: Masking against side-channel attacks: a formal security proof. In: Johansson, T., Nguyen, P.Q. (eds.) EUROCRYPT 2013. LNCS, vol. 7881, pp. 142–159. Springer, Heidelberg (2013)

37. Przydatek, B., Wullschleger, J.: Error-tolerant combiners for oblivious primitives. In: Aceto, L., Damgård, I., Goldberg, L.A., Halldórsson, M.M., Ingólfsdóttir, A., Walukiewicz, I. (eds.) ICALP 2008, Part II. LNCS, vol. 5126, pp. 461–472. Springer, Heidelberg (2008)

38. Vadhan, S.P.: Constructing locally computable extractors and cryptosystems in the bounded-storage model. J. Cryptol. 17(1), 43–77 (2004). http://dx.doi.org/10.1007/s00145-003-0237-x

39. Yao, A.C.C.: How to generate and exchange secrets (extended abstract). In: FOCS, pp. 162–167 (1986)

Efficient Multi-party Computation:
From Passive to Active Security
via Secure SIMD Circuits

Daniel Genkin[1,2], Yuval Ishai[1(✉)], and Antigoni Polychroniadou[3]

[1] Technion, Haifa, Israel
{danielg3,yuvali}@cs.technion.ac.il
[2] Tel Aviv University, Tel Aviv, Israel
[3] Aarhus University, Aarhus, Denmark
antigoni@cs.au.dk

Abstract. A central problem in cryptography is that of converting protocols that offer security against passive (or semi-honest) adversaries into ones that offer security against active (or malicious) adversaries. This problem has been the topic of a large body of work in the area of secure multiparty computation (MPC). Despite these efforts, there are still big efficiency gaps between the best protocols in these two settings. In two recent works, Genkin et al. (STOC 2014) and Ikarashi et al. (ePrint 2014) suggested the following new paradigm for efficiently transforming passive-secure MPC protocols into active-secure ones. They start by observing that in several natural information-theoretic MPC protocols, an arbitrary active attack on the protocol can be perfectly simulated in an ideal model that allows for *additive* attacks on the arithmetic circuit being evaluated. That is, the simulator is allowed to (blindly) modify the original circuit by adding an arbitrary field element to each wire. To protect against such attacks, the original circuit is replaced by a so-called *AMD circuit*, which can offer protection against such attacks with constant multiplicative overhead to the size.

Our motivating observation is that in the most efficient known information-theoretic MPC protocols, which are based on packed secret sharing, it is *not* the case that general attacks reduce to additive attacks. Instead, the corresponding ideal attack can include limited forms of linear combinations of wire values. We extend the AMD circuit methodology to so-called *secure SIMD circuits*, which offer protection against this more general class of attacks.

We apply secure SIMD circuits to obtain several asymptotic and concrete efficiency improvements over the current state of the art. In particular, we improve the additive per-layer overhead of the current best protocols from $O(n^2)$ to $O(n)$, where n is the number of parties, and obtain the first protocols based on packed secret sharing that "natively" achieve near-optimal security without incurring the high concrete cost of Bracha's committee-based security amplification method.

Our analysis is based on a new modular framework for proving reductions from general attacks to algebraic attacks. This framework allows us to reprove previous results in a conceptually simpler and more unified way, as well as obtain our new results.

© International Association for Cryptologic Research 2015
R. Gennaro and M. Robshaw (Eds.): CRYPTO 2015, Part II, LNCS 9216, pp. 721–741, 2015.
DOI: 10.1007/978-3-662-48000-7_35

1 Introduction

1.1 Overview

Secure multiparty computation (MPC) is a central research area in cryptography. An MPC protocol allows $n \geq 2$ parties to compute a function of their inputs without compromising the privacy of the inputs or the correctness of the outputs. This should hold even if some of the parties are corrupted by an adversary. Since its introduction in the 1980s [2,7,12,20], there has been a rich body of work dealing with many aspects of the problem, with a major focus on efficiency.

The difficulty of designing MPC protocols depends largely on the power of the adversary. An important distinction is between MPC protocols that offer security against *passive* (or semi-honest) adversaries, who follow the protocol's specification but try to learn information from messages they receive, and security against *active* (or malicious) adversaries, who are allowed to deviate from the protocol's specification in arbitrary ways. The security guarantees in the passive case are weaker, but the protocols are simpler and more efficient.

A common paradigm for designing actively secure MPC protocols (namely, ones that are secure against active adversaries) is to start with a passively secure protocol and then convert it into an actively secure protocol. Some relevant techniques include general-purpose "GMW-style" compilers that employ zero-knowledge proofs [6,12], ad-hoc protocols for verifying the correct execution of subprotocols [2,7], cut-and-choose techniques [18], or "MPC in the head" [15,16]. These techniques typically involve a significant overhead.

A different technique, which in some cases provides better results, was recently proposed independently by Genkin et al. [11] and Ikarashi et al. [14]. These works observe that in several known passively secure protocols for evaluating arithmetic circuits, the effect of any active adversary is limited to an *additive attack* on the circuit wires. That is, everything that an adversary can achieve by attacking the real protocol for evaluating C he could have also achieved by attacking an ideal circuit evaluation process in which he can *blindly* add a field element of his choice to each wire in C. In the following, we refer to such protocols as *additively corruptible protocols*. To secure such a protocol against active adversaries, it is enough to run it on a so-called *AMD circuit* $\overline{C}$ – a randomized circuit which is functionally equivalent to C but additionally offers resistance against additive attacks.[1] The results of [11,14] simplify feasibility results in the information-theoretic setting and obtain efficiency improvements, closing some previous asymptotic efficiency gaps between passively secure and actively secure protocols. This applies to the best known protocols that tolerate an optimal number of corrupted parties (i.e., $t < n/2$ parties using secure point-to-point channels or $t < n$ parties in a suitable hybrid model).

Our motivating observation is that the best information-theoretic MPC protocols that tolerate a slightly sub-optimal number of corrupted parties (e.g.,

---

[1] The work of [14] does not explicitly construct AMD circuits, but implicitly relies on a simple construction of AMD circuits that tolerate a restricted class of additive attacks which suffices in some cases.

$t < 0.49n$) are *not* additively corruptible. These protocols replace the standard secret sharing used in optimally resilient protocols by a more efficient *packed secret sharing* technique, and as a result provide better asymptotic efficiency. The ideal attack corresponding to an active adversary attacking these protocols can include a limited form of linear combinations that combine multiple wire values.[2] As a result, the techniques of [11,14] do not apply to such protocols. In the following, we refer to such protocols *linearly corruptible protocols*.

A second disadvantage of the techniques of [11,14] is that they are tailored to specific protocols. In particular, the part of the analysis that maps general attacks to additive attacks is done in an ad-hoc way per protocol without a unified framework that captures all additively corruptible protocols.

1.2 Our Contribution

In this paper we address both issues outlined above. First, we present a new general framework for proving that a passively secure protocol is additively or linearly corruptible. This framework is used to reprove previous results from [11] in a more unified way, and is also used to prove our new results. Second, we extend the AMD circuit constructions from [11] to offer security against linear attacks. We use these two types of results to close previous efficiency gaps between passively secure and actively secure information-theoretic protocols based on packed secret sharing.

We consider two regimes for such protocols: the *single input, single circuit* regime and the *Franklin and Yung (FY)* [10] regime for simultaneously evaluating ℓ copies of the circuit on different inputs. Notice that the latter is a special case of the former that allows for simpler and more efficient solutions. Currently, all actively secure protocols that rely on packed secret sharing (in both regimes) employ verification methods that introduce at least a quadratic overhead in the number of parties n, for each circuit layer. We reduce this overhead to quasi-linear (or linear in the FY regime), as in the best previous passively secure protocols. In the FY regime, by evaluating the circuit on $\ell = \Omega(n)$ inputs, the amortized per-layer overhead is reduced to constant, leading to the first actively secure protocols whose amortized communication complexity is only $O(|C| + n)$ even for circuits that are very narrow and deep. See Table 1 for a more detailed account of our results and a comparison with previous results.

In addition, we point out that the concrete efficiency of DIK-style protocols [1,8] (see [17]c entry of Table 1) involves prohibitively large constants when applied with near-optimal security threshold. Indeed, the threshold obtained directly by [8] is $t < n/4$ which is quite far from the optimal bound of $n/2$. To improve on this threshold, a general technique due to Bracha [5] is applied for boosting the resilience. The basic idea is that a constant-size committee runs an optimally resilient protocol to emulate the role of each server in the low-threshold protocol. While this technique can be implemented with a constant multiplicative overhead, this constant is very large. Our actively secure protocols natively

---

[2] In the full version we illustrate the necessity of extending the attack model to linear.

achieve a near-optimal security threshold with a low overhead, inheriting this feature from the passively secure protocols on which they are based.

Table 1. Comparison of information-theoretic MPC protocols for arithmetic circuits. Below, n is the number of parties, ϵ is an arbitrary small positive constant, C is an arithmetic circuit or an SIMD circuit, d_C is the multiplicative depth of C, and T is the set of corrupted parties such that $|T| \leq t$. The copies column indicates the number of simultaneously evaluated circuit copies. Passively secure protocols achieve perfect security while actively secure protocols realize C (with abort) with at most $O(1/|\mathbb{F}|)$ simulation error. The communication complexity column counts the total number of field elements exchanged between the parties. For the case of simultaneous evaluation of multiple copies, we count the amortized cost for evaluating a single copy of C. The protocols having resilience $|T| < n$ are constructed on the OT or OLE hybrid model. Note that the $\widetilde{O}$ notation suppresses logarithmic factors.

Ref.	Adv.	Copies	Resilience	Communication complexity
[12]	passive	1	$|T| < n$	$O(n^2|C|)$ for boolean circuits
[17]	passive	1	$|T| < n$	$O(n^2|C|)$
[2]	passive	1	$|T| < n/2$	$O(n^2|C|)$
[9]	passive	1	$|T| < n/2$	$O(n|C| + n^2)$
[10]	passive	$\Theta(n)$	$|T| < (1/2 - \epsilon)n$	$O(n|C|)$
[8]a	passive	$\Theta(n)$	$|T| < (1/2 - \epsilon)n$	$O(|C| + n)$
[8]b	passive	1	$|T| < (1/2 - \epsilon)n$	$\widetilde{O}(|C| + n \cdot d_C)$
[17]a	active	1	$|T| < n$	$O(n^2|C| + \log|\mathbb{F}| \cdot d_C)$
[11]	active	1	$|T| < n$	$O(n^2|C|)$
this work	active	1	$|T| < n$	$O(n^2|C|)$
[3]	active	1	$|T| < n/2$	$O(n|C| + n^2 \log n \cdot d_C) + \text{poly}(n)$
[11]	active	1	$|T| < n/2$	$O(n|C| + n^2)$
this work	active	1	$|T| < n/2$	$O(n|C| + n^2)$
[17]b	active	$\Theta(n)$	$|T| < (1/2 - \epsilon)n$	$O(|C| + n \cdot d_C)$
[17]c	active	$\Theta(n)$	$|T| < (1/2 - \epsilon)n$	$O(|C| + m \cdot d_C)^a$
this work	**active**	$\Theta(n)$	$|T| < (1/2 - \epsilon)n$	$O(|C| + n)$
[8]c	active	1	$|T| < (1/2 - \epsilon)n$	$\widetilde{O}(|C| + n^2 \cdot d_C)$
this work	**active**	1	$|T| < (1/2 - \epsilon)n$	$\widetilde{O}(|C| + n \cdot d_C + n^2)$

[a]In the client-server model, where m (n) is the number of clients (servers).

A key ingredient in our results is an extension of the additive attacks model considered in [11,14], which we now explain in more detail. Protocols that utilize packed secret sharing typically operate on SIMD *circuits*. An SIMD circuit is a generalization of arithmetic circuits, composed by ℓ-gates which get as input two wire *bundles* of size ℓ output a wire output *bundle* of size ℓ obtained by performing ℓ *point-wise* multiplications, additions and subtractions in parallel. Thus, SIMD circuits simultaneously evaluate ℓ copies of the same arithmetic circuit, on different inputs. Next, for protocols based on packed secret sharing, the ideal attack corresponding to deviations made by an active adversary can include a limited form of linear combinations of wire values. Thus, we extend the additive attacks considered in [11] to capture a stronger class of attacks, called *linear* attacks, applied to SIMD circuits.

A *linear* attack on an SIMD circuit changes the computation of a *multiplication* ℓ-gate by adding to the gate's output bundle a linear function $f : \mathbb{F}^{2\ell} \to \mathbb{F}^{\ell}$ of all the wires in the gate's two input bundles. In addition, we also allow a linear attack to specify an *additive* attack on all wire bundles inside the SIMD circuit. We note that for the case where $\ell = 1$ linear attacks are equivalent to additive attacks (see Sect. 2.2 for details).

In the sequel, we prove that for natural protocols based on packed secret sharing, any deviation made by an active adversary actually corresponds to a *linear* attack on the underlying SIMD circuit.

2 Detailed Overview of Results

2.1 Actively Secure MPC Protocols from AMD/SIMD Circuits

Our approach for constructing actively secure MPC protocols is as follows. We present a general *framework* and prove that any passively secure protocol π, satisfying the framework's requirements is indeed additively or linearly corruptible depending on whether π uses packed secret sharing or not. Next, in order to transform any passively secure protocol for evaluating a circuit C, which meets the framework's requirement, into an actively secure protocol, we apply the *same* passive protocol on a different circuit $\overline{C}^{\wedge UG}$ which is essentially the secure version of C. We thus transfer the responsibility of handling the consequences resulting from an active adversary deviating from the protocol, to $\overline{C}^{AUG}$.

We now describe different applications of our framework for existing MPC protocols. See Table 1 for a concise summary.

Applying our framework to an arithmetic version of the passively secure GMW protocol [12,17], in Theorem 10 we match the results of [11, Theorem 1.5] obtaining an actively secure protocol for computing a circuit C, without an honest majority, using $O(n^2|C|)$ calls to an OLE-oracle.[3] In the honest majority setting, applying our framework to the passively secure DN protocol [9], we match the results of [11, Theorem 1.4] and [14] obtaining an actively secure protocol with communication complexity of $O(n|C| + n^2)$ field elements.

Next, in the FY regime, by applying our framework to the passively secure DIK protocol ([8]a), we improve the result of [17]b by eliminating the dependence of the additive term on the depth of the circuit.

---

[3] In fact, we slightly improve the construction of [11] by reducing the statistical simulation error from $O(|C|/|\mathbb{F}|)$ to $O(1/|\mathbb{F}|)$.

Theorem 1. *Let n, t, ℓ be positive integers such that $n = 2t + 2\ell - 1$ and let C be an n-party SIMD ℓ-circuit over a finite field $\mathbb{F}$. Then, there exists a protocol π, in the FY regime, that (t, ϵ)-securely computes C with abort for $\epsilon = O(\ell \log \ell / |\mathbb{F}|)$ and with communication complexity of $O(n|\mathsf{C}| + n^2)$ field elements. Setting $\ell = \Theta(n)$ yields an amortized communication complexity of $O(|\mathsf{C}| + n)$ field elements.*

Finally, applying our framework to the passively secure DIK protocol in the single input single circuit regime ([8]b), we improve the actively secure protocol of [8]c by reducing its additive term from $\widetilde{O}(n^2 \cdot d_\mathsf{C})$ to $\widetilde{O}(n \cdot d_\mathsf{C} + n^2)$.

Theorem 2. *Let n, t, ℓ be positive integers such that $n = 2t + 2\ell - 1$ and let C be an n-party circuit over a finite field $\mathbb{F}$. Then there exists an n-party protocol π, in the single circuit single input regime, that (t, ϵ)-securely computes C with abort for $\epsilon = O(\ell \log \ell / |\mathbb{F}|)$ and with communication complexity $\widetilde{O}\left((|\mathsf{C}|n + n^2 \cdot d_\mathsf{C})/\ell + n^2\right)$ field elements. By setting $\ell = \Theta(n)$ we obtain that the communication complexity of π is $\widetilde{O}\left(|\mathsf{C}| + n \cdot d_\mathsf{C} + n^2\right)$ field elements.*

2.2 Additive and Linear Attack Secure AMD/SIMD Circuits

We now define the notion of linear-attack security. Let C be a circuit to be computed. We say that a randomized SIMD circuit $\overline{\mathsf{C}}$ is an ϵ-linear-attack secure implementation of C if $\overline{\mathsf{C}}$ correctly computes C, when not attacked, and moreover any linear attack on $\overline{\mathsf{C}}$ has the same effect on the outputs of $\overline{\mathsf{C}}$ (up to ϵ statical error) as applying some *additive attack* on the inputs and outputs of $\overline{\mathsf{C}}$ alone.

Definition 1 (Linear-attack and *additive-attack* security). *A randomized SIMD circuit $\overline{\mathsf{C}}$ is said to be an ϵ-linear-attack secure implementation of a (possibly randomized) circuit $\mathsf{C} : (\mathbb{F}^\ell)^n \to (\mathbb{F}^\ell)^k$ if the following holds:*

- *Completeness. For all $\mathbf{x} \in (\mathbb{F}^\ell)^n$ it holds that $\overline{\mathsf{C}}(\mathbf{x}) \equiv \mathsf{C}(\mathbf{x})$.*
- *Linear-Attack security. For any circuit $\overline{\mathsf{C}}^{\mathbf{L}}$ obtained by subjecting $\overline{\mathsf{C}}$ to a linear attack $\mathbf{L}$, there exists $\mathbf{a}^{\mathsf{in}} \in (\mathbb{F}^\ell)^n$ and a distribution $\mathcal{A}^{\mathsf{out}}$ over $(\mathbb{F}^\ell)^k$ such that for any $\mathbf{x} \in (\mathbb{F}^\ell)^n$ it holds that $SD(\overline{\mathsf{C}}^{\mathbf{L}}(\mathbf{x}), \mathsf{C}(\mathbf{x} + \mathbf{a}^{\mathsf{in}}) + \mathcal{A}^{\mathsf{out}}) \leq \epsilon$, where SD denotes statistical distance between two distributions.*

Finally, we say that $\overline{\mathsf{C}}$ is an additive-attack-secure implementation of C if $\overline{\mathsf{C}}$ has the same completeness property as above as well as the same security property with the linear attack $\mathbf{L}$ replaced by an additive attack $\mathbf{A}$.

Notice that restricting Definition 1 for $\ell = 1$ yields exactly the model considered in [11]. This is the case since for non-SIMD circuits, any additive attack can be converted into a linear attack. Conversely, we notice that a linear attack on the output of a multiplication gate can be easily converted to an additive attack on its two inputs. Notice that this equivalence does not hold when $\ell > 1$.

In Sect. 5, we present a construction for securing circuits against additive attacks. While our construction has the same asymptotic efficiency as the construction of [11], it has much better *concrete* efficiency, as well as an improved soundness error of $O(1/|\mathbb{F}|)$ (compared to $O(|C|/|\mathbb{F}|)$ in [11]).

Theorem 3 (Cf. Theorem 6). *For any arithmetic circuit* $\mathsf{C} : \mathbb{F}^n \to \mathbb{F}^k$ *there exists a randomized circuit* $\overline{\mathsf{C}} : \mathbb{F}^n \to \mathbb{F}^k$ *such that* $\overline{\mathsf{C}}$ *is an* ϵ-*additive-attack secure implementation of* C *where* $\epsilon = O(1/|\mathbb{F}|)$ *and* $|\overline{\mathsf{C}}| = O(|\mathsf{C}|)$.

Next, departing from the case of $\ell = 1$, in the full version we present a construction for securing SIMD circuits against linear attacks.

Theorem 4. *For any* SIMD *circuit* $\mathsf{C} : \left(\mathbb{F}^\ell\right)^n \to \left(\mathbb{F}^\ell\right)^k$ *there exists a randomized* SIMD *circuit* $\overline{\mathsf{C}}$ *such that* $\overline{\mathsf{C}}$ *is an* ϵ-*linear-attack secure implementation of* C *where* $\epsilon = O\left(\ell \log \ell / |\mathbb{F}|\right)$ *and* $|\overline{\mathsf{C}}| = O(|\mathsf{C}| + \log \ell)$.

3   Our Techniques

3.1   Constructing Actively Secure MPC Protocols

Our framework for proving that a passively secure protocol π is in fact additively or linearly corruptible, consists of three steps. We point out that while these steps modify the original protocols, are only a *thought-experiment* used to prove the main claim about the effect of an active adversary on the underlying circuit that parties try to evaluate. The only *real* modification required to the protocol in order to transform it to an actively secure protocol, is to execute it on an additive-attack or linear-attack secure circuit (see below).

Step 1: Protocol Randomization. In order to convert an active adversary controlling a set of parties $\mathcal{T}$ to an additive attack, we first transform a protocol π to another protocol $\pi^{\mathcal{T}}$ such that all the messages $m_{\overline{\mathcal{T}},\mathcal{T}}$ sent by the parties in $\mathcal{T}$ to the parties in $\mathcal{T}$ (except during the last communication round) syntactically depend only on the randomness of π. In particular, we require that $m_{\overline{\mathcal{T}},\mathcal{T}}$ does not depend on the inputs $\mathbf{x}_{\overline{\mathcal{T}}}$ of the parties in $\overline{\mathcal{T}}$ or on the messages that the parties in $\overline{\mathcal{T}}$ receive during the protocol. In such case we say that $\pi^{\mathcal{T}}$ is $\mathcal{T}$-randomized.

We first show that for many natural MPC protocols, for any set of parties $\mathcal{T}$, such that $|\mathcal{T}|$ is below the privacy threshold of a protocol, it is possible to construct a $\mathcal{T}$-randomized protocol, $\pi^{\mathcal{T}}$, such that any deviation from π made by an active adversary has the same effect as performing the same deviation from $\pi^{\mathcal{T}}$. In this case we say that $\pi^{\mathcal{T}}$ is $\mathcal{T}$-equivalent to π. See Definition 3.

Notice that $\mathcal{T}$-randomization requirement is stronger than privacy. This is since $\mathcal{T}$-randomization requires that the *values* of $m_{\overline{\mathcal{T}},\mathcal{T}}$ do not depend on the inputs of the parties in $\overline{\mathcal{T}}$ or on messages that parties in $\overline{\mathcal{T}}$ received as opposed to privacy which makes a similar requirement on the *distribution* of $m_{\overline{\mathcal{T}},\mathcal{T}}$. See Step 2 for the necessity of the $\mathcal{T}$-randomization requirement.

Step 2: From General Adversaries to Additive Attacks. We now reduce any adversary controlling a set of parties $\mathcal{T}$, attacking a $\mathcal{T}$-randomized protocol π, to an *additive* attack on the protocol circuit C_π where C_π is a direct implementation of the arithmetic operations performed by π. C_π gets as input the inputs $\mathbf{x}$ of the parties in π as well the randomness $\mathbf{r}$ used in π. It then evaluates π on inputs $(\mathbf{x}; \mathbf{r})$ and outputs the outputs of all the parties following an execution of $\pi(\mathbf{x}; \mathbf{r})$.

Since π is $\mathcal{T}$-randomized we can simulate from the randomness $\mathbf{r}$ for π and from the inputs $\mathbf{x}_\mathcal{T}$, the view $\widetilde{u}_\mathcal{T}$ (except during the last round) of the parties in $\mathcal{T}$. Next, we determine the additive attack on C_π corresponding to an adversary Adv controlling the parties in $\mathcal{T}$ as follows. We first honestly simulate the parties in $\mathcal{T}$ on their view $\widetilde{u}_\mathcal{T}$ and obtain the messages $\widetilde{m}_{\mathcal{T},\overline{\mathcal{T}}}$ sent by the parties in $\mathcal{T}$ to the parties in $\overline{\mathcal{T}}$ during an honest execution of π. Next, we invoke Adv on the view $\widetilde{u}_\mathcal{T}$ and obtain the messages $\widetilde{m}_{\mathcal{T},\overline{\mathcal{T}}}^{\mathsf{Adv}}$ sent by the parties in $\mathcal{T}$ to the parties in $\overline{\mathcal{T}}$ during a real execution of π in the presence of Adv. Finally we determine the additive attack $\mathbf{A}$ on C_π by computing $\mathbf{A} \leftarrow \widetilde{m}_{\mathcal{T},\overline{\mathcal{T}}}^{\mathsf{Adv}} - \widetilde{m}_{\mathcal{T},\overline{\mathcal{T}}}$.

Since π is $\mathcal{T}$-randomized, it is the case that inside C_π under the additive attack $\mathbf{A}$ it holds that $\widetilde{m}_{\mathcal{T},\overline{\mathcal{T}}}^{\mathsf{Adv}} = \widetilde{m}_{\mathcal{T},\overline{\mathcal{T}}} + \mathbf{A}$, for any input $\mathbf{x}_{\overline{\mathcal{T}}}$ of the parties in $\overline{\mathcal{T}}$ as well as for any messages that these parties receive during π. We thus correctly simulate, inside C_π, the effect of Adv on π. Notice that this is not necessary true in case π is $\mathcal{T}$-private since for any selection of the randomness $\mathbf{r}$, the specific values of the messages sent by the parties in $\overline{\mathcal{T}}$ to Adv might depend on their inputs $\mathbf{x}_{\overline{\mathcal{T}}}$ to π. Since $\mathbf{x}_{\overline{\mathcal{T}}}$ is not known to the simulator, it cannot generate the correct view $\widetilde{u}_\mathcal{T}$ required in order to compute $\widetilde{m}_{\mathcal{T},\overline{\mathcal{T}}}^{\mathsf{Adv}}$ and $\widetilde{m}_{\mathcal{T},\overline{\mathcal{T}}}$.

Step 3: Translate Attacks on C_π to Attacks on C. We translate additive attacks on C_π to equivalent attack on C. In Sect. 7, we present the notion of *homomorphic circuits* and prove that if a circuit C' is homomorphic to a circuit C then for any additive attack $\mathbf{A}'$ on C' there exists an equivalent additive attack $\mathbf{A}$ on C such that $C^{\mathbf{A}}(\mathbf{x}) = C'^{\mathbf{A}'}(\mathbf{x})$, for all $\mathbf{x}$. Next, extending the notion of circuit homomorphism to SIMD circuits, in the full version we define the notion of ℓ-homomorphic circuits and prove that if a circuit C' is ℓ-homomorphic to an SIMD circuit C, then for any additive attack $\mathbf{A}'$ on C' there exists an equivalent *linear* attack on C such that $C^{\mathbf{L}}(\mathbf{x}) = C'^{\mathbf{A}'}(\mathbf{x})$ for all $\mathbf{x}$.

Application to Natural MPC Protocols. In Sect. 8 we apply the above transformations on the arithmetic version of the passively secure GMW protocol, proving that it is additively corruptible. Next, in the full version we apply the above transformations to the passively secure DN and DIK protocols, proving that these protocols are additively and linear corruptible, respectively.

MPC Protocols Using Linear or Additive Attack Secure Circuits. The notions of linear and additive-attack security only require that any attack will be equivalent to an additive attack on the inputs and the outputs of the evaluated circuit. Thus, directly executing an additively or linearly corruptible MPC protocol over an additive-attack secure or linear-attack secure circuit C still leaves the inputs and the outputs of C unprotected. Instead, before securing C against additive or linear attacks, we first modify C to C^{AUG} which gets as inputs an AMD encoding of C's inputs and produces an encoding of C's outputs. We then transform C^{AUG} to an additive-attack or linear-attack secure circuit $\overline{C}^{\mathsf{AUG}}$ and evaluate $\overline{C}^{AUG}$ using a passively secure protocol, asking each party to locally compute an AMD encoding of the inputs as well as locally decode the outputs.

3.2 Securing Circuits Against Additive and Linear Attacks

We first present our techniques for additive-attack security (see Sect. 5). Given a circuit C, in the additive-attack secure version $\overline{C}$ of C, every wire of C is paired with a wire that carries a corresponding MAC tag. Next, each gate in C is replaced by a small gadget computing the gate's result as well as its corresponding MAC tag. In addition, this gadget also gets as inputs the MAC tags corresponding to the gate's inputs. Using these tags, the gadget verifies that the gate's result was computed correctly. Notice that the MAC tag verification itself is also vulnerable to additive (and later linear) attacks. However, we construct the verification circuit in such a way that even in the presence of attacks, it outputs a random value if the MAC computation or MAC verification fails for some gate.

The Basic Additive-Attack Secure Circuit Compiler. Similar to [4,11] we use a multiplicative MAC in order to additive-attack secure the output of each gate. Concretely, for each input gate a, its corresponding MAC tag will be $a' = a \cdot v$ where v is a randomly selected field element acting as the MAC key (fixed to be the same value for all gates). Next, for every addition gate $c = a + b$ with inputs a, b and associated MAC tags a', b', we compute the MAC tag c' associated with c directly by computing $c' = a' + b'$.

For every multiplication gate $c = a \cdot b$ with inputs a, b and associated MAC tags a', b', we need to ensure the correct computation of $c = a \cdot b$. Given a MAC tag of the expected result of c and the MAC tags of a, b, we could have verified that under an additive attack indeed $c \cdot v = a \cdot b \cdot v$. Thus, we must somehow combine the (assumed to be correct) MAC tag values $a' = a \cdot v$ and $b' = b \cdot v$ in order to generate the tag of the expected result $a \cdot b \cdot v$. Moreover, this tag generation must be done in such a way that ensures that no combination of attacks on the tag generation circuit and on the multiplication gate's actual computation, can produce an incorrect result without being detected.

In [11], this was solved by setting the MAC tag c' to be $c' = a' \cdot b' = a \cdot b \cdot v^2$. The construction of [11] was based on the fact that an additive attack $\mathbf{A}$ on the computation of $c'^{\mathbf{A}} = (a' + \mathbf{A}_{a',c'})(b' + \mathbf{A}_{b',c'})$ introduces additional monomials of the form $\mathbf{A}_{b',c'} \cdot a' = \mathbf{A}_{b',c'} \cdot av$ or $\mathbf{A}_{a',c'} \cdot b' = \mathbf{A}_{a',c'} \cdot bv$ and it cannot introduce additional monomials of the form $a' \cdot b' = a \cdot b \cdot v^2$, where $\mathbf{A}_{a',c'}$ denotes the attack $\mathbf{A}$ restricted to the wire connecting the gates a', c' inside C. Next, in [11] it was shown that in case these additional monomials are present, they cannot be canceled out by any other combination of additive attacks, thereby making $\overline{C}$ abort the computation by masking its results with a completely random value.

The main problem with the basic construction of [11] is that even when no attacks are present, the degree of the MAC key v inside c' increases from v to v^2. This limits the construction of [11] to only low-degree circuits as well as requiring complicated ad-hoc gadgets to support addition and subtraction gates with MAC tags having different degrees of v. Finally, in order to additive-attack secure arbitrary-degree circuits, [11] employs a degree reduction procedure vastly increasing the concrete overhead of the overall construction.

An Efficient MAC Combination Gadget. In Construction 1, we solve the problem of combining MAC tags in a different way. Let $a' = a \cdot v$ and $b' = b \cdot v$. We first compute c' as $c' = a' \cdot b = (a \cdot v) \cdot b$. Moreover, we also compute $c'' = a \cdot b' = a \cdot (b \cdot v)$. Therefore, if no additive attack is present, it is always the case that $c' - c'' = 0$. However, notice that an additive attack on c' can only produce monomials of the form $A_{a',c'} \cdot b$ or of the from $A_{b,c'} \cdot (a \cdot v)$. In contrast, notice that any additive attack on c'' can only produce monomials of the form $A_{a,c'} \cdot (b \cdot v)$ and $A_{b,c'} \cdot a$ which cannot be canceled out by any of the monomials produced by the attack on c'. Thus, by checking that $c' - c'' = 0$, we either obtain that $c' - c''$ is non-zero with high probability (making the entire circuit to abort) or that no attack was mounted on the circuits computing c' and c''. In the latter case, we obtain that $c' = a \cdot b \cdot v$, which is the correct MAC tag of the expected result of the multiplication gate $c = a \cdot b$ under the key v.

Computing Multiplication Gates. Next, we use the MAC tag c' computed previously in order to verify the correct computation of c. We achieve this by computing the output of c and then MAC it by multiplying with the MAC key v. Next, we check that the above result matches the *known-good* MAC tag c'. This last check is implemented by computing $c \cdot v - c'$ and having $\overline{C}$ abort in case $c \cdot v - c' \neq 0$. Notice that any additive attack on c can only introduce (after the multiplication by the MAC key) monomials of the from $a \cdot v$ or $b \cdot v$ which cannot be canceled-out by the MAC tag $a \cdot b \cdot v$. Hence, we conclude that in the presence of an additive attack the gate output check fails, making $\overline{C}$ abort.

Computing Addition Gates. Notice that in the above described construction, the degree of the used MAC key v is always 1 and in particular it does not increase after the computation of multiplication gates. Therefore, given an addition gate $c = a + b$, we can compute the MAC tag for c by computing $c' = a' + b'$. This avoids the ad-hoc gadget of [11] for additive-attack securely computing MAC tags where the inputs of the addition gate are of different degrees. Eliminating this gadget also simplifies the circuit randomization process (see below).

Avoiding Degree-Reduction. Next, since the degree of the key does not increase after the execution of each multiplication, this allows us to directly additive-attack secure arbitrary circuits without the need to reduce the degree (as opposed to the construction of [11]). This, together with a simplified circuit randomization process (described below), induces a big improvement in the concrete overhead of the construction compared to the construction of [11].

Circuit Randomization Process. The above described construction only achieves additive-attack security for the case where the inputs to each multiplication gate are *almost* random (See Definition 5). Moreover, it is also required that each input of C is also almost random (individually). We force the inputs of a multiplication gate $c = a \cdot b$ to be almost random as follows. First, we additively secret share a and b to $(a - r_1, r_1)$ and $(b - r_2, r_2)$. We then compute the output of c by $c = (a - r_1)(b - r_2) + (a - r_1) \cdot r_2 + r_1 \cdot (b - r_2) + r_1 \cdot r_2 = a \cdot b$. Notice that in this case, the inputs of every multiplication gate are uniformly random. Randomizing the inputs of C is done similarly, see full version for details.

Protecting SIMD Circuits Against Linear Attacks. As described above, a linear attack $\mathbf{L}$ on a multiplication gate c with input gates a and b specifies a linear function $f : \mathbb{F}^{2\ell} \to \mathbb{F}^\ell$ (in the gate's input *bundles*) to be added to the gate's output *bundle*. We specify f using two $\ell \times \ell$ matrices $\mathbf{L}_{a,c}$ and $\mathbf{L}_{b,c}$, changing the computation performed by c to be $c = a \odot b + \mathbf{L}_{a,c} \cdot a + \mathbf{L}_{b,c} \cdot b$. Notice that $\mathbf{L}$ only introduces monomials of the form $\mathbf{L}_{a,c} \cdot a$, $\mathbf{L}_{b,c} \cdot b$ but not of the form $a \odot b$, where $\odot$ denotes ℓ-wide point-wise multiplication of two wire bundles. In the full version, we extend the high-level ideas of the above described construction to handle SIMD circuits and linear attacks.

Next, our basic construction for transforming an SIMD circuit C to a functionally-equivalent linear-attack secure SIMD circuit $\overline{C}$ guarantees that every linear attack on $\overline{C}$ is either equivalent to an *additive* attack on the inputs and outputs of C, or some wire in a special bundle $\mathbf{f}$, which denotes an error flag inside $\overline{C}$, becomes non-zero. In such a case, we would like another bundle $\mathbf{f}'$ to be almost random. In the full version, we design a special-purpose gadget, called $\overline{\text{Mix}}$ circuit, which satisfies the above property, even in the presence of linear attacks.

4 Preliminaries

Arithmetic Circuits. Following [11], an *arithmetic circuit* C is a directed acyclic graph whose vertices are called *gates* and whose edges are called *wires*. Every in-degree 0 gate in C is labeled by a variable from a set of variables $X = \{x_1, \cdots, x_n\}$ and is referred to as an input gate. All other gates have in-degree 2, are labeled by elements from $\{+, -, \times\}$ and referred to as add, sub and mult gates, respectively. Every gate of out-degree 0 is called an output gate. We assume that the output gates are ordered. In some cases we also allow in-degree 0 gates labeled by rand referred to as *randomness* gates. A circuit containing rand gates is called a *randomized* circuit. For a (possibly randomized) circuit C and for a gate g of C we denote by g_x the distribution of the output value of g (defined in a natural way) when C is being evaluated on an input x.

SIMD Circuits. An SIMD circuit with bundle size ℓ is defined similar to arithmetic circuits. We refer to the edges of an SIMD circuit C as *wire bundles* or *bundles* and to vertices of an SIMD circuits as ℓ-*gates*. We write $C : \left(\mathbb{F}^\ell\right)^n \to \left(\mathbb{F}^\ell\right)^k$ to indicate that C is an SIMD circuit with n input bundles and k output bundles. Each multiplication, addition or subtraction ℓ-gate of an SIMD circuit gets as input two wire bundles of size ℓ and outputs a bundle of size ℓ obtained by performing ℓ *point-wise* multiplications, additions or subtractions in parallel.[4]

---

[4] Notice that for the case of SIMD circuits, the notion of in-degree of a gate corresponds to the number of its input wire-bundles (as opposed to individual wires). Thus the in-degree of $\{\times, +, -\}$ gates is 2. The notion of out-degree is defined similarly.

We also allow SIMD circuits to contain an additional type of ℓ-gates with in-degree and out-degree 1, referred to as routing ℓ-gates. Each routing ℓ-gate is labeled by a function $\rho : [\ell] \rightarrow [\ell]$. We shall sometimes refer to these routing ℓ-gates as ρ-gates. A ρ-gate on an input bundle $\mathbf{a} = (a_1, \cdots, a_\ell)$ outputs a bundle $\mathbf{b} = (b_1, \cdots, b_\ell)$ such that $b_i = a_{\rho(i)}$ for all $1 \leq i \leq \ell$.

Additive Attacks. An additive attack $\mathbf{A}$ changes the computation performed by a circuit C by specifying for every wire in C, connecting gates a and b, a value to be added to the output of a. The derived value is then used for the computation of b. In addition, $\mathbf{A}$ specifies values to be added to the outputs of C. Note that an additive attack on a circuit C is a fixed vector of field elements which is independent from the inputs and internal values of C.

Linear Attacks. A *linear attack* $\mathbf{L}$ on an SIMD circuit changes the computation of a multiplication ℓ-gate by adding to each wire in the gate's output bundle a linear function of all the wires in the gate's two input bundles. In particular, for any multiplication ℓ-gate c with input bundles a and b, a linear attack $\mathbf{L}$ specifies a linear function $f : \mathbb{F}^{2\ell} \rightarrow \mathbb{F}^\ell$ such that the output bundle of c is equal to $\mathsf{c} = \mathsf{a} \odot \mathsf{b} + f(\mathsf{a}, \mathsf{b})$, where $\odot$ denotes point-wise multiplication of two wire bundles. In addition, similar to additive attacks, we allow a linear attack $\mathbf{L}$ to specify an additive attack $\mathbf{L}^{\mathsf{out}}$ on the outputs of the SIMD circuit C.

Attacks on Addition and Subtraction Gates. We do not allow linear attacks on addition and subtraction gates. This is since mounting an attack of the form $f(\mathsf{a}, \mathsf{b}) = -\mathsf{a} - \mathsf{b}$ the adversary is able to fix an output of an addition gate $\mathsf{c} = \mathsf{a} + \mathsf{b} - \mathsf{a} - \mathsf{b}$ to be always zero. Therefore, allowing for such attacks means that it is possible to override the output of these gates to be an arbitrary value. Such attacks are not supported by our constructions.[5]

Additive Attacks on SIMD Circuits. Note that allowing additive attacks on wire bundles of SIMD circuits (in addition to linear attacks) will not provide the adversary with additional capabilities in modifying the circuit's computation. This is since for any pair of attacks $(\mathbf{A}, \mathbf{L})$ on an SIMD circuit C where $\mathbf{A}$ is an additive attack and $\mathbf{L}$ is a linear attack there exists a functionally-equivalent linear attack $\mathbf{L}'$. The linear attack $\mathbf{L}'$ can be constructed as follows. First, the additive attacks specified by $\mathbf{A}$ can be pushed "downstream" through the circuit till the inputs of the multiplication gates and the outputs of the output gates. Next, additive attacks on inputs of a multiplication gate c, can be added to the diagonal of the appropriate matrices as specified by $\mathbf{L}$, yielding $\mathbf{L}'$.

Additive Attacks in Secure Multi-party Computation. In the following we define the notion of additively corruptible versions of a functionality. Without loss of generality, we only consider functionalities where only P_1 gets an output. That is, functionalities of the form $f : \mathbb{F}^{I_1} \times \cdots \times \mathbb{F}^{I_n} \rightarrow \mathbb{F}^{O_1}$ where $(I_1, \cdots, I_n, O_1)$ are positive integers. Note that we can move to individual outputs using a standard transformation (See [13, Sect. 2.5.2]).

---

[5] Note that linear attacks on multiplication gates suffice to achieve MPC tasks.

Definition 2. *Let C be an n-party circuit. We define the* additively corruptible *version of C to be an n-party functionality f_C^A that takes additional input from the adversary representing an additive attack, A, on C. For an input x and additive attack A, f_C^A outputs $C^A(x)$. The notion of a linearly corruptible circuit is defined similarly, replacing the additive attack A with a linear attack L.*

Next, we define the notion of T-equivalent protocols.

Definition 3. *Let π and π' be two protocols for computing an n-party circuit C in the f and f' hybrid models respectively. We say that π is T-equivalent to π' if for any adversary Adv controlling a set of parties $T \subseteq P$ and for any input x it holds that $\text{Real}_{\pi,T}^{\text{Adv},f}(x) \equiv \text{Real}_{\pi',T}^{\text{Adv},f'}(x)$.*

5 Additive Security for Arithmetic Circuits

In this section we simplify the construction of [11] improving its additive-attack security from $O(|C|/|\mathbb{F}|)$ to $O(1/|\mathbb{F}|)$, as well as improving its concrete efficiency. Following the approach of [11], we first present a simpler construction whose security holds only when the circuit's wire values satisfy some local randomness property (Construction 1). In the full version, we show how to eliminate this assumption by applying general transformations to the circuit.

We begin by defining additive-attack security for specific input distributions.

Definition 4. *Let $\mathbb{F}$ be a finite field, $C : \mathbb{F}^n \to \mathbb{F}^k$ an arithmetic circuit, and I a distribution over $\mathbb{F}^n$. We say that a circuit $\overline{C} : \mathbb{F}^n \to \mathbb{F}^{k+1}$ is an ϵ-additive-attack secure implementation of C with respect to I if the following holds:*

- *Completeness. For all $x \in \mathbb{F}^n$, $\overline{C}(x) \equiv C(x)$.*
- *Security with respect to I. For any additive attack A, there exists $a^{\text{in}} \in \mathbb{F}^n$ and a distribution A^{out} over $\mathbb{F}^k$ such that $SD(\overline{C}^A(I), C(I + a^{\text{in}}) + A^{\text{out}}) \leq \epsilon$.*

The construction guarantees security as defined in Definition 1 with $\epsilon = O(1/|\mathbb{F}|)$, under the assumption that the inputs of the circuit as well as the inputs of each multiplication gate are sufficiently random. Unlike the basic construction of [11], the construction described in this section does not require the randomization of the inputs of addition and subtraction gates. Thus, below we define a weaker notion of locally random circuits compared to the one used in [11], by not imposing any requirement about the inputs of addition and subtraction gates. This also greatly simplifies the construction of such circuits.

Definition 5 (Locally Random Circuits). *Let $\mathbb{F}$ be a finite field, C be a randomized arithmetic circuit. We say that C is locally ϵ-random with respect to a distribution I if the following two properties hold.*

1. **Local Randomization of Input Gates.** *For any $y \in \mathbb{F}$ and for any $1 \leq i \leq n$ the probability over selecting $x \leftarrow I$ that $x_i = y$ is at most $|\mathbb{F}| \cdot \epsilon$.*

2. **Local Randomization of Multiplication Gates.** *For any $(y, z) \in \mathbb{F}^2$ and any pair of gates (a, b), whose outputs are the inputs to some multiplication gate in C, it holds that the probability, over the internal randomness of C and the selection $\mathbf{x} \leftarrow I$, that $(\mathsf{a}_{\mathbf{x}}, \mathsf{b}_{\mathbf{x}}) = (y, z)$ is at most ϵ.*

We now present our basic construction for constructing additive-attack circuits.

Construction 1. *Let $\mathsf{C} : \mathbb{F}^n \to \mathbb{F}^k$ be a circuit. Define a circuit $\overline{\mathsf{C}}$ that on input $\mathbf{x}$ computes $\mathbf{z} = \mathsf{C}(\mathbf{x})$ and then performs the following:*

MAC Generation Circuit:

1. *Generate a random elements $\mathbf{r}, \mathbf{v} \in \mathbb{F}$ and compute $\mathbf{r}' \leftarrow \mathbf{r} \cdot \mathbf{v}$.*
2. *For each input gate c, compute the value $\mathsf{c}' \leftarrow \mathsf{c} \cdot \mathbf{v}$.*
3. *For each non-input gate c let a, b be its inputs and let a', b' be the MAC tags corresponding to a and b. Compute the MAC tag c' as follows:*
   (a) *If c is a multiplication gate, let $\mathsf{c}' \leftarrow \mathsf{a}' \cdot \mathsf{b}$ and let $\mathsf{c}'' \leftarrow \mathsf{a} \cdot \mathsf{b}'$.*
   (b) *If c is an addition gate let $\mathsf{c}' \leftarrow \mathsf{a}' + \mathsf{b}'$. Similarly, if c is a subtraction gate let $\mathsf{c}' \leftarrow \mathsf{a}' - \mathsf{b}'$.*

MAC Checking Circuit:

4. *For every input gate c in C, generate a random element t^{c} and compute*
$$\mathsf{g}^{\mathsf{c}} \leftarrow \mathsf{c} + \mathsf{r}, \qquad \mathsf{h}'^{\mathsf{c}} \leftarrow \mathsf{c}' + \mathsf{r}', \qquad \mathsf{g}'^{\mathsf{c}} \leftarrow \mathsf{g}^{\mathsf{c}} \cdot \mathbf{v}, \qquad \mathsf{f}^{\mathsf{c}} \leftarrow \mathsf{h}'^{\mathsf{c}} - \mathsf{g}'^{\mathsf{c}}.$$
5. *Compute $\mathsf{f}_1 \leftarrow \sum_{\mathsf{c} \in \mathsf{inpt}_{\mathsf{C}}} \mathsf{t}^{\mathsf{c}} \cdot \mathsf{f}^{\mathsf{c}}$ where $\mathsf{inpt}_{\mathsf{C}}$ is the set of the input gates of C.*
6. *For every multiplication gate c, generate two random field elements $\mathsf{t}^{\mathsf{c}}, \mathsf{w}^{\mathsf{c}}$ and compute $\mathsf{f}^{\mathsf{c}} \leftarrow \mathsf{c}' - \mathsf{c}'', \qquad \mathsf{g}^{\mathsf{c}} \leftarrow \mathsf{c} \cdot \mathbf{v}, \qquad \mathsf{h}^{\mathsf{c}} \leftarrow \mathsf{g}^{\mathsf{c}} - \mathsf{c}'.$*
7. *Let $\mathsf{mul}_{\mathsf{C}}$ be the set of all multiplication gates in C, compute $\mathsf{f}_2 \leftarrow \sum_{\mathsf{c} \in \mathsf{mul}_{\mathsf{C}}} \mathsf{w}^{\mathsf{c}} \cdot \mathsf{f}^{\mathsf{c}}$ and $\mathsf{f}_3 \leftarrow \sum_{\mathsf{c} \in \mathsf{mul}_{\mathsf{C}}} \mathsf{t}^{\mathsf{c}} \cdot \mathsf{h}^{\mathsf{c}}$.*
8. *Compute $\mathsf{f} \leftarrow \mathsf{f}_1 \cdot \mathsf{s}_1 + \mathsf{f}_2 \cdot \mathsf{s}_2 + \mathsf{f}_3 \cdot \mathsf{s}_3$ where $\mathsf{s}_1, \mathsf{s}_2, \mathsf{s}_3$ are random field elements.*

Output Generation: *Output $\mathbf{z} + \mathbf{f} \cdot \mathbf{r}$ where $\mathbf{r}$ is a random vector from $\mathbb{F}^k$.*

In the full version we prove the following theorems.

Theorem 5. *Let $\mathsf{C} : \mathbb{F}^n \to \mathbb{F}^k$ be a randomized arithmetic circuit which is locally ϵ-random with respect to and input distribution I. Then the circuit $\overline{\mathsf{C}}$ obtained by applying Construction 1 to C is a $(|\mathbb{F}| \cdot \epsilon + 1/|\mathbb{F}|)$-additive-attack secure implementation of C with respect to I. Moreover, $|\overline{\mathsf{C}}| = O(|\mathsf{C}|)$.*

Theorem 6 (Additive-attack Security). *For any arithmetic circuit $\mathsf{C} : \mathbb{F}^n \to \mathbb{F}^k$ there exists a randomized circuit $\overline{\mathsf{C}} : \mathbb{F}^n \to \mathbb{F}^k$ such that $\overline{\mathsf{C}}$ is an ϵ-additive-attack secure implementation of C where $\epsilon = O(1/|\mathbb{F}|)$. Moreover, $|\overline{\mathsf{C}}| = O(|\mathsf{C}|)$.*

Notice that unlike the work of [11], the error parameter of the construction is $O(1/|\mathbb{F}|)$. This matches the result of [14], but in a stronger attack model.

6   From General Adversaries to Additive Attacks

In this section we reduce any general adversary attacking a randomized protocol π to an additive attack on the protocol circuit C_π defined as follows. We compile a protocol π into a circuit, C_π, by writing all local computations performed by the parties as circuits and whenever a party P_i sends a message to P_j, we connect the corresponding parts of the circuits representing P_i and P_j using wires. Notice that for every input $\mathbf{x}$ and randomness $\mathbf{r}$, it holds that $\pi(\mathbf{x}; \mathbf{r}) = C_\pi(\mathbf{x}, \mathbf{r})$.

We now define the notion of a *last-round-private* protocol.

Definition 6. *Let T be a set of corrupted parties and let π be a T-randomized n-party protocol for computing an n-input circuit $C : \mathbb{F}^{I_1} \times \cdots \times \mathbb{F}^{I_n} \to \mathbb{F}^{O_1}$. We say that π is T-last-round-private if the following hold.*

1. **Structure of the Last Round.** *During the last round, only P_1 computes the output vector $\mathbf{z}$, in the following way. Let $\overline{T}' \subseteq \overline{T}$ be the set of parties from $\overline{T}$ sending messages to P_1 during the last round. Each output $\{z_i\}_{1 \leq i \leq O_1}$ is computed by P_1 evaluating two linear functions F_T and $F_{\overline{T}'}$ such that $z_i = F_T(l^i_{T,P_1}) + F_{\overline{T}'}(l^i_{\overline{T}',P_1})$ where the messages $l^i_{T,P_1}, l^i_{\overline{T}',P_1}$ are the shares corresponding to z_i received by P_1 from the parties in T and $\overline{T}'$, respectively.*

2. **Privacy of the Last Round.** *Fix an input $\mathbf{x}_T$ and randomness $\mathbf{r}_T$ to the circuit C_π for the parties in T. In addition, fix an additive attack $\mathbf{A}$ on C_π and fix a view $\widehat{u}_T$ of the parties in T during an execution of $C_\pi^\mathbf{A}$ on $(\mathbf{x}_T, \mathbf{r}_T)$. Let $\mathbf{Z}$ be the distribution of outputs in $C_\pi^\mathbf{A}$ conditioned on $(\mathbf{x}_T, \mathbf{r}_T, \mathbf{A}, \widehat{u}_T)$ and fix $\mathbf{z}$ from the support of $\mathbf{Z}$. Finally, let $\widehat{l}_{T,P_1}$ be the messages received by P_1 from the parties in T during the last round of $C_\pi^\mathbf{A}$ as uniquely defined by $(\mathbf{x}_T, \mathbf{r}_T, \widehat{u}_T)$. We require that the distribution of the messages $\widehat{l}_{\overline{T}',P_1}$, over the unfixed randomness $\mathbf{r}_{\overline{T}}$ is uniform conditioned on $F_{\overline{T}'}(\widehat{l}_{\overline{T}',P_1}) = \mathbf{z} - F_T(\widehat{l}_{T,P_1})$.*

In the full version we prove the following theorem.

Theorem 7. *Let π be a T-last-round-private and T-randomized protocol. Then for any active adversary Adv controlling the parties in T there exists a simulator Sim such that for any input $\mathbf{x}$ it holds that $\mathsf{Ideal}^{\mathsf{Sim}}_{f_{C_\pi^\mathbf{A}}, T}(\mathbf{x}) \equiv \mathsf{Real}^{\mathsf{Adv}}_{\pi, T}(\mathbf{x})$.*

7   Homomorphism for Standard Circuits

In this section we prove that if two circuits C and C' meet certain properties, then for any additive attack on C' there exists an equivalent additive attack on C. Applying this approach to C_π, we prove that any additive attack on C_π corresponds to an additive attack on C.

Without loss of generality, we express every multiplication gate as a product of its inputs where each input is an arbitrary fixed linear combination of the preceding addition and subtraction gates up to the depth of the preceding multiplication gate.

Definition 7. *Let* C *be a randomized circuit and let* c *be an in-degree 2 multiplication gate inside* C. *We define two ordered sets* left_c *and* right_c, *as follows.*

$$\mathsf{left}_c = \left\{ \mathsf{a} \in \{\times, \mathsf{input}\} \ : \ \begin{array}{l} \exists \text{path } \textit{from } \mathsf{a} \textit{ to the first input of } \mathsf{c} \\ \textit{which only contains gates from the set } \{+,-\} \end{array} \right\}$$

$$\mathsf{right}_c = \left\{ \mathsf{a} \in \{\times, \mathsf{input}\} \ : \ \begin{array}{l} \exists \text{path } \textit{from } \mathsf{a} \textit{ to the second input of } \mathsf{c} \\ \textit{which only contains gates from the set } \{+,-\} \end{array} \right\}$$

The ordered sets left_c *and* right_c *naturally define two linear functions* $l^c : \mathbb{F}^{|\mathsf{left}_c|} \to \mathbb{F}$ *and* $r^c : \mathbb{F}^{|\mathsf{right}_c|} \to \mathbb{F}$ *representing the output of* c *as a function of the outputs of the preceding* mult *and* input *gates. More specifically, for any input* $\mathbf{x}$ *to* C *it holds that* $c_{\mathbf{x}} = l^c(\mathbf{a}_{\mathbf{x}}) \cdot r^c(\mathbf{b}_{\mathbf{x}})$ *where* $\mathbf{a} = \mathsf{left}_c$ *and* $\mathbf{b} = \mathsf{right}_c$.

We now express every output gate which is an addition or subtraction gate as a fixed linear combination of the output of the proceeding multiplication gates.

Definition 8. *Let* C *be a deterministic circuit and let* c *be an output gate that is an* add *or* sub *gate. We define the ordered set* in_c *as follows.*

$$\mathsf{in}_c = \left\{ \mathsf{a} \in \{\times, \mathsf{input}\} \ : \ \begin{array}{l} \exists \text{path } \textit{from } \mathsf{a} \textit{ to either of the two inputs of } \mathsf{c} \\ \textit{which only contains gates from the set } \{+,-\} \end{array} \right\}$$

The set in_c *naturally defines a linear function* $f^c : \mathbb{F}^{|\mathsf{in}_c|} \to \mathbb{F}$ *representing the output of* c *as a function of the outputs of the preceding* mult *and* input *gates. More specifically, for input* $\mathbf{x}$ *to* C *it holds that* $c_{\mathbf{x}} = f^c(\mathbf{a}_{\mathbf{x}})$ *where* $\mathbf{a} = \mathsf{in}_c$.

We now define the notion of circuit homomorphism. Later, we prove that if a circuit C′ is homomorphic to a circuit C then any additive attack on C′ can be simulated by an additive attack on C. Applying the above on MPC protocols, as long as the circuit C_π of a protocol π is homomorphic to C, then any additive attack on C_π can be simulated by an additive attack on C. Combining this with the result of Sect. 6, we obtain that for any protocol π computing a circuit C, which is $\mathcal{T}$-*randomized*, $\mathcal{T}$-*last-round-private* and *homomorphic* to C, any attack mounted by an active adversary is equivalent to an additive attack on C.

Definition 9 (Circuit Homomorphism). *Let* C *be a deterministic circuit. A circuit* C′ *is said to be homomorphic to* C *if there exists a mapping* $\mathcal{H}$ *from the* input *and* mult *gates of* C *to the gates of* C′ *such that the following properties hold. Below, for any gate* c *of* C *we denote the output of* $\mathcal{H}(c)$ *by* c′.

1. **Input.** *For any* input *gate* c *of* C *and for any input* $\mathbf{x}$ *it holds that* $c_{\mathbf{x}} = c'_{\mathbf{x}}$.
2. **Multiplications.** *For any* mult *gate* c *we require that there exists constant* $\lambda^c \in \mathbb{F}$ *with the following properties for any input* $\mathbf{x}$:
   (a) *It holds that* $c'_{\mathbf{x}} + \lambda^c = l^c((a'_{\mathbf{x}} + \lambda^a)_{a \in \mathsf{left}_c}) \cdot r^c((b'_{\mathbf{x}} + \lambda^b)_{b \in \mathsf{right}_c})$.
   (b) *For every* mult *gate used for the computation of the output of* c′ *inside* C′, *the left input is a linear function of* $l^c((a'_{\mathbf{x}})_{a \in \mathsf{left}_c})$ *and the right input is a linear function of* $r^c((b'_{\mathbf{x}})_{b \in \mathsf{right}_c})$.

3. **Outputs.** *We first require that both* C *and* C' *have the same number of output gates. Let* c *be the* i*-th gate of* C*, we distinguish two different cases.*

(a) *Let* o' *be the* i*-th output gate of* C'*. If* c *is a mult gate, then* $o'_{\mathbf{x}} = c'_{\mathbf{x}} + \lambda^c$.[6]

(b) *If* c *is an add, or sub gate then the* i*-th output of* C'*,* $o'_{\mathbf{x}}$ *is equal to*
$$o'_{\mathbf{x}} = f^c\left((a'_{\mathbf{x}} + \lambda^a)_{a \in in_c}\right) \text{ for all input } \mathbf{x}.$$

Moreover, we require that the recovery of the output from the gates o' *of* C' *is performed without computing any* mult *gates.*

Remark 1 *Given two circuits* C, C'*, a mapping* $\mathcal{H}$*, a constant* λ^c *and functions* l^c *and* r^c *for every* mult *gate* c *in* C*, it is possible to decide in polynomial time if* C' *is homomorphic to* C*. Checking that the requirements of Definition 9 hold can be done symbolically using the gate's output as variables.*

For simplicity of exposition, Definition 9 is tailored to protocols working on additive secret sharing such as the GMW protocol. A simple generalization of Definition 9 captures protocols working on any linear secret sharing scheme, such as the DN and DIK. See full version for details.

Lemma 1 *Let* C *be a deterministic circuit and let* C' *be a circuit homomorphic to* C*. Then for any additive attack* $\mathbf{A}'$ *on* C' *there exists an additive attack* $\mathbf{A}$ *on* C *such that for any input* $\mathbf{x}$ *it holds that* $C'^{\mathbf{A}'}(\mathbf{x}) = C^{\mathbf{A}}(\mathbf{x})$.

We now extend Lemma 1 to handle n-party circuits computed during an MPC protocol. We begin by defining the notion of $\mathcal{T}$-homomorphic circuits.

Definition 10 *Let* π *be an* n*-party protocol,* C *be an* n*-party circuit and let* $\mathcal{T}$ *be a set of parties. We say that* C_π *is* $\mathcal{T}$*-homomorphic to* C *if for any input* $\mathbf{x}_{\mathcal{T}}$ *for the parties in* $\mathcal{T}$ *and for every randomness* $\mathbf{r}$*, the circuit* $C_\pi((\mathbf{x}_{\mathcal{T}}, \cdot), \mathbf{r})$ *obtained by fixing the inputs* $\mathbf{x}_{\mathcal{T}}$ *and* $\mathbf{r}$ *inside* C_π *is homomorphic to* $C(\mathbf{x}_{\mathcal{T}}, \cdot)$.

In the full version we prove the following theorem.

Theorem 8 *Let* π *be an* n*-party protocol for computing a circuit* $C : \mathbb{F}^{I_1} \times \cdots \times \mathbb{F}^{I_n} \to \mathbb{F}^{O_1}$ *in the* f*-hybrid model and let* $\mathcal{T}$ *be a set of parties such that* π *is* $\mathcal{T}$*-randomized,* $\mathcal{T}$*-last-round-private and* C_π *is* $\mathcal{T}$*-homomorphic to* C*. Then for any active adversary* Adv *controlling the parties in* $\mathcal{T}$ *there exists a simulator* Sim *such that for any input* $\mathbf{x}$ *it holds that* $\mathsf{Ideal}^{\mathsf{Sim}}_{f_C^A, \mathcal{T}}(\mathbf{x}) \equiv \mathsf{Real}^{\mathsf{Adv}, f}_{\pi, \mathcal{T}}(\mathbf{x})$.

8 The GMW Protocol

In this section we prove that an arithmetic generalization of the passively secure GMW protocol [12] is additively corruptible. We first extend the GMW protocol to the arithmetic setting [17], where the OT oracle is replaced by oblivious linear function evaluation (OLE) [19].

---

[6] Notice that here we do not require that $\mathcal{H}(c) = o'$. This is since already $\mathcal{H}(c) = c'$ and moreover there exists a gate o' such that $o' = c'_{\mathbf{x}} + \lambda^c$.

Definition 11 (The OLE functionality). *Let $\mathbb{F}$ be a finite field. We define the functionality f_{OLE} that on inputs $(a, b) \in \mathbb{F}^2$ from the sender and $x \in \mathbb{F}$ from the receiver outputs $\perp$ to the sender and $a \cdot x + b$ to the receiver.*

We now proceed describing an arithmetic version of the GMW protocol in the OLE-hybrid model [12,17]. We begin by describing the Input-Share$_{\mathsf{GMW}}$ and Mult$_{\mathsf{GMW}}$ protocols used to evaluate input and multiplication gates.

Construction 2 (Subprotocol Input-Share$_{\mathsf{GMW}}$). *The subprotocol Input-Share$_{\mathsf{GMW}}$ is defined as follows. Each party P_i on input x computes a random additive sharing of x, denoted by $[\mathbf{x}]_{\mathsf{add}} = (x_1, \ldots, x_n)$, and deals it among all the parties.*

Construction 3 (Subprotocol Mult$_{\mathsf{GMW}}$). *The subprotocol Mult$_{\mathsf{GMW}}$ gets as input additive sharings $[\mathbf{a}]_{\mathsf{add}}$, $[\mathbf{b}]_{\mathsf{add}}$ and outputs an additive sharing $[\mathbf{c}]_{\mathsf{add}}$ such that $\mathbf{c} = \mathbf{a} \cdot \mathbf{b}$. The protocol proceeds as follows.*

1. *Each ordered pair of parties P_i, P_j, such that $i \neq j$, performs the following.*
   (a) *P_i generates a random value $r_{i,j}$ and acting as a sender sends $(a_i, r_{i,j})$ to the OLE oracle. P_j acting as a receiver sends b_j to the OLE oracle.*
   (b) *The OLE oracle responds with $s_{i,j} = a_i \cdot b_j + r_{i,j}$ to P_j.*
2. *Each party P_i computes $c_i \leftarrow a_i \cdot b_i + \sum_{\substack{j=1 \\ j \neq i}}^{n} (s_{j,i} - r_{i,j})$.*

We now proceed in describing the passively secure GMW protocol.

Construction 4 (Passively secure GMW protocol). *Let $\mathsf{C} : \mathbb{F}^{I_1} \times \cdots \times \mathbb{F}^{I_n} \to \mathbb{F}^{O_1}$ be an n-party circuit. The protocol GMW_C for C proceeds as follows:*

1. ***Input sharing phase.*** *For each input gate associated to party P_i, party P_i executes the protocol Input-Share$_{\mathsf{GMW}}$ described in Construction 2.*
2. ***Circuit evaluation phase.*** *For each gate c in C with input sharings $[\mathbf{a}]_{\mathsf{add}} = (a_1, \ldots, a_n)$ and $[\mathbf{b}]_{\mathsf{add}} = (b_1, \ldots, b_n)$ proceed as follows:*

   Evaluating addition and subtraction gates. *For the case of addition gates, all parties locally compute $[\mathbf{c}]_{\mathsf{add}} \leftarrow [\mathbf{a}]_{\mathsf{add}} + [\mathbf{b}]_{\mathsf{add}}$. Similarly, for subtraction gates, all parties locally compute $[\mathbf{c}]_{\mathsf{add}} \leftarrow [\mathbf{a}]_{\mathsf{add}} - [\mathbf{b}]_{\mathsf{add}}$.*

   Evaluating multiplication gates. *All the parties execute the Mult$_{\mathsf{GMW}}$ protocol described in Construction 3 on inputs $[\mathbf{a}]_{\mathsf{add}}$ and $[\mathbf{b}]_{\mathsf{add}}$.*
3. ***Output recovery phase.*** *At the end of the computation, for each output gate c of C all the parties hold a sharing $[\mathbf{c}]_{\mathsf{add}}$ corresponding to its value. For each output gate c, the parties generate a random sharing $[\mathbf{z}]_{\mathsf{add}}$ of 0 and compute $[\mathbf{c}']_{\mathsf{add}} \leftarrow [\mathbf{c}]_{\mathsf{add}} + [\mathbf{z}]_{\mathsf{add}}$. Parties $\{P_2, \cdots, P_n\}$ send their shares of $[\mathbf{c}']_{\mathsf{add}}$ to P_1. Then P_1 recovers the output c by computing $c \leftarrow \sum_{i=1}^{n} c_i'$.*

The works of [12,17] analyzed the passively secure GMW protocol.

Theorem 9 ([12,17]). *For any n-party circuit $\mathsf{C} : \mathbb{F}^{I_1} \times \cdots \times \mathbb{F}^{I_n} \to \mathbb{F}^{O_1}$, the protocol GMW_C in the OLE hybrid model is passively secure against any adversary controlling at most $n - 1$ parties. Moreover, the communication complexity (in field elements) as well as the number of oracle calls of GMW_C is $O(n^2 |C|)$.*

8.1   Randomizing the GMW Protocol

Note that the protocol Input-Share$_{\mathsf{GMW}}$ is already randomized. This is since additive secret sharing is done by having the party P_i, holding the input x, send random shares r_j to all other parties and then compute his share to be $x - \sum_j r_j$. Therefore, the messages exchanged during the input sharing phase are already input-independent. We now describe how to randomize the evaluation of multiplication gates in GMW protocol. In the Mult$_{\mathsf{GMW}}$ protocol, all messages received by the parties are sent by the f_{OLE} oracle. We thus construct the $f_{\mathsf{OLE}}^{\mathcal{T}}$ oracle which sends messages to the parties in $\mathcal{T}$ which only depend syntacticly on the randomness of the protocol and not on the inputs of the parties in $\overline{\mathcal{T}}$.

Construction 5 (The $f_{\mathsf{OLE}}^{\mathcal{T}}$ Functionality). *Let $\mathcal{T}$ be a set of parties. We define the functionality $f_{\mathsf{OLE}}^{\mathcal{T}}$ that on inputs (a, b) from a party P_i acting as a sender and $x \in \mathbb{F}$ from a party P_j acting as a receiver performs the following.*

1. $P_j \in \mathcal{T}$ *and* $P_i \in \overline{\mathcal{T}}$. *Let* P_h *be the first party not in* $\mathcal{T}$. $f_{\mathsf{OLE}}^{\mathcal{T}}$ *generates a random value* e, *sends* $\perp$ *to* P_i *and* e *to* P_j *and* $ax + b - e$ *to* P_h.
2. **Otherwise.** *In this case* $f_{\mathsf{OLE}}^{\mathcal{T}}$ *sends* $\perp$ *to* P_i *and* $ax + b$ *to* P_j.

In the following we describe the Mult$_{\mathsf{GMW}}^{\mathcal{T}}$ protocol in the $f_{\mathsf{OLE}}^{\mathcal{T}}$ hybrid model.

Construction 6 (Subprotocol Mult$_{\mathsf{GMW}}^{\mathcal{T}}$). *Let $\mathcal{T}$ be a set of parties and let P_h be the first party not in $\mathcal{T}$. The subprotocol Mult$_{\mathsf{GMW}}^{\mathcal{T}}$, in the $f_{\mathsf{OLE}}^{\mathcal{T}}$ hybrid model, gets as input additive sharings of $[\mathbf{a}]_{\mathsf{add}}$, $[\mathbf{b}]_{\mathsf{add}}$ and outputs an additive sharing $[\mathbf{c}]_{\mathsf{add}}$ such that $\mathbf{c} = \mathbf{a} \cdot \mathbf{b}$. The protocol proceeds as follows.*

1. *Each ordered pair of parties P_i, P_j, such that $i \neq j$, performs the following.*
   (a) P_i *generates a random value $r_{i,j}$ and acting as a sender sends $(a_i, r_{i,j})$ to the $f_{\mathsf{OLE}}^{\mathcal{T}}$ oracle. P_j acting as a receiver sends b_j to the $f_{\mathsf{OLE}}^{\mathcal{T}}$ oracle.*
   (b) *The $f_{\mathsf{OLE}}^{\mathcal{T}}$ oracle responds with $s_{i,j}$ to P_j, and with $s'_{i,j}$ to P_h in case that $P_j \in \mathcal{T}$ and $P_i \in \overline{\mathcal{T}}$.*
2. *Each party $P_i \in \mathcal{T}$ computes $c_i \leftarrow a_i \cdot b_i + \sum_{\substack{j=1 \\ j \neq i}}^{n} (s_{j,i} - r_{i,j})$.*
3. *Each party $P_i \in \overline{\mathcal{T}}$, such that $P_i \neq P_h$, generates his share c_i of c uniformly at random, computes $d_i \leftarrow a_i \cdot b_i + \sum_{\substack{j=1 \\ j \neq i}}^{n} (s_{j,i} - r_{i,j})$ and sends (c_i, d_i) to P_h.*
4. *Party P_h computes $c_h \leftarrow a_h \cdot b_h + \sum_{\substack{P_i \in \overline{\mathcal{T}} \\ P_i \neq P_h}} (d_i - c_i) + \sum_{\substack{P_i \in \overline{\mathcal{T}} \\ P_j \in \mathcal{T}}} s'_{i,j}$.*

Next, we describe the GMW$_{\mathsf{C}}^{\mathcal{T}}$ protocol. In the full version we prove that GMW$_{\mathsf{C}}^{\mathcal{T}}$ is $\mathcal{T}$-randomized and $\mathcal{T}$-equivalent to GMW$_{\mathsf{C}}$.

Construction 7 (GMW$_{\mathsf{C}}^{\mathcal{T}}$ Protocol). *Let $\mathsf{C} : \mathbb{F}^{I_1} \times \cdots \times \mathbb{F}^{I_n} \to \mathbb{F}^{O_1}$ be an n-party circuit and let $\mathcal{T}$ be a set of parties such that $|\mathcal{T}| < n$. The protocol GMW$_{\mathsf{C}}^{\mathcal{T}}$ for C is defined to be the same as the GMW$_{\mathsf{C}}$ protocol form Construction 4 except that the parties execute the Mult$_{\mathsf{GMW}}^{\mathcal{T}}$ protocol instead of Mult$_{\mathsf{GMW}}$.*

Lemma 2. *Let C be an n-party circuit. For any set of parties $\mathcal{T}$ such that $|\mathcal{T}| < n$ the protocol GMW$_{\mathsf{C}}^{\mathcal{T}}$ is $\mathcal{T}$-randomized and is $\mathcal{T}$-equivalent to GMW$_{\mathsf{C}}$.*

8.2   The GMW Protocol in the Presence of an Active Adversary

In this section we prove that the execution of the passively secure GMW protocol is additively corruptible. We begin by stating that $\mathsf{GMW}_\mathsf{C}^\mathcal{T}$ defined in Construction 4 is $\mathcal{T}$-last-round-private as well as $\mathcal{T}$-homomorphic to C.

Lemma 3. *Let n be positive integer and let C be an n-party circuit. Then for any set of parties $\mathcal{T}$ such that $|\mathcal{T}| < n$ it holds that the protocol $\mathsf{GMW}_\mathsf{C}^\mathcal{T}$ for computing C is $\mathcal{T}$-last-round-private as well as $\mathcal{T}$-homomorphic to C.*

Proof (sketch). The $\mathcal{T}$-last-round-private property follows from the fact that during the output recovery phase of the $\mathsf{GMW}_\mathsf{C}^\mathcal{T}$, all the parties locally re-randomize their shares with random sharings of 0. We now prove that $\mathsf{C}_{\mathsf{GMW}_\mathsf{C}^\mathcal{T}}$ is indeed $\mathcal{T}$-homomorphic to C. Fix randomness $\mathbf{r}$ for $\mathsf{C}_{\mathsf{GMW}_\mathsf{C}^\mathcal{T}}$. Next, for any input gate c of C, we set the homomorphism $\mathcal{H}$ to map c to the corresponding input gate in $\mathsf{C}_{\mathsf{GMW}_\mathsf{C}^\mathcal{T}}$. Finally, for every multiplication gate c of C, we set $\mathcal{H}$ to map c to a wire in $\mathsf{C}_{\mathsf{GMW}_\mathsf{C}^\mathcal{T}}$ corresponding to the share c_h, held by the party P_h in step 4 of the $\mathsf{Mult}_{\mathsf{GMW}}^\mathcal{T}$ protocol. Finally, we set λ^c to be the sum of all the shares c_i generated during steps 2 and 3 of $\mathsf{Mult}_{\mathsf{GMW}}^\mathcal{T}$. Notice that since $\mathsf{Mult}_{\mathsf{GMW}}^\mathcal{T}$ is $\mathcal{T}$-randomized, λ^c can be uniquely determined from $\mathbf{r}$. It can be easily verified that for every choice of $\mathbf{r}$ the homomorphism $\mathcal{H}$ as well as the constants λ^c, where c is a multiplication gate, satisfy all the requirements of Definition 9.   □

Combining the results of Lemmas 2 and 3 and Theorem 8 with additive-attack constructions in Sect. 5 we obtain the following theorem.

Theorem 10 (Cf. Theorem 1.5 in [11]). *For any n-party circuit $\mathsf{C} : \mathbb{F}^{I_1} \times \cdots \times \mathbb{F}^{I_n} \to \mathbb{F}^{O_1}$ there exists a protocol π for $O(1/|\mathbb{F}|)$-securely computing C with abort in the OLE hybrid model. Moreover π invokes the OLE oracle $O(n^2|\mathsf{C}|)$ times and has a total communication complexity of $O(n^2|\mathsf{C}|)$ field elements.*

Acknowledgments. We thank Manoj Prabhakaran, Amit Sahai and Eran Tromer for helpful discussions.

This research was supported by the European Union's Tenth Framework Programme (FP10/2010-2016) under grant agreement no. 259426 ERC-CaC. The first author was also supported by the Check Point Institute for Information Security; the Israeli Centers of Research Excellence I-CORE program (center 4/11); the Israeli Ministry of Science and Technology; the Leona M. & Harry B. Helmsley Charitable Trust. The second author was also supported by ISF grant 1709/14, and BSF grant 2012378. The third author was also supported by the Danish National Research Foundation; the National Science Foundation of China (grant no. 61061130540) for the Sino-Danish CTIC; the CFEM supported by the Danish Strategic Research Council.

References

1. Baron, J., El-Defrawy, K., Lampkins, J., Ostrovsky, R.: How to withstand mobile virus attacks, revisited. In: PODC. pp. 293–302 (2014)

2. Ben-Or, M., Goldwasser, S., Wigderson, A.: Completeness theorems for non-cryptographic fault-tolerant distributed computation (extended abstract). In: STOC. pp. 1–10 (1988)
3. Ben-Sasson, E., Fehr, S., Ostrovsky, R.: Near-linear unconditionally-secure multiparty computation with a dishonest minority. In: Safavi-Naini, R., Canetti, R. (eds.) CRYPTO 2012. LNCS, vol. 7417, pp. 663–680. Springer, Heidelberg (2012)
4. Bendlin, R., Damgård, I., Orlandi, C., Zakarias, S.: Semi-homomorphic encryption and multiparty computation. In: Paterson, K.G. (ed.) EUROCRYPT 2011. LNCS, vol. 6632, pp. 169–188. Springer, Heidelberg (2011)
5. Bracha, G.: An O(log n) expected rounds randomized byzantine generals protocol. J. ACM **34**(4), 910–920 (1987)
6. Canetti, R., Lindell, Y., Ostrovsky, R., Sahai, A.: Universally composable two-party and multi-party secure computation. In: STOC. pp. 494–503 (2002)
7. Chaum, D., Crépeau, C., Damgård, I.: Multiparty unconditionally secure protocols (extended abstract). In: STOC. pp. 11–19 (1988)
8. Damgård, I., Ishai, Y., Krøigaard, M.: Perfectly secure multiparty computation and the computational overhead of cryptography. In: Gilbert, H. (ed.) EUROCRYPT 2010. LNCS, vol. 6110, pp. 445–465. Springer, Heidelberg (2010)
9. Damgård, I.B., Nielsen, J.B.: Scalable and unconditionally secure multiparty computation. In: Menezes, A. (ed.) CRYPTO 2007. LNCS, vol. 4622, pp. 572–590. Springer, Heidelberg (2007)
10. Franklin, M.K., Yung, M.: Communication complexity of secure computation (extended abstract). In: STOC. pp. 699–710 (1992)
11. Genkin, D., Ishai, Y., Prabhakaran, M., Sahai, A., Tromer, E.: Circuits resilient to additive attacks with applications to secure computation. In: STOC. pp. 495–504 (2014)
12. Goldreich, O., Micali, S., Wigderson, A.: How to play any mental game or a completeness theorem for protocols with honest majority. In: STOC. pp. 218–229 (1987)
13. Hazay, C., Lindell, Y.: Efficient Secure Two-Party Protocols - Techniques and Constructions. Information Security and Cryptography. Springer, Heidelberg (2010)
14. Ikarashi, D., Kikuchi, R., Hamada, K., Chida, K.: Actively private and correct mpc scheme in t<n/2 from passively secure schemes with small overhead. IACR Cryptology ePrint Archive **2014**, 304 (2014)
15. Ishai, Y., Kushilevitz, E., Ostrovsky, R., Sahai, A.: Zero-knowledge from secure multiparty computation. In: STOC. pp. 21–30 (2007)
16. Ishai, Y., Prabhakaran, M., Sahai, A.: Founding cryptography on oblivious transfer – efficiently. In: Wagner, D. (ed.) CRYPTO 2008. LNCS, vol. 5157, pp. 572–591. Springer, Heidelberg (2008)
17. Ishai, Y., Prabhakaran, M., Sahai, A.: Secure arithmetic computation with no honest majority. In: Reingold, O. (ed.) TCC 2009. LNCS, vol. 5444, pp. 294–314. Springer, Heidelberg (2009)
18. Lindell, Y., Pinkas, B.: An efficient protocol for secure two-party computation in the presence of malicious adversaries. In: Naor, M. (ed.) EUROCRYPT 2007. LNCS, vol. 4515, pp. 52–78. Springer, Heidelberg (2007)
19. Naor, M., Pinkas, B.: Oblivious polynomial evaluation. SIAM J. Comput. **35**(5), 1254–1281 (2006)
20. Yao, A.C.: Protocols for secure computations (extended abstract). In: FOCS. pp. 160–164 (1982)

Large-Scale Secure Computation: Multi-party Computation for (Parallel) RAM Programs

Elette Boyle[1](✉), Kai-Min Chung[2], and Rafael Pass[3]

[1] Technion Israel, Haifa, Israel
eboyle@alum.mit.edu
[2] Academica Sinica, Taipei, Taiwan
kmchung@iis.sinica.edu.tw
[3] Cornell University, Ithaca, USA
rafael@cs.cornell.edu

Abstract. We present the first efficient (i.e., polylogarithmic overhead) method for securely and privately processing large data sets over multiple parties with *parallel, distributed algorithms*. More specifically, we demonstrate load-balanced, statistically secure computation protocols for computing Parallel RAM (PRAM) programs, handling $(1/3 - \epsilon)$ fraction malicious players, while preserving up to polylogarithmic factors the computation, parallel time, and memory complexities of the PRAM program, aside from a one-time execution of a broadcast protocol per party. Additionally, our protocol has polylog communication locality—that is, each of the n parties speaks only with polylog(n) other parties.

1 Introduction

Large data sets, such as medical data, genetic data, transaction data, the web and web access logs, and network traffic data, are now in abundance. Much of the data is stored or made accessible in a distributed fashion, having necessitated the development of efficient distributed protocols that compute over such data. In particular, novel programming models for processing large data sets with *parallel, distributed algorithms*, such as MapReduce (and its implementation Hadoop) are emerging as crucial tools for leveraging this data in important ways.

But these methods require that the data itself is revealed to the participating servers performing the computation—and thus blatantly violate the privacy of

E. Boyle—The research of the first author has received funding from the European Union's Tenth Framework Programme (FP10/ 2010-2016) under grant agreement no. 259426 ERC-CaC, and ISF grant 1709/14.

R. Pass—Pass is supported in part by a Alfred P. Sloan Fellowship, Microsoft New Faculty Fellowship, NSF Award CNS-1217821, NSF CAREER Award CCF-0746990, NSF Award CCF-1214844, AFOSR YIP Award FA9550-10-1-0093, and DARPA and AFRL under contract FA8750-11-2- 0211. The views and conclusions contained in this document are those of the authors and should not be interpreted as representing the official policies, either expressed or implied, of the Defense Advanced Research Projects Agency or the US Government.

© International Association for Cryptologic Research 2015
R. Gennaro and M. Robshaw (Eds.): CRYPTO 2015, Part II, LNCS 9216, pp. 742–762, 2015.
DOI: 10.1007/978-3-662-48000-7_36

potentially sensitive data. As a consequence, such methods cannot be used in many critical applications (e.g., discovery of causes or treatments of diseases using genetic or medical data).

In contrast, methods such as secure multi-party computation (MPC), introduced in the seminal works of Yao [Yao86] and Goldreich, Micali and Wigderson [GMW87], enable securely and privately performing any computation on individuals private inputs (assuming some fraction of the parties are honest). However, despite great progress in developing these techniques, there are no MPC protocols whose efficiency and communication requirements scale to the modern regime of large-scale distributed, parallel data processing.

We are concerned with merging these two approaches. In particular,

We seek MPC protocols that efficiently (technically, with polylogarithmic overhead) enable secure and private processing of large data sets with parallel, distributed algorithms.

Explicitly, in this large-scale regime, the following properties are paramount:

1. *Exploiting Random Access.* Computations on large data sets are frequently "lightweight": accessing a small number of dynamically chosen data items, relying on conditional branching, and/or maintaining small memory. This means that converting a program first into a circuit to enable its secure computation, which immediately obliterates these gains, will not be a feasible option.
2. *Exploiting Parallelism.* In fact, as mentioned, to effectively solve large-scale problems, modern programming models heavily leverage parallelism. The notion of a Parallel RAM (PRAM) better captures such computing models. In the PRAM model of computation, several (polynomially many) CPUs run simultaneously, potentially communicating with one another, while accessing the same shared external memory. We consider a PRAM model with a variable number of CPUs but with a fixed activation structure (i.e., what processors are activated at which time steps is fixed). Note that such a model simultaneously captures RAMs (a single CPU) and circuits (the circuit topology dictates the CPU activation structure).
3. *Exploiting Plurality of Users.* In the setting of MPC we would like to leverage not only parallelism within a single party (i.e., if a party has multiple CPUs that may run in parallel), but also that we have a large number of parties that can run in parallel. So, if we have n parties, each with k processors, we ideally would like to securely compute PRAMs that use nk CPUs (as opposed to just k CPUs).

Additionally, the following desiderata are often of importance:

4. *Load balancing.* When the data set contains tens or hundreds of thousands of users' data, it is often unreasonable to assume that any single user can provide memory, computation, or communication resources on the order of the data of *all users*. Rather, we would like to *balance* the load across nodes.

5. *Communication Locality.* In many cases, establishing a secure communication channel with a large number of distinct parties may be costly, and thus we would like to minimize the *locality of communication* [BGT13]: that is, the number of total parties that each party must send and receive message to during the course of the protocol.

To date, no existing work addresses secure computation of Parallel RAM programs. Indeed, nearly all results in MPC require a *circuit* model for the function being evaluated (including the line of work on scalable MPC [DI06, DIK+08, DKMS12, ZMS14]), and thus inherit resource requirements that are linear in the circuit size. Even for (sequential) RAM, the only known protocols either only handle two parties [OS97, GKK+11, LO13, GGH+13], or in the context of multiparty computation require all parties to store *all inputs* [DMN11], rendering the protocol useless in a large-scale setting (even forgetting about computation load balancing and locality).

1.1   Our Results

We present a statistically secure MPC for (any sequence of) PRAMs handling $(1/3 - \epsilon)$ fraction static corruptions in a synchronous communication network, with secure point-to-point channels. In addition, our protocol is strongly *load balanced* and *communication local* (i.e., $\mathsf{polylog}(n)$ locality). We state our theorem assuming each party itself is a k-processor PRAM, for parameter k.

Theorem 1 (Informal – Main Theorem). *For any constant $\epsilon > 0$ and polynomial parallelism parameter $k = k(n)$, there exists an n-party statistically secure (with error negligible in n) protocol for computing any adaptively chosen sequence of PRAM programs Π_j with fixed CPU activation structures (and that may have bounded shared state), handling $(1/3 - \epsilon)$ fraction static corruptions with the following complexities, where each party is a k-processor PRAM (and where $|x|, |y|$ denote per-party input and output size,[1] $\mathsf{space}(\Pi)$, $\mathsf{comp}(\Pi)$, and $\mathsf{time}(\Pi)$ denote the worst-case space, computation, and (parallel) runtime of Π, and $CPUs(\Pi)$ denotes the number of CPUs of Π):*

- *Computation per party, per Π_j: $\tilde{O}(\mathsf{comp}(\Pi_j)/n + |y|)$.*
- *Time steps, per Π_j: $\tilde{O}\left(\mathsf{time}(\Pi_j) \cdot \max\left\{1, \frac{CPUs(\Pi)}{nk}\right\}\right)$.*
- *Memory per party: $\tilde{O}\left(|x| + |y| + \max_{j=1}^{N} \mathsf{space}(\Pi_j)/n\right)$.*
- *Communication Locality: $\tilde{O}(1)$.*

given a one-time preprocessing phase with complexity:

- *Computation per party: $\tilde{O}(|x|)$, plus single broadcast of $\tilde{O}(1)$ bits.*
- *Time steps: $\tilde{O}\left(\max\left\{1, \frac{|x|}{k}\right\}\right)$.*

---

[1] For simplicity of exposition, we assume all parties have the same input size and receive the same output.

Additionally, our protocol achieves a strong "online" load-balancing guarantee: at all times during the protocol, all parties' communication and computation loads vary by at most a constant multiplicative factor (up to a $\mathsf{polylog}(n)$ additive term).

Remark 1 (Round complexity). As is the case with all general MPC protocols in the information-theoretic setting to date, the round complexity of our protocol corresponds directly with the time complexity (as when restricted to circuits, parallel complexity corresponds to circuit depth). That is, for each evaluated PRAM program Π_j, the protocol runs in $\tilde{O}(\mathsf{time}(\Pi_j))$ sequential communication rounds to securely evaluate Π_j.

Remark 2 (On the achieved parameters). Note that in terms of memory, each party only stores her input, output, and her "fair" share of the required space complexity, up to polylogarithmic factors. In terms of computation (up to polylogarithmic factors), each party does her "fair" share of the computation, receives her outputs, and in addition is required to read her entire input at an initial preprocessing stage (even though the computations may only involve a subset of the input bits; this additional overhead of "touching" the whole input once is necessary to achieve security).[2] Finally, the time complexity corresponds to the parallel complexity of the PRAM being computed, as long as the combined number of available processors nk from all parties matches or exceeds the number of required parallel processes of the program (and degrades with the corresponding deficit).

Remark 3 (Instantiating the single-use broadcast). The broadcast channel can be instantiated either by the $O(\sqrt{n})$-locality broadcast protocol of King *et al.* [KSSV06], or the $\mathsf{polylog}(n)$-*average* locality protocol of [BSGH13] at the expense of a cost of a one-time per-party computational cost of $O(\sqrt{n})$, or average cost of $\mathsf{polylog}(n)$, respectively. We separate the broadcast cost from our protocol complexity measures to emphasize that any (existing or future) broadcast protocol can be directly plugged in, yielding associated desirable properties.[3]

1.2   Construction Overview

Our starting point is an *Oblivious PRAM (OPRAM)* compiler [BCP14b, GO96], a tool that compiles any PRAM program into one whose memory access patterns

---

[2] For general secure computation, and even if we restrict to functionalities that only access a few parties' inputs, and only a few bits of their data, essentially all parties must perform computation at least $\Omega(|x|)$. To see this, consider secure computation of a "multi-party Private Information Retrieval (PIR)" functionality: each party $i > 1$ has as input some "big data" x_i, and party 1 has as input a party index i and an index j into their data x_i. The functionality returns $x_i[j]$ (i.e., the j'th bit of party i's data) to party 1 and nothing to everyone else. We claim that each party $i > 1$ must access every bit of x_i; if not, it learns that particular bit of its data was not requested, which it cannot learn in an ideal execution of the functionality.

[3] For instance, it remains open to achieve statistically secure broadcast with worst-case $\mathsf{polylog}(n)$ locality.

are independent of the data (i.e., "oblivious"). Such a compiler (with polyloga-rithmic overhead) was recently attained by [BCP14b].

Indeed, it is no surprise that such a tool will be useful toward our goal. It has been demonstrated in the sequential setting that Oblivious (sequential) RAM (ORAM) compilers can be used to builds secure 2-party protocols for RAM programs [OS97, GKK+11, LO13, GGH+13]. Taking a similar approach, building upon the OPRAM compiler of [BCP14b] directly yields 2-party protocols for PRAMs.

However, OPRAM on its own does not directly provide a solution for *multi-party* computation (when there are many parties). While this approach gives protocols whose complexities scale well with the RAM (or PRAM) complexity of the programs, the complexities grow poorly with the number of parties. Indeed, the only current technique for securely evaluating a RAM program on *multiple* parties' inputs [DMN11] is for all parties to hold secret shares of all parties' inputs, and then jointly execute (using standard MPC for circuits) the trusted CPU instructions of the ORAM-compiled version of the program. This means each party must communicate and maintain information of size equivalent to *all parties'* inputs, and everyone must talk to everyone else for every time step of the RAM program evaluation.

One may attempt to improve the situation by first electing a small $\mathsf{polylog}(n)$-size representative committee of parties, and then only performing the above steps within this committee. This approach drops the total communication and computation of the protocol to reasonable levels. However, this approach does not save the subset of elected parties from carrying the burden of the entire computation. In particular, each elected party must memory storage equal to the size of *all parties' inputs combined*, making the protocol unusable for "large-scale" computation.

In this paper, we provide a new approach for dealing with this issue. We show how to use an OPRAM in a way that achieves balancing of memory, computation, and communication across all parties.

Our MPC construction proceeds in the following steps:

1. **From OPRAM to MPC.** Given an OPRAM, we begin by considering MPC in a "benign" adversarial setting, which we refer to as *oblivious multi-party computation*, where all parties are assumed to be honest, and we only require that an external attacker that views communication and activation (including memory and computation usages) patterns does not learn anything about the inputs. We show:
   (a) OPRAM yields efficient *memory-balanced* oblivious MPC for PRAM.
   (b) Using committee election techniques (à la [KLST11, DKMS12, BGT13]), any oblivious multi-party computation can be compiled into a standard secure MPC with only $\mathsf{polylog}$ overhead (and a one-time use of a broadcast channel per party).
2. **Load Balancing & Communication Locality.** We next show semi-generic compilers for "nice" (formally defined) *oblivious multi-party* protocols, each introducing only $\mathsf{polylog}(n)$ overhead:

(a) From any "nice" protocol to one whose computation and communication are *load-balanced*.
(b) From any "nice" protocol to one that is both *load-balanced* and *communication local* (i.e., polylog(n) locality).

Our final result is obtained by combining the above steps and observing that Step 1(b) preserves load-balancing and communication locality (and thus can be applied *after* Step 2). Let us mention that *just* Step 1 (together with existing construction of ORAMs) already yields the first MPC protocol for (sequential) RAM programs in which no party must store all parties' inputs. Additionally, *just* Step 1 (together with the OPRAM construction of [BCP14b]) yields the first MPC for PRAMs.

We now expand upon each of these steps.

MPC from OPRAM. Recall that our construction proceeds via an intermediate notion of *oblivious* security, in which we do not require security against corrupted parties, but rather against an external adversary who sees the activation patterns (i.e., accessed memory addresses and computation times) and communication patterns (i.e., sender/receiver ids and message lengths) of parties throughout the protocol.

Oblivious MPC from OPRAM. At a high level, our protocol will emulate a *distributed* OPRAM[4] structure, where the CPUs and memory cells in the OPRAM are *each associated with parties*. (Recall that we need only achieve "oblivious" security, and thus can trust individual parties with these tasks). The "CPU" parties will control the evaluation flow of the (OPRAM-compiled) program, communicating with the parties emulating the role of the appropriate memory cells for each address to be accessed in the (OPRAM-compiled) database.

The distributed OPRAM structure will enable us to evenly spread the memory burden across parties, incurring only polylog(n) overhead in total memory and computation, and while guaranteeing that the communication patterns between committees (corresponding to data access patterns) do not reveal information on the underlying secret values.

This framework shares a similar flavor to the protocols of [DKMS12, BGJK12], which assign committees to each of the gates of a circuit being evaluated, and to [BGT13], which uses CPU and input committees to direct program execution and distributedly store parties' inputs. The distributed OPRAM idea improves and conceptually simplifies the input storage handling of Boyle *et al.* [BGT13], in which n committees holding the n parties' inputs execute a distributed "oblivious input shuffling" procedure to break the link between which committees are communicating and which inputs are being accessed in the computation.

---

[4] We remark that the term "distributed ORAM" was used with a different meaning in [LO13], in regard to an ORAM that was split across two users.

Compiling from "Oblivious" Security to Malicious Security. We next present a general compiler taking an oblivious protocol to one that is secure against $(1/3 - \epsilon)n$ statically corrupted malicious parties. (This step can be viewed as a refinement and generalization of ideas from [KLST11, DKMS12, BGT13].) We ensure the compiler tightly preserves the computation, memory, load-balancing, and communication locality of the original protocol, up to polylog(n) factors (modulo a one-time broadcast per party). This enables us to apply the transformation to any of the oblivious protocols resulting from the intermediate steps in our progression.

At a high level, the compiler takes the following form: (1) First, the parties collectively elect a large number of "good" committees, each of size polylog(n), where "good" means each committee is composed of at least $2/3$ honest parties, and that parties are spread roughly evenly across committees. (2) Each party will verifiably secret share his input among the corresponding committee C_i. (3) From this point on, the role of each party P_i in the original protocol will be emulated by the corresponding committee C_i. That is, each local P_i computation will be executed via a small-scale MPC among C_i, and each communication from P_i to P_j will be performed via an MPC among committees C_i and C_j.

The primary challenge in this step is how to elect such committees while incurring only polylog(n) locality and computation per party. To do so, we build atop the "almost-everywhere" scalable committee election protocol of King *et al.* [KSSV06] to elect a single good committee, and then show that one may use a polylog(n)-wise independent function family $\{F_s\}_{s \in S}$ to elect the remaining committees with small description size (in the fashion of [KLST11, BGT13], for the case of combinatorial samplers and computational pseudorandom functions), with committee i defined as $C_i := F_s(i)$ for fixed random seed s.

We remark that, aside from the one-time broadcast, this compiler preserves load balancing and polylog(n) locality. Indeed, load balancing is maintained since the committee setup procedure is computationally inexpensive, and each party appears in roughly the same number of "worker" committees. The locality of the resulting protocol increases by an additive polylog(n) for the committee setup, and a multiplicative polylog(n) term since all communications are now performed among polylog(n)-size *committees* instead of individual parties.

Load Balancing Distributed Protocols

Load-Balancing (Without Locality). We now show how to modify our protocol such that the total computational complexity and memory balancing are preserved, while additionally achieving a strong *computation* load balancing property—with high probability, at all times throughout the protocol execution, every party performs close to $1/n$ fraction of current total work, up to an additive polylog(n) amount of work. This will hold simultaneously for both computation and communication.[5]

---

[5] Note that while our current protocol is memory balanced, it is currently rather imbalanced in computation: e.g., the parties emulating OPRAM CPUs are required to perform computation that is proportional to the whole PRAM computation.

We present and analyze our load-balancing solution in the intermediate *oblivious* MPC security setting (recall that one can then apply the compiler from Step 2(b) above to obtain malicious MPC with analogous load-balancing). Let us mention that there is a huge literature on "load-balanced distributed computation" (e.g., [ACMR95, MPS02, MR98, AAK08]): As far as we can tell, our setting differs from the typical studied scenarios in that we must load balance an underlying distributed protocol, as opposed to a collection of independent "non-communicating jobs". Indeed, the main challenge in our setting is to deal with the fact that "jobs" talk to one another, and this communication must remain efficient also be made load balanced. Furthermore, we seek a load-balanced solution with communication locality.

We consider a large class of arbitrary (potentially load-unbalanced and large-locality) distributed protocols Π, where we view each party in this underlying protocol as a "job". Our goal is to load-balance Π by passing "jobs" between "workers" (which will be the actual parties in the new protocols). More precisely, we start off with any protocol Π that satisfies the following (natural) "nice" properties:

- Each "job" has polylog(n) size state;
- In each round, each "job" performs at most polylog(n) computation and communication;
- In each round, each "job" communicates (either sending or receiving a message) to at most one other "job".

It can be verified that these properties hold for our oblivious MPC for PRAM protocol.

Our load-balanced version of such a protocol first randomly[6] efficiently assigns "workers" (i.e., parties) to "jobs". Next, whenever a worker W has performed "enough" work for a particular job J, it randomly selects a replacement worker W' and passes the job over to it (that is, it passes over the state of the job J—which is "small" by assumption). The key obstacle in our setting is that the job J may later communicate with many other jobs, and all the workers responsible for those jobs need to be informed of the switch (and in particular, who the new worker responsible for the job J is). Since the number of jobs is $\Omega(n)$, workers cannot afford to store a complete directory of which worker is currently responsible for each job.

We overcome this obstacle by first modifying Π to ensure that it has small locality—this enables each job to only maintain a short list of the workers currently responsible for the *"neighboring"* jobs. We achieve this locality by requiring that parties (i.e., jobs) in the original protocol Π route their messages along the hypercube. Now, whenever a worker W for a job J is being replaced by some worker W', W informs all J's neighboring jobs (i.e., the workers responsible for them) of this change. We use the Valiant-Brebner [VB81] routing procedure to implement the hypercube routing because it ensures a desirable "low-congestion

---

[6] In the actual analysis, we show that it also suffices to use polylog(n)-wise independent randomness to pick this and subsequent assignments.

property," which in our setting translates to ensuring that the overhead of routing is not too high for any individual worker.

The above description has not yet mentioned what it means for a worker to have done "enough" work for a job J. Each round a job is active (i.e., performing some computation), its "cost" increases by 1—we refer to this as an *emulation cost*. Additionally, each time a worker W is switched out from a job J, then J's and each of J's neighboring jobs' costs are increased by 1—we refer to this as a *switch cost*. Finally, once a job's (total) cost has reached a particular threshold τ, its cost is reset to 1 and the worker responsible for the job is switched out. The threshold τ is set to $2 \log M + 1$ where M is the number of jobs.

We show: (1) This switching does not introduce too much overhead. We, in fact, show that the total induced switching cost is bounded above by the emulation cost. (2) The resulting total work is load balanced across workers—we show this by first demonstrating that the protocol is load-balanced in expectation, and then using concentration to argue our stronger online load-balancing property.

Finally, note that although communication between *jobs* is being routed through the hypercube, and thus the job communication protocol has small locality, the final load-balanced protocol, being run by *workers*, does *not* have small locality. This is because workers are assigned the role of many different jobs over time, and may possibly speak to a new set of neighbors for each position. (Indeed, over time, each worker will eventually need to speak to every other worker). We next show how to modify this protocol to achieve *locality*, while preserving load-balancing.

Achieving Both Load-Balancing and Locality. In our final step, we show how to modify the above-mentioned protocol to also achieve locality. We modify the protocol to also let *workers* route messages through a low-degree network (on top of the routing in the previous step). This immediately ensures locality. But, we must be careful to ensure that the additional message passing does not break load-balancing.

A natural idea is to again simply pass messages between *workers* along a low-degree hypercube network via Valiant-Brebner (VB) routing [VB81]. Indeed, the low-congestion property will ensure (as before) that routing does not incur too large an overhead for each worker.

However, when analyzing the overall load balance (for workers), we see an inherent distinction between this case and the previous. Previously, the nodes of the hypercube corresponded to *jobs*, each emulated by workers who swap in and out over time. When the underlying jobs protocol required job s to send a message to job t, the resulting message routing induced a cost along a path of neighboring jobs (that is, the workers emulating them), *independent of which workers are currently emulating them*. This independence, together with the fact that a worker passes his job after performing "enough" work for it, enabled us to obtain concentration bounds on overall load balancing over the random assignment of workers to jobs.

Now, the nodes correspond directly to *workers*. When the underlying jobs protocol requires a message transferred from job s to job t, routing along the workers' graph must traverse a path from the *worker currently emulating job s* to the *worker currently emulating job t*, removing the crucial independence property from above. Even worse, workers along the routing path can now incur costs *even if they are not assigned to any job*. In this case, it is not even clear that job passing in of itself will be sufficient to ensure balancing.

To get around these issues, we add an extra step in the VB routing procedure (itself inspired by [VB81]) to break potential bad correlations. The idea is as follows: To route from the worker W_s emulating job s to the worker W_t emulating job t, we first route (as usual) from W_s to a *random* worker W_u, and then from W_u to W_t; i.e., travel from W_s to W_t by "walking into the woods" and back. We may now partition the cost of routing into these two sub-parts, each associated with a single active job (s or t). Now, although workers along the worker-routing path will still incur costs from this routing (even though their jobs may be completely unrelated), the *distribution* of these costs on workers depends only on the identity of the initiating worker (W_s or W_t). We may thus generalize the previous analysis to argue that if the expectation of work is load-balanced, then it still has concentration in this case.

For a modular analysis, we formalize the required properties of the underlying communication network and routing algorithm (to be used for the s-to-u and u-to-t routing) as a *local load-balanced routing network*, and show that the hypercube network together with VB routing satisfies these conditions.

1.3   Discussion and Future Work

With the explosive growth of data made available in a distributed fashion, and the growth of efficient parallel, distributed algorithms (such as those enabled by MapReduce) to compute on this data, ensuring privacy and security in such large-scale parallel settings is of fundamental importance. We have taken the first steps in addressing this problem by presenting the first protocols for secure multi-party computation, that with only polylogarithmic overhead, enable evaluating PRAM programs on a (large) number of parties' inputs. Our work leaves open several interesting open problems:

Honest Majority. We have assumed that 2/3 of the players are honest. In the absence of a broadcast channel,[7] it is known that this is optimal. But if we assume the existence of a broadcast channel, it may suffice to assume 1/2 fraction honest players.

Asynchrony. Our protocol assumes a synchronous communication network. We leave open the handling of asynchronous communication.

Trading efficiency for security. An interesting avenue to pursue are various tradeoffs between boosted efficiency and partial sacrifices in security. For example, in some settings, it is not detrimental to leak which parties' inputs were used within the computation; in such scenarios, one could then hope

---

[7] While the statement of our result makes use of a broadcast channel, as we mention, this channel can also be instantiated with known protocols.

to remove the one-time $\Theta(n|x|)$ input preprocessing cost. Similarly, it may be acceptable to reveal the input-specific resources (runtime, space) required by the program on parties inputs; in such cases, we may modify the protocol to take only input-specific runtime and use input-specific memory.

In this work we focus only on achieving standard "full" security. However, we remark that our protocol can serve as a solid basis for achieving such tradeoffs (e.g., a straightforward tweak to our protocol results in input-specific resource use).

Communication complexity. As with all existing generic multi-party computation protocols in the information-theoretic setting, the communication complexity of our protocol is equal to its computation complexity. In contrast, in the computational setting (based on cryptographic assumptions), protocols with communication complexity below the complexity of the evaluated function have been constructed by relying on *fully homomorphic encryption (FHE)* [Gen09] (e.g., [Gen09, AJLA+12, MSS13]). We leave as an interesting open question whether FHE-style techniques can be applied also to our protocol to improve the communication complexity, based on computational assumptions.

1.4 Overview of the Paper

Section 2 contains preliminaries. In Sect. 3 we provide our ultimate theorem, and the sequence of intermediate notions and theorems which combine to yield this final result. We refer the reader to the full version of this work [BCP14a] for a complete descriptions and proofs.

2 Preliminaries

2.1 Multi-party Computation (MPC)

Protocol Syntax. We model parties as (parallel) RAM machines. An n-party protocol Φ is described as a collection of n (parallel) RAM programs $(P_i)_{i \in [n]}$, to be executed by the respective parties, containing additional special communication instructions $\mathsf{Comm}(i, \mathsf{msg})$, indicating for the executing party to send message msg to party i.

The per-party space, computation, and time complexities of the protocol $\Phi = (P_i)_{i \in [n]}$ are defined directly with respect to the corresponding party's PRAM program P_i, where each Comm is charged as a single computation time step. (See Sect. 2.2 for a definition of $CPUs(P)$, $\mathsf{space}(P)$, $\mathsf{comp}(P)$, $\mathsf{time}(P)$ for PRAM P). The analogous total protocol complexities are defined as expected: Namely, $\mathsf{space}(\Phi)$ and $\mathsf{comp}(\Phi)$ are the *sums*, $\mathsf{space}(\Phi) = \sum_{i \in [n]} \mathsf{space}(P_i)$, $\mathsf{comp}(\Phi) = \sum_{i \in [n]} \mathsf{comp}(P_i)$, and $\mathsf{time}(\Phi)$ is the *maximum*, $\mathsf{time}(\Phi) = \max_{i \in [n]} \mathsf{time}(P_i)$.

MPC Security. We consider the standard notion of (statistical) MPC security. We refer the reader to e.g. [BGW88] for more a more complete description of MPC security within this setting.

2.2   Parallel RAM (PRAM) Programs

A Concurrent Read Concurrent Write (CRCW) m-processor *parallel random-access machine (PRAM)* with memory size n consists of numbered processors $CPU_1, \ldots, CPU_m$, each with local memory registers of size $\log n$, which operate synchronously in parallel and can make access to shared "external" memory of size n.

A PRAM program Π (given m, n, and some input x stored in shared memory) provides CPU-specific execution instructions, which can access the shared data via commands $\mathsf{Access}(r, v)$, where $r \in [n]$ is an index to a memory location, and v is a word (of size $\log n$) or $\perp$. Each $\mathsf{Access}(r, v)$ instruction is executed as:

1. **Read** from shared memory cell address r; denote value by v_{old}.
2. **Write** value $v \neq \perp$ to address r (if $v = \perp$, then take no action).
3. **Return** v_{old}.

In the case that two or more processors simultaneously initiate $\mathsf{Access}(r, v_i)$ with the same address r, then all requesting processors receive the previously existing memory value v_{old}, and the memory is rewritten with the value v_i corresponding to the lowest-numbered CPU i for which $v_i \neq \perp$.

We more generally support PRAM programs with a dynamic number of processors (i.e., m_i processors required for each time step i of the computation), as long as this sequence of processor numbers $m_1, m_2, \ldots$ is fixed, public information. The complexity of our OPRAM solution will scale with the number of required processors in each round, instead of the maximum number of required processors.

We consider the following *worst-case* metrics of a PRAM (over all inputs):

- $CPUs(\Pi)$: number of parallel processors required by Π.
- $\mathsf{space}(\Pi)$: largest database address accessed by Π.
- $\mathsf{time}(\Pi)$: maximum number of time steps taken by any processor to evaluate Π (where each Access is charged as a single step).[8]
- $\mathsf{comp}(\Pi)$: the total sum of all computation steps of active CPUs evaluating Π (which, for programs with fixed activation schedules as we consider, is a fixed value).

3   Local, Load-Balanced MPC for PRAM

Ultimately, we construct a protocol that securely realizes the ideal functionality $\mathcal{F}_{\mathsf{PRAMs}}$ (Fig. 1) for evaluating a sequence of PRAM programs (with bounded state maintained between program) on parties' fixed inputs. For simplicity of exposition, we assume each party has equal input size and receives the same

---

[8] We remark that the PRAM time complexity of any function f is bounded above by its circuit depth complexity (where the PRAM complexity of f is defined as the minimal value of $\mathsf{time}(\Pi)$ of any PRAM Π which evaluates f).

output. We further assume the total remnant state from one program execution to the next is bounded in size by the combined input size of all parties.[9]

Theorem 2 (Main Theorem). *For any constant $\epsilon > 0$ and polynomial parallelism parameter $k = k(n)$, there exists an n-party statistically secure (with error negligible in n) protocol realizing the functionality $\mathcal{F}_{\mathsf{PRAMs}}$, handling $(1/3 - \epsilon)$ fraction static corruptions with the following complexities, where each party is a k-processor PRAM (and where $|x|, |y|$ denote per-party input and output size, $\mathsf{space}(\Pi)$, $\mathsf{comp}(\Pi)$, and $\mathsf{time}(\Pi)$ denote the worst-case space, computation, and (parallel) runtime of Π, and $CPUs(\Pi)$ denotes the number of CPUs of Π):*

- *Computation per party, per Π_j: $\tilde{O}\big(\mathsf{comp}(\Pi_j)/n + |y|\big)$.*
- *Time steps, per Π_j: $\tilde{O}\left(\mathsf{time}(\Pi_j) \cdot \max\left\{1, \frac{CPUs(\Pi)}{nk}\right\}\right)$.*
- *Memory per party: $\tilde{O}\left(|x| + |y| + \max_{j=1}^{N} \mathsf{space}(\Pi_j)/n\right)$.*
- *Communication Locality: $\tilde{O}(1)$.*

given a one-time preprocessing phase with complexity:

- *Computation per party: $\tilde{O}(|x|)$, plus single broadcast of $\tilde{O}(1)$ bits.*
- *Time steps: $\tilde{O}\left(\max\left\{1, \frac{|x|}{k}\right\}\right)$.*

Additionally, the protocol achieves $\mathsf{polylog}(n)$ communication locality, and a strong "online" load-balancing guarantee:

  Online Load Balancing: *For every constant $\delta > 0$, with all but negligible probability in n, the following holds at all times during the protocol: Let cc and $\mathsf{cc}(W_j)$ denote the total communication complexity and communication complexity of party P_j, comp and $\mathsf{comp}(P_j)$ denote the total computation complexity and computation complexity of party P_j, we have*

$$\frac{(1-\delta)}{n}\mathsf{cc} - \mathsf{polylog}(n) \leq \mathsf{cc}(P_j) \leq \frac{(1+\delta)}{n}\mathsf{cc} + \mathsf{polylog}(n)$$

$$\frac{(1-\delta)}{n}\mathsf{comp} - \mathsf{polylog}(n) \leq \mathsf{comp}(P_j) \leq \frac{(1+\delta)}{n}\mathsf{comp} + \mathsf{polylog}(n).$$

3.1   Proof of Main Theorem

At a very high level, the proof takes three steps: We first obtain MPC realizing $\mathcal{F}_{\mathsf{PRAMs}}$ with a weaker notion of *oblivious* security. We then show how to attain communication locality and load balancing, while preserving oblivious security. (This combines two steps described within the introduction). Finally, we convert the obliviously secure protocol to one secure in the malicious setting. We now proceed to describe these steps in greater technical detail.

---

[9] To support larger shared state size $\mathsf{space}^{\mathsf{Remnant}}$, the memory requirements of the protocol must grow with an extra additive $\tilde{O}(\mathsf{space}^{\mathsf{Remnant}})$.

Ideal Functionality $\mathcal{F}_{\mathsf{PRAMs}}$:

$\mathcal{F}_{\mathsf{PRAMs}}$ running with parties $P_1, \ldots, P_n$ and an adversary proceeds as follows. The functionality maintains longterm storage of parties' inputs $\{x_i\}_{i \in [n]}$ (each of equal size $|x|$), per-CPU state information state_i, and remnant memory $\mathsf{data}^{\mathsf{Remnant}}$ of total size $\mathsf{space}^{\mathsf{Remnant}} \in O(n \cdot |x|)$ transferred from computation to computation.

- Initialize $\mathsf{data}^{\mathsf{Remnant}} \leftarrow \emptyset$ and $\mathsf{state}_i \leftarrow \emptyset$ for each processor $i \in [m]$.
- Input Submission: Upon receiving an input $(commit, sid, input, x_i)$ from party P_i, record the value x_i as the input of P_i.
- Computation: Upon receiving a tuple $(compute, sid, \Pi, \mathsf{space}, \mathsf{time})$ consisting of an m-processor PRAM program Π, a space bound space, and a time bound time, execute Π as $(\mathsf{output}, \mathsf{state}_1, \ldots, \mathsf{state}_m, \mathsf{data}^{\mathsf{Remnant}}) \leftarrow \Pi(x_1, \ldots, x_n, \mathsf{state}_1, \ldots, \mathsf{state}_m, \mathsf{data}^{\mathsf{Remnant}})$ with the current value of state_i for each CPU $i \in [m]$. Send output to all parties.

Fig. 1. The ideal functionality $\mathcal{F}_{\mathsf{PRAMs}}$, corresponding to secure computation of a sequence of adaptively chosen PRAMs on parties' inputs.

Step 1: Oblivious-Secure MPC for PRAM. Intuitively, an adversary in the oblivious model is not allowed to corrupt any parties, and instead is restricted to seeing the "externally measurable" properties of the protocol (e.g., party response times, communication patterns, etc.).

Definition 1 Oblivious Secure MPC). *Secure realization of a functionality F by a protocol in the oblivious model is defined by the following real-ideal world scenario:*

Ideal World: Same as standard MPC without corrupted parties. That is, the adversary learns only public outputs of the functionality F evaluated on honest-party inputs.

Real World: Instead of corrupting parties, viewing their states, and controlling their actions (as in the standard malicious adversarial setting), the adversary is now limited as an external observer, and is given access only to the following information:

- *Activation Patterns: Complete list of tuples of the form*
  - *(timestep, party-id, compute-time): Specifying all local computation times of parties.*
  - *(timestep, party-id, local-mem-addr): Specifying all memory access patterns of parties.*
- *Communication Patterns: Complete list of tuples of the form*
  - *(timestep, sndr-id, rcvr-id, msg-len): Specifying all sender-receiver pairs, in addition to the corresponding communicated message bit-length.*

The output of the real-world experiment consists of the outputs of the (honest) parties, in addition to an arbitrary PPT function of the adversary's view at the conclusion of the protocol.

(Statistical) Security: For every PPT adversary $\mathcal{A}$ in the real-world execution, there exists a PPT ideal-world adversary § for which for every environment $\mathcal{Z}$, we have $\mathsf{output}_{\mathsf{Real}}(1^k, \mathcal{A}, \mathcal{Z}) \overset{s}{\cong} \mathsf{output}_{\mathsf{Ideal}}(1^k, §, \mathcal{Z})$.

Toward our result, it will be advantageous to think of computations as composed of several sub-parts, or *"jobs,"* that each maintain and compute on small polylogarithmic-size state (Note that this is natural in the PRAM setting, where each CPU has polylogarithmic-size local memory). Later, to achieve load balancing, jobs will be assigned to and passed around between *"workers,"* so that each worker roughly performs the same amount of work. (The small state requirement per job will guarantee that "job passing" is not too expensive). Then, to obtain malicious security, each worker will ultimately be emulated by a committee of parties via small-scale MPCs; because of the polynomial overhead in the underlying MPC protocol, it will be important that this is only done for computations of $\mathsf{polylog}(n)$ size on $\mathsf{polylog}(n)$-size memory.

We now define the notion of a protocol in the *jobs model*.

Definition 2 (Jobs Model). *Let n be a security parameter. A jobs protocol consists of a $\mathsf{poly}(n)$-size set Jobs of agents (called jobs), and a distributed protocol description $\Pi_{\mathcal{J}}$, instructing each job to perform local computations and to communicate over a synchronized network (via point-to-point communication), with the following properties:*

- *Bounded memory: each job's space complexity is $w \in \mathsf{polylog}(n)$.*
- *Bounded per-round computation and communication: the computation and communication complexity of each job at each round is upper bounded by $w \in \mathsf{polylog}(n)$.*

A job is active in a round if it performs computation within this round.

A jobs protocol is further said to have injective communication if the following property is satisfied:

- *Injective communication: each round, a set of jobs are activated, and each sends a single $\mathsf{polylog}(n)$-sized message to a distinct job.*

By convention, we assume the first m_{in} jobs of a jobs protocol are *input jobs*, the last m_{out} are *output jobs*, and the remaining jobs are *helper jobs*. Each input job J_i holds a single-word input $x_i \in \{0,1\}^w$ (for $w \in \mathsf{polylog}(n)$); output and helper jobs have no input. We then have a canonical correspondence between functionalities in the standard n-party setting and the equivalent functionalities in the Worker-Jobs Model:

- Functionality $\mathcal{F}$: In the n-party setting. Accepts inputs x_i from each party P_i, evaluates $y \leftarrow F(x_1||\cdots||x_n)$, outputs the resulting value y to all parties P_i.
- Functionality $\mathcal{F}^{\mathsf{Jobs}}$: In the Jobs Model. Accepts (short) inputs x_u^i from each Input Job, evaluates $y \leftarrow F(x_1||\cdots||x_\ell)$, and distributes the resulting value y (in short pieces) to the Output Jobs.

We may analogously define *oblivious security* of a jobs protocol (where jobs are honest and the adversary sees only "externally measurable" properties of the protocol, as in Definition 1). Within the jobs model, we thus wish to securely realize the functionality $\mathcal{F}_{\text{PRAMs}}^{\text{Jobs}}$, equivalent to $\mathcal{F}_{\text{PRAMs}}$ with the above syntactic change. Note that in the regime of oblivious security, a jobs protocol yields a *memory*-balanced protocol in the standard n-party model, by simply assigning jobs to the n parties evenly.

Theorem 3. *There exists an oblivious-secure protocol in the Jobs Model realizing the functionality $\mathcal{F}_{\text{PRAMs}}^{\text{Jobs}}$ for securely computing a sequence of N adaptively chosen PRAM programs Π_j, with the following complexities (where $n \cdot |x|, |y|$ denote the total input and output size, and $\mathsf{space}(\Pi)$, comp, and $\mathsf{time}(\Pi)$ denote the worst-case space, computation, and (parallel) runtime of Π over all inputs):*

- *Number of jobs: $\tilde{O}\left(n \cdot |x| + |y| + \max_{j \in [N]} \mathsf{space}(\Pi_j)\right)$.*
- *Computation complexity, per Π_j: $\tilde{O}(\mathsf{comp}(\Pi_j))$.*
- *Time steps, per Π_j: $\tilde{O}(\mathsf{time}(\Pi_j))$.*
- *The number of active jobs in each round is $O(\max_{j \in [N]} CPUs(\Pi_j))$.*

given a one-time preprocessing phase with complexity

- *Computation complexity: $\tilde{O}(n \cdot |x|)$.*
- *Time steps: $\tilde{O}(1)$.*

Further, the protocol has injective communication: in each round, each activated job sends a single $\mathsf{polylog}(n)$-size message to a distinct job.

Recall within the Jobs Model each job is limited to maintaining state of size $\mathsf{polylog}(n)$; thus the *memory requirement* of the above protocol is

$$\tilde{O}\left(n \cdot |x| + |y| + \max_{j \in [N]} \mathsf{space}(\Pi_j)\right),$$

based on the number of required jobs.

Idea of proof. The result builds upon the existence of an Oblivious PRAM compiler with $\mathsf{polylog}(n)$ time and space overhead that is *collision-free* (i.e., where no two CPUs must access the same memory address in the same timestep), which is guaranteed to exist unconditionally based on [BCP14b]. In addition to the standard Input and Output jobs, our protocol will have one Helper job for each of the CPUs and each memory cell in the database of the OPRAM-compiled program. The CPU jobs store the local state and perform the computations of their corresponding CPU. In each round that the ith CPU's instructions dictate a memory access at location $\mathsf{addr}^{(i)}$, the CPU job i will communicate with the Memory job $\mathsf{addr}^{(i)}$ to perform the access. (Thus, in each round, at most $2 \cdot CPUs(\mathsf{OPRAM}(\Pi))$ jobs are active, where $\mathsf{OPRAM}(\Pi)$ denotes the OPRAM-compilation of Π). Activation and communication patterns in the resulting protocol are simulatable directly by the OPRAM security. The preprocessing phase of the protocol corresponds to inserting all inputs into the OPRAM-protected database in parallel (i.e., emulating the OPRAM-compiled input insertion program that simply inserts each input x_i into address i of the database).

Step 2: Locality and Load Balancing. This step attains polylog(n) communication *locality*,[10] and computation *load balancing* from any jobs protocol Π_J with injective communication. We do so by emulating Π_J by a fixed set of parties (which we sometimes refer to as "workers"), where each worker is assigned several jobs, and will pass jobs to other workers once he has performed a certain amount of work. This yields a standard N-party protocol with a special *decomposable state* structure: i.e., parties' memory can be decomposed into separate polylog(n)-size memory blocks, which are only ever computed on independently or in pairs, in steps of polylog(n) computation per round. This is because parties' computation is limited to individual jobs to which it was assigned.[11]

Definition 3 (Decomposable State). *An N-party protocol Π is said to have decomposable state if for every party P, the local memory* mem *of P can be decomposed into* polylog(n)-*size blocks* mem $= ($mem$_1,$ mem$_2, \ldots,$ mem$_m)$ *such that: In each round of Π, the (parallel) local computation performed by party P is described as a list $\{(i, j, f_{i,j})\}_{(i,j)\in I}$ for some $I \subseteq [m] \times [m]$, such that each $f_{i,j}$ has complexity* polylog(n). *For each $(i, j) \in I$, party P executes $($mem$_i,$ mem$_j) \leftarrow f_{i,j}($mem$_i,$ mem$_j)$.[12] By convention, received communication messages are stored in local memory.*

We achieve the following "fully load-balanced" properties. Note that the first two properties correspond directly to our final load-balancing goal. The final property will be used to ensure that no individual worker is ever assigned drastically more than the expected number of simultaneous parallel computation tasks; this is important since workers will eventually be emulated by (technically, committees of) parties, who themselves may have bounded parallelism capability (i.e., small number of CPUs).

Definition 4 (Fully Load Balanced). *An N-party protocol Π is said to be fully load balanced with respect to security parameter n if the following properties hold:*

- *Memory load balancing: Let* space(Π) *denote the total space complexity of protocol Π. For every constant $\delta > 0$, with all but negligible probability in n, every party P_j has space complexity*

$$\text{space}(P_j) \leq \frac{(1+\delta)}{N} \text{space}(\Pi) + \text{polylog}(n).$$

- *Online computation/communication load balancing: For every constant $\delta > 0$, with all but negligible probability in n, the following holds at all times during*

---

[10] Recall a protocol has (communication) locality $\ell(n)$ if during the course of the protocol every party communicates with at most $\ell(n)$ other parties.

[11] Looking ahead, pairwise computation will be used when emulating job-to-job communication, and will be sufficient when the original jobs protocol has *injective communication*, so that each job communicates with at most one other job per round.

[12] With some canonical resolution for write conflicts. (In our constructions, the sets (i, j) will be disjoint).

the protocol: Let cc *and* $cc(P_j)$ *denote the total communication complexity and communication complexity of party* P_j, comp *and* $comp(P_j)$ *denote the total computation complexity and computation complexity of party* P_j, *we have*

$$\frac{(1-\delta)}{N}cc - polylog(n) \leq cc(P_j) \leq \frac{(1+\delta)}{N}cc + polylog(n)$$

$$\frac{(1-\delta)}{N}comp - polylog(n) \leq comp(P_j) \leq \frac{(1+\delta)}{N}comp + polylog(n).$$

- *Per-round per-party efficiency:*[13] *Let* A *be an upper bound on the number of active jobs at each round in* $\Pi_{\mathcal{J}}$. *With all but negligible probability in* n, *the per-round per-party computation complexity is upper bounded by* $\tilde{O}(1+(A/N))$.

Theorem 4. *Let* $\Pi_{\mathcal{J}}$ *be an* M-*job protocol with computation complexity* comp *and injective communication, realizing functionality* $\mathcal{F}^{Jobs}$. *Then there exists a fully load-balanced (Definition 4)* $\tilde{O}(n)$-*party protocol* $\Pi_{\mathcal{W}}$ *with decomposable states (Definition 3) that realizes* $\mathcal{F}$ *with total computation* $\tilde{O}(comp)$, *space complexity* $\tilde{O}(M)$, *and* polylog(n) *locality. If* $\Pi_{\mathcal{J}}$ *satisfies oblivious security, so does* $\Pi_{\mathcal{W}}$.

Idea of proof. Recall that in our construction of $\Pi_{\mathcal{W}}$ (in the introduction), at any point of the protocol execution, each job is assigned to a random worker[14] and is stored in at most 2 workers. This is sufficient to imply memory load balancing by standard concentration and union bounds. Online computation/communication load balancing follows by observing that (i) the job-passing pattern is independent of the worker-job assignment, and (ii) jobs are passed frequently enough before accumulating large cost. This allows us to think of the execution as partitioned into "job chunks" each of which is assigned to a random worker, thus amenable to concentration bounds. The last load-balanced property follows again by the fact that each job is independently assigned to a random worker and that each job only performs polylog(n) amount of work per round. To obtain locality, we consider a fixed low-degree communication network between workers, and pass messages using a load-balanced routing algorithm. Load balancing of this modified scheme follows by similar, but more delicate analysis.

The resulting protocol has *decomposable state*, since parties' memory and computation are completely local to individual jobs, or pairs of jobs in the case of emulating job-to-job communication (since the starting jobs protocol has injective communication).

Step 3: From Oblivious to Malicious Security. Finally, we present a general transformation that produces an n-party MPC protocol securely realizing a functionality $\mathcal{F}$ against $(1/3 - \epsilon)n$ static corruptions, given any $\tilde{\Theta}(n)$-party protocol

---

[13] We note that the last two properties are related but incomparable. The online load balancing property focuses on accumulated work, whereas the per-round per-party efficiency concerns upper bounds on per-round work, which is used to bound the required amount of parallelism to execute the protocol with efficient parallel time.

[14] Technically, the initial job-worker assignment is only K-wise independent for $K = \log^3 n$. Nevertheless, this is sufficient for concentration bounds to go through.

with decomposable states (see Definition 3) realizing the corresponding jobs-model functionality $\mathcal{F}^{\text{jobs}}$ with only *oblivious* security. This step can be viewed as a refinement and generalization of ideas from [KLST11, DKMS12, BGT13].

Theorem 5 (From Oblivious Security to Malicious Security). *Suppose there exists an $N \in \Theta(n \cdot \text{polylog}(n))$-party oblivious protocol with decomposable state, realizing functionality $\mathcal{F}^{\text{jobs}}$ in space, computation, and (parallel) time complexity* space, comp, time. *Then for any constant $\epsilon > 0$ there exists an n-party MPC protocol (with error negligible in n) securely realizing the corresponding functionality $\mathcal{F}$ against $(1/3 - \epsilon)n$ static corruptions, with the following complexities (where each party is a PRAM with possibly many processors), given a one-time preprocessing phase with a single broadcast of $\tilde{O}(1)$ bits per party:*

- *Per-party memory: $\tilde{O}(\text{space}/n)$.*
- *Total computation: $\tilde{O}(\text{comp})$.*
- *Time complexity: $\tilde{O}(\text{time})$.*

In addition, if the original protocol has $\tilde{O}(1)$ locality and is fully load-balanced (i.e., satisfying all properties of Definition 4), then the resulting protocol additionally possesses the following properties:

- *Communication locality $\tilde{O}(1)$.*
- *Online computation load balancing, as in Definition 4(c).*
- *Time complexity $\tilde{O}\left(\text{time} \cdot \max\left\{1, \frac{A}{nk}\right\}\right)$ when each party is limited to being a k-processor PRAM, where A denotes the maximum per-round per-party computation complexity of any party in the original oblivious-secure protocol.*[15]

Idea of Proof. The compiler takes the following form: First, parties collectively elect a large number of "good" committees, each of size $\text{polylog}(n)$, where "good" means each committee is composed of at least 2/3 honest parties, and that parties are spread roughly evenly across committees. The one-time broadcast is used to reach full agreement on the first committee. These committees will then emulate each of the *decomposable sub-computations* of the original protocol Π (see Definition 3), via small-scale MPCs. That is, committees are initialized with inputs by having the parties in Π' split their inputs into $\text{polylog}(n)$-size pieces and verifiably secret share them to the appropriate committee(s). Each local computation (and communication) in Π decomposes as a collection of $f_{i,j}$, each affecting only two committees (emulating mem_i and mem_j). Since committees are only size $\text{polylog}(n)$, and *each small-scale MPC has only $\text{polylog}(n)$ memory and computation* (because of decomposability), the memory, computation, and time complexity overhead is small. Since parties are spread across committees, the protocol remains load balanced. Finally, by using a perfectly secure underlying MPC protocol (such as [BGW88]), the only information revealed corresponds directly to the "observable" properties (communication patterns, etc.), thus reducing directly to oblivious security (as per Definition 1).

---

[15] In particular, for our MPC for PRAMs protocol formed by combining Steps 1 and 2, the parameter A will correspond to the number of CPUs required in the evaluated PRAM Π, with polylog overhead.

References

[AAK08] Awerbuch, B., Azar, Y., Khandekar, R.: Fast load balancing via bounded best response. In: SODA 2008, pp. 314–322 (2008)

[ACMR95] Adler, M., Chakrabarti, S., Mitzenmacher, M., Rasmussen, L.E.: Parallel randomized load balancing (preliminary version). In: STOC 1995, pp. 238–247 (1995)

[AJLA+12] Asharov, G., Jain, A., López-Alt, A., Tromer, E., Vaikuntanathan, V., Wichs, D.: Multiparty computation with low communication, computation and interaction via threshold FHE. In: Pointcheval, D., Johansson, T. (eds.) EUROCRYPT 2012. LNCS, vol. 7237, pp. 483–501. Springer, Heidelberg (2012)

[BCP14a] Boyle, E., Chung, K.-M., Pass, R.: Large-scale secure computation. Cryptology ePrint Archive, Report 2014/404 (2014)

[BCP14b] Boyle, E., Chung, K.-M., Pass, R.: Oblivious parallel ram. Cryptology ePrint Archive, Report 2014/594 (2014)

[BGJK12] Boyle, E., Goldwasser, S., Jain, A., Kalai, Y.T.: Multiparty computation secure against continual memory leakage. In: STOC, pp. 1235–1254 (2012)

[BGT13] Boyle, E., Goldwasser, S., Tessaro, S.: Communication locality in secure multi-party computation. In: Sahai, A. (ed.) TCC 2013. LNCS, vol. 7785, pp. 356–376. Springer, Heidelberg (2013)

[BGW88] Ben-Or, M., Goldwasser, S., Wigderson, A.: Completeness theorems for non-cryptographic fault-tolerant distributed computation (extended abstract). In: STOC, pp. 1–10 (1988)

[BSGH13] Braud-Santoni, N., Guerraoui, R., Huc, F.: Fast byzantine agreement. In: PODC, pp. 57–64 (2013)

[DI06] Damgård, I.B., Ishai, Y.: Scalable secure multiparty computation. In: Dwork, C. (ed.) CRYPTO 2006. LNCS, vol. 4117, pp. 501–520. Springer, Heidelberg (2006)

[DIK+08] Damgård, I., Ishai, Y., Krøigaard, M., Nielsen, J.B., Smith, A.: Scalable multiparty computation with nearly optimal work and resilience. In: Wagner, D. (ed.) CRYPTO 2008. LNCS, vol. 5157, pp. 241–261. Springer, Heidelberg (2008)

[DKMS12] Dani, V., King, V., Movahedi, M., Saia, J.: Breaking the o(nm) bit barrier: Secure multiparty computation with a static adversary. CoRR, abs/1203.0289 (2012)

[DMN11] Damgård, I., Meldgaard, S., Nielsen, J.B.: Perfectly secure oblivious RAM without random oracles. In: Ishai, Y. (ed.) TCC 2011. LNCS, vol. 6597, pp. 144–163. Springer, Heidelberg (2011)

[Gen09] Gentry, C.: Fully homomorphic encryption using ideal lattices. In: STOC, pp. 169–178 (2009)

[GGH+13] Gentry, C., Goldman, K.A., Halevi, S., Julta, C., Raykova, M., Wichs, D.: Optimizing ORAM and using it efficiently for secure computation. In: De Cristofaro, E., Wright, M. (eds.) PETS 2013. LNCS, vol. 7981, pp. 1–18. Springer, Heidelberg (2013)

[GKK+11] Gordon, S.D., Katz, J., Kolesnikov, V., Malkin, T., Raykova, M., Vahlis, Y.: Secure computation with sublinear amortized work. IACR Cryptology ePrint Archive, 2011:482 (2011)

[GMW87] Goldreich, O., Micali, S., Wigderson, A.: How to play any mental game or a completeness theorem for protocols with honest majority. In: STOC, pp. 218–229 (1987)

[GO96]   Goldreich, O., Ostrovsky, R.: Software protection and simulation on oblivious rams. J. ACM **43**(3), 431–473 (1996)

[KLST11] King, V., Lonargan, S., Saia, J., Trehan, A.: Load balanced scalable byzantine agreement through quorum building, with full information. In: Aguilera, M.K., Yu, H., Vaidya, N.H., Srinivasan, V., Choudhury, R.R. (eds.) ICDCN 2011. LNCS, vol. 6522, pp. 203–214. Springer, Heidelberg (2011)

[KSSV06] King, V., Saia, J., Sanwalani, V., Vee, E.: Scalable leader election. In: SODA, pp. 990–999 (2006)

[LO13]   Lu, S., Ostrovsky, R.: Distributed oblivious RAM for secure two-party computation. In: Sahai, A. (ed.) TCC 2013. LNCS, vol. 7785, pp. 377–396. Springer, Heidelberg (2013)

[MPS02] Mitzenmacher, M., Prabhakar, B., Shah, D.: Load balancing with memory. In: Proceedings, (FOCS), 16–19 November 2002, Vancouver, BC, Canada, pp. 799–808 (2002)

[MR98]   Muthukrishnan, S., Rajaraman, R.: An adversarial model for distributed dynamic load balancing. In: SPAA 1998, pp. 47–54 (1998)

[MSS13] Myers, S., Sergi, M., Shelat, A.: Black-box proof of knowledge of plaintext and multiparty computation with low communication overhead. In: Sahai, A. (ed.) TCC 2013. LNCS, vol. 7785, pp. 397–417. Springer, Heidelberg (2013)

[OS97]   Ostrovsky, R., Shoup, V.: Private information storage (extended abstract).In: STOC, pp. 294–303 (1997)

[VB81]   Valiant, L.G., Brebner, G.J.: Universal schemes for parallel communication. In: STOC, pp. 263–277 (1981)

[Yao86] Yao, A.C.-C.: How to generate and exchange secrets (extended abstract). In: FOCS, pp. 162–167 (1986)

[ZMS14] Zamani, M., Movahedi, M., Saia, J.: Millions of millionaires: Multiparty computation in large networks. Cryptology ePrint Archive, Report 2014/149 (2014)

Incoercible Multi-party Computation and Universally Composable Receipt-Free Voting

Joël Alwen[1]([✉]), Rafail Ostrovsky[2], Hong-Sheng Zhou[3], and Vassilis Zikas[4]

[1] IST Austria, Klosterneuburg, Austria
jalwen@ist.ac.at
[2] UCLA, Los Angeles, USA
rafail@cs.ucla.edu
[3] Virginia Commonwealth University, Richmond, USA
hszhou@vcu.edu
[4] ETH Zurich, Zurich, Switzerland
vzikas@inf.ethz.ch

Abstract. Composable notions of incoercibility aim to forbid a coercer from using anything beyond the coerced parties' inputs and outputs to catch them when they try to deceive him. Existing definitions are restricted to weak coercion types, and/or are not universally composable. Furthermore, they often make too strong assumptions on the knowledge of coerced parties—e.g., they assume they known the identities and/or the strategies of other coerced parties, or those of corrupted parties— which makes them unsuitable for applications of incoercibility such as e-voting, where colluding adversarial parties may attempt to coerce honest voters, e.g., by offering them money for a promised vote, and use their own view to check that the voter keeps his end of the bargain.

In this work we put forward the first universally composable notion of incoercible multi-party computation, which satisfies the above intuition and does not assume collusions among coerced parties or knowledge of the corrupted set. We define natural notions of UC incoercibility corresponding to standard coercion-types, i.e., receipt-freeness and resistance to full-active coercion. Importantly, our suggested notion has the unique property that it builds *on top* of the well studied UC framework by Canetti instead of modifying it. This guarantees backwards compatibility, and allows us to inherit results from the rich UC literature.

We then present MPC protocols which realize our notions of UC incoercibility given access to an arguably minimal setup—namely honestly generate tamper-proof hardware performing a very simple cryptographic operation—e.g., a smart card. This is, to our knowledge, the first proposed construction of an MPC protocol (for more than two parties) that is incoercibly secure *and* universally composable, and therefore the first construction of a universally composable receipt-free e-voting protocol.

Keywords: Multi-party computation · Universal composition · Receipt-freeness

V. Zikas—Research partly done while the author was at UCLA.

© International Association for Cryptologic Research 2015
R. Gennaro and M. Robshaw (Eds.): CRYPTO 2015, Part II, LNCS 9216, pp. 763–780, 2015.
DOI: 10.1007/978-3-662-48000-7_37

1   Introduction

Secure multi-party computation (MPC) allows n mutually distrustful parties to securely perform some joint computation on their inputs even in the presence of cheating parties. To capture worst-case (collaborative) cheating, a central adversary is assumed who gets to corrupt parties and uses them to attack the MPC protocol. Roughly speaking, security requires that the computation leaks no information to the adversarial parties about the inputs and outputs of uncorrupted, aka honest, parties (privacy) and that the corrupted parties cannot affect the output any more than choosing their own inputs (correctness).

The seminal works on MPC [3,12,18,36] established feasibility for arbitrary functions and started a rich and still evolving literature. Along the way, additional desired properties of MPC were investigated. Among these, *universal composability* guarantees that the protocol preserve its security even when executed within an online adversarial environment, e.g., along-side other (potentially insecure) protocols. Various frameworks for defining universal composability have been suggested [2,30], with Canneti's UC framework [6] being the most common.

The above frameworks make use of the so called *simulation-based* paradigm for defining security which, in a nutshell, can be described as follows: Let f denote a specification of the task that the parties wish to perform. Security of a protocol Π for f is defined by comparing its execution with an ideal scenario in which the parties have access to a fully trusted third party, the *functionality*, which takes their inputs, locally computes f, and returns to the parties their respective outputs. More concretely, a protocol Π is secure if for any adversary $\mathcal{A}$ attacking Π, there exists an ideal adversary $\mathcal{S}$ attacking the above ideal evaluation scenario, which simulates the attack (and view) of $\mathcal{A}$ towards any environment $\mathcal{Z}$ that gets to choose the parties' inputs and see their outputs.[1]

Arguably, UC security captures most security guarantees that one would expect from a multi-party protocol. Nonetheless, it does not capture *incoercibility* a property which is highly relevant for one of a prototypical application of MPC, namely secure e-voting. Intuitively, incoercibility ensures that even when some party is forced (or *coerced*) by some external entity into executing a strategy other than its originally intender, e.g., coerced to use a different input or even a different protocol, then the party can disobey (i.e., deceive) its coercer, e.g., use its originally intended input, without the coercer being able to detect it.

In the special case of e-voting, where parties are voters, this would mean that a coercer, e.g., a vote buyer that offers a voter money in exchange of his vote for some candidate c, is not able to verify whether the voter indeed voted for c or for some other candidate. In other words, the voter cannot use his transcript as a receipt that he voted for c, which is why in the context of voting the above type of incoercibility is often referred to as *receipt-freeness*.

---

[1] In strong (UC) definitions, it is required that this simulation is sound even in an on-line manner, i.e., $\mathcal{S}$ is not only required to simulated the view of $\mathcal{A}$ but has to do so against an online environment that might talk to the adversary at any point.

Which guarantees can we expect from a general definition of incoercibility? Clearly, if the coercer can use the outputs of the function to be computed to check upon the coerced party it is impossible to deceive him. Considering our voting scenario (concretely, majority election) if there are two candidates c_1 and c_2 and a set V of voters with $|V| = 2m + 1$ for some m, and the coercer coercing $v_i \in V$ knows that half of the parties in $V \setminus \{v_i\}$ voted for c_1 and the other half voted for c_2, then v_i cannot deceive its coercer, as his input uniquely defines the outcome of the election. Therefore, composable notions of incoercibility [9,35] aim for the next best thing, namely allow the parties to deceive their coercer within the "space of doubt" that the computed function allows them. In other words, an informal description of incoercibility requires that the parties can deceive their coercer when they are executing the protocol as good as they can deceive someone who only observes the inputs and outputs of the computation.

Of course, the above intuition becomes tricky to formulate when the protocol is supposed to be incoercible and simultaneously tolerate malicious adversaries. There are several parameters to take into account when designing such a definition. In the following we sketch those that, in our opinion, are most relevant.

Coercion Type. This specifies the power that the coercer has on the coerced party. Here we one can distinguish several types of coercion: *I/O-coercion* allows the coercer to provide an input to the party and only use its output. This is the simplest (and weakest) form of coercion as it is implied by UC security. A stronger type is *receipt-freeness* or *semi-honest* coercion; here, the coercer gets to provide an input to the coerced party, but expects to see a transcript which is consistent to this input. This type corresponds to the notion of coercion introduced in [9,10] and abstracts the receipt-freeness requirement in the voting literature [19,20,23,27,28,31–33].[2] Finally, *active coercion* is the strongest notion of coercion, where the adversary instructs the coerced party which messages to send in the protocol and expects to see all messages he receives (also in an online fashion). This type of coercion has been considered, explicitly or implicitly, in the stand-alone setting (i.e., without universal composition) by Moran and Naor [32] and more recently in the UC setting by Unruh and Müller-Quade [35].

Adaptive vs. Static. As with corruption, we can consider coercers who choose the set of parties to coerce at the beginning of the protocol execution, i.e., in a *static* manner, or *adaptively* during the protocol execution depending on their view so far—e.g., by observing the views of other coerced parties.

Coercer/Deceiver-Collusions. The vast majority of works in the multi-party literature assumes a so called *monolithic* adversary who coordinates the actions of corrupted parties. This naturally captures the worst-case scenario in which cheaters work together to attack the protocol. Analogously, works on incoercible computation [9,10,32,35] assume a monolithic coercer, i.e., a single entity which is in charge of coordinating coerced parties. This has the following counter-intuitive side-effect: in order for a coerced party to be able to deceive any such

---

[2] For the special case of encryption, resiliency to semi-honest coercion corresponds to the well-known concept of *deniability* [8].

a monolithic coercer it needs to coordinate its deception strategy with other coerced (or with honest) parties. In fact, in recent universally composable notions of incoercibility this deceiver coordination is explicit. For example, in [35] an even stronger requirement is assumed: the coerced parties which attempt a deception know the identities and deception strategies of other coerced parties, and even the identities of all corrupted parties. This is an unrealistic assumption in scenarios such as e-voting, where a potential vote-seller is most likely oblivious to who is cheating or to who else is selling its vote.

In order to avoid the above counter intuitive situation, in this work we assume that deception (therefore also coercion) is local to each coerced party, i.e., coercers of different parties are not *by default* colluding. Alas, casting our definition in the UC framework makes coercer collusion explicit: Although coercers are local, they can still be coordinated via an external channel, e.g., through the environment. In fact, in our definition the worst-case environment implicitly specifies such a worst-case coercion scenario.

Informants and Dependency between Corruption and Deception. Another question which is highly relevant for incoercibility, is whether or not coerced parties know the identities of the cheaters/adversaries. In particular, a worst case coercion scenario is the one in which the coercer and the adversary work together to check on the coerced parties—stated differently, the coercer uses corrupted parties as informants against coerced parties to detect if they are attempting to deceive him. (In the context of receipt-free voting, this corresponds to checking the view/receipt of vote sellers against the corresponding views of malicious parties.) Clearly, if a coerced party knows who are the informants then it is easier to deceive its coercer. (This is the approach taken in [35], where the identities of corrupted parties are accessible to the deceivers via a special register.) Arguably, however, this is not a realistic assumption as it reduces the effect of using informants—a vote buyer is unlikely to tell the vote seller how he can check upon him. The modeling approach taken in this work implies that real-world deceivers have no information on who is corrupted (or coerced).

Our Contributions. In this work we provide the first security definition of incoercible multi-party computation which is universally composable (UC) and makes minimal assumptions on the coerced parties' ability to deceive their coercer. Our definition offers the same flexibility on addressing different classes of coercion as standard security notion offers for corruptions. Indicatively, by instantiating it with different types of coercion we devise definitions of UC incoercibility against *semi-honest* coercions—corresponding to the classical notion of receipt-freeness—as well as of the more powerful *active coercions* corresponding to the strong receipt-freeness notion introduced in [32]. As a sanity check, we show that if the coercers only see the output of coerced parties (a notion which we call I/O-incoercibility), then any UC secure protocol is also incoercible.

In addition to flexibility, our definition has the following intuitive properties:

Universal composability and compatibility with standard UC. We prove universal composition theorems for all the suggested types of incoercibility, which imply that an incoercible protocol can be composed with any other incoercible protocol. Because our definition builds on top of the UC framework instead of modifying is (e.g., as in [10,35]), our protocols are automatically also universally composable with standard (coercible) UC protocols, at the cost, of course, of giving up incoercibility; that is, when composing an incoercibly UC protocol with a standard (coercible) UC protocol, we still get a UC secure protocol. We note in passing that defining incoercibility in UC has the additional advantage that it protects even against on-line coercer, e.g., vote-buyer that expect the receipt to be transmitted to them while the party is voting.

Minimal-knowledge deceptions. The deceivers in the real-world have no knowledge of who is coerced or corrupted, nor do they know which strategy other coerced parties will follow. Thus they need to deceive assuming that any party might be an informant.

Last but not least, we present a UC incoercible protocol for arbitrary multi-party computation which tolerates any number of actively corrupted and any number of coerced parties (for both semi-honest and active coercion), as long as there is at least one honest party. Our protocols make use of an arguably minimal and realistic assumption (see the discussion below), i.e., access to a simple honestly generated hardware token. To our knowledge, ours is the first protocol construction, which implements *any* functionality in the *multi-party* ($n > 2$ parties) setting. In fact, our construction can be seen a compiler, in this token-hybrid model, of UC secure to incoercible UC secure protocols. Therefore, when instantiated with a fast UC secure protocol it yields a realistic candidate for construction for UC secure incoercible e-voting.

Our protocol is proved secure against static coercion/corruption, but our proofs carry through (with minimal modifications) to the adaptive setting. In fact, our protocols realize an even stronger security definition in which the coercers, but *not* the coerced parties (i.e., the deceivers), might coordinate their strategies. However, we chose to keep the definition somewhat weaker, to leave space for more solutions or possibly different assumptions.[3]

The Ideal Token Assumption. Our protocols assume that each party has access to a hardware token which might perform fresh encryptions with to some hidden keys that are shared among the parties. The goal of the token is to offer the parties a source of hidden randomness that allows them to deceive their coercer. A setup of this type seems to be necessary for such a strong incoercibility notion when nearly everyone might be corrupted, since if the coerced parties have no external form of hidden randomness, then it seems impossible for them to deceive—the coercer might request their entire state and compare it with messages received by its informants, which would require the coerced party to align its lie with message it sends to the informants (whose the identities are unknown).

---

[3] Recall that our definition *does* allow coercer coordination through the environment.

On top of being minimal in the above sense, our encryption token assumption is also very easy to implement in reality for a system with a bounded number of participants—this is typically the case in elections: Let N be an upper bound on the voters; the voting registration authority (i.e., the token creator and distributor) computes N keys $k_1, \ldots, k_N$, one for every potential voter; every p_i who registers receives his ith token along with a vector of N random strings $(k_{1i}, \ldots, k_{Ni})$, corresponding to his keys-shares; the last p_i who registers (i.e., the last to be in the registration desk before it closes) receives his token, say the n-th token, along with the vector $(k_{1n}, \ldots, k_{Nn}) = (k_1, \ldots, k_n) \oplus \bigoplus_{i=1}^{n}(k_{1i}, \ldots, k_{Ni})$, where $\oplus$ denotes the component-wise application of the bit-wise xor operation. Note that the assumption of a hardware token (capturing pre-distributed smart cards) has been used extensively in practice, e.g., in the university elections in Austria and even the national elections in Finland and Estonia [14].

Related Literature. The incoercibility literature can roughly be split in two classes: works that look at the special case of receipt-free voting [1,4,15,16,19, 20,22,23,26–28,31,33,34] and works that look at the more general problem of incoercible realization of arbitrary multi-party functions [9,10,32,35]. Below, we focus on the second class which is closer to our goal and refer the reader to the full version of this work for a short survey of the voting-specific literature.

The first to consider incoercibility in the setting of general MPC were Canneti and Gennaro [9]. They put forth a notion of incoercibility for static off-line semi-honest coercions. Unfortunately their notion is only known to be sequentially composable and moreover the definition is not compatible with the more general setting of computing reactive functionalities. On the positive side, deception strategies are both split and oblivious of other deceivers, and [9] does provide a construction realizing a large class of (non-reactive) functions f.

Building on the idea of [9], Moran and Naor [32] define a stronger version of incoercibility against adaptive active coercions using split oblivious deception strategies. They go on to provide a construction implementing a voting functionality. Their model of communication and execution is based on that of [5] and, thus, provides sequential (but not concurrent or universal) composability [6]; also, similarly to [9], it is not clear how to extend the model in [32] to reactive functionalities (such as say a commitment scheme).

More recently Unruh and Müller-Quade [35] provided the first universally composable notion of incoercibility. Due to similarity in goals with our work, we provide a comparison with our definition and results. In a nutshell, the definition in [35] specifies the deception strategy D as an extra form of adversary-like machine. The requirement is that for any such deceiver D in the ideal world, there exists a corresponding real-world deceiver D_S (in [35] D_S is called deceiver simulator) such that for any (real-world) adversary $\mathcal{A}$ there exists and (ideal-world) simulator $\mathcal{S}$ that makes the ideal world where D controls the coerced and $\mathcal{S}$ the corrupted parties, indistinguishable from the real world where D_S controls the coerced and $\mathcal{A}$ the corrupted parties, in the presence of any environment $\mathcal{Z}$.[4] Importantly, in [35] it is explicitly assumed that the deceiver has

---

[4] In fact, the model of [35] builds on the externalized UC (EUC) model of Canetti et al. [7] which is designed to allow for deniable protocols.

access to a public register indicating which parties are corrupted and which are deceiving. As already mentioned, the above modelling choices of [35] have the following side-effects: (1) the real-world deceiving parties are explicitly allowed out-of-band communication (since deception is coordinated by the monolithic D_S) and (2) they know the identities of the corrupted parties, i.e., of the potentially informants. As discussed above these assumptions are not realistic for e-voting. Furthermore, the model of execution in [35] considerably deviates from the GUC model, e.g., it modifies the number of involved ITMs and the corruption mechanism, which can lead to syntactical incompatibilities with GUC protocols and issues with composition with (coercible) GUC protocols.[5]

An alternative approach to universally composable incoercibility was taken in the most recent revision of Canetti's UC paper, and adopted in [10] for the two-party setting. This definition builds on the idea from [9] and is for semi-honest coercions. Furthermore, the coercion mechanism in the multi-party setting is unspecified and no composition theorem is proved.[6]

In terms of protocols, in [10] a two-party protocol in the semi-honest coercion and corruption model is suggested assuming indistinguishability obfuscation [17]. Their approach is based on Yao's garble circuits and is specifically tailored to the two party setting; as they argue, their protocols are not universally composable under active corruption. On the other hand, in [35] a two-party protocol for computing a restricted class of two-party functionalities was suggested; also here it is unclear whether or not this approach can yield a protocol in the multi-party setting or for a wider class of two-party functionalities. Thus ours is the first UC secure incoercible multi-party protocol, which can be, for example, used for receipt-free voting— an inherently multi-party functionality.[7]

Outline of the Remainder of the Paper. In Sect. 2 we present our UC incoercibility definition. Subsequently, in Sect. 3 we describe instances of our definitions corresponding to the three standard coercion types, namely, I/O, receipt-freeness, and active coercion and corresponding composition theorems. Following that, in Sect. 4 we provide our UC receipt-free protocol for computing any given function. Our protocol is simple enough to be considered a good starting point for an alternative approach to existing e-voting protocol. Finally, in Sect. 5 we prove that our receipt-free protocol can withhold even active coercion attacks. Due to space limitation, the proofs have been moved to the full version of this work.

Preliminaries and Notation. Our definition of incoercibility builds on the Universal Composition framework of Canetti [6] from which we inherit the protocol execution model along with the (adaptive) corruption mechanism. We assume the reader has some familiarity with the UC framework [6] but in the following we recall some basic notation and terminology. We denote by ITM the set

---

[5] For example, the corruption mechanism as described in [35] does not specify that (let alone how) the deceiver simulates deception towards the corresponding adversary.

[6] Note that the Definition in [10] also changes the underlying model of computation, which makes it necessary to re-prove composition.

[7] Our protocol uses the CLOS protocol [11] in a black-box manner, and it remains secure even when CLOS is replaced by more efficient protocols, e.g., the IPS protocol [21] in the pre-processing model.

of efficient (e.g. poly-time) ITMs and by $[n]$ the set of integers $\{1, \ldots, n\}$. For simplicity, we use the notations "p_i" and "party i" interchangeably to refer to the party with identity i. For a set $\mathcal{J} \subseteq [n]$ if for each $i \in \mathcal{J}$ the ITM π_i is a protocol machine for party i then we use the shorthand $\pi_{\mathcal{J}}$ to refer to the $|\mathcal{J}|$-tuple $(\pi_{i_1}, \ldots, \pi_{i_{|\mathcal{J}|}})$. In particular we simply write π to denote $\pi_{[n]}$.

A protocol π UC emulates ρ if π can replace ρ in any environment in which it is executed; similarly, a protocol UC realizes a given functionality $\mathcal{F}$ if it UC emulated the the *dummy $\mathcal{F}$-hybrid protocol* ϕ, which simply relays inputs from the environment to $\mathcal{F}$ and vice versa. In [6] protocols come with their hybrids (so the hybrids are not written in the protocol notation); but for sake of clarity in order to make the hybrid-functionality explicit, we at times write it as a superscript of the protocol, e.g., we might denote a $\mathcal{G}$-hybrid protocol π as $\pi^{\mathcal{G}}$.

Finally, we use the following standard UC terminology: we say that a party (or functionality) P *issues a delayed message* x for another party P' (where x can be an input or an output for some functionality) to refer to the process in which P prepares x to be sent to P', but requests for the simulator's approval before actually sending it. Depending on whether or not this approval request includes the actual message, we refer to the delayed output as *public* or *private*, respectively. For details on delayed messages we refer to [6].

2 Our UC Incoercibility Definition

Our security notion aims to capture the intuition that deceiving one's coercer is as easy as the function we are computing allows it to be. Intuitively, this means that for any *(ideal-world) deception* strategy that the coerced party would follow in the ideal world—where the functionality takes care of the computation—there exists a corresponding *(real-world) deception* strategy that he can play in the real world which satisfies the following property:

> The distinguishing advantage of any set of coercers in distinguishing between executions in which parties deceive and ones where they do not deceive is the same in the ideal world (where coerced parties follow their ideal deception strategy DI) as it is in the real world (where parties follow their corresponding real-world deception strategy DR).

To capture worst-case incoercibility (and get composition) we let the environment play the role of the coercer. This makes the ability of coercers to collude explicit while capturing worst-case and on-line coercion strategies. However, in order to provide a flexible definition, which for example captures the standard notion of receipt-freeness, where coerced parties follow their protocol, we define the effect of a party's coercion as a transformation applied on its protocol, which specifies the control the environment/coercer has on a corrupted party. For example, in the case of receipt-freeness this transformation internally logs the state of the coerced party, and upon reception of a special message from $\mathcal{Z}$ requesting a "receipt" it hands $\mathcal{Z}$ this state. We refer to the next section for a detailed definition of different coercion types.

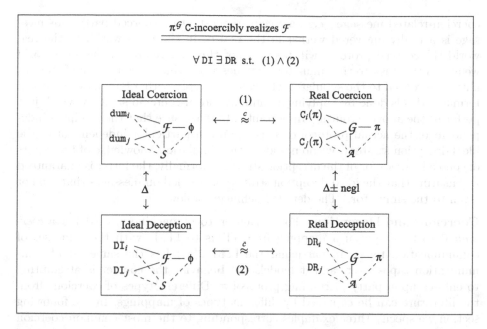

Fig. 1. The incoercibility definition. For clarity we explicitly write the hybrids $\mathcal{F}$ and $\mathcal{G}$. The interfaces on the right of $\mathcal{F}$ (resp. $\mathcal{G}$) correspond to interfaced of honest parties. Parties i and j are coerced according to the coercer C. Dashed lines denote communication tapes to the adversary implicitly modeling a network of insecure channels.

The above idea is demonstrated in Fig. 1, were the following four worlds are illustrated: the ideal world where coerced parties follow their coercer's instructions (top left), the ideal world where coerced parties attempt a deception (bottom left), the real world where coerced parties follow their coercer's instructions (top right), and the real world where the coerced parties attempt a deception (bottom right). As sketched above, incoercibility requires that if the advantage of the best environment (i.e., the one that maximizes its advantage) in distinguishing the top-left world from the bottom-left world is $0 \leq \Delta \leq 1$, then the advantage of the best environment in distinguishing the top-right world from the bottom-right world is also $\Delta' = \Delta$ (plus/minus some negligible quantity).

The above paradigm captures the intuition of incoercibility, but in order to get a more meaningful statement we need the incoercible protocol to also be secure, i.e., implements its specification. This means that when parties do follow their coercion instructions, the protocol should be a secure implementation of the given functionality. In the above terminology, there should be a simulator which makes the top-right world indistinguishable from the top-left world. This has two implications: First, together with the previous requirement, i.e., that $\Delta' = \Delta \pm negl.$ it implies that the bottom-right world should also be indistinguishable from the bottom-left world for the same simulator.

Second, to ensure that the top two worlds are indistinguishable for natural coercions, e.g., for receipt-freeness, we need that when the environment sends a coercion-related message—e.g., a receipt-request—to a coerced party, this message is actually answered wether in the real or in the ideal world. In the real world the coerced protocol will take care of this. Therefore, in the ideal world we assign this task to the simulator: any messages which is not for the functionality is re-routed to the simulator who can then reply with a (simulated) receipt; formally, this is done by applying a "dummy" ideal-coercion strategy which just performs the above rerouting. Importantly, to make sure that the receipt is independent of the actual protocol execution, and in particular independent of the ideal deception strategy, we do not allow the simulation knowledge of the inputs of coerced parties, or of the deception strategy (formally, the latter is guaranteed by ensuring that the ideal deception strategy is applied on messages that are not given to the simulator.) The detailed definition follows.

Coercions and Deceptions. For a given protocol machine π_i we define a *coercion* C to be a special a mapping from ITMs to ITMs with the same set of communication tapes. In particular the ITM $C(\pi_i)$ has the same set of communication tapes as π_i and it models the behavior the coercer is attempting to enforce upon party p_i running protocol π. Different types of coercions from the literature can be captured by different types of mappings. In the following section we specify three examples corresponding to the most common coercion types in the literature.

To model the ideal-world behavior (intuitively the "effective" behavior) of a coerced party when obeying its coercer, we use the protocol ITM dum called the *dummy coercion* (we at times refer to dum as the *extended dummy* protocol). As sketched above, dum ensures that the simulator handles all messages that are not intended for the functionality. More concretely, the following describes the behaviour of dum upon receiving a message from various parties.

From $\mathcal{Z}$: If the message has the form (x, fid) intended for delivery to functionality $\mathcal{F}$, dum forwards x to $\mathcal{F}$ using a private delayed input (c.f. Page 7). All other messages from $\mathcal{Z}$ are forwarded to the simulator.
From $\mathcal{F}$: Any message from $\mathcal{F}$ is delivered to the simulator.
From $\mathcal{S}$: If the message has the form (x, fid) then dum forwards x to $\mathcal{F}$. Otherwise it forwards the message to $\mathcal{Z}$.

An ideal-world *deception strategy* corresponds to an attempt of a coerced party to lie to the environment about its interaction with the ideal functionality $\mathcal{F}$. Thus, it can be described as a mapping applied on the messages that the deceiving party exchanges with the functionality and with the environment. To keep our assumptions minimal, we require the real-world (protocol) deception strategy to also have the same structure, i.e., be described it as mappings applied on the messages that the deceiving party p_i running a protocol exchanges with its hybrids and with the environment.[8]

Thus, to capture deception by party p_i running ITM π_i, we define a *deception strategy*, denoted by $D_i(\pi_i)$, to be an ITM which can be described via a triple

---

[8] A more liberal, but weaker, definition could allow the real-world deception strategy to be an arbitrary Turing machine with the same hybrids as p_i.

(D_i^1, π_i, D_i^2) of interconnected ITMs behaving as follows: π_i's messages to/from the adversary are not changed, but we place D_i^1 between π_i and $\mathcal{Z}$ while we place D_i^2 between π_i and its hybrids. For notational simplicity we, at times, omit the argument from $D_i(\cdot)$ and write D_i instead of $D_i(\pi_i)$ when the argument is already clear from the context.

Using these concepts we can now somewhat sharpen the above intuition on our definition. Informally, a protocol π UC *incoercibly* realized a functionality $\mathcal{F}$ with respect to a coercion C (in short: π C-IUC *realizes* $\mathcal{F}$) if the following two conditions are satisfied: (1) for any set $\mathcal{J} \subseteq [n]$ of coerced parties, when replacing the honest protocol π_i with the wrapped protocol $\widehat{\pi}_i = C_i(\pi)$ for all $i \in \mathcal{J}$ the resulting network UC realizes the $\mathcal{F}$-dummy protocol ϕ, where the parties $i \in \mathcal{J}$ use C_i instead of ϕ; and (2) for any player and their ideal deception $DI_i = D_i(\mathtt{dum}_i)$ there exists a real deceiving strategy $DR_i = D_i'(C_i(\pi))$ such that no environment can catch coerced parties lying with DI_i in ρ with probability better than catching them lying with DR_i in π.

To make the above intuition formal, we need the following notation. Let $\mathcal{J} \subseteq [n]$ denote the set of coerced parties. (To avoid unnecessarily complicated statements, we restrict to static coercion, so the set $\mathcal{J}$ is chosen by $\mathcal{Z}$ at the beginning of the protocol execution.) The execution of protocol π with coercion C corresponds to executing, *in the UC model of execution*, the protocol which results by replacing for each party $j \in \mathcal{J}$ it's protocol machine π_j with the above described $C_j(\pi)$. Much like UC, we write $\{\text{EXEC}_{\pi, C, \mathcal{A}, \mathcal{Z}}(\lambda, z)\}_{\lambda \in \mathbb{N}, z \in \{0,1\}^*}$ to denote the ensemble of the outputs of the environment $\mathcal{Z}$ when executing protocol π with the above modifications, in the presence of adversary $\mathcal{A}$. Consistently with the UC literature, we often write $\text{EXEC}_{\pi, C, \mathcal{A}, \mathcal{Z}}$ instead of $\{\text{EXEC}_{\pi, C, \mathcal{A}, \mathcal{Z}}(\lambda, z)\}_{\lambda \in \mathbb{N}, z \in \{0,1\}^*}$. We also use the notation $\text{UC-EXEC}_{\pi, \mathcal{A}, \mathcal{Z}}$ to denote the analogous ensemble of outputs for a standard UC execution. For clarity, for the dummy $\mathcal{F}$-hybrid protocol ϕ we might write $\text{EXEC}_{\mathcal{F}, C, \mathcal{S}, \mathcal{Z}}$ and $\text{UC-EXEC}_{\mathcal{F}, \mathcal{S}, \mathcal{Z}}$ instead of $\text{EXEC}_{\phi, C, \mathcal{A}, \mathcal{Z}}$ and $\text{UC-EXEC}_{\phi, \mathcal{A}, \mathcal{Z}}$, respectively.

Definition 1 (UC Incoercibility). *Let π be an n-party protocol and for an n-party functionality $\mathcal{F}$ let ϕ denote the dummy $\mathcal{F}$-hybrid protocol, and let C be a coercion. We say that π C-IUC realizes $\mathcal{F}$ if for every $i \in [n]$ and every ideal deception strategy DI_i there exists an real deception strategy DR_i with the following property. For every adversary $\mathcal{A}$ there exists a simulator S such that for any set $DI_{\mathcal{J}} = \{DI_i : i \in \mathcal{J}\}$ and every environments $\mathcal{Z}$:*

$$\text{EXEC}_{\phi, \mathtt{dum}_{\mathcal{J}}, \mathcal{S}, \mathcal{Z}} \overset{c}{\approx} \text{EXEC}_{\pi, C_{\mathcal{J}}, \mathcal{A}, \mathcal{Z}} \tag{1}$$

$$\text{EXEC}_{\phi, DI_{\mathcal{J}}, \mathcal{S}, \mathcal{Z}} \overset{c}{\approx} \text{EXEC}_{\pi, DR_{\mathcal{J}}, \mathcal{A}, \mathcal{Z}} \tag{2}$$

where dum *denotes the dummy coercer described above.*

We observe that when no party is coerced, i.e., $\mathcal{J} = \emptyset$, then the definition coincides with UC security which shows that incoercibility with respect to any type of coercion also implies standard UC security.

3   Types of Coercion

Using our definition we can capture the types of coercion previously considered (mainly in the e-voting) literature. These types are specified in this section, where we also prove the composability of the corresponding definitions.

I/O Coercion. As a sanity check we look at a particularly weak form of coercion called *input/output (I/O) coercion*. Intuitively, this corresponds to a setting where a party is being coerced to use a particular input and must return the output of the protocol to the coercer as evidence of it's actions. We capture this formally by defining the I/O coercion C^{io} to be identical to the dummy coercion; that is for any protocol machine π_i $\mathsf{C}^{io}(\pi_i) = \mathsf{dum}(\pi_i) = \pi_i$. In particular it faithfully uses the input to π_i supplied by $\mathcal{Z}$ and follows the code of π_i during the entire execution, and eventually returns the output back to $\mathcal{Z}$.

Not surprisingly, we already have I/O-incoercible protocols for a wide variety of functionalities since standard UC realization is equivalent to I/O-incoercible realization.

Theorem 1. *In the static corruption model protocol π UC realizes functionality $\mathcal{F}$ with static corruptions if and only if π C^{io}-IUC realizes $\mathcal{F}$.*

An immediate consequence of Theorem 1 and the UC composition theorem in [6] is that I/O-incoercibility is a composable notion.

Semi-honest Coercion (Receipt-Freeness). The type of incoercibility that has been mostly considered in the literature is the so-called receipt-freeness. The idea there is that the coercer expects to be provided with additional evidence of that a specific input was used. In the most severe case such a proof could, for example, be the entire view of a coerced party in the protocol execution.

In the following, we define the semi-honest coercion C^{sh}, which captures receipt freeness: at a high-level, for a given protocol machine π_i the ITM $\mathsf{C}^{sh}(\pi_i)$ behaves identically to π_i with the only difference that upon being asked by $\mathcal{Z}$, ITM $\mathsf{C}^{sh}(\pi_i)$ outputs all messages it has received from the adversary and it hybrids as well as it's random coins (i.e., the contents of his random tape). Note that, as $\mathcal{Z}$ already knows the messages it previously sent to $\mathsf{C}^{sh}(\pi_i)$, it can now reconstruct the entire view of p_i in the protocol.

Intuitively, the output of $\mathsf{C}^{sh}(\pi_i)$ can be used as a receipt that p_i is running π_i on the inputs chosen by $\mathcal{Z}$ as follows. On the one hand, any message p_i claims to have received over the insecure channels can be confirmed to $\mathcal{Z}$ by the informant. On the other hand, for any prefix of receipt causing π_i to send a message over the insecure channel, $\mathcal{Z}$ can check with it's informant if indeed exactly that message was sent by p_i at that point.

Theorem 2. *Let C be a semi-honest coercer, i.e., $\mathsf{C} = \mathsf{C}^{sh}$. If protocol π C-IUC-realizes functionality $\mathcal{F}$, and protocol σ C-IUC-realizes functionality $\mathcal{H}$ in the $\mathcal{F}$-hybrid world, then the composed protocol σ^π C-IUC-realizes functionality $\mathcal{H}$.*

Active Coercion. We next turn to defining active coercion. Here, instead of simply requiring a receipt, the coercer takes complete control over the actions of coerced parties. We capture this by introducing the fully-invasive—also referred to as *active* coercion C^A which allows the environment full control over the coerced party's interfaces. Formally, for any (set of) functionalities $\mathcal{G}$ and any $\mathcal{G}$-hybrid protocol π $C(\pi_i) = \bar{\phi}_i$ where $\bar{\phi}_i$ is the $\mathcal{G}$-hybrid dummy coercer's protocol, i.e., $\bar{\phi}_i = \mathtt{dum}_i^{\mathcal{G}}$. A universal composition theorem for incoercibility against active coercion can be proved along the lines of Theorem 2.

4 Receipt-Free-Incoercible Multi-party Computation

In this section we describe a protocol for IUC realizing any (well-formed [11]) n-party functionalities $\mathcal{F}$ in the presence of semi-honest (i.e. receipt-free) coercions. Our construction makes black-box use of the UC secure protocol by Canetti et al. [11] but it can be instantiated also with other (faster) UC secure protocols. In fact, our construction can be seen as a compiler of UC protocols to IUC secure protocols in the honestly generated hardware-token setting. Thus, by replacing the call to the protocol in [11] with a call to a faster UC secure protocol we obtain a reasonably efficient candidate for universally composable receipt-free voting.

Our protocol (compiler) assumes access to *honestly-generated* tamper resistant hardware tokens that perform encryption under a key which is secret shared among the parties.

Intuitively, the receipt-freeness of the protocol $\Pi_{\mathcal{F}}$ can be argued as follows: because the token does not reveal the encryption keys to anyone, the CPA security of the encryption scheme ensures that the adversary cannot distinguish encryptions of some x_i from encryption of an $x_i' \neq x_i$. Thus the real deceiver $\mathtt{DR}_i$ for a coerced p_i can simply change the input it provides the token according to the ideal deceiver $\mathtt{DI}_i$ and report back to $\mathcal{Z}$ (as part of the receipt) the actual reply to the token. Since we assume $t + t' < n$ there is at least one share of the decryption key unknown to $\mathcal{Z}$ and so it can not immediately detect that the ciphertext given in the receipt doesn't encrypt x_i. At this point $\mathtt{DR}_i$ can follow the rest of the protocol honestly and can report the remainder of it's view honestly in the receipt. A formal theorem and proof follow.

The Construction. For simplicity we restrict ourselves to non-reactive functionalities, also known as *secure function evaluation (SFE)*. (The general case can be reduced to this case by using a suitable form of secret sharing for maintaining the secret state of the reactive functionality.) Moreover, we describe our protocols as *synchronous protocols*, i.e., round-based protocols where messages sent in some round are delivered by the beginning of the next round; such protocols can be executed in UC as demonstrated in [24,25]. We point out that the protocols in [24] assume a global synchronizing clock; however, as noted in [24,25], when we do not require guaranteed termination, e.g., in fully asynchronous environments, the clock can be emulated by the parties exchanging dummy synchronization messages. We further assume that the parties have access to a broadcast channel.

Without loss of generality, we assume that the functionality $\mathcal{F}$ being computed has a global output obtained by evaluating the function f on the vector of inputs. The case of local (a.k.a. private) and/or randomized functionalities can be dealt with by using standard techniques (c.f. [29].) Furthermore, as is usual with UC functionalities, we assume that $\mathcal{F}$ delivers its outputs in a delayed manner—whenever an output is ready for some party the simulator $\mathcal{S}$ is notified and $\mathcal{F}$ waits for $\mathcal{S}$'s permission to deliver the output.[9] Finally, to ensure properly synchronized simulation, we need to allow $\mathcal{S}$ to know when honest parties hand their input to the functionality. Thus we assume that the functionality $\mathcal{F}$ informs the simulator upon reception of any input x_i from an honest party p_i. We point out that as we allow a dishonest majority, we are restricted to security with abort, i.e., upon receiving a special message (abort) from the simulator, the functionality $\mathcal{F}$ sets all outputs of honest parties to a special symbol $\bot$.

Finally, our protocols makes use of an *authenticated additive n-out-of-n secret sharing*. Informally, this is an additive secret sharing where each share is authenticated by a digital signature for which every party knows the verification key, buy no party knows the signing key. We refer to the full version for a formal specification of our scheme.

In the remainder of this section we present our protocol and prove its security. We start by describing the hardware token that our protocol needs. The token functionality $\mathcal{T}_{\mathsf{ThEnc}}$ captures a threshold authenticated encryption token and is described in Fig. 2. The token is parameterized by an IND-CPA secure symmetric key encryption scheme (Gen, Enc, Dec) and an existentially unforgeable signature scheme (Gen', Sign, Ver). Initially the token generates signature key pair (sk, vk). Then upon request from any party i (or the adversary when p_i is corrupt) it generates a random encryption key k_i for p_i and uses vk to compute an n-out-of-n authenticated sharing k_i. Each party $j \in [n]$ requests it's share $\langle k_i \rangle_j$. Subsequently, whenever p_i requests an encryption of some message m from the token, $\mathcal{T}_{\mathsf{ThEnc}}$ computes a *fresh* encryption of m under key k_i and hands the result to p_i.

Given hybrid access to $\mathcal{T}_{\mathsf{ThEnc}}$ $(\mathcal{P})$, our protocol $\pi_{\mathcal{F}}$ for C^{sh}-incoercibly (UC) securely realizing any given functionality $\mathcal{F}$ proceeds in three sequential phases:

1. In the setup phase, for each player (at their behest) an encryption key is generated and shared with an n-out-of-n authenticated secret sharing. Formally, for each $i \in [n]$ a message (keygen, i) is sent by p_i to the token.[10] Shares are then delivered to parties when they send a keyShare to $\mathcal{T}_{\mathsf{ThEnc}}$.
2. In the second phase, each p_i asks the token to encrypt its inputs x_i under key k_i, i.e., inputs (encrypt, x_i, i) to the token $\mathcal{T}_{\mathsf{ThEnc}}$ $(\mathcal{P})$.
3. Finally, in a third phase, the parties invoke a UC secure SFE protocol, e.g., the one from [11] denoted by Π_{CLOS}, to implement the functionality $\widehat{\mathcal{F}}$. Roughly

---

[9] Because we restrict to public-output functions, we can wlog assume that the output is issued in a *public* delayed manner (c.f., Sect. 1).

[10] Presumably in a real world setting this phase will be executed on behalf of the players by the authority in charge of running the election. Then the tokens with an initialized state can be distributed to the players.

speaking, $\widehat{\mathcal{F}}$ receives from each player as input a ciphertext and one key-share for each of the decryption keys $k_1, \ldots, k_n$, reconstructs the decryption keys from the shares, and uses them to decrypt the ciphertexts to obtain plaintexts $\{x_i\}_{i \in [n]}$. If either reconstruction (i.e. signature verification) or decryption fails then $\mathcal{F}$ outputs $\perp$. Otherwise it computes and outputs a fresh n-out-of-n authenticated sharing of the value $f(x_1, \ldots, x_n)$. We refer to the full version for a formal description of protocol $\Pi_{\mathcal{F}}$ and functionality $\widehat{\mathcal{F}}$.

Functionality $\mathcal{T}_{\text{ThEnc}}$ ($\mathcal{P}$)

The token functionality is parameterized by a symmetric-key CPA encryption scheme (Gen, Enc, Dec), an existentially unforgeable signature scheme (Gen', Sign, Ver), and a security parameter λ.

- Upon receiving message (keygen, i) from party $i \in \mathcal{P}$ (or the adversary) if p_i is honest and a message (keygen, i) has already been received from p_i then ignore it; otherwise execute the following steps:
    1. Sample $k_i \leftarrow \text{Gen}(1^\lambda)$.
    2. Sample an n-out-of-n authenticated sharing $\{\langle k_i \rangle_j\}_{j \in [n]}$ of k_i.
- Upon receiving message (keyShare, i, j) from party $p_i \in \mathcal{P}$, if share $\langle k_i \rangle_j$ has not already been sampled then send $\perp$ to p_i. Otherwise send it to party i.
- Upon receiving message (encrypt, m, i) from party p_i for some $m \in \{0, 1\}^*$, if a key k_i has not already been sampled then send $\perp$ to p_i. Otherwise compute a fresh encryption $e_i = \text{Enc}_{k_i}(m)$ and send it to p_i.

Fig. 2. Threshold encryption token functionality

The security of protocol $\Pi_{\mathcal{F}}$ is argued as follows: As long as there is at least one honest party, the adversary will not get information about any of the encryption keys k_i. This follows from the security of the encryption scheme (Gen, Enc, Dec) used by the token and the privacy of the protocol Π_{CLOS}. Thus the simulator can simulate the entire protocol execution by simply using encryptions of random messages to simulate the tokens responses and storing local (simulated) copied of coerced parties. The unforgeability of the signatures used by the token to authenticate the shares will guarantee that the adversary cannot alter the input of honest or coerced parties by giving faulty inputs to the execution of Π_{CLOS}.

Theorem 3. *Let $\mathcal{F}$ be a n-party well-formed functionality as above. Further let* (Gen, Enc, Dec) *be an encryption scheme secure against chosen plaintext attacks (IND-CPA) and* (Gen', Sign, Ver) *be an existentially unforgeable signature scheme. Then the protocol $\Pi_{\mathcal{F}}$ C^{sh}-incoercibly (UC) securely realizes the functionality $\mathcal{F}$ in the static corruption model in the presence of any t corrupted and t' coerced parties where $t + t' < n$.*

5 Active-Incoercible Multi-party Computation

In this section we consider the strongest form of coercion, namely active coercions. Recall that these essentially turn a coerced party into a dummy party with all interaction driven by $\mathcal{Z}$. It turns out that the protocol from the previous section achieving semi-honest-incoercibility can also be shown to achieve full active-incoercibility. There are two key differences between the two security notions which must be addressed in the proof.

1. In a simulation for a semi-honest coerced party p_i, the simulator $\mathcal{S}$ must maintain a simulated internal state of p_i so that it can always respond to a receipt request from $\mathcal{Z}$. However no such requirement is placed on $\mathcal{S}$ for active coercions making the job of $\mathcal{S}$ easier in this respect.
2. On the other hand, say $\mathcal{Z}$ instructs a coerced (non-deceiving) party p_i to give input x to $\mathcal{F}$. In both the semi-honest and active case in the ideal worlds p_i will forward x to $\mathcal{F}$. Moreover in the semi-honest case p_i would use x as input to the honest protocol. However case of an active coercion $\mathcal{Z}$ is essentially running the protocol on behalf of p_i as it wishes. Thus their is no guarantee that x will be the effective input of p_i in such a protocol execution. So $\mathcal{S}$ must now extract the effective input of p_i during the protocol execution and force p_i to submit that as input to $\mathcal{F}$ in place of x. (Indeed, this is where $\mathcal{S}$ uses the property that parties have *delayed* input to $\mathcal{F}$.) Otherwise the two world would, in general, be distinguishable.

Theorem 4 (Active-Incoercibility). *Let $\mathcal{F}$ be a n-party well-formed functionality as above. Further let $(\mathsf{Gen}, \mathsf{Enc}, \mathsf{Dec})$ be an encryption scheme secure against chosen plaintext attacks (IND-CPA) and $(\mathsf{Gen}', \mathsf{Sign}, \mathsf{Ver})$ be an existentially unforgeable signature scheme. Then the protocol $\Pi_{\mathcal{F}}$ C^A-incoercibly (UC) securely realizes the functionality $\mathcal{F}$ in the static corruption model in the presence of any t corrupted and t' coerced parties where $t + t' < n$.*

Acknowledgements. Joël Alwen was supported by the ERC starting grant (259668-PSPC). Rafail Ostrovsky was supported in part by NSF grants 09165174, 1065276, 1118126 and 1136174, US-Israel BSF grant 2008411, OKAWA Foundation Research Award, IBM Faculty Research Award, Xerox Faculty Research Award, B. John Garrick Foundation Award, Teradata Research Award, Lockheed-Martin Corporation Research Award, and the Defense Advanced Research Projects Agency through the U.S. Office of Naval Research under Contract N00014 -11 -1-0392. The views expressed are those of the author and do not reflect the official policy or position of the Department of Defense or the U.S. Government. Vassilis Zikas was supported in part by the Swiss National Science Foundation (SNF) via the Ambizione grant PZ00P-2142549.

References

1. Backes, M., Hritcu, C., Maffei, M.: Automated verification of remote electronic voting protocols in the applied pi-calculus. In: CSF, pp. 195–209. IEEE Computer Society (2008)

2. Backes, M., Pfitzmann, B., Waidner, M.: The reactive simulatability (RSIM) framework for asynchronous systems. Inf. Comput. **205**(12), 1685–1720 (2007)
3. Ben-Or, M., Goldwasser, S., Wigderson, A.: Completeness theorems for non-cryptographic fault-tolerant distributed computation (extended abstract). In: 20th ACM STOC, pp. 1–10. ACM Press, May 1988
4. Benaloh, J.C., Tuinstra, D.: Receipt-free secret-ballot elections (extended abstract). In: 26th ACM STOC, pp. 544–553. ACM Press, May 1994
5. Canetti, R.: Security and composition of multiparty cryptographic protocols. J. Cryptology **13**(1), 143–202 (2000)
6. Canetti, R.: Universally composable security: A new paradigm for cryptographic protocols. In: 42nd FOCS, pp. 136–145. IEEE Computer Society Press, October 2001
7. Canetti, R., Dodis, Y., Pass, R., Walfish, S.: Universally composable security with global setup. In: Vadhan, S.P. (ed.) TCC 2007. LNCS, vol. 4392, pp. 61–85. Springer, Heidelberg (2007)
8. Canetti, R., Dwork, C., Naor, M., Ostrovsky, R.: Deniable encryption. In: Kaliski Jr., B.S. (ed.) CRYPTO 1997. LNCS, vol. 1294, pp. 90–104. Springer, Heidelberg (1997)
9. Canetti, R., Gennaro, R.: Incoercible multiparty computation (extended abstract). In: FOCS, pp. 504–513. IEEE Computer Society (1996)
10. Canetti, R., Goldwasser, S., Poburinnaya, O.: Adaptively secure two-party computation from indistinguishability obfuscation. In: Dodis, Y., Nielsen, J.B. (eds.) TCC 2015, Part II. LNCS, vol. 9015, pp. 557–585. Springer, Heidelberg (2015)
11. Canetti, R., Lindell, Y., Ostrovsky, R., Sahai, A.: Universally composable two-party and multi-party secure computation. In: 34th ACM STOC, pp. 494–503. ACM Press, May 2002
12. Chaum, D., Crépeau, C., Damgård, I.: Multiparty unconditionally secure protocols (extended abstract). In: 20th ACM STOC, pp. 11–19. ACM Press, May 1988
13. Chaum, D., Jakobsson, M., Rivest, R.L., Ryan, P.Y.A., Benaloh, J., Kutylowski, M., Adida, B. (eds.): Towards Trustworthy Elections. LNCS, vol. 6000. Springer, Heidelberg (2010)
14. Commision, E.E.: Internet voting in estonia, October 2013
15. Delaune, S., Kremer, S., Ryan, M.: Coercion-resistance and receipt-freeness in electronic voting. In: CSFW, pp. 28–42. IEEE Computer Society (2006)
16. Delaune, S., Kremer, S., Ryan, M.: Verifying privacy-type properties of electronic voting protocols: a taster. In: Chaum et al. [13], pp. 289–309
17. Garg, S., Gentry, C., Halevi, S., Raykova, M., Sahai, A., Waters, B.: Candidate indistinguishability obfuscation and functional encryption for all circuits. In: 54th FOCS, pp. 40–49. IEEE Computer Society Press, October 2013
18. Goldreich, O., Micali, S., Wigderson, A.: How to play any mental game or a completeness theorem for protocols with honest majority. In: Aho, A. (ed.) 19th ACM STOC, pp. 218–229. ACM Press, May 1987
19. Heather, J., Schneider, S.: A formal framework for modelling coercion resistance and receipt freeness. In: Giannakopoulou, D., Méry, D. (eds.) FM 2012. LNCS, vol. 7436, pp. 217–231. Springer, Heidelberg (2012)
20. Hirt, M., Sako, K.: Efficient receipt-free voting based on homomorphic encryption. In: Preneel, B. (ed.) EUROCRYPT 2000. LNCS, vol. 1807, p. 539. Springer, Heidelberg (2000)
21. Ishai, Y., Prabhakaran, M., Sahai, A.: Founding cryptography on oblivious transfer – efficiently. In: Wagner, D. (ed.) CRYPTO 2008. LNCS, vol. 5157, pp. 572–591. Springer, Heidelberg (2008)

22. Jonker, H.L., de Vink, E.P.: Formalising receipt-freeness. In: Katsikas, S.K., López, J., Backes, M., Gritzalis, S., Preneel, B. (eds.) ISC 2006. LNCS, vol. 4176, pp. 476–488. Springer, Heidelberg (2006)

23. Juels, A., Catalano, D., Jakobsson, M.: Coercion-resistant electronic elections. In: Chaum et al. [13], pp. 37–63

24. Katz, J., Maurer, U., Tackmann, B., Zikas, V.: Universally composable synchronous computation. In: Sahai, A. (ed.) TCC 2013. LNCS, vol. 7785, pp. 477–498. Springer, Heidelberg (2013)

25. Kushilevitz, E., Lindell, Y., Rabin, T.: Information-theoretically secure protocols and security under composition. In: Kleinberg, J.M. (ed.) 38th ACM STOC, pp. 109–118. ACM Press, May 2006

26. Küsters, R., Truderung, T.: An epistemic approach to coercion-resistance for electronic voting protocols. In: 2009 IEEE Symposium on Security and Privacy, pp. 251–266. IEEE Computer Society Press, May 2009

27. Küsters, R., Truderung, T., Vogt, A.: Verifiability, privacy, and coercion-resistance: New insights from a case study. In: IEEE Symposium on Security and Privacy, pp. 538–553. IEEE Computer Society (2011)

28. Küsters, R., Truderung, T., Vogt, A.: A game-based definition of coercion-resistance and its applications. J. Comput. Secur. (special issue of selected CSF 2010 papers) 20(6/2012), 709–764 (2012)

29. Lindell, Y., Pinkas, B.: A proof of security of Yao's protocol for two-party computation. J. Cryptology 22(2), 161–188 (2009)

30. Maurer, U., Renner, R.: Abstract cryptography. In: Chazelle, B. (ed.) ICS 2011, pp. 1–21. Tsinghua University Press, January 2011

31. Michels, M., Horster, P.: Some remarks on a receipt-free and universally verifiable mix-typevoting scheme. In: Kim, K., Matsumoto, T. (eds.) ASIACRYPT 1996. LNCS, vol. 1163, pp. 125–132. Springer, Heidelberg (1996)

32. Moran, T., Naor, M.: Receipt-free universally-verifiable voting with everlasting privacy. In: Dwork, C. (ed.) CRYPTO 2006. LNCS, vol. 4117, pp. 373–392. Springer, Heidelberg (2006)

33. Okamoto, T.: Receipt-free electronic voting schemes for large scale elections. In: Christianson, B., Lomas, M., Crispo, B., Roe, M. (eds.) Security Protocols 1997. LNCS, vol. 1361, pp. 25–35. Springer, Heidelberg (1998)

34. Sako, K., Kilian, J.: Receipt-free mix-type voting scheme. In: Guillou, L.C., Quisquater, J.-J. (eds.) EUROCRYPT 1995. LNCS, vol. 921, pp. 393–403. Springer, Heidelberg (1995)

35. Unruh, D., Müller-Quade, J.: Universally composable incoercibility. In: Rabin, T. (ed.) CRYPTO 2010. LNCS, vol. 6223, pp. 411–428. Springer, Heidelberg (2010)

36. Yao, A.C.-C.: Protocols for secure computations (extended abstract). In: 23rd FOCS, pp. 160–164. IEEE Computer Society Press, November 1982

Author Index

Printed in the United States
By Bookmasters